Math for Your World

A Brief Guide to Getting **the Most** from This Book

Feature	Description	Benefit	Page
Section-Opening Scenarios	Every section opens with a scenario presenting a unique application of mathematics in your life outside the classroom.	Realizing that mathematics is everywhere will help motivate your learning.	229
Learning Objectives	Every section begins with a list of objectives. Each objective is restated in the margin where the objective is covered.	The objectives focus your reading by emphasizing what is most important and where to find it.	515
Detailed Worked-Out Examples	Examples are clearly written and provide step-by-step solutions. No steps are omitted, and each step is thoroughly explained to the right of the mathematics.	The blue annotations will help you understand the solutions by providing the reason why every mathematics step is true and necessary.	221
Applications Using Real-World Data	Interesting applications from nearly every discipline, supported by up-to-date real world data, are included in every section.	Ever wonder how you'll use mathematics? These applications will show you how mathematics can solve real problems.	166
Explanatory Voice Balloons Round to the nearest dollar.	Voice Balloons help to demystify mathematics. They translate mathematical language into plain English, clarify problem solving procedures, and present alternative ways of understanding concepts.	Does math ever look foreign to you? This feature often translates math into everyday English. If you understand what you're reading, it makes problem solving a lot easier.	222
Study Tips	The book's Study Tip boxes offer suggestions for problem solving, point out common errors to avoid, and provide informal hints and suggestions.	By seeing common mistakes, you'll be able to avoid them.	233
Achieving Success	The book's Achieving Success boxes offer strategies for success in learning math, as well as suggestions for future college coursework and training programs.	Follow these suggestions to help achieve your full academic potential in mathematics and beyond.	155
Brief Reviews	Brief Reviews cover skills you already learned but may have forgotten.	Having these refresher boxes easily accessible will help ease anxiety about skills you may have forgotten.	17

2 Work the Problems

Feature	Description	Benefit	Page
Checkpoint Examples ✓	Each example is followed by a similar matched problem, called a Checkpoint, that offers you the opportunity to work a similar exercise. The answers to the Checkpoints are provided in the answer section.	You learn best by doing. You'll solidify your understanding of worked examples if you try a similar problem right away to be sure you understand what you've just read.	470
Extensive and Varied Exercise Sets	An abundant collection of exercises is included in an exercise set at the end of each section. Exercises are organized within categories. The exercises in the second category, Practice Exercises, follow the same order as the section's worked examples.	The parallel order of the Practice Exercises lets you refer to the worked examples and use them as models for solving these problems.	156

3 Review for Quizzes and Tests

Feature	Description	Benefit	Page
Detailed Chapter Summaries	Each chapter contains a detailed review chart that summarizes the definitions and concepts in every section of the chapter. Examples that illustrate these key concepts are also included in the review.	Study each detailed summary and you'll know the most important material in the chapter!	42
Chapter Review Exercises	After the detailed summary, a comprehensive collection of review exercises for each of the chapter's sections provides the opportunity to practice the skills you just reviewed.	Practice makes perfect. These exercises contain the most significant problems for each of the chapter's sections.	493
Chapter Tests	Each chapter contains two practice tests with approximately 25 problems each that cover the important concepts in the chapter.	You can use the chapter tests to determine whether you have mastered the material covered in the chapter. Try taking them under test conditions: Time yourself and don't peek at the answers until you're done.	445
Video Resources on DVD	Available to students on DVD or on YouTube, these videos contain worked-out solutions to every Checkpoint problem in the text.	The videos let you visually review the solution to every Checkpoint problem so you can see the teacher walk through all the steps.	470

Math for Your World

Robert Blitzer

Miami Dade College

Prentice Hall
Boston • Columbus • Indianapolis • New York • San Francisco • Upper Saddle River
Amsterdam • Cape Town • Dubai • London • Madrid • Milan • Munich • Paris • Montréal • Toronto
Delhi • Mexico City • São Paulo • Sydney • Hong Kong • Seoul • Singapore • Taipei • Tokyo

Editor in Chief: *Anne Kelly*
Acquisitions Editor: *Marnie Greenhut*
Sponsoring Editor: *Dawn Murrin*
Editorial Assistant: *Elle Driska*
Senior Managing Editor: *Karen Wernholm*
Senior Production Project Manager: *Kathleen A. Manley*
Digital Assets Manager: *Marianne Groth*
Media Producer: *Audra Walsh*
Software Development: *Edward Chappell, MathXL; and Marty Wright, TestGen*
Senior Author Support/Technology Specialist: *Joe Vetere*
Senior Prepress Supervisor: *Caroline Fell*
Manufacturing Buyer: *Carol Melville and Debra Rossi*
Senior Media Buyer: *Ginny Michaud*
Production Coordination and Composition: *Preparé, Inc.*
Illustrations: *Scientific Illustrators/Laserwords*
Rights and Permissions Advisor: *Michael Joyce*
Image Manager: *Rachel Youdelman*
Photo Researcher: *Sheila Norman*
Design Manager: *Andrea Nix*
Art Director: *Christina Gleason*
Text Design: *Ellen Pettengell Design*
Cover Design: *Studio Montage*
Cover Photo: *Shutterstock/Stephen Coburn (suitcase); Shutterstock/Vaclav Volrab (giraffes); AFP/Getty Images (airport)*

Library of Congress Cataloging-in-Publication Data

Blitzer, Robert.
    Math for your world/Robert Blitzer.—1st ed.
        p. cm.
    Includes bibliographical references and index.
    SE ISBN 0-13-210351-6 (alk. paper)
    1. Mathematics—Textbooks.   I. Title.
    QA11.2.B55 2012
    510—dc22

                                            2010029922

4 5 6 7 8 9 10—CKV—14 13 12

Prentice Hall
is an imprint of

www.PearsonSchool.com

ISBN 10: 0-13-210351-6
ISBN 13: 978-0-13-210351-0

Table of Contents

To the Student

The bar graph shows some of the qualities that teens say make a great teacher.

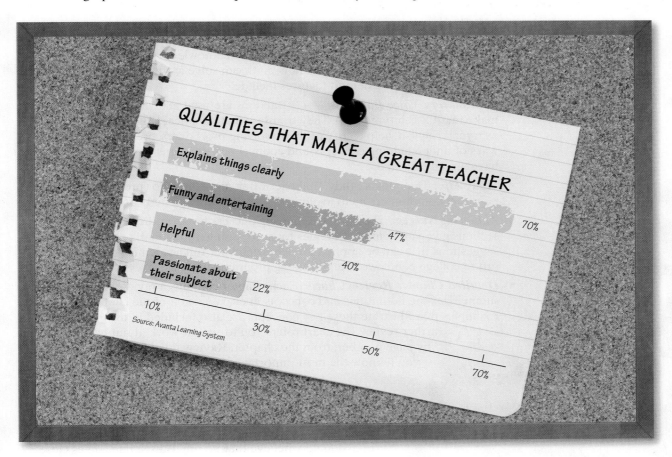

QUALITIES THAT MAKE A GREAT TEACHER

Explains things clearly — 70%

Funny and entertaining — 47%

Helpful — 40%

Passionate about their subject — 22%

10%   30%   50%   70%

Source: Avanta Learning System

It was my goal to incorporate each of the qualities that make a great teacher throughout the pages of this book.

Explains Things Clearly

I understand that your primary purpose in reading *Math for Your World* is to acquire a solid understanding of the required topics in your math class. In order to achieve this goal, I've carefully explained each topic. Important definitions and procedures are set off in boxes, and worked-out examples that present solutions in a step-by-step manner appear in every section. Each example is followed by a similar matched problem, called a Checkpoint, for you to try so that you can actively participate in the learning process as you read the book. (Answers to all Checkpoints appear in the back of the book.)

Funny/Entertaining

Who says that a math textbook can't be entertaining? From our quirky cover to the photos in the chapter and section openers, prepare to expect the unexpected. I hope some of the book's enrichment essays, called Blitzer Bonuses, will put a smile on your face from time to time.

Helpful

I designed the book's features to help you acquire knowledge of fundamental mathematics, as well as to show you how math can solve authentic problems that apply to your life. These helpful features include:

- *Explanatory Voice Balloons:* Voice Balloons are used in a variety of ways to make math less intimidating. They translate mathematical language into everyday English, help clarify problem-solving procedures, present alternative ways of understanding concepts, and connect new concepts to concepts you have already learned.
- *Study Tips:* The book's Study Tip boxes offer suggestions for problem solving, point out common errors to avoid, and provide informal hints and suggestions.
- *Achieving Success:* The book's Achieving Success boxes at the end of most sections give you helpful strategies for success in learning math, as well as suggestions that can be applied for achieving your full academic potential in future college coursework and training programs.
- *Detailed Chapter Review Charts:* Each chapter contains a review chart that summarizes the definitions and concepts in every section of the chapter. Examples that illustrate these key concepts are also included in the chart. For further review, the chart refers you to similar worked-out examples, by page number, from the chapter. Review these summaries and you'll know the most important material in the chapter!

Passionate about Their Subject

I passionately believe that no other discipline comes close to math in offering a more extensive set of tools for application and development of your mind. I wrote the book in Point Reyes National Seashore, 40 miles north of San Francisco. The park consists of 75,000 acres with miles of pristine surf-washed beaches, forested ridges, and bays bordered by white cliffs. It was my hope to convey the beauty and excitement of mathematics using nature's unspoiled beauty as a source of inspiration and creativity. Enjoy the pages that follow as you empower yourself with the mathematics needed to succeed in your class now, and later in college, training programs, your career, and your life.

Regards,

Bob

Robert Blitzer

Preface

Math for Your World provides a general survey of mathematical topics that are useful in our contemporary world. My general purpose in writing the book is to help high school students gain mathematical literacy by showing them how mathematics can be applied to their lives in interesting, enjoyable, and meaningful ways. The book has four major goals:

1. To help students acquire knowledge of fundamental mathematics that can be applied in future coursework and training programs.
2. To show students how mathematics can solve authentic problems that apply to their lives.
3. To develop mathematically proficient students who reason with quantitative issues and mathematical ideas they are likely to encounter in postsecondary education, career, and life.
4. To enable students to develop problem-solving skills, while fostering critical thinking, within an interesting setting.

The book's variety of topics and flexibility of sequence make it appropriate for a one-semester or one-year course for high school juniors and seniors that integrates topics from previous math courses and looks ahead to topics in future higher-education coursework. The text addresses the gamut of core content skills and Mathematical Practices covered in the Common Core State Standards for Mathematics. Students completing a course based on the topics in *Math for Your World* should be prepared for nonremedial or credit-bearing college mathematics courses, as well as for training programs for career-level jobs.

Pedagogical Features

The ability to read textbooks to gather and assimilate information is a key component to success in any coursework or training program.

One major obstacle in the way of achieving success in mathematics is the fact that many students do not read their textbook. Anecdotal evidence gathered over years highlights two basic reasons why students do not take advantage of math textbooks:

"I'll never use this information."
"I can't follow the explanations."

I've written every page of *Math for Your World* with the intent of eliminating these two objections. The ideas and tools I've used to do so are described in the following features:

- ***Learning Objectives.*** Learning objectives are clearly stated at the beginning of each section. These objectives help students recognize and focus on the section's most important ideas. The objectives are restated in the margin at their point of use.
- ***Chapter-Opening and Section-Opening Scenarios.*** Every chapter and every section open with a scenario presenting a unique application of mathematics in students' lives outside the classroom. These scenarios are revisited in the course of the chapter or section in an example, discussion, or exercise.

- *Applications Using Real-World Data.* A wide variety of interesting and innovative applications supported by up-to-date real world data are included in every section. Students will be able to relate to these applications, which look ahead to college and life beyond high school.

- *Detailed Worked-Out Examples.* Each example is titled, making the purpose of the example clear. Examples are clearly written and provide students with detailed step-by-step solutions. No steps are omitted and each step is thoroughly explained to the right of the mathematics.

- *Explanatory Voice Balloons.* Voice Balloons are used in a variety of ways to demystify mathematics. They translate mathematical language into everyday English, help clarify problem-solving procedures, present alternative ways of understanding concepts, and connect problem solving to concepts students have already learned.

- *Checkpoint Examples.* Each example is followed by a similar matched problem, called a Checkpoint, offering students the opportunity to test for conceptual understanding by working a similar exercise. The answers to the Checkpoints are provided in the answer section in the back of the book.

- *Extensive and Varied Exercise Sets.* An abundant collection of exercises is included in an exercise set at the end of each section. Exercises are organized into seven categories, a format that makes it easy to create well-rounded homework assignments.

  - Concept and Vocabulary Exercises are intended for classroom discussion in order to engage participation in the learning process. Exercises include fill-in-the-blank, true/false, open-ended questions, and items asking students to determine whether statements make sense. These exercises promote familiarity with basic vocabulary and concepts, prerequisites for students' success.

  - Practice Exercises follow the same order as the section's worked examples. This parallel order enables students to refer to the titled examples and their detailed explanations to successfully achieve each section's objectives.

  - Practice Plus Exercises contain more challenging practice exercises that often require students to combine several skills or concepts. These exercises provide the option of creating assignments that take practice exercises to a more challenging level.

  - Application Exercises give the option of assigning realistic, relevant, and unique applications consistent with your students' needs and interests.

  - Critical Thinking Exercises require students to employ analytic skills that go beyond applying each section's basic objectives. The exercises ask students to make sense of complex problems and persevere in solving them.

  - Technology Exercises enable students to use technological tools to explore and deepen their understanding of concepts.

  - Group Exercises contain projects and collaborative activities that give students the opportunity to work cooperatively as they think and talk about mathematics.

- *Brief Reviews.* The book's Brief Review boxes summarize mathematical skills that students should have learned previously, but which many students still need to review. This feature appears whenever a particular skill is first needed and eliminates the need for you to reteach that skill.

- *Study Tips.* The book's Study Tip boxes offer suggestions for problem solving, point out common errors to avoid, and provide informal hints and suggestions.

- *Achieving Success.* The book's Achieving Success boxes at the end of most sections offer strategies for success in math courses, as well as suggestions for achieving one's full academic potential.

- ***Blitzer Bonuses.*** These enrichment essays provide historical, interdisciplinary, and otherwise interesting connections to the mathematics under study, showing students that math is an interesting and dynamic discipline.

- ***Detailed Chapter Review Charts.*** Each chapter contains a review chart that summarizes the definitions and concepts in every section of the chapter. Examples that illustrate these key concepts are also included in the chart. For further review, the chart refers students to similar worked-out examples, by page number, from the chapter.

- ***End-of-Chapter Materials.*** A comprehensive collection of review exercises for each of the chapter's sections follows the review chart. This is followed by two equivalent versions of a chapter test that enables students to test their understanding of the material covered in the chapter.

I hope that my love for learning, as well as my respect for the diversity of students I have taught and learned from over the years, is apparent throughout *Math for Your World*. By connecting mathematics to the whole spectrum of learning, it is my intent to show high school students that their world is profoundly mathematical, and indeed, π is in the sky.

Robert Blitzer

Supplements List

Student Resources

The following supplements are available for purchase.

Video Resources on DVD with Optional Captioning

The Video Resources on DVD show teachers working through every Checkpoint problem in the text, making it easy for students to go back to something they learned in class. The DVD is ideal for distance learning, missed classes, or supplemental instruction. Videos include an optional window to display subtitles in both English and Spanish.

Teacher Resources

Annotated Teacher's Edition

The Annotated Teacher's Edition contains answers to most exercises on the page where they appear. Longer answers are in the back of the book.

Teacher's Solutions Manual

The Teacher's Solutions Manual includes detailed, worked-out solutions to all textbook exercises and all Checkpoints.

TestGen® (download only)

TestGen® enables teachers to build, edit, and print tests using a computerized test bank of questions developed to cover all the objectives of the text. TestGen® is algorithmically based, allowing teachers to create multiple but equivalent versions of the same question or test with the click of the button. Teachers can also modify test bank questions or add new questions.

Teacher's Testing Manual (download only)

The Teacher's Testing Manual includes pre-made tests derived from TestGen® that allow the teacher to assess students at the chapter, section, and objective levels.

PowerPoint Lecture Slides (download only)

These fully editable slides include definitions, key concepts, and examples. Slides can be projected in class, posted in an online class, or made available to students by the teacher for the purpose of review.

Image Resource Library (download only)

JPG or GIF files for all the art in the text is available for use in your own PowerPoint slides or to insert in worksheets or tests.

MathXL® for School (optional, for purchase only—access code required)

MathXL® for School is a powerful online homework, tutorial, and assessment system that accompanies Pearson Education's textbooks in mathematics or statistics.

With MathXL® for School, teachers can:

- Create, edit, and assign auto-graded online homework and tests correlated at the objective level to the textbook.

- Maintain records of all student and class performance tracked in MathXL® for School's online gradebook.

- Deliver quality, effective instruction regardless of experience level.

With MathXL® for School students can:

- Take chapter tests in MathXL® for School and receive personalized study plans based on their test results.

- Use the study plan to link directly to tutorial exercises for the objectives they need to study and retest.

- Access supplemental animations, video clips, and stepped out guided solutions directly from selected exercises.

For more information, visit our Web site at www.mathxlforschool.com, or contact your Pearson sales representative.

Acknowledgments

I would like to express my appreciation to all the reviewers, survey participants, and focus group attendees whose collective insights form the backbone of this text. In particular, I would like to thank the following people for reviewing *Math for Your World*.

Tom Beatini, *Glen Rock High School*

Jeremy Beckman, *Naches Valley High School*

Alexis Bergeven, *Milton-Freewater Unified School District #7*

Susan Blackwell, *First Flight High School*

Susan Bothman, *Ooltewah High School*

Patricia Buttolph, *Greater Southern Tier BOCES*

Veronica Carlson, *Moon Valley High School*

Kris Clester, *Boca Ciega High School*

Lauren Duff, *Montgomery County Public Schools*

Blake Edmondson, *Springville High School*

Peggy Foos, *Valley High School*

John Gianotti, *Argo High School*

Thomas Haver, *Winchester High School*

Beverly Kimes, *Birmingham City Schools*

André Mathurin, *Bellarmine College Preparatory*

Ron Millard, *Shawnee Mission South High School*

Donna Miller, *Eagle Grove High School*

Barbara Riebau, *Custer High School*

Leo Straughn, *Murphy High School*

Eileen Verde, *Ocean County Vocational-Technical School*

Additional acknowledgments are extended to Brad Davis, for preparing the answer section and annotated answers and serving as accuracy checker; Teri Lovelace, for helping to develop applications of special interest to high school students; Dan Miller and Kelly Barber, for preparing the solutions manuals; the Preparé, Inc. formatting team for the book's brilliant paging; Brian Morris at Scientific Illustrators, for superbly illustrating the book; Sheila Norman, photo researcher, for obtaining the book's photographs; and Rebecca Dunn, project manager, and Kathleen Manley, production editor, whose collective talents kept every aspect of this complex project moving through its many stages.

I would like to thank my editors at Pearson, Marnie Greenhut and Dawn Murrin, who guided and coordinated the book from manuscript through production. Thanks to Christina Gleason for the cover and interior design. Finally, thanks to the entire Pearson sales force, for your confidence and enthusiasm about the book.

Robert Blitzer

About the Author

Bob Blitzer is a native of Manhattan and received a Bachelor of Arts degree with dual majors in mathematics and psychology (minor: English literature) from the City College of New York. His unusual combination of academic interests led him toward a Master of Arts in mathematics from the University of Miami and a doctorate in behavioral sciences from Nova University. Bob's love for teaching mathematics was nourished for nearly 30 years at Miami Dade College, where he received numerous teaching awards, including Innovator of the Year from the League for Innovations in the Community College and an endowed chair based on excellence in the classroom. In addition to *Math for Your World*, Bob has written textbooks covering introductory algebra, college algebra, algebra and trigonometry, precalculus, and liberal arts mathematics, all published by Pearson Prentice Hall. When not secluded in his Northern California writer's cabin, Bob can be found hiking the beaches and trails of Point Reyes National Seashore, and tending to the chores required by his beloved entourage of horses, chickens, and irritable roosters.

Bob and his horse Jerid

Problem Solving and Critical Thinking

How would your lifestyle change if a gallon of gas cost $9.15? Or if the price of a staple such as milk was $15? That's how much those products would cost if their prices had increased at the same rate college tuition has increased since 1980.

Tuition and Fees at Four-Year Colleges

	School Year Ending 2000	School Year Ending 2009
Public	$3362	$6585
Private	$15,518	$25,143

Source: The College Board

If these trends continue, what can we expect in the 2010s and beyond? We can answer this question by using estimation techniques that allow us to represent the data mathematically. With such representations, called *mathematical models,* we can gain insights and predict what might occur in the future on a variety of issues, ranging from college costs to climate change.

Mathematical models involving college costs are developed in Example 8 and Checkpoint 8 of Section 1.2. In Exercise 16 of each chapter test, you will approach our climate crisis mathematically by developing models for data related to climate change.

1.1 Inductive and Deductive Reasoning

1. Understand and use inductive reasoning.

2. Understand and use deductive reasoning.

One of the newer frontiers of mathematics suggests that there is an underlying order in things that appear to be random, such as the hiss and crackle of background noises as you tune a radio. Irregularities in the heartbeat, some of them severe enough to cause a heart attack, or irregularities in our sleeping patterns, such as insomnia, are examples of chaotic behavior. Chaos in the mathematical sense does not mean a complete lack of form or arrangement. In mathematics, chaos is used to describe something that appears to be random but is not actually random. The patterns of chaos appear in images like the one shown on the right, called the Mandelbrot set. Magnified portions of this image yield repetitions of the original structure, as well as new and unexpected patterns. The Mandelbrot set transforms the hidden structure of chaotic events into a source of wonder and inspiration.

A magnification of the Mandelbrot set

Many people associate mathematics with tedious computation, meaningless algebraic procedures, and intimidating sets of equations. The truth is that mathematics is the most powerful means we have of exploring our world and describing how it works. The word *mathematics* comes from the Greek word *mathematikos*, which means "inclined to learn." To be mathematical literally means to be inquisitive, open-minded, and interested in a lifetime of pursuing knowledge!

Mathematics and Your Life

A major goal of this book is to show you how mathematics can be applied to your life in interesting, enjoyable, and meaningful ways. The ability to think mathematically and reason with quantitative issues will help you so that you can:

- apply techniques for solving mathematical problems to problems encountered in school and in every day life. (Chapter 1, Problem Solving and Critical Thinking)

- order and arrange your world by using sets to sort and classify information. (Chapter 2, Set Theory)

- put the numbers you encounter in the news, ranging from the national debt to costs for President Obama's 2009 economic stimulus package, into perspective. (Chapter 3, Number Theory and the Real Number System)

- use mathematical models to gain insights into a variety of issues, including the positive benefits that humor and laughter can have on your life. (Chapter 4, Algebra: Equations and Inequalities)

- establish a healthy weight range for your age and height. (Chapter 5, Algebra: Functions, Linear Functions, and Linear Systems)

- use basic ideas about savings, loans, and investments to achieve your financial goals. (Chapter 8, Personal Finance: Saving, Investing, and Spending)

- use geometry to study the shape of your world, enhancing your appreciation of nature's patterns and beauty. (Chapter 10, Geometry)

- develop an understanding of the fundamentals of statistics and how these numbers are used to make decisions. (Chapter 12, Statistics)
- use logic to evaluate the arguments of others and become a more effective advocate for your own beliefs. (Chapter 13, Logic)

Mathematics and Your Career

Generally speaking, the income of an occupation is related to the amount of education required. This, in turn, is usually related to the skill level required in language and mathematics. With our increasing reliance on technology, the more mathematics you know, the more career choices you will have.

Mathematics and Your World

Mathematics is a science that helps us recognize, classify, and explore the hidden patterns of our universe. Focusing on areas as different as planetary motion, animal markings, shapes of viruses, aerodynamics of figure skaters, and the very origin of the universe, mathematics is the most powerful tool available for revealing the underlying structure of our world. Within the last 30 years, mathematicians have even found order in chaotic events such as the uncontrolled storm of noise in the nerve cells of the brain during an epileptic seizure.

Inductive Reasoning

"It is better to take what may seem to be too much math rather than too little. Career plans change, and one of the biggest roadblocks in undertaking new educational or training goals is poor preparation in mathematics. Furthermore, not only do people qualify for more jobs with more math, they are also better able to perform their jobs."

Occupational
Outlook Quarterly

1 Understand and use inductive reasoning.

Mathematics involves the study of patterns. In everyday life, we frequently rely on patterns and routines to draw conclusions. Here is an example:

The last six times I went to the beach, the traffic was light on Wednesdays and heavy on Sundays. My conclusion is that weekdays have lighter traffic than weekends.

This type of reasoning process is referred to as *inductive reasoning*, or *induction*.

Inductive Reasoning

Inductive reasoning is the process of arriving at a general conclusion based on observations of specific examples.

Although inductive reasoning is a powerful method of drawing conclusions, we can never be absolutely certain that these conclusions are true. For this reason, the conclusions are called **conjectures**, **hypotheses**, or educated guesses. A strong inductive argument does not guarantee the truth of the conclusion, but rather provides strong support for the conclusion. If there is just one case for which the conjecture does not hold, then the conjecture is false. Such a case is called a **counterexample**.

Example 1 Finding a Counterexample

The ten symbols that we use to write numbers, namely 0, 1, 2, 3, 4, 5, 6, 7, 8, and 9, are called **digits**. In each example shown below, the sum of two two-digit numbers is a three-digit number.

$$\begin{array}{r} 47 \\ +73 \\ \hline 120 \end{array} \qquad \begin{array}{r} 56 \\ +46 \\ \hline 102 \end{array}$$

Two-digit numbers

Three-digit sums

Is the sum of two two-digit numbers always a three-digit number? Find a counter-example to show that the statement

The sum of two two-digit numbers is a three-digit number

is false.

Solution

There are many counterexamples, but we only need to find one. Here is an example that makes the statement false:

$$
\begin{array}{r}
\text{Two-digit} \\
\text{numbers}
\end{array}
\begin{array}{r}
56 \\
+43 \\
\hline
99
\end{array}
\quad \text{This is a two-digit sum, not a three-digit sum.}
$$

This example is a counterexample that shows the statement

The sum of two two-digit numbers is a three-digit number

is false.

> **Checkpoint 1** Find a counterexample to show that the statement
>
> The product of two two-digit numbers is a three-digit number
>
> is false.

Here are two examples of inductive reasoning:

- **Strong Inductive Argument** In a random sample of 1203 parents of children under 18, 15% of the parents said they were willing to pay for all of their child's college education. (*Source:* Sallie Mae/Gallup survey) We can conclude that there is a 95% probability that between 12.1% and 17.9% of all parents of children under 18 are willing to pay for all of their child's college education.

  In Chapter 12, you will learn how observations from a randomly selected group, one in which each member of the population has an equal chance of being selected, can provide probabilities of what is true about an entire population.

- **Weak Inductive Argument** Men have difficulty expressing their feelings. Neither my dad nor my boyfriend has ever cried in front of me.

  When generalizing from observations about your own circumstances and experiences, avoid jumping to hasty conclusions based on a few observations. Psychologists theorize that we do this - that is, place everyone in a neat category - to feel more secure about ourselves and our relationships to others.

Inductive reasoning is extremely important to mathematicians. Discovery in mathematics often begins with an examination of individual cases to reveal patterns about numbers.

Example 2 Using Inductive Reasoning

Identify a pattern in each list of numbers. Then use this pattern to find the next number.

a. 3, 12, 21, 30, 39, _____   **b.** 3, 12, 48, 192, 768, _____

c. 3, 4, 6, 9, 13, 18, _____   **d.** 3, 6, 18, 36, 108, 216, _____

Solution

a. Because 3, 12, 21, 30, 39, _____ is increasing relatively slowly, let's use addition as the basis for our individual observations.

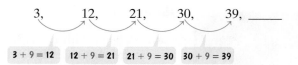

Generalizing from these observations, we conclude that each number after the first is obtained by adding 9 to the previous number. Using this pattern, the next number is 39 + 9, or 48.

b. Because 3, 12, 48, 192, 768, _____ is increasing relatively rapidly, let's use multiplication as the basis for our individual observations.

Generalizing from these observations, we conclude that each number after the first is obtained by multiplying the previous number by 4. Using this pattern, the next number is 768 × 4, or 3072.

c. Because 3, 4, 6, 9, 13, 18, _____ is increasing relatively slowly, let's use addition as the basis for our individual observations.

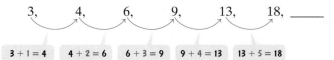

Generalizing from these observations, we conclude that each number after the first is obtained by adding a counting number to the previous number. The additions begin with 1 and continue through each successive counting number. Using this pattern, the next number is 18 + 6, or 24.

d. Because 3, 6, 18, 36, 108, 216, _____ is increasing relatively rapidly, let's use multiplication as the basis for our individual observations.

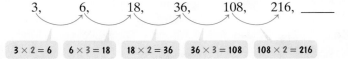

Generalizing from these observations, we conclude that each number after the first is obtained by multiplying the previous number by 2 or by 3. The multiplications begin with 2 and then alternate, multiplying by 2, then 3, then 2, then 3, and so on. Using this pattern, the next number is 216 × 3, or 648.

✔ **Checkpoint 2** Identify a pattern in each list of numbers. Then use this pattern to find the next number.

a. 3, 9, 15, 21, 27, _____   **b.** 2, 10, 50, 250, _____

c. 3, 6, 18, 72, 144, 432, 1728, _____   **d.** 1, 9, 17, 3, 11, 19, 5, 13, 21, _____

In our next example, the patterns are a bit more complex than the additions and multiplications we encountered in Example 2.

Example 3   Using Inductive Reasoning

Identify a pattern in each list of numbers. Then use this pattern to find the next number.

a. 1, 1, 2, 3, 5, 8, 13, 21, _____   **b.** 23, 54, 95, 146, 117, 98, _____

As this tree branches, the number of branches forms the Fibonacci sequence.

Solution

a. We begin with $1, 1, 2, 3, 5, 8, 13, 21$. Starting with the third number in the list, let's form our observations by comparing each number with the two numbers that immediately precede it.

1, 1, 2, 3, 5, 8, 13, 21, _____

preceded by 1 and 1:	preceded by 1 and 2:	preceded by 2 and 3:	preceded by 3 and 5:	preceded by 5 and 8:	preceded by 8 and 13:
$1 + 1 = 2$	$1 + 2 = 3$	$2 + 3 = 5$	$3 + 5 = 8$	$5 + 8 = 13$	$8 + 13 = 21$

Generalizing from these observations, we conclude that the first two numbers are 1. Each number thereafter is the sum of the two preceding numbers. Using this pattern, the next number is $13 + 21$, or 34. (The numbers 1, 1, 2, 3, 5, 8, 13, 21, and 34 are the first nine terms of the *Fibonacci sequence*, discussed in Chapter 3, Section 3.7.)

b. Now, we consider 23, 54, 95, 146, 117, 98. Let's use the digits that form each number as the basis for our individual observations. Focus on the sum of the digits, as well as the final digit increased by 1.

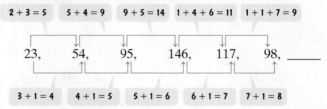

$2 + 3 = 5$	$5 + 4 = 9$	$9 + 5 = 14$	$1 + 4 + 6 = 11$	$1 + 1 + 7 = 9$

23, 54, 95, 146, 117, 98, _____

$3 + 1 = 4$	$4 + 1 = 5$	$5 + 1 = 6$	$6 + 1 = 7$	$7 + 1 = 8$

Generalizing from these observations, we conclude that for each number after the first, we obtain the first digit or the first two digits by adding the digits of the previous number. We obtain the last digit by adding 1 to the final digit of the preceding number. Applying this pattern to find the number that follows 98, the first two digits are $9 + 8$, or 17. The last digit is $8 + 1$, or 9. Thus, the next number in the list is 179.

Study Tip

The illusion in **Figure 1.1** is an ambiguous figure containing two patterns, where it is not clear which pattern should predominate. Do you see a goblet or two faces looking at each other? Like this ambiguous figure, some lists of numbers can display more than one pattern, particularly if only a few numbers are given. Inductive reasoning can result in more than one probable next number in a list.

Example: $1, 2, 4,$ _____

Pattern: Each number after the first is obtained by multiplying the previous number by 2. The missing number is 4×2, or 8.

Pattern: Each number after the first is obtained by adding successive counting numbers, starting with 1, to the previous number. The second number is $1 + 1$, or 2. The third number is $2 + 2$, or 4. The missing number is $4 + 3$, or 7.

Inductive reasoning can also result in different patterns that produce the same probable next number in a list.

Figure 1.1

Example: $1, 4, 9, 16, 25,$ _____

Pattern: Start by adding 3 to the first number. Then add successive odd numbers, 5, 7, 9, and so on. The missing number is $25 + 11$, or 36.

Pattern: Each number is obtained by squaring its position in the list: The first number is $1^2 = 1 \times 1 = 1$, the second number is $2^2 = 2 \times 2 = 4$, the third number is $3^2 = 3 \times 3 = 9$, and so on. The missing sixth number is $6^2 = 6 \times 6$, or 36.

The numbers that we found in Examples 2 and 3 are probable numbers. Perhaps you found patterns other than the ones we pointed out that might have resulted in different answers.

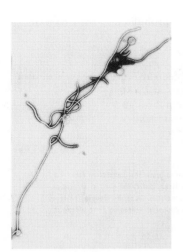

This electron microscope photograph shows the knotty shape of the Ebola virus.

Checkpoint 3 Identify a pattern in each list of numbers. Then use this pattern to find the next number.

▶ **a.** $1, 3, 4, 7, 11, 18, 29, 47,$ _____   **b.** $2, 3, 5, 9, 17, 33, 65, 129,$ _____

Mathematics is more than recognizing number patterns. It is about the patterns that arise in the world around us. For example, by describing patterns formed by various kinds of knots, mathematicians are helping scientists investigate the knotty shapes and patterns of viruses. One of the weapons used against viruses is based on recognizing visual patterns in the possible ways that knots can be tied.

Our next example deals with recognizing visual patterns.

Example 4   **Finding the Next Figure in a Visual Sequence**

Describe two patterns in this sequence of figures. Use the patterns to draw the next figure in the sequence.

 , , , , _____

Solution

The more obvious pattern is that the figures alternate between circles and squares. We conclude that the next figure will be a circle. We can identify the second pattern in the four regions containing no dots, one dot, two dots, and three dots. The dots are placed in order (no dots, one dot, two dots, three dots) in a clockwise direction. However, the entire pattern of dots rotates counterclockwise as we follow the figures from left to right.

This means that the next figure should be a circle with a single dot in the right-hand region, two dots in the bottom region, three dots in the left-hand region, and no dots in the top region. This figure is drawn in **Figure 1.2**.

Figure 1.2

Checkpoint 4 Describe two patterns in this sequence of figures. Use the patterns to draw the next figure in the sequence.

 , , , , _____

2 Understand and use deductive reasoning.

Deductive Reasoning

We use inductive reasoning in everyday life. Many of the conjectures that come from this kind of thinking seem highly likely, although we can never be absolutely certain that they are true. Another method of reasoning, called *deductive reasoning*, or *deduction*, can be used to prove that some conjectures are true.

Deductive Reasoning

Deductive reasoning is the process of proving a specific conclusion from one or more general statements. A conclusion that is proved to be true by deductive reasoning is called a **theorem**.

Deductive reasoning allows us to draw a specific conclusion from one or more general statements. On the next page are two examples of deductive reasoning. Notice that in both everyday situations, the general statement from which the conclusion is drawn is implied rather than directly stated.

Everyday Situation	Deductive Reasoning
One player to another in a Scrabble game: "You have to remove those five letters. You can't use TEXAS as a word."	• All proper names are prohibited in Scrabble. *general statement* TEXAS is a proper name. Therefore, TEXAS is prohibited in Scrabble. *conclusion*
Advice to college freshmen on choosing classes: "Never sign up for a 7 A.M. class. Yes, you did it in high school, but Mom was always there to keep waking you up, and if by some miracle you do make it to an early class, you will sleep through the lecture when you get there." (*Source: How to Survive Your Freshman Year*, Hundreds of Heads Books, 2004)	• All people need to sleep at 7 A.M. *general statement* You sign up for a class at 7 A.M. Therefore, you'll sleep through the lecture or not even make it to class. *conclusion* In Chapter 13, you'll learn how to prove this conclusion from the general statement in the first line. But is the general statement really true? Can we make assumptions about the sleeping patterns of all people, or are we using deductive reasoning to reinforce an untrue reality assumption?

Our next example illustrates the difference between inductive and deductive reasoning. The first part of the example involves reasoning that moves from specific examples to a general statement, illustrating inductive reasoning. The second part of the example begins with the general case rather than specific examples and illustrates deductive reasoning. To begin the general case, we use a letter to represent any one of various numbers. A letter used to represent any number in a collection of numbers is called a **variable**.

A Brief Review

In case you have forgotten some basic terms of arithmetic, the following list should be helpful.

Sum:	the result of addition
Difference:	the result of subtraction
Product:	the result of multiplication
Quotient:	the result of division

Example 5   Using Inductive and Deductive Reasoning

Consider the following procedure:

Select a number. Multiply the number by 6. Add 8 to the product. Divide this sum by 2. Subtract 4 from the quotient.

a. Repeat this procedure for at least four different numbers. Write a conjecture that relates the result of this process to the original number selected.

b. Use the variable n to represent the original number and use deductive reasoning to prove the conjecture in part (a).

Solution

a. First, let us pick our starting numbers. We will use 4, 7, 11, and 100, but we could pick any four numbers. Next we will apply the procedure given in this example to 4, 7, 11, and 100, four individual cases, in **Table 1.1**.

Table 1.1 Applying the Procedure in Four Individual Cases

Select a number.	4	7	11	100
Multiply the number by 6.	$4 \times 6 = 24$	$7 \times 6 = 42$	$11 \times 6 = 66$	$100 \times 6 = 600$
Add 8 to the product.	$24 + 8 = 32$	$42 + 8 = 50$	$66 + 8 = 74$	$600 + 8 = 608$
Divide this sum by 2.	$\dfrac{32}{2} = 16$	$\dfrac{50}{2} = 25$	$\dfrac{74}{2} = 37$	$\dfrac{608}{2} = 304$
Subtract 4 from the quotient.	$16 - 4 = 12$	$25 - 4 = 21$	$37 - 4 = 33$	$304 - 4 = 300$

Because we are asked to write a conjecture that relates the result of this process to the original number selected, let us focus on the result of each case.

Original number selected	4	7	11	100
Result of the process	12	21	33	300

Do you see a pattern? Our conjecture is that the result of the process is three times the original number selected. We have used inductive reasoning.

b. Now we begin with the general case rather than specific examples. We use the variable n to represent any number.

Select a number.	n
Multiply the number by 6.	$6n$ (This means 6 times n.)
Add 8 to the product.	$6n + 8$
Divide this sum by 2.	$\dfrac{6n + 8}{2} = \dfrac{6n}{2} + \dfrac{8}{2} = 3n + 4$
Subtract 4 from the quotient.	$3n + 4 - 4 = 3n$

Using the variable n to represent any number, the result is $3n$, or three times the number n. This proves that the result of the procedure is three times the original number selected for any number. We have used deductive reasoning.

Checkpoint 5 Consider the following procedure:

Select a number. Multiply the number by 4. Add 6 to the product. Divide this sum by 2. Subtract 3 from the quotient.

a. Repeat this procedure for at least four different numbers. Write a conjecture that relates the result of this process to the original number selected.

b. Use the variable n to represent the original number and use deductive reasoning to prove the conjecture in part (a).

Exercise Set 1.1

Concept and Vocabulary Exercises

In Exercises 1–3, fill in each blank so that the resulting statement is true.

1. The statement $3 + 3 = 6$ serves as a/an _____ to the conjecture that the sum of two odd numbers is an odd number.

2. Arriving at a specific conclusion from one or more general statements is called _____ reasoning.

3. Arriving at a general conclusion based on observations of specific examples is called _____ reasoning.

In Exercises 4–6, determine whether each statement is true or false. If the statement is false, make the necessary change(s) to produce a true statement.

4. Deductive reasoning is used to determine possible patterns when examining a list of numbers.

5. Some conjectures can be proved using deductive reasoning.

6. A theorem cannot have counterexamples.

Respond to Exercises 7–8 using verbal or written explanations.

7. The word *induce* comes from a Latin term meaning "to lead." Explain what leading has to do with inductive reasoning.

8. Describe what is meant by deductive reasoning. Give an example.

In Exercises 9–12, determine whether each statement makes sense or does not make sense, and explain your reasoning.

9. My 90-year-old grandfather smoked a pack of cigarettes a day his whole life, so I can conclude that cigarettes are not harmful.

10. Neither my parents nor my grandparents attended college, so I can conclude that I'm not college material.

11. I used the data shown in the bar graph, which summarizes a random sample of 1203 parents of children under 18, to conclude with certainty that 31% of all parents of children under 18 are willing to pay for half of their child's college education.

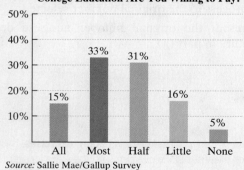

For How Much of Your Child's College Education Are You Willing to Pay?

Source: Sallie Mae/Gallup Survey

12. I used the data shown in the bar graph for Exercise 11, which summarizes a random sample of 1203 parents of children under 18, to conclude inductively that a greater percentage of all parents of children under 18 are willing to pay for all of their child's college education than for none of their child's college education.

Practice Exercises

In Exercises 13–20, find a counterexample to show that each of the statements is false.

13. No U.S. president has been younger than 65 at the time of his inauguration.

14. No singers appear in movies.

15. If a number is multiplied by itself, the result is even.

16. The sum of two three-digit numbers is a four-digit number.

17. Adding the same number to both the numerator and the denominator (top and bottom) of a fraction does not change the fraction's value.

18. If the difference between two numbers is odd, then the two numbers are both odd.

19. If a number is added to itself, the sum is greater than the original number.

20. If 1 is divided by a number, the quotient is less than that number.

In Exercises 21–50, identify a pattern in each list of numbers. Then use this pattern to find the next number. (More than one pattern might exist, so it is possible that there is more than one correct answer.)

21. 8, 12, 16, 20, 24, _____
22. 19, 24, 29, 34, 39, _____
23. 37, 32, 27, 22, 17, _____
24. 33, 29, 25, 21, 17, _____
25. 3, 9, 27, 81, 243, _____
26. 2, 8, 32, 128, 512, _____
27. 1, 2, 4, 8, 16, _____
28. 1, 5, 25, 125, _____
29. 1, 4, 1, 8, 1, 16, 1, _____
30. 1, 4, 1, 7, 1, 10, 1, _____
31. 4, 2, 0, −2, −4, _____
32. 6, 3, 0, −3, −6, _____
33. $\frac{1}{2}, \frac{1}{6}, \frac{1}{10}, \frac{1}{14}, \frac{1}{18}$, _____
34. $1, \frac{1}{2}, \frac{1}{3}, \frac{1}{4}, \frac{1}{5}$, _____
35. $1, \frac{1}{3}, \frac{1}{9}, \frac{1}{27}$, _____
36. $1, \frac{1}{2}, \frac{1}{4}, \frac{1}{8}$, _____
37. 3, 7, 12, 18, 25, 33, _____
38. 2, 5, 9, 14, 20, 27, _____
39. 3, 6, 11, 18, 27, 38, _____
40. 2, 5, 10, 17, 26, 37, _____
41. 3, 7, 10, 17, 27, 44, _____
42. 2, 5, 7, 12, 19, 31, _____
43. 2, 7, 12, 5, 10, 15, 8, 13, _____
44. 3, 9, 15, 5, 11, 17, 7, 13, _____
45. 3, 6, 5, 10, 9, 18, 17, 34, _____
46. 2, 6, 5, 15, 14, 42, 41, 123, _____
47. 64, −16, 4, −1, _____
48. 125, −25, 5, −1, _____
49. $(6, 2), (0, −4), \left(7\frac{1}{2}, 3\frac{1}{2}\right), (2, −2), (3, \underline{\quad})$
50. $\left(\frac{2}{3}, \frac{4}{9}\right), \left(\frac{1}{5}, \frac{1}{25}\right), (7, 49), \left(-\frac{5}{6}, \frac{25}{36}\right), \left(-\frac{4}{7}, \underline{\quad}\right)$

In Exercises 51–54, identify a pattern in each sequence of figures. Then use the pattern to find the next figure in the sequence.

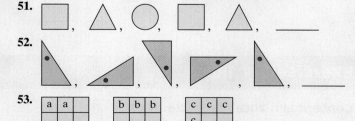

51.
52.
53.
54.

Exercises 55–58 describe procedures that are to be applied to numbers. In each exercise,

- **a.** *Repeat the procedure for four numbers of your choice. Write a conjecture that relates the result of the process to the original number selected.*

- **b.** *Use the variable n to represent the original number and use deductive reasoning to prove the conjecture in part (a).*

55. Select a number. Multiply the number by 4. Add 8 to the product. Divide this sum by 2. Subtract 4 from the quotient.

56. Select a number. Multiply the number by 3. Add 6 to the product. Divide this sum by 3. Subtract the original selected number from the quotient.

57. Select a number. Add 5. Double the result. Subtract 4. Divide by 2. Subtract the original selected number.

58. Select a number. Add 3. Double the result. Add 4. Divide by 2. Subtract the original selected number.

In Exercises 59–64, use inductive reasoning to predict the next line in each sequence of computations. Then use a calculator or perform the arithmetic by hand to determine whether your conjecture is correct.

59.
$$1 + 2 = \frac{2 \times 3}{2}$$
$$1 + 2 + 3 = \frac{3 \times 4}{2}$$
$$1 + 2 + 3 + 4 = \frac{4 \times 5}{2}$$
$$1 + 2 + 3 + 4 + 5 = \frac{5 \times 6}{2}$$

60.
$$3 + 6 = \frac{6 \times 3}{2}$$
$$3 + 6 + 9 = \frac{9 \times 4}{2}$$
$$3 + 6 + 9 + 12 = \frac{12 \times 5}{2}$$
$$3 + 6 + 9 + 12 + 15 = \frac{15 \times 6}{2}$$

61.
$$1 + 3 = 2 \times 2$$
$$1 + 3 + 5 = 3 \times 3$$
$$1 + 3 + 5 + 7 = 4 \times 4$$
$$1 + 3 + 5 + 7 + 9 = 5 \times 5$$

62.
$$\frac{1}{1 \times 2} + \frac{1}{2 \times 3} = \frac{2}{3}$$
$$\frac{1}{1 \times 2} + \frac{1}{2 \times 3} + \frac{1}{3 \times 4} = \frac{3}{4}$$
$$\frac{1}{1 \times 2} + \frac{1}{2 \times 3} + \frac{1}{3 \times 4} + \frac{1}{4 \times 5} = \frac{4}{5}$$

63.
$$9 \times 9 + 7 = 88$$
$$98 \times 9 + 6 = 888$$
$$987 \times 9 + 5 = 8888$$
$$9876 \times 9 + 4 = 88,888$$

64.
$$1 \times 9 - 1 = 8$$
$$21 \times 9 - 1 = 188$$
$$321 \times 9 - 1 = 2888$$
$$4321 \times 9 - 1 = 38,888$$

Practice Plus

In Exercises 65–66, use inductive reasoning to predict the next line in each sequence of computations. Then use a calculator or perform the arithmetic by hand to determine whether your conjecture is correct.

65.
$$33 \times 3367 = 111,111$$
$$66 \times 3367 = 222,222$$
$$99 \times 3367 = 333,333$$
$$132 \times 3367 = 444,444$$

66.
$$1 \times 8 + 1 = 9$$
$$12 \times 8 + 2 = 98$$
$$123 \times 8 + 3 = 987$$
$$1234 \times 8 + 4 = 9876$$
$$12,345 \times 8 + 5 = 98,765$$

67. Study the pattern in these examples:
$$a^2 \,\#\, a^4 = a^{10} \quad a^3 \,\#\, a^2 = a^7 \quad a^5 \,\#\, a^3 = a^{11}.$$

Select the equation that describes the pattern.

- **a.** $a^x \,\#\, a^y = a^{2x+y}$
- **b.** $a^x \,\#\, a^y = a^{x+2y}$
- **c.** $a^x \,\#\, a^y = a^{x+y+4}$
- **d.** $a^x \,\#\, a^y = a^{xy+2}$

68. Study the pattern in these examples:
$$a^5 * a^3 * a^2 = a^5 \quad a^3 * a^7 * a^2 = a^6 \quad a^2 * a^4 * a^8 = a^7.$$

Select the equation that describes the pattern.

- **a.** $a^x * a^y * a^z = a^{x+y+z}$
- **b.** $a^x * a^y * a^z = a^{\frac{xyz}{2}}$
- **c.** $a^x * a^y * a^z = a^{\frac{x+y+z}{2}}$
- **d.** $a^x * a^y * a^z = a^{\frac{xy}{2}+z}$

Application Exercises

In Exercises 69–72, identify the reasoning process, induction or deduction, in each example. Explain your answer.

69. It can be shown that
$$1 + 2 + 3 + \cdots + n = \frac{n(n+1)}{2}.$$

I can use this formula to conclude that the sum of the first one hundred counting numbers, $1 + 2 + 3 + \cdots + 100$, is
$$\frac{100(100+1)}{2} = \frac{100(101)}{2} = 50(101), \text{ or } 5050.$$

70. An HMO does a follow-up study on 200 randomly selected patients given a flu shot. None of these people became seriously ill with the flu. The study concludes that all HMO patients should be urged to get a flu shot in order to prevent a serious case of the flu.

71. The data in the graph are from a random sample of 1200 full-time four-year undergraduate college students on 100 U.S. campuses.

The Greatest Problems on Campus

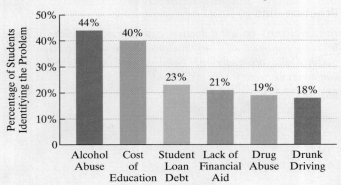

Source: Student Monitor LLC

We can conclude that there is a high probability that approximately 44% of all full-time four-year college students in the United States believe that alcohol abuse is one of the greatest problem on campus.

72. The course policy states that work turned in late will be marked down a grade. I turned in my report a day late, so it was marked down from B to C.

73. The ancient Greeks studied **figurate numbers**, so named because of their representations as geometric arrangements of points.

Triangular Numbers

1   3   6   10   15   21

Square Numbers

1   4   9   16   25

Pentagonal Numbers

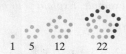

1   5   12   22

a. Use inductive reasoning to write the five triangular numbers that follow 21.

b. Use inductive reasoning to write the five square numbers that follow 25.

c. Use inductive reasoning to write the five pentagonal numbers that follow 22.

d. Use inductive reasoning to complete this statement: If a triangular number is multiplied by 8 and then 1 is added to the product, a _____ number is obtained.

74. The triangular arrangement of numbers shown below is known as **Pascal's triangle**, credited to French mathematician Blaise Pascal (1623–1662). Use inductive reasoning to find the six numbers designated by question marks.

$$\begin{array}{ccccccc} & & & 1 & & & \\ & & 1 & & 1 & & \\ & & 1 & 2 & 1 & & \\ & 1 & 3 & & 3 & 1 & \\ 1 & & 4 & 6 & 4 & & 1 \\ ? & ? & ? & ? & ? & ? \end{array}$$

Critical Thinking Exercises

75. If $(6-2)^2 = 36 - 24 + 4$ and $(8-5)^2 = 64 - 80 + 25$, use inductive reasoning to write a compatible expression for $(11-7)^2$.

76. The rectangle shows an array of nine numbers represented by combinations of the variables a, b, and c.

$a+b$	$a-b-c$	$a+c$
$a-b+c$	a	$a+b-c$
$a-c$	$a+b+c$	$a-b$

a. Determine the nine numbers in the array for $a = 10$, $b = 6$, and $c = 1$. What do you observe about the sum of the numbers in all rows, all columns, and the two diagonals?

b. Repeat part (a) for $a = 12$, $b = 5$, and $c = 2$.

c. Repeat part (a) for values of a, b, and c of your choice.

d. Use the results of parts (a) through (c) to make an inductive conjecture about the rectangular array of nine numbers represented by a, b, and c.

e. Use deductive reasoning to prove your conjecture in part (d).

77. Write a list of numbers that has two patterns so that the next number in the list can be 15 or 20.

78. a. Repeat the following procedure with at least five people. Write a conjecture that relates the result of the procedure to each person's birthday.

Take the number of the month of your birthday (January = 1, February = 2, ..., December = 12), multiply by 5, add 6, multiply this sum by 4, add 9, multiply this new sum by 5, and add the number of the day on which you were born. Finally, subtract 165.

b. Let M represent the month number and let D represent the day number of any person's birthday. Use deductive reasoning to prove your conjecture in part (a).

Technology Exercises

79. a. Use a calculator to find $6 \times 6, 66 \times 66, 666 \times 666$, and 6666×6666.

b. Describe a pattern in the numbers being multiplied and the resulting products.

c. Use the pattern to write the next two multiplications and their products. Then use your calculator to verify these results.

d. Is this process an example of inductive or deductive reasoning? Explain your answer.

80. a. Use a calculator to find $3367 \times 3, 3367 \times 6, 3367 \times 9$, and 3367×12.

b. Describe a pattern in the numbers being multiplied and the resulting products.

c. Use the pattern to write the next two multiplications and their products. Then use your calculator to verify these results.

d. Is this process an example of inductive or deductive reasoning? Explain your answer.

Group Exercise

81. Stereotyping refers to classifying people, places, or things according to common traits. Prejudices and stereotypes can function as assumptions in our thinking, appearing in inductive and deductive reasoning. For example, it is not difficult to find inductive reasoning that results in generalizations such as these, as well as deductive reasoning in which these stereotypes serve as assumptions:

> School has nothing to do with life.
>
> Intellectuals are nerds.
>
> People on welfare are lazy.

Each group member should find one example of inductive reasoning and one example of deductive reasoning in which stereotyping occurs. Upon returning to the group, present each example and then describe how the stereotyping results in faulty conjectures or prejudging situations and people.

1.2 Estimation, Graphs, and Mathematical Models

Objectives

1. Use estimation techniques to arrive at an approximate answer to a problem.

2. Apply estimation techniques to information given by graphs.

3. Develop mathematical models that estimate relationships between variables.

If present trends continue, is it possible that our descendants could live to be 200 years of age? To answer this question, we need to examine data for life expectancy and develop estimation techniques for representing the data mathematically. In this section, you will learn estimation methods that will enable you to obtain mathematical representations of data displayed by graphs, using these representations to predict what might occur in the future.

Estimation

1. **Use estimation techniques to arrive at an approximate answer to a problem.**

Estimation is the process of arriving at an approximate answer to a question. For example, companies estimate the amount of their products consumers are likely to use, and economists estimate financial trends. If you are about to cross a street, you may estimate the speed of oncoming cars so that you know whether or not to wait before crossing. Rounding numbers is also an estimation method. You might round a number without even being aware that you are doing so. You may say that you are 16 years old, rather than 16 years 5 months, or that you will be home in about a half-hour, rather than 25 minutes.

You will find estimation to be equally valuable in your work for this class. Making mistakes with a calculator or a computer is easy. Estimation can tell us whether the answer displayed for a computation makes sense.

In this section, we demonstrate several estimation methods. In the second part of the section, we apply these techniques to information given by graphs.

Performing computations using rounded numbers is one way to check whether an answer displayed by a calculator or a computer is reasonable. Rounding whole numbers depends on knowing the place values of the digits. (The digits that we use in base ten are 0, 1, 2, 3, 4, 5, 6, 7, 8, and 9.) The place that a digit occupies in a number tells us its value in that number. Here is an example using world population at 8:35 A.M. Eastern Time on January 4, 2009.

Place values

billion, hundred million, ten million, million, hundred thousand, ten thousand, thousand, hundred, ten, one

6 , 7   5   1 , 5   9   3 , 1   0   3

This number is read "six billion, seven hundred fifty-one million, five hundred ninety-three thousand, one hundred three."

Rounding Whole Numbers

1. Look at the digit to the right of the digit where rounding is to occur.
2. **a.** If the digit to the right is 5 or greater, add 1 to the digit to be rounded. Replace all digits to the right with zeros.
   b. If the digit to the right is less than 5, do not change the digit to be rounded. Replace all digits to the right with zeros.

The symbol ≈ means *is approximately equal to*. We will use this symbol when rounding numbers.

Example 1   Rounding a Whole Number

Round world population (6,751,593,103) as follows:

a. to the nearest million
b. to the nearest thousand.

Solution

a.   6,751,593,103   ≈   6,752,000,000

Millions digit, where rounding is to occur | Digit to the right is 5. | Add 1 to the digit to be rounded. | Replace all digits to the right with zeros.

World population to the nearest million is six billion, seven hundred fifty-two million.

b.   6,751,593,103   ≈   6,751,593,000

Thousands digit, where rounding is to occur | Digit to the right is less than 5. | Do not change the digit to be rounded. | Replace all digits to the right with zeros.

World population to the nearest thousand is six billion, seven hundred fifty-one million, five hundred ninety-three thousand.

✓ **Checkpoint 1**   Round world population (6,751,593,103) as follows:

a. to the nearest billion
b. to the nearest hundred million
▸ **c.** to the nearest ten thousand.

Rounding can also be applied to decimal notation, used to denote a part of a whole. Once again, the place that a digit occupies tells us its value. Here's an example using the first seven digits of the number π (pi). (We'll have more to say about π, whose digits extend endlessly with no repeating pattern, in Chapter 3.)

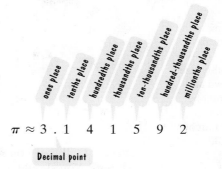

$$\pi \approx 3 \,.\, 1 \quad 4 \quad 1 \quad 5 \quad 9 \quad 2$$

Decimal point

We round the decimal part of a decimal number in nearly the same way that we round whole numbers. The only difference is that we drop the digits to the right of the rounding place, rather than replacing these digits with zeros.

Example 2 Rounding the Decimal Part of a Number

Round 3.141592, the first seven digits of π, as follows:

a. to the nearest hundredth
b. to the nearest thousandth.

Solution

a.
$$3.141592 \quad \approx \quad 3.14$$

| Hundredths digit, where rounding is to occur | Digit to the right is less than 5. | Do not change the digit to be rounded. | Drop all digits to the right. |

The number π to the nearest hundredth is three and fourteen hundredths.

b.
$$3.141592 \quad \approx \quad 3.142$$

| Thousandths digit, where rounding is to occur | Digit to the right is 5. | Add 1 to the digit to be rounded. | Drop all digits to the right. |

The number π to the nearest thousandth is three and one hundred forty-two thousandths.

Checkpoint 2   Round 3.141592, the first seven digits of π, as follows:

a. to the nearest tenth
b. to the nearest ten-thousandth.

Example 3 Estimation by Rounding

You purchased bread for $2.59, detergent for $2.17, a sandwich for $3.65, an apple for $0.47, and coffee for $5.79. The total bill was given as $18.67. Is this amount reasonable?

Solution

If you are in the habit of carrying a calculator to the store, you can answer the question by finding the exact cost of the purchase. However, estimation can be

used to determine if the bill is reasonable even if you do not have a calculator. We will round the cost of each item to the nearest dollar.

Round to the nearest dollar. → Use digits in the tenths place to do the rounding.

Bread	$2.59	≈ $3.00
Detergent	$2.17	≈ $2.00
Sandwich	$3.65	≈ $4.00
Apple	$0.47	≈ $0.00
Coffee	$5.79	≈ $6.00
		$15.00

The total bill that you were given, $18.67, seems a bit high compared to the $15.00 estimate. You should check the bill before paying it. Adding the prices of all seven items gives the true total bill of $14.67.

✓ **Checkpoint 3** You and a friend ate lunch at Ye Olde Cafe. The check for the meal showed soup for $2.40, tomato juice for $1.25, a roast beef sandwich for $4.60, a chicken salad sandwich for $4.40, two coffees totaling $1.40, apple pie for $1.85, and chocolate cake for $2.95.

a. Round the cost of each item to the nearest dollar and obtain an estimate for the food bill.

b. The total bill before tax was given as $21.85. Is this amount reasonable?

Example 4   Estimation by Rounding

A carpenter who works full time earns $28 per hour.

a. Estimate the carpenter's weekly salary.

b. Estimate the carpenter's annual salary.

Solution

a. In order to simplify the calculation, we can round the hourly rate of $28 to $30. Be sure to write out the units for each number in the calculation. The work week is 40 hours per week and the rounded salary is $30 per hour. We express this as

$$\frac{40 \text{ hours}}{\text{week}} \quad \text{and} \quad \frac{\$30}{\text{hour}}.$$

The word *per* is represented by the division bar. We multiply these two numbers to estimate the carpenter's weekly salary. We cancel out units that are identical if they are above and below the division bar.

$$\frac{40 \text{ \cancel{hours}}}{\text{week}} \times \frac{\$30}{\cancel{\text{hour}}} = \frac{\$1200}{\text{week}}$$

Thus, the carpenter earns approximately $1200 per week, written ≈$1200.

b. For the estimate of annual salary, we may round 52 weeks to 50 weeks. The annual salary is approximately the product of $1200 per week and 50 weeks per year:

$$\frac{\$1200}{\cancel{\text{week}}} \times \frac{50 \text{ \cancel{weeks}}}{\text{year}} = \frac{\$60,000}{\text{year}}.$$

Thus, the carpenter earns approximately $60,000 per year, or $60,000 annually, written ≈$60,000.

Checkpoint 4   A landscape architect who works full time earns $52 per hour.

a. Estimate the landscape architect's weekly salary. $\approx 40 \times 50 = \$2000$ per wk

b. Estimate the landscape architect's annual salary. $\approx 50 \times 2000 = \$100,000$ per yr

Estimation with Graphs

2 Apply estimation techniques to information given by graphs.

Magazines, newspapers, and websites often display information using circle, bar, and line graphs. The following examples illustrate how rounding and other estimation techniques can be applied to data displayed in each of these types of graphs.

Circle graphs, also called **pie charts**, show how a whole quantity is divided into parts. Circle graphs are divided into pieces, called **sectors**. **Figure 1.3** shows a circle graph that indicates how Americans disagree as to when "old age" begins.

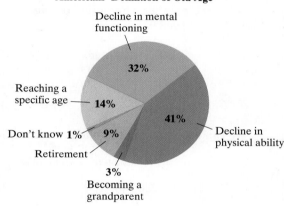

Figure 1.3
Source: American Demographics

A Brief Review • Percents

- **Percents** are the result of expressing numbers as part of 100. The word *percent* means *per hundred*. For example, the circle graph in **Figure 1.3** shows that 41% of Americans define old age by a decline in physical ability. Thus, 41 out of every 100 Americans define old age in this manner: $41\% = \frac{41}{100}$.

- To convert a number from percent form to decimal form, move the decimal point two places to the left and drop the percent sign. Example:

$$41\% = 41.\% = 0.41\%$$

Thus, $41\% = 0.41$.

- Many applications involving percent are based on the following formula:

$$\boxed{A} \text{ is } \boxed{P \text{ percent}} \text{ of } \boxed{B.}$$
$$A = P \cdot B.$$

Note that the word *of* implies multiplication.

A more thorough review of percents can be found in Section 7.1.

In our next example, we will use the information in the circle graph in **Figure 1.3** to estimate a quantity. Although different rounding results in different estimates, the whole idea behind the rounding process is to make calculations simple.

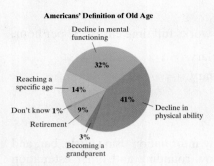

Americans' Definition of Old Age

Decline in mental functioning 32%

Reaching a specific age 14%

Don't know 1%

Retirement 9%

Becoming a grandparent 3%

Decline in physical ability 41%

Figure 1.3 (repeated)

Example 5 **Applying Estimation Techniques to a Circle Graph**

According to the U.S. Census Bureau, in 2008, there were 220,948,915 Americans 20 years and older. Assuming the circle graph in **Figure 1.3** is representative of this age group,

a. Use the appropriate information displayed by the graph to determine a calculation that shows the number of Americans 20 years and older who define old age by a decline in physical ability.

b. Use rounding to find a reasonable estimate for this calculation.

Solution

a. The circle graph in **Figure 1.3** indicates that 41% of Americans define old age by a decline in physical ability. Among the 220,948,915 Americans 20 years and older, the number who define old age in this manner is determined by finding 41% of 220,948,915.

The number of Americans 20 and older who define old age by a decline in physical ability	is	41%	of	the number of Americans 20 and older.

$$= 0.41 \times 220{,}948{,}915$$

b. We can use rounding to obtain a reasonable estimate of $0.41 \times 220{,}948{,}915$.

Round to the nearest ten million.

$$0.41 \times 220{,}948{,}915 \approx 0.4 \times 220{,}000{,}000 = 88{,}000{,}000$$

Round to the nearest tenth.

$$\begin{array}{r} 220{,}000{,}000 \\ \times \quad 0.4 \\ \hline 88{,}000{,}000.0 \end{array}$$

Our answer indicates that approximately 88,000,000 (88 million) Americans 20 years and older define old age by a decline in physical ability.

✓ **Checkpoint 5** Being aware of which appliances and activities in your home use the most energy can help you make sound decisions that allow you to decrease energy consumption and increase savings. The circle graph in **Figure 1.4** shows how energy consumption is distributed throughout a typical home.

Suppose that last year your family spent $2148.72 on natural gas and electricity. Assuming the circle graph in **Figure 1.4** is representative of your family's energy consumption,

a. Use the appropriate information displayed by the graph to determine a calculation that shows the amount your family spent on heating and cooling for the year.

b. Use rounding to find a reasonable estimate for this calculation.

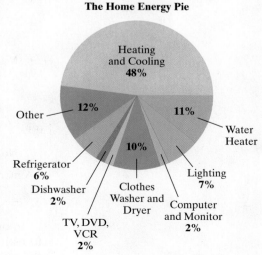

The Home Energy Pie

Heating and Cooling 48%

Other 12%

Water Heater 11%

Refrigerator 6%

Dishwasher 2%

TV, DVD, VCR 2%

Clothes Washer and Dryer 10%

Computer and Monitor 2%

Lighting 7%

Figure 1.4
Source: Natural Home and Garden

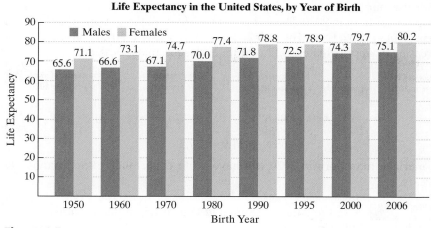

Figure 1.5
Source: National Center for Health Statistics

Bar **graphs** are convenient for comparing some measurable attribute of various items. The bars may be either horizontal or vertical, and their heights or lengths are used to show the amounts of different items. **Figure 1.5** is an example of a typical bar graph. The graph shows life expectancy for American men and American women born in various years from 1950 through 2006.

Example 6 Applying Estimation and Inductive Reasoning to Data in a Bar Graph

Use the data for men in **Figure 1.5** to estimate each of the following:

a. a man's increased life expectancy, rounded to the nearest hundredth of a year, for each subsequent birth year

b. the life expectancy of a man born in 2020.

Solution

a. One way to estimate increased life expectancy for each subsequent birth year is to generalize from the information given for 1950 (male life expectancy: 65.6 years) and for 2006 (male life expectancy: 75.1 years). The average yearly increase in life expectancy is the change in life expectancy from 1950 to 2006 divided by the change in time from 1950 to 2006.

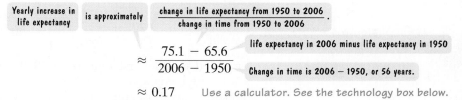

$$\approx \frac{75.1 - 65.6}{2006 - 1950}$$

$$\approx 0.17 \quad \text{Use a calculator. See the technology box below.}$$

For each subsequent birth year, a man's life expectancy is increasing by approximately 0.17 year.

Technology

Here is the calculator keystroke sequence needed to perform the computation in Example 6(a).

$$(\; 75.1 \; - \; 65.6 \;) \; \div \; (\; 2006 \; - \; 1950 \;).$$

Press $=$ on a scientific calculator or ENTER on a graphing calculator to display the answer. As specified, we round to the nearest hundredth.

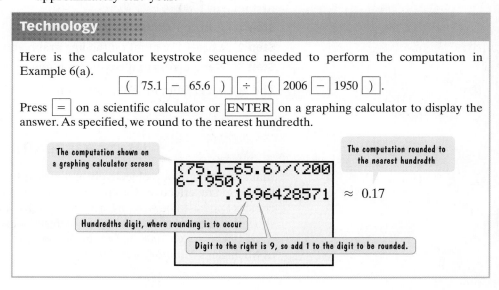

b. We can use our computation in part (a) to estimate the life expectancy of an American man born in 2020. The bar graph indicates that men born in 1950 had a life expectancy of 65.6 years. The year 2020 is 70 years after 1950 and life expectancy is increasing by approximately 0.17 year for each subsequent birth year.

Life expectancy for a man born in 2020	is approximately	life expectancy for a man born in 1950	plus	yearly increase in life expectancy	times	the number of years from 1950 to 2020.

$$\approx 65.6 \quad + \quad 0.17 \quad \times \quad 70$$

$$= 65.6 + 11.9 = 77.5$$

An American man born in 2020 will have a life expectancy of approximately 77.5 years.

Checkpoint 6   Use the data for women in **Figure 1.5** on the previous page to estimate each of the following:

a. a woman's increased life expectancy, rounded to the nearest hundredth of a year, for each subsequent birth year

b. the life expectancy, to the nearest tenth of a year, of a woman born in 2050.

Women's Average Age of First Marriage

Figure 1.6
Source: U.S. Census Bureau

Line graphs are often used to illustrate trends over time. Some measure of time, such as months or years, frequently appears on the horizontal axis. Amounts are generally listed on the vertical axis. Points are drawn to represent the given information. The graph is formed by connecting the points with line segments.

Figure 1.6 is an example of a typical line graph. The graph shows the average age at which women in the United States married for the first time from 1890 through 2008. The years are listed on the horizontal axis and the ages are listed on the vertical axis. The symbol ⤳ on the vertical axis shows that there is a break in values between 0 and 20. Thus, the first tick mark on the vertical axis represents an average age of 20.

Figure 1.6 shows how to find the average age at which women married for the first time in 1980.

Step 1   Locate 1980 on the horizontal axis.

Step 2   Locate the point on the line graph above 1980.

Step 3   Read across to the corresponding age on the vertical axis.

The age is 22. Thus, in 1980, women in the United States married for the first time at an average age of 22.

Example 7   Using a Line Graph

Math is a way to look objectively at issues in society to decide if campaigns to change behavior are effective. Campaigns urging high school students to make healthier choices appear to be working. The line graphs in **Figure 1.7** at the top of the next page show the decreasing percentage of high school seniors who used alcohol or cigarettes from 1980 through 2008 during the 30 days prior to being surveyed for the University of Michigan's *Monitoring the Future* study.

a. Find an estimate for the percentage of seniors who smoked cigarettes in 1995.

b. In which five-year period did the percentage of seniors who smoked cigarettes decrease at the greatest rate?

c. In which year labeled on the horizontal axis did 50% of the seniors use alcohol?

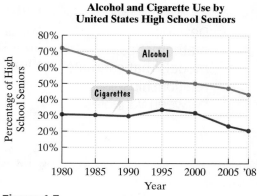

Figure 1.7
Source: University of Michigan

Solution

a. Estimating the Percentage Smoking Cigarettes in 1995

b. Identifying the Period of the Greatest Rate of Decreasing Cigarette Use

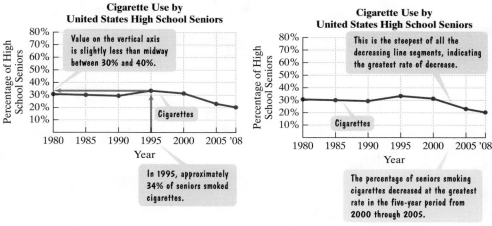

c. Identifying the Year When 50% of Seniors Used Alcohol

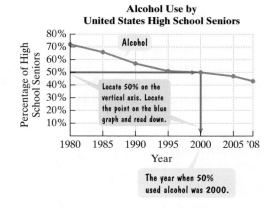

Checkpoint 7   Use the line graphs in **Figure 1.7** to solve this exercise.

a. Find an estimate for the percentage of seniors who used alcohol in 1985.

b. In which five-year period did the percentage of seniors who smoked cigarettes increase?

c. In which year labeled on the horizontal axis did approximately 57% of the seniors use alcohol?

Mathematical Models

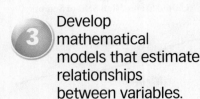

Develop mathematical models that estimate relationships between variables.

We have seen that American men born in 1950 have a life expectancy of 65.6 years, increasing by approximately 0.17 year for each subsequent birth year. We can use variables to express the life expectancy, E, for American men born x years after 1950.

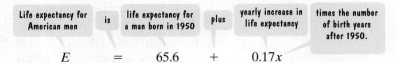

| Life expectancy for American men | is | life expectancy for a man born in 1950 | plus | yearly increase in life expectancy | times the number of birth years after 1950. |

$$E \quad = \quad 65.6 \quad + \quad 0.17x$$

A **formula** is a statement of equality that uses letters to express a relationship between two or more variables. Thus, $E = 65.6 + 0.17x$ is a formula describing life expectancy, E, for American men born x years after 1950. Be aware that this formula provides *estimates* of life expectancy, as shown in **Table 1.2**.

Table 1.2 Comparing Given Data with Estimates Determined by a Formula

Birth Year	Life Expectancy: Given Data	Life Expectancy: Formula Estimate $E = 65.6 + 0.17x$
1950	65.6	$E = 65.6 + 0.17(0) = 65.6 + 0 = 65.6$
1960	66.6	$E = 65.6 + 0.17(10) = 65.6 + 1.7 = 67.3$
1970	67.1	$E = 65.6 + 0.17(20) = 65.6 + 3.4 = 69.0$
1980	70.0	$E = 65.6 + 0.17(30) = 65.6 + 5.1 = 70.7$
1990	71.8	$E = 65.6 + 0.17(40) = 65.6 + 6.8 = 72.4$
1995	72.5	$E = 65.6 + 0.17(45) = 65.6 + 7.65 = 73.25$
2000	74.3	$E = 65.6 + 0.17(50) = 65.6 + 8.5 = 74.1$
2006	75.1	$E = 65.6 + 0.17(56) = 65.6 + 9.52 = 75.12$

In each row, we substitute the number of years after 1950 for x. The better estimates occur in 1950, 2000, and 2006.

The process of finding formulas to describe real-world phenomena is called **mathematical modeling**. Such formulas, together with the meaning assigned to the variables, are called **mathematical models**. We often say that these formulas model, or describe, the relationships among the variables.

Example 8 Modeling the Cost of Attending a Public College

The bar graph in **Figure 1.8** at the top of the next page shows the average cost of tuition and fees for public four-year colleges, adjusted for inflation.

a. Estimate the yearly increase in tuition and fees. Round to the nearest dollar.

b. Write a mathematical model that estimates the average cost of tuition and fees, T, at public four-year colleges for the school year ending x years after 2000.

c. Use the mathematical model from part (b) to project the average cost of tuition and fees at public four-year colleges for the school year ending in 2014.

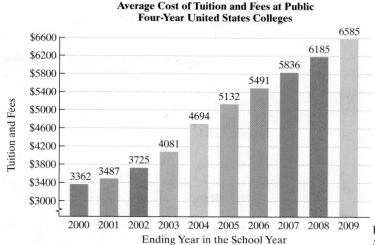

Average Cost of Tuition and Fees at Public Four-Year United States Colleges

Figure 1.8
Source: The College Board

Solution

a. We can use the data from 2000 and 2009 to estimate the yearly increase in tuition and fees.

> | Yearly increase in tuition and fees | is approximately | change in tuition and fees from 2000 to 2009 / change in time from 2000 to 2009 |

$$\approx \frac{6585 - 3362}{2009 - 2000}$$

$$= \frac{3223}{9} \approx 358$$

Each year the average cost of tuition and fees for public four-year colleges is increasing by approximately $358.

b. Now we can use variables to obtain a mathematical model that estimates the average cost of tuition and fees, T, for the school year ending x years after 2000.

> | The average cost of tuition and fees | is | tuition and fees in 2000 | plus | yearly increase in tuition and fees | times the number of years after 2000. |

$$T = 3362 + 358x$$

The mathematical model $T = 3362 + 358x$ estimates the average cost of tuition and fees, T, at public four-year colleges for the school year ending x years after 2000.

c. Now let's use the mathematical model to project the average cost of tuition and fees for the school year ending in 2014. Because 2014 is 14 years after 2000, we substitute 14 for x.

$T = 3362 + 358x$	This is the mathematical model from part (b).
$T = 3362 + 358(14)$	Substitute 14 for x.
$= 3362 + 5012$	Multiply: 358(14) = 5012.
$= 8374$	Add. On a calculator, enter 3362 + 358 × 14 and press = or ENTER.

Our model projects that the average cost of tuition and fees at public four-year colleges for the school year ending in 2014 will be $8374.

✓ Checkpoint 8   The bar graph in **Figure 1.9** shows the average cost of tuition and fees for private four-year colleges, adjusted for inflation.

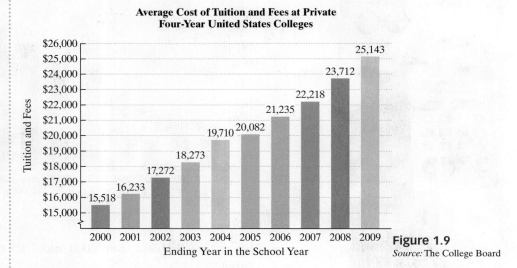

Average Cost of Tuition and Fees at Private Four-Year United States Colleges

Figure 1.9
Source: The College Board

a. Estimate the yearly increase in tuition and fees. Round to the nearest dollar.

b. Write a mathematical model that estimates the average cost of tuition and fees, T, at private four-year colleges for the school year ending x years after 2000.

c. Use the mathematical model from part (b) to project the average cost of tuition and fees at private four-year colleges for the school year ending in 2012.

Sometimes a mathematical model gives an estimate that is not a good approximation or is extended to include values of the variable that do not make sense. In these cases, we say that **model breakdown** has occurred. Models that accurately describe data for the past ten years might not serve as reliable predictions for what can reasonably be expected to occur in the future. Model breakdown can occur when formulas are extended too far into the future.

Achieving Success

Check! Check! Check!

After completing each Checkpoint or odd-numbered exercise, compare your answer with the one given in the answer section at the back of the book. To make this process more convenient, place a Post-it® or some other marker at the appropriate page of the answer section. If your answer is different from the one given in the answer section, try to figure out your mistake. Then correct the error. If you cannot determine what went wrong, show your work to your teacher. **By recording each step neatly and using as much paper as you need**, your teacher will find it easier to determine where you had trouble.

Exercise Set 1.2

Concept and Vocabulary Exercises

In Exercises 1–3, fill in each blank so that the resulting statement is true.

1. The process of arriving at an approximate answer to a computation such as 0.79×403 is called _____.

2. A graph that shows how a whole quantity is divided into parts is called a/an _____.

3. A formula that approximates real-world phenomena is called a/an _____.

In Exercises 4–6, determine whether each statement is true or false. If the statement is false, make the necessary change(s) to produce a true statement.

4. Decimal numbers are rounded by using the digit to the right of the digit where rounding is to occur.

5. Line graphs are often used to illustrate trends over time.

6. Mathematical modeling results in formulas that give exact values of real-world phenomena over time.

Respond to Exercises 7–18 using verbal or written explanations.

7. What is estimation? When is it helpful to use estimation?

8. Explain how to round 218,543 to the nearest thousand and to the nearest hundred thousand.

9. Explain how to round 14.26841 to the nearest hundredth and to the nearest thousandth.

10. What does the $\approx$ symbol mean?

11. In this era of calculators and computers, why is there a need to develop estimation skills?

12. Describe a circle graph.

13. Describe a bar graph.

14. Describe a line graph.

15. What does it mean when we say that a formula models real-world phenomena?

16. In 1999, the average movie ticket price was $5.08. This has increased by approximately $0.24 per year. Describe how to use this information to write a mathematical model that estimates the average movie ticket price, T, in dollars, x years after 1999.

17. Explain how to use the mathematical model described in Exercise 16 to project the average movie ticket price in 2020.

18. Describe one way in which you use estimation in a nonacademic area of your life.

In Exercises 19–22, determine whether each statement makes sense or does not make sense, and explain your reasoning.

19. When buying several items at the market, I use estimation before going to the cashier to be sure I have enough money to pay for the purchase.

20. It's not necessary to use estimation skills when using my calculator.

21. Being able to compute an exact answer requires a different ability than estimating the reasonableness of the answer.

22. My mathematical model estimates the data for the past ten years extremely well, so it will serve as an accurate prediction for what will occur in 2050.

Practice Exercises

The bar graph gives the 2008 populations of the ten most populous states in the United States. Use the appropriate information displayed by the graph to solve Exercises 23–24.

Population by State of the Ten Most Populace States

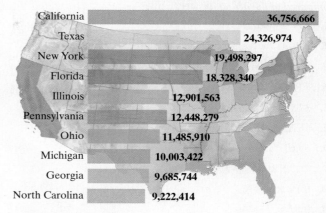

State	Population
California	36,756,666
Texas	24,326,974
New York	19,498,297
Florida	18,328,340
Illinois	12,901,563
Pennsylvania	12,448,279
Ohio	11,485,910
Michigan	10,003,422
Georgia	9,685,744
North Carolina	9,222,414

Source: U.S. Census Bureau

23. Round the population of New York to the nearest **a.** hundred, **b.** thousand, **c.** ten thousand, **d.** hundred thousand, **e.** million, **f.** ten million.

24. Select any two states other than New York. For each state selected, round the population to the nearest **a.** hundred, **b.** thousand, **c.** ten thousand, **d.** hundred thousand, **e.** million, **f.** ten million.

> *Pi goes on and on and on …*
> *And e is just as cursed.*
> *I wonder: Which is larger*
> *When their digits are reversed?*
>
> *Martin Gardner*

Although most people are familiar with π, the number e is more significant in mathematics, showing up in problems involving population growth and compound interest, and at the heart of the statistical bell curve. One way to think of e is the dollar amount you would have in a savings account at the end of the year if you invested $1 at the beginning of the year and the bank paid an annual interest rate of 100% compounded continuously (compounding interest every trillionth of a second, every quadrillionth of a second, etc.). Although continuous compounding sounds terrific, at the end of the year your $1 would have grown to a mere $e, or $2.72, rounded to the nearest cent. Here is a better approximation for e.

$$e \approx 2.718281828459045$$

In Exercises 25–30, use this approximation to round e as specified.

25. to the nearest thousandth

26. to the nearest ten-thousandth

27. to the nearest hundred-thousandth

28. to the nearest millionth

29. to nine decimal places

30. to ten decimal places

In Exercises 31–56, because different rounding results in different estimates, there is not one single, correct answer to each exercise.

In Exercises 31–44, obtain an estimate for each computation by rounding the numbers so that the resulting arithmetic can easily be performed by hand or in your head. Then use a calculator to perform the computation. How reasonable is your estimate when compared to the actual answer?

31. 359 + 596

32. 248 + 797

33. 8.93 + 1.04 + 19.26

34. 7.92 + 3.06 + 24.36

35. 32.15 − 11.239

36. 46.13 − 15.237

37. 39.67 × 5.5

38. 78.92 × 6.5

39. 0.79 × 414

40. 0.67 × 211

41. 47.83 ÷ 2.9

42. 54.63 ÷ 4.7

43. 32% of 187,253

44. 42% of 291,506

In Exercises 45–56, determine each estimate without using a calculator. Then use a calculator to perform the computation necessary to obtain an exact answer. How reasonable is your estimate when compared to the actual answer?

45. Estimate the total cost of six grocery items if their prices are $3.47, $5.89, $19.98, $2.03, $11.85, and $0.23.

46. Estimate the total cost of six grocery items if their prices are $4.23, $7.79, $28.97, $4.06, $13.43, and $0.74.

47. A full-time employee who works 40 hours per week earns $19.50 per hour. Estimate that person's annual income.

48. A full-time employee who works 40 hours per week earns $29.85 per hour. Estimate that person's annual income.

49. You lease a car at $605 per month for three years. Estimate the total cost of the lease.

50. You lease a car at $415 per month for four years. Estimate the total cost of the lease.

51. A raise of $310,000 is evenly distributed among 294 teachers. Estimate the amount each teacher receives.

52. A raise of $310,000 is evenly distributed among 196 teachers. Estimate the amount each teacher receives.

53. If a person who works 40 hours per week earns $61,500 per year, estimate that person's hourly wage.

54. If a person who works 40 hours per week earns $38,950 per year, estimate that person's hourly wage.

55. The average life expectancy in Canada is 80.1 years. Estimate the country's life expectancy in hours.

56. The average life expectancy in Mozambique is 40.3 years. Estimate the country's life expectancy in hours.

Practice Plus

In Exercises 57–58, obtain an estimate for each computation without using a calculator. Then use a calculator to perform the computation. How reasonable is your estimate when compared to the actual answer?

57. $\dfrac{0.19996 \times 107}{0.509}$

58. $\dfrac{0.47996 \times 88}{0.249}$

59. Ten people ordered calculators. The least expensive was $19.95 and the most expensive was $39.95. Half ordered a $29.95 calculator. Select the best estimate of the amount spent on calculators.
   a. $240   **b.** $310   **c.** $345   **d.** $355

60. Ten people ordered calculators. The least expensive was $4.95 and the most expensive was $12.95. Half ordered a $6.95 calculator. Select the best estimate of the amount spent on calculators.
   a. $160   **b.** $105   **c.** $75   **d.** $55

61. Traveling at an average rate of between 60 and 70 miles per hour for 3 to 4 hours, select the best estimate for the distance traveled.
   a. 90 miles   **b.** 190 miles   **c.** 225 miles   **d.** 275 miles

62. Traveling at an average rate of between 40 and 50 miles per hour for 3 to 4 hours, select the best estimate for the distance traveled.
   a. 120 miles   **b.** 160 miles   **c.** 195 miles   **d.** 210 miles

63. Imagine that you counted 60 numbers per minute and continued to count nonstop until you reached 10,000. Determine a reasonable estimate of the number of hours it would take you to complete the counting.

64. Imagine that you counted 60 numbers per minute and continued to count nonstop until you reached one million. Determine a reasonable estimate of the number of days it would take you to complete the counting.

Application Exercises

The circle graph shows the percentage of the 221,730,462 American adults who drink caffeinated beverages on a daily basis and the number of cups consumed per day. Use this information to solve Exercises 65–66.

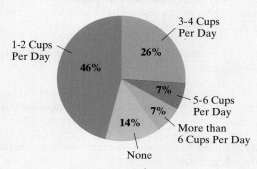

A Wired Nation: Percentage of American Adults Drinking Caffeinated Beverages

Source: Harris Interactive

65. Without using a calculator, estimate the number of American adults who drink from one to two cups of caffeinated beverages per day.

66. Without using a calculator, estimate the number of American adults who drink from three to four cups of caffeinated beverages per day.

An online test of English spelling looked at how well people spelled difficult words. The bar graph shows how many people per 100 spelled each word correctly. Use this information to solve Exercises 67–68.

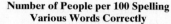
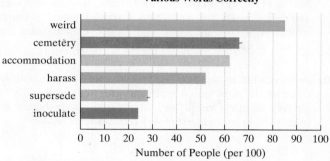

Number of People per 100 Spelling Various Words Correctly

Source: Vivian Cook, *Accomodating Brocolli in the Cemetary or Why Can't Anybody Spell?*, Simon and Schuster, 2004.

67. a. Estimate the number of people per 100 who spelled *weird* correctly.

    b. In a group consisting of 8729 randomly selected people, estimate how many more people can correctly spell *weird* than *inoculate*.

68. a. Estimate the number of people per 100 who spelled *cemetery* correctly.

    b. In a group consisting of 7219 randomly selected people, estimate how many more people can correctly spell *cemetery* than *supersede*.

Americans are getting married later in life, or not getting married at all. In 2008, more than half of Americans ages 25 through 29 were unmarried. The bar graph shows the percentage of never-married men and women in this age group. Use this information to solve Exercises 69–70.

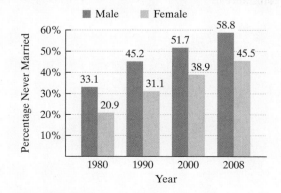

Percentage of United States Population Never Married, Ages 25–29

Source: U.S. Census Bureau

69. a. Estimate the average yearly increase in the percentage of never-married American males ages 25–29. Round the percent to the nearest tenth.

    b. Estimate the percentage of never-married American males ages 25–29 in 2010.

70. a. Estimate the average yearly increase in the percentage of never-married American females ages 25–29. Round the percent to the nearest tenth.

    b. Estimate the percentage of never-married American females ages 25–29 in 2010.

The graph shows the number, in millions, of men and women participating in high school athletics for selected years from 1976 through 2008. Use this information to solve Exercises 71–72.

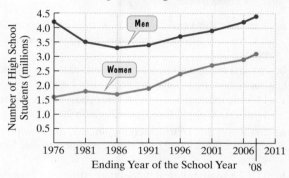

Participation in High School Athletics

Source: National Federation of State High School Associations Athletics Participation

71. a. For which year did the fewest number of men participate in high school athletics? Estimate the number of men who participated for that year. Express the answer in millions, rounded to one decimal place.

    b. In which five-year period did the number of women participating in high school athletics increase at the greatest rate?

    c. For which year on the graph did 3.5 million men participate in high school athletics?

    d. For which year on the graph was there the greatest difference between men's and women's participation?

72. a. For which year did the greatest number of men participate in high school athletics? Estimate the number of men who participated for that year. Express the answer in millions, rounded to one decimal place.

    b. In which five-year period did the number of men participating in high school athletics decrease at the greatest rate?

    c. For which year on the graph did approximately 2.4 million women participate in high school athletics?

    d. Approximately how many more women participated in high school athletics in 2008 than in 1976? Express the answer in millions, rounded to one decimal place.

The bar graph shows the population of the United States, in millions, for five selected years. Use this information to solve Exercises 73–74.

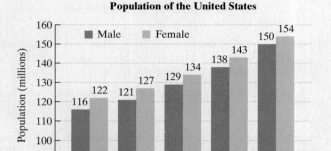

Population of the United States

Source: U.S. Census Bureau

73. a. Estimate the yearly increase in the female population. Express the answer in millions, rounded to one decimal place.

   b. Write a mathematical model that estimates the female U.S. population, F, in millions, x years after 1985.

   c. Use the mathematical model from part (b) to project the female U.S. population, in millions, in 2020.

74. a. Estimate the yearly increase in the male population. Express the answer in millions, rounded to one decimal place.

   b. Write a mathematical model that estimates the male U.S. population, M, in millions, x years after 1985.

   c. Use the mathematical model from part (b) to project the male U.S. population, in millions, in 2020.

Critical Thinking Exercises

In Exercises 75–78, match the story with the correct graph. The graphs are labeled (a), (b), (c), and (d).

75. As the blizzard got worse, the snow fell harder and harder.

76. The snow fell more and more softly.

77. It snowed hard, but then it stopped. After a short time, the snow started falling softly.

78. It snowed softly, and then it stopped. After a short time, the snow started falling hard.

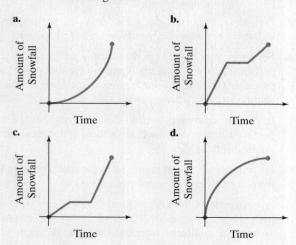

79. American children ages 2 to 17 spend 19 hours 40 minutes per week watching television. (*Source:* TV-Turnoff Network) From ages 2 through 17, inclusive, estimate the number of days an American child spends watching television. How many years, to the nearest tenth of a year, is that?

80. If you spend $1000 each day, estimate how long it will take to spend a billion dollars.

81. Take a moment to read the verse preceding Exercises 25–30 that mentions the numbers π and e, whose decimal representations continue infinitely with no repeating patterns. The verse was written by the American mathematician (and accomplished amateur magician!) Martin Gardner (1914–2010), author of more than 60 books, and best known for his "Mathematical Games" column, which ran in *Scientific American* for 25 years. Explain the humor in Gardner's question.

82. A forecaster at the National Hurricane Center needs to estimate the time until a hurricane with high probability of striking South Florida will hit Miami. Is it better to overestimate or underestimate? Explain your answer.

Group Exercises

83. Group members should devise an estimation process that can be used to answer each of the following questions. Use input from all group members to describe the best estimation process possible.

   a. Is it possible to walk from San Francisco to New York in a year?

   b. How much money is spent on ice cream in the United States each year?

84. Group members should begin by consulting an almanac, newspaper, magazine, or the Internet to find two graphs that show "intriguing" data changing from year to year. In one graph, the data values should be increasing relatively steadily. In the second graph, the data values should be decreasing relatively steadily. For each graph selected, write a mathematical model that estimates the changing variable x years after the graph's starting date. Then use each mathematical model to make predictions about what might occur in the future. Are there circumstances that might affect the accuracy of the prediction? List some of these circumstances.

1.3 : Problem Solving

Critical thinking and problem solving are essential skills in both school and work. A model for problem solving was established by the charismatic teacher and mathematician George Polya (1887–1985) in *How to Solve It* (Princeton University Press, Princeton, NJ, 1957). This book, first published in 1945, has sold more than one million copies and is available in 17 languages. Using a four-step procedure for problem solving, Polya's book demonstrates how to think clearly in any field.

"If you don't know where you're going, you'll probably end up some place else."

Yogi Berra

Polya's Four Steps in Problem Solving

Step 1   Understand the problem. Read the problem several times. The first reading can serve as an overview. In the second reading, write down what information is given and determine exactly what it is that the problem requires you to find.

Step 2   Devise a plan. The plan for solving the problem might involve one or more of these suggested problem-solving strategies:

- Use inductive reasoning to look for a pattern.
- Make a systematic list or a table.
- Use estimation to make an educated guess at the solution. Check the guess against the problem's conditions and work backward to eventually determine the solution.
- Try expressing the problem more simply and solve a similar simpler problem.
- Use trial and error.
- List the given information in a chart or table.
- Try making a sketch or a diagram to illustrate the problem.
- Relate the problem to a similar problem that you have seen before. Try applying the procedures used to solve the similar problem to the new one.
- Look for a "catch" if the answer seems too obvious. Perhaps the problem involves some sort of trick question deliberately intended to lead the problem solver in the wrong direction.
- Use the given information to eliminate possibilities.
- Use common sense.

Step 3   Carry out the plan and solve the problem.

Step 4   Look back and check the answer. The answer should satisfy the conditions of the problem. The answer should make sense and be reasonable. If this is not the case, recheck the method and any calculations. Perhaps there is an alternate way to arrive at a correct solution.

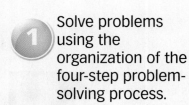

Solve problems using the organization of the four-step problem-solving process.

"For every complex problem, there is a solution that is simple, neat, and wrong."

H. L. Mencken

The very first step in problem solving involves evaluating the given information in a deliberate manner. Is there enough given to solve the problem? Is the information relevant to the problem's solution or are some facts not necessary to arrive at a solution?

Example 1   Finding What Is Missing

Which necessary piece of information is missing and prevents you from solving the following problem?

A man purchased five shirts, each at the same discount price. How much did he pay for them?

Solution

Step 1   Understand the problem. Here's what is given:

Number of shirts purchased: 5.

We must find how much the man paid for the five shirts.

Step 2   Devise a plan. The amount that the man paid for the five shirts is the number of shirts, 5, times the cost of each shirt. The discount price of each shirt is not given. This missing piece of information makes it impossible to solve the problem.

Checkpoint 1   Which necessary piece of information is missing and prevents you from solving the following problem?

The bill for your meal totaled $20.36, including the tax. How much change should you receive from the cashier?

Example 2   Finding What Is Unnecessary

In the following problem, one more piece of information is given than is necessary for solving the problem. Identify this unnecessary piece of information. Then solve the problem.

A roll of E-Z Wipe paper towels contains 100 sheets and costs $1.38. A comparable brand, Kwik-Clean, contains five dozen sheets per roll and costs $1.23. If you need three rolls of paper towels, which brand is the better value?

Solution

Step 1   Understand the problem. Here's what is given:

E-Z Wipe: 100 sheets per roll; $1.38
Kwik-Clean: 5 dozen sheets per roll; $1.23
Needed: 3 rolls.

We must determine which brand offers the better value.

Step 2   Devise a plan. The brand with the better value is the one that has the lower price per sheet. Thus, we can compare the two brands by finding the cost for one sheet of E-Z Wipe and one sheet of Kwik-Clean. The price per sheet, or the *unit price*, is the price of a roll divided by the number of sheets in the roll. The fact that three rolls are required is not relevant to the problem. This unnecessary piece of information is not needed to find which brand is the better value.

Step 3   Carry out the plan and solve the problem.

E-Z Wipe:

$$\text{price per sheet} = \frac{\text{price of a roll}}{\text{number of sheets per roll}}$$

$$= \frac{\$1.38}{100 \text{ sheets}} = \$0.0138 \approx \$0.01$$

Kwik-Clean:

$$\text{price per sheet} = \frac{\text{price of a roll}}{\text{number of sheets per roll}}$$

$$= \frac{\$1.23}{60 \text{ sheets}} = \$0.0205 \approx \$0.02$$

> 5 dozen = 5 × 12, or 60 sheets

By comparing unit prices, we see that E-Z Wipe, at approximately \$0.01 per sheet, is the better value.

Step 4   Look back and check the answer. We can double-check the arithmetic in each of our unit-price computations. We can also see if these unit prices satisfy the problem's conditions. The product of each brand's price per sheet and the number of sheets per roll should result in the given price for a roll.

E-Z Wipe: Check \$0.0138.   Kwik-Clean: Check \$0.0205.

$0.0138 \times 100 = \$1.38$ $0.0205 \times 60 = \$1.23$

> These are the given prices for a roll of each respective brand.

The unit prices satisfy the problem's conditions.

A generalization of our work in Example 2 allows you to compare different brands and make a choice between various products of different sizes. When shopping at the supermarket, a useful number to keep in mind is a product's *unit price*. The **unit price** is the total price divided by the total units. Among comparable brands, the best value is the product with the lowest unit price, assuming that the units are kept uniform.

The word *per* is used to state unit prices. For example, if a 12-ounce box of cereal sells for \$3.00, its unit price is determined as follows:

$$\text{Unit price} = \frac{\text{total price}}{\text{total units}} = \frac{\$3.00}{12 \text{ ounces}} = \$0.25 \text{ per ounce.}$$

Checkpoint 2   Solve the following problem. If the problem contains information that is not relevant to its solution, identify this unnecessary piece of information.

A manufacturer packages its apple juice in bottles and boxes. A 128-ounce bottle costs \$5.39 and a 9-pack of 6.75-ounce boxes costs \$3.15. Which packaging option is the better value?

Example 3   Applying the Four-Step Procedure

By paying \$100 cash up front and the balance at \$20 a week, how long will it take to pay for a bicycle costing \$680?

Solution

Step 1   Understand the problem. Here's what is given:

Cost of the bicycle: \$680

Amount paid in cash: \$100

Weekly payments: \$20.

Cost of the bicycle: $680
Amount paid in cash: $100
Weekly payments: $20

The given information (repeated)

If necessary, consult a dictionary to look up any unfamiliar words. The word *balance* means the amount still to be paid. We must find the balance to determine the number of weeks required to pay off the bicycle.

Step 2   Devise a plan. Subtract the amount paid in cash from the cost of the bicycle. This results in the amount still to be paid. Because weekly payments are $20, divide the amount still to be paid by 20. This will give the number of weeks required to pay for the bicycle.

Step 3   Carry out the plan and solve the problem. Begin by finding the balance, the amount still to be paid for the bicycle.

$680   cost of the bicycle
−$100   amount paid in cash
$580   amount still to be paid

Now divide the $580 balance by $20, the payment per week. The result of the division is the number of weeks needed to pay off the bicycle.

$$\frac{\$580}{\frac{\$20}{week}} = \$580 \times \frac{week}{\$20} = \frac{580\,weeks}{20} = 29\,weeks$$

It will take 29 weeks to pay for the bicycle.

Step 4   Look back and check the answer. We can certainly double-check the arithmetic either by hand or with a calculator. We can also see if the answer, 29 weeks to pay for the bicycle, satisfies the condition that the bicycle costs $680.

$20   weekly payments
× 29   number of weeks
$580   total of weekly payments

$580   total of weekly payments
+$100   amount paid in cash
$680   cost of bicycle

The answer of 29 weeks satisfies the condition that the cost of the bicycle is $680.

Checkpoint 3   By paying $350 cash up front and the balance at $45 per month, how long will it take to pay for a computer costing $980?

Making lists is a useful strategy in problem solving.

Study Tip

An effective way to see if you understand a problem is to restate the problem in your own words.

Example 4   Solving a Problem by Making a List

Suppose you are an engineer programming the automatic gate for a 50-cent toll. The gate should accept exact change only. It should not accept pennies. How many coin combinations must you program the gate to accept?

Solution

Step 1   Understand the problem. The total change must always be 50 cents. One possible coin combination is two quarters. Another is five dimes. We need to count all such combinations.

Step 2   Devise a plan. Make a list of all possible coin combinations. Begin with the coins of larger value and work toward the coins of smaller value.

Step 3   Carry out the plan and solve the problem. First we must find all of the coins that are not pennies but can combine to form 50 cents. This includes half-dollars, quarters, dimes, and nickels. Now we can set up a table. We will use these coins as table headings.

Half-Dollars	Quarters	Dimes	Nickels

Each row in the table will represent one possible combination for exact change. We start with the largest coin, the half-dollar. Only one half-dollar is needed to make exact change. No other coins are needed. Thus, we put a 1 in the half-dollars column and 0s in the other columns to represent the first possible combination.

Half-Dollars	Quarters	Dimes	Nickels
1	0	0	0

Likewise, two quarters are also exact change for 50 cents. We put a 0 in the half-dollars column, a 2 in the quarters column, and 0s in the columns for dimes and nickels.

Half-Dollars	Quarters	Dimes	Nickels
1	0	0	0
0	2	0	0

In this manner, we can find all possible combinations for exact change for the 50-cent toll. These combinations are shown in **Table 1.3**.

Table 1.3 Exact Change for 50 Cents: No Pennies

Half-Dollars	Quarters	Dimes	Nickels
1	0	0	0
0	2	0	0
0	1	2	1
0	1	1	3
0	1	0	5
0	0	5	0
0	0	4	2
0	0	3	4
0	0	2	6
0	0	1	8
0	0	0	10

Count the coin combinations shown in **Table 1.3**. How many coin combinations must the gate accept? You must program the gate to accept 11 coin combinations.

Step 4   Look back and check the answer. Double-check **Table 1.3** to make sure that no possible combinations have been omitted and that the total in each row is 50 cents. Double-check your count of the number of combinations.

Checkpoint 4   Suppose you are an engineer programming the automatic gate for a 30-cent toll. The gate should accept exact change only. It should not accept pennies. How many coin combinations must you program the gate to accept?

Sketches and diagrams are sometimes useful in problem solving.

Example 5 Solving a Problem by Using a Diagram

Four runners are in a one-mile race: Maria, Aretha, Thelma, and Debbie. Points are awarded only to the women finishing first or second. The first-place winner gets more points than the second-place winner. How many different arrangements of first- and second-place winners are possible?

Solution

Step 1 Understand the problem. Three possibilities for first and second position are

Maria-Aretha

Maria-Thelma

Aretha-Maria.

Notice that Maria finishing first and Aretha finishing second is a different outcome than Aretha finishing first and Maria finishing second. Order makes a difference because the first-place winner gets more points than the second-place winner. We must count all possibilities for first and second position.

Step 2 Devise a plan. If Maria finishes first, then each of the other three runners could finish second:

First place	Second place	Possibilities for first and second place
Maria	Aretha	Maria-Aretha
	Thelma	Maria-Thelma
	Debbie	Maria-Debbie

Similarly, we can list each woman as the possible first-place runner. Then we will list the other three women as possible second-place runners. Next we will determine the possibilities for first and second place. This diagram will show how the runners can finish first or second.

Step 3 Carry out the plan and solve the problem. Now we complete the diagram started in step 2. The diagram is shown in **Figure 1.10**.

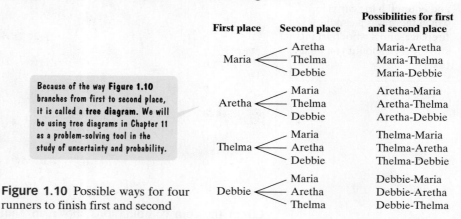

Because of the way **Figure 1.10** branches from first to second place, it is called a **tree diagram**. We will be using tree diagrams in Chapter 11 as a problem-solving tool in the study of uncertainty and probability.

Figure 1.10 Possible ways for four runners to finish first and second

Count the number of possibilities shown under the third column, "Possibilities for first and second place." Can you see that there are 12 possibilities? Therefore, 12 different arrangements of first- and second-place winners are possible.

Step 4 Look back and check the answer. Check the diagram in **Figure 1.10** to make sure that no possible first- and second-place outcomes have been left out. Double-check your count for the winning pairs of runners.

▶ **Checkpoint 5** Your "school wardrobe" is rather limited—just two pairs of jeans to choose from (one blue, one black) and three T-shirts to choose from (one beige, one yellow, and one blue). How many different outfits can you form?

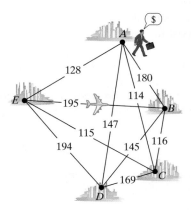

Figure 1.11

Example 6 Using a Reasonable Option to Solve a Problem with More Than One Solution

You live in city A and are visiting four prospective colleges in cities B, C, D, and E. Other than starting and ending the trip in city A, there are no restrictions as to the order in which the other four cities are visited.

The one-way fares between each of the cities are given in **Table 1.4**. A diagram that illustrates this information is shown in **Figure 1.11**.

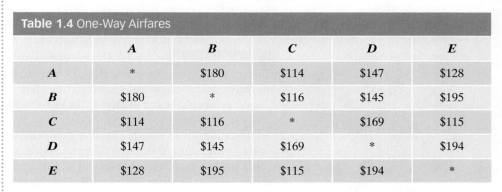

Table 1.4 One-Way Airfares					
	A	B	C	D	E
A	*	$180	$114	$147	$128
B	$180	*	$116	$145	$195
C	$114	$116	*	$169	$115
D	$147	$145	$169	*	$194
E	$128	$195	$115	$194	*

Your parents have asked you to suggest an order for visiting cities B, C, D, and E once, returning home to city A, for less than $750.

Solution

Step 1 Understand the problem. There are many ways to visit cities B, C, D, and E once, and return home to A. One route is

$$A, E, D, C, B, A.$$ Fly from A to E to D to C to B and then back to A.

The cost of this trip involves the sum of five costs, shown in both **Table 1.4** and **Figure 1.11**:

$$\$128 + \$194 + \$169 + \$116 + \$180 = \$787.$$

We must find a route that costs less than $750.

Step 2 Devise a plan. You start at city A. From there, fly to the city to which the airfare is cheapest. Then from there fly to the next city to which the airfare is cheapest, and so on. From the last of the cities, fly home to city A. Compute the cost of this trip to see if it is less than $750. If it is not, use trial and error to find other possible routes and select an order (if there is one) whose cost is less than $750.

Step 3 Carry out the plan and solve the problem. See **Figure 1.12**. The route is indicated using red lines with arrows.

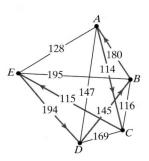

Figure 1.12

- Start at A.
- Choose the line segment with the smallest number: 114. Fly from A to C. (cost: $114)
- From C, choose the line segment with the smallest number that does not lead to A: 115. Fly from C to E. (cost: $115)
- From E, choose the line segment with the smallest number that does not lead to a city already visited: 194. Fly from E to D. (cost: $194)
- From D, there is little choice but to fly to B, the only city not yet visited. (cost: $145)
- From B, return home to A. (cost: $180)

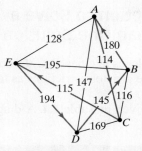

Figure 1.12 (repeated)

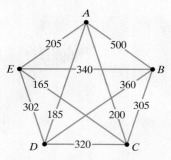

Figure 1.13

The route that we are considering is

$$A, C, E, D, B, A.$$

Let's see if the cost is less than $750. The cost is

$$\$114 + \$115 + \$194 + \$145 + \$180 = \$748.$$

Because the cost is less than $750, the sales director can follow the order A, C, E, D, B, A.

Step 4 Look back and check the answer. Use **Table 1.4** on the previous page or **Figure 1.12** to verify that the five numbers used in the sum shown above are correct. Use estimation to verify that $748 is a reasonable cost for the trip.

✓ **Checkpoint 6**   As in Example 6, you live in city A and are visiting four prospective colleges in cities $B, C, D,$ and E. The diagram in **Figure 1.13** shows the one-way airfares between any two cities. Give an order for visiting cities $B, C, D,$ and ▶ E once, returning home to city A, for less than $1460.

Becoming more confident as a problem solver will help you as you study the topics throughout this book. Furthermore, you can often apply problem-solving techniques to real-life situations in school and at work. Some people even use these techniques for dealing with personal problems in need of creative solutions.

Example 7   Problem Solving in School

You have a math test on Chapter 1 this Friday. Your organized math notebook contains class notes, homework, and a list of examples your teacher gave the class to review. Panic at 8 P.M. on Thursday: As you open the textbook to start studying, you remember that you left your notebook in your school locker. What can you do? Should you forget about studying for this test?

Solution

Step 1 Understand the problem. The test is tomorrow. At this hour of the evening, the school is locked, so you do not have access to your math notebook or the list of examples to study. You do have the textbook. If you decide not to study, you will probably fail the test.

Step 2 Devise a plan. You can study the information in the textbook and not give up. Each chapter contains a review chart that summarizes the definitions and concepts in every section of the chapter, complete with examples. There are also two practice tests that cover the important concepts in the chapter. Although it's late, you decide to use the review chart and one of the chapter tests to prepare for tomorrow's math test.

Step 3 Carry out the plan and solve the problem. You study the chapter review chart. Even without your math notebook, the summary of concepts and worked examples in the chart quickly remind you of the most important material in the chapter. Then you take one of the chapter tests and check your answers. Although it's now late and you feel exhausted, the test determines that you have mastered the material covered in the chapter. You are prepared for Friday's test.

Step 4 Look back. You realize that this problem could have been avoided if you had not waited until the night before the test to start studying. In the future, you decide to prepare for tests days in advance. The night before the exam you plan to do last-minute tune-ups and not exhaust yourself by cramming a chapter's worth of material into your brain. It also occurs to you that you did not do everything possible to solve the problem. You could have called a friend from class to get the list of examples your teacher gave the class to review.

✓ **Checkpoint 7** What problems have you had in studying for tests? Describe how the four steps in problem solving can be applied to find a solution to one of the problems.

Achieving Success

Ask! Ask! Ask!

Do not be afraid to ask questions in class. Your teacher may not realize that a concept is unclear until you raise your hand and ask a question. Other students who have problems asking questions in class will be appreciative that you have spoken up. Be polite and professional, but ask as many questions as required.

Exercise Set 1.3

Concept and Vocabulary Exercises

In Exercises 1–2, fill in each blank so that the resulting statement is true.

1. The first step in problem solving is to read the problem several times in order to _____ the problem.

2. The second step in problem solving is to _____ for solving the problem.

In Exercises 3–4, determine whether each statement is true or false. If the statement is false, make the necessary change(s) to produce a true statement.

3. Polya's four steps in problem solving make it possible to obtain answers to problems even if necessary pieces of information are missing.

4. When making a choice between various sizes of a product, the best value is the size with the lowest price.

Respond to Exercises 5–8 using verbal or written explanations.

5. Briefly describe Polya's four steps in problem solving.

In Exercises 6–8, explain the plan needed to solve the problem.

6. If you know how much was paid for several pounds of steak, find the cost of one pound.

7. If you know a person's age, find the year in which that person was born.

8. If you know how much you earn each hour, find your yearly income.

In Exercises 9–12, determine whether each statement makes sense or does not make sense, and explain your reasoning.

9. Polya's four steps in problem solving make it possible for me to solve any mathematical problem easily and quickly.

10. I used Polya's four steps in problem solving to deal with a personal problem in need of a creative solution.

11. I find it helpful to begin the problem-solving process by restating the problem in my own words.

12. When I get bogged down with a problem, there's no limit to the amount of time I should spend trying to solve it.

Practice and Application Exercises

Everyone can become a better, more confident problem solver. As in learning any other skill, learning problem solving requires hard work and patience. Work as many problems as possible in this exercise set. You may feel confused once in a while, but do not be discouraged. Thinking about a particular problem and trying different methods can eventually lead to new insights. Be sure to check over each answer carefully!

In Exercises 13–16, what necessary piece of information is missing that prevents solving the problem?

13. If a student saves $35 per week, how long will it take to save enough money to buy a computer?

14. If a steak sells for $8.15, what is the cost per pound?

15. If it takes you four minutes to read a page in a book, how many words can you read in one minute?

16. By paying $1500 cash and the balance in equal monthly payments, how many months would it take to pay for a car costing $12,495?

In Exercises 17–20, one more piece of information is given than is necessary for solving the problem. Identify this unnecessary piece of information. Then solve the problem.

17. A salesperson receives a weekly salary of $350. In addition, $15 is paid for every item sold in excess of 200 items. How much extra is received from the sale of 212 items?

18. You have $250 to spend and you need to purchase four new tires. If each tire weighs 21 pounds and costs $42 plus $2.50 tax, how much money will you have left after buying the tires?

19. A parking garage charges $2.50 for the first hour and $0.50 for each additional hour. If a customer gave the parking attendant $20.00 for parking from 10 A.M. to 3 P.M., how much did the garage charge?

20. An architect is designing a house. The scale on the plan is 1 inch = 6 feet. If the house is to have a length of 90 feet and a width of 30 feet, how long will the line representing the house's length be on the blueprint?

Use Polya's four-step method in problem solving to solve Exercises 21–52.

21. **a.** Which is the better value: a 15.3-ounce box of cereal for $3.37 or a 24-ounce box of cereal for $4.59?

    b. The supermarket displays the unit price for the 15.3-ounce box in terms of cost per ounce, but displays the unit price for the 24-ounce box in terms of cost per pound. What are the unit prices, to the nearest cent, given by the supermarket?

    c. Based on your work in parts (a) and (b), does the better value always have the lower displayed unit price? Explain your answer.

22. **a.** Which is the better value: a 12-ounce jar of honey for $2.25 or an 18-ounce jar of honey for $3.24?

    b. The supermarket displays the unit price for the 12-ounce jar in terms of cost per ounce, but displays the unit price for the 18-ounce jar in terms of cost per quart. Assuming 32 ounces in a quart, what are the unit prices, to the nearest cent, given by the supermarket?

    c. Based on your work in parts (a) and (b), does the better value always have the lower displayed unit price? Explain your answer.

23. One person earns $48,000 per year. Another earns $3750 per month. How much more does the first person earn in a year than the second?

24. At the beginning of a year, the odometer on a car read 25,124 miles. At the end of the year, it read 37,364 miles. If the car averaged 24 miles per gallon, how many gallons of gasoline did it use during the year?

25. A television sells for $750. Instead of paying the total amount at the time of the purchase, the same television can be bought by paying $100 down and $50 a month for 14 months. How much is saved by paying the total amount at the time of the purchase?

26. In a basketball game, the Bulldogs scored 34 field goals, each counting 2 points, and 13 foul goals, each counting 1 point. The Panthers scored 38 field goals and 8 foul goals. Which team won? By how many points did it win?

27. Calculators were purchased at $65 per dozen and sold at $20 for three calculators. Find the profit on six dozen calculators.

28. Pens are bought at 95¢ per dozen and sold in groups of four for $2.25. Find the profit on 15 dozen pens.

29. Each day a small business owner sells 200 pizza slices at $1.50 per slice and 85 sandwiches at $2.50 each. If business expenses come to $60 per day, what is the owner's profit for a ten-day period?

30. A tutoring center pays math tutors $5.15 per hour. Tutors earn an additional $1.20 per hour for each hour over 40 hours per week. A math tutor worked 42 hours one week and 45 hours the second week. How much did the tutor earn in this two-week period?

31. A car rents for $220 per week plus $0.25 per mile. Find the rental cost for a two-week trip of 500 miles for a group of three people.

32. A college graduate receives a salary of $2750 a month for her first job. During the year she plans to spend $4800 for rent, $8200 for food, $3750 for clothing, $4250 for household expenses, and $3000 for other expenses. With the money that is left, she expects to buy as many shares of stock at $375 per share as possible. How many shares will she be able to buy?

33. Charlene decided to ride her bike from her home to visit her friend Danny. Three miles away from home, her bike got a flat tire and she had to walk the remaining two miles to Danny's home. She could not repair the tire and had to walk all the way back home. How many more miles did Charlene walk than she rode?

34. A store received 200 containers of juice to be sold by April 1. Each container cost the store $0.75 and sold for $1.25. The store signed a contract with the vendor in which the vendor agreed to a $0.50 refund for every container not sold by April 1. If 150 containers were sold by April 1, how much profit did the store make?

35. A storeowner ordered 25 calculators that cost $30 each. The storeowner can sell each calculator for $35. The storeowner sold 22 calculators to customers. He had to return 3 calculators and pay a $2 charge for each returned calculator. Find the storeowner's profit.

36. New York City and Washington, D.C. are about 240 miles apart. A car leaves New York City at noon traveling directly south toward Washington, D.C. at 55 miles per hour. At the same time and along the same route, a second car leaves Washington, D.C. bound for New York City traveling directly north at 45 miles per hour. How far has each car traveled when the drivers meet for lunch at 2:24 P.M.?

37. An automobile purchased for $23,000 is worth $2700 after 7 years. Assuming that the car's value depreciated steadily from year to year, what was it worth at the end of the third year?

38. An automobile purchased for $34,800 is worth $8550 after 7 years. Assuming that the car's value depreciated steadily from year to year, what was it worth at the end of the third year?

39. A vending machine accepts nickels, dimes, and quarters. Exact change is needed to make a purchase. How many ways can a person with five nickels, three dimes, and two quarters make a 45-cent purchase from the machine?

40. How many ways can you make change for a quarter using only pennies, nickels, and dimes?

41. The members of the Student Activity Council at your high school are meeting to select two speakers for a month-long event celebrating artists and entertainers. The choices are Johnny Depp, Taylor Lautner, Jon Stewart, and Taylor Swift. How many different ways can the two speakers be selected?

42. The members of the Student Activity Council at your high school are meeting to select two speakers for a month-long event exploring why some people are most likely to succeed. The choices are Bill Clinton, Sean Combs, Donald Trump, Oprah Winfrey, and Simon Cowell. How many different ways can the two speakers be selected?

43. If you spend $4.79, in how many ways can you receive change from a five-dollar bill?

44. If you spend $9.74, in how many ways can you receive change from a ten-dollar bill?

45. You throw three darts at the board shown. Each dart hits the board and scores a 1, 5, or 10. How many different total scores can you make?

46. Suppose that you throw four darts at the board shown. With these four darts, there are 16 ways to hit four different numbers whose sum is 100. Describe one way you can hit four different numbers on the board that total 100.

47. Five friends (A, B, C, D, and E) agreed to share the expenses of a surprise birthday party equally. If A spent $42, B spent $10, C spent $26, D spent $32, and E spent $30, who owes money after the party and how much do they owe? To whom is money owed and how much should they receive? In order to resolve these discrepancies, who should pay how much to whom?

48. Six houses are spaced equally around a circular road. If it takes 10 minutes to walk from the first house to the third house, how long would it take to walk all the way around the road?

49. If a test has four true/false questions, in how many ways can there be three answers that are false and one answer that is true?

50. There are five people in a room. Each person shakes the hand of every other person exactly once. How many handshakes are exchanged?

51. Five runners, Andy, Beth, Caleb, Darnell, and Ella, are in a one-mile race. Andy finished the race seven seconds before Caleb. Caleb finished the race two seconds before Beth. Beth finished the race six seconds after Darnell. Ella finished the race eight seconds after Darnell. In which order did the runners finish the race?

52. Eight teams are competing in a volleyball tournament. Any team that loses a game is eliminated from the tournament. How many games must be played to determine the tournament winner?

In Exercises 53–54, you have three errands to run around town, although in no particular order. You plan to start and end at home.

You must go to the bank, the post office, and the dry cleaners. Distances, in miles, between any two of these locations are given in the diagram.

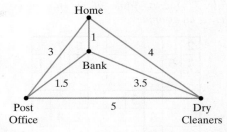

53. Determine a route whose distance is less than 12 miles for running the errands and returning home.

54. Determine a route whose distance exceeds 12 miles for running the errands and returning home.

55. The map shows five western states. Trace a route on the map that crosses each common state border exactly once.

56. The layout of a city with land masses and bridges is shown. Trace a route that shows people how to walk through the city so as to cross each bridge exactly once.

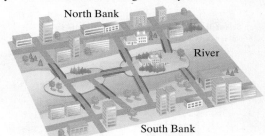

57. Jose, Bob, and Tony are college students living in adjacent dorm rooms. Bob lives in the middle dorm room. Their majors are business, psychology, and biology, although not necessarily in that order. The business major frequently uses the new computer in Bob's dorm room when Bob is in class. The psychology major and Jose both have 8 A.M. classes, and the psychology major knocks on Jose's wall to make sure he is awake. Determine Bob's major.

58. The figure represents a map of 13 countries. If countries that share a common border cannot be the same color, what is the minimum number of colors needed to color the map?

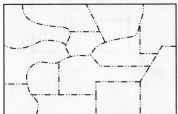

The sudoku (pronounced: sue-DOE-koo) craze, a number puzzle popular in Japan, hit the United States in 2005. A sudoku ("single number") puzzle consists of a 9-by-9 grid of 81 boxes subdivided into nine 3-by-3 squares. Some of the square boxes contain numbers. Here is an example:

2						6	7	
4	5							1
	9		7					
			3					
		4				2	8	
	1			2	4			
	8	3	9		7			
6	2			5		4	1	
							9	

The objective is to fill in the remaining squares so that every row, every column, and every 3-by-3 square contains each of the digits from 1 through 9 exactly once. (You can work this puzzle in Exercise 70, perhaps consulting one of the dozens of sudoku books in which the numerals 1 through 9 have created a cottage industry for publishers. There's even a Sudoku for Dummies.*)*

Trying to slot numbers into small checkerboard grids is not unique to sudoku. In Exercises 59–62, we explore some of the intricate patterns in other arrays of numbers, including magic squares. A **magic square** *is a square array of numbers arranged so that the numbers in all rows, all columns, and the two diagonals have the same sum. Here is an example of a magic square in which the sum of the numbers in each row, each column, and each diagonal is 15:*

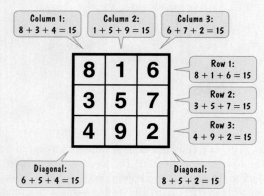

Exercises 59–60 are based on magic squares.

59. **a.** Use the properties of a magic square to fill in the missing numbers.

5		18
	15	
		25

   b. Show that the number of letters in the word for each number in the square in part (a) generates another magic square.

60. **a.** Use the properties of a magic square to fill in the missing numbers.

96		37	45
	43		
		25	57
23		82	78

   b. Show that if you reverse the digits for each number in the square in part (a), another magic square is generated.

   (*Source* for the *alphamagic square* in Exercise 59 and the *mirrormagic square* in Exercise 60: Clifford A. Pickover, *A Passion for Mathematics*, John Wiley & Sons, Inc., 2005.)

61. As in sudoku, fill in the missing numbers in the 3-by-3 square so that it contains each of the digits from 1 through 9 exactly once. Furthermore, in this *antimagic square*, the rows, the columns, and the two diagonals must have *different sums*.

9		7
	1	
3		5

62. The missing numbers in the 4-by-4 array are one-digit numbers. The sums for each row, each column, and one diagonal are listed in the voice balloons outside the array. Find the missing numbers.

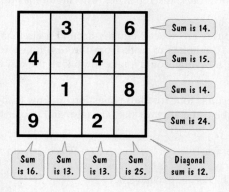

63. Some numbers in the printing of a division problem have become illegible. They are designated below by *. Fill in the blanks.

```
          1 * *
     * * ) 4 * * *
          2 8
          * 5 6
          * * *
          * * *
          * * *
              0
```

Critical Thinking Exercises

64. Gym lockers are to be numbered from 1 through 99 using metal numbers to be nailed onto each locker. How many 7s are needed?

65. Your family is on vacation in an isolated town. Everyone in the town was born there and has never left. You develop a toothache and check out the two dentists in town. One dentist has gorgeous teeth and one has teeth that show the effects of poor dental work. Which dentist should you choose and why?

66. India Jones is standing on a large rock in the middle of a square pool filled with hungry, man-eating piranhas. The edge of the pool is 20 feet away from the rock. India's mom wants to rescue her son, but she is standing on the edge of the pool with only two planks, each $19\frac{1}{2}$ feet long. How can India be rescued using the two planks?

67. One person tells the truth on Monday, Tuesday, Wednesday, and Thursday, but lies on all other days. A second person lies on Tuesday, Wednesday, and Thursday, but tells the truth on all other days. If both people state "I lied yesterday," then what day of the week is it today?

68. (This logic problem dates back to the eighth century.) A farmer needs to take his goat, wolf, and cabbage across a stream. His boat can hold him and one other passenger (the goat, wolf, or cabbage). If he takes the wolf with him, the goat will eat the cabbage. If he takes the cabbage, the wolf will eat the goat. Only when the farmer is present are the cabbage and goat safe from their respective predators. How does the farmer get everything across the stream?

69. As in sudoku, fill in the missing numbers along the sides of the triangle so that it contains each of the digits from 1 through 9 exactly once. Furthermore, each side of the triangle should contain four digits whose sum is 17.

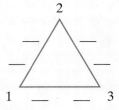

70. Solve the sudoku puzzle at the top of the left column on page 40.

71. A version of this problem, called the *missing dollar problem*, first appeared in 1933. Three people eat at a restaurant and receive a total bill for $30. They divide the amount equally and pay $10 each. The waiter gives the bill and the $30 to the manager, who realizes there is an error: The correct charge should be only $25. The manager gives the waiter five $1 bills to return to the customers, with the restaurant's apologies. However, the waiter is dishonest, keeping $2 and giving back only $3 to the customers. In conclusion, each of the three customers has paid $9 and the waiter has stolen $2, giving a total of $29. However, the original bill was $30. Where has the missing dollar gone?

72. A firefighter spraying water on a fire stood on the middle rung of a ladder. When the smoke became less thick, the firefighter moved up 4 rungs. However it got too hot, so the firefighter backed down 6 rungs. Later, the firefighter went up 7 rungs and stayed until the fire was out. Then, the firefighter climbed the remaining 4 rungs and entered the building. How many rungs does the ladder have?

73. The Republic of Teenville is composed of four states: A, B, C, and D. According to the country's constitution, the congress will have 30 seats, divided among the four states according to their respective populations. The table shows each state's population.

Population of Teenville by State

State	A	B	C	D	Total
Population (in thousands)	275	383	465	767	1890

Allocate the 30 congressional seats among the four states in a fair manner.

74. Two identical barrels are filled with water. The water temperature in the first barrel is 65°F. The water temperature in the second barrel is 15°F. A golf ball is dropped into each barrel. If the golf balls are identical, positioned at the same height, and dropped at the same time, which ball will touch the bottom of the barrel first?

Group Exercise

75. Listed below are common problems for high school students. Each problem has many potential solutions. Group members should select at least two problems in the list. For each selection, devise a plan for solving the problem and explain how the plan might be carried out.

- preparing for school in the morning
- being on time for classes or school appointments
- doing or finishing homework
- preparing for and taking exams
- managing time: balancing personal time, class time, and study time
- asking questions in class
- organizing a math notebook
- talking to a teacher or talking in front of a group
- deciding which school assignments should be completed first
- setting and achieving goals

(Source for the list of common problems: Cynthia Benjamin, *Problem Solving in School*, Pearson Learning Group, 1996.)

Chapter 1 Summary

1.1 Inductive and Deductive Reasoning

Definitions and Concepts

Inductive reasoning is the process of arriving at a general conclusion based on observations of specific examples. The conclusion is called a conjecture or a hypothesis. A case for which a conjecture is false is called a counterexample.

Examples

Find the next number, computation, or figure.

- $0, 4, 8, 12,$ Add 4: $12 + 4 = 16.$

- $\frac{1}{2}, \frac{1}{6}, \frac{1}{18}, \frac{1}{54},$ Multiply the denominator by 3: $\frac{1}{54 \times 3} = \frac{1}{162}.$

- $2! = 2 \cdot 1 = 2$
  $3! = 3 \cdot 2 \cdot 1 = 6$
  $4! = 4 \cdot 3 \cdot 2 \cdot 1 = 24$
  $5! = 5 \cdot 4 \cdot 3 \cdot 2 \cdot 1 = 120$

  The numbers before the ! symbol are increasing by 1. Each product begins with the number before the ! symbol and continues to decrease by 1 down through 1.
  $6! = 6 \cdot 5 \cdot 4 \cdot 3 \cdot 2 \cdot 1 = 720$

- , , , ,

  Figures alternate between triangles and rectangles. Numbers of appendages: 1, 2, 3, 1, 2, 3, etc.

Additional Examples to Review

Example 1, page 3; Example 2, page 4; Example 3, page 5; Example 4, page 7

Definitions and Concepts

Deductive reasoning is the process of proving a specific conclusion from one or more general statements. The statement that is proved is called a theorem.

Example

Select a number. Multiply the number by 6. Add 10 to the product. Divide the sum by 2. Subtract 5 from the quotient.

Select a number	2	3	4	n
Multiply the number by 6.	$2 \cdot 6 = 12$	$3 \cdot 6 = 18$	$4 \cdot 6 = 24$	$6n$
Add 10 to the product.	$12 + 10 = 22$	$18 + 10 = 28$	$24 + 10 = 34$	$6n + 10$
Divide the sum by 2.	$\frac{22}{2} = 11$	$\frac{28}{2} = 14$	$\frac{34}{2} = 17$	$\frac{6n + 10}{2} = 3n + 5$
Subtract 5 from the quotient.	$11 - 5 = 6$	$14 - 5 = 9$	$17 - 5 = 12$	$3n + 5 - 5 = 3n$
	Inductive conjecture: The result is three times the original number.			Deductive conclusion: The result, $3n$, is three times the original number n.

Additional Example to Review

Example 5, page 8

1.2 Estimation, Graphs, and Mathematical Models

Definitions and Concepts

Rounding Whole Numbers

1. Look at the digit to the right of the digit where rounding is to occur.
2. a. If the digit to the right is 5 or greater, add 1 to the digit to be rounded. Replace all digits to the right with zeros.
   b. If the digit to the right is less than 5, do not change the digit to be rounded. Replace all digits to the right with zeros.

The symbol ≈ means *is approximately equal to*.

Examples

Round 8,725,317 to
- the nearest million

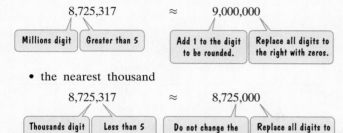

- the nearest thousand

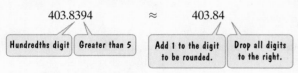

Additional Example to Review
Example 1, page 14

Definitions and Concepts

Decimal parts of numbers are rounded in nearly the same way as whole numbers. However, digits to the right of the rounding place are dropped.

Examples

Round 403.8394 to
- the nearest hundredth

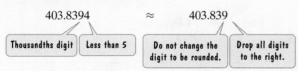

- the nearest thousandth

| 403.8394 | ≈ | 403.839 |

Thousandths digit · Less than 5 · Do not change the digit to be rounded. · Drop all digits to the right.

Additional Example to Review
Example 2, page 15

Definitions and Concepts

Estimation is the process of arriving at an approximate answer to a question. Computations can be estimated by using rounding that results in simplified arithmetic.

Examples

- $53.76 + $911.24 + $691.72 ≈ $50 + $900 + $700 = $1650
- 42% of 197,236 ≈ 40% of 200,000 = 0.4 × 200,000 = 80,000
- $793,165 × 19 ≈ $800,000 × 20 = $16,000,000
- 37.891 ÷ 5.023 ≈ 40 ÷ 5 = 8

Additional Examples to Review

Example 3, page 15; Example 4, page 16

Definitions and Concepts

Estimation is useful when interpreting information given by circle, bar, or line graphs.

Example

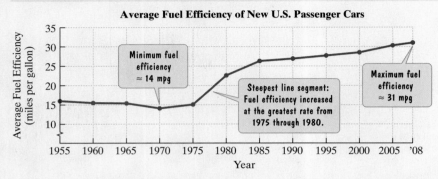

Average Fuel Efficiency of New U.S. Passenger Cars

Source: U.S. Department of Transportation

Additional Examples to Review

Example 5, page 18; Example 6, page 19; Example 7, page 20

Definitions and Concepts

The process of finding formulas to describe real-world phenomena is called mathematical modeling. Such formulas, together with the meaning assigned to the variables, are called mathematical models.

Example

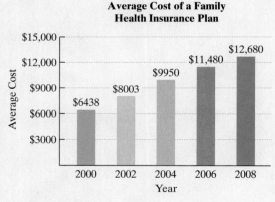

Average Cost of a Family Health Insurance Plan

Source: Kaiser Family Foundation

Yearly increase in average cost

$$\approx \frac{\text{change in cost from 2000 to 2008}}{\text{change in time from 2000 to 2008}}$$

$$= \frac{\$12,680 - \$6438}{2008 - 2000}$$

$$= \frac{\$6242}{8} \approx \$780$$

Mathematical model estimating average cost, C, in dollars, x years after 2000:

$$C = 6438 + 780x$$

Cost in 2000   Yearly increase

Additional Example to Review

Example 8, page 22

1.3 Problem Solving

Definitions and Concepts

Polya's Four Steps in Problem Solving

1. Understand the problem.
2. Devise a plan.
3. Carry out the plan and solve the problem.
4. Look back and check the answer.

Example

Phone Plan A offers a flat rate of $0.10 per minute. Phone Plan B offers a flat rate of $0.12 per minute, but the first 30 minutes each month are free. If you talk for 150 minutes in a month, which plan is the better deal and by how much?

- Understand the problem.
  Plan A: $0.10 per minute   Plan B: 30 free minutes, then $0.12 per minute
  We need to determine which plan is cheaper for 150 minutes of phone calls and by how much.

- Devise a plan.
  Find the cost of Plan A: Multiply the number of minutes, 150, by $0.10. Find the cost of
  Plan B: 30 minutes are free, so multiply the number of minutes, 150 − 30, or 120, by $0.12. The better deal is the cheaper plan. To determine by how much, subtract the cost of the cheaper plan from the cost of the more expensive plan.

- Carry out the plan and solve the problem.
  Cost of Plan A: $150 \times \$0.10 = \$15.00$
  Cost of Plan B: $120 \times \$0.12 = \$14.40$
  Plan B is the better deal by $15.00 − $14.40, or by $0.60.

- Look back and check the answer.
  Double-check the arithmetic by hand or with a calculator: $150 \boxed{\times} .10 \boxed{-} 120 \boxed{\times} .12 \boxed{=}$.
  You should get .6, the amount saved using Plan B.

Additional Examples to Review

Example 1, page 30; Example 2, page 30; Example 3, page 31; Example 4, page 32;
Example 5, page 34; Example 6, page 35

Review Exercises

Section 1.1 Inductive and Deductive Reasoning

1. Which reasoning process is shown in the following example? Explain your answer.

   All books by Stephen King have made the best-seller list. *Carrie* is a novel by Stephen King. Therefore, *Carrie* was on the best-seller list.

2. Which reasoning process is shown in the following example? Explain your answer.

   All books by Stephen King have made the best-seller list. Therefore, it is highly probable that the novel King is currently working on will make the best-seller list.

In Exercises 3–10, identify a pattern in each list of numbers. Then use this pattern to find the next number.

3. $4, 9, 14, 19,$ _____

4. $7, 14, 28, 56,$ _____

5. $1, 3, 6, 10, 15,$ _____

6. $\dfrac{3}{4}, \dfrac{3}{5}, \dfrac{1}{2}, \dfrac{3}{7},$ _____

7. $40, -20, 10, -5,$ _____

8. $40, -20, -80, -140,$ _____

9. $2, 2, 4, 6, 10, 16, 26,$ _____

10. $2, 6, 12, 36, 72, 216,$ _____

11. Identify a pattern in the following sequence of figures. Then use the pattern to find the next figure in the sequence.

, , , , _____

In Exercises 12–13, use inductive reasoning to predict the next line in each sequence of computations. Then perform the arithmetic to determine whether your conjecture is correct.

12.
$$2 = 4 - 2$$
$$2 + 4 = 8 - 2$$
$$2 + 4 + 8 = 16 - 2$$
$$2 + 4 + 8 + 16 = 32 - 2$$
_____

13.
$$111 \div 3 = 37$$
$$222 \div 6 = 37$$
$$333 \div 9 = 37$$
_____

14. Consider the following procedure:

   Select a number. Double the number. Add 4 to the product. Divide the sum by 2. Subtract 2 from the quotient.

   a. Repeat the procedure for four numbers of your choice. Write a conjecture that relates the result of the process to the original number selected.

   b. Use the variable *n* to represent the original number and use deductive reasoning to prove the conjecture in part (a).

Section 1.2 Estimation, Graphs, and Mathematical Models

15. In 2008, there were 4,257,473 live births in the United States.
   (*Source*: National Center for Health Statistics)

   Round the number of live births to the nearest

   a. hundred.   b. thousand.

   c. hundred thousand.   d. million.

16. A magnified view of the boundary of this black "buglike" shape, called the Mandelbrot set, was illustrated in the Section 1.1 opener on page 2.

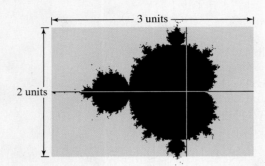

The area of the yellow rectangular region is the product of its length, 3 units, and its width, 2 units, or 6 square units. It is conjectured that the area of the black buglike region representing the Mandelbrot set is

$$\sqrt{6\pi - 1} - e \approx 1.5065916514855 \text{ square units.}$$

(*Source*: Robert P. Munafo, *Mandelbrot Set Glossary and Encyclopedia*)

Round the area of the Mandelbrot set to

  a. the nearest tenth.

  b. the nearest hundredth.

  c. the nearest thousandth.

  d. seven decimal places.

In Exercises 17–20, obtain an estimate for each computation by rounding the numbers so that the resulting arithmetic can easily be performed by hand or in your head. Then use a calculator to perform the computation. How reasonable is your estimate when compared to the actual answer?

17. 1.57 + 4.36 + 9.78

18. 8.83 × 49

19. 19.894 ÷ 4.179

20. 62.3% of 3847.6

In Exercises 21–24, determine each estimate without using a calculator. Then use a calculator to perform the computation necessary to obtain an exact answer. How reasonable is your estimate when compared to the actual answer?

21. Estimate the total cost of six grocery items if their prices are $8.47, $0.89, $2.79, $0.14, $1.19, and $4.76.

22. Estimate the salary of a worker who works for 78 hours at $7.85 per hour.

23. At a yard sale, a person bought 21 books at $0.85 each, two chairs for $11.95 each, and a ceramic plate for $14.65. Estimate the total amount spent.

24. The circle graph shows the most important issues for college students in 2008.

Most Important Issues for United States College Students

Source: Mother Jones, September/October 2008

Assume the circle graph is representative of college students who received bachelor's degrees in 2008. Among the 1,585,326 college students who received bachelor's degrees in 2008, estimate the number for whom the most important issue was global warming.

25. A small private school employs ten teachers with salaries ranging from $817 to $992 per week. Which of the following is the best estimate of the monthly payroll for the teachers?

  a. $30,000   **b.** $36,000

  c. $42,000   **d.** $50,000

26. Select the best estimate for the number of seconds in a day.

  a. 1500   **b.** 15,000

  c. 86,000   **d.** 100,000

27. Imagine the entire global population as a village of precisely 200 people. The bar graph shows some numeric observations based on this scenario.

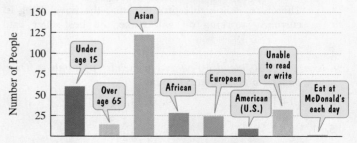

Earth's Population as a Village of 200 People

Source: Gary Rimmer, *Number Freaking*, The Disinformation Company Ltd.

  a. Which group in the village has a population that exceeds 100? Estimate this group's population.

  b. World population is approximately 33 million times the population of the village of 200 people. Use this observation to estimate the number of people in the world, in millions, unable to read or write.

28. The bar graph shows the percentage of people 25 years of age and older who were college graduates in the United States for seven selected years.

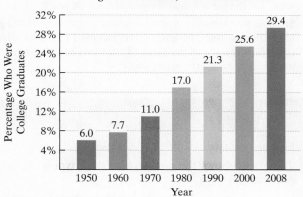

Percentage of College Graduates, among People Ages 25 and Older, in the United States

Source: U.S. Census Bureau

a. Estimate the average yearly increase in the percentage of college graduates. Round to the nearest tenth of a percent.

b. If the trend shown by the graph continues, estimate the percentage of people 25 years of age and older who will be college graduates in 2020.

29. During a diagnostic evaluation, a 33-year-old woman experienced a panic attack a few minutes after she had been asked to relax her whole body. The graph shows the rapid increase in heart rate during the panic attack.

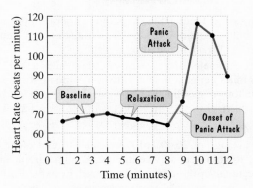

Heart Rate before and during a Panic Attack

Source: Davis and Palladino, *Psychology*, Fifth Edition, Prentice Hall, 2007.

a. Use the graph to estimate the woman's maximum heart rate during the first 12 minutes of the diagnostic evaluation. After how many minutes did this occur?

b. Use the graph to estimate the woman's minimum heart rate during the first 12 minutes of the diagnostic evaluation. After how many minutes did this occur?

c. During which time period did the woman's heart rate increase at the greatest rate?

d. After how many minutes was the woman's heart rate approximately 75 beats per minute?

30. In the years after warning labels were put on cigarette packs, the number of smokers dropped from approximately two in five adults to one in five. The bar graph shows the percentage of American adults who smoked cigarettes for selected years from 1965 through 2009.

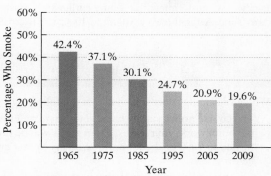

Percentage of American Adults Who Smoke Cigarettes

Source: Centers for Disease Control and Prevention

a. Estimate the yearly decrease in the percentage of American adults who smoke cigarettes. Round to the nearest tenth of a percent.

b. Write a mathematical model that estimates the percentage of American adults who smoked cigarettes, C, x years after 1965.

c. Use your mathematical model from part (b) to project the percentage of American adults who will smoke cigarettes in 2015.

Section 1.3 Problem Solving

31. What necessary piece of information is missing that prevents solving the following problem?

If 3 milligrams of a medicine is given for every 20 pounds of body weight, how many milligrams should be given to a six-year-old child?

32. In the following problem, there is one more piece of information given than is necessary for solving the problem. Identify this unnecessary piece of information. Then solve the problem.

A taxicab charges $3.00 for the first mile and 50 cents for each additional half-mile. After a six-mile trip, a customer handed the taxi driver a $20 bill. Find the cost of the trip.

Use the four-step method in problem solving to solve Exercises 33–38.

33. If there are seven frankfurters in one pound, how many pounds would you buy for a picnic to supply 28 people with two frankfurters each?

34. A car rents for $175 per week plus $0.30 per mile. Find the rental cost for a three-week trip of 1200 miles.

35. You are choosing between two plans at a discount warehouse. Plan A offers an annual membership fee of $100 and you pay 80% of the manufacturer's recommended list price. Plan B offers an annual membership fee of $40 and you pay 90% of the manufacturer's recommended list price. If you anticipate purchasing $1500 of merchandise in a year, which plan offers the better deal? By how much?

36. Miami is on Eastern Standard Time and San Francisco is on Pacific Standard Time, three hours earlier than Eastern Standard Time. A flight leaves Miami at 10 A.M. Eastern Standard Time, stops for 45 minutes in Houston, Texas, and arrives in San Francisco at 1:30 P.M. Pacific time. What is the actual flying time from Miami to San Francisco?

37. An automobile purchased for $37,000 is worth $2600 after eight years. Assuming that the value decreased steadily each year, what was the car worth at the end of the fifth year?

38. Suppose you are an engineer programming the automatic gate for a 35-cent toll. The gate is programmed for exact change only and will not accept pennies. How many coin combinations must you program the gate to accept?

Chapter 1 Test A

1. Which reasoning process is shown in the following example?

The course policy states that if you turn in at least 80% of the homework, your lowest exam grade will be dropped. I turned in 90% of the homework, so my lowest grade will be dropped.

2. Which reasoning process is shown in the following example?

We examine the fingerprints of 1000 people. No two individuals in this group of people have identical fingerprints. We conclude that for all people, no two people have identical fingerprints.

In Exercises 3–6, find the next number, computation, or figure, as appropriate.

3. 0, 5, 10, 15, _____

4. $\frac{1}{6}, \frac{1}{12}, \frac{1}{24}, \frac{1}{48}$, _____

5.
$3367 \times 3 = 10,101$
$3367 \times 6 = 20,202$
$3367 \times 9 = 30,303$
$3367 \times 12 = 40,404$
_____

6.

7. Consider the following procedure:

Select a number. Multiply the number by 4. Add 8 to the product. Divide the sum by 2. Subtract 4 from the quotient.

a. Repeat this procedure for three numbers of your choice. Write a conjecture that relates the result of the process to the original number selected.

b. Use the variable n to represent the original number and use deductive reasoning to prove the conjecture in part (a).

8. Round 3,279,425 to the nearest hundred thousand.

9. Round 706.3849 to the nearest hundredth.

In Exercises 10–14, determine each estimate without using a calculator. Different rounding results in different estimates, so there is not one single correct answer to each exercise. Use rounding to make the resulting calculations simple.

10. For a trip visiting prospective colleges, a student needs to spend $47.00 for gas, $311.00 for food, and $405.00 for a hotel room. If the student takes $681.79 from savings, estimate how much more money is needed for the trip.

11. The cost for opening a restaurant is $485,000. If 19 people decide to share equally in the business, estimate the amount each must contribute.

12. Find an estimate of 0.48992×121.976.

13. The graph shows the composition of a typical American community's trash.

Types of Trash in an American Community by Percentage of Total Weight

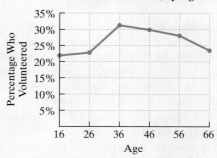

Paper 35%
Yard waste 12%
Food waste 12%
Plastic 11%
Metal 8%
Glass 5%
Other 17%

Source: U.S. Environmental Protection Agency

Across the United States, people generate approximately 512 billion pounds of trash per year. Estimate the number of pounds of trash in the form of plastic.

14. If the odometer of a car reads 71,911.5 miles and it averaged 28.9 miles per gallon, select the best estimate for the number of gallons of gasoline used.

a. 2400   **b.** 3200   **c.** 4000
d. 4800   **e.** 5600

15. An average of 26.4% of Americans 16 and older devoted time to volunteer service in 2008. The line graph shows the percentage of those who volunteered, by age.

Volunteer Activity in the United States, by Age

Source: Corporation for National and Community Services

a. For which age did the percentage of those who volunteered reach a maximum? Find a reasonable estimate of the percentage who volunteered for that age.

b. For which age was the percentage of volunteer activity at a minimum? Find a reasonable estimate of the percentage who volunteered for that age.

c. Between which two ages did the percentage of those who volunteered decrease at the greatest rate.

d. For which age did approximately 30% of Americans volunteer?

16. The amount of carbon dioxide in the atmosphere, measured in parts per million, has been increasing as a result of the burning of oil and coal. The buildup of gases and particles is believed to trap heat and raise the planet's temperature. The bar graph shows the average atmospheric concentration of carbon dioxide for seven selected years.

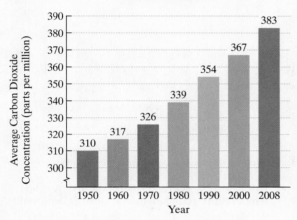

Average Atmospheric Concentration of Carbon Dioxide

Source: National Oceanic and Atmospheric Administration

a. Estimate the yearly increase in the average atmospheric concentration of carbon dioxide. Express the answer in parts per million, rounded to the nearest hundredth.

b. Write a mathematical model that estimates the average atmospheric concentration of carbon dioxide, C, in parts per million, x years after 1950.

c. If the trend shown by the data continues, use your mathematical model from part (b) to project the average atmospheric concentration of carbon dioxide in 2050.

17. The cost of renting a boat from Estes Rental is $9 per 15 minutes. The cost from Ship and Shore Rental is $20 per half-hour. If you plan to rent the boat for three hours, which business offers the better deal and by how much?

18. A bus operates between Miami International Airport and Miami Beach, 10 miles away. It makes 20 round trips per day carrying 32 passengers per trip. If the fare each way is $11.00, how much money is taken in from one day's operation?

19. By paying $50 cash up front and the balance at $35 a week, how long will it take to pay for a computer costing $960?

20. In 2000, the population of Greece was 10,600,000, with projections of a population decrease of 28,000 people per year. In the same year, the population of Belgium was 10,200,000, with projections of a population decrease of 12,000 people per year. (*Source:* United Nations) According to these projections, which country will have the greater population in 2035 and by how many more people?

Chapter 1 Test B

1. Which reasoning process is shown in the following example?

   In a Marist Poll survey of 938 randomly selected American adults, 47% of the respondents were annoyed by the word "whatever" in conversation. We conclude that there is a high probability that approximately 47% of all American adults find "whatever" annoying in conversation.

2. Which reasoning process is shown in the following example?

   In *I Can't Keep My Own Secrets* (Harper Teen, 2009), over 600 teens tell stories about their lives in the exact same length: six words. Three examples: "Born in the wrong decade, man." "Rather be alone in my room." "I fulfilled my awkwardness quota today."

   My friend told me that she was one of the authors represented in *I Can't Keep My Own Secrets*. I concluded that my friend told a story about her life in half a dozen words.

In Exercises 3–6, find the next number, computation, or figure, as appropriate.

3. 5, 7, 10, 14, _____

4. $\frac{1}{2}, \frac{1}{6}, \frac{1}{18}, \frac{1}{54},$ _____

5.
$$1 + 3 = 2^2$$
$$1 + 3 + 5 = 3^2$$
$$1 + 3 + 5 + 7 = 4^2$$
$$1 + 3 + 5 + 7 + 9 = 5^2$$

6.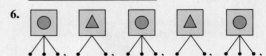

7. Consider the following procedure:

   Select a number. Multiply the number by 20. Add 12 to the product. Divide the sum by 4. Subtract 3 from the quotient.

   a. Repeat this procedure for three numbers of your choice. Write a conjecture that relates the result of the process to the original number selected.

   b. Use the variable n to represent the original number and use deductive reasoning to prove the conjecture in part (a).

8. Round 3,279,425 to the nearest million.

9. Round 706.38493 to the nearest thousandth.

In Exercises 10–14, determine each estimate without using a calculator. Different rounding results in different estimates, so there is not one single correct answer to each exercise. Use rounding to make the resulting calculations simple.

10. You purchase a sandwich for $4.85, juice for $1.79, a banana for $0.85, and two magazines costing $7.18. If you hand the clerk a $20 bill, estimate the amount of change you should receive.

11. A carpenter works 38 hours per week at an hourly rate of $22.50. Find a reasonable estimate of the carpenter's annual wages.

12. Find an estimate of $\dfrac{496.11}{19.7}$.

13. The circle graph shows the most important problems for high school teenagers.

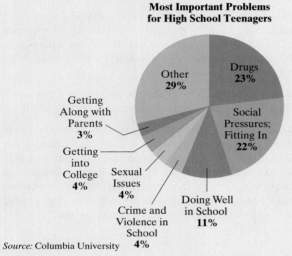

Most Important Problems for High School Teenagers

Source: Columbia University

Among the total high school teen population of 16,503,611 students, estimate the number for whom doing well in school is the most important problem.

14. If the odometer of a car reads 98,511.6 miles and it averaged 24.5 miles per gallon, estimate the number of gallons of gasoline used.

15. Contrary to popular belief, older people do not need less sleep than younger adults. However, the line graphs show that they awaken more often during the night. The numerous awakenings are one reason why some elderly individuals report that sleep is less restful than it had been in the past.

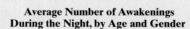

Average Number of Awakenings During the Night, by Age and Gender

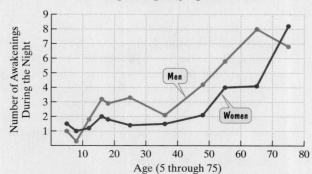

Source: Stephen Davis and Joseph Palladino, *Psychology*, Fifth Edition, Prentice Hall, 2007.

a. At which age, estimated to the nearest year, do women have the least number of awakenings during the night? What is the average number of awakenings at that age?

b. At which age do men have the greatest number of awakenings during the night? What is the average number of awakenings at that age?

c. Estimate, to the nearest tenth, the average number of awakenings during the night for 18-year-old men.

16. Our planet has been heating up for more than a century. Many experts have concluded that the increase in the amount of carbon dioxide in the atmosphere has been at least partly responsible for the increase in global surface temperature. The bar graph shows the average global temperature for selected years from 1900 through 2007.

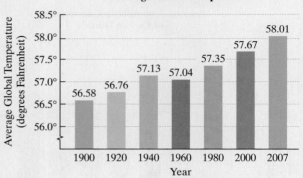

Average Global Temperature

Source: National Oceanic and Atmospheric Administration

a. Estimate the yearly increase in the average global temperature, rounded to the nearest hundredth of a degree.

b. Write a mathematical model that estimates the average global temperature, T, in degrees Fahrenheit, x years after 1900.

c. If the trend shown by the data continues, use your mathematical model from part (b) to project the average global temperature in 2050.

17. A midsize car can be rented for $23 per day plus $0.20 per mile. If you rent a midsize car for five days and drive 657 miles, what is the rental fee?

18. A razor can be used six times before it gets too dull to do a good job. The razors come in packs of ten. Each pack costs $5.00. If a man shaves three days each week, how much will he spend on razors for the year?

19. By paying $70 cash up front and the balance at $85 a month, how long will it take to pay for a bike costing $1600?

20. Company A pays $23,000 yearly with raises of $1200 per year. Company B pays $26,000 yearly with raises of $800 per year. Both companies begin paying raises in year 2. Which company will pay more in year 10? How much more?

Set Theory

Our bodies are fragile and complex, vulnerable to disease and easily damaged. The sequencing of the human genome in 2003—all 140,000 genes—should lead to rapid advances in treating heart disease, cancer, depression, Alzheimer's, and AIDS. Neural stem cell research could make it possible to repair brain damage and even re-create whole parts of the brain. There appears to be no limit to the parts of our bodies that can be replaced. By contrast, at the start of the twentieth century, we lacked even a basic understanding of the different types of human blood. The discovery of blood types, organized into collections called sets and illustrated by a special set diagram, rescued surgery patients from random, often lethal, transfusions. In this sense, the set diagram for blood types that you will encounter in this chapter reinforces our optimism that life does improve and that we are better off today than we were one hundred years ago.

Organizing and visually representing sets of human blood types is presented in the Blitzer Bonus on page 91. The vital role that this representation plays in blood transfusions is developed in Exercises 103–106 and Exercise 115 of Exercise Set 2.4.

2.1 Basic Set Concepts

1. Use three methods to represent sets.
2. Define and recognize the empty set.
3. Use the symbols ∈ and ∉.
4. Apply set notation to sets of natural numbers.
5. Determine a set's cardinal number.
6. Recognize equivalent sets.
7. Distinguish between finite and infinite sets.
8. Recognize equal sets.

We tend to place things in categories, which allows us to order and structure the world. For example, to which populations do you belong? Do you categorize yourself as a high school student? What about your gender? What about your special interests or your ethnic background? Our minds cannot find order and meaning without creating collections. Mathematicians call such collections *sets*. A **set** is a collection of objects whose contents can be clearly determined. The objects in a set are called the **elements**, or **members**, of the set.

A set must be **well defined**, meaning that its contents can be clearly determined. Using this criterion, the collection of actors who have won an Academy Award is a set. We can always determine whether or not a particular actor is an element of this collection. By contrast, consider the collection of great actors. Whether or not a person belongs to this collection is a matter of how we interpret the word *great*. In this text, we will only consider collections that form well-defined sets.

Methods for Representing Sets

1 Use three methods to represent sets.

An example of a set is the set of the days of the week, whose elements are Monday, Tuesday, Wednesday, Thursday, Friday, Saturday, and Sunday.

Capital letters are generally used to name sets. Let's use W to represent the set of the days of the week.

Three methods are commonly used to designate a set. One method is a **word description**. We can describe set W as the set of the days of the week. A second method is the **roster method**. This involves listing the elements of a set inside a pair of braces, { }. The braces at the beginning and end indicate that we are representing a set. The roster form uses commas to separate the elements of the set. Thus, we can designate the set W by listing its elements:

W = {Monday, Tuesday, Wednesday, Thursday, Friday, Saturday, Sunday}.

Grouping symbols such as parentheses, ( ), and square brackets, [ ], are not used to represent sets. Only commas are used to separate the elements of a set. Separators such as colons or semicolons are not used. Finally, the order in which the elements are listed in a set is not important. Thus, another way of expressing the set of the days of the week is

W = {Saturday, Sunday, Monday, Tuesday, Wednesday, Thursday, Friday}.

Example 1 Representing a Set Using a Description

Write a word description of the set

P = {Washington, Adams, Jefferson, Madison, Monroe}.

Solution

Set *P* is the set of the first five presidents of the United States.

 **Checkpoint 1** Write a word description of the set

$$L = \{a, b, c, d, e, f\}.$$

Example 2 Representing a Set Using the Roster Method

Set *C* is the set of U.S. coins with a value of less than a dollar. Express this set using the roster method.

Solution

$$C = \{penny, nickel, dime, quarter, half\text{-}dollar\}$$

 **Checkpoint 2** Set *M* is the set of months beginning with the letter A. Express this set using the roster method.

The third method for representing a set is with **set-builder notation**. Using this method, the set of the days of the week can be expressed as

$$W = \{x \mid x \text{ is a day of the week}\}.$$

Set *W*　is　the set of　all elements *x*　such that

We read this notation as "Set *W* is the set of all elements *x* such that *x* is a day of the week." Before the vertical line is the variable *x*, which represents an element in general. After the vertical line is the condition *x* must meet in order to be an element of the set.

Table 2.1 contains two examples of sets, each represented with a word description, the roster method, and set-builder notation.

The Beatles climbed to the top of the British music charts in 1963, conquering the United States a year later. In 2009, the successful videogame *The Beatles RockBand* allowed teens to meet the Beatles by actually *being* the Beatles.

Table 2.1 Sets Using Three Designations

Word Description	Roster Method	Set-Builder Notation
B is the set of members of the Beatles in 1963.	*B* = {George Harrison, John Lennon, Paul McCartney, Ringo Starr}	*B* = {*x*|*x* was a member of the Beatles in 1963}
S is the set of states whose names begin with the letter A.	*S* = {Alabama, Alaska, Arizona, Arkansas}	*S* = {*x*|*x* is a U.S. state whose name begins with the letter A}

Example 3 Converting from Set-Builder to Roster Notation

Express the set

$$A = \{x \mid x \text{ is a month that begins with the letter M}\}$$

using the roster method.

Solution

Set A, $\{x|x \text{ is a month that begins with the letter } M\}$, is the set of all elements x such that x is a month beginning with the letter M. There are two such months, namely March and May. Thus,

$$A = \{\text{March, May}\}.$$

 **Checkpoint 3** Express the set

$$O = \{x|x \text{ is a positive odd number less than 10}\}$$

using the roster method.

The representation of some sets by the roster method can be rather long, or even impossible, if we attempt to list every element. For example, consider the set of all lowercase letters of the English alphabet. If L is chosen as a name for this set, we can use set-builder notation to represent L as follows:

$$L = \{x|x \text{ is a lowercase letter of the English alphabet}\}.$$

A complete listing using the roster method is rather tedious:

$$L = \{\text{a, b, c, d, e, f, g, h, i, j, k, l, m, n, o, p, q, r, s, t, u, v, w, x, y, z}\}.$$

We can shorten the listing in set L by writing

$$L = \{\text{a, b, c, d, } \ldots, \text{ z}\}.$$

The three dots after the element d, called an *ellipsis*, indicate that the elements in the set continue in the same manner up to and including the last element z.

Blitzer Bonus

The Loss of Sets

Have you ever considered what would happen if we suddenly lost our ability to recall categories and the names that identify them? This is precisely what happened to Alice, the heroine of Lewis Carroll's *Through the Looking Glass*, as she walked with a fawn in "the woods with no names."

So they walked on together through the woods, Alice with her arms clasped lovingly round the soft neck of the Fawn, till they came out into another open field, and here the Fawn gave a sudden bound into the air, and shook itself free from Alice's arm. "I'm a Fawn!" it cried out in a voice of delight. "And, dear me! you're a human child!" A sudden look of alarm came into its beautiful brown eyes, and in another moment it had darted away at full speed.

By realizing that Alice is a member of the set of human beings, which in turn is part of the set of dangerous things, the fawn is overcome by fear. Thus, the fawn's experience is determined by the way it structures the world into sets with various characteristics.

John Tenniel, colored by Fritz Kredel

The Empty Set

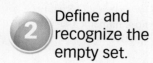 **2** Define and recognize the empty set.

Consider the following sets:

$$\{x|x \text{ is a fawn that speaks}\}$$
$$\{x|x \text{ is a number greater than 10 and less than 4}\}.$$

Can you see what these sets have in common? They both contain no elements. There are no fawns that speak. There are no numbers that are both greater than 10 and also less than 4. Sets such as these that contain no elements are called the *empty set*, or the *null set*.

> **The Empty Set**
>
> The **empty set**, also called the **null set**, is the set that contains no elements. The empty set is represented by { } or ∅.

Notice that { } and ∅ **have the same meaning**. However, **the empty set is not represented by {∅}**. This notation represents a set containing the element ∅.

Example 4   Recognizing the Empty Set

Which one of the following is the empty set?

a. {0}   **b.** 0

c. $\{x \mid x$ is a number less than 4 or greater than 10$\}$

d. $\{x \mid x$ is a square with exactly three sides$\}$

Solution

a. {0} is a set containing one element, 0. Because this set contains an element, it is not the empty set.

b. 0 is a number, not a set, so it cannot possibly be the empty set. It does, however, represent the number of members of the empty set.

c. $\{x \mid x$ is a number less than 4 or greater than 10$\}$ contains all numbers that are either less than 4, such as 3, or greater than 10, such as 11. Because some elements belong to this set, it cannot be the empty set.

d. $\{x \mid x$ is a square with exactly three sides$\}$ contains no elements. There are no squares with exactly three sides. This set is the empty set.

 Checkpoint 4   Which one of the following is the empty set?

a. $\{x \mid x$ is a number less than 3 or greater than 5$\}$

b. $\{x \mid x$ is a number less than 3 and greater than 5$\}$

▶ **c.** nothing   **d.** {∅}

Notations for Set Membership

We now consider two special notations that indicate whether or not a given object belongs to a set.

3 Use the symbols ∈ and ∉.

The Notations ∈ and ∉

The symbol ∈ is used to indicate that an object is an element of a set. The symbol ∈ is used to replace the words "is an element of."

The symbol ∉ is used to indicate that an object is *not* an element of a set. The symbol ∉ is used to replace the words "is not an element of."

Example 5   Using the Symbols ∈ and ∉

Determine whether each statement is true or false:

a. $r \in \{a, b, c, \ldots, z\}$ **b.** $7 \notin \{1, 2, 3, 4, 5\}$ **c.** $\{a\} \in \{a, b\}$.

Solution

a. Because r is an element of the set $\{a, b, c, \ldots, z\}$, the statement

$$r \in \{a, b, c, \ldots, z\}$$

is true.

Observe that an element can belong to a set in roster notation when three dots appear even though the element is not listed.

b. Because 7 is not an element of the set $\{1, 2, 3, 4, 5\}$, the statement

$$7 \notin \{1, 2, 3, 4, 5\}$$

is true.

c. We now consider the statement $\{a\} \in \{a, b\}$. Because $\{a\}$ is a set and the set $\{a\}$ is not an element of the set $\{a, b\}$, the statement

$$\{a\} \in \{a, b\}$$

is false.

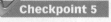 **Checkpoint 5**   Determine whether each statement is true or false:

a. $8 \in \{1, 2, 3, \ldots, 10\}$

b. $r \notin \{a, b, c, z\}$

c. $\{\text{Monday}\} \in \{x \mid x \text{ is a day of the week}\}$.

Sets of Natural Numbers

 4 Apply set notation to sets of natural numbers.

For much of the remainder of this section, we will focus on the set of numbers used for counting:

$$\{1, 2, 3, 4, 5, 6, 7, 8, 9, 10, 11, \ldots\}.$$

The set of counting numbers is also called the set of **natural numbers**. We represent this set by the boldface letter **N**.

> **The Set of Natural Numbers**
>
> $$\mathbf{N} = \{1, 2, 3, 4, 5, \ldots\}$$

The three dots, or ellipsis, after the 5 indicate that there is no final element and that the listing goes on forever.

Example 6   Representing Sets of Natural Numbers

Express each of the following sets using the roster method:

a. Set A is the set of natural numbers less than 5.

b. Set B is the set of natural numbers greater than or equal to 25.

c. $E = \{x \mid x \in \mathbf{N} \text{ and } x \text{ is even}\}$.

Solution

a. The natural numbers less than 5 are 1, 2, 3, and 4. Thus, set A can be expressed using the roster method as

$$A = \{1, 2, 3, 4\}.$$

b. The natural numbers greater than or equal to 25 are 25, 26, 27, 28, and so on. Set B in roster form is

$$B = \{25, 26, 27, 28, \ldots\}.$$

The three dots show that the listing goes on forever.

c. The set-builder notation

$$E = \{x \mid x \in \mathbf{N} \text{ and } x \text{ is even}\}$$

indicates that we want to list the set of all x such that x is an element of the set of natural numbers and x is even. The set of numbers that meets both conditions is the set of even natural numbers. The set in roster form is

$$E = \{2, 4, 6, 8, \ldots\}.$$

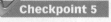 **Checkpoint 6**   Express each of the following sets using the roster method:

a. Set A is the set of natural numbers less than or equal to 3.

b. Set B is the set of natural numbers greater than 14.

c. $O = \{x \mid x \in \mathbf{N} \text{ and } x \text{ is odd}\}$.

A Brief Review • Inequality Notation

Inequality symbols are frequently used to describe sets of natural numbers. **Table 2.2** reviews basic inequality notation.

Table 2.2 Inequality Notation and Sets

Inequality Symbol and Meaning	Example — Set-Builder Notation	Roster Method
$x < a$ — *x* is less than *a*.	$\{x \mid x \in \mathbf{N} \text{ and } x < 4\}$ — *x* is a natural number less than 4.	$\{1, 2, 3\}$
$x \leq a$ — *x* is less than or equal to *a*.	$\{x \mid x \in \mathbf{N} \text{ and } x \leq 4\}$ — *x* is a natural number less than or equal to 4.	$\{1, 2, 3, 4\}$
$x > a$ — *x* is greater than *a*.	$\{x \mid x \in \mathbf{N} \text{ and } x > 4\}$ — *x* is a natural number greater than 4.	$\{5, 6, 7, 8, \ldots\}$
$x \geq a$ — *x* is greater than or equal to *a*.	$\{x \mid x \in \mathbf{N} \text{ and } x \geq 4\}$ — *x* is a natural number greater than or equal to 4.	$\{4, 5, 6, 7, \ldots\}$
$a < x < b$ — *x* is greater than *a* and less than *b*.	$\{x \mid x \in \mathbf{N} \text{ and } 4 < x < 8\}$ — *x* is a natural number greater than 4 and less than 8.	$\{5, 6, 7\}$
$a \leq x \leq b$ — *x* is greater than or equal to *a* and less than or equal to *b*.	$\{x \mid x \in \mathbf{N} \text{ and } 4 \leq x \leq 8\}$ — *x* is a natural number greater than or equal to 4 and less than or equal to 8.	$\{4, 5, 6, 7, 8\}$
$a \leq x < b$ — *x* is greater than or equal to *a* and less than *b*.	$\{x \mid x \in \mathbf{N} \text{ and } 4 \leq x < 8\}$ — *x* is a natural number greater than or equal to 4 and less than 8.	$\{4, 5, 6, 7\}$
$a < x \leq b$ — *x* is greater than *a* and less than or equal to *b*.	$\{x \mid x \in \mathbf{N} \text{ and } 4 < x \leq 8\}$ — *x* is a natural number greater than 4 and less than or equal to 8.	$\{5, 6, 7, 8\}$

Example 7   Representing Sets of Natural Numbers

Express each of the following sets using the roster method:

a. $\{x \mid x \in \mathbf{N} \text{ and } x \leq 100\}$ **b.** $\{x \mid x \in \mathbf{N} \text{ and } 70 \leq x < 100\}$.

Solution

a. $\{x \mid x \in \mathbf{N} \text{ and } x \leq 100\}$ represents the set of natural numbers less than or equal to 100. This set can be expressed using the roster method as

$$\{1, 2, 3, 4, \ldots, 100\}.$$

b. $\{x \mid x \in \mathbf{N} \text{ and } 70 \leq x < 100\}$ represents the set of natural numbers greater than or equal to 70 and less than 100. This set in roster form is $\{70, 71, 72, 73, \ldots, 99\}$.

✓ **Checkpoint 7** Express each of the following sets using the roster method:

a. $\{x \mid x \in \mathbf{N} \text{ and } x < 200\}$

b. $\{x \mid x \in \mathbf{N} \text{ and } 50 < x \leq 200\}$.

Blitzer Bonus
●●●●●●●●●●●●●●●
Math, Internet, and Texting Lingo

It's easy to feel overwhelmed by mathematical language and notation. However, Internet and texting lingo, also filled with symbols, appears not to intimidate American teens, who send and receive an average of 2300 text messages per month. (*Source*: Nielsen) FWIW (For what it's worth), :-) (smile), :–D (laugh), and :–((frown) are more complex and incomprehensible to math teachers than $\varnothing$, $\in$, $<$, or $>$. (It's also likely that your teachers believe 2 much txting is bad 4u.) Becoming familiar with mathematical symbols and notations is no more difficult than learning and using the special Internet/texting notations that keep you connected. TAFN (That's all for now.)

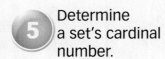

Determine a set's cardinal number.

Cardinality and Equivalent Sets

The number of elements in a set is called the **cardinal number**, or **cardinality**, of the set. For example, the set $\{a, e, i, o, u\}$ contains five elements and therefore has the cardinal number 5. We can also say that the set has a cardinality of 5.

> **Definition of a Set's Cardinal Number**
>
> The **cardinal number** of set A, represented by $n(A)$, is the number of distinct elements in set A. The symbol $n(A)$ is read "n of A."

Notice that the cardinal number of a set refers to the number of *distinct*, or different, elements in the set. **Repeating elements in a set neither adds new elements to the set nor changes its cardinality.** For example, $A = \{3, 5, 7\}$ and $B = \{3, 5, 5, 7, 7, 7\}$ represent the same set with three distinct elements, 3, 5, and 7. Thus, $n(A) = 3$ and $n(B) = 3$.

Example 8 ⬛ Determining a Set's Cardinal Number

Find the cardinal number of each of the following sets:

a. $A = \{7, 9, 11, 13\}$ **b.** $B = \{0\}$

c. $C = \{13, 14, 15, \ldots, 22, 23\}$ **d.** $\varnothing$.

Solution

The cardinal number for each set is found by determining the number of elements in the set.

a. $A = \{7, 9, 11, 13\}$ contains four distinct elements. Thus, the cardinal number of set A is 4. We also say that set A has a cardinality of 4, or $n(A) = 4$.

b. $B = \{0\}$ contains one element, namely 0. The cardinal number of set B is 1. Therefore, $n(B) = 1$.

c. Set $C = \{13, 14, 15, \ldots, 22, 23\}$ lists only five elements. However, the three dots indicate that the natural numbers from 16 through 21 are also in the set.

Counting the elements in the set, we find that there are 11 natural numbers in set C. The cardinality of set C is 11, and $n(C) = 11$.

d. The empty set, $\varnothing$, contains no elements. Thus, $n(\varnothing) = 0$.

 **Checkpoint 8** Find the cardinal number of each of the following sets:

a. $A = \{6, 10, 14, 15, 16\}$ **b.** $B = \{872\}$

c. $C = \{9, 10, 11, \ldots, 15, 16\}$ **d.** $D = \{\ \}$.

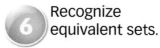 **6** Recognize equivalent sets.

Sets that contain the same number of elements are said to be *equivalent*.

> **Definition of Equivalent Sets**
>
> Set A is **equivalent** to set B means that set A and set B contain the same number of elements. For equivalent sets, $n(A) = n(B)$.

Here is an example of two equivalent sets:

$$n(A) = n(B) = 5$$

$$A = \{x \mid x \text{ is a vowel}\} = \{a, e, i, o, u\}$$
$$\updownarrow \updownarrow \updownarrow \updownarrow \updownarrow$$
$$B = \{x \mid x \in \mathbf{N} \text{ and } 3 \leq x \leq 7\} = \{3, 4, 5, 6, 7\}.$$

It is not necessary to count elements and arrive at 5 to determine that these sets are equivalent. The lines with arrowheads, $\updownarrow$, indicate that each element of set A can be paired with exactly one element of set B and each element of set B can be paired with exactly one element of set A. We say that the sets can be placed in a **one-to-one correspondence**.

> **One-to-One Correspondences and Equivalent Sets**
>
> **1.** If set A and set B can be placed in a one-to-one correspondence, then A is equivalent to B: $n(A) = n(B)$.
>
> **2.** If set A and set B cannot be placed in a one-to-one correspondence, then A is not equivalent to B: $n(A) \neq n(B)$.

Example 9 **Determining If Sets Are Equivalent**

The bar graph in **Figure 2.1** shows the average hours spent per day in leisure and sports activities for teens (ages 15–19) and the elderly (ages 75+) in the United States.

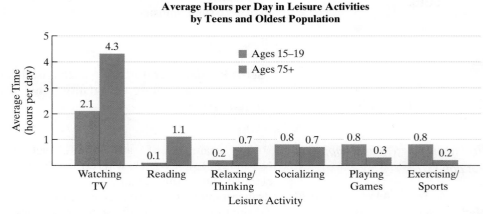

Figure 2.1

Source: Bureau of Labor Statistics

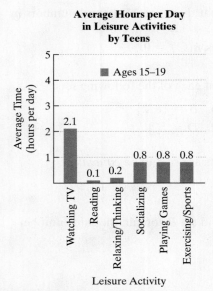

Average Hours per Day in Leisure Activities by Teens

Figure 2.1 (partly repeated)

Let

A = the set of the activities shown in **Figure 2.1**.

B = the set of the average hours per day teens ages 15–19 spend on these activities.

Are these sets equivalent? Explain.

Solution

Let's begin by expressing each set in roster form.

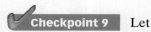

$A = \{\text{TV, reading, relaxing, socializing, games, exercising}\}$

$B = \{\ 2.1,\quad 0.1,\quad\quad 0.2,\quad\quad\quad\quad\quad 0.8\ \}$

> Do not write 0.8 three times. We are interested in each set's <u>distinct</u> elements.

There are two ways to determine that these sets are not equivalent.

Method 1. Trying to Set Up a One-to-One Correspondence
The lines with arrowheads between the sets in roster form indicate that the correspondence between the sets is not one-to-one. The elements socializing, games, and exercising from set A are each paired with the element 0.8 from set B. These sets are not equivalent.

Method 2. Counting Elements
Set A contains six distinct elements: $n(A) = 6$. Set B contains four distinct elements: $n(B) = 4$. Because the sets do not contain the same number of elements, they are not equivalent.

Checkpoint 9 Let

A = the set of activities shown in **Figure 2.1** on page 59.

B = the set of the average hours per day people 75 and older spend on these activities.

Are these sets equivalent? Explain.

Finite and Infinite Sets

7 Distinguish between finite and infinite sets.

Example 9 illustrated that to compare the cardinalities of two sets, pair off their elements. If there is not a one-to-one correspondence, the sets have different cardinalities and are not equivalent. Although this idea is obvious in the case of *finite sets*, some unusual conclusions emerge when dealing with *infinite sets*.

> **Finite Sets and Infinite Sets**
>
> Set A is a **finite set** if $n(A) = 0$ (that is, A is the empty set) or $n(A)$ is a natural number. A set whose cardinality is not 0 or a natural number is called an **infinite set**.

"Infinity is where things happen that don't."

W. W. Sawyer, *Prelude to Mathematics*, Penguin Books, 1960.

An example of an infinite set is the set of natural numbers, $\mathbf{N} = \{1, 2, 3, 4, 5, 6, \ldots\}$, where the ellipsis indicates that there is no last, or final, element. Does this set have a cardinality? The answer is yes, albeit one of the strangest numbers you've ever seen. The set of natural numbers is assigned the infinite cardinal number $\aleph_0$ (read: "aleph-null," aleph being the first letter of the Hebrew alphabet). What follows is a succession of mind-boggling results, including a hierarchy of different infinite numbers in which $\aleph_0$ is the smallest infinity:

$$\aleph_0 < \aleph_1 < \aleph_2 < \aleph_3 < \aleph_4 < \aleph_5\ldots.$$

Equal Sets

⑧ Recognize equal sets.

We conclude this section with another important concept of set theory, equality of sets.

Definition of Equality of Sets

Set A is **equal** to set B means that set A and set B contain exactly the same elements, regardless of order or possible repetition of elements. We symbolize the equality of sets A and B using the statement $A = B$.

Study Tip

In English, the words *equal* and *equivalent* often mean the same thing. This is not the case in set theory. **Equal sets** contain the **same elements**. **Equivalent sets** contain the **same number of elements**. If two sets are equal, then they must be equivalent. However, if two sets are equivalent, they are not necessarily equal.

For example, if $A = \{w, x, y, z\}$ and $B = \{z, y, w, x\}$, then $A = B$ because the two sets contain exactly the same elements.

Because equal sets contain the same elements, they also have the same cardinal number. For example, the equal sets $A = \{w, x, y, z\}$ and $B = \{z, y, w, x\}$ have four elements each. Thus, both sets have the same cardinal number: 4. Notice that a possible one-to-one correspondence between the equal sets A and B can be obtained by pairing each element with itself:

$$A = \{w, \ x, \ y, \ z\}$$
$$B = \{z, \ y, \ w, \ x\}$$

This illustrates an important point: **If two sets are equal, then they must be equivalent.**

Example 10   Determining Whether Sets Are Equal

Determine whether each statement is true or false:

  a. $\{4, 8, 9\} = \{8, 9, 4\}$ **b.** $\{1, 3, 5\} = \{0, 1, 3, 5\}.$

Solution

  a. The sets $\{4, 8, 9\}$ and $\{8, 9, 4\}$ contain exactly the same elements. Therefore, the statement

$$\{4, 8, 9\} = \{8, 9, 4\}$$

    is true.

  b. As we look at the given sets, $\{1, 3, 5\}$ and $\{0, 1, 3, 5\}$, we see that 0 is an element of the second set, but not the first. The sets do not contain exactly the same elements. Therefore, the sets are not equal. This means that the statement

$$\{1, 3, 5\} = \{0, 1, 3, 5\}$$

    is false.

✓ **Checkpoint 10**   Determine whether each statement is true or false:

▶   **a.** $\{O, L, D\} = \{D, O, L\}$ **b.** $\{4, 5\} = \{5, 4, \varnothing\}.$

Exercise Set 2.1

Concept and Vocabulary Exercises

In Exercises 1–6, fill in each blank so that the resulting statement is true.

1. A set that contains no elements is called the _____ set and is represented by _____.

2. The symbol $\in$ is used to indicate that _____.

3. The set $\{1, 2, 3, 4, 5, \ldots\}$ is called the set of _____.

4. The number of distinct elements in a set is called the _____ of the set.

5. Two sets that contain the same number of elements are called _____.

6. Two sets that contain the same elements are called _____.

In Exercises 7–12, determine whether each statement is true or false. If the statement is false, make the necessary change(s) to produce a true statement.

7. Two sets can be equal but not equivalent.

8. Any set in roster notation that contains three dots must be an infinite set.

9. $n(\varnothing) = 1$

10. Some sets that can be written in set-builder notation cannot be written in roster form.

11. If the elements in a set cannot be counted in a trillion years, the set is an infinite set.

12. Because 0 is not a natural number, it can be deleted from any set without changing the set's cardinality.

Respond to Exercises 13–18 using verbal or written explanations.

13. What is a set?

14. Describe the three methods used to represent a set. Give an example of a set represented by each method.

15. What is the empty set?

16. Explain what is meant by *equivalent sets*.

17. Explain what is meant by *equal sets*.

18. Use cardinality to describe the difference between a finite set and an infinite set.

In Exercises 19–22, determine whether each statement makes sense or does not make sense, and explain your reasoning.

19. I used the roster method to express the set of states that I have visited.

20. I used the roster method and natural numbers to express the set of average daily Fahrenheit temperatures throughout the month of July in Vostok Station, Antarctica, the coldest month in one of the coldest locations in the world.

21. Using this bar graph that shows the average number of hours that Americans sleep per day, I can see that there is a one-to-one correspondence between the set of six ages on the horizontal axis and the set of the average number of hours that men sleep per day.

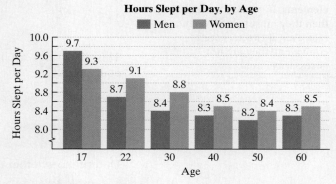

Hours Slept per Day, by Age

Source: ATUS, Bureau of Labor Statistics

22. Using the bar graph in Exercise 21, I can see that there is a one-to-one correspondence between the set of the average number of hours that men sleep per day and the set of the average number of hours that women sleep per day.

Practice Exercises

In Exercises 23–28, determine which collections are not well defined and therefore not sets.

23. The collection of U.S. presidents

24. The collection of juniors and seniors currently attending your high school

25. The collection of the five worst U.S. presidents

26. The collection of talented juniors and seniors currently attending your high school

27. The collection of natural numbers greater than one million

28. The collection of even natural numbers greater than 100

In Exercises 29–36, write a word description of each set. (More than one correct description may be possible.)

29. {Mercury, Venus, Earth, Mars, Jupiter, Saturn, Uranus, Neptune}

30. {Saturday, Sunday}

31. {January, June, July}

32. {April, August}

33. $\{6, 7, 8, 9, \ldots\}$

34. $\{9, 10, 11, 12, \ldots\}$

35. $\{6, 7, 8, 9, \ldots, 20\}$

36. $\{9, 10, 11, 12, \ldots, 25\}$

In Exercises 37–54, express each set using the roster method.

37. The set of the four seasons in a year

38. The set of months of the year that have exactly 30 days

39. $\{x | x$ is a month that ends with the letters b-e-r$\}$

40. $\{x | x$ is a lowercase letter of the alphabet that follows d and comes before j$\}$

41. The set of natural numbers less than 4

42. The set of natural numbers less than or equal to 6

43. The set of odd natural numbers less than 13

44. The set of even natural numbers less than 10

45. $\{x | x \in \mathbf{N}$ and $x \le 5\}$

46. $\{x | x \in \mathbf{N}$ and $x \le 4\}$

47. $\{x | x \in \mathbf{N}$ and $x > 5\}$

48. $\{x | x \in \mathbf{N}$ and $x > 4\}$

49. $\{x | x \in \mathbf{N}$ and $6 < x \le 10\}$

50. $\{x | x \in \mathbf{N}$ and $7 < x \le 11\}$

51. $\{x | x \in \mathbf{N}$ and $10 \le x < 80\}$

52. $\{x | x \in \mathbf{N}$ and $15 \le x < 60\}$

53. $\{x | x + 5 = 7\}$

54. $\{x | x + 3 = 9\}$

In Exercises 55–68, determine which sets are the empty set.

55. $\{\varnothing, 0\}$

56. $\{0, \varnothing\}$

57. $\{x | x$ is a woman who served as U.S. president before 2000$\}$

58. $\{x | x$ is a living U.S. president born before 1200$\}$

59. $\{x | x$ is the number of women who served as U.S. president before 2000$\}$

60. $\{x | x$ is the number of living U.S. presidents born before 1200$\}$

61. $\{x | x$ is a U.S. state whose name begins with the letter X$\}$

62. $\{x | x$ is a month of the year whose name begins with the letter X$\}$

63. $\{x | x < 2$ and $x > 5\}$

64. $\{x | x < 3$ and $x > 7\}$

65. $\{x | x \in \mathbf{N}$ and $2 < x < 5\}$

66. $\{x | x \in \mathbf{N}$ and $3 < x < 7\}$

67. $\{x | x$ is a number less than 2 or greater than 5$\}$

68. $\{x | x$ is a number less than 3 or greater than 7$\}$

In Exercises 69–88, determine whether each statement is true or false.

69. $3 \in \{1, 3, 5, 7\}$

70. $6 \in \{2, 4, 6, 8, 10\}$

71. $12 \in \{1, 2, 3, \ldots, 14\}$

72. $10 \in \{1, 2, 3, \ldots, 16\}$

73. $5 \in \{2, 4, 6, \ldots, 20\}$

74. $8 \in \{1, 3, 5, \ldots, 19\}$

75. $11 \notin \{1, 2, 3, \ldots, 9\}$

76. $17 \notin \{1, 2, 3, \ldots, 16\}$

77. $37 \notin \{1, 2, 3, \ldots, 40\}$

78. $26 \notin \{1, 2, 3, \ldots, 50\}$

79. $4 \notin \{x | x \in \mathbf{N}$ and x is even$\}$

80. $2 \in \{x | x \in \mathbf{N}$ and x is odd$\}$

81. $13 \notin \{x | x \in \mathbf{N}$ and $x < 13\}$

82. $20 \notin \{x | x \in \mathbf{N}$ and $x < 20\}$

83. $16 \notin \{x | x \in \mathbf{N}$ and $15 \le x < 20\}$

84. $19 \notin \{x | x \in \mathbf{N}$ and $16 \le x < 21\}$

85. $\{3\} \in \{3, 4\}$

86. $\{7\} \in \{7, 8\}$

87. $-1 \notin \mathbf{N}$

88. $-2 \notin \mathbf{N}$

In Exercises 89–102, find the cardinal number for each set.

89. $A = \{17, 19, 21, 23, 25\}$

90. $A = \{16, 18, 20, 22, 24, 26\}$

91. $B = \{2, 4, 6, \ldots, 30\}$

92. $B = \{1, 3, 5, \ldots, 21\}$

93. $C = \{x | x$ is a day of the week that begins with the letter A$\}$

94. $C = \{x | x$ is a month of the year that begins with the letter W$\}$

95. $D = \{$five$\}$

96. $D = \{$six$\}$

97. $A = \{x | x$ is a letter in the word *five*$\}$

98. $A = \{x | x$ is a letter in the word *six*$\}$

99. $B = \{x | x \in \mathbf{N}$ and $2 \le x < 7\}$

100. $B = \{x | x \in \mathbf{N}$ and $3 \le x < 10\}$

101. $C = \{x | x < 4$ and $x \ge 12\}$

102. $C = \{x | x < 5$ and $x \ge 15\}$

In Exercises 103–112,

    a. *Are the sets equivalent? Explain.*

    b. *Are the sets equal? Explain.*

103. A is the set of students at your high school. B is the set of students on the football team at your high school.

104. A is the set of states in the United States. B is the set of people who are now governors of the states in the United States.

105. $A = \{1, 2, 3, 4, 5\}$
$B = \{0, 1, 2, 3, 4\}$

106. $A = \{1, 3, 5, 7, 9\}$
$B = \{2, 4, 6, 8, 10\}$

107. $A = \{1, 1, 1, 2, 2, 3, 4\}$
$B = \{4, 3, 2, 1\}$

108. $A = \{0, 1, 1, 2, 2, 2, 3, 3, 3, 3\}$
$B = \{3, 2, 1, 0\}$

109. $A = \{x \mid x \in \mathbf{N}$ and $6 \le x < 10\}$
$B = \{x \mid x \in \mathbf{N}$ and $9 < x \le 13\}$

110. $A = \{x \mid x \in \mathbf{N}$ and $12 < x \le 17\}$
$B = \{x \mid x \in \mathbf{N}$ and $20 \le x < 25\}$

111. $A = \{x \mid x \in \mathbf{N}$ and $100 \le x \le 105\}$
$B = \{x \mid x \in \mathbf{N}$ and $99 < x < 106\}$

112. $A = \{x \mid x \in \mathbf{N}$ and $200 \le x \le 206\}$
$B = \{x \mid x \in \mathbf{N}$ and $199 < x < 207\}$

In Exercises 113–118, determine whether each set is finite or infinite.

113. $\{x \mid x \in \mathbf{N}$ and $x \ge 100\}$

114. $\{x \mid x \in \mathbf{N}$ and $x \ge 50\}$

115. $\{x \mid x \in \mathbf{N}$ and $x \le 1{,}000{,}000\}$

116. $\{x \mid x \in \mathbf{N}$ and $x \le 2{,}000{,}000\}$

117. The set of natural numbers less than 1

118. The set of natural numbers less than 0

Practice Plus

In Exercises 119–122, express each set using set-builder notation. Use inequality notation to express the condition x must meet in order to be a member of the set. (More than one correct inequality may be possible.)

119. $\{61, 62, 63, 64, \ldots\}$

120. $\{36, 37, 38, 39, \ldots\}$

121. $\{61, 62, 63, 64, \ldots, 89\}$

122. $\{36, 37, 38, 39, \ldots, 59\}$

Application Exercises

The bar graph shows the average number of minutes needed to work to pay for a Big Mac in selected cities. In Exercises 123–130, use the information given by the graph to represent each set by the roster method, or use the appropriate notation to indicate that the set is the empty set.

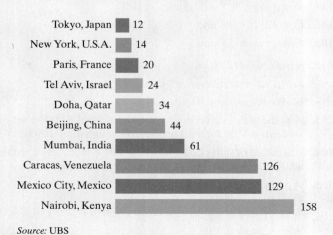

Earning a Big Mac:
Average Number of Minutes Needed
to Work to Pay for a Big Mac in Selected Cities

Source: UBS

123. The set of selected cities in which the number of minutes needed to work to pay for a Big Mac exceeds 100

124. The set of selected cities in which the number of minutes needed to work to pay for a Big Mac exceeds 61

125. The set of selected cities in which the number of minutes needed to work to pay for a Big Mac is at least 44 and at most 129

126. The set of selected cities in which the number of minutes needed to work to pay for a Big Mac is at least 34 and at most 126

127. $\{x \mid x$ is a selected city in which $12 \le$ number of working minutes to pay for a Big Mac $< 24\}$

128. $\{x \mid x$ is a selected city in which $12 <$ number of working minutes to pay for a Big Mac $\le 24\}$

129. $\{x \mid x$ is a selected city in which the number of working minutes to pay for a Big Mac $> 158\}$

130. $\{x \mid x$ is a selected city in which the number of working minutes to pay for a Big Mac $\ge 158\}$

A study of 900 working women in Texas showed that their feelings changed throughout the day. The line graph below shows 15 different times in a day and the average level of happiness for the women at each time. Based on the information given by the graph, represent each of the sets in Exercises 131–134 using the roster method.

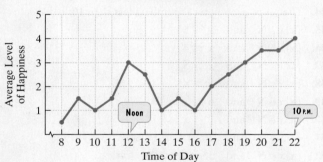

Average Level of Happiness
at Different Times of Day

Source: D. Kahneman et al. "A Survey Method for Characterizing Daily Life Experience," *Science.*

131. $\{x \mid x$ is a time of the day when the average level of happiness was 3$\}$

132. $\{x \mid x$ is a time of the day when the average level of happiness was 1$\}$

133. $\{x \mid x$ is a time of the day when
$3 <$ average level of happiness $< 4\}$

134. $\{x \mid x$ is a time of the day when
$3 <$ average level of happiness $\le 4\}$

135. Do the results of Exercise 131 or 132 indicate a one-to-one correspondence between the set representing the time of day and the set representing average level of happiness? Are these sets equivalent?

Critical Thinking Exercises

In Exercises 136–139, give examples of two sets that meet the given conditions. If the conditions are impossible to satisfy, explain why.

136. The two sets are equivalent but not equal.

137. The two sets are equivalent and equal.

138. The two sets are equal but not equivalent.

139. The two sets are neither equivalent nor equal.

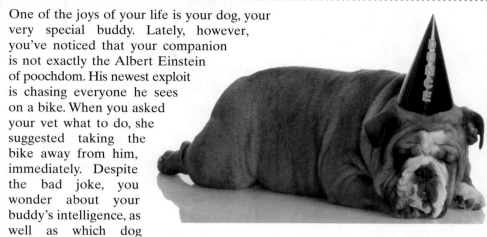

2.2 | Subsets

One of the joys of your life is your dog, your very special buddy. Lately, however, you've noticed that your companion is not exactly the Albert Einstein of poochdom. His newest exploit is chasing everyone he sees on a bike. When you asked your vet what to do, she suggested taking the bike away from him, immediately. Despite the bad joke, you wonder about your buddy's intelligence, as well as which dog breeds are considered the smartest. In this section, you will see an example involving sets, *subsets*, and the intelligence of dogs. We open the section with a discussion of subsets.

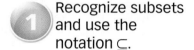 Recognize subsets and use the notation ⊆.

Subsets

In the twentieth century, all U.S. presidents were male. All the elements of the set of twentieth-century U.S. presidents are elements of the set of males. Situations like this, in which all the elements of one set are also elements of another set, are described by the following definition:

> **Definition of a Subset of a Set**
>
> Set A is a **subset** of set B, expressed as
>
> $$A \subseteq B,$$
>
> if every element in set A is also an element in set B.

Because every twentieth-century U.S. president was male, this means that the set of twentieth-century U.S. presidents is a subset of the set of males:

$$\{x \mid x \text{ is a twentieth-century U.S. president}\} \subseteq \{x \mid x \text{ is a male}\}.$$

Every person in this set, to the left of the subset symbol,

is also a member of this set, to the right of the subset symbol.

Observe that a subset is itself a set.

The notation $A \nsubseteq B$ means that A **is not a subset** of B. Set A is not a subset of set B if there is at least one element of set A that is not an element of set B. For example, consider the following sets:

$$A = \{1, 2, 3\} \quad \text{and} \quad B = \{1, 2\}.$$

Can you see that 3 is an element of set A that is not in set B? Thus, set A is not a subset of set B: $A \nsubseteq B$.

We can show that $A \subseteq B$ by showing that every element of set A also occurs as an element of set B. We can show that $A \nsubseteq B$ by finding one element of set A that is not in set B.

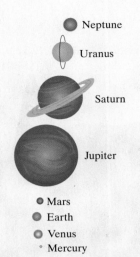

Neptune

Uranus

Saturn

Jupiter

Mars

Earth

Venus

Mercury

The eight planets in Earth's solar system

No, we did not forget Pluto. In 2006, based on the requirement that a planet must dominate its own orbit (Pluto is slave to Neptune's orbit), the International Astronomical Union removed Pluto from the list of planets and decreed that it belongs to a new category of heavenly body, a "dwarf planet."

Example 1   Using the Symbols $\subseteq$ and $\nsubseteq$

Write $\subseteq$ or $\nsubseteq$ in each blank to form a true statement:

a. $A = \{1, 3, 5, 7\}$
$B = \{1, 3, 5, 7, 9, 11\}$
A_____B

b. $A = \{x | x$ is a letter in the word *proof*$\}$
$B = \{y | y$ is a letter in the word *roof*$\}$
A_____B

c. $A = \{x | x$ is a planet of Earth's solar system$\}$
$B = \{$Mercury, Venus, Earth, Mars, Jupiter, Saturn, Uranus, Neptune$\}$
A_____B.

Solution

a. All the elements of $A = \{1, 3, 5, 7\}$ are also contained in $B = \{1, 3, 5, 7, 9, 11\}$. Therefore, set A is a subset of set B:

$$A \subseteq B.$$

b. Let's write the set of letters in the word *proof* and the set of letters in the word *roof* in roster form. In each case, we consider only the distinct elements, so there is no need to repeat the o.

$$A = \{p, r, o, f\} \qquad B = \{r, o, f\}$$

> The element p is in set A but not in set B.

Because there is an element in set A that is not in set B, set A is not a subset of set B:

$$A \nsubseteq B.$$

c. All the elements of

$$A = \{x | x \text{ is a planet of Earth's solar system}\}$$

are contained in

$$B = \{\text{Mercury, Venus, Earth, Mars, Jupiter, Saturn, Uranus, Neptune}\}.$$

Because all elements in set A are also in set B, set A is a subset of set B:

$$A \subseteq B.$$

Furthermore, the sets are equal $(A = B)$.

✓ Checkpoint 1   Write $\subseteq$ or $\nsubseteq$ in each blank to form a true statement:

a. $A = \{1, 3, 5, 6, 9, 11\}$
$B = \{1, 3, 5, 7\}$
A_____B

b. $A = \{x | x$ is a letter in the word *roof*$\}$
$B = \{y | y$ is a letter in the word *proof*$\}$
A_____B

c. $A = \{x | x$ is a day of the week$\}$
$B = \{$Monday, Tuesday, Wednesday, Thursday, Friday, Saturday, Sunday$\}$
A_____B.

Proper Subsets

2 Recognize proper subsets and use the notation $\subset$.

In Example 1(c) and Check Point 1(c), the given sets are equal and illustrate that **every set is a subset of itself**. If A is any set, then $A \subseteq A$ because it is obvious that each element of A is a member of A.

If we know that set A is a subset of set B and we exclude the possibility of the sets being equal, then set A is called a *proper subset* of set B, written $A \subset B$.

Definition of a Proper Subset of a Set

Set A is a **proper subset** of set B, expressed as $A \subset B$, if set A is a subset of set B and sets A and B are not equal ($A \neq B$).

Try not to confuse the symbols for subset, $\subseteq$, and proper subset, $\subset$. In some subset examples, both symbols can be placed between sets:

Set A Set B $\qquad\qquad$ Set A Set B

$$\{1,3\} \subseteq \{1,3,5\} \quad and \quad \{1,3\} \subset \{1,3,5\}.$$

A is a subset of *B*. Every element in *A* is also an element in *B*. $\qquad$ *A* is a proper subset of *B* because *A* and *B* are not equal sets.

By contrast, there are subset examples where only the symbol $\subseteq$ can be placed between sets:

Set A Set B

$$\{1,3,5\} \subseteq \{1,3,5\}.$$

A is a subset of *B*. Every element in *A* is also an element in *B*. *A* is not a proper subset of *B* because $A = B$. The symbol $\subset$ should not be placed between the sets.

Because the lower part of the subset symbol in $A \subseteq B$ suggests an equal sign, it is *possible* that sets A and B are equal, although they do not have to be. By contrast, the missing lower line for the proper subset symbol in $A \subset B$ indicates that sets A and B *cannot* be equal.

Study Tip

- The notation for "is a subset of," $\subseteq$, is similar to the notation for "is less than or equal to," $\leq$. Because the notations share similar ideas, $A \subseteq B$ applies to finite sets only if the cardinal number of set A is less than or equal to the cardinal number of set B.

- The notation for "is a proper subset of," $\subset$, is similar to the notation for "is less than," $<$. Because the notations share similar ideas, $A \subset B$ applies to finite sets only if the cardinal number of set A is less than the cardinal number of set B.

Example 2   Using the Symbols $\subseteq$ and $\subset$

Write $\subseteq$, $\subset$, or both in each blank to form a true statement:

a. $A = \{x \mid x$ is a person and x lives in San Francisco$\}$

$B = \{x \mid x$ is a person and x lives in California$\}$

A____B

b. $A = \{2,4,6,8\}$

$B = \{2,8,4,6\}$

A____B.

Solution

a. We begin with $A = \{x | x$ is a person and x lives in San Francisco$\}$ and $B = \{x | x$ is a person and x lives in California$\}$. Every person living in San Francisco is also a person living in California. Because each person in set A is contained in set B, set A is a subset of set B:

$$A \subseteq B.$$

Can you see that the two sets do not contain exactly the same elements and, consequently, are not equal? A person living in California outside San Francisco is in set B, but not in set A. Because the sets are not equal, set A is a proper subset of set B:

$$A \subset B.$$

The symbols $\subseteq$ and $\subset$ can both be placed in the blank to form a true statement.

b. We now consider $A = \{2, 4, 6, 8\}$ and $B = \{2, 8, 4, 6\}$. Every number in $A = \{2, 4, 6, 8\}$ is contained in $B = \{2, 8, 4, 6\}$, so set A is a subset of set B:

$$A \subseteq B.$$

Because the sets contain exactly the same elements and are equal, set A is *not* a proper subset of set B. The symbol $\subset$ cannot be placed in the blank if we want to form a true statement. (Because set A is not a proper subset of set B, it is correct to write $A \not\subset B$.)

Checkpoint 2   Write $\subseteq$, $\subset$, or both in each blank to form a true statement:

a. $A = \{2, 4, 6, 8\}$
$B = \{2, 8, 4, 6, 10\}$
A____B

b. $A = \{x | x$ is a person and x lives in Atlanta$\}$
$B = \{x | x$ is a person and x lives in Georgia$\}$
A____B.

Study Tip

Do not confuse the symbols $\in$ and $\subseteq$. The symbol $\in$ means "is an element of" and the symbol $\subseteq$ means "is a subset of." Notice the differences among the following true statements:

$4 \in \{4, 8\}$ $\{4\} \subseteq \{4, 8\}$ $\{4\} \notin \{4, 8\}$.

4 is an element of the set $\{4, 8\}$.

The set containing 4 is a subset of the set $\{4, 8\}$.

The set containing 4 is not an element of $\{4, 8\}$, although $\{4\} \in \{\{4\}, \{8\}\}$.

Subsets and the Empty Set

The meaning of $A \subseteq B$ leads to some interesting properties of the empty set.

Example 3   **The Empty Set as a Subset**

Let $A = \{\ \}$ and $B = \{1, 2, 3, 4, 5\}$. Is $A \subseteq B$?

Solution

A is not a subset of B ($A \not\subseteq B$) if there is at least one element of set A that is not an element of set B. Because A represents the empty set, there are no elements in set A, period, much less elements in A that do not belong to B. Because we cannot find an element in $A = \{\ \}$ that is not contained in $B = \{1, 2, 3, 4, 5\}$, this means that $A \subseteq B$. Equivalently, $\varnothing \subseteq B$.

> **Checkpoint 3**   Let $A = \{\ \}$ and $B = \{6, 7, 8\}$. Is $A \subseteq B$?

Example 3 illustrates the principle that **the empty set is a subset of every set**. Furthermore, the empty set is a proper subset of every set except itself.

> **The Empty Set as a Subset**
> 1. For any set B, $\varnothing \subseteq B$.
> 2. For any set B other than the empty set, $\varnothing \subset B$.

The Number of Subsets of a Given Set

③ Determine the number of subsets of a set.

If a set contains n elements, how many subsets can be formed? Let's observe some special cases, namely sets with 0, 1, 2, and 3 elements. We can use inductive reasoning to arrive at a general conclusion. We begin by listing subsets and counting the number of subsets in our list. This is shown in **Table 2.3**.

Table 2.3 The Number of Subsets: Some Special Cases

Set	Number of Elements	List of All Subsets	Number of Subsets
$\{\ \}$	0	$\{\ \}$	1
$\{a\}$	1	$\{a\}, \{\ \}$	2
$\{a, b\}$	2	$\{a, b\}, \{a\}, \{b\}, \{\ \}$	4
$\{a, b, c\}$	3	$\{a, b, c\},$ $\{a, b\}, \{a, c\}, \{b, c\},$ $\{a\}, \{b\}, \{c\}, \{\ \}$	8

Table 2.3 suggests that when we increase the number of elements in the set by one, the number of subsets doubles. The number of subsets appears to be a power of 2.

Number of elements	0	1	2	3
Number of subsets	$1 = 2^0$	$2 = 2^1$	$4 = 2 \times 2 = 2^2$	$8 = 2 \times 2 \times 2 = 2^3$

The power of 2 is the same as the number of elements in the set. Using inductive reasoning, if the set contains n elements, then the number of subsets that can be formed is 2^n.

> **Number of Subsets**
> The number of subsets of a set with n elements is 2^n.

For a given set, we know that every subset except the set itself is a proper subset. In **Table 2.3**, we included the set itself when counting the number of subsets. If we want to find the number of proper subsets, we must exclude counting the given set, thereby decreasing the number by 1.

A Brief Review • Powers of 2

If powers of 2 have you in an exponentially increasing state of confusion, here's a list of values that should be helpful. Observe how rapidly these values are increasing.

$2^0 = 1$

$2^1 = 2$

$2^2 = 2 \times 2 = 4$

$2^3 = 2 \times 2 \times 2 = 8$

$2^4 = 2 \times 2 \times 2 \times 2 = 16$

$2^5 = 2 \times 2 \times 2 \times 2 \times 2 = 32$

$2^6 = 64$

$2^7 = 128$

$2^8 = 256$

$2^9 = 512$

$2^{10} = 1024$

$2^{11} = 2048$

$2^{12} = 4096$

$2^{15} = 32,768$

$2^{20} = 1,048,576$

$2^{25} = 33,554,432$

$2^{30} = 1,073,741,824$

Number of Proper Subsets

The number of proper subsets of a set with n elements is $2^n - 1$.

Example 4 · Finding the Number of Subsets and Proper Subsets

Find the number of subsets and the number of proper subsets for each set:

a. $\{a, b, c, d, e\}$

b. $\{x \mid x \in \mathbf{N}$ and $9 \le x \le 15\}$.

Solution

a. A set with n elements has 2^n subsets. Because the set $\{a, b, c, d, e\}$ contains 5 elements, there are $2^5 = 2 \times 2 \times 2 \times 2 \times 2 = 32$ subsets. Of these, we must exclude counting the given set as a proper subset, so there are $2^5 - 1 = 32 - 1 = 31$ proper subsets.

b. We can write $\{x \mid x \in \mathbf{N}$ and $9 \le x \le 15\}$ in roster form as $\{9, 10, 11, 12, 13, 14, 15\}$. Because this set contains 7 elements, there are $2^7 = 2 \times 2 \times 2 \times 2 \times 2 \times 2 \times 2 = 128$ subsets. Of these, there are $2^7 - 1 = 128 - 1 = 127$ proper subsets.

✓ **Checkpoint 4** Find the number of subsets and the number of proper subsets for each set:

a. $\{a, b, c, d\}$

▶ **b.** $\{x \mid x \in \mathbf{N}$ and $3 \le x \le 8\}$.

Example 5 · Finding the Number of Subsets and Proper Subsets

According to *The Intelligence of Dogs* by Stanley Coren, the four smartest breeds of dogs are

1. Border collie

2. Poodle

3. German shepherd

4. Golden retriever.

A man is trying to decide whether to buy a dog. He can afford up to four dogs, but may buy fewer or none at all. However, if the man makes a purchase, he will buy one or more of the four smartest breeds. The kennel has only one dog from each breed—that is, one border collie, one poodle, one German shepherd, and one golden retriever.

a. Consider the set that represents these breeds:

$$\{\text{border collie, poodle, German shepherd, golden retriever}\}.$$

Find the number of subsets of this set. What does this number represent in practical terms?

b. List all the subsets and describe what they represent for the prospective dog owner.

c. How many of the subsets are proper subsets?

**Smartest Dogs
Rank/Breed**

1. Border collie

2. Poodle

3. German shepherd

4. Golden retriever.

Source: The Intelligence of Dogs

Solution

a. A set with n elements has 2^n subsets. Because the set of breeds contains four elements, the number of subsets is $2^4 = 2 \times 2 \times 2 \times 2 = 16$. In practical terms, there are 16 different ways to make a selection.

b. Purchase no dogs: $\varnothing$

Purchase one dog: {border collie}, {poodle},
{German shepherd}, {golden retriever}

Purchase two dogs: {border collie, poodle},
{border collie, German shepherd},
{border collie, golden retriever},
{poodle, German shepherd},
{poodle, golden retriever},
{German shepherd, golden retriever}

Purchase three dogs: {border collie, poodle, German shepherd},
{border collie, poodle, golden retriever},
{border collie, German shepherd, golden retriever},
{poodle, German shepherd, golden retriever}

Purchase four dogs: {border collie, poodle, German shepherd, golden retriever}

As predicted in part(a), 16 different selections are possible.

c. Every set except

{border collie, poodle, German shepherd, golden retriever}

is a proper subset of the given set. There are $16 - 1$, or 15, proper subsets. Although we obtained this number by counting, we can also use the fact that the number of proper subsets of a set with n elements is $2^n - 1$. Because the set of breeds contains four elements, there are $2^4 - 1$ proper subsets:

$$2^4 - 1 = (2 \times 2 \times 2 \times 2) - 1 = 16 - 1 = 15 \text{ proper subsets.}$$

Checkpoint 5   According to the American Film Institute, the set of the three best movies of all time consists of

{*Citizen Kane* (1941), *The Godfather* (1972), *Casablanca* (1942)}.

Suppose that you have all three films on DVD and decide to view some, all, or none of these films.

a. Find the number of subsets of the given set. What does this number represent in practical terms?

b. List all the subsets and describe what they represent in terms of viewing options.

c. How many of the subsets are proper subsets?

Achieving Success

Use index cards to help learn new terms.

Many of the terms, notations, and formulas used in this book may be new to you. Buy a pack of 3×5 index cards. On each card, list a new vocabulary word, symbol, or title of a formula. On the other side of the card, put the definition or formula. Here is an example:

Effective Index Card

Formula for the number of subsets	A set with n elements has 2^n subsets.
Front	Back

Review these cards frequently. Use the cards to quiz yourself and prepare for exams.

Exercise Set 2.2

Concept and Vocabulary Exercises

In Exercises 1–4, fill in each blank so that the resulting statement is true.

1. Set A is a subset of set B, expressed as _____ , means that _____ .

2. Set A is a proper subset of set B, expressed as _____ , means that set A is a subset of set B and _____ .

3. The number of subsets of a set with n elements is _____ .

4. The number of proper subsets of a set with n elements is _____ .

In Exercises 5–8, determine whether each statement is true or false. If the statement is false, make the necessary change(s) to produce a true statement.

5. The set $\{3\}$ has 2^3, or eight, subsets.

6. All sets have subsets.

7. Every set has a proper subset.

8. The set $\{3, \{1, 4\}\}$ has eight subsets.

Respond to Exercises 9–14 using verbal or written explanations.

9. Explain what is meant by a subset.

10. What is the difference between a subset and a proper subset?

11. Explain why the empty set is a subset of every set.

12. Describe the difference between the symbols $\in$ and $\subseteq$. Explain how each symbol is used.

13. Describe the formula for finding the number of subsets for a given set. Give an example.

14. Describe how to find the number of proper subsets for a given set. Give an example.

In Exercises 15–18, determine whether each statement makes sense or does not make sense, and explain your reasoning.

15. The set of my parents' six mortgage payments from January through June is a subset of the set of their 12 cable television payments from January through December.

16. Every time I increase the number of elements in a set by one, I double the number of subsets.

17. Because Example 5 (the smartest breeds of dogs) and Checkpoint 5 (the best movies of all time) involve different situations, I cannot solve them by the same method.

18. I recently purchased a set of books and am deciding which books, if any, to take on vacation. The number of subsets of my set of books gives me the number of different combinations of the books that I can take.

Practice Exercises

In Exercises 19–36, write $\subseteq$ or $\nsubseteq$ in each blank so that the resulting statement is true.

19. $\{1, 2, 5\}$ _____ $\{1, 2, 3, 4, 5, 6, 7\}$

20. $\{2, 3, 7\}$ _____ $\{1, 2, 3, 4, 5, 6, 7\}$

21. $\{-3, 0, 3\}$ _____ $\{-3, -1, 1, 3\}$

22. $\{-4, 0, 4\}$ _____ $\{-4, -3, -1, 1, 3, 4\}$

23. $\{$Monday, Friday$\}$ _____ $\{$Saturday, Sunday, Monday, Tuesday, Wednesday$\}$

24. $\{$Mercury, Venus, Earth$\}$ _____ $\{$Venus, Earth, Mars, Jupiter$\}$

25. $\{x \,|\, x$ is a cat$\}$ _____ $\{x \,|\, x$ is a black cat$\}$

26. $\{x \,|\, x$ is a dog$\}$ _____ $\{x \,|\, x$ is a purebred dog$\}$

27. $\{$c, o, n, v, e, r, s, a, t, i, o, n$\}$ _____ $\{$v, o, i, c, e, s, r, a, n, t, o, n$\}$

28. $\{$r, e, v, o, l, u, t, i, o, n$\}$ _____ $\{$t, o, l, o, v, e, r, u, i, n$\}$

29. $\left\{\frac{4}{7}, \frac{9}{13}\right\}$ _____ $\left\{\frac{7}{4}, \frac{13}{9}\right\}$

30. $\left\{\frac{1}{2}, \frac{1}{3}\right\}$ _____ $\{2, 3, 5\}$

31. $\varnothing$ _____ $\{2, 4, 6\}$

32. $\varnothing$ _____ $\{1, 3, 5\}$

33. $\{2, 4, 6\}$ _____ $\varnothing$

34. $\{1, 3, 5\}$ _____ $\varnothing$

35. $\{\ \}$ _____ $\varnothing$

36. $\varnothing$ _____ $\{\ \}$

In Exercises 37–58, determine whether $\subseteq$, $\subset$, both, or neither can be placed in each blank to form a true statement.

37. $\{$V, C, R$\}$ _____ $\{$V, C, R, S$\}$

38. $\{$F, I, N$\}$ _____ $\{$F, I, N, K$\}$

39. $\{0, 2, 4, 6, 8\}$ _____ $\{8, 0, 6, 2, 4\}$

40. $\{9, 1, 7, 3, 4\}$ _____ $\{1, 3, 4, 7, 9\}$

41. $\{x \,|\, x$ is a man$\}$ _____ $\{x \,|\, x$ is a woman$\}$

42. $\{x \,|\, x$ is a woman$\}$ _____ $\{x \,|\, x$ is a man$\}$

43. $\{x \,|\, x$ is a man$\}$ _____ $\{x \,|\, x$ is a person$\}$

44. $\{x \,|\, x$ is a woman$\}$ _____ $\{x \,|\, x$ is a person$\}$

45. $\{x \,|\, x$ is a man or a woman$\}$ _____ $\{x \,|\, x$ is a person$\}$

46. $\{x \,|\, x$ is a woman or a man$\}$ _____ $\{x \,|\, x$ is a person$\}$

47. $A = \{x \,|\, x \in \mathbf{N}$ and $5 < x < 12\}$
    $B =$ the set of natural numbers between 5 and 12
    A ___ B

48. $A = \{x \,|\, x \in \mathbf{N}$ and $3 < x < 10\}$
    $B =$ the set of natural numbers between 3 and 10
    A ___ B

49. $A = \{x \,|\, x \in \mathbf{N}$ and $5 < x < 12\}$
    $B =$ the set of natural numbers between 3 and 17
    A ___ B

50. $A = \{x \,|\, x \in \mathbf{N}$ and $3 < x < 10\}$
    $B =$ the set of natural numbers between 2 and 16
    A ___ B

51. $A = \{x \,|\, x \in \mathbf{N}$ and $5 < x < 12\}$
    $B = \{x \,|\, x \in \mathbf{N}$ and $2 \le x \le 11\}$
    A ___ B

52. $A = \{x \,|\, x \in \mathbf{N}$ and $3 < x < 10\}$
    $B = \{x \,|\, x \in \mathbf{N}$ and $2 \le x \le 8\}$
    A ___ B

53. $\varnothing$ ____ $\{7, 8, 9, \ldots, 100\}$

54. $\varnothing$ ____ $\{101, 102, 103, \ldots, 200\}$

55. $\{7, 8, 9, \ldots\}$ ____ $\varnothing$

56. $\{101, 102, 103, \ldots\}$ ____ $\varnothing$

57. $\varnothing$ ____ $\{\ \}$

58. $\{\ \}$ ____ $\varnothing$

In Exercises 59–72, determine whether each statement is true or false. If the statement is false, explain why.

59. Homer $\in$ {Homer, Marge, Bart, Lisa}

60. Canada $\in$ {Mexico, United States, Canada}

61. Homer $\subseteq$ {Homer, Marge, Bart, Lisa}

62. Canada $\subseteq$ {Mexico, United States, Canada}

63. {Homer} $\subseteq$ {Homer, Marge, Bart, Lisa}

64. {Canada} $\subseteq$ {Mexico, United States, Canada}

65. $\varnothing \in$ {Harry, James, Lily, Vernon}

66. $\varnothing \subseteq$ {Tina Fey, Miley Cyrus, Steve Carell}

67. $\{5\} \in \{\{5\}, \{9\}\}$

68. $\{1\} \in \{\{1\}, \{3\}\}$

69. $\{1, 4\} \not\subseteq \{4, 1\}$

70. $\{1, 4\} \not\subset \{4, 1\}$

71. $0 \notin \varnothing$

72. $\{0\} \not\subseteq \varnothing$

In Exercises 73–78, list all the subsets of the given set.

73. {border collie, poodle}   74. {Romeo, Juliet}

75. {t, a, b}   76. {I, II, III}

77. {0}   78. $\varnothing$

In Exercises 79–86, calculate the number of subsets and the number of proper subsets for each set.

79. $\{2, 4, 6, 8\}$

80. $\left\{\frac{1}{2}, \frac{1}{3}, \frac{1}{4}, \frac{1}{5}\right\}$

81. $\{2, 4, 6, 8, 10, 12\}$

82. $\{a, b, c, d, e, f\}$

83. $\{x \mid x \text{ is a day of the week}\}$

84. $\{x \mid x \text{ is a U.S. coin worth less than a dollar}\}$

85. $\{x \mid x \in \mathbf{N} \quad \text{and} \quad 2 < x < 6\}$

86. $\{x \mid x \in \mathbf{N} \quad \text{and} \quad 2 \leq x \leq 6\}$

Practice Plus

In Exercises 87–96, determine whether each statement is true or false. If the statement is false, make the necessary change(s) to produce a true statement.

87. The set $\{1, 2, 3, \ldots, 1000\}$ has 2^{1000} proper subsets.

88. The set $\{1, 2, 3, \ldots, 10,000\}$ has $2^{10,000}$ proper subsets.

89. $\{x \mid x \in \mathbf{N} \text{ and } 30 < x < 50\} \subseteq \{x \mid x \in \mathbf{N} \text{ and } 30 \leq x \leq 50\}$

90. $\{x \mid x \in \mathbf{N} \text{ and } 20 \leq x \leq 60\} \not\subseteq \{x \mid x \in \mathbf{N} \text{ and } 20 < x < 60\}$

91. $\varnothing \not\subseteq \{\varnothing, \{\varnothing\}\}$

92. $\{\varnothing\} \not\subseteq \{\varnothing, \{\varnothing\}\}$

93. $\varnothing \in \{\varnothing, \{\varnothing\}\}$

94. $\{\varnothing\} \in \{\varnothing, \{\varnothing\}\}$

95. If $A \subseteq B$ and $d \in A$, then $d \in B$.

96. If $A \subseteq B$ and $B \subseteq C$, then $A \subseteq C$.

Application Exercises

97. Houses in Euclid Estates are all identical. However, a person can purchase a new house with some, all, or none of a set of options. This set includes {pool, screened-in balcony, lake view, alarm system, upgraded landscaping}. How many options are there for purchasing a house in this community?

98. A cheese pizza can be ordered with some, all, or none of the following set of toppings: {beef, ham, mushrooms, sausage, peppers, pepperoni, olives, prosciutto, onion}. How many different variations are available for ordering a pizza?

99. Here's a sampler of Will Smith's films from 2002 through 2008, including box office grosses:

    Men in Black II (2002: $442 million), *Shark Tale* (2004: $367 million), *Hitch* (2005: $368 million), *The Pursuit of Happyness* (2006: $307 million), *I am Legend* (2007: $585 million), *Hancock* (2008: $624 million).
    (*Source: Entertainment Weekly*)

    Suppose that you have all six films on DVD and decide to view some, all, or none of these films. How many viewing options do you have?

100. A small town has four police cars. If a radio dispatcher receives a call, depending on the nature of the situation, no cars, one car, two cars, three cars, or all four cars can be sent. How many options does the dispatcher have for sending the police cars to the scene of the caller?

101. According to the U.S. Census Bureau, the most ethnically diverse U.S. cities are New York City, Los Angeles, Miami, Chicago, Washington, D.C., Houston, San Diego, and Seattle. If you decide to visit some, all, or none of these cities, how many travel options do you have?

102. First unveiled in 1998 based on more than 1500 ballots sent to film notables, the American Film Institute rated the top U.S. movies. Updated in 2007, the Institute selected *Citizen Kane* (1941), *The Godfather* (1972), *Casablanca* (1942), *Raging Bull* (1980), *Singin' in the Rain* (1952), *Gone with the Wind* (1939), *Lawrence of Arabia* (1962), *Schindler's List* (1993), *Vertigo* (1958), and *The Wizard of Oz* (1939). Suppose that you have all ten films on DVD and decide to view some, all, or none of these films. How many viewing options do you have?

Critical Thinking Exercises

103. Suppose that a nickel, a dime, and a quarter are on a table. You may select some, all, or none of the coins. Specify all of the different amounts of money that can be selected.

104. If a set has 127 proper subsets, how many elements are there in the set?

2.3 | Venn Diagrams and Set Operations

1. Understand the meaning of a universal set.

2. Understand the basic ideas of a Venn diagram.

3. Use Venn diagrams to visualize relationships between two sets.

4. Find the complement of a set.

5. Find the intersection of two sets.

6. Find the union of two sets.

7. Perform operations with sets.

8. Determine sets involving set operations from a Venn diagram.

9. Understand the meaning of *and* and *or*.

Sí TV, a 24-hour cable channel targeted to young U.S. Latinos, was launched in 2004 and is now in more than 18 million households. Its motto: "Speak English. Live Latin." As Latino spending power steadily rises, corporate America has discovered that Hispanic Americans, particularly young spenders between the ages of 14 and 34, want to be spoken to in English, even as they stay true to their Latino identity.

What is the primary language spoken at home by U.S. Hispanics? In this section, we use sets to analyze the answer to this question. By doing so, you will see how sets and their visual representations provide precise ways of organizing, classifying, and describing a wide variety of data.

① Understand the meaning of a universal set.

Universal Sets and Venn Diagrams

The circle graph in **Figure 2.2** categorizes America's 46 million Hispanics by the primary language spoken at home. The graph's sectors define four sets:

- the set of U.S. Hispanics who speak Spanish at home
- the set of U.S. Hispanics who speak English at home
- the set of U.S. Hispanics who speak both Spanish and English at home
- the set of U.S. Hispanics who speak neither Spanish nor English at home.

In discussing sets, it is convenient to refer to a general set that contains all elements under discussion. This general set is called the *universal set*. A **universal set**, symbolized by U, is a set that contains all the elements being considered in a given discussion or problem. Thus, a convenient universal set for the sets described above is

$$U = \text{the set of U.S. Hispanics.}$$

Notice how this universal set restricts our attention so that we can divide it into the four subsets shown by the circle graph in **Figure 2.2**.

We can obtain a more thorough understanding of sets and their relationship to a universal set by considering diagrams that allow visual analysis. **Venn diagrams**, named for the British logician John Venn (1834–1923), are used to show the visual relationship among sets.

Figure 2.3 is a Venn diagram. The universal set is represented by a region inside a rectangle. Subsets within the universal set are depicted by circles, or sometimes by ovals or other shapes. In this Venn diagram, set A is represented by the light blue region inside the circle.

The dark blue region in **Figure 2.3** represents the set of elements in the universal set U that are not in set A. By combining the regions shown by the light blue shading and the dark blue shading, we obtain the universal set, U.

Languages Spoken at Home by U.S. Hispanics

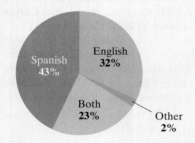

Spanish 43%
English 32%
Both 23%
Other 2%

Figure 2.2
Source: Time

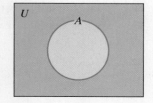

Figure 2.3

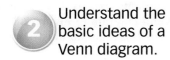

Understand the basic ideas of a Venn diagram.

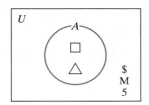

Figure 2.4

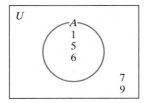

Figure 2.5

Use Venn diagrams to visualize relationships between two sets.

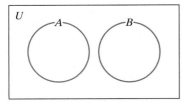

Figure 2.6

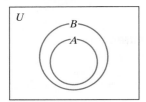

Figure 2.7

Example 1   Determining Sets from a Venn Diagram

Use the Venn diagram in **Figure 2.4** to determine each of the following sets:

a. U **b.** A **c.** the set of elements in U that are not in A.

Solution

a. Set U, the universal set, consists of all the elements within the rectangle. Thus, $U = \{\Box, \triangle, \$, M, 5\}$.

b. Set A consists of all the elements within the circle. Thus, $A = \{\Box, \triangle\}$.

c. The set of elements in U that are not in A, shown by the set of all the elements outside the circle, is $\{\$, M, 5\}$.

 Checkpoint 1   Use the Venn diagram in **Figure 2.5** to determine each of the following sets:

a. U

b. A

▶ **c.** the set of elements in U that are not in A.

Representing Two Sets in a Venn Diagram

There are a number of different ways to represent two subsets of a universal set in a Venn diagram. To help understand these representations, consider the following scenario:

Your high school math class has been asked to determine whether there is sufficient support at the local community college to have a blood drive. You take a survey to obtain information, asking students from the college

Would you be willing to donate blood?

Would you be willing to help serve a free breakfast to blood donors?

Set A represents the set of students willing to donate blood. Set B represents the set of students willing to help serve breakfast to donors. Possible survey results include the following:

• No students willing to donate blood are willing to serve breakfast, and vice versa.
• All students willing to donate blood are willing to serve breakfast.
• The same students who are willing to donate blood are willing to serve breakfast.
• Some of the students willing to donate blood are willing to serve breakfast.

We begin by using Venn diagrams to visualize these results. To do so, we consider four basic relationships and their visualizations.

Relationship 1: Disjoint Sets Two sets that have no elements in common are called **disjoint sets**. Two disjoint sets, A and B, are shown in the Venn diagram in **Figure 2.6**. Disjoint sets are represented as circles that do not overlap. No elements of set A are elements of set B, and vice versa.

Since set A represents the set of students willing to donate blood and set B represents the set of students willing to serve breakfast to donors, the set diagram illustrates

No students willing to donate blood are willing to serve breakfast, and vice versa.

Relationship 2: Proper Subsets If set A is a proper subset of set B $(A \subset B)$, the relationship is shown in the Venn diagram in **Figure 2.7**. All elements of set A are elements of set B. If an x representing an element is placed inside circle A, it automatically falls inside circle B.

Since set A represents the set of students willing to donate blood and set B represents the set of students willing to serve breakfast to donors, the set diagram illustrates

All students willing to donate blood are willing to serve breakfast.

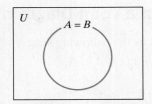

Figure 2.8

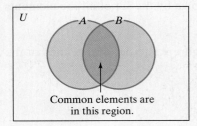

Common elements are
in this region.

Figure 2.9

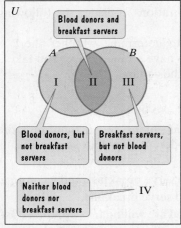

A: Set of blood donors
B: Set of breakfast servers

Figure 2.10

Relationship 3: Equal Sets If $A = B$, then set A contains exactly the same elements as set B. This relationship is shown in the Venn diagram in **Figure 2.8**. Because all elements in set A are in set B, and vice versa, this diagram illustrates that when $A = B$, then $A \subseteq B$ and $B \subseteq A$.

Since set A represents the set of students willing to donate blood and set B represents the set of students willing to serve breakfast to donors, the set diagram illustrates

The same students who are willing to donate blood are willing to serve breakfast.

Relationship 4: Sets with Some Common Elements In mathematics, the word *some* means *there exists at least one*. If set A and set B have at least one element in common, then the circles representing the sets must overlap. This is illustrated in the Venn diagram in **Figure 2.9**.

Since set A represents the set of students willing to donate blood and set B represents the set of students willing to serve breakfast to donors, the presence of at least one student in the dark blue region in **Figure 2.9** illustrates

Some students willing to donate blood are willing to serve breakfast.

In **Figure 2.10**, we've numbered each of the regions in the Venn diagram in **Figure 2.9**. Let's make sure we understand what these regions represent in terms of the campus blood drive scenario. Remember that A is the set of blood donors and B is the set of breakfast servers.

In **Figure 2.10**, we'll start with the innermost region, region II, and work outward to region IV.

Region II	This region represents the set of students willing to donate blood and serve breakfast. The elements that belong to both set A and set B are in this region.
Region I	This region represents the set of students willing to donate blood but not serve breakfast. The elements that belong to set A but not to set B are in this region.
Region III	This region represents the set of students willing to serve breakfast but not donate blood. The elements that belong to set B but not to set A are in this region.
Region IV	This region represents the set of students surveyed who are not willing to donate blood and are not willing to serve breakfast. The elements that belong to the universal set U that are not in sets A or B are in this region.

Example 2   Determining Sets from a Venn Diagram

Use the Venn diagram in **Figure 2.11** to determine each of the following sets:

a. U

b. B

c. the set of elements in A but not B

d. the set of elements in U that are not in B

e. the set of elements in both A and B.

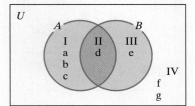

Figure 2.11

Solution

a. Set U, the universal set, consists of all elements within the rectangle. Taking the elements in regions I, II, III, and IV, we obtain $U = \{a, b, c, d, e, f, g\}$.

b. Set B consists of the elements in regions II and III. Thus, $B = \{d, e\}$.

c. The set of elements in A but not B, found in region I, is $\{a, b, c\}$.

d. The set of elements in U that are not in B, found in regions I and IV, is $\{a, b, c, f, g\}$.

e. The set of elements in both A and B, found in region II, is $\{d\}$.

> **Checkpoint 2**   Use the Venn diagram in **Figure 2.11** to determine each of the following sets:

a. A

b. the set of elements in B but not A

c. the set of elements in U that are not in A

d. the set of elements in U that are not in A or B.

Blitzer Bonus

A Venn Diagram for the Dogs

In the Venn diagram shown below, the universal set is the set of purebred dogs. Within the universal set, two subsets are shown.

U: Purebred Dogs

Set of Smartest Dogs
{Border collie, Poodle, German shepherd, Golden retriever, Doberman pinscher, Shetland sheepdog, Labrador retriever, Papillon, Rottweiler, Australian cattle dog}

Set of Least Intelligent Dogs
{Afghan hound, Basenji, Bulldog, Chow Chow, Borzoi, Bloodhound, Pekingese, Beagle, Mastiff, Basset hound}

If Afghan hounds could read, they'd probably disagree with being an element of the set of least intelligent dogs.

Source: Stanley Coren, The Intelligence of Dogs

The Complement of a Set

4 Find the complement of a set.

In arithmetic, we use operations such as addition and multiplication to combine numbers. We now turn to three set operations, called *complement, intersection*, and *union*. We begin by defining a set's complement.

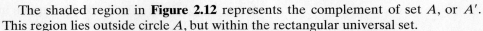

Definition of the Complement of a Set

The **complement** of set A, symbolized by A', is the set of all elements in the universal set that are *not* in A. This idea can be expressed in set-builder notation as follows:

$$A' = \{x \mid x \in U \quad \text{and} \quad x \notin A\}.$$

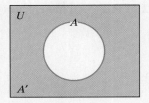

Figure 2.12

The shaded region in **Figure 2.12** represents the complement of set A, or A'. This region lies outside circle A, but within the rectangular universal set.

In order to find A', a universal set U must be given. A fast way to find A' is to cross out the elements in U that are given to be in set A. The set that remains is A'.

Example 3 **Finding a Set's Complement**

Let $U = \{1, 2, 3, 4, 5, 6, 7, 8, 9\}$ and $A = \{1, 3, 4, 7\}$. Find A'.

Solution

Set A' contains all the elements of set U that are not in set A. Because set A contains the elements $1, 3, 4,$ and 7, these elements cannot be members of set A':

$$\{\cancel{1}, 2, \cancel{3}, \cancel{4}, 5, 6, \cancel{7}, 8, 9\}.$$

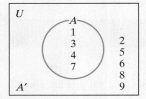

Figure 2.13

Thus, set A' contains $2, 5, 6, 8,$ and 9:

$$A' = \{2, 5, 6, 8, 9\}.$$

A Venn diagram illustrating A and A' is shown in **Figure 2.13**.

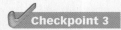 **Checkpoint 3** Let $U = \{a, b, c, d, e\}$ and $A = \{a, d\}$. Find A'.

The Intersection of Sets

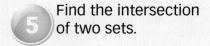 Find the intersection of two sets.

If A and B are sets, we can form a new set consisting of all elements that are in both A and B. This set is called the *intersection* of the two sets.

Definition of the Intersection of Sets

The **intersection** of sets A and B, written $A \cap B$, is the set of elements common to both set A and set B. This definition can be expressed in set-builder notation as follows:

$$A \cap B = \{x \mid x \in A \quad \text{and} \quad x \in B\}.$$

In Example 4, we are asked to find the intersection of two sets. This is done by listing the common elements of both sets. Because the intersection of two sets is also a set, we enclose these elements with braces.

Example 4 **Finding the Intersection of Two Sets**

Find each of the following intersections:

a. $\{7, 8, 9, 10, 11\} \cap \{6, 8, 10, 12\}$

b. $\{1, 3, 5, 7, 9\} \cap \{2, 4, 6, 8\}$

c. $\{1, 3, 5, 7, 9\} \cap \varnothing$.

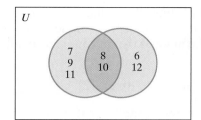

Figure 2.14 The numbers 8 and 10 belong to both sets.

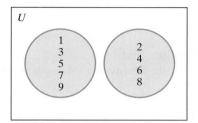

Figure 2.15 These disjoint sets have no common elements.

Solution

a. The elements common to $\{7, 8, 9, 10, 11\}$ and $\{6, 8, 10, 12\}$ are 8 and 10. Thus,

$$\{7, 8, 9, 10, 11\} \cap \{6, 8, 10, 12\} = \{8, 10\}.$$

The Venn diagram in **Figure 2.14** illustrates this situation.

b. The sets $\{1, 3, 5, 7, 9\}$ and $\{2, 4, 6, 8\}$ have no elements in common. Thus,

$$\{1, 3, 5, 7, 9\} \cap \{2, 4, 6, 8\} = \varnothing.$$

The Venn diagram in **Figure 2.15** illustrates this situation. The sets are disjoint.

c. There are no elements in $\varnothing$, the empty set. This means that there can be no elements belonging to both $\{1, 3, 5, 7, 9\}$ and $\varnothing$. Therefore,

$$\{1, 3, 5, 7, 9\} \cap \varnothing = \varnothing.$$

 **Checkpoint 4**   Find each of the following intersections:

a. $\{1, 3, 5, 7, 10\} \cap \{6, 7, 10, 11\}$

b. $\{1, 2, 3\} \cap \{4, 5, 6, 7\}$

► c. $\{1, 2, 3\} \cap \varnothing.$

The Union of Sets

 Find the union of two sets.

Another set that we can form from sets A and B consists of elements that are in A or B or in both sets. This set is called the *union* of the two sets.

> **Definition of the Union of Sets**
>
> The **union** of sets A and B, written $A \cup B$, is the set of elements that are members of set A or of set B or of both sets. This definition can be expressed in set-builder notation as follows:
>
> $$A \cup B = \{x \mid x \in A \quad \text{or} \quad x \in B\}.$$

We can find the union of set A and set B by listing the elements of set A. Then, we include any elements of set B that have not already been listed. Enclose all elements that are listed with braces. This shows that the union of two sets is also a set.

Example 5   **Finding the Union of Two Sets**

Find each of the following unions:

a. $\{7, 8, 9, 10, 11\} \cup \{6, 8, 10, 12\}$

b. $\{1, 3, 5, 7, 9\} \cup \{2, 4, 6, 8\}$

c. $\{1, 3, 5, 7, 9\} \cup \varnothing.$

Solution

This example uses the same sets as in Example 4. However, this time we are finding the unions of the sets, rather than their intersections.

a. To find $\{7, 8, 9, 10, 11\} \cup \{6, 8, 10, 12\}$, start by listing all the elements from the first set, namely 7, 8, 9, 10, and 11. Now list all the elements from the second set that are not in the first set, namely 6 and 12. The union is the set consisting of all these elements. Thus,

$$\{7, 8, 9, 10, 11\} \cup \{6, 8, 10, 12\} = \{6, 7, 8, 9, 10, 11, 12\}.$$

b. To find $\{1, 3, 5, 7, 9\} \cup \{2, 4, 6, 8\}$, list the elements from the first set, namely 1, 3, 5, 7, and 9. Now add to the list the elements in the second set that are not in the first set. This includes every element in the second set, namely 2, 4, 6, and 8. The union is the set consisting of all these elements, so

$$\{1, 3, 5, 7, 9\} \cup \{2, 4, 6, 8\} = \{1, 2, 3, 4, 5, 6, 7, 8, 9\}.$$

c. To find $\{1, 3, 5, 7, 9\} \cup \varnothing$, list the elements from the first set, namely 1, 3, 5, 7, and 9. Because there are no elements in $\varnothing$, the empty set, there are no additional elements to add to the list. Thus,

$$\{1, 3, 5, 7, 9\} \cup \varnothing = \{1, 3, 5, 7, 9\}.$$

Examples 4 and 5 illustrate the role that the empty set plays in intersection and union.

> **The Empty Set in Intersection and Union**
>
> For any set A,
>
> **1.** $A \cap \varnothing = \varnothing$
> **2.** $A \cup \varnothing = A$.

Checkpoint 5 Find each of the following unions:

a. $\{1, 3, 5, 7, 10\} \cup \{6, 7, 10, 11\}$
b. $\{1, 2, 3\} \cup \{4, 5, 6, 7\}$
c. $\{1, 2, 3\} \cup \varnothing$.

Performing Set Operations

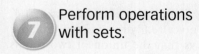 Perform operations with sets.

Some problems involve more than one set operation. The set notation specifies the order in which we perform these operations. **Always begin by performing any operations inside parentheses.** Here are two examples involving sets we will find in Example 6.

- Finding $(A \cup B)'$

  Step 1. Parentheses indicate to first find the union of A and B.

  Step 2. Find the complement of $A \cup B$.

- Finding $A' \cap B'$

  Step 1. Find the complement of A.

  Step 2. Find the complement of B.

  Step 3. Find the intersection of A' and B'.

Example 6   Performing Set Operations

Given

$$U = \{1, 2, 3, 4, 5, 6, 7, 8, 9, 10\}$$
$$A = \{1, 3, 7, 9\}$$
$$B = \{3, 7, 8, 10\},$$

find each of the following sets:

a. $(A \cup B)'$ **b.** $A' \cap B'$.

Solution

a. To find $(A \cup B)'$ we will first work inside the parentheses and determine $A \cup B$. Then we'll find the complement of $A \cup B$, namely $(A \cup B)'$.

$$A \cup B = \{1, 3, 7, 9\} \cup \{3, 7, 8, 10\} \quad \text{These are the given sets.}$$

$$= \{1, 3, 7, 8, 9, 10\} \quad \text{Join (unite) the elements, listing the common elements (3 and 7) only once.}$$

Now find $(A \cup B)'$, the complement of $A \cup B$.

$$(A \cup B)' = \{1, 3, 7, 8, 9, 10\}'$$

$$= \{2, 4, 5, 6\} \quad \text{List the elements in the universal set that are not listed in } \{1, 3, 7, 8, 9, 10\}: \{\cancel{1}, 2, \cancel{3}, 4, 5, 6, \cancel{7}, \cancel{8}, \cancel{9}, \cancel{10}\}.$$

b. To find $A' \cap B'$, we must first identify the elements in A' and B'. Set A' is the set of elements of U that are not in set A:

$$A' = \{2, 4, 5, 6, 8, 10\}. \quad \text{List the elements in the universal set that are not listed in } A = \{1, 3, 7, 9\}: \{\cancel{1}, 2, \cancel{3}, 4, 5, 6, \cancel{7}, 8, \cancel{9}, 10\}.$$

Set B' is the set of elements of U that are not in set B:

$$B' = \{1, 2, 4, 5, 6, 9\}. \quad \text{List the elements in the universal set that are not listed in } B = \{3, 7, 8, 10\}: \{1, 2, \cancel{3}, 4, 5, 6, \cancel{7}, \cancel{8}, 9, \cancel{10}\}.$$

Now we can find $A' \cap B'$, the set of elements belonging both to A' and to B':

$$A' \cap B' = \{2, 4, 5, 6, 8, 10\} \cap \{1, 2, 4, 5, 6, 9\}$$

$$= \{2, 4, 5, 6\}. \quad \text{The numbers 2, 4, 5, and 6 are common to both sets.}$$

✓ **Checkpoint 6**   Given $U = \{a, b, c, d, e\}$, $A = \{b, c\}$, and $B = \{b, c, e\}$, find each of the following sets:

a. $(A \cup B)'$ **b.** $A' \cap B'$.

Example 7   Determining Sets from a Venn Diagram

⑧ Determine sets involving set operations from a Venn diagram.

The Venn diagram in **Figure 2.16** percolates with interesting numbers. Use the diagram to determine each of the following sets:

a. $A \cup B$ **b.** $(A \cup B)'$
c. $A \cap B$ **d.** $(A \cap B)'$
e. $A' \cap B$ **f.** $A \cup B'$.

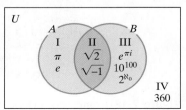

Figure 2.16

Solution

Refer to **Figure 2.16**, repeated below.

Set to Determine	Description of Set	Regions in Venn Diagram in Figure 2.16	Set in Roster Form
a. $A \cup B$	set of elements in A or B or both	I, II, III	$\{\pi, e, \sqrt{2}, \sqrt{-1}, e^{\pi i}, 10^{100}, 2^{\aleph_0}\}$
b. $(A \cup B)'$	set of elements in U that are not in $A \cup B$	IV	$\{360\}$
c. $A \cap B$	set of elements in both A and B	II	$\{\sqrt{2}, \sqrt{-1}\}$
d. $(A \cap B)'$	set of elements in U that are not in $A \cap B$	I, III, IV	$\{\pi, e, e^{\pi i}, 10^{100}, 2^{\aleph_0}, 360\}$
e. $A' \cap B$	set of elements that are not in A and are in B	III	$\{e^{\pi i}, 10^{100}, 2^{\aleph_0}\}$
f. $A \cup B'$	set of elements that are in A or not in B or both	I, II, IV	$\{\pi, e, \sqrt{2}, \sqrt{-1}, 360\}$

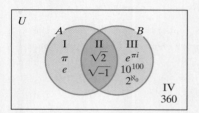

Figure 2.16 (repeated)

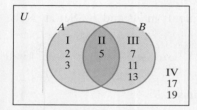

Figure 2.17

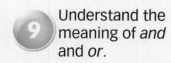
Understand the meaning of *and* and *or*.

✓ **Checkpoint 7**   Use the Venn diagram in **Figure 2.17** to determine each of the following sets:

a. $A \cap B$ **b.** $(A \cap B)'$ **c.** $A \cup B$
d. $(A \cup B)'$ **e.** $A' \cup B$ **f.** $A \cap B'$.

Sets and Precise Use of Everyday English

Set operations and Venn diagrams provide precise ways of organizing, classifying, and describing the vast array of sets and subsets we encounter every day. Let's see how this applies to the sets from the beginning of this section:

U = the set of U.S. Hispanics
S = the set of U.S. Hispanics who speak Spanish at home
E = the set of U.S. Hispanics who speak English at home.

When describing collections in everyday English, the word **or** refers to the **union** of sets. Thus, U.S. Hispanics who speak Spanish or English at home means those who speak Spanish or English or both. The word **and** refers to the **intersection** of sets. Thus, U.S. Hispanics who speak Spanish and English at home means those who speak both languages.

In **Figure 2.18**, we revisit the circle graph showing languages spoken at home by U.S. Hispanics. To the right of the circle graph, we've organized the data using a Venn diagram. The voice balloons indicate how the Venn diagram provides a more accurate understanding of the subsets and their data.

Languages Spoken at Home by U.S. Hispanics

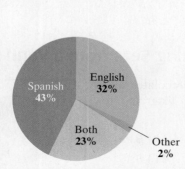

Figure 2.18 Comparing a circle graph and a Venn diagram
Source: Time

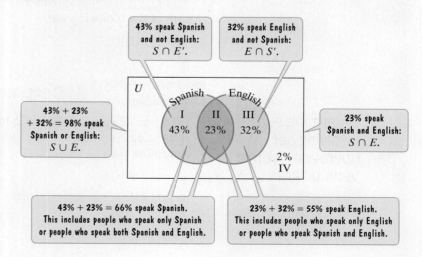

Exercise Set 2.3

Concept and Vocabulary Exercises

In Exercises 1–4, fill in each blank so that the resulting statement is true.

1. Visual relationships among sets are shown by ———————— .

2. The set of all elements in the universal set that are not in set A is called the ———————— of set A, and is symbolized by ———————— .

3. The set of elements common to both set A and set B is called the ———————— of sets A and B, and is symbolized by ———————— .

4. The set of elements that are members of set A or set B or of both sets is called the ———————— of sets A and B, and is symbolized by ———————— .

In Exercises 5–8, determine whether each statement is true or false. If the statement is false, make the necessary change(s) to produce a true statement.

5. Disjoint sets are represented by circles that do not overlap.

6. If set A is a proper subset of set B, the sets are represented by two circles where circle A is drawn outside of circle B.

7. Equal sets are represented by the same circle.

8. If two sets are represented by circles that overlap, then they have at least two elements in common.

Respond to Exercises 9–18 using verbal or written explanations.

9. Describe what is meant by a universal set. Provide an example.

10. What is a Venn diagram and how is it used?

11. Describe the Venn diagram for two disjoint sets. How does this diagram illustrate that the sets have no common elements?

12. Describe the Venn diagram for proper subsets. How does this diagram illustrate that the elements of one set are also in the second set?

13. Describe the Venn diagram for two equal sets. How does this diagram illustrate that the sets are equal?

14. Describe the Venn diagram for two sets with common elements. How does the diagram illustrate this relationship?

15. Describe what is meant by the complement of a set.

16. Is it possible to find a set's complement if a universal set is not given? Explain your answer.

17. Describe what is meant by the intersection of two sets. Give an example.

18. Describe what is meant by the union of two sets. Give an example.

In Exercises 19–22, determine whether each statement makes sense or does not make sense, and explain your reasoning.

19. Set A and set B share only one element, so I don't need to use overlapping circles to visualize their relationship.

20. Even if I'm not sure how mathematicians define irrational and complex numbers, telling me how these sets are related, I can construct a Venn diagram illustrating their relationship.

21. If I am given sets A and B, the set $(A \cup B)'$ indicates I should take the union of the complement of A and the complement of B.

22. I used a Venn diagram to represent the set of smartest dogs and the set of least intelligent dogs as overlapping circles.

Practice Exercises

In Exercises 23–26, describe a universal set U that includes all elements in the given sets. Answers may vary.

23. $A = \{$Bach, Mozart, Beethoven$\}$
    $B = \{$Brahms, Schubert$\}$

24. $A = \{$William Shakespeare, Charles Dickens$\}$
    $B = \{$Mark Twain, Robert Louis Stevenson$\}$

25. $A = \{$Pepsi, Sprite$\}$
    $B = \{$Coca Cola, Seven-Up$\}$

26. $A = \{$Acura RSX, Toyota Camry, Mitsubishi Lancer$\}$
    $B = \{$Dodge Ram, Chevrolet Impala$\}$

In Exercises 27–30, let $U = \{a, b, c, d, e, f, g\}$, $A = \{a, b, f, g\}$, $B = \{c, d, e\}$, $C = \{a, g\}$, and $D = \{a, b, c, d, e, f\}$. Use the roster method to write each of the following sets.

27. A' 28. B'

29. C' 30. D'

In Exercises 31–34, let $U = \{1, 2, 3, 4, \ldots, 20\}$, $A = \{1, 2, 3, 4, 5\}$, $B = \{6, 7, 8, 9\}$, $C = \{1, 3, 5, 7, \ldots, 19\}$, and $D = \{2, 4, 6, 8, \ldots, 20\}$. Use the roster method to write each of the following sets.

31. A' 32. B'

33. C' 34. D'

In Exercises 35–38, let U = {1, 2, 3, 4, …},
A = {1, 2, 3, 4, …, 20}, B = {1, 2, 3, 4, …, 50},
C = {2, 4, 6, 8, …}, and D ={1, 3, 5, 7, …}.
Use the roster method to write each of the following sets.

35. A' **36.** B'

37. C' **38.** D'

In Exercises 39–62, let

$$U = \{1, 2, 3, 4, 5, 6, 7\}$$
$$A = \{1, 3, 5, 7\}$$
$$B = \{1, 2, 3\}$$
$$C = \{2, 3, 4, 5, 6\}.$$

Find each of the following sets.

39. $A \cap B$ **40.** $B \cap C$

41. $A \cup B$ **42.** $B \cup C$

43. A' **44.** B'

45. $A' \cap B'$ **46.** $B' \cap C$

47. $A \cup C'$ **48.** $B \cup C'$

49. $(A \cap C)'$ **50.** $(A \cap B)'$

51. $A' \cup C'$ **52.** $A' \cup B'$

53. $(A \cup B)'$ **54.** $(A \cup C)'$

55. $A \cup \varnothing$ **56.** $C \cup \varnothing$

57. $A \cap \varnothing$ **58.** $C \cap \varnothing$

59. $A \cup U$ **60.** $B \cup U$

61. $A \cap U$ **62.** $B \cap U$

In Exercises 63–88, let

$$U = \{a, b, c, d, e, f, g, h\}$$
$$A = \{a, g, h\}$$
$$B = \{b, g, h\}$$
$$C = \{b, c, d, e, f\}.$$

Find each of the following sets.

63. $A \cap B$ **64.** $B \cap C$

65. $A \cup B$ **66.** $B \cup C$

67. A' **68.** B'

69. $A' \cap B'$ **70.** $B' \cap C$

71. $A \cup C'$ **72.** $B \cup C'$

73. $(A \cap C)'$ **74.** $(A \cap B)'$

75. $A' \cup C'$ **76.** $A' \cup B'$

77. $(A \cup B)'$ **78.** $(A \cup C)'$

79. $A \cup \varnothing$ **80.** $C \cup \varnothing$

81. $A \cap \varnothing$ **82.** $C \cap \varnothing$

83. $A \cup U$ **84.** $B \cup U$

85. $A \cap U$ **86.** $B \cap U$

87. $(A \cap B) \cup B'$ **88.** $(A \cup B) \cap B'$

In Exercises 89–100, use the Venn diagram to represent each set in roster form.

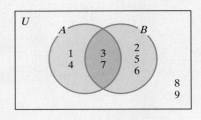

89. A **90.** B

91. U **92.** $A \cup B$

93. $A \cap B$ **94.** A'

95. B' **96.** $(A \cap B)'$

97. $(A \cup B)'$ **98.** $A' \cap B$

99. $A \cap B'$ **100.** $A \cup B'$

In Exercises 101–114, use the Venn diagram to determine each set or cardinality.

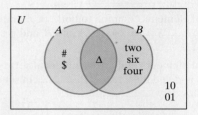

101. B **102.** A

103. $A \cup B$ **104.** $A \cap B$

105. $n(A \cup B)$ **106.** $n(A \cap B)$

107. $n(A')$ **108.** $n(B')$

109. $(A \cap B)'$ **110.** $(A \cup B)'$

111. $A' \cap B$ **112.** $A \cap B'$

113. $n(U) - n(B)$ **114.** $n(U) - n(A)$

Practice Plus

In Exercises 115–122, let

$$U = \{x | x \in \mathbf{N} \quad \text{and} \quad x < 9\}$$
$$A = \{x | x \text{ is an odd natural number} \quad \text{and} \quad x < 9\}$$
$$B = \{x | x \text{ is an even natural number} \quad \text{and} \quad x < 9\}$$
$$C = \{x | x \in \mathbf{N} \quad \text{and} \quad 1 < x < 6\}.$$

Find each of the following sets.

115. $A \cup B$ **116.** $B \cup C$

117. $A \cap U$ **118.** $A \cup U$

119. $A \cap C'$ **120.** $A \cap B'$

121. $(B \cap C)'$ **122.** $(A \cap C)'$

Application Exercises

A math tutor working with a small group of students asked each student when he or she had studied for class the previous weekend. Their responses are shown in the Venn diagram.

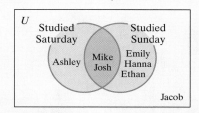

In Exercises 123–130, use the Venn diagram to list the elements of each set in roster form.

123. The set of students who studied Saturday

124. The set of students who studied Sunday

125. The set of students who studied Saturday or Sunday

126. The set of students who studied Saturday and Sunday

127. The set of students who studied Saturday and not Sunday

128. The set of students who studied Sunday and not Saturday

129. The set of students who studied neither Saturday nor Sunday

130. The set of students surveyed by the math tutor

The bar graph shows the percentage of Americans with gender preferences for various jobs.

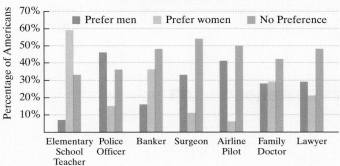

Source: Pew Research Center

In Exercises 131–136, use the information in the graph to place the indicated job in the correct region of the following Venn diagram.

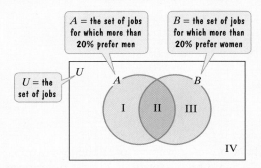

131. elementary school teacher

132. police officer

133. surgeon

134. banker

135. family doctor

136. lawyer

A **palindromic number** *is a natural number whose value does not change if its digits are reversed. Examples of palindromic numbers are 11, 454, and 261,162. In Exercises 137–146, use this definition to place the indicated natural number in the correct region of the following Venn diagram.*

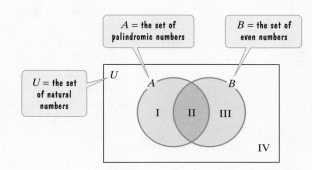

137. 11   **138.** 22

139. 15   **140.** 17

141. 454   **142.** 101

143. 9558   **144.** 9778

145. 9559   **146.** 9779

Critical Thinking Exercises

In Exercises 147–153, determine whether each statement is true or false. If the statement is false, make the necessary change(s) to produce a true statement.

147. $A \cap A' = \varnothing$

148. $(A \cup B) \subseteq A$

149. If $A \subseteq B$, then $A \cap B = B$.

150. $A \cap U = U$

151. $A \cup \varnothing = \varnothing$

152. If $A \subseteq B$, then $A \cap B = \varnothing$.

153. If $B \subseteq A$, then $A \cap B = B$.

In Exercises 154–157, assume $A \neq B$. Draw a Venn diagram that correctly illustrates the relationship between the sets.

154. $A \cap B = A$.

155. $A \cap B = B$.

156. $A \cup B = A$.

157. $A \cup B = B$.

2.4 Set Operations and Venn Diagrams with Three Sets

1. Perform set operations with three sets.

2. Use Venn diagrams with three sets.

3. Use Venn diagrams to prove equality of sets.

Should your blood type determine what you eat? The blood-type diet, developed by naturopathic physician Peter D'Adamo, is based on the theory that people with different blood types require different diets for optimal health. D'Adamo gives very detailed recommendations for what people with each type should and shouldn't eat. For example, he says shitake mushrooms are great for type B's, but bad for type O's. Type B? Type O? In this section, we present a Venn diagram with three sets that will give you a unique perspective on the different types of human blood. Despite this perspective, we'll have nothing to say about shitakes, avoiding the question as to whether or not the blood-type diet really works.

1 Perform set operations with three sets.

Set Operations with Three Sets

We now know how to find the union and intersection of two sets. We also know how to find a set's complement. In Example 1, we apply set operations to situations containing three sets.

Example 1   Set Operations with Three Sets

Given

$$U = \{1, 2, 3, 4, 5, 6, 7, 8, 9\}$$
$$A = \{1, 2, 3, 4, 5\}$$
$$B = \{1, 2, 3, 6, 8\}$$
$$C = \{2, 3, 4, 6, 7\},$$

find each of the following sets:

a. $A \cup (B \cap C)$

b. $(A \cup B) \cap (A \cup C)$

c. $A \cap (B \cup C')$.

Solution

Before determining each set, let's be sure we perform the operations in the correct order. Remember that we begin by performing any set operations inside parentheses.

- Finding $A \cup (B \cap C)$

  Step 1. Find the intersection of B and C.

  Step 2. Find the union of A and $(B \cap C)$.

- Finding $(A \cup B) \cap (A \cup C)$

  Step 1. Find the union of A and B.   **Step 2.** Find the union of A and C.

  Step 3. Find the intersection of $(A \cup B)$ and $(A \cup C)$.

- Finding $A \cap (B \cup C')$ **Step 1.** Find the complement of C.

  Step 2. Find the union of B and C'.

  Step 3. Find the intersection of A and $(B \cup C')$.

$U = \{1, 2, 3, 4, 5, 6, 7, 8, 9\}$

$A = \{1, 2, 3, 4, 5\}$

$B = \{1, 2, 3, 6, 8\}$

$C = \{2, 3, 4, 6, 7\}$

The given sets (repeated for your convenience)

a. To find $A \cup (B \cap C)$, first find the set within the parentheses, $B \cap C$:

$$B \cap C = \{1, 2, 3, 6, 8\} \cap \{2, 3, 4, 6, 7\} = \{2, 3, 6\}.$$

Common elements are **2, 3,** and **6.**

Now finish the problem by finding $A \cup (B \cap C)$:

$$A \cup (B \cap C) = \{1, 2, 3, 4, 5\} \cup \{2, 3, 6\} = \{1, 2, 3, 4, 5, 6\}.$$

List all elements in A and then add the only unlisted element in $B \cap C$, namely **6.**

b. To find $(A \cup B) \cap (A \cup C)$, first find the sets within parentheses. Start with $A \cup B$:

$$A \cup B = \{1, 2, 3, 4, 5\} \cup \{1, 2, 3, 6, 8\} = \{1, 2, 3, 4, 5, 6, 8\}.$$

List all elements in A and then add the unlisted elements in B, namely **6** and **8.**

Now find $A \cup C$:

$$A \cup C = \{1, 2, 3, 4, 5\} \cup \{2, 3, 4, 6, 7\} = \{1, 2, 3, 4, 5, 6, 7\}.$$

List all elements in A and then add the unlisted elements in C, namely **6** and **7.**

Now finish the problem by finding $(A \cup B) \cap (A \cup C)$:

$$(A \cup B) \cap (A \cup C) = \{1, 2, 3, 4, 5, 6, 8\} \cap \{1, 2, 3, 4, 5, 6, 7\} = \{1, 2, 3, 4, 5, 6\}.$$

Common elements are **1, 2, 3, 4, 5,** and **6.**

c. As in parts (a) and (b), to find $A \cap (B \cup C')$, begin with the set in parentheses. First we must find C', the set of elements in U that are not in C:

$$C' = \{1, 5, 8, 9\}. \quad \text{List the elements in } U \text{ that are not in}$$
$$C = \{2, 3, 4, 6, 7\}: \{1, \cancel{2}, \cancel{3}, \cancel{4}, 5, \cancel{6}, \cancel{7}, 8, 9\}.$$

$U = \{1, 2, 3, 4, 5, 6, 7, 8, 9\}$

$A = \{1, 2, 3, 4, 5\}$

$B = \{1, 2, 3, 6, 8\}$

$C = \{2, 3, 4, 6, 7\}$

The given sets (repeated)

Using $C' = \{1, 5, 8, 9\}$, we now can identify the elements of $B \cup C'$:

$$B \cup C' = \{1, 2, 3, 6, 8\} \cup \{1, 5, 8, 9\} = \{1, 2, 3, 5, 6, 8, 9\}.$$

> List all elements in B and then add the unlisted elements in C', namely 5 and 9.

Now finish the problem by finding $A \cap (B \cup C')$:

$$A \cap (B \cup C') = \{1, 2, 3, 4, 5\} \cap \{1, 2, 3, 5, 6, 8, 9\} = \{1, 2, 3, 5\}.$$

> Common elements are 1, 2, 3, and 5.

Checkpoint 1   Given $U = \{a, b, c, d, e, f\}$, $A = \{a, b, c, d\}$, $B = \{a, b, d, f\}$, and $C = \{b, c, f\}$, find each of the following sets:

    a. $A \cup (B \cap C)$

    b. $(A \cup B) \cap (A \cup C)$

    c. $A \cap (B \cup C')$.

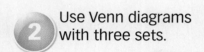 Use Venn diagrams with three sets.

Venn Diagrams with Three Sets

Venn diagrams can contain three or more sets, such as the diagram in **Figure 2.19**. The three sets in the figure separate the universal set, U, into eight regions. The numbering of these regions is arbitrary—that is, we can number any region as I, any region as II, and so on. Here is a description of each region, starting with the innermost region, region V, and working outward to region VIII.

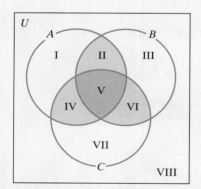

Figure 2.19 Three intersecting sets separate the universal set into eight regions.

The Region Shown in Dark Blue

Region V   This region represents elements that are common to sets A, B, and C: $A \cap B \cap C$.

The Regions Shown in Light Blue

Region II   This region represents elements in both sets A and B that are not in set C: $(A \cap B) \cap C'$.

Region IV   This region represents elements in both sets A and C that are not in set B: $(A \cap C) \cap B'$.

Region VI   This region represents elements in both sets B and C that are not in set A: $(B \cap C) \cap A'$.

The Regions Shown in White

Region I   This region represents elements in set A that are in neither sets B nor C: $A \cap (B' \cap C')$.

Region III   This region represents elements in set B that are in neither sets A nor C: $B \cap (A' \cap C')$.

Region VII   This region represents elements in set C that are in neither sets A nor B: $C \cap (A' \cap B')$.

Region VIII   This region represents elements in the universal set U that are not in sets A, B, or C: $A' \cap B' \cap C'$.

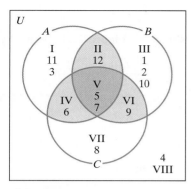

Figure 2.20

Example 2 | **Determining Sets from a Venn Diagram with Three Intersecting Sets**

Use the Venn diagram in **Figure 2.20** to determine each of the following sets:

a. A **b.** $A \cup B$ **c.** $B \cap C$

d. C' **e.** $A \cap B \cap C$.

Solution

Set to Determine	Description of Set	Regions in Venn Diagram	Set in Roster Form
a. A	set of elements in A	I, II, IV, V	$\{11, 3, 12, 6, 5, 7\}$
b. $A \cup B$	set of elements in A or B or both	I, II, III, IV, V, VI	$\{11, 3, 12, 1, 2, 10, 6, 5, 7, 9\}$
c. $B \cap C$	set of elements in both B and C	V, VI	$\{5, 7, 9\}$
d. C'	set of elements in U that are not in C	I, II, III, VIII	$\{11, 3, 12, 1, 2, 10, 4\}$
e. $A \cap B \cap C$	set of elements in A and B and C	V	$\{5, 7\}$

✓ **Checkpoint 2** Use the Venn diagram in **Figure 2.20** to determine each of the following sets:

a. C

b. $B \cup C$

c. $A \cap C$

d. B'

▶ **e.** $A \cup B \cup C$.

Proving the Equality of Sets

③ Use Venn diagrams to prove equality of sets.

Throughout Section 2.3, you were given two sets A and B and their universal set U and asked to find $(A \cap B)'$ and $A' \cup B'$. In each example, $(A \cap B)'$ and $A' \cup B'$ resulted in the same set. This occurs regardless of which sets we choose for A and B in a universal set U. Examining these individual cases and applying inductive reasoning, a conjecture (or educated guess) is that $(A \cap B)' = A' \cup B'$.

We can apply deductive reasoning to *prove* the statement $(A \cap B)' = A' \cup B'$ for *all* sets A and B in any universal set U. To prove that $(A \cap B)'$ and $A' \cup B'$ are equal, we use a Venn diagram. If both sets are represented by the same regions in this general diagram, then this proves that they are equal. Example 3 shows how this is done.

A Brief Review

In summary, here are the two forms of reasoning discussed in Chapter 1.

- **Inductive Reasoning:** Starts with individual observations and works to a general conjecture (or educated guess)

- **Deductive Reasoning:** Starts with general cases and works to the proof of a specific statement (or theorem)

Example 3 | **Proving the Equality of Sets**

Use a Venn diagram to prove that

$$(A \cap B)' = A' \cup B'.$$

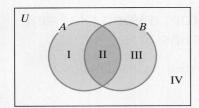

Figure 2.21

Solution

We use the Venn diagram with sets A and B, as shown in **Figure 2.21**. Begin by identifying the regions representing $(A \cap B)'$.

Set	Regions in the Venn Diagram
A	I, II
B	II, III
$A \cap B$	II (This is the region common to A and B.)
$(A \cap B)'$	I, III, IV (These are the regions in U that are not in $A \cap B$.)

Next, find the regions in **Figure 2.21** representing $A' \cup B'$.

Set	Regions in the Venn Diagram
A'	III, IV (These are the regions not in A.)
B'	I, IV (These are the regions not in B.)
$A' \cup B'$	I, III, IV (These are the regions obtained by uniting the regions representing A' and B'.)

Both $(A \cap B)'$ and $A' \cup B'$ are represented by the same regions, I, III, and IV, of the Venn diagram. This result proves that

$$(A \cap B)' = A' \cup B'$$

for all sets A and B in any universal set U.

Can you see how we applied deductive reasoning in Example 3? We started with the two general sets A and B in the Venn diagram in **Figure 2.21** and worked to the specific conclusion that $(A \cap B)'$ and $A' \cup B'$ represent the same regions in the diagram. Thus, the statement $(A \cap B)' = A' \cup B'$ is a theorem.

✓ **Checkpoint 3** Use the Venn diagram in **Figure 2.21** to solve this exercise.

a. Which region represents $(A \cup B)'$?

b. Which region represents $A' \cap B'$?

c. Based on parts (a) and (b), what can you conclude?

Example 4 Proving the Equality of Sets

Use a Venn diagram to prove that

$$A \cup (B \cap C) = (A \cup B) \cap (A \cup C).$$

Solution

Use a Venn diagram with three sets A, B, and C, as shown in **Figure 2.22**. Begin by identifying the regions representing $A \cup (B \cap C)$.

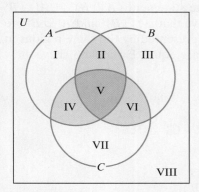

Figure 2.22

Set	Regions in the Venn Diagram
A	I, II, IV, V
$B \cap C$	V, VI (These are the regions common to B and C.)
$A \cup (B \cap C)$	I, II, IV, V, VI (These are the regions obtained by uniting the regions representing A and $B \cap C$.)

Next, find the regions representing $(A \cup B) \cap (A \cup C)$.

Set	Regions in the Venn Diagram
A	I, II, IV, V
B	II, III, V, VI
C	IV, V, VI, VII
$A \cup B$	I, II, III, IV, V, VI (Unite the regions representing A and B.)
$A \cup C$	I, II, IV, V, VI, VII (Unite the regions representing A and C.)
$(A \cup B) \cap (A \cup C)$	I, II, IV, V, VI (These are the regions common to $A \cup B$ and $A \cup C$.)

Both $A \cup (B \cap C)$ and $(A \cup B) \cap (A \cup C)$ are represented by the same regions, I, II, IV, V, and VI, of the Venn diagram. This result proves that

$$A \cup (B \cap C) = (A \cup B) \cap (A \cup C)$$

for all sets A, B, and C in any universal set U. Thus, the statement is a theorem.

 **Checkpoint 4**   Use the Venn diagram in **Figure 2.22** to solve this exercise.

a. Which regions represent $A \cap (B \cup C)$?

b. Which regions represent $(A \cap B) \cup (A \cap C)$?

▶ **c.** Based on parts (a) and (b), what can you conclude?

Blitzer Bonus

Blood Types and Venn Diagrams

In the early 1900s, the Austrian immunologist Karl Landsteiner discovered that all blood is not the same. Blood serum drawn from one person often clumped when mixed with the blood cells of another. The clumping was caused by different antigens, proteins, and carbohydrates that trigger antibodies and fight infection. Landsteiner classified blood types based on the presence or absence of the antigens A, B, and Rh in red blood cells. The Venn diagram in **Figure 2.23** contains eight regions representing the eight common blood groups.

In the Venn diagram, blood with the Rh antigen is labeled positive and blood lacking the Rh antigen is labeled negative. The region where the three circles intersect represents type AB⁺, indicating that a person with this blood type has the antigens A, B, and Rh. Observe that type O blood (both positive and negative) lacks A and B antigens. Type O⁻ lacks all three antigens, A, B, and Rh.

In blood transfusions, the recipient must have all or more of the antigens present in the donor's blood. This discovery rescued surgery patients from random, often lethal, transfusions. This knowledge made the massive blood drives during World War I possible. Eventually, it made the modern blood bank possible as well.

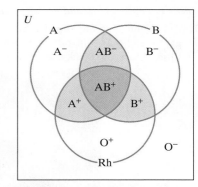

Figure 2.23 Human blood types

Achieving Success

Willpower! Willpower! Willpower!

Successful people are willing to suck it up from time to time and do things they don't like doing. They realize that these things will lead them to their goals. Do you think that a student who is committed to his or her education is entertained by proving theorems about sets? Do high school athletes always enjoy hours of practice each day? Sometimes you just have to exercise that special tool called willpower to get things done, whether you feel like it or not.

"I hated every minute of training. But I said, 'Don't quit. Suffer now and live the rest of your life as a champion.'"

—Muhammad Ali, boxing champion

Exercise Set 2.4

Concept and Vocabulary Exercises

In Exercises 1–2, fill in each blank so that the resulting statement is true.

1. In order to perform set operations such as $(A \cup B) \cap (A \cup C)$, begin by performing any set operations _____.

2. The three sets in the Venn diagrams that appeared throughout this section separate the universal set into _____ regions.

In Exercises 3–4, determine whether each statement is true or false. If the statement is false, make the necessary change(s) to produce a true statement.

3. Inductive reasoning is used to prove the equality of sets.

4. In this section we proved that $(A \cap B)' = A' \cup B'$ is a theorem, so this means that the complement of the intersection of two sets is the union of the complements of those sets.

Respond to Exercises 5–6 using verbal or written explanations.

5. If you are given three sets, A, B, and C in a universal set U, describe what is involved in determining $(A \cup B)' \cap C$. Be as specific as possible in your description.

6. Describe how a Venn diagram can be used to prove that $(A \cup B)'$ and $A' \cap B'$ are equal sets.

In Exercises 7–10, determine whether each statement makes sense or does not make sense, and explain your reasoning.

7. This Venn diagram showing color combinations from red, green, and blue illustrates that white is a combination of all three colors and black uses none of the colors.

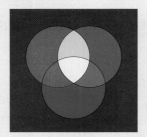

8. I used a Venn diagram to prove that $(A \cup B)'$ and $A' \cup B'$ are not equal.

9. I found 50 examples of two sets, A and B, for which $(A \cup B)'$ and $A' \cap B'$ resulted in the same set, so this proves that $(A \cup B)' = A' \cap B'$.

10. I used the innermost region of a Venn diagram with three sets to determine the common elements of the sets.

Practice Exercises

In Exercises 11–22, let

$$U = \{1, 2, 3, 4, 5, 6, 7\}$$
$$A = \{1, 3, 5, 7\}$$
$$B = \{1, 2, 3\}$$
$$C = \{2, 3, 4, 5, 6\}.$$

Find each of the following sets.

11. $A \cup (B \cap C)$
12. $A \cap (B \cup C)$
13. $(A \cup B) \cap (A \cup C)$
14. $(A \cap B) \cup (A \cap C)$
15. $A' \cap (B \cup C')$
16. $C' \cap (A \cup B')$
17. $(A' \cap B) \cup (A' \cap C')$
18. $(C' \cap A) \cup (C' \cap B')$
19. $(A \cup B \cup C)'$
20. $(A \cap B \cap C)'$
21. $(A \cup B)' \cap C$
22. $(B \cup C)' \cap A$

In Exercises 23–34, let

$$U = \{a, b, c, d, e, f, g, h\}$$
$$A = \{a, g, h\}$$
$$B = \{b, g, h\}$$
$$C = \{b, c, d, e, f\}.$$

Find each of the following sets.

23. $A \cup (B \cap C)$
24. $A \cap (B \cup C)$
25. $(A \cup B) \cap (A \cup C)$
26. $(A \cap B) \cup (A \cap C)$
27. $A' \cap (B \cup C')$
28. $C' \cap (A \cup B')$
29. $(A' \cap B) \cup (A' \cap C')$
30. $(C' \cap A) \cup (C' \cap B')$
31. $(A \cup B \cup C)'$
32. $(A \cap B \cap C)'$
33. $(A \cup B)' \cap C$
34. $(B \cup C)' \cap A$

In Exercises 35–42, use the Venn diagram shown to answer each question.

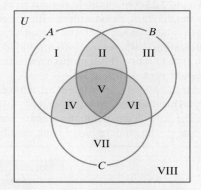

35. Which regions represent set B?
36. Which regions represent set C?
37. Which regions represent $A \cup C$?
38. Which regions represent $B \cup C$?
39. Which regions represent $A \cap B$?
40. Which regions represent $A \cap C$?
41. Which regions represent B'?
42. Which regions represent C'?

In Exercises 43–54, use the Venn diagram to represent each set in roster form.

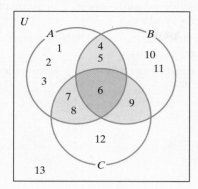

43. *A*

44. *B*

45. $A \cup B$

46. $B \cup C$

47. $(A \cup B)'$

48. $(B \cup C)'$

49. $A \cap B$

50. $A \cap C$

51. $A \cap B \cap C$

52. $A \cup B \cup C$

53. $(A \cap B \cap C)'$

54. $(A \cup B \cup C)'$

Use the Venn diagram shown to solve Exercises 55–58.

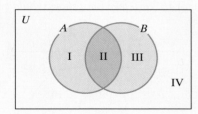

55. **a.** Which region represents $A \cap B$?

   b. Which region represents $B \cap A$?

   c. Based on parts (a) and (b), what can you conclude?

56. **a.** Which regions represent $A \cup B$?

   b. Which regions represent $B \cup A$?

   c. Based on parts (a) and (b), what can you conclude?

57. **a.** Which region(s) represents $(A \cap B)'$?

   b. Which region(s) represents $A' \cap B'$?

   c. Based on parts (a) and (b), are $(A \cap B)'$ and $A' \cap B'$ equal for all sets A and B? Explain your answer.

58. **a.** Which region(s) represents $(A \cup B)'$?

   b. Which region(s) represents $A' \cup B'$?

   c. Based on parts (a) and (b), are $(A \cup B)'$ and $A' \cup B'$ equal for all sets A and B? Explain your answer.

In Exercises 59–64, use the Venn diagram for Exercises 55–58 to determine whether the given sets are equal for all sets A and B.

59. $A' \cup B, A \cap B'$

60. $A' \cap B, A \cup B'$

61. $(A \cup B)', (A \cap B)'$

62. $(A \cup B)', A' \cap B$

63. $(A' \cap B)', A \cup B'$

64. $(A \cup B')', A' \cap B$

Use the Venn diagram shown to solve Exercises 65–68.

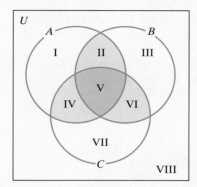

65. **a.** Which regions represent $(A \cap B) \cup C$?

   b. Which regions represent $(A \cup C) \cap (B \cup C)$?

   c. Based on parts (a) and (b), what can you conclude?

66. **a.** Which regions represent $(A \cup B) \cap C$?

   b. Which regions represent $(A \cap C) \cup (B \cap C)$?

   c. Based on parts (a) and (b), what can you conclude?

67. **a.** Which regions represent $A \cap (B \cup C)$?

   b. Which regions represent $A \cup (B \cap C)$?

   c. Based on parts (a) and (b), are $A \cap (B \cup C)$ and $A \cup (B \cap C)$ equal for all sets A, B, and C? Explain your answer.

68. **a.** Which regions represent $C \cup (B \cap A)$?

   b. Which regions represent $C \cap (B \cup A)$?

   c. Based on parts (a) and (b), are $C \cup (B \cap A)$ and $C \cap (B \cup A)$ equal for all sets A, B, and C? Explain your answer.

In Exercises 69–74, use the Venn diagram shown above to determine which statements are true for all sets A, B, and C, and, consequently, are theorems.

69. $A \cap (B \cup C) = (A \cap B) \cup C$

70. $A \cup (B \cap C) = (A \cup B) \cap C$

71. $B \cup (A \cap C) = (A \cup B) \cap (B \cup C)$

72. $B \cap (A \cup C) = (A \cap B) \cup (B \cap C)$

73. $A \cap (B \cup C)' = A \cap (B' \cap C')$

74. $A \cup (B \cap C)' = A \cup (B' \cup C')$

Practice Plus

75. **a.** Let $A = \{c\}, B = \{a, b\}, C = \{b, d\}$, and $U = \{a, b, c, d, e, f\}$. Find $A \cup (B' \cap C')$ and $(A \cup B') \cap (A \cup C')$.

   b. Let $A = \{1, 3, 7, 8\}, B = \{2, 3, 6, 7\}, C = \{4, 6, 7, 8\}$, and $U = \{1, 2, 3, \ldots, 8\}$. Find $A \cup (B' \cap C')$ and $(A \cup B') \cap (A \cup C')$.

c. Based on your results in parts (a) and (b), use inductive reasoning to write a conjecture that relates $A \cup (B' \cap C')$ and $(A \cup B') \cap (A \cup C')$.

d. Use deductive reasoning to determine whether your conjecture in part (c) is a theorem.

76. a. Let $A = \{3\}$, $B = \{1, 2\}$, $C = \{2, 4\}$, and $U = \{1, 2, 3, 4, 5, 6\}$. Find $(A \cup B)' \cap C$ and $A' \cap (B' \cap C)$.

b. Let $A = \{d, f, g, h\}$, $B = \{a, c, f, h\}$, $C = \{c, e, g, h\}$, and $U = \{a, b, c, \ldots, h\}$. Find $(A \cup B)' \cap C$ and $A' \cap (B' \cap C)$.

c. Based on your results in parts (a) and (b), use inductive reasoning to write a conjecture that relates $(A \cup B)' \cap C$ and $A' \cap (B' \cap C)$.

d. Use deductive reasoning to determine whether your conjecture in part (c) is a theorem.

Application Exercises

A math tutor working with a small study group has classified students in the group by whether or not they scored 90% or above on each of three tests. The results are shown in the Venn diagram.

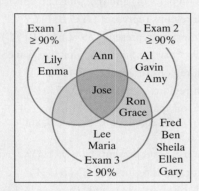

In Exercises 77–88, use the Venn diagram to represent each set in roster form.

77. The set of students who scored 90% or above on exam 2

78. The set of students who scored 90% or above on exam 3

79. The set of students who scored 90% or above on exam 1 and exam 3

80. The set of students who scored 90% or above on exam 1 and exam 2

81. The set of students who scored 90% or above on exam 1 and not on exam 2

82. The set of students who scored 90% or above on exam 3 and not on exam 1

83. The set of students who scored 90% or above on exam 1 or not on exam 2

84. The set of students who scored 90% or above on exam 3 or not on exam 1

85. The set of students who scored 90% or above on *exactly one* test

86. The set of students who scored 90% or above on *at least two* tests

87. The set of students who scored 90% or above on exam 2 and not on exam 1 and exam 3

88. The set of students who scored 90% or above on exam 1 and not on exam 2 and exam 3

89. Use the Venn diagram to describe a set of students that is the empty set.

90. Use the Venn diagram to describe the set {Fred, Ben, Sheila, Ellen, Gary}.

The chart shows the most popular shows on television in 2007, 2008, and 2009.

Most Popular Television Shows

2007	**2008**	**2009**
1. *American Idol*	**1.** *American Idol*	**1.** *American Idol*
2. *Dancing with the Stars*	**2.** *Dancing with the Stars*	**2.** *Dancing with the Stars*
3. *CSI*	**3.** *NBC Sunday Night Football*	**3.** *NBC Sunday Night Football*
4. *Grey's Anatomy*	**4.** *CSI*	**4.** *NCIS*
5. *House*	**5.** *Grey's Anatomy*	**5.** *CSI*
6. *Desperate Housewives*	**6.** *Samantha Who?*	**6.** *The Mentalist*

Source: Nielsen Media Research

In Exercises 91–96, use the Venn diagram to indicate in which region, I through VIII, each television show should be placed.

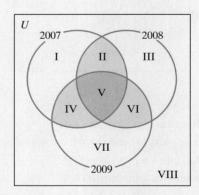

91. *NBC Sunday Night Football*

92. *Samantha Who?*

93. *CSI*

94. *Grey's Anatomy*

95. *House*

96. *60 Minutes*

The chart shows the top single recordings of all time.

Top Single Recordings

Title	Artist or Group	Sales	Year Released
"Candle in the Wind"	Elton John	37 million	1997
"White Christmas"	Bing Crosby	30 million	1942
"Rock Around the Clock"	Bill Haley and His Comets	17 million	1954
"I Want to Hold Your Hand"	The Beatles	12 million	1963
"It's Now or Never"	Elvis Presley	10 million	1960
"Hey Jude"	The Beatles	10 million	1968
"I Will Always Love You"	Whitney Houston	10 million	1992
"Hound Dog"	Elvis Presley	9 million	1956
"Diana"	Paul Anka	9 million	1957
"I'm a Believer"	The Monkees	8 million	1966

Source: RIAA

In Exercises 97–102, use the Venn diagram to indicate in which region, I through VIII, each recording should be placed.

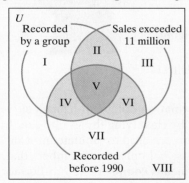

97. "Candle in the Wind"
98. "White Christmas"
99. "I Want to Hold Your Hand"
100. "Hey Jude"
101. "Diana"
102. "I'm a Believer"

Critical Thinking Exercises

The eight blood types discussed in the Blitzer Bonus on page 91 are shown once again in the Venn diagram at the top of the next column. In blood transfusions, the set of antigens in a donor's blood must be a subset of the set of antigens in a recipient's blood. Thus, the recipient must have all or more of the antigens present in the donor's blood. Use this information to solve Exercises 103–106.

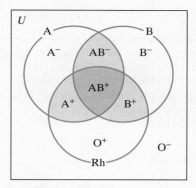

Human blood types

103. What is the blood type of a universal recipient?
104. What is the blood type of a universal donor?
105. Can an A^+ person donate blood to an A^- person?
106. Can an A^- person donate blood to an A^+ person?

In Exercises 107–114, use the symbols $A, B, C, \cap, \cup$, and $'$, as necessary, to describe each shaded region. More than one correct symbolic description may be possible.

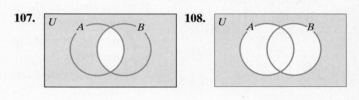

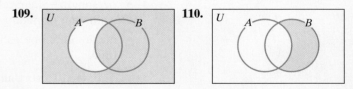

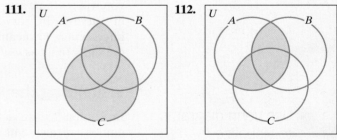

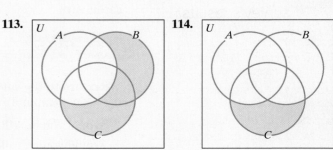

115. Each group member should find out his or her blood type. (If you cannot obtain this information, select a blood type that you find appealing!) Read the introduction to Exercises 103–106. Referring to the Venn diagram for these exercises, each group member should determine all other group members to whom blood can be donated and from whom it can be received.

116. The group should define three sets, each of which categorizes *U*, the set of students in the group, in different ways. Examples include the set of students with blonde hair, the set of students no more than 16 years old, and the set of students who spend more than 10 hours per week playing video games. Once you have defined the sets, construct a Venn diagram with three intersecting sets and eight regions. Each student should determine which region he or she belongs to. Illustrate the sets by writing each first name in the appropriate region.

2.5 : Survey Problems

Objectives

1. Use Venn diagrams to visualize a survey's results.

2. Use survey results to complete Venn diagrams and answer questions about the survey.

Seven out of ten teens ages 16 and 17 own cellphones. What they do with them is very different from what adults do. According to a Pew Research Center survey, teens are more likely than older adults to

- use their cellphone for text messaging.
- use their cellphone to take pictures.
- use their cellphone to play games.
- use their cellphone to access the Internet.
- personalize their cellphone by changing wallpaper or adding ring tones.
- make cellphone calls to fill up their free time.

The need to sort and organize information obtained through surveys is related to our need to find order and meaning by classifying things into sets. In this section, you will see how sets and Venn diagrams are used to tabulate information collected in surveys. In survey problems, it is helpful to remember that **and** means **intersection**, **or** means **union**, and **not** means **complement**. Furthermore, *but* means the same thing as *and*. Thus, **but** means **intersection**.

Visualizing the Results of a Survey

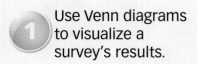
Use Venn diagrams to visualize a survey's results.

In Section 2.1, we defined the cardinal number of set *A*, denoted by $n(A)$, as the number of elements in set *A*. Venn diagrams are helpful in determining a set's cardinality.

Example 1   Using a Venn Diagram to Visualize the Results of a Survey

We return to the survey in which your high school math class asked students from the local community college two questions:

Would you be willing to donate blood?

Would you be willing to help serve a free breakfast to blood donors?

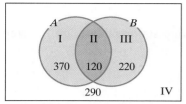

A: Set of students willing to
donate blood
B: Set of students willing to
serve breakfast to donors

Figure 2.24 Results of a survey

Set A represents the set of students willing to donate blood. Set B represents the set of students willing to help serve breakfast to donors. The survey results are summarized in **Figure 2.24**. Use the diagram to answer the following questions:

a. How many students are willing to donate blood?
b. How many students are willing to help serve a free breakfast to blood donors?
c. How many students are willing to donate blood and serve breakfast?
d. How many students are willing to donate blood or serve breakfast?
e. How many students are willing to donate blood but not serve breakfast?
f. How many students are willing to serve breakfast but not donate blood?
g. How many students are neither willing to donate blood nor serve breakfast?
h. How many students were surveyed?

Solution

a. The number of students willing to donate blood can be determined by adding the numbers in regions I and II. Thus, $n(A) = 370 + 120 = 490$. There are 490 students willing to donate blood.

b. The number of students willing to help serve a free breakfast to blood donors can be determined by adding the numbers in regions II and III. Thus, $n(B) = 120 + 220 = 340$. There are 340 students willing to help serve breakfast.

c. The number of students willing to donate blood and serve breakfast appears in region II, the region representing the intersection of the two sets. Thus, $n(A \cap B) = 120$. There are 120 students willing to donate blood and serve breakfast.

d. The number of students willing to donate blood or serve breakfast is found by adding the numbers in regions I, II, and III, representing the union of the two sets. We see that $n(A \cup B) = 370 + 120 + 220 = 710$. Therefore, 710 students in the survey are willing to donate blood or serve breakfast.

e. The region representing students who are willing to donate blood but not serve breakfast, $A \cap B'$, is region I. We see that 370 of the students surveyed are willing to donate blood but not serve breakfast.

f. Region III represents students willing to serve breakfast but not donate blood: $B \cap A'$. We see that 220 students surveyed are willing to help serve breakfast but not donate blood.

g. Students who are neither willing to donate blood nor serve breakfast, $A' \cap B'$, fall within the universal set, but outside circles A and B. These students fall in region IV, where the Venn diagram indicates that there are 290 elements. There are 290 students in the survey who are neither willing to donate blood nor serve breakfast.

h. We can find the number of students surveyed by adding the numbers in regions I, II, III, and IV. Thus, $n(U) = 370 + 120 + 220 + 290 = 1000$. There were 1000 students surveyed.

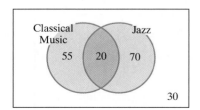

Figure 2.25

Checkpoint 1 In a survey on musical tastes, respondents were asked: Do you listen to classical music? Do you listen to jazz? The survey results are summarized in **Figure 2.25**. Use the diagram to answer the following questions:

a. How many respondents listened to classical music?
b. How many respondents listened to jazz?
c. How many respondents listened to both classical music and jazz?
d. How many respondents listened to classical music or jazz?
e. How many respondents listened to classical music but not jazz?
f. How many respondents listened to jazz but not classical music?
g. How many respondents listened to neither classical music nor jazz?
h. How many people were surveyed?

2 Use survey results to complete Venn diagrams and answer questions about the survey.

Solving Survey Problems

Venn diagrams are used to solve problems involving surveys. Here are the steps needed to solve survey problems:

> ### Solving Survey Problems
>
> 1. Use the survey's description to define sets and draw a Venn diagram.
> 2. Use the survey's results to determine the cardinality for each region in the Venn diagram. **Start with the intersection of the sets, the innermost region, and work outward.**
> 3. Use the completed Venn diagram to answer the problem's questions.

Example 2   Teens and Cellphones

A survey was taken that asked 150 randomly selected teens ages 16 and 17 what they do with their cellphones. Results of the survey showed that 30 teens used their cellphones to take pictures and play games, 100 teens used their cellphones to take pictures, and 70 teens used their cellphones to play games.

  a. Use a Venn diagram to illustrate the survey's results.

  b. How many of those surveyed used their cellphones to take pictures but not play games?

  c. How many of those surveyed did not use their cellphones to take pictures or play games?

Solution

  a. We begin with a Venn diagram that illustrates the survey's results.

    Step 1   Define sets and draw a Venn diagram. The Venn diagram in **Figure 2.26** shows two sets. Set P is the set of teens who use their cellphones to take pictures. Set G is the set of teens who use their cellphones to play games.

    Step 2   Determine the cardinality for each region in the Venn diagram, starting with the innermost region and working outward. We are given the following cardinalities:

    • There were 150 teens surveyed: $n(U) = 150$.

    • 30 teens used their cellphones to take pictures and play games: $n(P \cap G) = 30$.

    • 100 teens used their cellphones to take pictures: $n(P) = 100$.

    • 70 teens used their cellphones to play games: $n(G) = 70$.

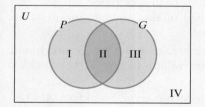

Figure 2.26

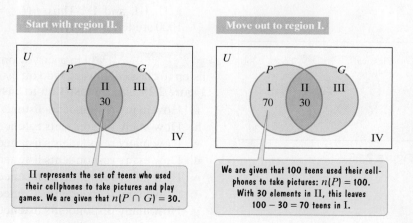

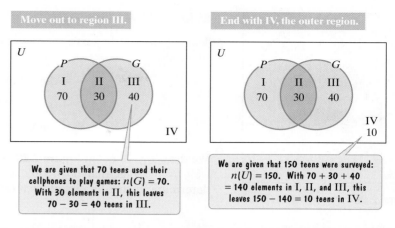

Move out to region III.

We are given that 70 teens used their cellphones to play games: $n(G) = 70$. With 30 elements in II, this leaves $70 - 30 = 40$ teens in III.

End with IV, the outer region.

We are given that 150 teens were surveyed: $n(U) = 150$. With $70 + 30 + 40 = 140$ elements in I, II, and III, this leaves $150 - 140 = 10$ teens in IV.

Step 3 Use the completed Venn diagram to answer the problem's questions. The completed Venn diagram is shown in **Figure 2.27**.

b. Those surveyed who used their cellphones to take pictures but not play games are shown in region I. This means that 70 teens used their cellphones to take pictures but not play games.

c. Those surveyed who did not use their cellphones to take pictures or play games are shown in region IV. This means that 10 teens did not use their cellphones to take pictures or play games.

Figure 2.27

Checkpoint 2   A survey was taken that asked 150 randomly selected teens ages 16 and 17 what they do with their cellphones. Results of the survey showed that 110 teens used their cellphones for text messaging and accessing the Internet, 140 teens used their cellphones for text messaging, and 117 teens used their cellphones to access the Internet.

a. Use a Venn diagram to illustrate the survey's results.

b. How many of those surveyed used their cellphones to access the Internet but not for text messaging?

c. How many of those surveyed did not use their cellphones for text messaging or accessing the Internet?

In the next example, we create a Venn diagram with three intersecting sets to illustrate a survey's results. In our final example, we use this Venn diagram to answer questions about the survey.

Example 3   Constructing a Venn Diagram for a Survey

Sixty people were contacted and responded to a movie survey. The following information was obtained:

a. 6 people liked comedies, dramas, and science fiction.

b. 13 people liked comedies and dramas.

c. 10 people liked comedies and science fiction.

d. 11 people liked dramas and science fiction.

e. 26 people liked comedies.

f. 21 people liked dramas.

g. 25 people liked science fiction.

Use a Venn diagram to illustrate the survey's results.

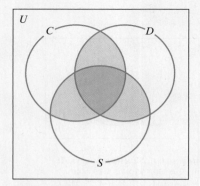

Figure 2.28

Study Tip

After entering the cardinality of the innermost region, work outward and use subtraction to obtain subsequent cardinalities. The phrase

    SURVEY SUBTRACT

is a helpful reminder of these repeated subtractions.

Solution

The set of people surveyed is a universal set with 60 elements containing three subsets:

$$C = \text{the set of those who like comedies}$$
$$D = \text{the set of those who like dramas}$$
$$S = \text{the set of those who like science fiction.}$$

We draw these sets in **Figure 2.28**. Now let's use the numbers in (a) through (g), as well as the fact that 60 people were surveyed, which we call condition (h), to determine the cardinality of each region in the Venn diagram.

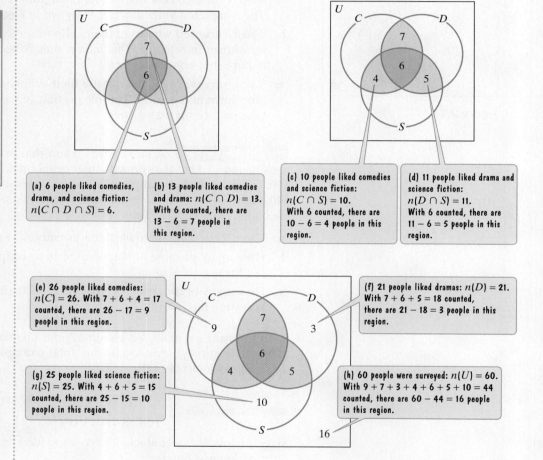

(a) 6 people liked comedies, drama, and science fiction: $n(C \cap D \cap S) = 6$.

(b) 13 people liked comedies and drama: $n(C \cap D) = 13$. With 6 counted, there are $13 - 6 = 7$ people in this region.

(c) 10 people liked comedies and science fiction: $n(C \cap S) = 10$. With 6 counted, there are $10 - 6 = 4$ people in this region.

(d) 11 people liked drama and science fiction: $n(D \cap S) = 11$. With 6 counted, there are $11 - 6 = 5$ people in this region.

(e) 26 people liked comedies: $n(C) = 26$. With $7 + 6 + 4 = 17$ counted, there are $26 - 17 = 9$ people in this region.

(f) 21 people liked dramas: $n(D) = 21$. With $7 + 6 + 5 = 18$ counted, there are $21 - 18 = 3$ people in this region.

(g) 25 people liked science fiction: $n(S) = 25$. With $4 + 6 + 5 = 15$ counted, there are $25 - 15 = 10$ people in this region.

(h) 60 people were surveyed: $n(U) = 60$. With $9 + 7 + 3 + 4 + 6 + 5 + 10 = 44$ counted, there are $60 - 44 = 16$ people in this region.

With a cardinality in each region, we have completed the Venn diagram that illustrates the survey's results.

Checkpoint 3 A survey of 250 memorabilia collectors showed the following results: 108 collected baseball cards. 92 collected comic books. 62 collected stamps. 29 collected baseball cards and comic books. 5 collected baseball cards and stamps. 2 collected comic books and stamps. 2 collected all three types of memorabilia. Use a Venn diagram to illustrate the survey's results.

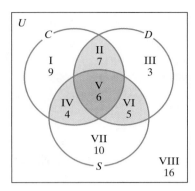

Figure 2.29

Example 4 Using a Survey's Venn Diagram

The Venn diagram in **Figure 2.29** shows the results of the movie survey in Example 3. How many of those surveyed liked

a. comedies, but neither dramas nor science fiction?

b. dramas and science fiction, but not comedies?

c. dramas or science fiction, but not comedies?

d. exactly one movie style?

e. at least two movie styles?

f. none of the movie styles?

Solution

$C \cap (D' \cap S')$

a. Those surveyed who liked comedies, but neither dramas nor science fiction, are represented in region I. There are 9 people in this category.

$(D \cap S) \cap C'$

b. Those surveyed who liked dramas and science fiction, but not comedies, are represented in region VI. There are 5 people in this category.

c. We are interested in those surveyed who liked dramas or science fiction, but not comedies:

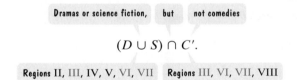

$$(D \cup S) \cap C'.$$

Regions II, III, IV, V, VI, VII   Regions III, VI, VII, VIII

The intersection of the regions in the voice balloons consists of the common regions shown in red, III, VI, and VII. There are 3 + 5 + 10 = 18 elements in these regions. There are 18 people who liked dramas or science fiction, but not comedies.

d. Those surveyed who liked exactly one movie style are represented in regions I, III, and VII. There are 9 + 3 + 10 = 22 elements in these regions. Thus, 22 people liked exactly one movie style.

e. Those surveyed who liked at least two movie styles are people who liked two or more types of movies. People who liked two movie styles are represented in regions II, IV, and VI. Those who liked three movie styles are represented in region V. Thus, we add the number of elements in regions II, IV, V, and VI: 7 + 4 + 6 + 5 = 22. Thus, 22 people liked at least two movie styles.

f. Those surveyed who liked none of the movie styles are represented in region VIII. There are 16 people in this category.

✓ **Checkpoint 4** Use the Venn diagram you constructed in Checkpoint 3 to determine how many of those surveyed collected

a. comic books, but neither baseball cards nor stamps.

b. baseball cards and stamps, but not comic books.

c. baseball cards or stamps, but not comic books.

d. exactly two types of memorabilia.

e. at least one type of memorabilia.

f. none of the types of memorabilia.

Exercise Set 2.5

Concept and Vocabulary Exercises

In Exercises 1–3, fill in each blank so that the resulting statement is true.

1. In survey problems, the word _____ means "intersection."

2. In survey problems, the word _____ means "union."

3. In survey problems, the word _____ means "complement."

In Exercises 4–5, determine whether each statement is true or false. If the statement is false, make the necessary change(s) to produce a true statement.

4. When filling in cardinalities for regions in a two-set Venn diagram, the innermost region, the intersection of the two sets, should be the last region to be filled in.

5. $n(A')$ can be obtained by subtracting $n(A)$ from $n(U)$.

6. In your own words, describe how to solve a survey problem.

In Exercises 7–8, determine whether each statement makes sense or does not make sense, and explain your reasoning.

7. I surveyed seniors in my high school who use the Internet and used a Venn diagram with two intersecting circles to show the number of seniors visiting social networking sites, as well as the number of seniors who have blogs.

8. A Venn diagram with three intersecting circles can be used to show American participation in the country's top three extreme sports: inline skating, skateboarding, and mountain biking.

Practice Exercises

Use the Venn diagram at the top of the next column, which shows the number of elements in regions I through IV, to answer the questions in Exercises 9–16.

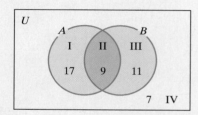

9. How many elements belong to set A?
10. How many elements belong to set B?
11. How many elements belong to set A but not set B?
12. How many elements belong to set B but not set A?
13. How many elements belong to set A or set B?
14. How many elements belong to set A and set B?
15. How many elements belong to neither set A nor set B?
16. How many elements are there in the universal set?

Use the accompanying Venn diagram, which shows the number of elements in region II, to answer Exercises 17–18.

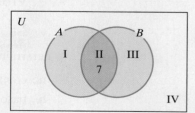

17. If $n(A) = 21$, $n(B) = 29$, and $n(U) = 48$, find the number of elements in each of regions I, III, and IV.

18. If $n(A) = 23$, $n(B) = 27$, and $n(U) = 53$, find the number of elements in each of regions I, III, and IV.

Use the accompanying Venn diagram, which shows the cardinality of each region, to answer Exercises 19–34.

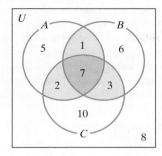

19. How many elements belong to set B?
20. How many elements belong to set A?
21. How many elements belong to set A but not set C?
22. How many elements belong to set B but not set A?
23. How many elements belong to set A or set C?
24. How many elements belong to set A or set B?
25. How many elements belong to set A and set C?
26. How many elements belong to set A and set B?
27. How many elements belong to set B and set C, but not to set A?
28. How many elements belong to set A and set C, but not to set B?
29. How many elements belong to set B or set C, but not to set A?
30. How many elements belong to set A or set C, but not to set B?
31. Considering sets A, B, and C, how many elements belong to exactly one of these sets?
32. Considering sets A, B, and C, how many elements belong to exactly two of these sets?
33. Considering sets A, B, and C, how many elements belong to at least one of these sets?
34. Considering sets A, B, and C, how many elements belong to at least two of these sets?

The accompanying Venn diagram shows the number of elements in region V. In Exercises 35–36, use the given cardinalities to determine the number of elements in each of the other seven regions.

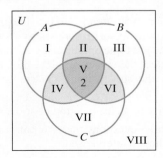

35. $n(U) = 30, n(A) = 11, n(B) = 8, n(C) = 14,$
$n(A \cap B) = 3, n(A \cap C) = 5, n(B \cap C) = 3$
36. $n(U) = 32, n(A) = 21, n(B) = 15, n(C) = 14,$
$n(A \cap B) = 6, n(A \cap C) = 7, n(B \cap C) = 8$

Practice Plus

In Exercises 37–40, use the Venn diagram and the given conditions to determine the number of elements in each region, or explain why the conditions are impossible to meet.

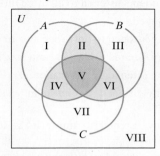

37. $n(U) = 38, n(A) = 26, n(B) = 21, n(C) = 18,$
$n(A \cap B) = 17, n(A \cap C) = 11, n(B \cap C) = 8,$
$n(A \cap B \cap C) = 7$
38. $n(U) = 42, n(A) = 26, n(B) = 22, n(C) = 25,$
$n(A \cap B) = 17, n(A \cap C) = 11, n(B \cap C) = 9,$
$n(A \cap B \cap C) = 5$
39. $n(U) = 40, n(A) = 10, n(B) = 11, n(C) = 12,$
$n(A \cap B) = 6, n(A \cap C) = 9, n(B \cap C) = 7,$
$n(A \cap B \cap C) = 2$
40. $n(U) = 25, n(A) = 8, n(B) = 9, n(C) = 10,$
$n(A \cap B) = 6, n(A \cap C) = 9, n(B \cap C) = 8,$
$n(A \cap B \cap C) = 5$

Application Exercises

In Exercises 41–46, construct a Venn diagram and determine the cardinality for each region. Use the completed Venn diagram to answer the questions.

41. A survey of 75 high school seniors was taken to determine where they got the news about what's going on in the world. Of those surveyed, 29 students got the news from the Internet, 43 from television, and 7 from both the Internet and television.

Of those surveyed,

a. How many got the news from only the Internet?
b. How many got the news from only television?
c. How many got the news from the Internet or television?
d. How many did not get the news from either the Internet or television?

42. A survey of 120 high school seniors who used the Internet indicated that 85 students visited social networking sites, 37 students had blogs, and 6 students visited social networking sites and had blogs.

Of those surveyed,

a. How many students only visited social networking sites?
b. How many students only had blogs?
c. How many students visited social networking sites or had blogs?
d. How many students neither visited social networking sites nor had blogs?

43. A survey of 80 students was taken to determine the musical styles they listened to. Forty-two students listened to rock, 34 to classical, and 27 to jazz. Twelve students listened to rock and jazz, 14 to rock and classical, and 10 to classical and jazz. Seven students listened to all three musical styles.

Of those surveyed,

a. How many listened to only rock music?

b. How many listened to classical and jazz, but not rock?

c. How many listened to classical or jazz, but not rock?

d. How many listened to music in exactly one of the musical styles?

e. How many listened to music in at least two of the musical styles?

f. How many did not listen to any of the musical styles?

44. A survey of 180 high school students was taken to determine participation in various extracurricular activities. Forty-three students were in the band, 52 participated in sports, and 35 participated in theater. Thirteen students participated in the band and sports, 14 in sports and theater, and 12 in band and theater. Five students participated in all three activities.

Of those surveyed,

a. How many participated in only sports?

b. How many participated in the band and sports, but not theater?

c. How many participated in the band or sports, but not theater?

d. How many participated in exactly one of these activities?

e. How many participated in at least two of these activities?

f. How many did not participate in any of the three activities?

45. In the August 2005 issue of *Consumer Reports*, readers suffering from depression reported that alternative treatments were less effective than prescription drugs. Suppose that 550 readers felt better taking prescription drugs, 220 felt better through meditation, and 45 felt better taking St. John's wort. Furthermore, 95 felt better using prescription drugs and meditation, 17 felt better using prescription drugs and St. John's wort, 35 felt better using meditation and St. John's wort, 15 improved using all three treatments, and 150 improved using none of these treatments. (Hypothetical results are partly based on percentages given in *Consumer Reports*.)

a. How many readers suffering from depression were included in the report?

Of those included in the report,

b. How many felt better using prescription drugs or meditation?

c. How many felt better using St. John's wort only?

d. How many improved using prescription drugs and meditation, but not St. John's wort?

e. How many improved using prescription drugs or St. John's wort, but not meditation?

f. How many improved using exactly two of these treatments?

g. How many improved using at least one of these treatments?

46. A survey of American adults was taken to determine the use of cellphones, digital cameras, and laptop computers. The results were as follows: 86 people used cellphones, 60 used digital cameras, 36 used laptop computers, 22 used cellphones and digital cameras, 17 used cellphones and laptop computers, 15 used digital cameras and laptop computers, 10 used all three technologies, and 30 used none of these technologies.

a. How many people were surveyed?

Of those surveyed,

b. How many used cellphones or digital cameras?

c. How many used laptop computers only?

d. How many used cellphones and digital cameras, but not laptop computers?

e. How many used cellphones or laptop computers, but not digital cameras?

f. How many used exactly two of these technologies?

g. How many people used at least one of these technologies?

Critical Thinking Exercises

47. In a survey of 150 students, 90 were taking mathematics and 30 were taking Spanish.

a. What is the least number of students who could have been taking both courses?

b. What is the greatest number of students who could have been taking both courses?

c. What is the greatest number of students who could have been taking neither course?

48. A person applying for the position of college registrar submitted the following report to the college president on 90 students: 31 take math; 28 take chemistry; 42 take psychology; 9 take math and chemistry; 10 take chemistry and psychology; 6 take math and psychology; 4 take all three subjects; and 20 take none of these courses. The applicant was not hired. Explain why.

Group Exercise

49. This group activity is intended to provide practice in the use of Venn diagrams to sort responses to a survey. The group will determine the topic of the survey. Although you will not actually conduct the survey, it might be helpful to imagine carrying out the survey using the students in your high school.

a. In your group, decide on a topic for the survey.

b. Devise three questions that the pollster will ask to the people who are interviewed.

c. Construct a Venn diagram that will assist the pollster in sorting the answers to the three questions. The Venn diagram should contain three intersecting circles within a universal set and eight regions.

d. Describe what each of the regions in the Venn diagram represents in terms of the questions in your poll.

Chapter 2 Summary

2.1 Basic Set Concepts

Definitions and Concepts

A set is a collection of objects whose contents can be clearly determined. The objects in a set are called the elements, or members, of the set.

Sets can be designated by word descriptions, the roster method (a listing within braces, separating elements with commas), or set-builder notation:

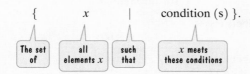

Examples

Word Description	Roster Method	Set-Builder Notation
The set of weekend days	{Saturday, Sunday}	$\{x \mid x$ is a weekend day$\}$
The set of letters in the word *rear*	{r, e, a}	$\{x \mid x$ is a letter in the word *rear*$\}$

Additional Examples to Review
Example 1, page 52; Example 2, page 53; Example 3, page 53

Definitions and Concepts

The empty set, or the null set, represented by { } or $\varnothing$, is a set that contains no elements.

Examples

- $\{x \mid x$ is a number less than 3 and greater than 5$\}$ is $\varnothing$ because no numbers are both less than 3 and greater than 5.
- $\{0\}$ is not $\varnothing$. The set contains one element, namely 0.

Additional Example to Review
Example 4, page 55

Definitions and Concepts

The symbol $\in$ means that an object is an element of a set. The symbol $\notin$ means that an object is not an element of a set.

Examples

- $7 \in \{1, 2, 3, \ldots, 9, 10\}$: 7 is an element of the set $\{1, 2, 3, \ldots, 9, 10\}$.
- $\{7\} \notin \{1, 2, 3, \ldots, 9, 10\}$: The set $\{7\}$ is not an element of the set $\{1, 2, 3, \ldots, 9, 10\}$.

Additional Example to Review
Example 5, page 55

Definitions and Concepts

The set of natural numbers is $\mathbf{N} = \{1, 2, 3, 4, 5, \ldots\}$. Inequality symbols are frequently used to describe sets of natural numbers.

Meanings of Inequality Symbols

$x < a$: x is less than a.
$x \leq a$: x is less than or equal to a.
$x > a$: x is greater than a.
$x \geq a$: x is greater than or equal to a.
$a < x < b$: x is greater than a and less than b.
$a \leq x \leq b$: x is greater than or equal to a and less than or equal to b.
$a \leq x < b$: x is greater than or equal to a and less than b.
$a < x \leq b$: x is greater than a and less than or equal to b.

Examples

Set-Builder Notation	Word Description	Roster Method
$\{x \mid x \in \mathbf{N} \text{ and } x \leq 60\}$	The set of natural numbers less than or equal to 60	$\{1, 2, 3, \ldots, 60\}$
$\{x \mid x \in \mathbf{N} \text{ and } 7 < x < 13\}$	The set of natural numbers greater than 7 and less than 13	$\{8, 9, 10, 11, 12\}$

Additional Examples to Review

Example 6, page 56; Example 7, page 57

Definitions and Concepts

The cardinal number of a set A, $n(A)$, is the number of distinct elements in set A. Repeating elements in a set neither adds new elements to the set nor changes its cardinality.

Examples

- $A = \{x \mid x \text{ is a day of the week}\}$: $n(A) = 7$
- $B = \{x \mid x \text{ is a letter in the word } rear\}$: $n(B) = 3$
- $C = \{5, 6, 7, \ldots, 20\}$: $n(C) = 16$

Additional Example to Review

Example 8, page 58

Definitions and Concepts

Equivalent sets have the same number of elements, or the same cardinality. A one-to-one correspondence between sets A and B means that each element in A can be paired with exactly one element in B, and vice versa. If two sets can be placed in a one-to-one correspondence, then they are equivalent.

Example

$\{x \mid x \text{ is a month beginning with the letter J}\}$ and $\{x \mid x \in \mathbf{N} \text{ and } x < 4\}$ are equivalent sets:

$$\{\text{January, June, July}\}$$
$$\updownarrow \quad \updownarrow \quad \updownarrow$$
$$\{\quad 1, \quad 2, \quad 3\ \}$$

Both sets have the same cardinality, namely 3.

Additional Example to Review

Example 9, page 59

Definitions and Concepts

Set A is a finite set if $n(A) = 0$ or if $n(A)$ is a natural number. A set that is not finite is an infinite set. Equal sets have exactly the same elements, regardless of order or possible repetition of elements. If two sets are equal, then they must be equivalent.

Examples

- Infinite Set

  $\{x | x \in \mathbf{N} \quad \text{and} \quad x \text{ is odd}\} = \{1, 3, 5, \ldots\}$

- Finite Set

  $\{x | x \in \mathbf{N} \quad \text{and} \quad x < 200\} = \{1, 2, 3, \ldots, 199\}$

- Equivalent Sets That Are Not Equal

  $A = \{1, 3, 5, 7\}$ and $B = \{2, 4, 6, 8\}$

The sets are equivalent: $n(A) = n(B) = 4$.
The sets are not equal: They do not contain exactly the same elements.

Additional Example to Review

Example 10, page 61

2.2 Subsets

Definitions and Concepts

Set A is a subset of set B, expressed as $A \subseteq B$, if every element in set A is also in set B. The notation $A \nsubseteq B$ means that set A is not a subset of set B, so there is at least one element of set A that is not an element of set B.

Examples

- $\{5, 6, 7\} \subseteq \{x | x \in \mathbf{N} \quad \text{and} \quad x < 8\}$ because every element of $\{5, 6, 7\}$ is an element of $\{1, 2, 3, 4, 5, 6, 7\}$.

- $\{a, b, c, e\} \nsubseteq \{a, b, e, f, g\}$ because at least one element of $\{a, b, c, e\}$ is not an element of $\{a, b, e, f, g\}$: $c \notin \{a, b, e, f, g\}$.

Additional Example to Review

Example 1, page 66

Definitions and Concepts

Set A is a proper subset of set B, expressed as $A \subset B$, if A is a subset of B and $A \neq B$.

Examples

- $\{1, 2\} \subseteq \{1, 2, 6\}$. Because the sets are not equal, it is also true that $\{1, 2\} \subset \{1, 2, 6\}$.

- $\{1, 2\} \subseteq \{x | x \in \mathbf{N} \quad \text{and} \quad x \leq 2\}$. Because the sets are equal, it is not correct to place the symbol $\subset$ between them: $\{1, 2\} \not\subset \{x | x \in \mathbf{N} \quad \text{and} \quad x \leq 2\}$.

Additional Example to Review

Example 2, page 67

Definitions and Concepts

The empty set is a subset of every set.

Examples

- $\varnothing \subseteq \{1, 3, 5\}$ and $\varnothing \subset \{1, 3, 5\}$.
- $\varnothing \subseteq \{ \ \}$, but $\varnothing \not\subset \{ \ \}$.

Additional Example to Review

Example 3, page 68

Definitions and Concepts

A set with n elements has 2^n subsets and $2^n - 1$ proper subsets.

Examples

- $\{1, 3, 5, 7, 9, 11\}$, a set with 6 elements, has $2^6 = 2 \times 2 \times 2 \times 2 \times 2 \times 2 = 64$ subsets.
- $\{1, 3, 5, 7, 9, 11\}$ has $64 - 1 = 63$ proper subsets.

Additional Examples to Review

Example 4, page 70; Example 5, page 70

2.3 Venn Diagrams and Set Operations

Definitions and Concepts

A universal set, symbolized by U, is a set that contains all the elements being considered in a given discussion or problem.

Examples

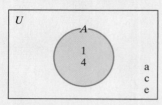

- $U = \{1, 4, a, c, e\}$
- $A = \{1, 4\}$
- $\{a, c, e\}$ is the set of elements in U that are not in A.

Additional Example to Review

Example 1, page 75

Definitions and Concepts

Venn Diagrams: Representing Two Subsets of a Universal Set

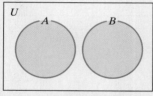

No A are B.
A and B are disjoint.

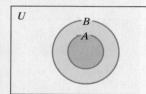

All A are B.
$A \subset B$

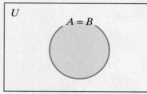

A and B are
equal sets.

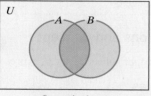

Some (at least
one) A are B.

Examples

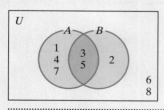

- $A = \{1, 3, 4, 5, 7\}$
- $B = \{2, 3, 5\}$
- Set of elements in A and $B = \{3, 5\}$
- $U = \{1, 2, 3, 4, 5, 6, 7, 8\}$

Additional Example to Review

Example 2, page 76

Definitions and Concepts

A' (the complement of set A), which can be read A prime or **not** A, is the set of all elements in the universal set that are not in A.

Example

If $U = \{a, b, c, d\}$ and $A = \{b, d\}$, then $A' = \{a, c\}$.

Additional Example to Review
Example 3, page 78

Definitions and Concepts

$A \cap B$ (A intersection B), which can be read set A **and** set B, is the set of elements common to both set A and set B.

Examples

- $\{a, b, c, d\} \cap \{c, d, e, f\} = \{c, d\}$
- $\{a, b, c, d\} \cap \{e, f\} = \varnothing$

Additional Example to Review
Example 4, page 78

Definitions and Concepts

$A \cup B$ (A union B), which can be read set A **or** set B, is the set of elements that are members of set A or of set B or of both sets.

Examples

- $\{a, b, c, d\} \cup \{c, d, e, f\} = \{a, b, c, d, e, f\}$
- $\{a, b, c, d\} \cup \{e, f\} = \{a, b, c, d, e, f\}$

Additional Example to Review
Example 5, page 79

Definitions and Concepts

Some problems involve more than one set operation. Begin by performing any operations inside parentheses.

Examples

$U = \{a, b, c, d, e, f\}$, $A = \{a, c, d\}$, $B = \{c, d, f\}$

- $(A \cup B)' = \{a, c, d, f\}' = \{b, e\}$
- $A' \cap B' = \{b, e, f\} \cap \{a, b, e\} = \{b, e\}$

Additional Example to Review
Example 6, page 81

Definitions and Concepts

Elements of sets involving a variety of set operations can be determined from Venn diagrams.

Examples

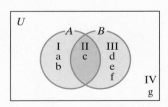

- $A \cup B$ (regions I, II, III) $= \{a, b, c, d, e, f\}$
- $(A \cup B)'$ (region IV) $= \{g\}$
- $A \cap B$ (region II) $= \{c\}$
- $(A \cap B)'$ (regions I, III, IV) $= \{a, b, d, e, f, g\}$
- $A' \cap B$ (region III) $= \{d, e, f\}$
- $A \cup B'$ (regions I, II, IV) $= \{a, b, c, g\}$

Additional Example to Review
Example 7, page 81

2.4 Set Operations and Venn Diagrams with Three Sets

Definitions and Concepts

When using set operations involving three sets, begin by performing operations within parentheses.

Examples

$U = \{1, 2, 3, 4, 5, 6\}, A = \{2, 4\}, B = \{2, 5\}, C = \{1, 5, 6\}$
- $(A \cap B) \cup C = \{2\} \cup \{1, 5, 6\} = \{1, 2, 5, 6\}$
- $A \cup (B \cap C') = \{2, 4\} \cup (\{2, 5\} \cap \{2, 3, 4\}) = \{2, 4\} \cup \{2\} = \{2, 4\}$

Additional Example to Review

Example 1, page 86

Definitions and Concepts

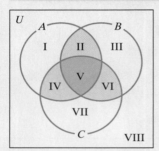

The figure on the left shows a Venn diagram with three intersecting sets that separate the universal set, U, into eight regions. Elements of sets involving a variety of set operations can be determined from this Venn diagram.

Examples

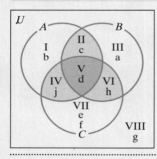

- B' (regions I, IV, VII, VIII) $= \{b, j, e, f, g\}$
- $A \cap C$ (regions IV, V) $= \{j, d\}$
- $B \cap C$ (regions V, VI) $= \{d, h\}$
- $(A \cap C) \cup (B \cap C) = \{j, d\} \cup \{d, h\} = \{j, d, h\}$
- $A \cap B \cap C$ (region V) $= \{d\}$
- $(A \cup B \cup C)'$ (region VIII) $= \{g\}$

Additional Example to Review

Example 2, page 89

Definitions and Concepts

If two specific sets represent the same regions of a general Venn diagram, then this deductively proves that the two sets are equal.

Example

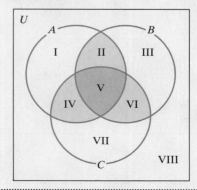

$(A \cap B) \cap C$ is represented by region V.

| Regions II, V | Regions IV, V, VI, VII |

$A \cap (B \cap C)$ is represented by region V.

| Regions I, II, IV, V | Regions V, VI |

Conclusion: $(A \cap B) \cap C = A \cap (B \cap C)$

Additional Examples to Review
Example 3, page 89; Example 4, page 90

2.5 Survey Problems

Definitions and Concepts

Venn diagrams can be used to organize information collected in surveys. When interpreting cardinalities in such diagrams, *and* and *but* mean intersection, *or* means union, and *not* means complement.

To solve a survey problem,

1. Define sets and draw a Venn diagram.
2. Fill in the cardinality of each region, starting with the innermost region and working outward.
3. Use the completed diagram to answer the problem's questions.

Example

Seventy people were contacted and responded to a television survey. The following information was obtained:

 a. 5 people liked reality shows, sitcoms, and dramas.

 b. 12 people liked reality shows and sitcoms.

 c. 13 people liked reality shows and dramas.

 d. 8 people liked sitcoms and dramas.

 e. 32 people liked reality shows.

 f. 25 people liked sitcoms.

 g. 25 people liked dramas.

Use a Venn diagram to illustrate the survey's results.

Let R = the set of those who liked reality shows, S = the set of those who liked sitcoms, and D = the set of those who liked dramas.

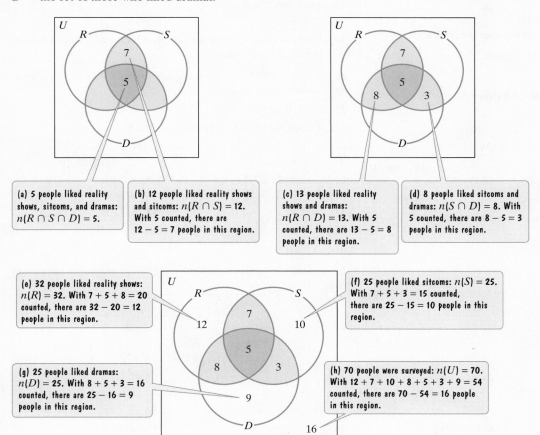

(a) 5 people liked reality shows, sitcoms, and dramas: $n(R \cap S \cap D) = 5$.

(b) 12 people liked reality shows and sitcoms: $n(R \cap S) = 12$. With 5 counted, there are $12 - 5 = 7$ people in this region.

(c) 13 people liked reality shows and dramas: $n(R \cap D) = 13$. With 5 counted, there are $13 - 5 = 8$ people in this region.

(d) 8 people liked sitcoms and dramas: $n(S \cap D) = 8$. With 5 counted, there are $8 - 5 = 3$ people in this region.

(e) 32 people liked reality shows: $n(R) = 32$. With $7 + 5 + 8 = 20$ counted, there are $32 - 20 = 12$ people in this region.

(f) 25 people liked sitcoms: $n(S) = 25$. With $7 + 5 + 3 = 15$ counted, there are $25 - 15 = 10$ people in this region.

(g) 25 people liked dramas: $n(D) = 25$. With $8 + 5 + 3 = 16$ counted, there are $25 - 16 = 9$ people in this region.

(h) 70 people were surveyed: $n(U) = 70$. With $12 + 7 + 10 + 8 + 5 + 3 + 9 = 54$ counted, there are $70 - 54 = 16$ people in this region.

Some Conclusions Based on the Venn Diagram

- 12 people liked reality shows, but neither sitcoms nor dramas.
- 3 people liked sitcoms and dramas, but not reality shows.
- 12 + 10 + 9 = 31 people liked exactly one style of entertainment.
- 16 people liked none of the entertainment styles.

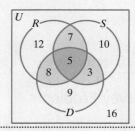

Additional Examples to Review

Example 1, page 96; Example 2, page 98; Example 3, page 99; Example 4, page 101

Review Exercises

Section 2.1 Basic Set Concepts

In Exercises 1–2, write a word description of each set. (More than one correct description may be possible.)

1. {Tuesday, Thursday}
2. {1, 2, 3, . . . , 10}

In Exercises 3–5, express each set using the roster method.

3. $\{x \mid x$ is a letter in the word *miss*$\}$
4. $\{x \mid x \in \mathbf{N}$ and $8 \le x < 13\}$
5. $\{x \mid x \in \mathbf{N}$ and $x \le 30\}$

In Exercises 6–7, determine which sets are the empty set.

6. $\{\varnothing\}$ 7. $\{x \mid x < 4$ and $x \ge 6\}$

In Exercises 8–9, fill in the blank with either $\in$ or $\notin$ to make each statement true.

8. 93 _____ {1, 2, 3, 4, . . . , 99, 100}
9. {d} _____ {a, b, c, d, e}

In Exercises 10–11, find the cardinal number for each set.

10. $A = \{x \mid x$ is a month of the year$\}$
11. $B = \{18, 19, 20, . . . , 31, 32\}$

In Exercises 12–13, fill in the blank with either $=$ or $\ne$ to make each statement true.

12. {0, 2, 4, 6, 8} _____ {8, 2, 6, 4}
13. $\{x \mid x \in \mathbf{N}$ and $x > 7\}$ _____ {8, 9, 10, . . . , 100}

In Exercises 14–15, determine if the pairs of sets are equivalent, equal, both, or neither.

14. $A = \{x \mid x$ is a lowercase letter that comes before f in the English alphabet$\}$
    $B = \{2, 4, 6, 8, 10\}$
15. $A = \{x \mid x \in \mathbf{N}$ and $3 < x < 7\}$
    $B = \{4, 5, 6\}$

In Exercises 16–17, determine whether each set is finite or infinite.

16. $\{x \mid x \in \mathbf{N}$ and $x < 50{,}000\}$
17. $\{x \mid x \in \mathbf{N}$ and x is even$\}$

Section 2.2 Subsets

In Exercises 18–20, write $\subseteq$ or $\nsubseteq$ in each blank so that the resulting statement is true.

18. {penny, nickel, dime}
    _____ {half-dollar, quarter, dime, nickel, penny}

19. $\{-1, 0, 1\}$ _____ $\{-3, -2, -1, 1, 2, 3\}$
20. $\varnothing$ _____ $\{x \mid x$ is an odd natural number$\}$

In Exercises 21–22, determine whether $\subseteq$, $\subset$, both, or neither can be placed in each blank to form a true statement.

21. $\{1, 2\}$ _____ $\{1, 1, 2, 2\}$
22. $\{x \mid x$ is a person living in the United States$\}$
    _____ $\{y \mid y$ is a person living on planet Earth$\}$

In Exercises 23–29, determine whether each statement is true or false. If the statement is false, explain why.

23. Texas $\in$ {Oklahoma, Louisiana, Georgia, South Carolina}
24. $4 \subseteq \{2, 4, 6, 8, 10, 12\}$
25. $\{e, f, g\} \subset \{d, e, f, g, h, i\}$
26. $\{\ominus, \varnothing\} \subset \{\varnothing, \ominus\}$
27. $\{3, 7, 9\} \subseteq \{9, 7, 3, 1\}$
28. {six} has 2^6 subsets.
29. $\varnothing \subseteq \{\ \}$
30. List all subsets for the set $\{1, 5\}$. Which one of these subsets is not a proper subset?

In Exercises 31–32, find the number of subsets and the number of proper subsets for each set.

31. $\{2, 4, 6, 8, 10\}$
32. $\{x \mid x$ is a month that begins with the letter J$\}$

Section 2.3 Venn Diagrams and Set Operations

In Exercises 33–37, let $U = \{1, 2, 3, 4, 5, 6, 7, 8\}$, $A = \{1, 2, 3, 4\}$, and $B = \{1, 2, 4, 5\}$. Find each of the following sets.

33. $A \cap B$ 34. $A \cup B'$
35. $A' \cap B$ 36. $(A \cup B)'$
37. $A' \cap B'$

In Exercises 38–45, use the Venn diagram to represent each set in roster form.

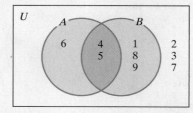

38. A 39. B'
40. $A \cup B$ 41. $A \cap B$

42. $(A \cap B)'$

43. $(A \cup B)'$

44. $A \cap B'$

45. U

Section 2.4 Set Operations and Venn Diagrams with Three Sets

In Exercises 46–47, let

$$U = \{1, 2, 3, 4, 5, 6, 7, 8\}$$
$$A = \{1, 2, 3, 4\}$$
$$B = \{1, 2, 4, 5\}$$
$$C = \{1, 5\}.$$

Find each of the following sets.

46. $A \cup (B \cap C)$

47. $(A \cap C)' \cup B$

In Exercises 48–53, use the Venn diagram to represent each set in roster form.

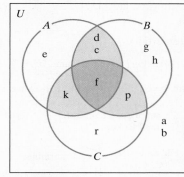

48. $A \cup C$

49. $B \cap C$

50. $(A \cap B) \cup C$

51. $A \cap C'$

52. $(A \cap C)'$

53. $A \cap B \cap C$

54. Use a Venn diagram with two intersecting circles to prove that
$$(A \cup B)' = A' \cap B'.$$

55. Use a Venn diagram with three intersecting circles to determine whether the following statement is a theorem:
$$A \cap (B \cup C) = A \cup (B \cap C).$$

56. *The Penguin Atlas of Women in the World* uses maps and graphics to present data on how women live across continents and cultures. The table is based on data from the atlas.

Women in the World: School, Work, and Literacy

	Percentage of College Students Who Are Women	Percentage of Women Working for Pay	Percentage of Women Who Are Illiterate
United States	57%	59%	1%
Italy	57%	37%	2%
Turkey	43%	28%	20%
Norway	60%	63%	1%
Pakistan	46%	33%	65%
Iceland	65%	71%	1%
Mexico	51%	40%	10%

Source: The Penguin Atlas of Women in the World, 2009

The data can be organized in the following Venn diagram:

Women in the World

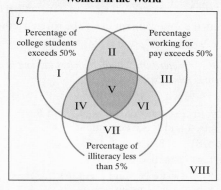

Use the data to determine the region in the Venn diagram into which each of the seven countries should be placed.

Section 2.5 Survey Problems

In Exercises 57–58, construct a Venn diagram and determine the cardinality for each region. Use the completed Venn diagram to answer the questions.

57. A survey of 1000 American adults was taken to analyze their investments. Of those surveyed, 650 had invested in stocks, 550 in bonds, and 400 in both stocks and bonds. Of those surveyed,

  a. How many invested in only stocks?

  b. How many invested in stocks or bonds?

  c. How many did not invest in either stocks or bonds?

58. A survey of 200 students at a nonresidential college was taken to determine how the students got to campus during the fall term. Of those surveyed, 118 used cars, 102 used public transportation, and 70 used bikes. Forty-eight students used cars and public transportation, 38 used cars and bikes, and 26 used public transportation and bikes. Twenty-two students used all three modes of transportation.

Of those surveyed,

  a. How many used only public transportation?

  b. How many used cars and public transportation, but not bikes?

  c. How many used cars or public transportation, but not bikes?

  d. How many used exactly two of these modes of transportation?

  e. How many did not use any of the three modes of transportation to get to campus?

Chapter 2 Test A

1. Express the following set using the roster method:
$\{x \mid x \in \mathbf{N} \text{ and } 17 < x \leq 24\}$.

In Exercises 2–9, determine whether each statement is true or false. If the statement is false, explain why.

2. $\{6\} \in \{1, 2, 3, 4, 5, 6, 7\}$

3. If $A = \{x \mid x \text{ is a day of the week}\}$ and $B = \{2, 4, 6, \ldots, 14\}$, then sets A and B are equivalent.

4. $\{2, 4, 6, 8\} = \{8, 8, 6, 6, 4, 4, 2\}$

5. $\{d, e, f, g\} \subseteq \{a, b, c, d, e, f\}$

6. $\{3, 4, 5\} \subset \{x \mid x \in \mathbf{N} \text{ and } x < 6\}$

7. $14 \notin \{1, 2, 3, 4, \ldots, 39, 40\}$

8. $\{a, b, c, d, e\}$ has 25 subsets.

9. The empty set is a proper subset of any set, including itself.

10. List all subsets for the set $\{6, 9\}$. Which of these subsets is not a proper subset?

In Exercises 11–15, let
$$U = \{a, b, c, d, e, f, g\}$$
$$A = \{a, b, c, d\}$$
$$B = \{c, d, e, f\}$$
$$C = \{a, e, g\}.$$

Find each of the following sets or cardinalities.

11. $A \cup B$ **12.** $(B \cap C)'$

13. $A \cap C'$ **14.** $(A \cup B) \cap C$

15. $n(A \cup B')$

In Exercises 16–18, use the Venn diagram to represent each set in roster form.

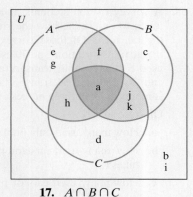

16. A' **17.** $A \cap B \cap C$

18. $(A \cap B) \cup (A \cap C)$

19. Use the Venn diagram shown to determine whether the following statement is a theorem:
$$A' \cap (B \cup C) = (A' \cap B) \cup (A' \cap C).$$

Show work clearly as you develop the regions representing each side of the statement.

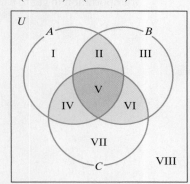

20. Here is a list of some famous people on whom the FBI kept files:

Famous Person	Length of FBI File
Bud Abbott (entertainer)	14 pages
Charlie Chaplin (entertainer)	2063 pages
Albert Einstein (scientist)	1800 pages
Martin Luther King, Jr. (civil rights leader)	17,000 pages
Elvis Presley (entertainer)	663 pages
Jackie Robinson (athlete)	131 pages
Eleanor Roosevelt (first lady; U.N. representative)	3000 pages
Frank Sinatra (entertainer)	1275 pages

Source: Paul Grobman, *Vital Statistics*, Plume, 2005.

The data can be organized in the following Venn diagram:

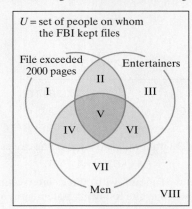

Use the data to determine the region in the Venn diagram into which each of the following people should be placed.

a. Chaplin   **b.** Einstein

c. King   **d.** Roosevelt

e. Sinatra

21. A winter resort took a poll of its 350 visitors to see which winter activities people enjoyed. The results were as follows: 178 liked to ski, 154 liked to snowboard, 57 liked to ice skate, 49 liked to ski and snowboard, 15 liked to ski and ice skate, 2 liked to snowboard and ice skate, and 2 liked all three activities.

a. Use a Venn diagram to illustrate the survey's results.

Use the Venn diagram to determine how many of those surveyed enjoyed

b. exactly one of these activities.

c. none of these activities.

d. at least two of these activities.

e. snowboarding and ice skating, but not skiing.

f. snowboarding or ice skating, but not skiing.

g. only skiing.

Chapter 2 Test B

1. Express the following set using the roster method:

   $\{x \mid x \in \mathbf{N} \quad \text{and} \quad 8 \le x < 14\}.$

In Exercises 2–9, determine whether each statement is true or false. If the statement is false, explain why.

2. $\{d\} \in \{a, b, c, d, e, f\}$

3. If $A = \{x \mid x \text{ is a month of the year}\}$ and $B = \{1, 3, 5, \ldots, 23\}$, then sets A and B are equivalent.

4. $\{1, 3, 5, 7\} = \{7, 5, 3, 3, 1\}$

5. $\{6, 7, 8, 9\} \subseteq \{3, 4, 5, 6, 7, 9, 10\}$

6. $\{2, 4, 6\} \subset \{x \mid x \in \mathbf{N} \quad \text{and} \quad x < 7\}$

7. $33 \notin \{1, 2, 3, \ldots, 99, 100\}$

8. $\{28, 29, 30, 31, 32\}$ has 32 subsets.

9. $\{ \ \} \subseteq \varnothing$

10. List all subsets for the set $\{6, 8, 10\}$. Which of these subsets is not a proper subset?

In Exercises 11–15, let

$$U = \{1, 2, 3, 4, 5, 6, 7\}$$
$$A = \{1, 2, 3, 4\}$$
$$B = \{3, 4, 5, 6\}$$
$$C = \{1, 5, 7\}.$$

Find each of the following sets or cardinalities.

11. $A \cup C$

12. $(A \cap B)'$

13. $B \cap C'$

14. $(B \cup C) \cap A$

15. $n(C \cup B')$

In Exercises 16–18, use the Venn diagram to represent each set in roster form.

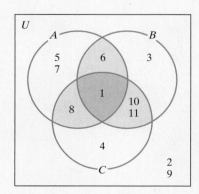

16. C'

17. $(C \cap B)'$

18. $(A \cup B) \cap (A \cup C)$

19. Use the Venn diagram shown to determine whether the following statement is a theorem:

$$A \cap (B \cup C) = A \cup (B \cap C).$$

Show work clearly as you develop the regions representing each side of the statement.

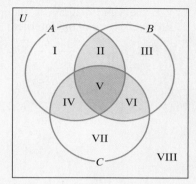

20. The chart shows three health indicators for seven countries or regions.

Worldwide Health Indicators

Country/Region	Male Life Expectancy	Female Life Expectancy	Persons per Doctor
United States	74.8	80.1	360
Italy	77.6	83.2	180
Russia	59.9	73.3	240
East Africa	46.9	48.2	13,620
Japan	78.6	85.6	530
England	75.9	81.0	720
Iran	68.6	71.4	1200

Source: Time Almanac 2009

The data can be organized in the following Venn diagram:

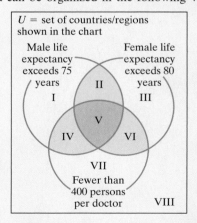

Use the data to determine the region in the Venn diagram into which each of the following countries or regions should be placed.

a. Japan
b. United States
c. East Africa
d. Russia
e. Italy

21. A survey of 120 American adults was taken to determine where they accessed the Internet. The results were as follows: 68 people accessed the Internet at school, 70 at home, 31 at work, 45 at school and at home, 14 at school and at work, 12 at home and at work, and 5 at school and at home and at work.

    a. Use a Venn diagram to illustrate the survey's results.

Use the Venn diagram to determine how many of those surveyed accessed the Internet

    b. in exactly one location.

    c. in none of these locations.

    d. in at least two locations.

    e. at school and at home, but not at work.

    f. at school or at home, but not at work.

    g. only at school.

Number Theory and the Real Number System

3

When you are picked up from school, talk radio is on again. Just as you are about to ask to change the station, you hear politicians discussing the problem of the national debt that exceeds $12 trillion. They state that it's more than the gross domestic product of China, the world's second-richest nation, and four times greater than the combined net worth of America's 691 billionaires. They make it seem like the national debt is a real problem, but later you realize that you don't really know what a number like 12 trillion means. If the national debt were evenly divided among all citizens of the country, how much would every man, woman, and child have to pay? Is economic doomsday about to arrive?

Literacy with numbers, called numeracy, is a prerequisite for functioning in a meaningful way personally, professionally, and as a citizen. In this chapter, our focus is on understanding numbers, their properties, and their applications. The problem of placing the $12.3 trillion national debt in perspective appears as Example 9 in Section 3.6.

Number Theory: Prime and Composite Numbers

3.1

Objectives

1. Determine divisibility.

2. Write the prime factorization of a composite number.

3. Find the greatest common divisor of two numbers.

4. Solve problems using the greatest common divisor.

5. Find the least common multiple of two numbers.

6. Solve problems using the least common multiple.

Number Theory and Divisibility

You are organizing an athletics program at your school. You need to divide 40 boys and 24 girls into all-male and all-female teams so that each team has the same number of people. The boys' teams should have the same number of players as the girls' teams. What is the largest number of people that can be placed on a team?

This problem can be solved using a branch of mathematics called **number theory**. Number theory is primarily concerned with the properties of numbers used for counting, namely 1, 2, 3, 4, 5, and so on. The set of counting numbers is also called the set of **natural numbers**. As we saw in Chapter 2, we represent this set by the letter **N**.

> **The Set of Natural Numbers**
> $$\mathbf{N} = \{1, 2, 3, 4, 5, 6, 7, 8, 9, 10, 11, \ldots\}$$

We can solve the school athletics program problem. However, to do so we must understand the concept of divisibility. For example, there are a number of different ways to divide the 24 girls into teams, including

1 team with all 24 girls:	$1 \times 24 = 24$
2 teams with 12 girls per team:	$2 \times 12 = 24$
3 teams with 8 girls per team:	$3 \times 8 = 24$
4 teams with 6 girls per team:	$4 \times 6 = 24$
6 teams with 4 girls per team:	$6 \times 4 = 24$
8 teams with 3 girls per team:	$8 \times 3 = 24$
12 teams with 2 girls per team:	$12 \times 2 = 24$
24 teams with 1 girl per team:	$24 \times 1 = 24.$

The natural numbers that are multiplied together resulting in a product of 24 are called *factors* of 24. Any natural number can be expressed as a product of two or more natural numbers. The natural numbers that are multiplied are called the **factors** of the product. Notice that a natural number may have many factors.

$$2 \times 12 = 24 \qquad 3 \times 8 = 24 \qquad 6 \times 4 = 24$$

Factors of 24   Factors of 24   Factors of 24

The numbers 1, 2, 3, 4, 6, 8, 12, and 24 are all factors of 24. Each of these numbers divides 24 without a remainder.

In general, let a and b represent natural numbers. We say that a is **divisible** by b if the operation of dividing a by b leaves a remainder of 0.

A natural number is divisible by all of its factors. Thus, 24 is divisible by 1, 2, 3, 4, 6, 8, 12, and 24. Using the factor 8, we can express this divisibility in a number of ways:

24 is **divisible** by 8.

8 is a **divisor** of 24.

8 **divides** 24.

Mathematicians use a special notation to indicate divisibility.

> **Divisibility**
>
> If a and b are natural numbers, a is **divisible** by b if the operation of dividing a by b leaves a remainder of 0. This is the same as saying that b is a **divisor** of a, or b **divides** a. All three statements are symbolized by writing
>
> $$b|a.$$

Using this new notation, we can write

$$12|24.$$

Twelve divides 24 because 24 divided by 12 leaves a remainder of 0. By contrast, 13 does not divide 24 because 24 divided by 13 does not leave a remainder of 0. The notation

$$13\nmid24$$

means that 13 does not divide 24.

 **Determine divisibility.**

Table 3.1 shows some common rules for divisibility. Divisibility rules for 7 and 11 are difficult to remember and are not included in the table.

Table 3.1 Rules of Divisibility

Divisible By	Test	Example
2	The last digit is 0, 2, 4, 6, or 8.	5,892,796 is divisible by 2 because the last digit is 6.
3	The sum of the digits is divisible by 3.	52,341 is divisible by 3 because the sum of the digits is $5 + 2 + 3 + 4 + 1 = 15$, and 15 is divisible by 3.
4	The last two digits form a number divisible by 4.	3,947,136 is divisible by 4 because 36 is divisible by 4.
5	The number ends in 0 or 5.	28,160 and 72,805 end in 0 and 5, respectively. Both are divisible by 5.
6	The number is divisible by both 2 and 3. (In other words, the number is even and the sum of its digits is divisible by 3.)	954 is divisible by 2 because it ends in 4. 954 is also divisible by 3 because the digit sum is 18, which is divisible by 3. Because 954 is divisible by both 2 and 3, it is divisible by 6.
8	The last three digits form a number that is divisible by 8.	593,777,832 is divisible by 8 because 832 is divisible by 8.
9	The sum of the digits is divisible by 9.	5346 is divisible by 9 because the sum of the digits, 18, is divisible by 9.
10	The last digit is 0.	998,746,250 is divisible by 10 because the number ends in 0.
12	The number is divisible by both 3 and 4. (In other words, the sum of the digits is divisible by 3 and the last two digits form a number divisible by 4.)	614,608,176 is divisible by 3 because the digit sum is 39, which is divisible by 3. It is also divisible by 4 because the last two digits form 76, which is divisible by 4. Because 614,608,176 is divisible by both 3 and 4, it is divisible by 12.

Calculators and Divisibility

You can use a calculator to verify each result in Example 1. Consider part (a):

$$4 \mid 3,754,086.$$

Divide 3,754,086 by 4 using the following keystrokes:

3754086 ÷ 4.

Press = or ENTER . The number displayed is 938521.5. This is not a natural number. The 0.5 shows that the division leaves a nonzero remainder. Thus, 4 does not divide 3,754,086. The given statement is false.

Now consider part (b):

$$9 \nmid 4,119,706,413.$$

Use your calculator to divide the number on the right by 9:

4119706413 ÷ 9.

Press = or ENTER . The display is 457745157. This is a natural number. The remainder of the division is 0, so 9 does divide 4,119,706,413. The given statement is false.

The number 2 is the only even prime number. Every other even number has at least three factors: 1, 2, and the number itself.

Write the prime factorization of a composite number.

Example 1   Using the Rules of Divisibility

Which one of the following statements is true?

a. $4 \mid 3,754,086$ **b.** $9 \nmid 4,119,706,413$ **c.** $8 \mid 677,840$

Solution

a. $4 \mid 3,754,086$ states that 4 divides 3,754,086. **Table 3.1** indicates that for 4 to divide a number, the last two digits must form a number that is divisible by 4. Because 86 is not divisible by 4, the given statement is false.

b. $9 \nmid 4,119,706,413$ states that 9 does *not* divide 4,119,706,413. Based on **Table 3.1**, if the sum of the digits is divisible by 9, then 9 does indeed divide this number. The sum of the digits is $4 + 1 + 1 + 9 + 7 + 0 + 6 + 4 + 1 + 3 = 36$, which is divisible by 9. Because 4,119,706,413 is divisible by 9, the given statement is false.

c. $8 \mid 677,840$ states that 8 divides 677,840. **Table 3.1** indicates that for 8 to divide a number, the last three digits must form a number that is divisible by 8. Because 840 is divisible by 8, then 8 divides 677,840, and the given statement is true.

The statement given in part (c) is the only true statement.

Checkpoint 1   Which one of the following statements is true?

a. $8 \mid 48,324$ **b.** $6 \mid 48,324$ **c.** $4 \nmid 48,324$

Prime Factorization

By developing some other ideas of number theory, we will be able to solve the school athletics program problem. We begin with the definition of a prime number.

Prime Numbers

A **prime number** is a natural number greater than 1 that has only itself and 1 as factors.

Using this definition, we see that the number 7 is a prime number because it has only 1 and 7 as factors. Said in another way, 7 is prime because it is divisible by only 1 and 7. The first ten prime numbers are 2, 3, 5, 7, 11, 13, 17, 19, 23, and 29. Each number in this list has exactly two divisors, itself and 1. By contrast, 9 is not a prime number; in addition to being divisible by 1 and 9, it is also divisible by 3. The number 9 is an example of a *composite number*.

Composite Numbers

A **composite number** is a natural number greater than 1 that is divisible by a number other than itself and 1.

Using this definition, the first ten composite numbers are 4, 6, 8, 9, 10, 12, 14, 15, 16, and 18. Each number in this list has at least three distinct divisors.

By the definitions above, both prime numbers and composite numbers must be natural numbers *greater than* 1, so **the natural number 1 is neither prime nor composite**.

Every composite number can be expressed as the product of prime numbers. For example, the composite number 45 can be expressed as

$$45 = 3 \times 3 \times 5.$$

Note that 3 and 5 are prime numbers. Expressing a composite number as the product of prime numbers is called **prime factorization**. The prime factorization of 45 is $3 \times 3 \times 5$. The order in which we write these factors does not matter. This means that

$$45 = 3 \times 3 \times 5$$
$$\text{or} \quad 45 = 5 \times 3 \times 3$$
$$\text{or} \quad 45 = 3 \times 5 \times 3.$$

In Chapter 1, we defined a **theorem** as a statement that can be proved using deductive reasoning. The ancient Greeks proved that if the order of the factors is disregarded, there is only one prime factorization possible for any given composite number. This statement is called the **Fundamental Theorem of Arithmetic**.

> **The Fundamental Theorem of Arithmetic**
>
> Every composite number can be expressed as a product of prime numbers in one and only one way (if the order of the factors is disregarded).

One method used to find the prime factorization of a composite number is called a **factor tree**. To use this method, begin by selecting any two numbers, other than 1, whose product is the number to be factored. One or both of the factors may not be prime numbers. Continue to factor composite numbers. Stop when all numbers are prime.

Example 2 Prime Factorization Using a Factor Tree

Find the prime factorization of 700.

Solution

Start with any two numbers, other than 1, whose product is 700, such as 7 and 100. This forms the first branch of the tree. Continue factoring the composite number or numbers that result (in this case 100), branching until each branch ends with a prime number.

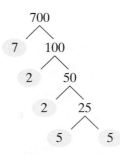

The prime factors are shown on light blue ovals. Thus, the prime factorization of 700 is

$$700 = 7 \times 2 \times 2 \times 5 \times 5.$$

We can use exponents to show the repeated prime factors:

$$700 = 7 \times 2^2 \times 5^2.$$

Using a dot to indicate multiplication and arranging the factors from least to greatest, we can write

$$700 = 2^2 \cdot 5^2 \cdot 7.$$

 **Checkpoint 2** Find the prime factorization of 120.

A Brief Review • Exponents

- If n is a natural number,

$$b^n = \underbrace{b \cdot b \cdot b \cdot \cdots \cdot b.}_{\text{Exponent or Power}}$$

Base b appears as a factor n times.

- b^n is read "the nth power of b" or "b to the nth power." Thus, the nth power of b is defined as the product of n factors of b. The expression b^n is called an **exponential expression**. Furthermore, $b^1 = b$.

Exponential Expression	Read	Evaluation
8^1	8 to the first power	$8^1 = 8$
5^2	5 to the second power or 5 squared	$5^2 = 5 \cdot 5 = 25$
6^3	6 to the third power or 6 cubed	$6^3 = 6 \cdot 6 \cdot 6 = 216$
10^4	10 to the fourth power	$10^4 = 10 \cdot 10 \cdot 10 \cdot 10 = 10,000$
2^5	2 to the fifth power	$2^5 = 2 \cdot 2 \cdot 2 \cdot 2 \cdot 2 = 32$

Greatest Common Divisor

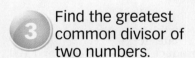

Find the greatest common divisor of two numbers.

The greatest common divisor (GCD) of two or more natural numbers is the largest number that is a divisor (or factor) of all the numbers. For example, 8 is the greatest common divisor of 32 and 40 because it is the largest natural number that divides both 32 and 40. Some pairs of numbers have 1 as their greatest common divisor. Such number pairs are said to be **relatively prime**. For example, the greatest common divisor of 5 and 26 is 1. Thus, 5 and 26 are relatively prime.

The greatest common divisor can be found using prime factorizations.

Finding the Greatest Common Divisor Using Prime Factorizations

To find the greatest common divisor of two or more numbers,

1. Write the prime factorization of each number.
2. Select each prime factor with the smallest exponent that is common to each of the prime factorizations.
3. Form the product of the numbers from step 2. The greatest common divisor is the product of these factors.

Example 3   Finding the Greatest Common Divisor

Find the greatest common divisor of 216 and 234.

Solution

Step 1 **Write the prime factorization of each number.** Begin by writing the prime factorizations of 216 and 234.

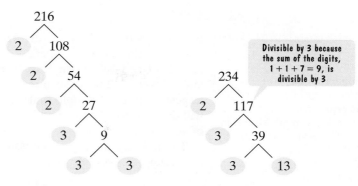

The factor tree on the left indicates that

$$216 = 2^3 \times 3^3.$$

The factor tree on the right indicates that

$$234 = 2 \times 3^2 \times 13.$$

Step 2   Select each prime factor with the smaller exponent that is common to each of the prime factorizations. Look at the factorizations of 216 and 234 from step 1. Can you see that 2 is a prime number common to the factorizations of 216 and 234? Likewise, 3 is also a prime number common to the two factorizations. By contrast, 13 is a prime number that is not common to both factorizations.

$$216 = 2^3 \times 3^3$$
$$234 = 2 \times 3^2 \times 13$$

2 is a prime number common to both factorizations.	3 is a prime number common to both factorizations.

Now we need to use these prime factorizations to determine which exponent is appropriate for 2 and which exponent is appropriate for 3. The appropriate exponent is the smaller exponent associated with the prime number in the factorizations. The exponents associated with 2 in the factorizations are 1 and 3, so we select 1. Therefore, one factor for the greatest common divisor is 2^1, or 2. The exponents associated with 3 in the factorizations are 2 and 3, so we select 2. Therefore, another factor for the greatest common divisor is 3^2.

$$216 = 2^3 \times 3^3$$

The smaller exponent on 2 is 1.	The smaller exponent on 3 is 2.

$$234 = 2^1 \times 3^2 \times 13$$

Step 3   Form the product of the numbers from step 2. The greatest common divisor is the product of these factors.

$$\text{Greatest common divisor} = 2 \times 3^2 = 2 \times 9 = 18$$

The greatest common divisor of 216 and 234 is 18.

✓ **Checkpoint 3**   Find the greatest common divisor of 225 and 825.

 ④ Solve problems using the greatest common divisor.

Example 4   Solving a Problem Using the Greatest Common Divisor

For a school athletics program, you need to divide 40 boys and 24 girls into all-male and all-female teams so that each team has the same number of people. What is the largest number of people that can be placed on a team?

Solution

Because 40 boys are to be divided into teams, the number of boys on each team must be a divisor of 40. Because 24 girls are to be divided into teams, the number of girls placed on a team must be a divisor of 24. Although the teams are all-male and all-female, the same number of people must be placed on each team. The largest number of people that can be placed on a team is the largest number that will divide into 40 and 24 without a remainder. This is the greatest common divisor of 40 and 24.

To find the greatest common divisor of 40 and 24, begin with their prime factorizations.

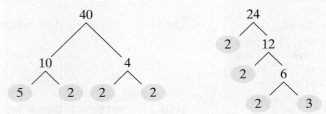

The factor trees indicate that

$$40 = 2^3 \times 5 \qquad \text{and} \qquad 24 = 2^3 \times 3.$$

We see that 2 is a prime number common to both factorizations. The exponents associated with 2 in the factorizations are 3 and 3, so we select 3.

$$\text{Greatest common divisor} = 2^3 = 2 \times 2 \times 2 = 8$$

The largest number of people that can be placed on a team is 8. Thus, the 40 boys can form five teams with 8 boys per team. The 24 girls can form three teams with 8 girls per team.

Checkpoint 4   A chorus teacher needs to divide 192 boys and 288 girls into all-male and all-female singing groups so that each group has the same number of people. What is the largest number of people that can be placed in each singing group?

Least Common Multiple

5 Find the least common multiple of two numbers.

The **least common multiple** (LCM) of two or more natural numbers is the smallest natural number that is divisible by all of the numbers. One way to find the least common multiple is to make a list of the numbers that are divisible by each number. This list represents the **multiples** of each number. For example, if we wish to find the least common multiple of 15 and 20, we can list the sets of multiples of 15 and multiples of 20.

$$\begin{cases} \text{Numbers Divisible by 15:} \\ \quad \text{Multiples of 15:} \qquad \{15, 30, 45, 60, 75, 90, 105, 120, \dots\} \end{cases}$$

$$\begin{cases} \text{Numbers Divisible by 20:} \\ \quad \text{Multiples of 20:} \qquad \{20, 40, 60, 80, 100, 120, 140, 160, \dots\} \end{cases}$$

Two common multiples of 15 and 20 are 60 and 120. The least common multiple is 60. Equivalently, 60 is the smallest number that is divisible by both 15 and 20.

Sometimes a partial list of the multiples for each of two numbers does not reveal the smallest number that is divisible by both given numbers. A more efficient method for finding the least common multiple is to use prime factorizations.

6 Solve problems using the least common multiple.

Finding the Least Common Multiple Using Prime Factorizations

To find the least common multiple of two or more numbers,

1. Write the prime factorization of each number.
2. Select every prime factor that occurs, raised to the greatest power to which it occurs, in these factorizations.
3. Form the product of the numbers from step 2. The least common multiple is the product of these factors.

Example 5 Finding the Least Common Multiple

Find the least common multiple of 144 and 300.

Solution

Step 1 Write the prime factorization of each number. Write the prime factorizations of 144 and 300.

$$144 = 2^4 \times 3^2$$
$$300 = 2^2 \times 3 \times 5^2$$

Step 2 Select every prime factor that occurs, raised to the greater power to which it occurs, in these factorizations. The prime factors that occur are 2, 3, and 5. The greater exponent that appears on 2 is 4, so we select 2^4. The greater exponent that appears on 3 is 2, so we select 3^2. The only exponent that occurs on 5 is 2, so we select 5^2. Thus, we have selected 2^4, 3^2, and 5^2.

Step 3 Form the product of the numbers from step 2. The least common multiple is the product of these factors.

$$\text{Least common multiple} = 2^4 \times 3^2 \times 5^2 = 16 \times 9 \times 25 = 3600$$

The least common multiple of 144 and 300 is 3600. The smallest natural number divisible by both 144 and 300 is 3600.

 **Checkpoint 5** Find the least common multiple of 18 and 30.

Example 6 Solving a Problem Using the Least Common Multiple

A movie theater runs its films continuously. One movie runs for 80 minutes and a second runs for 120 minutes. Both movies begin at 4:00 P.M. When will the movies begin again at the same time?

Solution

The shorter movie lasts 80 minutes, or 1 hour, 20 minutes. It begins at 4:00, so it will be shown again at 5:20. The longer movie lasts 120 minutes, or 2 hours. It begins at 4:00, so it will be shown again at 6:00. We are asked to find when the movies will begin again at the same time. Therefore, we are looking for the least common multiple of 80 and 120. Find the least common multiple and then add this number of minutes to 4:00 P.M.

Begin with the prime factorizations of 80 and 120:

$$80 = 2^4 \times 5$$
$$120 = 2^3 \times 3 \times 5.$$

Study Tip

Example 6 can also be solved by making a partial list of starting times for each movie.

Shorter Movie (Runs 1 hour, 20 minutes):

4:00, 5:20, 6:40, 8:00,...

Longer Movie (Runs 2 hours):

4:00, 6:00, 8:00,...

The list reveals that both movies start together again at 8:00 P.M.

Using $80 = 2^4 \times 5$ and $120 = 2^3 \times 3 \times 5$, we select each prime factor, with the greater exponent from each factorization.

$$\text{Least common multiple} = 2^4 \times 3 \times 5 = 16 \times 3 \times 5 = 240$$

Therefore, it will take 240 minutes, or 4 hours, for the movies to begin again at the same time. By adding 4 hours to 4:00 P.M., they will start together again at 8:00 P.M.

✓ **Checkpoint 6** A movie theater runs two documentary films continuously. One documentary runs for 40 minutes and a second documentary runs for 60 minutes. Both movies begin at 3:00 P.M. When will the movies begin again at the same time?

Achieving Success

As you work the assigned exercises, use the "Ten Minutes of Frustration" rule. If you have exhausted every possible means for solving a problem and you are still bogged down, stop after ten minutes. Put a question mark by the exercise and move on. When you return to class, ask your teacher for assistance.

Exercise Set 3.1

Concept and Vocabulary Exercises

In Exercises 1–4, fill in each blank so that the resulting statement is true.

1. A natural number greater than 1 that has only itself and 1 as factors is called a/an _____ number.

2. A natural number greater than 1 that is divisible by a number other than itself and 1 is called a/an _____ number.

3. The largest number that is a factor of two or more natural numbers is called their _____.

4. The smallest number that is divisible by two or more natural numbers is called their _____.

In Exercises 5–8, determine whether each statement is true or false. If the statement is false, make the necessary change(s) to produce a true statement.

5. The notation $b|a$ means that b is divisible by a.

6. $b \nmid a$ means that b does not divide a.

7. The words *factor* and *divisor* have opposite meanings.

8. A number can only be divisible by exactly one number.

Respond to Exercises 9–18 using verbal or written explanations.

9. If a is a factor of c, what does this mean?

10. How do you know that 45 is divisible by 5?

11. What does "a is divisible by b" mean?

12. Describe the difference between a prime number and a composite number.

13. What does the Fundamental Theorem of Arithmetic state?

14. What is the greatest common divisor of two or more natural numbers?

15. Describe how to find the greatest common divisor of two natural numbers.

16. What is the least common multiple of two or more natural numbers?

17. Describe how to find the least common multiple of two natural numbers.

18. The process of finding the greatest common divisor of two natural numbers is similar to finding the least common multiple of the numbers. Describe how the two processes differ.

In Exercises 19–22, determine whether each statement makes sense or does not make sense, and explain your reasoning.

19. I'm working with a prime number that intrigues me because it has three natural number factors.

20. When I find the greatest common factor, I select common prime factors with the greatest exponent and when I find the least common multiple, I select common prime factors with the smallest exponent.

21. I need to separate 70 boys and 175 girls into all-male or all-female teams with the same number of people on each team. By finding the least common multiple of 70 and 175, I can determine the largest number of people that can be placed on a team.

22. The number 12171330203317121 and the sentence "we panic in a pew" have something in common: They read the same from left to right and from right to left.

Practice Exercises

Use rules of divisibility to determine whether each number given in Exercises 23–32 is divisible by

a. 2   **b.** 3   **c.** 4   **d.** 5   **e.** 6
f. 8   **g.** 9   **h.** 10   **i.** 12.

23. 6944   **24.** 7245

25. 21,408   **26.** 25,025

27. 26,428   **28.** 89,001

29. 374,832   **30.** 347,712

31. 6,126,120   **32.** 5,941,221

In Exercises 33–46, use a calculator to determine whether each statement is true or false. If the statement is true, explain why this is so using one of the rules of divisibility in **Table 3.1** *on page 119.*

33. $3|5958$ **34.** $3|8142$ **35.** $4|10{,}612$

36. $4|15{,}984$ **37.** $5|38{,}814$ **38.** $5|48{,}659$

39. $6|104{,}538$ **40.** $6|163{,}944$ **41.** $8|20{,}104$

42. $8|28{,}096$ **43.** $9|11{,}378$ **44.** $9|23{,}772$

45. $12|517{,}872$ **46.** $12|785{,}172$

In Exercises 47–66, find the prime factorization of each composite number.

47. 75   **48.** 45   **49.** 56

50. 48   **51.** 105   **52.** 180

53. 500   **54.** 360   **55.** 663

56. 510   **57.** 885   **58.** 999

59. 1440   **60.** 1280   **61.** 1996

62. 1575   **63.** 3675   **64.** 8316

65. 85,800   **66.** 30,600

In Exercises 67–78, find the greatest common divisor of the numbers.

67. 42 and 56   **68.** 25 and 70   **69.** 16 and 42

70. 66 and 90   **71.** 60 and 108   **72.** 96 and 212

73. 72 and 120   **74.** 220 and 400   **75.** 342 and 380

76. 224 and 430   **77.** 240 and 285   **78.** 150 and 480

In Exercises 79–90, find the least common multiple of the numbers.

79. 42 and 56   **80.** 25 and 70   **81.** 16 and 42

82. 66 and 90   **83.** 60 and 108   **84.** 96 and 212

85. 72 and 120   **86.** 220 and 400   **87.** 342 and 380

88. 224 and 430   **89.** 240 and 285   **90.** 150 and 480

Practice Plus

In Exercises 91–96, determine all values of d that make each statement true.

91. $9|12{,}34d$ **92.** $9|23{,}42d$ **93.** $8|76{,}523{,}45d$

94. $8|88{,}888{,}82d$ **95.** $4|963{,}23d$ **96.** $4|752{,}67d$

A **perfect number** *is a natural number that is equal to the sum of its factors, excluding the number itself. In Exercises 97–100, determine whether or not each number is perfect.*

97. 28   **98.** 6   **99.** 20   **100.** 50

A prime number is an **emirp** *("prime" spelled backward) if it becomes a different prime number when its digits are reversed. In Exercises 101–104, determine whether or not each prime number is an emirp.*

101. 41   **102.** 43   **103.** 107   **104.** 113

A prime number p such that $2p + 1$ is also a prime number is called a **Germain prime**, *named after the German mathematician Sophie Germain (1776–1831), who made major contributions to number theory. In Exercises 105–108, determine whether or not each prime number is a Germain prime.*

105. 13   **106.** 11   **107.** 241   **108.** 97

109. Find the product of the greatest common divisor of 24 and 27 and the least common multiple of 24 and 27. Compare this result to the product of 24 and 27. Write a conjecture based on your observation.

110. Find the product of the greatest common divisor of 48 and 72 and the least common multiple of 48 and 72. Compare this result to the product of 48 and 72. Write a conjecture based on your observation.

Application Exercises

111. In Carl Sagan's novel *Contact*, Ellie Arroway, the book's heroine, has been working at SETI, the Search for Extraterrestrial Intelligence, listening to the crackle of the cosmos. One night, as the radio telescopes are turned toward Vega, they suddenly pick up strange pulses through the background noise. Two pulses are followed by a pause, then three pulses, five, seven,

$$11, \quad 13, \quad 17, \quad 19, \quad 23, \quad 29, \quad 31, \ldots$$

  2 3 5   7   11   13   17   19   23   29   31   37

continuing through 97. Then it starts all over again. Ellie is convinced that only intelligent life could generate the structure in the sequence of pulses. "It's hard to imagine some radiating plasma sending out a regular set of mathematical signals like this." What is it about the structure of the pulses that the book's heroine recognizes as the sign of intelligent life? Asked in another way, what is significant about the numbers of pulses?

112. There are two species of insects, *Magicicada septendecim* and *Magicicada tredecim*, that live in the same environment. They have a life cycle of exactly 17 and 13 years, respectively. For all but their last year, they remain in the ground feeding on the sap of tree roots. Then, in their last year, they emerge en masse from the ground as fully formed cricketlike insects, taking over the forest in a single night. They chirp loudly, mate, eat, lay eggs, then die six weeks later.

(*Source:* Marcus du Sautoy, *The Music of the Primes*, HarperCollins, 2003.)

   a. Suppose that the two species have life cycles that are not prime, say 18 and 12 years, respectively. List the set of multiples of 18 that are less than or equal to 216. List the set of multiples of 12 that are less than or equal to 216. Over a 216-year period, how many times will the two species emerge in the same year and compete to share the forest?

   b. Recall that both species have evolved prime-number life cycles, 17 and 13 years, respectively. Find the least common multiple of 17 and 13. How often will the two species have to share the forest?

   c. Compare your answers to parts (a) and (b). What explanation can you offer for each species having a prime number of years as the length of its life cycle?

113. A relief worker needs to divide 300 bottles of water and 144 cans of food into groups that each contain the same number of items. Also, each group must have the same type of item (bottled water or canned food). What is the largest number of relief supplies that can be put in each group?

114. A chorus teacher needs to divide 180 boys and 144 girls into all-male and all-female singing groups so that each group has the same number of people. What is the largest number of people that can be placed in each singing group?

115. You have in front of you 310 five-dollar bills and 460 ten-dollar bills. Your problem: Place the five-dollar bills and the ten-dollar bills in stacks so that each stack has the same number of bills, and each stack contains only one kind of bill (five-dollar or ten-dollar). What is the largest number of bills that you can place in each stack?

116. Harley collects sports cards. He has 360 football cards and 432 baseball cards. Harley plans to arrange his cards in stacks so that each stack has the same number of cards. Also, each stack must have the same type of card (football or baseball). Every card in Harley's collection is to be placed in one of the stacks. What is the largest number of cards that can be placed in each stack?

117. You and your brother both work the 4:00 P.M. to 8:00 P.M. shift. You have every sixth night off. Your brother has every tenth night off. Both of you were off on June 1. Your brother would like to see a bargain 5:00 movie with you. When will the two of you have the same night off again?

118. A movie theater runs its films continuously. One movie is a short documentary that runs for 40 minutes. The other movie is a full-length feature that runs for 100 minutes. Each film is shown in a separate theater. Both movies begin at noon. When will the movies begin again at the same time?

119. Two people are jogging around the sidewalk surrounding the high school campus in the same direction. One person can run completely around the track in 15 minutes. The second person takes 18 minutes. If they both start running in the same place at the same time, how long will it take them to be together at this place if they continue to run?

120. Two people are in a bicycle race around a circular track. One rider can race completely around the track in 40 seconds. The other rider takes 45 seconds. If they both begin the race at a designated starting point, how long will it take them to be together at this starting point again if they continue to race around the track?

Critical Thinking Exercises

121. Write a four-digit natural number that is divisible by 4 and not by 8.

122. Find the greatest common divisor and the least common multiple of $2^{17} \cdot 3^{25} \cdot 5^{31}$ and $2^{14} \cdot 3^{37} \cdot 5^{30}$. Express answers in the same form as the numbers given.

123. A middle-aged man observed that his present age was a prime number. He also noticed that the number of years in which his age would again be prime was equal to the number of years ago in which his age was prime. How old is the man?

124. A movie theater runs its films continuously. One movie runs for 85 minutes and a second runs for 100 minutes. The theater has a 15-minute intermission after each movie, at which point the movie is shown again. If both movies start at noon, when will the two movies start again at the same time?

125. The difference between consecutive prime numbers is always an even number, except for two particular prime numbers. What are those numbers?

Technology Exercises

*Use the divisibility rules listed in **Table 3.1** on page 119 to answer the questions in Exercises 126–128. Then, using a calculator, perform the actual division to determine whether your answer is correct.*

126. Is 67,234,096 divisible by 4?

127. Is 12,541,750 divisible by 3?

128. Is 48,201,651 divisible by 9?

Group Exercises

The following topics from number theory are appropriate for either individual or group research projects. A report should be given to the class on the researched topic. Useful references include books about numbers and number theory, books whose purpose is to excite the reader about mathematics, history of mathematics books, encyclopedias, and the Internet.

129. Euclid and Number Theory

130. An Unsolved Problem from Number Theory

131. Perfect Numbers

132. Deficient and Abundant Numbers

133. Formulas That Yield Primes

134. The Sieve of Eratosthenes

3.2 The Integers; Order of Operations

1. Define the integers.

2. Graph integers on a number line.

3. Use the symbols $<$ and $>$.

4. Find the absolute value of an integer.

5. Perform operations with integers.

6. Use the order of operations agreement.

---

1 Define the integers.

In 2009, the United States faced the worst economic crisis since the Great Depression. From the political left and right, watchdogs of the federal budget warned of continued fiscal trouble. President Obama's $787 billion economic stimulus package of spending increases and tax cuts contributed to a record projected federal deficit of $1.8 trillion. In this section, we use operations on a set of numbers called the *integers* to describe numerically the sad state of the nation's finances.

Defining the Integers

In Section 3.1, we applied some ideas of number theory to the set of natural, or counting, numbers:

$$\text{Natural numbers} = \{1, 2, 3, 4, 5, \ldots\}.$$

When we combine the number 0 with the natural numbers, we obtain the set of **whole numbers**:

$$\text{Whole numbers} = \{0, 1, 2, 3, 4, 5, \ldots\}.$$

The whole numbers do not allow us to describe certain everyday situations. For example, if the balance in your checking account is $30 and you write a check for $35, your checking account is overdrawn by $5. We can write this as -5, read *negative* 5. The set consisting of the natural numbers, 0, and the negatives of the natural numbers is called the set of **integers**.

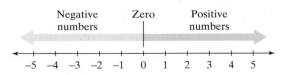

$$\text{Integers} = \{\ldots, -4, -3, -2, -1, 0, 1, 2, 3, 4, \ldots\}$$

Negative integers   Positive integers

Notice that the term *positive integers* is another name for the natural numbers. The positive integers can be written in two ways:

1. Use a "+" sign. For example, +4 is "positive four."

2. Do not write any sign. For example, 4 is assumed to be "positive four."

The Number Line; The Symbols $<$ and $>$

The **number line** is a graph we use to visualize the set of integers, as well as other sets of numbers. The number line is shown in **Figure 3.1**.

Negative numbers   Zero   Positive numbers

$$\begin{array}{ccccccccccc} -5 & -4 & -3 & -2 & -1 & 0 & 1 & 2 & 3 & 4 & 5 \end{array}$$

Figure 3.1 The number line

The number line extends indefinitely in both directions, shown by the arrows on the left and the right. Zero separates the positive numbers from the negative

2 Graph integers on a number line.

numbers on the number line. The positive integers are located to the right of 0 and the negative integers are located to the left of 0. **Zero is neither positive nor negative.** For every positive integer on a number line, there is a corresponding negative integer on the opposite side of 0.

Integers are graphed on a number line by placing a dot at the correct location for each number.

Example 1   **Graphing Integers on a Number Line**

Graph:

 a. −3   **b.** 4   **c.** 0.

Solution

Place a dot at the correct location for each integer.

Checkpoint 1 Graph:

 a. −4   **b.** 0   **c.** 3.

We will use the following symbols for comparing two integers:

  < means "is less than."

  > means "is greater than."

3 Use the symbols < and >.

On the number line, the integers increase from left to right. The *lesser* of two integers is the one farther to the *left* on a number line. The *greater* of two integers is the one farther to the *right* on a number line.

Look at the number line in **Figure 3.2**. The integers −4 and −1 are graphed.

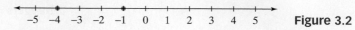

Figure 3.2

Observe that −4 is to the left of −1 on the number line. This means that −4 is less than −1.

$$-4 < -1$$ −4 is less than −1 because −4 is to the **left** of −1 on the number line.

In **Figure 3.2**, we can also observe that −1 is to the right of −4 on the number line. This means that −1 is greater than −4.

$$-1 > -4$$ −1 is greater than −4 because −1 is to the **right** of −4 on the number line.

The symbols < and > are called **inequality symbols**. These symbols always point to the lesser of the two integers when the inequality statement is true.

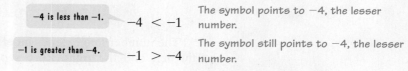

Example 2   Using the Symbols $<$ and $>$

Insert either $<$ or $>$ in the shaded area between the integers to make each statement true:

a. $-4 \blacksquare 3$ **b.** $-1 \blacksquare -5$ **c.** $-5 \blacksquare -2$ **d.** $0 \blacksquare -3$.

Solution

The solution is illustrated by the number line in **Figure 3.3**.

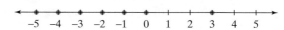

-5 -4 -3 -2 -1 0 1 2 3 4 5   **Figure 3.3**

a. $-4 < 3$ (negative 4 is less than 3) because -4 is to the left of 3 on the number line.

b. $-1 > -5$ (negative 1 is greater than negative 5) because -1 is to the right of -5 on the number line.

c. $-5 < -2$ (negative 5 is less than negative 2) because -5 is to the left of -2 on the number line.

d. $0 > -3$ (zero is greater than negative 3) because 0 is to the right of -3 on the number line.

Study Tip

You can think of negative integers as amounts of money that you *owe*. It's better to owe less, so

$$-1 > -5.$$

✔ **Checkpoint 2**   Insert either $<$ or $>$ in the shaded area between the integers to make each statement true:

a. $6 \blacksquare -7$ **b.** $-8 \blacksquare -1$

c. $-25 \blacksquare -2$ **d.** $-14 \blacksquare 0$.

The symbols $<$ and $>$ may be combined with an equal sign, as shown in the following table:

	Symbols	Meaning	Examples	Explanation
This inequality is true if either the $<$ part or the $=$ part is true.	$a \leq b$	a is less than or equal to b.	$2 \leq 9$ $9 \leq 9$	Because $2 < 9$ Because $9 = 9$
This inequality is true if either the $>$ part or the $=$ part is true.	$b \geq a$	b is greater than or equal to a.	$9 \geq 2$ $2 \geq 2$	Because $9 > 2$ Because $2 = 2$

Absolute Value

④ Find the absolute value of an integer.

Absolute value describes distance from 0 on a number line. If a represents an integer, the symbol $|a|$ represents its absolute value, read "the absolute value of a." For example,

$$|-5| = 5.$$

The absolute value of -5 is 5 because -5 is 5 units from 0 on a number line.

> **Absolute Value**
>
> The **absolute value** of an integer a, denoted by $|a|$, is the distance from 0 to a on the number line. Because absolute value describes a distance, it is never negative.

Example 3   Finding Absolute Value

Find the absolute value:

a. $|-3|$ **b.** $|5|$ **c.** $|0|$.

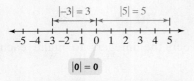

$$|0| = 0$$

Figure 3.4 Absolute value describes distance from 0 on a number line.

Solution

The solution is illustrated in **Figure 3.4**.

a. $|-3| = 3$ *The absolute value of −3 is 3 because −3 is 3 units from 0.*

b. $|5| = 5$ *5 is 5 units from 0.*

c. $|0| = 0$ *0 is 0 units from itself.*

Example 3 illustrates that the absolute value of a positive integer or 0 is the number itself. The absolute value of a negative integer, such as −3, is the number without the negative sign. Zero is the only real number whose absolute value is 0: $|0| = 0$. **The absolute value of any integer other than 0 is always positive.**

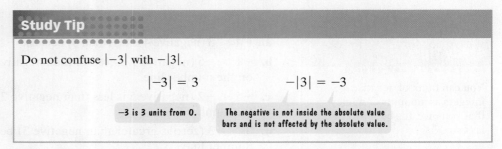

> **Study Tip**
>
> Do not confuse $|-3|$ with $-|3|$.
>
> $$|-3| = 3 \qquad\qquad -|3| = -3$$
>
> −3 is 3 units from 0.   The negative is not inside the absolute value bars and is not affected by the absolute value.

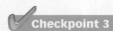

Checkpoint 3   Find the absolute value:

a. $|-8|$ **b.** $|6|$ **c.** $-|8|$.

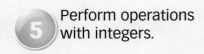

⑤   Perform operations with integers.

Addition of Integers

It has not been a good day! First, you lost your wallet with $30 in it. Then, you borrowed $10 to get through the day, which you somehow misplaced. Your loss of $30 followed by a loss of $10 is an overall loss of $40. This can be written

$$-30 + (-10) = -40.$$

The result of adding two or more numbers is called the **sum** of the numbers. The sum of −30 and −10 is −40.

You can think of gains and losses of money to find sums. For example, to find $17 + (-13)$, think of a gain of $17 followed by a loss of $13. There is an overall gain of $4. Thus, $17 + (-13) = 4$. In the same way, to find $-17 + 13$, think of a loss of $17 followed by a gain of $13. There is an overall loss of $4, so $-17 + 13 = -4$.

Using gains and losses, we can develop the following rules for adding integers:

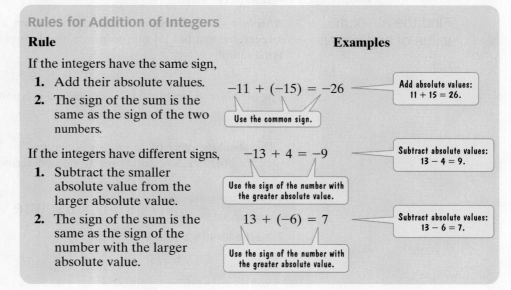

Rules for Addition of Integers

Rule	Examples
If the integers have the same sign, **1.** Add their absolute values. **2.** The sign of the sum is the same as the sign of the two numbers.	$-11 + (-15) = -26$ — Add absolute values: $11 + 15 = 26$. Use the common sign.
If the integers have different signs, **1.** Subtract the smaller absolute value from the larger absolute value. **2.** The sign of the sum is the same as the sign of the number with the larger absolute value.	$-13 + 4 = -9$ — Subtract absolute values: $13 - 4 = 9$. Use the sign of the number with the greater absolute value. $13 + (-6) = 7$ — Subtract absolute values: $13 - 6 = 7$. Use the sign of the number with the greater absolute value.

Calculators and Adding Integers

You can use a calculator to add integers. Here are the keystrokes for finding $-11 + (-15)$:

Scientific Calculator

Graphing Calculator

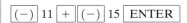

Here are the keystrokes for finding $-13 + 4$:

Scientific Calculator

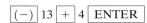

Graphing Calculator

$(-)$ 13 $+$ 4 $\boxed{\text{ENTER}}$

In addition to gains and losses of money, another good analogy for adding integers is temperatures above and below zero on a thermometer. Think of the thermometer as a number line standing straight up. For example,

$$-11 + (-15) = -26$$

If it's 11 below zero and the temperature falls 15 degrees, it will then be 26 below zero.

$$-13 + 4 = -9$$

If it's 13 below zero and the temperature rises 4 degrees, the new temperature will be 9 below zero.

$$13 + (-6) = 7.$$

If it's 13 above zero and the temperature falls 6 degrees, it will then be 7 above zero.

Using the analogies of gains and losses of money or temperatures can make the formal rules for addition of integers easy to use.

Can you guess what number is displayed if you use a calculator to find a sum such as $18 + (-18)$? If you gain 18 and then lose 18, there is neither an overall gain nor loss. Thus,

$$18 + (-18) = 0.$$

We call 18 and -18 **additive inverses**. Additive inverses have the same absolute value, but lie on opposite sides of zero on the number line. Thus, -7 is the additive inverse of 7, and 5 is the additive inverse of -5. In general, the sum of any integer and its additive inverse is 0:

$$a + (-a) = 0.$$

Subtraction of Integers

Suppose that a computer that normally sells for $1500 has a price reduction of $600. The computer's reduced price, $900, can be expressed in two ways:

$$1500 - 600 = 900 \quad \text{or} \quad 1500 + (-600) = 900.$$

This means that

$$1500 - 600 = 1500 + (-600).$$

To subtract 600 from 1500, we add 1500 and the additive inverse of 600. Generalizing from this situation, we define subtraction as follows:

> **Definition of Subtraction**
>
> For all integers a and b,
>
> $$a - b = a + (-b).$$
>
> In words, to subtract b from a, add the additive inverse of b to a. The result of subtraction is called the **difference**.

Example 4   Subtracting Integers

Subtract:

a. $17 - (-11)$ **b.** $-18 - (-5)$ **c.** $-18 - 5$.

Solution

a. $17 - (-11) = 17 + 11 = 28$

> Change the subtraction to addition.   Replace −11 with its additive inverse.

b. $-18 - (-5) = -18 + 5 = -13$

> Change the subtraction to addition.   Replace −5 with its additive inverse.

c. $-18 - 5 = -18 + (-5) = -23$

> Change the subtraction to addition.   Replace 5 with its additive inverse.

✓ **Checkpoint 4** Subtract:

a. $30 - (-7)$ **b.** $-14 - (-10)$ **c.** $-14 - 10$.

Subtraction is used to solve problems in which the word *difference* appears. The difference between integers a and b is expressed as $a - b$.

Example 5   An Application of Subtraction Using the Word *Difference*

The bar graph in **Figure 3.5** shows the budget surplus or deficit for the United States government from 2000 through 2008. What is the difference between the 2000 surplus and the 2008 deficit?

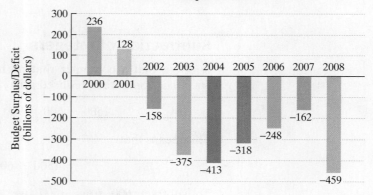

U.S. Government Budget Surplus/Deficit

Figure 3.5
Source: Budget of the U.S. Government

Solution

> The difference   is   the 2000 surplus   minus   the 2008 deficit.

$$= 236 - (-459)$$
$$= 236 + 459 = 695$$

The difference between the 2000 surplus and the 2008 deficit is $695 billion.

✓ **Checkpoint 5** Use **Figure 3.5** to find the difference between the 2007 deficit and the 2008 deficit.

Multiplication of Integers

The result of multiplying two or more numbers is called the **product** of the numbers. You can think of multiplication as repeated addition or subtraction that starts at 0. For example,

$$3(-4) = 0 + (-4) + (-4) + (-4) = -12$$

> The numbers have different signs and the product is negative.

and

$$(-3)(-4) = 0 - (-4) - (-4) - (-4) = 0 + 4 + 4 + 4 = 12.$$

> The numbers have the same sign and the product is positive.

These observations give us the following rules for multiplying integers:

Rules for Multiplying Integers

Rule	Examples
1. The product of two integers with different signs is found by multiplying their absolute values. The product is negative.	• $7(-5) = -35$
2. The product of two integers with the same sign is found by multiplying their absolute values. The product is positive.	• $(-6)(-11) = 66$
3. The product of 0 and any integer is 0: $a \cdot 0 = 0$ and $0 \cdot a = 0$.	• $-17(0) = 0$
4. If no number is 0, a product with an odd number of negative factors is found by multiplying absolute values. The product is negative.	• $-2(-3)(-5) = -30$ Three (odd) negative factors
5. If no number is 0, a product with an even number of negative factors is found by multiplying absolute values. The product is positive.	• $-2(3)(-5) = 30$ Two (even) negative factors

Exponential Notation

Because exponents indicate repeated multiplication, rules for multiplying real numbers can be used to evaluate exponential expressions.

Example 6   Evaluating Exponential Expressions

Evaluate:

a. $(-6)^2$ **b.** -6^2 **c.** $(-5)^3$ **d.** $(-2)^4$.

Solution

a. $(-6)^2 = (-6)(-6) = 36$ **b.** $-6^2 = -(6 \cdot 6) = -36$

> Base is −6.   Same signs give positive product.

> Base is 6. The negative is not inside parentheses and is not taken to the second power.

c. $(-5)^3 = (-5)(-5)(-5) = -125$ **d.** $(-2)^4 = (-2)(-2)(-2)(-2) = 16$

> An odd number of negative factors gives a negative product.

> An even number of negative factors gives a positive product.

✓ Checkpoint 6   Evaluate:

a. $(-5)^2$ **b.** -5^2 **c.** $(-4)^3$ **d.** $(-3)^4$.

Blitzer Bonus

Integers, Karma, and Exponents

On Friday the 13th, are you a bit more careful crossing the street even if you don't consider yourself superstitious? Numerology, the belief that certain integers have greater significance and can be lucky or unlucky, is widespread in many cultures.

Integer	Connotation	Culture	Origin	Example
4	Negative	Chinese	The word for the number 4 sounds like the word for death.	Many buildings in China have floor-numbering systems that skip 40–49.
7	Positive	United States	In dice games, this prime number is the most frequently rolled number with two dice.	There was a spike in the number of couples getting married on 7/7/07.
8	Positive	Chinese	It's considered a sign of prosperity.	The Beijing Olympics began at 8 P.M. on 8/8/08.
13	Negative	Various	Various reasons, including the number of people at the Last Supper	Many buildings around the world do not label any floor "13."
18	Positive	Jewish	The Hebrew letters spelling *chai*, or living, are the 8th and 10th in the alphabet, adding up to 18	Monetary gifts for celebrations are often given in multiples of 18.
666	Negative	Christian	The New Testament's Book of Revelation identifies 666 as the "number of the beast," which some say refers to Satan.	In 2008, Reeves, Louisiana, eliminated 666 as the prefix of its phone numbers.

Source: The New York Times

Although your author is not a numerologist, he is intrigued by curious exponential representations for 666:

$$666 = 6 + 6 + 6 + 6^3 + 6^3 + 6^3$$
$$666 = 1^3 + 2^3 + 3^3 + 4^3 + 5^3 + 6^3 + 5^3 + 4^3 + 3^3 + 2^3 + 1^3$$
$$666 = 2^2 + 3^2 + 5^2 + 7^2 + 11^2 + 13^2 + 17^2$$

> Sum of the squares of the first seven prime numbers

$$666 = 1^6 - 2^6 + 3^6.$$

Division of Integers

The result of dividing the integer a by the nonzero integer b is called the **quotient** of the numbers. We can write this quotient as $a \div b$ or $\frac{a}{b}$.

A relationship exists between multiplication and division. For example,

$$\frac{-12}{4} = -3 \text{ means that } 4(-3) = -12.$$

$$\frac{-12}{-4} = 3 \text{ means that } -4(3) = -12.$$

Because of the relationship between multiplication and division, the rules for obtaining the sign of a quotient are the same as those for obtaining the sign of a product.

Rules for Dividing Integers

Rule	Examples
1. The quotient of two integers with different signs is found by dividing their absolute values. The quotient is negative.	• $\dfrac{80}{-4} = -20$ • $\dfrac{-15}{5} = -3$
2. The quotient of two integers with the same sign is found by dividing their absolute values. The quotient is positive.	• $\dfrac{27}{9} = 3$ • $\dfrac{-45}{-3} = 15$
3. Zero divided by any nonzero integer is zero.	• $\dfrac{0}{-5} = 0$ (because $-5(0) = 0$)
4. Division by 0 is undefined.	• $\dfrac{-8}{0}$ is undefined (because 0 cannot be multiplied by an integer to obtain -8).

6 Use the order of operations agreement.

Order of Operations

Suppose that you want to find the value of $3 + 7 \cdot 5$. Which procedure shown below is correct?

$$3 + 7 \cdot 5 = 3 + 35 = 38 \qquad \text{or} \qquad 3 + 7 \cdot 5 = 10 \cdot 5 = 50$$

If you know the answer, you probably know certain rules, called the **order of operations**, that make sure there is only one correct answer. One of these rules states that if a problem contains no parentheses, perform multiplication before addition. Thus, the procedure on the left is correct because the multiplication of 7 and 5 is done first. Then the addition is performed. The correct answer is 38.

Here are the rules for determining the order in which operations should be performed:

Order of Operations

1. Perform all operations within grouping symbols.
2. Evaluate all exponential expressions.
3. Do all multiplications and divisions in the order in which they occur, working from left to right.
4. Finally, do all additions and subtractions in the order in which they occur, working from left to right.

In the third step, be sure to do all multiplications and divisions *as they occur* from left to right. For example,

$$8 \div 4 \cdot 2 = 2 \cdot 2 = 4$$ Do the division first because it occurs first.

$$8 \cdot 4 \div 2 = 32 \div 2 = 16.$$ Do the multiplication first because it occurs first.

Example 7 Using the Order of Operations

Simplify: $6^2 - 24 \div 2^2 \cdot 3 + 1$.

Solution

There are no grouping symbols. Thus, we begin by evaluating exponential expressions. Then we multiply or divide. Finally, we add or subtract.

$$6^2 - 24 \div 2^2 \cdot 3 + 1$$

$= 36 - 24 \div 4 \cdot 3 + 1$ Evaluate exponential expressions: $6^2 = 6 \cdot 6 = 36$ and $2^2 = 2 \cdot 2 = 4$.

$= 36 - 6 \cdot 3 + 1$ Perform the multiplications and divisions from left to right. Start with $24 \div 4 = 6$.

$= 36 - 18 + 1$ Now do the multiplication: $6 \cdot 3 = 18$.

$= 18 + 1$ Finally, perform the additions and subtractions from left to right. Subtract: $36 - 18 = 18$.

$= 19$ $18 + 1 = 19$.

▶ **Checkpoint 7** Simplify: $7^2 - 48 \div 4^2 \cdot 5 + 2$.

Example 8 Using the Order of Operations

Simplify: $(-6)^2 - (5 - 7)^2(-3)$.

Solution

Because grouping symbols appear, we perform the operation within parentheses first.

$$(-6)^2 - (5 - 7)^2(-3)$$

$= (-6)^2 - (-2)^2(-3)$ Work inside parentheses first: $5 - 7 = 5 + (-7) = -2$.

$= 36 - 4(-3)$ Evaluate exponential expressions: $(-6)^2 = (-6)(-6) = 36$ and $(-2)^2 = (-2)(-2) = 4$.

$= 36 - (-12)$ Multiply: $4(-3) = -12$.

$= 48$ Subtract: $36 - (-12) = 36 + 12 = 48$.

▶ **Checkpoint 8** Simplify: $(-8)^2 - (10 - 13)^2(-2)$.

Achieving Success

Learn from your mistakes. Being human means making mistakes. By finding and understanding your errors, you will become a better math student.

Source of Error	Remedy
Not Understanding a Concept	Review the concept by finding a similar example in your textbook or class notes. Ask your teacher questions to help clarify the concept.
Skipping Steps	Show clear step-by-step solutions. Detailed solution procedures help organize your thoughts and enhance understanding. Doing too many steps mentally often results in preventable mistakes.
Carelessness	Write neatly. Not being able to read your own math writing leads to errors. Avoid writing in pen so you won't have to put huge marks through incorrect work.

"You can achieve your goal if you persistently pursue it."

—Cha Sa-Soon, a 68-year-old South Korean woman who passed her country's written driver's-license exam on her 950th try (*Source: Newsweek*)

Exercise Set 3.2

Concept and Vocabulary Exercises

In Exercises 1–4, fill in each blank so that the resulting statement is true.

1. The integers are defined by the set _____.

2. If $a < b$, then a is located to the _____ of b on a number line.

3. On a number line, the absolute value of a, denoted $|a|$, represents _____.

4. Two integers that have the same absolute value, but lie on opposite sides of zero on a number line, are called _____.

In Exercises 5–8, determine whether each statement is true or false. If the statement is false, make the necessary change(s) to produce a true statement.

5. The sum of a positive integer and a negative integer is always a positive integer.

6. The difference between 0 and a negative integer is always a positive integer.

7. The product of a positive integer and a negative integer is never a positive integer.

8. The quotient of 0 and a negative integer is undefined.

Respond to Exercises 9–16 using verbal or written explanations.

9. How does the set of integers differ from the set of whole numbers?

10. Explain how to graph an integer on a number line.

11. Explain how to add integers.

12. Explain how to subtract integers.

13. Explain how to multiply integers.

14. Explain how to divide integers.

15. Describe what it means to raise a number to a power. In your description, include a discussion of the difference between -5^2 and $(-5)^2$.

16. Why is $\frac{0}{4}$ equal to 0, but $\frac{4}{0}$ undefined?

In Exercises 17–20, determine whether each statement makes sense or does not make sense, and explain your reasoning.

17. Without adding integers, I can see that the sum of -227 and 319 is greater than the sum of 227 and -319.

18. I found the variation in U.S. temperature by subtracting the record low temperature, a negative integer, from the record high temperature, a positive integer.

19. I've noticed that the sign rules for dividing integers are slightly different than the sign rules for multiplying integers.

20. The rules for the order of operations avoid the confusion of obtaining different results when I simplify the same expression.

Practice Exercises

In Exercises 21–24, start by drawing a number line that shows integers from −5 to 5. Then graph each of the following integers on your number line.

21. 3   **22.** 5   **23.** −4   **24.** −2

In Exercises 25–32, insert either < or > in the shaded area between the integers to make the statement true.

25. −2 ■ 7   **26.** −1 ■ 13

27. −13 ■ −2   **28.** −1 ■ −13

29. 8 ■ −50   **30.** 7 ■ −9

31. −100 ■ 0   **32.** 0 ■ −300

In Exercises 33–38, find the absolute value.

33. $|-14|$ **34.** $|-16|$

35. $|14|$ **36.** $|16|$

37. $|-300,000|$ **38.** $|-1,000,000|$

In Exercises 39–50, find each sum.

39. $-7 + (-5)$ **40.** $-3 + (-4)$

41. $12 + (-8)$ **42.** $13 + (-5)$

43. $6 + (-9)$ **44.** $3 + (-11)$

45. $-9 + (+4)$ **46.** $-7 + (+3)$

47. $-9 + (-9)$ **48.** $-13 + (-13)$

49. $9 + (-9)$ **50.** $13 + (-13)$

In Exercises 51–62, perform the indicated subtraction.

51. $13 - 8$ **52.** $14 - 3$

53. $8 - 15$ **54.** $9 - 20$

55. $4 - (-10)$ **56.** $3 - (-17)$

57. $-6 - (-17)$ **58.** $-4 - (-19)$

59. $-12 - (-3)$ **60.** $-19 - (-2)$

61. $-11 - 17$ **62.** $-19 - 21$

In Exercises 63–72, find each product.

63. $6(-9)$ **64.** $5(-7)$

65. $(-7)(-3)$ **66.** $(-8)(-5)$

67. $(-2)(6)$ **68.** $(-3)(10)$

69. $(-13)(-1)$ **70.** $(-17)(-1)$

71. $0(-5)$ **72.** $0(-8)$

In Exercises 73–86, evaluate each exponential expression.

73. 5^2 **74.** 6^2

75. $(-5)^2$ **76.** $(-6)^2$

77. 4^3 **78.** 2^3

79. $(-5)^3$ **80.** $(-4)^3$

81. $(-5)^4$ **82.** $(-4)^4$

83. -3^4 **84.** -1^4

85. $(-3)^4$ **86.** $(-1)^4$

In Exercises 87–100, find each quotient, or, if applicable, state that the expression is undefined.

87. $\frac{-12}{4}$ **88.** $\frac{-40}{5}$

89. $\frac{21}{-3}$ **90.** $\frac{60}{-6}$

91. $\frac{-90}{-3}$ **92.** $\frac{-66}{-6}$

93. $\frac{0}{-7}$ **94.** $\frac{0}{-8}$

95. $\frac{-7}{0}$ **96.** $\frac{0}{0}$

97. $(-480) \div 24$ **98.** $(-300) \div 12$

99. $(465) \div (-15)$ **100.** $(-594) \div (-18)$

In Exercises 101–120, use the order of operations to find the value of each expression.

101. $7 + 6 \cdot 3$ **102.** $-5 + (-3) \cdot 8$

103. $(-5) - 6(-3)$ **104.** $-8(-3) - 5(-6)$

105. $6 - 4(-3) - 5$ **106.** $3 - 7(-1) - 6$

107. $3 - 5(-4 - 2)$ **108.** $3 - 9(-1 - 6)$

109. $(2 - 6)(-3 - 5)$ **110.** $9 - 5(6 - 4) - 10$

111. $3(-2)^2 - 4(-3)^2$ **112.** $5(-3)^2 - 2(-2)^3$

113. $(2 - 6)^2 - (3 - 7)^2$

114. $(4 - 6)^2 - (5 - 9)^3$

115. $6(3 - 5)^3 - 2(1 - 3)^3$

116. $-3(-6 + 8)^3 - 5(-3 + 5)^3$

117. $8^2 - 16 \div 2^2 \cdot 4 - 3$

118. $10^2 - 100 \div 5^2 \cdot 2 - (-3)$

119. $24 \div [3^2 \div (8 - 5)] - (-6)$

120. $30 \div [5^2 \div (7 - 12)] - (-9)$

Practice Plus

In Exercises 121–130, use the order of operations to find the value of each expression.

121. $8 - 3[-2(2 - 5) - 4(8 - 6)]$

122. $8 - 3[-2(5 - 7) - 5(4 - 2)]$

123. $-2^2 + 4[16 \div (3 - 5)]$

124. $-3^2 + 2[20 \div (7 - 11)]$

125. $4|10 - (8 - 20)|$

126. $-5|7 - (20 - 8)|$

127. $[-5^2 + (6 - 8)^3 - (-4)] - [|-2|^3 + 1 - 3^2]$

128. $[-4^2 + (7 - 10)^3 - (-27)] - [|-2|^5 + 1 - 5^2]$

129. $\dfrac{12 \div 3 \cdot 5|2^2 + 3^2|}{7 + 3 - 6^2}$

130. $\dfrac{-3 \cdot 5^2 + 89}{(5 - 6)^2 - 2|3 - 7|}$

In Exercises 131–134, express each sentence as a single numerical expression. Then use the order of operations to simplify the expression.

131. Cube −2. Subtract this exponential expression from −10.

132. Cube −5. Subtract this exponential expression from −100.

133. Subtract 10 from 7. Multiply this difference by 2. Square this product.

134. Subtract 11 from 9. Multiply this difference by 2. Raise this product to the fourth power.

Application Exercises

135. The peak of Mount McKinley, the highest point in the United States, is 20,320 feet above sea level. Death Valley, the lowest point in the United States, is 282 feet below sea level. What is the difference in elevation between the peak of Mount McKinley and Death Valley?

136. The peak of Mount Kilimanjaro, the highest point in Africa, is 19,321 feet above sea level. Qattara Depression, Egypt, the lowest point in Africa, is 436 feet below sea level. What is the difference in elevation between the peak of Mount Kilimanjaro and the Qattara Depression?

Life expectancy for the average American man is 75.2 years; for a woman, it's 80.4 years. The number line, with points representing eight integers, indicates factors, many within our control, that can stretch or shrink one's probable life span. Use this information to solve Exercises 137–146.

Stretching or Shrinking One's Life Span

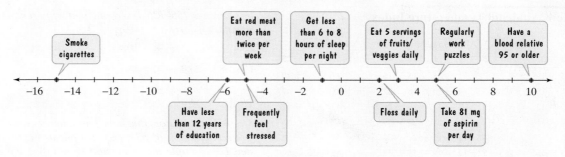

Years of Life Gained or Lost

Source: Newsweek

137. If you have a blood relative 95 or older and you smoke cigarettes, do you stretch or shrink your life span? By how many years?

138. If you floss daily and eat red meat more than twice per week, do you stretch or shrink your life span? By how many years?

139. If you frequently feel stressed and have less than 12 years of education, do you stretch or shrink your life span? By how many years?

140. If you get less than 6 to 8 hours of sleep per night and smoke cigarettes, do you stretch or shrink your life span? By how many years?

141. What is the difference in the life span between a person who regularly works puzzles and a person who eats red meat more than twice per week?

142. What is the difference in the life span between a person who eats 5 servings of fruits/veggies daily and a person who frequently feels stressed?

143. What happens to the life span for a person who takes 81 mg of aspirin per day and eats red meat more than twice per week?

144. What happens to the life span for a person who regularly works puzzles and a person who frequently feels stressed?

145. What is the difference in the life span between a person with less than 12 years of education and a person who smokes cigarettes?

146. What is the difference in the life span between a person who gets less than 6 to 8 hours of sleep per night and a person who smokes cigarettes?

The bar graph shows the amount of money, in billions of dollars, collected and spent by the U.S. government from 2001 through 2008. Use the information from the graph to solve Exercises 147–150. Express answers in billions of dollars.

Money Collected and Spent by the United States Government

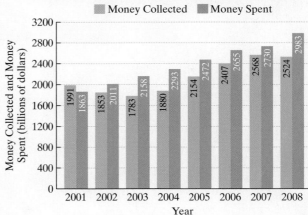

Source: Budget of the U.S. Government

147. In 2003, what was the difference between the amount of money collected and the amount spent? Was there a budget surplus or deficit in 2003?

(In Exercises 148–150, refer to the bar graph at the bottom of page 141.)

148. In 2004, what was the difference between the amount of money collected and the amount spent? Was there a budget surplus or deficit in 2004?

149. What is the difference between the 2001 surplus and the 2008 deficit?

150. What is the difference between the 2001 surplus and the 2007 deficit?

The way that we perceive the temperature on a cold day depends on both air temperature and wind speed. The windchill is what the air temperature would have to be with no wind to achieve the same chilling effect on the skin. In 2002, the National Weather Service issued new windchill temperatures, shown in the table below. Use the information from the table to solve Excercise 151–154.

New Windchill Temperature Index

Air Temperature (°F)

Wind Speed (miles per hour)	30	25	20	15	10	5	0	−5	−10	−15	−20	−25
5	25	19	13	7	1	−5	−11	−16	−22	−28	−34	−40
10	21	15	9	3	−4	−10	−16	−22	−28	−35	−41	−47
15	19	13	6	0	−7	−13	−19	−26	−32	−39	−45	−51
20	17	11	4	−2	−9	−15	−22	−29	−35	−42	−48	−55
25	16	9	3	−4	−11	−17	−24	−31	−37	−44	−51	−58
30	15	8	1	−5	−12	−19	−26	−33	−39	−46	−53	−60
35	14	7	0	−7	−14	−21	−27	−34	−41	−48	−55	−62
40	13	6	−1	−8	−15	−22	−29	−36	−43	−50	−57	−64
45	12	5	−2	−9	−16	−23	−30	−37	−44	−51	−58	−65
50	12	4	−3	−10	−17	−24	−31	−38	−45	−52	−60	−67
55	11	4	−3	−11	−18	−25	−32	−39	−46	−54	−61	−68
60	10	3	−4	−11	−19	−26	−33	−40	−48	−55	−62	−69

Frostbite occurs in 15 minutes or less.

Source: National Weather Service

151. What is the difference between how cold the temperature feels with winds at 10 miles per hour and 25 miles per hour when the air temperature is 15 °F?

152. What is the difference between how cold the temperature feels with winds at 5 miles per hour and 30 miles per hour when the air temperature is 10 °F?

153. What is the difference in the windchill temperature between an air temperature of 5 °F with winds at 50 miles per hour and an air temperature of −10 °F with winds at 5 miles per hour?

154. What is the difference in the windchill temperature between an air temperature of 5 °F with winds at 55 miles per hour and an air temperature of −5 °F with winds at 10 miles per hour?

Critical Thinking Exercises

In Exercises 155–156, insert one pair of parentheses to make each calculation correct.

155. $8 - 2 \cdot 3 - 4 = 10$

156. $8 - 2 \cdot 3 - 4 = 14$

Technology Exercises

Scientific calculators that have parentheses keys allow for the entry and computation of relatively complicated expressions in a single step. For example, the expression $15 + (10 - 7)^2$ can be evaluated by entering the following keystrokes:

15 $+$ $($ 10 $-$ 7 $)$ y^x 2 $=$.

Find the value of each expression in Exercises 157–159 in a single step on your scientific calculator.

157. $8 - 2 \cdot 3 - 9$

158. $(8 - 2) \cdot (3 - 9)$

159. $5^3 + 4 \cdot 9 - (8 + 9 \div 3)$

3.3   The Rational Numbers

You are making eight dozen chocolate chip cookies for a large neighborhood block party. The recipe lists the ingredients needed to prepare five dozen cookies, such as $\frac{3}{4}$ cup sugar. How do you adjust the amount of sugar, as well as the amounts of each of the other ingredients, given in the recipe?

Adapting a recipe to suit a different number of portions usually involves working with numbers that are not integers. For example, the number describing the amount of sugar, $\frac{3}{4}$ (cup), is not an integer, although it consists of the quotient of two integers, 3 and 4. Before returning to the problem of changing the size of a recipe, we study a new set of numbers consisting of the quotients of integers.

Defining the Rational Numbers

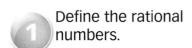

If two integers are added, subtracted, or multiplied, the result is always another integer. This, however, is not always the case with division. For example, 10 divided by 5 is the integer 2. By contrast, 5 divided by 10 is $\frac{1}{2}$, and $\frac{1}{2}$ is not an integer. To permit divisions such as $\frac{5}{10}$, we enlarge the set of integers, calling the new collection the *rational numbers*. The set of **rational numbers** consists of all the numbers that can be expressed as a quotient of two integers, with the denominator not 0.

> ### The Rational Numbers
> The set of **rational numbers** is the set of all numbers which can be expressed in the form $\frac{a}{b}$, where a and b are integers and b is not equal to 0. The integer a is called the **numerator**, and the integer b is called the **denominator**.

The following numbers are examples of rational numbers:

$$\tfrac{1}{2}, \ \tfrac{-3}{4}, \ 5, \ 0.$$

The integer 5 is a rational number because it can be expressed as the quotient of integers: $5 = \frac{5}{1}$. Similarly, 0 can be written as $\frac{0}{1}$.

In general, every integer a is a rational number because it can be expressed in the form $\frac{a}{1}$.

Reducing Rational Numbers

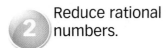

A rational number is **reduced to its lowest terms**, or **simplified**, when the numerator and denominator have no common divisors other than 1. Reducing rational numbers to lowest terms is done using the **Fundamental Principle of Rational Numbers**.

> ### The Fundamental Principle of Rational Numbers
>
> If $\frac{a}{b}$ is a rational number and c is any number other than 0,
>
> $$\frac{a \cdot c}{b \cdot c} = \frac{a}{b}.$$
>
> The rational numbers $\frac{a}{b}$ and $\frac{a \cdot c}{b \cdot c}$ are called **equivalent fractions**.

When using the Fundamental Principle to reduce a rational number, the simplification can be done in one step by finding the greatest common divisor of the numerator and the denominator, and using it for c. Thus, **to reduce a rational number to its lowest terms, divide both the numerator and the denominator by their greatest common divisor.**

For example, consider the rational number $\frac{12}{100}$. The greatest common divisor of 12 and 100 is 4. We reduce to lowest terms as follows:

$$\frac{12}{100} = \frac{3 \cdot 4}{25 \cdot 4} = \frac{3}{25} \quad \text{or} \quad \frac{12}{100} = \frac{12 \div 4}{100 \div 4} = \frac{3}{25}.$$

Example 1   Reducing a Rational Number

Reduce $\frac{130}{455}$ to lowest terms.

Solution

Begin by finding the greatest common divisor of 130 and 455.

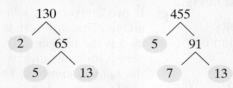

Thus, $130 = 2 \cdot 5 \cdot 13$ and $455 = 5 \cdot 7 \cdot 13$. The greatest common divisor is $5 \cdot 13$, or 65. Divide the numerator and the denominator of the given rational number by $5 \cdot 13$ or by 65.

$$\frac{130}{455} = \frac{2 \cdot 5 \cdot 13}{5 \cdot 7 \cdot 13} = \frac{2}{7} \quad \text{or} \quad \frac{130}{455} = \frac{130 \div 65}{455 \div 65} = \frac{2}{7}$$

There are no common divisors of 2 and 7 other than 1. Thus, the rational number $\frac{2}{7}$ is in its lowest terms.

Checkpoint 1   Reduce $\frac{72}{90}$ to lowest terms.

Mixed Numbers and Improper Fractions

③ Convert between mixed numbers and improper fractions.

A **mixed number** consists of the sum of an integer and a rational number, expressed without the use of an addition sign. Here is an example of a mixed number:

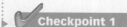

$$3\frac{4}{5}.$$

The integer is 3 and the rational number is $\frac{4}{5}$. $3\frac{4}{5}$ means $3 + \frac{4}{5}$.

The mixed number $3\frac{4}{5}$ is read "three and four-fifths."

An **improper fraction** is a rational number for which the numerator is greater than the denominator. An example of an improper fraction is $\frac{19}{5}$.

The mixed number $3\frac{4}{5}$ can be converted to the improper fraction $\frac{19}{5}$ using the boxed procedure at the top of the next page.

Converting a Positive Mixed Number to an Improper Fraction

1. Multiply the denominator of the rational number by the integer and add the numerator to this product.
2. Place the sum in step 1 over the denominator in the mixed number.

Example 2   **Converting from a Mixed Number to an Improper Fraction**

Convert $3\frac{4}{5}$ to an improper fraction.

Solution

$$3\frac{4}{5} = \frac{5 \cdot 3 + 4}{5}$$

Multiply the denominator by the integer and add the numerator.

Place the sum over the mixed number's denominator.

$$= \frac{15 + 4}{5} = \frac{19}{5}$$

Checkpoint 2   Convert $2\frac{5}{8}$ to an improper fraction.

When converting a negative mixed number to an improper fraction, copy the negative sign and then follow the previous procedure. For example,

$$-2\frac{3}{4} = -\frac{4 \cdot 2 + 3}{4} = -\frac{8 + 3}{4} = -\frac{11}{4}.$$

Copy the negative sign from step to step and convert $2\frac{3}{4}$ to an improper fraction.

A positive improper fraction can be converted to a mixed number using the following procedure:

Study Tip

$-2\frac{3}{4}$ means
$$-\left(2\frac{3}{4}\right) \quad \text{or} \quad -\left(2 + \frac{3}{4}\right).$$
$-2\frac{3}{4}$ does not mean
$$-2 + \frac{3}{4}.$$

Converting a Positive Improper Fraction to a Mixed Number

1. Divide the denominator into the numerator. Record the quotient and the remainder.
2. Write the mixed number using the following form:

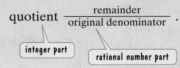

$$\text{quotient } \frac{\text{remainder}}{\text{original denominator}}.$$

integer part   *rational number part*

Example 3   **Converting from an Improper Fraction to a Mixed Number**

Convert $\frac{42}{5}$ to a mixed number.

Solution

Step 1   Divide the denominator into the numerator.

$$\begin{array}{r} 8 \\ 5\overline{)42} \\ 40 \\ \hline 2 \end{array}$$

quotient

remainder

Step 2   Write the mixed number using quotient $\dfrac{\text{remainder}}{\text{original denominator}}$. We use

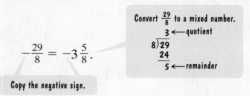

, and obtain $\dfrac{42}{5} = 8\dfrac{2}{5}$.

✓ **Checkpoint 3**   Convert $\frac{5}{3}$ to a mixed number.

When converting a negative improper fraction to a mixed number, copy the negative sign and then follow the previous procedure. For example,

$$-\frac{29}{8} = -3\frac{5}{8}.$$

Copy the negative sign.

Convert $\frac{29}{8}$ to a mixed number.

$$\begin{array}{r} 3 \leftarrow \text{quotient} \\ 8\overline{)29} \\ \underline{24} \\ 5 \leftarrow \text{remainder} \end{array}$$

④ Express rational numbers as decimals.

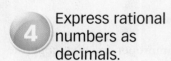

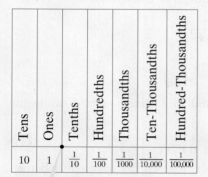

Rational Numbers and Decimals

We have seen that a rational number is the quotient of integers. Rational numbers can also be expressed as decimals. As shown in the place-value chart in the margin, it is convenient to represent rational numbers with denominators of 10, 100, 1000, and so on as decimals. For example,

$$\frac{7}{10} = 0.7, \quad \frac{3}{100} = 0.03, \quad \text{and} \quad \frac{8}{1000} = 0.008.$$

Any rational number $\frac{a}{b}$ can be expressed as a decimal by dividing the denominator, b, into the numerator, a.

Example 4   **Expressing Rational Numbers as Decimals**

Express each rational number as a decimal:

a. $\frac{5}{8}$ **b.** $\frac{7}{11}$.

Solution

In each case, divide the denominator into the numerator.

a.
$$\begin{array}{r} 0.625 \\ 8\overline{)5.000} \\ \underline{4\,8} \\ 20 \\ \underline{16} \\ 40 \\ \underline{40} \\ 0 \end{array}$$
$\dfrac{5}{8} = 0.625$

b.
$$\begin{array}{r} 0.6363\ldots \\ 11\overline{)7.0000\ldots} \\ \underline{6\,6} \\ 40 \\ \underline{33} \\ 70 \\ \underline{66} \\ 40 \\ \underline{33} \\ 70 \\ \vdots \end{array}$$
$\dfrac{7}{11} = 0.6363\ldots$

In Example 4, the decimal for $\frac{5}{8}$, namely 0.625, stops and is called a **terminating decimal**. Other examples of terminating decimals are

$$\frac{1}{4} = 0.25, \quad \frac{2}{5} = 0.4, \quad \text{and} \quad \frac{7}{8} = 0.875.$$

By contrast, the division process for $\frac{7}{11}$ results in $0.6363\ldots$, with the digits 63 repeating over and over indefinitely. To indicate this, write a bar over the digits that repeat. Thus,

$$\frac{7}{11} = 0.\overline{63}.$$

The decimal for $\frac{7}{11}, 0.\overline{63}$, is called a **repeating decimal**. Other examples of repeating decimals are

$$\frac{1}{3} = 0.333\ldots = 0.\overline{3} \quad \text{and} \quad \frac{2}{3} = 0.666\ldots = 0.\overline{6}.$$

> **Rational Numbers and Decimals**
>
> Any rational number can be expressed as a decimal. The resulting decimal will either terminate (stop), or it will have a digit that repeats or a block of digits that repeats.

✓ **Checkpoint 4** Express each rational number as a decimal:

▶ **a.** $\frac{3}{8}$ **b.** $\frac{5}{11}$.

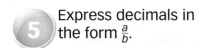

5 Express decimals in the form $\frac{a}{b}$.

Tens	Ones	Tenths	Hundredths	Thousandths	Ten-Thousandths	Hundred-Thousandths
10	1	$\frac{1}{10}$	$\frac{1}{100}$	$\frac{1}{1000}$	$\frac{1}{10,000}$	$\frac{1}{100,000}$

decimal point

Reversing Directions: Expressing Decimals as Quotients of Two Integers

Terminating decimals can be expressed with denominators of 10, 100, 1000, 10,000, and so on. Use the place-value chart shown in the margin. The digits to the right of the decimal point are the numerator of the rational number. To find the denominator, observe the last digit to the right of the decimal point. The place value of this digit will indicate the denominator.

Example 5 **Expressing Terminating Decimals in $\frac{a}{b}$ Form**

Express each terminating decimal as a quotient of integers:

a. 0.7 **b.** 0.49 **c.** 0.048.

Solution

a. $0.7 = \frac{7}{10}$ because the 7 is in the tenths position.

b. $0.49 = \frac{49}{100}$ because the last digit on the right, 9, is in the hundredths position.

c. $0.048 = \frac{48}{1000}$ because the digit on the right, 8, is in the thousandths position.

Reducing to lowest terms, $\frac{48}{1000} = \frac{48 \div 8}{1000 \div 8} = \frac{6}{125}$.

✓ **Checkpoint 5** Express each terminating decimal as a quotient of integers, reduced to lowest terms:

▶ **a.** 0.9 **b.** 0.86 **c.** 0.053.

A Brief Review • Solving One-Step Equations

- Solving an equation involves determining all values that result in a true statement when substituted into the equation. Such values are solutions of the equation.

  Example
  The solution of $x - 4 = 10$ is 14 because $14 - 4 = 10$ is a true statement.

- Two basic rules can be used to solve equations:

  1. We can add or subtract the same quantity on both sides of an equation.

  2. We can multiply or divide both sides of an equation by the same quantity, as long as we do not multiply or divide by zero.

Examples of Equations That Can Be Solved in One Step

Equation	How to Solve	Solving the Equation	The Equation's Solution
$x - 4 = 10$	Add 4 to both sides.	$x - 4 + 4 = 10 + 4$ $x = 14$	14
$y + 12 = 17$	Subtract 12 from both sides.	$y + 12 - 12 = 17 - 12$ $y = 5$	5
$99n = 53$	Divide both sides by 99.	$\frac{99n}{99} = \frac{53}{99}$ $n = \frac{53}{99}$	$\frac{53}{99}$
$\frac{z}{5} = 9$	Multiply both sides by 5.	$5 \cdot \frac{z}{5} = 5 \cdot 9$ $z = 45$	45

Equations whose solutions require more than one step are discussed in Chapter 4.

Why have we provided this brief review of equations that can be solved in one step? If you are given a rational number as a repeating decimal, there is a technique for expressing the number as a quotient of integers that requires solving a one-step equation. We begin by illustrating the technique with an example. Then we will summarize the steps in the procedure and apply them to another example.

Example 6 Expressing a Repeating Decimal in $\frac{a}{b}$ Form

Express $0.\overline{6}$ as a quotient of integers.

Solution

Step 1 Let n equal the repeating decimal. Let $n = 0.\overline{6}$, so that $n = 0.66666\ldots$.

Step 2 If there is one repeating digit, multiply both sides of the equation in step 1 by 10.

$$n = 0.66666\ldots \qquad \text{This is the equation from step 1.}$$
$$10n = 10(0.66666\ldots) \qquad \text{Multiply both sides by 10.}$$
$$10n = 6.66666\ldots \qquad \text{Multiplying by 10 moves the decimal point one place to the right.}$$

Step 3 Subtract the equation in step 1 from the equation in step 2.

Remember from algebra that n means $1n$. Thus, $10n - 1n = 9n$.

$$
\begin{aligned}
10n &= 6.66666\ldots \qquad \text{This is the equation from step 2.}\\
- \quad n &= 0.66666\ldots \qquad \text{This is the equation from step 1.}\\
\hline
9n &= 6
\end{aligned}
$$

Step 4   Divide both sides of the equation in step 3 by the number in front of *n* and solve for *n*. We solve $9n = 6$ for *n* by dividing both sides by 9.

$$9n = 6 \qquad \text{This is the equation from step 3.}$$

$$\frac{9n}{9} = \frac{6}{9} \qquad \text{Divide both sides by 9.}$$

$$n = \frac{6}{9} = \frac{2}{3} \qquad \text{Reduce } \tfrac{6}{9} \text{ to lowest terms:}$$

$$\frac{6}{9} = \frac{2 \cdot \cancel{3}}{3 \cdot \cancel{3}} = \frac{2}{3}.$$

We began the solution process with $n = 0.\overline{6}$, and now we have $n = \frac{2}{3}$. Therefore,

$$0.\overline{6} = \frac{2}{3}.$$

Here are the steps for expressing a repeating decimal as a quotient of integers. Assume that the repeating digit or digits begin directly to the right of the decimal point.

> **Expressing a Repeating Decimal as a Quotient of Integers**
>
> **Step 1** Let *n* equal the repeating decimal.
>
> **Step 2** Multiply both sides of the equation in step 1 by 10 if one digit repeats, by 100 if two digits repeat, by 1000 if three digits repeat, and so on.
>
> **Step 3** Subtract the equation in step 1 from the equation in step 2.
>
> **Step 4** Divide both sides of the equation in step 3 by the number in front of *n* and solve for *n*.

 **Checkpoint 6**   Express $0.\overline{2}$ as a quotient of integers.

Example 7   Expressing a Repeating Decimal in $\frac{a}{b}$ Form

Express $0.\overline{53}$ as a quotient of integers.

Solution

Step 1   Let *n* equal the repeating decimal. Let $n = 0.\overline{53}$, so that $n = 0.535353\ldots$.

Step 2   If there are two repeating digits, multiply both sides of the equation in step 2 by 100.

$$n = 0.535353\ldots \qquad \text{This is the equation from step 1.}$$

$$100n = 100(0.535353\ldots) \qquad \text{Multiply both sides by 100.}$$

$$100n = 53.535353\ldots \qquad \begin{array}{l}\text{Multiplying by 100 moves the decimal}\\\text{point two places to the right.}\end{array}$$

Step 3   Subtract the equation in step 1 from the equation in step 2.

$$100n = 53.535353\ldots \qquad \text{This is the equation from step 2.}$$
$$-\quad n = 0.535353\ldots \qquad \text{This is the equation from step 1.}$$
$$\overline{99n = 53}$$

Step 4   Divide both sides of the equation in step 3 by the number in front of *n* and solve for *n*. We solve $99n = 53$ for *n* by dividing both sides by 99.

$$99n = 53 \quad \text{This is the equation from step 3.}$$

$$\frac{99n}{99} = \frac{53}{99} \quad \text{Divide both sides by 99.}$$

$$n = \frac{53}{99}$$

Because *n* equals $0.\overline{53}$ and *n* equals $\frac{53}{99}$,

$$0.\overline{53} = \frac{53}{99}.$$

✓ **Checkpoint 7** Express $0.\overline{79}$ as a quotient of integers.

Multiplying and Dividing Rational Numbers

The product of two rational numbers is found as follows:

6 Multiply and divide rational numbers.

Multiplying Rational Numbers

The product of two rational numbers is the product of their numerators divided by the product of their denominators.

If $\frac{a}{b}$ and $\frac{c}{d}$ are rational numbers, then $\frac{a}{b} \cdot \frac{c}{d} = \frac{a \cdot c}{b \cdot d}$.

Example 8   Multiplying Rational Numbers

Multiply. If possible, reduce the product to its lowest terms:

a. $\frac{3}{8} \cdot \frac{5}{11}$ **b.** $\left(-\frac{2}{3}\right)\left(-\frac{9}{4}\right)$ **c.** $\left(3\frac{2}{3}\right)\left(1\frac{1}{4}\right)$.

Solution

a. $\frac{3}{8} \cdot \frac{5}{11} = \frac{3 \cdot 5}{8 \cdot 11} = \frac{15}{88}$

b. $\left(-\frac{2}{3}\right)\left(-\frac{9}{4}\right) = \frac{(-2)(-9)}{3 \cdot 4} = \frac{18}{12} = \frac{3 \cdot 6}{2 \cdot 6} = \frac{3}{2}$ or $1\frac{1}{2}$

c. $\left(3\frac{2}{3}\right)\left(1\frac{1}{4}\right) = \frac{11}{3} \cdot \frac{5}{4} = \frac{11 \cdot 5}{3 \cdot 4} = \frac{55}{12}$ or $4\frac{7}{12}$

Study Tip

You can divide numerators and denominators by common factors *before* performing multiplication. Then multiply the remaining factors in the numerators and multiply the remaining factors in the denominators. For example,

$$\frac{7}{15} \cdot \frac{20}{21} = \frac{\overset{1}{\cancel{7}}}{\underset{3}{\cancel{15}}} \cdot \frac{\overset{4}{\cancel{20}}}{\underset{3}{\cancel{21}}} = \frac{1 \cdot 4}{3 \cdot 3} = \frac{4}{9}.$$

✓ **Checkpoint 8** Multiply. If possible, reduce the product to its lowest terms:

a. $\frac{4}{11} \cdot \frac{2}{3}$ **b.** $\left(-\frac{3}{7}\right)\left(-\frac{14}{4}\right)$ **c.** $\left(3\frac{2}{5}\right)\left(1\frac{1}{2}\right)$.

Two numbers whose product is 1 are called **reciprocals**, or **multiplicative inverses**, of each other. Thus, the reciprocal of 2 is $\frac{1}{2}$ and the reciprocal of $\frac{1}{2}$ is 2 because $2 \cdot \frac{1}{2} = 1$. In general, if $\frac{c}{d}$ is a nonzero rational number, its reciprocal is $\frac{d}{c}$ because $\frac{c}{d} \cdot \frac{d}{c} = 1$.

Reciprocals are used to find the quotient of two rational numbers.

Dividing Rational Numbers

The quotient of two rational numbers is the product of the first number and the reciprocal of the second number.

If $\frac{a}{b}$ and $\frac{c}{d}$ are rational numbers and $\frac{c}{d}$ is not 0, then $\frac{a}{b} \div \frac{c}{d} = \frac{a}{b} \cdot \frac{d}{c} = \frac{a \cdot d}{b \cdot c}$.

Example 9   Dividing Rational Numbers

Divide. If possible, reduce the quotient to its lowest terms:

a. $\frac{4}{5} \div \frac{1}{10}$ **b.** $-\frac{3}{5} \div \frac{7}{11}$ **c.** $4\frac{3}{4} \div 1\frac{1}{2}$.

Solution

a. $\frac{4}{5} \div \frac{1}{10} = \frac{4}{5} \cdot \frac{10}{1} = \frac{4 \cdot 10}{5 \cdot 1} = \frac{40}{5} = 8$

b. $-\frac{3}{5} \div \frac{7}{11} = -\frac{3}{5} \cdot \frac{11}{7} = \frac{-3(11)}{5 \cdot 7} = -\frac{33}{35}$

c. $4\frac{3}{4} \div 1\frac{1}{2} = \frac{19}{4} \div \frac{3}{2} = \frac{19}{4} \cdot \frac{2}{3} = \frac{19 \cdot 2}{4 \cdot 3} = \frac{38}{12} = \frac{19 \cdot \cancel{2}}{6 \cdot \cancel{2}} = \frac{19}{6}$ or $3\frac{1}{6}$

Checkpoint 9   Divide. If possible, reduce the quotient to its lowest terms:

a. $\frac{9}{11} \div \frac{5}{4}$ **b.** $-\frac{8}{15} \div \frac{2}{5}$ **c.** $3\frac{3}{8} \div 2\frac{1}{4}$.

Adding and Subtracting Rational Numbers

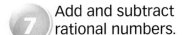

Add and subtract rational numbers.

Rational numbers with identical denominators are added and subtracted using the following rules:

> **Adding and Subtracting Rational Numbers with Identical Denominators**
>
> The sum or difference of two rational numbers with identical denominators is the sum or difference of their numerators over the common denominator.
> If $\frac{a}{b}$ and $\frac{c}{b}$ are rational numbers, then $\frac{a}{b} + \frac{c}{b} = \frac{a + c}{b}$ and $\frac{a}{b} - \frac{c}{b} = \frac{a - c}{b}$.

Example 10   Adding and Subtracting Rational Numbers with Identical Denominators

Perform the indicated operations:

a. $\frac{3}{7} + \frac{2}{7}$ **b.** $\frac{11}{12} - \frac{5}{12}$ **c.** $-5\frac{1}{4} - \left(-2\frac{3}{4}\right)$.

Solution

a. $\frac{3}{7} + \frac{2}{7} = \frac{3 + 2}{7} = \frac{5}{7}$

b. $\frac{11}{12} - \frac{5}{12} = \frac{11 - 5}{12} = \frac{6}{12} = \frac{1 \cdot \cancel{6}}{2 \cdot \cancel{6}} = \frac{1}{2}$

c. $-5\frac{1}{4} - \left(-2\frac{3}{4}\right) = -\frac{21}{4} - \left(-\frac{11}{4}\right) = -\frac{21}{4} + \frac{11}{4} = \frac{-21 + 11}{4} = \frac{-10}{4} = -\frac{5}{2}$ or $-2\frac{1}{2}$

Checkpoint 10   Perform the indicated operations:

a. $\frac{5}{12} + \frac{3}{12}$ **b.** $\frac{7}{4} - \frac{1}{4}$ **c.** $-3\frac{3}{8} - \left(-1\frac{1}{8}\right)$.

   If the rational numbers to be added or subtracted have different denominators, we use the least common multiple of their denominators to rewrite the rational numbers. The least common multiple of the denominators is called the **least common denominator (LCD)**.

Rewriting rational numbers with a least common denominator is done using the Fundamental Principle of Rational Numbers, discussed at the beginning of this section. Recall that if $\frac{a}{b}$ is a rational number and c is a nonzero number, then

$$\frac{a}{b} = \frac{a}{b} \cdot \frac{c}{c} = \frac{a \cdot c}{b \cdot c}.$$

Multiplying the numerator and the denominator of a rational number by the same nonzero number is equivalent to multiplying by 1, resulting in an equivalent fraction.

Example 11 Adding Rational Numbers with Unlike Denominators

Find the sum: $\frac{3}{4} + \frac{1}{6}$.

Solution

The smallest number divisible by both 4 and 6 is 12. Therefore, 12 is the least common multiple of 4 and 6, and will serve as the least common denominator. To obtain a denominator of 12, multiply the denominator and the numerator of the first rational number, $\frac{3}{4}$, by 3. To obtain a denominator of 12, multiply the denominator and the numerator of the second rational number, $\frac{1}{6}$, by 2.

$$\frac{3}{4} + \frac{1}{6} = \frac{3}{4} \cdot \frac{3}{3} + \frac{1}{6} \cdot \frac{2}{2}$$

Rewrite each rational number as an equivalent fraction with a denominator of 12; $\frac{3}{3} = 1$ and $\frac{2}{2} = 1$, and multiplying by 1 does not change a number's value.

$$= \frac{9}{12} + \frac{2}{12}$$

Multiply.

$$= \frac{11}{12}$$

Add numerators and put this sum over the least common denominator.

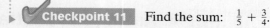 **Checkpoint 11** Find the sum: $\frac{1}{5} + \frac{3}{4}$.

If the least common denominator cannot be found by inspection, use prime factorizations of the denominators and the method for finding their least common multiple, discussed in Section 3.1.

Example 12 Subtracting Rational Numbers with Unlike Denominators

Perform the indicated operation: $\frac{1}{15} - \frac{7}{24}$.

Solution

We need to first find the least common denominator, which is the least common multiple of 15 and 24. What is the smallest number divisible by both 15 and 24? The answer is not obvious, so we begin with the prime factorization of each number.

$$15 = 5 \cdot 3$$
$$24 = 8 \cdot 3 = 2^3 \cdot 3$$

The different factors are 5, 3, and 2. Using the greater number of times each factor appears in either factorization, the least common multiple is $5 \cdot 3 \cdot 2^3 = 5 \cdot 3 \cdot 8 = 120$. We will now express each rational number with a denominator of 120, which is the

Technology

Here is a possible keystroke sequence on a graphing calculator for the subtraction problem in Example 12:

$$1 \;\boxed{\div}\; 15 \;\boxed{-}\; 7 \;\boxed{\div}\; 24$$

$$\boxed{\blacktriangleright \text{Frac}} \;\; \boxed{\text{ENTER}}.$$

```
1/15-7/24▶Frac
          -9/40
```

The calculator display reads $-9/40$, serving as a check for our answer in Example 12.

least common denominator. For the first rational number, $\frac{1}{15}$, 120 divided by 15 is 8. Thus, we will multiply the numerator and the denominator by 8. For the second rational number, $\frac{7}{24}$, 120 divided by 24 is 5. Thus, we will multiply the numerator and the denominator by 5.

$$\frac{1}{15} - \frac{7}{24} = \frac{1}{15} \cdot \frac{8}{8} - \frac{7}{24} \cdot \frac{5}{5} \qquad \text{Rewrite each rational number as an equivalent fraction with a denominator of 120.}$$

$$= \frac{8}{120} - \frac{35}{120} \qquad \text{Multiply.}$$

$$= \frac{8 - 35}{120} \qquad \text{Subtract the numerators and put this difference over the least common denominator.}$$

$$= \frac{-27}{120} \qquad \text{Perform the subtraction.}$$

$$= \frac{-9 \cdot 3}{40 \cdot 3} \qquad \text{Reduce to lowest terms.}$$

$$= -\frac{9}{40}$$

Checkpoint 12 Perform the indicated operation: $\frac{3}{10} - \frac{7}{12}$.

Order of Operations with Rational Numbers

In the previous section, we presented rules for determining the order in which operations should be performed: operations in grouping symbols; exponential expressions; multiplication/division (left to right); addition/subtraction (left to right). In our next example, we apply the order of operations to an expression with rational numbers.

8 Use the order of operations agreement with rational numbers.

Example 13   Using the Order of Operations

Simplify: $\left(\frac{1}{2}\right)^3 - \left(\frac{1}{2} - \frac{3}{4}\right)^2 (-4)$.

Solution

Because grouping symbols appear, we perform the operation within parentheses first.

$$\left(\frac{1}{2}\right)^3 - \left(\frac{1}{2} - \frac{3}{4}\right)^2 (-4)$$

$$= \left(\frac{1}{2}\right)^3 - \left(-\frac{1}{4}\right)^2 (-4) \qquad \text{Work inside parentheses first:} \quad \frac{1}{2} - \frac{3}{4} = \frac{2}{4} - \frac{3}{4} = \frac{2}{4} + \left(-\frac{3}{4}\right) = -\frac{1}{4}.$$

$$= \frac{1}{8} - \frac{1}{16}(-4) \qquad \text{Evaluate exponential expressions:} \quad \left(\frac{1}{2}\right)^3 = \frac{1}{2} \cdot \frac{1}{2} \cdot \frac{1}{2} = \frac{1}{8} \text{ and } \left(-\frac{1}{4}\right)^2 = \left(-\frac{1}{4}\right)\left(-\frac{1}{4}\right) = \frac{1}{16}.$$

$$= \frac{1}{8} - \left(-\frac{1}{4}\right) \qquad \text{Multiply:} \quad \frac{1}{16} \cdot \left(\frac{-4}{1}\right) = -\frac{4}{16} = -\frac{1}{4}.$$

$$= \frac{3}{8} \qquad \text{Subtract:} \quad \frac{1}{8} - \left(-\frac{1}{4}\right) = \frac{1}{8} + \frac{1}{4} = \frac{1}{8} + \frac{2}{8} = \frac{3}{8}.$$

Checkpoint 13 Simplify:  $\left(-\frac{1}{2}\right)^2 - \left(\frac{7}{10} - \frac{8}{15}\right)^2 (-18)$.

9 Apply the density property of rational numbers.

Density of Rational Numbers

It is always possible to find a rational number between any two distinct rational numbers. Mathematicians express this idea by saying that the set of rational numbers is **dense**.

> **Density of the Rational Numbers**
>
> If r and t represent rational numbers, with $r < t$, then there is a rational number s such that s is between r and t:
>
> $$r < s < t.$$

One way to find a rational number between two given rational numbers is to find the rational number halfway between them. Add the given rational numbers and divide their sum by 2, thereby finding the average of the numbers.

Example 14   Illustrating the Density Property

Find the rational number halfway between $\frac{1}{2}$ and $\frac{3}{4}$.

Solution

First, add $\frac{1}{2}$ and $\frac{3}{4}$.

$$\frac{1}{2} + \frac{3}{4} = \frac{2}{4} + \frac{3}{4} = \frac{5}{4}$$

Next, divide this sum by 2.

$$\frac{5}{4} \div \frac{2}{1} = \frac{5}{4} \cdot \frac{1}{2} = \frac{5}{8}$$

The number $\frac{5}{8}$ is halfway between $\frac{1}{2}$ and $\frac{3}{4}$. Thus,

$$\frac{1}{2} < \frac{5}{8} < \frac{3}{4}.$$

Study Tip

The inequality $\frac{1}{2} < \frac{5}{8} < \frac{3}{4}$ is more obvious if all denominators are changed to 8:

$$\frac{4}{8} < \frac{5}{8} < \frac{6}{8}.$$

We can repeat the procedure of Example 14 and find the rational number halfway between $\frac{1}{2}$ and $\frac{5}{8}$. Repeated application of this procedure implies the following surprising result.

Between any two distinct rational numbers are *infinitely many* rational numbers.

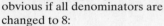 **Checkpoint 14**   Find the rational number halfway between $\frac{1}{3}$ and $\frac{1}{2}$.

Problem Solving with Rational Numbers

10 Solve problems involving rational numbers.

A common application of rational numbers involves preparing food for a different number of servings than what the recipe gives. The amount of each ingredient can be found as follows:

Amount of ingredient needed

$$= \frac{\text{desired serving size}}{\text{recipe serving size}} \times \text{ingredient amount in the recipe.}$$

Example 15   Adjusting the Size of a Recipe

A chocolate-chip cookie recipe for five dozen cookies requires $\frac{3}{4}$ cup sugar. If you want to make eight dozen cookies, how much sugar is needed?

Solution

Amount of sugar needed

$$= \frac{\text{desired serving size}}{\text{recipe serving size}} \times \text{sugar amount in recipe}$$

$$= \frac{8 \text{ dozen}}{5 \text{ dozen}} \times \frac{3}{4} \text{ cup}$$

The amount of sugar needed, in cups, is determined by multiplying the rational numbers:

$$\frac{8}{5} \times \frac{3}{4} = \frac{8 \cdot 3}{5 \cdot 4} = \frac{24}{20} = \frac{6 \cdot \cancel{4}}{5 \cdot \cancel{4}} = 1\frac{1}{5}.$$

Thus, $1\frac{1}{5}$ cups of sugar is needed. (Depending on the measuring cup you are using, you may need to round the sugar amount to $1\frac{1}{4}$ cups.)

✓ **Checkpoint 15** A chocolate-chip cookie recipe for five dozen cookies requires two eggs. If you want to make seven dozen cookies, exactly how many eggs are needed? Now round your answer to a realistic number that does not involve a fractional part of an egg.

Blitzer Bonus

NUMB3RS: Solving Crime with Mathematics

NUMB3RS is a prime-time TV crime series. The show's hero, Charlie Epps, is a brilliant mathematician who uses his powerful skills to help the FBI identify and catch criminals. The episodes are entertaining and the basic premise shows how math is a powerful weapon in the never-ending fight against crime. *NUMB3RS* is significant because it is the first popular weekly drama that revolves around mathematics. A team of mathematician advisors ensure that the equations seen in the scripts are real and relevant to the episodes. The mathematical content of the show includes many topics from this book, ranging from prime numbers, probability theory, and basic geometry.

Throughout its run, *NUMB3RS* has been the most watched series in its time slot on Friday nights. Episodes begin with a spoken tribute about the importance of mathematics:

"We all use math everywhere. To tell time, to predict the weather, to handle money . . . Math is more than formulas and equations. Math is more than numbers. It is logic. It is rationality. It is using your mind to solve the biggest mysteries we know."

Achieving Success

Read the textbook. There is a big difference between high school math classes and college courses. In high school, teachers generally cover all material for tests in class through lectures and/or activities. In college, students are responsible for information in their textbook, whether it is covered in class or not. The ability to read textbooks and learn from their pages is an essential skill for achieving success in college mathematics. Begin practicing this skill now. Read this book even if your teacher covers everything in class. Not only will you reinforce what you've learned in class, but you'll also be developing a skill that is a key component for success in higher education.

Exercise Set 3.3

Concept and Vocabulary Exercises

In Exercises 1–4, fill in each blank so that the resulting statement is true.

1. The set of _____ is the set of all numbers which can be expressed in the form $\frac{a}{b}$, where a and b are _____ and b is not equal to _____.

2. The number $\frac{17}{5}$ is an example of _____ because _____.

3. Numbers in the form $\frac{a}{b}$ (see Exercise 1) can be expressed as decimals. The decimals either _____ or _____.

4. The quotient of two fractions is the product of the first number and the _____ of the second number.

In Exercises 5–8, determine whether each statement is true or false. If the statement is false, make the necessary change(s) to produce a true statement.

5. $\frac{1}{2} + \frac{1}{5} = \frac{2}{7}$

6. $\frac{1}{2} \div 4 = 2$

7. Every fraction has infinitely many equivalent fractions.

8. $\dfrac{3+7}{30} = \dfrac{\overset{1}{3}+7}{\underset{10}{\cancel{30}}} = \dfrac{8}{10} = \dfrac{4}{5}$

Respond to Exercises 9–18 using verbal or written explanations.

9. What is a rational number?

10. Explain how to reduce a rational number to its lowest terms.

11. Explain how to convert from a mixed number to an improper fraction. Use $7\frac{2}{3}$ as an example.

12. Explain how to convert from an improper fraction to a mixed number. Use $\frac{47}{5}$ as an example.

13. Explain how to write a rational number as a decimal.

14. Explain how to write $0.\overline{9}$ as a quotient of integers.

15. Explain how to multiply rational numbers. Use $\frac{5}{6} \cdot \frac{1}{2}$ as an example.

16. Explain how to divide rational numbers. Use $\frac{5}{6} \div \frac{1}{2}$ as an example.

17. Explain how to add rational numbers with different denominators. Use $\frac{5}{6} + \frac{1}{2}$ as an example.

18. What does it mean when we say that the set of rational numbers is dense?

In Exercises 19–22, determine whether each statement makes sense or does not make sense, and explain your reasoning.

19. I saved money by buying a computer for $\frac{3}{2}$ of its original price.

20. I find it easier to multiply $\frac{1}{5}$ and $\frac{3}{4}$ than to add them.

21. My calculator shows the decimal form for the rational number $\frac{3}{11}$ as 0.2727273, so $\frac{3}{11} = 0.2727273$.

22. It took me approximately one hour to count all the rational numbers between 0 and 1.

Practice Exercises

In Exercises 23–34, reduce each rational number to its lowest terms.

23. $\frac{10}{15}$ 24. $\frac{18}{45}$ 25. $\frac{15}{18}$ 26. $\frac{16}{64}$
27. $\frac{24}{42}$ 28. $\frac{32}{80}$ 29. $\frac{60}{108}$ 30. $\frac{112}{128}$
31. $\frac{342}{380}$ 32. $\frac{210}{252}$ 33. $\frac{308}{418}$ 34. $\frac{144}{300}$

In Exercises 35–40, convert each mixed number to an improper fraction.

35. $2\frac{3}{8}$ 36. $2\frac{7}{9}$ 37. $-7\frac{3}{5}$
38. $-6\frac{2}{5}$ 39. $12\frac{7}{16}$ 40. $11\frac{5}{16}$

In Exercises 41–46, convert each improper fraction to a mixed number.

41. $\frac{23}{5}$ 42. $\frac{47}{8}$ 43. $-\frac{76}{9}$
44. $-\frac{59}{9}$ 45. $\frac{711}{20}$ 46. $\frac{788}{25}$

In Exercises 47–58, express each rational number as a decimal.

47. $\frac{3}{4}$ 48. $\frac{3}{5}$ 49. $\frac{7}{20}$ 50. $\frac{3}{20}$
51. $\frac{7}{8}$ 52. $\frac{5}{16}$ 53. $\frac{9}{11}$ 54. $\frac{3}{11}$
55. $\frac{22}{7}$ 56. $\frac{20}{3}$ 57. $\frac{2}{7}$ 58. $\frac{5}{7}$

In Exercises 59–70, express each terminating decimal as a quotient of integers. If possible, reduce to lowest terms.

59. 0.3   60. 0.9   61. 0.4
62. 0.6   63. 0.39   64. 0.59
65. 0.82   66. 0.64   67. 0.725
68. 0.625   69. 0.5399   70. 0.7006

In Exercises 71–78, express each repeating decimal as a quotient of integers. If possible, reduce to lowest terms.

71. $0.\overline{7}$ 72. $0.\overline{1}$ 73. $0.\overline{9}$ 74. $0.\overline{3}$
75. $0.\overline{36}$ 76. $0.\overline{81}$ 77. $0.\overline{257}$ 78. $0.\overline{529}$

In Exercises 79–126, perform the indicated operations. If possible, reduce the answer to its lowest terms.

79. $\frac{3}{8} \cdot \frac{7}{11}$ 80. $\frac{5}{8} \cdot \frac{3}{11}$ 81. $\left(-\frac{1}{10}\right)\left(\frac{7}{12}\right)$

82. $\left(-\frac{1}{8}\right)\left(\frac{5}{9}\right)$ 83. $\left(-\frac{2}{3}\right)\left(-\frac{9}{4}\right)$ 84. $\left(-\frac{5}{4}\right)\left(-\frac{6}{7}\right)$

85. $\left(3\frac{3}{4}\right)\left(1\frac{3}{5}\right)$ 86. $\left(2\frac{4}{5}\right)\left(1\frac{1}{4}\right)$ 87. $\frac{5}{4} \div \frac{3}{8}$

88. $\frac{5}{8} \div \frac{4}{3}$ 89. $-\frac{7}{8} \div \frac{15}{16}$ 90. $-\frac{13}{20} \div \frac{4}{5}$

91. $6\frac{3}{5} \div 1\frac{1}{10}$ 92. $1\frac{3}{4} \div 2\frac{5}{8}$ 93. $\frac{2}{11} + \frac{3}{11}$

94. $\frac{5}{13} + \frac{2}{13}$ 95. $\frac{5}{6} - \frac{1}{6}$ 96. $\frac{7}{12} - \frac{5}{12}$

97. $\frac{7}{12} - \left(-\frac{1}{12}\right)$ 98. $\frac{5}{16} - \left(-\frac{5}{16}\right)$ 99. $\frac{1}{2} + \frac{1}{5}$

100. $\frac{1}{3} + \frac{1}{5}$ 101. $\frac{3}{4} + \frac{3}{20}$ 102. $\frac{2}{5} + \frac{2}{15}$

103. $\frac{5}{24} + \frac{7}{30}$ 104. $\frac{7}{108} + \frac{55}{144}$ 105. $\frac{13}{18} - \frac{2}{9}$

106. $\frac{13}{15} - \frac{2}{45}$ 107. $\frac{4}{3} - \frac{3}{4}$ 108. $\frac{3}{2} - \frac{2}{3}$

109. $\frac{1}{15} - \frac{27}{50}$ 110. $\frac{4}{15} - \frac{1}{6}$ 111. $2\frac{2}{3} + 1\frac{3}{4}$

112. $2\frac{1}{8} + 3\frac{3}{4}$

113. $3\frac{2}{3} - 2\frac{1}{2}$

114. $3\frac{3}{4} - 2\frac{1}{3}$

115. $-5\frac{2}{3} + 3\frac{1}{6}$

116. $-2\frac{1}{2} + 1\frac{3}{4}$

117. $-1\frac{4}{7} - \left(-2\frac{5}{14}\right)$

118. $-1\frac{4}{9} - \left(-2\frac{5}{18}\right)$

119. $\left(\frac{1}{2} - \frac{1}{3}\right) \div \frac{5}{8}$

120. $\left(\frac{1}{2} + \frac{1}{4}\right) \div \left(\frac{1}{2} + \frac{1}{3}\right)$

121. $-\frac{9}{4}\left(\frac{1}{2}\right) + \frac{3}{4} \div \frac{5}{6}$

122. $\left[-\frac{4}{7} - \left(-\frac{2}{5}\right)\right]\left[-\frac{3}{8} + \left(-\frac{1}{9}\right)\right]$

123. $\dfrac{\frac{7}{9} - 3}{\frac{5}{6}} \div \frac{3}{2} + \frac{3}{4}$

124. $\dfrac{\frac{17}{25}}{\frac{3}{5} - 4} \div \frac{1}{5} + \frac{1}{2}$

125. $\frac{1}{4} - 6(2 + 8) \div \left(-\frac{1}{3}\right)\left(-\frac{1}{9}\right)$

126. $\frac{3}{4} - 4(2 + 7) \div \left(-\frac{1}{2}\right)\left(-\frac{1}{6}\right)$

In Exercises 127–132, find the rational number halfway between the two numbers in each pair.

127. $\frac{1}{4}$ and $\frac{1}{3}$ **128.** $\frac{2}{3}$ and $\frac{5}{6}$ **129.** $\frac{1}{2}$ and $\frac{2}{3}$

130. $\frac{3}{5}$ and $\frac{2}{3}$ **131.** $-\frac{2}{3}$ and $-\frac{5}{6}$ **132.** -4 and $-\frac{7}{2}$

Different operations with the same rational numbers usually result in different answers. Exercises 133–134 illustrate some curious exceptions.

133. Show that $\frac{13}{4} + \frac{13}{9}$ and $\frac{13}{4} \times \frac{13}{9}$ give the same answer.

134. Show that $\frac{169}{30} + \frac{13}{15}$ and $\frac{169}{30} \div \frac{13}{15}$ give the same answer.

Practice Plus

In Exercises 135–138, perform the indicated operations. Leave denominators in prime factorization form.

135. $\dfrac{5}{2^2 \cdot 3^2} - \dfrac{1}{2 \cdot 3^2}$

136. $\dfrac{7}{3^2 \cdot 5^2} - \dfrac{1}{3 \cdot 5^3}$

137. $\dfrac{1}{2^4 \cdot 5^3 \cdot 7} + \dfrac{1}{2 \cdot 5^4} - \dfrac{1}{2^3 \cdot 5^2}$

138. $\dfrac{1}{2^3 \cdot 17^8} + \dfrac{1}{2 \cdot 17^9} - \dfrac{1}{2^2 \cdot 3 \cdot 17^8}$

In Exercises 139–142, express each rational number as a decimal. Then insert either $<$ or $>$ in the shaded area between the rational numbers to make the statement true.

139. $\dfrac{6}{11}$ ▪ $\dfrac{7}{12}$

140. $\dfrac{29}{36}$ ▪ $\dfrac{28}{35}$

141. $-\dfrac{5}{6}$ ▪ $-\dfrac{8}{9}$

142. $-\dfrac{1}{125}$ ▪ $-\dfrac{3}{500}$

Application Exercises

A study of teens under the age of 18 revealed how they dealt with stress. The circle graphs at the top of the next column show the breakdown of the number of men and women who responded to stress in four different ways. Use this information to solve Exercises 143–144.

How Teens Deal with Stress

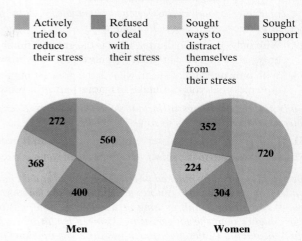

Actively tried to reduce their stress

Refused to deal with their stress

Sought ways to distract themselves from their stress

Sought support

Men   Women

Source: John Hopkins Bloomberg School of Public Health

143. a. What fractional part of the men actively tried to reduce their stress? Reduce this fraction to its lowest terms.

   b. Express the rational number in part (a) as a decimal. What percentage of teen males actively tried to reduce their stress?

   c. What fractional part of the women actively tried to reduce their stress? Reduce this fraction to its lowest terms.

   d. Express the rational number in part (c) as a decimal. What percentage of teen females actively tried to reduce their stress?

   e. What is the difference between the percentage of teen females and teen males who actively tried to reduce their stress?

144. a. What fractional part of the men refused to deal with their stress? Reduce this fraction to its lowest terms.

   b. Express the rational number in part (a) as a decimal. What percentage of teen males refused to deal with their stress?

   c. What fractional part of the women refused to deal with their stress? Reduce this fraction to its lowest terms.

   d. Express the rational number in part (c) as a decimal. What percentage of teen females refused to deal with their stress?

   e. What is the difference between the percentage of teen males and teen females who refused to deal with their stress?

Use the following list of ingredients for chocolate brownies to solve Exercises 145–150.

Ingredients for 16 Brownies

$\frac{2}{3}$ *cup butter, 5 ounces unsweetened chocolate, $1\frac{1}{2}$ cups sugar, 2 teaspoons vanilla, 2 eggs, 1 cup flour*

145. How much of each ingredient is needed to make 8 brownies?

146. How much of each ingredient is needed to make 12 brownies?

147. How much of each ingredient is needed to make 20 brownies?

148. How much of each ingredient is needed to make 24 brownies?

Ingredients for 16 Brownies (*repeated for Exercises 149–150*)
$\frac{2}{3}$ *cup butter, 5 ounces unsweetened chocolate,* $1\frac{1}{2}$ *cups sugar,*
2 teaspoons vanilla, 2 eggs, 1 cup flour

149. With only one cup of butter, what is the greatest number of brownies that you can make? (Ignore part of a brownie.)

150. With only one cup of sugar, what is the greatest number of brownies that you can make? (Ignore part of a brownie.)

A mix for eight servings of instant potatoes requires $2\frac{2}{3}$ *cups of water. Use this information to solve Exercises 151–152.*

151. If you want to make 11 servings, how much water is needed?

152. If you want to make six servings, how much water is needed?

The sounds created by plucked or bowed strings of equal diameter and tension produce various notes depending on the lengths of the strings. If a string is half as long as another, its note will be an octave higher than the longer string. Using a length of 1 unit to represent middle C, the diagram shows different fractions of the length of this unit string needed to produce the notes D, E, F, G, A, B, and c one octave higher than middle C.

For many of the strings, the length is $\frac{8}{9}$ *of the length of the previous string. For example, the A string is* $\frac{8}{9}$ *of the length of the string needed to produce the note for G:* $\frac{8}{9} \cdot \frac{2}{3} = \frac{16}{27}.$
Use this information to solve Exercises 153–154.

153. a. Which strings from D through c are $\frac{8}{9}$ of the length of the preceding string?

   b. How is your answer to part (a) shown on this one-octave span of the piano keyboard?

One Octave

154. a. Which strings from D through c are not $\frac{8}{9}$ of the length of the preceding string?

   b. How is your answer to part (a) shown on the one-octave span on the piano keyboard in Exercise 153(b)?

155. A board $7\frac{1}{2}$ inches long is cut from a board that is 2 feet long. If the width of the saw cut is $\frac{1}{16}$ inch, what is the length of the remaining piece?

156. A board that is $7\frac{1}{4}$ inches long is cut from a board that is 3 feet long. If the width of the saw cut is $\frac{1}{16}$ inch, what is the length of the remaining piece?

157. A franchise is owned by three people. The first owns $\frac{5}{12}$ of the business and the second owns $\frac{1}{4}$ of the business. What fractional part of the business is owned by the third person?

158. At a workshop on enhancing creativity, $\frac{1}{4}$ of the participants are musicians, $\frac{2}{5}$ are artists, $\frac{1}{10}$ are actors, and the remaining participants are writers. What fraction of the people attending the workshop are writers?

159. If you walk $\frac{3}{4}$ mile and then jog $\frac{2}{5}$ mile, what is the total distance covered? How much farther did you walk than jog?

160. Many companies pay people extra when they work more than a regular 40-hour work week. The overtime pay is often $1\frac{1}{2}$ times the regular hourly rate. This is called time and a half. A summer job for students pays $12 an hour and offers time and a half for the hours worked over 40. If a student works 46 hours during one week, what is the student's total pay before taxes?

161. A will states that $\frac{3}{5}$ of the estate is to be divided among relatives. Of the remaining estate, $\frac{1}{4}$ goes to charity. What fraction of the estate goes to charity?

162. The legend of a map indicates that 1 inch = 16 miles. If the distance on the map between two cities is $2\frac{3}{8}$ inches, how far apart are the cities?

Critical Thinking Exercises

163. Shown below is a short excerpt from "The Star-Spangled Banner." The time is $\frac{3}{4}$, which means that each measure must contain notes that add up to $\frac{3}{4}$. The values of the different notes tell musicians how long to hold each note.

$$\circ = 1 \qquad \text{♩} = \frac{1}{2} \qquad \text{♩} = \frac{1}{4} \qquad \text{♪} = \frac{1}{8}$$

Use vertical lines to divide this line of "The Star-Spangled Banner" into measures.

say does that   Star-span-gled   Ban-ner   yet   wave   O'er the

164. Use inductive reasoning to predict the addition problem and the sum that will appear in the fourth row. Then perform the arithmetic to verify your conjecture.

$$\frac{1}{1 \cdot 2} + \frac{1}{2 \cdot 3} = \frac{2}{3}$$

$$\frac{1}{1 \cdot 2} + \frac{1}{2 \cdot 3} + \frac{1}{3 \cdot 4} = \frac{3}{4}$$

$$\frac{1}{1 \cdot 2} + \frac{1}{2 \cdot 3} + \frac{1}{3 \cdot 4} + \frac{1}{4 \cdot 5} = \frac{4}{5}$$

Technology Exercises

165. Use a calculator to express the following rational numbers as decimals.

a. $\frac{197}{800}$ **b.** $\frac{4539}{3125}$ **c.** $\frac{7}{6250}$

166. Some calculators have a fraction feature. This feature allows you to perform operations with fractions and displays the answer as a fraction reduced to its lowest terms. If your calculator has this feature, use it to verify any five of the answers that you obtained in Exercises 79–126.

Group Exercise

167. Each member of the group should present an application of rational numbers. The application can be based on research or on how the group member uses rational numbers in his or her life. If you are not sure where to begin, ask yourself how your life would be different if fractions and decimals were concepts unknown to our civilization.

3.4 The Irrational Numbers

1. Define the irrational numbers.

2. Simplify square roots.

3. Perform operations with square roots.

4. Rationalize denominators.

For the followers of the Greek mathematician Pythagoras in the sixth century B.C., numbers took on a life-and-death importance. The "Pythagorean Brotherhood" was a secret group whose members were convinced that properties of whole numbers were the key to understanding the universe. Members of the Brotherhood (which admitted women) thought that all numbers that were not whole numbers could be represented as the ratio of whole numbers. A crisis occurred for the Pythagoreans when they discovered the existence of a number that was not rational. Because the

Pythagoras

Shown here is Renaissance artist Raphael Sanzio's (1483–1520) image of Pythagoras from *The School of Athens* mural. Detail of left side.

Stanza della Segnatura, Vatican Palace, Vatican State. Scala/Art Resource, NY.

Pythagoreans viewed numbers with reverence and awe, the punishment for speaking about this number was death. However, a member of the Brotherhood revealed the secret of the number's existence. When he later died in a shipwreck, his death was viewed as punishment from the gods.

The triangle in **Figure 3.6** led the Pythagoreans to the discovery of a number that could not be expressed as the quotient of integers. Based on their understanding of the relationship among the sides of this triangle, they knew that the length of the side shown in red had to be a number that, when squared, is equal to 2. The Pythagoreans discovered that this number seemed to be close to the rational numbers

$$\frac{14}{10}, \frac{141}{100}, \frac{1414}{1000}, \frac{14{,}141}{10{,}000}, \text{ and so on.}$$

However, they were shocked to find that there is no quotient of integers whose square is equal to 2.

The positive number whose square is equal to 2 is written $\sqrt{2}$. We read this "the square root of 2," or "radical 2." The symbol $\sqrt{}$ is called the **radical sign**. The number under the radical sign, in this case 2, is called the **radicand**. The entire symbol $\sqrt{2}$ is called a **radical**.

Using deductive reasoning, mathematicians have proved that $\sqrt{2}$ cannot be represented as a quotient of integers. This means that there is no terminating or repeating decimal that can be multiplied by itself to give 2. We can, however, give a decimal approximation for $\sqrt{2}$. We use the symbol $\approx$, which means "is approximately equal to." Thus,

$$\sqrt{2} \approx 1.414214.$$

We can verify that this is only an approximation by multiplying 1.414214 by itself. The product is not exactly 2:

$$1.414214 \times 1.414214 = 2.000001237796.$$

A number like $\sqrt{2}$, whose decimal representation does not come to an end and does not have a block of repeating digits, is an example of an **irrational number**.

Length = ?

Length: 1 unit

Length: 1 unit

Figure 3.6

Define the irrational numbers.

The Irrational Numbers

The set of **irrational numbers** is the set of numbers whose decimal representations are neither terminating nor repeating.

Perhaps the best known of all the irrational numbers is π (pi). This irrational number represents the distance around a circle (its circumference) divided by the diameter of the circle. In the *Star Trek* episode "Wolf in the Fold," Spock foils an evil computer by telling it to "compute the last digit in the value of π." Because π is an irrational number, there is no last digit in its decimal representation:

$$\pi = 3.14159265358979323846264433832795\ldots.$$

The nature of the irrational number π has fascinated mathematicians for centuries. Amateur and professional mathematicians have taken up the challenge of calculating π to more and more decimal places. Although such an exercise may seem pointless, it serves as the ultimate stress test for new high-speed computers and also as a test for the long-standing, but still unproven, conjecture that the distribution of digits in π is completely random.

Technology

You can obtain decimal approximations for irrational numbers using a calculator. For example, to approximate $\sqrt{2}$, use the following keystrokes:

Scientific Calculator	Graphing Calculator
2 √ or 2 2ND INV x^2	√ 2 ENTER or 2ND INV x^2 2 ENTER

> Some graphing calculators show an open parenthesis after displaying √. In this case, enter a closed parenthesis,), after **2**.

The display may read 1.41421356237, although your calculator may show more or fewer digits. Between which two integers would you graph $\sqrt{2}$ on a number line?

Square Roots

The United Nations Building in New York was designed to represent its mission of promoting world harmony. Viewed from the front, the building looks like three rectangles stacked upon each other. In each rectangle, the width divided by the height is $\sqrt{5} + 1$ to 2, approximately 1.618 to 1. The ancient Greeks believed that such a rectangle, called a **golden rectangle**, was the most pleasing of all rectangles. The comparison 1.618 to 1 is approximate because $\sqrt{5}$ is an irrational number.

The **principal square root** of a nonnegative number n, written $\sqrt{n}$, is the positive number that when multiplied by itself gives n. Thus,

$$\sqrt{36} = 6 \text{ because } 6 \cdot 6 = 36$$

and

$$\sqrt{81} = 9 \text{ because } 9 \cdot 9 = 81.$$

The U.N. building is designed with three golden rectangles.

Notice that both $\sqrt{36}$ and $\sqrt{81}$ are rational numbers because 6 and 9 are terminating decimals. Thus, **not all square roots are irrational**.

Numbers such as 36 and 81 are called *perfect squares*. A **perfect square** is a number that is the square of a whole number. The first few perfect squares are listed below.

0 $= 0^2$	**16** $= 4^2$	**64** $= 8^2$	**144** $= 12^2$
1 $= 1^2$	**25** $= 5^2$	**81** $= 9^2$	**169** $= 13^2$
4 $= 2^2$	**36** $= 6^2$	**100** $= 10^2$	**196** $= 14^2$
9 $= 3^2$	**49** $= 7^2$	**121** $= 11^2$	**225** $= 15^2$

The principal square root of a perfect square is a whole number. For example,

$$\sqrt{0} = 0, \sqrt{1} = 1, \sqrt{4} = 2, \sqrt{9} = 3, \sqrt{16} = 4, \sqrt{25} = 5, \sqrt{36} = 6,$$

and so on.

Simplifying Square Roots

2 Simplify square roots.

A rule for simplifying square roots can be generalized by comparing $\sqrt{25 \cdot 4}$ and $\sqrt{25} \cdot \sqrt{4}$. Notice that

$$\sqrt{25 \cdot 4} = \sqrt{100} = 10 \quad \text{and} \quad \sqrt{25} \cdot \sqrt{4} = 5 \cdot 2 = 10.$$

Because we obtain 10 in both situations, the original radicals must be equal. That is,

$$\sqrt{25 \cdot 4} = \sqrt{25} \cdot \sqrt{4}.$$

This result is a particular case of the **product rule for square roots** that can be generalized as follows:

> **The Product Rule for Square Roots**
>
> If a and b represent nonnegative numbers, then
>
> $$\sqrt{ab} = \sqrt{a} \cdot \sqrt{b} \quad \text{and} \quad \sqrt{a} \cdot \sqrt{b} = \sqrt{ab}.$$
>
> The square root of a product is the product of the square roots.

Example 1 shows how the product rule is used to remove from the square root any perfect squares that occur as factors.

Study Tip

There are no addition or subtraction rules for square roots:

$$\sqrt{a + b} \neq \sqrt{a} + \sqrt{b}$$
$$\sqrt{a - b} \neq \sqrt{a} - \sqrt{b}.$$

For example, if $a = 9$ and $b = 16$,

$$\sqrt{9 + 16} = \sqrt{25} = 5$$

and

$$\sqrt{9} + \sqrt{16} = 3 + 4 = 7.$$

Thus,

$$\sqrt{9 + 16} \neq \sqrt{9} + \sqrt{16}.$$

Example 1 Simplifying Square Roots

Simplify, if possible:

a. $\sqrt{75}$ **b.** $\sqrt{500}$ **c.** $\sqrt{17}$.

Solution

a. $\sqrt{75} = \sqrt{25 \cdot 3}$ *25 is the greatest perfect square that is a factor of 75.*

       $= \sqrt{25} \cdot \sqrt{3}$ *$\sqrt{ab} = \sqrt{a} \cdot \sqrt{b}$*

       $= 5\sqrt{3}$ *Write $\sqrt{25}$ as 5.*

b. $\sqrt{500} = \sqrt{100 \cdot 5}$ *100 is the greatest perfect square factor of 500.*

       $= \sqrt{100} \cdot \sqrt{5}$ *$\sqrt{ab} = \sqrt{a} \cdot \sqrt{b}$*

       $= 10\sqrt{5}$ *Write $\sqrt{100}$ as 10.*

c. Because 17 has no perfect square factors (other than 1), $\sqrt{17}$ cannot be simplified.

✓ **Checkpoint 1** Simplify, if possible:

a. $\sqrt{12}$ **b.** $\sqrt{60}$ **c.** $\sqrt{55}$.

3 Perform operations with square roots.

Multiplying Square Roots

If a and b are nonnegative, then we can use the product rule

$$\sqrt{a} \cdot \sqrt{b} = \sqrt{a \cdot b}$$

to multiply square roots. The product of the square roots is the square root of the product. Once the square roots are multiplied, simplify the square root of the product when possible.

Example 2 Multiplying Square Roots

Multiply:

  a. $\sqrt{2} \cdot \sqrt{5}$ **b.** $\sqrt{7} \cdot \sqrt{7}$ **c.** $\sqrt{6} \cdot \sqrt{12}$.

Solution

  a. $\sqrt{2} \cdot \sqrt{5} = \sqrt{2 \cdot 5} = \sqrt{10}$

  b. $\sqrt{7} \cdot \sqrt{7} = \sqrt{7 \cdot 7} = \sqrt{49} = 7$

  c. $\sqrt{6} \cdot \sqrt{12} = \sqrt{6 \cdot 12} = \sqrt{72} = \sqrt{36 \cdot 2} = \sqrt{36} \cdot \sqrt{2} = 6\sqrt{2}$

> It is possible to multiply irrational numbers and obtain a rational number for the product.

 **Checkpoint 2** Multiply:

  a. $\sqrt{3} \cdot \sqrt{10}$ **b.** $\sqrt{10} \cdot \sqrt{10}$ **c.** $\sqrt{6} \cdot \sqrt{2}$.

Dividing Square Roots

Another property for square roots involves division.

The Quotient Rule for Square Roots

If a and b represent nonnegative numbers and $b \neq 0$, then

$$\frac{\sqrt{a}}{\sqrt{b}} = \sqrt{\frac{a}{b}} \quad \text{and} \quad \sqrt{\frac{a}{b}} = \frac{\sqrt{a}}{\sqrt{b}}.$$

The quotient of two square roots is the square root of the quotient.

Once the square roots are divided, simplify the square root of the quotient when possible.

Example 3 Dividing Square Roots

Find the quotient:

  a. $\dfrac{\sqrt{75}}{\sqrt{3}}$ **b.** $\dfrac{\sqrt{90}}{\sqrt{2}}$.

Solution

  a. $\dfrac{\sqrt{75}}{\sqrt{3}} = \sqrt{\dfrac{75}{3}} = \sqrt{25} = 5$

  b. $\dfrac{\sqrt{90}}{\sqrt{2}} = \sqrt{\dfrac{90}{2}} = \sqrt{45} = \sqrt{9 \cdot 5} = \sqrt{9} \cdot \sqrt{5} = 3\sqrt{5}$

 **Checkpoint 3** Find the quotient:

  a. $\dfrac{\sqrt{80}}{\sqrt{5}}$ **b.** $\dfrac{\sqrt{48}}{\sqrt{6}}$.

Adding and Subtracting Square Roots

The number that multiplies a square root is called the square root's **coefficient**. For example, in $3\sqrt{5}$, 3 is the coefficient of the square root.

Square roots with the same radicand can be added or subtracted by adding or subtracting their coefficients:

$$a\sqrt{c} + b\sqrt{c} = (a + b)\sqrt{c} \qquad a\sqrt{c} - b\sqrt{c} = (a - b)\sqrt{c}.$$

Sum of coefficients times the common square root

Difference of coefficients times the common square root

Example 4   Adding and Subtracting Square Roots

Add or subtract as indicated:

a. $7\sqrt{2} + 5\sqrt{2}$ **b.** $2\sqrt{5} - 6\sqrt{5}$ **c.** $3\sqrt{7} + 9\sqrt{7} - \sqrt{7}.$

Solution

a. $7\sqrt{2} + 5\sqrt{2} = (7 + 5)\sqrt{2}$
$$= 12\sqrt{2}$$

b. $2\sqrt{5} - 6\sqrt{5} = (2 - 6)\sqrt{5}$
$$= -4\sqrt{5}$$

c. $3\sqrt{7} + 9\sqrt{7} - \sqrt{7} = 3\sqrt{7} + 9\sqrt{7} - 1\sqrt{7}$ Write $\sqrt{7}$ as $1\sqrt{7}$.
$$= (3 + 9 - 1)\sqrt{7}$$
$$= 11\sqrt{7}$$

✔ **Checkpoint 4**   Add or subtract as indicated:

a. $8\sqrt{3} + 10\sqrt{3}$
b. $4\sqrt{13} - 9\sqrt{13}$
▶ **c.** $7\sqrt{10} + 2\sqrt{10} - \sqrt{10}.$

In some situations, it is possible to add and subtract square roots that do not contain a common square root by first simplifying.

Example 5   Adding and Subtracting Square Roots by First Simplifying

Add or subtract as indicated:

a. $\sqrt{2} + \sqrt{8}$ **b.** $4\sqrt{50} - 6\sqrt{32}.$

Solution

a. $\sqrt{2} + \sqrt{8}$

$\qquad = \sqrt{2} + \sqrt{4 \cdot 2}$ Split 8 into two factors such that one factor is a perfect square.

$\qquad = 1\sqrt{2} + 2\sqrt{2}$ $\sqrt{4 \cdot 2} = \sqrt{4} \cdot \sqrt{2} = 2\sqrt{2}$

$\qquad = (1 + 2)\sqrt{2}$ Add coefficients and retain the common square root.

$\qquad = 3\sqrt{2}$ Simplify.

Study Tip

Sums or differences of square roots that cannot be simplified and that do not contain a common square root cannot be combined into one term by adding or subtracting coefficients. Some examples:

- $5\sqrt{3} + 3\sqrt{5}$ cannot be combined by adding coefficients. The square roots, $\sqrt{3}$ and $\sqrt{5}$, are different.
- $28 + 7\sqrt{3}$, or $28\sqrt{1} + 7\sqrt{3}$, cannot be combined by adding coefficients. The square roots, $\sqrt{1}$ and $\sqrt{3}$, are different.

b. $4\sqrt{50} - 6\sqrt{32}$

$= 4\sqrt{25 \cdot 2} - 6\sqrt{16 \cdot 2}$ *25 is the greatest perfect square factor of 50 and 16 is*
the greatest perfect square factor of 32.

$= 4 \cdot 5\sqrt{2} - 6 \cdot 4\sqrt{2}$ $\sqrt{25 \cdot 2} = \sqrt{25}\sqrt{2} = 5\sqrt{2}$ *and*
$\sqrt{16 \cdot 2} = \sqrt{16}\sqrt{2} = 4\sqrt{2}$

$= 20\sqrt{2} - 24\sqrt{2}$ *Multiply.*

$= (20 - 24)\sqrt{2}$ *Subtract coefficients and retain the common square root.*

$= -4\sqrt{2}$ *Simplify.*

✓ **Checkpoint 5**   Add or subtract as indicated:

a. $\sqrt{3} + \sqrt{12}$ **b.** $4\sqrt{8} - 7\sqrt{18}$.

Rationalizing Denominators

4 **Rationalize denominators.**

The calculator screen in **Figure 3.7** shows approximate values for $\dfrac{1}{\sqrt{3}}$ and $\dfrac{\sqrt{3}}{3}$. The two approximations are the same. This is not a coincidence:

$$\frac{1}{\sqrt{3}} = \frac{1}{\sqrt{3}} \cdot \boxed{\frac{\sqrt{3}}{\sqrt{3}}} = \frac{\sqrt{3}}{\sqrt{9}} = \frac{\sqrt{3}}{3}$$

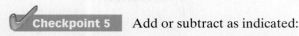

Any nonzero number divided by itself is 1. Multiplication by 1 does not change the value of $\dfrac{1}{\sqrt{3}}$.

```
1/√(3)
         .5773502692
√(3)/3
         .5773502692
```

Figure 3.7 The calculator screen shows approximate values for $\dfrac{1}{\sqrt{3}}$ and $\dfrac{\sqrt{3}}{3}$.

This process involves rewriting a radical expression as an equivalent expression in which the denominator no longer contains any radicals. The process is called **rationalizing the denominator**. If the denominator contains the square root of a natural number that is not a perfect square, **multiply the numerator and the denominator by the smallest number that produces the square root of a perfect square in the denominator**.

Example 6   Rationalizing Denominators

Rationalize the denominator:

a. $\dfrac{15}{\sqrt{6}}$ **b.** $\sqrt{\dfrac{3}{5}}$ **c.** $\dfrac{12}{\sqrt{8}}$.

Solution

a. If we multiply the numerator and the denominator of $\dfrac{15}{\sqrt{6}}$ by $\sqrt{6}$, the denominator becomes $\sqrt{6} \cdot \sqrt{6} = \sqrt{36} = 6$. Therefore, we multiply by 1, choosing $\dfrac{\sqrt{6}}{\sqrt{6}}$ for 1.

$$\frac{15}{\sqrt{6}} = \frac{15}{\sqrt{6}} \cdot \frac{\sqrt{6}}{\sqrt{6}} = \frac{15\sqrt{6}}{\sqrt{36}} = \frac{15\sqrt{6}}{6} = \frac{5\sqrt{6}}{2}$$

Multiply by 1.   *Simplify:* $\dfrac{15}{6} = \dfrac{5 \cdot 3}{2 \cdot 3} = \dfrac{5}{2}$.

b. $\sqrt{\dfrac{3}{5}} = \dfrac{\sqrt{3}}{\sqrt{5}} = \dfrac{\sqrt{3}}{\sqrt{5}} \cdot \dfrac{\sqrt{5}}{\sqrt{5}} = \dfrac{\sqrt{15}}{\sqrt{25}} = \dfrac{\sqrt{15}}{5}$

Multiply by 1.

c. The *smallest* number that will produce a perfect square in the denominator of $\dfrac{12}{\sqrt{8}}$ is $\sqrt{2}$, because $\sqrt{8} \cdot \sqrt{2} = \sqrt{16} = 4$. We multiply by 1, choosing $\dfrac{\sqrt{2}}{\sqrt{2}}$ for 1.

$$\frac{12}{\sqrt{8}} = \frac{12}{\sqrt{8}} \cdot \frac{\sqrt{2}}{\sqrt{2}} = \frac{12\sqrt{2}}{\sqrt{16}} = \frac{12\sqrt{2}}{4} = 3\sqrt{2}$$

 **Checkpoint 6**   Rationalize the denominator:

a. $\dfrac{25}{\sqrt{10}}$ **b.** $\sqrt{\dfrac{2}{7}}$ **c.** $\dfrac{5}{\sqrt{18}}$.

Blitzer Bonus

Golden Rectangles

The early Greeks believed that the most pleasing of all rectangles were **golden rectangles**, whose ratio of width to height is

$$\frac{w}{h} = \frac{\sqrt{5} + 1}{2}.$$

The Parthenon at Athens fits into a golden rectangle once the triangular pediment is reconstructed.

Irrational Numbers and Other Kinds of Roots

Irrational numbers appear in the form of roots other than square roots. The symbol $\sqrt[3]{}$ represents the **cube root** of a number. For example,

$$\sqrt[3]{8} = 2 \text{ because } 2 \cdot 2 \cdot 2 = 8 \quad \text{and} \quad \sqrt[3]{64} = 4 \text{ because } 4 \cdot 4 \cdot 4 = 64.$$

Although these cube roots are rational numbers, most cube roots are not. For example,

$$\sqrt[3]{217} \approx 6.0092 \text{ because } (6.0092)^3 \approx 216.995, \text{ not exactly } 217.$$

There is no end to the kinds of roots for numbers. For example, $\sqrt[4]{}$ represents the **fourth root** of a number. Thus, $\sqrt[4]{81} = 3$ because $3 \cdot 3 \cdot 3 \cdot 3 = 81$. Although the fourth root of 81 is rational, most fourth roots, fifth roots, and so on tend to be irrational.

Achieving Success

Warm up your brain before starting the assigned homework. Researchers say the mind can be strengthened, just like your muscles, with regular training and rigorous practice. Think of the book's exercise sets as brain calisthenics. If you're feeling a bit sluggish before any of your mental workouts, try this warmup:

In the list below say the color the word is printed in, not the word itself. Once you can do this in 15 seconds without an error, the warmup is over and it's time to move on to the assigned exercises.

Blue Yellow Red Green **Yellow Green Blue** Red **Yellow Red**

Blitzer Bonus
● ● ● ● ● ● ● ● ● ● ● ● ● ● ● ● ●

A Radical Idea: Time Is Relative

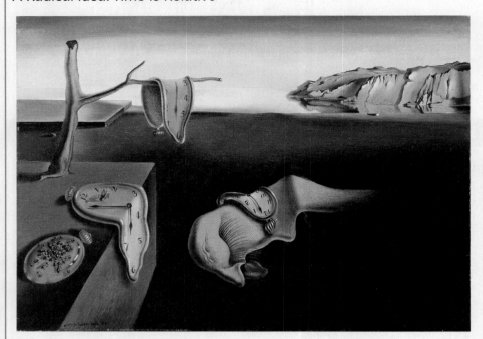

Salvador Dali "The Persistence of Memory" 1931, oil on canvas, $9\frac{1}{2} \times 13$ in. (24.1 × 33 cm). The Museum of Modern Art/Licensed by Scala-Art Resource. N.Y. © 1999 Demart Pro Arte. Geneva/Artists Rights Society (ARS). New York.

What does travel in space have to do with square roots? Imagine that in the future we will be able to travel at velocities approaching the speed of light (approximately 186,000 miles per second). According to Einstein's theory of special relativity, time would pass more quickly on Earth than it would in the moving spaceship. The special-relativity equation

$$R_a = R_f \sqrt{1 - \left(\frac{v}{c}\right)^2}$$

gives the aging rate of an astronaut, R_a, relative to the aging rate of a friend, R_f, on Earth. In this formula, v is the astronaut's speed and c is the speed of light. As the astronaut's speed approaches the speed of light, we can substitute c for v.

$R_a = R_f \sqrt{1 - \left(\frac{v}{c}\right)^2}$ Einstein's equation gives the aging rate of an astronaut, R_a, relative to the aging rate of a friend, R_f, on Earth.

$R_a = R_f \sqrt{1 - \left(\frac{c}{c}\right)^2}$ The velocity, v, is approaching the speed of light, c, so let $v = c$.

$\quad = R_f \sqrt{1 - 1}$ $\left(\frac{c}{c}\right)^2 = 1^2 = 1 \cdot 1 = 1$

$\quad = R_f \sqrt{0}$ Simplify the radicand: $1 - 1 = 0$.

$\quad = R_f \cdot 0$ $\sqrt{0} = 0$

$\quad = 0$ Multiply: $R_f \cdot 0 = 0$.

Close to the speed of light, the astronaut's aging rate, R_a, relative to that of a friend, R_f, on Earth is nearly 0. What does this mean? As we age here on Earth, the space traveler would barely get older. The space traveler would return to an unknown futuristic world in which friends and loved ones would be long gone.

Exercise Set 3.4

Concept and Vocabulary Exercises

In Exercises 1–4, fill in each blank so that the resulting statement is true.

1. The set of irrational numbers is the set of numbers whose decimal representations are neither _____ nor _____.

2. The irrational number _____ represents the circumference of a circle divided by the diameter of the circle.

3. The square root of n, represented by _____, is the positive number that when multiplied by itself gives _____.

4. The number that multiplies a square root is called the square root's _____.

In Exercises 5–8, determine whether each statement is true or false. If the statement is false, make the necessary change(s) to produce a true statement.

5. The product of any two irrational numbers is always an irrational number.

6. $\sqrt{9} + \sqrt{16} = \sqrt{25}$

7. $\sqrt{\sqrt{16}} = 2$

8. $\dfrac{\sqrt{64}}{2} = \sqrt{32}$

Respond to Exercises 9–14 using verbal or written explanations.

9. Describe the difference between a rational number and an irrational number.

10. Using $\sqrt{50}$, explain how to simplify a square root.

11. Describe how to multiply square roots.

12. Explain how to add square roots with the same radicand.

13. Explain how to add $\sqrt{3} + \sqrt{12}$.

14. Describe what it means to rationalize a denominator. Use $\dfrac{2}{\sqrt{5}}$ in your explanation.

In Exercises 15–18, determine whether each statement makes sense or does not make sense, and explain your reasoning.

15. The humor in this cartoon is based on the fact that the football will never be hiked.

FOXTROT © 2003 Bill Amend.

16. The humor in this cartoon is based on the fact that the enemy will never be charged.

HAGAR THE HORRIBLE © 1995 Johnny Hart

17. The joke in this Peanuts cartoon would be more effective if Woodstock had rationalized the denominator correctly in the last frame.

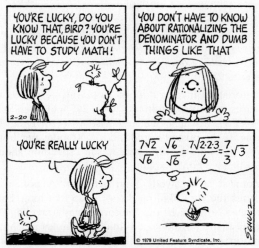

PEANUTS © United Feature Syndicate, Inc.

18. The joke in this FoxTrot cartoon would be more effective if Paige and Jason had skated the same number.

FOXTROT © 2003 Bill Amend.

Practice Exercises

Evaluate each expression in Exercises 19–28.

19. $\sqrt{9}$ **20.** $\sqrt{16}$

21. $\sqrt{25}$ **22.** $\sqrt{49}$

23. $\sqrt{64}$ **24.** $\sqrt{100}$

25. $\sqrt{121}$ **26.** $\sqrt{144}$

27. $\sqrt{169}$ **28.** $\sqrt{225}$

In Exercises 29–34, use a calculator with a square root key to find a decimal approximation for each square root. Round the number displayed to the nearest **a.** *tenth,* **b.** *hundredth,* **c.** *thousandth.*

29. $\sqrt{173}$

30. $\sqrt{3176}$

31. $\sqrt{17{,}761}$

32. $\sqrt{779{,}264}$

33. $\sqrt{\pi}$

34. $\sqrt{2\pi}$

In Exercises 35–42, simplify the square root.

35. $\sqrt{20}$ **36.** $\sqrt{50}$

37. $\sqrt{80}$ **38.** $\sqrt{12}$

39. $\sqrt{250}$ **40.** $\sqrt{192}$

41. $7\sqrt{28}$ **42.** $3\sqrt{52}$

In Exercises 43–74, perform the indicated operation. Simplify the answer when possible.

43. $\sqrt{7} \cdot \sqrt{6}$ **44.** $\sqrt{19} \cdot \sqrt{3}$

45. $\sqrt{6} \cdot \sqrt{6}$ **46.** $\sqrt{5} \cdot \sqrt{5}$

47. $\sqrt{3} \cdot \sqrt{6}$ **48.** $\sqrt{12} \cdot \sqrt{2}$

49. $\sqrt{2} \cdot \sqrt{26}$ **50.** $\sqrt{5} \cdot \sqrt{50}$

51. $\dfrac{\sqrt{54}}{\sqrt{6}}$ **52.** $\dfrac{\sqrt{75}}{\sqrt{3}}$

53. $\dfrac{\sqrt{90}}{\sqrt{2}}$ **54.** $\dfrac{\sqrt{60}}{\sqrt{3}}$

55. $\dfrac{-\sqrt{96}}{\sqrt{2}}$ **56.** $\dfrac{-\sqrt{150}}{\sqrt{3}}$

57. $7\sqrt{3} + 6\sqrt{3}$ **58.** $8\sqrt{5} + 11\sqrt{5}$

59. $4\sqrt{13} - 6\sqrt{13}$ **60.** $6\sqrt{17} - 8\sqrt{17}$

61. $\sqrt{5} + \sqrt{5}$ **62.** $\sqrt{3} + \sqrt{3}$

63. $4\sqrt{2} - 5\sqrt{2} + 8\sqrt{2}$ **64.** $6\sqrt{3} + 8\sqrt{3} - 16\sqrt{3}$

65. $\sqrt{5} + \sqrt{20}$ **66.** $\sqrt{3} + \sqrt{27}$

67. $\sqrt{50} - \sqrt{18}$ **68.** $\sqrt{63} - \sqrt{28}$

69. $3\sqrt{18} + 5\sqrt{50}$ **70.** $4\sqrt{12} + 2\sqrt{75}$

71. $\dfrac{1}{4}\sqrt{12} - \dfrac{1}{2}\sqrt{48}$ **72.** $\dfrac{1}{5}\sqrt{300} - \dfrac{2}{3}\sqrt{27}$

73. $3\sqrt{75} + 2\sqrt{12} - 2\sqrt{48}$ **74.** $2\sqrt{72} + 3\sqrt{50} - \sqrt{128}$

In Exercises 75–84, rationalize the denominator.

75. $\dfrac{5}{\sqrt{3}}$ **76.** $\dfrac{12}{\sqrt{5}}$

77. $\dfrac{21}{\sqrt{7}}$ **78.** $\dfrac{30}{\sqrt{5}}$

79. $\dfrac{12}{\sqrt{30}}$ **80.** $\dfrac{15}{\sqrt{50}}$

81. $\dfrac{15}{\sqrt{12}}$ **82.** $\dfrac{13}{\sqrt{40}}$

83. $\sqrt{\dfrac{2}{5}}$ **84.** $\sqrt{\dfrac{5}{7}}$

Practice Plus

In Exercises 85–92, perform the indicated operations. Simplify the answer when possible.

85. $3\sqrt{8} - \sqrt{32} + 3\sqrt{72} - \sqrt{75}$

86. $3\sqrt{54} - 2\sqrt{24} - \sqrt{96} + 4\sqrt{63}$

87. $3\sqrt{7} - 5\sqrt{14} \cdot \sqrt{2}$

88. $4\sqrt{2} - 8\sqrt{10} \cdot \sqrt{5}$

89. $\dfrac{\sqrt{32}}{5} + \dfrac{\sqrt{18}}{7}$ **90.** $\dfrac{\sqrt{27}}{2} + \dfrac{\sqrt{75}}{7}$

91. $\dfrac{\sqrt{2}}{\sqrt{3}} + \dfrac{\sqrt{3}}{\sqrt{2}}$ **92.** $\dfrac{\sqrt{2}}{\sqrt{7}} + \dfrac{\sqrt{7}}{\sqrt{2}}$

Application Exercises

The formula

$$d = \sqrt{\dfrac{3h}{2}}$$

models the distance, d, in miles, that a person h feet high can see to the horizon. Use this formula to solve Exercises 93–94.

93. The pool deck on a cruise ship is 72 feet above the water. How far can passengers on the pool deck see? Write the answer in simplified radical form. Then use the simplified radical form and a calculator to express the answer to the nearest tenth of a mile.

94. The captain of a cruise ship is on the star deck, which is 120 feet above the water. How far can the captain see? Write the answer in simplified radical form. Then use the simplified radical form and a calculator to express the answer to the nearest tenth of a mile.

Police use the formula $v = 2\sqrt{5L}$ to estimate the speed of a car, v, in miles per hour, based on the length, L, in feet, of its skid marks upon sudden braking on a dry asphalt road. Use the formula to solve Exercises 95–96.

95. A motorist is involved in an accident. A police officer measures the car's skid marks to be 245 feet long. Estimate the speed at which the motorist was traveling before braking. If the posted speed limit is 50 miles per hour and the motorist tells the officer he was not speeding, should the officer believe him? Explain.

96. A motorist is involved in an accident. A police officer measures the car's skid marks to be 45 feet long. Estimate the speed at which the motorist was traveling before braking. If the posted speed limit is 35 miles per hour and the motorist tells the officer she was not speeding, should the officer believe her? Explain.

97. The graph shows the median heights for boys of various ages in the United States from birth through 60 months, or five years old.

Boys' Heights

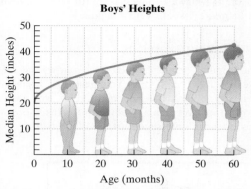

Source: Laura Walther Nathanson, *The Portable Pediatrician for Parents*

a. Use the graph to estimate the median height, to the nearest inch, of boys who are 50 months old. 41 in.

b. The formula $h = 2.9\sqrt{x} + 20.1$ models the median height, h, in inches, of boys who are x months of age. According to the formula, what is the median height of boys who are 50 months old? Use a calculator and round to the nearest tenth of an inch. How well does your estimate from part (a) describe the median height obtained from the formula?

98. The graph shows the median heights for girls of various ages in the United States from birth through 60 months, or five years old.

Girls' Heights

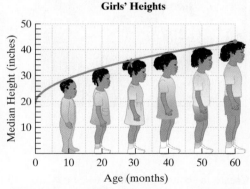

Source: Laura Walther Nathanson, *The Portable Pediatrician for Parents*

a. Use the graph to estimate the median height, to the nearest inch, of girls who are 50 months old.

b. The formula $h = 3.1\sqrt{x} + 19$ models the median height, h, in inches, of girls who are x months of age. According to the formula, what is the median height of girls who are 50 months old? Use a calculator and round to the nearest tenth of an inch. How well does your estimate from part (a) describe the median height obtained from the formula?

Autism is a neurological disorder that impedes language and derails social and emotional development. New findings suggest that the condition is not a sudden calamity that strikes children at the age of 2 or 3, but a developmental problem linked to *abnormally rapid brain growth during infancy. The graphs show that the heads of severely autistic children start out smaller than average and then go through a period of explosive growth. Exercises 99–100 involve mathematical models for the data shown by the graphs.*

Developmental Differences between Healthy Children and Severe Autistics

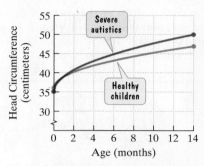

Source: The Journal of the American Medical Association

99. The data for one of the two groups shown by the graphs can be modeled by

$$H = 2.9\sqrt{x} + 36,$$

where H is the head circumference, in centimeters, at age x months, $0 \le x \le 14$.

a. According to the model, what is the head circumference at birth?

b. According to the model, what is the head circumference at 9 months?

c. According to the model, what is the head circumference at 14 months? Use a calculator and round to the nearest tenth of a centimeter.

d. Use the values that you obtained in parts (a) through (c) and the graphs shown above to determine if the given model describes healthy children or severe autistics.

100. The data for one of the two groups shown by the graphs can be modeled by

$$H = 4\sqrt{x} + 35,$$

where H is the head circumference, in centimeters, at age x months, $0 \le x \le 14$.

a. According to the model, what is the head circumference at birth?

b. According to the model, what is the head circumference at 9 months?

c. According to the model, what is the head circumference at 14 months? Use a calculator and round to the nearest centimeter.

d. Use the values that you obtained in parts (a) through (c) and the graphs shown above to determine if the given model describes healthy children or severe autistics.

The Blitzer Bonus on page 166 gives Einstein's special-relativity equation

$$R_a = R_f \sqrt{1 - \left(\frac{v}{c}\right)^2}$$

for the aging rate of an astronaut, R_a, relative to the aging rate of a friend on Earth, R_f, where v is the astronaut's speed and c is the speed of light. Take a few minutes to read the essay and then solve Exercises 101–102.

101. You are moving at 80% of the speed of light. Substitute $0.8c$ in the equation shown above. What is your aging rate relative to a friend on Earth? If 100 weeks have passed for your friend, how long were you gone?

102. You are moving at 90% of the speed of light. Substitute $0.9c$ in the equation shown on the left. What is your aging rate, correct to two decimal places, relative to a friend on Earth? If 100 weeks have passed for your friend, how long, to the nearest week, were you gone?

Critical Thinking Exercises

103. The popular comic strip *FoxTrot* follows the off-the-wall lives of the Fox family. Youngest son Jason is forever obsessed by his love of math. Use this math-themed *FoxTrot* to solve as many of the problems as possible. Then decode Jason's message shown on the first line.

FOXTROT © 2009 Bill Amend

104. Read the Blitzer Bonus on page 166. The future is now: You have the opportunity to explore the cosmos in a starship traveling near the speed of light. The experience will enable you to understand the mysteries of the universe in deeply personal ways, transporting you to unimagined levels of knowing and being. The down side: You return from your two-year journey to a futuristic world in which friends and loved ones are long gone. Do you explore space or stay here on Earth? What are the reasons for your choice?

In Exercises 105–107, insert either $<$ or $>$ in the shaded area between the numbers to make each statement true.

105. $\sqrt{2}$ ■ 1.5

106. $-\pi$ ■ -3.5

107. $-\dfrac{3.14}{2}$ ■ $-\dfrac{\pi}{2}$

108. How does doubling a number affect its square root?

109. Between which two consecutive integers is $-\sqrt{47}$?

110. Simplify: $\sqrt{2} + \sqrt{\dfrac{1}{2}}$.

111. Create a counterexample to show that the following statement is false: The difference between two distinct irrational numbers is always an irrational number.

Group Exercises

The following topics related to irrational numbers are appropriate for either individual or group research projects. A report should be given to the class on the researched topic.

112. A History of How Irrational Numbers Developed

113. Pi: Its History, Applications, and Curiosities

114. Proving That $\sqrt{2}$ Is Irrational

115. Imaginary Numbers: Their History, Applications, and Curiosities

116. The Golden Rectangle in Art and Architecture

3.5 : Real Numbers and Their Properties

Objectives

1. Recognize subsets of the real numbers.

2. Recognize properties of real numbers.

1 Recognize subsets of the real numbers.

The Set of Real Numbers

The vampire legend is death as seducer; he/she sucks our blood to take us to a perverse immortality. The vampire resembles us, but appears only at night, hidden among mortals. In this section, you will find vampires in the world of numbers. Mathematicians even use the labels *vampire* and *weird* to describe sets of numbers. However, the label that appears most frequently is *real*. The union of the rational numbers and the irrational numbers is the set of **real numbers**.

The sets that make up the real numbers are summarized in **Table 3.2**. We refer to these sets as **subsets** of the real numbers, meaning that all elements in each subset are also elements in the set of real numbers.

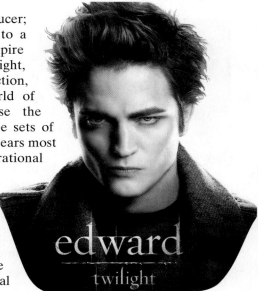

edward

twilight

Real numbers

```
┌─────────────────────────┬──────────────┐
│ Rational                │ Irrational   │
│ numbers                 │ numbers      │
│  ┌────────────────────┐ │              │
│  │ Integers           │ │              │
│  │  ┌───────────────┐ │ │              │
│  │  │ Whole         │ │ │              │
│  │  │ numbers       │ │ │              │
│  │  │  ┌──────────┐ │ │ │              │
│  │  │  │ Natural  │ │ │ │              │
│  │  │  │ numbers  │ │ │ │              │
│  │  │  └──────────┘ │ │ │              │
│  │  └───────────────┘ │ │              │
│  └────────────────────┘ │              │
└─────────────────────────┴──────────────┘
```

This diagram shows that every real number is rational or irrational.

Table 3.2 Important Subsets of the Real Numbers

Name	Description	Examples
Natural numbers	$\{1, 2, 3, 4, 5, \ldots\}$ These are the numbers that we use for counting.	$2, 3, 5, 17$
Whole numbers	$\{0, 1, 2, 3, 4, 5, \ldots\}$ The set of whole numbers includes 0 and the natural numbers.	$0, 2, 3, 5, 17$
Integers	$\{\ldots, -5, -4, -3, -2, -1, 0, 1, 2, 3, 4, 5, \ldots\}$ The set of integers includes the whole numbers and the negatives of the natural numbers.	$-17, -5, -3, -2, 0, 2, 3, 5, 17$
Rational numbers	$\left\{\dfrac{a}{b}\,\middle|\, a \text{ and } b \text{ are integers and } b \neq 0\right\}$ The set of rational numbers is the set of all numbers that can be expressed as a quotient of two integers, with the denominator not 0. Rational numbers can be expressed as terminating or repeating decimals.	$-17 = \frac{-17}{1}, -5 = \frac{-5}{1}, -3, -2,$ $0, 2, 3, 5, 17,$ $\frac{2}{5} = 0.4,$ $\frac{-2}{3} = -0.6666\ldots = -0.\overline{6}$
Irrational numbers	The set of irrational numbers is the set of all numbers whose decimal representations are neither terminating nor repeating. Irrational numbers cannot be expressed as a quotient of integers.	$\sqrt{2} \approx 1.414214$ $-\sqrt{3} \approx -1.73205$ $\pi \approx 3.142$ $-\frac{\pi}{2} \approx -1.571$

Example 1 ::: Classifying Real Numbers

Consider the following set of numbers:

$$\left\{-7, -\frac{3}{4}, 0, 0.\overline{6}, \sqrt{5}, \pi, 7.3, \sqrt{81}\right\}.$$

List the numbers in the set that are

a. natural numbers.　　**b.** whole numbers.　　**c.** integers.

d. rational numbers.　　**e.** irrational numbers.　　**f.** real numbers.

Solution

a. Natural numbers: The natural numbers are the numbers used for counting. The only natural number in the set is $\sqrt{81}$ because $\sqrt{81} = 9$. (9 multiplied by itself, or 9^2, is 81.)

b. Whole numbers: The whole numbers consist of the natural numbers and 0. The elements of the set that are whole numbers are 0 and $\sqrt{81}$.

c. Integers: The integers consist of the natural numbers, 0, and the negatives of the natural numbers. The elements of the set that are integers are $\sqrt{81}$, 0, and -7.

d. Rational numbers: All numbers in the set that can be expressed as the quotient of integers are rational numbers. These include $-7\left(-7 = \frac{-7}{1}\right)$, $-\frac{3}{4}$, $0\left(0 = \frac{0}{1}\right)$, and $\sqrt{81}\left(\sqrt{81} = \frac{9}{1}\right)$. Furthermore, all numbers in the set that are terminating or repeating decimals are also rational numbers. These include $0.\overline{6}$ and 7.3.

e. Irrational numbers: The irrational numbers in the set are $\sqrt{5}(\sqrt{5} \approx 2.236)$ and $\pi(\pi \approx 3.14)$. Both $\sqrt{5}$ and π are only approximately equal to 2.236 and 3.14, respectively. In decimal form, $\sqrt{5}$ and π neither terminate nor have blocks of repeating digits.

f. Real numbers: All the numbers in the given set are real numbers.

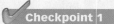 **Checkpoint 1** Consider the following set of numbers:

$$\left\{-9, -1.3, 0, 0.\overline{3}, \frac{\pi}{2}, \sqrt{9}, \sqrt{10}\right\}.$$

List the numbers in the set that are

a. natural numbers.　　　　　**b.** whole numbers.

c. integers.　　　　　　　　**d.** rational numbers.

▸ **e.** irrational numbers.　　　　**f.** real numbers.

Properties of the Real Numbers

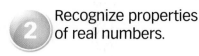 **Recognize properties of real numbers.**

When you use your calculator to add two real numbers, you can enter them in either order. The fact that two real numbers can be added in either order is called the **commutative property of addition**. You probably use this property, as well as other properties of the real numbers listed in **Table 3.3**, without giving it much thought. The properties of the real numbers are especially useful in algebra, as we shall see in Chapter 4.

Table 3.3 Properties of the Real Numbers

Name	Meaning	Examples
Closure Property of Addition	The sum of any two real numbers is a real number.	$4\sqrt{2}$ is a real number and $5\sqrt{2}$ is a real number, so $4\sqrt{2} + 5\sqrt{2}$, or $9\sqrt{2}$, is a real number.
Closure Property of Multiplication	The product of any two real numbers is a real number.	10 is a real number and $\frac{1}{2}$ is a real number, so $10 \cdot \frac{1}{2}$, or 5, is a real number.
Commutative Property of Addition	Changing order when adding does not affect the sum. $a + b = b + a$	• $13 + 7 = 7 + 13$ • $\sqrt{2} + \sqrt{5} = \sqrt{5} + \sqrt{2}$
Commutative Property of Multiplication	Changing order when multiplying does not affect the product. $ab = ba$	• $13 \cdot 7 = 7 \cdot 13$ • $\sqrt{2} \cdot \sqrt{5} = \sqrt{5} \cdot \sqrt{2}$
Associative Property of Addition	Changing grouping when adding does not affect the sum. $(a + b) + c = a + (b + c)$	$(7 + 2) + 5 = 7 + (2 + 5)$ $9 + 5 = 7 + 7$ $14 = 14$
Associative Property of Multiplication	Changing grouping when multiplying does not affect the product. $(ab)c = a(bc)$	$(7 \cdot 2) \cdot 5 = 7 \cdot (2 \cdot 5)$ $14 \cdot 5 = 7 \cdot 10$ $70 = 70$
Distributive Property of Multiplication over Addition	Multiplication distributes over addition. $a \cdot (b + c) = a \cdot b + a \cdot c$	$7(4 + \sqrt{3}) = 7 \cdot 4 + 7 \cdot \sqrt{3}$ $= 28 + 7\sqrt{3}$
Identity Property of Addition	Zero can be deleted from a sum. $a + 0 = a$ $0 + a = a$	• $\sqrt{3} + 0 = \sqrt{3}$ • $0 + \pi = \pi$
Identity Property of Multiplication	One can be deleted from a product. $a \cdot 1 = a$ $1 \cdot a = a$	• $\sqrt{3} \cdot 1 = \sqrt{3}$ • $1 \cdot \pi = \pi$
Inverse Property of Addition	The sum of a real number and its additive inverse gives 0, the additive identity. $a + (-a) = 0$ $(-a) + a = 0$	• $\sqrt{3} + \left(-\sqrt{3}\right) = 0$ • $-\pi + \pi = 0$
Inverse Property of Multiplication	The product of a nonzero real number and its multiplicative inverse gives 1, the multiplicative identity. $a \cdot \dfrac{1}{a} = 1, a \neq 0$ $\dfrac{1}{a} \cdot a = 1, a \neq 0$	• $\sqrt{3} \cdot \dfrac{1}{\sqrt{3}} = 1$ • $\dfrac{1}{\pi} \cdot \pi = 1$

Example 2   Identifying Properties of the Real Numbers

Name the property illustrated:

a. $\sqrt{3}\cdot 7 = 7\cdot\sqrt{3}$ **b.** $(4+7)+6 = 4+(7+6)$

c. $2(3+\sqrt{5}) = 6+2\sqrt{5}$ **d.** $\sqrt{2}+(\sqrt{3}+\sqrt{7}) = \sqrt{2}+(\sqrt{7}+\sqrt{3})$

e. $17+(-17)=0$ **f.** $\sqrt{2}\cdot 1 = \sqrt{2}.$

Solution

a. $\sqrt{3}\cdot 7 = 7\cdot\sqrt{3}$ Commutative property of multiplication

b. $(4+7)+6 = 4+(7+6)$ Associative property of addition

c. $2(3+\sqrt{5}) = 6+2\sqrt{5}$ Distributive property of multiplication over addition

d. $\sqrt{2}+(\sqrt{3}+\sqrt{7}) = \sqrt{2}+(\sqrt{7}+\sqrt{3})$ The only change between the left and the right sides is in the order that $\sqrt{3}$ and $\sqrt{7}$ are added. The order is changed from $\sqrt{3}+\sqrt{7}$ to $\sqrt{7}+\sqrt{3}$ using the commutative property of addition.

e. $17+(-17)=0$ Inverse property of addition

f. $\sqrt{2}\cdot 1 = \sqrt{2}$ Identity property of multiplication

✓ **Checkpoint 2**   Name the property illustrated:

a. $(4\cdot 7)\cdot 3 = 4\cdot(7\cdot 3)$

b. $3(\sqrt{5}+4) = 3(4+\sqrt{5})$

c. $3(\sqrt{5}+4) = 3\sqrt{5}+12$

d. $2(\sqrt{3}+\sqrt{7}) = (\sqrt{3}+\sqrt{7})2$

e. $1+0=1$

f. $-4\left(-\dfrac{1}{4}\right) = 1.$

Although the entire set of real numbers is closed with respect to addition and multiplication, some of the subsets of the real numbers do not satisfy the closure property for a given operation. If an operation on a set results in just one number that is not in that set, then the set is not closed for that operation.

Example 3   Verifying Closure

a. Are the integers closed with respect to multiplication?

b. Are the irrational numbers closed with respect to multiplication?

c. Are the natural numbers closed with respect to division?

Solution

a. Consider some examples of the multiplication of integers:

$$3\cdot 2 = 6 \qquad 3(-2) = -6 \qquad -3(-2) = 6 \qquad -3\cdot 0 = 0.$$

The product of any two integers is always a positive integer, a negative integer, or zero, which is an integer. Thus, the integers are closed under the operation of multiplication.

b. If we multiply two irrational numbers, must the product always be an irrational number? The answer is no. Here is an example:

$$\sqrt{7} \cdot \sqrt{7} = \sqrt{49} = 7.$$

Both irrational   Not an irrational number

This means that the irrational numbers are not closed under the operation of multiplication.

c. If we divide any two natural numbers, must the quotient always be a natural number? The answer is no. Here is an example:

$$4 \div 8 = \tfrac{1}{2}.$$

Both natural numbers   Not a natural number

Thus, the natural numbers are not closed under the operation of division.

 **Checkpoint 3**

a. Are the natural numbers closed with respect to multiplication?

▶ **b.** Are the integers closed with respect to division?

The commutative property involves a change in order with no change in the final result. However, changing the order in which we subtract and divide real numbers can produce different answers. For example,

$$7 - 4 \neq 4 - 7 \quad \text{and} \quad 6 \div 2 \neq 2 \div 6.$$

Because the real numbers are not commutative with respect to subtraction and division, it is important that you enter numbers in the correct order when using a calculator to perform these operations.

The associative property does not hold for the operations of subtraction and division. The examples below show that if we change groupings when subtracting or dividing three numbers, the answer may change.

$$(6 - 1) - 3 \neq 6 - (1 - 3) \qquad (8 \div 4) \div 2 \neq 8 \div (4 \div 2)$$
$$5 - 3 \neq 6 - (-2) \qquad\qquad 2 \div 2 \neq 8 \div 2$$
$$2 \neq 8 \qquad\qquad\qquad\qquad 1 \neq 4$$

Blitzer Bonus

Beyond the Real Numbers

Only real numbers greater than or equal to zero have real number square roots. The square root of -1, $\sqrt{-1}$, is not a real number. This is because there is no real number that can be multiplied by itself that results in -1. Multiplying any real number by itself can never give a negative product. In the sixteenth century, mathematician Girolamo Cardano (1501–1576) wrote that square roots of negative numbers would cause "mental tortures." In spite of these "tortures," mathematicians invented a new number, called i, to represent $\sqrt{-1}$. The number i is not a real number; it is called an **imaginary number**. Thus, $\sqrt{9} = 3$, $-\sqrt{9} = -3$, but $\sqrt{-9}$ is not a real number. However, $\sqrt{-9}$ is an imaginary number, represented by $3i$. The adjective *real* as a way of describing what we now call the real numbers was first used by the French mathematician and philosopher René Descartes (1596–1650) in response to the concept of imaginary numbers.

THE KID WHO LEARNED ABOUT MATH
ON THE STREET

If you divide 6,973 by 0, you die

Once, this guy tried to find the square root of -9, and his eyeballs turned black.

This girl my brother knows found out exactly what π equals, but she went nuts.

Source: © 2000 Roz Chast from Cartoonbank.com. All rights reserved.

Achieving Success

Manage your time. Use a day planner such as the one shown below. (Go online and search "day planner" or "day scheduler" to find a schedule grid that you can print and use.)

Sample Day Planner

	Monday	Tuesday	Wednesday	Thursday	Friday	Saturday	Sunday
5:00 A.M.							
6:00 A.M.							
7:00 A.M.							
8:00 A.M.							
9:00 A.M.							
10:00 A.M.							
11:00 A.M.							
12:00 P.M.							
1:00 P.M.							
2:00 P.M.							
3:00 P.M.							
4:00 P.M.							
5:00 P.M.							
6:00 P.M.							
7:00 P.M.							
8:00 P.M.							
9:00 P.M.							
10:00 P.M.							
11:00 P.M.							
Midnight							

- On the Sunday before the week begins, fill in the time slots with fixed items such as school, extra curricular activities, work, etc.
- Because your education should be a top priority, decide what times you would like to study and do homework. Fill in these activities on the day planner.
- Plan other flexible activities such as exercise, socializing, etc. around the times already established.
- Be flexible. Things may come up that are unavoidable or some items may take longer than planned. However, be honest with yourself: Playing video games with friends is not "unavoidable."
- Stick to your schedule. At the end of the week, you will look back and be impressed at all the things you have accomplished.

Exercise Set 3.5

Concept and Vocabulary Exercises

In Exercises 1–8, fill in each blank so that the resulting statement is true.

1. Every real number is either _____ or _____.

2. The _____ property of addition states that the sum of any two real numbers is a real number.

3. If a and b are real numbers, the commutative property of multiplication states that _____.

4. If $a, b,$ and c are real numbers, the associative property of addition states that _____.

5. If $a, b,$ and c are real numbers, the distributive property states that _____.

6. The _____ property of addition states that zero can be deleted from a sum.

7. The _____ property of multiplication states that _____ can be deleted from a product.

8. The product of a nonzero real number and its _____ gives 1, the _____.

In Exercises 9–16, determine whether each statement is true or false. If the statement is false, make the necessary change(s) to produce a true statement.

9. Every rational number is an integer.

10. Some whole numbers are not integers.

11. Some rational numbers are not positive.

12. Irrational numbers cannot be negative.

13. Subtraction is a commutative operation.

14. $(24 \div 6) \div 2 = 24 \div (6 \div 2)$

15. $7 \cdot a + 3 \cdot a = a \cdot (7 + 3)$

16. $2 \cdot a + 5 = 5 \cdot a + 2$

Respond to Exercises 17–24 using verbal or written explanations.

17. What does it mean when we say that the rational numbers are a subset of the real numbers?

18. What does it mean if we say that a set is closed under a given operation?

19. State the commutative property of addition and give an example.

20. State the commutative property of multiplication and give an example.

21. State the associative property of addition and give an example.

22. State the associative property of multiplication and give an example.

23. State the distributive property of multiplication over addition and give an example.

24. Does $7 \cdot (4 \cdot 3) = 7 \cdot (3 \cdot 4)$ illustrate the commutative property or the associative property? Explain your answer.

In Exercises 25–28, determine whether each statement makes sense or does not make sense, and explain your reasoning.

25. The humor in this cartoon is based on the fact that "rational" and "real" have different meanings in mathematics and in everyday speech.

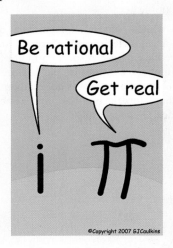

©Copyright 2007 GJCaulkins

26. The number of pages in this book is a real number.

27. The book that I'm reading on the history of π appropriately contains an irrational number of pages.

28. I took the phrase "one plus twelve," used commutative and associative ideas by changing both the order and grouping of the letters, and arrived at the phrase "two plus eleven."

Practice Exercises

In Exercises 29–32, list all numbers from the given set that are

   a. *natural numbers.*   **b.** *whole numbers.*

   c. *integers.*   **d.** *rational numbers.*

   e. *irrational numbers.*   **f.** *real numbers.*

29. $\{-9, -\frac{4}{5}, 0, 0.25, \sqrt{3}, 9.2, \sqrt{100}\}$

30. $\{-7, -0.\overline{6}, 0, \sqrt{49}, \sqrt{50}\}$

31. $\{-11, -\frac{5}{6}, 0, 0.75, \sqrt{5}, \pi, \sqrt{64}\}$

32. $\{-5, -0.\overline{3}, 0, \sqrt{2}, \sqrt{4}\}$

33. Give an example of a whole number that is not a natural number.

34. Give an example of an integer that is not a whole number.

35. Give an example of a rational number that is not an integer.

36. Give an example of a rational number that is not a natural number.

37. Give an example of a number that is an integer, a whole number, and a natural number.

38. Give an example of a number that is a rational number, an integer, and a real number.

39. Give an example of a number that is an irrational number and a real number.

40. Give an example of a number that is a real number, but not an irrational number.

Complete each statement in Exercises 41–43 to illustrate the commutative property.

41. $3 + (4 + 5) = 3 + (5 + \underline{\quad})$

42. $\sqrt{5} \cdot 4 = 4 \cdot \underline{\quad}$

43. $9 \cdot (6 + 2) = 9 \cdot (2 + \underline{\quad})$

Complete each statement in Exercises 44–45 to illustrate the associative property.

44. $(3 + 7) + 9 = \underline{\quad} + (7 + \underline{\quad})$

45. $(4 \cdot 5) \cdot 3 = \underline{\quad} \cdot (5 \cdot \underline{\quad})$

Complete each statement in Exercises 46–48 to illustrate the distributive property.

46. $3 \cdot (6 + 4) = 3 \cdot 6 + 3 \cdot \underline{\quad}$

47. $\underline{\quad} \cdot (4 + 5) = 7 \cdot 4 + 7 \cdot 5$

48. $2 \cdot (\underline{\quad} + 3) = 2 \cdot 7 + 2 \cdot 3$

Use the distributive property to simplify the radical expressions in Exercises 49–56.

49. $5(6 + \sqrt{2})$ **50.** $4(3 + \sqrt{5})$

51. $\sqrt{7}(3 + \sqrt{2})$ **52.** $\sqrt{6}(7 + \sqrt{5})$

53. $\sqrt{3}(5 + \sqrt{3})$ **54.** $\sqrt{7}(9 + \sqrt{7})$

55. $\sqrt{6}(\sqrt{2} + \sqrt{6})$ **56.** $\sqrt{10}(\sqrt{2} + \sqrt{10})$

In Exercises 57–72, state the name of the property illustrated.

57. $6 + (-4) = (-4) + 6$

58. $11 \cdot (7 + 4) = 11 \cdot 7 + 11 \cdot 4$

59. $6 + (2 + 7) = (6 + 2) + 7$

60. $6 \cdot (2 \cdot 3) = 6 \cdot (3 \cdot 2)$

61. $(2 + 3) + (4 + 5) = (4 + 5) + (2 + 3)$

62. $7 \cdot (11 \cdot 8) = (11 \cdot 8) \cdot 7$

63. $2(-8 + 6) = -16 + 12$

64. $-8(3 + 11) = -24 + (-88)$

65. $(2\sqrt{3}) \cdot \sqrt{5} = 2(\sqrt{3} \cdot \sqrt{5})$

66. $\sqrt{2}\pi = \pi\sqrt{2}$

67. $\sqrt{17} \cdot 1 = \sqrt{17}$

68. $\sqrt{17} + 0 = \sqrt{17}$

69. $\sqrt{17} + (-\sqrt{17}) = 0$

70. $\sqrt{17} \cdot \dfrac{1}{\sqrt{17}} = 1$

71. $\dfrac{1}{\sqrt{2} + \sqrt{7}} (\sqrt{2} + \sqrt{7}) = 1$

72. $(\sqrt{2} + \sqrt{7}) + -(\sqrt{2} + \sqrt{7}) = 0$

In Exercises 73–77, use two numbers to show that

73. the natural numbers are not closed with respect to subtraction.

74. the natural numbers are not closed with respect to division.

75. the integers are not closed with respect to division.

76. the irrational numbers are not closed with respect to subtraction.

77. the irrational numbers are not closed with respect to multiplication.

Practice Plus

In Exercises 78–81, determine if each statement is true or false. Do not use a calculator.

78. $468(787 + 289) = 787 + 289(468)$

79. $468(787 + 289) = 787(468) + 289(468)$

80. $58 \cdot 9 + 32 \cdot 9 = (58 + 32) \cdot 9$

81. $58 \cdot 9 \cdot 32 \cdot 9 = (58 \cdot 32) \cdot 9$

Application Exercises

In Exercises 82–85, use the definition of vampire numbers from the Blitzer Bonus on page 172 to determine which products are vampires.

82. $15 \times 93 = 1395$

83. $80 \times 86 = 6880$

84. $20 \times 51 = 1020$

85. $146 \times 938 = 136{,}948$

A **narcissistic number** *is an n-digit number equal to the sum of each of its digits raised to the nth power. Here's an example:*

$$153 = 1^3 + 5^3 + 3^3.$$

Three digits, so exponents are 3

In Exercises 86–89, determine which real numbers are narcissistic.

86. 370

87. 371

88. 372

89. 9474

90. The algebraic expressions

$$\frac{D(A + 1)}{24} \quad \text{and} \quad \frac{DA + D}{24}$$

describe the drug dosage for children between the ages of 2 and 13. In each algebraic expression, D stands for an adult dose and A represents the child's age.

a. Name the property that explains why these expressions are equal for all values of D and A.

b. If an adult dose of ibuprofen is 200 milligrams, what is the proper dose for a 12-year-old child? Use both forms of the algebraic expressions to answer the question. Which form is easier to use?

Critical Thinking Exercises

In Exercises 91–92, name the property used to go from step to step each time that (why?) occurs.

91. $7 + 2(x + 9)$

$= 7 + (2x + 18)$ (why?)

$= 7 + (18 + 2x)$ (why?)

$= (7 + 18) + 2x$ (why?)

$= 25 + 2x$

$= 2x + 25$ (why?)

92. $5(x + 4) + 3x$

$= (5x + 20) + 3x$ (why?)

$= (20 + 5x) + 3x$ (why?)

$= 20 + (5x + 3x)$ (why?)

$= 20 + (5 + 3)x$ (why?)

$= 20 + 8x$

$= 8x + 20$ (why?)

93. Closure illustrates that a characteristic of a set is not necessarily a characteristic of all of its subsets. The real numbers are closed with respect to multiplication, but the irrational numbers, a subset of the real numbers, are not. Give an example of a set that is not mathematical that has a particular characteristic, but which has a subset without this characteristic.

3.6 Exponents and Scientific Notation

Bigger than the biggest thing ever and then some. Much bigger than that in fact, really amazingly immense, a totally stunning size, real 'wow, that's big', time . . . Gigantic multiplied by colossal multiplied by staggeringly huge is the sort of concept we're trying to get across here.

Douglas Adams, *The Restaurant at the End of the Universe*

Although Adams's description may not quite apply to our $12.4 trillion national debt, exponents can be used to explore the meaning of this

"staggeringly huge" number. In this section, you will learn to use exponents to provide a way of putting large and small numbers in perspective.

Properties of Exponents

1 Use properties of exponents.

We have seen that exponents are used to indicate repeated multiplication. Now consider the multiplication of two exponential expressions, such as $b^4 \cdot b^3$. We are multiplying 4 factors of b and 3 factors of b. We have a total of 7 factors of b:

$$b^4 \cdot b^3 = (b \cdot b \cdot b \cdot b)(b \cdot b \cdot b) = b^7.$$

4 factors of b 3 factors of b Total: 7 factors of b

The product is exactly the same if we add the exponents:

$$b^4 \cdot b^3 = b^{4+3} = b^7.$$

Properties of exponents allow us to perform operations with exponential expressions without having to write out long strings of factors. Three such properties are given in **Table 3.4**.

Table 3.4 Properties of Exponents		
Property	**Meaning**	**Examples**
The Product Rule $b^m \cdot b^n = b^{m+n}$	When multiplying exponential expressions with the same base, add the exponents. Use this sum as the exponent of the common base.	$9^6 \cdot 9^{12} = 9^{6+12} = 9^{18}$
The Power Rule $(b^m)^n = b^{m \cdot n}$	When an exponential expression is raised to a power, multiply the exponents. Place the product of the exponents on the base and remove the parentheses.	$(3^4)^5 = 3^{4 \cdot 5} = 3^{20}$ $(5^3)^8 = 5^{3 \cdot 8} = 5^{24}$
The Quotient Rule $\dfrac{b^m}{b^n} = b^{m-n}$	When dividing exponential expressions with the same base, subtract the exponent in the denominator from the exponent in the numerator. Use this difference as the exponent of the common base.	$\dfrac{5^{12}}{5^4} = 5^{12-4} = 5^8$ $\dfrac{9^{40}}{9^5} = 9^{40-5} = 9^{35}$

The third property in **Table 3.4**, the quotient rule, can lead to a zero exponent when subtracting exponents. Here is an example:

$$\frac{4^3}{4^3} = 4^{3-3} = 4^0.$$

We can see what this zero exponent means by evaluating 4^3 in the numerator and the denominator:

$$\frac{4^3}{4^3} = \frac{4 \cdot 4 \cdot 4}{4 \cdot 4 \cdot 4} = \frac{64}{64} = 1.$$

This means that 4^0 must equal 1. This example illustrates the zero exponent rule.

The Zero Exponent Rule

If b is any real number other than 0,

$$b^0 = 1.$$

Example 1 **Using the Zero Exponent Rule**

Use the zero exponent rule to simplify:

a. 7^0 b. π^0 c. $(-5)^0$ d. $-5^0.$

Solution

a. $7^0 = 1$ b. $\pi^0 = 1$ c. $(-5)^0 = 1$ d. $-5^0 = -1.$

Only 5 is raised to the 0 power.

Checkpoint 1 Use the zero exponent rule to simplify:

a. 19^0 b. $(3\pi)^0$ c. $(-14)^0$ d. $-14^0.$

The quotient rule can result in a negative exponent. Consider, for example, $4^3 \div 4^5$:

$$\frac{4^3}{4^5} = 4^{3-5} = 4^{-2}.$$

We can see what this negative exponent means by evaluating the numerator and the denominator:

$$\frac{4^3}{4^5} = \frac{4 \cdot 4 \cdot 4}{4 \cdot 4 \cdot 4 \cdot 4 \cdot 4} = \frac{1}{4^2}.$$

Notice that $\frac{4^3}{4^5}$ equals both 4^{-2} and $\frac{1}{4^2}$. This means that 4^{-2} must equal $\frac{1}{4^2}$. This example is a particular case of the negative exponent rule.

Study Tip

$\frac{4^3}{4^5}$ and $\frac{4^5}{4^3}$ represent different numbers:

$$\frac{4^3}{4^5} = 4^{3-5} = 4^{-2} = \frac{1}{4^2} = \frac{1}{16}$$

$$\frac{4^5}{4^3} = 4^{5-3} = 4^2 = 16.$$

The Negative Exponent Rule

If b is any real number other than 0 and m is a natural number,

$$b^{-m} = \frac{1}{b^m}.$$

Example 2	Using the Negative Exponent Rule

Use the negative exponent rule to simplify:

a. 8^{-2} **b.** 5^{-3} **c.** 7^{-1}.

Solution

a. $8^{-2} = \dfrac{1}{8^2} = \dfrac{1}{8 \cdot 8} = \dfrac{1}{64}$ **b.** $5^{-3} = \dfrac{1}{5^3} = \dfrac{1}{5 \cdot 5 \cdot 5} = \dfrac{1}{125}$ **c.** $7^{-1} = \dfrac{1}{7^1} = \dfrac{1}{7}$

✓ **Checkpoint 2**   Use the negative exponent rule to simplify:

a. 9^{-2} **b.** 6^{-3} **c.** 12^{-1}.

Powers of Ten

Exponents and their properties allow us to represent and compute with numbers that are large or small. For example, one billion, or 1,000,000,000 can be written as 10^9. In terms of exponents, 10^9 might not look very large, but consider this: If you can count to 200 in one minute and decide to count for 12 hours a day at this rate, it would take you in the region of 19 years, 9 days, 5 hours, and 20 minutes to count to 10^9!

Powers of ten follow two basic rules:

1. **A positive exponent tells how many 0s follow the 1.** For example, 10^9 (one billion) is a 1 followed by nine zeros: 1,000,000,000. A googol, 10^{100}, is a 1 followed by one hundred zeros. (A googol far exceeds the number of protons, neutrons, and electrons in the universe.) A googol is a veritable pipsqueak compared to the googolplex, 10 raised to the googol power, or $10^{10^{100}}$; that's a 1 followed by a googol zeros. (If each zero in a googolplex were no larger than a grain of sand, there would not be enough room in the universe to represent the number.)

2. **A negative exponent tells how many places there are to the right of the decimal point.** For example, 10^{-9} (one billionth) has nine places to the right of the decimal point. The nine places include eight 0s and the 1.

$$10^{-9} = 0.\underbrace{000000001}_{\text{nine places}}$$

John Scott *"Invoking the Googolplex"* 1981, water color on paper, 24 × 18 in.
Source: Photo courtesy of The Gallery, Stratford, Ontario.

Blitzer Bonus

Earthquakes and Powers of Ten

The earthquake that ripped through northern California on October 17, 1989, measured 7.1 on the Richter scale, killed more than 60 people, and injured more than 2400. Shown here is San Francisco's Marina district, where shock waves tossed houses off their foundations and into the street.

The Richter scale is misleading because it is not actually a 1 to 8, but rather a 1 to 10 million scale. Each level indicates a tenfold increase in magnitude from the previous level, making a 7.0 earthquake a million times greater than a 1.0 quake.

The following is a translation of the Richter scale:

Richter Number (R)	Magnitude (10^{R-1})
1	$10^{1-1} = 10^0 = 1$
2	$10^{2-1} = 10^1 = 10$
3	$10^{3-1} = 10^2 = 100$
4	$10^{4-1} = 10^3 = 1000$
5	$10^{5-1} = 10^4 = 10,000$
6	$10^{6-1} = 10^5 = 100,000$
7	$10^{7-1} = 10^6 = 1,000,000$
8	$10^{8-1} = 10^7 = 10,000,000$

Table 3.5 Names of Large Numbers	
10^2	hundred
10^3	thousand
10^6	million
10^9	billion
10^{12}	trillion
10^{15}	quadrillion
10^{18}	quintillion
10^{21}	sextillion
10^{24}	septillion
10^{27}	octillion
10^{30}	nonillion
10^{100}	googol
10^{googol}	googolplex

 Convert from scientific notation to decimal notation.

Scientific Notation

In 2009, the United States government spent more than it had collected in taxes, resulting in a record budget deficit of $1.35 trillion. Put into perspective, the $1.35 trillion could pay for 40,000 players like Alex Rodriguez, whose $33 million salary in 2009 made him baseball's richest man. Because a trillion is 10^{12} (see **Table 3.5**), the 2009 budget deficit can be expressed as

$$\$1.35 \times 10^{12}.$$

The number 1.35×10^{12} is written in a form called *scientific notation*.

> ### Scientific Notation
>
> A positive number is written in **scientific notation** when it is expressed in the form
>
> $$a \times 10^n,$$
>
> where a is a number greater than or equal to 1 and less than 10 ($1 \leq a < 10$) and n is an integer.

It is customary to use the multiplication symbol, $\times$, rather than a dot, when writing a number in scientific notation.

Here are two examples of numbers in scientific notation:

- Each day, 2.6×10^7 pounds of dust from the atmosphere settle on Earth.
- The length of the AIDS virus is 1.1×10^{-4} millimeter.

We can use n, the exponent on the 10 in $a \times 10^n$, to change a number in scientific notation to decimal notation. If n is **positive**, move the decimal point in a to the **right** n places. If n is **negative**, move the decimal point in a to the **left** $|n|$ places.

Example 3 ▦ **Converting from Scientific to Decimal Notation**

Write each number in decimal notation:

a. 2.6×10^7 **b.** 1.1×10^{-4}.

Solution

In each case, we use the exponent on the 10 to move the decimal point. In part (a), the exponent is positive, so we move the decimal point to the right. In part (b), the exponent is negative, so we move the decimal point to the left.

a. $2.6 \times 10^7 = 26,000,000$

$n = 7$

Move the decimal point 7 places to the right.

b. $1.1 \times 10^{-4} = 0.00011$

$n = -4$

Move the decimal point $|-4|$ places, or 4 places, to the left.

✓ 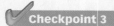 Write each number in decimal notation:

a. 7.4×10^9 **b.** 3.017×10^{-6}.

Convert from decimal notation to scientific notation.

To convert a positive number from decimal notation to scientific notation, we reverse the procedure of Example 3.

> **Converting from Decimal to Scientific Notation**
>
> Write the number in the form $a \times 10^n$.
>
> - Determine a, the numerical factor. Move the decimal point in the given number to obtain a number greater than or equal to 1 and less than 10.
> - Determine n, the exponent on 10^n. The absolute value of n is the number of places the decimal point was moved. The exponent n is positive if the given number is greater than 10 and negative if the given number is between 0 and 1.

Example 4   Converting from Decimal Notation to Scientific Notation

Write each number in scientific notation:

a. 4,600,000   **b.** 0.000023.

Solution

$$\textbf{a.}\quad 4{,}600{,}000 \;=\; 4.6 \;\times\; 10^6$$

This number is greater than 10, so n is positive in $a \times 10^n$.	Move the decimal point in 4,600,000 to get $1 \le a < 10$.	The decimal point moved 6 places from 4,600,000 to 4.6.

$$\textbf{b.}\quad 0.000023 \;=\; 2.3 \;\times\; 10^{-5}$$

This number is less than 1, so n is negative in $a \times 10^n$.	Move the decimal point in 0.000023 to get $1 \le a < 10$.	The decimal point moved 5 places from 0.000023 to 2.3.

✓ **Checkpoint 4**   Write each number in scientific notation:

a. 7,410,000,000   **b.** 0.000000092.

Example 5   Expressing the Cost of the 2009 Economic Stimulus Spending Package in Scientific Notation

The cost of President Obama's 2009 economic stimulus package was $787 billion. Express the cost in scientific notation.

Solution

Because a billion is 10^9, the cost of the bill can be expressed as

$$787 \times 10^9.$$

> This factor is not between 1 and 10, so the number is not in scientific notation.

The voice balloon indicates that we need to convert 787 to scientific notation.

$$787 \times 10^9 = (7.87 \times 10^2) \times 10^9 = 7.87 \times 10^{2+9} = 7.87 \times 10^{11}$$

> $787 = 7.87 \times 10^2$

▶ In scientific notation, the cost of the economic stimulus package was $\$7.87 \times 10^{11}$.

✓ **Checkpoint 5** As of December 2009, the population of the United States was approximately 307 million.

   a. Express the population in scientific notation.

▸   **b.** Complete the following statement: The population exceeds $\frac{3}{10}$ of a _____.

Computations with Scientific Notation

④ Perform computations using scientific notation.

We use the product rule for exponents to multiply numbers in scientific notation:

$$(a \times 10^n) \times (b \times 10^m) = (a \times b) \times 10^{n+m}.$$

Add the exponents on 10 and multiply the other parts of the numbers separately.

Example 6 | Multiplying Numbers in Scientific Notation

Multiply: $(3.4 \times 10^9)(2 \times 10^{-5})$. Write the product in decimal notation.

Solution

$$\begin{aligned}
(3.4 \times 10^9)(2 \times 10^{-5}) &= (3.4 \times 2) \times (10^9 \times 10^{-5}) && \text{Regroup factors.}\\
&= 6.8 \times 10^{9+(-5)} && \text{Add the exponents on 10}\\
&&& \text{and multiply the other parts.}\\
&= 6.8 \times 10^4 && \text{Simplify.}\\
&= 68{,}000 && \text{Write the product in decimal}\\
&&& \text{notation.}
\end{aligned}$$

✓ **Checkpoint 6** Multiply: $(1.3 \times 10^7)(4 \times 10^{-2})$. Write the product in
▸ decimal notation.

We use the quotient rule for exponents to divide numbers in scientific notation:

$$\frac{a \times 10^n}{b \times 10^m} = \left(\frac{a}{b}\right) \times 10^{n-m}.$$

Subtract the exponents on 10 and divide the other parts of the numbers separately.

Example 7 | Dividing Numbers in Scientific Notation

Divide: $\dfrac{8.4 \times 10^{-7}}{4 \times 10^{-4}}$. Write the quotient in decimal notation.

Solution

$$\frac{8.4 \times 10^{-7}}{4 \times 10^{-4}} = \left(\frac{8.4}{4}\right) \times \left(\frac{10^{-7}}{10^{-4}}\right)$$ Regroup factors.

$$= 2.1 \times 10^{-7-(-4)}$$ Subtract the exponents on 10 and divide the other parts.

$$= 2.1 \times 10^{-3}$$ Simplify: $-7 - (-4) = -7 + 4 = -3$.

$$= 0.0021$$ Write the quotient in decimal notation.

✔ **Checkpoint 7** Divide: $\dfrac{6.9 \times 10^{-8}}{3 \times 10^{-2}}$. Write the quotient in decimal notation.

Multiplication and division involving very large or very small numbers can be performed by first converting each number to scientific notation.

Example 8   Using Scientific Notation to Multiply

Multiply: $0.00064 \times 9{,}400{,}000{,}000$. Express the product in **a.** scientific notation and **b.** decimal notation.

Solution

a. $0.00064 \times 9{,}400{,}000{,}000$

$$= 6.4 \times 10^{-4} \times 9.4 \times 10^{9}$$ Write each number in scientific notation.

$$= (6.4 \times 9.4) \times (10^{-4} \times 10^{9})$$ Regroup factors.

$$= 60.16 \times 10^{-4+9}$$ Add the exponents on 10 and multiply the other parts.

$$= 60.16 \times 10^{5}$$ Simplify.

$$= (6.016 \times 10) \times 10^{5}$$ Express 60.16 in scientific notation.

$$= 6.016 \times 10^{6}$$ Add exponents on 10: $10^{1} \times 10^{5} = 10^{1+5} = 10^{6}$.

b. The answer in decimal notation is obtained by moving the decimal point in 6.016 six places to the right. The product is 6,016,000.

✔ **Checkpoint 8**   Multiply: $0.0036 \times 5{,}200{,}000$. Express the product in **a.** scientific notation and **b.** decimal notation.

Applications: Putting Numbers in Perspective

5 Solve applied problems using scientific notation.

Due to tax cuts and spending increases, the United States began accumulating large deficits in the 1980s. To finance the deficit, the government had borrowed $12.3 trillion as of December 2009. The graph in **Figure 3.8** shows the national debt increasing over time.

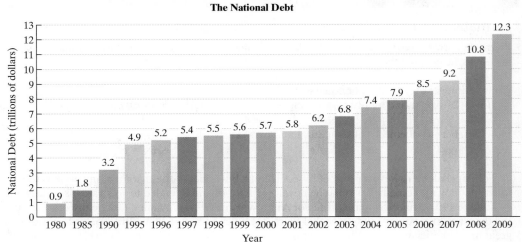

Figure 3.8
Source: Office of Management and Budget

Example 9 shows how we can use scientific notation to comprehend the meaning of a number such as 12.3 trillion.

Example 9 — The National Debt

As of December 2009, the national debt was \$12.3 trillion, or 12.3×10^{12} dollars. At that time, the U.S. population was approximately 307,000,000 (307 million), or 3.07×10^8. If the national debt was evenly divided among every individual in the United States, how much would each citizen have to pay?

Solution

Although it is not necessary to do so, let's express the total debt, 12.3×10^{12} dollars in scientific notation as 1.23×10^{13}. The amount each citizen must pay is the total debt, 1.23×10^{13} dollars, divided by the number of citizens, 3.07×10^8.

$$\frac{1.23 \times 10^{13}}{3.07 \times 10^8} = \left(\frac{1.23}{3.07}\right) \times \left(\frac{10^{13}}{10^8}\right)$$
$$\approx 0.401 \times 10^{13-8}$$
$$= 0.401 \times 10^5$$
$$= (4.01 \times 10^{-1}) \times 10^5$$
$$= 4.01 \times 10^4$$
$$= 40{,}100$$

Every U.S. citizen would have to pay approximately \$40,100 to the federal government to pay off the national debt.

If a number is written in scientific notation, $a \times 10^n$, the digits in a are called **significant digits**.

National Debt: 1.23×10^{13} U.S. Population: 3.07×10^8

Three significant digits   Three significant digits

Because these were the given numbers in Example 9, we rounded the answer, 4.01×10^4, to three significant digits. When multiplying or dividing in scientific notation where rounding is necessary and rounding instructions are not given, **round the scientific notation answer to the least number of significant digits found in any of the given numbers**.

✓ **Checkpoint 9**   In Example 5, we saw that the cost of the 2009 economic stimulus package was \$787 billion, or 7.87×10^{11} dollars. If this cost was evenly divided among every individual in the United States (approximately 3.07×10^8 people), how much would each citizen have to pay?

Blitzer Bonus

Fermat's Last Theorem

Pierre de Fermat (1601–1665) was a lawyer who enjoyed studying mathematics. In a margin of one of his books, he claimed that no natural numbers satisfy
$$a^n + b^n = c^n$$
if n is an integer greater than or equal to 3.

If $n = 2$, we can find numbers satisfying $a^n + b^n = c^n$, or $a^2 + b^2 = c^2$. For example,
$$3^2 + 4^2 = 5^2.$$
However, Fermat claimed that no natural numbers satisfy
$$a^3 + b^3 = c^3, \qquad a^4 + b^4 = c^4, \qquad a^5 + b^5 = c^5,$$
and so on.

Fermat claimed to have a proof of his conjecture, but added, "The margin of my book is too narrow to write it down." Some believe that he never had a proof and intended to frustrate his colleagues.

In June 1993, Princeton math professor Andrew Wiles (1953–) claimed that he discovered a proof of the theorem. Subsequent study revealed flaws, but Wiles corrected them. In 1995, his final proof served as a classic example of how great mathematicians accomplish great things: a combination of genius, hard work, frustration, and trial and error.

Andrew Wiles

Exercise Set 3.6

Concept and Vocabulary Exercises

In Exercises 1–5, fill in each blank so that the resulting statement is true.

1. When multiplying expressions with the same base, _____ the exponents.

2. When an exponential expression is raised to a power, _____ the exponents.

3. When dividing exponential expressions with the same base, _____ the exponents.

4. Any nonzero real number raised to the zero power is equal to _____.

5. A positive number is written in scientific notation when the first factor is _____ and the second factor is _____.

In Exercises 6–10, determine whether each statement is true or false. If the statement is false, make the necessary change(s) to produce a true statement.

6. $2^3 \cdot 2^5 = 4^8$ 7. $\dfrac{10^8}{10^4} = 10^2$ 8. $5^{-2} = -5^2$

9. A trillion is one followed by 12 zeros.

10. According to *Mother Jones* magazine, sending all 2009 U.S. high school graduates to private colleges would cost $347 billion. Because a billion is 10^9, the cost in scientific notation is 347×10^9 dollars.

Respond to Exercises 11–20 using verbal or written explanations.

11. Explain the product rule for exponents. Use $2^3 \cdot 2^5$ in your explanation.

12. Explain the power rule for exponents. Use $(3^2)^4$ in your explanation.

13. Explain the quotient rule for exponents. Use $\dfrac{5^8}{5^2}$ in your explanation.

14. Explain the zero exponent rule and give an example.

15. Explain the negative exponent rule and give an example.

16. How do you know if a number is written in scientific notation?

17. Explain how to convert from scientific to decimal notation and give an example.

18. Explain how to convert from decimal to scientific notation and give an example.

19. Suppose you are looking at a number in scientific notation. Describe the size of the number you are looking at if the exponent on ten is **a.** positive, **b.** negative, **c.** zero.

20. Describe one advantage of expressing a number in scientific notation over decimal notation.

In Exercises 21–24, determine whether each statement makes sense or does not make sense, and explain your reasoning.

21. If 5^{-2} is raised to the third power, the result is a number between 0 and 1.

22. The expression $\dfrac{a^n}{b^0}$ is undefined because division by 0 is undefined.

23. For a recent year, total tax collections in the United States were 2.02×10^5.

24. I just finished reading a book that contained approximately 1.04×10^5 words.

Practice Exercises

In Exercises 25–36, use properties of exponents to simplify each expression. First express the answer in exponential form. Then evaluate the expression.

25. $2^2 \cdot 2^3$ **26.** $3^3 \cdot 3^2$ **27.** $4 \cdot 4^2$

28. $5 \cdot 5^2$ **29.** $(2^2)^3$ **30.** $(3^3)^2$

31. $(1^4)^5$ **32.** $(1^3)^7$ **33.** $\dfrac{4^7}{4^5}$

34. $\dfrac{6^7}{6^5}$ **35.** $\dfrac{2^8}{2^4}$ **36.** $\dfrac{3^8}{3^4}$

In Exercises 37–48, use the zero and negative exponent rules to simplify each expression.

37. 3^0 **38.** 9^0 **39.** $(-3)^0$

40. $(-9)^0$ **41.** -3^0 **42.** -9^0

43. 2^{-2} **44.** 3^{-2} **45.** 4^{-3}

46. 2^{-3} **47.** 2^{-5} **48.** 2^{-6}

In Exercises 49–54, use properties of exponents to simplify each expression. First express the answer in exponential form. Then evaluate the expression.

49. $3^4 \cdot 3^{-2}$ **50.** $2^5 \cdot 2^{-2}$ **51.** $3^{-3} \cdot 3$

52. $2^{-3} \cdot 2$ **53.** $\dfrac{2^3}{2^7}$ **54.** $\dfrac{3^4}{3^7}$

In Exercises 55–70, express each number in decimal notation.

55. 2.7×10^2 **56.** 4.7×10^3

57. 9.12×10^5 **58.** 8.14×10^4

59. 8×10^7 **60.** 7×10^6

61. 1×10^5 **62.** 1×10^8

63. 7.9×10^{-1} **64.** 8.6×10^{-1}

65. 2.15×10^{-2} **66.** 3.14×10^{-2}

67. 7.86×10^{-4} **68.** 4.63×10^{-5}

69. 3.18×10^{-6} **70.** 5.84×10^{-7}

In Exercises 71–90, express each number in scientific notation.

71. 370 **72.** 530

73. 3600 **74.** 2700

75. 32,000 **76.** 64,000

77. 220,000,000 **78.** 370,000,000,000

79. 0.027 **80.** 0.014

81. 0.0037 **82.** 0.00083

83. 0.00000293 **84.** 0.000000647

85. 820×10^5 **86.** 630×10^8

87. 0.41×10^6 **88.** 0.57×10^9

89. 2100×10^{-9} **90.** $97,000 \times 10^{-11}$

In Exercises 91–104, perform the indicated operation and express each answer in decimal notation.

91. $(2 \times 10^3)(3 \times 10^2)$ **92.** $(5 \times 10^2)(4 \times 10^4)$

93. $(2 \times 10^9)(3 \times 10^{-5})$ **94.** $(4 \times 10^8)(2 \times 10^{-4})$

95. $(4.1 \times 10^2)(3 \times 10^{-4})$ **96.** $(1.2 \times 10^3)(2 \times 10^{-5})$

97. $\dfrac{12 \times 10^6}{4 \times 10^2}$ **98.** $\dfrac{20 \times 10^{20}}{10 \times 10^{15}}$

99. $\dfrac{15 \times 10^4}{5 \times 10^{-2}}$ **100.** $\dfrac{18 \times 10^2}{9 \times 10^{-3}}$

101. $\dfrac{6 \times 10^3}{2 \times 10^5}$ **102.** $\dfrac{8 \times 10^4}{2 \times 10^7}$

103. $\dfrac{6.3 \times 10^{-6}}{3 \times 10^{-3}}$ **104.** $\dfrac{9.6 \times 10^{-7}}{3 \times 10^{-3}}$

In Exercises 105–114, perform the indicated operation by first expressing each number in scientific notation. Write the answer in scientific notation.

105. $(82,000,000)(3,000,000,000)$

106. $(94,000,000)(6,000,000,000)$

107. $(0.0005)(6,000,000)$

108. $(0.000015)(0.004)$

109. $\dfrac{9,500,000}{500}$ **110.** $\dfrac{30,000}{0.0005}$

111. $\dfrac{0.00008}{200}$ **112.** $\dfrac{0.0018}{0.0000006}$

113. $\dfrac{480,000,000,000}{0.00012}$ **114.** $\dfrac{0.000000096}{16,000}$

Practice Plus

In Exercises 115–118, perform the indicated operations. Express each answer as a fraction reduced to its lowest terms.

115. $\dfrac{2^4}{2^5} + \dfrac{3^3}{3^5}$ **116.** $\dfrac{3^5}{3^6} + \dfrac{2^3}{2^6}$

117. $\dfrac{2^6}{2^4} - \dfrac{5^4}{5^6}$ **118.** $\dfrac{5^6}{5^4} - \dfrac{2^4}{2^6}$

In Exercises 119–122, perform the indicated computations. Express answers in scientific notation.

119. $(5 \times 10^3)(1.2 \times 10^{-4}) \div (2.4 \times 10^2)$

120. $(2 \times 10^2)(2.6 \times 10^{-3}) \div (4 \times 10^3)$

121. $\dfrac{(1.6 \times 10^4)(7.2 \times 10^{-3})}{(3.6 \times 10^8)(4 \times 10^{-3})}$

122. $\dfrac{(1.2 \times 10^6)(8.7 \times 10^{-2})}{(2.9 \times 10^6)(3 \times 10^{-3})}$

Application Exercises

We have seen that in 2009, the United States government spent more than it had collected in taxes, resulting in a record budget deficit of $1.35 trillion. In Exercises 123–126, you will use scientific notation to put a number like 1.35 trillion in perspective.

123. a. Express 1.35 trillion in scientific notation.

b. Express the 2009 U.S. population, 307 million, in scientific notation.

c. Use your scientific notation answers from parts (a) and (b) to answer this question: If the 2009 budget deficit was evenly divided among every individual in the United States, how much would each citizen have to pay? Express the answer in scientific and decimal notations.

124. a. Express 1.35 trillion in scientific notation.

b. A trip around the world at the Equator is approximately 25,000 miles. Express this number in scientific notation.

c. Use your scientific notation answers from parts (a) and (b) to answer this question: How many times can you circle the world at the Equator by traveling 1.35 trillion miles?

125. If there are approximately 3.2×10^7 seconds in a year, approximately how many years is 1.35 trillion seconds? (*Note:* 1.35 trillion seconds would take us back in time to a period when Neanderthals were using stones to make tools.)

126. The Washington Monument, overlooking the U.S. Capitol, stands about 555 feet tall. Stacked end to end, how many monuments would it take to reach 1.35 trillion feet? (*Note:* That's more than twice the distance from Earth to the sun.)

The bar graph shows the U.S. student population in public schools, in millions, and the amount the U.S. government spent on public education, in billions of dollars, for five selected years. Exercises 127–128 are based on the numbers displayed by the graph.

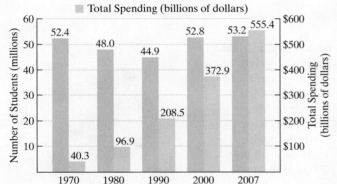

United States Public Schools: Students and Spending

Source: U.S. Department of Education

127. a. Express the 2007 student population in scientific notation.

b. Express the amount that the U.S. government spent on public education in 2007 in scientific notation.

c. Use your scientific notation answers from parts (a) and (b) to determine the amount that the U.S. government spent per pupil on public education in 2007. Express the answer in scientific and decimal notations.

128. a. Express the 2000 student population in scientific notation.

b. Express the amount that the U.S. government spent on public education in 2000 in scientific notation.

c. Use your scientific notation answers from parts (a) and (b) to determine the amount that the U.S. government spent per pupil on public education in 2000. Express the answer in scientific and decimal notations.

The bar graph at the top of the next column quantifies our love for movies by showing the number of tickets sold, in millions, and the average price per ticket for five selected years. Exercises 129–130 are based on the numbers displayed by the graph. Do not round the answers.

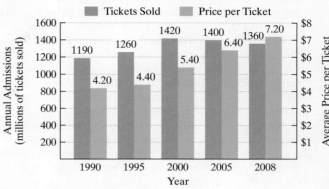

United States Film Admissions and Admission Charges

Source: National Association of Theater Owners

129. Use scientific notation to compute the amount of money that the motion picture industry made from box-office receipts in 2008. Express the answer in scientific notation.

130. Use scientific notation to compute the amount of money that the motion picture industry made from box office receipts in 2005. Express the answer in scientific notation.

131. The mass of one oxygen molecule is 5.3×10^{-23} gram. Find the mass of 20,000 molecules of oxygen. Express the answer in scientific notation.

132. The mass of one hydrogen atom is 1.67×10^{-24} gram. Find the mass of 80,000 hydrogen atoms. Express the answer in scientific notation.

Critical Thinking Exercises

In Exercises 133–139, determine whether each statement is true or false. If the statement is false, make the necessary change(s) to produce a true statement.

133. $4^{-2} < 4^{-3}$

134. $5^{-2} > 2^{-5}$

135. $5^2 \cdot 5^{-2} > 2^5 \cdot 2^{-5}$

136. $534.7 = 5.347 \times 10^3$

137. $\dfrac{8 \times 10^{30}}{4 \times 10^{-5}} = 2 \times 10^{25}$

138. $(7 \times 10^5) + (2 \times 10^{-3}) = 9 \times 10^2$

139. $(4 \times 10^3) + (3 \times 10^2) = 43 \times 10^2$

140. Give an example of a number for which there is no advantage to using scientific notation instead of decimal notation. Explain why this is the case.

141. The mad Dr. Frankenstein has gathered enough bits and pieces (so to speak) for $2^{-1} + 2^{-2}$ of his creature-to-be. Write a fraction that represents the amount of his creature that must still be obtained.

Technology Exercises

142. Use a calculator in fraction mode to check your answers in Exercises 43–48.

143. Use a calculator to check any three of your answers in Exercises 55–70.

144. Use a calculator to check any three of your answers in Exercises 71–90.

145. Use a calculator with an $\boxed{\text{EE}}$ or $\boxed{\text{EXP}}$ key to check any four of your computations in Exercises 91–114. Display the result of the computation in scientific notation and in decimal notation.

Group Exercises

146. Putting Numbers into Perspective. A large number can be put into perspective by comparing it with another number. For example, we put the $12.3 trillion national debt (Example 9) into perspective by comparing this number to the number of U.S. citizens. In Exercises 123–126, we put the $1.35 trillion budget deficit into perspective by comparing 1.35 trillion to the number of U.S. citizens, the distance around the world, the number of seconds in a year, and the height of the Washington Monument.

For this project, each group member should consult an almanac, a newspaper, or the Internet to find a number greater than one million. Explain to other members of the group the context in which the large number is used. Express the number in scientific notation. Then put the number into perspective by comparing it with another number.

147. (Refer to the Blitzer Bonus on page 186.) Fermat's most notorious theorem baffled the greatest minds for more than three centuries. In 1995, after ten years of work, Princeton University's Andrew Wiles proved Fermat's Last Theorem. *People* magazine put him on its list of "the 25 most intriguing people of the year," the Gap asked him to model jeans, and Barbara Walters chased him for an interview. "Who's Barbara Walters?" asked the bookish Wiles, who had somehow gone through life without a television.

Using the 1993 PBS documentary *Solving Fermat: Andrew Wiles* or information about Andrew Wiles on the Internet, research and present a group seminar on what Wiles did to prove Fermat's Last Theorem, the problems he encountered along the way, and how Wiles overcame those problems.

3.7 Arithmetic and Geometric Sequences

Objectives

1. Write terms of an arithmetic sequence.

2. Use the formula for the general term of an arithmetic sequence.

3. Write terms of a geometric sequence.

4. Use the formula for the general term of a geometric sequence.

Blitzer Bonus

Fibonacci Numbers on the Piano Keyboard

One Octave

Numbers in the Fibonacci sequence can be found in an octave on the piano keyboard. The octave contains 2 black keys in one cluster, 3 black keys in another cluster, a total of 5 black keys, 8 white keys, and a total of 13 keys altogether. The numbers 2, 3, 5, 8, and 13 are the third through seventh terms of the Fibonacci sequence.

Sequences

Many creations in nature involve intricate mathematical designs, including a variety of spirals. For example, the arrangement of the individual florets in the head of a sunflower forms spirals. In some species, there are 21 spirals in the clockwise direction and 34 in the counterclockwise direction. The precise numbers depend on the species of sunflower: 21 and 34, or 34 and 55, or 55 and 89, or even 89 and 144.

This observation becomes even more interesting when we consider a sequence of numbers investigated by Leonardo of Pisa, also known as Fibonacci, an Italian mathematician of the thirteenth century. The **Fibonacci sequence** of numbers is an infinite sequence that begins as follows:

$$1, 1, 2, 3, 5, 8, 13, 21, 34, 55, 89, 144, 233, \ldots.$$

The first two terms are 1. Every term thereafter is the sum of the two preceding terms. For example, the third term, 2, is the sum of the first and second terms: $1 + 1 = 2$. The fourth term, 3, is the sum of the second and third terms: $1 + 2 = 3$, and so on. Did you know that the number of spirals in a daisy or a sunflower, 21 and 34, are two Fibonacci numbers? The number of spirals in a pinecone, 8 and 13, and a pineapple, 8 and 13, are also Fibonacci numbers.

We can think of a **sequence** as a list of numbers that are related to each other by a rule. The numbers in a sequence are called its **terms**. The letter a with a subscript is used to represent the terms of a sequence. Thus, a_1 represents the first term of the sequence, a_2 represents the second term, a_3 the third term, and so on. This notation is shown for the first six terms of the Fibonacci sequence:

$$1, \quad 1, \quad 2, \quad 3, \quad 5, \quad 8.$$

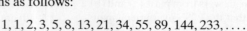

$a_1 = 1$ $a_2 = 1$ $a_3 = 2$ $a_4 = 3$ $a_5 = 5$ $a_6 = 8$

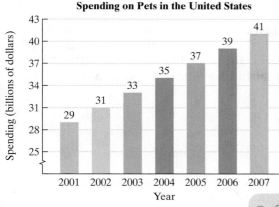

Spending on Pets in the United States

Figure 3.9
Source: American Pet Products Manufacturers Association

Arithmetic Sequences

The bar graph in **Figure 3.9** shows how much Americans spent on their pets, rounded to the nearest billion dollars, each year from 2001 through 2007.

The graph illustrates that each year spending increased by $2 billion. The sequence of annual spending

$$29, 31, 33, 35, 37, 39, 41, \ldots$$

shows that each term after the first, 29, differs from the preceding term by a constant amount, namely 2. This sequence is an example of an *arithmetic sequence*.

> **Definition of an Arithmetic Sequence**
>
> An **arithmetic sequence** is a sequence in which each term after the first differs from the preceding term by a constant amount. The difference between consecutive terms is called the **common difference** of the sequence.

The common difference, d, is found by subtracting any term from the term that directly follows it. In the following examples, the common difference is found by subtracting the first term from the second term: $a_2 - a_1$.

Arithmetic Sequence	Common Difference
$29, 31, 33, 35, 37, \ldots$	$d = 31 - 29 = 2$
$-5, -2, 1, 4, 7, \ldots$	$d = -2 - (-5) = -2 + 5 = 3$
$8, 3, -2, -7, -12, \ldots$	$d = 3 - 8 = -5$

If the first term of an arithmetic sequence is a_1, each term after the first is obtained by adding d, the common difference, to the previous term.

1 Write terms of an arithmetic sequence.

Example 1 **Writing the Terms of an Arithmetic Sequence**

Write the first six terms of the arithmetic sequence with first term 6 and common difference 4.

Solution

The first term is 6. The second term is $6 + 4$, or 10. The third term is $10 + 4$, or 14, and so on. The first six terms are

$$6, 10, 14, 18, 22, \text{ and } 26.$$

✓ **Checkpoint 1** Write the first six terms of the arithmetic sequence with first term 100 and common difference 20.

Example 2 **Writing the Terms of an Arithmetic Sequence**

Write the first six terms of the arithmetic sequence with $a_1 = 5$ and $d = -2$.

Solution

The first term, a_1, is 5. The common difference, d, is -2. To find the second term, we add -2 to 5, giving 3. For the next term, we add -2 to 3, and so on. The first six terms are
$$5, 3, 1, -1, -3, \text{ and } -5.$$

✓ **Checkpoint 2** Write the first six terms of the arithmetic sequence with $a_1 = 8$ and $d = -3$.

2 Use the formula for the general term of an arithmetic sequence.

The General Term of an Arithmetic Sequence

Consider an arithmetic sequence whose first term is a_1 and whose common difference is d. We are looking for a formula for the general term, a_n. Let's begin by writing the first six terms. The first term is a_1. The second term is $a_1 + d$. The third term is $a_1 + d + d$, or $a_1 + 2d$. Thus, we start with a_1 and add d to each successive term. The first six terms are

$$a_1, \quad a_1 + d, \quad a_1 + 2d, \quad a_1 + 3d, \quad a_1 + 4d, \quad a_1 + 5d.$$

| a_1, first term | a_2, second term | a_3, third term | a_4, fourth term | a_5, fifth term | a_6, sixth term |

Applying inductive reasoning to the pattern of the terms results in the following formula for the general term, or the nth term, of an arithmetic sequence:

> ### General Term of an Arithmetic Sequence
> The nth term (the general term) of an arithmetic sequence with first term a_1 and common difference d is
> $$a_n = a_1 + (n - 1)d.$$

Example 3 — Using the Formula for the General Term of an Arithmetic Sequence

Find the eighth term of the arithmetic sequence whose first term is 4 and whose common difference is -7.

Solution

To find the eighth term, a_8, we replace n in the formula with 8, a_1 with 4, and d with -7.
$$a_n = a_1 + (n - 1)d$$
$$a_8 = 4 + (8 - 1)(-7) = 4 + 7(-7) = 4 + (-49) = -45$$

The eighth term is -45. We can check this result by writing the first eight terms of the sequence:

$$4, -3, -10, -17, -24, -31, -38, -45.$$

✓ **Checkpoint 3** Find the ninth term of the arithmetic sequence whose first term is 6 and whose common difference is -5.

In Chapter 1, we saw that the process of finding formulas to describe real-world phenomena is called mathematical modeling. Such formulas, together with the meaning assigned to the variables, are called mathematical models. Example 4 illustrates how the formula for the general term of an arithmetic sequence can be used to develop a mathematical model.

Example 4 — Using an Arithmetic Sequence to Model the Shifting Gender of U.S. Car Drivers

The graph in **Figure 3.10** shows that in 2002, 49.8% of car drivers in the United States were women. On average, the percentage of female drivers has increased by approximately 0.08 each year.

a. Write a formula for the nth term of the arithmetic sequence that models the percentage of drivers who were women n years after 2001.

b. Use the model from part (a) to project the percentage of drivers who will be women in 2021.

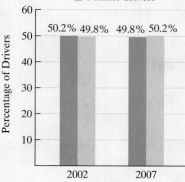

Percentage of United States Car Drivers, by Gender
- ■ Male drivers
- ■ Female drivers

50.2% 49.8% 49.8% 50.2%

Figure 3.10
Source: Federal Highway Administration

Solution

a. We can express the percentage of drivers who were women with the following arithmetic sequence:

49.8,   49.88,   49.96, … .

a_1: 49.8% of drivers were women in 2002, 1 year after 2001.	a_2: 49.88% (49.8 + 0.08) of drivers were women in 2003, 2 years after 2001.	a_3: 49.96% (49.88 + 0.08) of drivers were women in 2004, 3 years after 2001.

In this sequence, a_1, the first term, represents the percentage of drivers who were women in 2002. Each subsequent year this percentage increased by 0.08, so $d = 0.08$. We use the formula for the general term of an arithmetic sequence to write the nth term of the sequence that models the percentage of drivers who were women n years after 2001.

$a_n = a_1 + (n - 1)d$ This is the formula for the general term of an arithmetic sequence.

$a_n = 49.8 + (n - 1)0.08$ $a_1 = 49.8$ and $d = 0.08$.

$a_n = 49.8 + 0.08n - 0.08$ Distribute 0.08 to each term in parentheses.

$a_n = 0.08n + 49.72$ Simplify: $49.8 - 0.08 = 49.72$.

Thus, the percentage of drivers who were women n years after 2001 can be modeled by $a_n = 0.08n + 49.72$.

b. Now we need to project the percentage of drivers who will be women in 2021. The year 2021 is 20 years after 2001: that is, $2021 - 2001 = 20$. We substitute 20 for in $a_n = 0.08n + 49.72$.

$$a_{20} = 0.08(20) + 49.72 = 1.6 + 49.72 = 51.32$$

The 20th term of the sequence is 51.32. Therefore, the model projects that 51.32% of drivers will be women in 2021.

Checkpoint 4   The bar graph in **Figure 3.10** shows that in 2002, 50.2% of car drivers in the United States were men. On average, the percentage of male drivers has decreased by approximately 0.08 each year.

a. Write a formula for the nth term of the arithmetic sequence that models the percentage of drivers who were men n years after 2001.

b. Use the model from part (a) to project the percentage of drivers who will be men in 2011.

Geometric Sequences

Figure 3.11 shows a sequence in which the number of squares is increasing. From left to right, the number of squares is 1, 5, 25, 125, and 625. In this sequence, each term after the first, 1, is obtained by multiplying the preceding term by a constant amount, namely 5. This sequence of increasing numbers of squares is an example of a *geometric sequence*.

Figure 3.11 A geometric sequence of squares

> **Definition of a Geometric Sequence**
>
> A **geometric sequence** is a sequence in which each term after the first is obtained by multiplying the preceding term by a fixed nonzero constant. The amount by which we multiply each time is called the **common ratio** of the sequence.

The common ratio, r, is found by dividing any term after the first term by the term that directly precedes it. In the examples below, the common ratio is found by dividing the second term by the first term: $\dfrac{a_2}{a_1}$.

Geometric Sequence	**Common Ratio**
$1, 5, 25, 125, 625, \ldots$	$r = \frac{5}{1} = 5$
$4, 8, 16, 32, 64, \ldots$	$r = \frac{8}{4} = 2$
$6, -12, 24, -48, 96, \ldots$	$r = \frac{-12}{6} = -2$
$9, -3, 1, -\frac{1}{3}, \frac{1}{9}, \ldots$	$r = \frac{-3}{9} = -\frac{1}{3}$

> **Study Tip**
>
> When the common ratio of a geometric sequence is negative, the signs of the terms alternate.

3 Write terms of a geometric sequence.

How do we write out the terms of a geometric sequence when the first term and the common ratio are known? We multiply the first term by the common ratio to get the second term, multiply the second term by the common ratio to get the third term, and so on.

Example 5   Writing the Terms of a Geometric Sequence

Write the first six terms of the geometric sequence with first term 6 and common ratio $\frac{1}{3}$.

Solution

The first term is 6. The second term is $6 \cdot \frac{1}{3}$, or 2. The third term is $2 \cdot \frac{1}{3}$, or $\frac{2}{3}$. The fourth term is $\frac{2}{3} \cdot \frac{1}{3}$, or $\frac{2}{9}$, and so on. The first six terms are

$$6, 2, \tfrac{2}{3}, \tfrac{2}{9}, \tfrac{2}{27}, \text{ and } \tfrac{2}{81}.$$

✓ **Checkpoint 5**   Write the first six terms of the geometric sequence with first term 12 and common ratio $-\frac{1}{2}$.

The General Term of a Geometric Sequence

4 Use the formula for the general term of a geometric sequence.

Consider a geometric sequence whose first term is a_1 and whose common ratio is r. We are looking for a formula for the general term, a_n. Let's begin by writing the first six terms. The first term is a_1. The second term is $a_1 r$. The third term is $a_1 r \cdot r$, or $a_1 r^2$. The fourth term is $a_1 r^2 \cdot r$, or $a_1 r^3$, and so on. Starting with a_1 and multiplying each successive term by r, the first six terms are

$$a_1, \qquad a_1 r, \qquad a_1 r^2, \qquad a_1 r^3, \qquad a_1 r^4, \qquad a_1 r^5.$$

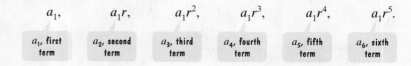

Applying inductive reasoning to the pattern of the terms results in the following formula for the general term, or the nth term, of a geometric sequence:

> **General Term of a Geometric Sequence**
>
> The nth term (the general term) of a geometric sequence with first term a_1 and common ratio r is
>
> $$a_n = a_1 r^{n-1}.$$

Example 6 — Using the Formula for the General Term of a Geometric Sequence

Find the eighth term of the geometric sequence whose first term is -4 and whose common ratio is -2.

Solution

To find the eighth term, a_8, we replace n in the formula with 8, a_1 with -4, and r with -2.

$$a_n = a_1 r^{n-1}$$
$$a_8 = -4(-2)^{8-1} = -4(-2)^7 = -4(-128) = 512$$

The eighth term is 512. We can check this result by writing the first eight terms of the sequence: $-4, 8, -16, 32, -64, 128, -256, 512$.

✓ **Checkpoint 6** Find the seventh term of the geometric sequence whose first term is 5 and whose common ratio is -3.

Example 7 — Geometric Population Growth

The table shows the population of the United States in 2000, with estimates given by the Census Bureau for 2001 through 2006.

Year	2000	2001	2002	2003	2004	2005	2006
Population (millions)	281.4	284.5	287.6	290.8	294.0	297.2	300.5

a. Show that the population is increasing geometrically.

b. Write the general term for the geometric sequence modeling the population of the United States, in millions, n years after 1999.

c. Project the U.S. population, in millions, for the year 2019.

Solution

a. First, we use the sequence of population growth, 281.4, 284.5, 287.6, 290.8, and so on, to divide the population for each year by the population in the preceding year.

$$\frac{284.5}{281.4} \approx 1.011, \quad \frac{287.6}{284.5} \approx 1.011, \quad \frac{290.8}{287.6} \approx 1.011$$

Continuing in this manner, we will keep getting approximately 1.011. This means that the population is increasing geometrically with $r \approx 1.011$. In this situation, the common ratio is the growth rate, indicating that the population of the United States in any year shown in the sequence is approximately 1.011 times the population the year before.

b. The sequence of the U.S. population growth is

$$281.4, 284.5, 287.6, 290.8, 294.0, 297.2, 300.5, \ldots .$$

Because the population is increasing geometrically, we can find the general term of this sequence using

$$a_n = a_1 r^{n-1}.$$

In this sequence, $a_1 = 281.4$ and [from part (a)] $r \approx 1.011$. We substitute these values into the formula for the general term. This gives the general term for the geometric sequence describing the U.S. population, in millions, n years after 1999.

$$a_n = 281.4(1.011)^{n-1}$$

c. We can use the formula for the general term, a_n, in part (b) to project the U.S. population for the year 2019. The year 2019 is 20 years after 1999— that is, $2019 - 1999 = 20$. Thus, $n = 20$. We substitute 20 for n in $a_n = 281.4(1.011)^{n-1}$.

$$a_{20} = 281.4(1.011)^{20-1} = 281.4(1.011)^{19} \approx 346.4$$

The model projects that the United States will have a population of approximately 346.4 million in the year 2019.

 **Checkpoint 7** Write the general term for the geometric sequence

$$3, 6, 12, 24, 48, \ldots .$$

▶ Then use the formula for the general term to find the eighth term.

Achieving Success

Assuming that you have done very well preparing for an exam, **there are certain things you can do that will make you a better test taker**.

- Get a good sleep the night before the exam.
- Have a good breakfast that balances protein, carbohydrates, and fruit.
- Briefly review the material in the chapter summary or any summary you may have created.
- Bring everything you need to the exam, including two pencils, an eraser, scratch paper (if permitted), a calculator (if you're allowed to use one), water, and a watch.
- Survey the entire exam quickly to get an idea of its length.
- Read the directions to each problem carefully. Make sure that you have answered the specific question asked.
- Work the easy problems first. Then return to the hard problems you are not sure of. Doing the easy problems first will build your confidence. If you get bogged down on any one problem, you may not be able to complete the exam and receive credit for the questions you can easily answer.
- Attempt every problem. There may be partial credit even if you do not obtain the correct answer.
- Work carefully. Show your step-by-step solutions neatly. Check your work and answers.
- Watch the time. Pace yourself and be aware of when half the time is up. Determine how much of the exam you have completed. This will indicate if you're moving at a good pace or need to speed up. Prepare to spend more time on problems worth more points.
- Never turn in a test early. Use every available minute you are given for the test. If you have extra time, double check your arithmetic and look over your solutions.

Exercise Set 3.7

Concept and Vocabulary Exercises

In Exercises 1–4, fill in each blank so that the resulting statement is true.

1. A sequence in which each term after the first differs from the preceding term by a constant amount is called a/an _____ sequence. The difference between consecutive terms is called the _____ of the sequence.

2. The nth term of the sequence described in Exercise 1 is given by the formula _____, where a_1 is _____ and d is _____.

3. A sequence in which each term after the first is obtained by multiplying the preceding term by a fixed nonzero number is called a/an _____ sequence. The amount by which we multiply each time is called the _____ of the sequence.

4. The nth term of the sequence described in Exercise 3 is given by the formula _____, where a_1 is _____ and r is _____.

In Exercises 5–11, determine whether each statement is true or false. If the statement is false, make the necessary change(s) to produce a true statement.

5. The common difference for the arithmetic sequence given by $1, -1, -3, -5, \ldots$ is 2.

6. The sequence $1, 4, 8, 13, 19, 26, \ldots$ is an arithmetic sequence.

7. The nth term of an arithmetic sequence whose first term is a_1 and whose common difference is d is $a_n = a_1 + nd$.

8. The sequence $2, 6, 24, 120, \ldots$ is an example of a geometric sequence.

9. Adjacent terms in a geometric sequence have a common difference.

10. A sequence that is not arithmetic must be geometric.

11. If a sequence is geometric, we can write as many terms as we want by repeatedly multiplying by the common ratio.

Respond to Exercises 12–18 using verbal or written explanations.

12. What is a sequence? Give an example with your description.

13. What is an arithmetic sequence? Give an example with your description.

14. What is the common difference in an arithmetic sequence?

15. What is a geometric sequence? Give an example with your description.

16. What is the common ratio in a geometric sequence?

17. If you are given a sequence that is arithmetic or geometric, how can you determine which type of sequence it is?

18. For the first 30 days of a flu outbreak, the number of students at your school who become ill is increasing. Which is worse: The number of students with the flu is increasing arithmetically or is increasing geometrically? Explain your answer.

In Exercises 19–22, determine whether each statement makes sense or does not make sense, and explain your reasoning.

19. Now that I've studied sequences, I realize that the joke in this cartoon is based on the fact that you can't have a negative number of sheep.

WHEN MATHEMATICIANS CAN'T SLEEP

20. The sequence for the number of seats per row in our movie theater as the rows move toward the back is arithmetic with $d = 1$ so people don't block the view of those in the row behind them.

21. There's no end to the number of geometric sequences that I can generate whose first term is 5 if I pick nonzero numbers r and multiply 5 by each value of r repeatedly.

22. I've noticed that the big difference between arithmetic and geometric sequences is that arithmetic sequences are based on addition and geometric sequences are based on multiplication.

Practice Exercises

In Exercises 23–42, write the first six terms of the arithmetic sequence with the first term, a_1, and common difference, d.

23. $a_1 = 8, d = 2$

24. $a_1 = 5, d = 3$

25. $a_1 = 200, d = 20$

26. $a_1 = 300, d = 50$

27. $a_1 = -7, d = 4$

28. $a_1 = -8, d = 5$

29. $a_1 = -400, d = 300$

30. $a_1 = -500, d = 400$

31. $a_1 = 7, d = -3$

32. $a_1 = 9, d = -5$

33. $a_1 = 200, d = -60$

34. $a_1 = 300, d = -90$

35. $a_1 = \frac{5}{2}, d = \frac{1}{2}$

36. $a_1 = \frac{3}{4}, d = \frac{1}{4}$

37. $a_1 = \frac{3}{2}, d = \frac{1}{4}$

38. $a_1 = \frac{3}{2}, d = -\frac{1}{4}$

39. $a_1 = 4.25, d = 0.3$

40. $a_1 = 6.3, d = 0.25$

41. $a_1 = 4.5, d = -0.75$

42. $a_1 = 3.5, d = -1.75$

In Exercises 43–62, find the indicated term for the arithmetic sequence with first term, a_1, and common difference, d.

43. Find a_6, when $a_1 = 13, d = 4$.

44. Find a_{16}, when $a_1 = 9, d = 2$.

45. Find a_{50}, when $a_1 = 7, d = 5$.

46. Find a_{60}, when $a_1 = 8, d = 6$.

47. Find a_9, when $a_1 = -5, d = 9$.

48. Find a_{10}, when $a_1 = -8, d = 10$.

49. Find a_{200}, when $a_1 = -40, d = 5$.

50. Find a_{150}, when $a_1 = -60, d = 5$.

51. Find a_{10}, when $a_1 = 8, d = -10$.

52. Find a_{11}, when $a_1 = 10, d = -6$.

53. Find a_{60}, when $a_1 = 35, d = -3$.

54. Find a_{70}, when $a_1 = -32, d = 4$.

55. Find a_{12}, when $a_1 = 12, d = -5$.

56. Find a_{20}, when $a_1 = -20, d = -4$.

57. Find a_{90}, when $a_1 = -70, d = -2$.

58. Find a_{80}, when $a_1 = 106, d = -12$.

59. Find a_{12}, when $a_1 = 6, d = \frac{1}{2}$.

60. Find a_{14}, when $a_1 = 8, d = \frac{1}{4}$.

61. Find a_{50}, when $a_1 = 14, d = -0.25$.

62. Find a_{110}, when $a_1 = -12, d = -0.5$.

In Exercises 63–70, write a formula for the general term (the nth term) of each arithmetic sequence. Then use the formula for a_n to find a_{20}, the 20th term of the sequence.

63. $1, 5, 9, 13, \ldots$

64. $2, 7, 12, 17, \ldots$

65. $7, 3, -1, -5, \ldots$

66. $6, 1, -4, -9, \ldots$

67. $a_1 = 9, d = 2$

68. $a_1 = 6, d = 3$

69. $a_1 = -20, d = -4$

70. $a_1 = -70, d = -5$

In Exercises 71–92, write the first six terms of the geometric sequence with the first term, a_1, and common ratio, r.

71. $a_1 = 4, r = 2$

72. $a_1 = 2, r = 3$

73. $a_1 = 1000, r = 1$

74. $a_1 = 5000, r = 1$

75. $a_1 = 3, r = -2$

76. $a_1 = 2, r = -3$

77. $a_1 = 10, r = -4$

78. $a_1 = 20, r = -4$

79. $a_1 = 2000, r = -1$

80. $a_1 = 3000, r = -1$

81. $a_1 = -2, r = -3$

82. $a_1 = -4, r = -2$

83. $a_1 = -6, r = -5$

84. $a_1 = -8, r = -5$

85. $a_1 = \frac{1}{4}, r = 2$

86. $a_1 = \frac{1}{2}, r = 2$

87. $a_1 = \frac{1}{4}, r = \frac{1}{2}$

88. $a_1 = \frac{1}{5}, r = \frac{1}{2}$

89. $a_1 = -\frac{1}{16}, r = -4$

90. $a_1 = -\frac{1}{8}, r = -2$

91. $a_1 = 2, r = 0.1$

92. $a_1 = -1000, r = 0.1$

In Exercises 93–112, find the indicated term for the geometric sequence with first term, a_1, and common ratio, r.

93. Find a_7, when $a_1 = 4, r = 2$.

94. Find a_5, when $a_1 = 4, r = 3$.

95. Find a_{20}, when $a_1 = 2, r = 3$.

96. Find a_{20}, when $a_1 = 2, r = 2$.

97. Find a_{100}, when $a_1 = 50, r = 1$.

98. Find a_{200}, when $a_1 = 60, r = 1$.

99. Find a_7, when $a_1 = 5, r = -2$.

100. Find a_4, when $a_1 = 4, r = -3$.

101. Find a_{30}, when $a_1 = 2, r = -1$.

102. Find a_{40}, when $a_1 = 6, r = -1$.

103. Find a_6, when $a_1 = -2, r = -3$.

104. Find a_5, when $a_1 = -5, r = -2$.

105. Find a_8, when $a_1 = 6, r = \frac{1}{2}$.

106. Find a_8, when $a_1 = 12, r = \frac{1}{2}$.

107. Find a_6, when $a_1 = 18, r = -\frac{1}{3}$.

108. Find a_4, when $a_1 = 9, r = -\frac{1}{3}$.

109. Find a_{40}, when $a_1 = 1000, r = -\frac{1}{2}$.

110. Find a_{30}, when $a_1 = 8000, r = -\frac{1}{2}$.

111. Find a_8, when $a_1 = 1,000,000, r = 0.1$.

112. Find a_8, when $a_1 = 40,000, r = 0.1$.

In Exercises 113–120, write a formula for the general term (the nth term) of each geometric sequence. Then use the formula for a_n to find a_7, the seventh term of the sequence.

113. $3, 12, 48, 192, \ldots$

114. $3, 15, 75, 375, \ldots$

115. $18, 6, 2, \frac{2}{3}, \ldots$

116. $12, 6, 3, \frac{3}{2}, \ldots$

117. $1.5, -3, 6, -12, \ldots$

118. $5, -1, \frac{1}{5}, -\frac{1}{25}, \ldots$

119. $0.0004, -0.004, 0.04, -0.4, \ldots$

120. $0.0007, -0.007, 0.07, -0.7, \ldots$

Determine whether each sequence in Exercises 121–136 is arithmetic or geometric. Then find the next two terms.

121. $2, 6, 10, 14, \ldots$

122. $3, 8, 13, 18, \ldots$

123. $5, 15, 45, 135, \ldots$

124. $15, 30, 60, 120, \ldots$

125. $-7, -2, 3, 8, \ldots$

126. $-9, -5, -1, 3, \ldots$

127. $3, \frac{3}{2}, \frac{3}{4}, \frac{3}{8}, \ldots$

128. $6, 3, \frac{3}{2}, \frac{3}{4}, \ldots$

129. $\frac{1}{2}, 1, \frac{3}{2}, 2, \ldots$

130. $\frac{2}{3}, 1, \frac{4}{3}, \frac{5}{3}, \ldots$

131. $7, -7, 7, -7, \ldots$

132. $6, -6, 6, -6, \ldots$

133. $7, -7, -21, -35, \ldots$

134. $6, -6, -18, -30, \ldots$

135. $\sqrt{5}, 5, 5\sqrt{5}, 25, \ldots$

136. $\sqrt{3}, 3, 3\sqrt{3}, 9, \ldots$

Practice Plus

The sum, S_n, of the first n terms of an arithmetic sequence is given by

$$S_n = \frac{n}{2}(a_1 + a_n),$$

in which a_1 is the first term and a_n is the nth term. The sum, S_n, of the first n terms of a geometric sequence is given by

$$S_n = \frac{a_1(1 - r^n)}{1 - r},$$

in which a_1 is the first term and r is the common ratio ($r \neq 1$). In Exercises 137–144, determine whether each sequence is arithmetic or geometric. Then use the appropriate formula to find S_{10}, the sum of the first ten terms.

137. $4, 10, 16, 22, \ldots$

138. $7, 19, 31, 43, \ldots$

139. $2, 6, 18, 54, \ldots$

140. $3, 6, 12, 24, \ldots$

141. $3, -6, 12, -24, \ldots$

142. $4, -12, 36, -108, \ldots$

143. $-10, -6, -2, 2, \ldots$

144. $-15, -9, -3, 3, \ldots$

145. Use the appropriate formula shown above to find $1 + 2 + 3 + 4 + \cdots + 100$, the sum of the first 100 natural numbers.

146. Use the appropriate formula shown above to find $2 + 4 + 6 + 8 + \cdots + 200$, the sum of the first 100 positive even integers.

Application Exercises

The bar graphs at the top of the next column show changes in educational attainment for Americans ages 25 and older from 1970 to 2007. Exercises 147–148 involve developing arithmetic sequences that model the data.

Educational Attainment for Americans Ages 25 and Older

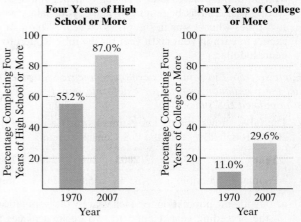

Source: U.S. Census Bureau

147. In 1970, 11.0% of Americans ages 25 and older had completed four years of college or more. On average, this percentage has increased by approximately 0.5 each year.

    a. Write a formula for the *n*th term of the arithmetic sequence that models the percentage of Americans ages 25 and older who completed four years of college or more *n* years after 1969.

    b. Use the model from part (a) to project the percentage of Americans ages 25 and older who will have completed four years of college or more by 2019.

148. In 1970, 55.2% of Americans ages 25 and older had completed four years of high school or more. On average, this percentage has increased by approximately 0.86 each year.

    a. Write a formula for the *n*th term of the arithmetic sequence that models the percentage of Americans ages 25 and older who completed four years of high school or more *n* years after 1969.

    b. Use the model from part (a) to project the percentage of Americans ages 25 and older who will have completed four years of high school or more by 2019.

149. Company A pays $24,000 yearly with raises of $1600 per year. Company B pays $28,000 yearly with raises of $1000 per year. Which company will pay more in year 10? How much more?

150. Company A pays $23,000 yearly with raises of $1200 per year. Company B pays $26,000 yearly with raises of $800 per year. Which company will pay more in year 10? How much more?

In Exercises 151–152, suppose you save $1 the first day of a month, $2 the second day, $4 the third day, and so on. That is, each day you save twice as much as you did the day before.

151. What will you put aside for savings on the fifteenth day of the month?

152. What will you put aside for savings on the thirtieth day of the month?

153. A professional baseball player signs a contract with a beginning salary of $3,000,000 for the first year and an annual increase of 4% per year beginning in the second year. That is, beginning in year 2, the athlete's salary will be 1.04 times what it was in the previous year. What is the athlete's salary for year 7 of the contract? Round to the nearest dollar.

154. You are offered a job that pays $30,000 for the first year with an annual increase of 5% per year beginning in the second year. That is, beginning in year 2, your salary will be 1.05 times what it was in the previous year. What can you expect to earn in your sixth year on the job? Round to the nearest dollar.

In Exercises 155–156, you will develop geometric sequences that model the population growth for California and Texas, the two most-populated U.S. states.

155. The table shows population estimates for California from 2003 through 2006 from the U.S. Census Bureau.

Year	2003	2004	2005	2006
Population in millions	35.48	35.89	36.13	36.46

a. Divide the population for each year by the population in the preceding year. Round to two decimal places and show that California has a population increase that is approximately geometric.

b. Write the general term of the geometric sequence modeling California's population, in millions, *n* years after 2002.

c. Use your model from part (b) to project California's population, in millions, for the year 2010. Round to two decimal places.

156. The table shows population estimates for Texas from 2003 through 2006 from the U.S. Census Bureau.

Year	2003	2004	2005	2006
Population in millions	22.12	22.49	22.86	23.41

a. Divide the population for each year by the population in the preceding year. Round to two decimal places and show that Texas has a population increase that is approximately geometric.

b. Write the general term of the geometric sequence modeling Texas's population, in millions, *n* years after 2002.

c. Use your model from part (b) to project Texas's population, in millions, for the year 2010. Round to two decimal places.

Critical Thinking Exercises

157. A person is investigating two employment opportunities. They both have a beginning salary of $20,000 per year. Company A offers an increase of $1000 per year. Company B offers 5% more than during the preceding year. Which company will pay more in the sixth year?

158. Would you rather have $10,000,000 and a brand new BMW, or would you rather have what you would receive on days 28, 29, and 30 combined if we give you 1¢ today, 2¢ tomorrow, 4¢ on day 3, 8¢ on day 4, 16¢ on day 5, and so on? Explain.

Group Exercise

159. Enough curiosities involving the Fibonacci sequence exist to warrant a flourishing Fibonacci Association. It publishes a quarterly journal. Do some research on the Fibonacci sequence in the library or on the Internet, and find one property that interests you. After doing this research, get together with your group to share these intriguing properties.

Chapter 3 Summary

3.1 Number Theory: Prime and Composite Numbers

Definitions and Concepts

The set of natural numbers is $\{1, 2, 3, 4, 5, \dots\}$.
$b \mid a$ (b divides a: a is divisible by b) for natural numbers a and b if the operation of dividing a by b leaves a remainder of 0.

Rules of Divisibility

Divisible By	Test	Divisible By	Test
2	The last digit is 0, 2, 4, 6, or 8.	8	The last three digits form a number that is divisible by 8.
3	The sum of the digits is divisible by 3.	9	The sum of the digits is divisible by 9.
4	The last two digits form a number divisible by 4.	10	The last digit is 0.
5	The number ends in 0 or 5.	12	The number is divisible by both 3 and 4. (In other words, the sum of the digits is divisible by 3 and the last two digits form a number divisible by 4.)
6	The number is divisible by both 2 and 3. (In other words, the number is even and the sum of its digits is divisible by 3.)		

Example

- Which of the numbers 2, 3, 4, 5, 6, 8, 9, 10, and 12 divide 614,608,176?

  $2|614,608,176$ because the last digit is 6.

  $3|614,608,176$ because the sum of the digits is $6 + 1 + 4 + 6 + 0 + 8 + 1 + 7 + 6 = 39$, and 39 is divisible by 3.

  $4|614,608,176$ because the last two digits form the number 76, which is divisible by 4.

  $5\nmid614,608,176$ because the number does not end in 0 or 5.

  $6|614,608,176$ because the number is divisible by both 2 and 3.

  $8|614,608,176$ because the last three digits form the number 176, which is divisible by 8.

  $9\nmid614,608,176$ because the sum of the digits, 39, is not divisible by 9.

  $10\nmid614,608,176$ because the last digit is not 0.

  $12|614,608,176$ because the number is divisible by both 3 and 4.

Additional Example to Review

Example 1, page 120

Definitions and Concepts

A prime number is a natural number greater than 1 that has only itself and 1 as factors. A composite number is a natural number greater than 1 that is divisible by a number other than itself and 1.
The Fundamental Theorem of Arithmetic: Every composite number can be expressed as a product of prime numbers in one and only one way (if the order of the factors is disregarded).

Example

- Find the prime factorization of 300.

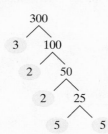

$300 = 3 \cdot 2^2 \cdot 5^2 = 2^2 \cdot 3 \cdot 5^2$

Additional Example to Review

Example 2, page 121

Definitions and Concepts

The greatest common divisor of two or more natural numbers is the largest number that is a divisor (or factor) of all the numbers. To find the greatest common divisor of two or more numbers,

1. Write the prime factorization of each number.
2. Select each prime factor with the smallest exponent that is common to each of the prime factorizations.
3. Form the product of the numbers from step 2. The greatest common divisor is the product of these factors.

Example

- Find the greatest common divisor of 168 and 180.

  $168 = 2 \cdot 2 \cdot 2 \cdot 3 \cdot 7 = 2^3 \cdot 3 \cdot 7$
  $180 = 2 \cdot 2 \cdot 3 \cdot 3 \cdot 5 = 2^2 \cdot 3^2 \cdot 5$

  Greatest common divisor $= 2^2 \cdot 3 = 4 \cdot 3 = 12$

Additional Examples to Review

Example 3, page 122; Example 4, page 123

Definitions and Concepts

The least common multiple of two or more natural numbers is the smallest natural number that is divisible by all of the numbers. To find the least common multiple of two or more numbers,

1. Write the prime factorization of each number.
2. Select every prime factor that occurs, raised to the greatest power to which it occurs, in these factorizations.
3. Form the product of the numbers from step 2. The least common multiple is the product of these factors.

Example

- Find the least common multiple of 168 and 180.

  $168 = 2^3 \cdot 3 \cdot 7$ and $180 = 2^2 \cdot 3^2 \cdot 5$

  Least common multiple $= 2^3 \cdot 3^2 \cdot 5 \cdot 7 = 8 \cdot 9 \cdot 5 \cdot 7 = 2520$

  The smallest natural number divisible by both 168 and 180 is 2520.

Additional Examples to Review

Example 5, page 125; Example 6, page 125

3.2 The Integers; Order of Operations

Definitions and Concepts

The set of whole numbers is $\{0, 1, 2, 3, 4, 5, \ldots\}$. The set of integers is $\{\ldots, -3, -2, -1, 0, 1, 2, 3, \ldots\}$. Integers are graphed on a number line by placing a dot at the correct location for each number.
$a < b$ (a is less than b) means a is to the left of b on a number line.
$a > b$ (a is greater than b) means a is to the right of b on a number line.

Example

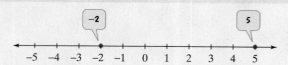

- $-2 < 5$ and $5 > -2$.

Additional Examples to Review

Example 1, page 130; Example 2, page 131

Definitions and Concepts

$|a|$, the absolute value of a, is the distance of a from 0 on a number line. The absolute value of a positive number is the number itself. The absolute value of 0 is 0: $|0| = 0$. The absolute value of a negative number is the number without the negative sign. For example, $|-8| = 8$.

Examples

- $|-1000| = 1000$ - $|1| = 1$ - $|0| = 0$

Additional Example to Review

Example 3, page 131

Definitions and Concepts

Addition of Integers

If the integers have the same sign,

1. Add their absolute values.
2. The sign of the sum is the same as the sign of the two numbers.

If the integers have different signs,
1. Subtract the smaller absolute value from the larger absolute value.
2. The sign of the sum is the same as the sign of the number with the larger absolute value.

The result of addition is called the sum.

Examples

- $12 + 17 = 29$ • $-12 + (-17) = -29$ • $12 + (-17) = -5$ • $-12 + 17 = 5$

Additional Examples to Review

See the box on page 132.

Definitions and Concepts

Subtraction of Integers

Additive inverses have the same absolute value, but lie on opposite sides of zero on a number line.
For all integers a and b,

$$a - b = a + (-b).$$

In words, to subtract b from a, add the additive inverse of b to a. The result of subtraction is called the difference.

Examples

- $25 - (-13) = 25 + 13 = 38$ • $-25 - (-13) = -25 + 13 = -12$ • $-25 - 13 = -25 + (-13) = -38$

Additional Examples to Review

Example 4, page 133; Example 5, page 134

Definitions and Concepts

Multiplication of Integers
1. The product of two integers with different signs is found by multiplying their absolute values. The product is negative.
2. The product of two integers with the same sign is found by multiplying their absolute values. The product is positive.
3. The product of 0 and any integer is 0: $a \cdot 0 = 0$ and $0 \cdot a = 0$.
4. If no number is 0, a product with an odd number of negative factors is found by multiplying absolute values. The product is negative.
5. If no number is 0, a product with an even number of negative factors is found by multiplying absolute values. The product is positive.

Examples

- $8(-10) = -80$ • $-8(10) = -80$ • $8(10) = 80$ • $(-8)(-10) = 80$
- $(-8)(-10)(-2) = -160$ • $-8(10)(-2) = 160$

  Three (odd) negative factors   Two (even) negative factors

- $(-2)^3 = (-2)(-2)(-2) = -8$ • $(-10)^2 = (-10)(-10) = 100$

Additional Examples to Review

See the box on page 135; Example 6, page 135

Definitions and Concepts

Division of Integers
1. The quotient of two integers with different signs is found by dividing their absolute values. The quotient is negative.

2. The quotient of two integers with the same sign is found by dividing their absolute values. The quotient is positive.

3. Zero divided by any nonzero integer is zero.

4. Division by 0 is undefined.

Examples

- $\dfrac{-20}{5} = -4$ • $\dfrac{20}{-5} = -4$ • $\dfrac{20}{5} = 4$

- $\dfrac{-20}{-5} = 4$ • $\dfrac{0}{-5} = 0$ • $\dfrac{-5}{0}$ is undefined.

Additional Examples to Review

See the box near the top of page 137.

Definitions and Concepts

Order of Operations

1. Perform all operations within grouping symbols.

2. Evaluate all exponential expressions.

3. Do all multiplications and divisions from left to right.

4. Do all additions and subtractions from left to right.

Examples

- $-10 - (8 - 14) = -10 - (-6) = -10 + 6 = -4$
- $30 \div (10 - 5^2) = 30 \div (10 - 25) = 30 \div (-15) = -2$
- $(9 - 12)^3(3 - 7)^2 = (-3)^3(-4)^2 = -27(16) = -432$

Additional Examples to Review

Example 7, page 138; Example 8, page 138

3.3 The Rational Numbers

Definitions and Concepts

The set of rational numbers is the set of all numbers which can be expressed in the form $\frac{a}{b}$, where a and b are integers and b is not equal to 0.

A rational number is reduced to its lowest terms, or simplified, by dividing both the numerator and the denominator by their greatest common divisor.

Example

- $\dfrac{375}{1000} = \dfrac{5 \cdot 75}{5 \cdot 200} = \dfrac{\cancel{5} \cdot \cancel{25} \cdot 3}{\cancel{5} \cdot \cancel{25} \cdot 8} = \dfrac{3}{8}$

Additional Example to Review

Example 1, page 144

Definitions and Concepts

A mixed number consists of the sum of an integer and a rational number, expressed without the use of an addition sign. An improper fraction is a rational number whose numerator is greater than its denominator.

Converting a Positive Mixed Number to an Improper Fraction

1. Multiply the denominator of the rational number by the integer and add the numerator to this product.

2. Place the sum in step 1 over the denominator in the mixed number.

Converting a Positive Improper Fraction to a Mixed Number

1. Divide the denominator into the numerator. Record the quotient and the remainder.

2. Write the mixed number using the following form:

$$\text{quotient} \; \frac{\text{remainder}}{\text{original denominator}}.$$

Examples

- Convert $4\frac{2}{3}$ to an improper fraction.

$$4\frac{2}{3} = \frac{3 \cdot 4 + 2}{3} = \frac{12 + 2}{3} = \frac{14}{3}$$

- Convert $\frac{37}{5}$ to a mixed number.

$$\begin{array}{r} 7 \\ 5\overline{)37} \\ \underline{35} \\ 2 \end{array} \qquad \frac{37}{5} = 7\frac{2}{5}$$

Additional Examples to Review

Example 2, page 145; Example 3, page 145

Definitions and Concepts

Any rational number can be expressed as a decimal. The resulting decimal will either terminate (stop), or it will have a digit that repeats or a block of digits that repeat. The rational number $\frac{a}{b}$ is expressed as a decimal by dividing b into a.

Examples

- $\dfrac{3}{8} = 0.375$

  $$\begin{array}{r} 0.375 \\ 8\overline{)3.000} \end{array}$$

- $\dfrac{7}{12} = 0.58\overline{3}$

  $$\begin{array}{r} 0.58333\cdots \\ 12\overline{)7.00000\cdots} \end{array}$$

Additional Example to Review

Example 4, page 146

Definitions and Concepts

To express a terminating decimal as a quotient of integers, the digits to the right of the decimal point are the numerator. The place value of the last digit to the right of the decimal point determines the denominator.

Examples

- $0.3 = \dfrac{3}{10}$
- $0.37 = \dfrac{37}{100}$
- $0.019 = \dfrac{19}{1000}$

Additional Example to Review

Example 5, page 147

Definitions and Concepts

To express a repeating decimal as a quotient of integers, use the following procedure:

Step 1 Let n equal the repeating decimal.

Step 2 Multiply both sides of the equation in step 1 by 10 if one digit repeats, by 100 if two digits repeat, by 1000 if three digits repeat, and so on.

Step 3 Subtract the equation in step 1 from the equation in step 2.

Step 4 Divide both sides of the equation in step 3 by the number in front of n and solve for n.

Example

- Express $0.\overline{24}$ as a quotient of integers.

  1. $n = 0.\overline{24} = 0.242424\ldots$

  2. $100n \qquad = 24.242424\ldots$

  3. $\begin{aligned} 100n &= 24.242424\ldots \\ -n &= 0.242424\ldots \\ \hline 99n &= 24 \end{aligned}$

  4. $\dfrac{99n}{99} = \dfrac{24}{99}$

  $n = \dfrac{24}{99}$ \quad Conclusion: $0.\overline{24} = \dfrac{24}{99}$

Additional Examples to Review

Example 6, page 148; Example 7, page 149

Definitions and Concepts

The product of two rational numbers is the product of their numerators divided by the product of their denominators.

Examples

- $\left(-\dfrac{3}{5}\right)\left(\dfrac{2}{7}\right) = \dfrac{-3(2)}{5 \cdot 7} = \dfrac{-6}{35} = -\dfrac{6}{35}$

- $\dfrac{3}{20}\left(6\dfrac{1}{4}\right) = \dfrac{3}{20} \cdot \dfrac{25}{4} = \dfrac{3}{\underset{4}{\cancel{20}}} \cdot \dfrac{\overset{5}{\cancel{25}}}{4} = \dfrac{3 \cdot 5}{4 \cdot 4} = \dfrac{15}{16}$

Additional Example to Review

Example 8, page 150

Definitions and Concepts

Two numbers whose product is 1 are called reciprocals, or multiplicative inverses, of each other. The quotient of two rational numbers is the product of the first number and the reciprocal of the second number.

Example

- $\dfrac{3}{5} \div 2\dfrac{1}{5} = \dfrac{3}{5} \div \dfrac{11}{5} = \dfrac{3}{5} \cdot \dfrac{5}{11} = \dfrac{3}{\underset{1}{\cancel{5}}} \cdot \dfrac{\cancel{5}^{1}}{11} = \dfrac{3 \cdot 1}{1 \cdot 11} = \dfrac{3}{11}$

Additional Example to Review

Example 9, page 151

Definitions and Concepts

The sum or difference of two rational numbers with identical denominators is the sum or difference of their numerators over the common denominator.

Examples

- $\dfrac{3}{5} + \dfrac{1}{5} = \dfrac{3 + 1}{5} = \dfrac{4}{5}$

- $\dfrac{11}{12} - \dfrac{1}{12} = \dfrac{11 - 1}{12} = \dfrac{10}{12} = \dfrac{2 \cdot 5}{2 \cdot 6} = \dfrac{5}{6}$

- $-3\dfrac{1}{4} - 5\dfrac{3}{4} = -\dfrac{13}{4} - \dfrac{23}{4} = \dfrac{-13 - 23}{4} = \dfrac{-36}{4} = -9$

Additional Example to Review

Example 10, page 151

Definitions and Concepts

Add or subtract rational numbers with unlike denominators by first expressing each rational number with the least common denominator and then following the procedure at the bottom of the previous page.

Example

- $\dfrac{3}{10} - \dfrac{7}{12}$

  $10 = 2 \cdot 5$ and $12 = 2^2 \cdot 3$, so the LCD is $2^2 \cdot 3 \cdot 5 = 60$.

  $\dfrac{3}{10} - \dfrac{7}{12} = \dfrac{3}{10} \cdot \dfrac{6}{6} - \dfrac{7}{12} \cdot \dfrac{5}{5} = \dfrac{18}{60} - \dfrac{35}{60} = \dfrac{18 - 35}{60} = \dfrac{-17}{60} = -\dfrac{17}{60}$

Additional Examples to Review

Example 11, page 152; Example 12, page 152

Definitions and Concepts

The order of operations can be applied to an expression with rational numbers.

Example

- $\dfrac{5}{6} - 3\left(\dfrac{1}{2} + \dfrac{2}{3}\right) = \dfrac{5}{6} - 3\left(\dfrac{1}{2} \cdot \dfrac{3}{3} + \dfrac{2}{3} \cdot \dfrac{2}{2}\right) = \dfrac{5}{6} - 3\left(\dfrac{3}{6} + \dfrac{4}{6}\right)$

  The least common denominator is 6.

  $= \dfrac{5}{6} - 3\left(\dfrac{3 + 4}{6}\right) = \dfrac{5}{6} - \dfrac{3}{1} \cdot \dfrac{7}{6} = \dfrac{5}{6} - \dfrac{3 \cdot 7}{6} = \dfrac{5}{6} - \dfrac{21}{6}$

  $= \dfrac{5 - 21}{6} = \dfrac{-16}{6} = \dfrac{-2 \cdot 8}{2 \cdot 3} = \dfrac{-8}{3} = -\dfrac{8}{3} \text{ or } -2\dfrac{2}{3}$

Additional Example to Review

Example 13, page 153

Definitions and Concepts

Density of the Rational Numbers

Given any two distinct rational numbers, there is always a rational number between them. To find the rational number halfway between two rational numbers, add the rational numbers and divide their sum by 2.

Example

- Find the rational number halfway between $\frac{1}{5}$ and $\frac{1}{3}$.

  $\left(\dfrac{1}{5} + \dfrac{1}{3}\right) \div 2 = \left(\dfrac{1}{5} \cdot \dfrac{3}{3} + \dfrac{1}{3} \cdot \dfrac{5}{5}\right) \div 2 = \left(\dfrac{3}{15} + \dfrac{5}{15}\right) \div 2$

  The least common denominator is 15.

  $= \dfrac{8}{15} \div \dfrac{2}{1} = \dfrac{\overset{4}{8}}{15} \cdot \dfrac{1}{\underset{1}{2}} = \dfrac{4}{15}$

  Thus, $\dfrac{1}{5} < \dfrac{4}{15} < \dfrac{1}{3}$.

Additional Example to Review

Example 14, page 154

3.4 The Irrational Numbers

Definitions and Concepts

The set of irrational numbers is the set of numbers whose decimal representations are neither terminating nor repeating. Examples of irrational numbers are $\sqrt{2} \approx 1.414$ and $\pi \approx 3.142$.

Simplifying square roots: Use the product rule, $\sqrt{ab} = \sqrt{a} \cdot \sqrt{b}$, to remove from the square root any perfect squares that occur as factors.

Example

- $\sqrt{50} = \sqrt{25 \cdot 2} = \sqrt{25} \cdot \sqrt{2} = 5\sqrt{2}$

Additional Example to Review

Example 1, page 161

Definitions and Concepts

Multiplying square roots: $\sqrt{a} \cdot \sqrt{b} = \sqrt{ab}$. The product of square roots is the square root of the product.

Example

- $\sqrt{10} \cdot \sqrt{8} = \sqrt{10 \cdot 8} = \sqrt{80} = \sqrt{16 \cdot 5} = \sqrt{16} \cdot \sqrt{5} = 4\sqrt{5}$

Additional Example to Review

Example 2, page 162

Definitions and Concepts

Dividing square roots: $\dfrac{\sqrt{a}}{\sqrt{b}} = \sqrt{\dfrac{a}{b}}$. The quotient of square roots is the square root of the quotient.

Example

- $\dfrac{\sqrt{500}}{\sqrt{2}} = \sqrt{\dfrac{500}{2}} = \sqrt{250} = \sqrt{25 \cdot 10} = \sqrt{25} \cdot \sqrt{10} = 5\sqrt{10}$

Additional Example to Review

Example 3, page 162

Definitions and Concepts

Adding and subtracting square roots: If the radicals have the same radicand, add or subtract their coefficients. The answer is the sum or difference of the coefficients times the common square root. Addition or subtraction is sometimes possible by first simplifying the square roots.

Examples

- $2\sqrt{13} - 9\sqrt{13} = (2 - 9)\sqrt{13} = -7\sqrt{13}$
- $2\sqrt{32} + 4\sqrt{50} = 2\sqrt{16 \cdot 2} + 4\sqrt{25 \cdot 2} = 2 \cdot 4\sqrt{2} + 4 \cdot 5\sqrt{2} = 8\sqrt{2} + 20\sqrt{2}$
  $= (8 + 20)\sqrt{2} = 28\sqrt{2}$

Additional Examples to Review

Example 4, page 163; Example 5, page 163

Definitions and Concepts

Rationalizing denominators: Multiply the numerator and the denominator by the smallest number that produces a perfect square radicand in the denominator.

Example

• $\dfrac{15}{\sqrt{3}} = \dfrac{15}{\sqrt{3}} \cdot \dfrac{\sqrt{3}}{\sqrt{3}} = \dfrac{15\sqrt{3}}{\sqrt{9}} = \dfrac{15\sqrt{3}}{3} = \dfrac{\overset{5}{\cancel{15}}\sqrt{3}}{\underset{1}{\cancel{3}}} = 5\sqrt{3}$

Additional Example to Review

Example 6, page 164

3.5 Real Numbers and Their Properties

Definitions and Concepts

The set of real numbers is obtained by combining the rational numbers with the irrational numbers.

Real numbers

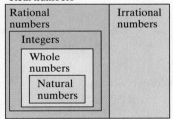

This diagram shows that every real number is rational or irrational.

Subsets of the Real Numbers

Name	Description
Natural numbers	$\{1, 2, 3, 4, 5, \ldots\}$ These are the numbers that we use for counting.
Whole numbers	$\{0, 1, 2, 3, 4, 5, \ldots\}$ The set of whole numbers includes 0 and the natural numbers.
Integers	$\{\ldots, -5, -4, -3, -2, -1, 0, 1, 2, 3, 4, 5, \ldots\}$ The set of integers includes the whole numbers and the negatives of the natural numbers.
Rational numbers	$\left\{ \dfrac{a}{b} \,\middle|\, a \text{ and } b \text{ are integers and } b \neq 0 \right\}$ The set of rational numbers is the set of all numbers that can be expressed as a quotient of two integers, with the denominator not 0. Rational numbers can be expressed as terminating or repeating decimals.
Irrational numbers	The set of irrational numbers is the set of all numbers whose decimal representations are neither terminating nor repeating. Irrational numbers cannot be expressed as a quotient of integers.

Example

• Consider the set $\left\{-6, -\frac{2}{7}, 0, 0.\overline{3}, \sqrt{5}, \sqrt{16}, 5\pi\right\}$

Natural numbers: $\left\{\sqrt{16}\ (\text{or } 4)\right\}$

Whole numbers: $\left\{0, \sqrt{16}\right\}$

Integers: $\left\{-6, 0, \sqrt{16}\right\}$

Rational numbers: $\left\{-6, -\frac{2}{7}, 0, 0.\overline{3}\left(\text{or } \frac{1}{3}\right), \sqrt{16}\right\}$

Irrational numbers: $\left\{\sqrt{5}, 5\pi\right\}$

Real numbers: $\left\{-6, -\frac{2}{7}, 0, 0.\overline{3}, \sqrt{5}, \sqrt{16}, 5\pi\right\}$

Additional Example to Review

Example 1, page 172

Definitions and Concepts

Properties of the Real Numbers

Name	Meaning
Closure Property of Addition	The sum of any two real numbers is a real number.
Closure Property of Multiplication	The product of any two real numbers is a real number.
Commutative Property of Addition	Changing order when adding does not affect the sum. $a + b = b + a$
Commutative Property of Multiplication	Changing order when multiplying does not affect the product. $ab = ba$
Associative Property of Addition	Changing grouping when adding does not affect the sum. $(a + b) + c = a + (b + c)$
Associative Property of Multiplication	Changing grouping when multiplying does not affect the product. $(ab)c = a(bc)$
Distributive Property of Multiplication over Addition	Multiplication distributes over addition. $a \cdot (b + c) = a \cdot b + a \cdot c$
Identity Property of Addition	Zero can be deleted from a sum. $a + 0 = a$ $0 + a = a$
Identity Property of Multiplication	One can be deleted from a product. $a \cdot 1 = a$ $1 \cdot a = a$
Inverse Property of Addition	The sum of a real number and its additive inverse gives 0, the additive identity. $a + (-a) = 0$ $(-a) + a = 0$
Inverse Property of Multiplication	The product of a nonzero real number and its multiplicative inverse gives 1, the multiplicative identity. $a \cdot \dfrac{1}{a} = 1, a \neq 0$ $\dfrac{1}{a} \cdot a = 1, a \neq 0$

Examples

- $7(4 \cdot 3) = 7(3 \cdot 4)$ Commutative property of multiplication
- $7(4 + 3) = 7 \cdot 4 + 7 \cdot 3$ Distributive property
- $(7 + 4) + 3 = 7 + (4 + 3)$ Associative property of addition
- $7 \cdot \frac{1}{7} = 1$ Inverse property of multiplication

Additional Examples to Review

Example 2, page 174; Example 3, page 174

3.6 Exponents and Scientific Notation

Definitions and Concepts

Properties of Exponents

- Product rule: $b^m \cdot b^n = b^{m+n}$
- Power rule: $\left(b^m\right)^n = b^{m \cdot n}$

- Quotient rule: $\dfrac{b^m}{b^n} = b^{m-n}, b \neq 0$
- Zero exponent rule: $b^0 = 1, b \neq 0$
- Negative exponent rule: $b^{-m} = \dfrac{1}{b^m}, b \neq 0$

Examples

- $2^3 \cdot 2^5 = 2^{3+5} = 2^8 = 2 \cdot 2 \cdot 2 \cdot 2 \cdot 2 \cdot 2 \cdot 2 \cdot 2 = 256$
- $\dfrac{5^6}{5^3} = 5^{6-3} = 5^3 = 5 \cdot 5 \cdot 5 = 125$
- $9^{-2} = \dfrac{1}{9^2} = \dfrac{1}{9 \cdot 9} = \dfrac{1}{81}$
- $4^0 + 4^{-3} = 1 + \dfrac{1}{4^3} = 1 + \dfrac{1}{4 \cdot 4 \cdot 4} = 1 + \dfrac{1}{64} = \dfrac{64}{64} + \dfrac{1}{64} = \dfrac{65}{64}$

Additional Examples to Review

Table 3.4, page 179; Example 1, page 180; Example 2, page 181

Definitions and Concepts

A positive number in scientific notation is expressed as $a \times 10^n$, where $1 \leq a < 10$ and n is an integer.

Changing from Scientific to Decimal Notation: If n is positive, move the decimal point in a to the right n places. If n is negative, move the decimal point in a to the left $|n|$ places.

Examples

- $3.4 \times 10^5 = 340,000$

  $n = 5$

  Move the decimal point 5 places to the right.

- $6.1 \times 10^{-3} = 0.0061$

  $n = -3$

  Move the decimal point $|-3| = 3$ places to the left.

Additional Example to Review

Example 3, page 182

Definitions and Concepts

Changing from Decimal to Scientific Notation: Move the decimal point in the given number to obtain a, where $1 \leq a < 10$. The number of places the decimal point moves gives the absolute value of n in $a \times 10^n$; n is positive if the number is greater than 10 and negative if the number is less than 1.

Examples

- $2800 = 2.8 \times 10^3$ The decimal point moved 3 places from 2800 to 2.8.
- $0.000019 = 1.9 \times 10^{-5}$ The decimal point moved 5 places from 0.000019 to 1.9.

Additional Examples to Review

Example 4, page 183; Example 5, page 183

Definitions and Concepts

The product and quotient rules for exponents are used to multiply and divide numbers in scientific notation. If a number is written in scientific notation, $a \times 10^n$, the digits in a are called significant digits. If rounding is necessary, round the scientific notation answer to the least number of significant digits found in any of the given numbers.

Examples

• Write the product in scientific notation and decimal notation.

$$(8 \times 10^9)(6 \times 10^{-4}) = (8 \times 6) \times (10^9 \times 10^{-4}) = 48 \times 10^{9+(-4)} = 48 \times 10^5$$

$$= (4.8 \times 10) \times 10^5 = 4.8 \times 10^6 = 4,800,000$$

Not in scientific notation

Scientific notation

Decimal notation

• Use scientific notation to divide and write the answer in scientific notation.

$$\frac{50,000,000}{0.0025} = \frac{5 \times 10^7}{2.5 \times 10^{-3}} = \left(\frac{5}{2.5}\right) \times \left(\frac{10^7}{10^{-3}}\right) = 2 \times 10^{7-(-3)} = 2 \times 10^{7+3} = 2 \times 10^{10}$$

Additional Examples to Review

Example 6, page 184; Example 7, page 184; Example 8, page 185; Example 9, page 186

3.7 Arithmetic and Geometric Sequences

Definitions and Concepts

In an arithmetic sequence, each term after the first differs from the preceding term by a constant, the common difference. Subtract any term from the term that directly follows it to find the common difference.

Example

• Write the first six terms of the arithmetic sequence with first term 6 and common difference -2.

$$6, 6 + (-2) = 4, 4 + (-2) = 2, 2 + (-2) = 0, 0 + (-2) = -2, -2 + (-2) = -4$$

The first six terms are $6, 4, 2, 0, -2$, and -4.

Additional Examples to Review

Example 1, page 191; Example 2, page 191

Definitions and Concepts

The general term, or the nth term, of an arithmetic sequence is

$$a_n = a_1 + (n - 1)d,$$

where a_1 is the first term and d is the common difference.

Example

• Find the eighth term of the arithmetic sequence whose first term is 5 and whose common difference is -4.

Given: $a_1 = 5, d = -4$;   Find: a_8.

Use $a_n = a_1 + (n - 1)d$, with $n = 8$.

$$a_8 = 5 + (8 - 1)(-4) = 5 + 7(-4) = 5 + (-28) = -23$$

Additional Examples to Review
Example 3, page 192; Example 4, page 192

Definitions and Concepts

In a geometric sequence, each term after the first is obtained by multiplying the preceding term by a nonzero constant, the common ratio. Divide any term after the first by the term that directly precedes it to find the common ratio.

Example

• Write the first five terms of the geometric sequence with first term 20 and common ratio $-\frac{1}{2}$.

$$20, \quad 20\left(-\frac{1}{2}\right) = -10, \quad -10\left(-\frac{1}{2}\right) = 5, \quad 5\left(-\frac{1}{2}\right) = -\frac{5}{2}, \quad -\frac{5}{2}\left(-\frac{1}{2}\right) = \frac{5}{4}$$

The first five terms are $20, -10, 5, -\frac{5}{2}$, and $\frac{5}{4}$.

Additional Example to Review
Example 5, page 194

Definitions and Concepts

The general term, or the nth term, of a geometric sequence is

$$a_n = a_1 r^{n-1},$$

where a_1 is the first term and r is the common ratio.

Example

• Find the sixth term of the geometric sequence whose first term is -10 and whose common ratio is 2.

Given: $a_1 = -10, r = 2$;   Find: a_6.

Use $a_n = a_1 r^{n-1}$ with $n = 6$.

$$a_6 = -10 \cdot 2^{6-1} = -10 \cdot 2^5 = -10(32) = -320$$

Additional Examples to Review
Example 6, page 195; Example 7, page 195

Review Exercises

Section 3.1 Number Theory: Prime and Composite Numbers

In Exercises 1 and 2, determine whether the number is divisible by each of the following numbers: 2, 3, 4, 5, 6, 8, 9, 10, and 12. If you are using a calculator, explain the divisibility shown by your calculator using one of the rules of divisibility.

1. 238,632

2. 421,153,470

In Exercises 3–5, find the prime factorization of each composite number.

3. 705

4. 960

5. 6825

In Exercises 6–8, find the greatest common divisor and the least common multiple of the numbers.

6. 30 and 48

7. 36 and 150

8. 216 and 254

9. For a school athletics program, you need to divide 24 boys and 60 girls into all-male and all-female teams so that each team has the same number of people. What is the largest number of people that can be placed on a team?

10. The library at a high school runs videotapes of two study skill programs continuously. One videotape runs for 42 minutes and a second runs for 56 minutes. Both videotapes begin at 9:00 A.M. When will the videos of the two programs begin again at the same time?

Section 3.2 The Integers; Order of Operations

In Exercises 11–12, insert either $<$ or $>$ in the shaded area between the integers to make the statement true.

11. $-93 \blacksquare 17$ **12.** $-2 \blacksquare -200$

In Exercises 13–15, find the absolute value.

13. $|-860|$ **14.** $|53|$ **15.** $|0|$

Perform the indicated operations in Exercises 16–28.

16. $8 + (-11)$ **17.** $-6 + (-5)$

18. $-7 - 8$ **19.** $-7 - (-8)$

20. $(-9)(-11)$ **21.** $5(-3)$

22. $\dfrac{-36}{-4}$ **23.** $\dfrac{20}{-5}$

24. $-40 \div 5 \cdot 2$ **25.** $-6 + (-2) \cdot 5$

26. $6 - 4(-3 + 2)$ **27.** $28 \div (2 - 4^2)$

28. $36 - 24 \div 4 \cdot 3 - 1$

29. **a.** In 2007, the U.S. government collected \$2568 billion and spent \$2730 billion. Express the budget deficit as a negative integer, in billions of dollars.

     b. In 2001, the Congressional Budget Office projected a budget surplus of \$573 billion for 2007. What is the difference between the 2007 projected surplus and the actual 2007 deficit that you determined in part (a)?

Section 3.3 The Rational Numbers

In Exercises 30–32, reduce each rational number to its lowest terms.

30. $\dfrac{40}{75}$ **31.** $\dfrac{36}{150}$ **32.** $\dfrac{165}{180}$

In Exercises 33–34, convert each mixed number to an improper fraction.

33. $5\frac{9}{11}$ **34.** $-3\frac{2}{7}$

In Exercises 35–36, convert each improper fraction to a mixed number.

35. $\dfrac{27}{5}$ **36.** $-\dfrac{17}{9}$

In Exercises 37–40, express each rational number as a decimal.

37. $\dfrac{4}{5}$ **38.** $\dfrac{3}{7}$ **39.** $\dfrac{5}{8}$ **40.** $\dfrac{9}{16}$

In Exercises 41–44, express each terminating decimal as a quotient of integers in lowest terms.

41. 0.6 - **42.** 0.68 **43.** 0.588 **44.** 0.0084

In Exercises 45–47, express each repeating decimal as a quotient of integers in lowest terms.

45. $0.\overline{5}$ **46.** $0.\overline{34}$ **47.** $0.1\overline{13}$

In Exercises 48–58, perform the indicated operations. Where possible, reduce the answer to lowest terms.

48. $\dfrac{3}{5} \cdot \dfrac{7}{10}$ **49.** $\left(3\frac{1}{3}\right)\left(1\frac{3}{4}\right)$

50. $\dfrac{4}{5} \div \dfrac{3}{10}$ **51.** $-1\frac{2}{3} \div 6\frac{2}{3}$

52. $\dfrac{2}{9} + \dfrac{4}{9}$ **53.** $\dfrac{7}{9} + \dfrac{5}{12}$

54. $\dfrac{3}{4} - \dfrac{2}{15}$ **55.** $\dfrac{1}{3} + \dfrac{1}{2} \cdot \dfrac{4}{5}$

56. $\dfrac{3}{8}\left(\dfrac{1}{2} + \dfrac{1}{3}\right)$ **57.** $\dfrac{1}{2} - \dfrac{2}{3} \div \dfrac{5}{9} + \dfrac{3}{10}$

58. $\left(\dfrac{1}{2} + \dfrac{1}{3}\right) \div \left(\dfrac{1}{4} - \dfrac{3}{8}\right)$

In Exercises 59–60, find the rational number halfway between the two numbers in each pair.

59. $\dfrac{1}{7}$ and $\dfrac{1}{8}$ **60.** $\dfrac{3}{4}$ and $\dfrac{3}{5}$

61. A recipe for chicken enchiladas is meant for six people and requires $4\frac{1}{2}$ pounds of chicken. If you want to serve 15 people, how much chicken is needed?

62. The gas tank of a car is filled to its capacity. The first day, $\frac{1}{4}$ of the tank's gas is used for travel. The second day, $\frac{1}{3}$ of the tank's original amount of gas is used for travel. What fraction of the tank is filled with gas at the end of the second day?

Section 3.4 The Irrational Numbers

In Exercises 63–66, simplify the square root.

63. $\sqrt{28}$ **64.** $\sqrt{72}$

65. $\sqrt{150}$ **66.** $\sqrt{300}$

In Exercises 67–75, perform the indicated operation. Simplify the answer when possible.

67. $\sqrt{6} \cdot \sqrt{8}$ **68.** $\sqrt{10} \cdot \sqrt{5}$

69. $\dfrac{\sqrt{24}}{\sqrt{2}}$ **70.** $\dfrac{\sqrt{27}}{\sqrt{3}}$

71. $\sqrt{5} + 4\sqrt{5}$ **72.** $7\sqrt{11} - 13\sqrt{11}$

73. $\sqrt{50} + \sqrt{8}$ **74.** $\sqrt{3} - 6\sqrt{27}$

75. $2\sqrt{18} + 3\sqrt{8}$

In Exercises 76–77, rationalize the denominator.

76. $\dfrac{30}{\sqrt{5}}$ **77.** $\sqrt{\dfrac{2}{3}}$

78. Paleontologists use the mathematical model $W = 4\sqrt{2x}$ to estimate the walking speed of a dinosaur, W, in feet per second, where x is the length, in feet, of the dinosaur's leg. What is the walking speed of a dinosaur whose leg length is 6 feet? Express the answer in simplified radical form. Then use your calculator to estimate the walking speed to the nearest tenth of a foot per second.

Section 3.5 Real Numbers and Their Properties

79. Consider the set

$$\left\{-17, -\tfrac{9}{13}, 0, 0.75, \sqrt{2}, \pi, \sqrt{81}\right\}.$$

List all numbers from the set that are **a.** natural numbers, **b.** whole numbers, **c.** integers, **d.** rational numbers, **e.** irrational numbers, **f.** real numbers.

80. Give an example of an integer that is not a natural number.

81. Give an example of a rational number that is not an integer.

82. Give an example of a real number that is not a rational number.

In Exercises 83–90, state the name of the property illustrated.

83. $3 + 17 = 17 + 3$

84. $(6 \cdot 3) \cdot 9 = 6 \cdot (3 \cdot 9)$

85. $\sqrt{3}(\sqrt{5} + \sqrt{3}) = \sqrt{15} + 3$

86. $(6 \cdot 9) \cdot 2 = 2 \cdot (6 \cdot 9)$

87. $\sqrt{3}(\sqrt{5} + \sqrt{3}) = (\sqrt{5} + \sqrt{3})\sqrt{3}$

88. $(3 \cdot 7) + (4 \cdot 7) = (4 \cdot 7) + (3 \cdot 7)$

89. $-3\left(-\dfrac{1}{3}\right) = 1$

90. $\sqrt{7} \cdot 1 = \sqrt{7}$

In Exercises 91–92, give an example to show that

91. The natural numbers are not closed with respect to division.

92. The whole numbers are not closed with respect to subtraction.

Section 3.6 Exponents and Scientific Notation

In Exercises 93–103, evaluate each expression.

93. $6 \cdot 6^2$

94. $2^3 \cdot 2^3$

95. $(2^2)^2$

96. $(3^3)^2$

97. $\dfrac{5^6}{5^4}$

98. 7^0

99. $(-7)^0$

100. 6^{-3}

101. 2^{-4}

102. $\dfrac{7^4}{7^6}$

103. $3^5 \cdot 3^{-2}$

In Exercises 104–107, express each number in decimal notation.

104. 4.6×10^2

105. 3.74×10^4

106. 2.55×10^{-3}

107. 7.45×10^{-5}

In Exercises 108–113, express each number in scientific notation.

108. 7520

109. 3,590,000

110. 0.00725

111. 0.000000409

112. 420×10^{11}

113. 0.97×10^{-4}

In Exercises 114–117, perform the indicated operation and express each answer in decimal notation.

114. $(3 \times 10^7)(1.3 \times 10^{-5})$

115. $(5 \times 10^3)(2.3 \times 10^2)$

116. $\dfrac{6.9 \times 10^3}{3 \times 10^5}$

117. $\dfrac{2.4 \times 10^{-4}}{6 \times 10^{-6}}$

In Exercises 118–121, perform the indicated operation by first expressing each number in scientific notation. Write the answer in scientific notation.

118. $(60,000)(540,000)$

119. $(91,000)(0.0004)$

120. $\dfrac{8,400,000}{4000}$

121. $\dfrac{0.000003}{0.00000006}$

In Exercises 122–123, use 10^6 for one million and 10^9 for one billion to rewrite the number in each statement in scientific notation.

122. The 2009 economic stimulus package allocated \$53.6 billion for grants to states for education.

123. The population of the United States at the time the economic stimulus package was voted into law was approximately 307 million.

124. Use your scientific notation answers from Exercises 122 and 123 to answer this question:

If the cost for grants to states for education was evenly divided among every individual in the United States, how much would each citizen have to pay?

125. The human body contains approximately 3.2×10^4 microliters of blood for every pound of body weight. Each microliter of blood contains approximately 5×10^6 red blood cells. Express in scientific notation the approximate number of red blood cells in the body of a 180-pound person.

Section 3.7 Arithmetic and Geometric Sequences

In Exercises 126–128, write the first six terms of the arithmetic sequence with the first term, a_1, and common difference, d.

126. $a_1 = 7, d = 4$

127. $a_1 = -4, d = -5$

128. $a_1 = \frac{3}{2}, d = -\frac{1}{2}$

In Exercises 129–131, find the indicated term for the arithmetic sequence with first term, a_1, and common difference, d.

129. Find a_6, when $a_1 = 5, d = 3$.

130. Find a_{12}, when $a_1 = -8, d = -2$.

131. Find a_{14}, when $a_1 = 14, d = -4$.

In Exercises 132–133, write a formula for the general term (the nth term) of each arithmetic sequence. Then use the formula for a_n to find a_{20}, the 20th term of the sequence.

132. $-7, -3, 1, 5, \ldots$

133. $a_1 = 200, d = -20$

In Exercises 134–136, write the first six terms of the geometric sequence with the first term, a_1, and common ratio, r.

134. $a_1 = 3, r = 2$

135. $a_1 = \frac{1}{2}, r = \frac{1}{2}$

136. $a_1 = 16, r = -\frac{1}{2}$

In Exercises 137–139, find the indicated term for the geometric sequence with first term, a_1, and common ratio, r.

137. Find a_4, when $a_1 = 2, r = 3$.

138. Find a_6, when $a_1 = 16, r = \frac{1}{2}$.

139. Find a_5, when $a_1 = -3, r = 2$.

In Exercises 140–141, write a formula for the general term (the nth term) of each geometric sequence. Then use the formula for a_n to find a_8, the eighth term of the sequence.

140. $1, 2, 4, 8, \ldots$

141. $100, 10, 1, \frac{1}{10}, \ldots$

Determine whether each sequence in Exercises 142–145 is arithmetic or geometric. Then find the next two terms.

142. $4, 9, 14, 19, \ldots$

143. $2, 6, 18, 54, \ldots$

144. $1, \frac{1}{4}, \frac{1}{16}, \frac{1}{64}, \ldots$

145. $0, -7, -14, -21, \ldots$

146. The graphic indicates that there are more eyes at school.

Percentage of United States Students Ages 12–18 Seeing Security Cameras at School

Source: Department of Education

In 2001, 39% of students ages 12–18 reported seeing one or more security cameras at their school. On average, this percentage has increased by approximately 4.75 each year since then.

a. Write a formula for the *n*th term of the arithmetic sequence that describes the percentage of students ages 12–18 who reported seeing security cameras at school *n* years after 2000.

b. Use the model to predict the percentage of students ages 12–18 who will report seeing security cameras at school by the year 2013.

147. Projections for the U.S. population, ages 85 and older, are shown in the following table.

Year	2000	2010	2020	2030	2040	2050
Projected Population in millions	4.2	5.9	8.3	11.6	16.2	22.7

Actual 2000 population

Source: U.S. Census Bureau

a. Show that the U.S. population, ages 85 and older, is projected to increase geometrically.

b. Write the general term of the geometric sequence describing the U.S. population, ages 85 and older, in millions, *n* decades after 1990.

c. Use the model in part (b) to project the U.S. population, ages 85 and older, in 2080.

Chapter 3 Test A

1. Which of the numbers 2, 3, 4, 5, 6, 8, 9, 10, and 12 divide 391,248?

2. Find the prime factorization of 252.

3. Find the greatest common divisor and the least common multiple of 48 and 72.

Perform the indicated operations in Exercises 4–6.

4. $-6 - (5 - 12)$

5. $(-3)(-4) \div (7 - 10)$

6. $(6 - 8)^2 (5 - 7)^3$

7. Express $\frac{7}{12}$ as a decimal.

8. Express $0.\overline{64}$ as a quotient of integers in lowest terms.

In Exercises 9–11, perform the indicated operations. Where possible, reduce the answer to its lowest terms.

9. $\left(-\frac{3}{7}\right) \div \left(-2\frac{1}{7}\right)$

10. $\frac{19}{24} - \frac{7}{40}$

11. $\frac{1}{2} - 8\left(\frac{1}{4} + 1\right)$

12. Find the rational number halfway between $\frac{1}{2}$ and $\frac{2}{3}$.

13. Multiply and simplify: $\sqrt{10} \cdot \sqrt{5}$.

14. Add: $\sqrt{50} + \sqrt{32}$.

15. Rationalize the denominator: $\dfrac{6}{\sqrt{2}}$.

16. List all the rational numbers in this set:

$$\left\{-7, -\frac{4}{5}, 0, 0.25, \sqrt{3}, \sqrt{4}, \frac{22}{7}, \pi\right\}.$$

In Exercises 17–18, state the name of the property illustrated.

17. $3(2 + 5) = 3(5 + 2)$

18. $6(7 + 4) = 6 \cdot 7 + 6 \cdot 4$

In Exercises 19–21, evaluate each expression.

19. $3^3 \cdot 3^2$ **20.** $\dfrac{4^6}{4^3}$ **21.** 8^{-2}

22. Multiply and express the answer in decimal notation.

$$(3 \times 10^8)(2.5 \times 10^{-5})$$

23. Divide by first expressing each number in scientific notation. Write the answer in scientific notation.

$$\frac{49,000}{0.007}$$

24. One way economists measure economic well-being is by calculating a nation's gross domestic product (GDP), the total value of all final goods and services produced in an economy. In 2009, the GDP of the United States was $14.4 trillion.

   a. Express the GDP in scientific notation.

   b. If the GDP was evenly distributed among all Americans (3.07×10^8 people), how much would each citizen receive?

25. In 2009, the population of the United States was approximately 3.1×10^8 and on average each person consumed approximately 6000 ounces of soda. Express the total consumption of soda for that year in scientific notation.

(*Source: Upfront*)

26. Write the first six terms of the arithmetic sequence with first term, a_1, and common difference, d.

$$a_1 = 1, d = -5$$

27. Find a_9, the ninth term of the arithmetic sequence with the first term, a_1, and common difference, d.

$$a_1 = -2, d = 3$$

28. Write the first six terms of the geometric sequence with first term, a_1, and common ratio, r.

$$a_1 = 16, r = \frac{1}{2}$$

29. Find a_7, the seventh term of the geometric sequence with the first term, a_1, and common ratio, r.

$$a_1 = 5, r = 2$$

Chapter 3 Test B

1. Which of the numbers 2, 3, 4, 5, 6, 8, 9, 10, and 12 divide 97,128?

2. Find the prime factorization of 360.

3. Find the greatest common divisor and the least common multiple of 72 and 108.

Perform the indicated operations in Exercises 4–6.

4. $-12 - (14 - 24)$

5. $(-8)(-3) \div (9 - 13)$

6. $(7 - 10)^2(5 - 15)^3$

7. Express $\frac{11}{12}$ as a decimal.

8. Express $0.\overline{51}$ as a quotient of integers in lowest terms.

In Exercises 9–11, perform the indicated operations. Where possible, reduce the answer to its lowest terms.

9. $\left(-\frac{5}{11}\right) \div \left(-2\frac{2}{9}\right)$

10. $\frac{17}{24} - \frac{5}{36}$

11. $\frac{1}{3} - 5\left(\frac{1}{2} + 1\right)$

12. Find the rational number halfway between $\frac{1}{5}$ and $\frac{3}{4}$.

13. Multiply and simplify: $\sqrt{30} \cdot \sqrt{3}$.

14. Add: $\sqrt{75} + \sqrt{12}$.

15. Rationalize the denominator: $\frac{15}{\sqrt{5}}$

16. List all the rational numbers in this set:

$$\left\{-100, -\frac{2}{3}, 0, 1.57283, \sqrt{5}, 2\pi, \sqrt{100}, 12.\overline{7}\right\}.$$

In Exercises 17–18, state the name of the property illustrated.

17. $3(2 + 5) = (2 + 5)3$

18. $3(2 \cdot 5) = (3 \cdot 2)5$

In Exercises 19–21, evaluate each expression.

19. $2^4 \cdot 2^2$ **20.** $\frac{3^{40}}{3^{35}}$ **21.** 5^{-3}

22. Multiply and express the answer in decimal notation.

$$(8 \times 10^{14})(3.5 \times 10^{-6})$$

23. Divide by first expressing each number in scientific notation. Write the answer in scientific notation.

$$\frac{36,000,000}{0.0006}$$

24. One way economists measure economic well-being is by calculating a nation's gross domestic product (GDP), the total value of all final goods and services produced in an economy. In 2009, the GDP of China was $7.8 trillion.

  a. Express the GDP in scientific notation.

  b. If the GDP was evenly distributed among China's 1.3 billion people, how much would each person receive?

25. In 2009, the population of the United States was approximately 3.1×10^8 and on average each person received 14,000 e-mails. Express the total number of e-mails received for that year in scientific notation. Do not round the answer.

  (*Source: Time*)

26. Write the first six terms of the arithmetic sequence with first term, a_1, and common difference, d.

$$a_1 = 7, d = -3$$

27. Find a_{10}, the tenth term of the arithmetic sequence with first term, a_1, and common difference, d.

$$a_1 = -9, d = 5$$

28. Write the first six terms of the geometric sequence with first term, a_1, and common ratio, r.

$$a_1 = 25, r = -\frac{1}{5}$$

29. Find a_8, the eighth term of the geometric sequence with first term, a_1, and common ratio, r.

$$a_1 = 10, r = 2$$

Algebra: Equations and Inequalities

4

The belief that humor and laughter can have positive effects on our lives is not new.

The Bible tells us, "A merry heart doeth good like a medicine, but a broken spirit drieth the bones." (Proverbs 17:22)

Some random humor factoids:
• The average adult laughs 15 times each day. (Newhouse News Service)
• Forty-six percent of people who are telling a joke laugh more than the people they are telling it to. (*U.S. News and World Report*)
• Algebra can be used to model the influence that humor plays in our responses to negative life events. (Bob Blitzer, *Math for Your World*)

That last tidbit that your author threw into the list is true. Based on our sense of humor, there is actually a formula that predicts how we will respond to difficult life events.

Formulas can be used to explain what is happening in the present and to make predictions about what might occur in the future. In this chapter, you will learn to use formulas and mathematical models in new ways that will help you to recognize patterns, logic, and order in a world that can appear chaotic to the untrained eye.

Humor opens Section 4.2, and the advantage of having a sense of humor becomes laughingly evident in the models in Example 6 on page 236.

4.1 Algebraic Expressions and Formulas

Objectives

1 Evaluate algebraic expressions.

2 Use mathematical models.

3 Understand the vocabulary of algebraic expressions.

4 Simplify algebraic expressions.

Feeling attractive with a suntan that gives you a "healthy glow"? Think again. Direct sunlight is known to promote skin cancer. Although sunscreens protect you from burning, dermatologists are concerned with the long-term damage that results from the sun even without sunburn.

Algebraic Expressions

Let's see what this child's "healthy glow" has to do with algebra. The biggest difference between arithmetic and algebra is the use of *variables* in algebra. A **variable** is a letter that represents a variety of different numbers. For example, we can let x represent the number of minutes that a person can stay in the sun without burning with no sunscreen. With a number 6 sunscreen, exposure time without burning is six times as long, or 6 times x. This can be written $6 \cdot x$, but it is usually expressed as $6x$. Placing a number and a letter next to one another indicates multiplication.

Notice that $6x$ combines the number 6 and the variable x using the operation of multiplication. A combination of variables and numbers using the operations of addition, subtraction, multiplication, or division, as well as powers or roots, is called an **algebraic expression**. Here are some examples of algebraic expressions:

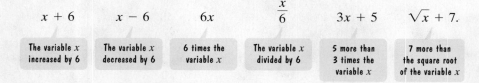

$$x + 6 \qquad x - 6 \qquad 6x \qquad \frac{x}{6} \qquad 3x + 5 \qquad \sqrt{x} + 7.$$

| The variable x increased by 6 | The variable x decreased by 6 | 6 times the variable x | The variable x divided by 6 | 5 more than 3 times the variable x | 7 more than the square root of the variable x |

Evaluating Algebraic Expressions

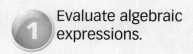

1 Evaluate algebraic expressions.

Evaluating an algebraic expression means finding the value of the expression for a given value of the variable. For example, we can evaluate $6x$ (from the sunscreen example) when $x = 15$. We substitute 15 for x. We obtain $6 \cdot 15$, or 90. This means that if you can stay in the sun for 15 minutes without burning when you don't put on any lotion, then with a number 6 lotion, you can "cook" for 90 minutes without burning.

Many algebraic expressions contain more than one operation. Evaluating an algebraic expression correctly involves carefully applying the order of operations agreement that we studied in Chapter 3.

The Order of Operations Agreement

1. Perform operations within the innermost parentheses and work outward. If the algebraic expression involves a fraction, treat the numerator and the denominator as if they were each enclosed in parentheses.

2. Evaluate all exponential expressions.

3. Perform multiplications and divisions as they occur, working from left to right.

4. Perform additions and subtractions as they occur, working from left to right.

Example 1 **Evaluating an Algebraic Expression**

Evaluate $7 + 5(x - 4)^3$ for $x = 6$.

Solution

$$7 + 5(x - 4)^3 = 7 + 5(6 - 4)^3 \quad \text{Replace x with 6.}$$
$$= 7 + 5(2)^3 \quad \text{First work inside parentheses: } 6 - 4 = 2.$$
$$= 7 + 5(8) \quad \text{Evaluate the exponential expression:} \\ 2^3 = 2 \cdot 2 \cdot 2 = 8.$$
$$= 7 + 40 \quad \text{Multiply: } 5(8) = 40.$$
$$= 47 \quad \text{Add.}$$

▶ ✓ **Checkpoint 1** Evaluate $8 + 6(x - 3)^2$ for $x = 13$.

Study Tip

Notice the difference between the following evaluations:

- x^2 for $x = -6$

  $x^2 = (-6)^2$

  $= (-6)(-6) = 36$

- $-x^2$ for $x = 6$

  $-x^2 = -6^2 = -6 \cdot 6 = -36$

  > The negative is not inside parentheses and is not taken to the second power.

Work carefully when evaluating algebraic expressions with exponents and negatives.

Example 2 **Evaluating an Algebraic Expression**

Evaluate $x^2 + 5x - 3$ for $x = -6$.

Solution

We substitute -6 for each of the two occurrences of x. Then we use the order of operations to evaluate the algebraic expression.

$$x^2 + 5x - 3 \quad \text{This is the given algebraic expression.}$$
$$= (-6)^2 + 5(-6) - 3 \quad \text{Substitute } -6 \text{ for each x.}$$
$$= 36 + 5(-6) - 3 \quad \text{Evaluate the exponential expression:} \\ (-6)^2 = (-6)(-6) = 36.$$
$$= 36 + (-30) - 3 \quad \text{Multiply: } 5(-6) = -30.$$
$$= 6 - 3 \quad \text{Add and subtract from left to right.} \\ \text{First add: } 36 + (-30) = 6.$$
$$= 3 \quad \text{Subtract.}$$

▶ ✓ **Checkpoint 2** Evaluate $x^2 + 4x - 7$ for $x = -5$.

Example 3 **Evaluating an Algebraic Expression**

Evaluate $-2x^2 + 5xy - y^3$ for $x = 4$ and $y = -2$.

Solution

We substitute 4 for each x and -2 for each y. Then we use the order of operations to evaluate the algebraic expression.

$$-2x^2 + 5xy - y^3 \quad \text{This is the given algebraic expression.}$$
$$= -2 \cdot 4^2 + 5 \cdot 4(-2) - (-2)^3 \quad \text{Substitute 4 for x and } -2 \text{ for y.}$$
$$= -2 \cdot 16 + 5 \cdot 4(-2) - (-8) \quad \text{Evaluate the exponential expressions:} \\ 4^2 = 4 \cdot 4 = 16 \text{ and} \\ (-2)^3 = (-2)(-2)(-2) = -8.$$
$$= -32 + (-40) - (-8) \quad \text{Multiply: } -2 \cdot 16 = -32 \text{ and} \\ 5(4)(-2) = 20(-2) = -40.$$
$$= -72 - (-8) \quad \text{Add and subtract from left to right. First add:} \\ -32 + (-40) = -72.$$
$$= -64 \quad \text{Subtract: } -72 - (-8) = -72 + 8 = -64.$$

▶ ✓ **Checkpoint 3** Evaluate $-3x^2 + 4xy - y^3$ for $x = 5$ and $y = -1$.

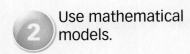

Use mathematical models.

Formulas and Mathematical Models

An **equation** is formed when an equal sign is placed between two algebraic expressions. One aim of algebra is to provide a compact, symbolic description of the world. These descriptions involve the use of *formulas*. A **formula** is an equation that uses variables to express a relationship between two or more quantities.

Here are two examples of formulas related to heart rate and exercise.

Couch-Potato Exercise

$$H = \frac{1}{5}(220 - a)$$

Heart rate, in beats per minute,   is $\frac{1}{5}$ of   the difference between 220 and your age.

Working It

$$H = \frac{9}{10}(220 - a)$$

Heart rate, in beats per minute,   is $\frac{9}{10}$ of   the difference between 220 and your age.

These important definitions are repeated from earlier chapters.

The process of finding formulas to describe real-world phenomena is called **mathematical modeling**. Such formulas, together with the meaning assigned to the variables, are called **mathematical models**. We often say that these formulas model, or describe, the relationships among the variables.

Example 4   Modeling Caloric Needs

The bar graph in **Figure 4.1** shows the estimated number of calories per day needed to maintain energy balance for various gender and age groups for moderately active lifestyles. (Moderately active means a lifestyle that includes physical activity equivalent to walking 1.5 to 3 miles per day at 3 to 4 miles per hour, in addition to the light physical activity associated with typical day-to-day life.)

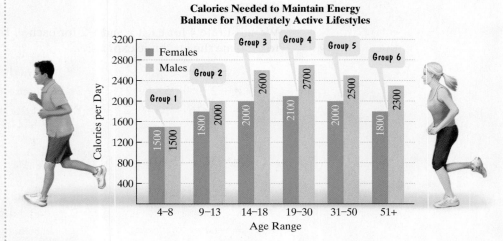

Calories Needed to Maintain Energy Balance for Moderately Active Lifestyles

Figure 4.1
Source: U.S.D.A.

The mathematical model

$$F = -66x^2 + 526x + 1030$$

describes the number of calories needed per day, F, by females in age group x with moderately active lifestyles. According to the model, how many calories per day are needed by females between the ages of 19 and 30, inclusive, with this lifestyle? Does this underestimate or overestimate the number shown by the graph in **Figure 4.1**? By how much?

Solution

Because the 19–30 age range is designated as group 4, we substitute 4 for x in the given model. Then we use the order of operations to find F, the number of calories needed per day by females between the ages of 19 and 30.

$F = -66x^2 + 526x + 1030$	This is the given mathematical model.
$F = -66 \cdot 4^2 + 526 \cdot 4 + 1030$	Replace each occurrence of x with 4.
$F = -66 \cdot 16 + 526 \cdot 4 + 1030$	Evaluate the exponential expression: $4^2 = 4 \cdot 4 = 16$.
$F = -1056 + 2104 + 1030$	Multiply from left to right: $-66 \cdot 16 = -1056$ and $526 \cdot 4 = 2104$.
$F = 2078$	Add.

The formula indicates that females in the 19–30 age range with moderately active lifestyles need 2078 calories per day. **Figure 4.1** indicates that 2100 calories are needed. Thus, the mathematical model underestimates caloric needs by $2100 - 2078$ calories, or by 22 calories per day.

 **Checkpoint 4** The mathematical model

$$M = -120x^2 + 998x + 590$$

describes the number of calories needed per day, M, by males in age group x with moderately active lifestyles. According to the model, how many calories per day are needed by males between the ages of 19 and 30, inclusive, with this lifestyle? Does this underestimate or overestimate the number shown by the graph in **Figure 4.1**? ▶ By how much?

The Vocabulary of Algebraic Expressions

3 Understand the vocabulary of algebraic expressions.

We have seen that an algebraic expression combines numbers and variables. Here is another example of an algebraic expression:

$$7x - 9y - 3.$$

The **terms** of an algebraic expression are those parts that are separated by addition. For example, we can rewrite $7x - 9y - 3$ as

$$7x + (-9y) + (-3).$$

This expression contains three terms, namely $7x$, $-9y$, and -3.

The numerical part of a term is called its **coefficient**. In the term $7x$, the 7 is the coefficient. In the term $-9y$, the -9 is the coefficient.

Coefficients of 1 and -1 are not written. Thus, the coefficient of x, meaning $1x$, is 1. Similarly, the coefficient of $-y$, meaning $-1y$, is -1.

A term that consists of just a number is called a **numerical term** or a **constant**. The numerical term of $7x - 9y - 3$ is -3.

The parts of each term that are multiplied are called the **factors** of the term. The factors of the term $7x$ are 7 and x.

Like terms are terms that have the same variable factors. For example, $3x$ and $7x$ are like terms.

Simplify algebraic
expressions.

Simplifying Algebraic Expressions

The properties of real numbers that we discussed in Chapter 3 can be applied to algebraic expressions.

Properties of Real Numbers

Property	Example
Commutative Property of Addition $a + b = b + a$	$13x^2 + 7x = 7x + 13x^2$
Commutative Property of Multiplication $ab = ba$	$x \cdot 6 = 6x$
Associative Property of Addition $(a + b) + c = a + (b + c)$	$3 + (8 + x) = (3 + 8) + x = 11 + x$
Associative Property of Multiplication $(ab)c = a(bc)$	$-2(3x) = (-2 \cdot 3)x = -6x$
Distributive Property $a(b + c) = ab + ac$	$5(3x + 7) = 5 \cdot 3x + 5 \cdot 7 = 15x + 35$
$a(b - c) = ab - ac$	$4(2x - 5) = 4 \cdot 2x - 4 \cdot 5 = 8x - 20$

Study Tip

To combine like terms mentally, add or subtract the coefficients of the terms. Use this result as the coefficient of the terms' variable factor(s).

The distributive property in the form

$$ba + ca = (b + c)a$$

enables us to add or subtract like terms. For example,

$$3x + 7x = (3 + 7)x = 10x$$
$$7y^2 - y^2 = 7y^2 - 1y^2 = (7 - 1)y^2 = 6y^2.$$

This process is called **combining like terms**.

An algebraic expression is **simplified** when parentheses have been removed and like terms have been combined.

Example 5 Simplifying an Algebraic Expression

Simplify: $5(3x - 7) - 6x$.

Solution

$$5(3x - 7) - 6x$$

$= 5 \cdot 3x - 5 \cdot 7 - 6x$ Use the distributive property to remove the parentheses.

$= 15x - 35 - 6x$ Multiply.

$= (15x - 6x) - 35$ Group like terms.

$= 9x - 35$ Combine like terms: $15x - 6x = (15 - 6)x = 9x$.

▶  **Checkpoint 5**   Simplify: $7(2x - 3) - 11x$.

Example 6   Simplifying an Algebraic Expression

Simplify: $6(2x^2 + 4x) + 10(4x^2 + 3x)$.

Solution

> $52x^2$ and $54x$ are not like terms. They contain different variable factors, x^2 and x, and cannot be combined.

$6(2x^2 + 4x) + 10(4x^2 + 3x)$

$= 6 \cdot 2x^2 + 6 \cdot 4x + 10 \cdot 4x^2 + 10 \cdot 3x$ Use the distributive property to remove the parentheses.

$= 12x^2 + 24x + 40x^2 + 30x$ Multiply.

$= (12x^2 + 40x^2) + (24x + 30x)$ Group like terms.

$= 52x^2 + 54x$ Combine like terms:
$12x^2 + 40x^2 = (12 + 40)x^2 = 52x^2$
and $24x + 30x = (24 + 30)x = 54x$.

Checkpoint 6   Simplify: $7(4x^2 + 3x) + 2(5x^2 + x)$.

It is not uncommon to see algebraic expressions with parentheses preceded by a negative sign or subtraction. An expression of the form $-(a + b)$ can be simplified as follows:

$$-(a + b) = -1(a + b) = (-1)a + (-1)b = -a + (-b) = -a - b.$$

Do you see a fast way to obtain the simplified expression on the right? **If a negative sign or a subtraction symbol appears outside parentheses, drop the parentheses and change the sign of every term within the parentheses.** For example,

$$-(3x^2 - 7x - 4) = -3x^2 + 7x + 4.$$

Example 7   Simplifying an Algebraic Expression

Simplify: $8x + 2[5 - (x - 3)]$.

Solution

$8x + 2[5 - (x - 3)]$

$= 8x + 2[5 - x + 3]$ Drop parentheses and change the sign of each term in parentheses: $-(x - 3) = -x + 3$.

$= 8x + 2[8 - x]$ Simplify inside brackets: $5 + 3 = 8$.

$= 8x + 16 - 2x$ Apply the distributive property:

$2[8 - x] = 2 \cdot 8 - 2x = 16 - 2x$.

$= (8x - 2x) + 16$ Group like terms.

$= 6x + 16$ Combine like terms: $8x - 2x = (8 - 2)x = 6x$.

Checkpoint 7   Simplify: $6x + 4[7 - (x - 2)]$.

Blitzer Bonus

Using Algebra to Solve Problems in Your Everyday Life

In *Geek Logik* (Workman Publishing, 2006), humorist Garth Sundem presents formulas covering dating, romance, career, finance, everyday decisions, and health. On the right is a sample of one of his formulas that "takes the guesswork out of life, providing easier living through algebra."

Should You Apologize to Your Friend?

In the formula on the right, each variable is assigned a number from 1 (low) to 10 (high).

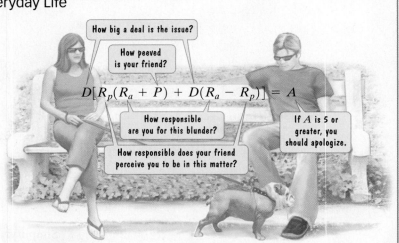

How big a deal is the issue?

How peeved is your friend?

$$D[R_p(R_a + P) + D(R_a - R_p)] = A$$

How responsible are you for this blunder?

If A is 5 or greater, you should apologize.

How responsible does your friend perceive you to be in this matter?

Achieving Success

Algebra is cumulative. This means that the topics build on one another. Understanding each topic depends on understanding the previous material. Do not let yourself fall behind.

Exercise Set 4.1

Concept and Vocabulary Exercises

In Exercises 1–6, fill in each blank so that the resulting statement is true.

1. Finding the value of an algebraic expression for a given value of the variable is called _____ the expression.

2. When an equal sign is placed between two algebraic expressions, an _____ is formed.

3. The parts of an algebraic expression that are separated by addition are called the _____ of the expression.

4. In the algebraic expression $7x$, 7 is called the _____ because it is the numerical part.

5. In the algebraic expression $7x$, 7 and x are called _____ because they are multiplied together.

6. The algebraic expressions $3x$ and $7x$ are called _____ because they contain the same variable to the same power.

In Exercises 7–14, determine whether each statement is true or false. If the statement is false, make the necessary change(s) to produce a true statement.

7. The term x has no coefficient.

8. $5 + 3(x - 4) = 8(x - 4) = 8x - 32$

9. $-x - x = -x + (-x) = 0$

10. $x - 0.02(x + 200) = 0.98x - 4$

11. $3 + 7x = 10x$

12. $b \cdot b = 2b$

13. $(3y - 4) - (8y - 1) = -5y - 3$

14. $-4y + 4 = -4(y + 4)$

Respond to Exercises 15–20 using verbal or written explanations.

15. What is an algebraic expression? Provide an example with your description.

16. What does it mean to evaluate an algebraic expression? Provide an example with your description.

17. What is a term? Provide an example with your description.

18. What are like terms? Provide an example with your description.

19. Explain how to add like terms. Give an example.

20. What does it mean to simplify an algebraic expression?

In Exercises 21–24, determine whether each statement makes sense or does not make sense, and explain your reasoning.

21. I did not use the distributive property to simplify $3(2x + 5x)$.

22. The terms $13x^2$ and $10x$ both contain the variable x, so I can combine them to obtain $23x^3$.

23. Regardless of what real numbers I substitute for x and y, I will always obtain zero when evaluating $2x^2y - 2yx^2$.

24. A model that describes the percentage of U.S. teens who used social networks x years after 2008 cannot be used to estimate the percentage in 2008.

Practice Exercises

In Exercises 25–58, evaluate the algebraic expression for the given value or values of the variables.

25. $5x + 7$; $x = 4$
26. $9x + 6$; $x = 5$
27. $-7x - 5$; $x = -4$
28. $-6x - 13$; $x = -3$
29. $x^2 + 4$; $x = 5$
30. $x^2 + 9$; $x = 3$
31. $x^2 - 6$; $x = -2$
32. $x^2 - 11$; $x = -3$
33. $-x^2 + 4$; $x = 5$
34. $-x^2 + 9$; $x = 3$
35. $-x^2 - 6$; $x = -2$
36. $-x^2 - 11$; $x = -3$
37. $x^2 + 4x$; $x = 10$
38. $x^2 + 6x$; $x = 9$
39. $8x^2 + 17$; $x = 5$
40. $7x^2 + 25$; $x = 3$
41. $x^2 - 5x$; $x = -11$
42. $x^2 - 8x$; $x = -5$
43. $x^2 + 5x - 6$; $x = 4$
44. $x^2 + 7x - 4$; $x = 6$
45. $4 + 5(x - 7)^3$; $x = 9$
46. $6 + 5(x - 6)^3$; $x = 8$
47. $x^2 - 3(x - y)$; $x = 2, y = 8$
48. $x^2 - 4(x - y)$; $x = 3, y = 8$
49. $2x^2 - 5x - 6$; $x = -3$
50. $3x^2 - 4x - 9$; $x = -5$
51. $-5x^2 - 4x - 11$; $x = -1$
52. $-6x^2 - 11x - 17$; $x = -2$
53. $3x^2 + 2xy + 5y^2$; $x = 2, y = 3$
54. $4x^2 + 3xy + 2y^2$; $x = 3, y = 2$
55. $-x^2 - 4xy + 3y^3$; $x = -1, y = -2$
56. $-x^2 - 3xy + 4y^3$; $x = -3, y = -1$
57. $\dfrac{2x + 3y}{x + 1}$; $x = -2, y = 4$
58. $\dfrac{2x + y}{xy - 2x}$; $x = -2, y = 4$

The formula

$$C = \frac{5}{9}(F - 32)$$

expresses the relationship between Fahrenheit temperature, F, and Celsius temperature, C. In Exercises 59–60, use the formula to convert the given Fahrenheit temperature to its equivalent temperature on the Celsius scale.

59. $50°F$
60. $86°F$

A football was kicked vertically upward from a height of 4 feet with an initial speed of 60 feet per second. The formula

$$h = 4 + 60t - 16t^2$$

describes the ball's height above the ground, h, in feet, t seconds after it was kicked. Use this formula to solve Exercises 61–62.

61. What was the ball's height 2 seconds after it was kicked?
62. What was the ball's height 3 seconds after it was kicked?

In Exercises 63–84, simplify each algebraic expression.

63. $7x + 10x$
64. $5x + 13x$
65. $5x^2 - 8x^2$
66. $7x^2 - 10x^2$
67. $3(x + 5)$
68. $4(x + 6)$
69. $4(2x - 3)$
70. $3(4x - 5)$
71. $5(3x + 4) - 4$
72. $2(5x + 4) - 3$
73. $5(3x - 2) + 12x$
74. $2(5x - 1) + 14x$
75. $7(3y - 5) + 2(4y + 3)$
76. $4(2y - 6) + 3(5y + 10)$
77. $5(3y - 2) - (7y + 2)$
78. $4(5y - 3) - (6y + 3)$
79. $3(-4x^2 + 5x) - (5x - 4x^2)$
80. $2(-5x^2 + 3x) - (3x - 5x^2)$
81. $7 - 4[3 - (4y - 5)]$
82. $6 - 5[8 - (2y - 4)]$
83. $8x - 3[5 - (7 - 6x)]$
84. $7x - 4[6 - (8 - 5x)]$

Practice Plus

In Exercises 85–88, simplify each algebraic expression.

85. $18x^2 + 4 - [6(x^2 - 2) + 5]$
86. $14x^2 + 5 - [7(x^2 - 2) + 4]$
87. $2(3x^2 - 5) - [4(2x^2 - 1) + 3]$
88. $4(6x^2 - 3) - [2(5x^2 - 1) + 1]$

Application Exercises

The maximum heart rate, in beats per minute, that you should achieve during exercise is 220 minus your age:

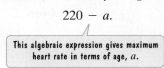

$$220 - a.$$

This algebraic expression gives maximum heart rate in terms of age, a.

The bar graph shows the target heart rate ranges for four types of exercise goals. The lower and upper limits of these ranges are fractions of the maximum heart rate, 220 – a. Exercises 89–90 are based on the information in the graph.

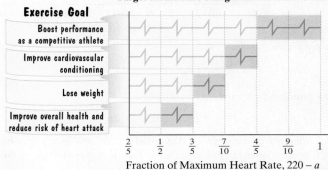

Target Heart Rate Ranges for Exercise Goals

Fraction of Maximum Heart Rate, 220 – a

89. If your exercise goal is to improve cardiovascular conditioning, the graph shows the following range for target heart rate, H, in beats per minute:

Lower limit of range $H = \dfrac{7}{10}(220 - a)$

Upper limit of range $H = \dfrac{4}{5}(220 - a).$

a. What is the lower limit of the heart rate range, in beats per minute, for a 20-year-old with this exercise goal?
b. What is the upper limit of the heart rate range, in beats per minute, for a 20-year-old with this exercise goal?

90. If your exercise goal is to improve overall health, the graph on the previous page shows the following range for target heart rate, *H*, in beats per minute:

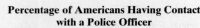

Lower limit of range $H = \dfrac{1}{2}(220 - a)$

Upper limit of range $H = \dfrac{3}{5}(220 - a)$.

   a. What is the lower limit of the heart rate range, in beats per minute, for a 30-year-old with this exercise goal?

   b. What is the upper limit of the heart rate range, in beats per minute, for a 30-year-old with this exercise goal?

Hello, officer! The bar graph shows the percentage of Americans in various age groups who had contact with a police officer, for anything from an arrest to asking directions, in a recent year.

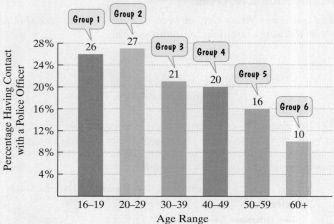

Percentage of Americans Having Contact with a Police Officer

Source: Bureau of Justice Statistics

The mathematical model

$$P = -0.5x^2 + 0.1x + 26.9$$

describes the percentage of Americans, P, in age group x who had contact with a police officer. Use this formula to solve Exercises 91–92.

91. According to the formula, what percentage of Americans ages 60 and older had contact with a police officer? Does this underestimate or overestimate the percentage shown by the graph? By how much?

92. According to the formula, what percentage of Americans between the ages of 20 and 29, inclusive, had contact with a police officer? Does this underestimate or overestimate the percentage shown by the graph? By how much?

The line graph at the top of the next column shows the cost of inflation. What cost $10,000 in 1967 would cost the amount shown by the graph in subsequent years.

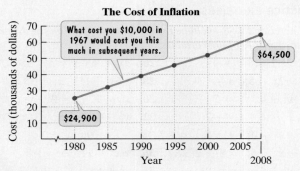

The Cost of Inflation

What cost you $10,000 in 1967 would cost you this much in subsequent years.

$64,500

$24,900

Source: Bureau of Labor Statistics

Here are two mathematical models for the data shown by the graph. In each formula, C represents the cost x years after 1980 of what cost $10,000 in 1967.

Model 1 $C = 1388x + 24{,}963$

Model 2 $C = 3x^2 + 1308x + 25{,}268$

Use these models to solve Exercises 93–96.

93. **a.** Use the graph to estimate the cost in 2000, to the nearest thousand dollars, of what cost $10,000 in 1967.

   b. Use model 1 to determine the cost in 2000. How well does this describe your estimate from part (a)?

   c. Use model 2 to determine the cost in 2000. How well does this describe your estimate from part (a)?

94. **a.** Use the graph to estimate the cost in 1990, to the nearest thousand dollars, of what cost $10,000 in 1967.

   b. Use model 1 to determine the cost in 1990. How well does this describe your estimate from part (a)?

   c. Use model 2 to determine the cost in 1990. How well does this describe your estimate from part (a)?

95. Which model is a better description for the cost in 1980 of what cost $10,000 in 1967? Does this model underestimate or overestimate the cost shown by the graph? By how much?

96. Which model is a better description for the cost in 2008 of what cost $10,000 in 1967? Does this model underestimate or overestimate the cost shown by the graph? By how much?

Critical Thinking Exercises

97. A business that manufactures small alarm clocks has weekly fixed costs of $5000. The average cost per clock for the business to manufacture *x* clocks is described by

$$\frac{0.5x + 5000}{x}.$$

   a. Find the average cost when *x* = 100, 1000, and 10,000.

   b. Like all other businesses, the alarm clock manufacturer must make a profit. To do this, each clock must be sold for at least 50¢ more than what it costs to manufacture. Due to competition from a larger company, the clocks can be sold for $1.50 each and no more. Our small manufacturer can only produce 2000 clocks weekly. Does this business have much of a future? Explain.

98. Think of a situation where you either apologized or did not apologize for a blunder you committed. Use the formula in the Blitzer Bonus on page 226 to determine whether or not you should have apologized. How accurately does the formula model what you actually did?

4.2 Linear Equations in One Variable

Objectives

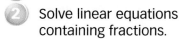

1. Solve linear equations.

2. Solve linear equations containing fractions.

3. Identify equations with no solution or infinitely many solutions.

Sense of Humor and Depression

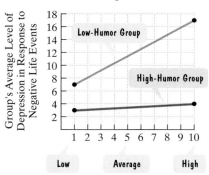

Figure 4.2

Source: Steven Davis and Joseph Palladino, *Psychology*, Fifth Edition. Prentice Hall, 2007.

The belief that humor and laughter can have positive benefits on our lives is not new. The graphs in **Figure 4.2** indicate that persons with a low sense of humor have higher levels of depression in response to negative life events than those with a high sense of humor. These graphs can be modeled by the following formulas:

Low-Humor Group	High-Humor Group
$D = \dfrac{10}{9}x + \dfrac{53}{9}$	$D = \dfrac{1}{9}x + \dfrac{26}{9}.$

In each formula, x represents the intensity of a negative life event (from 1, low, to 10, high) and D is the level of depression in response to that event.

Suppose that the low-humor group averages a level of depression of 10 in response to a negative life event. We can determine the intensity of that event by substituting 10 for D in the low-humor model, $D = \dfrac{10}{9}x + \dfrac{53}{9}$:

$$10 = \frac{10}{9}x + \frac{53}{9}.$$

The two sides of an equation can be reversed. So, we can also express this equation as

$$\frac{10}{9}x + \frac{53}{9} = 10.$$

Notice that the highest exponent on the variable is 1. Such an equation is called a *linear equation in one variable*. In this section, we will study how to solve such equations.

Solving Linear Equations in One Variable

1. Solve linear equations.

We begin with the general definition of a linear equation in one variable.

Definition of a Linear Equation

A **linear equation in one variable** x is an equation that can be written in the form

$$ax + b = 0,$$

where a and b are real numbers, and $a \neq 0$.

An example of a linear equation in one variable is

$$4x + 12 = 0.$$

Solving an equation in x involves determining all values of x that result in a true statement when substituted into the equation. Such values are **solutions**, or **roots**, of the equation. For example, substitute -3 for x in $4x + 12 = 0$. We obtain

$$4(-3) + 12 = 0, \quad \text{or} \quad -12 + 12 = 0.$$

This simplifies to the true statement $0 = 0$. Thus, -3 is a solution of the equation $4x + 12 = 0$. We also say that -3 **satisfies** the equation $4x + 12 = 0$, because when we substitute -3 for x, a true statement results. The set of all such solutions is called the equation's **solution set**. For example, the solution set of the equation $4x + 12 = 0$ is $\{-3\}$.

Two or more equations that have the same solution set are called **equivalent equations**. For example, the equations

$$4x + 12 = 0 \quad \text{and} \quad 4x = -12 \quad \text{and} \quad x = -3$$

are equivalent equations because the solution set for each is $\{-3\}$. To solve a linear equation in x, we transform the equation into an equivalent equation one or more times. Our final equivalent equation should be of the form

$$x = \text{a number}.$$

The solution set of this equation is the set consisting of the number.

To generate equivalent equations, we will use the following properties:

The Addition and Multiplication Properties of Equality

The Addition Property of Equality

The same real number or algebraic expression may be added to both sides of an equation without changing the equation's solution set.

$$a = b \text{ and } a + c = b + c \text{ are equivalent equations.}$$

The Multiplication Property of Equality

The same nonzero real number may multiply both sides of an equation without changing the equation's solution set.

$$a = b \text{ and } ac = bc \text{ are equivalent equations as long as } c \neq 0.$$

Because subtraction is defined in terms of addition, the addition property also lets us subtract the same number from both sides of an equation without changing the equation's solution set. Similarly, because division is defined in terms of multiplication, the multiplication property of equality can be used to divide both sides of an equation by the same nonzero number to obtain an equivalent equation.

Table 4.1 illustrates how these properties are used to isolate x to obtain an equation of the form $x = $ a number.

Table 4.1 Using Properties of Equality to Solve Equations

Equation	How to Isolate x	Solving the Equation	The Equation's Solution Set
$x - 3 = 8$	Add 3 to both sides.	$x - 3 + 3 = 8 + 3$ $x = 11$	$\{11\}$
$x + 7 = -15$	Subtract 7 from both sides.	$x + 7 - 7 = -15 - 7$ $x = -22$	$\{-22\}$
$6x = 30$	Divide both sides by 6 (or multiply both sides by $\frac{1}{6}$).	$\dfrac{6x}{6} = \dfrac{30}{6}$ $x = 5$	$\{5\}$
$\dfrac{x}{5} = 9$	Multiply both sides by 5.	$5 \cdot \dfrac{x}{5} = 5 \cdot 9$ $x = 45$	$\{45\}$

These equations are solved using the **Addition Property of Equality.**

These equations are solved using the **Multiplication Property of Equality.**

Example 1   Using Properties of Equality to Solve an Equation

Solve and check: $2x + 3 = 17$.

Solution

Our goal is to obtain an equivalent equation with x isolated on one side and a number on the other side.

$$2x + 3 = 17 \quad \text{This is the given equation.}$$
$$2x + 3 - 3 = 17 - 3 \quad \text{Subtract 3 from both sides.}$$
$$2x = 14 \quad \text{Simplify.}$$
$$\frac{2x}{2} = \frac{14}{2} \quad \text{Divide both sides by 2.}$$
$$x = 7 \quad \text{Simplify: } \frac{2x}{2} = 1x = x \text{ and } \frac{14}{2} = 7.$$

Now we check the proposed solution, 7, by replacing x with 7 in the original equation.

$$2x + 3 = 17 \quad \text{This is the original equation.}$$
$$2 \cdot 7 + 3 \stackrel{?}{=} 17 \quad \text{Substitute 7 for x. The question mark indicates that we do not yet know if the two sides are equal.}$$
$$14 + 3 \stackrel{?}{=} 17 \quad \text{Multiply: } 2 \cdot 7 = 14.$$
$$17 = 17 \quad \text{Add: } 14 + 3 = 17.$$

This statement is true.

Because the check results in a true statement, we conclude that the solution set of the given equation is $\{7\}$.

Checkpoint 1 Solve and check: $4x + 5 = 29$.

Here is a step-by-step procedure for solving a linear equation in one variable. Not all of these steps are necessary to solve every equation.

> **Solving a Linear Equation**
>
> 1. Simplify the algebraic expression on each side by removing grouping symbols and combining like terms.
> 2. Collect all the variable terms on one side and all the constants, or numerical terms, on the other side.
> 3. Isolate the variable and solve.
> 4. Check the proposed solution in the original equation.

Example 2   Solving a Linear Equation

Solve and check: $2(x - 4) - 5x = -5$.

Solution

Step 1   Simplify the algebraic expression on each side.

$$2(x - 4) - 5x = -5 \qquad \text{This is the given equation.}$$
$$2x - 8 - 5x = -5 \qquad \text{Use the distributive property.}$$
$$-3x - 8 = -5 \qquad \text{Combine like terms: } 2x - 5x = -3x.$$

Step 2   Collect variable terms on one side and constants on the other side. The only variable term in $-3x - 8 = -5$ is $-3x$, and $-3x$ is already on the left side. We will collect constants on the right side by adding 8 to both sides.

$$-3x - 8 + 8 = -5 + 8 \qquad \text{Add 8 to both sides.}$$
$$-3x = 3 \qquad \text{Simplify.}$$

Step 3   Isolate the variable and solve. We isolate the variable, x, by dividing both sides of $-3x = 3$ by -3.

$$\frac{-3x}{-3} = \frac{3}{-3} \qquad \text{Divide both sides by } -3.$$

$$x = -1 \qquad \text{Simplify: } \frac{-3x}{-3} = 1x = x \text{ and } \frac{3}{-3} = -1.$$

Step 4   Check the proposed solution in the original equation. Substitute -1 for x in the original equation.

$$2(x - 4) - 5x = -5 \qquad \text{This is the original equation.}$$
$$2(-1 - 4) - 5(-1) \overset{?}{=} -5 \qquad \text{Substitute } -1 \text{ for } x.$$
$$2(-5) - 5(-1) \overset{?}{=} -5 \qquad \begin{array}{l}\text{Simplify inside parentheses:}\\ -1 - 4 = -1 + (-4) = -5.\end{array}$$
$$-10 - (-5) \overset{?}{=} -5 \qquad \text{Multiply: } 2(-5) = -10 \text{ and } 5(-1) = -5.$$

$$\underset{\text{This statement is true.}}{-5 = -5} \qquad -10 - (-5) = -10 + 5 = -5$$

Because the check results in a true statement, we conclude that the solution set of the given equation is $\{-1\}$.

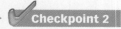 **Checkpoint 2**   Solve and check: $6(x - 3) - 10x = -10$.

Study Tip

We simplify algebraic expressions. We solve algebraic equations. Although basic rules of algebra are used in both procedures, notice the differences between the procedures:

Simplifying an Algebraic Expression	**Solving an Algebraic Equation**

Simplify: $3(x - 7) - (5x - 11)$.   Solve: $3(x - 7) - (5x - 11) = 14$.

> This is not an equation. There is no equal sign.

> This is an equation. There is an equal sign.

Solution $3(x - 7) - (5x - 11)$
$$= 3x - 21 - 5x + 11$$
$$= (3x - 5x) + (-21 + 11)$$
$$= -2x + (-10)$$
$$= -2x - 10$$

> Stop! Further simplification is not possible. Avoid the common error of setting $-2x - 10$ equal to 0.

Solution $3(x - 7) - (5x - 11) = 14$
$$3x - 21 - 5x + 11 = 14$$
$$-2x - 10 = 14$$

> Add 10 to both sides.

$$-2x - 10 + 10 = 14 + 10$$
$$-2x = 24$$

> Divide both sides by -2.

$$\frac{-2x}{-2} = \frac{24}{-2}$$
$$x = -12$$

The solution set is $\{-12\}$.

Example 3   Solving a Linear Equation

Solve and check: $5x - 12 = 8x + 24$.

Solution

Step 1   Simplify the algebraic expression on each side. Neither side contains grouping symbols or like terms that can be combined. Therefore, we can skip this step.

Step 2   Collect variable terms on one side and constants on the other side. One way to do this is to collect variable terms on the left and constants on the right. This is accomplished by subtracting $8x$ from both sides and adding 12 to both sides.

$5x - 12 = 8x + 24$	This is the given equation.
$5x - 12 - 8x = 8x + 24 - 8x$	Subtract $8x$ from both sides.
$-3x - 12 = 24$	Simplify: $5x - 8x = -3x$.
$-3x - 12 + 12 = 24 + 12$	Add 12 to both sides and collect constants on the right side.
$-3x = 36$	Simplify.

Step 3   Isolate the variable and solve. We isolate the variable, x, by dividing both sides of $-3x = 36$ by -3.

$\dfrac{-3x}{-3} = \dfrac{36}{-3}$	Divide both sides by -3.
$x = -12$	Simplify.

Study Tip

You can solve
$$5x - 12 = 8x + 24$$
by collecting variable terms on the right and numbers on the left. To collect variable terms on the right, subtract $5x$ from both sides:
$$5x - 12 - 5x = 8x + 24 - 5x$$
$$-12 = 3x + 24.$$
To collect numbers on the left, subtract 24 from both sides:
$$-12 - 24 = 3x + 24 - 24$$
$$-36 = 3x.$$
Now isolate x by dividing both sides by 3:
$$\frac{-36}{3} = \frac{3x}{3}$$
$$-12 = x.$$
This is the same equation that we obtained in Example 3.

Step 4   Check the proposed solution in the original equation. Because we isolated the variable and obtained $x = -12$, substitute -12 for x in the original equation.

$$5x - 12 = 8x + 24 \qquad \text{This is the original equation.}$$

$$5(-12) - 12 \overset{?}{=} 8(-12) + 24 \qquad \text{Substitute } -12 \text{ for x.}$$

$$-60 - 12 \overset{?}{=} -96 + 24 \qquad \text{Multiply: } 5(-12) = -60 \text{ and } 8(-12) = -96.$$

> This statement is true. $-72 = -72$ Add: $-60 + (-12) = -72$ and $-96 + 24 = -72.$

Because the check results in a true statement, we conclude that the solution set of the given equation is $\{-12\}$.

> **Checkpoint 3**   Solve the equation: $2x + 9 = 8x - 3$.

Example 4   **Solving a Linear Equation**

Solve and check: $2(x - 3) - 17 = 13 - 3(x + 2)$.

Solution

Step 1   Simplify the algebraic expression on each side.

> Do not begin with $13 - 3$. Multiplication (the distributive property) is applied before subtraction.

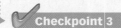

$$2(x - 3) - 17 = 13 - 3(x + 2) \qquad \text{This is the given equation.}$$

$$2x - 6 - 17 = 13 - 3x - 6 \qquad \text{Use the distributive property.}$$

$$2x - 23 = -3x + 7 \qquad \text{Combine like terms.}$$

Step 2   Collect variable terms on one side and constants on the other side. We will collect variable terms of $2x - 23 = -3x + 7$ on the left by adding $3x$ to both sides. We will collect the numbers on the right by adding 23 to both sides.

$$2x - 23 + 3x = -3x + 7 + 3x \qquad \text{Add 3x to both sides.}$$

$$5x - 23 = 7 \qquad \text{Simplify: } 2x + 3x = 5x.$$

$$5x - 23 + 23 = 7 + 23 \qquad \text{Add 23 to both sides.}$$

$$5x = 30 \qquad \text{Simplify.}$$

Step 3   Isolate the variable and solve. We isolate the variable, x, by dividing both sides of $5x = 30$ by 5.

$$\frac{5x}{5} = \frac{30}{5} \qquad \text{Divide both sides by 5.}$$

$$x = 6 \qquad \text{Simplify.}$$

Step 4   Check the proposed solution in the original equation. Substitute 6 for x in the original equation.

$$2(x - 3) - 17 = 13 - 3(x + 2) \qquad \text{This is the original equation.}$$

$$2(6 - 3) - 17 \overset{?}{=} 13 - 3(6 + 2) \qquad \text{Substitute 6 for x.}$$

$$2(3) - 17 \overset{?}{=} 13 - 3(8) \qquad \text{Simplify inside parentheses.}$$

$$6 - 17 \overset{?}{=} 13 - 24 \qquad \text{Multiply.}$$

$$-11 = -11 \qquad \text{Subtract.}$$

The true statement $-11 = -11$ verifies that the solution set is $\{6\}$.

> **Checkpoint 4**   Solve and check: $4(2x + 1) = 29 + 3(2x - 5)$.

Linear Equations with Fractions

2 Solve linear equations containing fractions.

Equations are easier to solve when they do not contain fractions. How do we remove fractions from an equation? We begin by multiplying both sides of the equation by the least common denominator of any fractions in the equation. The least common denominator is the smallest number that all denominators will divide into. Multiplying every term on both sides of the equation by the least common denominator will eliminate the fractions in the equation. Example 5 shows how we "clear an equation of fractions."

Example 5   Solving a Linear Equation Involving Fractions

Solve and check: $\dfrac{3x}{2} = \dfrac{8x}{5} - 4$.

Solution

The denominators are 2 and 5. The smallest number that is divisible by both 2 and 5 is 10. We begin by multiplying both sides of the equation by 10, the least common denominator.

$$\frac{3x}{2} = \frac{8x}{5} - 4 \qquad \text{This is the given equation.}$$

$$10 \cdot \frac{3x}{2} = 10\left(\frac{8x}{5} - 4\right) \qquad \text{Multiply both sides by 10.}$$

$$10 \cdot \frac{3x}{2} = 10 \cdot \frac{8x}{5} - 10 \cdot 4 \qquad \text{Use the distributive property. Be sure to multiply all terms by 10.}$$

$$\overset{5}{\cancel{10}} \cdot \frac{3x}{\underset{1}{\cancel{2}}} = \overset{2}{\cancel{10}} \cdot \frac{8x}{\underset{1}{\cancel{5}}} - 40 \qquad \text{Divide out common factors in the multiplications.}$$

$$15x = 16x - 40 \qquad \begin{array}{l}\text{Complete the multiplications: } 5 \cdot 3x = 15x \text{ and} \\ 2 \cdot 8x = 16. \text{ The fractions are now cleared.}\end{array}$$

At this point, we have an equation similar to those we have previously solved. Collect the variable terms on one side and the constants on the other side.

$$15x - 16x = 16x - 40 - 16x \qquad \begin{array}{l}\text{Subtract 16x from both sides to get the variable} \\ \text{terms on the left.}\end{array}$$

$$-x = -40 \qquad \text{Simplify.}$$

> **We're not finished. A negative sign should not precede x.**

Isolate x by multiplying or dividing both sides of this equation by -1.

$$\frac{-x}{-1} = \frac{-40}{-1} \qquad \text{Divide both sides by } -1.$$

$$x = 40 \qquad \text{Simplify.}$$

Check the proposed solution. Substitute 40 for x in the original equation. You should obtain $60 = 60$. This true statement verifies that the solution set is $\{40\}$.

▶ **Checkpoint 5**   Solve and check: $\dfrac{2x}{3} = 7 - \dfrac{x}{2}$.

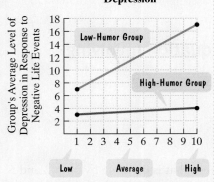

Sense of Humor and Depression

Group's Average Level of Depression in Response to Negative Life Events

Low-Humor Group

High-Humor Group

Low   Average   High

Intensity of Negative Life Event

Figure 4.2 (repeated)

| Example 6 | An Application: Responding to Negative Life Events |

In the section opener, we introduced line graphs, repeated in **Figure 4.2**, indicating that persons with a low sense of humor have higher levels of depression in response to negative life events than those with a high sense of humor. These graphs can be modeled by the following formulas:

Low-Humor Group   High-Humor Group

$$D = \frac{10}{9}x + \frac{53}{9} \qquad D = \frac{1}{9}x + \frac{26}{9}.$$

In each formula, x represents the intensity of a negative life event (from 1, low, to 10, high) and D is the average level of depression in response to that event. If the high-humor group averages a level of depression of 3.5, or $\frac{7}{2}$, in response to a negative life event, what is the intensity of that event? How is the solution shown on the red line graph in **Figure 4.2**?

Solution

We are interested in the intensity of a negative life event with an average level of depression of $\frac{7}{2}$ for the high-humor group. We substitute $\frac{7}{2}$ for D in the high-humor model and solve for x, the intensity of the negative life event.

$$D = \frac{1}{9}x + \frac{26}{9}$$
    This is the given formula for the high-humor group.

$$\frac{7}{2} = \frac{1}{9}x + \frac{26}{9}$$
    Replace D with $\frac{7}{2}$.

$$18 \cdot \frac{7}{2} = 18\left(\frac{1}{9}x + \frac{26}{9}\right)$$
    Multiply both sides by 18, the least common denominator.

$$18 \cdot \frac{7}{2} = 18 \cdot \frac{1}{9}x + 18 \cdot \frac{26}{9}$$
    Use the distributive property.

$$\overset{9}{\cancel{18}} \cdot \frac{7}{2} = \overset{2}{\cancel{18}} \cdot \frac{1}{\underset{1}{9}}x + \overset{2}{\cancel{18}} \cdot \frac{26}{\underset{1}{9}}$$
    Divide out common factors in the multiplications.

$$63 = 2x + 52$$
    Complete the multiplications. The fractions are now cleared.

$$63 - 52 = 2x + 52 - 52$$
    Subtract 52 from both sides to get constants on the left.

$$11 = 2x$$
    Simplify.

$$\frac{11}{2} = \frac{2x}{2}$$
    Divide both sides by 2.

$$\frac{11}{2} = x$$
    Simplify.

The formula indicates that if the high-humor group averages a level of depression of 3.5 in response to a negative life event, the intensity of that event is $\frac{11}{2}$, or 5.5. This is illustrated on the line graph for the high-humor group in **Figure 4.3**.

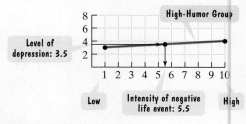

Level of depression: 3.5

High-Humor Group

Low   Intensity of negative life event: 5.5   High

Figure 4.3

✓ **Checkpoint 6** Use the model for the low-humor group given in Example 6 on the previous page to solve this problem. If the low-humor group averages a level of depression of 10 in response to a negative life event, what is the intensity of that event? How is the solution shown on the blue line graph in **Figure 4.2** on the previous page?

Equations with No Solution or Infinitely Many Solutions

3 Identify equations with no solution or infinitely many solutions.

Thus far, each equation that we have solved has had a single solution. However, some equations are not true for even one real number. By contrast, other equations are true for all real numbers.

If you attempt to solve an equation with no solution, you will eliminate the variable and obtain a false statement, such as $2 = 5$. If you attempt to solve an equation that is true for every real number, you will eliminate the variable and obtain a true statement, such as $4 = 4$.

Example 7 **Attempting to Solve an Equation with No Solution**

Solve: $2x + 6 = 2(x + 4)$.

Solution

$$2x + 6 = 2(x + 4) \quad \text{This is the given equation.}$$
$$2x + 6 = 2x + 8 \quad \text{Use the distributive property.}$$
$$2x + 6 - 2x = 2x + 8 - 2x \quad \text{Subtract 2x from both sides.}$$

Keep reading. 6 = 8 is not the solution.
$$6 = 8 \quad \text{Simplify.}$$

The original equation, $2x + 6 = 2(x + 4)$, is equivalent to the statement $6 = 8$, which is false for every value of x. The equation has no solution. The solution set is $\varnothing$, the empty set.

✓ **Checkpoint 7** Solve: $3x + 7 = 3(x + 1)$.

Example 8 **Solving an Equation for Which Every Real Number Is a Solution**

Solve: $4x + 6 = 6(x + 1) - 2x$.

Solution

$$4x + 6 = 6(x + 1) - 2x \quad \text{This is the given equation.}$$
$$4x + 6 = 6x + 6 - 2x \quad \text{Apply the distributive property on the right side.}$$
$$4x + 6 = 4x + 6 \quad \text{Combine like terms on the right side: } 6x - 2x = 4x.$$

Can you see that the equation $4x + 6 = 4x + 6$ is true for every value of x? Let's continue solving the equation by subtracting $4x$ from both sides.

$$4x + 6 - 4x = 4x + 6 - 4x$$

Keep reading. 6 = 6 is not the solution.
$$6 = 6$$

The original equation, $4x + 6 = 6(x + 1) - 2x$, is equivalent to the statement $6 = 6$, which is true for every value of x. Thus, the solution set consists of the set of all real numbers, expressed in set-builder notation as $\{x \mid x$ is a real number$\}$. Try substituting any real number of your choice for x in the original equation. You will obtain a true statement.

✓ **Checkpoint 8** Solve: $7x + 9 = 9(x + 1) - 2x$.

Achieving Success

"I've already had algebra, so I can coast through this part of the course."

Don't use this "been there done that" mind set as an excuse for laziness and bad study habits.

Colleges are not interested in whether you've *had* algebra. What they want to determine is whether you've *learned* algebra. College placement exams are designed to see if you are ready to incorporate algebra as a prerequisite in courses such as business, statistics, biology, chemistry, and precalculus. Placement exams contain algebra problems you need to be able to address in college courses. You will be placed in a math class according to your performance. Learn algebra now and you can reap the benefits throughout college.

Exercise Set 4.2

Concept and Vocabulary Exercises

In Exercises 1–4, fill in each blank so that the resulting statement is true.

1. An equation that can be written in the form $ax + b = 0$, where a and b are _____ and $a \neq 0$, is called a/an _____.

2. If $a = b$, then $a + c =$ _____, which is called the _____ property of equality.

3. If $a = b$ and $c \neq 0$, then _____ $= bc$, which is called the _____ property of equality.

4. The algebraic expression $7(x - 4) + 2x$ can be _____, whereas the algebraic equation $7(x - 4) + 2x = 35$ can be _____.

In Exercises 5–8, determine whether each statement is true or false. If the statement is false, make the necessary change(s) to produce a true statement.

5. The equation $2x + 5 = 0$ is equivalent to $2x = 5$.

6. The equation $x + \frac{1}{3} = \frac{1}{2}$ is equivalent to $x + 2 = 3$.

7. The equation $3x = 2x$ has no solution.

8. The equation $3(x + 4) = 3(4 + x)$ has precisely one solution.

Respond to Exercises 9–14 using verbal or written explanations.

9. What is the solution set of an equation?

10. State the addition property of equality and give an example.

11. State the multiplication property of equality and give an example.

12. How do you know if an equation has one solution, no solution, or infinitely many solutions?

13. What is the difference between solving an equation such as $2(x - 4) + 5x = 34$ and simplifying an algebraic expression such as $2(x - 4) + 5x$? If there is a difference, which topic should be taught first? Why?

14. Suppose that you solve $\frac{x}{5} - \frac{x}{2} = 1$ by multiplying both sides by 20, rather than the least common denominator of 5 and 2 (namely, 10). Describe what happens. If you get the correct solution, why do you think we clear the equation of fractions by multiplying by the *least* common denominator?

In Exercises 15–18, determine whether each statement makes sense or does not make sense, and explain your reasoning.

15. Although I can solve $3x + \frac{1}{5} = \frac{1}{4}$ by first subtracting $\frac{1}{5}$ from both sides, I find it easier to begin by multiplying both sides by 20, the least common denominator.

16. Because I know how to clear an equation of fractions, I decided to clear the equation $0.5x + 8.3 = 12.4$ of decimals by multiplying both sides by 10.

17. The number 3 satisfies the equation $7x + 9 = 9(x + 1) - 2x$, so $\{3\}$ is the equation's solution set.

18. I can solve $-2x = 10$ using the addition property of equality.

Practice Exercises

In Exercises 19–76, solve and check each equation.

19. $x - 7 = 3$
20. $x - 3 = -17$
21. $x + 5 = -12$
22. $x + 12 = -14$
23. $\frac{x}{3} = 4$
24. $\frac{x}{5} = 3$
25. $5x = 45$
26. $6x = 18$
27. $8x = -24$
28. $5x = -25$
29. $-8x = 2$
30. $-6x = 3$
31. $5x + 3 = 18$
32. $3x + 8 = 50$
33. $6x - 3 = 63$
34. $5x - 8 = 72$
35. $4x - 14 = -82$
36. $9x - 14 = -77$
37. $14 - 5x = -41$
38. $25 - 6x = -83$
39. $9(5x - 2) = 45$
40. $10(3x + 2) = 70$
41. $5x - (2x - 10) = 35$
42. $11x - (6x - 5) = 40$
43. $3x + 5 = 2x + 13$
44. $2x - 7 = 6 + x$
45. $8x - 2 = 7x - 5$
46. $13x + 14 = -5 + 12x$
47. $7x + 4 = x + 16$
48. $8x + 1 = x + 43$
49. $8y - 3 = 11y + 9$
50. $5y - 2 = 9y + 2$
51. $2(4 - 3x) = 2(2x + 5)$
52. $3(5 - x) = 4(2x + 1)$
53. $8(y + 2) = 2(3y + 4)$
54. $3(3y - 1) = 4(3 + 3y)$
55. $3(x + 1) = 7(x - 2) - 3$
56. $5x - 4(x + 9) = 2x - 3$
57. $5(2x - 8) - 2 = 5(x - 3) + 3$
58. $7(3x - 2) + 5 = 6(2x - 1) + 24$
59. $6 = -4(1 - x) + 3(x + 1)$
60. $100 = -(x - 1) + 4(x - 6)$
61. $10(z + 4) - 4(z - 2) = 3(z - 1) + 2(z - 3)$
62. $-2(z - 4) - (3z - 2) = -2 - (6z - 2)$
63. $\frac{2x}{3} - 5 = 7$
64. $\frac{3x}{4} - 9 = -6$
65. $\frac{x}{3} + \frac{x}{2} = \frac{5}{6}$
66. $\frac{x}{4} - \frac{x}{5} = 1$
67. $20 - \frac{z}{3} = \frac{z}{2}$
68. $\frac{z}{5} - \frac{1}{2} = \frac{z}{6}$
69. $\frac{y}{3} + \frac{2}{5} = \frac{y}{5} - \frac{2}{5}$
70. $\frac{y}{12} + \frac{1}{6} = \frac{y}{2} - \frac{1}{4}$
71. $\frac{3x}{4} - 3 = \frac{x}{2} + 2$
72. $\frac{3x}{5} - \frac{2}{5} = \frac{x}{3} + \frac{2}{5}$
73. $\frac{3x}{5} - x = \frac{x}{10} - \frac{5}{2}$
74. $2x - \frac{2x}{7} = \frac{x}{2} + \frac{17}{2}$
75. $\frac{x - 3}{5} - 1 = \frac{x - 5}{4}$
76. $\frac{x - 2}{3} - 4 = \frac{x + 1}{4}$

In Exercises 77–96, solve each equation. Use set notation to express solution sets for equations with no solution or equations that are true for all real numbers.

77. $3x - 7 = 3(x + 1)$
78. $2(x - 5) = 2x + 10$
79. $2(x + 4) = 4x + 5 - 2x + 3$
80. $3(x - 1) = 8x + 6 - 5x - 9$
81. $7 + 2(3x - 5) = 8 - 3(2x + 1)$
82. $2 + 3(2x - 7) = 9 - 4(3x + 1)$
83. $4x + 1 - 5x = 5 - (x + 4)$
84. $5x - 5 = 3x - 7 + 2(x + 1)$
85. $4(x + 2) + 1 = 7x - 3(x - 2)$
86. $5x - 3(x + 1) = 2(x + 3) - 5$
87. $3 - x = 2x + 3$
88. $5 - x = 4x + 5$
89. $\frac{x}{3} + 2 = \frac{x}{3}$
90. $\frac{x}{4} + 3 = \frac{x}{4}$
91. $\frac{x}{3} = \frac{x}{2}$
92. $\frac{x}{4} = \frac{x}{3}$
93. $\frac{x - 2}{5} = \frac{3}{10}$
94. $\frac{x + 4}{8} = \frac{3}{16}$
95. $\frac{x}{2} - \frac{x}{4} + 4 = x + 4$
96. $\frac{x}{2} + \frac{2x}{3} + 3 = x + 3$

Practice Plus

97. Evaluate $x^2 - x$ for the value of x satisfying $4(x - 2) + 2 = 4x - 2(2 - x)$.
98. Evaluate $x^2 - x$ for the value of x satisfying $2(x - 6) = 3x + 2(2x - 1)$.
99. Evaluate $x^2 - (xy - y)$ for x satisfying $\frac{x}{5} - 2 = \frac{x}{3}$ and y satisfying $-2y - 10 = 5y + 18$.
100. Evaluate $x^2 - (xy - y)$ for x satisfying $\frac{3x}{2} + \frac{3x}{4} = \frac{x}{4} - 4$ and y satisfying $5 - y = 7(y + 4) + 1$.

In Exercises 101–108, solve each equation.

101. $[(3 + 6)^2 \div 3] \cdot 4 = -54x$
102. $2^3 - [4(5 - 3)^3] = -8x$
103. $5 - 12x = 8 - 7x - [6 \div 3(2 + 5^3) + 5x]$
104. $2(5x + 58) = 10x + 4(21 \div 3.5 - 11)$
105. $0.7x + 0.4(20) = 0.5(x + 20)$
106. $0.5(x + 2) = 0.1 + 3(0.1x + 0.3)$
107. $4x + 13 - \{2x - [4(x - 3) - 5]\} = 2(x - 6)$
108. $-2\{7 - [4 - 2(1 - x) + 3]\} = 10 - [4x - 2(x - 3)]$

Application Exercises

The latest guidelines, which apply to both men and women, give healthy weight ranges, rather than specific weights, for your height. The further you are above the upper limit of your range, the greater are the risks of developing weight-related health problems. The bar graph shows these ranges for various heights for people between the ages of 19 and 34, inclusive.

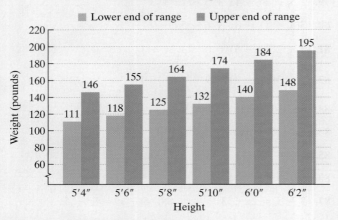

Healthy Weight Ranges for Men and Women, Ages 19 to 34

Source: U.S. Department of Health and Human Services

The mathematical model

$$\frac{W}{2} - 3H = 53$$

describes a weight, W, in pounds, that lies within the healthy weight range for a person whose height is H inches over 5 feet. Use this information to solve Exercises 109–110.

109. Use the formula to find a healthy weight for a person whose height is 5′6″. (*Hint:* H = 6 because this person's height is 6 inches over 5 feet.) How many pounds is this healthy weight below the upper end of the range shown by the bar graph?

110. Use the formula to find a healthy weight for a person whose height is 6′0″. (*Hint:* H = 12 because this person's height is 12 inches over 5 feet.) How many pounds is this healthy weight below the upper end of the range shown by the bar graph?

The formula

$$p = 15 + \frac{5d}{11}$$

describes the pressure of sea water, p, in pounds per square inch, at a depth of d feet below the surface. Use the formula to solve Exercises 111–112.

111. The record depth for breath-held diving, by Francisco Ferreras (Cuba) off Grand Bahama Island, on November 14, 1993, involved pressure of 201 pounds per square inch. To what depth did Ferreras descend on this ill-advised venture? (He was underwater for 2 minutes and 9 seconds!)

112. At what depth is the pressure 20 pounds per square inch?

Critical Thinking Exercises

113. Suppose you are an algebra teacher grading the following solution on an examination:

Solve: $-3(x - 6) = 2 - x$.
Solution: $-3x - 18 = 2 - x$
$-2x - 18 = 2$
$-2x = -16$
$x = 8.$

You should note that 8 checks, and the solution set is {8}. The student who worked the problem therefore wants full credit. Can you find any errors in the solution? If full credit is 10 points, how many points would you give the student? Justify your position.

114. Although the formulas in Example 6 on page 236 are correct, some people object to representing the variables with numbers, such as a 1-to-10 scale for the intensity of a negative life event. What might be their objection to quantifying the variables in this situation?

115. Write three equations whose solution set is {5}.

116. If x represents a number, write an English sentence about the number that results in an equation with no solution.

117. A woman's height, h, is related to the length of the femur, f (the bone from the knee to the hip socket), by the formula $f = 0.432h - 10.44$. Both h and f are measured in inches. A partial skeleton is found of a woman in which the femur is 16 inches long. Police find the skeleton in an area where a woman slightly over 5 feet tall has been missing for over a year. Could the partial skeleton be that of the missing woman? Explain.

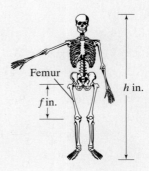

4.3 Applications of Linear Equations

① Use linear equations to solve problems.

② Solve a formula for a variable.

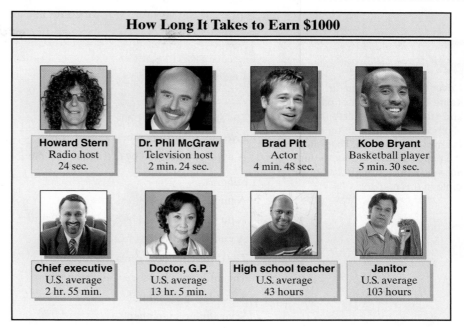

How Long It Takes to Earn $1000

Howard Stern
Radio host
24 sec.

Dr. Phil McGraw
Television host
2 min. 24 sec.

Brad Pitt
Actor
4 min. 48 sec.

Kobe Bryant
Basketball player
5 min. 30 sec.

Chief executive
U.S. average
2 hr. 55 min.

Doctor, G.P.
U.S. average
13 hr. 5 min.

High school teacher
U.S. average
43 hours

Janitor
U.S. average
103 hours

Source: Time

In this section, you'll see examples and exercises focused on how much money Americans earn. These situations illustrate a step-by-step strategy for solving problems. As you become familiar with this strategy, you will learn to solve a wide variety of problems.

Problem Solving with Linear Equations

① Use linear equations to solve problems.

We have seen that a model is a mathematical representation of a real-world situation. In this section, we will be solving problems that are presented in English. This means that we must obtain models by translating from the ordinary language of English into the language of algebraic equations. To translate, however, we must understand the English prose and be familiar with the forms of algebraic language. Below are some general steps we will follow in solving word problems.

Strategy for Solving Word Problems

Step 1 Read the problem carefully several times until you can state in your own words what is given and what the problem is looking for. Let x (or any variable) represent one of the unknown quantities in the problem.

Step 2 If necessary, write expressions for any other unknown quantities in the problem in terms of x.

Step 3 Write an equation in x that models the verbal conditions of the problem.

Step 4 Solve the equation and answer the problem's question.

Step 5 Check the solution *in the original wording* of the problem, not in the equation obtained from the words.

The most difficult step in the strategy for solving word problems involves translating verbal conditions into an algebraic equation. Translations of some commonly used English phrases are listed in **Table 4.2**. We choose to use x to represent the variable, but we could use any letter.

Table 4.2 Algebraic Translations of English Phrases

English Phrase	Algebraic Expression
Addition	
The sum of a number and 7	$x + 7$
Five more than a number; a number plus 5	$x + 5$
A number increased by 6; 6 added to a number	$x + 6$
Subtraction	
A number minus 4	$x - 4$
A number decreased by 5	$x - 5$
A number subtracted from 8	$8 - x$
The difference between a number and 6	$x - 6$
The difference between 6 and a number	$6 - x$
Seven less than a number	$x - 7$
Seven minus a number	$7 - x$
Nine fewer than a number	$x - 9$
Multiplication	
Five times a number	$5x$
The product of 3 and a number	$3x$
Two-thirds of a number (used with fractions)	$\frac{2}{3}x$
Seventy-five percent of a number (used with decimals)	$0.75x$
Thirteen multiplied by a number	$13x$
A number multiplied by 13	$13x$
Twice a number	$2x$
Division	
A number divided by 3	$\frac{x}{3}$
The quotient of 7 and a number	$\frac{7}{x}$
The quotient of a number and 7	$\frac{x}{7}$
The reciprocal of a number	$\frac{1}{x}$
More than one operation	
The sum of twice a number and 7	$2x + 7$
Twice the sum of a number and 7	$2(x + 7)$
Three times the sum of 1 and twice a number	$3(1 + 2x)$
Nine subtracted from 8 times a number	$8x - 9$
Twenty-five percent of the sum of 3 times a number and 14	$0.25(3x + 14)$
Seven times a number, increased by 24	$7x + 24$
Seven times the sum of a number and 24	$7(x + 24)$

Example 1 **Education Pays Off**

The graph in **Figure 4.4** shows average yearly earnings in the United States by highest educational attainment.

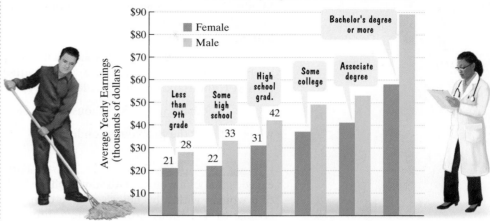

Figure 4.4
Source: U.S. Census Bureau

The average yearly salary of a man with an associate degree exceeds that of a man with some college by $4 thousand. The average yearly salary of a man with a bachelor's degree or more exceeds that of a man with some college by $40 thousand. Combined, three men with each of these educational attainments earn $191 thousand. Find the average yearly salary of men with each of these levels of education.

Solution

Step 1   Let x represent one of the unknown quantities. We know something about salaries of men with associate degrees and bachelor's degrees or more: They exceed the salary of a man with some college by $4 thousand and $40 thousand, respectively. We will let

x = the average yearly salary of a man with some college
(in thousands of dollars).

Step 2   Represent other unknown quantities in terms of x. Because a man with an associate degree earns $4 thousand more than a man with some college, let

$x + 4$ = the average yearly salary of a man with an associate degree.

Because a man with a bachelor's degree or more earns $40 thousand more than a man with some college, let

$x + 40$ = the average yearly salary of a man with a bachelor's degree or more.

Step 3   Write an equation in x that models the conditions. Combined, three men with each of these educational attainments earn $191 thousand.

Salary: some college	plus	salary: associate degree	plus	salary: bachelor's degree or more	equal	$191 thousand.
x	$+$	$(x + 4)$	$+$	$(x + 40)$	$=$	191

Step 4   Solve the equation and answer the question.

$$x + (x + 4) + (x + 40) = 191$$ This is the equation that models the problem's conditions.

$$3x + 44 = 191$$ Remove parentheses, regroup, and combine like terms.

$$3x = 147$$ Subtract 44 from both sides.

$$x = 49$$ Divide both sides by 3.

Because we isolated the variable in the model and obtained $x = 49$,

$$\text{average salary with some college} = x = 49$$
$$\text{average salary with an associate degree} = x + 4 = 49 + 4 = 53$$
$$\text{average salary with a bachelor's degree or more} = x + 40 = 49 + 40 = 89.$$

Men with some college average $49 thousand per year, men with associate degrees average $53 thousand per year, and men with bachelor's degrees or more average $89 thousand per year.

Step 5   Check the proposed solution in the original wording of the problem. The problem states that combined, three men with each of these educational attainments earn $191 thousand. Using the salaries we determined in step 4, the sum is

$$\$49 \text{ thousand} + \$53 \text{ thousand} + \$89 \text{ thousand, or } \$191 \text{ thousand,}$$

which satisfies the problem's conditions.

✔ **Checkpoint 1**   The average yearly salary of a woman with an associate degree exceeds that of a woman with some college by $4 thousand. The average yearly salary of a woman with a bachelor's degree or more exceeds that of a woman with some college by $21 thousand. Combined, three women with each of these educational attainments earn $136 thousand. Find the average yearly salary of women with each of these levels of education. (These salaries are illustrated by the bar graph on the previous page.)

Example 2   Modeling Attitudes of College Freshmen

Researchers have surveyed college freshmen every year since 1969. **Figure 4.5** shows that attitudes about some life goals have changed dramatically. **Figure 4.5** shows that the freshmen class of 2009 was more interested in making money than the freshmen of 1969 had been. In 1969, 42% of first-year college students considered "being well-off financially" essential or very important. For the period from 1969 through 2009, this percentage increased by approximately 0.9 each year. If this trend continues, by which year will all college freshmen consider "being well-off financially" essential or very important?

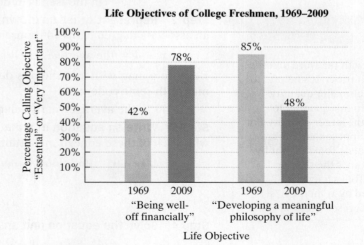

Life Objectives of College Freshmen, 1969–2009

Figure 4.5
Source: Higher Education Research Institute

Solution

Step 1   Let x represent one of the unknown quantities. We are interested in the year when all college freshmen, or 100% of the freshmen, will consider this life objective essential or very important. Let

$x=$ the number of years after 1969 when all freshmen will consider "being well-off financially" essential or very important.

Step 2   Represent other unknown quantities in terms of *x*. There are no other unknown quantities to find, so we can skip this step.

Step 3   Write an equation in *x* that models the conditions.

The 1969 percentage	increased by	0.9 each year for x years	equals	100% of the freshmen.
42	+	0.9x	=	100

Step 4   Solve the equation and answer the question.

$$42 + 0.9x = 100 \quad \text{This is the equation that models the problem's conditions.}$$
$$42 - 42 + 0.9x = 100 - 42 \quad \text{Subtract 42 from both sides.}$$
$$0.9x = 58 \quad \text{Simplify.}$$
$$\frac{0.9x}{0.9} = \frac{58}{0.9} \quad \text{Divide both sides by 0.9.}$$
$$x = 64.\overline{4} \approx 64 \quad \text{Simplify and round to the nearest whole number.}$$

Using current trends, by approximately 64 years after 1969, or in 2033, all freshmen will consider "being well-off financially" essential or very important.

Step 5   Check the proposed solution in the original wording of the problem. The problem states that all freshmen (100%, represented by 100 using the model) will consider the objective essential or very important. Does this approximately occur if we increase the 1969 percentage, 42%, by 0.9 each year for 64 years, our proposed solution?

$$42 + 0.9(64) = 42 + 57.6 = 99.6 \approx 100$$

This verifies that using trends shown in **Figure 4.5**, all first-year college students will consider the objective essential or very important approximately 64 years after 1969.

A Brief Review • Clearing an Equation of Decimals

- You can clear an equation of decimals by multiplying each side by a power of 10. The exponent on 10 will be equal to the greatest number of digits to the right of any decimal point in the equation.
- Multiplying a decimal number by 10^n has the effect of moving the decimal point n places to the right.

Example

$$42 + 0.9x = 100$$

The greatest number of digits to the right of any decimal point in the equation is 1. Multiply each side by 10^1, or 10.

$$10(42 + 0.9x) = 10(100)$$
$$10(42) + 10(0.9x) = 10(100)$$
$$420 + 9x = 1000$$
$$420 - 420 + 9x = 1000 - 420$$
$$9x = 580$$
$$\frac{9x}{9} = \frac{580}{9}$$
$$x = 64.\overline{4} \approx 64$$

It is not a requirement to clear decimals before solving an equation. Compare this solution to the one in step 4 of Example 2. Which method do you prefer?

Your author teaching high
school math in 1969

✓ Checkpoint 2 **Figure 4.5** on page 244 shows that the freshmen class of 2009 was less interested in developing a philosophy of life than the freshmen of 1969 had been. In 1969, 85% of the freshmen considered this objective essential or very important. Since then, this percentage has decreased by approximately 0.9 each year. If this trend continues, by which year will only 25% of college freshmen consider "developing a meaningful philosophy of life" essential or very important?

Blitzer Bonus

Forty Years of Change

1969

2009

Compared to 1969, college freshmen in 2009 had making money on their minds. Here are some other random factoids and statistics about life then versus life today.

	1969	2009
President	Richard Nixon	Barack Obama
U.S. Population	203 million	308 million
Life Expectancy	Male: 66.6 years Female: 73.1 years	Male: 75.1 years Female: 80.2 years
Cost per Gallon of Premium Gas	$0.36	$2.95
Average Cost of a New Car	$3270	$28,400
Average Income	$8550	$39,750
Unemployment	3.5%	10%
Minimum Voting Age	21	18
Technology	Teens listen to music on record players and stereos. People talk on phones through land lines.	Most teens listen to music digitally on iPods and mp3 players. Cellphones, smartphones, and laptops revolutionize global communication.
Percentage of Teens Enjoying Parents' Company	57%	64%
Percentage of Teens Stating the Importance of Patriotism	55%	32%

Sources: Channel One News, infoplease.com, *Scholastic Scope*

Example 3   Selecting a Rental Car

Acme Car rental agency charges $4 a day plus $0.15 per mile, whereas Interstate rental agency charges $20 a day and $0.05 per mile. For how many miles of driving in a day will the costs for the two rentals be the same?

Solution

Step 1   Let x represent one of the unknown quantities. Let

x = the number of miles driven in a day for which the two rentals cost the same.

Step 2   Represent other unknown quantities in terms of x. There are no other unknown quantities, so we can skip this step.

Step 3   Write an equation in x that models the conditions. The daily cost for Acme is the daily charge, $4, plus the per-mile charge, $0.15, times the number of miles driven, x. The daily cost for Interstate is the daily charge, $20, plus the per-mile charge, $0.05, times the number of miles driven, x.

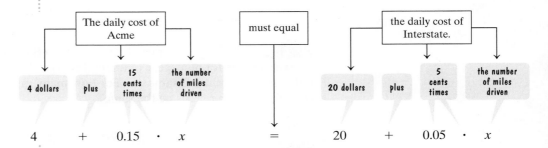

$$4 \quad + \quad 0.15 \cdot x \quad = \quad 20 \quad + \quad 0.05 \cdot x$$

Step 4   Solve the equation and answer the question.

$$4 + 0.15x = 20 + 0.05x \qquad \text{This is the equation that models the problem's conditions.}$$

$$4 + 0.15x - 0.05x = 20 + 0.05x - 0.05x \qquad \text{Subtract 0.05x from both sides.}$$

$$4 + 0.1x = 20 \qquad \text{Simplify.}$$

$$4 + 0.1x - 4 = 20 - 4 \qquad \text{Subtract 4 from both sides.}$$

$$0.1x = 16 \qquad \text{Simplify.}$$

$$\frac{0.1x}{0.1} = \frac{16}{0.1} \qquad \text{Divide both sides by 0.1.}$$

$$x = 160 \qquad \text{Simplify.}$$

Because x represents the number of miles driven in a day for which the two rentals cost the same, the costs will be the same with 160 miles of driving.

Step 5   Check the proposed solution in the original wording of the problem. The problem states that the costs for the two rentals should be the same. Let's see if they are with 160 miles of driving.

$$\text{Cost for Acme} = \$4 + \$0.15(160) = \$4 + \$24 = \$28$$

Daily charge   Per-mile charge

$$\text{Cost for Interstate} = \$20 + \$0.05(160) = \$20 + \$8 = \$28$$

With 160 miles driven in a day, both rentals cost $28. Thus, the proposed solution, 160 miles per day, satisfies the problem's conditions.

If you prefer, you can solve $4 + 0.15x = 20 + 0.05x$, the equation that modeled the conditions in Example 3, by first clearing the decimals. The greatest number of digits to the right of any decimal point in the equation is 2. Multiply each side by 10^2, or 100.

$$4 + 0.15x = 20 + 0.05x$$
$$100(4 + 0.15x) = 100(20 + 0.05x)$$
$$100(4) + 100(0.15x) = 100(20) + 100(0.05x)$$
$$400 + 15x = 2000 + 5x$$
$$400 + 15x - 5x = 2000 + 5x - 5x$$
$$400 + 10x = 2000$$
$$400 + 10x - 400 = 2000 - 400$$
$$10x = 1600$$
$$\frac{10x}{10} = \frac{1600}{10}$$
$$x = 160$$

Checkpoint 3 A car can be rented from Basic Rental for $260 per week with no extra charge for mileage. Continental charges $80 per week plus 25 cents for each mile driven to rent the same car. For how many miles of driving in a week will the costs for the two rentals be the same?

Example 4 A Price Reduction on a Digital Camera

Your local computer store is having a terrific sale on digital cameras. After a 40% price reduction, you purchase a digital camera for $276. What was the camera's price before the reduction?

Solution

Step 1 Let x represent one of the unknown quantities. We will let

x = the original price of the digital camera prior to the reduction.

Step 2 Represent other unknown quantities in terms of x. There are no other unknown quantities to find, so we can skip this step.

Step 3 Write an equation in x that models the conditions. The camera's original price minus the 40% reduction is the reduced price, $276.

$$x - 0.4x = 276$$

Step 4 Solve the equation and answer the question.

$x - 0.4x = 276$ This is the equation that models the problem's conditions.

$0.6x = 276$ Combine like terms: $x - 0.4x = 1x - 0.4x = 0.6x$.

$\frac{0.6x}{0.6} = \frac{276}{0.6}$ Divide both sides by 0.6.

$x = 460$ Simplify: $0.6\overline{)276.0}$ = 460.

The digital camera's price before the reduction was $460.

Step 5   Check the proposed solution in the original wording of the problem. The price before the reduction, $460, minus the 40% reduction should equal the reduced price given in the original wording, $276:

$$460 - 40\% \text{ of } 460 = 460 - 0.4(460) = 460 - 184 = 276.$$

This verifies that the digital camera's price before the reduction was $460.

✓ **Checkpoint 4**   After a 30% price reduction, you purchase a new computer
▶ for $840. What was the computer's price before the reduction?

Solving a Formula for One of Its Variables

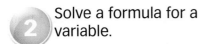
Solve a formula for a variable.

We know that solving an equation is the process of finding the number (or numbers) that make the equation a true statement. All of the equations we have solved contained only one letter, x.

By contrast, formulas contain two or more letters, representing two or more variables. An example is the formula for the perimeter of a rectangle:

$$P = 2l + 2w.$$

> A rectangle's perimeter is the sum of twice its length and twice its width.

We say that this formula is solved for the variable P because P is alone on one side of the equation and the other side does not contain a P.

Solving a formula for a variable means rewriting the formula so that the variable is isolated on one side of the equation. It does not mean obtaining a numerical value for that variable.

To solve a formula for one of its variables, treat that variable as if it were the only variable in the equation. Think of the other variables as if they were numbers. Isolate all terms with the specified variable on one side of the equation and all terms without the specified variable on the other side. Then divide both sides by the same nonzero quantity to get the specified variable alone. The next two examples show how to do this.

Example 5   Solving a Formula for a Variable

Solve the formula $P = 2l + 2w$ for l.

Solution

First, isolate $2l$ on the right by subtracting $2w$ from both sides. Then solve for l by dividing both sides by 2.

> We need to isolate l.

$$P = 2l + 2w \qquad \text{This is the given formula.}$$

$$P - 2w = 2l + 2w - 2w \qquad \text{Isolate 2l by subtracting 2w from both sides.}$$

$$P - 2w = 2l \qquad \text{Simplify.}$$

$$\frac{P - 2w}{2} = \frac{2l}{2} \qquad \text{Solve for l by dividing both sides by 2.}$$

$$\frac{P - 2w}{2} = l \qquad \text{Simplify.}$$

Equivalently, $l = \dfrac{P - 2w}{2}$.

✓ **Checkpoint 5**   Solve the formula $P = 2l + 2w$ for w.

Example 6 Solving a Formula for a Variable

The total price of an article purchased on a monthly deferred payment plan is described by the following formula:

$$T = D + pm.$$

In this formula, T is the total price, D is the down payment, p is the monthly payment, and m is the number of months one pays. Solve the formula for p.

Solution

First, isolate pm on the right by subtracting D from both sides. Then, isolate p from pm by dividing both sides of the formula by m.

We need to isolate p.

$T = D + pm$ This is the given formula. We want p alone.

$T - D = D - D + pm$ Isolate pm by subtracting D from both sides.

$T - D = pm$ Simplify.

$\dfrac{T - D}{m} = \dfrac{pm}{m}$ Now isolate p by dividing both sides by m.

$\dfrac{T - D}{m} = p$ Simplify: $\dfrac{pm}{m} = \dfrac{p\cancel{m}}{\cancel{m}} = \dfrac{p}{1} = p.$

✓ **Checkpoint 6** Solve the formula $T = D + pm$ for m.

Achieving Success

To prepare for further coursework and training programs, read *Math Study Skills for College Students* by Alan Bass (Pearson, 2008). Many of the successful strategies suggested by your author are based on Professor Bass's discussions of what students need to know to achieve success in college math courses. The book even includes an entire chapter on overcoming math anxiety. The following excerpt punctuates the example that opened this section, "Education Pays Off."

"When employers see a college degree, they know the candidate has been exposed to and successfully assimilated a general, well-rounded body of knowledge. A college degree shows discipline, character, and sharp thinking skills. Apparently, that's pretty important. According to the U.S. Census Bureau, college graduates will earn $2.1 million more over the course of their life than high school graduates. Assuming that your degree takes four years, that's like getting paid $2625 for each day you attend college! Work hard!"

—Alan Bass

Exercise Set 4.3

Concept and Vocabulary Exercises

In Exercises 1–4, fill in each blank so that the resulting statement is true.

1. According to a survey by the Kaiser Family Foundation, the number of minutes per day that teens spend watching TV exceeds the number of minutes listening to music by 120. If x represents the number of minutes each day listening to music, the number of minutes spent watching TV can be represented by _____.

2. In 2006, 55% of teens used social networking sites. For the period from 2006 through 2009, this percentage increased by approximately 6 each year. The percentage of teens using social networking sites x years after 2006 can be represented by _____.

3. I can rent a car for $125 per week plus $0.15 for each mile driven. If I drive x miles in a week, the cost for the rental can be represented by _____.

4. I purchased a computer after a 35% price reduction. If x represents the computer's original price, the reduced price can be represented by _____.

In Exercises 5–8, determine whether each equation correctly models the given conditions. If the model is incorrect, make the necessary change(s) to produce an accurate model.

5. The top two extreme sports in the United States are inline skating and skateboarding. Participation in inline skating exceeds that of skateboarding by 2.4 million. Combined, 19.2 million Americans participate in these sports. (*Source*: Sporting Goods Manufacturers Association) I can find the number of skateboarders by solving $x + 2.4 = 19.2$ for x.

6. In 2006, 28% of teens had blogs. For the period from 2006 through 2009, this percentage decreased by approximately 4 per year. If this trend continues, I can find the year in which no teens will have blogs by solving $28 - 4x = 0$ for x.

7. A car rental company charges $125 per week plus $0.20 per mile to rent one of its cars. If I spent $335 for the week, I can find the number of miles driven by solving $125 - 0.20x = 335$ for x.

8. After a 40% reduction, I spent $12 for a dictionary. I can find the dictionary's price before the reduction by solving $x - 0.4 = 12$ for x.

Respond to Exercises 9–10 using verbal or written explanations.

9. In your own words, describe a step-by-step approach for solving algebraic word problems.

10. Explain what it means to solve a formula for a variable.

In Exercises 11–14, determine whether each statement makes sense or does not make sense, and explain your reasoning.

11. By modeling attitudes of college freshmen from 1969 through 2009, I can make precise predictions about the attitudes of the freshman class of 2050.

12. I find the hardest part in solving a word problem is writing the equation that models the verbal conditions.

13. I solved a formula for one of its variables, so now I have a numerical value for that variable.

14. After solving $P = 2l + 2w$ for l, I obtained $2l = P - 2w$.

Practice Exercises

Use the five-step strategy for solving word problems to find the number or numbers described in Exercises 15–24.

15. When five times a number is decreased by 4, the result is 26. What is the number?

16. When two times a number is decreased by 3, the result is 11. What is the number?

17. When a number is decreased by 20% of itself, the result is 20. What is the number?

18. When a number is decreased by 30% of itself, the result is 28. What is the number?

19. When 60% of a number is added to the number, the result is 192. What is the number?

20. When 80% of a number is added to the number, the result is 252. What is the number?

21. 70% of what number is 224?

22. 70% of what number is 252?

23. One number exceeds another by 26. The sum of the numbers is 64. What are the numbers?

24. One number exceeds another by 24. The sum of the numbers is 58. What are the numbers?

Practice Plus

In Exercises 25–32, write each English phrase as an algebraic expression. Then simplify the expression. Let x represent the number.

25. A number decreased by the sum of the number and four

26. A number decreased by the difference between eight and the number

27. Six times the product of negative five and a number

28. Ten times the product of negative four and a number

29. The difference between the product of five and a number and twice the number

30. The difference between the product of six and a number and negative two times the number

31. The difference between eight times a number and six more than three times the number

32. Eight decreased by three times the sum of a number and six

Application Exercises

The bar graph shows the ten most popular college majors with median, or middlemost, salaries for those with a bachelor's degree only. Exercises 33–38 involve some of the median salaries represented by the graph.

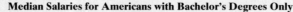

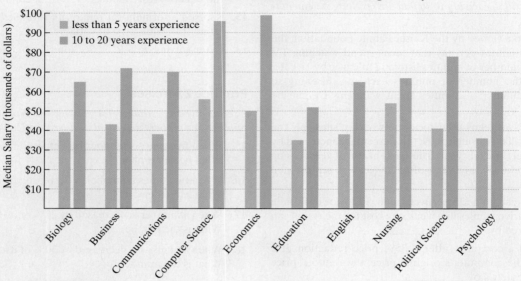

Median Salaries for Americans with Bachelor's Degrees Only

Source: PayScale (2008 data)

33. For those with less than 5 years experience, the median salary of a computer science major exceeds that of a psychology major by $20 thousand. Combined, two people with each of these majors and less than 5 years experience earn $92 thousand. Find the median salary of a computer science major with less than 5 years experience and the median salary of a psychology major with less than 5 years experience.

34. The median salary of a computer science major with less than 5 years experience exceeds that of an education major with 10 to 20 years experience by $4 thousand. Combined, two people with each of these majors and specified years of experience earn $108 thousand. Find the median salary of a computer science major with less than 5 years experience and the median salary of an education major with 10 to 20 years experience.

35. For those with 10 to 20 years experience, the median salary of a computer science major exceeds that of a political science major by $18 thousand and the median salary of an economics major exceeds that of a political science major by $21 thousand. Combined, three people with each of these majors and 10 to 20 years experience earn $273 thousand. Find the median salary for people with each major and 10 to 20 years experience.

36. For those with less than 5 years experience, the median salary of a biology major exceeds that of a communications major by $1 thousand and the median salary of a business major exceeds that of a communications major by $5 thousand. Combined, three people with each of these majors and less than 5 years experience earn $120 thousand. Find the median salary for people with each major and less than 5 years experience.

37. The median salary of a business major with 10 to 20 years experience is $14 thousand less than twice that of a business major with less than 5 years experience. Combined, two business majors with these levels of experience earn $115 thousand. Find the median salary of a business major with less than 5 years experience and a business major with 10 to 20 years experience.

38. The median salary of a nursing major with 10 to 20 years experience is $3 thousand less than twice that of an education major with less than 5 years experience. Combined, two people with each of these majors and specified years of experience earn $102 thousand. Find the median salary of a nursing major with 10 to 20 years experience and an education major with less than 5 years experience.

On average, every minute of every day, 150 babies are born. The bar graph represents the results of a single day of births, deaths, and population increase worldwide. Exercises 39–40 are based on the information displayed by the graph.

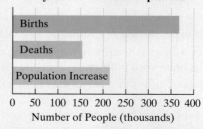

Daily Growth of World Population

Source: James Henslin, *Sociology,* Ninth Edition, Pearson, 2008.

39. Each day, the number of births in the world is 92 thousand less than three times the number of deaths.

  a. If the population increase in a single day is 214 thousand, determine the number of births and deaths per day.

b. If the population increase in a single day is 214 thousand, by how many millions of people does the worldwide population increase each year? Round to the nearest million.

c. Based on your answer to part (b), approximately how many years does it take for the population of the world to increase by an amount greater than the entire U.S. population (308 million)?

40. Each day, the number of births in the world exceeds twice the number of deaths by 61 thousand.

a. If the population increase in a single day is 214 thousand, determine the number of births and deaths per day.

b. If the population increase in a single day is 214 thousand, by how many millions of people does the worldwide population increase each year? Round to the nearest million.

c. Based on your answer to part (b), approximately how many years does it take for the population of the world to increase by an amount greater than the entire U.S. population (308 million)?

Even as Americans increasingly view a college education as essential for success, many believe that a college education is becoming less available to qualified students. Exercises 41–42 are based on the data displayed by the graph.

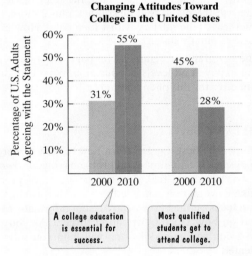

Changing Attitudes Toward College in the United States

Source: Public Agenda

41. In 2000, 31% of U.S. adults viewed a college education as essential for success. For the period from 2000 through 2010, the percentage viewing a college education as essential for success increased on average by approximately 2.4 each year. If this trend continues, by which year will 67% of all American adults view a college education as essential for success?

42. In 2000, 45% of U.S. adults believed that most qualified students get to attend college. For the period from 2000 through 2010, the percentage who believed that a college education is available to most qualified students decreased by approximately 1.7 each year. If this trend continues, by which year will only 11% of all American adults believe that most qualified students get to attend college?

43. A new car worth $24,000 is depreciating in value by $3000 per year. After how many years will the car's value be $9000?

44. A new car worth $45,000 is depreciating in value by $5000 per year. After how many years will the car's value be $10,000?

45. You are choosing between two health clubs. Club A offers membership for a fee of $40 plus a monthly fee of $25. Club B offers membership for a fee of $15 plus a monthly fee of $30. After how many months will the total cost at each health club be the same? What will be the total cost for each club?

46. Video Store A charges $9 to rent a video game for one week. Although only members can rent from the store, membership is free. Video Store B charges only $4 to rent a video game for one week. Only members can rent from the store and membership is $50 per year. After how many video-game rentals will the total amount spent at each store be the same? What will be the total amount spent at each store?

47. The bus fare in a city is $1.25. People who use the bus have the option of purchasing a monthly discount pass for $15.00. With the discount pass, the fare is reduced to $0.75. Determine the number of times in a month the bus must be used so that the total monthly cost without the discount pass is the same as the total monthly cost with the discount pass.

48. A discount pass for a bridge costs $30 per month. The toll for the bridge is normally $5.00, but it is reduced to $3.50 for people who have purchased the discount pass. Determine the number of times in a month the bridge must be crossed so that the total monthly cost without the discount pass is the same as the total monthly cost with the discount pass.

49. You are choosing between two plans at a discount warehouse. Plan A offers an annual membership fee of $100 and you pay 80% of the manufacturer's recommended list price. Plan B offers an annual membership fee of $40 and you pay 90% of the manufacturer's recommended list price. How many dollars of merchandise would you have to purchase in a year to pay the same amount under both plans? What will be the cost for each plan?

50. You are choosing between two plans at a discount warehouse. Plan A offers an annual membership fee of $300 and you pay 70% of the manufacturer's recommended list price. Plan B offers an annual membership fee of $40 and you pay 90% of the manufacturer's recommended list price. How many dollars of merchandise would you have to purchase in a year to pay the same amount under both plans? What will be the cost for each plan?

51. In 2008, there were 13,300 students at college A, with a projected enrollment increase of 1000 students per year. In the same year, there were 26,800 students at college B, with a projected enrollment decline of 500 students per year. According to these projections, when will the colleges have the same enrollment? What will be the enrollment in each college at that time?

52. In 2000, the population of Greece was 10,600,000, with projections of a population decrease of 28,000 people per year. In the same year, the population of Belgium was 10,200,000, with projections of a population decrease of 12,000 people per year. (*Source:* United Nations) According to these projections, when will the two countries have the same population? What will be the population at that time?

53. After a 20% reduction, you purchase a television for $336. What was the television's price before the reduction?

54. After a 30% reduction, you purchase a dictionary for $30.80. What was the dictionary's price before the reduction?

55. Including 8% sales tax, an inn charges $162 per night. Find the inn's nightly cost before the tax is added.

56. Including 5% sales tax, an inn charges $252 per night. Find the inn's nightly cost before the tax is added.

Exercises 57–58 involve markup, the amount added to the dealer's cost of an item to arrive at the selling price of that item.

57. The selling price of a refrigerator is $584. If the markup is 25% of the dealer's cost, what is the dealer's cost of the refrigerator?

58. The selling price of a scientific calculator is $15. If the markup is 25% of the dealer's cost, what is the dealer's cost of the calculator?

In Exercises 59–76, solve each formula for the specified variable. Do you recognize the formula? If so, what does it describe?

59. $A = LW$ for L

60. $D = RT$ for R

61. $A = \frac{1}{2}bh$ for b

62. $V = \frac{1}{3}Bh$ for B

63. $I = Prt$ for P

64. $C = 2\pi r$ for r

65. $E = mc^2$ for m

66. $V = \pi r^2 h$ for h

67. $y = mx + b$ for m

68. $P = C + MC$ for M

69. $A = \frac{1}{2}h(a + b)$ for a

70. $A = \frac{1}{2}h(a + b)$ for b

71. $S = P + Prt$ for r

72. $S = P + Prt$ for t

73. $Ax + By = C$ for x

74. $Ax + By = C$ for y

75. $a_n = a_1 + (n - 1)d$ for n

76. $a_n = a_1 + (n - 1)d$ for d

Critical Thinking Exercises

77. The price of a dress is reduced by 40%. When the dress still does not sell, it is reduced by 40% of the reduced price. If the price of the dress after both reductions is $72, what was the original price?

78. In a film, the actor Charles Coburn plays an elderly "uncle" character criticized for marrying a woman when he is 3 times her age. He wittily replies, "Ah, but in 20 years time I shall only be twice her age." How old is the "uncle" and the woman?

79. Suppose that we agree to pay you 8¢ for every problem in this chapter that you solve correctly and fine you 5¢ for every problem done incorrectly. If at the end of 26 problems we do not owe each other any money, how many problems did you solve correctly?

80. It was wartime when the Ricardos found out Mrs. Ricardo was pregnant. Ricky Ricardo was drafted and made out a will, deciding that $14,000 in a savings account was to be divided between his wife and his child-to-be. Rather strangely, and certainly with gender bias, Ricky stipulated that if the child were a boy, he would get twice the amount of the mother's portion. If it were a girl, the mother would get twice the amount the girl was to receive. We'll never know what Ricky was thinking of, for (as fate would have it) he did not return from war. Mrs. Ricardo gave birth to twins—a boy and a girl. How was the money divided?

81. A thief steals a number of rare plants from a nursery. On the way out, the thief meets three security guards, one after another. To each security guard, the thief is forced to give one-half the plants that he still has, plus 2 more. Finally, the thief leaves the nursery with 1 lone palm. How many plants were originally stolen?

82. The bar graph in **Figure 4.4** on page 243 indicates a gender gap in pay at all levels of education. What explanations can you offer for this phenomenon?

Group Exercise

83. One of the best ways to learn how to *solve* a word problem in algebra is to *design* word problems of your own. Creating a word problem makes you very aware of precisely how much information is needed to solve the problem. You must also focus on the best way to present information to a reader and on how much information to give. As you write your problem, you gain skills that will help you solve problems created by others.

The group should design five different word problems that can be solved using linear equations. All of the problems should be on different topics. For example, the group should not have more than one problem on price reduction. The group should turn in both the problems and their algebraic solutions.

(If you're not sure where to begin, consider using some of the data in the Blitzer Bonus on page 246 comparing life in 1969 and 2009.)

4.4 : Modeling with Proportions

Objectives

Objectives

1 Solve proportions.

2 Solve problems using proportions.

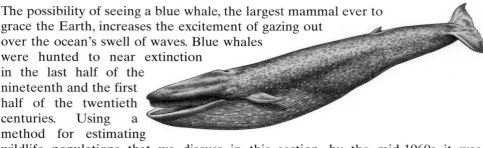

The possibility of seeing a blue whale, the largest mammal ever to grace the Earth, increases the excitement of gazing out over the ocean's swell of waves. Blue whales were hunted to near extinction in the last half of the nineteenth and the first half of the twentieth centuries. Using a method for estimating wildlife populations that we discuss in this section, by the mid-1960s it was determined that the world population of blue whales was less than 1000. This led the International Whaling Commission to ban the killing of blue whales to prevent their extinction. A dramatic increase in blue whale sightings indicates an ongoing increase in their population and the success of the killing ban.

Proportions

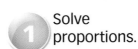

1 Solve proportions.

A **ratio** compares quantities by division. For example, a group contains 60 women and 30 men. The ratio of women to men is $\frac{60}{30}$. We can express this ratio as a fraction reduced to lowest terms:

$$\frac{60}{30} = \frac{2 \cdot \cancel{30}}{1 \cdot \cancel{30}} = \frac{2}{1}.$$

This ratio can be expressed as 2:1, or 2 to 1.

A **proportion** is a statement that says that two ratios are equal. If the ratios are $\frac{a}{b}$ and $\frac{c}{d}$, then the proportion is

$$\frac{a}{b} = \frac{c}{d}.$$

We can clear this equation of fractions by multiplying both sides by bd:

$$\frac{a}{b} = \frac{c}{d} \qquad \text{This is the given proportion.}$$

$$bd \cdot \frac{a}{b} = bd \cdot \frac{c}{d} \qquad \text{Multiply both sides by } bd \ (b \neq 0 \text{ and } d \neq 0). \text{ Then simplify.}$$

On the left, $\frac{\cancel{b}d}{1} \cdot \frac{a}{\cancel{b}} = da = ad$. On the right, $\frac{b\cancel{d}}{1} \cdot \frac{c}{\cancel{d}} = bc$.

$$ad = bc.$$

We see that the following principle is true for any proportion:

bc

$$\frac{a}{b} \diagup\diagdown \frac{c}{d}$$

ad

The cross-products principle: $ad = bc$

> **The Cross-Products Principle for Proportions**
>
> $$\text{If } \ \frac{a}{b} = \frac{c}{d}, \quad \text{then} \quad ad = bc. \quad (b \neq 0 \text{ and } d \neq 0)$$
>
> The cross products ad and bc are equal.

For example, since $\frac{2}{3} = \frac{6}{9}$, we see that $2 \cdot 9 = 3 \cdot 6$, or $18 = 18$. We can also use $\frac{2}{3} = \frac{6}{9}$ and conclude that $3 \cdot 6 = 2 \cdot 9$. When using the cross-products principle, it does not matter on which side of the equation each product is placed.

If three of the numbers in a proportion are known, the value of the missing quantity can be found by using the cross-products principle. This idea is illustrated in Example 1(a).

Example 1 ⬛ Solving Proportions

Solve each proportion and check:

a. $\dfrac{63}{x} = \dfrac{7}{5}$ **b.** $\dfrac{20}{x - 10} = \dfrac{30}{x}$.

Solution

Cross products

a.

$\dfrac{63}{x} = \dfrac{7}{5}$ *This is the given proportion.*

$63 \cdot 5 = 7x$ *Apply the cross-products principle.*

$315 = 7x$ *Simplify.*

$\dfrac{315}{7} = \dfrac{7x}{7}$ *Divide both sides by 7.*

$45 = x$ *Simplify.*

The solution set is $\{45\}$.

Check

$\dfrac{63}{45} \overset{?}{=} \dfrac{7}{5}$ *Substitute 45 for x in $\dfrac{63}{x} = \dfrac{7}{5}$.*

$\dfrac{7 \cdot \cancel{9}}{5 \cdot \cancel{9}} \overset{?}{=} \dfrac{7}{5}$ *Reduce $\dfrac{63}{45}$ to lowest terms.*

$\dfrac{7}{5} = \dfrac{7}{5}$ *This true statement verifies that the solution set is $\{45\}$.*

b.

$\dfrac{20}{x - 10} = \dfrac{30}{x}$ *This is the given proportion.*

$20x = 30(x - 10)$ *Apply the cross-products principle.*

$20x = 30x - 30 \cdot 10$ *Use the distributive property.*

$20x = 30x - 300$ *Simplify.*

$20x - 30x = 30x - 300 - 30x$ *Subtract 30x from both sides.*

$-10x = -300$ *Simplify.*

$\dfrac{-10x}{-10} = \dfrac{-300}{-10}$ *Divide both sides by −10.*

$x = 30$ *Simplify.*

The solution set is $\{30\}$.

Check

$\dfrac{20}{30 - 10} \overset{?}{=} \dfrac{30}{30}$ *Substitute 30 for x in $\dfrac{20}{x - 10} = \dfrac{30}{x}$.*

$\dfrac{20}{20} \overset{?}{=} \dfrac{30}{30}$ *Subtract: $30 - 10 = 20$.*

$1 = 1$ *This true statement verifies that the solution is 30.*

✓ **Checkpoint 1**   Solve each proportion and check:

a. $\dfrac{10}{x} = \dfrac{2}{3}$ **b.** $\dfrac{22}{60 - x} = \dfrac{2}{x}$.

Applications of Proportions

2 Solve problems using proportions.

We now turn to practical application problems that can be solved by modeling with proportions. Here is a procedure for solving these problems:

> **Solving Applied Problems Using Proportions**
> 1. Read the problem and represent the unknown quantity by x (or any letter).
> 2. Set up a proportion that models the problem's conditions. List the given ratio on one side and the ratio with the unknown quantity on the other side. Each respective quantity should occupy the same corresponding position on each side of the proportion.
> 3. Drop units and apply the cross-products principle.
> 4. Solve for x and answer the question.

Example 2 Applying Proportions: Calculating Taxes

The property tax on a house with an assessed value of $480,000 is $5760. Determine the property tax on a house with an assessed value of $600,000, assuming the same tax rate.

Solution

Step 1 Represent the unknown by x. Let $x =$ the tax on the $600,000 house.

Step 2 Set up a proportion. We will set up a proportion comparing taxes to assessed value.

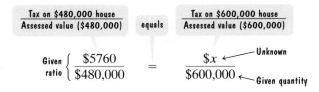

Step 3 Drop the units and apply the cross-products principle. We drop the dollar signs and begin to solve for x.

$$\frac{5760}{480,000} = \frac{x}{600,000}$$ This is the proportion that models the problem's conditions.

$$480,000x = (5760)(600,000)$$ Apply the cross-products principle.

$$480,000x = 3,456,000,000$$ Multiply.

Step 4 Solve for x and answer the question.

$$\frac{480,000x}{480,000} = \frac{3,456,000,000}{480,000}$$ Divide both sides by 480,000.

$$x = 7200$$ Simplify.

The property tax on the $600,000 house is $7200.

Checkpoint 2 The property tax on a house with an assessed value of $250,000 is $3500. Determine the property tax on a house with an assessed value of $420,000, assuming the same tax rate.

Sampling in Nature

The method that was used to estimate the blue whale population described in the section opener is called the **capture-recapture method**. Because it is impossible to count each individual animal within a population, wildlife biologists randomly catch and tag a given number of animals. Sometime later they capture a second sample of animals and count the number of recaptured tagged animals. The total size of the wildlife population is then estimated using the following proportion:

$$\left.\underset{\substack{\text{Initially} \\ \text{unknown} \\ (x) \longrightarrow}}{\frac{\text{Original number of}}{\text{tagged animals}}\atop \frac{}{\substack{\text{Total number} \\ \text{of animals in the} \\ \text{population}}}} = \frac{\frac{\text{Number of recaptured}}{\text{tagged animals}}}{\substack{\text{Number of animals} \\ \text{in second sample}}}\right\} \begin{array}{c} \text{Known} \\ \text{ratio} \end{array}$$

Although this is called the capture-recapture method, it is not necessary to recapture animals in order to observe whether or not they are tagged. This could be done from a distance, with binoculars for instance.

Example 3 Applying Proportions: Estimating Wildlife Population

Wildlife biologists catch, tag, and then release 135 deer back into a wildlife refuge. Two weeks later they observe a sample of 140 deer, 30 of which are tagged. Assuming the ratio of tagged deer in the sample holds for all deer in the refuge, approximately how many deer are in the refuge?

Solution

Step 1 Represent the unknown by x. Let x = the total number of deer in the refuge.

Step 2 Set up a proportion.

$$\underset{\text{Unknown} \longrightarrow}{\frac{\substack{\text{Original number} \\ \text{of tagged deer}}}{\substack{\text{Total number} \\ \text{of deer}}}} \quad \underset{\text{equals}}{=} \quad \left.\frac{\substack{\text{Number of tagged deer} \\ \text{in the observed sample}}}{\substack{\text{Total number of deer} \\ \text{in the observed sample}}}\right\} \begin{array}{c} \text{Known} \\ \text{ratio} \end{array}$$

$$\frac{135}{x} = \frac{30}{140}$$

Steps 3 and 4 Apply the cross-products principle, solve, and answer the question.

$$\frac{135}{x} = \frac{30}{140} \quad \text{This is the proportion that models the problem's conditions.}$$

$$(135)(140) = 30x \quad \text{Apply the cross-products principle.}$$

$$18,900 = 30x \quad \text{Multiply.}$$

$$\frac{18,900}{30} = \frac{30x}{30} \quad \text{Divide both sides by 30.}$$

$$630 = x \quad \text{Simplify.}$$

There are approximately 630 deer in the refuge.

Checkpoint 3 Wildlife biologists catch, tag, and then release 120 deer back into a wildlife refuge. Two weeks later they observe a sample of 150 deer, 25 of which are tagged. Assuming the ratio of tagged deer in the sample holds for all deer in the refuge, approximately how many deer are in the refuge?

Exercise Set 4.4

Concept and Vocabulary Exercises

In Exercises 1–2, fill in each blank so that the resulting statement is true.

1. A statement that two ratios are equal is called a/an
   _____.

2. The cross-products principle states that if
   $\frac{a}{b} = \frac{c}{d}$ ($b \neq 0$ and $d \neq 0$), then _____.

In Exercises 3–4, determine whether each statement is true or false. If the statement is false, make the necessary change(s) to produce a true statement.

3. A statement that two proportions are equal is called a ratio.

4. The ratio of 1.25 to 2 is equal to the ratio 5:8.

Respond to Exercises 5–8 using verbal or written explanations.

5. What is a proportion? Give an example with your description.

6. Explain the difference between a ratio and a proportion.

7. Explain how to solve a proportion. Illustrate your explanation with an example.

8. Explain the meaning of this statement: A company's monthly sales are proportional to its advertising budget.

In Exercises 9–12, determine whether each statement makes sense or does not make sense, and explain your reasoning.

9. I can solve $\frac{x}{9} = \frac{4}{6}$ by using the cross-products principle or by multiplying both sides by 18, the least common denominator.

10. I made an error using the cross-products principle and put each product on the wrong side of the equation.

11. The capture-recapture method may be inaccurate if some of the animals are hiding.

12. At the rate shown in the cartoon, five potatoes would cost $36.67.

REAL LIFE ADVENTURES by Gary Wise and Lance Aldrich

If the people who own the shops at the airport owned other things.

Universal Press Syndicate

Practice Exercises

In Exercises 13–26, solve each proportion and check.

13. $\frac{24}{x} = \frac{12}{7}$

14. $\frac{56}{x} = \frac{8}{7}$

15. $\frac{x}{6} = \frac{18}{4}$

16. $\frac{x}{32} = \frac{3}{24}$

17. $\frac{-3}{8} = \frac{x}{40}$

18. $\frac{-3}{8} = \frac{6}{x}$

19. $\frac{x}{12} = -\frac{3}{4}$

20. $\frac{x}{64} = -\frac{9}{16}$

21. $\frac{x-2}{12} = \frac{8}{3}$

22. $\frac{x-4}{10} = \frac{3}{5}$

23. $\frac{x}{7} = \frac{x+14}{5}$

24. $\frac{x}{5} = \frac{x-3}{2}$

25. $\frac{y+10}{10} = \frac{y-2}{4}$

26. $\frac{2}{y-5} = \frac{3}{y+6}$

Practice Plus

In Exercises 27–32, solve each proportion for x.

27. $\dfrac{x}{a} = \dfrac{b}{c}$

28. $\dfrac{a}{x} = \dfrac{b}{c}$

29. $\dfrac{a+b}{c} = \dfrac{x}{d}$

30. $\dfrac{a-b}{c} = \dfrac{x}{d}$

31. $\dfrac{x+a}{a} = \dfrac{b+c}{c}$

32. $\dfrac{ax-b}{b} = \dfrac{c-d}{d}$

Application Exercises

33. The property tax on a house with an assessed value of $520,000 is $7280. Determine the property tax on a house with an assessed value of $650,000, assuming the same tax rate.

34. The property tax on a house with an assessed value of $350,000 is $4200. Determine the property tax on a house with an assessed value of $720,000, assuming the same tax rate.

35. An alligator's tail length is proportional to its body length. An alligator with a body length of 4 feet has a tail length of 3.6 feet. What is the tail length of an alligator whose body length is 6 feet?

|← Body length →|← Tail length →|

36. An object's weight on the moon is proportional to its weight on Earth. Neil Armstrong, the first person to step on the moon on July 20, 1969, weighed 360 pounds on Earth (with all of his equipment on) and 60 pounds on the moon. What is the moon weight of a person who weighs 186 pounds on Earth?

37. St. Paul Island in Alaska has 12 fur seal rookeries (breeding places). In 1961, to estimate the fur seal pup population in the Gorbath rookery, 4963 fur seal pups were tagged in early August. In late August, a sample of 900 pups was observed and 218 of these were found to have been previously tagged. Estimate the total number of fur seal pups in this rookery.

38. To estimate the number of bass in a lake, wildlife biologists tagged 50 bass and released them in the lake. Later they netted 108 bass and found that 27 of them were tagged. Approximately how many bass are in the lake?

Critical Thinking Exercises

39. The front sprocket on a bicycle has 60 teeth and the rear sprocket has 20 teeth. For mountain biking, an owner needs a 5-to-1 front-to-rear ratio. If only one of the sprockets is to be replaced, describe the two ways in which this can be done.

40. A baseball player's batting average is the ratio of the number of hits to the number of times at bat. A baseball player has 10 hits out of 20 times at bat. How many consecutive times at bat must a hit be made to raise the player's batting average to 0.600?

Technology Exercises

Use a calculator to solve Exercises 41–42. Round your answer to two decimal places.

41. Solve: $\dfrac{7.32}{x} = \dfrac{-19.03}{28}$

42. On a map, 2 inches represent 13.47 miles. How many miles does a person plan to travel if the distance on the map is 9.85 inches?

4.5 Modeling Using Variation

Objectives

1. Solve direct variation problems.

2. Solve inverse variation problems.

3. Solve combined variation problems.

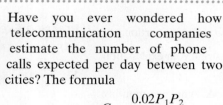

Have you ever wondered how telecommunication companies estimate the number of phone calls expected per day between two cities? The formula

$$C = \dfrac{0.02 P_1 P_2}{d^2}$$

shows that the daily number of phone calls, C, increases as the populations of the cities, P_1 and P_2, in thousands, increase and decreases as the distance, d, between the cities increases.

Certain formulas occur so frequently in applied situations that they are given special names. Variation formulas show how one quantity changes in relation to other quantities. Quantities can vary *directly* or *inversely*. In this section, we look at situations that can be modeled by each of these kinds of variation.

Solve direct
variation problems.

Direct Variation

When you swim underwater, the pressure in your ears depends on the depth at which you are swimming. The formula

$$p = 0.43d$$

describes the water pressure, p, in pounds per square inch, at a depth of d feet. We can use this formula to determine the pressure in your ears at various depths:

If $d = 20$, $p = 0.43(20) = 8.6$. At a depth of 20 feet, water pressure is 8.6 pounds per square inch.

Doubling the depth doubles the pressure.

If $d = 40$, $p = 0.43(40) = 17.2$. At a depth of 40 feet, water pressure is 17.2 pounds per square inch.

Doubling the depth doubles the pressure.

If $d = 80$, $p = 0.43(80) = 34.4$. At a depth of 80 feet, water pressure is 34.4 pounds per square inch.

The formula $p = 0.43d$ illustrates that water pressure is a constant multiple of your underwater depth. If your depth is doubled, the pressure is doubled; if your depth is tripled, the pressure is tripled; and so on. Because of this, the pressure in your ears is said to **vary directly** as your underwater depth. The **equation of variation** is

$$p = 0.43d.$$

Generalizing our discussion of pressure and depth, we obtain the following statement:

Direct Variation

If a situation is described by an equation in the form

$$y = kx,$$

where k is a nonzero constant, we say that y **varies directly as** x. The number k is called the **constant of variation**.

Problems involving direct variation can be solved using the following procedure. This procedure applies to direct variation problems, as well as to the other kinds of variation problems that we will discuss.

Solving Variation Problems

1. Write an equation that models the given English statement.
2. Substitute the given pair of values into the equation in step 1 and solve for k, the constant of variation.
3. Substitute the value of k into the equation in step 1.
4. Use the equation from step 3 to answer the problem's question.

Example 1   Solving a Direct Variation Problem

The volume of blood, B, in a person's body varies directly as body weight, W. A person who weighs 160 pounds has approximately 5 quarts of blood. Estimate the number of quarts of blood in a person who weighs 200 pounds.

Solution

Step 1   Write an equation. We know that *y varies directly as x* is expressed as

$$y = kx.$$

By changing letters, we can write an equation that models the following English statement: The volume of blood, B, varies directly as body weight, W.

$$B = kW$$

Step 2   Use the given values to find *k*. A person who weighs 160 pounds has approximately 5 quarts of blood. Substitute 160 for W and 5 for B in the direct variation equation. Then solve for k.

$B = kW$	The volume of blood varies directly as body weight.
$5 = k \cdot 160$	Substitute 160 for W and 5 for B.
$\dfrac{5}{160} = \dfrac{k \cdot 160}{160}$	Divide both sides by 160.
$0.03125 = k$	Express $\frac{5}{160}$, or $\frac{1}{32}$, in decimal form.

Step 3   Substitute the value of *k* into the equation.

$B = kW$	Use the equation from step 1.
$B = 0.03125W$	Replace k, the constant of variation, with 0.03125.

Step 4   Answer the problem's question. We are interested in estimating the number of quarts of blood in a person who weighs 200 pounds. Substitute 200 for W in $B = 0.03125W$ and solve for B.

$B = 0.03125W$	This is the equation from step 3.
$B = 0.03125(200)$	Substitute 200 for W.
$= 6.25$	Multiply.

A person who weighs 200 pounds has approximately 6.25 quarts of blood.

Checkpoint 1   The number of gallons of water, W, used when taking a shower varies directly as the time, t, in minutes, in the shower. A shower lasting 5 minutes uses 30 gallons of water. How much water is used in a shower lasting 11 minutes?

A direct variation situation can involve variables to higher powers. For example, y can vary directly as x^2 ($y = kx^2$) or as x^3 ($y = kx^3$).

Direct Variation with Powers

y **varies directly as the *n*th power of** *x* if there exists some nonzero constant k such that

$$y = kx^n.$$

Example 2 Solving a Direct Variation Problem

The distance, s, that a body falls from rest varies directly as the square of the time, t, of the fall. If skydivers fall 64 feet in 2 seconds, how far will they fall in 4.5 seconds?

Solution

Step 1 Write an equation. We know that *y varies directly as the square of x* is expressed as

$$y = kx^2.$$

By changing letters, we can write an equation that models the following English statement: Distance, s, varies directly as the square of time, t, of the fall.

$$s = kt^2$$

Step 2 Use the given values to find k. Skydivers fall 64 feet in 2 seconds. Substitute 64 for s and 2 for t in the direct variation equation. Then solve for k.

$$s = kt^2 \qquad \text{Distance varies directly as the square of time.}$$
$$64 = k \cdot 2^2 \qquad \text{Skydivers fall 64 feet in 2 seconds.}$$
$$64 = 4k \qquad \text{Simplify: } 2^2 = 4.$$
$$\frac{64}{4} = \frac{4k}{4} \qquad \text{Divide both sides by 4.}$$
$$16 = k \qquad \text{Simplify.}$$

Step 3 Substitute the value of k into the equation.

$$s = kt^2 \qquad \text{Use the equation from step 1.}$$
$$s = 16t^2 \qquad \text{Replace k, the constant of variation, with 16.}$$

Step 4 Answer the problem's question. How far will the skydivers fall in 4.5 seconds? Substitute 4.5 for t in $s = 16t^2$ and solve for s.

$$s = 16(4.5)^2 = 16(20.25) = 324$$

Thus, in 4.5 seconds, the skydivers will fall 324 feet.

✓ **Checkpoint 2** The weight of a great white shark varies directly as the cube of its length. A great white shark caught off Catalina Island, California, was 15 feet long and weighed 2025 pounds. What was the weight of the 25-foot-long shark in the novel *Jaws*?

Inverse Variation

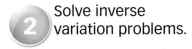

② Solve inverse variation problems.

The distance from San Francisco to Los Angeles is 420 miles. The time that it takes to drive from San Francisco to Los Angeles depends on the rate at which one drives and is given by

$$\text{Time} = \frac{420}{\text{Rate}}.$$

For example, if you average 30 miles per hour, the time for the drive is

$$\text{Time} = \frac{420}{30} = 14,$$

or 14 hours. If you average 50 miles per hour, the time for the drive is

$$\text{Time} = \frac{420}{50} = 8.4,$$

or 8.4 hours. As your rate (or speed) increases, the time for the trip decreases and vice versa.

We can express the time for the 420-mile San Francisco–Los Angeles trip using t for time and r for rate:

$$t = \frac{420}{r}.$$

This equation is an example of an **inverse variation** equation. Time, t, **varies inversely** as rate, r. When two quantities vary inversely, one quantity increases as the other decreases and vice versa.

Generalizing, we obtain the following statement:

> **Inverse Variation**
>
> If a situation is described by an equation in the form
>
> $$y = \frac{k}{x},$$
>
> where k is a nonzero constant, we say that **y varies inversely as x**. The number k is called the **constant of variation**.

We use the same procedure to solve inverse variation problems as we did to solve direct variation problems. Example 3 illustrates this procedure.

Example 3   Solving an Inverse Variation Problem

When you use a spray can and press the valve at the top, you decrease the pressure of the gas in the can. This decrease of pressure causes the volume of the gas in the can to increase. Because the gas needs more room than is provided in the can, it expands in spray form through the small hole near the valve. In general, if the temperature is constant, the pressure, P, of a gas in a container varies inversely as the volume, V, of the container. The pressure of a gas sample in a container whose volume is 8 cubic inches is 12 pounds per square inch. If the sample expands to a volume of 22 cubic inches, what is the new pressure of the gas?

Solution

Step 1   Write an equation. We know that *y varies inversely as x* is expressed as

$$y = \frac{k}{x}.$$

By changing letters, we can write an equation that models the following English statement: The pressure, P, of a gas in a container varies inversely as the volume, V.

$$P = \frac{k}{V}$$

Step 2   Use the given values to find k. The pressure of a gas sample in a container whose volume is 8 cubic inches is 12 pounds per square inch. Substitute 12 for P and 8 for V in the inverse variation equation. Then solve for k.

$$P = \frac{k}{V} \qquad \text{Pressure varies inversely as volume.}$$

$$12 = \frac{k}{8} \qquad \text{The pressure in an 8 cubic-inch container is 12 pounds per square inch.}$$

$$12 \cdot 8 = \frac{k}{8} \cdot 8 \qquad \text{Multiply both sides by 8.}$$

$$96 = k \qquad \text{Simplify.}$$

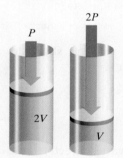

Doubling the pressure halves the volume.

Step 3 Substitute the value of k into the equation.

$$P = \frac{k}{V} \quad \text{Use the equation from step 1.}$$

$$P = \frac{96}{V} \quad \text{Replace } k, \text{ the constant of variation, with 96.}$$

Step 4 Answer the problem's question. We need to find the pressure when the volume expands to 22 cubic inches. Substitute 22 for V and solve for P.

$$P = \frac{96}{V} = \frac{96}{22} = 4\frac{4}{11}$$

When the volume is 22 cubic inches, the pressure of the gas is $4\frac{4}{11}$ pounds per square inch.

> ✓ **Checkpoint 3**   The length of a violin string varies inversely as the frequency of its vibrations. A violin string 8 inches long vibrates at a frequency of 640 cycles per second. What is the frequency of a 10-inch string?

Combined Variation

 Solve combined variation problems.

In **combined variation**, direct and inverse variation occur at the same time. For example, as the advertising budget, A, of a company increases, its monthly sales, S, also increase. Monthly sales vary directly as the advertising budget:

$$S = kA.$$

By contrast, as the price of the company's product, P, increases, its monthly sales, S, decrease. Monthly sales vary inversely as the price of the product:

$$S = \frac{k}{P}.$$

We can combine these two variation equations into one combined equation:

$$S = \frac{kA}{P}.$$

> Monthly sales , S, vary directly as the advertising budget, A, and inversely as the price of the product, P.

The following example illustrates an application of combined variation.

Example 4   **Solving a Combined Variation Problem**

The owners of Rollerblades Plus determine that the monthly sales, S, of its skates vary directly as its advertising budget, A, and inversely as the price of the skates, P. When $60,000 is spent on advertising and the price of the skates is $40, the monthly sales are 12,000 pairs of rollerblades.

a. Write an equation of variation that describes this situation.

b. Determine monthly sales if the amount of the advertising budget is increased to $70,000.

Solution

a. Write an equation.

$$S = \frac{kA}{P}$$

> Translate "sales vary directly as the advertising budget and inversely as the skates' price."

Use the given values to find k.

$$S = \frac{kA}{P}$$

Sales vary directly as the advertising budget, A, and inversely as the skates' price, P.

$$12{,}000 = \frac{k(60{,}000)}{40}$$

When \$60,000 is spent on advertising $(A = 60{,}000)$ and the price is \$40 $(P = 40)$, monthly sales are 12,000 units $(S = 12{,}000)$.

$$12{,}000 = k \cdot 1500$$

Divide 60,000 by 40.

$$\frac{12{,}000}{1500} = \frac{k \cdot 1500}{1500}$$

Divide both sides of the equation by 1500.

$$8 = k$$

Simplify.

Therefore, the equation of variation that models monthly sales is

$$S = \frac{8A}{P}.$$

Substitute 8 for k in $S = \frac{kA}{P}$.

b. The advertising budget is increased to \$70,000, so $A = 70{,}000$. The skates' price is still \$40, so $P = 40$.

$$S = \frac{8A}{P}$$

This is the equation that models monthly sales from part (a).

$$S = \frac{8(70{,}000)}{40}$$

Substitute 70,000 for A and 40 for P.

$$S = 14{,}000$$

Simplify.

With a \$70,000 advertising budget and \$40 price, the company can expect to sell 14,000 pairs of rollerblades in a month (up from 12,000).

Checkpoint 4 The number of minutes needed to solve an exercise set of variation problems varies directly as the number of problems and inversely as the number of people working to solve the problems. It takes 4 people 32 minutes to solve 16 problems. How many minutes will it take 8 people to solve 24 problems?

Achieving Success

Avoid asking for help on a problem that you have not thought about. This is basically asking your teacher to do the work for you. First try solving the problem on your own!

Exercise Set 4.5

Concept and Vocabulary Exercises

In Exercises 1–4, fill in each blank so that the resulting statement is true.

1. y varies directly as x can be modeled by the equation _____.

2. y varies directly as the nth power of x can be modeled by the equation _____.

3. y varies inversely as x can be modeled by the equation _____.

4. y varies directly as x and inversely as z can be modeled by the equation _____.

In Exercises 5–8, determine whether each statement is true or false. If the statement is false, make the necessary change(s) to produce a true statement.

5. A man's weight, W, varies directly as the cube of his height, H, so

$$W = k^3 H.$$

6. Radiation machines, used to treat tumors, produce an intensity of radiation, R, that varies inversely as the square of the distance from the machine, D, so

$$R = k\sqrt{D}.$$

7. Body-mass index, BMI, varies directly as one's weight, w, in pounds, and inversely as the square of one's height, h, in inches, so,

$$\text{BMI} = \frac{kh^2}{w}.$$

8. The average number of daily phone calls, C, between two cities varies directly as the product of their populations, P_1 and P_2, and inversely as the square of the distance, d, between them, so

$$C = \frac{kP_1P_2}{d^2}.$$

Respond to Exercises 9–12 using verbal or written explanations.

9. What does it mean if two quantities vary directly?

10. In your own words, explain how to solve a variation problem.

11. What does it mean if two quantities vary inversely?

12. Explain what is meant by combined variation. Give an example with your explanation.

In Exercises 13–16, determine whether each statement makes sense or does not make sense, and explain your reasoning.

13. The time it takes me to get to campus varies directly as my rate of travel.

14. It seems reasonable that a student's grade in a course varies directly as the number of hours spent studying.

15. It seems reasonable that a student's grade in a course varies directly as the number of hours spent watching TV.

16. It seems reasonable that a student's exam score varies inversely as the number of missed assignments.

Practice Exercises

Use the four-step procedure for solving variation problems given on page 261 to solve Exercises 17–28.

17. y varies directly as x. $y = 65$ when $x = 5$. Find y when $x = 12$.

18. y varies directly as x. $y = 45$ when $x = 5$. Find y when $x = 13$.

19. y varies directly as x^2. $y = 24$ when $x = 2$. Find y when $x = 5$.

20. y varies directly as x^2. $y = 45$ when $x = 3$. Find y when $x = 10$.

21. y varies inversely as x. $y = 12$ when $x = 5$. Find y when $x = 2$.

22. y varies inversely as x. $y = 6$ when $x = 3$. Find y when $x = 9$.

23. y varies inversely as $\sqrt{x}$. $y = 10$ when $x = 4$. Find y when $x = 100$.

24. y varies inversely as $\sqrt{x}$. $y = 20$ when $x = 25$. Find y when $x = 4$.

25. y varies directly as x and inversely as z. $y = 100$ when $x = 5$ and $z = 10$. Find y when $x = 3$ and $z = 60$.

26. y varies directly as x and inversely as z. $y = 20$ when $x = 5$ and $z = 100$. Find y when $x = 8$ and $z = 16$.

27. y varies directly as x and inversely as the square of z. $y = 20$ when $x = 50$ and $z = 5$. Find y when $x = 3$ and $z = 6$.

28. a varies directly as b and inversely as the square of c. $a = 7$ when $b = 9$ and $c = 6$. Find a when $b = 4$ and $c = 8$.

Practice Plus

In Exercises 29–34, write an equation that expresses each relationship. Then solve the equation for y.

29. x varies directly as y.

30. x^2 varies directly as y.

31. x varies inversely as y.

32. x^2 varies inversely as y.

33. x varies directly as y and inversely as z.

34. x^2 varies directly as y and inversely as z.

Application Exercises

Use the four-step procedure for solving variation problems given on page 261 to solve Exercises 35–44.

35. The height that a ball bounces varies directly as the height from which it was dropped. A tennis ball dropped from 12 inches bounces 8.4 inches. From what height was the tennis ball dropped if it bounces 56 inches?

36. The distance that a spring will stretch varies directly as the force applied to the spring. A force of 12 pounds is needed to stretch a spring 9 inches. What force is required to stretch the spring 15 inches?

37. If all men had identical body types, their weight would vary directly as the cube of their height. Shown below is Robert Wadlow, who reached a record height of 8 feet 11 inches (107 inches) before his death at age 22. If a man who is 5 feet 10 inches tall (70 inches) with the same body type as Mr. Wadlow weighs 170 pounds, what was Robert Wadlow's weight shortly before his death?

38. The number of houses that can be served by a water pipe varies directly as the square of the diameter of the pipe. A water pipe that has a 10-centimeter diameter can supply 50 houses.

   a. How many houses can be served by a water pipe that has a 30-centimeter diameter?

   b. What size water pipe is needed for a new subdivision of 1250 houses?

39. The figure shows that a bicyclist tips the cycle when making a turn. The angle B, formed by the vertical direction and the bicycle, is called the banking angle. The banking angle varies inversely as the cycle's turning radius. When the turning radius is 4 feet, the banking angle is 28°. What is the banking angle when the turning radius is 3.5 feet?

$B°$

40. The water temperature of the Pacific Ocean varies inversely as the water's depth. At a depth of 1000 meters, the water temperature is 4.4° Celsius. What is the water temperature at a depth of 5000 meters?

41. Radiation machines, used to treat tumors, produce an intensity of radiation that varies inversely as the square of the distance from the machine. At 3 meters, the radiation intensity is 62.5 milliroentgens per hour. What is the intensity at a distance of 2.5 meters?

42. The illumination provided by a car's headlight varies inversely as the square of the distance from the headlight. A car's headlight produces an illumination of 3.75 footcandles at a distance of 40 feet. What is the illumination when the distance is 50 feet?

43. Body-mass index, or BMI, takes both weight and height into account when assessing whether an individual is underweight or overweight. BMI varies directly as one's weight, in pounds, and inversely as the square of one's height, in inches. In adults, normal values for the BMI are between 20 and 25, inclusive. Values below 20 indicate that an individual is underweight and values above 30 indicate that an individual is obese. A person who weighs 180 pounds and is 5 feet, or 60 inches, tall has a BMI of 35.15. What is the BMI, to the nearest tenth, for a 170-pound person who is 5 feet 10 inches tall. Is this person overweight?

44. One's intelligence quotient, or IQ, varies directly as a person's mental age and inversely as that person's chronological age. A person with a mental age of 25 and a chronological age of 20 has an IQ of 125. What is the chronological age of a person with a mental age of 40 and an IQ of 80?

Heart rates and life spans of most mammals can be modeled using inverse variation. The bar graph shows the average heart rate and the average life span of five mammals. You will use the data to solve Exercises 45–46.

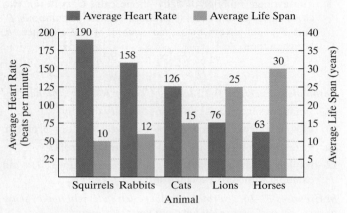

Heart Rate and Life Span

Source: The Handy Science Answer Book, Visible Ink Press, 2003.

45. a. A mammal's average life span, L, in years, varies inversely as its average heart rate, R, in beats per minute. Use the data shown for horses to write the equation that models this relationship.

b. Is the inverse variation equation in part (a) an exact model or an approximate model for the data shown for lions?

c. Elephants have an average heart rate of 27 beats per minute. Determine their average life span.

46. a. A mammal's average life span, L, in years, varies inversely as its average heart rate, R, in beats per minute. Use the data shown for cats to write the equation that models this relationship.

b. Is the inverse variation equation in part (a) an exact model or an approximate model for the data shown for squirrels?

c. Mice have an average heart rate of 634 beats per minute. Determine their average life span, rounded to the nearest year.

Critical Thinking Exercises

47. Sound intensity varies inversely as the square of the distance from the sound source. If you are in a movie theater and you change your seat to one that is twice as far from the speakers, how does the new sound intensity compare to that of your original seat?

48. Many people claim that as they get older, time seems to pass more quickly. Suppose that the perceived length of a period of time is inversely proportional to your age. How long will a year seem to be when you are three times as old as you are now?

49. In a hurricane, the wind pressure varies directly as the square of the wind velocity. If wind pressure is a measure of a hurricane's destructive capacity, what happens to this destructive power when the wind speed doubles?

50. The heat generated by a stove element varies directly as the square of the voltage and inversely as the resistance. If the voltage remains constant, what needs to be done to triple the amount of heat generated?

51. Galileo's telescope brought about revolutionary changes in astronomy. A comparable leap in our ability to observe the universe took place as a result of the Hubble Space Telescope. The space telescope can see stars and galaxies whose brightness is $\frac{1}{50}$ of the faintest objects now observable using ground-based telescopes. Use the fact that the brightness of a point source, such as a star, varies inversely as the square of its distance from an observer to show that the space telescope can see about seven times farther than a ground-based telescope.

Group Exercise

52. Begin by deciding on a product that interests the group because you are now in charge of advertising this product. Demand for the product varies directly as the amount spent on advertising and inversely as the price of the product. However, as more money is spent on advertising, the price of your product rises. Under what conditions would members recommend an increased expense in advertising? Once you've determined what your product is, write formulas for the given conditions and experiment with hypothetical numbers. What other factors might you take into consideration in terms of your recommendation? How do these factor affect the demand for your product?

4.6 Linear Inequalities in One Variable

Objectives

① Graph subsets of real numbers on a number line.

② Solve linear inequalities.

③ Identify linear inequalities with no solution or all real numbers as solutions.

④ Solve applied problems using linear inequalities.

Rent-a-Heap, a car rental company, charges $125 per week plus $0.20 per mile to rent one of their cars. Suppose you are limited by how much money you can spend for the week: You can spend at most $335. If we let x represent the number of miles you drive the heap in a week, we can write an inequality that models the given conditions:

The weekly charge of $125	plus	the charge of $0.20 per mile for x miles	must be less than or equal to	$335.
125	+	0.20x	≤	335.

Notice that the highest exponent on the variable is 1. Such an inequality is called a *linear inequality in one variable*. The symbol between the two sides of an inequality can be ≤ (is less than or equal to), < (is less than), ≥ (is greater than or equal to), or > (is greater than).

In this section, we will study how to solve linear inequalities such as $125 + 0.20x \le 335$. **Solving an inequality** is the process of finding the set of numbers that makes the inequality a true statement. These numbers are called the **solutions** of the inequality and we say that they **satisfy** the inequality. The set of all solutions is called the **solution set** of the inequality. We begin by discussing how to represent these solution sets, which are subsets of real numbers, on a number line.

Graphing Subsets of Real Numbers on a Number Line

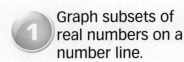

Graph subsets of real numbers on a number line.

Table 4.3 shows how to represent various subsets of real numbers on a number line. Open dots (○) indicate that a number is not included in a set. Closed dots (●) indicate that a number is included in a set.

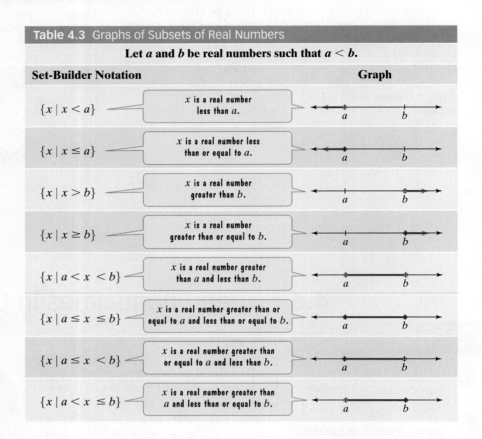

Table 4.3 Graphs of Subsets of Real Numbers

Let a and b be real numbers such that $a < b$.

Set-Builder Notation		Graph
$\{x \mid x < a\}$	x is a real number less than a.	
$\{x \mid x \leq a\}$	x is a real number less than or equal to a.	
$\{x \mid x > b\}$	x is a real number greater than b.	
$\{x \mid x \geq b\}$	x is a real number greater than or equal to b.	
$\{x \mid a < x < b\}$	x is a real number greater than a and less than b.	
$\{x \mid a \leq x \leq b\}$	x is a real number greater than or equal to a and less than or equal to b.	
$\{x \mid a \leq x < b\}$	x is a real number greater than or equal to a and less than b.	
$\{x \mid a < x \leq b\}$	x is a real number greater than a and less than or equal to b.	

Example 1   Graphing Subsets of Real Numbers

Graph each set:

a. $\{x \mid x < 3\}$ **b.** $\{x \mid x \geq -1\}$ **c.** $\{x \mid -1 < x \leq 3\}$.

Solution

a. $\{x \mid x < 3\}$ — x is a real number less than 3.

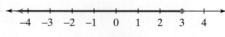

b. $\{x \mid x \geq -1\}$ — x is a real number greater than or equal to −1.

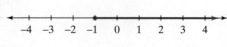

c. $\{x \mid -1 < x \leq 3\}$ — x is a real number greater than −1 and less than or equal to 3.

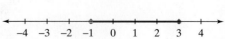

✓ **Checkpoint 1**   Graph each set:

a. $\{x \mid x < 4\}$ **b.** $\{x \mid x \geq -2\}$ **c.** $\{x \mid -4 \leq x < 1\}$.

Solving Linear Inequalities in One Variable

We know that a linear equation in x can be expressed as $ax + b = 0$. A **linear inequality in x** can be written in one of the following forms:

$$ax + b < 0, \quad ax + b \leq 0, \quad ax + b > 0, \quad ax + b \geq 0.$$

In each form, $a \neq 0$.

Back to our question that opened this section: How many miles can you drive your Rent-a-Heap car if you can spend at most $335? We answer the question by solving

$$0.20x + 125 \leq 335$$

for x. The solution procedure is nearly identical to that for solving

$$0.20x + 125 = 335.$$

Our goal is to get x by itself on the left side. We do this by subtracting 125 from both sides to isolate $0.20x$:

$$0.20x + 125 \leq 335 \qquad \text{This is the given inequality.}$$
$$0.20x + 125 - 125 \leq 335 - 125 \qquad \text{Subtract 125 from both sides.}$$
$$0.20x \leq 210. \qquad \text{Simplify.}$$

Finally, we isolate x from $0.20x$ by dividing both sides of the inequality by 0.20:

$$\frac{0.20x}{0.20} \leq \frac{210}{0.20} \qquad \text{Divide both sides by 0.20.}$$
$$x \leq 1050. \qquad \text{Simplify.}$$

With at most $335 per week to spend, you can travel at most 1050 miles.

We started with the inequality $0.20x + 125 \leq 335$ and obtained the inequality $x \leq 1050$ in the final step. Both of these inequalities have the same solution set, namely $\{x \mid x \leq 1050\}$. Inequalities such as these, with the same solution set, are said to be **equivalent**.

We isolated x from $0.20x$ by dividing both sides of $0.20x \leq 210$ by 0.20, a positive number. Let's see what happens if we divide both sides of an inequality by a negative number. Consider the inequality $10 < 14$. Divide 10 and 14 by -2:

$$\frac{10}{-2} = -5 \quad \text{and} \quad \frac{14}{-2} = -7.$$

Because -5 lies to the right of -7 on the number line, -5 is greater than -7:

$$-5 > -7.$$

Notice that the direction of the inequality symbol is reversed:

$$10 < 14$$

Dividing by -2 changes the direction of the inequality symbol.

$$-5 > -7.$$

In general, **when we multiply or divide both sides of an inequality by a negative number, the direction of the inequality symbol is reversed**. When we reverse the direction of the inequality symbol, we say that we change the *sense* of the inequality.

---

2 Solve linear inequalities.

Study Tip

English phrases such as "at least" and "at most" can be modeled by inequalities.

English Sentence	Inequality
x is at least 5.	$x \geq 5$
x is at most 5.	$x \leq 5$
x is between 5 and 7.	$5 < x < 7$
x is no more than 5.	$x \leq 5$
x is no less than 5.	$x \geq 5$

We can summarize the discussion on the previous page with the following statement:

> **Solving Linear Inequalities**
>
> The procedure for solving linear inequalities is the same as the procedure for solving linear equations, with one important exception: When multiplying or dividing both sides of the inequality by a negative number, reverse the direction of the inequality symbol, changing the sense of the inequality.

Example 2   Solving a Linear Inequality

Solve and graph the solution set: $4x - 7 \geq 5$.

Solution

Our goal is to get x by itself on the left side. We do this by first getting $4x$ by itself, adding 7 to both sides.

$$4x - 7 \geq 5 \qquad \text{This is the given inequality.}$$
$$4x - 7 + 7 \geq 5 + 7 \qquad \text{Add 7 to both sides.}$$
$$4x \geq 12 \qquad \text{Simplify.}$$

Next, we isolate x from $4x$ by dividing both sides by 4. The inequality symbol stays the same because we are dividing by a positive number.

$$\frac{4x}{4} \geq \frac{12}{4} \qquad \text{Divide both sides by 4.}$$
$$x \geq 3 \qquad \text{Simplify.}$$

The solution set consists of all real numbers that are greater than or equal to 3, expressed in set-builder notation as $\{x \mid x \geq 3\}$. The graph of the solution set is shown as follows:

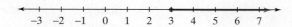

We cannot check all members of an inequality's solution set, but we can take a few values to get an indication of whether or not it is correct. In Example 2, we found that the solution set of $4x - 7 \geq 5$ is $\{x \mid x \geq 3\}$. Show that 3 and 4 satisfy the inequality, whereas 2 does not.

✓ **Checkpoint 2**   Solve and graph the solution set: $5x - 3 \leq 17$.

Example 3   Solving Linear Inequalities

Solve and graph the solution set:

a. $\frac{1}{3}x < 5$ **b.** $-3x < 21$.

Solution

In each case, our goal is to isolate x. In the first inequality, this is accomplished by multiplying both sides by 3. In the second inequality, we can do this by dividing both sides by -3.

a. $\dfrac{1}{3}x < 5$ *This is the given inequality.*

$3 \cdot \dfrac{1}{3}x < 3 \cdot 5$ *Isolate x by multiplying by 3 on both sides.*

The symbol $<$ stays the same because we are multiplying both sides by a positive number.

$x < 15$ *Simplify.*

The solution set is $\{x \mid x < 15\}$. The graph of the solution set is shown as follows:

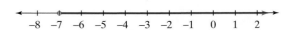

b. $-3x < 21$ *This is the given inequality.*

$\dfrac{-3x}{-3} > \dfrac{21}{-3}$ *Isolate x by dividing by -3 on both sides.*

The symbol $<$ must be reversed because we are dividing both sides by a negative number.

$x > -7$ *Simplify.*

The solution set is $\{x \mid x > -7\}$. The graph of the solution set is shown as follows:

✔ **Checkpoint 3**   Solve and graph the solution set:

▶ **a.** $\dfrac{1}{4}x < 2$ **b.** $-6x < 18.$

Example 4   Solving a Linear Inequality

Solve and graph the solution set: $6x - 12 > 8x + 2.$

Solution

We will get x by itself on the left side. We begin by subtracting $8x$ from both sides so that the variable term appears on the left.

$6x - 12 > 8x + 2$ *This is the given inequality.*

$6x - 8x - 12 > 8x - 8x + 2$ *Subtract 8x on both sides with the goal of isolating x on the left.*

$-2x - 12 > 2$ *Simplify.*

Next, we get $-2x$ by itself, adding 12 to both sides.

$-2x - 12 + 12 > 2 + 12$ *Add 12 to both sides.*

$-2x > 14$ *Simplify.*

In order to solve $-2x > 14$, we isolate x from $-2x$ by dividing both sides by -2. The direction of the inequality symbol must be reversed because we are dividing by a negative number.

$$\frac{-2x}{-2} < \frac{14}{-2}$$ Divide both sides by -2 and change the sense of the inequality.

$$x < -7$$ Simplify.

The solution set is $\{x|x < -7\}$. The graph of the solution set is shown as follows:

▶ **Checkpoint 4** Solve and graph the solution set: $7x - 3 > 13x + 33$.

Example 5 Solving a Linear Inequality

Solve and graph the solution set:

$$2(x - 3) + 5x \le 8(x - 1).$$

Solution

Begin by simplifying the algebraic expression on each side.

$$2(x - 3) + 5x \le 8(x - 1)$$ This is the given inequality.

$$2x - 6 + 5x \le 8x - 8$$ Use the distributive property.

$$7x - 6 \le 8x - 8$$ Add like terms on the left: $2x + 5x = 7x$.

We will get x by itself on the left side. Subtract $8x$ from both sides.

$$7x - 8x - 6 \le 8x - 8x - 8$$
$$-x - 6 \le -8$$

Next, we get $-x$ by itself, adding 6 to both sides.

$$-x - 6 + 6 \le -8 + 6$$
$$-x \le -2$$

To isolate x, we must eliminate the negative sign in front of the x. Because $-x$ means $-1x$, we can do this by dividing both sides of the inequality by -1. This reverses the direction of the inequality symbol.

$$\frac{-x}{-1} \ge \frac{-2}{-1}$$ Divide both sides by -1 and change the sense of the inequality.

$$x \ge 2$$ Simplify.

The solution set is $\{x|x \ge 2\}$. The graph of the solution set is shown as follows:

▶ **Checkpoint 5** Solve and graph the solution set:

$$2(x - 3) - 1 \le 3(x + 2) - 14.$$

Study Tip

You can solve

$$7x - 6 \le 8x - 8$$

by isolating x on the right side. Subtract $7x$ from both sides and add 8 to both sides:

$$7x - 6 - 7x \le 8x - 8 - 7x$$
$$-6 \le x - 8$$
$$-6 + 8 \le x - 8 + 8$$
$$2 \le x.$$

This last inequality means the same thing as

$$x \ge 2.$$

Solution sets, in this case $\{x|x \ge 2\}$, are expressed with the variable on the left and the constant on the right.

In our next example, the inequality has three parts:

$$-3 < 2x + 1 \le 3.$$

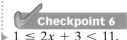

2x + 1 is greater than −3
and less than or equal to 3.

By performing the same operation on all three parts of the inequality, our goal is to **isolate x in the middle**.

Example 6   Solving a Three-Part Inequality

Solve and graph the solution set:

$$-3 < 2x + 1 \le 3.$$

Solution

We would like to isolate x in the middle. We can do this by first subtracting 1 from all three parts of the inequality. Then we isolate x from $2x$ by dividing all three parts of the inequality by 2.

$-3 < 2x + 1 \le 3$	This is the given inequality.
$-3 - 1 < 2x + 1 - 1 \le 3 - 1$	Subtract 1 from all three parts.
$-4 < 2x \le 2$	Simplify.
$\dfrac{-4}{2} < \dfrac{2x}{2} \le \dfrac{2}{2}$	Divide each part by 2.
$-2 < x \le 1$	Simplify.

The solution set consists of all real numbers greater than -2 and less than or equal to 1, represented by $\{x \mid -2 < x \le 1\}$. The graph is shown as follows:

```
─┼──┼──┼──○──┼──┼──●──┼──┼──┼──┼──→
 −5 −4 −3 −2 −1  0  1  2  3  4  5
```

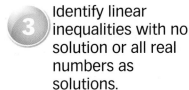 **Checkpoint 6**   Solve and graph the solution set on a number line:
$1 \le 2x + 3 < 11$.

Inequalities with Unusual Solution Sets

3 Identify linear inequalities with no solution or all real numbers as solutions.

We have seen that some equations have no solution. This is also true for some inequalities. An example of such an inequality is

$$x > x + 1.$$

There is no number that is greater than itself plus 1. This inequality has no solution. Its solution set is $\varnothing$, the empty set.

By contrast, some inequalities are true for all real numbers. An example of such an inequality is

$$x < x + 1.$$

Every real number is less than itself plus 1. The solution set is expressed in set-builder notation as $\{x \mid x \text{ is a real number}\}$.

Recognizing Inequalities with No Solution or All Real Numbers as Solutions

If you attempt to solve an inequality with no solution or one that is true for every real number, you will eliminate the variable.

- An inequality with no solution results in a false statement, such as $0 > 1$. The solution set is $\varnothing$, the empty set.

- An inequality that is true for every real number results in a true statement, such as $0 < 1$. The solution set is $\{x \mid x \text{ is a real number}\}$.

Example 7   Solving a Linear Inequality

Solve: $3(x + 1) > 3x + 5$.

Solution

$$3(x + 1) > 3x + 5 \qquad \text{This is the given inequality.}$$

$$3x + 3 > 3x + 5 \qquad \text{Apply the distributive property.}$$

$$3x + 3 - 3x > 3x + 5 - 3x \qquad \text{Subtract 3x from both sides.}$$

> Keep reading. 3 > 5 is not the solution.

$$3 > 5 \qquad \text{Simplify.}$$

The original inequality is equivalent to the statement $3 > 5$, which is false for every value of x. The inequality has no solution. The solution set is $\varnothing$, the empty set.

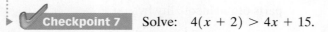 **Checkpoint 7**   Solve: $4(x + 2) > 4x + 15$.

Example 8   Solving a Linear Inequality

Solve: $2(x + 5) \leq 5x - 3x + 14$.

Solution

$$2(x + 5) \leq 5x - 3x + 14 \qquad \text{This is the given inequality.}$$

$$2x + 10 \leq 5x - 3x + 14 \qquad \text{Apply the distributive property.}$$

$$2x + 10 \leq 2x + 14 \qquad \text{Combine like terms.}$$

$$2x + 10 - 2x \leq 2x + 14 - 2x \qquad \text{Subtract 2x from both sides.}$$

> Keep reading. 10 ≤ 14 is not the solution.

$$10 \leq 14 \qquad \text{Simplify.}$$

The original inequality is equivalent to the statement $10 \leq 14$, which is true for every value of x. The solution is the set of all real numbers, expressed in set-builder notation as $\{x \mid x \text{ is a real number}\}$.

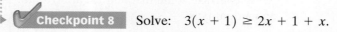 **Checkpoint 8**   Solve: $3(x + 1) \geq 2x + 1 + x$.

 Solve applied problems using linear inequalities.

As you continue your higher education, you will discover that different professors may use different grading systems to determine your final course grade. Some professors require a final examination; others do not. In our next example, a final exam is required *and* it counts as two grades.

Example 9   An Application: Final Course Grade

To earn an A in a course, you must have a final average of at least 90%. On the first four examinations, you have grades of 86%, 88%, 92%, and 84%. If the final examination counts as two grades, what must you get on the final to earn an A in the course?

Solution

We will use our five-step strategy for solving algebraic word problems.

Steps 1 and 2   Represent unknown quantities in terms of *x*. Let

$$x = \text{your grade on the final examination.}$$

Step 3   Write an inequality in *x* that models the conditions. The average of the six grades is found by adding the grades and dividing the sum by 6.

$$\text{Average} = \frac{86 + 88 + 92 + 84 + x + x}{6}$$

Because the final counts as two grades, the *x* (your grade on the final examination) is added twice. This is also why the sum is divided by 6.

To get an A, your average must be at least 90. This means that your average must be greater than or equal to 90.

Your average	must be greater than or equal to	90.
$\dfrac{86 + 88 + 92 + 84 + x + x}{6}$	$\geq$	90

Step 4   Solve the inequality and answer the problem's question.

$\dfrac{86 + 88 + 92 + 84 + x + x}{6} \geq 90$	This is the inequality that models the given conditions.
$\dfrac{350 + 2x}{6} \geq 90$	Combine like terms in the numerator.
$6\left(\dfrac{350 + 2x}{6}\right) \geq 6(90)$	Multiply both sides by 6, clearing the fraction.
$350 + 2x \geq 540$	Multiply.
$350 + 2x - 350 \geq 540 - 350$	Subtract 350 from both sides.
$2x \geq 190$	Simplify.
$\dfrac{2x}{2} \geq \dfrac{190}{2}$	Divide both sides by 2.
$x \geq 95$	Simplify.

You must get at least 95% on the final examination to earn an A in the course.

Step 5   Check. We can perform a partial check by computing the average with any grade that is at least 95. We will use 96. If you get 96% on the final examination, your average is

$$\frac{86 + 88 + 92 + 84 + 96 + 96}{6} = \frac{542}{6} = 90\frac{1}{3}.$$

Because $90\frac{1}{3} > 90$, you earn an A in the course.

Checkpoint 9   To earn a B in a course, you must have a final average of at least 80%. On the first three examinations, you have grades of 82%, 74%, and 78%. If the final examination counts as two grades, what must you get on the final to earn a B in the course?

Exercise Set 4.6

Concept and Vocabulary Exercises

In Exercises 1–6, fill in each blank so that the resulting statement is true.

1. On a number line, an open dot indicates that a number _____ in a solution set, and a closed dot indicates that a number _____ in a solution set.

2. If an inequality's solution set consists of all real numbers, x, that are less than a, the solution set is represented in set-builder notation as _____.

3. If an inequality's solution set consists of all real numbers, x, that are greater than a and less than or equal to b, the solution set is represented in set-builder notation as _____.

4. When multiplying or dividing both sides of an inequality by a negative number, _____ of the inequality symbol.

5. If the solution of an inequality results in a false statement such as $0 > 1$, the solution set is _____.

6. If the solution of an inequality in x results in a true statement such as $0 < 1$, the solution set is _____.

In Exercises 7–10, determine whether each statement is true or false. If the statement is false, make the necessary change(s) to produce a true statement.

7. The inequality $x - 3 > 0$ is equivalent to $x < 3$.

8. The statement "x is at most 5" is written $x < 5$.

9. The inequality $-4x < -20$ is equivalent to $x > -5$.

10. The statement "the sum of x and 6% of x is at least 80" is modeled by $x + 0.06x \geq 80$.

Respond to Exercises 11–14 using verbal or written explanations.

11. When graphing the solutions of an inequality, what is the difference between an open dot and a closed dot?

12. When solving an inequality, when is it necessary to change the direction of the inequality symbol? Give an example.

13. Describe ways in which solving a linear inequality is similar to solving a linear equation.

14. Describe ways in which solving a linear inequality is different than solving a linear equation.

In Exercises 15–18, determine whether each statement makes sense or does not make sense, and explain your reasoning.

15. I can check inequalities by substituting 0 for the variable: When 0 belongs to the solution set, I should obtain a true statement, and when 0 does not belong to the solution set, I should obtain a false statement.

16. In an inequality such as $5x + 4 < 8x - 5$, I can avoid division by a negative number depending on which side I collect the variable terms and on which side I collect the constant terms.

17. I solved $-2x + 5 \geq 13$ and concluded that -4 is the greatest integer in the solution set.

18. I began the solution of $5 - 3(x + 2) > 10x$ by simplifying the left side, obtaining $2x + 4 > 10x$.

Practice Exercises

In Exercises 19–30, graph each set of real numbers on a number line.

19. $\{x \mid x > 6\}$　　　　20. $\{x \mid x > -2\}$

21. $\{x \mid x < -4\}$　　　　22. $\{x \mid x < 0\}$

23. $\{x \mid x \geq -3\}$　　　　24. $\{x \mid x \geq -5\}$

25. $\{x \mid x \leq 4\}$　　　　26. $\{x \mid x \leq 7\}$

27. $\{x \mid -2 < x \leq 5\}$　　　　28. $\{x \mid -3 \leq x < 7\}$

29. $\{x \mid -1 < x < 4\}$　　　　30. $\{x \mid -7 \leq x \leq 0\}$

In Exercises 31–84, solve each inequality and graph the solution set on a number line.

31. $x - 3 > 2$　　　　32. $x + 1 < 5$

33. $x + 4 \leq 9$　　　　34. $x - 5 \geq 1$

35. $x - 3 < 0$　　　　36. $x + 4 \geq 0$

37. $4x < 20$　　　　38. $6x \geq 18$

39. $3x \geq -15$ **40.** $7x < -21$

41. $2x - 3 > 7$ **42.** $3x + 2 \leq 14$

43. $3x + 3 < 18$ **44.** $8x - 4 > 12$

45. $\frac{1}{2}x < 4$ **46.** $\frac{1}{2}x > 3$

47. $\frac{x}{3} > -2$ **48.** $\frac{x}{4} < -1$

49. $-3x < 15$ **50.** $-7x > 21$

51. $-3x \geq -15$ **52.** $-7x \leq -21$

53. $3x + 4 \leq 2x + 7$ **54.** $2x + 9 \leq x + 2$

55. $5x - 9 < 4x + 7$ **56.** $3x - 8 < 2x + 11$

57. $-2x - 3 < 3$ **58.** $14 - 3x > 5$

59. $3 - 7x \leq 17$ **60.** $5 - 3x \geq 20$

61. $-x < 4$ **62.** $-x > -3$

63. $5 - x \leq 1$ **64.** $3 - x \geq -3$

65. $2x - 5 > -x + 6$ **66.** $6x - 2 \geq 4x + 6$

67. $2x - 5 < 5x - 11$ **68.** $4x - 7 > 9x - 2$

69. $3(x + 1) - 5 < 2x + 1$ **70.** $4(x + 1) + 2 \geq 3x + 6$

71. $8x + 3 > 3(2x + 1) - x + 5$

72. $7 - 2(x - 4) < 5(1 - 2x)$

73. $\frac{x}{4} - \frac{3}{2} \leq \frac{x}{2} + 1$ **74.** $\frac{3x}{10} + 1 \geq \frac{1}{5} - \frac{x}{10}$

75. $1 - \frac{x}{2} > 4$ **76.** $7 - \frac{4}{5}x < \frac{3}{5}$

77. $6 < x + 3 < 8$ **78.** $7 < x + 5 < 11$

79. $-3 \leq x - 2 < 1$ **80.** $-6 < x - 4 \leq 1$

81. $-11 < 2x - 1 \leq -5$ **82.** $3 \leq 4x - 3 < 19$

83. $-3 \leq \frac{2}{3}x - 5 < -1$ **84.** $-6 \leq \frac{1}{2}x - 4 < -3$

In Exercises 85–94, solve each inequality.

85. $4x - 4 < 4(x - 51)$ **86.** $3x - 5 < 3(x - 2)$

87. $x + 3 < x + 7$ **88.** $x + 4 < x + 10$

89. $7x \leq 7(x - 2)$ **90.** $3x + 1 \leq 3(x - 2)$

91. $2(x + 3) > 2x + 1$ **92.** $5(x + 4) > 5x + 10$

93. $5x - 4 \leq 4(x - 1)$ **94.** $6x - 3 \leq 3(x - 1)$

Practice Plus

In Exercises 95–98, write an inequality with x isolated on the left side that is equivalent to the given inequality.

95. $Ax + By > C$; Assume $A > 0$.

96. $Ax + By \leq C$; Assume $A > 0$.

97. $Ax + By > C$; Assume $A < 0$.

98. $Ax + By \leq C$; Assume $A < 0$.

In Exercises 99–104, use set-builder notation to describe all real numbers satisfying the given conditions.

99. A number increased by 5 is at least two times the number.

100. A number increased by 12 is at least four times the number.

101. Twice the sum of four and a number is at most 36.

102. Three times the sum of five and a number is at most 48.

103. If the quotient of three times a number and five is increased by four, the result is no more than 34.

104. If the quotient of three times a number and four is decreased by three, the result is no less than 9.

Application Exercises

The graphs show that the three components of love, namely passion, intimacy, and commitment, progress differently over time. Passion peaks early in a relationship and then declines. By contrast, intimacy and commitment build gradually. Use the graphs to solve Exercises 105–112. Assume that x represents years in a relationship.

The Course of Love Over Time

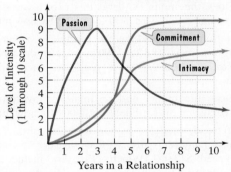

Source: R. J. Sternberg, A Triangular Theory of Love, *Psychological Review*, 93, 119–135.

105. Use set-builder notation to write an inequality that expresses for which years in a relationship intimacy is greater than commitment.

106. Use set-builder notation to write an inequality that expresses for which years in a relationship passion is greater than or equal to intimacy.

107. What is the relationship between passion and intimacy for $\{x | 5 \leq x < 7\}$?

108. What is the relationship between intimacy and commitment for $\{x | 4 \leq x < 7\}$?

109. What is the relationship between passion and commitment for $\{x | 6 < x < 8\}$?

110. What is the relationship between passion and commitment for $\{x | 7 < x < 9\}$?

111. What is the maximum level of intensity for passion? After how many years in a relationship does this occur?

112. After approximately how many years do levels of intensity for commitment exceed the maximum level of intensity for passion?

The bar graph shows that the average credit-card debt per U.S. household more than tripled over the 16-year period from 1991 through 2007. The data shown by the graph can be modeled by

$$d = 417x + 3085,$$

where d is the average credit-card debt per U.S. household x years after 1990. Use this information to solve Exercises 113–114.

Average Credit-Card Debt per U.S. Household

Source: CardTrak.com

113. a. Does the model underestimate or overestimate the average credit-card debt for 2007? By how much?

b. If trends modeled by the formula continue, describe how many years after 1990 average credit-card debt will exceed $12,259. Which years are included in your description?

114. a. Does the model underestimate or overestimate the average credit-card debt for 2006? By how much?

b. If trends modeled by the formula continue, describe how many years after 1990 average credit-card debt will exceed $13,093. Which years are included in your description?

115. On two examinations, you have grades of 86 and 88. There is an optional final examination, which counts as one grade. You decide to take the final in order to get a course grade of A, meaning a final average of at least 90.

a. What must you get on the final to earn an A in the course?

b. By taking the final, if you do poorly, you might risk the B that you have in the course based on the first two exam grades. If your final average is less than 80, you will lose your B in the course. Describe the grades on the final that will cause this to happen.

116. On three examinations, you have grades of 88, 78, and 86. There is still a final examination, which counts as one grade.

a. In order to get an A, your average must be at least 90. If you get 100 on the final, compute your average and determine if an A in the course is possible.

b. To earn a B in the course, you must have a final average of at least 80. What must you get on the final to earn a B in the course?

117. A car can be rented from Continental Rental for $80 per week plus 25 cents for each mile driven. How many miles can you travel if you can spend at most $400 for the week?

118. A car can be rented from Basic Rental for $60 per week plus 50 cents for each mile driven. How many miles can you travel if you can spend at most $600 for the week?

119. An elevator at a construction site has a maximum capacity of 3000 pounds. If the elevator operator weighs 245 pounds and each cement bag weighs 95 pounds, up to how many bags of cement can be safely lifted on the elevator in one trip?

120. An elevator at a construction site has a maximum capacity of 2800 pounds. If the elevator operator weighs 265 pounds and each cement bag weighs 65 pounds, up to how many bags of cement can be safely lifted on the elevator in one trip?

121. A basic cellphone plan costs $20 per month for 60 calling minutes. Additional time costs $0.40 per minute. The formula

$$C = 20 + 0.40(x - 60)$$

gives the monthly cost for this plan, C, for x calling minutes, where $x > 60$. How many calling minutes are possible for a monthly cost of at least $28 and at most $40?

122. The formula for converting Fahrenheit temperature, F, to Celsius temperature, C, is

$$C = \frac{5}{9}(F - 32).$$

If Celsius temperature ranges from 15° to 35°, inclusive, what is the range for the Fahrenheit temperature?

Critical Thinking Exercises

123. A car can be rented from Basic Rental for $260 per week with no extra charge for mileage. Continental charges $80 per week plus 25 cents for each mile driven to rent the same car. How many miles should be driven in a week to make the rental cost for Basic Rental a better deal than Continental's?

124. Membership in a fitness club costs $500 yearly plus $1 per hour spent working out. A competing club charges $440 yearly plus $1.75 per hour for use of their equipment. How many hours must a person work out yearly to make membership in the first club cheaper than membership in the second club?

125. A company manufactures and sells personalized stationery. The weekly fixed cost is $3000 and it cost $3.00 to produce each package of stationery. The selling price is $5.50 per package. How many packages of stationery must be produced and sold each week for the company to generate a profit?

126. What's wrong with this argument? Suppose x and y represent two real numbers, where $x > y$.

$2 > 1$	This is a true statement.
$2(y - x) > 1(y - x)$	Multiply both sides by $y - x$.
$2y - 2x > y - x$	Use the distributive property.
$y - 2x > -x$	Subtract y from both sides.
$y > x$	Add 2x to both sides.

The final inequality, $y > x$, is impossible because we were initially given $x > y$.

Chapter 4 Summary

4.1 Algebraic Expressions and Formulas

Definitions and Concepts

An algebraic expression combines variables and numbers using addition, subtraction, multiplication, division, powers, or roots.

Evaluating an algebraic expression means finding its value for a given value of the variable or for given values of the variables. Once these values are substituted, follow the order of operations agreement:

1. Perform operations within the innermost parentheses and work outward. If the algebraic expression involves a fraction, treat the numerator and the denominator as if they were each enclosed in parentheses.
2. Evaluate all exponential expressions.
3. Perform multiplications and divisions as they occur, working from left to right.
4. Perform additions and subtractions as they occur, working from left to right.

Example

- Evaluate $x^3 - 5(x - 2)^2$ when $x = -4$.
$$x^3 - 5(x - 2)^2 = (-4)^3 - 5(-4 - 2)^2 = (-4)^3 - 5(-6)^2$$
$$= -64 - 5(36) = -64 - 180 = -244$$

Additional Examples to Review
Example 1, page 221; Example 2, page 221; Example 3, page 221

Definitions and Concepts

An equation is a statement that two expressions are equal. Formulas are equations that express relationships among two or more variables. Mathematical modeling is the process of finding formulas to describe real-world phenomena. Such formulas, together with the meaning assigned to the variables, are called mathematical models. The formulas are said to model, or describe, the relationships among the variables.

Example

- The formula
$$h = -16t^2 + 200t + 4$$

models the height, h, in feet, of fireworks t seconds after launch. What is the height after 2 seconds?

$$h = -16(2)^2 + 200(2) + 4$$
$$= -16(4) + 200(2) + 4$$
$$= -64 + 400 + 4 = 340$$

The height after 2 seconds is 340 feet.

Additional Example to Review
Example 4, page 222

Definitions and Concepts

Terms of an algebraic expression are separated by addition. Like terms have the same variables with the same exponents on the variables. To add or subtract like terms, add or subtract the coefficients and copy the common variable.

An algebraic expression is simplified when parentheses have been removed and like terms have been combined.

Example

- Simplify: $7(3x - 4) - (10x - 5)$.

$$7(3x - 4) - (10x - 5) = 21x - 28 - 10x + 5 = 21x - 10x - 28 + 5 = 11x - 23$$

Additional Examples to Review

Example 5, page 224; Example 6, page 225; Example 7, page 225

4.2 Linear Equations in One Variable

Definitions and Concepts

A linear equation in one variable can be written in the form $ax + b = 0$, where a and b are real numbers, and $a \neq 0$.

Solving a linear equation is the process of finding the set of numbers that makes the equation a true statement. These numbers are the solutions. The set of all such solutions is the solution set.

Equivalent equations have the same solution set. The addition and multiplication properties are used to generate equivalent equations.

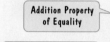 **Addition Property of Equality** — The same real number or algebraic expression may be added to both sides of an equation without changing the equation's solution set.

Multiplication Property of Equality — The same nonzero real number may multiply both sides of an equation without changing the equation's solution set.

Solving a Linear Equation

1. Simplify the algebraic expression on each side by removing grouping symbols and combining like terms.
2. Collect all the variable terms on one side and all the constants, or numerical terms, on the other side.
3. Isolate the variable and solve.
4. Check the proposed solution in the original equation.

Example

- Solve: $4(x - 5) = 2x - 14$.

$$4x - 20 = 2x - 14$$
$$4x - 2x - 20 = 2x - 2x - 14$$
$$2x - 20 = -14$$
$$2x - 20 + 20 = -14 + 20$$
$$2x = 6$$
$$\frac{2x}{2} = \frac{6}{2}$$
$$x = 3 \quad \text{Checking gives } -8 = -8, \text{ so } \{3\} \text{ is the}$$
$$\text{solution set.}$$

Additional Examples to Review

Example 1, page 231; Example 2, page 232; Example 3, page 233; Example 4, page 234

Definitions and Concepts

If an equation contains fractions, begin by multiplying both sides of the equation by the least common denominator of the fractions in the equation, thereby clearing fractions.

Example

- Solve: $\dfrac{x}{5} + \dfrac{1}{2} = \dfrac{x}{2} - 1.$

$$10\left(\dfrac{x}{5} + \dfrac{1}{2}\right) = 10\left(\dfrac{x}{2} - 1\right)$$

$$10 \cdot \dfrac{x}{5} + 10 \cdot \dfrac{1}{2} = 10 \cdot \dfrac{x}{2} - 10 \cdot 1$$

$$2x + 5 = 5x - 10$$

$$-3x = -15$$

$$x = 5 \quad \text{Checking gives } \tfrac{3}{2} = \tfrac{3}{2}, \text{ so } \{5\} \text{ is the solution set.}$$

Additional Examples to Review

Example 5, page 235; Example 6, page 236

Definitions and Concepts

If a false statement (such as $-6 = 7$) is obtained in solving an equation, the equation has no solution. The solution set is $\varnothing$, the empty set.

If a true statement (such as $-6 = -6$) is obtained in solving an equation, the equation has infinitely many solutions. The solution set is the set of all real numbers, written $\{x \mid x \text{ is a real number}\}$.

Examples

- Solve: $3x + 2 = 3(x + 5).$

$$3x + 2 = 3x + 15$$

$$3x + 2 - 3x = 3x + 15 - 3x$$

$$2 = 15$$

No solution: $\varnothing$

- Solve: $2(x + 4) = x + x + 8.$

$$2x + 8 = 2x + 8$$

$$2x + 8 - 2x = 2x + 8 - 2x$$

$$8 = 8$$

Solution set: $\{x \mid x \text{ is a real number}\}$

Additional Examples to Review

Example 7, page 237; Example 8, page 237

4.3 Applications of Linear Equations

Definitions and Concepts

A Strategy for Solving Word Problems Using Linear Equations

Step 1 Read the problem carefully several times until you can state in your own words what is given and what the problem is looking for. Let x (or any variable) represent one of the unknown quantities in the problem.

Step 2 If necessary, write expressions for any other unknown quantities in the problem in terms of x.

Step 3 Write an equation in x that models the verbal conditions of the problem.

Step 4   Solve the equation and answer the problem's question.

Step 5   Check the solution *in the original wording* of the problem, not in the equation obtained from the words.

Example

- After a 60% reduction, a graphing calculator sold for $32. What was the original price?

  Let x = the original price.

Original price	minus	60% reduction	=	reduced price
x	$-$	$0.6x$	$=$	32

$$0.4x = 32$$

$$\frac{0.4x}{0.4} = \frac{32}{0.4}$$

$$x = 80$$

The original price was $80. Check this amount using the first sentence in the problem's conditions.

Additional Examples to Review

Example 1, page 243; Example 2, page 244; Example 3, page 247; Example 4, page 248

Definitions and Concepts

Solving a formula for a variable means rewriting the formula so that the variable is isolated on one side of the equation.

Example

- Solve for r: $E = I(R + r)$.

$$E = IR + Ir \quad \boxed{\text{We need to isolate } r.}$$

$$E - IR = Ir$$

$$\frac{E - IR}{I} = r$$

Additional Examples to Review

Example 5, page 249; Example 6, page 250

4.4 Modeling with Proportions

Definitions and Concepts

The ratio of a to b is written $\dfrac{a}{b}$, or $a{:}b$. A proportion is a statement in the form $\dfrac{a}{b} = \dfrac{c}{d}$.

The cross-products principle states that if $\dfrac{a}{b} = \dfrac{c}{d}$, then $ad = bc$.

Example

- Solve: $\dfrac{-3}{4} = \dfrac{x}{12}$.

$$4x = -3(12)$$

$$4x = -36$$

$$x = -9 \quad \text{Checking gives } -\frac{3}{4} = -\frac{3}{4}, \text{ so } \{-9\} \text{ is the solution set.}$$

Additional Example to Review
Example 1, page 256

Definitions and Concepts

Solving Applied Problems Using Proportions

1. Read the problem and represent the unknown quantity by x (or any letter).
2. Set up a proportion that models the problem's conditions. List the given ratio on one side and the ratio with the unknown quantity on the other side. Each respective quantity should occupy the same corresponding position on each side of the proportion.
3. Drop units and apply the cross-products principle.
4. Solve for x and answer the question.

Example

- 30 elk are tagged and released. Sometime later, a sample of 80 elk are observed and 10 are tagged. How many elk are there?

$$x = \text{number of elk}$$

$$\frac{\text{Tagged}}{\text{Total}} \quad \frac{30}{x} = \frac{10}{80}$$

$$10x = 30 \cdot 80$$
$$10x = 2400$$
$$x = 240$$

There are 240 elk.

Additional Examples to Review
Example 2, page 257; Example 3, page 258

4.5 Modeling Using Variation

Definitions and Concepts

Statements of Variation

In each statement, the number k is the constant of variation.

y varies directly as x: $\quad y = kx$

y varies directly as x^n: $\quad y = kx^n$

y varies inversely as x: $\quad y = \dfrac{k}{x}$

y varies inversely as x^n: $\quad y = \dfrac{k}{x^n}$

y varies directly as x and inversely as z: $\quad y = \dfrac{kx}{z}$

Solving Variation Problems

1. Write an equation that models the given English statement.
2. Substitute the given pair of values into the equation in step 1 and solve for k, the constant of variation.
3. Substitute the value of k into the equation in step 1.
4. Use the equation from step 3 to answer the problem's question.

Example

- The time that it takes you to drive a certain distance varies inversely as your driving rate. Averaging 40 miles per hour, it takes you 10 hours to drive the distance. How long would the trip take averaging 50 miles per hour?

1. $t = \dfrac{k}{r}$ Time, t, varies inversely as rate, r.

2. It takes 10 hours at 40 miles per hour.

$$10 = \frac{k}{40}$$
$$k = 10(40) = 400$$

3. $t = \dfrac{400}{r}$

4. How long does it take at 50 miles per hour? Substitute 50 for r.

$$t = \frac{400}{50} = 8$$

It takes 8 hours at 50 miles per hour.

Additional Examples to Review

Example 1, page 262; Example 2, page 263; Example 3, page 264; Example 4, page 265

4.6 Linear Inequalities in One Variable

Definitions and Concepts

Graphs of Subsets of Real Numbers

Set-Builder Notation	Graph
$\{x \mid x < a\}$	
$\{x \mid x \le a\}$	
$\{x \mid x > b\}$	
$\{x \mid x \ge b\}$	

Set-Builder Notation	Graph
$\{x \mid a < x < b\}$	
$\{x \mid a \le x \le b\}$	
$\{x \mid a \le x < b\}$	
$\{x \mid a < x \le b\}$	

Examples

- $\{x \mid x \le 1\}$

- $\{x \mid x > -1\}$

- $\{x \mid -2 \le x < 3\}$

Additional Example to Review

Example 1, page 270

Definitions and Concepts

A linear inequality in one variable can be written in one of the following forms, where $a \neq 0$:

$$ax + b < 0, \quad ax + b \leq 0, \quad ax + b > 0, \quad ax + b \geq 0.$$

The procedure for solving linear inequalities is the same as the procedure for solving linear equations, with one important exception: When multiplying or dividing both sides of the inequality by a negative number, reverse the direction of the inequality symbol, changing the sense of the inequality.

Example

• Solve:
$$x + 4 \geq 6x - 16.$$
$$x + 4 - 6x \geq 6x - 16 - 6x$$
$$-5x + 4 \geq -16$$
$$-5x + 4 - 4 \geq -16 - 4$$
$$-5x \geq -20$$
$$\frac{-5x}{-5} \leq \frac{-20}{-5}$$
$$x \leq 4$$

Solution set: $\{x \mid x \leq 4\}$

Additional Examples to Review

Example 2, page 272; Example 3, page 272; Example 4, page 273; Example 5, page 274

Definitions and Concepts

An inequality in x with three parts is solved by isolating x in the middle.

Example

• Solve:
$$-9 < 5x + 1 \leq 16.$$
$$-9 - 1 < 5x + 1 - 1 \leq 16 - 1$$
$$-10 < 5x \leq 15$$
$$\frac{-10}{5} < \frac{5x}{5} \leq \frac{15}{5}$$
$$-2 < x \leq 3$$

Solution set: $\{x \mid -2 < x \leq 3\}$

Additional Example to Review

Example 6, page 275

Definitions and Concepts

If a false statement (such as $3 > 5$) is obtained in solving an inequality, the inequality has no solution. The solution set is $\varnothing$, the empty set.

If a true statement (such as $3 \leq 5$) is obtained in solving an inequality, the inequality has infinitely many solutions. The solution set is the set of all real numbers, written $\{x \mid x \text{ is a real number}\}$.

Examples

• Solve:
$$2(2x + 5) > 4x + 13.$$
$$4x + 10 > 4x + 13$$
$$4x - 4x + 10 > 4x - 4x + 13$$
$$10 > 13$$

No solution: $\varnothing$

- Solve: $3(x - 1) + 6 \geq 7x - 4x$.

$$3x - 3 + 6 \geq 3x$$
$$3x + 3 \geq 3x$$
$$3x - 3x + 3 \geq 3x - 3x$$
$$3 \geq 0$$

Solution set: $\{x | x \text{ is a real number}\}$

Additional Examples to Review

Example 7, page 276; Example 8, page 276

Review Exercises

Section 4.1 Algebraic Expressions and Formulas

In Exercises 1–3, evaluate the algebraic expression for the given value of the variable.

1. $6x + 9; x = 4$
2. $7x^2 + 4x - 5; x = -2$
3. $6 + 2(x - 8)^3; x = 5$
4. The bar graph shows the percentage of U.S. adults who have been tested for HIV, by age.

Percentage of Adults in the United States Tested for HIV, by Age

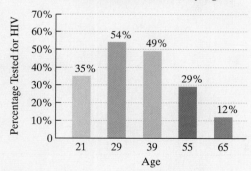

Source: National Center for Health Statistics

The data in the graph can be modeled by the formula

$$P = -0.05x^2 + 3.6x - 15,$$

where P represents the percentage of U.S. adults tested for HIV at age x. According to the formula, what percentage of U.S. adults who are 21 years old have been tested for HIV? Does the model underestimate or overestimate the percent displayed by the bar graph? By how much?

In Exercises 5–7, simplify each algebraic expression.

5. $5(2x - 3) + 7x$
6. $3(4y - 5) - (7y - 2)$
7. $2(x^2 + 5x) + 3(4x^2 - 3x)$

Section 4.2 Linear Equations in One Variable

In Exercises 8–14, solve each equation.

8. $4x + 9 = 33$
9. $5x - 3 = x + 5$
10. $3(x + 4) = 5x - 12$
11. $2(x - 2) + 3(x + 5) = 2x - 2$
12. $\dfrac{2x}{3} = \dfrac{x}{6} + 1$
13. $7x + 5 = 5(x + 3) + 2x$
14. $7x + 13 = 2(2x - 5) + 3x + 23$

Section 4.3 Applications of Linear Equations

15. The number of TV channels is increasing. The bar graph shows the total channels available in the average U.S. household each year from 2000 through 2006.

Number of TV Channels in the Average United States Household

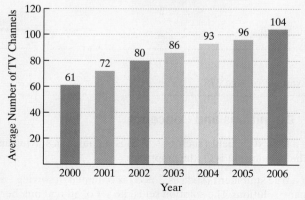

Source: Nielsen Media Research

Here are two mathematical models for the data shown by the graph. In each formula, N represents the number of TV channels in the average U.S. household x years after 2000.

Model 1 ⟶ $N = 6.8x + 64$

Model 2 ⟶ $N = -0.5x^2 + 9.5x + 62$

a. Which model better describes the data for 2000?

b. Does model 2 underestimate or overestimate the number of channels for 2006? By how many channels?

c. Use model 1 to project in which year we will have 166 TV channels.

16. Compared with peers a decade ago, young people spend 79 more minutes of free time each day listening to music, watching TV and movies, playing video games, and hanging out online. The bar graph shows daily media consumption for U.S. children and teens ages 8 to 18.

Average Number of Minutes per Day U.S. Children and Teens, Ages 8 to 18, Spend with Various Media

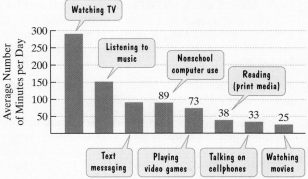

Source: Kaiser Family Foundation

Time spent per day watching TV exceeds time spent text messaging by 180 minutes. Time spent listening to music exceeds time spent text messaging by 61 minutes. Combined, these three activities consume 511 minutes (more than $8\frac{1}{2}$ hours) each day. Find the average number of minutes per day that young people ages 8 to 18 spend on each of these activities.

17. The line graph indicates that in 2007, 15% of households in the United States had only cellphones. For the period from 2007 through 2009, this percentage increased by 3 each year. If this trend continues, by which year will 51% of U.S. households have only cellphones?

Percentage of Cellphone-Only Homes in the U.S.

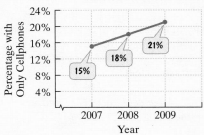

Source: Nielsen

18. You are choosing between two cellphone plans. One plan has a monthly fee of $15 with a charge of $0.05 per minute. The other plan has a monthly fee of $5 with a charge of $0.07 per minute. For how many minutes of phone calls will the costs for the two plans be the same?

19. After a 20% price reduction, a cordless phone sold for $48. What was the phone's price before the reduction?

20. A salesperson earns $300 per week plus 5% commission of sales. How much must be sold to earn $800 in a week?

In Exercises 21–24, solve each formula for the specified variable.

21. $Ax - By = C$ for x

22. $A = \frac{1}{2}bh$ for h

23. $A = \dfrac{B + C}{2}$ for B

24. $vt + gt^2 = s$ for g

Section 4.4 Modeling with Proportions

In Exercises 25–28, solve each proportion.

25. $\dfrac{3}{x} = \dfrac{15}{25}$

26. $\dfrac{-7}{5} = \dfrac{91}{x}$

27. $\dfrac{x + 2}{3} = \dfrac{4}{5}$

28. $\dfrac{5}{x + 7} = \dfrac{3}{x + 3}$

29. If a school board determines that there should be 3 teachers for every 50 students, how many teachers are needed for an enrollment of 5400 students?

30. To determine the number of trout in a lake, a conservationist catches 112 trout, tags them, and returns them to the lake. Later, 82 trout are caught, and 32 of them are found to be tagged. How many trout are in the lake?

Section 4.5 Modeling Using Variation

31. Many areas of Northern California depend on the snowpack of the Sierra Nevada mountain range for their water supply. The volume of water produced from melting snow varies directly as the volume of snow. Meteorologists have determined that 250 cubic centimeters of snow will melt to 28 cubic centimeters of water. How much water does 1200 cubic centimeters of melting snow produce?

32. The distance that a body falls from rest varies directly as the square of the time of the fall. If skydivers fall 144 feet in 3 seconds, how far will they fall in 10 seconds?

33. The pitch of a musical tone varies inversely as its wavelength. A tone has a pitch of 660 vibrations per second and a wavelength of 1.6 feet. What is the pitch of a tone that has a wavelength of 2.4 feet?

34. The loudness of a stereo speaker, measured in decibels, varies inversely as the square of your distance from the speaker. When you are 8 feet from the speaker, the loudness is 28 decibels. What is the loudness when you are 4 feet from the speaker?

35. The time required to assemble computers varies directly as the number of computers assembled and inversely as the number of workers. If 30 computers can be assembled by 6 workers in 10 hours, how long would it take 5 workers to assemble 40 computers?

Section 4.6 Linear Inequalities in One Variable

In Exercises 36–44, solve each inequality and graph the solution set on a number line. It is not necessary to provide graphs if the inequality has no solution or is true for all real numbers.

36. $2x - 5 < 3$

37. $\frac{x}{2} > -4$

38. $3 - 5x \leq 18$

39. $4x + 6 < 5x$

40. $6x - 10 \geq 2(x + 3)$

41. $4x + 3(2x - 7) \leq x - 3$

42. $-1 < 4x + 2 \leq 6$

43. $2(2x + 4) > 4(x + 2) - 6$

44. $-2(x - 4) \leq 3x + 1 - 5x$

45. To pass a course, a student must have an average on three examinations of at least 60. If a student scores 42 and 74 on the first two tests, what must be earned on the third test to pass the course?

Chapter 4 Test A

1. Evaluate $x^3 - 4(x - 1)^2$ when $x = -2$.

2. Simplify: $5(3x - 2) - (x - 6)$.

3. The formula

$$T = 15,395 + 988x - 2x^2$$

models the average cost of tuition and fees, T, at private four-year U.S. colleges for the school year ending x years after 2000. Use the formula to determine the average cost of tuition and fees at private four-year colleges for the school year ending in 2010.

In Exercises 4–7, solve each equation.

4. $12x + 4 = 7x - 21$

5. $3(2x - 4) = 9 - 3(x + 1)$

6. $3(x - 4) + x = 2(6 + 2x)$

7. $\frac{x}{5} - 2 = \frac{x}{3}$

8. Solve for y: $By - Ax = A$.

9. The formula $T = 383x + 3136$ models the average cost of tuition and fees, T, at public four-year U.S. colleges for the school year ending x years after 2000. Use the model to project when tuition and fees at public four-year colleges will average $8498.

In Exercises 10–11, solve each proportion.

10. $\dfrac{5}{8} = \dfrac{x}{12}$

11. $\dfrac{x + 5}{8} = \dfrac{x + 2}{5}$

12. How will you spend your average life expectancy of 78 years? According to the American Bureau of Labor Statistics, you will devote 62 years to sleeping, working, and watching TV. The number of years sleeping will exceed the number of years watching TV by 19. The number of years working will exceed the number of years watching TV by 16. Over your lifetime, how many years will you spend on each of these activities?

13. You bought a new car for $50,750. Its value is decreasing by $5500 per year. After how many years will its value be $12,250?

14. Photo Shop A charges $1.60 to develop a roll of film plus $0.11 for each print. Photo Shop B charges $1.20 to develop a roll of film plus $0.13 per print. For how many prints will the amount spent at each photo shop be the same? What will be that amount?

15. After a 60% reduction, a jacket sold for $20. What was the jacket's price before the reduction?

16. Park rangers catch, tag, and release 200 tule elk back into a wildlife refuge. Two weeks later they observe a sample of 150 elk, of which 5 are tagged. Assuming that the ratio of tagged elk in the sample holds for all elk in the refuge, how many elk are there in the park?

17. The Mach number is a measurement of speed named after the man who suggested it, Ernst Mach (1838–1916). The speed of an aircraft varies directly as its Mach number. Shown here are two aircraft. Use the figures for the Concorde to determine the Blackbird's speed.

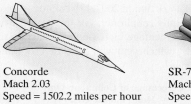

Concorde
Mach 2.03
Speed = 1502.2 miles per hour

SR-71 Blackbird
Mach 3.3
Speed = ?

18. The intensity of light received at a source varies inversely as the square of the distance from the source. A particular light has an intensity of 20 foot-candles at 15 feet. What is the light's intensity at 10 feet?

In Exercises 19–21, solve each inequality and graph the solution set on a number line.

19. $6 - 9x \geq 33$

20. $4x - 2 > 2(x + 6)$

21. $-3 \leq 2x + 1 < 6$

22. A student has grades on three examinations of 76, 80, and 72. What must the student earn on a fourth examination in order to have an average of at least 80?

Chapter 4 Test B

1. Evaluate $x^3 - 5(x - 2)^2$ when $x = -4$.

2. Simplify: $7(2x - 3) - (x - 10)$.

3. The bar graph shows the percentage of U.S. adults who received a flu shot each year from 2005 through 2009. The data can be modeled by the formula

$$F = 28 + 6x - 0.6x^2,$$

where F is the percentage of U.S adults who got a flu shot x years after 2005. Does the model underestimate or overestimate the percentage shown for 2009? By how much?

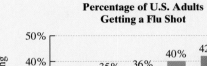

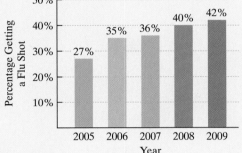

Source: Harris Interactive

In Exercises 4–7, solve each equation.

4. $3 - x = 24 - 4x$

5. $7 - 3(1 - 2x) = 4 + 2x$

6. $5(x - 2) + x = 3(2x - 1) - 7$

7. $\dfrac{x}{3} = 7 - \dfrac{x}{4}$

8. Solve for x: $B = Ax + C$.

9. The bar graph shows that Americans increasingly view a college education as essential for success.

Percentage of Americans Believing College Is Essential for Success

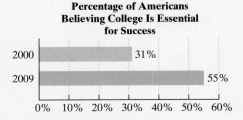

Source: Public Agenda, 2010

The formula $S = 2.7x + 31$ models the percentage of Americans believing college is essential for success, S, x years after 2000. Use the model to project by which year 85% of Americans will view a college education as essential for success.

In Exercises 10–11, solve each proportion.

10. $\dfrac{3}{4} = \dfrac{x}{10}$

11. $\dfrac{x + 4}{5} = \dfrac{x - 2}{7}$

12. The bar graph shows average yearly earnings in the United States for people with a college education, by final degree earned.

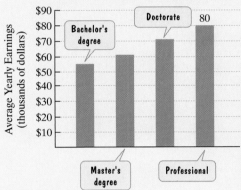

Source: U.S. Census Bureau

The average yearly salary of an American whose final degree is a master's exceeds that of an American whose final degree is a bachelor's by $6 thousand. The average yearly salary of an American whose final degree is a doctorate exceeds that of an American whose final degree is a bachelor's by $16 thousand. Combined, three people with each of these educational attainments earn $187 thousand. Find the average yearly salary of Americans with each of these final degrees.

13. You bought a new car for $42,620. Its value is decreasing by $4500 per year. After how many years will its value be $15,620?

14. A discount pass for a bridge costs $27 per month. The toll for the bridge is normally $4.00, but it is reduced to $2.50 for people who have purchased the discount pass. Determine the number of times in a month the bridge must be crossed so that the total monthly cost without the discount pass is the same as the total monthly cost with the discount pass.

15. After a 30% reduction, a camera sold for $602. What was the camera's price before the reduction?

16. Park rangers catch, tag, and then release 318 deer back into a state park. Two weeks later, they select a sample of 168 deer, 56 of which are tagged. Assuming the ratio of tagged deer in the sample holds for all deer in the park, approximately how many deer are in the park?

17. The distance that a body falls from rest varies directly as the square of the time of the fall. If skydivers fall 400 feet in 5 seconds, how far will they fall in 9 seconds?

18. The amount of current flowing in an electrical circuit varies inversely as the resistance in the circuit. When the resistance in a particular circuit is 5 ohms, the current is 42 amperes. What is the current when the resistance is 4 ohms?

In Exercises 19–21, solve each inequality and graph the solution set on a number line.

19. $7 - 8x \leq 47$

20. $2(x + 7) < 8 - 4x$

21. $-3 < 5x + 2 \leq 27$

22. A student has grades on three examinations of 88, 90, and 84. What must the student earn on a fourth examination in order to have an average of at least 90?

Algebra: Graphs, Functions, Linear Functions, and Linear Systems

5

Television, movies, and magazines place great emphasis on physical beauty. Our culture emphasizes physical appearance to such an extent that it is a central factor in the perception and judgment of others. The modern emphasis on thinness as the ideal body shape has been suggested as a major cause of eating disorders among adolescent girls.

Cultural values of physical attractiveness change over time. During the 1950s, actress Jayne Mansfield embodied the postwar ideal: curvy, buxom, and big-hipped. Men, too, have been caught up in changes of how they "ought" to look. The 1960s' ideal was the soft and scrawny hippie. Today's ideal man is tough and muscular.

Given the importance of culture in setting standards of attractiveness, how can you establish a healthy weight range for your age and height? In this chapter, we will use systems of inequalities to explore these skin-deep issues.

You'll find a weight for various ages using the models (mathematical, not fashion) in Example 4 of Section 5.5 and Exercises 63–66 in Exercise Set 5.5. Exercises 69–70 use graphs and a formula for body-mass index to indicate whether you are obese, overweight, borderline overweight, normal weight, or underweight.

5.1 Graphing and Functions

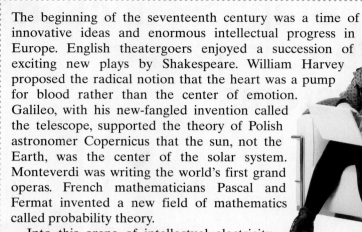

The beginning of the seventeenth century was a time of innovative ideas and enormous intellectual progress in Europe. English theatergoers enjoyed a succession of exciting new plays by Shakespeare. William Harvey proposed the radical notion that the heart was a pump for blood rather than the center of emotion. Galileo, with his new-fangled invention called the telescope, supported the theory of Polish astronomer Copernicus that the sun, not the Earth, was the center of the solar system. Monteverdi was writing the world's first grand operas. French mathematicians Pascal and Fermat invented a new field of mathematics called probability theory.

Into this arena of intellectual electricity stepped French aristocrat René Descartes (1596–1650). Descartes (pronounced "day cart"), propelled by the creativity surrounding him, developed a new branch of mathematics that brought together algebra and geometry in a unified way—a way that visualized numbers as points on a graph, equations as geometric figures, and geometric figures as equations. This new branch of mathematics, called *analytic geometry*, established Descartes as one of the founders of modern thought and among the most original mathematicians and philosophers of any age. We begin this section by looking at Descartes's deceptively simple idea, called the **rectangular coordinate system** or (in his honor) the **Cartesian coordinate system**.

Points and Ordered Pairs

Descartes used two number lines that intersect at right angles at their zero points, as shown in **Figure 5.1**. The horizontal number line is the **x-axis**. The vertical number line is the **y-axis**. The point of intersection of these axes is their zero points, called the **origin**. Positive numbers are shown to the right and above the origin. Negative numbers are shown to the left and below the origin. The axes divide the plane into four quarters, called **quadrants**. The points located on the axes are not in any quadrant.

Each point in the rectangular coordinate system corresponds to an **ordered pair** of real numbers, (x, y). Examples of such pairs are $(-5, 3)$ and $(3, -5)$. The first number in each pair, called the **x-coordinate**, denotes the distance and direction from the origin along the x-axis. The second number in each pair, called the **y-coordinate**, denotes vertical distance and direction along a line parallel to the y-axis or along the y-axis itself.

Figure 5.2 shows how we **plot**, or locate, the points corresponding to the ordered pairs $(-5, 3)$ and $(3, -5)$. We plot $(-5, 3)$ by going 5 units from 0 to the left along the x-axis. Then we go 3 units up parallel to the y-axis. We plot $(3, -5)$ by going 3 units from 0 to the right along the x-axis and 5 units down parallel to the y-axis. The phrase "the points corresponding to the ordered pairs $(-5, 3)$ and $(3, -5)$" is often abbreviated as "the points $(-5, 3)$ and $(3, -5)$."

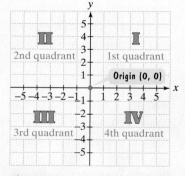

Figure 5.1 The rectangular coordinate system

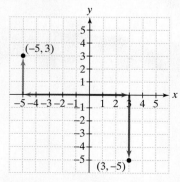

Figure 5.2 Plotting $(-5, 3)$ and $(3, -5)$

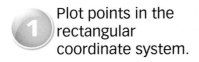

Plot points in the rectangular coordinate system.

Example 1 Plotting Points in the Rectangular Coordinate System

Plot the points: $A(-3,5), B(2,-4), C(5,0), D(-5,-3), E(0,4),$ and $F(0,0)$.

Solution

See **Figure 5.3**. We move from the origin and plot the points in the following way:

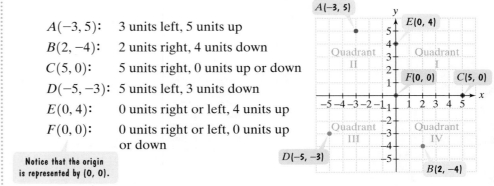

$A(-3,5)$:	3 units left, 5 units up
$B(2,-4)$:	2 units right, 4 units down
$C(5,0)$:	5 units right, 0 units up or down
$D(-5,-3)$:	5 units left, 3 units down
$E(0,4)$:	0 units right or left, 4 units up
$F(0,0)$:	0 units right or left, 0 units up or down

Notice that the origin is represented by (0, 0).

Figure 5.3 Plotting points

Checkpoint 1   Plot the points: $A(-2,4), B(4,-2), C(-3,0),$ and $D(0,-3)$.

Graphs of Equations

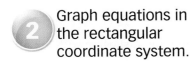

Graph equations in the rectangular coordinate system.

A relationship between two quantities can sometimes be expressed as an **equation in two variables**, such as

$$y = 4 - x^2.$$

A **solution of an equation in two variables**, x and y, is an ordered pair of real numbers with the following property: When the x-coordinate is substituted for x and the y-coordinate is substituted for y in the equation, we obtain a true statement. For example, consider the equation $y = 4 - x^2$ and the ordered pair $(3, -5)$. When 3 is substituted for x and -5 is substituted for y, we obtain the statement $-5 = 4 - 3^2$, or $-5 = 4 - 9$, or $-5 = -5$. Because this statement is true, the ordered pair $(3, -5)$ is a solution of the equation $y = 4 - x^2$. We also say that $(3, -5)$ **satisfies** the equation.

We can generate as many ordered-pair solutions as desired to $y = 4 - x^2$ by substituting numbers for x and then finding the corresponding values for y. For example, suppose we let $x = 3$:

Start with x.   *Compute y.*   *Form the ordered pair (x, y).*

x	$y = 4 - x^2$	Ordered Pair (x, y)
3	$y = 4 - 3^2 = 4 - 9 = -5$	$(3, -5)$

Let x = 3.   *(3, -5) is a solution of $y = 4 - x^2$.*

The **graph of an equation in two variables** is the set of all points whose coordinates satisfy the equation. One method for graphing such equations is the **point-plotting method**. First, we find several ordered pairs that are solutions of the equation. Next, we plot these ordered pairs as points in the rectangular coordinate system. Finally, we connect the points with a smooth curve or line. This often gives us a picture of all ordered pairs that satisfy the equation.

Example 2   Graphing an Equation Using the Point-Plotting Method

Graph $y = 4 - x^2$. Select integers for x, starting with -3 and ending with 3.

Solution

For each value of x, we find the corresponding value for y.

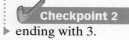

Start with x.   Compute y.   Form the ordered pair (x, y).

We selected integers from -3 to 3, inclusive, to include three negative numbers, 0, and three positive numbers. We also wanted to keep the resulting computations for y relatively simple.

x	$y = 4 - x^2$	Ordered Pair (x, y)
-3	$y = 4 - (-3)^2 = 4 - 9 = -5$	$(-3, -5)$
-2	$y = 4 - (-2)^2 = 4 - 4 = 0$	$(-2, 0)$
-1	$y = 4 - (-1)^2 = 4 - 1 = 3$	$(-1, 3)$
0	$y = 4 - 0^2 = 4 - 0 = 4$	$(0, 4)$
1	$y = 4 - 1^2 = 4 - 1 = 3$	$(1, 3)$
2	$y = 4 - 2^2 = 4 - 4 = 0$	$(2, 0)$
3	$y = 4 - 3^2 = 4 - 9 = -5$	$(3, -5)$

Now we plot the seven points and join them with a smooth curve, as shown in **Figure 5.4**. The graph of $y = 4 - x^2$ is a curve where the part of the graph to the right of the y-axis is a reflection of the part to the left of it and vice versa. The arrows on the left and the right of the curve indicate that it extends indefinitely in both directions.

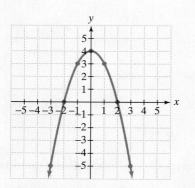

Figure 5.4 The graph of $y = 4 - x^2$

✓ **Checkpoint 2**   Graph $y = 4 - x$. Select integers for x, starting with -3 and ending with 3.

    Part of the beauty of the rectangular coordinate system is that it allows us to "see" formulas and visualize the solution to a problem. This idea is demonstrated in Example 3.

Example 3   An Application Using Graphs of Equations

The toll to a bridge costs $2.50. Commuters who use the bridge frequently have the option of purchasing a monthly discount pass for $21.00. With the discount pass, the toll is reduced to $1.00. The monthly cost, y, of using the bridge x times can be described by the following formulas:

Without the discount pass:

$$y = 2.5x$$ The monthly cost, y, is $2.50 times the number of times, x, that the bridge is used.

With the discount pass:

$$y = 21 + 1 \cdot x$$ The monthly cost, y, is $21 for the discount pass plus $1 times
$$y = 21 + x.$$ the number of times, x, that the bridge is used.

a. Let $x = 0, 2, 4, 10, 12, 14,$ and 16. Make a table of values for each equation showing seven solutions for the equation.

b. Graph the equations in the same rectangular coordinate system.

c. What are the coordinates of the intersection point for the two graphs? Interpret the coordinates in practical terms.

Solution

a. Tables of values showing seven solutions for each equation follow.

Without the Discount Pass

x	$y = 2.5x$	(x, y)
0	$y = 2.5(0) = 0$	$(0, 0)$
2	$y = 2.5(2) = 5$	$(2, 5)$
4	$y = 2.5(4) = 10$	$(4, 10)$
10	$y = 2.5(10) = 25$	$(10, 25)$
12	$y = 2.5(12) = 30$	$(12, 30)$
14	$y = 2.5(14) = 35$	$(14, 35)$
16	$y = 2.5(16) = 40$	$(16, 40)$

With the Discount Pass

x	$y = 21 + x$	(x, y)
0	$y = 21 + 0 = 21$	$(0, 21)$
2	$y = 21 + 2 = 23$	$(2, 23)$
4	$y = 21 + 4 = 25$	$(4, 25)$
10	$y = 21 + 10 = 31$	$(10, 31)$
12	$y = 21 + 12 = 33$	$(12, 33)$
14	$y = 21 + 14 = 35$	$(14, 35)$
16	$y = 21 + 16 = 37$	$(16, 37)$

b. Now we are ready to graph the two equations. Because the x- and y-coordinates are nonnegative, it is only necessary to use the origin, the positive portions of the x- and y-axes, and the first quadrant of the rectangular coordinate system. The x-coordinates begin at 0 and end at 16. We will let each tick mark on the x-axis represent two units. However, the y-coordinates begin at 0 and get as large as 40 in the formula that describes the monthly cost without the coupon book. So that our y-axis does not get too long, we will let each tick mark on the y-axis represent five units. Using this setup and the two tables of values, we construct the graphs of $y = 2.5x$ and $y = 21 + x$, shown in **Figure 5.5**.

c. The graphs intersect at $(14, 35)$. This means that if the bridge is used 14 times in a month, the total monthly cost without the discount pass is the same as the total monthly cost with the discount pass, namely $35.

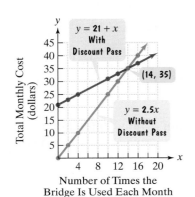

Figure 5.5 Options for a toll

In **Figure 5.5**, look at the two graphs to the right of the intersection point $(14, 35)$. The red graph of $y = 21 + x$ lies below the blue graph of $y = 2.5x$. This means that if the bridge is used more than 14 times in a month $(x > 14)$, the (red) monthly cost, y, with the discount pass is less than the (blue) monthly cost, y, without the discount pass.

Checkpoint 3   The toll to a bridge costs $2.00. If you use the bridge x times in a month, the monthly cost, y, is $y = 2x$. With a $10 discount pass, the toll is reduced to $1.00. The monthly cost, y, of using the bridge x times in a month with the discount pass is $y = 10 + x$.

a. Let $x = 0, 2, 4, 6, 8, 10,$ and 12. Make tables of values showing seven solutions of $y = 2x$ and seven solutions of $y = 10 + x$.

b. Graph the equations in the same rectangular coordinate system.

c. What are the coordinates of the intersection point for the two graphs? Interpret the coordinates in practical terms.

Functions

3 Use function notation.

Reconsider one of the equations from Example 3, $y = 2.5x$. Recall that this equation describes the monthly cost, y, of using the bridge x times, with a toll of $2.50 each time the bridge is used. The monthly cost, y, depends on the number of times the bridge is used, x. For each value of x, there is one and only one value of y. **If an equation in two variables (x and y) yields precisely one value of y for each value of x, we say that y is a function of x.**

The notation $y = f(x)$ indicates that the variable y is a function of x. The notation $f(x)$ is read "f of x."

For example, the formula for the cost of the bridge

$$y = 2.5x$$

can be expressed in function notation as

$$f(x) = 2.5x.$$

We read this as "*f* of *x* is equal to $2.5x$." If, say, *x* equals 10 (meaning that the bridge is used 10 times), we can find the corresponding value of *y* (monthly cost) using the equation $f(x) = 2.5x$.

$$f(x) = 2.5x$$
$$f(10) = 2.5(10) \quad \text{To find } f(10), \text{ read "f of 10," replace x with 10.}$$
$$= 25$$

Because $f(10) = 25$ (*f* of 10 equals 25), this means that if the bridge is used 10 times in a month, the total monthly cost is \$25.

Table 5.1 compares our previous notation with the new notation of functions.

Table 5.1 Function Notation	
"y Equals" Notation	**"f(x) Equals" Notation**
$y = 2.5x$	$f(x) = 2.5x$
If $x = 10$, $y = 2.5(10) = 25.$	$f(10) = 2.5(10) = 25$ *f* of 10 equals 25.

In our next example, we will apply function notation to three different functions. It would be awkward to call all three functions *f*. We will call the first function *f*, the second function *g*, and the third function *h*. These are the letters most frequently used to name functions.

Example 4   Using Function Notation

Find each of the following:

a. $f(4)$ for $f(x) = 2x + 3$ **b.** $g(-2)$ for $g(x) = 2x^2 - 1$

c. $h(-5)$ for $h(r) = r^3 - 2r^2 + 5.$

Solution

a. $f(x) = 2x + 3$ This is the given function.

$\quad f(4) = 2 \cdot 4 + 3$ To find *f* of 4, replace *x* with 4.

$\quad\quad = 8 + 3$ Multiply: $2 \cdot 4 = 8$.

$\quad f(4) = 11$ *f* of 4 is 11.   Add.

b. $g(x) = 2x^2 - 1$ This is the given function.

$\quad g(-2) = 2(-2)^2 - 1$ To find *g* of -2, replace *x* with -2.

$\quad\quad = 2(4) - 1$ Evaluate the exponential expression: $(-2)^2 = 4$.

$\quad\quad = 8 - 1$ Multiply: $2(4) = 8$.

$\quad g(-2) = 7$ *g* of -2 is 7.   Subtract.

c. $h(r) = r^3 - 2r^2 + 5$ The function's name is *h* and *r* represents the function's input.

$\quad h(-5) = (-5)^3 - 2(-5)^2 + 5$ To find *h* of -5, replace each occurrence of *r* with -5.

$\quad\quad = -125 - 2(25) + 5$ Evaluate exponential expressions: $(-5)^3 = -125$ and $(-5)^2 = 25$.

$\quad\quad = -125 - 50 + 5$ Multiply: $2(25) = 50$.

$\quad h(-5) = -170$ *h* of -5 is -170. $-125 - 50 = -175$ and $-175 + 5 = -170$.

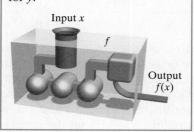

✓ **Checkpoint 4**   Find each of the following:

a. $f(6)$ for $f(x) = 4x + 5$

b. $g(-5)$ for $g(x) = 3x^2 - 10$

▶ **c.** $h(-4)$ for $h(r) = r^2 - 7r + 2$.

Example 5   An Application Involving Function Notation

Tailgaters beware: If your car is going 35 miles per hour on dry pavement, your required stopping distance is 160 feet, or the width of a football field. At 65 miles per hour, the distance required is 410 feet, or well over the length of a football field. **Figure 5.6** shows stopping distances for cars at various speeds on dry roads and on wet roads. **Figure 5.7** uses a line graph to represent stopping distances at various speeds on dry roads.

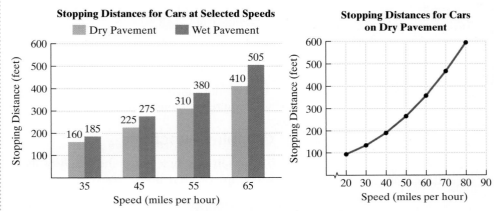

Figure 5.6

Source: National Highway Traffic Safety Administration

Figure 5.7

a. Use the line graph in **Figure 5.7** to estimate a car's required stopping distance at 60 miles per hour on dry pavement. Round to the nearest ten feet.

b. The function

$$f(x) = 0.0875x^2 - 0.4x + 66.6$$

models a car's required stopping distance, $f(x)$, in feet, on dry pavement traveling at x miles per hour. Use this function to find the required stopping distance at 60 miles per hour. Round to the nearest foot.

Solution

a. The required stopping distance at 60 miles per hour is estimated using the point shown in **Figure 5.8**. The second coordinate of this point extends slightly more than midway between 300 and 400 on the vertical axis. Thus, 360 is a reasonable estimate. We conclude that at 60 miles per hour on dry pavement, the required stopping distance is approximately 360 feet.

b. Now we use the given function to determine the required stopping

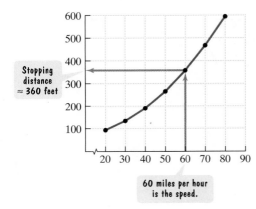

Figure 5.8

distance at 60 miles per hour. We need to find $f(60)$. The arithmetic gets somewhat "messy," so it is probably a good idea to use a calculator.

$f(x) = 0.0875x^2 - 0.4x + 66.6$	This function models stopping distance, $f(x)$, at x miles per hour.
$f(60) = 0.0875(60)^2 - 0.4(60) + 66.6$	Replace each x with 60.
$\quad = 0.0875(3600) - 0.4(60) + 66.6$	Use the order of operations, first evaluating the exponential expression.
$\quad = 315 - 24 + 66.6$	Perform the multiplications.
$\quad = 357.6$	Subtract and add as indicated.
$\quad \approx 358$	As directed, we've rounded to the nearest foot.

We see that $f(60) \approx 358$—that is, f of 60 is approximately 358. The model indicates that the required stopping distance on dry pavement at 60 miles per hour is approximately 358 feet.

Checkpoint 5

a. Use the line graph in **Figure 5.7** on the previous page to estimate a car's required stopping distance at 40 miles per hour on dry pavement. Round to the nearest ten feet.

b. Use the function in Example 5(b), $f(x) = 0.0875x^2 - 0.4x + 66.6$, to find the required stopping distance at 40 miles per hour. Round to the nearest foot.

Graphing Functions

④ Graph functions.

The **graph of a function** is the graph of its ordered pairs. In our next example, we will graph two functions.

Example 6 **Graphing Functions**

Graph the functions $f(x) = 2x$ and $g(x) = 2x + 4$ in the same rectangular coordinate system. Select integers for x from -2 to 2, inclusive.

Solution

For each function, we use the suggested values for x to create a table of some of the coordinates. These tables are shown below. Then, we plot the five points in each table and connect them, as shown in **Figure 5.9**. The graph of each function is a straight line. Do you see a relationship between the two graphs? The graph of g is the graph of f shifted vertically up 4 units.

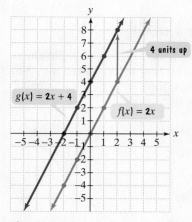

Figure 5.9

x	$f(x) = 2x$	(x, y) or $(x, f(x))$
-2	$f(-2) = 2(-2) = -4$	$(-2, -4)$
-1	$f(-1) = 2(-1) = -2$	$(-1, -2)$
0	$f(0) = 2 \cdot 0 = 0$	$(0, 0)$
1	$f(1) = 2 \cdot 1 = 2$	$(1, 2)$
2	$f(2) = 2 \cdot 2 = 4$	$(2, 4)$

x	$g(x) = 2x + 4$	(x, y) or $(x, g(x))$
-2	$g(-2) = 2(-2) + 4 = 0$	$(-2, 0)$
-1	$g(-1) = 2(-1) + 4 = 2$	$(-1, 2)$
0	$g(0) = 2 \cdot 0 + 4 = 4$	$(0, 4)$
1	$g(1) = 2 \cdot 1 + 4 = 6$	$(1, 6)$
2	$g(2) = 2 \cdot 2 + 4 = 8$	$(2, 8)$

Choose x. Compute $f(x)$ by evaluating f at x. Form the ordered pair.   Choose x. Compute $g(x)$ by evaluating g at x. Form the ordered pair.

Checkpoint 6 Graph the functions $f(x) = 2x$ and $g(x) = 2x - 3$ in the same rectangular coordinate system. Select integers for x from -2 to 2, inclusive. How is the graph of g related to the graph of f?

The Vertical Line Test

Not every graph in the rectangular coordinate system is the graph of a function. The definition of a function specifies that no value of x can be paired with two or more different values of y. Consequently, if a graph contains two or more different points with the same first coordinate, the graph cannot represent a function. This is illustrated in **Figure 5.10**. Observe that points sharing a common first coordinate are vertically above or below each other.

This observation is the basis of a useful test for determining whether a graph defines y as a function of x. The test is called the **vertical line test**.

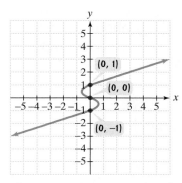

Figure 5.10 y is not a function of x because 0 is paired with three values of y, namely, 1, 0, and -1.

> **The Vertical Line Test for Functions**
>
> If any vertical line intersects a graph in more than one point, the graph does not define y as a function of x.

Technology

A graphing calculator is a powerful tool that quickly generates the graph of an equation in two variables. Here is the graph of $y = 4 - x^2$ that we drew by hand in **Figure 5.4** on page 296.

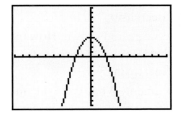

What differences do you notice between this graph and the graph we drew by hand? This graph seems a bit "jittery." Arrows do not appear on the left and right ends of the graph. Furthermore, numbers are not given along the axes. For the graph shown above, the x-axis extends from -10 to 10 and the y-axis also extends from -10 to 10. The distance represented by each consecutive tick mark is one unit. We say that the **viewing window** is $[-10, 10, 1]$ by $[-10, 10, 1]$.

To graph an equation in x and y using a graphing calculator, enter the equation, which must be solved for y, and specify the size of the viewing window. The size of the viewing window sets minimum and maximum values for both the x- and y-axes. Enter these values, as well as the values between consecutive tick marks, on the respective axes. The $[-10, 10, 1]$ by $[-10, 10, 1]$ viewing window used above is called the **standard viewing window**.

Example 7   Using the Vertical Line Test

Use the vertical line test to identify graphs in which y is a function of x.

a. 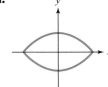   b. 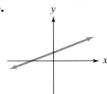   c. 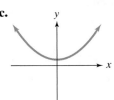   d.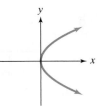

Solution

y is a function of x for the graphs in (b) and (c).

a. 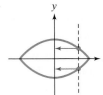   b. 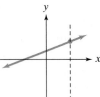   c. 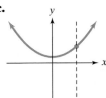   d.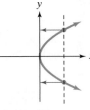

y **is not a function** of x. Two values of y correspond to one x-value.

y **is a function** of x.

y **is a function** of x.

y **is not a function** of x. Two values of y correspond to one x-value.

✓ **Checkpoint 7**   Use the vertical line test to identify graphs in which y is a function of x.

a. 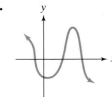   b. 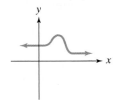   c.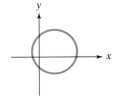

6 Obtain information about a function from its graph.

Obtaining Information from Graphs

Example 8 illustrates how to obtain information about a function from its graph.

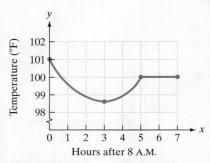

Figure 5.11 Body temperature from 8 A.M. through 3 P.M.

| Example 8 | **Analyzing the Graph of a Function** |

Too late for that flu shot now! It's only 8 A.M. and you're feeling lousy. Fascinated by the way that algebra models the world (your author is projecting a bit here), you construct a graph showing your body temperature from 8 A.M. through 3 P.M. You decide to let x represent the number of hours after 8 A.M. and y represent your body temperature at time x. The graph is shown in **Figure 5.11**. The symbol ╪ on the y-axis shows that there is a break in values between 0 and 98. Thus, the first tick mark on the y-axis represents a temperature of 98 °F.

a. What is your temperature at 8 A.M.?

b. During which period of time is your temperature decreasing?

c. Estimate your minimum temperature during the time period shown. How many hours after 8 A.M. does this occur? At what time does this occur?

d. During which period of time is your temperature increasing?

e. Part of the graph is shown as a horizontal line segment. What does this mean about your temperature and when does this occur?

f. Explain why the graph defines y as a function of x.

Solution

a. Because x is the number of hours after 8 A.M., your temperature at 8 A.M. corresponds to $x = 0$. Locate 0 on the horizontal axis and look at the point on the graph above 0. **Figure 5.12** shows that your temperature at 8 A.M. is 101°F.

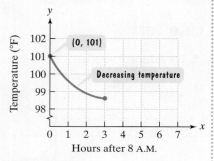

Figure 5.12

b. Your temperature is decreasing when the graph falls from left to right. This occurs between $x = 0$ and $x = 3$, also shown in **Figure 5.12**. Because x represents the number of hours after 8 A.M., your temperature is decreasing between 8 A.M. and 11 A.M.

c. Your minimum temperature can be found by locating the lowest point on the graph. This point lies above 3 on the horizontal axis, shown in **Figure 5.13**. The y-coordinate of this point falls more than midway between 98 and 99, at approximately 98.6. The lowest point on the graph, (3, 98.6), shows that your minimum temperature, 98.6 °F, occurs 3 hours after 8 A.M., at 11 A.M.

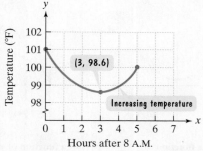

Figure 5.13

d. Your temperature is increasing when the graph rises from left to right. This occurs between $x = 3$ and $x = 5$, shown in **Figure 5.13**. Because x represents the number of hours after 8 A.M., your temperature is increasing between 11 A.M. and 1 P.M.

e. The horizontal line segment shown in **Figure 5.14** indicates that your temperature is neither increasing nor decreasing. Your temperature remains the same, 100 °F, between $x = 5$ and $x = 7$. Thus, your temperature is at a constant 100 °F between 1 P.M. and 3 P.M.

f. The complete graph of your body temperature from 8 A.M. through 3 P.M. is shown in **Figure 5.14**. No vertical line can be drawn that intersects this blue graph more than once. By the vertical line test, the graph defines y as a function of x. In practical terms, this means that your body temperature is a

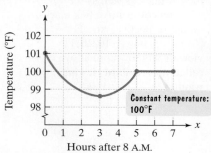

Figure 5.14

function of time. Each hour (or fraction of an hour) after 8 A.M., represented by x, yields precisely one body temperature, represented by y.

Checkpoint 8 When a physician injects a drug into a patient's muscle, the concentration of the drug in the body, measured in milligrams per 100 milliliters, depends on the time elapsed after the injection, measured in hours. **Figure 5.15** shows the graph of drug concentration over time, where x represents hours after the injection and y represents the drug concentration at time x.

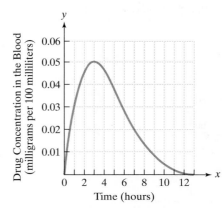

Figure 5.15

a. During which period of time is the drug concentration increasing?

b. During which period of time is the drug concentration decreasing?

c. What is the drug's maximum concentration and when does this occur?

d. What happens by the end of 13 hours?

▶ e. Explain why the graph defines y as a function of x.

Functions and Graphs for Cellphone Plans

A cellphone company offers the following plan:

- $20 per month buys 60 minutes.
- Additional time costs $0.40 per minute.

We can represent this plan mathematically by writing the total monthly cost, C, as a function of the number of calling minutes, t.

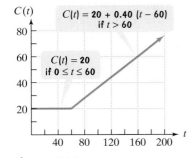

Figure 5.16

$$C(t) = \begin{cases} 20 & \text{if } 0 \le t \le 60 \\ 20 + 0.40(t - 60) & \text{if } t > 60 \end{cases}$$

The cost is $20 for up to and including 60 calling minutes.

The cost is $20 plus $0.40 per minute for additional time for more than 60 calling minutes.

$20 for first 60 minutes $0.40 per minute   times the number of calling minutes exceeding 60

A function that is defined by two (or more) equations is called a **piecewise function**. Many cellphone plans can be represented with piecewise functions. The graph of the piecewise function described above is shown in **Figure 5.16**.

Example 9 Using a Function for a Cellphone Plan

Use the function that describes the cellphone plan

$$C(t) = \begin{cases} 20 & \text{if } 0 \le t \le 60 \\ 20 + 0.40(t - 60) & \text{if } t > 60 \end{cases}$$

to find and interpret each of the following:

a. $C(30)$

b. $C(100)$.

$$C(t) = \begin{cases} 20 & \text{if } 0 \le t \le 60 \\ 20 + 0.40(t-60) & \text{if } t > 60 \end{cases}$$

The piecewise function describing a cellphone plan (repeated)

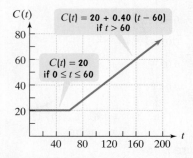

Figure 5.16 (repeated)

Solution

a. To find $C(30)$, we let $t = 30$. Because 30 lies between 0 and 60, we use the first line of the piecewise function.

$C(t) = 20$ This is the function's equation for $0 \le t \le 60$.

$C(30) = 20$ Replace t with 30. Regardless of this function's input, the constant output is 20.

This means that with 30 calling minutes, the monthly cost is $20. This can be visually represented by the point $(30, 20)$ on the first piece of the graph in **Figure 5.16**. Take a moment to identify this point on the graph.

b. To find $C(100)$, we let $t = 100$. Because 100 is greater than 60, we use the second line of the piecewise function.

$C(t) = 20 + 0.40(t - 60)$ This is the function's equation for $t > 60$.

$C(100) = 20 + 0.40(100 - 60)$ Replace t with 100.

$= 20 + 0.40(40)$ Subtract within parentheses: $100 - 60 = 40$.

$= 20 + 16$ Multiply: $0.40(40) = 16$.

$= 36$ Add: $20 + 16 = 36$.

Thus, $C(100) = 36$. This means that with 100 calling minutes, the monthly cost is $36. This can be visually represented by the point $(100, 36)$ on the second piece of the graph in **Figure 5.16**. Take a moment to identify this point on the graph.

 **Checkpoint 9** Use the function in Example 9 to find and interpret each of the following:

a. $C(40)$ **b.** $C(80)$.

▶ Identify your solutions by points on the graph in **Figure 5.16**.

Achieving Success

We have seen that **the best way to achieve success in math is through practice**. Keeping up with your homework, preparing for tests, asking questions of your teacher, reading your textbook, and attending all classes will help you learn the material and boost your confidence. Use language in a proactive way that reflects a sense of responsibility for your own successes and failures.

Reactive Language	Proactive Language
I'll try.	I'll do it.
That's just the way I am.	I can do better than that.
There's not a thing I can do.	I have options for improvement.
I have to.	I choose to.
I can't.	I can find a way.

Exercise Set 5.1

Concept and Vocabulary Exercises

In Exercises 1–4, fill in each blank so that the resulting statement is true.

1. The first number in an ordered pair is called the _____ and the second number is called the _____.

2. The ordered pair $(1, 3)$ is a _____ of the equation $y = 5x - 2$ because when 1 is substituted for x and 3 is substituted for y, we obtain a true statement. We also say that $(1, 3)$ _____ the equation.

3. If an equation in two variables (x and y) yields precisely one value of _____ for each value of _____, we say that y is a/an _____ of x.

4. If any vertical line intersects a graph _____, the graph does not define y as a/an _____ of x.

In Exercises 5–10, determine whether each statement is true or false. If the statement is false, make the necessary change(s) to produce a true statement.

5. If $(2, 5)$ satisfies an equation, then $(5, 2)$ also satisfies the equation.

6. The ordered pair $(3, 4)$ satisfies the equation
$$2y - 3x = -6.$$

7. No two ordered pairs of a function can have the same x-coordinate and different y-coordinates.

8. No two ordered pairs of a function can have the same y-coordinate and different x-coordinates.

9. A vertical line can intersect the graph of a function at more than one point.

10. A horizontal line can intersect the graph of a function at more than one point.

Respond to Exercises 11–16 using verbal or written explanations.

11. What is the rectangular coordinate system?

12. Explain how to plot a point in the rectangular coordinate system. Give an example with your explanation.

13. Explain why $(5, -2)$ and $(-2, 5)$ do not represent the same ordered pair.

14. Explain how to graph an equation in the rectangular coordinate system.

15. What is a function?

16. Explain how the vertical line test is used to determine whether a graph represents a function.

In Exercises 17–20, determine whether each statement makes sense or does not make sense, and explain your reasoning.

17. My body temperature is a function of the time of day.

18. Using $f(x) = 3x + 2$, I found $f(50)$ by applying the distributive property to $(3x + 2)50$.

19. I knew how to use point plotting to graph the equation $y = x^2 - 1$, so there was really nothing new to learn when I used the same technique to graph the function $f(x) = x^2 - 1$.

20. The graph of my function revealed aspects of its behavior that were not obvious by just looking at its equation.

Practice Exercises

In Exercises 21–40, plot the given point in a rectangular coordinate system.

21. $(1, 4)$ **22.** $(2, 5)$ **23.** $(-2, 3)$
24. $(-1, 4)$ **25.** $(-3, -5)$ **26.** $(-4, -2)$
27. $(4, -1)$ **28.** $(3, -2)$ **29.** $(-4, 0)$
30. $(-5, 0)$ **31.** $(0, -3)$ **32.** $(0, -4)$
33. $(0, 0)$ **34.** $\left(-3, -1\frac{1}{2}\right)$ **35.** $\left(-2, -3\frac{1}{2}\right)$
36. $(-5, -2.5)$ **37.** $(3.5, 4.5)$ **38.** $(2.5, 3.5)$
39. $(1.25, -3.25)$ **40.** $(2.25, -4.25)$

Graph each equation in Exercises 41–52. Select integers for x from -3 to 3, inclusive.

41. $y = x^2 - 2$ **42.** $y = x^2 + 2$
43. $y = x - 2$ **44.** $y = x + 2$
45. $y = 2x + 1$ **46.** $y = 2x - 4$
47. $y = -\frac{1}{2}x$ **48.** $y = -\frac{1}{2}x + 2$
49. $y = x^3$ **50.** $y = x^3 - 1$
51. $y = |x| + 1$ **52.** $y = |x| - 1$

In Exercises 53–66, evaluate each function at the given value of the variable.

53. $f(x) = x - 4$ **a.** $f(8)$ **b.** $f(1)$
54. $f(x) = x - 6$ **a.** $f(9)$ **b.** $f(2)$
55. $f(x) = 3x - 2$ **a.** $f(7)$ **b.** $f(0)$
56. $f(x) = 4x - 3$ **a.** $f(7)$ **b.** $f(0)$
57. $g(x) = x^2 + 1$ **a.** $g(2)$ **b.** $g(-2)$
58. $g(x) = x^2 + 4$ **a.** $g(3)$ **b.** $g(-3)$
59. $g(x) = -x^2 + 2$ **a.** $g(4)$ **b.** $g(-3)$
60. $g(x) = -x^2 + 1$ **a.** $g(5)$ **b.** $g(-4)$
61. $h(r) = 3r^2 + 5$ **a.** $h(4)$ **b.** $h(-1)$
62. $h(r) = 2r^2 - 4$ **a.** $h(5)$ **b.** $h(-1)$
63. $f(x) = 2x^2 + 3x - 1$ **a.** $f(3)$ **b.** $f(-4)$
64. $f(x) = 3x^2 + 4x - 2$ **a.** $f(2)$ **b.** $f(-1)$
65. $f(x) = \dfrac{x}{|x|}$ **a.** $f(6)$ **b.** $f(-6)$
66. $f(x) = \dfrac{|x|}{x}$ **a.** $f(5)$ **b.** $f(-5)$

In Exercises 67–74, evaluate $f(x)$ for the given values of x. Then use the ordered pairs $(x, f(x))$ from your table to graph the function.

67. $f(x) = x^2 - 1$ **68.** $f(x) = x^2 + 1$

x	$f(x) = x^2 - 1$
-2	
-1	
0	
1	
2	

x	$f(x) = x^2 + 1$
-2	
-1	
0	
1	
2	

69. $f(x) = x - 1$ **70.** $f(x) = x + 1$

x	$f(x) = x - 1$
-2	
-1	
0	
1	
2	

x	$f(x) = x + 1$
-2	
-1	
0	
1	
2	

71. $f(x) = (x - 2)^2$

x	$f(x) = (x - 2)^2$
0	
1	
2	
3	
4	

72. $f(x) = (x + 1)^2$

x	$f(x) = (x + 1)^2$
-3	
-2	
-1	
0	
1	

73. $f(x) = x^3 + 1$

x	$f(x) = x^3 + 1$
-3	
-2	
-1	
0	
1	

74. $f(x) = (x + 1)^3$

x	$f(x) = (x + 1)^3$
-3	
-2	
-1	
0	
1	

For Exercises 75–82, use the vertical line test to identify graphs in which y is a function of x.

75.

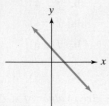

76.

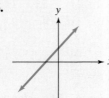

77.

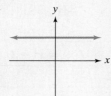

78.

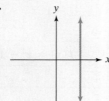

79.

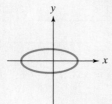

80.

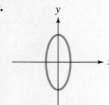

81.

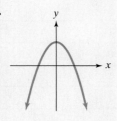

82.

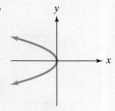

Practice Plus

In Exercises 83–84, let $f(x) = x^2 - x + 4$ and $g(x) = 3x - 5$.

83. Find $g(1)$ and $f(g(1))$.

84. Find $g(-1)$ and $f(g(-1))$.

In Exercises 85–86, let f and g be defined by the following table:

x	$f(x)$	$g(x)$
-2	6	0
-1	3	4
0	-1	1
1	-4	-3
2	0	-6

85. Find $\sqrt{f(-1) - f(0)} - [g(2)]^2 + f(-2) \div g(2) \cdot g(-1)$.

86. Find $|f(1) - f(0)| - [g(1)]^2 + g(1) \div f(-1) \cdot g(2)$.

In Exercises 87–90, write each English sentence as an equation in two variables. Then graph the equation.

87. The y-value is four more than twice the x-value.

88. The y-value is the difference between four and twice the x-value.

89. The y-value is three decreased by the square of the x-value.

90. The y-value is two more than the square of the x-value.

Application Exercises

A football is thrown by a quarterback to a receiver. The points in the figure show the height of the football, in feet, above the ground in terms of its distance, in yards, from the quarterback. Use this information to solve Exercises 91–96.

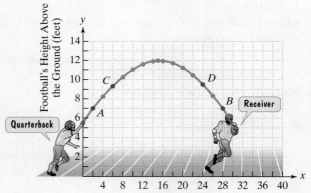

Distance of the Football from the Quarterback (yards)

91. Find the coordinates of point A. Then interpret the coordinates in terms of the information given.

92. Find the coordinates of point B. Then interpret the coordinates in terms of the information given.

93. Estimate the coordinates of point C.

94. Estimate the coordinates of point D.

95. What is the football's maximum height? What is its distance from the quarterback when it reaches its maximum height?

96. What is the football's height when it is caught by the receiver? What is the receiver's distance from the quarterback when he catches the football?

The male minority? The bar graph shows the number of bachelor's degrees, in thousands, awarded to men and women in the United States for selected years, with projections for 2010. The trend indicated by the graphs is among the hottest topics of debate among college-admissions officers. Some private liberal arts colleges have quietly begun special efforts to recruit men—including admissions preferences for them.

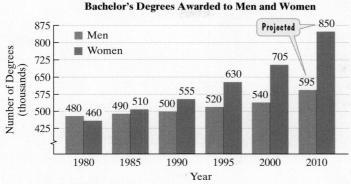

Bachelor's Degrees Awarded to Men and Women

Source: Department of Education

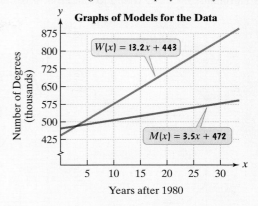

Graphs of Models for the Data

The function $M(x) = 3.5x + 472$ models the number of bachelor's degrees, in thousands, awarded to men x years after 1980. The function $W(x) = 13.2x + 443$ models the number of bachelor's degrees, in thousands, awarded to women x years after 1980. The graphs of functions M and W are shown to the right of the actual data. Use this information to solve Exercises 97–100.

97. a. Find and interpret $W(20)$. Identify this information as a point on the graph of the function for women.

   b. Does $W(20)$ overestimate or underestimate the actual data shown by the bar graph? By how much?

98. a. Find and interpret $M(20)$. Identify this information as a point on the graph of the function for men.

   b. Does $M(20)$ overestimate or underestimate the actual data shown by the bar graph? By how much?

99. a. Use the two functions to find and interpret $W(10) - M(10)$. Identify this information as an appropriate distance between the graphs of the functions for women and men.

   b. Does $W(10) - M(10)$ overestimate or underestimate the actual data shown by the bar graph? By how much?

100. a. Use the two functions to find and interpret $W(5) - M(5)$. Identify this information as an appropriate distance between the graphs of the functions for women and men.

   b. Does $W(5) - M(5)$ overestimate or underestimate the actual data shown by the bar graph? By how much?

The function $f(x) = 0.4x^2 - 36x + 1000$ models the number of accidents, $f(x)$, per 50 million miles driven as a function of a driver's age, x, in years, where x includes drivers from ages 16 through 74, inclusive. The graph of f is shown. Use the equation for f to solve Exercises 101–104.

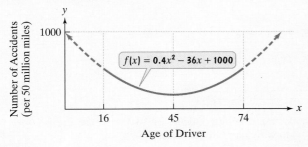

101. Find and interpret $f(20)$. Identify this information as a point on the graph of f at the bottom of the previous column.

102. Find and interpret $f(50)$. Identify this information as a point on the graph of f at the bottom of the previous column.

103. For what value of x does the graph reach its lowest point? Use the equation for f to find the minimum value of y. Describe the practical significance of this minimum value.

104. Use the graph to identify two different ages for which drivers have the same number of accidents. Use the equation for f to find the number of accidents for drivers at each of these ages.

Critical Thinking Exercises

A cellphone company offers the following plans. Also given are the piecewise functions that model these plans. Use this information to solve Exercises 105–106.

Plan A

- *$30 per month buys 120 minutes.*
- *Additional time costs $0.30 per minute.*

$$C(t) = \begin{cases} 30 & \text{if } 0 \le t \le 120 \\ 30 + 0.30(t - 120) & \text{if } t > 120 \end{cases}$$

Plan B

- *$40 per month buys 200 minutes.*
- *Additional time costs $0.30 per minute.*

$$C(t) = \begin{cases} 40 & \text{if } 0 \le t \le 200 \\ 40 + 0.30(t - 200) & \text{if } t > 200 \end{cases}$$

105. Simplify the algebraic expression in the second line of the piecewise function for plan A. Then use point-plotting to graph the function.

106. Simplify the algebraic expression in the second line of the piecewise function for plan B. Then use point-plotting to graph the function.

In Exercises 107–108, write a piecewise function that models each cellphone billing plan. Then graph the function.

107. $30 per month buys 500 minutes. Additional time costs $0.25 per minute.

108. $60 per month buys 2000 minutes. Additional time costs $0.10 per minute.

In Exercises 109–112, select the graph that best illustrates each story.

109. An airplane flew from Miami to San Francisco.

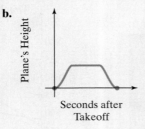

110. At noon, you begin to breathe in.

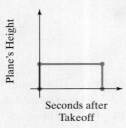

111. Measurements are taken of a person's height from birth to age 100.

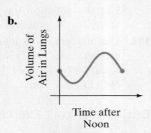

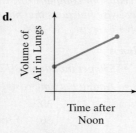

112. You begin your bike ride by riding down a hill. Then you ride up another hill. Finally, you ride along a level surface before coming to a stop.

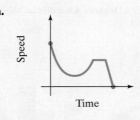
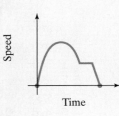
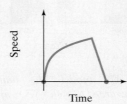
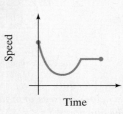

In Exercises 113–114, write a story, or description, to match each title and graph.

113. **Checking Account Balance**   **114.**   **Hair Length**

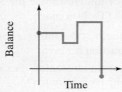
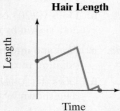

In Exercises 115–118, use the graphs of f and g to find each number.

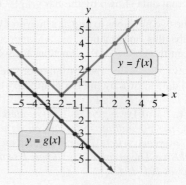

115. $f(-1) + g(-1)$ **116.** $f(1) + g(1)$

117. $f(g(-1))$ **118.** $f(g(1))$

Technology Exercise

119. Use a graphing calculator to verify the graphs that you drew by hand in Exercises 67–74.

Group Exercise

120. (For assistance with this exercise, refer to the discussion of cellphone plans beginning on page 303, as well as to Exercises 105–106.) Group members who have cellphone plans should describe the total monthly cost of the plan as follows:

$_____ per month buys _____ minutes. Additional time costs $_____ per minute.

(For simplicity, ignore other charges.) The group should select any three plans, from "basic" to "premier." For each plan selected, write a piecewise function that describes the plan and graph the function. Graph the three functions in the same rectangular coordinate system. Now examine the graphs. For any given number of calling minutes, the best plan is the one whose graph is lowest at that point. Compare the three calling plans. Is one plan always a better deal than the other two? If not, determine the interval of calling minutes for which each plan is the best deal. (You can check out cellphone plans by visiting www.point.com.)

5.2 Linear Functions and Their Graphs

Objectives

1. Use intercepts to graph a linear equation.

2. Calculate slope.

3. Use the slope and y-intercept to graph a line.

4. Graph horizontal or vertical lines.

5. Interpret slope as rate of change.

6. Use slope and y-intercept to model data.

It's hard to believe that this gas-guzzler, with its huge fins and overstated design, was available in 1957 for approximately $1800. Sadly, its elegance quickly faded, depreciating by $300 per year, often sold for scrap just six years after its glorious emergence from the dealer's showroom.

From these casual observations, we can obtain a mathematical model and its graph. The model is

$$y = -300x + 1800.$$

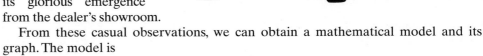
The car is depreciating by $300 per year for x years. The new car is worth $1800.

In this model, y is the car's value after x years. **Figure 5.17** shows the equation's graph. Using function notation, we can rewrite the equation as

$$f(x) = -300x + 1800.$$

A function such as this, whose graph is a straight line, is called a **linear function**. In this section, we will study linear functions and their graphs.

Graphing Using Intercepts

There is another way that we can write the equation

$$y = -300x + 1800.$$

We will collect the x- and y-terms on the left side. This is done by adding $300x$ to both sides:

$$300x + y = 1800.$$

All equations of the form $Ax + By = C$ are straight lines when graphed, as long as A and B are not both zero. Such equations are called **linear equations in two variables**. We can quickly obtain the graph for equations in this form when none of A, B, or C is zero by finding the points where the graph intersects the

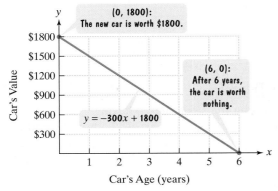

(0, 1800): The new car is worth $1800.

(6, 0): After 6 years, the car is worth nothing.

$y = -300x + 1800$

Car's Value

Car's Age (years)

Figure 5.17

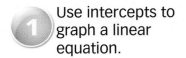
1 Use intercepts to graph a linear equation.

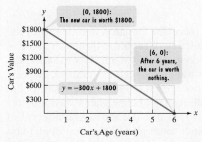

Figure 5.17 (repeated)

x-axis and the *y*-axis. The *x*-coordinate of the point where the graph intersects the *x*-axis is called the ***x*-intercept**. The *y*-coordinate of the point where the graph intersects the *y*-axis is called the ***y*-intercept**.

The graph of $300x + y = 1800$ in **Figure 5.17** intersects the *x*-axis at $(6, 0)$, so the *x*-intercept is 6. The graph intersects the *y*-axis at $(0, 1800)$, so the *y*-intercept is 1800.

> **Locating Intercepts**
>
> To locate the *x*-intercept, set $y = 0$ and solve the equation for *x*.
> To locate the *y*-intercept, set $x = 0$ and solve the equation for *y*.

An equation of the form $Ax + By = C$ as described above can be graphed by finding the *x*- and *y*-intercepts, plotting the intercepts, and drawing a straight line through these points. When graphing using intercepts, it is a good idea to use a third point, a checkpoint, before drawing the line. A checkpoint can be obtained by selecting a value for *x*, other than 0 or the *x*-intercept, and finding the corresponding value for *y*. The checkpoint should lie on the same line as the *x*- and *y*-intercepts. If it does not, recheck your work and find the error.

Study Tip

Note that $3x + 2y = 6$ is of the form $Ax + By = C$.

$$3x + 2y = 6$$

$A = 3$ $B = 2$ $C = 6$

In this case, none of A, B, or C is zero.

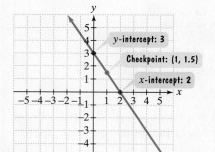

Figure 5.18 The graph of $3x + 2y = 6$

Example 1   Using Intercepts to Graph a Linear Equation

Graph: $3x + 2y = 6$.

Solution

Find the *x*-intercept by letting $y = 0$ and solving for *x*.	Find the *y*-intercept by letting $x = 0$ and solving for *y*.
$3x + 2y = 6$	$3x + 2y = 6$
$3x + 2 \cdot 0 = 6$	$3 \cdot 0 + 2y = 6$
$3x = 6$	$2y = 6$
$x = 2$	$y = 3$

The *x*-intercept is 2, so the line passes through the point $(2, 0)$. The *y*-intercept is 3, so the line passes through the point $(0, 3)$.

For our checkpoint, we choose a value for *x* other than 0 or the *x*-intercept, 2. We will let $x = 1$ and find the corresponding value for *y*.

$3x + 2y = 6$ This is the given equation.
$3 \cdot 1 + 2y = 6$ Substitute 1 for x.
$3 + 2y = 6$ Simplify.
$2y = 3$ Subtract 3 from both sides.
$y = \dfrac{3}{2}$ Divide both sides by 2.

The checkpoint is the ordered pair $\left(1, \frac{3}{2}\right)$, or $(1, 1.5)$.

The three points in **Figure 5.18** lie along the same line. Drawing a line through the three points results in the graph of $3x + 2y = 6$. The arrowheads at the ends of the line show that the line continues indefinitely in both directions.

✓ **Checkpoint 1**   Graph: $2x + 3y = 6$.

Slope

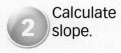

Mathematicians have developed a useful measure of the steepness of a line, called the *slope* of the line. Slope compares the vertical change (the **rise**) to the horizontal change (the **run**) when moving from one fixed point to another along the line. To calculate the slope of a line, we use a ratio that compares the change in *y* (the rise) to the change in *x* (the run).

Calculate slope.

Definition of Slope

The **slope** of the line through the distinct points (x_1, y_1) and (x_2, y_2) is

$$\frac{\text{Change in } y}{\text{Change in } x} = \frac{\text{Rise}}{\text{Run}}$$

Vertical change — Rise

Horizontal change — Run

$$= \frac{y_2 - y_1}{x_2 - x_1}$$

where $x_2 - x_1 \neq 0$.

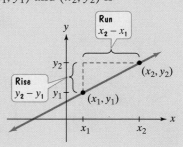

It is common notation to let the letter m represent the slope of a line. The letter m is used because it is the first letter of the French verb *monter*, meaning "to rise," or "to ascend."

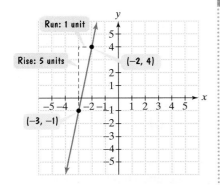

Figure 5.19 Visualizing a slope of 5

Example 2   Using the Definition of Slope

Find the slope of the line passing through each pair of points:

  a. $(-3, -1)$ and $(-2, 4)$ **b.** $(-3, 4)$ and $(2, -2)$.

Solution

a. Let $(x_1, y_1) = (-3, -1)$ and $(x_2, y_2) = (-2, 4)$. We obtain the slope as follows:

$$m = \frac{\text{Change in } y}{\text{Change in } x} = \frac{y_2 - y_1}{x_2 - x_1} = \frac{4 - (-1)}{-2 - (-3)} = \frac{5}{1} = 5.$$

The situation is illustrated in **Figure 5.19**. The slope of the line is 5, indicating that there is a vertical change, a rise, of 5 units for each horizontal change, a run, of 1 unit. The slope is positive and the line rises from left to right.

Study Tip

When computing slope, it makes no difference which point you call (x_1, y_1) and which point you call (x_2, y_2). If we let $(x_1, y_1) = (-2, 4)$ and $(x_2, y_2) = (-3, -1)$, the slope is still 5:

$$m = \frac{\text{Change in } y}{\text{Change in } x} = \frac{y_2 - y_1}{x_2 - x_1} = \frac{-1 - 4}{-3 - (-2)} = \frac{-5}{-1} = 5.$$

However, you should not subtract in one order in the numerator $(y_2 - y_1)$ and then in the opposite order in the denominator $(x_1 - x_2)$. The slope is *not* -5:

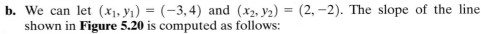

$$\frac{-1 - 4}{-2 - (-3)} = \frac{-5}{1} = -5. \quad \text{Incorrect}$$

b. We can let $(x_1, y_1) = (-3, 4)$ and $(x_2, y_2) = (2, -2)$. The slope of the line shown in **Figure 5.20** is computed as follows:

$$m = \frac{\text{Change in } y}{\text{Change in } x} = \frac{y_2 - y_1}{x_2 - x_1} = \frac{-2 - 4}{2 - (-3)} = \frac{-6}{5} = -\frac{6}{5}.$$

The slope of the line is $-\frac{6}{5}$. For every vertical change of -6 units (6 units down), there is a corresponding horizontal change of 5 units. The slope is negative and the line falls from left to right.

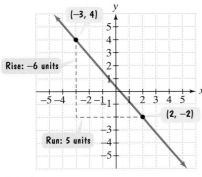

Figure 5.20 Visualizing a slope of $-\frac{6}{5}$

> **Checkpoint 2** Find the slope of the line passing through each pair of points:
>
> **a.** $(-3, 4)$ and $(-4, -2)$
> **b.** $(4, -2)$ and $(-1, 5)$.

Example 2 illustrates that a line with a positive slope is rising from left to right and a line with a negative slope is falling from left to right. By contrast, a horizontal line neither rises nor falls and has a slope of zero. A vertical line has no horizontal change, so $x_2 - x_1 = 0$ in the formula for slope. Because we cannot divide by zero, the slope of a vertical line is undefined. This discussion is summarized in **Table 5.2**.

Table 5.2 Possibilities for a Line's Slope

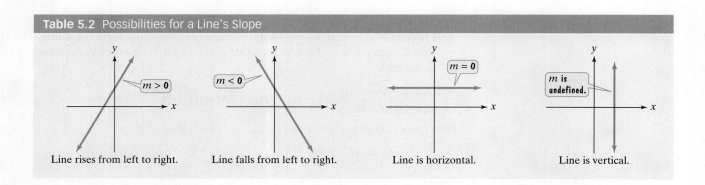

Line rises from left to right.	Line falls from left to right.	Line is horizontal.	Line is vertical.

The Slope-Intercept Form of the Equation of a Line

3 Use the slope and *y*-intercept to graph a line.

We can use the definition of slope to write the equation of any nonvertical line with slope m and y-intercept b. Because the y-intercept is b, the point $(0, b)$ lies on the line. Now, let (x, y) represent any other point on the line, shown in **Figure 5.21**. Keep in mind that the point (x, y) is arbitrary and is not in one fixed position. By contrast, the point $(0, b)$ is fixed.

Regardless of where the point (x, y) is located, the steepness of the line in **Figure 5.21** remains the same. Thus, the ratio for slope stays a constant m. This means that for all points along the line

$$m = \frac{\text{Change in } y}{\text{Change in } x} = \frac{y - b}{x - 0} = \frac{y - b}{x}.$$

We can clear the fraction by multiplying both sides by x, the denominator. Note that x is not zero since (x, y) is distinct from $(0, b)$, the only point on the line with first coordinate 0.

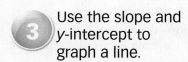
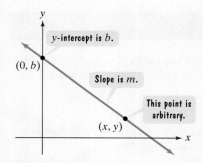

Figure 5.21 A line with slope m and y-intercept b

$$m = \frac{y - b}{x} \qquad \text{This is the slope of the line in Figure 5.21.}$$

$$mx = \frac{y - b}{x} \cdot x \qquad \text{Multiply both sides by x.}$$

$$mx = y - b \qquad \text{Simplify: } \frac{y - b}{x} \cdot x = y - b.$$

$$mx + b = y - b + b \qquad \text{Add b to both sides and solve for y.}$$

$$mx + b = y \qquad \text{Simplify.}$$

Now, if we reverse the two sides, we obtain the *slope-intercept form* of the equation of a line.

Slope-Intercept Form of the Equation of a Line

The **slope-intercept form of the equation** of a nonvertical line with slope m and y-intercept b is

$$y = mx + b.$$

The slope-intercept form of a line's equation, $y = mx + b$, can be expressed in function notation by replacing y with $f(x)$:

$$f(x) = mx + b.$$

We have seen that functions in the form $f(x) = mx + b$ are called **linear functions**. Thus, in the equation of a linear function, the x-coefficient is the line's slope and the constant term is the y-intercept. Here are two examples:

$$y = 2x - 4 \qquad\qquad f(x) = \frac{1}{2}x + 2.$$

The slope is 2.   The y-intercept is -4.   The slope is $\frac{1}{2}$.   The y-intercept is 2.

If a linear function's equation is in slope-intercept form, we can use the y-intercept and the slope to obtain its graph.

Graphing $y = mx + b$ Using the Slope and y-Intercept

1. Plot the point containing the y-intercept on the y-axis. This is the point $(0, b)$.
2. Obtain a second point using the slope, m. Write m as a fraction, and use rise over run, starting at the point containing the y-intercept, to plot this point.
3. Use a straightedge to draw a line through the two points. Draw arrowheads at the ends of the line to show that the line continues indefinitely in both directions.

Study Tip

Writing the slope, m, as a fraction allows you to identify the rise (the fraction's numerator) and the run (the fraction's denominator).

| **Example 3** | **Graphing by Using the Slope and y-Intercept** |

Graph the linear function $y = \dfrac{2}{3}x + 2$ by using the slope and y-intercept.

Solution

The equation of the linear function is in the form $y = mx + b$. We can find the slope, m, by identifying the coefficient of x. We can find the y-intercept, b, by identifying the constant term.

$$y = \frac{2}{3}x + 2$$

The slope is $\frac{2}{3}$.   The y-intercept is 2.

Now that we have identified the slope and the y-intercept, we use the three-step procedure in the box above to graph the equation.

Step 1 Plot the point containing the y-intercept on the y-axis. The y-intercept is 2. We plot $(0, 2)$, shown in **Figure 5.22**.

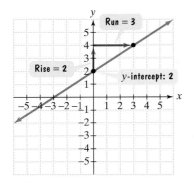

Figure 5.22 The graph of $y = \frac{2}{3}x + 2$

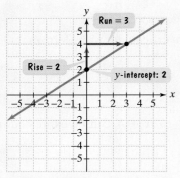

Figure 5.22 The graph of $y = \frac{2}{3}x + 2$ (repeated)

Step 2 Obtain a second point using the slope, *m*. Write *m* as a fraction, and use rise over run, starting at the point containing the *y*-intercept, to plot this point. The slope, $\frac{2}{3}$, is already written as a fraction:

$$m = \frac{2}{3} = \frac{\text{Rise}}{\text{Run}}.$$

We plot the second point on the line by starting at $(0, 2)$, the first point. Based on the slope, we move 2 units *up* (the rise) and 3 units to the *right* (the run). This puts us at a second point on the line, $(3, 4)$, shown in **Figure 5.22**.

Step 3 Use a straightedge to draw a line through the two points. The graph of $y = \frac{2}{3}x + 2$ is shown in **Figure 5.22**.

✓ **Checkpoint 3** Graph the linear function $y = \frac{3}{5}x + 1$ by using the slope and ▶ *y*-intercept.

Earlier in this section, we considered linear functions written in the form $Ax + By = C$. We used *x*- and *y*-intercepts, as well as checkpoints, to graph these functions. It is also possible to obtain the graphs by using the slope and *y*-intercept. To do this, begin by solving $Ax + By = C$ for *y*. This will put the equation in slope-intercept form. Then use the three-step procedure to graph the equation. This is illustrated in Example 4.

Example 4 Graphing by Using the Slope and *y*-Intercept

Graph the linear function $2x + 5y = 0$ by using the slope and *y*-intercept.

Solution

We put the equation in slope-intercept form by solving for *y*.

$2x + 5y = 0$	This is the given equation.
$2x - 2x + 5y = 0 - 2x$	Subtract 2x from both sides.
$5y = -2x + 0$	Simplify.
$\dfrac{5y}{5} = \dfrac{-2x + 0}{5}$	Divide both sides by 5.
$y = \dfrac{-2x}{5} + \dfrac{0}{5}$	Divide each term in the numerator by 5.
$y = -\dfrac{2}{5}x + 0$	Simplify. Equivalently, $f(x) = -\frac{2}{5}x + 0$.

Now that the equation is in slope-intercept form, we can use the slope and *y*-intercept to obtain its graph. Examine the slope-intercept form:

$$y = -\frac{2}{5}x + 0.$$

slope: $-\frac{2}{5}$ *y*-intercept: 0

Note that the slope is $-\frac{2}{5}$ and the *y*-intercept is 0. Use the *y*-intercept to plot $(0, 0)$ on the *y*-axis. Then locate a second point by using the slope.

$$m = -\frac{2}{5} = \frac{-2}{5} = \frac{\text{Rise}}{\text{Run}}$$

Because the rise is -2 and the run is 5, move *down* 2 units and to the *right* 5 units, starting at the point $(0, 0)$. This puts us at a second point on the line, $(5, -2)$. The ▶ graph of $2x + 5y = 0$ is the line drawn through these points, shown in **Figure 5.23**.

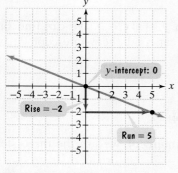

Figure 5.23 The graph of $2x + 5y = 0$, or $y = -\frac{2}{5}x$

The equation $2x + 5y = 0$ in Example 4 is of the form $Ax + By = C$ with $C = 0$. If you try graphing $2x + 5y = 0$ by using intercepts, you will find that the x-intercept is 0 and the y-intercept is 0. This means that the graph passes through the origin. A second point must be found to graph the line. In Example 4, the line's slope gave us the second point.

> ✓ **Checkpoint 4**   Graph the linear function $3x + 4y = 0$ by using the slope and y-intercept.

Equations of Horizontal and Vertical Lines

4 Graph horizontal or vertical lines.

If a line is horizontal, its slope is zero: $m = 0$. Thus, the equation $y = mx + b$ becomes $y = b$, where b is the y-intercept. All horizontal lines have equations of the form $y = b$.

Example 5   Graphing a Horizontal Line

Graph $y = -4$ in the rectangular coordinate system.

Solution

All ordered pairs that are solutions of $y = -4$ have a value of y that is always -4. Any value can be used for x. (Think of $y = -4$ as $0x + 1y = -4$.) In the table at the right, we have selected three of the possible values for x: -2, 0, and 3. The table shows that three ordered pairs that are solutions of $y = -4$ are $(-2, -4)$, $(0, -4)$, and $(3, -4)$. Drawing a line that passes through the three points gives the horizontal line shown in **Figure 5.24**.

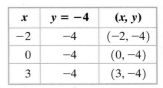

x	$y = -4$	(x, y)
-2	-4	$(-2, -4)$
0	-4	$(0, -4)$
3	-4	$(3, -4)$

For all choices of x, y is a constant -4.

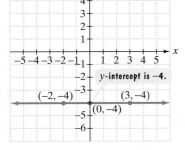

Figure 5.24 The graph of $y = -4$ or $f(x) = -4$

> ✓ **Checkpoint 5**   Graph $y = 3$ in the rectangular coordinate system.

Next, let's see what we can discover about the graph of an equation of the form $x = a$ by looking at an example.

Example 6   Graphing a Vertical Line

Graph $x = 2$ in the rectangular coordinate system.

Solution

All ordered pairs that are solutions of $x = 2$ have a value of x that is always 2. Any value can be used for y. (Think of $x = 2$ as $1x + 0y = 2$.) In the table on the right, we have selected three of the possible values for y: -2, 0, and 3. The table shows that three ordered pairs that are solutions of $x = 2$ are $(2, -2)$, $(2, 0)$, and $(2, 3)$. Drawing a line that passes through the three points gives the vertical line shown in **Figure 5.25**.

For all choices of y,

x is always 2.

$x = 2$	y	(x, y)
2	-2	$(2, -2)$
2	0	$(2, 0)$
2	3	$(2, 3)$

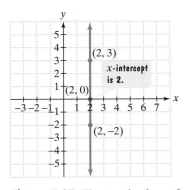

Figure 5.25 The graph of $x = 2$

Does a vertical line represent the graph of a linear function? No. Look at the graph of $x = 2$ in **Figure 5.25**. A vertical line drawn through $(2, 0)$ intersects the graph infinitely many times. This shows that infinitely many outputs are associated with the input 2. **No vertical line represents a linear function.** All other lines are graphs of functions.

> ### Horizontal and Vertical Lines
>
> The graph of $y = b$ or $f(x) = b$ is a horizontal line. The y-intercept is b.
>
> The graph of $x = a$ is a vertical line. The x-intercept is a.
>
>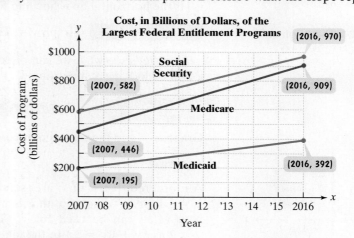

✓ **Checkpoint 6**   Graph $x = -2$ in the rectangular coordinate system.

Slope as Rate of Change

⑤ Interpret slope as rate of change.

Slope is defined as the ratio of a change in y to a corresponding change in x. Our next example shows how slope can be interpreted as a **rate of change** in an applied situation.

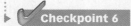 **Example 7**   **Slope as a Rate of Change**

In 2007, the U.S. government faced the prospect of paying out more and more in Social Security, Medicare, and Medicaid benefits. The line graphs in **Figure 5.26** show the costs of these entitlement programs, in billions of dollars, for 2007 with projections for years through 2016. Find the slope of the line segment representing Social Security. Round to one decimal place. Describe what the slope represents.

Figure 5.26
Source: Congressional Budget Office

Solution

We let x represent a year and y the cost, in billions of dollars, of Social Security in that year. The two points shown on the line segment for Social Security have the following coordinates:

$$(2007, 582) \quad \text{and} \quad (2016, 970).$$

| In 2007, the cost of Social Security was $582 billion. | In 2016, the cost of Social Security is projected to be $970 billion. |

Now we compute the slope.

$$m = \frac{\text{Change in } y}{\text{Change in } x} = \frac{970 - 582}{2016 - 2007}$$

The unit in the numerator is *billions of dollars.*

The unit in the denominator is *year.*

$$= \frac{388}{9} \approx 43.1$$

The slope indicates that for the period from 2007 through 2016, the cost of Social Security is projected to increase by approximately $43.1 billion per year. The rate of change is approximately $43.1 billion per year.

▶ **Checkpoint 7**   Find the slope of the line segment representing Medicare in **Figure 5.26**. Round to one decimal place. Describe what the slope represents.

Modeling Data with the Slope-Intercept Form of the Equation of a Line

Use slope and y-intercept to model data.

Linear functions are useful for modeling data that fall on or near a line. For example, **Figure 5.27** shows the average premium for employer-sponsored health insurance, usually divided between the worker and employer, from 1999 through 2008. Example 8 and Checkpoint 8 illustrate how we can use the equation $y = mx + b$ to develop models for the data. The red modifications below the x-axis show additions to the original display so that we can obtain these models.

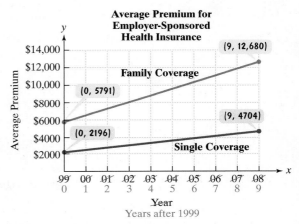

Figure 5.27
Source: Kaiser Family Foundation

Example 8   **Modeling with the Slope-Intercept Equation**

a. Use the two points for single coverage in **Figure 5.27** to find a function in the form $S(x) = mx + b$ that models the average premium, $S(x)$, for single coverage x years after 1999.

b. Use the model to project the average premium for single coverage in 2012.

Solution

a. We will use the line segment passing through the points $(0, 2196)$ and $(9, 4704)$ to obtain a model. We need values for m, the slope, and b, the y-intercept.

$$S(x) = mx + b$$

$m = \dfrac{\text{Change in } y}{\text{Change in } x}$

$= \dfrac{4704 - 2196}{9 - 0} \approx 279$

The point $(0, 2196)$ lies on the line segment, so the y-intercept is 2196: $b = 2196$.

Using $m \approx 279$ and $b = 2196$, the average premium, $S(x)$, for single-coverage employer-sponsored health insurance x years after 1999 can be modeled by the linear function

$$S(x) = 279x + 2196.$$

The slope, approximately 279, indicates that this premium is increasing at a rate of approximately \$279 per year.

b. Now let's use the model to project the average premium for single coverage in 2012. Because 2012 is 13 years after 1999, substitute 13 for x in $S(x) = 279x + 2196$ and evaluate the function at 13.

$$S(13) = 279(13) + 2196 = 3627 + 2196 = 5823$$

Our model projects an average premium of \$5823 for single coverage in 2012.

 Checkpoint 8

a. Use the two points for family coverage in **Figure 5.27** on the previous page to find a function in the form $F(x) = mx + b$ that models the average premium, $F(x)$, for family coverage x years after 1999. Round m to the nearest dollar.

▸ **b.** Use the model to project the average premium for family coverage in 2011.

Achieving Success

The Secret of Math Success

What's the secret of math success? The bar graph in **Figure 5.28** shows that Japanese teachers and students are more likely than their American counterparts to believe that the key to doing well in math is working hard. Americans tend to think that either you have mathematical intelligence or you don't. Alan Bass, author of *Math Study Skills* (Pearson Education, 2008), strongly disagrees with this American perspective:

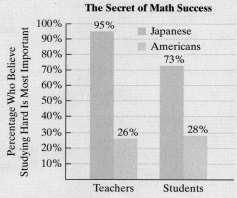

The Secret of Math Success

Figure 5.28
Source: Wade and Tavris, *Psychology*, Ninth Edition, Pearson, 2008.

"Human beings are easily intelligent enough to understand the basic principles of math. I cannot repeat this enough, but I'll try . . .

Poor performance in math is not due to a lack of intelligence! The fact is that the key to success in math is in taking an intelligent approach. Students come up to me and say, 'I'm just not good at math.' Then I ask to see their class notebooks and they show me a chaotic mess of papers jammed into a folder. In math, that's a lot like taking apart your car's engine, dumping the heap of disconnected parts back under the hood, and then going to a mechanic to ask why it won't run. Students come to me and say, 'I'm just not good at math.' Then I ask them about their study habits and they say, 'I have to do all my studying on the weekend.' In math, that's a lot like trying to do all your eating or sleeping on the weekends and wondering why you're so tired and hungry all the time. **How you approach math is much more important than how smart you are.**"

—Alan Bass

Exercise Set 5.2

Concept and Vocabulary Exercises

In Exercises 1–6, fill in each blank so that the resulting statement is true.

1. The x-coordinate of the point where a graph crosses the x-axis is called the _____.

2. The point $(0, 3)$ lies along a line, so 3 is the _____ of that line.

3. The slope of the line through the distinct points (x_1, y_1) and (x_2, y_2) is _____.

4. The slope-intercept form of the equation of a line is _____, where m represents the _____ and b represents the _____.

5. The graph of $y = b$ is a/an _____ line.

6. The graph of $x = a$ is a/an _____ line.

In Exercises 7–10, determine whether each statement is true or false. If the statement is false, make the necessary change(s) to produce a true statement.

7. The equation $y = mx + b$ shows that no line can have a y-intercept that is numerically equal to its slope.

8. Every line in the rectangular coordinate system has an equation that can be expressed in slope-intercept form.

9. The line $3x + 2y = 5$ has slope $-\frac{3}{2}$.

10. The line $2y = 3x + 7$ has a y-intercept of 7.

Respond to Exercises 11–18 using verbal or written explanations.

11. Describe how to find the x-intercept of a linear equation.

12. Describe how to find the y-intercept of a linear equation.

13. What is the slope of a line?

14. Describe how to calculate the slope of a line passing through two points.

15. Describe how to graph a line using the slope and y-intercept. Provide an original example with your description.

16. What does it mean if the slope of a line is 0?

17. What does it mean if the slope of a line is undefined?

18. Explain why the y-values can be any number for the equation $x = 5$. How is this shown in the graph of the equation?

In Exercises 19–22, determine whether each statement makes sense or does not make sense, and explain your reasoning.

19. When finding the slope of the line passing through $(-1, 5)$ and $(2, -3)$, I must let (x_1, y_1) be $(-1, 5)$ and (x_2, y_2) be $(2, -3)$.

20. A linear function that models tuition and fees at public four-year colleges from 2000 through 2008 has negative slope.

21. Because the variable m does not appear in $Ax + By = C$, equations in this form make it impossible to determine the line's slope.

22. If I drive m miles in a year, the function $c(x) = 0.25m + 3500$ models the annual cost, $c(x)$, in dollars, of operating my car, so the function shows that with no driving at all, the cost is $3500, and the rate of increase in this cost is $0.25 for each mile that I drive.

Practice Exercises

In Exercises 23–30, use the x- and y-intercepts to graph each linear equation.

23. $x - y = 3$

24. $x + y = 4$

25. $3x - 4y = 12$

26. $2x - 5y = 10$

27. $2x + y = 6$

28. $x + 3y = 6$

29. $5x = 3y - 15$

30. $3x = 2y + 6$

In Exercises 31–42, calculate the slope of the line passing through the given points. If the slope is undefined, so state. Then indicate whether the line rises, falls, is horizontal, or is vertical.

31. $(2, 6)$ and $(3, 5)$

32. $(4, 2)$ and $(3, 4)$

33. $(-2, 1)$ and $(2, 2)$

34. $(-1, 3)$ and $(2, 4)$

35. $(-2, 4)$ and $(-1, -1)$

36. $(6, -4)$ and $(4, -2)$

37. $(5, 3)$ and $(5, -2)$

38. $(3, -4)$ and $(3, 5)$

39. $(2, 0)$ and $(0, 8)$

40. $(3, 0)$ and $(0, -9)$

41. $(5, 1)$ and $(-2, 1)$

42. $(-2, 3)$ and $(1, 3)$

In Exercises 43–54, graph each linear function using the slope and y-intercept.

43. $y = 2x + 3$

44. $y = 2x + 1$

45. $y = -2x + 4$

46. $y = -2x + 3$

47. $y = \frac{1}{2}x + 3$

48. $y = \frac{1}{2}x + 2$

49. $f(x) = \frac{2}{3}x - 4$

50. $f(x) = \frac{3}{4}x - 5$

51. $y = -\frac{3}{4}x + 4$

52. $y = -\frac{2}{3}x + 5$

53. $f(x) = -\frac{5}{3}x$

54. $f(x) = -\frac{4}{3}x$

In Exercises 55–62,

a. *Put the equation in slope-intercept form by solving for y.*

b. *Identify the slope and the y-intercept.*

c. *Use the slope and y-intercept to graph the line.*

55. $3x + y = 0$

56. $2x + y = 0$

57. $3y = 4x$

58. $4y = 5x$

59. $2x + y = 3$

60. $3x + y = 4$

61. $7x + 2y = 14$

62. $5x + 3y = 15$

In Exercises 63–70, graph each horizontal or vertical line.

63. $y = 4$ **64.** $y = 2$

65. $y = -2$ **66.** $y = -3$

67. $x = 2$ **68.** $x = 4$

69. $x + 1 = 0$ **70.** $x + 5 = 0$

Practice Plus

In Exercises 71–74, find the slope of the line passing through each pair of points or state that the slope is undefined. Assume that all variables represent positive real numbers. Then indicate whether the line through the points rises, falls, is horizontal, or is vertical.

71. $(0, a)$ and $(b, 0)$

72. $(-a, 0)$ and $(0, -b)$

73. (a, b) and $(a, b + c)$

74. $(a - b, c)$ and $(a, a + c)$

In Exercises 75–76, find the slope and y-intercept of each line whose equation is given. Assume that $B \neq 0$.

75. $Ax + By = C$

76. $Ax = By - C$

In Exercises 77–78, find the value of y if the line through the two given points is to have the indicated slope.

77. $(3, y)$ and $(1, 4)$, $m = -3$

78. $(-2, y)$ and $(4, -4)$, $m = \frac{1}{3}$

Use the figure to make the lists in Exercises 79–80.

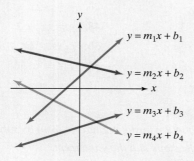

79. List the slopes m_1, m_2, m_3, and m_4 in order of decreasing size.

80. List the y-intercepts b_1, b_2, b_3, b_4 in order of decreasing size.

Application Exercises

The graph shows the percentage of sales of recorded music in the United States for rock and rap/hip hop. Use this information to solve Exercises 81–82.

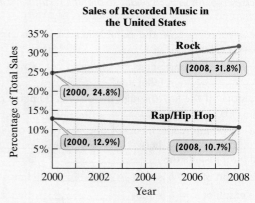

Source: RIAA

81. Find the slope of the line segment representing rap/hip hop. What does this represent in terms of the rate of change in the percentage of total sales of rap/hip hop for the period shown?

82. Find the slope of the line segment representing rock. What does this represent in terms of the rate of change in the percentage of total sales of rock for the period shown?

The bar graph shows that as online news has grown, traditional news media have slipped.

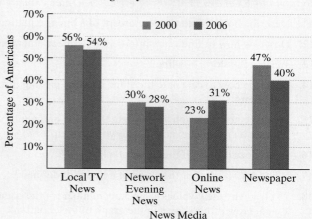

Source: Pew Research Center

In Exercises 83–84, find a linear function in slope-intercept form that models the given description. Each function should model the percentage of Americans, P(x), who regularly used the news outlet x years after 2000.

83. In 2000, 47% of Americans regularly used newspapers for getting news and this percentage has decreased at an average rate of approximately 1.2 each year since then.

84. In 2000, 23% of Americans regularly used online news for getting news and this percentage has increased at an average rate of approximately 1.3 each year since then.

The bar graph shows the racial and ethnic composition of the United States population in 2008, with projections for 2050. Exercises 85–86 involve the graphs shown in the rectangular coordinate system for two of the racial/ethnic groups.

Racial and Ethnic Composition of the United States Population

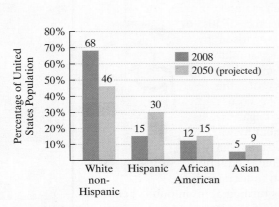

Source: Wikipedia

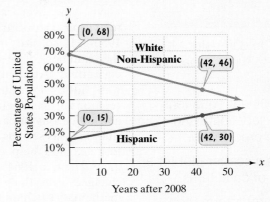

85. a. Use the two points for white non-Hispanics to find a function in the form $p(x) = mx + b$ that models the percentage of white non-Hispanics, $p(x)$, in the United States population x years after 2008. Round m to two decimal places.

b. Use the model from part (a) to project the percentage of white non-Hispanics in the United States in 2108.

86. a. Use the two points for Hispanics to find a function in the form $p(x) = mx + b$ that models the percentage of Hispanics, $p(x)$, in the United States population x years after 2008. Round m to two decimal places.

b. Use the model from part (a) to project the percentage of Hispanics in the United States in 2108.

Critical Thinking Exercises

In Exercises 87–88, find the coefficients that must be placed in each shaded area so that the function's graph will be a line satisfying the specified conditions.

87. ▪x + ▪y = 12; x-intercept = -2; y-intercept = 4

88. ▪x + ▪y = 12; y-intercept = -6; slope = $\frac{1}{2}$.

89. Prove that the equation of a line passing through $(a, 0)$ and $(0, b)(a \neq 0, b \neq 0)$ can be written in the form $\frac{x}{a} + \frac{y}{b} = 1$. Why is this called the *intercept form* of a line?

90. The relationship between Celsius temperature, C, and Fahrenheit temperature, F, can be described by a linear equation in the form $F = mC + b$. The graph of this equation contains the point $(0, 32)$: Water freezes at $0\,°C$ or at $32\,°F$. The line also contains the point $(100, 212)$: Water boils at $100\,°C$ or at $212\,°F$. Write the linear equation expressing Fahrenheit temperature in terms of Celsius temperature.

Technology Exercises

91. Use a graphing utility to verify any three of your hand-drawn graphs in Exercises 43–54.

92. Use a graphing utility to verify any three of your hand-drawn graphs in Exercises 55–62. Solve the equation for y before entering it.

5.3 The Point-Slope Form of the Equation of a Line; Scatter Plots and Regression Lines

1. Use the point-slope form to write equations of a line.

2. Write linear equations that model data and make predictions.

3. Make a scatter plot for a table of data items.

4. Interpret information given in a scatter plot.

Have you ever watched any movies or television shows from the 1940s or 1950s? If so, are you surprised by the number of people smoking cigarettes? At that time, there was little awareness of the relationship between tobacco use and numerous diseases. Cigarette smoking was seen as a healthy way to relax and help digest a hearty meal. Then, in 1964, a linear equation changed everything. To understand the mathematics behind this turning point in public health, we explore another form of a line's equation.

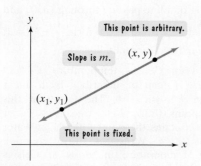

Figure 5.29 A line passing through (x_1, y_1) with slope m

Point-Slope Form

We can use the slope of a line to obtain another useful form of the line's equation. Consider a nonvertical line that has slope m and contains the point (x_1, y_1). Now let (x, y) represent any other point on the line, shown in **Figure 5.29**. Keep in mind that the point (x, y) is arbitrary and is not in one fixed position. By contrast, the point (x_1, y_1) is fixed.

Regardless of where the point (x, y) is located on the line, the steepness of the line in **Figure 5.29** remains the same. Thus, the ratio for slope stays a constant m. This means that for all points (x, y) along the line,

$$m = \frac{\text{Change in } y}{\text{Change in } x} = \frac{y - y_1}{x - x_1}.$$

We can clear the fraction by multiplying both sides by $x - x_1$, the least common denominator, where $x - x_1 \neq 0$.

$$m = \frac{y - y_1}{x - x_1} \qquad \text{This is the slope of the line in Figure 5.29.}$$

$$m(x - x_1) = \frac{y - y_1}{x - x_1} \cdot (x - x_1) \qquad \text{Multiply both sides by } x - x_1.$$

$$m(x - x_1) = y - y_1 \qquad \text{Simplify: } \frac{y - y_1}{x - x_1} \cdot (x - x_1) = y - y_1.$$

Now, if we reverse the two sides, we obtain the **point-slope form** of the equation of a line.

Study Tip

When writing the point-slope form of a line's equation, you will never substitute numbers for x and y. You will substitute values for x_1, y_1, and m.

Point-Slope Form of the Equation of a Line

The **point-slope form of the equation** of a nonvertical line with slope m that passes through the point (x_1, y_1) is

$$y - y_1 = m(x - x_1).$$

For example, the point-slope form of the equation of the line passing through $(1, 4)$ with slope 2 $(m = 2)$ is

$$y - 4 = 2(x - 1).$$

Using the Point-Slope Form to Write a Line's Equation

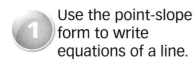

Use the point-slope form to write equations of a line.

If we know the slope of a line and a point not containing the y-intercept through which the line passes, the point-slope form is the equation that we should use. Once we have obtained this equation, it is customary to solve for y and write the equation in slope-intercept form. Examples 1 and 2 illustrate these ideas.

Example 1   **Writing the Point-Slope Form and the Slope-Intercept Form**

Write the point-slope form and the slope-intercept form of the equation of the line with slope 4 that passes through the point $(-1, 3)$.

Solution

We begin with the point-slope form of the equation of a line with $m = 4$, $x_1 = -1$, and $y_1 = 3$.

$$y - y_1 = m(x - x_1)$$ This is the point-slope form of the equation.

$$y - 3 = 4[x - (-1)]$$ Substitute: $(x_1, y_1) = (-1, 3)$ and $m = 4$.

$$y - 3 = 4(x + 1)$$ We now have the point-slope form of the equation of the given line.

Now we solve this equation for y and write an equivalent equation in slope-intercept form ($y = mx + b$).

We need to isolate y.

$$y - 3 = 4(x + 1)$$ This is the point-slope form of the equation.

$$y - 3 = 4x + 4$$ Use the distributive property.

$$y = 4x + 7$$ Add 3 to both sides.

The slope-intercept form of the line's equation is $y = 4x + 7$.

✓ **Checkpoint 1**   Write the point-slope form and the slope-intercept form of the equation of the line with slope 6 that passes through the point $(2, -5)$.

Example 2   **Writing the Point-Slope Form and the Slope-Intercept Form**

A line passes through the points $(4, -3)$ and $(-2, 6)$. (See **Figure 5.30**.) Find the equation of the line

a. in point-slope form.   **b.** in slope-intercept form.

Solution

a. To use the point-slope form, we need to find the slope. The slope is the change in the y-coordinates divided by the corresponding change in the x-coordinates.

$$m = \frac{6 - (-3)}{-2 - 4} = \frac{9}{-6} = -\frac{3}{2}$$ This is the definition of slope using $(4, -3)$ and $(-2, 6)$.

We can take either point on the line to be (x_1, y_1). Let's use $(x_1, y_1) = (4, -3)$. Now, we are ready to write the point-slope form of the equation.

$$y - y_1 = m(x - x_1)$$ This is the point-slope form of the equation.

$$y - (-3) = -\frac{3}{2}(x - 4)$$ Substitute: $(x_1, y_1) = (4, -3)$ and $m = -\frac{3}{2}$.

$$y + 3 = -\frac{3}{2}(x - 4)$$ Simplify.

This equation is one point-slope form of the equation of the line shown in **Figure 5.30**.

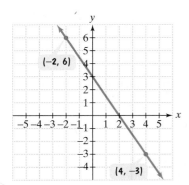

Figure 5.30

b. Now, we solve the equation $y + 3 = -\dfrac{3}{2}(x - 4)$ for y and write an equivalent equation in slope-intercept form ($y = mx + b$).

We need to isolate y.

$$y + 3 = -\frac{3}{2}(x - 4) \qquad \text{This is the point-slope form of the equation.}$$

$$y + 3 = -\frac{3}{2}x + 6 \qquad \text{Use the distributive property.}$$

$$y = -\frac{3}{2}x + 3 \qquad \text{Subtract 3 from both sides.}$$

This equation is the slope-intercept form of the equation of the line shown in **Figure 5.30** on the previous page.

✓ Checkpoint 2 A line passes through the points $(-2, -1)$ and $(-1, -6)$. Find the equation of the line

a. in point-slope form.

▶ **b.** in slope-intercept form.

The many forms of a line's equation can be a bit overwhelming. **Table 5.3** summarizes the various forms and contains the most important things you should remember about each form.

Table 5.3 Equations of Lines

Form	What You Should Know
$Ax + By = C$	• Graph equations in this form using intercepts (x-intercept: set $y = 0$; y-intercept: set $x = 0$) and a checkpoint.
$y = b$	• Graph equations in this form as horizontal lines with b as the y-intercept.
$x = a$	• Graph equations in this form as vertical lines with a as the x-intercept.
Slope-Intercept Form $y = mx + b$	• Graph equations in this form using the y-intercept, b, and the slope, m. • Start with this form when writing a linear equation if you know a line's slope and y-intercept.
Point-Slope Form $y - y_1 = m(x - x_1)$	• Start with this form when writing a linear equation if you know the slope of the line and a point on the line not containing the y-intercept or two points on the line, neither of which contains the y-intercept. Calculate the slope using $$m = \frac{\text{Change in } y}{\text{Change in } x} = \frac{y_2 - y_1}{x_2 - x_1}.$$ Although you begin with point-slope form, you usually solve for y and convert to slope-intercept form.

Applications

2 Write linear equations that model data and make predictions.

Linear equations are useful for modeling data that fall on or near a line. For example, the bar graph in **Figure 5.31(a)** gives the median age of the U.S. population in the indicated year. (The median age is the age in the middle when all the ages of the U.S. population are arranged from youngest to oldest.) The data are displayed as a set of five points in a rectangular coordinate system in **Figure 5.31(b)**.

The Graying of America: Median Age of the United States Population

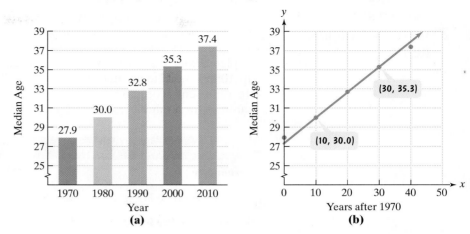

Figure 5.31
Source: U.S. Census Bureau

A set of points representing data is called a **scatter plot**. Also shown on the scatter plot in **Figure 5.31(b)** is a line that passes through or near the five points. By writing the equation of this line, we can obtain a model of the data and make predictions about the median age of the U.S. population in the future.

Example 3 **Modeling the Graying of America**

Write the slope-intercept form of the equation of the line shown in **Figure 5.31(b)**. Use the function to predict the median age of the U.S. population in 2020.

Solution

The line in **Figure 5.31(b)** passes through $(10, 30.0)$ and $(30, 35.3)$. We start by finding its slope.

$$m = \frac{\text{Change in } y}{\text{Change in } x} = \frac{35.3 - 30.0}{30 - 10} = \frac{5.3}{20} = 0.265$$

The slope indicates that each year the median age of the U.S. population is increasing by 0.265 year.

Now, we write the slope-intercept form of the equation of the line.

$y - y_1 = m(x - x_1)$	Begin with the point-slope form.
$y - 30.0 = 0.265(x - 10)$	Either ordered pair can be (x_1, y_1). Let $(x_1, y_1) = (10, 30.0)$. From above, $m = 0.265$.
$y - 30.0 = 0.265x - 2.65$	Apply the distributive property.
$y = 0.265x + 27.35$	Add 30 to both sides and solve for y.

A linear function that models the median age of the U.S. population, y, x years after 1970 is

$$y = 0.265x + 27.35 \quad \text{or} \quad f(x) = 0.265x + 27.35.$$

Now, let's use this equation to predict the median age in 2020. Because 2020 is 50 years after 1970, substitute 50 for x and compute y.

$$f(50) = 0.265(50) + 27.35 = 40.6$$

Our model predicts that the median age of the U.S. population in 2020 will be 40.6.

✓ **Checkpoint 3** Use the data points $(10, 30.0)$ and $(20, 32.8)$ from **Figure 5.31(b)** to write the slope-intercept form of an equation that models the median age of the U.S. population x years after 1970. Use this model to predict the median age in 2020.

Make a scatter plot for a table of data items.

Scatter Plots and Correlation

Is there a relationship between education and prejudice? With increased education, does a person's level of prejudice tend to increase, decrease, or stay the same? For each person in our sample, we will record the number of years of school completed and the score on a test measuring prejudice. Higher scores on this 1-to-10 test indicate greater prejudice. Using x to represent years of education and y to represent scores on a test measuring prejudice, **Table 5.4** shows these two quantities for a random sample of ten people.

Table 5.4 Recording Two Quantities in a Sample of Ten People										
Respondent	**A**	**B**	**C**	**D**	**E**	**F**	**G**	**H**	**I**	**J**
Years of education (x)	12	5	14	13	8	10	16	11	12	4
Score on prejudice test (y)	1	7	2	3	5	4	1	2	3	10

We can make a scatter plot of the data in **Table 5.4** by drawing a horizontal axis to represent years of education and a vertical axis to represent scores on a test measuring prejudice. We then represent each of the ten respondents with a single point on the graph. For example, the dot for respondent A is located to represent 12 years of education on the horizontal axis and 1 on the prejudice test on the vertical axis. Plotting each of the ten pairs of data in a rectangular coordinate system results in the scatter plot shown in **Figure 5.32**.

A scatter plot like the one in **Figure 5.32** can be used to determine whether two quantities are related. If there is a clear relationship, the quantities are said to be **correlated**. The scatter plot shows a downward trend among the data points, although there are a few exceptions. People with increased education tend to have a lower score on the test measuring prejudice. **Correlation** is used to determine if there is a relationship between two variables and, if so, the strength and direction of that relationship.

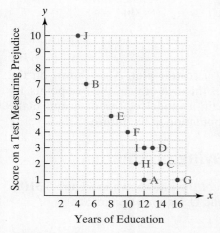

Figure 5.32 A scatter plot for education-prejudice data

Correlation and Causal Connections

Correlations can often be seen when data items are displayed on a scatter plot. Although the scatter plot in **Figure 5.32** indicates a correlation between education and prejudice, we cannot conclude that increased education causes a person's level of prejudice to decrease. There are at least three possible explanations:

1. The correlation between increased education and decreased prejudice is simply a coincidence.

2. Education usually involves classrooms with a variety of different kinds of people. Increased exposure to diversity in the classroom setting, which accompanies increased levels of education, might be an underlying cause for decreased prejudice.

3. Education, the process of acquiring knowledge, requires people to look at new ideas and see things in different ways. Thus, education causes one to be more tolerant and less prejudiced.

Establishing that one thing causes another is extremely difficult, even if there is a strong correlation between these things. For example, as the air temperature increases, there is an increase in the number of people stung by jellyfish at the beach. This does not mean that an increase in air temperature causes more people to be stung. It might mean that because it is hotter, more people go into the water. With an increased number of swimmers, more people are likely to be stung. In short, correlation is not necessarily causation.

> **Study Tip**
>
> The numbered list on the right represents three possibilities. Perhaps you can provide a better explanation about decreasing prejudice with increased education.

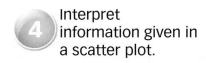

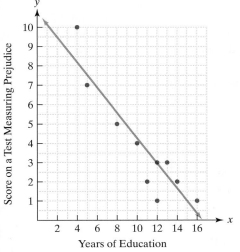

Figure 5.33 A scatter plot with a regression line

Regression Lines and Correlation Coefficients

Figure 5.33 shows the scatter plot for the education-prejudice data. Also shown is a straight line that seems to approximately "fit" the data points. Most of the data points lie either near or on this line. A line that best fits the data points in a scatter plot is called a **regression line**. The regression line is the particular line in which the spread of the data points around it is as small as possible.

A measure that is used to describe the strength and direction of a relationship between variables whose data points lie on or near a line is called the **correlation coefficient**, designated by r. **Figure 5.34** shows scatter plots and correlation coefficients. Variables are **positively correlated** if they tend to increase or decrease together, as in **Figure 5.34(a)**, **(b)**, and **(c)**. By contrast, variables are **negatively correlated** if one variable tends to decrease while the other increases, as in **Figure 5.34(e)**, **(f)**, and **(g)**. **Figure 5.34** illustrates that a correlation coefficient, r, is a number between -1 and 1, inclusive. **Figure 5.34(a)** shows a value of 1. This indicates a **perfect positive correlation** in which all points in the scatter plot lie precisely on the regression line that rises from left to right. **Figure 5.34(g)** shows a value of -1. This indicates a **perfect negative correlation** in which all points in the scatter plot lie precisely on the regression line that falls from left to right.

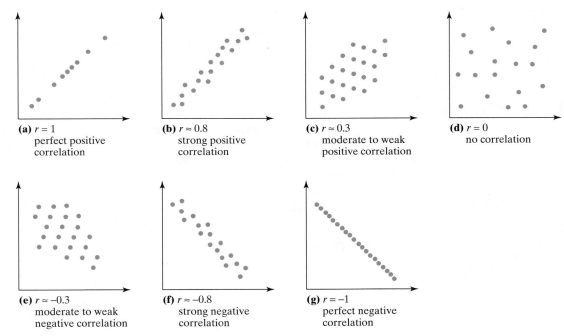

Figure 5.34 Scatter plots and correlation coefficients

Take another look at **Figure 5.34**. If r is between 0 and 1, as in **(b)** and **(c)**, the two variables are positively correlated, but not perfectly. Although all the data points will not lie on the regression line, as in **(a)**, an increase in one variable tends to be accompanied by an increase in the other.

Negative correlations are also illustrated in **Figure 5.34**. If r is between 0 and -1, as in **(e)** and **(f)**, the two variables are negatively correlated, but not perfectly. Although all the data points will not lie on the regression line, as in **(g)**, an increase in one variable tends to be accompanied by a decrease in the other.

Example 4

Math provides an objective way to look at social and political issues. For example, the points in the scatter plot in **Figure 5.35** show the number of firearms per 100 persons and the number of deaths per 100,000 persons for the ten industrialized countries with the highest death rates. The correlation coefficient for the data is 0.89. What does this correlation coefficient indicate about the strength and direction of the relationship between firearms per 100 persons and deaths per 100,000 persons?

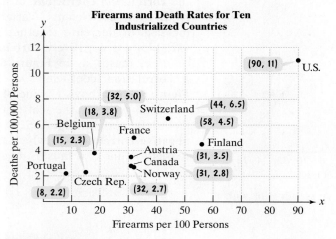

Figure 5.35
Source: International Action Network on Small Arms

Solution

The correlation coefficient $r = 0.89$ is close to 1, indicating a strong, but not perfect, positive correlation. There is a moderately strong positive relationship between the number of firearms per 100 persons and the number of deaths per 100,000 persons. As the number of firearms per 100 persons increases, the number of deaths per 100,000 persons also tends to increase.

> **Checkpoint 4** The points in the scatter plot in **Figure 5.36** show the number of hours per week spent watching TV, by annual income. The correlation coefficient for the data is −0.86. What does this correlation coefficient indicate about the strength and direction of the relationship between annual income and hours per week watching TV?

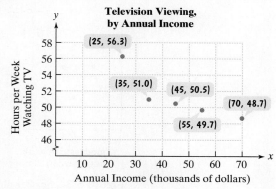

Figure 5.36
Source: Nielsen Media Research

How to Obtain the Correlation Coefficient and the Equation of the Regression Line

The easiest way to find the correlation coefficient and the equation of the regression line is to use a graphing or statistical calculator. Graphing calculators have statistical menus that enable you to enter the x and y data items for the variables. Based on this information, you can instruct the calculator to display a scatter plot, the equation of the regression line, and the correlation coefficient.

Shown below are the data involving the number of years of school, *x*, completed by ten randomly selected people and their scores on a test measuring prejudice, *y*. Recall that higher scores on the measure of prejudice (1 to 10) indicate greater levels of prejudice.

Respondent	A	B	C	D	E	F	G	H	I	J
Years of education (*x*)	12	5	14	13	8	10	16	11	12	4
Score on prejudice test (*y*)	1	7	2	3	5	4	1	2	3	10

After entering the *x* and *y* data items into a graphing calculator, we obtain the scatter plot of the data and the regression line, shown in **Figure 5.37**.

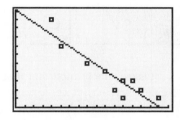

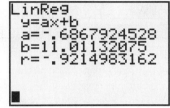

Figure 5.37	Figure 5.38

Also displayed by the calculator is the regression line's equation and the correlation coefficient, *r*, shown in **Figure 5.38**. The value of *r*, approximately −0.92, is fairly close to −1 and indicates a strong negative correlation. This means that the more education a person has, the less prejudiced that person is (based on scores on the test measuring levels of prejudice). Observe that the slope of the line is approximately −0.69. With a negative correlation coefficient, it makes sense that the slope of the regression line is negative. This line falls from left to right, indicating a negative correlation.

Blitzer Bonus

Cigarettes and Lung Cancer

This scatter plot shows a relationship between cigarette consumption among males and deaths due to lung cancer per million males. The data are from 11 countries and date back to a 1964 report by the U.S. Surgeon General. The scatter plot can be modeled by a line whose slope indicates an increasing death rate from lung cancer with increased cigarette consumption. At that time, the tobacco industry argued that in spite of this regression line, tobacco use is not the cause of cancer. Recent data do, indeed, show a causal effect between tobacco use and numerous diseases.

Source: Smoking and Health, Washington, D.C., 1964

Achieving Success

FoxTrot

FOXTROT © 2002 Bill Amend

Because concepts in mathematics build on each other, **it is extremely important that you complete all homework assignments**. This requires more than attempting a few of the assigned exercises. When it comes to assigned homework, you need to do four things and to do these things consistently throughout any math course:

1. Attempt to work *every* assigned problem.
2. Check your answers.
3. Correct your errors.
4. Ask for help with the problems you have attempted, but do not understand.

Exercise Set 5.3

Concept and Vocabulary Exercises

In Exercises 1–4, fill in each blank so that the resulting statement is true.

1. The point-slope form of the equation of a line is _____, where (x_1, y_1) represents _____ and m represents the line's _____.

2. A set of points representing data is called a/an _____.

3. The line that best fits a set of points is called a/an _____.

4. A measure that is used to describe the strength and direction of a relationship between variables whose data points lie on or near a line is called the _____, ranging from $r = $ _____ to $r = $ _____.

In Exercises 5–8, determine whether each statement is true or false. If the statement is false, make the necessary change(s) to produce a true statement.

5. The line whose equation is $y - 3 = 7(x + 2)$ passes through $(-3, 2)$.

6. The point-slope form can be applied to obtain the equation of the line through the points $(2, -5)$ and $(2, 6)$.

7. If $r = 0$, there is no correlation between two variables.

8. If $r = 1$, changes in one variable cause changes in the other variable.

Respond to Exercises 9–14 using verbal or written explanations.

9. Describe how to write the equation of a line if its slope and a point on the line are known.

10. Describe how to write the equation of a line if two points on the line are known.

11. What is a scatter plot?

12. How does a scatter plot indicate that two variables are correlated?

13. What is meant by a regression line?

14. When all points in a scatter plot fall on the regression line, what is the value of the correlation coefficient? Describe what this means.

In Exercises 15–18, determine whether each statement makes sense or does not make sense, and explain your reasoning.

15. I can use any two points in a scatter plot with different x-coordinates to write the point-slope form of the equation of the line through those points. However, the other data points in the scatter plot might not fall on, or even near, this line.

16. There is a strong negative correlation, -0.91, for data relating the percentage of people in various countries who are literate and the percentage who are undernourished. I can conclude that an increase in literacy causes a decrease in undernourishment.

17. I'm working with a data set for which the correlation coefficient and the slope of the regression line have opposite signs.

18. I read that there is a correlation of 0.72 between IQ scores of identical twins reared apart, so I would expect a significantly lower correlation, approximately 0.52, between IQ scores of identical twins reared together.

Practice Exercises

Write the point-slope form of the equation of the line satisfying each of the conditions in Exercises 19–46. Then use the point-slope form of the equation to write the slope-intercept form of the equation.

19. Slope = 3, passing through $(2, 5)$

20. Slope = 6, passing through $(3, 1)$

21. Slope = 5, passing through $(-2, 6)$

22. Slope = 7, passing through $(-4, 9)$

23. Slope = -8, passing through $(-3, -2)$

24. Slope = -4, passing through $(-5, -2)$

25. Slope = -12, passing through $(-8, 0)$

26. Slope = -11, passing through $(0, -3)$

27. Slope = -1, passing through $\left(-\frac{1}{2}, -2\right)$

28. Slope = -1, passing through $\left(-4, -\frac{1}{4}\right)$

29. Slope = $\frac{1}{2}$, passing through the origin

30. Slope = $\frac{1}{3}$, passing through the origin

31. Slope = $-\frac{2}{3}$, passing through $(6, -2)$

32. Slope = $-\frac{3}{5}$, passing through $(10, -4)$

33. Passing through $(1, 2)$ and $(5, 10)$

34. Passing through $(3, 5)$ and $(8, 15)$

35. Passing through $(-3, 0)$ and $(0, 3)$

36. Passing through $(-2, 0)$ and $(0, 2)$

37. Passing through $(-3, -1)$ and $(2, 4)$

38. Passing through $(-2, -4)$ and $(1, -1)$

39. Passing through $(-4, -1)$ and $(3, 4)$

40. Passing through $(-6, 1)$ and $(2, -5)$

41. Passing through $(-3, -1)$ and $(4, -1)$

42. Passing through $(-2, -5)$ and $(6, -5)$

43. Passing through $(2, 4)$ with x-intercept = -2

44. Passing through $(1, -3)$ with x-intercept = -1

45. x-intercept = $-\frac{1}{2}$ and y-intercept = 4

46. x-intercept = 4 and y-intercept = -2

In Exercises 47–50, make a scatter plot for the given data. Use the scatter plot to describe whether or not the variables appear to be related.

47.

x	1	6	4	3	7	2
y	2	5	3	3	4	1

48.

x	2	1	6	3	4
y	4	5	10	8	9

49.

x	8	6	1	5	4	10	3
y	2	4	10	5	6	2	9

50.

x	4	5	2	1
y	1	3	5	4

Use the scatter plots shown, labeled (a)–(f), to solve Exercises 51–54.

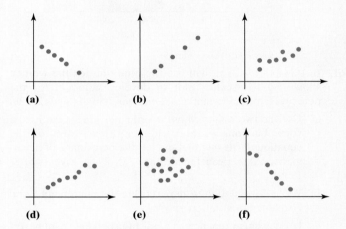

(a) **(b)** **(c)**

(d) **(e)** **(f)**

51. Which scatter plot indicates a perfect negative correlation?

52. Which scatter plot indicates a perfect positive correlation?

53. In which scatter plot is $r = 0.9$?

54. In which scatter plot is $r = 0.01$?

Practice Plus

Two nonintersecting lines that lie in the same rectangular coordinate system are parallel. If two lines are parallel, the ratio of the vertical change to the horizontal change is the same for both lines. Because two parallel lines have the same "steepness," they must have the same slope.

> If two nonvertical lines are parallel, then they have the same slope.

In Exercises 55–60, use this information to write equations in point-slope form and slope-intercept form of the line satisfying the given conditions.

55. The line passes through $(-3, 2)$ and is parallel to the line whose equation is $y = 4x + 1$.

56. The line passes through $(5, -3)$ and is parallel to the line whose equation is $y = 2x + 1$.

57. The line passes through $(-1, -5)$ and is parallel to the line whose equation is $3x + y = 6$.

58. The line passes through $(-4, -7)$ and is parallel to the line whose equation is $6x + y = 8$.

59. The line has an x-intercept at -4 and is parallel to the line containing $(3, 1)$ and $(2, 6)$.

60. The line has an x-intercept at -6 and is parallel to the line containing $(4, -3)$ and $(2, 2)$.

Application Exercises

Americans are getting married later in life, or not getting married at all. In 2008, nearly half of Americans ages 25 through 29 were unmarried. The bar graph shows the percentage of never-married men and women in this age group for four selected years. The data are displayed as two sets of four points each, one scatter plot for the percentage of never-married American men and one for the percentage of never-married American women. Also shown for each scatter plot is a line that passes through or near the four points. Use these lines to solve Exercises 61–62.

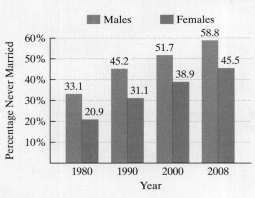

Percentage of United States Population Never Married, Ages 25–29

Source: U.S. Census Bureau

61. In this exercise, you will use the blue line for the women shown on the scatter plot to develop a model for the percentage of never-married American females ages 25–29.

   a. Use the two points whose coordinates are shown by the voice balloons to find the point-slope form of the equation of the line that models the percentage of never-married American females ages 25–29, y, x years after 1980.

   b. Write the equation from part (a) in slope-intercept form. Use function notation.

   c. Use the linear function to predict the percentage of never-married American females, ages 25–29, in 2020.

62. In this exercise, you will use the red line for the men shown on the scatter plot to develop a model for the percentage of never-married American males ages 25–29.

   a. Use the two points whose coordinates are shown by the voice balloons to find the point-slope form of the equation of the line that models the percentage of never-married American males ages 25–29, y, x years after 1980.

   b. Write the equation from part (a) in slope-intercept form. Use function notation.

   c. Use the linear function to predict the percentage of never-married American males, ages 25–29, in 2015.

Just as money doesn't buy happiness for individuals, the two don't necessarily go together for countries either. However, the scatter plot does show a relationship between a country's annual per capita income and the percentage of people in that country who call themselves "happy." Use the scatter plot to determine whether each of the statements in Exercises 63–70 is true or false.

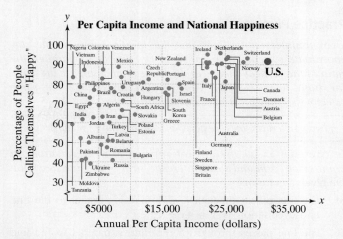

Source: Richard Layard, *Happiness: Lessons from a New Science*, Penguin, 2005.

63. There is no correlation between per capita income and the percentage of people who call themselves "happy."

64. There is an almost-perfect positive correlation between per capita income and the percentage of people who call themselves "happy."

65. There is a positive correlation between per capita income and the percentage of people who call themselves "happy."

66. As per capita income decreases, the percentage of people who call themselves "happy" also tends to decrease.

67. The country with the lowest per capita income has the least percentage of people who call themselves "happy."

68. The country with the highest per capita income has the greatest percentage of people who call themselves "happy."

69. A reasonable estimate of the correlation coefficient for the data is 0.8.

70. A reasonable estimate of the correlation coefficient for the data is -0.3.

Critical Thinking Exercises

For the pairs of quantities in Exercises 71–74, describe whether a scatter plot will show a positive correlation, a negative correlation, or no correlation. If there is a correlation, is it strong, moderate, or weak? Explain your answers.

71. Height and weight

72. Number of days absent and grade in a course

73. Height and grade in a course

74. Hours of video games played and grade in a course

75. Give an example of two variables with a strong positive correlation and explain why this is so.

76. Give an example of two variables with a strong negative correlation and explain why this is so.

77. Give an example of two variables with a strong correlation, where neither variable is the cause of the other.

Technology Exercise

78. Consider the following data:

Literacy and Hunger

| Country | Percentage Who Are | |
	Literate x	Undernourished y
Cuba	100	2
Egypt	71	4
Ethiopia	36	46
Grenada	96	7
Italy	98	2
Jamaica	80	9
Jordan	91	6
Pakistan	50	24
Russia	99	3
Togo	53	24
Uganda	67	19

Source: The Penguin State of the World Atlas, 2008

a. Make a scatter plot for the data.

b. Use the scatter plot to determine the correlation coefficient that best describes the relationship between the variables. Select from the following options:

$r = 1; r = 0.9; r = 0.4; r = 0; r = -0.4; r = -0.9; r = -1.$

c. Enter the x and y data items into a graphing calculator and obtain the scatter plot of the data and the graph of the regression line.

d. Use your calculator to obtain the regression line's equation and the correlation coefficient. Round all values to two decimal places. Is the value of r consistent with your answer from part (b)?

e. Use the equation of the regression line from part (d) to answer the following question: What percentage of people, to the nearest percent, can we anticipate are undernourished in a country where 60% of the people are literate?

Group Exercise

79. The group should select two variables, related to people in your high school, that it believes have a strong positive or negative correlation. Once these variables have been determined,

a. Collect at least 30 ordered pairs of data (x, y) from a sample of people in your school.

b. Draw a scatter plot for the data collected.

c. Does the scatter plot indicate a positive correlation, a negative correlation, or no relationship between the variables?

d. Use a graphing calculator to determine r. Does the value of r reinforce the impression conveyed by the scatter plot?

e. Use a graphing calculator to determine the equation of the regression line.

f. Use the regression line's equation to make a prediction about a y-value given an x-value.

g. Are the results of this project consistent with the group's original belief about the correlation between the variables, or are there some surprises in the data collected?

Systems of Linear Equations in Two Variables

5.4

Objectives

1. Decide whether an ordered pair is a solution of a linear system.

2. Solve linear systems by graphing.

3. Solve linear systems by substitution.

4. Solve linear systems by addition.

5. Identify systems that do not have exactly one ordered-pair solution.

6. Solve problems using systems of linear equations.

In high school, some students develop the habit of procrastinating. A study of college students showed that students who procrastinate typically have more symptoms of physical illness. Researchers identified college students who generally were procrastinators or non-procrastinators. The students were asked to report throughout the semester how many symptoms of physical illness they had experienced. **Figure 5.39** shows that by late in the semester, all students experienced increases in symptoms. Early in the semester, procrastinators reported fewer symptoms, but late in the semester, as work came due, they reported more symptoms than their nonprocrastinating peers.

The data in **Figure 5.39** can be analyzed using a pair of linear models in two variables. The figure shows that by week 6, both groups reported the same number of symptoms of illness, an average of approximately 3.5 symptoms per group. In this section, you will learn two algebraic methods, called

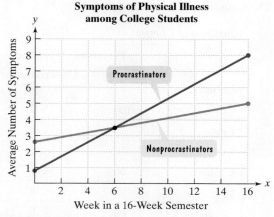

Figure 5.39
Source: Gerrig and Zimbardo, *Psychology and Life*, 18th Edition. Allyn and Bacon, 2008.

substitution and *addition*, that will reinforce this graphic observation, verifying (6, 3.5) as the point of intersection.

Systems of Linear Equations and Their Solutions

1. Decide whether an ordered pair is a solution of a linear system.

We have seen that all equations in the form $Ax + By = C$, A and B not both zero, are straight lines when graphed. Two such equations are called a **system of linear equations** or a **linear system**. A **solution to a system of linear equations in two variables** is an ordered pair that satisfies both equations in the system. For example, $(3, 4)$ satisfies the system

$$\begin{cases} x + y = 7 & \text{(3 + 4 is, indeed, 7.)} \\ x - y = -1. & \text{(3 - 4 is, indeed, -1.)} \end{cases}$$

Thus, $(3, 4)$ satisfies both equations and is a solution of the system. The solution can be described by saying that $x = 3$ and $y = 4$. The solution can also be described using set notation. The solution set of the system is $\{(3, 4)\}$—that is, the set consisting of the ordered pair $(3, 4)$.

Example 1 **Determining Whether an Ordered Pair Is a Solution of a System**

Determine whether $(1, 2)$ is a solution of the system:

$$\begin{cases} 2x - 3y = -4 \\ 2x + y = 4. \end{cases}$$

Solution

Because 1 is the x-coordinate and 2 is the y-coordinate of $(1, 2)$, we replace x with 1 and y with 2.

$$2x - 3y = -4 \qquad\qquad\qquad 2x + y = 4$$
$$2(1) - 3(2) \stackrel{?}{=} -4 \qquad\qquad 2(1) + 2 \stackrel{?}{=} 4$$
$$2 - 6 \stackrel{?}{=} -4 \qquad\qquad\qquad 2 + 2 \stackrel{?}{=} 4$$
$$-4 = -4, \quad \text{true} \qquad\qquad\qquad 4 = 4, \quad \text{true}$$

The pair $(1, 2)$ satisfies both equations: It makes each equation true. Thus, the pair is a solution of the system.

✓ **Checkpoint 1** Determine whether $(-4, 3)$ is a solution of the system:

$$\begin{cases} x + 2y = 2 \\ x - 2y = 6. \end{cases}$$

A system of linear equations can have exactly one solution, no solution, or infinitely many solutions. We begin with solving systems that have exactly one solution.

Solving Linear Systems by Graphing

2 Solve linear systems by graphing.

The solution to a system of linear equations can be found by graphing both of the equations in the same rectangular coordinate system. For a system with one solution, **the coordinates of the point of intersection of the lines is the system's solution**.

Example 2 **Solving a Linear System by Graphing**

Solve by graphing:

$$\begin{cases} x + 2y = 2 \\ x - 2y = 6. \end{cases}$$

Solution

We find the solution by graphing both $x + 2y = 2$ and $x - 2y = 6$ in the same rectangular coordinate system. We will use intercepts to graph each equation.

$$x + 2y = 2$$

x-intercept: Set $y = 0$.	**y-intercept: Set $x = 0$.**
$x + 2 \cdot 0 = 2$	$0 + 2y = 2$
$x = 2$	$2y = 2$
	$y = 1$

The line passes through $(2, 0)$.   The line passes through $(0, 1)$.

We graph $x + 2y = 2$ as a blue line in **Figure 5.40**.

$$x - 2y = 6$$

x-intercept: Set $y = 0$.	**y-intercept: Set $x = 0$.**
$x - 2 \cdot 0 = 6$	$0 - 2y = 6$
$x = 6$	$-2y = 6$
	$y = -3$

The line passes through $(6, 0)$.   The line passes through $(0, -3)$.

We graph $x - 2y = 6$ as a red line in **Figure 5.40**.

The system is graphed in **Figure 5.40**. To ensure that the graph is accurate, check the coordinates of the intersection point, $(4, -1)$, in both equations.

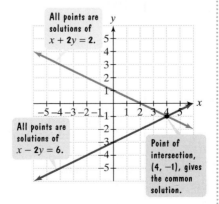

Figure 5.40 Visualizing a system's solution

Study Tip

When solving linear systems by graphing, neatly drawn graphs are essential for determining points of intersection.

- Use rectangular coordinate graph paper.
- Use a ruler or straightedge.
- Use a pencil with a sharp point.

We replace x with 4 and y with -1.

$$x + 2y = 2$$
$$4 + 2(-1) \overset{?}{=} 2$$
$$4 + (-2) \overset{?}{=} 2$$
$$2 = 2, \quad \text{true}$$

$$x - 2y = 6$$
$$4 - 2(-1) \overset{?}{=} 6$$
$$4 - (-2) \overset{?}{=} 6$$
$$4 + 2 \overset{?}{=} 6$$
$$6 = 6, \quad \text{true}$$

The pair $(4, -1)$ satisfies both equations—that is, it makes each equation true. This verifies that the system's solution set is $\{(4, -1)\}$.

Checkpoint 2 Solve by graphing:

$$\begin{cases} 2x + 3y = 6 \\ 2x + y = -2. \end{cases}$$

Solving Linear Systems by the Substitution Method

3 Solve linear systems by substitution.

Finding the solution to a linear system by graphing equations may not be easy to do. For example, a solution of $\left(-\frac{2}{3}, \frac{157}{29}\right)$ would be difficult to "see" as an intersection point on a graph.

Let's consider a method that does not depend on finding a system's solution visually: the substitution method. This method involves converting the system to one equation in one variable by an appropriate substitution.

Study Tip

In step 1, you can choose which variable to isolate in which equation. If possible, solve for a variable whose coefficient is 1 or -1 to avoid working with fractions.

Solving Linear Systems by Substitution

1. Solve either of the equations for one variable in terms of the other. (If one of the equations is already in this form, you can skip this step.)
2. Substitute the expression found in step 1 into the *other* equation. This will result in an equation in one variable.
3. Solve the equation containing one variable.
4. Back-substitute the value found in step 3 into the equation from step 1. Simplify and find the value of the remaining variable.
5. Check the proposed solution in both of the system's given equations.

Example 3 Solving a System by Substitution

Solve by the substitution method:

$$\begin{cases} y = -x - 1 \\ 4x - 3y = 24. \end{cases}$$

Solution

Step 1 Solve either of the equations for one variable in terms of the other. This step has already been done for us. The first equation, $y = -x - 1$, is solved for y in terms of x.

Step 2 Substitute the expression from step 1 into the other equation. We substitute the expression $-x - 1$ for y into the other equation:

$$y = \boxed{-x - 1} \qquad 4x - 3\,\boxed{y} = 24 \quad \text{Substitute } -x - 1 \text{ for } y.$$

This gives us an equation in one variable, namely
$$4x - 3(-x - 1) = 24.$$
The variable y has been eliminated.

Step 3   Solve the resulting equation containing one variable.

$4x - 3(-x - 1) = 24$	This is the equation containing one variable.
$4x + 3x + 3 = 24$	Apply the distributive property.
$7x + 3 = 24$	Combine like terms.
$7x = 21$	Subtract 3 from both sides.
$x = 3$	Divide both sides by 7.

Step 4   Back-substitute the obtained value into the equation from step 1. We now know that the x-coordinate of the solution is 3. To find the y-coordinate, we back-substitute the x-value into the equation from step 1.

$y = -x - 1$	This is the equation from step 1.

Substitute 3 for x.

$y = -3 - 1$	
$y = -4$	Simplify.

With $x = 3$ and $y = -4$, the proposed solution is $(3, -4)$.

Step 5   Check. Check the proposed solution, $(3, -4)$, in both of the system's given equations. Replace x with 3 and y with -4.

$y = -x - 1$	$4x - 3y = 24$
$-4 \overset{?}{=} -3 - 1$	$4(3) - 3(-4) \overset{?}{=} 24$
$-4 = -4,$ true	$12 + 12 \overset{?}{=} 24$
	$24 = 24,$ true

The pair $(3, -4)$ satisfies both equations. The system's solution set is $\{(3, -4)\}$.

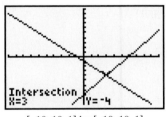
✓ **Checkpoint 3**   Solve by the substitution method:
$$\begin{cases} y = 3x - 7 \\ 5x - 2y = 8. \end{cases}$$

Example 4   Solving a System by Substitution

Solve by the substitution method:
$$\begin{cases} 5x - 4y = 9 \\ x - 2y = -3. \end{cases}$$

Solution

Step 1   Solve either of the equations for one variable in terms of the other. We begin by isolating one of the variables in either of the equations. By solving for x in the second equation, which has a coefficient of 1, we can avoid fractions.

$x - 2y = -3$	This is the second equation in the given system.
$x = 2y - 3$	Solve for x by adding 2y to both sides.

Step 2   Substitute the expression from step 1 into the other equation. We substitute $2y - 3$ for x in the first equation.

$$x = \boxed{2y - 3} \qquad 5\boxed{x} - 4y = 9$$

Substituting $2y - 3$ for x in $5x - 4y = 9$ gives us an equation in one variable, namely

$$5(2y - 3) - 4y = 9.$$

The variable x has been eliminated.

Step 3 Solve the resulting equation containing one variable.

$$5(2y - 3) - 4y = 9 \qquad \text{This is the equation containing one variable.}$$
$$10y - 15 - 4y = 9 \qquad \text{Apply the distributive property.}$$
$$6y - 15 = 9 \qquad \text{Combine like terms: } 10y - 4y = 6y.$$
$$6y = 24 \qquad \text{Add 15 to both sides.}$$
$$y = 4 \qquad \text{Divide both sides by 6.}$$

Step 4 Back-substitute the obtained value into the equation from step 1. Now that we have the y-coordinate of the solution, we back-substitute 4 for y in the equation $x = 2y - 3$.

$$x = 2y - 3 \qquad \text{Use the equation obtained in step 1.}$$
$$x = 2(4) - 3 \qquad \text{Substitute 4 for y.}$$
$$x = 8 - 3 \qquad \text{Multiply.}$$
$$x = 5 \qquad \text{Subtract.}$$

With $x = 5$ and $y = 4$, the proposed solution is $(5, 4)$.

Step 5 Check. Take a moment to show that $(5, 4)$ satisfies both given equations, $5x - 4y = 9$ and $x - 2y = -3$. The solution set is $\{(5, 4)\}$.

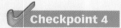 **Checkpoint 4**   Solve by the substitution method:

$$\begin{cases} 3x + 2y = -1 \\ x - y = 3. \end{cases}$$

Solving Linear Systems by the Addition Method

4 Solve linear systems by addition.

The substitution method is most useful if one of the given equations has an isolated variable. A third, and frequently the easiest, method for solving a linear system is the addition method. Like the substitution method, the addition method involves eliminating a variable and ultimately solving an equation containing only one variable. However, this time we eliminate a variable by adding the equations.

For example, consider the following system of linear equations:

$$\begin{cases} 3x - 4y = 11 \\ -3x + 2y = -7. \end{cases}$$

When we add these two equations, the x-terms are eliminated. This occurs because the coefficients of the x-terms, 3 and -3, are opposites (additive inverses) of each other:

$$\begin{cases} 3x - 4y = 11 \\ \underline{-3x + 2y = -7} \end{cases}$$

Add: $-2y = 4$ The sum is an equation in one variable.

$$y = -2 \qquad \text{Divide both sides by } -2 \text{ and solve for y.}$$

Now we can back-substitute -2 for y into one of the original equations to find x. It does not matter which equation we use: We will obtain the same value for x in either case. If we use either equation, we can show that $x = 1$ and the solution $(1, -2)$ satisfies both equations in the system.

When we use the addition method, we want to obtain two equations whose sum is an equation containing only one variable. The key step is to **obtain, for one of the variables, coefficients that differ only in sign**. To do this, we may need to multiply one or both equations by some nonzero number so that the coefficients of one of the variables, x or y, become opposites. Then when the two equations are added, this variable is eliminated.

Solving Linear Systems by Addition

1. If necessary, rewrite both equations in the form $Ax + By = C$.
2. If necessary, multiply either equation or both equations by appropriate nonzero numbers so that the sum of the x-coefficients or the sum of the y-coefficients is 0.
3. Add the equations in step 2. The sum is an equation in one variable.
4. Solve the equation in one variable.
5. Back-substitute the value obtained in step 4 into either of the given equations and solve for the other variable.
6. Check the solution in both of the original equations.

Example 5   Solving a System by the Addition Method

Solve by the addition method:

$$\begin{cases} 3x + 2y = 48 \\ 9x - 8y = -24. \end{cases}$$

Solution

Step 1   Rewrite both equations in the form $Ax + By = C$. Both equations are already in this form. Variable terms appear on the left and constants appear on the right.

Step 2   If necessary, multiply either equation or both equations by appropriate numbers so that the sum of the x-coefficients or the sum of the y-coefficients is 0. We can eliminate x or y. Let's eliminate x. Consider the terms in x in each equation, that is, $3x$ and $9x$. To eliminate x, we can multiply each term of the first equation by -3 and then add the equations.

$$\begin{cases} 3x + 2y = 48 \\ 9x - 8y = -24 \end{cases} \xrightarrow[\text{No change}]{\text{Multiply by } -3.} \begin{cases} -9x - 6y = -144 \\ \underline{9x - 8y = -24} \end{cases}$$

Step 3   Add the equations.   Add: $-14y = -168$

Step 4   Solve the equation in one variable. We solve $-14y = -168$ by dividing both sides by -14.

$$\frac{-14y}{-14} = \frac{-168}{-14} \qquad \text{Divide both sides by } -14.$$
$$y = 12 \qquad \text{Simplify.}$$

Step 5   Back-substitute and find the value for the other variable. We can back-substitute 12 for y into either one of the given equations. We'll use the first one.

$$\begin{aligned} 3x + 2y &= 48 && \text{This is the first equation in the given system.} \\ 3x + 2(12) &= 48 && \text{Substitute 12 for } y. \\ 3x + 24 &= 48 && \text{Multiply.} \\ 3x &= 24 && \text{Subtract 24 from both sides.} \\ x &= 8 && \text{Divide both sides by 3.} \end{aligned}$$

We found that $y = 12$ and $x = 8$. The proposed solution is $(8, 12)$.

Step 6   Check. Take a few minutes to show that $(8, 12)$ satisfies both of the original equations in the system. The solution set is $\{(8, 12)\}$.

Checkpoint 5 Solve by the addition method:

$$\begin{cases} 4x + 5y = 3 \\ 2x - 3y = 7. \end{cases}$$

Example 6   Solving a System by the Addition Method

Solve by the addition method:

$$\begin{cases} 7x = 5 - 2y \\ 3y = 16 - 2x. \end{cases}$$

Solution

Step 1   Rewrite both equations in the form $Ax + By = C$. We first arrange the system so that variable terms appear on the left and constants appear on the right. We obtain

$$7x + 2y = 5 \qquad \text{Add 2y to both sides of the first equation.}$$

$$2x + 3y = 16. \qquad \text{Add 2x to both sides of the second equation.}$$

Step 2   If necessary, multiply either equation or both equations by appropriate numbers so that the sum of the x-coefficients or the sum of the y-coefficients is 0. We can eliminate x or y. Let's eliminate y by multiplying the first equation by 3 and the second equation by -2.

$$\begin{cases} 7x + 2y = 5 \\ 2x + 3y = 16 \end{cases} \xrightarrow[\text{Multiply by } -2.]{\text{Multiply by 3.}} \begin{cases} 21x + 6y = 15 \\ \underline{-4x - 6y = -32} \end{cases}$$

Step 3   Add the equations.   Add: $17x + 0y = -17$

$$17x = -17$$

Step 4   Solve the equation in one variable. We solve $17x = -17$ by dividing both sides by 17.

$$\frac{17x}{17} = \frac{-17}{17} \qquad \text{Divide both sides by 17.}$$

$$x = -1 \qquad \text{Simplify.}$$

Step 5   Back-substitute and find the value for the other variable. We can back-substitute -1 for x into either one of the given equations. We'll use the second one.

$$3y = 16 - 2x \qquad \text{This is the second equation in the given system.}$$
$$3y = 16 - 2(-1) \qquad \text{Substitute } -1 \text{ for x.}$$
$$3y = 16 + 2 \qquad \text{Multiply.}$$
$$3y = 18 \qquad \text{Add.}$$
$$y = 6 \qquad \text{Divide both sides by 3.}$$

With $x = -1$ and $y = 6$, the proposed solution is $(-1, 6)$.

Step 6   Check. Take a moment to show that $(-1, 6)$ satisfies both given equations. The solution is $(-1, 6)$ and the solution set is $\{(-1, 6)\}$.

Checkpoint 6 Solve by the addition method:

$$\begin{cases} 3x = 2 - 4y \\ 5y = -1 - 2x. \end{cases}$$

Linear Systems Having No Solution or Infinitely Many Solutions

⑤ Identify systems that do not have exactly one ordered-pair solution.

We have seen that a system of linear equations in two variables represents a pair of lines. The lines either intersect at one point, are parallel, or are identical. Thus, there are three possibilities for the number of solutions to a system of two linear equations.

The Number of Solutions to a System of Two Linear Equations

The number of solutions to a system of two linear equations in two variables is given by one of the following. (See **Figure 5.41**.)

Number of Solutions	What This Means Graphically
Exactly one ordered-pair solution	The two lines intersect at one point.
No solution	The two lines are parallel.
Infinitely many solutions	The two lines are identical.

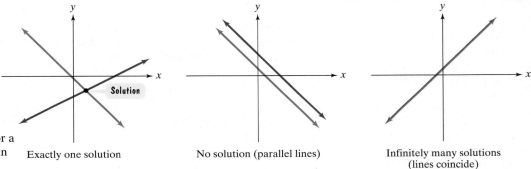

Figure 5.41 Possible graphs for a system of two linear equations in two variables

Exactly one solution   No solution (parallel lines)   Infinitely many solutions (lines coincide)

Example 7 A System with No Solution

Solve the system:

$$\begin{cases} 4x + 6y = 12 \\ 6x + 9y = 12. \end{cases}$$

Solution

Because no variable is isolated, we will use the addition method. To obtain coefficients of x that differ only in sign, we multiply the first equation by 3 and the second equation by -2.

$$\begin{cases} 4x + 6y = 12 \\ 6x + 9y = 12 \end{cases} \xrightarrow[\text{Multiply by } -2.}]{\text{Multiply by 3.}} \begin{cases} 12x + 18y = 36 \\ -12x - 18y = -24 \end{cases}$$

Add: $\quad 0 = 12$

There are no values of x and y for which $0 = 12$. No values of x and y satisfy $0x + 0y = 12$.

The false statement $0 = 12$ indicates that the system has no solution. The solution set is the empty set, $\varnothing$.

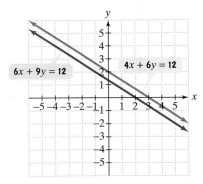

Figure 5.42 The graph of a system with no solution

$6x + 9y = 12$ $4x + 6y = 12$

The lines corresponding to the two equations in Example 7 are shown in **Figure 5.42**. The lines are parallel and have no point of intersection.

✓ **Checkpoint 7**   Solve the system:

$$\begin{cases} x + 2y = 4 \\ 3x + 6y = 13. \end{cases}$$

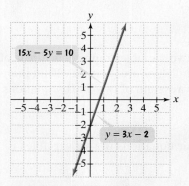

Figure 5.43 The graph of a system with infinitely many solutions

Although the system in Example 8 has infinitely many solutions, this does not mean that any ordered pair of numbers you can form will be a solution. The ordered pair (x, y) must satisfy one of the system's equations, $y = 3x - 2$ or $15x - 5y = 10$, and there are infinitely many such ordered pairs. Because the graphs are coinciding lines, the ordered pairs that are solutions of one of the equations are also solutions of the other equation.

6   Solve problems using systems of linear equations.

Example 8   A System with Infinitely Many Solutions

Solve the system:

$$\begin{cases} y = 3x - 2 \\ 15x - 5y = 10. \end{cases}$$

Solution

Because the variable y is isolated in $y = 3x - 2$, the first equation, we can use the substitution method. We substitute the expression for y into the second equation.

$$y = \boxed{3x - 2} \qquad 15x - 5\boxed{y} = 10 \quad \text{Substitute } 3x - 2 \text{ for } y.$$

$$15x - 5(3x - 2) = 10 \qquad \begin{array}{l}\text{The substitution results in an} \\ \text{equation in one variable.}\end{array}$$

$$15x - 15x + 10 = 10 \qquad \text{Apply the distributive property.}$$

$$10 = 10 \qquad \text{Simplify.}$$

This statement is true for all values of x and y.

In our final step, both variables have been eliminated and the resulting statement, $10 = 10$, is true. This true statement indicates that the system has infinitely many solutions. The solution set consists of all points (x, y) lying on either of the coinciding lines, $y = 3x - 2$ or $15x - 5y = 10$, as shown in **Figure 5.43**.

We express the solution set for the system in one of two equivalent ways:

$$\{(x, y) \mid y = 3x - 2\} \qquad \text{or} \qquad \{(x, y) \mid 15x - 5y = 10\}.$$

The set of all ordered pairs (x, y) such that $y = 3x - 2$

The set of all ordered pairs (x, y) such that $15x - 5y = 10$

✓ **Checkpoint 8**   Solve the system:

$$\begin{cases} y = 4x - 4 \\ 8x - 2y = 8. \end{cases}$$

Linear Systems Having No Solution or Infinitely Many Solutions

If both variables are eliminated when solving a system of linear equations by substitution or addition, one of the following is true:

1. There is no solution if the resulting statement is false.
2. There are infinitely many solutions if the resulting statement is true.

Modeling with Systems of Equations: Making Money (and Losing It)

What does every entrepreneur, from a kid selling lemonade to Donald Trump, want to do? Generate profit, of course. The profit made is the money taken in, or the revenue, minus the money spent, or the cost.

Revenue and Cost Functions

A company produces and sells x units of a product. Its **revenue** is the money generated by selling x units of the product. Its **cost** is the cost of producing x units of the product.

Revenue Function

$$R(x) = (\text{price per unit sold})x$$

Cost Function

$$C(x) = \text{fixed cost} + (\text{cost per unit produced})x$$

The point of intersection of the graphs of the revenue and cost functions is called the **break-even point**. The x-coordinate of the point reveals the number of units that a company must produce and sell so that money coming in, the revenue, is equal to money going out, the cost. The y-coordinate of the break-even point gives the amount of money coming in and going out. Example 9 illustrates the use of the substitution method in determining a company's break-even point.

Example 9   Finding a Break-Even Point

Technology is now promising to bring light, fast, and beautiful wheelchairs to millions of disabled people. A company is planning to manufacture these radically different wheelchairs. Fixed cost will be $500,000 and it will cost $400 to produce each wheelchair. Each wheelchair will be sold for $600.

a. Write the cost function, C, of producing x wheelchairs.

b. Write the revenue function, R, from the sale of x wheelchairs.

c. Determine the break-even point. Describe what it means.

Solution

a. The cost function is the sum of the fixed cost and variable cost.

$$\underset{\substack{\text{Fixed cost of}\\\text{\$500,000}}}{\boxed{}} \quad \text{plus} \quad \underset{\substack{\text{Variable cost: \$400 for}\\\text{each chair produced}}}{\boxed{}}$$

$$C(x) = 500{,}000 + 400x$$

b. The revenue function is the money generated from the sale of x wheelchairs. We are given that each wheelchair will be sold for $600.

$$\underset{\text{Revenue per chair, \$600, times}}{\boxed{}} \quad \underset{\text{the number of chairs sold}}{\boxed{}}$$

$$R(x) = 600x$$

c. The break-even point occurs where the graphs of C and R intersect. Thus, we find this point by solving the system

$$\begin{cases} C(x) = 500{,}000 + 400x \\ R(x) = 600x \end{cases} \quad \text{or} \quad \begin{cases} y = 500{,}000 + 400x \\ y = 600x. \end{cases}$$

Using substitution, we can substitute $600x$ for y in the first equation:

$$600x = 500{,}000 + 400x \qquad \text{Substitute 600x for y in } y = 500{,}000 + 400x.$$
$$200x = 500{,}000 \qquad \text{Subtract 400x from both sides.}$$
$$x = 2500 \qquad \text{Divide both sides by 200.}$$

Back-substituting 2500 for x in either of the system's equations (or functions), $C(x) = 500{,}000 + 400x$ or $R(x) = 600x$, we obtain

$$R(2500) = 600(2500) = 1{,}500{,}000.$$

$$\underset{\text{We used } R(x) = 600x.}{\boxed{}}$$

The break-even point is $(2500, 1{,}500{,}000)$. This means that the company will break even if it produces and sells 2500 wheelchairs. At this level, the money coming in is equal to the money going out: $1,500,000.

Figure 5.44 shows the graphs of the revenue and cost functions for the wheelchair business. Similar graphs and models apply no matter how small or large a business venture may be.

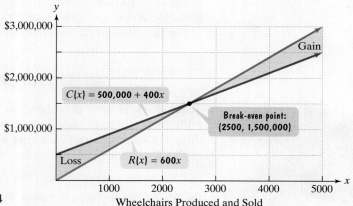

Figure 5.44

Wheelchairs Produced and Sold

The intersection point confirms that the company breaks even by producing and selling 2500 wheelchairs. Can you see what happens for $x < 2500$? The red cost graph lies above the blue revenue graph. The cost is greater than the revenue and the business is losing money. Thus, if they sell fewer than 2500 wheelchairs, the result is a *loss*. By contrast, look at what happens for $x > 2500$. The blue revenue graph lies above the red cost graph. The revenue is greater than the cost and the business is making money. Thus, if they sell more than 2500 wheelchairs, the result is a *gain*.

> **Checkpoint 9** A company that manufactures running shoes has a fixed cost of $300,000. Additionally, it costs $30 to produce each pair of shoes. They are sold at $80 per pair.
>
> **a.** Write the cost function, C, of producing x pairs of running shoes.
> **b.** Write the revenue function, R, from the sale of x pairs of running shoes.
> ▶ **c.** Determine the break-even point. Describe what it means.

The profit generated by a business is the money taken in (its revenue) minus the money spent (its cost). Thus, once a business has modeled its revenue and cost with a system of equations, it can determine its *profit function*, $P(x)$.

> ### The Profit Function
>
> The profit, $P(x)$, generated after producing and selling x units of a product is given by the **profit function**
>
> $$P(x) = R(x) - C(x),$$
>
> where R and C are the revenue and cost functions, respectively.

The profit function for the wheelchair business in Example 9 is

$$
\begin{aligned}
P(x) &= R(x) - C(x) \\
&= 600x - (500,000 + 400x) \\
&= 200x - 500,000.
\end{aligned}
$$

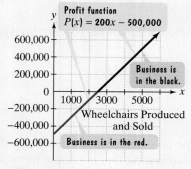

Figure 5.45

The graph of this profit function is shown in **Figure 5.45**. The red portion lies below the x-axis and shows a loss when fewer than 2500 wheelchairs are sold. The business is "in the red." The black portion lies above the x-axis and shows a gain when more than 2500 wheelchairs are sold. The wheelchair business is "in the black."

Achieving Success

Big Bucks for Teens

Can a teen start a business that is enjoyable and has a profit function in the black? We suggest reading *Quick Cash for Teens* (Sterling Publishing, 2009) by Peter G. Bielagus, a financial expert on teens and their money. Bielagus delivers the bottom line on becoming a teen entrepreneur, including plans for 101 businesses that are easy to start up with little money as you continue your education. While other teenagers were earning minimum wage, Bielagus enjoyed making $75 an hour as a teen entrepreneur. You can learn more by visiting www.quickcashforteens.com.

Exercise Set 5.4

Concept and Vocabulary Exercises

In Exercises 1–6, fill in each blank so that the resulting statement is true.

1. A solution to a system of linear equations in two variables is an ordered pair that _____.

2. When solving a system of linear equations by graphing, the system's solution is determined by using _____ of the lines representing the graphs of the system's equations.

3. Two methods for solving systems of linear equations without the use of graphs are _____ and _____.

4. If both variables are eliminated when solving a system of linear equations, there is _____ if the resulting statement is false and _____ if the resulting statement is true.

5. A company's _____ function is the money generated by selling x units of its product. The difference between this function and the company's cost function is called its _____ function.

6. A company has a graph that shows the money it generates by selling x units of its product. It also has a graph that shows its cost of producing x units of its product. The point of intersection of these graphs is called the company's _____.

In Exercises 7–10, determine whether each statement is true or false. If the statement is false, make the necessary change(s) to produce a true statement.

7. Unlike the graphing method, where solutions cannot be seen, the substitution method provides a way to visualize solutions as intersection points.

8. To solve the system

$$\begin{cases} 2x - y = 5 \\ 3x + 4y = 7 \end{cases}$$

by substitution, replace y in the second equation with $5 - 2x$.

9. If the two equations in a linear system are $5x - 3y = 7$ and $4x + 9y = 11$, multiplying the first equation by 4, the second by 5, and then adding equations will eliminate x.

10. If x can be eliminated by the addition method, y cannot be eliminated by using the original equations of the system.

Respond to Exercises 11–20 using verbal or written explanations.

11. What is a system of linear equations? Provide an example with your description.

12. What is the solution to a system of linear equations?

13. Explain how to solve a system of equations using graphing.

14. Explain how to solve a system of equations using the substitution method. Use $y = 3 - 3x$ and $3x + 4y = 6$ to illustrate your explanation.

15. Explain how to solve a system of equations using the addition method. Use $3x + 5y = -2$ and $2x + 3y = 0$ to illustrate your explanation.

16. What is the disadvantage to solving a system of equations using the graphing method?

17. When is it easier to use the addition method rather than the substitution method to solve a system of equations?

18. When using the addition or substitution method, how can you tell whether a system of linear equations has infinitely many solutions? What is the relationship between the graphs of the two equations?

19. When using the addition or substitution method, how can you tell whether a system of linear equations has no solution? What is the relationship between the graphs of the two equations?

20. Describe the break-even point for a business.

In Exercises 21–24, determine whether each statement makes sense or does not make sense, and explain your reasoning.

21. Even if a linear system has a solution set involving fractions, such as $\left\{ \left(\frac{8}{11}, \frac{43}{11} \right) \right\}$, I can use graphs to determine if the solution set is reasonable.

22. Each equation in a system of linear equations has infinitely many ordered-pair solutions.

23. Every system of linear equations has infinitely many ordered-pair solutions.

24. I find it easiest to use the addition method when one of the equations has a variable on one side by itself.

Practice Exercises

In Exercises 25–28, determine whether the given ordered pair is a solution of the system.

25. $(2, 3)$

$$\begin{cases} x + 3y = 11 \\ x - 5y = -13 \end{cases}$$

26. $(-3, 5)$

$$\begin{cases} 9x + 7y = 8 \\ 8x - 9y = -69 \end{cases}$$

27. $(2, 5)$
$$\begin{cases} 2x + 3y = 17 \\ x + 4y = 16 \end{cases}$$

28. $(8, 5)$
$$\begin{cases} 5x - 4y = 20 \\ 3y = 2x + 1 \end{cases}$$

In Exercises 29–36, solve each system by graphing. Check the coordinates of the intersection point in both equations.

29. $\begin{cases} x + y = 6 \\ x - y = 2 \end{cases}$

30. $\begin{cases} x + y = 2 \\ x - y = 4 \end{cases}$

31. $\begin{cases} 2x - 3y = 6 \\ 4x + 3y = 12 \end{cases}$

32. $\begin{cases} 4x + y = 4 \\ 3x - y = 3 \end{cases}$

33. $\begin{cases} y = x + 5 \\ y = -x + 3 \end{cases}$

34. $\begin{cases} y = x + 1 \\ y = 3x - 1 \end{cases}$

35. $\begin{cases} y = -x - 1 \\ 4x - 3y = 24 \end{cases}$

36. $\begin{cases} y = 3x - 4 \\ 2x + y = 1 \end{cases}$

In Exercises 37–48, solve each system by the substitution method. Be sure to check all proposed solutions.

37. $\begin{cases} x + y = 4 \\ y = 3x \end{cases}$

38. $\begin{cases} x + y = 6 \\ y = 2x \end{cases}$

39. $\begin{cases} x + 3y = 8 \\ y = 2x - 9 \end{cases}$

40. $\begin{cases} 2x - 3y = -13 \\ y = 2x + 7 \end{cases}$

41. $\begin{cases} x + 3y = 5 \\ 4x + 5y = 13 \end{cases}$

42. $\begin{cases} 3x + y = -18 \\ 5x - 2y = -8 \end{cases}$

43. $\begin{cases} 2x - y = -5 \\ x + 5y = 14 \end{cases}$

44. $\begin{cases} 2x + 3y = 11 \\ x - 4y = 0 \end{cases}$

45. $\begin{cases} 2x - y = 3 \\ 5x - 2y = 10 \end{cases}$

46. $\begin{cases} -x + 3y = 10 \\ 2x + 8y = -6 \end{cases}$

47. $\begin{cases} x + 8y = 6 \\ 2x + 4y = -3 \end{cases}$

48. $\begin{cases} -4x + y = -11 \\ 2x - 3y = 5 \end{cases}$

In Exercises 49–60, solve each system by the addition method. Be sure to check all proposed solutions.

49. $\begin{cases} x + y = 1 \\ x - y = 3 \end{cases}$

50. $\begin{cases} x + y = 6 \\ x - y = -2 \end{cases}$

51. $\begin{cases} 2x + 3y = 6 \\ 2x - 3y = 6 \end{cases}$

52. $\begin{cases} 3x + 2y = 14 \\ 3x - 2y = 10 \end{cases}$

53. $\begin{cases} x + 2y = 2 \\ -4x + 3y = 25 \end{cases}$

54. $\begin{cases} 2x - 7y = 2 \\ 3x + y = -20 \end{cases}$

55. $\begin{cases} 4x + 3y = 15 \\ 2x - 5y = 1 \end{cases}$

56. $\begin{cases} 3x - 7y = 13 \\ 6x + 5y = 7 \end{cases}$

57. $\begin{cases} 3x - 4y = 11 \\ 2x + 3y = -4 \end{cases}$

58. $\begin{cases} 2x + 3y = -16 \\ 5x - 10y = 30 \end{cases}$

59. $\begin{cases} 2x = 3y - 4 \\ -6x + 12y = 6 \end{cases}$

60. $\begin{cases} 5x = 4y - 8 \\ 3x + 7y = 14 \end{cases}$

In Exercises 61–68, solve by the method of your choice. Identify systems with no solution and systems with infinitely many solutions, using set notation to express their solution sets.

61. $\begin{cases} x = 9 - 2y \\ x + 2y = 13 \end{cases}$

62. $\begin{cases} 6x + 2y = 7 \\ y = 2 - 3x \end{cases}$

63. $\begin{cases} y = 3x - 5 \\ 21x - 35 = 7y \end{cases}$

64. $\begin{cases} 9x - 3y = 12 \\ y = 3x - 4 \end{cases}$

65. $\begin{cases} 3x - 2y = -5 \\ 4x + y = 8 \end{cases}$

66. $\begin{cases} 2x + 5y = -4 \\ 3x - y = 11 \end{cases}$

67. $\begin{cases} x + 3y = 2 \\ 3x + 9y = 6 \end{cases}$

68. $\begin{cases} 4x - 2y = 2 \\ 2x - y = 1 \end{cases}$

Practice Plus

Use the graphs of the linear functions to solve Exercises 69–70.

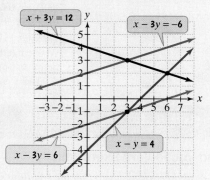

69. Write the linear system whose solution set is $\{(6, 2)\}$. Express each equation in the system in slope-intercept form.

70. Write the linear system whose solution set is $\varnothing$. Express each equation in the system in slope-intercept form.

In Exercises 71–72, solve each system for x and y, expressing either value in terms of a or b, if necessary. Assume that $a \neq 0$ and $b \neq 0$.

71. $\begin{cases} 5ax + 4y = 17 \\ ax + 7y = 22 \end{cases}$

72. $\begin{cases} 4ax + by = 3 \\ 6ax + 5by = 8 \end{cases}$

73. For the linear function $f(x) = mx + b$, $f(-2) = 11$ and $f(3) = -9$. Find m and b.

74. For the linear function $f(x) = mx + b$, $f(-3) = 23$ and $f(2) = -7$. Find m and b.

Application Exercises

The figure shows the graphs of the cost and revenue functions for a company that manufactures and sells small radios. Use the information in the figure to solve Exercises 75–80.

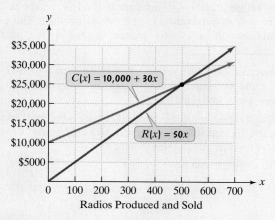

Radios Produced and Sold

75. How many radios must be produced and sold for the company to break even?

76. More than how many radios must be produced and sold for the company to have a profit?

77. Use the formulas shown in the voice balloons to find $R(200) - C(200)$. Describe what this means for the company.

78. Use the formulas shown in the voice balloons to find $R(300) - C(300)$. Describe what this means for the company.

79. a. Use the formulas shown in the voice balloons to write the company's profit function, P, from producing and selling x radios.

   b. Find the company's profit if 10,000 radios are produced and sold.

80. a. Use the formulas shown in the voice balloons to write the company's profit function, P, from producing and selling x radios.

   b. Find the company's profit if 20,000 radios are produced and sold.

Exercises 81–84 describe a number of business ventures. For each exercise,

  a. *Write the cost function, C.*

  b. *Write the revenue function, R.*

  c. *Determine the break-even point. Describe what it means.*

81. A company that manufactures small canoes has a fixed cost of $18,000. It costs $20 to produce each canoe. The selling price is $80 per canoe. (In solving this exercise, let x represent the number of canoes produced and sold.)

82. A company that manufactures bicycles has a fixed cost of $100,000. It costs $100 to produce each bicycle. The selling price is $300 per bike. (In solving this exercise, let x represent the number of bicycles produced and sold.)

83. You invest in a new play. The cost includes an overhead of $30,000, plus production costs of $2500 per performance. A sold-out performance brings in $3125. (In solving this exercise, let x represent the number of sold-out performances.)

84. You invested $30,000 and started a business writing greeting cards. Supplies cost 2¢ per card and you are selling each card for 50¢. (In solving this exercise, let x represent the number of cards produced and sold.)

85. We opened this section with a study showing that late in the semester, procrastinating students reported more symptoms of physical illness than their nonprocrastinating peers.

  a. At the beginning of the semester, procrastinators reported an average of 0.8 symptoms, increasing at a rate of 0.45 symptoms per week. Write a function that models the average number of symptoms, y, after x weeks.

  b. At the beginning of the semester, nonprocrastinators reported an average of 2.6 symptoms, increasing at a rate of 0.15 symptoms per week. Write a function that models the average number of symptoms, y, after x weeks.

  c. By which week in the semester did both groups report the same number of symptoms of physical illness? For that week, how many symptoms were reported by each group? How is this shown in **Figure 5.39** on page 334?

86. a. In 2004, approximately 140 million desktop PCs were sold worldwide, increasing at a rate of 4.8 million PCs per year. Write a function that models the number of desktop PCs sold worldwide, D, in millions, x years after 2004. (*Source:* iSuppli)

  b. In 2004, approximately 46 million laptop PCs were sold worldwide, increasing at a rate of 23.6 million PCs per year. Write a function that models the number of laptop PCs sold worldwide, L, in millions, x years after 2004. (*Source:* iSuppli)

  c. By which year did the sale of laptops equal the sale of desktops? How many PCs of each type were sold in that year?

  d. Use your answers from parts (a) through (c) to construct a graph that shows sales of desktop PCs and laptop PCs, in millions, from 2004 through 2010.

Use a system of linear equations to solve Exercises 87–92.

The graph shows the four candy bars with the highest fat content, representing grams of fat and calories in each bar. Exercises 87–90 are based on the graph.

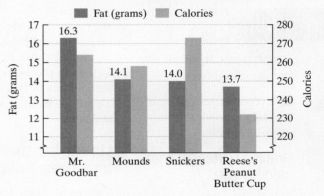

Candy Bars with the Highest Fat Content

Source: Krantz and Sveum, *The World's Worsts*, Harper Collins, 2005.

87. One Mr. Goodbar and two Mounds bars contain 780 calories. Two Mr. Goodbars and one Mounds bar contain 786 calories. Find the caloric content of each candy bar.

88. One Snickers bar and two Reese's Peanut Butter Cups contain 737 calories. Two Snickers bars and one Reese's Peanut Butter Cup contain 778 calories. Find the caloric content of each candy bar.

89. A collection of Halloween candy contains a total of five Mr. Goodbars and Mounds bars. Chew on this: The grams of fat in these candy bars exceed the daily maximum desirable fat intake of 70 grams by 7.1 grams. How many bars of each kind of candy are contained in the Halloween collection?

90. A collection of Halloween candy contains a total of 12 Snickers bars and Reese's Peanut Butter Cups. Chew on this: The grams of fat in these candy bars exceed twice the daily maximum desirable fat intake of 70 grams by 26.5 grams. How many bars of each kind of candy are contained in the Halloween collection?

91. A hotel has 200 rooms. Those with kitchen facilities rent for $100 per night and those without kitchen facilities rent for $80 per night. On a night when the hotel was completely occupied, revenues were $17,000. How many of each type of room does the hotel have?

92. A new restaurant is to contain two-seat tables and four-seat tables. Fire codes limit the restaurant's maximum occupancy to 56 customers. If the owners have hired enough servers to handle 17 tables of customers, how many of each kind of table should they purchase?

Critical Thinking Exercises

93. Write a system of equations having $\{(-2, 7)\}$ as a solution set. (More than one system is possible.)

94. One apartment is directly above a second apartment. The resident living downstairs calls his neighbor living above him and states, "If one of you is willing to come downstairs, we'll have the same number of people in both apartments." The upstairs resident responds, "We're all too tired to move. Why don't one of you come up here? Then we will have twice as many people up here as you've got down there." How many people are in each apartment?

95. A set of identical twins can only be distinguished by the characteristic that one always tells the truth and the other always lies. One twin tells you of a lucky number pair: "When I multiply my first lucky number by 3 and my second lucky number by 6, the addition of the resulting numbers produces a sum of 12. When I add my first lucky number and twice my second lucky number, the sum is 5." Which twin is talking?

5.5 Linear Inequalities in Two Variables

Objectives

1. Graph a linear inequality in two variables.

2. Use mathematical models involving linear inequalities.

3. Graph a system of linear inequalities.

We opened the chapter noting that the modern emphasis on thinness as the ideal body shape has been suggested as a major cause of eating disorders. In this section, as well as in the exercise set, we use systems of linear inequalities in two variables that will enable you to identify a healthy weight range for various heights and ages.

Linear Inequalities in Two Variables and Their Solutions

We have seen that equations in the form $Ax + By = C$, where A and B are not both zero, are straight lines when graphed. If we change the symbol $=$ to $>$, $<$, $\geq$, or $\leq$, we obtain a **linear inequality in two variables**. Some examples of linear inequalities in two variables are $x + y > 2$, $3x - 5y \leq 15$, and $2x - y < 4$.

A **solution of an inequality in two variables**, x and y, is an ordered pair of real numbers with the following property: When the x-coordinate is substituted for x and the y-coordinate is substituted for y in the inequality, we obtain a true statement. For example, $(3, 2)$ is a solution of the inequality $x + y > 1$. When 3 is substituted for x and 2 is substituted for y, we obtain the true statement $3 + 2 > 1$, or $5 > 1$. Because there are infinitely many pairs of numbers that have a sum greater than 1, the inequality $x + y > 1$ has infinitely many solutions. Each ordered-pair solution is said to **satisfy** the inequality. Thus, $(3, 2)$ satisfies the inequality $x + y > 1$.

The Graph of a Linear Inequality in Two Variables

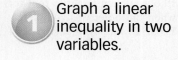

1. Graph a linear inequality in two variables.

We know that the graph of an equation in two variables is the set of all points whose coordinates satisfy the equation. Similarly, the **graph of an inequality in two variables** is the set of all points whose coordinates satisfy the inequality.

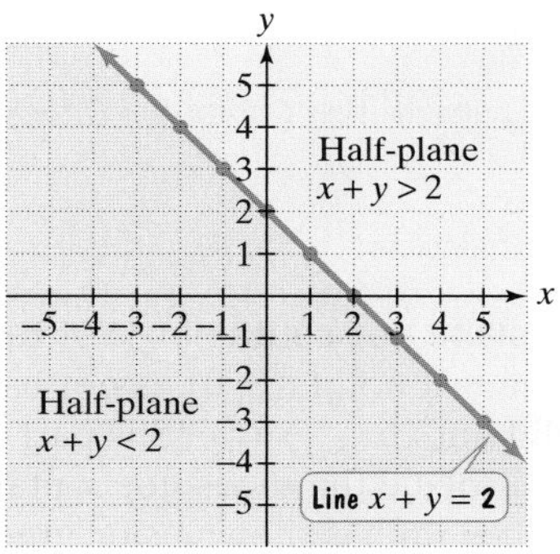

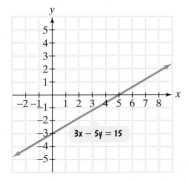

Figure 5.46

Let's use **Figure 5.46** to get an idea of what the graph of a linear inequality in two variables looks like. Part of the figure shows the graph of the linear equation $x + y = 2$. The line divides the points in the rectangular coordinate system into three sets. First, there is the set of points along the line satisfying $x + y = 2$. Next, there is the set of points in the green region above the line. Points in the green region satisfy the linear inequality $x + y > 2$. Finally, there is the set of points in the purple region below the line. Points in the purple region satisfy the linear inequality $x + y < 2$.

A **half-plane** is the set of all the points on one side of a line. In **Figure 5.46**, the green region is a half-plane. The purple region is also a half-plane. A half-plane is the graph of a linear inequality that involves $>$ or $<$. The graph of an inequality that involves $\geq$ or $\leq$ is a half-plane and a line. A solid line is used to show that a line is part of a graph. A dashed line is used to show that a line is not part of a graph.

Graphing a Linear Inequality in Two Variables

1. Replace the inequality symbol with an equal sign and graph the corresponding linear equation. Draw a solid line if the original inequality contains a $\leq$ or $\geq$ symbol. Draw a dashed line if the original inequality contains a $<$ or $>$ symbol.

2. Choose a test point from one of the half-planes. (Do not choose a point on the line.) Substitute the coordinates of the test point into the inequality.

3. If a true statement results, shade the half-plane containing this test point. If a false statement results, shade the half-plane not containing this test point.

Example 1   Graphing a Linear Inequality in Two Variables

Graph: $3x - 5y \geq 15$.

Solution

Step 1   Replace the inequality symbol by = and graph the linear equation. We need to graph $3x - 5y = 15$. We can use intercepts to graph this line.

We set $y = 0$ to find the x-intercept.	We set $x = 0$ to find the y-intercept.
$3x - 5y = 15$	$3x - 5y = 15$
$3x - 5 \cdot 0 = 15$	$3 \cdot 0 - 5y = 15$
$3x = 15$	$-5y = 15$
$x = 5$	$y = -3$

The x-intercept is 5, so the line passes through $(5, 0)$. The y-intercept is -3, so the line passes through $(0, -3)$. Using the intercepts, the line is shown in **Figure 5.47** as a solid line. This is because the inequality $3x - 5y \geq 15$ contains a $\geq$ symbol, in which equality is included.

Step 2   Choose a test point from one of the half-planes and not from the line. Substitute its coordinates into the inequality. The line $3x - 5y = 15$ divides the plane into three parts—the line itself and two half-planes. The points in one half-plane satisfy $3x - 5y > 15$. The points in the other half-plane satisfy $3x - 5y < 15$. We need to find which half-plane belongs to the solution of $3x - 5y \geq 15$. To do so, we test a point from either half-plane. The origin, $(0, 0)$, is the easiest point to test.

Figure 5.47 Preparing to graph $3x - 5y \geq 15$

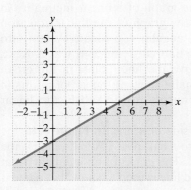

Figure 5.48 The graph of $3x - 5y \geq 15$

$$3x - 5y \geq 15 \qquad \text{This is the given inequality.}$$
$$3 \cdot 0 - 5 \cdot 0 \overset{?}{\geq} 15 \qquad \text{Test } (0, 0) \text{ by substituting 0 for } x \text{ and 0 for } y.$$
$$0 - 0 \overset{?}{\geq} 15 \qquad \text{Multiply.}$$
$$0 \geq 15 \qquad \text{This statement is false.}$$

Step 3 If a false statement results, shade the half-plane not containing the test point. Because 0 is not greater than or equal to 15, the test point, $(0, 0)$, is not part of the solution set. Thus, the half-plane below the solid line $3x - 5y = 15$ is part of the solution set. The solution set is the line and the half-plane that does not contain the point $(0, 0)$, indicated by shading this half-plane. The graph is shown using green shading and a blue line in **Figure 5.48**.

✓ Checkpoint 1 Graph: $2x - 4y \geq 8$.

When graphing a linear inequality, test a point that lies in one of the half-planes and *not on the line dividing the half-planes*. The test point $(0, 0)$ is convenient because it is easy to calculate when 0 is substituted for each variable. However, if $(0, 0)$ lies on the dividing line and not in a half-plane, a different test point must be selected.

Example 2 **Graphing a Linear Inequality in Two Variables**

Graph: $y > -\dfrac{2}{3}x$.

Solution

Step 1 Replace the inequality symbol by = and graph the linear equation. Because we are interested in graphing $y > -\frac{2}{3}x$, we begin by graphing $y = -\frac{2}{3}x$. We can use the slope and the y-intercept to graph this linear function.

$$y = -\frac{2}{3}x + 0$$

$$\text{Slope} = \frac{-2}{3} = \frac{\text{rise}}{\text{run}} \qquad y\text{-intercept} = 0$$

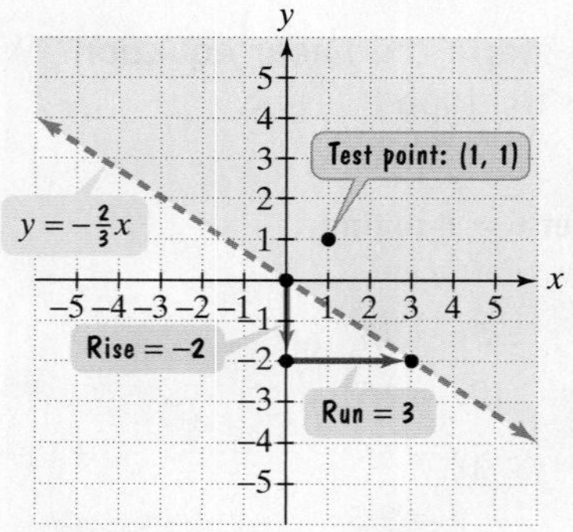

The y-intercept is 0, so the line passes through $(0, 0)$. Using the y-intercept and the slope, the line is shown in **Figure 5.49** as a dashed line. This is because the inequality $y > -\frac{2}{3}x$ contains a $>$ symbol, in which equality is not included.

Step 2 Choose a test point from one of the half-planes and not from the line. Substitute its coordinates into the inequality. We cannot use $(0, 0)$ as a test point because it lies on the line and not in a half-plane. Let's use $(1, 1)$, which lies in the half-plane above the line.

$$y > -\frac{2}{3}x \qquad \text{This is the given inequality.}$$
$$1 \overset{?}{>} -\frac{2}{3} \cdot 1 \qquad \text{Test } (1, 1) \text{ by substituting 1 for } x \text{ and 1 for } y.$$
$$1 > -\frac{2}{3} \qquad \text{This statement is true.}$$

Figure 5.49 The graph of $y > -\frac{2}{3}x$

Step 3 If a true statement results, shade the half-plane containing the test point. Because 1 is greater than $-\frac{2}{3}$, the test point, $(1, 1)$, is part of the solution set. All the points on the same side of the line $y = -\frac{2}{3}x$ as the point $(1, 1)$ are members of the solution set. The solution set is the half-plane that contains the point $(1, 1)$, indicated by shading this half-plane. The graph is shown using green shading and a dashed blue line in **Figure 5.49**.

Checkpoint 2   Graph: $y > -\dfrac{3}{4}x.$

Graphing Linear Inequalities without Using Test Points

You can graph inequalities in the form $y > mx + b$ or $y < mx + b$ without using test points. The inequality symbol indicates which half-plane to shade.

- If $y > mx + b$, shade the half-plane above the line $y = mx + b$.
- If $y < mx + b$, shade the half-plane below the line $y = mx + b$.

Turn back a page and observe how this is illustrated in **Figure 5.49**. The graph of $y > -\frac{2}{3}x$ is the half-plane above the line $y = -\frac{2}{3}x$.

It is also not necessary to use test points when graphing inequalities involving half-planes on one side of a vertical or a horizontal line.

Study Tip

Continue using test points to graph inequalities in the form $Ax + By > C$ or $Ax + By < C$. The graph of $Ax + By > C$ can lie above or below the line given by $Ax + By = C$, depending on the value of B. The same comment applies to the graph of $Ax + By < C$.

For the Vertical Line $x = a$:

- If $x > a$, shade the half-plane to the right of $x = a$.
- If $x < a$, shade the half-plane to the left of $x = a$.

For the Horizontal Line $y = b$:

- If $y > b$, shade the half-plane above $y = b$.
- If $y < b$, shade the half-plane below $y = b$.

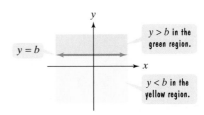

Example 3   **Graphing Inequalities without Using Test Points**

Graph each inequality in a rectangular coordinate system:

a. $y \leq -3$ **b.** $x > 2.$

Solution

a. $y \leq -3$ **b.** $x > 2$

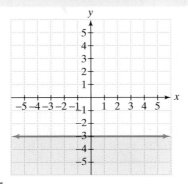

Graph $y = -3$, a horizontal line with y-intercept -3. The line is solid because equality is included in $y \leq -3$. Because of the less than part of $\leq$, shade the half-plane below the horizontal line.

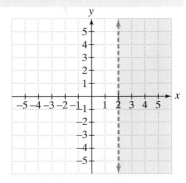

Graph $x = 2$, a vertical line with x-intercept 2. The line is dashed because equality is not included in $x > 2$. Because of $>$, the greater than symbol, shade the half-plane to the right of the vertical line.

Checkpoint 3   Graph each inequality in a rectangular coordinate system:

a. $y > 1$ **b.** $x \leq -2.$

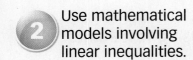

2 Use mathematical models involving linear inequalities.

Modeling with Systems of Linear Inequalities

Just as two or more linear equations make up a system of linear equations, two or more linear inequalities make up a **system of linear inequalities**. A **solution of a system of linear inequalities** in two variables is an ordered pair that satisfies each inequality in the system.

Example 4   **Does Your Weight Fit You?**

The latest guidelines, which apply to both men and women, give healthy weight ranges, rather than specific weights, for your height. **Figure 5.50** shows the healthy weight region for various heights for people between the ages of 19 and 34, inclusive.

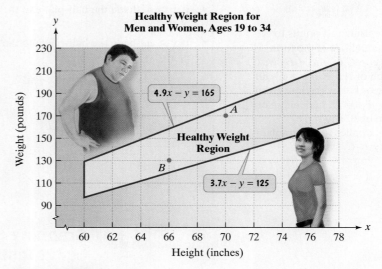

Figure 5.50
Source: U.S. Department of Health and Human Services

If x represents height, in inches, and y represents weight, in pounds, the healthy weight region in **Figure 5.50** can be modeled by the following system of linear inequalities:

$$\begin{cases} 4.9x - y \geq 165 \\ 3.7x - y \leq 125. \end{cases}$$

Show that point A in **Figure 5.50** is a solution of the system of inequalities that describes healthy weight.

Solution

Point A has coordinates $(70, 170)$. This means that if a person is 70 inches tall, or 5 feet 10 inches, and weighs 170 pounds, then that person's weight is within the healthy weight region. We can show that $(70, 170)$ satisfies the system of inequalities by substituting 70 for x and 170 for y in each inequality in the system.

$$4.9x - y \geq 165 \qquad\qquad 3.7x - y \leq 125$$
$$4.9(70) - 170 \geq 165 \qquad 3.7(70) - 170 \leq 125$$
$$343 - 170 \geq 165 \qquad 259 - 170 \leq 125$$
$$173 \geq 165, \quad \text{true} \qquad 89 \leq 125, \quad \text{true}$$

The coordinates $(70, 170)$ make each inequality true. Thus, $(70, 170)$ satisfies the system for the healthy weight region and is a solution of the system.

Checkpoint 4   Show that point B in **Figure 5.50** is a solution of the system of inequalities that describes healthy weight.

Graphing Systems of Linear Inequalities

3 Graph a system of linear inequalities.

The **solution set of a system of linear inequalities in two variables** is the set of all ordered pairs that satisfy each inequality in the system. Thus, to graph a system of inequalities in two variables, begin by graphing each individual inequality in the same rectangular coordinate system. Then find the region, if there is one, that is common to every graph in the system. This region of intersection gives a picture of the system's solution set.

Example 5   Graphing a System of Linear Inequalities

Graph the solution set of the system:

$$\begin{cases} x - y < 1 \\ 2x + 3y \geq 12. \end{cases}$$

Solution

Replacing each inequality symbol in $x - y < 1$ and $2x + 3y \geq 12$ with an equal sign indicates that we need to graph $x - y = 1$ and $2x + 3y = 12$. We can use intercepts to graph these lines.

$x - y = 1$

x-intercept: $x - 0 = 1$ *Set $y = 0$ in each equation.*
$\qquad\qquad\qquad x = 1$
The line passes through $(1, 0)$.

y-intercept: $0 - y = 1$ *Set $x = 0$ in each equation.*
$\qquad\qquad\qquad -y = 1$
$\qquad\qquad\qquad y = -1$
The line passes through $(0, -1)$.

$2x + 3y = 12$

x-intercept: $2x + 3 \cdot 0 = 12$
$\qquad\qquad\qquad 2x = 12$
$\qquad\qquad\qquad x = 6$
The line passes through $(6, 0)$.

y-intercept: $2 \cdot 0 + 3y = 12$
$\qquad\qquad\qquad 3y = 12$
$\qquad\qquad\qquad y = 4$
The line passes through $(0, 4)$.

Now we are ready to graph the solution set of the system of linear inequalities.

Graph $x - y < 1$. The blue line, $x - y = 1$, is dashed: Equality is not included in $x - y < 1$. Because $(0, 0)$ makes the inequality true $(0 - 0 < 1$, or $0 < 1$, is true), shade the half-plane containing $(0, 0)$ in yellow.

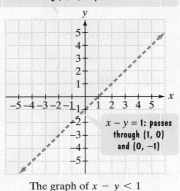

The graph of $x - y < 1$

Add the graph of $2x + 3y \geq 12$. The red line, $2x + 3y = 12$, is solid: Equality is included in $2x + 3y \geq 12$. Because $(0, 0)$ makes the inequality false $(2 \cdot 0 + 3 \cdot 0 \geq 12$, or $0 \geq 12$, is false), shade the half-plane not containing $(0, 0)$ using green vertical shading.

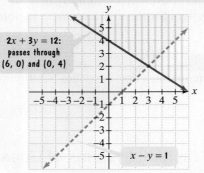

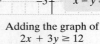

Adding the graph of $2x + 3y \geq 12$

The solution set of the system is graphed as the intersection (the overlap) of the two half-planes. This is the region in which the yellow shading and the green vertical shading overlap.

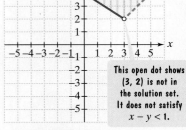

This open dot shows $(3, 2)$ is not in the solution set. It does not satisfy $x - y < 1$.

The graph of $x - y < 1$ and $2x + 3y \geq 12$

Checkpoint 5 Graph the solution set of the system:
$$\begin{cases} x + 2y > 4 \\ 2x - 3y \leq -6. \end{cases}$$

Example 6 Graphing a System of Linear Inequalities

Graph the solution set of the system:
$$\begin{cases} x \leq 4 \\ y > -2. \end{cases}$$

Solution

Graph $x \leq 4$. The blue vertical line, $x = 4$, is solid. Graph $x < 4$, the half-plane to the left of the blue line, using yellow shading.

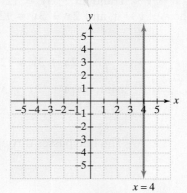

The graph of $x \leq 4$

Add the graph of $y > -2$. The red horizontal line, $y = -2$, is dashed. Graph $y > -2$, the half-plane above the dashed red line, using green vertical shading.

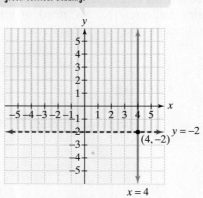

Adding the graph of $y > -2$

The solution set of the system is graphed as the intersection (the overlap) of the two half-planes. This is the region in which the yellow shading and the green vertical shading overlap.

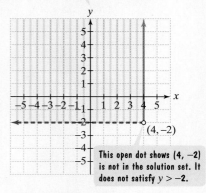

This open dot shows $(4, -2)$ is not in the solution set. It does not satisfy $y > -2$.

The graph of $x \leq 4$ and $y > -2$

Checkpoint 6 Graph the solution set of the system:
$$\begin{cases} x < 3 \\ y \geq -1. \end{cases}$$

Have you ever experienced learning *parts* of a subject with no idea at all of how those parts fit together? If you have had that experience, what you were missing was the logic of that subject matter. Thinking critically in math involves an understanding of how the parts fit together to form a coherent whole.

For example, the coherent whole in this chapter involved functions. We learned that y is a function of x if there is precisely one value of y for each value of x. We graphed functions and applied the vertical line test to identify graphs of functions. We used our understanding of functions and the definition of *slope* (change in y divided by change in x) to develop various forms for equations of lines:

$$Ax + By = C, \quad y = mx + b, \quad y - y_1 = m(x - x_1), \quad y = b, \quad x = a.$$

| Slope-Intercept Form | | Point-Slope Form | | Horizontal Line | | Vertical Line |

We used what we learned about these equations to solve systems of linear equations and systems of linear inequalities by graphing. We also used nongraphic methods, substitution and addition, to solve linear systems of equations.

As you study mathematics, be sure you understand how the various topics fit together with one another to form a coherent whole: a logical story.

Exercise Set 5.5

Concept and Vocabulary Exercises

In Exercises 1–4, fill in each blank so that the resulting statement is true.

1. The ordered pair $(5, 4)$ is a/an _____ of the inequality $x + y > 2$ because when 5 is substituted for _____ and 4 is substituted for _____, the true statement _____ is obtained.

2. The set of all points that satisfy an inequality is called the _____ of the inequality.

3. The set of all points on one side of a line is called a/an _____.

4. Two or more linear inequalities make up a/an _____ of linear inequalities.

In Exercises 5–8, determine whether each statement is true or false. If the statement is false, make the necessary change(s) to produce a true statement.

5. The ordered pair $(0, -3)$ satisfies $y > 2x - 3$.

6. The graph of $x < y + 1$ is the half-plane below the line $x = y + 1$.

7. In graphing $y \geq 4x$, a dashed line is used.

8. The graph of $x < 4$ is the half-plane to the left of the vertical line described by $x = 4$.

Respond to Exercises 9–14 using verbal or written explanations.

9. What is a half-plane?

10. What does a dashed line mean in the graph of an inequality?

11. What does a solid line mean in the graph of an inequality?

12. Explain how to graph $2x - 3y < 6$.

13. Compare the graphs of $3x - 2y > 6$ and $3x - 2y \leq 6$. Discuss similarities and differences between the graphs.

14. Describe how to solve a system of linear inequalities.

In Exercises 15–18, determine whether each statement makes sense or does not make sense, and explain your reasoning.

15. When graphing a linear inequality, I should always use $(0, 0)$ as a test point because it's easy to perform the calculations when 0 is substituted for each variable.

16. When graphing $3x - 4y < 12$, it's not necessary for me to graph the linear equation $3x - 4y = 12$ because the inequality contains a $<$ symbol, in which equality is not included.

17. The reason that systems of linear inequalities are appropriate for modeling healthy weight is because guidelines give healthy weight ranges, rather than specific weights, for various heights.

18. I graphed the solution set of $y \geq x + 2$ and $x \geq 1$ without using test points.

Practice Exercises

In Exercises 19–40, graph each linear inequality.

19. $x + y \geq 2$
20. $x - y \leq 1$
21. $3x - y \geq 6$
22. $3x + y \leq 3$
23. $2x + 3y > 12$
24. $2x - 5y < 10$
25. $5x + 3y \leq -15$
26. $3x + 4y \leq -12$
27. $2y - 3x > 6$
28. $2y - x > 4$
29. $y > \dfrac{1}{3}x$
30. $y > \dfrac{1}{4}x$

31. $y \leq 3x + 2$

32. $y \leq 2x - 1$

33. $y < -\dfrac{1}{4}x$

34. $y < -\dfrac{1}{3}x$

35. $x \leq 2$

36. $x \leq -4$

37. $y > -4$

38. $y > -2$

39. $y \geq 0$

40. $x \geq 0$

In Exercises 41–56, graph the solution set of each system of inequalities.

41. $\begin{cases} 3x + 6y \leq 6 \\ 2x + \ y \leq 8 \end{cases}$

42. $\begin{cases} x - y \geq 4 \\ x + y \leq 6 \end{cases}$

43. $\begin{cases} 2x + y < 3 \\ \ x - y > 2 \end{cases}$

44. $\begin{cases} x + \ y < 4 \\ 4x - 2y < 6 \end{cases}$

45. $\begin{cases} 2x + y < 4 \\ \ x - y > 4 \end{cases}$

46. $\begin{cases} 2x - y < 3 \\ \ x + y < 6 \end{cases}$

47. $\begin{cases} x \geq 2 \\ y \leq 3 \end{cases}$

48. $\begin{cases} x \geq 4 \\ y \leq 2 \end{cases}$

49. $\begin{cases} x \leq 5 \\ y > -3 \end{cases}$

50. $\begin{cases} x \leq 3 \\ y > -1 \end{cases}$

51. $\begin{cases} x - y \leq 1 \\ x \geq 2 \end{cases}$

52. $\begin{cases} 4x - 5y \geq -20 \\ x \geq -3 \end{cases}$

53. $\begin{cases} y > 2x - 3 \\ y < -x + 6 \end{cases}$

54. $\begin{cases} y < -2x + 4 \\ y < x - 4 \end{cases}$

55. $\begin{cases} x + 2y \leq 4 \\ y \geq x - 3 \end{cases}$

56. $\begin{cases} x + y \leq 4 \\ y \geq 2x - 4 \end{cases}$

Practice Plus

In Exercises 57–58, write each sentence as an inequality in two variables. Then graph the inequality.

57. The y-variable is at least 4 more than the product of -2 and the x-variable.

58. The y-variable is at least 2 more than the product of -3 and the x-variable.

In Exercises 59–60, write the given sentences as a system of inequalities in two variables. Then graph the system.

59. The sum of the x-variable and the y-variable is at most 4. The y-variable added to the product of 3 and the x-variable does not exceed 6.

60. The sum of the x-variable and the y-variable is at most 3. The y-variable added to the product of 4 and the x-variable does not exceed 6.

The graphs of solution sets of systems of inequalities involve finding the intersection of the solution sets of two or more inequalities. By contrast, in Exercises 61–62, you will be graphing the union of the solution sets of two inequalities.

61. Graph the union of $y > \frac{3}{2}x - 2$ and $y < 4$.

62. Graph the union of $x - y \geq -1$ and $5x - 2y \leq 10$.

Application Exercises

The figure shows the healthy weight region for various heights for people ages 35 and older.

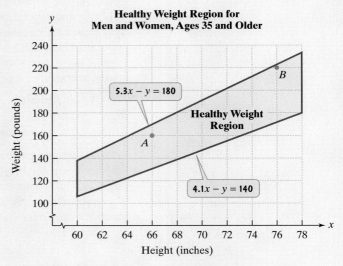

Healthy Weight Region for Men and Women, Ages 35 and Older

Source: U.S. Department of Health and Human Services

If x represents height, in inches, and y represents weight, in pounds, the healthy weight region can be modeled by the following system of linear inequalities:

$$\begin{cases} 5.3x - y \geq 180 \\ 4.1x - y \leq 140. \end{cases}$$

Use this information to solve Exercises 63–66.

63. Show that point A is a solution of the system of inequalities that describes healthy weight for this age group.

64. Show that point B is a solution of the system of inequalities that describes healthy weight for this age group.

65. Is a person in this age group who is 6 feet tall weighing 205 pounds within the healthy weight region?

66. Is a person in this age group who is 5 feet 8 inches tall weighing 135 pounds within the healthy weight region?

67. Many elevators have a capacity of 2000 pounds.

   a. If a child averages 50 pounds and an adult 150 pounds, write an inequality that describes when x children and y adults will cause the elevator to be overloaded.

   b. Graph the inequality. Because x and y must be positive, limit the graph to quadrant I only.

   c. Select an ordered pair satisfying the inequality. What are its coordinates and what do they represent in this situation?

68. A patient is not allowed to have more than 330 milligrams of cholesterol per day from a diet of eggs and meat. Each egg provides 165 milligrams of cholesterol. Each ounce of meat provides 110 milligrams.

   a. Write an inequality that describes the patient's dietary restrictions for x eggs and y ounces of meat.

   b. Graph the inequality. Because x and y must be positive, limit the graph to quadrant I only.

   c. Select an ordered pair satisfying the inequality. What are its coordinates and what do they represent in this situation?

The graph of an inequality in two variables is a region in the rectangular coordinate system. Regions in coordinate systems have numerous applications. For example, the regions in the following two graphs indicate whether a person is obese, overweight, borderline overweight, normal weight, or underweight.

Females

Males

Source: Centers for Disease Control and Prevention

Each horizontal axis shows a person's age. Each vertical axis shows that person's body-mass index (BMI), computed using the following formula:

$$\text{BMI} = \frac{703W}{H^2}.$$

The variable W represents weight, in pounds. The variable H represents height, in inches. Use this information and the graphs above to solve Exercises 69–70.

69. A man is 20 years old, 72 inches (6 feet) tall, and weighs 200 pounds.

    a. Compute the man's BMI. Round to the nearest tenth.

    b. Use the man's age and his BMI to locate this information as a point in the coordinate system for males. Is this person obese, overweight, borderline overweight, normal weight, or underweight?

70. A woman is 25 years old, 66 inches (5 feet, 6 inches) tall, and weighs 105 pounds.

    a. Compute the woman's BMI. Round to the nearest tenth.

    b. Use the woman's age and her BMI to locate this information as a point in the coordinate system for females. Is this person obese, overweight, borderline overweight, normal weight, or underweight?

Critical Thinking Exercises

In Exercises 71–72, write a system of inequalities for each graph.

71.

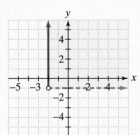

72.

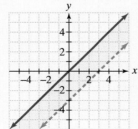

Without graphing, in Exercises 73–76, determine if each system has no solution or infinitely many solutions.

73. $\begin{cases} 3x + y < 9 \\ 3x + y > 9 \end{cases}$

74. $\begin{cases} 6x - y \leq 24 \\ 6x - y > 24 \end{cases}$

75. $\begin{cases} 3x + y \leq 9 \\ 3x + y \geq 9 \end{cases}$

76. $\begin{cases} 6x - y \leq 24 \\ 6x - y \geq 24 \end{cases}$

Chapter 5 Summary

5.1 Graphing and Functions

Definitions and Concepts

The rectangular coordinate system is formed using two number lines that intersect at right angles at their zero points. The horizontal line is the *x*-axis and the vertical line is the *y*-axis. Their point of intersection, $(0, 0)$, is the origin. Each point in the system corresponds to an ordered pair of real numbers, (x, y).

Example

- Plot: $(2, 3)$, $(-5, 4)$, $(-4\ -3)$, and $(5, -2)$.

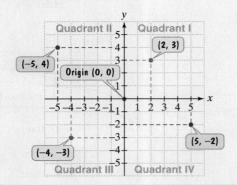

Additional Example to Review

Example 1, page 295

Definitions and Concepts

The graph of an equation in two variables is the set of all points whose coordinates satisfy the equation.

Example

- Graph: $y = x^2 - 1$.

x	$y = x^2 - 1$	Ordered Pair (x, y)
-2	$(-2)^2 - 1 = 3$	$(-2, -3)$
-1	$(-1)^2 - 1 = 0$	$(-1, 0)$
0	$0^2 - 1 = -1$	$(0, -1)$
1	$1^2 - 1 = 0$	$(1, 0)$
2	$2^2 - 1 = 3$	$(2, 3)$

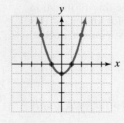

Additional Examples to Review

Example 2, page 296; Example 3, page 296

Definitions and Concepts

If an equation in x and y yields one value of y for each value of x, then y is a function of x, indicated by writing $f(x)$ for y.

Examples

- $f(x) = 5x - 2$; $f(-4) = 5(-4) - 2 = -20 - 2 = -22$
- $g(x) = 2x^2 - 5x - 1$; $g(-3) = 2(-3)^2 - 5(-3) - 1 = 2(9) - 5(-3) - 1$
  $$= 18 - (-15) - 1 = 18 + 15 - 1 = 32$$
- $h(x) = 6$; $h(-3) = 6, h(0) = 6, h(100) = 6$

Additional Examples to Review

Example 4, page 298; Example 5, page 299

Definitions and Concepts

The graph of a function is the graph of its ordered pairs.

Example

• Graph: $f(x) = |x| + 2$. Select integers for x from -2 to 2, inclusive.

x	$f(x) = |x| + 2$	(x, y)
-2	$f(-2) = |-2| + 2 = 2 + 2 = 4$	$(-2, 4)$
-1	$f(-1) = |-1| + 2 = 1 + 2 = 3$	$(-1, 3)$
0	$f(0) = |0| + 2 = 0 + 2 = 2$	$(0, 2)$
1	$f(2) = |1| + 2 = 1 + 2 = 3$	$(1, 3)$
2	$f(2) = |2| + 2 = 2 + 2 = 4$	$(2, 4)$

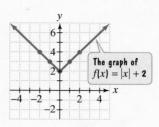

The graph of $f(x) = |x| + 2$

Additional Example to Review

Example 6, page 300

Definitions and Concepts

The Vertical Line Test: If any vertical line intersects a graph in more than one point, the graph does not define y as a function of x.

Examples

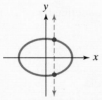

Not the graph of
a function

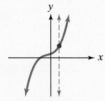

The graph of
a function

Additional Example to Review

Example 7, page 301

5.2 Linear Functions and Their Graphs

Definitions and Concepts

A function whose graph is a straight line is a linear function.

If a graph intersects the x-axis at $(a, 0)$, then a is an x-intercept.
If a graph intersects the y-axis at $(0, b)$, then b is a y-intercept.

The graph of $Ax + By = C$, a linear equation in two variables, is a straight line as long as A and B are not both zero. If none of A, B, and C is zero, the line can be graphed using intercepts and a checkpoint. To locate the x-intercept, set $y = 0$ and solve for x. To locate the y-intercept, set $x = 0$ and solve for y.

Example

Graph using intercepts: $4x + 3y = 12$.

x-intercept: $4x = 12$ Line passes through (3, 0).
(Set $y = 0$.) $x = 3$

y-intercept: $3y = 12$ Line passes through (0, 4).
(Set $x = 0$.) $y = 4$

Checkpoint: Let $x = 2$.

$$4 \cdot 2 + 3y = 12$$
$$8 + 3y = 12$$
$$3y = 4$$
$$y = \frac{4}{3}$$

Checkpoint is $\left(2, \frac{4}{3}\right)$, or $\left(2, 1\frac{1}{3}\right)$.

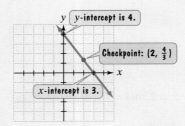

y-intercept is 4.

Checkpoint: $\left(2, \frac{4}{3}\right)$

x-intercept is 3.

Additional Example to Review

Example 1, page 310

Definitions and Concepts

The slope of the line through (x_1, y_1) and (x_2, y_2) is

$$m = \frac{\text{Rise}}{\text{Run}} = \frac{y_2 - y_1}{x_2 - x_1}.$$

If the slope is positive, the line rises from left to right. If the slope is negative, the line falls from left to right. The slope of a horizontal line is 0. The slope of a vertical line is undefined.

Examples

- For points $(-7, 2)$ and $(3, -4)$, the slope of the line through the points is:

$$m = \frac{\text{Change in } y}{\text{Change in } x} = \frac{-4 - 2}{3 - (-7)} = \frac{-6}{10} = -\frac{3}{5}.$$

The slope is negative, so the line falls.

- For points $(2, -5)$ and $(2, 16)$, the slope of the line through the points is:

$$m = \frac{\text{Change in } y}{\text{Change in } x} = \frac{16 - (-5)}{2 - 2} = \frac{21}{0}.$$

undefined

The slope is undefined, so the line is vertical.

Additional Examples to Review

Example 2, page 311; Example 7, page 316

Definitions and Concepts

The equation $y = mx + b$ is the slope-intercept form of the equation of a line, in which m is the slope and b is the y-intercept. Using function notation, the equation is $f(x) = mx + b$.

Example

- Graph: $f(x) = -\dfrac{3}{4}x + 1$.

Slope is $-\frac{3}{4}$. y-intercept is 1.

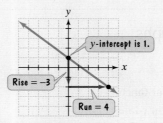

y-intercept is 1.

Rise = -3

Run = 4

Additional Examples to Review

Example 3, page 313; Example 4, page 314; Example 8, page 317

Definitions and Concepts

Horizontal and Vertical Lines

1. The graph of $y = b$ is a horizontal line that intersects the y-axis at $(0, b)$.
2. The graph of $x = a$ is a vertical line that intersects the x-axis at $(a, 0)$.

Examples

• Graph: $y = 3$.

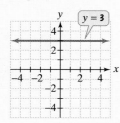

• Graph: $x = -2$.

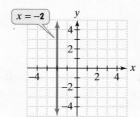

Additional Examples to Review

Example 5, page 315; Example 6, page 315

5.3 The Point-Slope Form of the Equation of a Line; Scatter Plots and Regression Lines

Definitions and Concepts

The point-slope form of the equation of a nonvertical line with slope m that passes through the point (x_1, y_1) is

$$y - y_1 = m(x - x_1).$$

Example

  Slope $= -4$, passing through $(-1, 5)$

$m = -4$ $x_1 = -1$ $y_1 = 5$

The point-slope form of the line's equation is

$$y - 5 = -4[x - (-1)].$$

Simplify:

$$y - 5 = -4(x + 1).$$

Additional Example to Review

Example 1, page 323

Definitions and Concepts

To write the point-slope form of the line passing through two points, begin by using the points to compute the slope, m. Use either given point as (x_1, y_1) and write the point-slope equation:

$$y - y_1 = m(x - x_1).$$

Solving this equation for y gives the slope-intercept form of the line's equation.

Example

• Write equations in point-slope form and in slope-intercept form of the line passing through $(4, 1)$ and $(3, -2)$.

$$m = \frac{-2 - 1}{3 - 4} = \frac{-3}{-1} = 3$$

Using $(4, 1)$ as (x_1, y_1), the point-slope form of the equation is

$$y - 1 = 3(x - 4).$$

Solve for y to obtain the slope-intercept form.

$$y - 1 = 3x - 12$$
$$y = 3x - 11$$

Additional Examples to Review
Example 2, page 323; Example 3, page 325

Definitions and Concepts

A plot of data points is called a scatter plot. If the points lie approximately along a line, the line that best fits the data is called a regression line.

A correlation coefficient, r, measures the strength and direction of a possible relationship between variables. If $r = 1$, there is a perfect positive correlation, and if $r = -1$, there is a perfect negative correlation. If $r = 0$, there is no relationship between the variables.

Examples

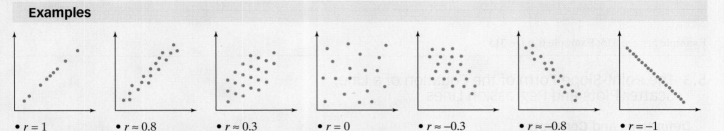

- $r = 1$
  perfect positive
  correlation

- $r \approx 0.8$
  strong positive
  correlation

- $r \approx 0.3$
  moderate to
  weak positive
  correlation

- $r = 0$
  no correlation

- $r \approx -0.3$
  moderate to
  weak negative
  correlation

- $r \approx -0.8$
  strong negative
  correlation

- $r = -1$
  perfect negative
  correlation

Additional Example to Review
Example 4, page 328

5.4 Systems of Linear Equations in Two Variables

Definitions and Concepts

Two equations in the form $Ax + By = C$ are called a system of linear equations. A solution of the system is an ordered pair that satisfies both equations in the system.

Example

- Determine whether $(3, -1)$ is a solution of

$$\begin{cases} 2x + 5y = 1 \\ 4x + y = 11. \end{cases}$$

Replace x with 3 and y with -1 in both equations.

$$2x + 5y = 1 \qquad\qquad 4x + y = 11$$
$$2 \cdot 3 + 5(-1) \stackrel{?}{=} 1 \qquad\qquad 4 \cdot 3 + (-1) \stackrel{?}{=} 11$$
$$6 + (-5) \stackrel{?}{=} 1 \qquad\qquad 12 + (-1) \stackrel{?}{=} 11$$
$$1 = 1, \ \text{true} \qquad\qquad 11 = 11, \ \text{true}$$

Thus, $(3, -1)$ is a solution of the system.

Additional Example to Review
Example 1, page 335

Definitions and Concepts

Linear systems with one solution can be solved by graphing. The coordinates of the point of intersection of the lines are the system's solution.

Example

- Solve by graphing:

$$\begin{cases} 2x + y = 4 \\ x + y = 2. \end{cases}$$

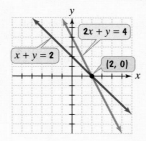

The solution set is $\{(2, 0)\}$.

Additional Example to Review

Example 2, page 335

Definitions and Concepts

Solving Linear Systems by the Substitution Method

1. Solve either of the equations for one variable in terms of the other. (If one of the equations is already in this form, you can skip this step.)
2. Substitute the expression found in step 1 into the *other* equation. This will result in an equation in one variable.
3. Solve the equation containing one variable.
4. Back-substitute the value found in step 3 into the equation from step 1. Simplify and find the value of the remaining variable.
5. Check the proposed solution in both of the system's given equations.

Example

- Solve by the substitution method:

$$\begin{cases} y = 2x + 3 \\ 7x - 5y = -18. \end{cases}$$

Substitute $2x + 3$ for y in the second equation.

$$7x - 5(2x + 3) = -18$$
$$7x - 10x - 15 = -18$$
$$-3x - 15 = -18$$
$$-3x = -3$$
$$x = 1$$

Find y. Substitute 1 for x in $y = 2x + 3$.

$$y = 2 \cdot 1 + 3 = 2 + 3 = 5$$

The solution, $(1, 5)$, checks. The solution set is $\{(1, 5)\}$.

Additional Examples to Review

Example 3, page 336; Example 4, page 337

Definitions and Concepts

Solving Linear Systems by the Addition Method

1. If necessary, rewrite both equations in the form $Ax + By = C$.
2. If necessary, multiply either equation or both equations by appropriate nonzero numbers so that the sum of the x-coefficients or the sum of the y-coefficients is 0.
3. Add the equations in step 2. The sum is an equation in one variable.
4. Solve the equation in one variable.
5. Back-substitute the value obtained in step 4 into either of the given equations and solve for the other variable.
6. Check the solution in both of the original equations.

Example

- Solve by the addition method:

$$\begin{cases} 2x + y = 10 \\ 3x + 4y = 25. \end{cases} \quad \xrightarrow[\text{No change}]{\text{Multiply by } -4.} \quad \begin{cases} -8x - 4y = -40 \\ \underline{3x + 4y = 25} \end{cases}$$

$$\text{Add:} \quad -5x = -15$$
$$x = 3$$

Find y. Back-substitute 3 for x. Use the first equation, $2x + y = 10$.

$$2(3) + y = 10$$
$$6 + y = 10$$
$$y = 4$$

The solution, $(3, 4)$, checks. The solution set is $\{(3, 4)\}$.

Additional Examples to Review

Example 5, page 339; Example 6, page 340

Definitions and Concepts

When solving by substitution or addition, if the variable is eliminated and a false statement results, the linear system has no solution. If the variable is eliminated and a true statement results, the system has infinitely many solutions.

Examples

- Solve:

$$\begin{cases} x + 2y = 4 \\ 3x + 6y = 13. \end{cases} \quad \xrightarrow[\text{No change}]{\text{Multiply by } -3.} \quad \begin{cases} -3x - 6y = -12 \\ \underline{3x + 6y = 13} \end{cases}$$

$$\text{Add:} \quad 0 = 1, \quad \text{false}$$

No solution: $\varnothing$

- Solve: Substitute $4x - 4$ for y in the second equation.

$$\begin{cases} y = 4x - 4 \\ 4x - y = 4. \end{cases}$$

$$4x - (4x - 4) = 4$$
$$4x - 4x + 4 = 4$$
$$4 = 4, \quad \text{true}$$

Solution set: $\{(x, y) \mid y = 4x - 4\}$ or $\{(x, y) \mid 4x - y = 4\}$

Additional Examples to Review

Example 7, page 341; Example 8, page 342

Definitions and Concepts

Functions of Business
A company produces and sells x units of a product.

Revenue Function: $R(x) = (\text{price per unit sold})x$
Cost Function: $C(x) = \text{fixed cost} + (\text{cost per unit produced})x$
Profit Function: $P(x) = R(x) - C(x)$

The point of intersection of the graphs of R and C is the break-even point. The x-coordinate of the point reveals the number of units that a company must produce and sell so that the money coming in, the revenue, is equal to the money going out, the cost. The y-coordinate gives the amount of money coming in and going out.

Example

- A company that manufactures lamps has a fixed cost of $80,000 and it costs $20 to produce each lamp. Lamps are sold for $70.

  a. Write the cost function.   **b.** Write the revenue function.

  $C(x) = 80,000 + 20x$ $R(x) = 70x$

  > Fixed cost | Variable cost: $20 per lamp

  > Revenue per lamp, $70, times number of lamps sold

  c. Find the break-even point.
  Solve
  $$\begin{cases} y = 80,000 + 20x \\ y = 70x \end{cases}$$
  by substitution. Solving
  $$70x = 80,000 + 20x$$
  yields $x = 1600$. Back-substituting yields $y = 112,000$. The break-even point is $(1600, 112,000)$: The company breaks even if it sells 1600 lamps. At this level, money coming in equals money going out: $112,000.

Additional Example to Review
Example 9, page 343

5.5 Linear Inequalities in Two Variables

Definitions and Concepts

A linear inequality in two variables can be written in the form
$Ax + By > C, Ax + By \geq C, Ax + By < C,$ or $Ax + By \leq C.$

Graphing a Linear Inequality in Two Variables

1. Replace the inequality symbol with an equal sign and graph the corresponding linear equation. Draw a solid line if the original inequality contains a $\leq$ or $\geq$ symbol. Draw a dashed line if the original inequality contains a $<$ or $>$ symbol.
2. Choose a test point from one of the half-planes. (Do not choose a point on the line.) Substitute the coordinates of the test point into the inequality.
3. If a true statement results, shade the half-plane containing this test point. If a false statement results, shade the half-plane not containing this test point.

Example

- Graph: $x - 2y \leq 4$.

  1. Graph $x - 2y = 4$. Use a solid line because the inequality symbol is $\leq$.
  2. Test $(0,0)$.
  $$x - 2y \leq 4$$
  $$0 - 2 \cdot 0 \overset{?}{\leq} 4$$
  $$0 \leq 4, \quad \text{true}$$

  3. The inequality is true. Shade the half-plane containing $(0,0)$.

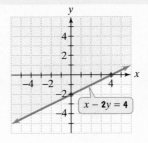

Additional Examples to Review
Example 1, page 349; Example 2, page 350

Definitions and Concepts

Some inequalities can be graphed without using test points, including $y > mx + b$ (the half-plane above $y = mx + b$), $y < mx + b$, $x > a$ (the half-plane to the right of $x = a$), $x < a$, $y > b$ (the half-plane above $y = b$), and $y < b$.

Examples

- Graph: $y < 2$.

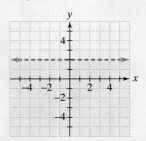

- Graph: $x \geq -1$.

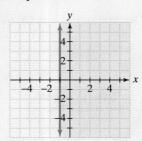

Additional Example to Review
Example 3, page 351

Definitions and Concepts

Two or more linear inequalities make up a system of linear inequalities. A solution is an ordered pair satisfying all inequalities in the system.

Graphing Systems of Linear Inequalities
1. Graph each inequality in the system in the same rectangular coordinate system.
2. Find the intersection of the individual graphs.

Example

- Graph the solution set of the system:

$$\begin{cases} y \leq -2x \\ x - y \geq 3. \end{cases}$$

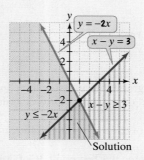

Additional Examples to Review
Example 4, page 352; Example 5, page 353; Example 6, page 354

Review Exercises

Section 5.1 Graphing and Functions

In Exercises 1–4, plot the given point in a rectangular coordinate system.

1. $(2, 5)$
2. $(-4, 3)$
3. $(-5, -3)$
4. $(2, -5)$

Graph each equation in Exercises 5–7. Let $x = -3, -2, -1, 0, 1, 2,$ and 3.

5. $y = 2x - 2$
6. $y = |x| + 2$
7. $y = x$
8. If $f(x) = 4x + 11$, find $f(-2)$.
9. If $f(x) = -7x + 5$, find $f(-3)$.

10. If $f(x) = 3x^2 - 5x + 2$, find $f(4)$.

11. If $f(x) = -3x^2 + 6x + 8$, find $f(-4)$.

In Exercises 12–13, evaluate $f(x)$ for the given values of x. Then use the ordered pairs $(x, f(x))$ from your table to graph the function.

12. $f(x) = \frac{1}{2}|x|$ **13.** $f(x) = x^2 - 2$

x	$f(x) = \frac{1}{2}\|x\|$
-6	
-4	
-2	
0	
2	
4	
6	

x	$f(x) = x^2 - 2$
-2	
-1	
0	
1	
2	

In Exercises 14–15, use the vertical line test to identify graphs in which y is a function of x.

14.
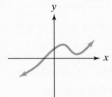

15.
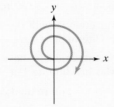

16. The line graph shows the top marginal income tax rates in the United States from 1925 through 2005.

Top United States Marginal Tax Rates, 1925–2005

Source: National Taxpayers Union

a. What are the coordinates of point A? What does this mean in terms of the information given by the graph?

b. Estimate the top marginal tax rate in 2005.

c. For the period shown, during which year did the United States have the highest marginal tax rate? Estimate, to the nearest percent, the tax rate for that year.

d. For the period from 1950 through 2005, during which year did the United States have the lowest marginal tax rate? Estimate, to the nearest percent, the tax rate for that year.

e. For the period shown, during which ten-year period did the top marginal tax rate remain constant? Estimate, to the nearest percent, the tax rate for that period.

f. For the period shown, during which five-year period did the top marginal tax rate increase most rapidly? Estimate, to the nearest percent, the increase in the top tax rate for that period.

Section 5.2 Linear Functions and Their Graphs

In Exercises 17–19, use the x- and y-intercepts to graph each linear equation.

17. $2x + y = 4$ **18.** $2x - 3y = 6$ **19.** $5x - 3y = 15$

In Exercises 20–23, calculate the slope of the line passing through the given points. If the slope is undefined, so state. Then indicate whether the line rises, falls, is horizontal, or is vertical.

20. $(3, 2)$ and $(5, 1)$ **21.** $(-1, 2)$ and $(-3, -4)$

22. $(-3, 4)$ and $(6, 4)$ **23.** $(5, 3)$ and $(5, -3)$

In Exercises 24–27, graph each linear function using the slope and y-intercept.

24. $y = 2x - 4$ **25.** $y = -\frac{2}{3}x + 5$

26. $f(x) = \frac{3}{4}x - 2$ **27.** $y = \frac{1}{2}x + 0$

In Exercises 28–30, **a.** *Write the equation in slope-intercept form;* **b.** *Identify the slope and the y-intercept;* **c.** *Use the slope and y-intercept to graph the line.*

28. $2x + y = 0$ **29.** $3y = 5x$ **30.** $3x + 2y = 4$

In Exercises 31–33, graph each horizontal or vertical line.

31. $x = 3$ **32.** $y = -4$ **33.** $x + 2 = 0$

34. The scatter plot indicates a relationship between the percentage of adult females in a country who are literate and the mortality of children under five. Also shown is a line that passes through or near the points.

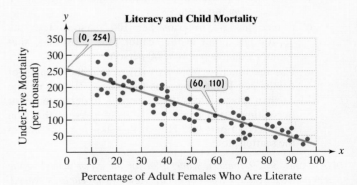

Literacy and Child Mortality

Source: United Nations

a. According to the graph, what is the y-intercept of the line? Describe what this represents in this situation.

b. Use the coordinates of the two points shown to compute the slope of the line. Describe what this means in terms of the rate of change.

c. Use the y-intercept from part (a) and the slope from part (b) to write a linear function that models child mortality, $f(x)$, per thousand, for children under five in a country where $x\%$ of adult women are literate.

d. Use the function from part (c) to predict the mortality rate of children under five in a country where 50% of adult females are literate.

Section 5.3 The Point-Slope Form of the Equation of a Line; Scatter Plots and Regression Lines

In Exercises 35–36, use the given conditions to write an equation for each line in point-slope form and in slope-intercept form.

35. Passing through $(-3, 2)$ with slope -6

36. Passing through $(1, 6)$ and $(-1, 2)$

37. Shown, again, is the scatter plot that indicates a relationship between the number of firearms per 100 persons and the number of deaths per 100,000 persons for industrialized countries with the highest death rates. Also shown is a line that passes through or near the points.

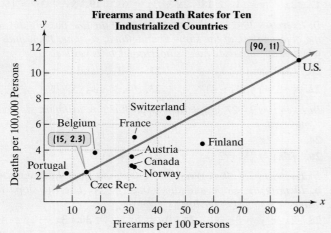

Firearms and Death Rates for Ten Industrialized Countries

Source: International Action Network on Small Arms

a. Use the two points whose coordinates are shown by the voice balloons to find an equation in point-slope form for the line that models deaths per 100,000 persons, y, as a function of firearms per 100 persons, x.

b. Write the equation in part (a) in slope-intercept form. Use function notation.

c. France has 32 firearms per 100 persons. Use the appropriate point in the scatter plot to estimate that country's deaths per 100,000 persons.

d. Use the function from part (b) to find the number of deaths per 100,000 persons for France. Round to one decimal place. Does the function underestimate or overestimate the deaths per 100,000 persons that you estimated in part (c)? How is this shown by the line in the scatter plot?

In Exercises 38–39, make a scatter plot for the given data. Use the scatter plot to describe whether or not the variables appear to be related.

38.

x	1	3	4	6	8	9
y	1	2	3	3	5	5

Shown, again, is the scatter plot that indicates a relationship between the percentage of adult females in a country who are literate and the mortality of children under five. Also shown is the regression line. Use this information to determine whether each of the statements in Exercises 40–46 is true or false.

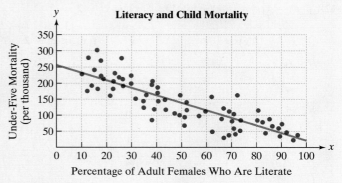

Literacy and Child Mortality

Source: United Nations

40. There is a perfect negative correlation between the percentage of adult females who are literate and under-five mortality.

41. As the percentage of adult females who are literate increases, under-five mortality tends to decrease.

42. The country with the least percentage of adult females who are literate has the greatest under-five mortality.

43. No two countries have the same percentage of adult females who are literate but different under-five mortalities.

44. There are more than 20 countries in this sample.

45. There is no correlation between the percentage of adult females who are literate and under-five mortality.

46. The country with the greatest percentage of adult females who are literate has an under-five mortality rate that is less than 50 children per thousand.

47. Which one of the following scatter plots indicates a correlation coefficient of approximately -0.9? (Scatter plots continue at the top of the next page.)

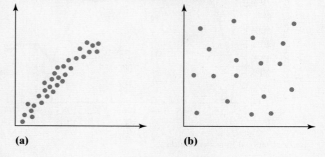

(a)   (b)

39.

Country	Canada	U.S.	Mexico	Brazil	Costa Rica	Denmark	China	Egypt	Pakistan	Bangladesh	Australia	Japan	Russia
Life expectancy in years, x	81	78	76	72	77	78	73	72	64	63	82	82	66
Infant deaths per 1000 births, y	5.1	6.3	19.0	23.3	9.0	4.4	21.2	28.4	66.9	57.5	4.8	2.8	10.8

Source: U.S. Bureau of the Census International Database

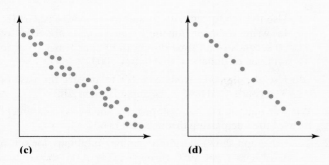

(c)   **(d)**

Section 5.4 Systems of Linear Equations in Two Variables

In Exercises 48–50, solve each system by graphing. Check the coordinates of the intersection point in both equations.

48. $\begin{cases} x + y = 5 \\ 3x - y = 3 \end{cases}$ **49.** $\begin{cases} 2x - y = -1 \\ x + y = -5 \end{cases}$

50. $\begin{cases} y = -x + 5 \\ 2x - y = 4 \end{cases}$

In Exercises 51–53, solve each system by the substitution method.

51. $\begin{cases} 2x + 3y = 2 \\ x = 3y + 10 \end{cases}$ **52.** $\begin{cases} y = 4x + 1 \\ 3x + 2y = 13 \end{cases}$

53. $\begin{cases} x + 4y = 14 \\ 2x - y = 1 \end{cases}$

In Exercises 54–56, solve each system by the addition method.

54. $\begin{cases} x + 2y = -3 \\ x - y = -12 \end{cases}$ **55.** $\begin{cases} 2x - y = 2 \\ x + 2y = 11 \end{cases}$

56. $\begin{cases} 5x + 3y = 1 \\ 3x + 4y = -6 \end{cases}$

In Exercises 57–59, solve by the method of your choice. Identify systems with no solution and systems with infinitely many solutions, using set notation to express their solution sets.

57. $\begin{cases} y = -x + 4 \\ 3x + 3y = -6 \end{cases}$ **58.** $\begin{cases} 3x + y = 8 \\ 2x - 5y = 11 \end{cases}$

59. $\begin{cases} 3x - 2y = 6 \\ 6x - 4y = 12 \end{cases}$

60. A company is planning to manufacture computer desks. The fixed cost will be $60,000 and it will cost $200 to produce each desk. Each desk will be sold for $450.

   a. Write the cost function, C, of producing x desks.

   b. Write the revenue function, R, from the sale of x desks.

   c. Determine the break-even point. Describe what it means.

61. The graph shows that from 2000 through 2006, Americans unplugged land lines and switched to cellphones.

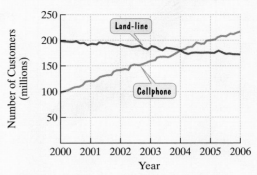

Number of Cellphone and Land-Line Customers in the United States

Source: Federal Communications Commission

   a. Use the graphs to estimate the point of intersection. In what year was the number of cellphone and land-line customers the same? How many millions of customers were there for each?

   b. In 2000, there were 98 million cellphone customers. For the period shown by the graph, this increased at an average rate of 19.8 million customers per year. Write a function that models the number of cellphone customers, y, in millions, x years after 2000.

   c. The function $4.3x + y = 198$ models the number of land-line customers, y, in millions, x years after 2000. Use this model and the model you obtained in part (b) to determine the year, rounded to the nearest year, when the number of cellphone and land-line customers was the same. According to the models, how many millions of customers, rounded to the nearest ten million, were there for each?

   d. How well do the models in parts (b) and (c) describe the point of intersection of the graphs that you estimated in part (a)?

Section 5.5 Linear Inequalities in Two Variables

In Exercises 62–68, graph each linear inequality.

62. $x - 3y \le 6$ **63.** $2x + 3y \ge 12$ **64.** $2x - 7y > 14$

65. $y > \frac{3}{5}x$ **66.** $y \le -\frac{1}{2}x + 2$ **67.** $x \le 2$

68. $y > -3$

In Exercises 69–74, graph the solution set of each system of linear inequalities.

69. $\begin{cases} 3x - y \le 6 \\ x + y \ge 2 \end{cases}$ **70.** $\begin{cases} x + y < 4 \\ x - y < 4 \end{cases}$ **71.** $\begin{cases} x \le 3 \\ y > -2 \end{cases}$

72. $\begin{cases} 4x + 6y \le 24 \\ y > 2 \end{cases}$ **73.** $\begin{cases} x + y \le 6 \\ y \ge 2x - 3 \end{cases}$ **74.** $\begin{cases} y < -x + 4 \\ y > x - 4 \end{cases}$

Chapter 5 Test A

1. Graph $y = |x| - 2$. Let $x = -3, -2, -1, 0, 1, 2,$ and 3.

2. If $f(x) = 3x^2 - 7x - 5$, find $f(-2)$.

In Exercises 3–4, use the vertical line test to identify graphs in which y is a function of x.

3.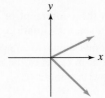

4.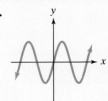

5. The graph shows the height, in meters, of an eagle in terms of its time, in seconds, in flight.

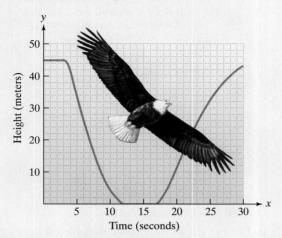

Time (seconds)

a. Is the eagle's height a function of time? Use the graph to explain why or why not.

b. Find $f(15)$. Describe what this means in practical terms.

c. What is a reasonable estimate of the eagle's maximum height?

d. During which period of time was the eagle descending?

6. Use the x- and y-intercepts to graph $4x - 2y = -8$.

7. Find the slope of the line passing through $(-3, 4)$ and $(-5, -2)$.

In Exercises 8–9, graph each linear function using the slope and y-intercept.

8. $y = \frac{2}{3}x - 1$ **9.** $f(x) = -2x + 3$

10. On Super Bowl Sunday, viewers of the big game expect to be entertained by commercials that offer an upbeat mix of punch lines, animal tricks, humor, and special effects. The price tag for a 30-second ad slot also follows tradition, up to an average of $3 million in 2009 from $2.7 million in 2008. The scatter plot shows the cost, in millions of dollars, of 30 seconds of ad time in the Super Bowl from 2003 through 2009. Also shown is a line that passes through or near the seven data points.

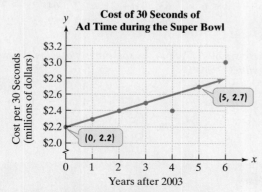

Cost of 30 Seconds of Ad Time during the Super Bowl

Source: TNS Media Intelligence

a. According to the graph, what is the y-intercept of the line? Describe what this represents in this situation.

b. Use the coordinates of the two points shown to compute the slope. What does this represent in terms of the rate of change in the cost of Super Bowl commercials?

c. Use the y-intercept shown and the slope from part (b) to write a linear function that models the cost of 30 seconds of ad time during the Super Bowl, $C(x)$, in millions of dollars, x years after 2003.

d. Use the function from part (c) to project the cost of 30 seconds of ad time in the Super Bowl in 2013.

11. Write an equation in point-slope form and in slope-intercept form of the line passing through $(2, 1)$ and $(-1, -8)$.

12. The bar graph shows world population, in billions, for seven selected years from 1950 through 2008. Also shown is a scatter plot with a line passing through two of the data points.

World Population, 1950–2008

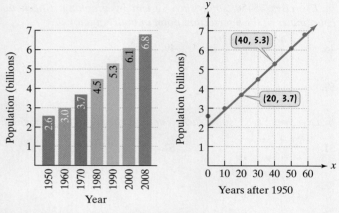

Source: U.S. Census Bureau, International Database

a. Use the two points whose coordinates are shown by the voice balloons to write the slope-intercept form of an equation that models world population, y, in billions, x years after 1950.

b. Use the model from part (a) to project world population in 2025.

13. Make a scatter plot for the given data. Use the scatter plot to describe whether or not the variables appear to be related.

x	1	4	3	5	2
y	5	2	2	1	4

The scatter plot shows the number of minutes each of 16 people exercise per week and the number of headaches per month each person experiences. Use the scatter plot to determine whether each of the statements in Exercises 14–16 is true or false.

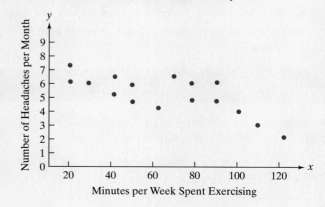

14. An increase in the number of minutes devoted to exercise causes a decrease in headaches.

15. There is a perfect negative correlation between time spent exercising and number of headaches.

16. The person who exercised most per week had the least number of headaches per month.

17. Solve by graphing:

$$\begin{cases} x + y = 6 \\ 4x - y = 4. \end{cases}$$

18. Solve by substitution:

$$\begin{cases} x = y + 4 \\ 3x + 7y = -18. \end{cases}$$

19. Solve by addition:

$$\begin{cases} 5x + 4y = 10 \\ 3x + 5y = -7. \end{cases}$$

20. A company is planning to produce and sell a new line of computers. The fixed cost will be $360,000 and it will cost $850 to produce each computer. Each computer will be sold for $1150.

  a. Write the cost function, C, of producing x computers.

  b. Write the revenue function, R, from the sale of x computers.

  c. Determine the break-even point. Describe what it means.

Graph each linear inequality in Exercises 21–23.
21. $3x - 2y < 6$ **22.** $y \le \frac{1}{2}x - 1$ **23.** $y > -1$

24. Graph the system of linear inequalities:

$$\begin{cases} 2x - y \le 4 \\ 2x - y > -1. \end{cases}$$

Chapter 5 Test B

1. Graph $y = |x| - 1$. Let $x = -3, -2, -1, 0, 1, 2,$ and 3.

2. If $f(x) = 4x^2 - 6x - 3$, find $f(-5)$.

In Exercises 3–4, use the vertical line test to identify graphs in which y is a function of x.

3. 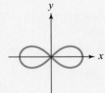   **4.**

5. With aging, body fat increases and muscle mass declines. The graphs show the average percent body fat in adult women and men as they age from 25 to 75 years.

Average Percent Body Fat in Adults

Source: Thompson et al., *The Science of Nutrition*, Benjamin Cummings, 2008.

  a. Is the average percent body fat in women a function of age? Use the red graph to explain why or why not.

  b. Between which ages is the average body fat in women increasing?

  c. At which age does the average body fat in men reach a maximum? What is the average percent body fat for that age?

  d. If $M(a)$ describes the average body fat in men at age a, use the appropriate graph to find $M(35)$. Describe what this means in practical terms.

6. Use the x- and y-intercepts to graph $2x - 4y = -8$.

7. Find the slope of the line passing through $(-5, 3)$ and $(-7, -5)$.

In Exercises 8–9, graph each linear function using the slope and y-intercept.
8. $y = \frac{1}{2}x - 1$ **9.** $f(x) = -2x + 4$

10. No matter who hosts, the Miss America pageant keeps losing viewers. The scatter plot shows the number of viewers of the pageant, in millions, along with some of the MCs from 1989 through 2009. Also shown is a line passing through two of the data points.

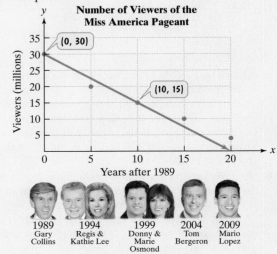

Years after 1989

Source: Entertainment Weekly

(In parts (a)-(d), refer to the graph at the bottom of the previous page.)

   a. According to the graph, what is the *y*-intercept of the line? Describe what this represents in this situation.

   b. Use the coordinates of the two points shown to compute the slope. What does this represent in terms of the rate of change in the number of viewers of the Miss America pageant?

   c. Use the *y*-intercept shown and the slope from part (b) to write a linear function that models the number of viewers of the Miss America pageant, $V(x)$, in millions, *x* years after 1989.

   d. Use the function from part (c) to determine the number of viewers of the pageant in 2009. Does this underestimate or overestimate the actual number of viewers shown by the scatter plot? By approximately how many millions of viewers?

11. Write an equation in point-slope form and in slope-intercept form of the line passing through $(2, 4)$ and $(4, -2)$.

12. Studies show that texting while driving is as risky as driving with a 0.08 blood alcohol level, the standard for drunk driving. The bar graph shows the number of fatalities in the United States involving distracted driving from 2004 through 2008. Although the distracted category involves such activities as talking on cellphones, conversing with passengers, and eating, experts at the National Highway Traffic Safety Administration claim that texting while driving is the clearest menace because it requires looking away from the road.

Number of Highway Fatalities in the United States Involving Distracted Driving

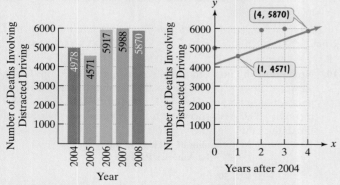

Source: National Highway Traffic Safety Administration

   a. Shown to the right of the bar graph is a scatter plot with a line passing through two of the data points. Use the two points whose coordinates are shown by the voice balloons to write the slope-intercept form of an equation that models the number of highway fatalities involving distracted driving, *y*, in the United States *x* years after 2004.

   b. In 2010, surveys showed overwhelming public support to ban texting while driving, although at that time only 19 states and Washington, D.C., outlawed the practice. Without additional laws that penalize texting drivers, use the model from part (a) to project the number of fatalities in the United States in 2014 involving distracted driving.

13. Make a scatter plot for the given data. Use the scatter plot to describe whether or not the variables appear to be related.

x	1	4	3	5	2
y	2	5	3	7	3

The scatter plot in the figure shows the relationship between the number of hours that students studied for a quiz and their quiz scores. The highest quiz score is 10 and the lowest is 1. Use the scatter plot to determine whether each of the statements in Exercises 14–16 is true or false.

Hours Studied and Quiz Score Results

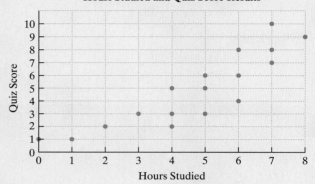

14. There is a strong positive correlation between hours studied and quiz scores.

15. The student who studied for one hour had a better quiz score than the student who did not study.

16. The student who studied the greatest amount of time had the best quiz score.

17. Solve by graphing:

$$\begin{cases} 2x + y = 8 \\ x - y = 1. \end{cases}$$

18. Solve by substitution:

$$\begin{cases} x = 7y - 22 \\ 5x + 2y = 1. \end{cases}$$

19. Solve by addition:

$$\begin{cases} 2x - 3y = -8 \\ 11x + 5y = -1. \end{cases}$$

20. A company is planning to manufacture PDAs (personal digital assistants). The fixed cost will be $400,000 and it will cost $20 to produce each PDA. Each PDA will be sold for $100.

   a. Write the cost function, *C*, of producing *x* PDAs.

   b. Write the revenue function, *R*, from the sale of *x* PDAs.

   c. Determine the break-even point. Describe what it means.

Graph each linear inequality in Exercises 21–23.

21. $2x - 3y \geq -6$ **22.** $y < -\frac{1}{2}x + 1$ **23.** $x > -2$

24. Graph the system of linear inequalities:

$$\begin{cases} x + y \geq 2 \\ x - y \geq 4. \end{cases}$$

Polynomials, Quadratic Equations, and Quadratic Functions

We are surrounded by evidence that the world is profoundly mathematical. After turning a somersault, a diver's path can be modeled by a function in the form $f(x) = ax^2 + bx + c$, called a *quadratic function,* as can the path of a football tossed from a quarterback to a receiver, or the path of a flipped coin. Even if you throw an object directly upward, although its path is straight and vertical, its changing height over time is described by a quadratic function. And tailgaters take note: whether you're driving a car or a truck on dry or wet roads, an array of quadratic functions that model your required stopping distances at various speeds are available to help you become a safer driver.

The quadratic functions surrounding our long history of objects that are thrown, kicked, or hit appear throughout the chapter, including Example 4 of Section 6.3 and Example 7 of Section 6.5. Tailgaters on motorcycles should pay close attention to Exercises 78-79 in the Chapter Review Exercises.

Operations with Polynomials; Polynomial Functions

6.1

Objectives

1. Understand the vocabulary of polynomials.

2. Add polynomials.

3. Subtract polynomials.

4. Multiply polynomials.

5. Use FOIL in polynomial multiplication.

6. Multiply the sum and difference of two terms.

7. Square binomials.

8. Use polynomial functions.

We're born. We die. **Figure 6.1** quantifies these statements by showing the numbers of births and deaths in the United States for six selected years.

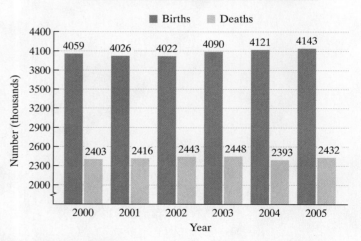

Numbers of Births and Deaths in the United States

Figure 6.1
Source: U.S. Department of Health and Human Services

Here are two functions that model the data in **Figure 6.1**:

$$B(x) = 7.4x^2 - 15x + 4046 \qquad D(x) = -3.5x^2 + 20x + 2405.$$

Number of births, $B(x)$, in thousands, x years after **2000**

Number of deaths, $D(x)$, in thousands, x years after **2000**

The algebraic expressions that appear on the right sides of the models are examples of *polynomials*. A **polynomial** is a single term or the sum of two or more terms containing variables with whole-number exponents. The polynomials above each contain three terms. In a **polynomial function**, the expression that defines the function is a polynomial. Polynomial functions are used in such diverse areas as science, business, medicine, psychology, and sociology. In this section, we consider basic ideas about polynomials and their operations.

Understand the
vocabulary of
polynomials.

How We Describe Polynomials

Consider the polynomial
$$7x^3 - 9x^2 + 13x - 6.$$
We can express this polynomial as
$$7x^3 + (-9x^2) + 13x + (-6).$$
The polynomial contains four terms. It is customary to write the terms in the order of descending powers of the variable. This is the **standard form** of a polynomial.

Some polynomials contain only one variable. Each term of such a polynomial in x is of the form ax^n. If $a \neq 0$, the **degree** of the term ax^n is n. For example, the degree of the term $7x^3$ is 3.

> **The Degree of ax^n**
>
> If $a \neq 0$, the degree of the term ax^n is n. The degree of a nonzero constant term is 0. The constant 0 has no defined degree.

Here is an example of a polynomial and the degree of each of its four terms:
$$6x^4 - 3x^3 + 2x - 5.$$

degree 4	degree 3	degree 1	degree of nonzero constant: 0

Notice that the exponent on x for the term $2x$, meaning $2x^1$, is understood to be 1. For this reason, the degree of $2x$ is 1. You can think of -5 as $-5x^0$; thus, its degree is 0.

A polynomial is simplified when it contains no grouping symbols and no like terms. A simplified polynomial that has exactly one term is called a **monomial**. A **binomial** is a simplified polynomial that has two terms. A **trinomial** is a simplified polynomial with three terms. Simplified polynomials with four or more terms have no special names.

The **degree of a polynomial** is the greatest of the degrees of all the terms of the polynomial. For example, $4x^2 + 3x$ is a binomial of degree 2 because the degree of the first term is 2, and the degree of the other term is less than 2. Also, $7x^5 - 2x^2 + 4$ is a trinomial of degree 5 because the degree of the first term is 5, and the degrees of the other terms are less than 5. In a simplified polynomial containing only one variable, the term of the greatest degree is called the **leading term**. Its coefficient is called the **leading coefficient**.

Up to now, we have used x to represent the variable in a polynomial. However, any letter can be used. For example,

- $7x^5 - 3x^3 + 8$ is a polynomial (in x) of degree 5. Because there are three terms, the polynomial is a trinomial. The leading coefficient is 7.
- $6y^3 + 4y^2 - y + 3$ is a polynomial (in y) of degree 3. Because there are four terms, the polynomial has no special name. The leading coefficient is 6.
- $z^7 + \sqrt{2}$ is a polynomial (in z) of degree 7. Because there are two terms, the polynomial is a binomial. The leading coefficient is 1.

Adding Polynomials

Add
polynomials.

Polynomials are added by combining like terms. For example, we can combine the monomials $-9x^3$ and $13x^3$ using addition as follows:
$$-9x^3 + 13x^3 = (-9 + 13)x^3 = 4x^3.$$

These like terms both contain x to the third power.	Add coefficients and keep the same variable factor, x^3.

Example 1 **Adding Polynomials**

Add: $(-6x^3 + 5x^2 + 4) + (2x^3 + 7x^2 - 10)$.

Solution

$$(-6x^3 + 5x^2 + 4) + (2x^3 + 7x^2 - 10)$$
$$= -6x^3 + 5x^2 + 4 + 2x^3 + 7x^2 - 10 \quad \text{Remove the parentheses. Like terms are shown in the same color.}$$
$$= \underline{-6x^3 + 2x^3} + \underline{5x^2 + 7x^2} + \underline{4 - 10} \quad \text{Rearrange the terms so that like terms are adjacent.}$$
$$= \quad -4x^3 \quad + 12x^2 \quad - 6 \quad \text{Combine like terms.}$$
$$= -4x^3 + 12x^2 - 6 \quad \text{This is the same sum as above, written more concisely.}$$

Polynomials can be added by arranging like terms in columns and combining like terms, column by column. Here's the solution to Example 1 using columns and a vertical format:

$$\begin{array}{r} -6x^3 + 5x^2 + 4 \\ 2x^3 + 7x^2 - 10 \\ \hline -4x^3 + 12x^2 - 6. \end{array}$$

✓ **Checkpoint 1** Add: $(-7x^3 + 4x^2 + 3) + (4x^3 + 6x^2 - 13)$.

Subtracting Polynomials

3 Subtract polynomials.

We subtract real numbers by adding the opposite, or additive inverse, of the number being subtracted. For example,

$$8 - 3 = 8 + (-3) = 5.$$

Similarly, we subtract one polynomial from another by adding the opposite of the polynomial being subtracted.

> **Subtracting Polynomials**
>
> To subtract two polynomials, change the sign of every term of the second polynomial. Add this result to the first polynomial.

Example 2 **Subtracting Polynomials**

Subtract: $(7x^3 - 8x^2 + 9x - 6) - (2x^3 - 6x^2 - 3x + 9)$.

Solution

$$(7x^3 - 8x^2 + 9x - 6) - (2x^3 - 6x^2 - 3x + 9)$$

Change the sign of each coefficient.

$$= (7x^3 - 8x^2 + 9x - 6) + (-2x^3 + 6x^2 + 3x - 9) \quad \text{Rewrite subtraction as addition of the additive inverse.}$$
$$= \underbrace{(7x^3 - 2x^3)} + \underbrace{(-8x^2 + 6x^2)} + \underbrace{(9x + 3x)} + \underbrace{(-6 - 9)} \quad \text{Group like terms.}$$
$$= \quad 5x^3 \quad + \quad (-2x^2) \quad + \quad 12x \quad + \quad (-15) \quad \text{Combine like terms.}$$
$$= 5x^3 - 2x^2 + 12x - 15 \quad \text{Simplify.}$$

You can also subtract polynomials using a vertical format. Here's the solution to Example 2 using a vertical format. Notice that you still distribute the negative sign, thereby obtaining the opposite.

$$7x^3 - 8x^2 + 9x - 6$$
$$-(2x^3 - 6x^2 - 3x + 9)$$

Change the sign of every term. →

$$7x^3 - 8x^2 + 9x - 6$$
$$-2x^3 + 6x^2 + 3x - 9$$
$$\overline{5x^3 - 2x^2 + 12x - 15}$$

 **Checkpoint 2**   Subtract: $(14x^3 - 5x^2 + x - 9) - (4x^3 - 3x^2 - 7x + 1)$.

Multiplying Polynomials

④ Multiply Polynomials.

The product of two monomials is obtained by using properties of exponents. For example,

$$(-8x^6)(5x^3) = -8 \cdot 5x^{6+3} = -40x^9.$$

Multiply coefficients and add exponents.

A Brief Review • Differences between Adding and Multiplying Monomials

- Don't confuse adding and multiplying monomials.

  Addition:

  $$5x^4 + 6x^4 = 11x^4$$

  Multiplication:

  $$(5x^4)(6x^4) = (5 \cdot 6)(x^4 \cdot x^4)$$
  $$= 30x^{4+4}$$
  $$= 30x^8$$

- Only like terms can be added or subtracted, but unlike terms may be multiplied.

  Addition:

  $5x^4 + 3x^2$ cannot be simplified.

  Multiplication:

  $$(5x^4)(3x^2) = (5 \cdot 3)(x^4 \cdot x^2)$$
  $$= 15x^{4+2}$$
  $$= 15x^6$$

Multiplying a Monomial and a Polynomial That Is Not a Monomial

We use the distributive property to multiply a monomial and a polynomial that is not a monomial. For example,

$$3x^2(2x^3 + 5x) = 3x^2 \cdot 2x^3 + 3x^2 \cdot 5x = 3 \cdot 2x^{2+3} + 3 \cdot 5x^{2+1} = 6x^5 + 15x^3.$$

Monomial   Binomial

Multiply coefficients and add exponents.

Multiplying a Monomial and a Polynomial That Is Not a Monomial

To multiply a monomial and a polynomial, use the distributive property to multiply each term of the polynomial by the monomial.

Example 3 Multiplying a Monomial and a Trinomial

Multiply: $4x^3(6x^5 - 2x^2 + 3)$.

Solution

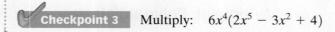

$$4x^3(6x^5 - 2x^2 + 3) = 4x^3 \cdot 6x^5 - 4x^3 \cdot 2x^2 + 4x^3 \cdot 3$$ Use the distributive property.
$$= 24x^8 - 8x^5 + 12x^3$$ Multiply coefficients and add exponents.

✓ **Checkpoint 3** Multiply: $6x^4(2x^5 - 3x^2 + 4)$

Multiplying Polynomials When Neither Is a Monomial

How do we multiply two polynomials if neither is a monomial? For example, consider

$$(2x + 3)(x^2 + 4x + 5).$$

Binomial   Trinomial

One way to perform $(2x + 3)(x^2 + 4x + 5)$ is to distribute $2x$ throughout the trinomial

$$2x(x^2 + 4x + 5)$$

and 3 throughout the trinomial

$$3(x^2 + 4x + 5).$$

Then combine the like terms that result.

> **Multiplying Polynomials When Neither Is a Monomial**
> Multiply each term of one polynomial by each term of the other polynomial. Then combine like terms.

Example 4 Multiplying a Binomial and a Trinomial

Multiply: $(2x + 3)(x^2 + 4x + 5)$.

Solution

$$(2x + 3)(x^2 + 4x + 5)$$

$$= 2x(x^2 + 4x + 5) + 3(x^2 + 4x + 5)$$ Multiply the trinomial by each term of the binomial.

$$= 2x \cdot x^2 + 2x \cdot 4x + 2x \cdot 5 + 3x^2 + 3 \cdot 4x + 3 \cdot 5$$ Use the distributive property.

$$= 2x^3 + 8x^2 + 10x + 3x^2 + 12x + 15$$ Multiply monomials: Multiply coefficients and add exponents.

$$= 2x^3 + 11x^2 + 22x + 15$$ Combine like terms: $8x^2 + 3x^2 = 11x^2$ and $10x + 12x = 22x$.

Another method for performing the multiplication is to use a vertical format similar to that used for multiplying whole numbers.

$$
\begin{array}{r}
x^2 + 4x + 5 \\
2x + 3 \\
\hline
3x^2 + 12x + 15 \\
2x^3 + 8x^2 + 10x \\
\hline
2x^3 + 11x^2 + 22x + 15
\end{array}
$$

$3(x^2 + 4x + 5)$

$2x(x^2 + 4x + 5)$ Write like terms in the same column.

Combine like terms.

 **Checkpoint 4**   Multiply: $(5x - 2)(3x^2 - 5x + 4)$.

The Product of Two Binomials: FOIL

⑤ Use FOIL in polynomial multiplication.

Frequently, we need to find the product of two binomials. One way to perform this multiplication is to distribute each term in the first binomial through the second binomial. For example, we can find the product of the binomials $3x + 2$ and $4x + 5$ as follows:

$$
\begin{aligned}
(3x + 2)(4x + 5) &= 3x(4x + 5) + 2(4x + 5) \\
&= 3x(4x) + 3x(5) + 2(4x) + 2(5) \\
&= 12x^2 + 15x + 8x + 10.
\end{aligned}
$$

Distribute $3x$ over $4x + 5$.

Distribute 2 over $4x + 5$.

We'll combine these like terms later. For now, our interest is in how to obtain *each* of these four terms.

We can also find the product of $3x + 2$ and $4x + 5$ using a method called FOIL, which is based on our work shown above. Any two binomials can be quickly multiplied by using the FOIL method, in which **F** represents the product of the **first** terms in each binomial, **O** represents the product of the **outside** terms, **I** represents the product of the **inside** terms, and **L** represents the product of the **last**, or second, terms in each binomial. For example, we can use the FOIL method to find the product of the binomials $3x + 2$ and $4x + 5$ as follows:

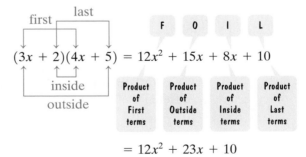

$$= 12x^2 + 23x + 10$$ Combine like terms.

In general, here's how to use the FOIL method to find the product of $ax + b$ and $cx + d$:

Using the FOIL Method to Multiply Binomials

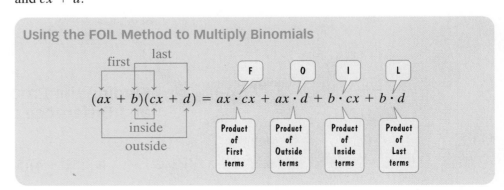

Example 5   Using the FOIL Method

Multiply:

a. $(3x + 4)(5x - 3)$ **b.** $(5x^3 - 6)(4x^3 - x)$.

Solution

a.

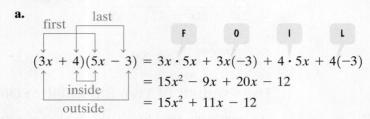

$$(3x + 4)(5x - 3) = 3x \cdot 5x + 3x(-3) + 4 \cdot 5x + 4(-3)$$
$$= 15x^2 - 9x + 20x - 12$$
$$= 15x^2 + 11x - 12$$

b.

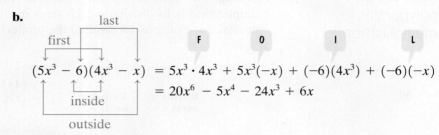

$$(5x^3 - 6)(4x^3 - x) = 5x^3 \cdot 4x^3 + 5x^3(-x) + (-6)(4x^3) + (-6)(-x)$$
$$= 20x^6 - 5x^4 - 24x^3 + 6x$$

✓ **Checkpoint 5** Multiply:

a. $(7x + 5)(4x - 3)$

▶ **b.** $(4x^3 - 5)(x^3 - 3x)$.

Multiplying the Sum and Difference of Two Terms

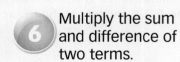

6 Multiply the sum and difference of two terms.

We can use the FOIL method to multiply $A + B$ and $A - B$ as follows:

$$\overset{\text{F} \quad \text{O} \quad \text{I} \quad \text{L}}{(A + B)(A - B) = A^2 - AB + AB - B^2 = A^2 - B^2}.$$

Notice that the outside and inside products have a sum of 0 and the terms cancel. The FOIL multiplication provides us with a quick rule for multiplying the sum and difference of two terms.

> **The Product of the Sum and Difference of Two Terms**
>
> $$(A + B)(A - B) = A^2 - B^2$$
>
> | The product of the sum and the difference of the same two terms | is | the square of the first term minus the square of the second term. |

Example 6   Finding the Product of the Sum and Difference of Two Terms

Multiply:

a. $(4y + 3)(4y - 3)$ **b.** $(5a^4 + 6)(5a^4 - 6)$.

Solution

Use the formula for the product of the sum and difference of two terms.

$$(A + B)(A - B) = A^2 - B^2$$

| | First term squared | − | Second term squared | = | Product |

a. $(4y + 3)(4y - 3)$ = $(4y)^2 - 3^2$ = $16y^2 - 9$

b. $(5a^4 + 6)(5a^4 - 6)$ = $(5a^4)^2 - 6^2$ = $25a^8 - 36$

✓ **Checkpoint 6** Multiply:

▶ **a.** $(7x + 8)(7x - 8)$ **b.** $(2y^3 - 5)(2y^3 + 5)$.

The Square of a Binomial

Square binomials.

Let us find $(A + B)^2$, the square of a binomial sum. To do so, we begin with the FOIL method and look for a general rule.

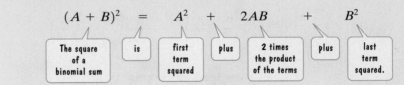

$$(A + B)^2 = (A + B)(A + B) = A \cdot A + A \cdot B + A \cdot B + B \cdot B$$
$$= A^2 + 2AB + B^2$$

This result implies the following rule:

> **The Square of a Binomial Sum**
>
> $$(A + B)^2 = A^2 + 2AB + B^2$$
>
> | The square of a binomial sum | is | first term squared | plus | 2 times the product of the terms | plus | last term squared. |

Example 7 Finding the Square of a Binomial Sum

Multiply:

a. $(x + 3)^2$ **b.** $(3x + 7)^2$.

Solution

Use the formula for the square of a binomial sum.

$$(A + B)^2 = A^2 + 2AB + B^2$$

	(First Term)²	+	2 · Product of the Terms	+	(Last Term)²	= Product
a. $(x + 3)^2 =$	x^2	+	$2 \cdot x \cdot 3$	+	3^2	$= x^2 + 6x + 9$
b. $(3x + 7)^2 =$	$(3x)^2$	+	$2(3x)(7)$	+	7^2	$= 9x^2 + 42x + 49$

✓ **Checkpoint 7** Multiply:

▶ **a.** $(x + 10)^2$ **b.** $(5x + 4)^2$.

A similar pattern occurs for $(A - B)^2$, the square of a binomial difference. Using the FOIL method on $(A - B)^2$, we obtain the following rule:

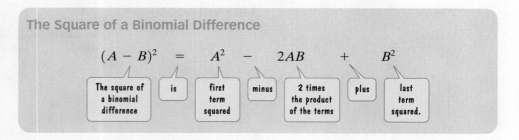

The Square of a Binomial Difference

$$(A - B)^2 = A^2 - 2AB + B^2$$

The square of a binomial difference | is | first term squared | minus | 2 times the product of the terms | plus | last term squared.

Example 8   Finding the Square of a Binomial Difference

Multiply:

a. $(x - 4)^2$ **b.** $(5y - 6)^2$.

Solution

Use the formula for the square of a binomial difference.

$$(A - B)^2 = A^2 - 2AB + B^2$$

	(First Term)^2	−	2 · Product of the Terms	+	(Last Term)^2	= Product
a. $(x - 4)^2 =$	x^2	−	$2 \cdot x \cdot 4$	+	4^2	$= x^2 - 8x + 16$
b. $(5y - 6)^2 =$	$(5y)^2$	−	$2(5y)(6)$	+	6^2	$= 25y^2 - 60y + 36$

Checkpoint 8   Multiply:

a. $(x - 9)^2$ **b.** $(7x - 3)^2$.

Polynomial Functions

8 Use polynomial functions.

We have seen that a polynomial function is defined by an expression that is a polynomial. In our final example, we apply operations with polynomials to polynomial functions.

Example 9   Modeling with Polynomial Functions

We opened the section with functions that model the number of births and deaths in the United States from 2000 through 2005:

$$B(x) = 7.4x^2 - 15x + 4046 \qquad D(x) = -3.5x^2 + 20x + 2405.$$

Number of births, $B(x)$, in thousands, x years after 2000

Number of deaths, $D(x)$, in thousands, x years after 2000

a. Write a function that models the change in U.S. population, $C(x)$, for each year from 2000 through 2005.

b. Use the function from part (a) to find the change in U.S. population in 2003.

c. Does the result in part (b) overestimate or underestimate the actual population change in 2003 obtained from the data in **Figure 6.1** on page 374? By how much?

Solution

a. The change in population, $C(x)$, is the number of births minus the number of deaths: $C(x) = B(x) - D(x)$.

$C(x)$

$= B(x) - D(x)$

$= (7.4x^2 - 15x + 4046) - (-3.5x^2 + 20x + 2405)$ Substitute the given functions.

$= 7.4x^2 - 15x + 4046 + 3.5x^2 - 20x - 2405$ Remove parentheses and change the sign of each term in the second set of parentheses.

$= (7.4x^2 + 3.5x^2) + (-15x - 20x) + (4046 - 2405)$ Group like terms.

$= 10.9x^2 - 35x + 1641$ Combine like terms.

The function

$$C(x) = 10.9x^2 - 35x + 1641$$

models the change in U.S. population, $C(x)$, in thousands, x years after 2000.

b. Because 2003 is 3 years after 2000, we substitute 3 for x in the change function $C(x)$.

$C(x) = 10.9x^2 - 35x + 1641$ Use the change-in-population function.

$C(3) = 10.9(3)^2 - 35(3) + 1641$ Substitute 3 for x.

$= 10.9(9) - 35(3) + 1641$ Evaluate the exponential expression: $3^2 = 9$.

$= 98.1 - 105 + 1641$ Perform the multiplications.

$= 1634.1$ Subtract and add from left to right.

We see that $C(3) = 1634.1$. The model indicates that there was a population increase of 1634.1 thousand, or 1,634,100 people, in 2003.

c. The data for 2003 in **Figure 6.1** on page 374 show 4090 thousand births and 2448 thousand deaths.

$$\text{population change} = \text{births} - \text{deaths}$$
$$= 4090 - 2448 = 1642$$

The actual population increase was 1642 thousand, or 1,642,000. Our model gave us an increase of 1634.1 thousand. Thus, the model underestimates the actual increase by $1642 - 1634.1$, or 7.9 thousand people.

> **Checkpoint 9**   Use the birth and death models from Example 9.

a. Write a function that models the total number of births and deaths in the United States, $T(x)$, for each year from 2000 through 2005.

b. Use the function from part (a) to find the total number of births and deaths in the United States in 2005.

c. Does the result in part (b) overestimate or underestimate the actual number of total births and deaths in 2005 obtained from the data in **Figure 6.1** on page 374? By how much?

Blitzer Bonus

Labrador Retrievers and Polynomial Multiplication

The color of a Labrador retriever is determined by its pair of genes. A single gene is inherited at random from each parent. The black-fur gene, B, is dominant. The yellow-fur gene, Y, is recessive. This means that labs with at least one black-fur gene (BB or BY) have black coats. Only labs with two yellow-fur genes (YY) have yellow coats.

Axl, your author's yellow lab, pictured on the right, inherited his genetic makeup from two black BY parents.

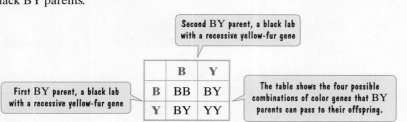

Second BY parent, a black lab with a recessive yellow-fur gene

First BY parent, a black lab with a recessive yellow-fur gene

	B	Y
B	BB	BY
Y	BY	YY

The table shows the four possible combinations of color genes that BY parents can pass to their offspring.

Old Dog... New Chicks

Because YY is one of four possible outcomes, the probability that a yellow lab like Axl will be the offspring of these black parents is $\frac{1}{4}$. The probabilities suggested by the table can be modeled by the expression $\left(\frac{1}{2}B + \frac{1}{2}Y\right)^2$.

$$\left(\frac{1}{2}B + \frac{1}{2}Y\right)^2 = \left(\frac{1}{2}B\right)^2 + 2\left(\frac{1}{2}B\right)\left(\frac{1}{2}Y\right) + \left(\frac{1}{2}Y\right)^2$$

$$= \frac{1}{4}BB + \frac{1}{2}BY + \frac{1}{4}YY$$

The probability of a black lab with two dominant black genes is $\frac{1}{4}$.

The probability of a black lab with a recessive yellow gene is $\frac{1}{2}$.

The probability of a yellow lab with two recessive yellow genes is $\frac{1}{4}$.

Achieving Success

Form a study group with other students in your class. Working in small groups often serves as an excellent way to learn and reinforce new material. Set up helpful procedures and guidelines for the group. "Talk" math by discussing and explaining the concepts and exercises to one another.

Exercise Set 6.1

Concept and Vocabulary Exercises

In Exercises 1–4, fill in each blank so that the resulting statement is true.

1. Exponents on the variables of a polynomial must be _____ numbers.

2. A polynomial with one term is called a/an _____, a polynomial with two terms is called a/an _____, and a polynomial with three terms is called a/an _____.

3. The degree of ax^n, $a \neq 0$, is _____.

4. The degree of a polynomial is _____.

In Exercises 5–8, determine whether each statement is true or false. If the statement is false, make the necessary change(s) to produce a true statement.

5. $4x^3 + 7x^2 - 5x + 2x^{-1}$ is a polynomial containing four terms.

6. If two polynomials of degree 2 are added, the sum must be a polynomial of degree 2.

7. $(x^2 - 7x) - (x^2 - 4x) = -11x$ for all values of x.

8. All terms of a polynomial are monomials.

Respond to Exercises 9–20 using verbal or written explanations.

9. What is a polynomial?

10. Explain how to determine the degree of each term of a polynomial.

11. Explain how to determine the degree of a polynomial.

12. What is a polynomial function?

13. Explain how to add polynomials.

14. Explain how to subtract polynomials.

15. Explain how to multiply a monomial and a polynomial that is not a monomial. Give an example.

16. Explain how to multiply a binomial and a trinomial.

17. What is the FOIL method and when is it used? Give an example of the method.

18. Explain how to find the product of the sum and difference of two terms. Give an example with your explanation.

19. Explain how to square a binomial sum. Give an example.

20. Explain how to square a binomial difference. Give an example.

In Exercises 21–24, determine whether each statement makes sense or does not make sense, and explain your reasoning.

21. Knowing the difference between factors and terms is important: In $(3x^2)^2$, I can distribute the exponent 2 on each factor, but in $(x^2 + 3)^2$, I cannot do the same thing on each term.

22. Many English words have prefixes with meanings similar to those used to describe polynomials, such as *monologue*, *binocular*, and *tricuspid*.

23. I used the FOIL method to find the product of $x + 5$ and $x^2 + 2x + 1$.

24. Instead of using the formula for the square of a binomial sum, I prefer to write the binomial sum twice and then apply the FOIL method.

Practice Exercises

In Exercises 25–28, is the algebraic expression a polynomial? If it is, write the polynomial in standard form.

25. $2x + 3x^2 - 5$

26. $2x + 3x^{-1} - 5$

27. $2x^2 + 3^{-1}x - 5$

28. $x^2 - x^3 + x^4 - 5$

In Exercises 29–32, find the degree of the polynomial.

29. $3x^2 - 5x + 4$

30. $-4x^3 + 7x^2 - 11$

31. $x^2 - 4x^3 + 9x - 12x^4 + 63$

32. $x^2 - 8x^3 + 15x^4 + 91$

In Exercises 33–38, perform the indicated operations. Write the resulting polynomial in standard form and indicate its degree.

33. $(-6x^3 + 5x^2 - 8x + 9) + (17x^3 + 2x^2 - 4x - 13)$

34. $(-7x^3 + 6x^2 - 11x + 13) + (19x^3 - 11x^2 + 7x - 17)$

35. $(17x^3 - 5x^2 + 4x - 3) - (5x^3 - 9x^2 - 8x + 11)$

36. $(18x^4 - 2x^3 - 7x + 8) - (9x^4 - 6x^3 - 5x + 7)$

37. $(5x^2 - 7x - 8) + (2x^2 - 3x + 7) - (x^2 - 4x - 3)$

38. $(8x^2 + 7x - 5) - (3x^2 - 4x) - (-6x^3 - 5x^2 + 3)$

In Exercises 39–80, find each product.

39. $4x^2(3x + 2)$

40. $5x^2(6x + 7)$

41. $2y(y^2 - 5y)$

42. $3y(y^2 - 4y)$

43. $5x^3(2x^5 - 4x^2 + 9)$

44. $6x^3(3x^5 - 5x^2 + 7)$

45. $(x + 1)(x^2 - x + 1)$

46. $(x + 5)(x^2 - 5x + 25)$

47. $(2x - 3)(x^2 - 3x + 5)$

48. $(2x - 1)(x^2 - 4x + 3)$

49. $(x + 7)(x + 3)$

50. $(x + 8)(x + 5)$

51. $(x - 5)(x + 3)$

52. $(x - 1)(x + 2)$

53. $(3x + 5)(2x + 1)$

54. $(7x + 4)(3x + 1)$

55. $(2x - 3)(5x + 3)$

56. $(2x - 5)(7x + 2)$

57. $(5x^2 - 4)(3x^2 - 7)$

58. $(7x^2 - 2)(3x^2 - 5)$

59. $(8x^3 + 3)(x^2 - 5)$

60. $(7x^3 + 5)(x^2 - 2)$

61. $(x + 3)(x - 3)$

62. $(x + 5)(x - 5)$

63. $(3x + 2)(3x - 2)$

64. $(2x + 5)(2x - 5)$

65. $(5 - 7x)(5 + 7x)$

66. $(4 - 3x)(4 + 3x)$

67. $(4x^2 + 5x)(4x^2 - 5x)$

68. $(3x^2 + 4x)(3x^2 - 4x)$

69. $(1 - y^5)(1 + y^5)$

70. $(2 - y^5)(2 + y^5)$

71. $(x + 2)^2$

72. $(x + 5)^2$

73. $(2x + 3)^2$

74. $(3x + 2)^2$

75. $(x - 3)^2$

76. $(x - 4)^2$

77. $(4x^2 - 1)^2$

78. $(5x^2 - 3)^2$

79. $(7 - 2x)^2$

80. $(9 - 5x)^2$

Practice Plus

In Exercises 81–88, perform the indicated operation or operations.

81. $(3x + 4)^2 - (3x - 4)^2$

82. $(5x + 2)^2 - (5x - 2)^2$

83. $(5x - 7)(3x - 2) - (4x - 5)(6x - 1)$

84. $(3x + 5)(2x - 9) - (7x - 2)(x - 1)$

85. $(2x + 5)(2x - 5)(4x^2 + 25)$

86. $(3x + 4)(3x - 4)(9x^2 + 16)$

87. $\dfrac{(2x - 7)^5}{(2x - 7)^3}$

88. $\dfrac{(5x - 3)^6}{(5x - 3)^4}$

Application Exercises

The bar graph shows the percentage of total U.S. dollar sales of recorded music for rock, country, and rap/hip-hop for five selected years from 2000 through 2008.

Sales of Recorded Music in the United States, by Genre

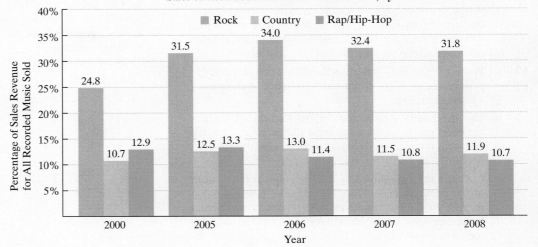

Source: Recording Industry Association of America

The data shown by the bar graph at the bottom of the previous page can be modeled by the following functions:

$$R(x) = -0.21x^2 + 2.6x + 24.7$$

$$C(x) = -0.09x^2 + 0.8x + 10.7$$

$$H(x) = -0.09x^2 + 0.4x + 13.0.$$

Each function gives the percentage of sales revenue, $R(x)$, $C(x)$, or $H(x)$, for all recorded music sold in the United States x years after 2000. Use this information to solve Exercises 89–92.

89. a. Write a function f that gives the combined percentage of total dollar sales for rock and country x years after 2000.

b. Use the function from part (a) to find the combined percentage of total dollar sales for rock and country in 2007.

c. Consider the data shown by the bar graph at the bottom of the previous page. Does the function value in part (b) overestimate or underestimate the actual percentage of total dollar sales for rock and country combined in 2007? By how much?

90. a. Write a function g that gives the combined percentage of total dollar sales for rock and rap/hip-hop x years after 2000.

b. Use the function from part (a) to find the combined percentage of total dollar sales for rock and rap/hip-hop in 2007.

c. Consider the data shown by the bar graph at the bottom of the previous page. Does the function value in part (b) overestimate or underestimate the actual percentage of total dollar sales for rock and rap/hip-hop combined in 2007? By how much?

91. a. Write a function F that gives the difference between the percentage of total dollar sales for rock and rap/hip-hop x years after 2000.

b. Use the function from part (a) to find the difference between the percentage of total dollar sales for rock and rap/hip-hop in 2008. Round to the nearest tenth of a percent.

c. Consider the data shown by the bar graph at the bottom of the previous page. Does the rounded function value in part (b) overestimate or underestimate the actual difference between the percentage of total dollar sales for rock and rap/hip-hop in 2008? By how much?

92. a. Write a function G that gives the difference between the percentage of total dollar sales for rock and country x years after 2000.

b. Use the function from part (a) to find the difference between the percentage of total dollar sales for rock and country in 2008. Round to the nearest tenth of a percent.

c. Consider the data shown by the bar graph at the bottom of the previous page. Does the rounded function value in part (b) overestimate or underestimate the actual difference between the percentage of total dollar sales for rock and country in 2008? By how much?

The area, A, of a rectangle with length l and width w is given by the formula $A = lw$. In Exercises 93–98, use this formula to write a polynomial in standard form that models, or represents, the area of each shaded region.

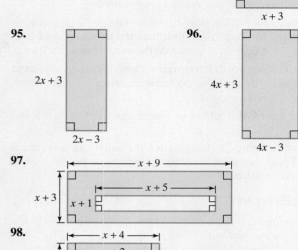

93. $x+1$, $x+1$

94. $x+3$, $x+3$

95. $2x+3$, $2x-3$

96. $4x+3$, $4x-3$

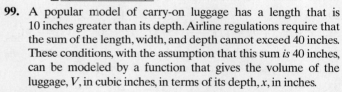

97. $x+9$, $x+5$, $x+3$, $x+1$

98. $x+4$, $x+2$, $x+3$, $x+1$

99. A popular model of carry-on luggage has a length that is 10 inches greater than its depth. Airline regulations require that the sum of the length, width, and depth cannot exceed 40 inches. These conditions, with the assumption that this sum *is* 40 inches, can be modeled by a function that gives the volume of the luggage, V, in cubic inches, in terms of its depth, x, in inches.

$$\underbrace{\text{Volume}}\ =\ \underbrace{\text{depth}}\ \cdot\ \underbrace{\text{length}}\ \cdot\ \underbrace{\text{width: } 40 - (\text{depth} + \text{length})}$$

$$V(x)\ =\ x\ \cdot\ (x+10)\ \cdot\ [40-(x+x+10)]$$

$$V(x)\ =\ x(x+10)(30-2x)$$

a. Perform the multiplications in the formula for $V(x)$ and express the formula in standard form.

b. Use the formula from part (a) to find $V(10)$. Describe what this means in practical terms.

c. The graph of the function modeling the volume of carry-on luggage is shown below. Identify your answer from part (b) as a point on the graph.

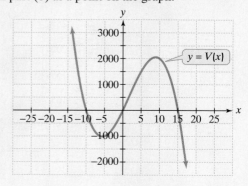

$y = V(x)$

d. Use the graph to describe realistic values for x for the volume function, where x represents the depth of the carry-on luggage. Use set-builder notation to express these realistic values.

100. Before working this exercise, be sure that you have read the Blitzer Bonus on page 384. The table shows the four combinations of color genes that a YY yellow lab and a BY black lab can pass to their offspring.

	B	**Y**
Y	BY	YY
Y	BY	YY

a. How many combinations result in a yellow lab with two recessive yellow genes? What is the probability of a yellow lab?

b. How many combinations result in a black lab with a recessive yellow gene? What is the probability of a black lab?

c. Find the product of Y and $\frac{1}{2}$B $+ \frac{1}{2}$Y. How does this product model the probabilities that you determined in parts (a) and (b)?

Critical Thinking Exercises

In Exercises 101–103, perform the indicated operations.

101. $[(7x^2 + 5) + 4x][(7x^2 + 5) - 4x]$

102. $[(3x^2 + x) + 1]^2$

103. $(x^n + 2)(x^n - 2) - (x^n - 3)^2$

104. Express the area of the plane figure shown as a polynomial in standard form.

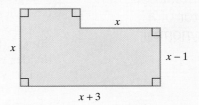

6.2 | Factoring Polynomials

Objectives

1 Factor out the greatest common factor of a polynomial.

2 Factor by grouping.

3 Factor a trinomial whose leading coefficient is 1.

4 Factor a trinomial whose leading coefficient is not 1.

5 Factor the difference of squares.

6 Factor perfect square trinomials.

7 Use a general strategy for factoring polynomials.

A two-year-old boy is asked, "Do you have a brother?" He answers, "Yes." "What is your brother's name?" "Tom." Asked if Tom has a brother, the two-year-old replies, "No." The child can go in the direction from self to brother, but he cannot reverse this direction and move from brother back to self.

As our intellects develop, we learn to reverse the direction of our thinking. Reversibility of thought is found throughout algebra. For example, we can multiply polynomials and show that

$$5x(2x + 3) = 10x^2 + 15x.$$

We can also reverse this process and express the resulting polynomial as

$$10x^2 + 15x = 5x(2x + 3).$$

Factoring a polynomial containing the sum of monomials means finding an equivalent expression that is a product.

Factoring $10x^2 + 15x$

Sum of monomials | Equivalent expression that is a product

$$10x^2 + 15x = 5x(2x + 3)$$

The factors of $10x^2 + 15x$ are $5x$ and $2x + 3$.

In this section, we will be **factoring over the set of integers**, meaning that the coefficients in the factors are integers. Polynomials that cannot be factored using integer coefficients are called **irreducible over the integers**, or **prime**.

The goal in factoring a polynomial is to use one or more factoring techniques until each of the polynomial's factors, except possibly for a monomial factor, is prime or irreducible. In this situation, the polynomial is said to be **factored completely**.

We will now discuss basic techniques for factoring polynomials.

Factoring Out the Greatest Common Factor

1 Factor out the greatest common factor of a polynomial.

In any factoring problem, the first step is to look for the *greatest common factor*. The **greatest common factor**, abbreviated GCF, is an expression with the greatest coefficient and of the highest degree that divides each term of the polynomial. Can you see that $7x$ is the greatest common factor of $21x^2 + 28x$? 7 is the greatest integer that divides both 21 and 28. Furthermore, x is the greatest power of x that divides x^2 and x.

The variable part of the greatest common factor always contains the *smallest* power of a variable that appears in all terms of the polynomial. For example, consider the polynomial

$$21x^2 + 28x.$$

x^1, or x, is the variable raised to the smallest exponent.

We see that x is the variable part of the greatest common factor, $7x$.

When factoring a monomial from a polynomial, determine the greatest common factor of all terms in the polynomial. Sometimes there may not be a GCF other than 1. When a GCF other than 1 exists, we use the following procedure:

> **Factoring a Monomial from a Polynomial**
>
> 1. Determine the greatest common factor of all terms in the polynomial.
> 2. Express each term as the product of the GCF and its other factor.
> 3. Use the distributive property to factor out the GCF.

Example 1   Factoring Out the Greatest Common Factor

Factor: $21x^2 + 28x$.

Solution

The GCF of the two terms of the polynomial is $7x$.

$$21x^2 + 28x$$
$$= 7x(3x) + 7x(4) \qquad \text{Express each term as the product of the GCF and its other factor.}$$
$$= 7x(3x + 4) \qquad \text{Factor out the GCF.}$$

We can check this factorization by multiplying $7x$ and $3x + 4$, obtaining the original polynomial as the answer.

▶    Factor: $20x^2 + 30x$.

Example 2   Factoring Out the Greatest Common Factor

Factor:

a. $18x^3 + 27x^2$ **b.** $x^2(x + 3) + 5(x + 3)$.

Solution

a. First, determine the greatest common factor.

> 9 is the greatest integer that divides 18 and 27.

$$18x^3 + 27x^2$$

> x^2 is the greatest expression that divides x^3 and x^2.

The GCF of the two terms of the polynomial is $9x^2$.

$$18x^3 + 27x^2$$
$$= 9x^2(2x) + 9x^2(3) \quad \text{Express each term as the product of the GCF}$$
$$\text{and its other factor.}$$
$$= 9x^2(2x + 3) \quad \text{Factor out the GCF.}$$

b. In this situation, the greatest common factor is the common binomial factor $(x + 3)$. We factor out this common factor as follows:

$$x^2(x + 3) + 5(x + 3) = (x + 3)(x^2 + 5). \quad \text{Factor out the common binomial factor.}$$

> ✓ **Checkpoint 2**   Factor:
>
> **a.** $10x^3 - 4x^2$ **b.** $2x(x - 7) + 3(x - 7)$.

Factoring by Grouping

2 Factor by grouping.

The terms of some polynomials have only a greatest common factor of 1. However, by a suitable grouping of the terms, it still may be possible to factor by factoring out common factors. This process, called **factoring by grouping**, is illustrated in Example 3.

Example 3   Factoring by Grouping

Factor: $x^3 + 4x^2 + 3x + 12$.

Solution

There is no factor other than 1 common to all terms. However, we can group terms that have a common factor:

$$\boxed{x^3 + 4x^2} \quad + \quad \boxed{3x + 12}.$$

> Common factor   Common factor
> is x^2.   is 3.

Study Tip

In Example 3, you can group the terms as follows:

$$(x^3 + 3x) + (4x^2 + 12).$$

Factor out the greatest common factor from each group and complete the factoring process. Describe what happens. What can you conclude?

We now factor the given polynomial as follows:

$$x^3 + 4x^2 + 3x + 12$$
$$= (x^3 + 4x^2) + (3x + 12) \quad \text{Group terms with common factors.}$$
$$= x^2(x + 4) + 3(x + 4) \quad \text{Factor out the greatest common factor from}$$
$$\text{the grouped terms. The remaining two terms}$$
$$\text{have } x + 4 \text{ as a common binomial factor.}$$
$$= (x + 4)(x^2 + 3). \quad \text{Factor out the GCF, } x + 4.$$

Thus, $x^3 + 4x^2 + 3x + 12 = (x + 4)(x^2 + 3)$.   Check   the   factorization   by multiplying the right side of the equation using the FOIL method. Because the factorization is correct, you should obtain the original polynomial.

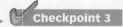 **Checkpoint 3**   Factor: $x^3 + 5x^2 + 2x + 10$.

> **Factoring by Grouping**
> 1. Group terms that have a common monomial factor. There will usually be two groups. Sometimes the terms must be rearranged.
> 2. Factor out the common monomial factor from each group.
> 3. Factor out the remaining common binomial factor (if one exists).

Factoring a Trinomial Whose Leading Coefficient Is 1

3 Factor a trinomial whose leading coefficient is 1.

In Section 6.1, we used the FOIL method to multiply two binomials. The product was often a trinomial. The following are some examples:

Factored Form		F O I L		Trinomial Form
$(x + 3)(x + 4)$	$=$	$x^2 + 4x + 3x + 12$	$=$	$x^2 + 7x + 12$
$(x - 3)(x - 4)$	$=$	$x^2 - 4x - 3x + 12$	$=$	$x^2 - 7x + 12$
$(x + 3)(x - 5)$	$=$	$x^2 - 5x + 3x - 15$	$=$	$x^2 - 2x - 15$

Observe that each trinomial is of the form $x^2 + bx + c$, where the coefficient of the squared term is 1. Our goal is to start with the trinomial form and, assuming that it is factorable, return to the factored form.

The first FOIL multiplication shown in our list indicates that

$$(x + 3)(x + 4) = x^2 + 7x + 12.$$

Let's reverse the sides of this equation:

$$x^2 + 7x + 12 = (x + 3)(x + 4).$$

We can make several important observations about the factors on the right side.

$x^2 + 7x + 12 = (x + 3)(x + 4)$ $x^2 + 7x + 12 = (x + 3)(x + 4)$ $x^2 + 7x + 12 = (x + 3)(x + 4)$

I: $3x$
O: $4x$

The first term of each factor is x. The product of the First terms is $x \cdot x = x^2$.

3 and 4 are factors of 12. The product of the Last terms is $3 \cdot 4 = 12$.

The sum of the Outside and Inside products is $4x + 3x = 7x$.

These observations provide us with a procedure for factoring $x^2 + bx + c$.

A Strategy for Factoring $x^2 + bx + c$

1. Enter x as the first term of each factor.

$$(x\quad)(x\quad) = x^2 + bx + c$$

2. List all pairs of factors of the constant c.

3. Try various combinations of these factors as the second term in each set of parentheses. Select the combination in which the sum of the Outside and Inside products is equal to bx.

$$(x + \square)(x + \square) = x^2 + bx + c$$

I
O
Sum of O + I

4. Check your work by multiplying the factors using the FOIL method. You should obtain the original trinomial.

If none of the possible combinations yield an Outside product and an Inside product whose sum is equal to bx, the trinomial cannot be factored using integers and is called **prime** over the set of integers.

Example 4 Factoring a Trinomial Whose Leading Coefficient Is 1

Factor: $x^2 + 6x + 8$.

Solution

Step 1 Enter x as the first term of each factor.

$$x^2 + 6x + 8 = (x\quad)(x\quad)$$

To find the second term of each factor, we must find two integers whose product is 8 and whose sum is 6.

Step 2 List all pairs of factors of the constant, 8.

Factors of 8	8, 1	4, 2	−8, −1	−4, −2

Step 3 Try various combinations of these factors. The correct factorization of $x^2 + 6x + 8$ is the one in which the sum of the Outside and Inside products is equal to $6x$. Here is a list of the possible factorizations:

Possible Factorizations of $x^2 + 6x + 8$	Sum of Outside and Inside Products (Should Equal $6x$)
$(x + 8)(x + 1)$	$x + 8x = 9x$
$(x + 4)(x + 2)$	$2x + 4x = 6x$
$(x − 8)(x − 1)$	$-x - 8x = -9x$
$(x − 4)(x − 2)$	$-2x - 4x = -6x$

This is the required middle term.

Thus, $x^2 + 6x + 8 = (x + 4)(x + 2)$.

Step 4 Check this result, $x^2 + 6x + 8 = (x + 4)(x + 2)$, by multiplying the right side using the FOIL method. You should obtain the original trinomial. Because of the commutative property, the factorization can also be expressed as

$$x^2 + 6x + 8 = (x + 2)(x + 4).$$

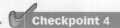 **Checkpoint 4** Factor: $x^2 + 5x + 6$.

Example 5 **Factoring a Trinomial Whose Leading Coefficient Is 1**

Factor: $x^2 + 2x - 35$.

Solution

Step 1 Enter x as the first term of each factor.

$$x^2 + 2x - 35 = (x \quad)(x \quad)$$

To find the second term of each factor, we must find two integers whose product is -35 and whose sum is 2.

Step 2 List all pairs of factors of the constant, -35.

Factors of -35	$35, -1$	$-35, 1$	$-7, 5$	$7, -5$

Step 3 Try various combinations of these factors. The correct factorization of $x^2 + 2x - 35$ is the one in which the sum of the Outside and Inside products is equal to $2x$. Here is a list of the possible factorizations:

Possible Factorizations of $x^2 + 2x - 35$	Sum of Outside and Inside Products (Should Equal $2x$)
$(x - 1)(x + 35)$	$35x - x = 34x$
$(x + 1)(x - 35)$	$-35x + x = -34x$
$(x - 7)(x + 5)$	$5x - 7x = -2x$
$(x + 7)(x - 5)$	$-5x + 7x = 2x$

> This is the required middle term.

Thus, $x^2 + 2x - 35 = (x + 7)(x - 5)$ or $(x - 5)(x + 7)$.

Step 4 Verify the factorization using the FOIL method.

$$\overset{\text{F} \quad \text{O} \quad \text{I} \quad \text{L}}{(x + 7)(x - 5) = x^2 - 5x + 7x - 35 = x^2 + 2x - 35}$$

Because the product of the factors is the original trinomial, the factorization is correct.

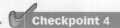 **Checkpoint 5** Factor: $x^2 + 3x - 10$.

> **Study Tip**
>
> To factor $x^2 + bx + c$ when c is negative, find two numbers with opposite signs whose sum is the coefficient of the middle term.
>
> $$x^2 + 2x - 35 = (x + 7)(x - 5)$$
>
> Negative   Opposite signs

Factoring a Trinomial Whose Leading Coefficient Is Not 1

④ Factor a trinomial whose leading coefficient is not 1.

How do we factor a trinomial such as $3x^2 - 20x + 28$? Notice that the leading coefficient is 3. We must find two binomials whose product is $3x^2 - 20x + 28$. The product of the First terms must be $3x^2$:

$$(3x \quad)(x \quad).$$

From this point on, the factoring strategy is exactly the same as the one we use to factor a trinomial whose leading coefficient is 1.

A Strategy for Factoring $ax^2 + bx + c$

Assume, for the moment, that the greatest common factor is 1.

1. Find two **First** terms whose product is ax^2:

$$(\square x + \quad)(\square x + \quad) = ax^2 + bx + c.$$

2. Find two **Last** terms whose product is c:

$$(\square x + \square)(\square x + \square) = ax^2 + bx + c.$$

3. By trial and error, perform steps 1 and 2 until the sum of the **Outside** product and **Inside** product is bx:

$$(\square x + \square)(\square x + \square) = ax^2 + bx + c.$$

$$\text{Sum of O + I}$$

If no such combination exists, the polynomial is prime.

Example 6   **Factoring a Trinomial Whose Leading Coefficient Is Not 1**

Factor: $3x^2 - 20x + 28$.

Solution

Step 1   Find two First terms whose product is $3x^2$.

$$3x^2 - 20x + 28 = (3x \quad)(x \quad)$$

Step 2   Find two Last terms whose product is 28. The number 28 has pairs of factors that are either both positive or both negative. Because the middle term, $-20x$, is negative, both factors must be negative. The negative factorizations of 28 are $-1(-28)$, $-2(-14)$, and $-4(-7)$.

Step 3   Try various combinations of these factors. The correct factorization of $3x^2 - 20x + 28$ is the one in which the sum of the Outside and Inside products is equal to $-20x$. Here is a list of the possible factorizations:

Possible Factorizations of $3x^2 - 20x + 28$	Sum of Outside and Inside Products (Should Equal $-20x$)
$(3x - 1)(x - 28)$	$-84x - x = -85x$
$(3x - 28)(x - 1)$	$-3x - 28x = -31x$
$(3x - 2)(x - 14)$	$-42x - 2x = -44x$
$(3x - 14)(x - 2)$	$-6x - 14x = -20x$
$(3x - 4)(x - 7)$	$-21x - 4x = -25x$
$(3x - 7)(x - 4)$	$-12x - 7x = -19x$

This is the required middle term.

Thus,

$$3x^2 - 20x + 28 = (3x - 14)(x - 2) \text{ or } (x - 2)(3x - 14).$$

Step 4 Verify the factorization, $3x^2 - 20x + 28 = (3x - 14)(x - 2)$, using the FOIL method.

$$(3x - 14)(x - 2) = 3x \cdot x + 3x(-2) + (-14) \cdot x + (-14)(-2)$$
$$= 3x^2 - 6x - 14x + 28$$
$$= 3x^2 - 20x + 28$$

Because this is the trinomial we started with, the factorization is correct.

▶ ✔ **Checkpoint 6** Factor: $5x^2 - 14x + 8$.

Example 7   Factoring a Trinomial Whose Leading Coefficient Is Not 1

Factor: $8y^2 - 10y - 3$.

Solution

Step 1   Find two First terms whose product is $8y^2$.

$$8y^2 - 10y - 3 \overset{?}{=} (8y \quad)(y \quad)$$
$$8y^2 - 10y - 3 \overset{?}{=} (4y \quad)(2y \quad)$$

Step 2   Find two Last terms whose product is -3. The possible factorizations are $1(-3)$ and $-1(3)$.

Step 3   Try various combinations of these factors. The correct factorization of $8y^2 - 10y - 3$ is the one in which the sum of the Outside and Inside products is equal to $-10y$. Here is a list of the possible factorizations:

Possible Factorizations of $8y^2 - 10y - 3$	Sum of Outside and Inside Products (Should Equal $-10y$)
$(8y + 1)(y - 3)$	$-24y + y = -23y$
$(8y - 3)(y + 1)$	$8y - 3y = 5y$
$(8y - 1)(y + 3)$	$24y - y = 23y$
$(8y + 3)(y - 1)$	$-8y + 3y = -5y$
$(4y + 1)(2y - 3)$	$-12y + 2y = -10y$
$(4y - 3)(2y + 1)$	$4y - 6y = -2y$
$(4y - 1)(2y + 3)$	$12y - 2y = 10y$
$(4y + 3)(2y - 1)$	$-4y + 6y = 2y$

These four factorizations are $(8y \quad)(y \quad)$ with $1(-3)$ and $-1(3)$ as factorizations of -3.

These four factorizations are $(4y \quad)(2y \quad)$ with $1(-3)$ and $-1(3)$ as factorizations of -3.

This is the required middle term.

Thus,

$$8y^2 - 10y - 3 = (4y + 1)(2y - 3) \text{ or } (2y - 3)(4y + 1).$$

Show that either of these factorizations is correct by multiplying the factors using the FOIL method. You should obtain the original trinomial.

▶ ✔ **Checkpoint 7** Factor: $6y^2 + 19y - 7$.

Not every trinomial can be factored. For example, consider

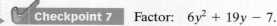

$$6x^2 + 14x + 7 = (6x + \square)(x + \square)$$
$$6x^2 + 14x + 7 = (3x + \square)(2x + \square).$$

The possible factors for the last term are 1 and 7. However, regardless of how these factors are placed in the boxes shown, the sum of the Outside and Inside products is not equal to $14x$. Thus, the trinomial $6x^2 + 14x + 7$ cannot be factored and is prime.

Factoring the Difference of Two Squares

⑤ Factor the difference of squares.

A method for factoring the difference of two squares is obtained by reversing the product formula for the sum and difference of two terms.

> ### The Difference of Two Squares
> If A and B are real numbers, variables, or algebraic expressions, then
> $$A^2 - B^2 = (A + B)(A - B).$$
> In words: The difference of the squares of two terms factors as the product of the sum and the difference of those terms.

Example 8 Factoring the Difference of Two Squares

Factor:

a. $x^2 - 4$ **b.** $81x^2 - 49.$

Solution

We must express each term as the square of some monomial. Then we use the formula for factoring $A^2 - B^2$.

a. $x^2 - 4 = x^2 - 2^2 = (x + 2)(x - 2)$

$$A^2 - B^2 = (A + B)(A - B)$$

b. $81x^2 - 49 = (9x)^2 - 7^2 = (9x + 7)(9x - 7)$

✓ **Checkpoint 8** Factor:

▶ **a.** $x^2 - 81$ **b.** $36x^2 - 25.$

We have seen that a polynomial is factored completely when it is written as the product of prime polynomials. To be sure that you have factored completely, check to see whether any factors with more than one term in the factored polynomial can be factored further. If so, continue factoring.

Example 9 A Repeated Factorization

Factor completely: $x^4 - 81.$

Solution

$x^4 - 81 = (x^2)^2 - 9^2$	Express as the difference of two squares.
$= (x^2 + 9)(x^2 - 9)$	The factors are the sum and the difference of the expressions being squared.
$= (x^2 + 9)(x^2 - 3^2)$	The factor $x^2 - 9$ is the difference of two squares and can be factored.
$= (x^2 + 9)(x + 3)(x - 3)$	The factors of $x^2 - 9$ are the sum and the difference of the expressions being squared.

Study Tip

Factoring $x^4 - 81$ as
$$(x^2 + 9)(x^2 - 9)$$
is not a complete factorization. The second factor, $x^2 - 9$, is itself a difference of two squares and can be factored.

✓ **Checkpoint 9** Factor completely: $81x^4 - 16.$

⑥ Factor perfect square trinomials.

Factoring Perfect Square Trinomials

Our next factoring technique is obtained by reversing the product formulas for squaring binomials. The trinomials that are factored using this technique are called **perfect square trinomials**.

Factoring Perfect Square Trinomials

Let A and B be real numbers, variables, or algebraic expressions.

1. $A^2 + 2AB + B^2 = (A + B)^2$ **2.** $A^2 - 2AB + B^2 = (A - B)^2$

Same sign   Same sign

The two items in the box show that perfect square trinomials, $A^2 + 2AB + B^2$ and $A^2 - 2AB + B^2$, come in two forms: one in which the coefficient of the middle term is positive and one in which the coefficient of the middle term is negative. Here's how to recognize a perfect square trinomial:

1. The first and last terms are squares of monomials or integers.

2. The middle term is twice the product of the expressions being squared in the first and last terms.

Example 10 **Factoring Perfect Square Trinomials**

Factor:

a. $x^2 + 6x + 9$ **b.** $25x^2 - 60x + 36$.

Solution

a. $x^2 + 6x + 9 = x^2 + 2 \cdot x \cdot 3 + 3^2 = (x + 3)^2$ The middle term has a positive sign.

$$A^2 \; + \; 2AB \; + \; B^2 \; = \; (A + B)^2$$

b. We suspect that $25x^2 - 60x + 36$ is a perfect square trinomial because $25x^2 = (5x)^2$ and $36 = 6^2$. The middle term can be expressed as twice the product of $5x$ and 6.

$$25x^2 - 60x + 36 = (5x)^2 - 2 \cdot 5x \cdot 6 + 6^2 = (5x - 6)^2$$

$$A^2 \; - \; 2AB \; + \; B^2 \; = \; (A - B)^2$$

 **Checkpoint 10** Factor:

a. $x^2 + 14x + 49$ **b.** $16x^2 - 56x + 49$.

A Strategy for Factoring Polynomials

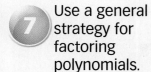 **⑦ Use a general strategy for factoring polynomials.**

It is important to practice factoring a wide variety of polynomials so that you can quickly select the appropriate technique. A polynomial is factored completely when all its polynomial factors, except possibly for monomial factors, are prime. Because of the commutative property, the order of the factors does not matter.

A Strategy for Factoring a Polynomial

1. If the terms have a common factor, factor out the GCF.
2. Determine the number of terms in the polynomial and try factoring as follows:
   a. If there are two terms, can the binomial be factored as the difference of two squares?
   $$A^2 - B^2 = (A + B)(A - B)$$
   b. If there are three terms, is the trinomial a perfect square trinomial? If so, factor by using one of the following formulas:
   $$A^2 + 2AB + B^2 = (A + B)^2$$
   $$A^2 - 2AB + B^2 = (A - B)^2.$$
   If the trinomial is not a perfect square trinomial, try factoring by trial and error.
   c. If there are four or more terms, try factoring by grouping.
3. Check to see if any factors with more than one term in the factored polynomial can be factored further. If so, factor completely.
4. Check by multiplying.

Example 11 Factoring a Polynomial

Factor: $3x^2 - 6x - 45$.

Solution

Step 1 If the terms have a common factor, factor out the GCF. Because 3 is common to all terms, we factor it out.

$$3x^2 - 6x - 45 = 3(x^2 - 2x - 15) \quad \text{Factor out the GCF.}$$

Step 2 Determine the number of terms and factor accordingly. The factor $x^2 - 2x - 15$ has three terms, but it is not a perfect square trinomial. We factor it using trial and error.

$$3x^2 - 6x - 45 = 3(x^2 - 2x - 15) = 3(x - 5)(x + 3)$$

Step 3 Check to see if factors can be factored further. In this case, they cannot, so we have factored completely.

Step 4 Check by multiplying.

$$3(x - 5)(x + 3) = 3(x^2 - 2x - 15) = 3x^2 - 6x - 45$$

FOIL

This is the original polynomial, so the factorization is correct.

> **Checkpoint 11** Factor: $4x^2 - 16x - 48$.

Example 12 Factoring a Polynomial

Factor: $7x^5 - 7x$.

Solution

Step 1 If the terms have a common factor, factor out the GCF. Because $7x$ is common to both terms, we factor it out.

$$7x^5 - 7x = 7x(x^4 - 1) \quad \text{Factor out the GCF.}$$

Step 2   Determine the number of terms and factor accordingly. Up to this point, we have $7x^5 - 7x = 7x(x^4 - 1)$. However, the factor $x^4 - 1$ has two terms. This binomial can be expressed as $(x^2)^2 - 1^2$, so it can be factored as the difference of two squares.

$$7x^5 - 7x = 7x(x^4 - 1) = 7x(x^2 + 1)(x^2 - 1)$$ Use $A^2 - B^2 = (A + B)(A - B)$ on $x^4 - 1$: $A = x^2$ and $B = 1$.

Step 3   Check to see if factors can be factored further. We note that $(x^2 - 1)$ is also the difference of two squares, $x^2 - 1^2$, so we continue factoring.

$$7x^5 - 7x = 7x(x^2 + 1)(x + 1)(x - 1)$$ Factor $x^2 - 1$ as the difference of two squares.

Step 4   Check by multiplying.

$$7x(x^2 + 1)(x + 1)(x - 1) = 7x(x^2 + 1)(x^2 - 1) = 7x(x^4 - 1) = 7x^5 - 7x$$

We obtain the original polynomial, so the factorization is correct.

> ✓ **Checkpoint 12**   Factor: $4x^5 - 64x$.

Example 13   Factoring a Polynomial

Factor: $x^3 - 5x^2 - 4x + 20$.

Solution

Step 1   If the terms have a common factor, factor out the GCF. Other than 1, there is no common factor.

Step 2   Determine the number of terms and factor accordingly. There are four terms. We try factoring by grouping.

$$x^3 - 5x^2 - 4x + 20$$
$$= (x^3 - 5x^2) + (-4x + 20)$$ Group terms with common factors.
$$= x^2(x - 5) - 4(x - 5)$$ Factor from each group.
$$= (x - 5)(x^2 - 4)$$ Factor out the common binomial factor, $x - 5$.

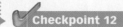

> Notice that it is necessary to factor out a negative number from the second grouping to obtain a common binomial factor, $x - 5$, for the two groupings.

Step 3   Check to see if factors can be factored further. We note that $(x^2 - 4)$ is the difference of two squares, $x^2 - 2^2$, so we continue factoring.

$$x^3 - 5x^2 - 4x + 20 = (x - 5)(x + 2)(x - 2)$$ Factor $x^2 - 4$ as the difference of two squares.

We have factored completely because no factor with more than one term can be factored further.

Step 4   Check by multiplying.

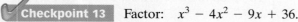

$$(x - 5)(x + 2)(x - 2) = (x - 5)(x^2 - 4) = x^3 - 4x - 5x^2 + 20$$
$$= x^3 - 5x^2 - 4x + 20$$

We obtain the original polynomial, so the factorization is correct.

> ✓ **Checkpoint 13**   Factor: $x^3 - 4x^2 - 9x + 36$.

Example 14 Factoring a Polynomial

Factor: $2x^3 - 24x^2 + 72x$.

Solution

Step 1 If the terms have a common factor, factor out the GCF. Because $2x$ is common to all terms, we factor it out.

$$2x^3 - 24x^2 + 72x = 2x(x^2 - 12x + 36) \quad \text{Factor out the GCF.}$$

Step 2 Determine the number of terms and factor accordingly. The factor $x^2 - 12x + 36$ has three terms. Is it a perfect square trinomial? Yes. The first term, x^2, is the square of a monomial. The last term, 36 or 6^2, is the square of an integer. The middle term involves twice the product of x and 6. We factor using $A^2 - 2AB + B^2 = (A - B)^2$.

$$2x^3 - 24x^2 + 72x = 2x(x^2 - 12x + 36)$$
$$= 2x(\underbrace{x^2 - 2 \cdot x \cdot 6 + 6^2}_{A^2 - 2\ A\ B + B^2}) \quad \text{The second factor is a perfect square trinomial.}$$
$$= 2x(x - 6)^2 \quad A^2 - 2AB + B^2 = (A - B)^2$$

Step 3 Check to see if factors can be factored further. In this problem, they cannot, so we have factored completely.

Step 4 Check by multiplying. Let's verify that $2x^3 - 24x^2 + 72x = 2x(x - 6)^2$.

$$2x(x - 6)^2 = 2x(x^2 - 12x + 36) = 2x^3 - 24x^2 + 72x$$

We obtain the original polynomial, so the factorization is correct.

▶ **Checkpoint 14** Factor: $3x^3 - 30x^2 + 75x$.

Blitzer Bonus

Candy and Quotients of Polynomials

A few candy factoids:

- Despite the fact that eating lots of candy has negative effects on health and teeth, the average American spends $84 per year on candy, consuming 23.9 pounds annually.
- 55% of candy sales are "impulse buys."
- The word *candy* originated with the Sanskrit word *khanda*, meaning "pieces of crystallized sugar."
- A function consisting of the quotient of two polynomials models what happens in your mouth after eating "pieces of crystallized sugar":

$$f(x) = \frac{6.5x^2 - 20.4x + 234}{x^2 + 36}.$$

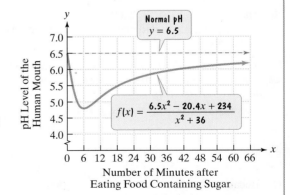

Figure 6.2

The function models the pH level, $f(x)$, of the human mouth after a person eats food containing sugar. The graph of this function is shown in **Figure 6.2**. The figure indicates that

- The normal pH level of the human mouth is 6.5.
- After eating food containing sugar, the pH level is the lowest, approximately 4.8, after 6 minutes.
- After quickly dropping below normal, the pH level of the mouth slowly begins to approach the normal level of 6.5 during the first hour.

Achieving Success

As you continue your study of mathematics, you will see how functions involving quotients of polynomials provide insights into phenomena as diverse as the aftereffects of candy on your mouth, the cost of environmental cleanup, and even our ongoing processes of learning and forgetting. Factoring is an essential skill for working with such functions. **Success in mathematics cannot be achieved without a complete understanding of factoring.** Now is the time to practice factoring a wide variety of polynomials. As you approach this exercise set, be sure you can apply each of the factoring techniques discussed in the section. This can be achieved by working the assigned problems from Exercises 23–74. Then it's time to work problems that require you to select the appropriate technique or techniques from Exercises 75–132. The more deeply you force your brain to think about factoring by working many exercises, the better will be your chances of achieving success in future algebra courses.

Exercise Set 6.2

Concept and Vocabulary Exercises

In Exercises 1–6, fill in each blank so that the resulting statement is true.

1. The polynomial $10x^2 + 50x$ can be factored by factoring out _____, called the _____.

2. The polynomial $x^3 + 5x^2 + 6x + 30$ can be factored using a technique called _____.

3. In order to factor $7x^2 - 23x + 6$, begin by finding two first terms whose product is _____ and two last terms whose product is _____.

4. The difference of the squares of two terms factors as a product of the _____ and the _____ of those terms.

5. The polynomial $x^2 + 6x + 9$ is a/an _____ trinomial because x^2 is the square of _____, 9 is the square of _____, and the middle term is twice the product of _____.

6. The trinomial $A^2 + 2AB + B^2$ can be factored as _____ and the trinomial $A^2 - 2AB + B^2$ can be factored as _____.

In Exercises 7–12, determine whether each statement is true or false. If the statement is false, make the necessary change(s) to produce a true statement.

7. One factor of $x^2 + x + 20$ is $x + 5$.

8. A trinomial can never have two identical factors.

9. One factor of $y^2 + 5y - 24$ is $y - 3$.

10. $x^2 + 4 = (x + 2)(x + 2)$

11. Because $x^2 - 25 = (x + 5)(x - 5)$, then $x^2 + 25 = (x - 5)(x + 5)$.

12. All perfect square trinomials are squares of binomials.

Respond to Exercises 13–18 using verbal or written explanations.

13. Using an example, explain how to factor out the greatest common factor of a polynomial.

14. Suppose that a polynomial contains four terms. Explain how to use factoring by grouping to factor the polynomial.

15. Explain how to factor $3x^2 + 10x + 8$.

16. Explain how to factor the difference of two squares. Provide an example with your explanation.

17. What is a perfect square trinomial and how is it factored?

18. What does it mean to factor completely?

In Exercises 19–22, determine whether each statement makes sense or does not make sense, and explain your reasoning.

19. Although $20x^3$ appears in both $20x^3 + 8x^2$ and $20x^3 + 10x$, I'll need to factor $20x^3$ in different ways to obtain each polynomial's factorization.

20. You grouped the polynomial's terms using different groupings than I did, yet we both obtained the same factorization.

21. Multiplying polynomials is relatively mechanical, but factoring often requires a great deal of thought.

22. When a factorization requires two factoring techniques, I'm less likely to make errors if I show one technique at a time rather than combining the two factorizations into one step.

Practice Exercises

In Exercises 23–32, factor out the greatest common factor.

23. $18x + 27$	24. $16x - 24$
25. $3x^2 + 6x$	26. $4x^2 - 8x$
27. $9x^4 - 18x^3 + 27x^2$	28. $6x^4 - 18x^3 + 12x^2$
29. $x(x + 5) + 3(x + 5)$	30. $x(2x + 1) + 4(2x + 1)$
31. $x^2(x - 3) + 12(x - 3)$	32. $x^2(2x + 5) + 17(2x + 5)$

In Exercises 33–38, factor by grouping. If necessary, factor out a negative number from the second grouping to obtain a common binomial factor.

33. $x^3 - 2x^2 + 5x - 10$	34. $x^3 - 3x^2 + 4x - 12$
35. $x^3 - x^2 + 2x - 2$	36. $x^3 + 6x^2 - 2x - 12$
37. $3x^3 - 2x^2 - 6x + 4$	38. $x^3 - x^2 - 5x + 5$

In Exercises 39–56, factor each trinomial, or state that the trinomial is prime.

39. $x^2 + 5x + 6$	40. $x^2 + 8x + 15$
41. $x^2 - 2x - 15$	42. $x^2 - 4x - 5$
43. $x^2 - 8x + 15$	44. $x^2 - 14x + 45$

45. $3x^2 - x - 2$ **46.** $2x^2 + 5x - 3$

47. $3x^2 - 25x - 28$ **48.** $3x^2 - 2x - 5$

49. $6x^2 - 11x + 4$ **50.** $6x^2 - 17x + 12$

51. $4x^2 + 16x + 15$ **52.** $8x^2 + 33x + 4$

53. $9x^2 - 9x + 2$ **54.** $9x^2 + 5x - 4$

55. $20x^2 + 27x - 8$ **56.** $15x^2 - 19x + 6$

In Exercises 57–66, factor the difference of two squares.

57. $x^2 - 100$ **58.** $x^2 - 144$

59. $36x^2 - 49$ **60.** $64x^2 - 81$

61. $9 - 25y^2$ **62.** $36 - 49y^2$

63. $x^4 - 16$ **64.** $x^4 - 1$

65. $16x^4 - 81$ **66.** $81x^4 - 1$

In Exercises 67–74, factor each perfect square trinomial.

67. $x^2 + 2x + 1$ **68.** $x^2 + 4x + 4$

69. $x^2 - 14x + 49$ **70.** $x^2 - 10x + 25$

71. $4x^2 + 4x + 1$ **72.** $25x^2 + 10x + 1$

73. $9x^2 - 6x + 1$ **74.** $64x^2 - 16x + 1$

In Exercises 75–132, factor completely, or state that the polynomial is prime.

75. $5x^3 - 20x$ **76.** $4x^3 - 100x$

77. $7x^3 + 7x$ **78.** $6x^3 + 24x$

79. $5x^2 - 5x - 30$ **80.** $5x^2 - 15x - 50$

81. $2x^4 - 162$ **82.** $7x^4 - 7$

83. $x^3 + 2x^2 - 9x - 18$ **84.** $x^3 + 3x^2 - 25x - 75$

85. $3x^3 - 24x^2 + 48x$ **86.** $5x^3 - 20x^2 + 20x$

87. $6x^2 + 8x$ **88.** $21x^2 - 35x$

89. $2y^2 - 2y - 112$ **90.** $6x^2 - 6x - 12$

91. $7y^4 + 14y^3 + 7y^2$ **92.** $2y^4 + 28y^3 + 98y^2$

93. $y^2 + 8y - 16$ **94.** $y^2 - 18y - 81$

95. $16y^2 - 4y - 2$ **96.** $32y^2 + 4y - 6$

97. $r^2 - 25r$ **98.** $3r^2 - 27r$

99. $4w^2 + 8w - 5$ **100.** $35w^2 - 2w - 1$

101. $x^3 - 4x$ **102.** $9x^3 - 9x$

103. $x^2 + 64$ **104.** $y^2 + 36$

105. $9y^2 + 13y + 4$ **106.** $20y^2 + 12y + 1$

107. $y^3 + 2y^2 - 4y - 8$ **108.** $y^3 + 2y^2 - y - 2$.

109. $16y^2 + 24y + 9$ **110.** $25y^2 + 20y + 4$

111. $4y^3 - 28y^2 + 40y$ **112.** $7y^3 - 21y^2 + 14y$

113. $y^5 - 81y$ **114.** $y^5 - 16y$

115. $20a^4 - 45a^2$ **116.** $48a^4 - 3a^2$

117. $9x^4 + 18x^3 + 6x^2$ **118.** $10x^4 + 20x^3 + 15x^2$

119. $12y^2 - 11y + 2$ **120.** $21x^2 - 25x - 4$

121. $9y^2 - 64$ **122.** $100y^2 - 49$

123. $9y^2 + 64$ **124.** $100y^2 + 49$

125. $2y^3 + 3y^2 - 50y - 75$ **126.** $12y^3 + 16y^2 - 3y - 4$

127. $2r^3 + 30r^2 - 68r$ **128.** $3r^3 - 27r^2 - 210r$

129. $8x^5 - 2x^3$ **130.** $y^9 - y^5$

131. $3x^2 + 243$ **132.** $27x^2 + 75$

Practice Plus

In Exercises 133–140, factor completely.

133. $10x^2(x + 1) - 7x(x + 1) - 6(x + 1)$

134. $12x^2(x - 1) - 4x(x - 1) - 5(x - 1)$

135. $6x^4 + 35x^2 - 6$ **136.** $7x^4 + 34x^2 - 5$

137. $(x - 7)^2 - 4a^2$ **138.** $(x - 6)^2 - 9a^2$

139. $x^2 + 8x + 16 - 25a^2$ **140.** $x^2 + 14x + 49 - 16a^2$

Application Exercises

141. Your computer store is having an incredible sale. The price on one model is reduced by 40%. Then the sale price is reduced by another 40%. If x is the computer's original price, the sale price can be modeled by

$$(x - 0.4x) - 0.4(x - 0.4x).$$

     a. Factor out $(x - 0.4x)$ from each term. Then simplify the resulting expression.

     b. Use the simplified expression from part (a) to answer these questions. With a 40% reduction followed by a 40% reduction, is the computer selling at 20% of its original price? If not, at what percentage of the original price is it selling?

142. Your local electronics store is having an end-of-the-year sale. The price on a plasma television had been reduced by 30%. Now the sale price is reduced by another 30%. If x is the television's original price, the sale price can be modeled by

$$(x - 0.3x) - 0.3(x - 0.3x).$$

     a. Factor out $(x - 0.3x)$ from each term. Then simplify the resulting expression.

     b. Use the simplified expression from part (a) to answer these questions. With a 30% reduction followed by a 30% reduction, is the television selling at 40% of its original price? If not, at what percentage of the original price is it selling?

In Exercises 143–146,

     a. *Write an expression for the area of the shaded region.*

     b. *Write the expression in factored form.*

143.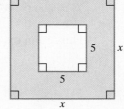

144.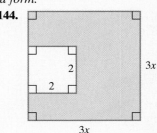

145.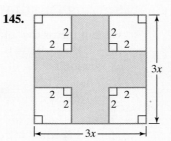

146.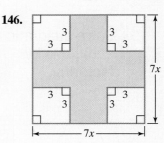

Critical Thinking Exercises

147. Where is the error in this "proof" that $2 = 0$?

$a = b$	Suppose that a and b are any equal real numbers.
$a^2 = b^2$	Square both sides of the equation.
$a^2 - b^2 = 0$	Subtract b^2 from both sides.
$2(a^2 - b^2) = 2 \cdot 0$	Multiply both sides by 2.
$2(a^2 - b^2) = 0$	On the right side, $2 \cdot 0 = 0$.
$2(a + b)(a - b) = 0$	Factor $a^2 - b^2$.
$2(a + b) = 0$	Divide both sides by $a - b$.
$2 = 0$	Divide both sides by $a + b$.

In Exercises 148–151, factor each polynomial.

148. $x^2 - y^2 + 3x + 3y$

149. $x^{2n} - 25y^{2n}$

150. $4x^{2n} + 12x^n + 9$

151. $(x + 3)^2 - 2(x + 3) + 1$

In Exercises 152–153, find all positive integers b so that the trinomial can be factored.

152. $x^2 + bx + 15$

153. $x^2 + 4x + b$

6.3 Solving Quadratic Equations by Factoring

Objectives

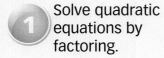 Solve quadratic equations by factoring.

② Solve higher-degree polynomial equations by factoring.

③ Solve problems using quadratic equations.

Motion and change are the very essence of life. Moving air brushes against our faces; rain falls on our heads; birds fly past us; plants spring from the earth, grow, and then die; and rocks thrown upward reach a maximum height before falling to the ground. In this section, you will use quadratic functions and factoring strategies to model and visualize motion. Analyzing the where and when of moving objects involves equations in which the highest exponent on the variable is 2, called *quadratic equations*.

The Standard Form of a Quadratic Equation

We begin by defining a quadratic equation.

> ### Definition of a Quadratic Equation
>
> A **quadratic equation** in x is an equation that can be written in the **standard form**
> $$ax^2 + bx + c = 0,$$
> where a, b, and c are real numbers, with $a \neq 0$. A quadratic equation in x is also called a **second-degree polynomial equation** in x.

Here is an example of a quadratic equation in standard form:

$$x^2 - 12x + 27 = 0.$$

$$\boxed{a = 1} \quad \boxed{b = -12} \quad \boxed{c = 27}$$

Solving Quadratic Equations by Factoring

① Solve quadratic equations by factoring.

We can factor the left side of the quadratic equation $x^2 - 12x + 27 = 0$. We obtain $(x - 3)(x - 9) = 0$. If a quadratic equation has zero on one side and a factored expression on the other side, it can be solved using the **zero-product principle**.

> ### The Zero-Product Principle
>
> If the product of two algebraic expressions is zero, then at least one of the factors is equal to zero.
>
> If $AB = 0$, then $A = 0$ or $B = 0$.

For example, consider the equation $(x - 3)(x - 9) = 0$. According to the zero-product principle, this product can be zero only if at least one of the factors is zero. We set each individual factor equal to zero and solve the resulting equations for x.

$$(x - 3)(x - 9) = 0$$
$$x - 3 = 0 \quad \text{or} \quad x - 9 = 0$$
$$x = 3 \qquad\qquad x = 9$$

The solutions of the original quadratic equation, $x^2 - 12x + 27 = 0$, are 3 and 9. The solution set is $\{3, 9\}$.

> **Solving a Quadratic Equation by Factoring**
>
> 1. If necessary, rewrite the equation in the standard form $ax^2 + bx + c = 0$, moving all terms to one side, thereby obtaining zero on the other side.
> 2. Factor completely.
> 3. Apply the zero-product principle, setting each factor containing a variable equal to zero.
> 4. Solve the equations in step 3.
> 5. Check the solutions in the original equation.

Example 1　　**Solving a Quadratic Equation by Factoring**

Solve:　$2x^2 - 5x = 12$.

Solution

Step 1　Move all terms to one side and obtain zero on the other side. Subtract 12 from both sides and write the equation in standard form.

$$2x^2 - 5x - 12 = 12 - 12$$
$$2x^2 - 5x - 12 = 0$$

Step 2　Factor.

$$(2x + 3)(x - 4) = 0$$

Steps 3 and 4　Set each factor equal to zero and solve the resulting equations.

$$2x + 3 = 0 \quad \text{or} \quad x - 4 = 0$$
$$2x = -3 \qquad\qquad x = 4$$
$$x = -\frac{3}{2}$$

Step 5　Check the solutions in the original equation.

Check $-\dfrac{3}{2}$:

$$2x^2 - 5x = 12$$
$$2\left(-\frac{3}{2}\right)^2 - 5\left(-\frac{3}{2}\right) \stackrel{?}{=} 12$$
$$2\left(\frac{9}{4}\right) - 5\left(-\frac{3}{2}\right) \stackrel{?}{=} 12$$
$$\frac{9}{2} + \frac{15}{2} \stackrel{?}{=} 12$$
$$\frac{24}{2} \stackrel{?}{=} 12$$
$$12 = 12, \quad \text{true}$$

Check 4:

$$2x^2 - 5x = 12$$
$$2(4)^2 - 5(4) \stackrel{?}{=} 12$$
$$2(16) - 5(4) \stackrel{?}{=} 12$$
$$32 - 20 \stackrel{?}{=} 12$$
$$12 = 12, \quad \text{true}$$

The solutions are $-\frac{3}{2}$ and 4, and the solution set is $\left\{-\frac{3}{2}, 4\right\}$.

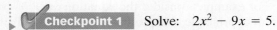 **Checkpoint 1** Solve: $2x^2 - 9x = 5$.

Study Tip

Do not confuse factoring a polynomial with solving a quadratic equation by factoring.

Factoring a Polynomial	**Solving a Quadratic Equation**
Factor: $2x^2 - 5x - 12$.	Solve: $2x^2 - 5x - 12 = 0$.

> This is not an equation. There is no equal sign.

> This is an equation. There is an equal sign.

Solution: $(2x + 3)(x - 4)$

Solution: $(2x + 3)(x - 4) = 0$
$$2x + 3 = 0 \quad \text{or} \quad x - 4 = 0$$
$$x = -\frac{3}{2} \qquad\qquad x = 4$$

> Stop! Avoid the common error of setting each factor equal to zero.

The solution set is $\left\{-\dfrac{3}{2}, 4\right\}$.

A polynomial function of the form $y = ax^2 + bx + c$ or $f(x) = ax^2 + bx + c$, $a \neq 0$, is called a **quadratic function**. An example of a quadratic function is $y = 2x^2 - 5x - 12$ or $f(x) = 2x^2 - 5x - 12$.

There is an important relationship between a quadratic equation in standard form, such as

$$2x^2 - 5x - 12 = 0$$

and a quadratic function, such as

$$y = 2x^2 - 5x - 12.$$

The solutions of $ax^2 + bx + c = 0$ correspond to the x-intercepts of the graph of the quadratic function $y = ax^2 + bx + c$.
For example, you can visualize the solutions of $2x^2 - 5x - 12 = 0$ by looking at the x-intercepts of the graph of the quadratic function $y = 2x^2 - 5x - 12$. The graph, shaped like a bowl, is shown in **Figure 6.3**. The solutions of the equation $2x^2 - 5x - 12 = 0$, $-\frac{3}{2}$ and 4, appear as the graph's x-intercepts.

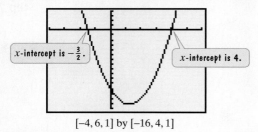

> x-intercept is $-\frac{3}{2}$.

> x-intercept is 4.

$[-4, 6, 1]$ by $[-16, 4, 1]$

Figure 6.3

Example 2 **Solving Quadratic Equations by Factoring**

Solve:

a. $5x^2 = 20x$ **b.** $x^2 + 4 = 8x - 12$ **c.** $(x - 7)(x + 5) = -20$.

Solution

a.
$$5x^2 = 20x \qquad \text{This is the given equation.}$$
$$5x^2 - 20x = 0 \qquad \text{Subtract 20x from both sides and write the equation in standard form.}$$
$$5x(x - 4) = 0 \qquad \text{Factor.}$$
$$5x = 0 \quad \text{or} \quad x - 4 = 0 \qquad \text{Set each factor equal to 0.}$$
$$x = 0 \qquad\qquad x = 4 \qquad \text{Solve the resulting equations.}$$

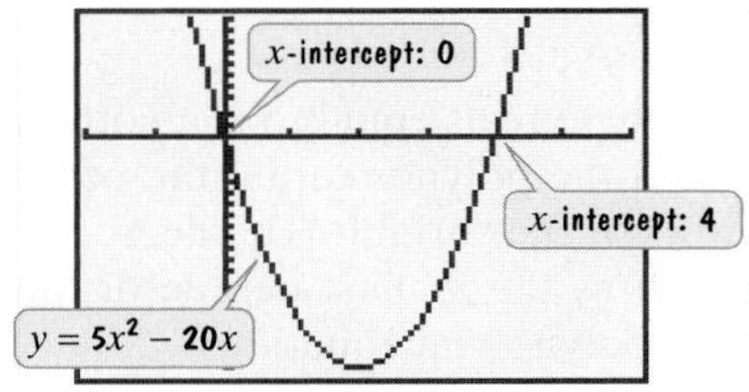

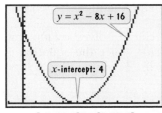

[−2, 6, 1] by [−20, 10, 1]

Figure 6.4 The solution set of $5x^2 = 20x$, or $5x^2 - 20x = 0$, is $\{0, 4\}$.

Check by substituting 0 and 4 into the given equation. The graph of $y = 5x^2 - 20x$, obtained with a graphing utility, is shown in **Figure 6.4**. The x-intercepts are 0 and 4. This verifies that the solutions are 0 and 4, and the solution set is $\{0, 4\}$.

b.
$$x^2 + 4 = 8x - 12 \qquad \text{This is the given equation.}$$

$$x^2 - 8x + 16 = 0 \qquad \text{Write the equation in standard form by}$$
$$\text{subtracting 5x and adding 12 on both sides.}$$

$$(x - 4)(x - 4) = 0 \qquad \text{Factor.}$$

$$x - 4 = 0 \quad \text{or} \quad x - 4 = 0 \qquad \text{Set each factor equal to 0.}$$

$$x = 4 \qquad\qquad x = 4 \qquad \text{Solve the resulting equations.}$$

Notice that there is only one solution (or, if you prefer, a repeated solution.) The trinomial $x^2 - 8x + 16$ is a perfect square trinomial that could have been factored as $(x - 4)^2$. The graph of $y = x^2 - 8x + 16$, obtained with a graphing utility, is shown in **Figure 6.5**. The graph has only one x-intercept at 4. This verifies that the equation's solution is 4 and the solution set is $\{4\}$.

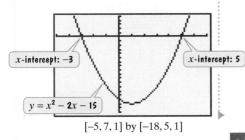

[−1, 10, 1] by [0, 20, 1]

Figure 6.5 The solution set of $x^2 + 4 = 8x - 12$, or $x^2 - 8x + 16 = 0$, is $\{4\}$.

c. Be careful! Although the left side of $(x - 7)(x + 5) = -20$ is factored, we cannot use the zero-product principle. Why not? The right side of the equation is not 0. So we begin by multiplying the factors on the left side of the equation. Then we add 20 to both sides to obtain 0 on the right side.

$$(x - 7)(x + 5) = -20 \qquad \text{This is the given equation.}$$

$$x^2 - 2x - 35 = -20 \qquad \text{Use the FOIL method to multiply on the left side.}$$

$$x^2 - 2x - 15 = 0 \qquad \text{Add 20 to both sides.}$$

$$(x + 3)(x - 5) = 0 \qquad \text{Factor.}$$

$$x + 3 = 0 \quad \text{or} \quad x - 5 = 0 \qquad \text{Set each factor equal to 0.}$$

$$x = -3 \qquad\qquad x = 5 \qquad \text{Solve the resulting equations.}$$

Check by substituting −3 and 5 into the given equation. The graph of $y = x^2 - 2x - 15$, obtained with a graphing utility, is shown in **Figure 6.6**. The x-intercepts are −3 and 5. This verifies that the solutions are −3 and 5, and the solution set is $\{-3, 5\}$.

Figure 6.6 The solution set of $(x - 7)(x + 5) = -20$, or $x^2 - 2x - 15 = 0$, is $\{-3, 5\}$.

Study Tip

Avoid the following errors:

$$5x^2 = 20x$$
$$\frac{5x^2}{x} = \frac{20x}{x}$$
$$5x = 20$$
$$x = 4$$

Never divide both sides of an equation by x. Division by zero is undefined and x may be zero. Indeed, the solutions for this equation (Example 2a) are 0 and 4. Dividing both sides by x does not permit us to find both solutions.

$$(x - 7)(x + 5) = -20$$
$$x - 7 = -20 \quad \text{or} \quad x + 5 = -20$$
$$x = -13 \quad \text{or} \quad x = -25$$

The zero-product principle cannot be used because the right side of the equation is not equal to 0.

✓ **Checkpoint 2**   Solve:

a. $3x^2 = 2x$

b. $x^2 + 7 = 10x - 18$

▶ **c.** $(x - 2)(x + 3) = 6.$

2 Solve higher-degree polynomial equations by factoring.

Polynomial Equations

A **polynomial equation** is the result of setting two polynomials equal to each other. The equation is in **standard form** if one side is 0 and the polynomial on the other side is in standard form, that is, in descending powers of the variable. The **degree of a polynomial equation** in standard form is the same as the highest degree of the terms in the polynomial. Here are examples of three polynomial equations:

$$2x^2 + 7x = 4 \qquad\qquad x^3 + x^2 = 4x + 4$$
$$3x - 21 = 0 \qquad 2x^2 + 7x - 4 = 0 \qquad x^3 + x^2 - 4x - 4 = 0.$$

This equation is of degree 1 because 1 is the highest degree.

This equation is of degree 2 because 2 is the highest degree.

This equation is of degree 3 because 3 is the highest degree.

Notice that a polynomial equation of degree 1 is a linear equation. A polynomial equation of degree 2 is a quadratic equation.

Some polynomial equations of degree 3 or higher can be solved by moving all terms to one side, thereby obtaining 0 on the other side. Once the equation is in standard form, factor and then set each factor equal to 0.

Example 3 Solving a Polynomial Equation by Factoring

Solve by factoring: $x^3 + x^2 = 4x + 4.$

Solution

Step 1 Move all terms to one side and obtain zero on the other side. Subtract $4x$ and subtract 4 from both sides.

$$x^3 + x^2 - 4x - 4 = 4x + 4 - 4x - 4$$
$$x^3 + x^2 - 4x - 4 = 0$$

Step 2 Factor. Use factoring by grouping. Group terms that have a common factor.

$$\boxed{x^3 + x^2} + \boxed{-4x - 4} = 0$$

Common factor is x^2.   Common factor is −4.

$$x^2(x + 1) - 4(x + 1) = 0 \qquad \text{Factor } x^2 \text{ from the first two terms and } -4 \text{ from the last two terms.}$$

$$(x + 1)(x^2 - 4) = 0 \qquad \text{Factor out the common binomial, } x + 1, \text{ from each term.}$$

$$(x + 1)(x + 2)(x - 2) = 0 \qquad \text{Factor completely by factoring } x^2 - 4 \text{ as the difference of two squares.}$$

Steps 3 and 4 Set each factor equal to zero and solve the resulting equation.

$$x + 1 = 0 \ \text{ or } \ x + 2 = 0 \ \text{ or } \ x - 2 = 0$$
$$x = -1 \qquad\quad x = -2 \qquad\quad x = 2$$

Step 5 Check the solutions in the original equation. Check the three solutions, $-1, -2$, and 2, by substituting them into the original equation. Can you verify that the solutions are $-1, -2$, and 2, and the solution set is $\{-2, -1, 2\}$?

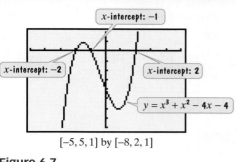

Technology

Graphic Connections

You can use a graphing utility to check the solutions to $x^3 + x^2 - 4x - 4 = 0$. Graph $y = x^3 + x^2 - 4x - 4$, as shown in **Figure 6.7**. The x-intercepts are $-2, -1$, and 2, corresponding to the equation's solutions.

$[-5, 5, 1]$ by $[-8, 2, 1]$

Figure 6.7

 **Checkpoint 3**   Solve by factoring: $2x^3 + 3x^2 = 8x + 12$.

Applications of Quadratic Equations

③ Solve problems using quadratic equations.

Solving quadratic equations by factoring can be used to answer questions about variables contained in mathematical models.

Example 4   **Modeling Motion**

You throw a ball straight up from a rooftop 384 feet high with an initial speed of 32 feet per second. The function

$$s(t) = -16t^2 + 32t + 384$$

describes the ball's height above the ground, $s(t)$, in feet, t seconds after you throw it. The ball misses the rooftop on its way down and eventually strikes the ground. How long will it take for the ball to hit the ground?

Solution

The ball hits the ground when $s(t)$, its height above the ground, is 0 feet. Thus, we substitute 0 for $s(t)$ in the given function and solve for t.

$s(t) = -16t^2 + 32t + 384$ This is the function that models the ball's height.

$0 = -16t^2 + 32t + 384$ Substitute 0 for $s(t)$.

$0 = -16(t^2 - 2t - 24)$ Factor out -16.

$0 = -16(t - 6)(t + 4)$ Factor $t^2 - 2t - 24$, the trinomial.

Do not set the constant, -16, equal to zero: $-16 \neq 0$.

$t - 6 = 0$ or $t + 4 = 0$ Set each variable factor equal to 0.

$t = 6$ $t = -4$ Solve for t.

Because we begin describing the ball's height at $t = 0$, we discard the solution $t = -4$. The ball hits the ground after 6 seconds.

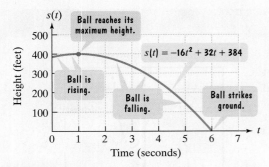

Figure 6.8

Figure 6.8 shows the graph of the quadratic function $s(t) = -16t^2 + 32t + 384$. The horizontal axis is labeled t, for the ball's time in motion. The vertical axis is labeled $s(t)$, for the ball's height above the ground at time t. Because time and height are both positive, the function is graphed in quadrant I only.

The graph visually shows what we discovered algebraically: The ball hits the ground after 6 seconds. The graph also reveals that the ball reaches its maximum height, 400 feet, after 1 second. Then the ball begins to fall.

Checkpoint 4 Use the function $s(t) = -16t^2 + 32t + 384$ to determine when the ball's height is 336 feet. Identify your meaningful solution as a point on the graph in **Figure 6.8**.

A Brief Review • Area and Volume

- The area of a two-dimensional figure is the number of square units, such as square inches (in.^{2}) or square feet (ft^2), that it takes to fill the interior of the figure.

  Example:

- The area, A, of a rectangle with length l and width w is given by the formula $A = lw$.

  Example:

  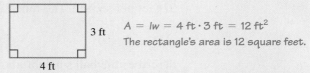

- The volume of a three-dimensional figure is the number of cubic units, such as cubic inches (in.^{3}) or cubic feet (ft^3), that it takes to fill the interior of the figure.

  Example:

  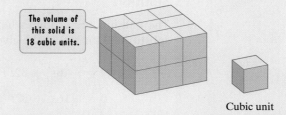

- The volume, V, of a rectangular solid with length l, width w, and height h is given by the formula $V = lwh$.

  Example:

  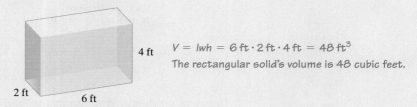

A complete discussion of area and volume can be found in Chapters 9 and 10.

In our next example, we use our five-step strategy from Section 4.3 for solving algebraic word problems.

Example 5   Solving a Problem Involving Landscape Design

A rectangular garden measures 80 feet by 60 feet. A large path of uniform width is to be added along both shorter sides and one longer side of the garden. The landscape designer doing the work wants to double the garden's area with the addition of this path. How wide should the path be?

Solution

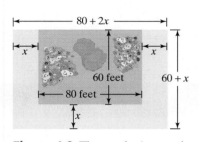

Figure 6.9 The garden's area is to be doubled by adding the path.

Step 1   Let x represent one of the unknown quantities. We will let

$$x = \text{the width of the path.}$$

The situation is illustrated in **Figure 6.9**. The figure shows the original 80-by-60 foot rectangular garden and the path of width x added along both shorter sides and one longer side.

Step 2   Represent other unknown quantities in terms of x. Because the path is added along both shorter sides and one longer side, **Figure 6.9** shows that

$$80 + 2x = \text{the length of the new, expanded rectangle}$$
$$60 + x = \text{the width of the new, expanded rectangle.}$$

Step 3   Write an equation in x that models the conditions. The area of the rectangle must be doubled by the addition of the path.

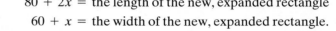

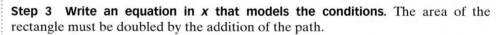

Step 4   Solve the equation and answer the question.

$(80 + 2x)(60 + x) = 2 \cdot 80 \cdot 60$	This is the equation that models the problem's conditions.
$4800 + 200x + 2x^2 = 9600$	Multiply. Use FOIL on the left side.
$2x^2 + 200x - 4800 = 0$	Subtract 9600 from both sides and write the equation in standard form.
$2(x^2 + 100x - 2400) = 0$	Factor out 2, the GCF.
$2(x - 20)(x + 120) = 0$	Factor the trinomial.
$x - 20 = 0 \quad \text{or} \quad x + 120 = 0$	Set each variable factor equal to 0.
$x = 20 \quad \text{or} \qquad\quad x = -120$	Solve for x.

The path cannot have a negative width. Because -120 is geometrically impossible, we use $x = 20$. The width of the path should be 20 feet.

Step 5   Check the proposed solution in the original wording of the problem. Has the landscape architect doubled the garden's area with the 20-foot-wide path? The area of the garden is 80 feet times 60 feet, or 4800 square feet. Because $80 + 2x$ and $60 + x$ represent the length and width of the expanded rectangle,

$$80 + 2x = 80 + 2 \cdot 20 = 120 \text{ feet is the expanded rectangle's length.}$$
$$60 + x = 60 + 20 \quad = 80 \text{ feet is the expanded rectangle's width.}$$

The area of the expanded rectangle is 120 feet times 80 feet, or 9600 square feet. This is double the area of the garden, 4800 square feet, as specified by the problem's conditions.

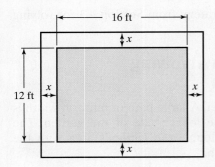

A rectangular garden measures 16 feet by 12 feet. A path of uniform width is to be added so as to surround the entire garden. The landscape artist doing the work wants the garden and path to cover an area of 320 square feet. ▶ How wide should the path be?

Achieving Success

In college, it is recommended that students study and do homework for at least two hours for each hour of class time. For high school students and college students, finding the necessary time to study is not always easy. In order to make education a top priority, efficient time management is essential. You can continue to improve your time management by identifying your biggest time-wasters. Do you really need to spend hours on the phone, update your Facebook page throughout the night, or watch that sitcom rerun? My biggest time-wasters: _____

Exercise Set 6.3

Concept and Vocabulary Exercises

In Exercises 1–4, fill in each blank so that the resulting statement is true.

1. A quadratic equation in x can be written in the standard form _____, where $a \neq 0$.

2. The zero-product principle states that if $AB = 0$, then _____.

3. The solutions of a quadratic equation correspond to the _____ of the graph of a quadratic function.

4. The polynomial equation $x^3 + x^2 - 4x - 4 = 0$ is written in standard form because one side is 0 and the polynomial on the other side is in _____ powers of the variable.

In Exercises 5–8, determine whether each statement is true or false. If the statement is false, make the necessary change(s) to produce a true statement.

5. If $(x + 3)(x - 4) = 2$, then $x + 3 = 0$ or $x - 4 = 0$.

6. The solutions of the equation $4(x - 5)(x + 3) = 0$ are $4, 5$, and -3.

7. Quadratic equations solved by factoring always have two different solutions.

8. Both 0 and $-\pi$ are solutions of the equation $x(x + \pi) = 0$.

Respond to Exercises 9–16 using verbal or written explanations.

9. What is a quadratic equation?

10. What is the zero-product principle?

11. Explain how to solve $x^2 - x = 6$.

12. Describe the relationship between the solutions of a quadratic equation and the graph of the corresponding quadratic function.

13. What is a polynomial equation? When is it in standard form?

14. What is the degree of a polynomial equation? What are polynomial equations of degree 1 and degree 2, respectively, called?

15. Explain how to solve $x^3 + x^2 = x + 1$.

16. A toy rocket is launched vertically upward. Using a quadratic equation, we find that the rocket will reach a height of 220 feet at 2.5 seconds and again at 5.5 seconds. How can this be?

In Exercises 17–20, determine whether each statement makes sense or does not make sense, and explain your reasoning.

17. When solving $4(x - 3)(x + 2) = 0$ and $4x(x + 2) = 0$, I can ignore the monomial factors.

18. I set the quadratic equation $2x^2 - 5x = 12$ equal to zero and obtained $2x^2 - 5x = 0$.

19. Because some trinomials are prime, some quadratic equations cannot be solved by factoring.

20. I'm looking at a graph with one x-intercept, so it must be the graph of a linear function or a vertical line.

Practice Exercises

In Exercises 21–56, use factoring to solve each quadratic equation. Check by substitution or by using a graphing utility and identifying x-intercepts.

21. $x^2 + x - 12 = 0$

22. $x^2 - 2x - 15 = 0$

23. $x^2 + 6x = 7$

24. $x^2 - 4x = 45$

25. $3x^2 + 10x - 8 = 0$

26. $2x^2 - 5x - 3 = 0$

27. $5x^2 = 8x - 3$

28. $7x^2 = 30x - 8$

29. $3x^2 = 2 - 5x$

30. $5x^2 = 2 + 3x$

31. $x^2 = 8x$

32. $x^2 = 4x$

33. $3x^2 = 5x$

34. $2x^2 = 5x$

35. $x^2 + 4x + 4 = 0$

36. $x^2 + 6x + 9 = 0$

37. $x^2 = 14x - 49$

38. $x^2 = 12x - 36$

39. $9x^2 = 30x - 25$

40. $4x^2 = 12x - 9$

41. $x^2 - 25 = 0$

42. $x^2 - 49 = 0$

43. $9x^2 = 100$

44. $4x^2 = 25$

45. $x(x - 3) = 18$

46. $x(x - 4) = 21$

47. $(x - 3)(x + 8) = -30$

48. $(x - 1)(x + 4) = 14$

49. $x(x + 8) = 16(x - 1)$

50. $x(x + 9) = 4(2x + 5)$

51. $(x + 1)^2 - 5(x + 2) = 3x + 7$

52. $(x + 1)^2 = 2(x + 5)$

53. $x(8x + 1) = 3x^2 - 2x + 2$

54. $2x(x + 3) = -5x - 15$

55. $\dfrac{x^2}{18} + \dfrac{x}{2} + 1 = 0$

56. $\dfrac{x^2}{4} - \dfrac{5x}{2} + 6 = 0$

In Exercises 57–66, use factoring to solve each polynomial equation. Check by substitution or by using a graphing utility and identifying x-intercepts.

57. $x^3 + 4x^2 - 25x - 100 = 0$ **58.** $x^3 - 2x^2 - x + 2 = 0$

59. $x^3 - x^2 = 25x - 25$

60. $x^3 + 2x^2 = 16x + 32$

61. $3x^4 - 48x^2 = 0$

62. $5x^4 - 20x^2 = 0$

63. $x^4 - 4x^3 + 4x^2 = 0$

64. $x^4 - 6x^3 + 9x^2 = 0$

65. $2x^3 + 16x^2 + 30x = 0$

66. $3x^3 - 9x^2 - 30x = 0$

In Exercises 67–70, determine the x-intercepts of the graph of each quadratic function. Then match the function with its graph, labeled (a)–(d). Each graph is shown in a $[-10, 10, 1]$ by $[-10, 10, 1]$ viewing rectangle.

67. $y = x^2 - 6x + 8$

68. $y = x^2 - 2x - 8$

69. $y = x^2 + 6x + 8$

70. $y = x^2 + 2x - 8$

a.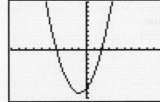

b.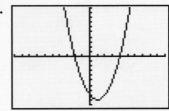

c.

d.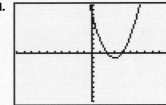

Practice Plus

In Exercises 71–74, solve each polynomial equation.

71. $x(x + 1)^3 - 42(x + 1)^2 = 0$

72. $x(x - 2)^3 - 35(x - 2)^2 = 0$

73. $-4x[x(3x - 2) - 8](25x^2 - 40x + 16) = 0$

74. $-7x[x(2x - 5) - 12](9x^2 + 30x + 25) = 0$

In Exercises 75–78, find all values of c satisfying the given conditions.

75. $f(x) = x^2 - 4x - 27$ and $f(c) = 5$.

76. $f(x) = 5x^2 - 11x + 6$ and $f(c) = 4$.

77. $f(x) = 2x^3 + x^2 - 8x + 2$ and $f(c) = 6$.

78. $f(x) = x^3 + 4x^2 - x + 6$ and $f(c) = 10$.

In Exercises 79–82, find all numbers satisfying the given conditions.

79. The product of the number decreased by 1 and increased by 4 is 24.

80. The product of the number decreased by 6 and increased by 2 is 20.

81. If 5 is subtracted from 3 times the number, the result is the square of 1 less than the number.

82. If the square of the number is subtracted from 61, the result is the square of 1 more than the number.

Application Exercises

A gymnast dismounts the uneven parallel bars at a height of 8 feet with an initial upward velocity of 8 feet per second. The function

$$s(t) = -16t^2 + 8t + 8$$

describes the height of the gymnast's feet above the ground, s(t), in feet, t seconds after dismounting. The graph of the function is shown, with unlabeled tick marks along the horizontal axis. Use the function to solve Exercises 83–84.

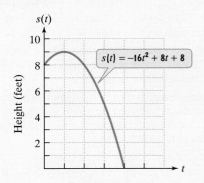

Time (seconds)

83. How long will it take the gymnast to reach the ground? Use this information to provide a number on each tick mark along the horizontal axis in the figure shown.

84. When will the gymnast be 8 feet above the ground? Identify the solution(s) as one or more points on the graph.

In a round-robin chess tournament, each player is paired with every other player once. The function

$$f(x) = \frac{x^2 - x}{2}$$

models the number of chess games, $f(x)$, that must be played in a round-robin tournament with x chess players. Use this function to solve Exercises 85–86.

85. In a round-robin chess tournament, 21 games were played. How many players were entered in the tournament?

86. In a round-robin chess tournament, 36 games were played. How many players were entered in the tournament?

The graph of the quadratic function in Exercises 85–86 is shown. Use the graph to solve Exercises 87–88.

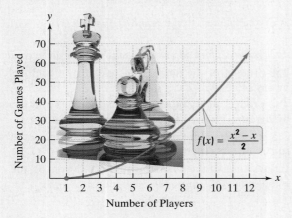

87. Identify your solution to Exercise 85 as a point on the graph.

88. Identify your solution to Exercise 86 as a point on the graph.

89. The length of a rectangular sign is 3 feet longer than the width. If the sign's area is 54 square feet, find its length and width.

90. A rectangular parking lot has a length that is 3 yards greater than the width. The area of the parking lot is 180 square yards. Find the length and the width.

91. Each side of a square is lengthened by 3 inches. The area of this new, larger square is 64 square inches. Find the length of a side of the original square.

92. Each side of a square is lengthened by 2 inches. The area of this new, larger square is 36 square inches. Find the length of a side of the original square.

93. A pool measuring 10 meters by 20 meters is surrounded by a path of uniform width, as shown in the figure. If the area of the pool and the path combined is 600 square meters, what is the width of the path?

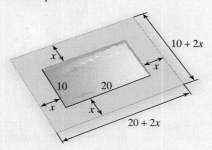

94. A vacant rectangular lot is being turned into a community vegetable garden measuring 15 meters by 12 meters. A path of uniform width is to surround the garden. If the area of the lot is 378 square meters, find the width of the path surrounding the garden.

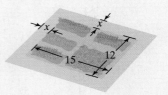

95. As part of a landscaping project, you put in a flower bed measuring 10 feet by 12 feet. You plan to surround the bed with a uniform border of low-growing plants.

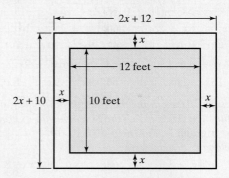

a. Write a polynomial that describes the area of the uniform border that surrounds your flower bed. *Hint:* The area of the border is the area of the large rectangle shown in the figure minus the area of the flower bed.

b. The low growing plants surrounding the flower bed require 1 square foot each when mature. If you have 168 of these plants, how wide a strip around the flower bed should you prepare for the border?

96. As part of a landscaping project, you put in a flower bed measuring 20 feet by 30 feet. To finish off the project, you are putting in a uniform border of pine bark around the outside of the rectangular garden. You have enough pine bark to cover 336 square feet. How wide should the border be?

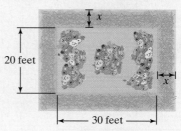

97. A machine produces open boxes using square sheets of metal. The figure at the top of the next page illustrates that the machine cuts equal-sized squares measuring 2 inches on a side from the corners and then shapes the metal into an open box by turning up the sides. If each box must have a volume of 200 cubic inches, find the length and width of the open box.

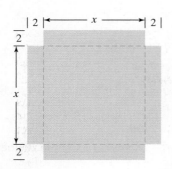

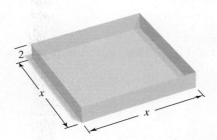

98. A machine produces open boxes using square sheets of metal. The machine cuts equal-sized squares measuring 3 inches on a side from the corners and then shapes the metal into an open box by turning up the sides. If each box must have a volume of 75 cubic inches, find the length and width of the open box.

Critical Thinking Exercises

In Exercises 99–100, solve each equation.

99. $3^{x^2-9x+20} = 1$

100. $(x^2 - 5x + 5)^3 = 1$

101. Write a quadratic equation in standard form whose solutions are -3 and 5.

Technology Exercises

In Exercises 102–105, use the x-intercepts of the graph of y in a $[-10, 10, 1]$ *by* $[-13, 10, 1]$ *viewing rectangle to solve the quadratic equation. Check by substitution.*

102. Use the graph of $y = x^2 + 3x - 4$ to solve
$$x^2 + 3x - 4 = 0.$$

103. Use the graph of $y = x^2 + x - 6$ to solve
$$x^2 + x - 6 = 0.$$

104. Use the graph of $y = (x - 2)(x + 3) - 6$ to solve
$$(x - 2)(x + 3) - 6 = 0.$$

105. Use the graph of $y = x^2 - 2x + 1$ to solve
$$x^2 - 2x + 1 = 0.$$

Solving Quadratic Equations by the Square Root Property and the Quadratic Formula

6.4

Objectives

1 Solve quadratic equations using the square root property.

2 Solve quadratic equations using the quadratic formula.

3 Determine the most efficient method to use when solving a quadratic equation.

4 Use the quadratic formula to solve problems.

Until fairly recently, many doctors believed that your blood pressure was theirs to know and yours to worry about. Today, however, people are encouraged to find out their blood pressure. That pumped-up cuff that squeezes against your upper arm measures blood pressure in millimeters (mm) of mercury (Hg). Blood pressure is given in two numbers: systolic pressure over diastolic pressure, such as 120 over 80. Systolic pressure is the pressure of blood against the artery walls when the heart contracts. Diastolic pressure is the pressure of blood against the artery walls when the heart is at rest.

In this section, we will present a formula that will enable you to solve a quadratic equation, $ax^2 + bx + c = 0$, even if the trinomial $ax^2 + bx + c$ cannot be factored. Using this formula, we will work with functions that model how systolic pressure for men and for women changes with age.

Solve quadratic equations using the square root property.

The Square Root Property

Let's begin with a relatively simple quadratic equation:

$$x^2 = 9.$$

The value of x must be a number whose square is 9. There are two such numbers:

$$x = \sqrt{9} = 3 \quad \text{or} \quad x = -\sqrt{9} = -3.$$

Thus, the solutions of $x^2 = 9$ are 3 and -3. This is an example of the **square root property**.

> **The Square Root Property**
>
> If u is an algebraic expression and d is a positive real number, then $u^2 = d$ is equivalent to $u = \sqrt{d}$ or $u = -\sqrt{d}$:
>
> $$\text{If } u^2 = d, \quad \text{then } u = \sqrt{d} \quad \text{or} \quad u = -\sqrt{d}.$$
>
> Equivalently,
>
> $$\text{If } u^2 = d, \quad \text{then } u = \pm\sqrt{d}.$$

Notice that $u = \pm\sqrt{d}$ is a shorthand notation to indicate that $u = \sqrt{d}$ or $u = -\sqrt{d}$. Although we usually read $u = \pm\sqrt{d}$ as "u equals plus or minus the square root of d," we actually mean that u is the positive square root of d or the negative square root of d.

Example 1   Solving a Quadratic Equation by the Square Root Property

Solve: $3x^2 = 18$.

Solution

To apply the square root property, we need a squared expression by itself on one side of the equation.

$$3x^2 = 18$$

We want x^2 by itself.

We can get x^2 by itself if we divide both sides by 3.

$3x^2 = 18$	This is the original equation.
$\dfrac{3x^2}{3} = \dfrac{18}{3}$	Divide both sides by 3.
$x^2 = 6$	Simplify.
$x = \sqrt{6} \quad \text{or} \quad x = -\sqrt{6}$	Apply the square root property.

Now let's check these proposed solutions in the original equation. Because the equation has an x^2-term and no x-term, we can check both values, $\pm\sqrt{6}$, at once.

Check $\sqrt{6}$ and $-\sqrt{6}$:

$3x^2 = 18$	This is the original equation.
$3(\pm\sqrt{6})^2 \overset{?}{=} 18$	Substitute the proposed solutions.
$3 \cdot 6 \overset{?}{=} 18$	$(\pm\sqrt{6})^2 = 6$
$18 = 18,$	true

The solutions are $-\sqrt{6}$ and $\sqrt{6}$. The solution set is $\{-\sqrt{6}, \sqrt{6}\}$ or $\{\pm\sqrt{6}\}$.

▶ ✓ **Checkpoint 1**   Solve: $4x^2 = 28$.

In this section, we will express irrational solutions in simplified radical form, rationalizing denominators when possible.

Example 2   **Solving a Quadratic Equation by the Square Root Property**

Solve: $2x^2 - 7 = 0$.

Solution

To solve by the square root property, we isolate the squared expression on one side of the equation.

$$2x^2 - 7 = 0$$

We want x^2 by itself.

$$
\begin{aligned}
2x^2 - 7 &= 0 && \text{This is the original equation.}\\
2x^2 &= 7 && \text{Add 7 to both sides.}\\
x^2 &= \frac{7}{2} && \text{Divide both sides by 2.}
\end{aligned}
$$

$$x = \sqrt{\frac{7}{2}} \quad \text{or} \quad x = -\sqrt{\frac{7}{2}} \qquad \text{Apply the square root property.}$$

Because the proposed solutions are opposites, we can rationalize both denominators at once:

$$\pm\sqrt{\frac{7}{2}} = \pm\frac{\sqrt{7}}{\sqrt{2}} \cdot \frac{\sqrt{2}}{\sqrt{2}} = \pm\frac{\sqrt{14}}{2}.$$

Substitute these values into the original equation and verify that the solutions are $-\dfrac{\sqrt{14}}{2}$ and $\dfrac{\sqrt{14}}{2}$. The solution set is $\left\{-\dfrac{\sqrt{14}}{2}, \dfrac{\sqrt{14}}{2}\right\}$ or $\left\{\pm\dfrac{\sqrt{14}}{2}\right\}$.

▶ ✓ **Checkpoint 2**   Solve: $3x^2 - 11 = 0$.

Can we solve an equation such as $(x - 1)^2 = 5$ using the square root property? Yes. The equation is in the form $u^2 = d$, where u^2, the squared expression, is by itself on the left side and d is positive.

$$(x - 1)^2 \quad = \quad 5$$

This is u^2 in $u^2 = d$ with $u = x - 1$.   This is d in $u^2 = d$ with $d = 5$.

Example 3   **Solving a Quadratic Equation by the Square Root Property**

Solve by the square root property: $(x - 1)^2 = 5$.

Solution

$$(x - 1)^2 = 5 \qquad \text{This is the original equation.}$$
$$x - 1 = \sqrt{5} \quad \text{or} \quad x - 1 = -\sqrt{5} \qquad \text{Apply the square root property.}$$
$$x = 1 + \sqrt{5} \qquad x = 1 - \sqrt{5} \qquad \text{Add 1 to both sides in each equation.}$$

Check $1 + \sqrt{5}$:
$$(x - 1)^2 = 5$$
$$(1 + \sqrt{5} - 1)^2 \overset{?}{=} 5$$
$$(\sqrt{5})^2 \overset{?}{=} 5$$
$$5 = 5, \quad \text{true}$$

Check $1 - \sqrt{5}$:
$$(x - 1)^2 = 5$$
$$(1 - \sqrt{5} - 1)^2 \overset{?}{=} 5$$
$$(-\sqrt{5})^2 \overset{?}{=} 5$$
$$5 = 5, \quad \text{true}$$

The solutions are $1 \pm \sqrt{5}$, and the solution set is $\{1 + \sqrt{5}, 1 - \sqrt{5}\}$ or $\{1 \pm \sqrt{5}\}$.

✓ **Checkpoint 3** Solve: $(x - 3)^2 = 10.$

The Quadratic Formula

② Solve quadratic equations using the quadratic formula.

The solutions of a quadratic equation cannot always be found by factoring. Some trinomials are difficult to factor, and others cannot be factored (that is, they are prime). Furthermore, not every quadratic equation is of the form $u^2 = d$, where solutions can be obtained by the square root property. However, there is a formula that can be used to solve all quadratic equations, whether or not they contain factorable trinomials or are of the form $u^2 = d$. The formula is called the *quadratic formula*.

Study Tip

The entire numerator of the quadratic formula must be divided by $2a$. Always write the fraction bar all the way across the numerator.

$$x = \frac{-b \pm \sqrt{b^2 - 4ac}}{2a}$$

The Quadratic Formula

The solutions of a quadratic equation in the form $ax^2 + bx + c = 0$, with $a \neq 0$, are given by the **quadratic formula**

$$x = \frac{-b \pm \sqrt{b^2 - 4ac}}{2a}.$$

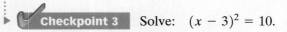 *x equals negative b plus or minus the square root of $b^2 - 4ac$, all divided by 2a.*

To use the quadratic formula, be sure that the quadratic equation is expressed with all terms on one side and zero on the other side. It may be necessary to begin by rewriting the equation in this form. Then determine the numerical values for a (the coefficient of the x^2-term), b (the coefficient of the x-term), and c (the constant term). Substitute the values of a, b, and c into the quadratic formula and evaluate the expression. The $\pm$ sign indicates that there are two solutions of the equation.

Example 4 Solving a Quadratic Equation Using the Quadratic Formula

Solve using the quadratic formula: $2x^2 + 9x - 5 = 0.$

Solution

The given equation is in the desired form, with all terms on one side and zero on the other side. Begin by identifying the values for a, b, and c.

$$2x^2 + 9x - 5 = 0$$

$a = 2$ $b = 9$ $c = -5$

Substituting these values into the quadratic formula and simplifying gives the equation's solutions.

$$x = \frac{-b \pm \sqrt{b^2 - 4ac}}{2a}$$
Use the quadratic formula.

$$x = \frac{-9 \pm \sqrt{9^2 - 4(2)(-5)}}{2(2)}$$
Substitute the values for a, b, and c:
$a = 2, b = 9$, and $c = -5$.

$$= \frac{-9 \pm \sqrt{81 + 40}}{4}$$
$9^2 - 4(2)(-5) = 81 - (-40) = 81 + 40$

$$= \frac{-9 \pm \sqrt{121}}{4}$$
Add under the radical sign.

$$= \frac{-9 \pm 11}{4}$$
$\sqrt{121} = 11$

Now we will evaluate this expression in two different ways to obtain the two solutions. On the left, we will *add* 11 to -9. On the right, we will *subtract* 11 from -9.

$$x = \frac{-9 + 11}{4} \quad \text{or} \quad x = \frac{-9 - 11}{4}$$

$$= \frac{2}{4} = \frac{1}{2} \qquad\qquad = \frac{-20}{4} = -5$$

Check the solutions, -5 and $\frac{1}{2}$, by substituting them into the original equation. The solution set is $\left\{ -5, \frac{1}{2} \right\}$.

 **Checkpoint 4**   Solve using the quadratic formula: $8x^2 + 2x - 1 = 0$.

The quadratic equation in Example 4 has rational solutions, namely -5 and $\frac{1}{2}$. The equation can also be solved by factoring. Take a few minutes to do this now and convince yourself that you will arrive at the same two solutions.

Any quadratic equation that has rational solutions can be solved by factoring or using the quadratic formula. However, quadratic equations with irrational solutions cannot be solved by factoring. These equations can be readily solved using the quadratic formula.

Example 5   **Solving a Quadratic Equation Using the Quadratic Formula**

Solve using the quadratic formula: $2x^2 = 4x + 1$.

Solution

The quadratic equation must have zero on one side to identify the values for a, b, and c. To move all terms to one side and obtain zero on the right, we subtract $4x + 1$ from both sides. Then we can identify the values for a, b, and c.

$$2x^2 = 4x + 1 \quad \text{This is the given equation.}$$

$$2x^2 - 4x - 1 = 0 \quad \text{Subtract } 4x + 1 \text{ from both sides.}$$

$a = 2$ $b = -4$ $c = -1$

Substituting $a = 2$, $b = -4$, and $c = -1$ into the quadratic formula and simplifying gives the equation's solutions.

$$x = \frac{-b \pm \sqrt{b^2 - 4ac}}{2a}$$

Use the quadratic formula.

$$x = \frac{-(-4) \pm \sqrt{(-4)^2 - 4(2)(-1)}}{2(2)}$$

Substitute the values for a, b, and c: $a = 2$, $b = -4$, and $c = -1$.

$$= \frac{4 \pm \sqrt{16 - (-8)}}{4}$$

$-(-4) = 4$, $(-4)^2 = (-4)(-4) = 16$, and $4(2)(-1) = -8$.

$$= \frac{4 \pm \sqrt{24}}{4}$$

$16 - (-8) = 16 + 8 = 24$

The solutions are $\dfrac{4 + \sqrt{24}}{4}$ and $\dfrac{4 - \sqrt{24}}{4}$. These solutions are irrational numbers.

You can use a calculator to obtain a decimal approximation for each solution. However, in situations such as this that do not involve applications, it is best to leave the irrational solutions in radical form as exact answers. In some cases, we can simplify this radical form. Using methods for simplifying square roots discussed in Section 3.4, we can simplify $\sqrt{24}$:

$$\sqrt{24} = \sqrt{4 \cdot 6} = \sqrt{4}\sqrt{6} = 2\sqrt{6}.$$

Now we can use this result to simplify the two solutions. First, use the distributive property to factor out 2 from both terms in the numerator. Then, divide the numerator and the denominator by 2.

$$x = \frac{4 \pm \sqrt{24}}{4} = \frac{4 \pm 2\sqrt{6}}{4} = \frac{\overset{1}{\cancel{2}}(2 \pm \sqrt{6})}{\underset{2}{\cancel{4}}} = \frac{2 \pm \sqrt{6}}{2}$$

In simplified radical form, the equation's solution set is

$$\left\{ \frac{2 + \sqrt{6}}{2}, \frac{2 - \sqrt{6}}{2} \right\}.$$

Examples 4 and 5 illustrate that the solutions of quadratic equations can be rational or irrational numbers. In Example 4, the expression under the square root was 121, a perfect square ($\sqrt{121} = 11$), and we obtained rational solutions. In Example 5, this expression was 24, which is not a perfect square (although we simplified $\sqrt{24}$ to $2\sqrt{6}$), and we obtained irrational solutions. If the expression under the square root simplifies to a negative number, then the quadratic equation has **no real solution**. The solution set consists of *imaginary numbers*, discussed in the Blitzer Bonus on page 175.

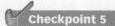 **Checkpoint 5** Solve using the quadratic formula: $2x^2 = 6x - 1$.

Determining Which Method to Use

 Determine the most efficient method to use when solving a quadratic equation.

All quadratic equations can be solved by the quadratic formula. However, if an equation is in the form $u^2 = d$, such as $x^2 = 5$ or $(2x + 3)^2 = 8$, it is faster to use the square root property, taking the square root of both sides. If the equation is not in the form $u^2 = d$, write the quadratic equation in standard form $(ax^2 + bx + c = 0)$. Try to solve the equation by factoring. If $ax^2 + bx + c$ cannot be factored, then solve the quadratic equation by the quadratic formula.

Table 6.1 summarizes our observations about which technique to use when solving a quadratic equation.

Table 6.1 Determining the Most Efficient Technique to Use When Solving a Quadratic Equation

Description and Form of the Quadratic Equation	Most Efficient Solution Method	Example
$ax^2 + bx + c = 0$ and $ax^2 + bx + c$ can be factored easily.	Factor and use the zero-product principle.	$3x^2 + 5x - 2 = 0$ $(3x - 1)(x + 2) = 0$ $3x - 1 = 0$ or $x + 2 = 0$ $x = \dfrac{1}{3}$ $x = -2$
$ax^2 + c = 0$ The quadratic equation has no x-term. ($b = 0$)	Solve for x^2 and apply the square root property.	$4x^2 - 7 = 0$ $4x^2 = 7$ $x^2 = \dfrac{7}{4}$ $x = \pm\dfrac{\sqrt{7}}{2}$
$u^2 = d$; u is a first-degree polynomial.	Use the square root property.	$(x + 4)^2 = 5$ $x + 4 = \pm\sqrt{5}$ $x = -4 \pm \sqrt{5}$
$ax^2 + bx + c = 0$ and $ax^2 + bx + c$ cannot be factored or the factoring is too difficult.	Use the quadratic formula: $x = \dfrac{-b \pm \sqrt{b^2 - 4ac}}{2a}.$	$x^2 - 2x - 6 = 0$ $\boxed{a = 1}\ \boxed{b = -2}\ \boxed{c = -6}$ $x = \dfrac{-(-2) \pm \sqrt{(-2)^2 - 4(1)(-6)}}{2(1)}$ $= \dfrac{2 \pm \sqrt{4 - 4(1)(-6)}}{2(1)}$ $= \dfrac{2 \pm \sqrt{28}}{2} = \dfrac{2 \pm \sqrt{4}\sqrt{7}}{2}$ $= \dfrac{2 \pm 2\sqrt{7}}{2} = \dfrac{2\left(1 \pm \sqrt{7}\right)}{2}$ $= 1 \pm \sqrt{7}$

④ Use the quadratic formula to solve problems.

Applications

Quadratic equations can be solved to answer questions about variables contained in quadratic functions.

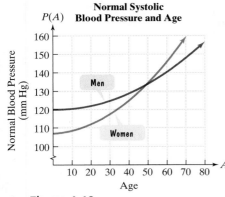

Normal Systolic Blood Pressure and Age

Figure 6.10

Example 6   **Blood Pressure and Age**

The graphs in **Figure 6.10** illustrate that a person's normal systolic blood pressure, measured in millimeters of mercury (mm Hg), depends on his or her age. The function

$$P(A) = 0.006A^2 - 0.02A + 120$$

models a man's normal systolic pressure, $P(A)$, at age A.

a. Find the age, to the nearest year, of a man whose normal systolic blood pressure is 125 mm Hg.

b. Use the graphs in **Figure 6.10** to describe the differences between the normal systolic blood pressures of men and women as they age.

Solution

a. We are interested in the age of a man with a normal systolic blood pressure of 125 millimeters of mercury. Thus, we substitute 125 for $P(A)$ in the given function for men. Then we solve for A, the man's age.

$P(A) = 0.006A^2 - 0.02A + 120$ This is the given function for men.

$125 = 0.006A^2 - 0.02A + 120$ Substitute 125 for $P(A)$.

$0 = 0.006A^2 - 0.02A - 5$ Subtract 125 from both sides and write the quadratic equation in standard form.

$$a = 0.006 \quad b = -0.02 \quad c = -5$$

Because the trinomial on the right side of the equation appears difficult to factor and is, in fact, prime, we solve using the quadratic formula.

Notice that the variable is A, rather than the usual x.

$A = \dfrac{-b \pm \sqrt{b^2 - 4ac}}{2a}$ Use the quadratic formula.

$= \dfrac{-(-0.02) \pm \sqrt{(-0.02)^2 - 4(0.006)(-5)}}{2(0.006)}$ Substitute the values for a, b, and c: $a = 0.006$, $b = -0.02$, and $c = -5$.

$= \dfrac{0.02 \pm \sqrt{0.1204}}{0.012}$ Use a calculator to simplify the expression under the square root.

$\approx \dfrac{0.02 \pm 0.347}{0.012}$ Use a calculator: $\sqrt{0.1204} \approx 0.347$.

$A \approx \dfrac{0.02 + 0.347}{0.012}$ or $A \approx \dfrac{0.02 - 0.347}{0.012}$

$A \approx 31$ $A \approx -27$ Use a calculator and round to the nearest integer.

Reject this solution. Age cannot be negative.

The positive solution, $A \approx 31$, indicates that 31 is the approximate age of a man whose normal systolic blood pressure is 125 mm Hg. This is illustrated by the black lines with the arrows on the red graph representing men in **Figure 6.11**.

b. Take a second look at the graphs in **Figure 6.11**. Before approximately age 50, the blue graph representing women's normal systolic blood pressure lies below the red graph representing men's normal systolic blood pressure. Thus, up to age 50, women's normal systolic blood pressure is lower than men's, although it is increasing at a faster rate. After age 50, women's normal systolic blood pressure is higher than men's.

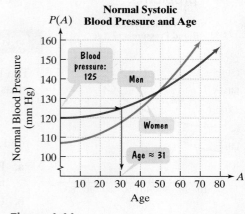

Figure 6.11

Checkpoint 6 The function $P(A) = 0.01A^2 + 0.05A + 107$ models a woman's normal systolic blood pressure, $P(A)$, at age A. Use this function to find the age, to the nearest year, of a woman whose normal systolic blood pressure is 115 mm Hg. Use the blue graph in **Figure 6.11** to verify your solution.

Blitzer Bonus

Art, Nature, and Quadratic Equations

A **golden rectangle** can be a rectangle of any size, but its long side must be Φ times as long as its short side, where $\Phi \approx 1.6$. Artists often use golden rectangles in their work because they are considered to be more visually pleasing than other rectangles.

In *The Bathers at Asnières*, by the French impressionist Georges Seurat (1859–1891), the artist positions parts of the painting as though they were inside golden rectangles.

Georges Seurat, "The Bathers at Asnières," 1883–1884. Oil on canvas, $79\frac{1}{2} \times 118\frac{1}{2}$ in. Erich Lessing/Art Resource, NY.

If a golden rectangle is divided into a square and a rectangle, as in **Figure 6.12(a)**, the smaller rectangle is a golden rectangle. If the smaller golden rectangle is divided again, the same is true of the yet smaller rectangle, and so on. The process of repeatedly dividing each golden rectangle in this manner is illustrated in **Figure 6.12(b)**. We've also created a spiral by connecting the opposite corners of all the squares with a smooth curve. This spiral matches the spiral shape of the chambered nautilus shell shown in **Figure 6.12(c)**. The shell spirals out at an ever-increasing rate that is governed by this geometry.

Golden Rectangle *A*

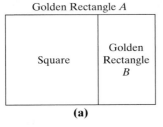

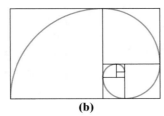

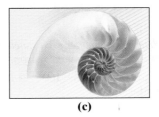

(a)   **(b)**   **(c)**

Figure 6.12

In the Exercise Set that follows, you will use the golden rectangles in **Figure 6.12(a)** to obtain an exact value for Φ, the ratio of the long side to the short side in a golden rectangle of any size. Your model will involve a quadratic equation that can be solved by the quadratic formula. (See Exercise 103.)

Achieving Success

Two Ways to Stay Sharp

- **Concentrate on one task at a time.** Do not multitask. Doing several things at once can cause confusion and can take longer to complete the tasks than tackling them sequentially.

- **Get enough sleep.** Fatigue impedes the ability to learn and do complex tasks.

Exercise Set 6.4

Concept and Vocabulary Exercises

In Exercises 1–4, fill in each blank so that the resulting statement is true.

1. The square root property states that if $u^2 = d$, then _____.

2. The solutions of $ax^2 + bx + c = 0, a \neq 0$, are given by _____, called the _____.

3. The most efficient method for solving $(x + 1)^2 = 7$ is _____.

4. The most efficient method for solving $x^2 - 2x - 10 = 0$ is _____.

In Exercises 5–8, determine whether each statement is true or false. If the statement is false, make the necessary change(s) to produce a true statement.

5. The solutions of $3x^2 - 5 = 0$ are $-\dfrac{\sqrt{5}}{3}$ and $\dfrac{\sqrt{5}}{3}$.

6. In using the quadratic formula to solve the quadratic equation $5x^2 = 2x - 7$, we have $a = 5, b = 2$, and $c = -7$.

7. The quadratic formula can be expressed as
$$x = -b \pm \frac{\sqrt{b^2 - 4ac}}{2a}.$$

8. The solutions $\dfrac{4 \pm \sqrt{3}}{2}$ can be simplified to $2 \pm \sqrt{3}$.

Respond to Exercises 9–12 using verbal or written explanations.

9. What is the square root property?

10. Explain how to solve $(x - 1)^2 = 16$ using the square root property.

11. What is the quadratic formula and why is it useful?

12. Explain how to solve $x^2 + 6x + 8 = 0$ using the quadratic formula.

In Exercises 13–16, determine whether each statement makes sense or does not make sense, and explain your reasoning.

13. Because I want to solve $25x^2 - 169 = 0$ fairly quickly, I'll use the quadratic formula.

14. I simplified $\dfrac{3 + 2\sqrt{3}}{2}$ to $3 + \sqrt{3}$ because 2 is a factor of $2\sqrt{3}$.

15. The fastest way for me to solve $x^2 - x - 2 = 0$ is to use the quadratic formula.

16. Because $4x - 8 = 0$ can be written as $0x^2 + 4x - 8 = 0$, I can use the quadratic formula with $a = 0$, $b = 4$, and $c = -8$ to solve the equation.

Practice Exercises

In Exercises 17–46, solve each quadratic equation by the square root property. If possible, simplify radicals or rationalize denominators.

17. $x^2 = 16$
18. $x^2 = 100$
19. $y^2 = 81$
20. $y^2 = 144$
21. $x^2 = 7$
22. $x^2 = 13$
23. $x^2 = 50$
24. $x^2 = 27$
25. $5x^2 = 20$
26. $3x^2 = 75$
27. $4y^2 = 49$
28. $16y^2 = 25$
29. $2x^2 + 1 = 51$
30. $3x^2 - 1 = 47$
31. $3x^2 - 2 = 0$
32. $3x^2 - 5 = 0$
33. $5z^2 - 7 = 0$
34. $5z^2 - 2 = 0$
35. $(x - 3)^2 = 16$
36. $(x - 2)^2 = 25$
37. $(x + 5)^2 = 121$
38. $(x + 6)^2 = 144$
39. $(3x + 2)^2 = 9$
40. $(2x + 1)^2 = 49$
41. $(x - 5)^2 = 3$
42. $(x - 3)^2 = 15$
43. $(y + 8)^2 = 11$
44. $(y + 7)^2 = 5$
45. $(z - 4)^2 = 18$
46. $(z - 6)^2 = 12$

Solve the equations in Exercises 47–66 using the quadratic formula.

47. $x^2 + 8x + 15 = 0$
48. $x^2 + 8x + 12 = 0$
49. $x^2 + 5x + 3 = 0$
50. $x^2 + 5x + 2 = 0$
51. $x^2 + 4x = 6$
52. $x^2 + 2x = 4$
53. $x^2 + 4x - 7 = 0$
54. $x^2 + 4x + 1 = 0$
55. $x^2 - 3x = 18$
56. $x^2 - 3x = 10$
57. $6x^2 - 5x - 6 = 0$
58. $9x^2 - 12x - 5 = 0$
59. $x^2 - 2x - 10 = 0$
60. $x^2 + 6x - 10 = 0$
61. $x^2 - x = 14$
62. $x^2 - 5x = 10$
63. $6x^2 + 6x + 1 = 0$
64. $3x^2 = 5x - 1$
65. $4x^2 = 12x - 9$
66. $9x^2 + 6x + 1 = 0$

In Exercises 67–86, solve each equation by the method of your choice. Simplify irrational solutions, if possible.

67. $2x^2 - x = 1$
68. $3x^2 - 4x = 4$
69. $5x^2 + 2 = 11x$
70. $5x^2 = 6 - 13x$
71. $3x^2 = 60$
72. $2x^2 = 250$
73. $x^2 - 2x = 1$
74. $2x^2 + 3x = 1$
75. $(2x + 3)(x + 4) = 1$
76. $(2x - 5)(x + 1) = 2$
77. $(3x - 4)^2 = 16$
78. $(2x + 7)^2 = 25$
79. $3x^2 - 12x + 12 = 0$
80. $9 - 6x + x^2 = 0$
81. $4x^2 - 16 = 0$
82. $3x^2 - 27 = 0$
83. $x^2 + 9x = 0$
84. $x^2 - 6x = 0$
85. $(3x - 2)^2 = 10$
86. $(4x - 1)^2 = 15$

Practice Plus

In Exercises 87–94, solve each equation by the method of your choice.

87. $\dfrac{3x^2}{4} - \dfrac{5x}{2} - 2 = 0$
88. $\dfrac{x^2}{3} - \dfrac{x}{2} - \dfrac{3}{2} = 0$

89. $(x - 1)(3x + 2) = -7(x - 1)$

90. $x(x + 1) = 4 - (x + 2)(x + 2)$

91. $(2x - 6)(x + 2) = 5(x - 1) - 12$

92. $7x(x - 2) = 3 - 2(x + 4)$

93. $2x^2 - 9x - 3 = 9 - 9x$

94. $3x^2 - 6x - 3 = 12 - 6x$

95. When the sum of 6 and twice a positive number is subtracted from the square of the number, 0 results. Find the number.

96. When the sum of 1 and twice a negative number is subtracted from twice the square of the number, 0 results. Find the number.

Application Exercises

The function $s(t) = 16t^2$ models the distance, $s(t)$, in feet, that an object falls in t seconds. Use this function and the square root property to solve Exercises 97–98. Express answers in simplified radical form. Then use your calculator to find a decimal approximation to the nearest tenth of a second.

97. A sky diver jumps from an airplane and falls for 4800 feet before opening a parachute. For how many seconds was the diver in a free fall?

98. A sky diver jumps from an airplane and falls for 3200 feet before opening a parachute. For how many seconds was the diver in a free fall?

A substantial percentage of the United States population is foreign-born. The bar graph shows the percentage of foreign-born Americans for selected years from 1920 through 2006.

Percentage of the United States Population That Was Foreign-Born, 1920–2006

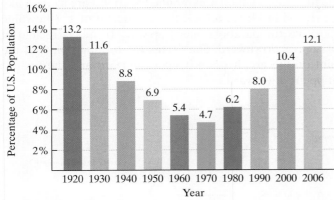

Source: U.S. Census Bureau

The percentage, $p(x)$, of the United States population that was foreign-born x years after 1920 can be modeled by the function

$$p(x) = 0.004x^2 - 0.37x + 14.1.$$

Use the function to solve Exercises 99–100.

99. a. According to the model, what percentage of the U.S. population was foreign-born in 2000? Does the model underestimate or overestimate the actual number displayed by the bar graph? By how much?

b. If trends shown by the model continue, in which year will 25% of the U.S. population be foreign-born? Round to the nearest year.

100. a. According to the model, what percentage of the U.S. population was foreign-born in 1990? Does the model underestimate or overestimate the actual number displayed by the bar graph? By how much?

b. If trends shown by the model continue, in which year will 17% of the U.S. population be foreign-born? Round to the nearest year.

101. The length of a rectangle is 3 meters longer than the width. If the area is 36 square meters, find the rectangle's dimensions. Round to the nearest tenth of a meter.

102. The length of a rectangle is 2 meters longer than the width. If the area is 10 square meters, find the rectangle's dimensions. Round to the nearest tenth of a meter.

103. If you have not yet done so, read the Blitzer Bonus on page 421. In this exercise, you will use the golden rectangles shown to obtain an exact value for Φ, the ratio of the long side to the short side in a golden rectangle of any size.

Golden Rectangle A

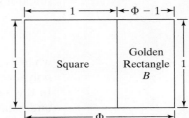

a. The golden ratio in rectangle A, or the ratio of the long side to the short side, can be modeled by $\dfrac{\Phi}{1}$. Write a fractional expression that models the golden ratio in rectangle B.

b. Set the expression for the golden ratio in rectangle A equal to the expression for the golden ratio in rectangle B. Solve the resulting proportion using the quadratic formula. Express Φ as an exact value in simplified radical form.

c. Use your solution from part (b) to complete this statement: The ratio of the long side to the short side in a golden rectangle of any size is _____ to 1.

Critical Thinking Exercises

104. Solve: $x^4 - 8x^2 + 15 = 0$.

105. Solve: $x^2 + 2\sqrt{3}x - 9 = 0$.

106. Solve for t: $s = -16t^2 + v_0t$.

107. The radicand of the quadratic formula, $b^2 - 4ac$, can be used to determine whether $ax^2 + bx + c = 0$ has solutions that are rational, irrational, or not real numbers. Explain how this works. Is it possible to determine the kinds of answers that one will obtain to a quadratic equation without actually solving the equation? Explain.

108. A rectangular vegetable garden is 5 feet wide and 9 feet long. The garden is to be surrounded by a tile border of uniform width. If there are 40 square feet of tile for the border, how wide, to the nearest tenth of a foot, should it be?

6.5 Quadratic Functions and Their Graphs

Many sports involve objects that are thrown, kicked, or hit, and then proceed with no additional force of their own. Such objects are called **projectiles**. Paths of projectiles, as well as their heights over time, can be modeled by quadratic functions. We have seen that a **quadratic function** is any function of the form

$$f(x) = ax^2 + bx + c,$$

where a, b, and c are real numbers, with $a \neq 0$. A quadratic function is a polynomial function whose greatest exponent is 2 when written in standard form. In this section, you will learn to use graphs of quadratic functions to gain a visual understanding of the algebra that describes football, baseball, basketball, the shot put, and other projectile sports.

Graphs of Quadratic Functions

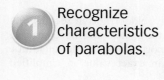

Recognize characteristics of parabolas.

The graph of any quadratic function is called a **parabola**. Parabolas are shaped like bowls or inverted bowls, as shown in **Figure 6.13**. If the coefficient of x^2 (the value of a in $ax^2 + bx + c$) is positive, the parabola opens upward. If the coefficient of x^2 is negative, the parabola opens downward. The **vertex** (or turning point) of the parabola is the lowest point on the graph when it opens upward and the highest point on the graph when it opens downward.

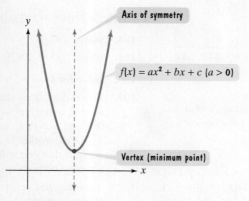

$a > 0$: Parabola opens upward.

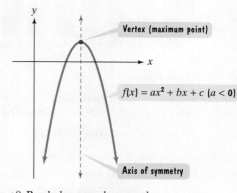

$a < 0$: Parabola opens downward.

Figure 6.13 Characteristics of graphs of quadratic functions

Look at the unusual image of the word *mirror* shown to the left. The artist, Scott Kim, has created the image so that the two halves of the whole are mirror images of each other. A parabola shares this kind of symmetry, in which a vertical line through the vertex divides the figure in half. Parabolas are symmetric with respect to this line, called the **axis of symmetry**. If a parabola is folded along its axis of symmetry, the two halves match exactly.

Example 1 Using Point Plotting to Graph a Parabola

Consider the quadratic function $f(x) = x^2 + 4x + 3$.

a. Is the graph a parabola that opens upward or downward?

b. Use point plotting to graph the parabola. Select integers from -5 to 1, inclusive, for x.

Solution

a. To determine whether a parabola opens upward or downward, we begin by identifying a, the coefficient of x^2. The following voice balloons show the values for a, b, and c in $f(x) = x^2 + 4x + 3$. Notice that we wrote x^2 as $1x^2$.

$$f(x) = 1x^2 + 4x + 3$$

| a, the coefficient of x^2, is 1. | b, the coefficient of x, is 4. | c, the constant term, is 3. |

When a is greater than 0, a parabola opens upward. When a is less than 0, a parabola opens downward. Because $a = 1$, which is greater than 0, the parabola opens upward.

b. To use point plotting to graph the parabola, we first make a table of x- and y-coordinates.

x	$f(x) = x^2 + 4x + 3$	(x, y)
-5	$f(-5) = (-5)^2 + 4(-5) + 3 = 8$	$(-5, 8)$
-4	$f(-4) = (-4)^2 + 4(-4) + 3 = 3$	$(-4, 3)$
-3	$f(-3) = (-3)^2 + 4(-3) + 3 = 0$	$(-3, 0)$
-2	$f(-2) = (-2)^2 + 4(-2) + 3 = -1$	$(-2, -1)$
-1	$f(-1) = (-1)^2 + 4(-1) + 3 = 0$	$(-1, 0)$
0	$f(0) = 0^2 + 4(0) + 3 = 3$	$(0, 3)$
1	$f(1) = 1^2 + 4(1) + 3 = 8$	$(1, 8)$

Then we plot the points and connect them with a smooth curve. The graph of $f(x) = x^2 + 4x + 3$ is shown in **Figure 6.14**.

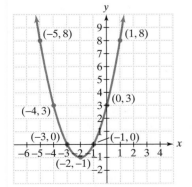

Figure 6.14 The graph of $f(x) = x^2 + 4x + 3$

Checkpoint 1 Consider the quadratic function $f(x) = x^2 - 6x + 8$.

a. Is the graph a parabola that opens upward or downward?

b. Use point plotting to graph the parabola. Select integers from 0 to 6, inclusive, for x.

Several points are important when graphing a quadratic function. These points, labeled in **Figure 6.15**, are the x-intercepts (although not every parabola has two x-intercepts), the y-intercept, and the vertex. Let's see how we can locate these points.

Finding a Parabola's *x*-Intercepts

At each point where a parabola crosses the x-axis, the value of y, or $f(x)$, equals 0. Thus, the x-intercepts can be found by replacing $f(x)$ with 0 in $f(x) = ax^2 + bx + c$. Use factoring or the quadratic formula to solve the resulting quadratic equation for x.

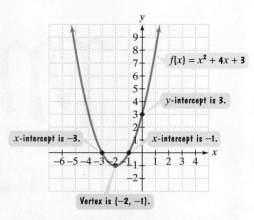

Figure 6.15 Useful points in graphing a parabola

2 Find a parabola's intercepts.

Example 2   Finding a Parabola's *x*-Intercepts

Find the x-intercepts for the parabola whose equation is $f(x) = x^2 + 4x + 3$.

Solution

Replace $f(x)$ with 0 in $f(x) = x^2 + 4x + 3$. We obtain $0 = x^2 + 4x + 3$, or $x^2 + 4x + 3 = 0$. We can solve this quadratic equation by factoring.

$$x^2 + 4x + 3 = 0$$
$$(x + 3)(x + 1) = 0$$
$$x + 3 = 0 \quad \text{or} \quad x + 1 = 0$$
$$x = -3 \qquad\qquad x = -1$$

Thus, the x-intercepts are -3 and -1. The parabola passes through $(-3, 0)$ and $(-1, 0)$, as shown in **Figure 6.15**.

✓ **Checkpoint 2**   Find the x-intercepts for the parabola whose equation is
▶ $f(x) = x^2 - 6x + 8$.

Finding a Parabola's *y*-Intercept

At the point where a parabola crosses the y-axis, the value of x equals 0. Thus, the y-intercept can be found by replacing x with 0 in $f(x) = ax^2 + bx + c$. Simple arithmetic will produce a value for $f(0)$, which is the y-intercept.

Example 3   Finding a Parabola's *y*-Intercept

Find the y-intercept for the parabola whose equation is $f(x) = x^2 + 4x + 3$.

Solution

Replace x with 0 in $f(x) = x^2 + 4x + 3$.

$$f(0) = 0^2 + 4 \cdot 0 + 3 = 0 + 0 + 3 = 3$$

The y-intercept is 3. The parabola passes through $(0, 3)$, as shown in **Figure 6.15**.

✓ **Checkpoint 3** Find the y-intercept for the parabola whose equation is $f(x) = x^2 - 6x + 8$.

Finding a Parabola's Vertex

3 Find a parabola's vertex.

Keep in mind that a parabola's vertex is its turning point. The x-coordinate of the vertex for the parabola in **Figure 6.15**, -2, is midway between the x-intercepts, -3 and -1. If a parabola has two x-intercepts, they are found by solving $ax^2 + bx + c = 0$. The solutions of this equation,

$$x = \frac{-b - \sqrt{b^2 - 4ac}}{2a} \quad \text{and} \quad x = \frac{-b + \sqrt{b^2 - 4ac}}{2a},$$

are the x-intercepts. The value of x midway between these intercepts is $x = -\dfrac{b}{2a}$.

This equation can be used to find the x-coordinate of the vertex even when no x-intercepts exist.

> **The Vertex of a Parabola**
>
> For a parabola whose equation is $f(x) = ax^2 + bx + c$,
>
> 1. The x-coordinate of the vertex is $-\dfrac{b}{2a}$.
> 2. The y-coordinate of the vertex is found by substituting the x-coordinate into the parabola's equation and evaluating the function at this value of x.

Example 4 **Finding a Parabola's Vertex**

Find the vertex for the parabola whose equation is $f(x) = x^2 + 4x + 3$.

Solution

We know that the x-coordinate of the vertex is $x = -\dfrac{b}{2a}$. Let's identify the numbers a, b, and c in the given equation, which is in the form $f(x) = ax^2 + bx + c$.

$$f(x) = x^2 + 4x + 3$$

$$a = 1 \qquad b = 4 \qquad c = 3$$

Substitute the values of a and b into the equation for the x-coordinate:

$$x = -\frac{b}{2a} = -\frac{4}{2 \cdot 1} = -\frac{4}{2} = -2$$

The x-coordinate of the vertex is -2. We substitute -2 for x into the equation of the function, $f(x) = x^2 + 4x + 3$, to find the y-coordinate:

$$f(-2) = (-2)^2 + 4(-2) + 3 = 4 + (-8) + 3 = -1.$$

The vertex is $(-2, -1)$, as shown in **Figure 6.15**.

✓ **Checkpoint 4** Find the vertex for the parabola whose equation is $f(x) = x^2 - 6x + 8$.

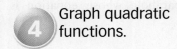

Graph quadratic functions.

A Strategy for Graphing Quadratic Functions

Here is a procedure to sketch the graph of the quadratic function $f(x) = ax^2 + bx + c$:

> **Graphing Quadratic Functions**
>
> The graph of $f(x) = ax^2 + bx + c$, called a parabola, can be graphed using the following steps:
>
> 1. Determine whether the parabola opens upward or downward. If $a > 0$, it opens upward. If $a < 0$, it opens downward.
>
> 2. Determine the vertex of the parabola. The x-coordinate is $-\dfrac{b}{2a}$. The y-coordinate is found by substituting the x-coordinate into the parabola's equation and evaluating the function at this value of x.
>
> 3. Find any x-intercepts by replacing $f(x)$ with 0. Solve the resulting quadratic equation for x. The real solutions are the x-intercepts.
>
> 4. Find the y-intercept by replacing x with 0 and computing $f(0)$. Because $f(0) = a \cdot 0^2 + b \cdot 0 + c$ simplifies to c, the y-intercept is c, the constant term, and the parabola passes through $(0, c)$.
>
> 5. Plot the intercepts and the vertex.
>
> 6. Connect these points with a smooth curve.

Example 5 **Graphing a Quadratic Function**

Graph the quadratic function: $f(x) = x^2 - 2x - 3$.

Solution

We can graph this function by following the steps in the box.

Step 1 Determine how the parabola opens. Note that a, the coefficient of x^2, is 1. Thus, $a > 0$; this positive value tells us that the parabola opens upward.

Step 2 Find the vertex. We know that the x-coordinate of the vertex is $-\dfrac{b}{2a}$. Let's identify the numbers a, b, and c in the given equation, which is in the form $f(x) = ax^2 + bx + c$.

$$f(x) = x^2 - 2x - 3$$

$$\boxed{a = 1} \qquad \boxed{b = -2} \quad \boxed{c = -3}$$

Now we substitute the values of a and b, $a = 1$ and $b = -2$, into the expression for the x-coordinate:

$$x\text{-coordinate of vertex} = -\frac{b}{2a} = -\frac{-2}{2(1)} = -\left(\frac{-2}{2}\right) = -(-1) = 1.$$

The x-coordinate of the vertex is 1. We can substitute 1 for x in the equation of the function, $f(x) = x^2 - 2x - 3$, to find the y-coordinate:

$$y\text{-coordinate of vertex} = f(1) = 1^2 - 2 \cdot 1 - 3 = 1 - 2 - 3 = -4.$$

The vertex is $(1, -4)$.

Step 3 Find the x-intercepts. Replace $f(x)$ with 0 in $f(x) = x^2 - 2x - 3$. We obtain $0 = x^2 - 2x - 3$ or $x^2 - 2x - 3 = 0$. We can solve this quadratic equation by factoring.

$$x^2 - 2x - 3 = 0$$
$$(x - 3)(x + 1) = 0$$
$$x - 3 = 0 \quad \text{or} \quad x + 1 = 0$$
$$x = 3 \qquad\qquad x = -1$$

The x-intercepts are 3 and -1. The parabola passes through $(3, 0)$ and $(-1, 0)$.

Step 4 Find the y-intercept. Replace x with 0 in $f(x) = x^2 - 2x - 3$:

$$f(0) = 0^2 - 2 \cdot 0 - 3 = 0 - 0 - 3 = -3.$$

The y-intercept is -3, which is the constant term in the function's equation. The parabola passes through $(0, -3)$.

Steps 5 and 6 Plot the intercepts and the vertex. Connect these points with a smooth curve. The intercepts and the vertex are shown as the four labeled points in **Figure 6.16**. Also shown is the graph of the quadratic function, obtained by connecting the points with a smooth curve.

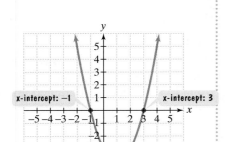

Figure 6.16 The graph of $f(x) = x^2 - 2x - 3$

▶ **Checkpoint 5** Graph the quadratic function: $f(x) = x^2 + 6x + 5$.

Example 6 Graphing a Quadratic Function

Graph the quadratic function: $f(x) = -x^2 + 4x - 1$.

Solution

Step 1 Determine how the parabola opens. Note that a, the coefficient of x^2, is -1. Thus, $a < 0$; this negative value tells us that the parabola opens downward.

Step 2 Find the vertex. The x-coordinate of the vertex is $-\dfrac{b}{2a}$.

$$f(x) = -x^2 + 4x - 1$$

$$x\text{-coordinate of vertex} = -\frac{b}{2a} = -\frac{4}{2(-1)} = -\left(\frac{4}{-2}\right) = -(-2) = 2$$

Substitute 2 for x in $f(x) = -x^2 + 4x - 1$ to find the y-coordinate:

$$y\text{-coordinate of vertex} = f(2) = -2^2 + 4 \cdot 2 - 1 = -4 + 8 - 1 = 3.$$

The vertex is $(2, 3)$.

Step 3 Find the x-intercepts. Replace $f(x)$ with 0 in $f(x) = -x^2 + 4x - 1$. We obtain $0 = -x^2 + 4x - 1$ or $-x^2 + 4x - 1 = 0$. This quadratic equation cannot be solved by factoring. We will use the quadratic formula to solve it with $a = -1, b = 4$, and $c = -1$.

$$x = \frac{-b \pm \sqrt{b^2 - 4ac}}{2a} = \frac{-4 \pm \sqrt{4^2 - 4(-1)(-1)}}{2(-1)} = \frac{-4 \pm \sqrt{16 - 4}}{-2}$$

> To locate the x-intercepts, we need decimal approximations. Thus, there is no need to simplify the radical form of the solutions.

$$x = \frac{-4 + \sqrt{12}}{-2} \approx 0.3 \quad \text{or} \quad x = \frac{-4 - \sqrt{12}}{-2} \approx 3.7$$

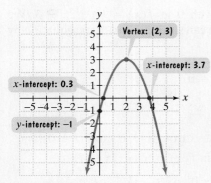

Figure 6.17 The graph of $f(x) = -x^2 + 4x - 1$

The x-intercepts are approximately 0.3 and 3.7. The parabola passes through $(0.3, 0)$ and $(3.7, 0)$.

Step 4 Find the y-intercept. Replace x with 0 in $f(x) = -x^2 + 4x - 1$:

$$f(0) = -0^2 + 4 \cdot 0 - 1 = -1.$$

The y-intercept is -1. The parabola passes through $(0, -1)$.

Steps 5 and 6   Plot the intercepts and the vertex. Connect these points with a smooth curve. The intercepts and the vertex are shown as the four labeled points in **Figure 6.17**. Also shown is the graph of the quadratic function, obtained by connecting the points with a smooth curve.

✓ **Checkpoint 6**   Graph the quadratic function: $f(x) = -x^2 - 2x + 5.$

Applications of Quadratic Functions

 5 Solve problems involving a quadratic function's minimum or maximum value.

Many applied problems involve finding the maximum or minimum value of a quadratic function, as well as where this value occurs.

Example 7   **The Parabolic Path of a Punted Football**

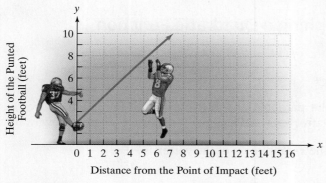

Figure 6.18

Figure 6.18 shows that when a football is kicked, the nearest defensive player is 6 feet from the point of impact with the kicker's foot. The height of the punted football, $f(x)$, in feet, can be modeled by

$$f(x) = -0.01x^2 + 1.18x + 2,$$

where x is the ball's horizontal distance, in feet, from the point of impact with the kicker's foot.

a. What is the maximum height of the punt and how far from the point of impact does this occur?

b. How far must the nearest defensive player, who is 6 feet from the kicker's point of impact, reach to block the punt?

c. If the ball is not blocked by the defensive player, how far down the field will it go before hitting the ground?

d. Graph the function that models the football's parabolic path.

Solution

a. We begin by identifying the numbers a, b, and c in the function's equation.

$$f(x) = -0.01x^2 + 1.18x + 2$$

$a = -0.01$ $b = 1.18$ $c = 2$

Because $a < 0$, the function has a maximum that occurs at $x = -\dfrac{b}{2a}$.

$$x = -\frac{b}{2a} = -\frac{1.18}{2(-0.01)} = -(-59) = 59$$

This means that the maximum height of the punt occurs 59 feet from the kicker's point of impact. The maximum height of the punt is

$$f(59) = -0.01(59)^2 + 1.18(59) + 2 = 36.81,$$

or 36.81 feet.

b. **Figure 6.18** shows that the defensive player is 6 feet from the kicker's point of impact. To block the punt, he must touch the football along its parabolic path. This means that we must find the height of the ball 6 feet from the kicker. Replace x with 6 in the given function, $f(x) = -0.01x^2 + 1.18x + 2$.

$$f(6) = -0.01(6)^2 + 1.18(6) + 2 = -0.36 + 7.08 + 2 = 8.72$$

The defensive player must reach 8.72 feet above the ground to block the punt.

c. Assuming that the ball is not blocked by the defensive player, we are interested in how far down the field it will go before hitting the ground. We are looking for the ball's horizontal distance, x, when its height above the ground, $f(x)$, is 0 feet. To find this x-intercept, replace $f(x)$ with 0 in $f(x) = -0.01x^2 + 1.18x + 2$. We obtain $0 = -0.01x^2 + 1.18x + 2$, or $-0.01x^2 + 1.18x + 2 = 0$. This quadratic equation cannot be solved by factoring. We will use the quadratic formula to solve it.

$-0.01x^2 + 1.18x + 2 = 0$

$a = -0.01$ $b = 1.18$ $c = 2$

The equation for determining the ball's maximum horizontal distance

Use a calculator to evaluate the radicand.

$$x = \frac{-b \pm \sqrt{b^2 - 4ac}}{2a} = \frac{-1.18 \pm \sqrt{(1.18)^2 - 4(-0.01)(2)}}{2(-0.01)} = \frac{-1.18 \pm \sqrt{1.4724}}{-0.02}$$

$$x = \frac{-1.18 + \sqrt{1.4724}}{-0.02} \quad \text{or} \quad x = \frac{-1.18 - \sqrt{1.4724}}{-0.02}$$

$x \approx -1.7$ $\qquad\qquad x \approx 119.7$ Use a calculator and round to the nearest tenth.

Reject this value. We are interested in the football's height corresponding to horizontal distances from its point of impact onward, or $x \geq 0$.

If the football is not blocked by the defensive player, it will go approximately 119.7 feet down the field before hitting the ground.

d. In terms of graphing the model for the football's parabolic path, $f(x) = -0.01x^2 + 1.18x + 2$, we have already determined the vertex and the approximate x-intercept.

vertex: $(59, 36.81)$

The ball's maximum height, 36.81 feet, occurs at a horizontal distance of 59 feet.

x-intercept: 119.7

The ball's maximum horizontal distance is approximately 119.7 feet.

Figure 6.18 indicates that the y-intercept is 2, meaning that the ball is kicked from a height of 2 feet. Let's verify this value by replacing x with 0 in $f(x) = -0.01x^2 + 1.18x + 2$.

$$f(0) = -0.01 \cdot 0^2 + 1.18 \cdot 0 + 2 = 0 + 0 + 2 = 2$$

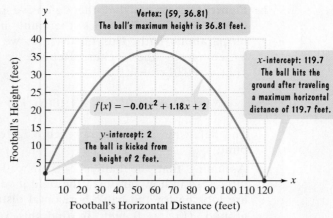

Using the vertex, $(59, 36.81)$, the x-intercept, 119.7, and the y-intercept, 2, the graph of the equation that models the football's parabolic path is shown in **Figure 6.19**. The graph is shown only for $x \geq 0$, indicating horizontal distances that begin at the football's impact with the kicker's foot and end with the ball hitting the ground.

Figure 6.19 The parabolic path of a punted football

✔ **Checkpoint 7**   An archer's arrow follows a parabolic path. The height of the arrow, $f(x)$, in feet, can be modeled by

$$f(x) = -0.005x^2 + 2x + 5,$$

where x is the arrow's horizontal distance from the archer, in feet.

 a. What is the maximum height of the arrow and how far from its release does this occur?

 b. Find the horizontal distance the arrow travels before it hits the ground. Round to the nearest foot.

▶ **c.** Graph the function that models the arrow's parabolic path.

Achieving Success

Address your stress. Stress levels can help or hinder performance. The parabola in **Figure 6.20** serves as a model that shows people under both low stress and high stress perform worse than their moderate-stress counterparts.

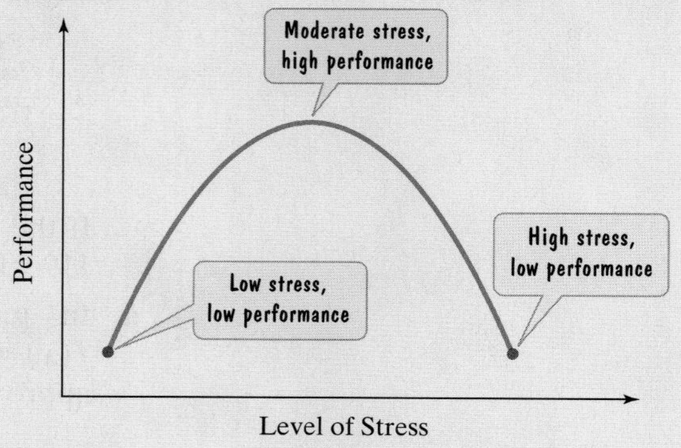

Figure 6.20
Source: Herbert Benson, *Your Maximum Mind,* Random House, 1987.

Exercise Set 6.5

Concept and Vocabulary Exercises

In Exercises 1–4, fill in each blank so that the resulting statement is true.

 1. The graph of $f(x) = ax^2 + bx + c$ opens upward if _____ and opens downward if _____.

 2. The x-intercepts for the graph of $f(x) = ax^2 + bx + c$ can be found by determining the real solutions of the equation _____.

 3. The y-intercept for the graph of $f(x) = ax^2 + bx + c$ can be determined by replacing x with _____ and computing _____.

 4. The x-coordinate of the vertex of the parabola whose equation is $f(x) = ax^2 + bx + c$ is _____. The y-coordinate of the vertex is found by substituting _____ into the parabola's equation and evaluating the function at this value of x.

In Exercises 5–8, determine whether each statement is true or false. If the statement is false, make the necessary change(s) to produce a true statement.

 5. The x-coordinate of the vertex of the parabola whose equation is $f(x) = ax^2 + bx + c$ is $\dfrac{b}{2a}$.

6. If a parabola has only one x-intercept, then the x-intercept is also the vertex.

7. There is no relationship between the graph of $f(x) = ax^2 + bx + c$ and the number of real solutions of the equation $ax^2 + bx + c = 0$.

8. If $f(x) = 4x^2 - 40x + 4$, then the vertex is the highest point on the graph.

Respond to Exercises 9–13 using verbal or written explanations.

9. What is a parabola? Describe its shape.

10. Explain how to decide whether a parabola opens upward or downward.

11. If a parabola has two x-intercepts, explain how to find them.

12. Explain how to find a parabola's y-intercept.

13. Describe how to find a parabola's vertex.

In Exercises 14–18, determine whether each statement makes sense or does not make sense, and explain your reasoning.

14. This work by artist Scott Kim (1955–) has the same kind of symmetry as the graph of a quadratic function.

"DYSLEXIA," 1981

15. I'm graphing a parabola that opens downward and whose vertex is a minimum point.

16. I'm graphing a quadratic function with two y-intercepts.

17. Parabolas that open up appear to form smiles ($a > 0$), while parabolas that open down frown ($a < 0$).

18. I threw a baseball vertically upward and its path was a parabola.

Practice Exercises

In Exercises 19–22, determine if the parabola whose equation is given opens upward or downward.

19. $f(x) = x^2 - 4x + 3$
20. $f(x) = x^2 - 6x + 5$
21. $f(x) = -2x^2 + x + 6$
22. $f(x) = -2x^2 - 4x + 6$

In Exercises 23–28, find the x-intercepts for the parabola whose equation is given. If the x-intercepts are irrational numbers, round your answers to the nearest tenth.

23. $f(x) = x^2 - 4x + 3$
24. $f(x) = x^2 - 6x + 5$
25. $f(x) = -x^2 + 8x - 12$
26. $f(x) = -x^2 - 2x + 3$
27. $f(x) = x^2 + 2x - 4$
28. $f(x) = x^2 + 8x + 14$

In Exercises 29–36, find the y-intercept for the parabola whose equation is given.

29. $f(x) = x^2 - 4x + 3$
30. $f(x) = x^2 - 6x + 5$
31. $f(x) = -x^2 + 8x - 12$
32. $f(x) = -x^2 - 2x + 3$
33. $f(x) = x^2 + 2x - 4$
34. $f(x) = x^2 + 8x + 14$
35. $f(x) = x^2 + 6x$
36. $f(x) = x^2 + 8x$

In Exercises 37–42, find the vertex for the parabola whose equation is given.

37. $f(x) = x^2 - 4x + 3$
38. $f(x) = x^2 - 6x + 5$
39. $f(x) = 2x^2 + 4x - 6$
40. $f(x) = -2x^2 - 4x - 2$
41. $f(x) = x^2 + 6x$
42. $f(x) = x^2 + 8x$

In Exercises 43–54, graph each quadratic function.

43. $f(x) = x^2 + 8x + 7$
44. $f(x) = x^2 + 10x + 9$
45. $f(x) = x^2 - 2x - 8$
46. $f(x) = x^2 + 4x - 5$
47. $f(x) = -x^2 + 4x - 3$
48. $f(x) = -x^2 + 2x + 3$
49. $f(x) = x^2 - 1$
50. $f(x) = x^2 - 4$
51. $f(x) = x^2 + 2x + 1$
52. $f(x) = x^2 - 2x + 1$
53. $f(x) = -2x^2 + 4x + 5$
54. $f(x) = -3x^2 + 6x - 2$

Practice Plus

In Exercises 55–61, find the vertex for the parabola whose equation is given by first writing the equation in the form $f(x) = ax^2 + bx + c$.

55. $f(x) = (x - 3)^2 + 2$
56. $f(x) = (x - 4)^2 + 3$
57. $f(x) = (x + 5)^2 - 4$
58. $f(x) = (x + 6)^2 - 5$
59. $f(x) = 2(x - 1)^2 - 3$
60. $f(x) = 2(x - 1)^2 - 4$
61. $f(x) = -3(x + 2)^2 + 5$

62. Generalize your work in Exercises 55–61 and complete the following statement: For a parabola whose equation is $f(x) = a(x - h)^2 + k$, the vertex is the point _____.

Application Exercises

An athlete whose event is the shot put releases the shot with the same initial velocity, but at different angles. The figure shows the parabolic paths for shots released at angles of 35° and 65°. Exercises 63–64 are based on the functions that model the parabolic paths.

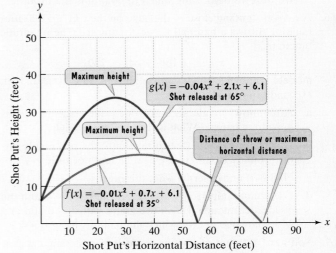

63. When the shot whose path is shown by the blue graph is released at an angle of 35°, its height, $f(x)$, in feet, can be modeled by
$$f(x) = -0.01x^2 + 0.7x + 6.1,$$
where x is the shot's horizontal distance, in feet, from its point of release. Use this model to solve parts (a) through (c) and verify your answers using the blue graph.

a. What is the maximum height of the shot and how far from its point of release does this occur?

b. What is the shot's maximum horizontal distance, to the nearest tenth of a foot, or the distance of the throw?

c. From what height was the shot released?

64. Refer to the red graph at the bottom of the previous page. When the shot whose path is shown by the red graph is released at an angle of 65°, its height, $g(x)$, in feet, can be modeled by

$$g(x) = -0.04x^2 + 2.1x + 6.1,$$

where x is the shot's horizontal distance, in feet, from its point of release. Use this model to solve parts (a) through (c) and verify your answers using the red graph.

  a. What is the maximum height, to the nearest tenth of a foot, of the shot and how far from its point of release does this occur?

  b. What is the shot's maximum horizontal distance, to the nearest tenth of a foot, or the distance of the throw?

  c. From what height was the shot released?

65. A person standing close to the edge on the top of a 160-foot building throws a baseball vertically upward. The quadratic function

$$s(t) = -16t^2 + 64t + 160$$

models the ball's height above the ground, $s(t)$, in feet, t seconds after it was thrown.

  a. After how many seconds does the ball reach its maximum height? What is the maximum height?

  b. How many seconds does it take until the ball finally hits the ground? Round to the nearest tenth of a second.

  c. Find $s(0)$ and describe what this means.

  d. Use your results from parts (a) through (c) to graph the quadratic function. Begin the graph with $t = 0$ and end with the value of t for which the ball hits the ground.

66. A person standing close to the edge on the top of a 200-foot building throws a baseball vertically upward. The quadratic function

$$s(t) = -16t^2 + 64t + 200$$

models the ball's height above the ground, $s(t)$, in feet, t seconds after it was thrown.

  a. After how many seconds does the ball reach its maximum height? What is the maximum height?

  b. How many seconds does it take until the ball finally hits the ground? Round to the nearest tenth of a second.

  c. Find $s(0)$ and describe what this means.

  d. Use your results from parts (a) through (c) to graph the quadratic function. Begin the graph with $t = 0$ and end with the value of t for which the ball hits the ground.

67. You have 120 feet of fencing to enclose a rectangular plot that borders on a river. If you do not fence the side along the river, find the length and width of the plot that will maximize the area. What is the largest area that can be enclosed?

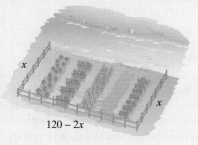

$120 - 2x$

68. You have 100 yards of fencing to enclose a rectangular area. Find the dimensions of the rectangle that maximize the enclosed area. What is the maximum area?

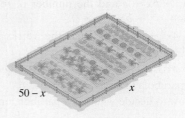

$50 - x$ x

Critical Thinking Exercises

69. Among all pairs of numbers whose sum is 16, find a pair whose product is as large as possible. What is the maximum product?

70. Among all pairs of numbers whose sum is 20, find a pair whose product is as large as possible. What is the maximum product?

71. Graph $x + y = 2$ and $y = x^2 - 4x + 4$ in the same rectangular coordinate system. What are the coordinates of the points of intersection? Show that each ordered pair is a solution of the system

$$\begin{cases} x + y = 2 \\ y = x^2 - 4x + 4. \end{cases}$$

72. You have 80 yards of fencing to enclose a rectangular region. Find the dimensions of the rectangle that maximize the enclosed area. What is the maximum area?

73. A parabola has x-intercepts at 3 and 7, a y-intercept at -21, and $(5, 4)$ for its vertex. Write the parabola's equation.

Technology Exercise

74. Use a graphing utility to verify any five of your hand drawn graphs in Exercises 43–54.

Group Exercise

75. Each group member should consult an almanac, newspaper, magazine, or the Internet to find data that initially increase and then decrease, or vice versa, and therefore can be modeled by a quadratic function. Group members should select the two sets of data that are most interesting and relevant. For each data set selected,

  a. Use the quadratic regression feature of a graphing utility to find the quadratic function that best fits the data.

  b. Use the equation of the quadratic function to make a prediction from the data. What circumstances might affect the accuracy of your prediction?

  c. Use the equation of the quadratic function to write and solve a problem involving maximizing or minimizing the function.

Chapter 6 Summary

6.1 Operations with Polynomials; Polynomial Functions

Definitions and Concepts

A polynomial is a single term or the sum of two or more terms containing variables with whole number exponents. A monomial is a polynomial with exactly one term; a binomial has exactly two terms; a trinomial has exactly three terms. The degree of a polynomial is the greatest of the powers of all the terms. The standard form of a polynomial is written in descending powers of the variable. In a polynomial function, the expression that defines the function is a polynomial.

To add polynomials, add like terms.

Example

- $(6x^3 + 5x^2 - 7x) + (-9x^3 + x^2 + 6x)$

  $= (6x^3 - 9x^3) + (5x^2 + x^2) + (-7x + 6x)$

  $= -3x^3 + 6x^2 - x$

Additional Example to Review

Example 1, page 376

Definitions and Concepts

To subtract two polynomials, change the sign of every term of the second polynomial. Add this result to the first polynomial.

Example

- $(5y^3 - 9y^2 - 4) - (3y^3 - 12y^2 - 5)$

  $= (5y^3 - 9y^2 - 4) + (-3y^3 + 12y^2 + 5)$

  $= (5y^3 - 3y^3) + (-9y^2 + 12y^2) + (-4 + 5)$

  $= 2y^3 + 3y^2 + 1$

Additional Example to Review

Example 2, page 376

Definitions and Concepts

To multiply a monomial and a polynomial that is not a monomial, use the distributive property to multiply each term of the polynomial by the monomial.

Example

- $2x^4(3x^2 - 6x + 5)$

  $= 2x^4 \cdot 3x^2 - 2x^4 \cdot 6x + 2x^4 \cdot 5$

  $= 6x^6 - 12x^5 + 10x^4$

Additional Example to Review

Example 3, page 378

Definitions and Concepts

To multiply polynomials when neither is a monomial, multiply each term of one polynomial by each term of the other polynomial. Then combine like terms.

Example

- $(2x + 3)(5x^2 - 4x + 2)$

$$= 2x(5x^2 - 4x + 2) + 3(5x^2 - 4x + 2)$$
$$= 10x^3 - 8x^2 + 4x + 15x^2 - 12x + 6$$
$$= 10x^3 + 7x^2 - 8x + 6$$

Additional Example to Review
Example 4, page 378

Definitions and Concepts

The FOIL method may be used when multiplying two binomials: First terms multiplied. Outside terms multiplied. Inside terms multiplied. Last terms multiplied.

Example

$$\boxed{F} \quad \boxed{O} \quad \boxed{I} \quad \boxed{L}$$

- $(3x + 7)(2x - 5) = 3x \cdot 2x + 3x(-5) + 7 \cdot 2x + 7(-5)$
$$= 6x^2 - 15x + 14x - 35$$
$$= 6x^2 - x - 35$$

Additional Example to Review
Example 5, page 380

Definitions and Concepts

The Product of the Sum and Difference of Two Terms
$$(A + B)(A - B) = A^2 - B^2$$

Example

- $(4x + 7)(4x - 7) = (4x)^2 - 7^2$
$$= 16x^2 - 49$$

Additional Example to Review
Example 6, page 380

Definitions and Concepts

The Square of a Binomial Sum
$$(A + B)^2 = A^2 + 2AB + B^2$$

Example

- $(x^2 + 6)^2 = (x^2)^2 + 2 \cdot x^2 \cdot 6 + 6^2$
$$= x^4 + 12x^2 + 36$$

Additional Example to Review
Example 7, page 381

Definitions and Concepts

The Square of a Binomial Difference
$$(A - B)^2 = A^2 - 2AB + B^2$$

Example

- $(9x - 3)^2 = (9x)^2 - 2 \cdot 9x \cdot 3 + 3^2$
  $$= 81x^2 - 54x + 9$$

Additional Example to Review
Example 8, page 382

6.2 Factoring Polynomials

Definitions and Concepts

Factoring a polynomial consisting of the sum of monomials means finding an equivalent expression that is a product. Polynomials that cannot be factored using integer coefficients are called prime polynomials over the integers. The greatest common factor, GCF, is an expression that divides every term of the polynomial. The GCF is the product of the largest common numerical factor and the variable of lowest degree common to every term of the polynomial. To factor a monomial from a polynomial, express each term as the product of the GCF and its other factor. Then use the distributive property to factor out the GCF.

Examples

- $20x^3 + 30x^2 = 10x^2(2x) + 10x^2(3) = 10x^2(2x + 3)$
- $x^2(x + 4) + 7(x + 4) = (x + 4)(x^2 + 7)$

Additional Examples to Review
Example 1, page 388; Example 2, page 389

Definitions and Concepts

To factor by grouping, factor out the GCF from each group. Then factor out the remaining common factor.

Example

- $x^3 + 2x^2 + 7x + 14 = (x^3 + 2x^2) + (7x + 14) = x^2(x + 2) + 7(x + 2) = (x + 2)(x^2 + 7)$

Additional Example to Review
Example 3, page 389

Definitions and Concepts

To factor a trinomial of the form $x^2 + bx + c$, find two numbers whose product is c and whose sum is b. The factorization is
$$(x + \text{one number})(x + \text{other number}).$$

Example

- Factor: $x^2 + 9x + 20$.
  Find two numbers whose product is 20 and whose sum is 9. The numbers are 4 and 5.
  $$x^2 + 9x + 20 = (x + 4)(x + 5)$$

Additional Examples to Review
Example 4, page 391; Example 5, page 392

Definitions and Concepts

To factor $ax^2 + bx + c$, try various combinations of factors of ax^2 and c until a middle term of bx is obtained for the sum of the outside and inside products.

Example

• Factor: $2x^2 + 7x - 15$.

> Factors of $2x^2$:
> $2x, x$

> Factors of -15: 1 and -15,
> -1 and 15, 3 and -5, -3 and 5

$$(2x - 3)(x + 5)$$

Sum of outside and inside products should equal $7x$.

$$10x - 3x = 7x$$

Thus, $2x^2 + 7x - 15 = (2x - 3)(x + 5)$.

Additional Examples to Review

Example 6, page 393; Example 7, page 394

Definitions and Concepts

The Difference of Two Squares

$$A^2 - B^2 = (A + B)(A - B)$$

Example

• $16x^2 - 25 = (4x)^2 - 5^2 = (4x + 5)(4x - 5)$

Additional Examples to Review

Example 8, page 395; Example 9, page 395

Definitions and Concepts

Perfect Square Trinomials

$$A^2 + 2AB + B^2 = (A + B)^2$$
$$A^2 - 2AB + B^2 = (A - B)^2$$

Examples

• $x^2 + 20x + 100 = x^2 + 2 \cdot x \cdot 10 + 10^2 = (x + 10)^2$
• $9x^2 - 30x + 25 = (3x)^2 - 2 \cdot 3x \cdot 5 + 5^2 = (3x - 5)^2$

Additional Example to Review

Example 10, page 396

Definitions and Concepts

A Factoring Strategy

1. Factor out the GCF.
2. **a.** If two terms, try
   $$A^2 - B^2 = (A + B)(A - B).$$
   b. If three terms, try
   $$A^2 + 2AB + B^2 = (A + B)^2$$
   $$A^2 - 2AB + B^2 = (A - B)^2.$$
   If not a perfect square trinomial, try factoring by trial and error.
   c. If four terms, try factoring by grouping.
3. See if any factors can be factored further.
4. Check by multiplying.

Example

• Factor: $3x^4 + 12x^3 - 3x^2 - 12x$.

The GCF is $3x$.

$$3x^4 + 12x^3 - 3x^2 - 12x$$
$$= 3x(x^3 + 4x^2 - x - 4)$$

> Four terms: Try grouping.

> Factor out a negative number from the second grouping to obtain a common binomial factor, $x + 4$, for the two groupings.

$$= 3x[x^2(x + 4) - 1(x + 4)]$$
$$= 3x(x + 4)(x^2 - 1)$$

> This can be factored further.

$$= 3x(x + 4)(x + 1)(x - 1)$$

Additional Examples to Review

Example 11, page 397; Example 12, page 397; Example 13, page 398; Example 14, page 399

6.3 Solving Quadratic Equations by Factoring

Definitions and Concepts

A quadratic equation in x can be written in the standard form

$$ax^2 + bx + c = 0, a \neq 0.$$

A polynomial equation is the result of setting two polynomials equal to each other. The equation is in standard form if one side is 0 and the polynomial on the other side is in standard form, that is, in descending powers of the variable. In standard form, its degree is the highest degree of any term in the equation. A polynomial equation of degree 1 is a linear equation and of degree 2 a quadratic equation. Some polynomial equations can be solved by writing the equation in standard form, factoring, and then using the zero-product principle: If a product is 0, then at least one of the factors is equal to 0.

Examples

• Solve: $5x^2 + 7x = 6$.

$$5x^2 + 7x - 6 = 0$$
$$(5x - 3)(x + 2) = 0$$
$$5x - 3 = 0 \quad \text{or} \quad x + 2 = 0$$
$$5x = 3 \qquad\qquad x = -2$$
$$x = \frac{3}{5}$$

The solutions are -2 and $\frac{3}{5}$, and the solution set is $\left\{-2, \frac{3}{5}\right\}$.

(The solutions are the x-intercepts of the graph of the quadratic function $f(x) = 5x^2 + 7x - 6$.)

• Solve: $x^3 - 9x^2 = x - 9$.

$$x^3 - 9x^2 - x + 9 = 0$$
$$x^2(x - 9) - 1(x - 9) = 0$$
$$(x - 9)(x^2 - 1) = 0$$
$$(x - 9)(x + 1)(x - 1) = 0$$
$$x - 9 = 0 \quad \text{or} \quad x + 1 = 0 \quad \text{or} \quad x - 1 = 0$$
$$x = 9 \qquad\qquad x = -1 \qquad\qquad x = 1$$

The solution set is $\{-1, 1, 9\}$.

Additional Examples to Review

Example 1, page 403; Example 2, page 404; Example 3, page 406; Example 4, page 407;
Example 5, page 409

6.4 Solving Quadratic Equations by the Square Root Property and the Quadratic Formula

Definitions and Concepts

The Square Root Property

If u is an algebraic expression and d is a positive real number, then

$$\text{If } u^2 = d, \quad \text{then } u = \sqrt{d} \quad \text{or} \quad u = -\sqrt{d}.$$

Equivalently,

$$\text{If } u^2 = d, \quad \text{then } u = \pm\sqrt{d}.$$

Examples

- Solve: $3x^2 = 7$

$$x^2 = \frac{7}{3}$$

$$x = \pm\sqrt{\frac{7}{3}} = \pm\frac{\sqrt{7}}{\sqrt{3}} = \pm\frac{\sqrt{7}}{\sqrt{3}}\cdot\frac{\sqrt{3}}{\sqrt{3}} = \pm\frac{\sqrt{21}}{3}$$

Solution set: $\left\{\pm\dfrac{\sqrt{21}}{3}\right\}$

- Solve: $(x - 6)^2 = 50.$

$$x - 6 = \pm\sqrt{50}$$

$$x = 6 \pm \sqrt{50} = 6 \pm \sqrt{25\cdot 2} = 6 \pm 5\sqrt{2}$$

Solution set: $\{6 \pm 5\sqrt{2}\}$

Additional Examples to Review

Example 1, page 414; Example 2, page 415; Example 3, page 415

Definitions and Concepts

The solutions of a quadratic equation in standard form

$$ax^2 + bx + c = 0, \quad a \neq 0,$$

are given by the quadratic formula

$$x = \frac{-b \pm \sqrt{b^2 - 4ac}}{2a}.$$

Example

- Solve using the quadratic formula:

$$2x^2 = 6x - 3.$$

First, write the equation in standard form by subtracting $6x$ and adding 3 on both sides.

$$2x^2 - 6x + 3 = 0$$

$$\boxed{a = 2} \quad \boxed{b = -6} \quad \boxed{c = 3}$$

$$x = \frac{-b \pm \sqrt{b^2 - 4ac}}{2a} = \frac{-(-6) \pm \sqrt{(-6)^2 - 4\cdot 2\cdot 3}}{2\cdot 2} = \frac{6 \pm \sqrt{36 - 24}}{4} = \frac{6 \pm \sqrt{12}}{4} = \frac{6 \pm \sqrt{4\cdot 3}}{4} = \frac{6 \pm 2\sqrt{3}}{4}$$

$$= \frac{2(3 \pm \sqrt{3})}{2\cdot 2} = \frac{3 \pm \sqrt{3}}{2}$$

Solution set: $\left\{\dfrac{3 \pm \sqrt{3}}{2}\right\}$

Additional Examples to Review
Example 4, page 416; Example 5, page 417; **Table 6.1**, page 419; Example 6, page 419

6.5 Quadratic Functions and Their Graphs

Definitions and Concepts

The graph of the quadratic function $f(x) = ax^2 + bx + c$, called a parabola, can be graphed using the following steps:

1. If $a > 0$, the parabola opens upward. If $a < 0$, it opens downward.

2. Find the vertex, the lowest point if the parabola opens upward and the highest point if it opens downward. The x-coordinate of the vertex is $-\dfrac{b}{2a}$. Substitute this value into the quadratic function's equation and evaluate to find the y-coordinate.

3. Find any x-intercepts by letting $f(x) = 0$ and solving the resulting equation.

4. Find the y-intercept by letting $x = 0$ and computing $f(0)$.

5. Plot the intercepts and the vertex.

6. Connect these points with a smooth curve.

Example

- Graph: $f(x) = x^2 - 2x - 8$.

  $\boxed{a = 1}$ $\boxed{b = -2}$ $\boxed{c = -8}$

- $a > 0$, so the parabola opens upward.

- Vertex: x-coordinate $= -\dfrac{b}{2a} = -\dfrac{(-2)}{2 \cdot 1} = 1$

  y-coordinate $= f(1) = 1^2 - 2 \cdot 1 - 8 = -9$

  Vertex is $(1, -9)$.

- x-intercepts: Let $f(x) = 0$.

  $x^2 - 2x - 8 = 0$

  $(x - 4)(x + 2) = 0$

  $x - 4 = 0$ or $x + 2 = 0$

  $x = 4$ $x = -2$

  The parabola passes through $(4, 0)$ and $(-2, 0)$.

- y-intercept: Let $x = 0$.

  $f(0) = 0^2 - 2 \cdot 0 - 8 = 0 - 0 - 8 = -8$

  The parabola passes through $(0, -8)$.

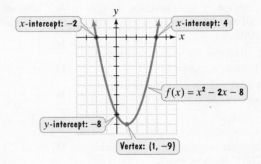

Additional Examples to Review
Example 5, page 428; Example 6, page 429; Example 7, page 430

Review Exercises

Section 6.1 Operations with Polynomials; Polynomial Functions

In Exercises 1–2, perform the indicated operations. Write the resulting polynomial in standard form and indicate its degree.

1. $(-6x^3 + 7x^2 - 9x + 3) + (14x^3 + 3x^2 - 11x - 7)$

2. $(13x^4 - 8x^3 + 2x^2) - (5x^4 - 3x^3 + 2x^2 - 6)$

In Exercises 3–8, find each product.

3. $8x^2(5x^3 - 2x - 1)$

4. $(3x - 2)(4x^2 + 3x - 5)$

5. $(3x - 5)(2x + 1)$

6. $(4x + 5)(4x - 5)$

7. $(2x + 5)^2$

8. $(3x - 4)^2$

9. The bar graph shows the percentage of total U.S. dollar sales of recorded music for two age groups for six selected years from 1990 through 2008.

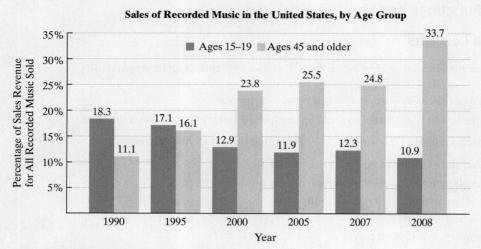

Sales of Recorded Music in the United States, by Age Group

Source: Recording Industry Association of America

The data can be modeled by the following functions:

| Teens ages 15–19 | $T(x) = 0.009x^2 - 0.6x + 18.7$ |

| Adults 45 and older | $F(x) = -0.004x^2 + 1.1x + 11.2.$ |

Each function gives the percentage of sales revenue by age group, $T(x)$ and $F(x)$, for all recorded music sold in the United Sates x years after 1990.

a. Write a function c that gives the combined percentage of total dollar sales for both age groups.

b. Use the function from part (a) to find the combined percentage of total dollar sales for both age groups in 2000.

c. Does the function value in part (b) overestimate or underestimate the actual percentage of total dollar sales for both age groups in 2000? By how much?

d. Write a function d that gives the difference between the percentage of total dollar sales for adults 45 and older and teens ages 15–19.

e. Use the function from part (d) to find the difference between total dollar sales for the two age groups in 1990.

f. Does the function value in part (e) overestimate or underestimate the actual difference in total dollar sales for the two age groups in 1990? By how much?

Section 6.2 Factoring Polynomials

In Exercises 10–11, factor out the greatest common factor.

10. $12x^3 + 16x^2 - 400x$

11. $x^2(x - 5) + 13(x - 5)$

In Exercises 12–13, factor by grouping.

12. $x^3 + 3x^2 + 2x + 6$

13. $x^3 + 5x + x^2 + 5$

In Exercises 14–25, factor each trinomial completely, or state that the trinomial is prime.

14. $x^2 - 3x + 2$

15. $x^2 - x - 20$

16. $x^2 + 19x + 48$

17. $x^2 - 6x + 8$

18. $x^2 + 5x - 9$

19. $3x^2 + 17x + 10$

20. $5y^2 - 17y + 6$

21. $4x^2 + 4x - 15$

22. $5y^2 + 11y + 4$

23. $8x^2 + 8x - 6$

24. $2x^3 + 7x^2 - 72x$

25. $12y^3 + 28y^2 + 8y$

In Exercises 26–28, factor the difference of two squares.

26. $4x^2 - 1$

27. $81 - 100y^2$

28. $z^4 - 16$

In Exercises 29–33, factor any perfect square trinomials, or state that the polynomial is prime.

29. $x^2 + 22x + 121$

30. $x^2 - 16x + 64$

31. $9y^2 + 48y + 64$

32. $16x^2 - 40x + 25$

33. $25x^2 + 15x + 9$

In Exercises 34–54, factor completely, or state that the polynomial is prime.

34. $x^3 - 8x^2 + 7x$

35. $10y^2 + 9y + 2$

36. $128 - 2y^2$

37. $9x^2 + 6x + 1$

38. $20x^7 - 36x^3$

39. $x^3 - 3x^2 - 9x + 27$

40. $y^2 + 16$

41. $2x^3 + 19x^2 + 35x$

42. $3x^3 - 30x^2 + 75x$

43. $4y^4 - 36y^2$

44. $5x^2 + 20x - 105$

45. $9x^2 + 8x - 3$

46. $10x^5 - 44x^4 + 16x^3$

47. $100y^2 - 49$

48. $9x^5 - 18x^4$

49. $x^4 - 1$

50. $6x^2 + 11x - 10$

51. $3x^4 - 12x^2$

52. $x^2 - x - 90$

53. $32y^3 + 32y^2 + 6y$

54. $2y^2 - 16y + 32$

Section 6.3 Solving Quadratic Equations by Factoring

In Exercises 55–59, use factoring to solve each polynomial equation.

55. $x^2 + 6x + 5 = 0$

56. $3x^2 = 22x - 7$

57. $(x + 3)(x - 2) = 50$

58. $3x^2 = 12x$

59. $x^3 + 5x^2 = 9x + 45$

60. The length of a rectangular sign is 3 feet longer than the width. If the sign has space for 54 square feet of advertising, find its length and its width.

61. A painting measuring 10 inches by 16 inches is surrounded by a frame of uniform width. If the combined area of the painting and frame is 280 square inches, determine the width of the frame.

Section 6.4 Solving Quadratic Equations by the Square Root Property and the Quadratic Formula

In Exercises 62–69, solve each quadratic equation by the square root property. If possible, simplify radicals or rationalize denominators.

62. $x^2 = 64$

63. $x^2 = 17$

64. $2x^2 = 150$

65. $(x - 3)^2 = 9$

66. $(y + 4)^2 = 5$

67. $3y^2 - 5 = 0$

68. $(2x - 7)^2 = 25$

69. $(x + 5)^2 = 12$

In Exercises 70–72, solve each equation using the quadratic formula. Simplify irrational solutions, if possible.

70. $2x^2 + 5x - 3 = 0$ **71.** $x^2 = 2x + 4$

72. $3x^2 + 5 = 9x$

In Exercises 73–77, solve each equation by the method of your choice. Simplify irrational solutions, if possible.

73. $2x^2 - 11x + 5 = 0$

74. $(3x + 5)(x - 3) = 5$

75. $3x^2 - 7x + 1 = 0$

76. $x^2 - 9 = 0$

77. $(2x - 3)^2 = 5$

78. The graph shows stopping distances for motorcycles at various speeds on dry roads and on wet roads.

Stopping Distances for Motorcycles at Selected Speeds

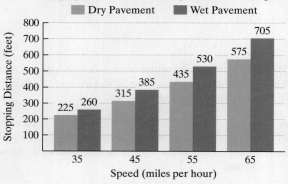

Source: National Highway Traffic Safety Administration

The functions

$$f(x) = 0.125x^2 - 0.8x + 99$$

Dry pavement

and

Wet pavement

$$g(x) = 0.125x^2 + 2.3x + 27$$

model a motorcycle's stopping distance, $f(x)$ or $g(x)$, in feet, traveling at x miles per hour. Function f models stopping distance on dry pavement and function g models stopping distance on wet pavement.

a. Use function g to find the stopping distance on wet pavement for a motorcycle traveling at 35 miles per hour. Round to the nearest foot. Does your rounded answer overestimate or underestimate the stopping distance shown by the graph? By how many feet?

b. Use function f to determine a motorcycle's speed requiring a stopping distance on dry pavement of 267 feet.

79. The graphs of the functions in Exercise 78 are shown at the top of the next column for $\{x | x \geq 30\}$.

a. How is your answer to Exercise 78(a) shown on the graph of g?

b. How is your answer to Exercise 78(b) shown on the graph of f?

Graphs of Models for a Motorcycle's Stopping Distances

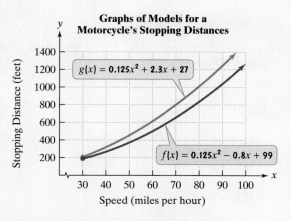

80. A baseball is hit by a batter. The function

$$s(t) = -16t^2 + 140t + 3$$

models the ball's height above the ground, $s(t)$, in feet, t seconds after it is hit. How long will it take for the ball to strike the ground? Round to the nearest tenth of a second.

Section 6.5 Quadratic Functions and Their Graphs

In Exercises 81–84, graph each quadratic function.

81. $f(x) = x^2 - 6x - 7$

82. $f(x) = -x^2 - 2x + 3$

83. $f(x) = -3x^2 + 6x + 1$

84. $f(x) = x^2 - 4x$

85. Fireworks are launched into the air. The function

$$f(t) = -16t^2 + 200t + 4$$

models the fireworks' height, $f(t)$, in feet, t seconds after they are launched. When should the fireworks explode so that they go off at the greatest height? What is that height?

86. A quarterback tosses a football to a receiver 40 yards downfield. The function

$$f(x) = -0.025x^2 + x + 6$$

models the football's height, $f(x)$, in feet, when it is x yards from the quarterback.

a. How many yards from the quarterback does the football reach its greatest height? What is that height?

b. If a defender is 38 yards from the quarterback, how far must he reach to deflect or catch the ball?

c. If the football is neither deflected by the defender nor caught by the receiver, how many yards will it go, to the nearest tenth of a yard, before hitting the ground?

d. Graph the quadratic function that models the football's parabolic path.

Chapter 6 Test A

In Exercises 1–8, perform the indicated operations.

1. $(7x^3 + 3x^2 - 5x - 11) + (6x^3 - 2x^2 + 4x - 13)$

2. $(9x^3 - 6x^2 - 11x - 4) - (4x^3 - 8x^2 - 13x + 5)$

3. $6x^2(8x^3 - 5x - 2)$

4. $(3x + 2)(x^2 - 4x - 3)$

5. $(3y + 7)(2y - 9)$

6. $(7x + 5)(7x - 5)$

7. $(x^2 + 3)^2$

8. $(5x - 3)^2$

In Exercises 9–17, factor completely, or state that the polynomial is prime.

9. $x^2 - 9x + 18$

10. $x^3 + 2x^2 + 3x + 6$

11. $15y^4 - 35y^3 + 10y^2$

12. $25x^2 - 9$

13. $x^2 + 4$

14. $36x^2 - 84x + 49$

15. $7x^2 - 50x + 7$

16. $4y^2 - 36$

17. $4x^2 + 12x + 9$

In Exercises 18–23, solve each equation by the method of your choice. Simplify irrational solutions, if possible.

18. $3x^2 + 5x + 1 = 0$

19. $(3x - 5)(x + 2) = -6$

20. $(2x + 1)^2 = 36$

21. $2x^2 = 6x - 1$

22. $2x^2 + 9x = 5$

23. $x^3 - 4x^2 - x + 4 = 0$

In Exercises 24–25, graph each quadratic function.

24. $f(x) = x^2 + 2x - 8$

25. $f(x) = -2x^2 + 16x - 24$

A baseball player hits a pop fly into the air. The function

$$s(t) = -16t^2 + 64t + 5$$

models the ball's height above the ground, $s(t)$, in feet, t seconds after it is hit. Use the function to solve Exercises 26–27.

26. When does the baseball reach its maximum height? What is that height?

27. After how many seconds does the baseball hit the ground? Round to the nearest tenth of a second.

28. The length of a rectangular sign is 3 feet longer than the width. If the sign has space for 40 square feet of advertising, find its length and its width.

Chapter 6 Test B

In Exercises 1–8, perform the indicated operations.

1. $(10x^3 + 7x^2 - 11x - 5) + (8x^3 - 6x^2 + 10x - 15)$

2. $(6x^3 - 9x^2 - 4x - 11) - (8x^3 - 4x^2 - 5x + 13)$

3. $8x^2(5x^3 - 6x - 3)$

4. $(2x + 3)(x^2 - 5x - 4)$

5. $(3y + 5)(4y - 7)$

6. $(5x + 7)(5x - 7)$

7. $(x^2 + 4)^2$

8. $(7x - 2)^2$

In Exercises 9–17, factor completely, or state that the polynomial is prime.

9. $x^2 - 9x + 20$

10. $x^3 + 4x^2 + 5x + 20$

11. $12y^4 - 27y^3 + 6y^2$

12. $9x^2 - 25$

13. $x^2 + 9$

14. $25x^2 - 90x + 81$

15. $5x^2 - 41x + 8$

16. $5y^2 - 80$

17. $9x^2 + 12x + 4$

In Exercises 18–23, solve each equation by the method of your choice. Simplify irrational solutions, if possible.

18. $2x^2 - 7x + 4 = 0$

19. $(5x + 4)(x - 1) = 2$

20. $(2x - 1)^2 = 25$

21. $3x^2 = 6x - 2$

22. $3x^2 - 5x = 2$

23. $x^3 + 3x^2 - x - 3 = 0$

In Exercises 24–25, graph each quadratic function.

24. $f(x) = x^2 - 6x + 5$

25. $f(x) = -2x^2 - 4x + 6$

A model rocket is launched upward from a platform 40 feet above the ground. The quadratic function

$$s(t) = -16t^2 + 400t + 40$$

models the rocket's height above the ground, $s(t)$, in feet, t seconds after it was launched. Use this function to solve Exercises 26–27.

26. When does the rocket reach its maximum height? What is that height?

27. After how many seconds does the rocket hit the ground? Round to the nearest tenth of a second.

28. An architect is allowed 15 square yards of floor space to add a small bedroom to a house. Because of the room's design in relationship to the existing structure, the width of the rectangular floor must be 7 yards less than two times the length. Find the length and width of the rectangular floor that the architect is permitted.

Personal Finance: Taxes and Interest

7

We all want a wonderful life with fulfilling work, good health, and loving relationships. And let's be honest: Financial security, or even a small fortune, wouldn't hurt! Achieving this goal depends on understanding basic ideas about savings, loans, and investments. A solid understanding of the topics in this chapter and the next can pay, literally, by making your financial goals a reality.

7.1 Percent, Sales Tax, and Discounts

Objectives

1. Express a fraction as a percent.

2. Express a decimal as a percent.

3. Express a percent as a decimal.

4. Solve applied problems involving sales tax and discounts.

5. Determine percent increase or decrease.

6. Investigate some of the ways percent can be abused.

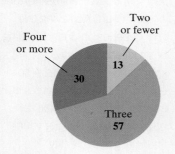

Figure 7.1 Number of bedrooms in privately owned single-family U.S. houses per 100 houses
Source: U.S. Census Bureau and HUD

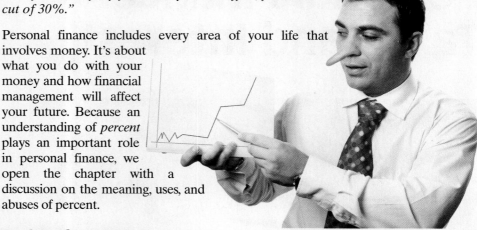

"And if elected, it is my solemn pledge to cut your taxes by 10% for each of my first three years in office, for a total cut of 30%."

Personal finance includes every area of your life that involves money. It's about what you do with your money and how financial management will affect your future. Because an understanding of *percent* plays an important role in personal finance, we open the chapter with a discussion on the meaning, uses, and abuses of percent.

Basics of Percent

Percents are the result of expressing numbers as a part of 100. The word *percent* means *per hundred*. For example, the circle graph in **Figure 7.1** shows that 57 out of every 100 single-family homes have three bedrooms. Thus, $\frac{57}{100} = 57\%$, indicating that 57% of the houses have three bedrooms. The percent sign, %, is used to indicate the number of parts out of one hundred parts.

A fraction can be expressed as a percent using the following procedure:

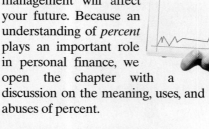

Expressing a Fraction as a Percent

1. Divide the numerator by the denominator.
2. Multiply the quotient by 100. This is done by moving the decimal point in the quotient two places to the right.
3. Attach a percent sign.

Express a fraction as a percent.

Example 1   Expressing a Fraction as a Percent

Express $\frac{5}{8}$ as a percent.

Solution

Step 1   Divide the numerator by the denominator.

$$5 \div 8 = 0.625$$

Step 2   Multiply the quotient by 100.

$$0.625 \times 100 = 62.5$$

Step 3   Attach a percent sign.

$$62.5\%$$

Thus, $\frac{5}{8} = 62.5\%$.

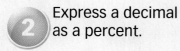 **Checkpoint 1**   Express $\frac{1}{8}$ as a percent.

Express a decimal as a percent.

Our work in Example 1 shows that $0.625 = 62.5\%$. This illustrates the procedure for expressing a decimal number as a percent.

Expressing a Decimal Number as a Percent
1. Move the decimal point two places to the right.
2. Attach a percent sign.

Example 2   Expressing a Decimal as a Percent

Express 0.47 as a percent.

Solution

Move decimal point two places right.

0.47 %   Attach a percent sign.

Thus, 0.47 = 47%.

Checkpoint 2   Express 0.023 as a percent.

We reverse the procedure of Example 2 to express a percent as a decimal number.

 Express a percent as a decimal.

Expressing a Percent as a Decimal Number
1. Move the decimal point two places to the left.
2. Remove the percent sign.

Example 3   Expressing Percents as Decimals

Express each percent as a decimal:
a. 19%   **b.** 180%.

Solution

Use the two steps in the box.

a.

$$19\% = 19.\% = 0.19\,\%$$

The percent sign is removed.

The decimal point starts at the far right.

The decimal point is moved two places to the left.

Thus, 19% = 0.19.

b. 180% = 1.80% = 1.80 or 1.8

Checkpoint 3   Express each percent as a decimal:
a. 67%   **b.** 250%.

If a fraction is part of a percent, as in $\frac{1}{4}\%$, begin by expressing the fraction as a decimal, retaining the percent sign. Then, express the percent as a decimal number. For example,

$$\frac{1}{4}\% = 0.25\% = 00.25\% = 0.0025.$$

Study Tip

Be careful with the zeros when expressing a small percent as a decimal number. For example,

$$\frac{1}{100}\% = 0.01\% = 00.01\% = 0.0001.$$

Solve applied problems involving sales tax and discounts.

Percent, Sales Tax, and Discounts

Many applications involving percent are based on the following formula:

A is P percent of B.

$$A = P \cdot B.$$

Note that the word *of* implies multiplication.

We can use this formula to determine the **sales tax** collected by states, counties, and cities on sales of items to customers. The sales tax is a percent of the cost of an item.

> Sales tax amount = tax rate × item's cost

Example 4 Percent and Sales Tax

Suppose that the local sales tax rate is 7.5% and you purchase a bicycle for $894.

a. How much tax is paid?

b. What is the bicycle's total cost?

Solution

a. Sales tax amount = tax rate × item's cost

$$= 7.5\% \times \$894 = 0.075 \times \$894 = \$67.05$$

> 7.5% of the item's cost, or 7.5% of $894

The tax paid is $67.05.

b. The bicycle's total cost is the purchase price, $894, plus the sales tax, $67.05.

$$\text{Total cost} = \$894.00 + \$67.05 = \$961.05$$

The bicycle's total cost is $961.05.

✓ **Checkpoint 4**　Suppose that the local sales tax rate is 6% and you purchase a computer for $1260.

a. How much tax is paid?

▶ **b.** What is the computer's total cost?

None of us is thrilled about sales tax, but we do like buying things that are *on sale*. Businesses reduce prices, or **discount**, to attract customers and to reduce inventory. The discount rate is a percent of the original price.

> Discount amount = discount rate × original price

Example 5 Percent and Sales Price

A computer with an original price of $1460 is on sale at 15% off.

a. What is the discount amount?

b. What is the computer's sale price?

Solution

a. Discount amount = discount rate × original price

$$= 15\% \times \$1460 = 0.15 \times \$1460 = \$219$$

15% of the original price, or 15% of $1460

The discount amount is $219.

b. The computer's sale price is the original price, $1460, minus the discount amount, $219.

$$\text{Sale price} = \$1460 - \$219 = \$1241$$

The computer's sale price is $1241.

✓ **Checkpoint 5**   A CD player with an original price of $380 is on sale at 35% off.

a. What is the discount amount?

▸ b. What is the CD player's sale price?

Percent and Change

5 Determine percent increase or decrease.

Percents are used for comparing changes, such as increases or decreases in sales, population, prices, and production. If a quantity changes, its **percent increase** or its **percent decrease** can be found as follows:

> **Finding Percent Increase or Percent Decrease**
>
> 1. Find the fraction for the percent increase or the percent decrease:
>
> $$\frac{\text{amount of increase}}{\text{original amount}} \quad \text{or} \quad \frac{\text{amount of decrease}}{\text{original amount}}.$$
>
> 2. Find the percent increase or the percent decrease by expressing the fraction in step 1 as a percent.

Example 6   **Finding Percent Increase and Decrease**

In 2000, world population was approximately 6 billion. **Figure 7.2** shows world population projections through the year 2150. The data are from the United Nations Family Planning Program and are based on optimistic or pessimistic expectations for successful control of human population growth.

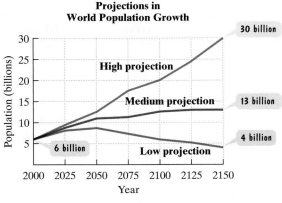

Figure 7.2
Source: United Nations

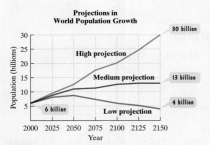

Figure 7.2 (repeated)

a. Find the percent increase in world population from 2000 to 2150 using the high projection data.

b. Find the percent decrease in world population from 2000 to 2150 using the low projection data.

Solution

a. Use the data shown on the blue, high-projection, graph.

$$\text{Percent increase} = \frac{\text{amount of increase}}{\text{original amount}}$$

$$= \frac{30 - 6}{6} = \frac{24}{6} = 4 = 400\%$$

The projected percent increase in world population is 400%.

b. Use the data shown on the green, low-projection, graph.

$$\text{Percent decrease} = \frac{\text{amount of decrease}}{\text{original amount}}$$

$$= \frac{6 - 4}{6} = \frac{2}{6} = \frac{1}{3} = 0.33\frac{1}{3} = 33\frac{1}{3}\%$$

The projected percent decrease in world population is $33\frac{1}{3}\%$.

In Example 6, we expressed the percent decrease as $33\frac{1}{3}\%$ because of the familiar conversion $\frac{1}{3} = 0.33\frac{1}{3}$. However, in many situations, rounding is needed. We suggest that you round to the nearest tenth of a percent. Carry the division in the fraction for percent increase or decrease to four places after the decimal point. Then round the decimal to three places, or to the nearest thousandth. Expressing this rounded decimal as a percent gives percent increase or decrease to the nearest tenth of a percent.

Study Tip

Notice the difference between the following examples:

- 2 is increased to 8.

$$\text{Percent increase} = \frac{\text{amount of increase}}{\text{original amount}} = \frac{6}{2} = 3 = 300\%$$

- 8 is decreased to 2.

$$\text{Percent decrease} = \frac{\text{amount of decrease}}{\text{original amount}} = \frac{6}{8} = \frac{3}{4} = 0.75 = 75\%$$

Although an increase from 2 to 8 is a 300% increase, a decrease from 8 to 2 is *not* a 300% decrease. **A percent decrease involving nonnegative quantities can never exceed 100%.** When a quantity is decreased by 100%, it is reduced to zero.

✓ **Checkpoint 6**

a. If 6 is increased to 10, find the percent increase.

b. If 10 is decreased to 6, find the percent decrease.

Example 7   **Finding Percent Decrease**

A jacket regularly sells for $135.00. The sale price is $60.75. Find the percent decrease of the sale price from the regular price.

Solution

$$\text{Percent decrease} = \frac{\text{amount of decrease}}{\text{original amount}}$$

$$= \frac{135.00 - 60.75}{135} = \frac{74.25}{135} = 0.55 = 55\%$$

The percent decrease of the sale price from the regular price is 55%. This means that the sale price of the jacket is 55% lower than the regular price.

Checkpoint 7 A television regularly sells for $940. The sale price is $611. Find the percent decrease of the sale price from the regular price.

Abuses of Percent

6 | Investigate some of the ways percent can be abused.

In our next examples, we look at a few of the many ways that percent can be used incorrectly. Confusion often arises when percent increase (or decrease) refers to a changing quantity that is itself a percent.

Example 8 Percents of Percents

John Tesh, while he was still coanchoring *Entertainment Tonight*, reported that the PBS series *The Civil War* had an audience of 13% versus the usual 4% PBS audience, "an increase of more than 300%." Did Tesh report the percent increase correctly?

Solution

We begin by finding the percent increase.

$$\text{Percent increase} = \frac{\text{amount of increase}}{\text{original amount}}$$

$$= \frac{13\% - 4\%}{4\%} = \frac{9\%}{4\%} = \frac{9}{4} = 2.25 = 225\%$$

The percent increase for PBS was 225%. This is not more than 300%, so Tesh did not report the percent increase correctly.

Checkpoint 8 An episode of a television series had an audience of 12% versus its usual 10%. What was the percent increase for this episode?

Example 9 Promises of a Politician

A politician states, "If you elect me to office, I promise to cut your taxes for each of my first three years in office by 10% each year, for a total reduction of 30%." Evaluate the accuracy of the politician's statement.

Solution

To make things simple, let's assume that a taxpayer paid $100 in taxes in the year previous to the politician's election. A 10% reduction during year 1 is 10% of $100.

$$10\% \text{ of previous year tax} = 10\% \text{ of } \$100 = 0.10 \times \$100 = \$10$$

With a 10% reduction the first year, the taxpayer will pay only $100 − $10, or $90, in taxes during the politician's first year in office.

The following table shows how we calculate the new, reduced tax for each of the first three years in office:

Year	Tax Paid the Year Before	10% Reduction	Taxes Paid This Year
1	$100	$0.10 \times \$100 = \$10$	$\$100 - \$10 = \$90$
2	$90	$0.10 \times \$90 = \$9$	$\$90 - \$9 = \$81$
3	$81	$0.10 \times \$81 = \$8.10$	$\$81 - \$8.10 = \$72.90$

Now, we determine the percent decrease in taxes over the three years.

$$\text{Percent decrease} = \frac{\text{amount of decrease}}{\text{original amount}}$$

$$= \frac{\$100 - \$72.90}{\$100} = \frac{\$27.10}{\$100} = \frac{27.1}{100} = 0.271 = 27.1\%$$

The taxes decline by 27.1%, not by 30%. The politician is ill-informed in saying that three consecutive 10% cuts add up to a total tax cut of 30%. In our calculation, which serves as a counterexample to the promise, the total tax cut is only 27.1%.

Checkpoint 9 Suppose you paid $1200 in taxes. During year 1, taxes decrease by 20%. During year 2, taxes increase by 20%.

a. What do you pay in taxes for year 2?

b. How do your taxes for year 2 compare with what you originally paid, namely $1200? If the taxes are not the same, find the percent increase or decrease.

Blitzer Bonus

High Schoolers' Financial Literacy

Scores have been falling on tests that measure financial literacy. Here are four items from a test given to high school seniors. Would you ace this one?

1. Which of the following is true about sales taxes?
   A. The national sales-tax percentage rate is 6%.
   B. The Federal Government will deduct it from your paycheck.
   C. You don't have to pay the tax if your income is very low.
   D. It makes things more expensive for you to buy.

58% of high school seniors answered incorrectly.

2. If you have caused an accident, which type of automobile insurance would cover damage to your own car?
   A. Comprehensive
   B. Liability
   C. Term
   D. Collision

63% of high school seniors answered incorrectly.

3. Which of the following types of investment would best protect the purchasing power of a family's savings in the event of a sudden increase in inflation?
   A. A 10-year bond issued by a corporation
   B. A certificate of deposit at a bank
   C. A 25-year corporate bond
   D. A house financed with a fixed-rate mortgage

64% of high school seniors answered incorrectly.

4. Sara and Joshua just had a baby. They received money as baby gifts and want to put it away for the baby's education. Which of the following tends to have the highest growth over periods of time as long as 18 years?
   A. A checking account
   B. Stocks
   C. A U.S. government savings bond
   D. A savings account

83% of high school seniors answered incorrectly.

Source: The JumpStart Coalition's 2008 Personal Financial Survey

Answers: 1. D; 2. D; 3. D; 4. B

Exercise Set 7.1

Concept and Vocabulary Exercises

In Exercises 1–8, fill in each blank so that the resulting statement is true.

1. Percents are the result of expressing numbers as part of _____.

2. To express $\frac{7}{8}$ as a percent, divide _____ by _____, multiply the quotient by _____, and attach _____.

3. To express 0.1 as a percent, move the decimal point _____ places to the _____ and attach _____.

4. To express 7.5% as a decimal, move the decimal point _____ places to the _____ and remove _____.

5. To find the sales tax amount, multiply the _____ and the _____.

6. To find the discount amount, multiply the _____ and the _____.

7. The numerator of the fraction for percent increase is _____ and the denominator of the fraction for percent increase is _____.

8. The numerator of the fraction for percent decrease is _____ and the denominator of the fraction for percent decrease is _____.

Exercises 9–10 are based on items from a financial literacy survey from the Center for Economic and Entrepreneurial Literacy. Determine whether each statement is true or false. If the statement is false, make the necessary change(s) to produce a true statement.

9. Santa had to lay off 25% of his eight reindeer because of the bad economy, so only seven reindeer remained. (65% answered this question incorrectly. Santa might consider leaving *Math for Your World* in stockings across the country.)

10. You spent 1% of your $50,000-per-year salary on gifts, so you spent $5000 on gifts for the year.

Respond to Exercises 11–16 using verbal or written explanations.

11. What is a percent?

12. Describe how to express a decimal number as a percent and give an example.

13. Describe how to express a percent as a decimal number and give an example.

14. Explain how to use the sales tax rate to determine an item's total cost.

15. Describe how to find percent increase and give an example.

16. Describe how to find percent decrease and give an example.

In Exercises 17–20, determine whether each statement makes sense or does not make sense, and explain your reasoning.

17. I have $100 and my restaurant bill comes to $80, which is not enough to leave a 20% tip.

18. I found the percent decrease in a jacket's price to be 120%.

19. My weight increased by 1% in January and 1% in February, so my increase in weight over the two months is 2%.

20. My rent increased from 20% to 30% of my income, so the percent increase is 10%.

Practice Exercises

In Exercises 21–30, express each fraction as a percent.

21. $\frac{2}{5}$ 22. $\frac{3}{5}$ 23. $\frac{1}{4}$ 24. $\frac{3}{4}$

25. $\frac{3}{8}$ 26. $\frac{7}{8}$ 27. $\frac{1}{40}$ 28. $\frac{3}{40}$

29. $\frac{9}{80}$ 30. $\frac{13}{80}$

In Exercises 31–40, express each decimal as a percent.

31. 0.59 32. 0.96 33. 0.3844

34. 0.003 35. 2.87 36. 9.83

37. 14.87 38. 19.63

39. 100 40. 95

In Exercises 41–54, express each percent as a decimal.

41. 72%   **42.** 38%   **43.** 43.6%

44. 6.25%   **45.** 130%   **46.** 260%

47. 2%   **48.** 6%   **49.** $\frac{1}{2}$%

50. $\frac{3}{4}$%   **51.** $\frac{5}{8}$%   **52.** $\frac{1}{8}$%

53. $62\frac{1}{2}$%   **54.** $87\frac{1}{2}$%

Use the percent formula, $A = PB$: A is P percent of B, to solve Exercises 55–58.

55. What is 3% of 200?   **56.** What is 8% of 300?

57. What is 18% of 40?   **58.** What is 16% of 90?

Practice Plus

There are three basic types of percent problems that can be solved using the percent formula $A = PB$.

Question	Given	Percent Formula
What is P percent of B?	P and B	Solve for A.
A is P percent of what?	A and P	Solve for B.
A is what percent of B?	A and B	Solve for P.

Exercises 55–58 involved using the formula to answer the first question. In Exercises 59–66, use the percent formula to answer the second or third question.

59. 3 is 60% of what?

60. 8 is 40% of what?

61. 24% of what number is 40.8?

62. 32% of what number is 51.2?

63. 3 is what percent of 15?

64. 18 is what percent of 90?

65. What percent of 2.5 is 0.3?

66. What percent of 7.5 is 0.6?

Application Exercises

67. Suppose that the local sales tax rate is 6% and you purchase a car for $32,800.

   a. How much tax is paid?

   b. What is the car's total cost?

68. Suppose that the local sales tax rate is 7% and you purchase a graphing calculator for $96.

   a. How much tax is paid?

   b. What is the calculator's total cost?

69. An exercise machine with an original price of $860 is on sale at 12% off.

   a. What is the discount amount?

   b. What is the exercise machine's sale price?

70. A dictionary that normally sells for $16.50 is on sale at 40% off.

   a. What is the discount amount?

   b. What is the dictionary's sale price?

The circle graph shows a breakdown of spending for the average U.S. household using 365 days worked as a basis of comparison. Use this information to solve Exercises 71–72. Round answers to the nearest tenth of a percent.

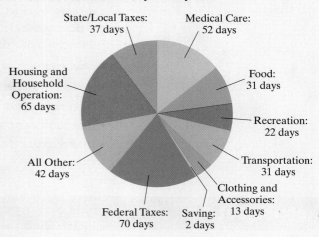

Spending for the Average American Household, by 365 Days Worked

State/Local Taxes: 37 days
Medical Care: 52 days
Food: 31 days
Recreation: 22 days
Transportation: 31 days
Clothing and Accessories: 13 days
Saving: 2 days
Federal Taxes: 70 days
All Other: 42 days
Housing and Household Operation: 65 days

Source: The Tax Foundation

71. What percentage of work time does the average U.S. household spend paying for federal taxes?

72. What percentage of work time does the average U.S. household spend paying for state and local taxes?

Although you want to choose a career that fits your interests and abilities, it is good to have an idea of what jobs pay when looking at career options. The bar graph shows the average yearly earnings of full-time employed college graduates with only a bachelor's degree based on their college major. Use this information to solve Exercises 73–74. Round all answers to the nearest tenth of a percent.

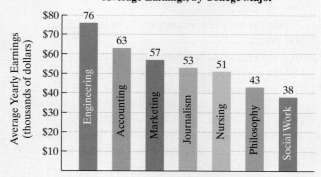

Average Earnings, by College Major

Average Yearly Earnings (thousands of dollars)

Engineering: 76
Accounting: 63
Marketing: 57
Journalism: 53
Nursing: 51
Philosophy: 43
Social Work: 38

Source: Arthur J. Keown, *Personal Finance*, Fourth Edition, Pearson, 2007.

73. Find the percent increase in the average yearly earnings from students majoring in social work to students majoring in engineering.

74. Find the percent increase for the average yearly earnings from students majoring in philosophy to students majoring in accounting.

75. A sofa regularly sells for $840. The sale price is $714. Find the percent decrease of the sale price from the regular price.

76. A FAX machine regularly sells for $380. The sale price is $266. Find the percent decrease of the sale price from the regular price.

77. Suppose that you have $10,000 in a rather risky investment recommended by your financial advisor. During the first year, your investment decreases by 30% of its original value. During the second year, your investment increases by 40% of its first-year value. Your advisor tells you that there must have been a 10% overall increase of your original $10,000 investment. Is your financial advisor using percentages properly? If not, what is your actual percent gain or loss of your original $10,000 investment?

78. The price of a color printer is reduced by 30% of its original price. When it still does not sell, its price is reduced by 20% of the reduced price. The salesperson informs you that there has been a total reduction of 50%. Is the salesperson using percentages properly? If not, what is the actual percent reduction from the original price?

Critical Thinking Exercises

79. What is the total cost of a $720 iPad that is on sale at 15% off if the local sales tax rate is 6%?

80. A condominium is taxed based on its $78,500 value. The tax rate is $3.40 for every $100 of value. If the tax is paid before March 1, 3% of the normal tax is given as a discount. How much tax is paid if the condominium owner takes advantage of the discount?

81. In January, each of 60 people purchased a $500 washing machine. In February, 10% fewer customers purchased the same washing machine that had increased in price by 20%. What was the change in sales from January to February?

82. When you buy something, it actually costs more than you may think—at least in terms of how much money you must earn to buy it. For example, if you pay 28% of your income in taxes, how much money would you have to earn to buy a used car for $7200?

7.2 Income Tax

Objectives

1. Determine gross income, adjustable gross income, and taxable income.

2. Calculate federal income tax.

3. Calculate FICA taxes.

4. Solve problems involving working teens and taxes.

"The Trouble with Trillions" episode of the *Simpsons* finds Homer frantically putting together his tax return two hours before the April 15th mailing deadline. In a frenzy, he shouts to his wife, "Marge, how many kids do we have, no time to count, I'll just estimate nine. If anyone asks, you need 24-hour nursing care, Lisa is a clergyman, Maggie is seven people, and Bart was wounded in Vietnam."

"Cool!" replies Bart.

It isn't only cartoon characters who are driven into states of frantic agitation over taxes. The average American pays over $10,000 per year in income tax. Yes, it's important to pay Uncle Sam what you owe, but not a penny more. People who do not understand the federal tax system often pay more than they have to. In this section, you will learn how income taxes are determined and calculated, reinforcing the role of tax planning in personal finance.

Paying Income Tax

Income tax is a percentage of your income collected by the government to fund its services and programs. The federal government collects income tax, and most, but not all, state governments do, too. (Alaska, Florida, Nevada, South Dakota, Texas, Washington, and Wyoming have no state income tax.) Tax revenue pays for our national defense, fire and police protection, road construction, schools, libraries, and parks. Without taxes, the government would not be able to conduct medical research, provide medical care for the elderly, or send astronauts into space.

Income tax is automatically withheld from your paycheck by your employer. The precise amount withheld for federal income tax depends on how you fill out your W-4 form, which you complete when you start a new job.

Although the United States Congress determines federal tax laws, the Internal Revenue Service (IRS) is the government body that enforces the laws and collects taxes. The IRS is a branch of the Treasury Department.

Determining Taxable Income

1 Determine gross income, adjustable gross income, and taxable income.

Federal income taxes are a percentage of your *taxable income*, which is based on your earnings in the calendar year—January to December. When the year is over, you have until April 15th to file your tax return.

Calculating your federal income tax begins with **gross income**, or total income for the year. This includes income from wages, tips, interest or dividends from investments, unemployment compensation, profits from a business, rental income, and even game-show winnings. It does not matter whether these winnings are in cash or in the form of items such as cars or vacations.

The next step in calculating your federal income tax is to determine your **adjusted gross income**. Adjusted gross income is figured by taking gross income and subtracting certain allowable amounts, called **adjustments**. These untaxed portions of gross income include contributions to certain retirement accounts and tax-deferred savings plans, interest paid on student loans, and alimony payments. In a traditional tax-deferred retirement plan, you get to deduct the full amount of your contribution from your gross income. You pay taxes on the money later, when you withdraw it at retirement.

$$\text{Adjusted gross income} = \text{Gross income} - \text{Adjustments}$$

IRS rules detail exactly what can be subtracted from gross income to determine your adjusted gross income.

Blitzer Bonus

Willie Nelson's Adjustments

In 1990, the IRS sent singer Willie Nelson a bill for $32 million. Egad! But as Willie said, "Thirty-two million ain't much if you say it fast." How did this happen? On bad advice, Willie got involved in a number of investments that he declared as adjustments. This reduced his adjusted gross income to a negligible amount that made paying taxes unnecessary. The IRS ruled these "adjustments" as blatant tax-avoidance schemes. Eventually, Willie and the IRS settled on a $9 million payment.

Source: Arthur J. Keown, *Personal Finance*, Fourth Edition, Pearson, 2007.

You are entitled to certain *exemptions* and *deductions*, subtracted from your adjusted gross income, before calculating your taxes. An **exemption** is a fixed amount on your return for each person supported by your income. You are entitled to this fixed amount ($3500 in 2008) for yourself and the same amount for each dependent. Until you reach the age of 18, you can be claimed as a dependent on your parent's or guardian's tax return. However, you are still required to file your own tax return if you are earning income.

A **standard deduction** is a lump-sum amount that you can subtract from your adjusted gross income. The IRS sets this amount. Most young people take the standard deduction because their financial situations are relatively simple and they are not eligible for numerous deductions associated with owning a home or making charitable contributions. **Itemized deductions** are deductions you list separately if you have incurred a large number of deductible expenses. Itemized deductions include interest on home mortgages, state income taxes, property taxes, charitable contributions, and medical expenses exceeding 7.5% of adjusted gross income. Taxpayers should choose the greater of a standard deduction or an itemized deduction.

Taxable income is figured by subtracting exemptions and deductions from adjusted gross income.

$$\text{Taxable income} = \text{Adjusted gross income} - (\text{Exemptions} + \text{Deductions})$$

Example 1   Gross Income, Adjusted Gross Income, and Taxable Income

A single man earned wages of $46,500, received $1850 in interest from a savings account, received $15,000 in winnings on a television game show, and contributed $2300 to a tax-deferred savings plan. He is entitled to a personal exemption of $3500 and a standard deduction of $5450. The interest on his home mortgage was $6500, he paid $2100 in property taxes and $1855 in state taxes, and he contributed $3000 to charity.

a. Determine the man's gross income.

b. Determine the man's adjusted gross income.

c. Determine the man's taxable income.

Solution

a. Gross income refers to this person's total income, which includes wages, interest from a savings account, and game-show winnings.

$$\text{Gross income} = \$46,500 + \$1850 + \$15,000 = \$63,350$$

| Wages | Earned interest | Game-show winnings |

The gross income is $63,350.

b. Adjusted gross income is gross income minus adjustments. The adjustment in this case is the contribution of $2300 to a tax-deferred savings plan. The full amount of this contribution is deducted from this year's gross income, although taxes will be paid on the money later when it is withdrawn, probably at retirement.

$$\text{Adjusted gross income} = \text{Gross income} - \text{Adjustments} = \$63,350 - \$2300 = \$61,050$$

| Contribution to a tax-deferred savings plan |

The adjusted gross income is $61,050.

c. We need to subtract exemptions and deductions from the adjusted gross income to determine the man's taxable income. This taxpayer is entitled to a personal exemption of $3500 and a standard deduction of $5450. However, a deduction greater than $5450 is obtained by itemizing deductions.

$$\text{Itemized deductions} = \$6500 + \$2100 + \$1855 + \$3000 = \$13,455$$

| Interest on home mortgage | Property taxes | State taxes | Charity |

We choose the itemized deductions of $13,455 because they are greater than the standard deduction of $5450.

$$\text{Taxable income} = \text{Adjusted gross income} - (\text{Exemptions} + \text{Deductions})$$
$$= \$61,050 - (\$3500 + \$13,455)$$
$$= \$61,050 - \$16,955$$
$$= \$44,095$$

The taxable income is $44,095.

> **Summary of Kinds of Income Associated with Federal Taxes**
>
> Gross income is total income for the year.
>
> Adjusted gross income = Gross income − Adjustments
>
> Taxable income = Adjusted gross income − (Exemptions + Deductions)

Checkpoint 1 A single woman earned wages of $87,200, received $2680 in interest from a savings account, and contributed $3200 to a tax-deferred savings plan. She is entitled to a personal exemption of $3500 and a standard deduction of $5450. The interest on her home mortgage was $11,700, she paid $4300 in property taxes and $5220 in state taxes, and contributed $15,000 to charity.

a. Determine the woman's gross income.

b. Determine the woman's adjusted gross income.

c. Determine the woman's taxable income.

Calculating Federal Income Tax

2 Calculate federal income tax.

A tax table is used to determine how much you owe based on your taxable income. However, you do not have to pay this much tax if you are entitled to any *tax credits*. **Tax credits** are sums of money that reduce the income tax owed by the full dollar-for-dollar amount of the credits.

Achieving Success

Taking a Bite Out of Taxes

A tax credit is not the same thing as a tax deduction. A tax deduction reduces taxable income, saving only a percentage of the deduction in taxes. A tax credit reduces the income tax owed on the full dollar amount of the credit. There are credits available for everything from donating a kidney to buying an energy efficient dishwasher. The American Opportunity Credit, included in the economic stimulus package of 2009, provides a tax credit of up to $2500 per student. The credit can be used to lower the costs of the first four years of college. You can claim the credit for up to 100% of the first $2000 in qualified college costs and 25% of the next $2000. Significantly, 40% of this credit is refundable. This means that even if you do not have any taxable income, you could receive a check from the government for up to $1000.

Credits are awarded for a variety of activities that the government wants to encourage. Because a tax credit represents a dollar-for-dollar reduction in your tax bill, it pays to know your tax credits. You can learn more about tax credits at www.irs.gov.

Most people pay part or all of their tax bill during the year. If you are employed, your employer deducts federal taxes through *withholdings* based on a percentage of your gross pay. If you are self-employed, you pay your tax bill through *quarterly estimated taxes.*

When you file your tax return, all you are doing is settling up with the IRS over the amount of taxes you paid during the year versus the federal income tax that you owe. Many people will have paid more during the year than they owe, in which case they receive a *tax refund.* Others will not have paid enough and need to send the rest to the IRS by the deadline.

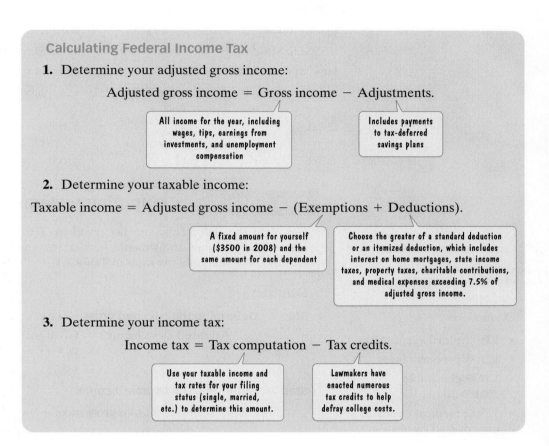

Calculating Federal Income Tax

1. Determine your adjusted gross income:

$$\text{Adjusted gross income} = \text{Gross income} - \text{Adjustments}.$$

All income for the year, including wages, tips, earnings from investments, and unemployment compensation

Includes payments to tax-deferred savings plans

2. Determine your taxable income:

$$\text{Taxable income} = \text{Adjusted gross income} - (\text{Exemptions} + \text{Deductions}).$$

A fixed amount for yourself ($3500 in 2008) and the same amount for each dependent

Choose the greater of a standard deduction or an itemized deduction, which includes interest on home mortgages, state income taxes, property taxes, charitable contributions, and medical expenses exceeding 7.5% of adjusted gross income.

3. Determine your income tax:

$$\text{Income tax} = \text{Tax computation} - \text{Tax credits}.$$

Use your taxable income and tax rates for your filing status (single, married, etc.) to determine this amount.

Lawmakers have enacted numerous tax credits to help defray college costs.

Tax Rate	Single
10%	up to $8025
15%	$8026 to $32,550

A portion of **Table 7.1** (The complete table appears on the next page.)

Table 7.1 on the next page shows 2008 tax rates, standard deductions, and exemptions for the four **filing status** categories described in the voice balloons. A portion of **Table 7.1** for a single filing status is given in the margin. The tax rates in the left column, called **marginal tax rates**, are assigned to various income ranges, called **margins**. For example, suppose you are single and your taxable income is $25,000. The singles column of the table shows that you must pay 10% tax on the first $8025, which is

$$10\% \text{ of } \$8025 = 0.10 \times \$8025 = \$802.50.$$

You must also pay 15% tax on the remaining $16,975 ($25,000 − $8025 = $16,975), which is

$$15\% \text{ of } \$16,975 = 0.15 \times \$16,975 = \$2546.25.$$

Your total tax is $802.50 + $2546.25 = $3348.75. In this scenario, your *marginal rate* is 15% and you are in the 15% *tax bracket.*

Table 7.1 2008 Marginal Tax Rates, Standard Deductions, and Exemptions

Tax Rate	Single *Unmarried, divorced, or legally separated*	Married Filing Separately *Married and each partner files a separate tax return*	Married Filing Jointly *Married and both partners file a single tax return*	Head of Household *Unmarried and paying more than half the cost of supporting a child or parent*
10%	up to $8025	up to $8025	up to $16,050	up to $11,450
15%	$8026 to $32,550	$8026 to $32,550	$16,051 to $65,100	$11,451 to $43,650
25%	$32,551 to $78,850	$32,551 to $65,725	$65,101 to $131,450	$43,651 to $112,650
28%	$78,851 to $164,550	$65,726 to $100,150	$131,451 to $200,300	$112,651 to $182,400
33%	$164,551 to $357,700	$100,151 to $178,850	$200,301 to $357,700	$182,401 to $357,700
35%	more than $357,700	more than $178,850	more than $357,700	more than $357,700
Standard Deduction	$5450	$5450	$10,900	$8000
Exemptions (per person)	$3500	$3500	$3500	$3500

Single Woman with No Dependents

Gross income: $62,000

Adjustments: $4000 paid to a tax-deferred IRA (Individual Retirement Account)

Deductions:

- $7500: mortgage interest
- $2200: property taxes
- $2400: charitable contributions
- $1500: medical expenses not covered by insurance

Tax credit: $500

Example 2 · Computing Federal Income Tax

Calculate the federal income tax owed by a single woman with no dependents whose gross income, adjustments, deductions, and credits are given in the margin. Use the 2008 marginal tax rates in **Table 7.1**.

Solution

Step 1 Determine the adjusted gross income.

$$\text{Adjusted gross income} = \text{Gross income} - \text{Adjustments}$$
$$= \$62,000 - \$4000$$
$$= \$58,000$$

Step 2 Determine the taxable income.

$$\text{Taxable income} = \text{Adjusted gross income} - (\text{Exemptions} + \text{Deductions})$$
$$= \$58,000 - (\$3500 + \text{Deductions})$$

The singles column in Table 7.1 shows a personal exemption of $3500.

The singles column in Table 7.1 shows a $5450 standard deduction. A greater deduction can be obtained by itemizing.

Itemized Deductions

$7500 : mortgage interest

$2200 : property taxes

$2400 : charitable contributions

$1500 : ~~medical expenses~~

Can only deduct amount in excess of 7.5% of adjusted gross income:
0.075 × $58,000 = $4350

$12,100 : total of deductible expenditures

We substitute $12,100 for deductions in the formula for taxable income.

$$\text{Taxable income} = \text{Adjusted gross income} - (\text{Exemptions} + \text{Deductions})$$
$$= \$58,000 - (\$3500 + \$12,100)$$
$$= \$58,000 - \$15,600$$
$$= \$42,400$$

Tax Rate	Single
10%	up to $8025
15%	$8026 to $32,550
25%	$32,551 to $78,850

A portion of **Table 7.1** (repeated)

Step 3   Determine the income tax.

$$\text{Income tax} = \text{Tax computation} - \text{Tax credits}$$
$$= \text{Tax computation} - \$500$$

We perform the tax computation using the singles rates in **Table 7.1**, partly repeated in the margin. Our taxpayer is in the 25% tax bracket because her taxable income, $42,400, is in the $32,551 to $78,850 income range. This means that she owes 10% on the first $8025 of her taxable income, 15% on her taxable income between $8026 and $32,550, inclusive, and 25% on her taxable income above $32,550.

> 10% marginal rate on first $8025 of taxable income

> 15% marginal rate on taxable income between $8026 and $32,550

> 25% marginal rate on taxable income above $32,550

$$\text{Tax computation} = 0.10 \times \$8025 + 0.15 \times (\$32,550 - \$8025) + 0.25 \times (\$42,400 - \$32,550)$$
$$= 0.10 \times \$8025 + 0.15 \times \$24,525 + 0.25 \times \$9850$$
$$= \$802.50 + \$3678.75 + \$2462.50$$
$$= \$6943.75$$

We substitute $6943.75 for the tax computation in the formula for income tax.

$$\text{Income tax} = \text{Tax computation} - \text{Tax credits}$$
$$= \$6943.75 - \$500$$
$$= \$6443.75$$

The federal income tax owed is $6443.75.

✓ **Checkpoint 2**   Use the 2008 marginal tax rates in **Table 7.1** to calculate the federal tax owed by a single man with no dependents whose gross income, adjustments, deductions, and credits are given as follows:

> Gross income: $40,000
> Adjustments: $1000
> Deductions: $3000: charitable contributions
> $1500: theft loss
> $300: cost of tax preparation
> Tax credit: none.

Blitzer Bonus

Teen Employment and Federal Taxes

Table 7.2 shows the percentage of teens who have jobs outside of school, by household income.

Table 7.2 Household Income and Teen Employment	
Household Income	**Teen Employment**
$20,000	14%
$60,000	26%
$80,000	32%
$120,000 – $150,000	33%
>$150,000	28%

Source: U.S. Center for Labor Market Studies, Northeastern University

Most working teens do not make more than $8000 per year, so their federal taxes are withheld at 10% of their gross pay.

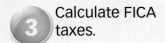

Calculate FICA taxes.

Social Security and Medicare (FICA)

In addition to income tax, we are required to pay the federal government **FICA** (Federal Insurance Contributions Act) taxes that are used for Social Security and Medicare benefits. **Social Security** provides payments to eligible retirees, people with health problems, eligible dependents of deceased persons, and disabled citizens. **Medicare** provides health care coverage mostly to Americans 65 and older.

For people who were not self-employed, the 2008 FICA tax rates were as follows:

- 7.65% on the first $102,000 from wages and tips
- 1.45% on income in excess of $102,000

The individual's employer must also pay matching amounts of FICA taxes. People who are self-employed must pay double the rates shown above.

Taxpayers are not permitted to subtract adjustments, exemptions, or deductions when determining FICA taxes.

Example 3   Computing FICA Tax

If you are not self-employed and earn $110,000, what are your FICA taxes?

Solution

The tax rates are 7.65% on the first $102,000 from wages and tips and 1.45% on income in excess of $102,000.

> 7.65% rate on the first $102,000 from wages and tips

> 1.45% rate on income in excess of $102,000

$$
\begin{aligned}
\text{FICA Tax} &= 0.0765 \times \$102{,}000 + 0.0145 \times (\$110{,}000 - \$102{,}000) \\
&= 0.0765 \times \$102{,}000 + 0.0145 \times \$8000 \\
&= \$7803 + \$116 \\
&= \$7919
\end{aligned}
$$

The FICA taxes are $7919.

✓ **Checkpoint 3**   If you are not self-employed and earn $200,000, what are your FICA taxes?

Example 4   Computing FICA Tax

If you are self-employed and earn $105,000, what are your FICA taxes?

Solution

People who are self-employed must pay double these rates: 7.65% on the first $102,000 from wages and tips and 1.45% on income in excess of $102,000.

> 7.65% rate on the first $102,000 from wages and tips

> Double

> Double

> 1.45% rate on income in excess of $102,000

$$
\begin{aligned}
\text{FICA Tax} &= 2 \times 0.0765 \times \$102{,}000 + 2 \times 0.0145 \times (\$105{,}000 - \$102{,}000) \\
&= 0.153 \times \$102{,}000 + 0.029 \times \$3000 \\
&= \$15{,}606 + \$87 \\
&= \$15{,}693
\end{aligned}
$$

The FICA taxes are $15,693.

Checkpoint 4 If you are self-employed and earn $125,000, what are your FICA taxes?

Working Teens and Taxes

For those of you who work part-time, getting paid is great. You finally have spending money. However, because employers withhold federal and state taxes, as well as FICA, your paychecks probably contain less spending money than you anticipated.

A pay stub attached to your paycheck provides a lot of information about the money you earned, including both your *gross pay* and your *net pay*. **Gross pay**, also known as **base pay**, is your salary prior to any withheld taxes for the pay period the check covers. Your gross pay is what you would receive if nothing were deducted. **Net pay** is the actual amount of your check after taxes have been withheld.

Example 5   Taxes for a Working Teen

You would like to have extra spending money and help save for college, so you decide to work part-time at the local gym. The job pays $10 per hour and you work 20 hours per week. Your employer withholds 10% of your gross pay for federal taxes, 7.65% for FICA taxes, and 3% for state taxes.

a. What is your weekly gross pay?
b. How much is withheld per week for federal taxes?
c. How much is withheld per week for FICA taxes?
d. How much is withheld per week for state taxes?
e. What is your weekly net pay?
f. What percentage of your gross pay is withheld for taxes? Round to the nearest tenth of a percent.

Solution

a. Your weekly gross pay is the number of hours worked, 20, times your hourly wage, $10 per hour.

$$\text{Gross pay} = 20 \text{ hours} \times \frac{\$10}{\text{hour}} = 20 \times \$10 = \$200$$

Your weekly gross pay, or what you would receive if nothing were deducted, is $200.

b. Your employer withholds 10% of your gross pay for federal taxes.

$$\text{Federal taxes} = 10\% \text{ of } \$200 = 0.10 \times \$200 = \$20$$

$20 is withheld per week for federal taxes.

c. Your employer withholds 7.65% of your gross pay for FICA taxes.

$$\text{FICA taxes} = 7.65\% \text{ of } \$200 = 0.0765 \times \$200 = \$15.30$$

$15.30 is withheld per week for FICA taxes.

d. Your employer withholds 3% of your gross pay for state taxes.

$$\text{State taxes} = 3\% \text{ of } \$200 = 0.03 \times \$200 = \$6$$

$6 is withheld per week for state taxes.

 ④ Solve problems involving working teens and taxes.

Blitzer Bonus

Percents and Tax Rates

In 1944 and 1945, the highest marginal tax rate in the United States was a staggering 94%. The tax rate on the highest-income Americans remained at approximately 90% throughout the 1950s, decreased to 70% in the 1960s and 1970s, and to 50% in the early 1980s before reaching a post–World War II low of 28% in 1998. (*Source:* IRS) In 2008, Denmark had the world's highest tax rate, starting at 42% and climbing to 68% for its wealthiest taxpayers.

e. Your weekly net pay is your gross pay minus the amounts withheld for federal, FICA, and state taxes.

Gross pay | Federal taxes | FICA taxes | State taxes

$$\begin{aligned}\text{Net pay} &= \$200 - (\$20 + \$15.30 + \$6)\\ &= \$200.00 - \$41.30\\ &= \$158.70\end{aligned}$$

Your weekly net pay is \$158.70. This is the actual amount of your paycheck.

f. Our work in part (e) shows that \$41.30 is withheld for taxes. The fractional part of your gross pay that is withheld for taxes is the amount that is withheld, \$41.30, divided by your gross pay, \$200.00. We then express this fraction as a percent.

Percent of gross pay withheld for taxes

$$= \frac{\$41.30}{\$200.00} = 0.2065 = 20.65\% \approx 20.7\%$$

Taxes / Gross pay

Your employer takes \$41.30 from your weekly gross salary and sends the money to the government. This represents approximately 20.7% of your gross pay.

Figure 7.3 contains a sample pay stub for the working teen in Example 5.

YOUR WORKPLACE 10 MAIN STREET ANY TOWN, STATE	YOUR NAME YOUR ADDRESS YOUR CITY, STATE, ZIP CODE SSN: 000-00-0000	Pay End Date: 0/00/10	
		Federal	**State**
		Single 1	Single 1

HOURS AND EARNINGS				
Current			**Year to Date**	
Base Rate ($10 per hour)	**Hours**	**Earnings**	**Hours**	**Earnings**
	20	$200	400	$4000

TAXES		
Description:	**Current:**	**Year to Date**
Federal	$20.00	$400.00
FICA	$15.30	$306.00
State	$ 6.00	$120.00
TOTAL	$41.30	$826.00
	Current:	**Year to Date**
TOTAL GROSS	$200.00	$4000.00
TOTAL DEDUCTIONS	$ 41.30	$ 826.00
NET PAY	$158.70	$3174.00

Figure 7.3 A teen's sample pay stub

Pay stubs are attached to your paycheck and usually contain four sections:

- Personal information about the employee: This may include your name, your address, your social security number, and your marital status.
- Information about earnings: This includes your hourly wage, the number of hours worked in the current pay period, and the number of hours worked year-to-date, which is the total number of hours worked since January 1 of the current year, and the amount earned during the current pay period and the amount earned year-to-date.
- Information about tax deductions, summarizing withholdings for the current pay period and the withholdings year-to-date.
- Gross pay, total deductions, and net pay for the current period, and the year-to-date total for each.

✓ **Checkpoint 5** You decide to work part-time at a local nursery. The job pays $12 per hour and you work 15 hours per week. Your employer withholds 10% of your gross pay for federal taxes, 7.65% for FICA taxes, and 4% for state taxes.

a. What is your weekly gross pay?

b. How much is withheld per week for federal taxes?

c. How much is withheld per week for FICA taxes?

d. How much is withheld per week for state taxes?

e. What is your weekly net pay?

f. What percentage of your gross pay is withheld for taxes? Round to the nearest tenth of a percent.

Exercise Set 7.2

Concept and Vocabulary Exercises

In Exercises 1–8, fill in each blank so that the resulting statement is true.

1. Your _____ income is your total income for the year.

2. Subtracting certain allowable amounts from the income in Exercise 1 results in your _____ income. These allowable amounts, or untaxed portions of income, are called _____.

3. A fixed amount deducted on your tax return for each person supported by your income, including yourself, is called a/an _____.

4. Your taxable income is your _____ income minus the sum of your _____ and _____.

5. Sums of money that reduce federal income tax by the full dollar-for-dollar amount are called _____.

6. Taxes used for Social Security and Medicare benefits are called _____ taxes.

7. Your base pay, or _____ pay, is your salary prior to any withheld taxes.

8. The actual amount of your paycheck after taxes have been withheld is called your _____ pay.

In Exercises 9–12, determine whether each statement is true or false. If the statement is false, make the necessary change(s) to produce a true statement.

9. Federal income tax is a percentage of your gross income.

10. If tax credits are equal, federal tax tables show that the greater your taxable income, the more you pay.

11. People in some states are not required to pay state income taxes.

12. FICA tax is a percentage of your gross income.

Respond to Exercises 13–23 using verbal or written explanations.

13. What is income tax? 14. What is gross income?

15. What is adjusted gross income?

16. What are exemptions? 17. What are deductions?

18. Under what circumstances should taxpayers itemize deductions?

19. How is taxable income determined?

20. What are tax credits?

21. What is the difference between a tax credit and a tax deduction?

22. What are FICA taxes?

23. How do you determine your net pay?

In Exercises 24–28, determine whether each statement makes sense or does not make sense, and explain your reasoning.

24. The only important thing to know about my taxes is whether I receive a refund or owe money on my return.

25. Because I am a teen with a part-time job, federal tax law does not allow me to itemize deductions.

26. I'm paying less federal tax on my first dollars of earnings and more federal tax on my last dollars of earnings.

27. My employer withholds the same amount of federal tax on my first dollars of earnings and my last dollars of earnings.

28. Once I begin college, I can choose a $4000 deduction or a $2500 credit to offset tuition and fees. I'll pay less federal taxes by selecting the $4000 deduction.

Practice and Application Exercises

In Exercises 29–30, find the gross income, the adjusted gross income, and the taxable income.

29. A taxpayer earned wages of $52,600, received $720 in interest from a savings account, and contributed $3200 to a tax-deferred retirement plan. He was entitled to a personal exemption of $3500 and had deductions totaling $7250.

30. A taxpayer earned wages of $23,500, received $495 in interest from a savings account, and contributed $1200 to a tax-deferred retirement plan. She was entitled to a personal exemption of $3500 and had deductions totaling $5450.

In Exercises 31–32, find the gross income, the adjusted gross income, and the taxable income. Base the taxable income on the greater of a standard deduction or an itemized deduction.

31. Suppose your neighbor earned wages of $86,250, received $1240 in interest from a savings account, and contributed $2200 to a tax-deferred retirement plan. She is entitled to a personal exemption of $3500 and a standard deduction of $5450. The interest on her home mortgage was $8900, she contributed $2400 to charity, and she paid $1725 in state taxes.

32. Suppose your neighbor earned wages of $319,150, received $1790 in interest from a savings account, and contributed $4100 to a tax-deferred retirement plan. He is entitled to a personal exemption of $3500 and the same exemption for each of his two children. He is also entitled to a standard deduction of $5450. The interest on his home mortgage was $51,235, he contributed $74,000 to charity, and he paid $12,760 in state taxes.

*In Exercises 33–42, use the 2008 marginal tax rates in **Table 7.1** on page 462 to compute the tax owed by each person or couple.*

33. a single man with a taxable income of $40,000

34. a single woman with a taxable income of $42,000

35. a married woman filing separately with a taxable income of $120,000

36. a married man filing separately with a taxable income of $110,000

37. a single man with a taxable income of $15,000 and a $2500 tax credit

38. a single woman with a taxable income of $12,000 and a $3500 tax credit

39. a married couple filing jointly with a taxable income of $250,000 and a $7500 tax credit

40. a married couple filing jointly with a taxable income of $400,000 and a $4500 tax credit

41. a head of household with a taxable income of $58,000 and a $6500 tax credit

42. a head of household with a taxable income of $46,000 and a $3000 tax credit

*In Exercises 43–46, use the 2008 marginal tax rates in **Table 7.1** on page 462 to calculate the income tax owed by each person.*

43. Single male, no dependents

    Gross income: $75,000

      Adjustments: $4000

        Deductions:

                $28,000 mortgage interest

                $4200 property taxes

                $3000 charitable contributions

      Tax credit: none

44. Single female, no dependents

    Gross income: $70,000

      Adjustments: $2000

        Deductions:

                $10,000 mortgage interest

                $2500 property taxes

                $1200 charitable contributions

      Tax credit: none

45. Unmarried head of household with two dependent children

    Gross income: $50,000

      Adjustments: none

        Deductions:

                $4500 state taxes

                $2000 theft loss

      Tax credit: $2000

46. Unmarried head of household with one dependent child

    Gross income: $40,000

      Adjustments: $1500

        Deductions:

                $3600 state taxes

                $800 charitable contributions

      Tax credit: $2500

For people who are not self-employed, the 2008 FICA tax rates were as follows:

- *7.65% on the first $102,000 from wages and tips*
- *1.45% on income in excess of $102,000.*

The individual's employer must also pay matching amounts of FICA taxes. People who are self-employed pay double the rates shown above. Use this information to solve Exercises 47–52.

47. If you are not self-employed and earn $120,000 what are your FICA taxes?

48. If you are not self-employed and earn $140,000 what are your FICA taxes?

49. If you are self-employed and earn $150,000, what are your FICA taxes?

50. If you are self-employed and earn $160,000, what are your FICA taxes?

51. To help pay for college, you worked part-time at a local restaurant, earning $20,000 in wages and tips.

  a. Calculate your FICA taxes.

  b. Use **Table 7.1** on page 462 to calculate your income tax. Assume you are single with no dependents, have no adjustments or tax credit, and you take the standard deduction.

  c. Including both FICA and income tax, what percentage of your gross income are your federal taxes? Round to the nearest tenth of a percent.

52. To help pay for college, you worked part-time at a local restaurant, earning $18,000 in wages and tips.

  a. Calculate your FICA taxes.

  b. Use **Table 7.1** on page 462 to calculate your income tax. Assume you are single with no dependents, have no adjustments or tax credit, and take the standard deduction.

  c. Including both FICA and income tax, what percentage of your gross income are your federal taxes? Round to the nearest tenth of a percent.

53. You decide to work part-time at a local supermarket. The job pays $8.50 per hour and you work 20 hours per week. Your employer withholds 10% of your gross pay for federal taxes, 7.65% for FICA taxes, and 3% for state taxes.

  a. What is your weekly gross pay?

  b. How much is withheld per week for federal taxes?

  c. How much is withheld per week for FICA taxes?

  d. How much is withheld per week for state taxes?

  e. What is your weekly net pay?

  f. What percentage of your gross pay is withheld for taxes? Round to the nearest tenth of a percent.

54. You decide to work part-time at a local veterinary hospital. The job pays $9.50 per hour and you work 20 hours per week. Your employer withholds 10% of your gross pay for federal taxes, 7.65% for FICA taxes, and 5% for state taxes.

  a. What is your weekly gross pay?

  b. How much is withheld per week for federal taxes?

  c. How much is withheld per week for FICA taxes?

  d. How much is withheld per week for state taxes?

  e. What is your weekly net pay?

  f. What percentage of your gross pay is withheld for taxes? Round to the nearest tenth of a percent.

Critical Thinking Exercises

55. Suppose you are in the 10% tax bracket. Once you begin college, you can choose a $4000 deduction or a $2500 credit to offset tuition and fees. Which option will reduce your tax bill by the greater amount? What is the difference in your savings between the two options?

56. A common complaint about income tax is "I can't afford to work more because it will put me in a higher tax bracket." Is it possible that being in a higher bracket means you actually lose money? Explain your answer.

57. Because of the mortgage interest tax deduction, is it possible to save money buying a house rather than renting, even though rent payments are lower than mortgage payments? Explain your answer.

Group Exercises

The following topics are appropriate for either individual or group research projects. Use the World Wide Web to investigate each topic.

58. Proposals to Simplify Federal Tax Laws and Filing Procedures

59. The Most Commonly Recommended Tax Saving Strategies

60. The Most Commonly Audited Tax Return Sections

61. Federal Tax Procedures Questioned over Issues of Fairness (Examples include the marriage penalty, the alternative minimum tax (AMT), and capital gains rates.)

7.3 Simple Interest

Objectives

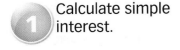

1. Calculate simple interest.

2. Use the future value formula.

3. Use the simple interest formula on discounted loans.

In 1626, Peter Minuit convinced the Wappinger Indians to sell him Manhattan Island for $24. If the native Americans had put the $24 into a bank account at a 5% interest rate compounded monthly, by the year 2010 there would have been well over $5 billion in the account!

Although you may not yet understand terms such as *interest rate* and *compounded monthly*, one thing seems clear: Money in certain savings accounts grows in remarkable ways. You, too, can take advantage of such accounts with astonishing results. In this section and the next, we will show you how.

Simple Interest

Interest is the amount of money that we get paid for lending or investing money, or that we pay for borrowing money. When we deposit money in a savings institution, the institution pays us interest for its use. When we borrow money, interest is the price we pay for the privilege of using the money until we repay it.

The amount of money that we deposit or borrow is called the **principal**. For example, if you deposit $2000 in a savings account, then $2000 is the principal. The amount of interest depends on the principal, the interest **rate**, which is given as a percent and varies from bank to bank, and the length of time for which the money is deposited. In this section, the rate is assumed to be annual (per year).

Simple interest involves interest calculated only on the principal. The following formula is used to find simple interest:

> **Calculating Simple Interest**
>
> $$\text{Interest} = \text{principal} \times \text{rate} \times \text{time}$$
> $$I = Prt$$
>
> The rate, r, is expressed as a decimal when calculating simple interest.

Study Tip

Throughout the remainder of this chapter and the next chapter, keep in mind that all given rates are assumed to be *per year*, unless otherwise stated.

Example 1 Calculating Simple Interest for One Year

Suppose that you deposit $2000 in a savings account at Hometown Bank, which has a rate of 6%. Find the interest at the end of the first year.

Solution

To find the interest at the end of the first year, we use the simple interest formula.

$$I = Prt = (2000)(0.06)(1) = 120$$

Principal, or amount deposited, is $2000.   Rate is 6% = 0.06.   Time is 1 year.

At the end of the first year, the interest is $120. You can withdraw the $120 interest and you still have $2000 in the savings account.

✓ Checkpoint 1   Suppose that you deposit $3000 in a savings account at Yourtown Bank, which has a rate of 5%. Find the interest at the end of the first year.

Example 2 Calculating Simple Interest for More Than One Year

A student took out a simple interest loan for $1800 for two years at a rate of 8% to purchase a used car. What is the interest on the loan?

Solution

To find the interest on the loan, we use the simple interest formula.

$$I = Prt = (1800)(0.08)(2) = 288$$

Principal, or amount borrowed, is $1800.   Rate is 8% = 0.08.   Time is 2 years.

The interest on the loan is $288.

✓ Checkpoint 2   A student took out a simple interest loan for $2400 for two years at a rate of 7%. What is the interest on the loan?

Simple interest is used for many short-term loans, including automobile and consumer loans. Imagine that a short-term loan is taken for 125 days. The time of the loan is $\frac{125}{365}$ because there are 365 days in a year. However, before the modern use of calculators and computers, the **Banker's rule** allowed financial institutions to use 360 in the denominator of such a fraction because this simplified the interest calculation. Using the Banker's rule, the time, t, for a 125-day short-term loan is

$$\frac{125 \text{ days}}{360 \text{ days}} = \frac{125}{360}.$$

Compare the values for time, t, for a 125-day short-term loan using denominators of 360 and 365.

$$\frac{125}{360} \approx 0.347 \qquad \frac{125}{365} \approx 0.342$$

The denominator of 360 benefits the bank by resulting in a greater period of time for the loan, and consequently more interest.

With the widespread use of calculators and computers, government agencies and the Federal Reserve Bank calculate simple interest using 365 days in a year, as do many credit unions and banks. However, there are still some financial institutions that use the Banker's rule with 360 days in a year because it produces a greater amount of interest.

Future Value: Principal Plus Interest

② Use the future value formula.

When a loan is repaid, the interest is added to the original principal to find the total amount due. In Example 2, at the end of two years, the student will have to repay

$$\text{principal} + \text{interest} = \$1800 + \$288 = \$2088.$$

In general, if a principal P is borrowed at a simple interest rate r, then after t years the amount due, A, can be determined as follows:

$$A = P + I = P + Prt = P(1 + rt).$$

The amount due, A, is called the **future value** of the loan. The principal borrowed now, P, is also known as the loan's **present value**.

> **Calculating Future Value for Simple Interest**
>
> The future value, A, of P dollars at simple interest rate r (as a decimal) for t years is given by
>
> $$A = P(1 + rt).$$

Example 3   Calculating Future Value

A loan of $1060 has been made at 6.5% for three months. Find the loan's future value.

Solution

The amount borrowed, or principal, P, is $1060. The rate, r, is 6.5%, or 0.065. The time, t, is given as 3 months. We need to express the time in years because the rate is understood to be 6.5% per year. Because 3 months is $\frac{3}{12}$ of a year, $t = \frac{3}{12} = \frac{1}{4} = 0.25$.

The loan's future value, or the total amount due after three months, is

$$A = P(1 + rt) = 1060[1 + (0.065)(0.25)] \approx \$1077.23.$$

Rounded to the nearest cent, the loan's future value is $1077.23.

Technology

$1060\,[1 + (0.065)(0.25)]$

On a Scientific Calculator:

$1060 \boxed{\times} \boxed{(} \boxed{1} \boxed{+} .065$

$\boxed{\times} .25 \boxed{)} \boxed{=}$

✓ **Checkpoint 3**   A loan of $2040 has been made at 7.5% for four months. Find the loan's future value.

| Example 4 | Earning Money by Putting Your Wallet Away Today |

Suppose you spend $4 each day, five days per week, on gourmet coffee.

a. How much do you spend on this item in a year?

b. If you invested your yearly spending on gourmet coffee in a savings account with a rate of 5%, how much would you have after one year?

Solution

a. Because you are spending $4 each day, five days per week, you are spending

$$\frac{\$4}{day} \times \frac{5 \text{ days}}{week} = \frac{\$4 \times 5}{week} = \frac{\$20}{week},$$

or $20 each week on gourmet coffee. Assuming that this continues throughout the 52 weeks in the year, you are spending

$$\frac{\$20}{week} \times \frac{52 \text{ weeks}}{year} = \frac{\$20 \times 52}{year} = \frac{\$1040}{year},$$

or $1040 each year on gourmet coffee.

b. Now suppose you invest $1040 in a savings account with a rate of 5%. To find your savings after one year, we use the future value formula for simple interest.

$$A = P(1 + rt) = 1040[1 + (0.05)(1)] = 1040(1.05) = 1092$$

Giving up the day-to-day expense of gourmet coffee can result in potential savings of $1092.

| Checkpoint 4 | In addition to jeopardizing your health, cigarette smoking is a costly addiction. Consider, for example, a person with a pack-a-day cigarette habit who spends $5 per day, seven days each week, on cigarettes.

a. How much is spent on this item in a year?

b. If this person invested the yearly spending on cigarettes in a savings account with a rate of 4%, how much would be saved after one year?

Achieving $uccess

Analyzing the Effects of Working While in School

An obvious way to have more money is by earning it and investing it. Working part-time while in high school or college has pros and cons.

Advantages

- General and career-specific experience
- Developing skills that can be useful after graduation
- Contacts who might be helpful in the future
- Time commitment that requires effective time management
- Extra money to spend and invest
- Confidence builder

Disadvantages

- Time commitment that reduces available study time
- Reduced opportunity for social and extracurricular activities
- Mentally shifting gears from work to classroom

The formula for future value, $A = P(1 + rt)$, has four variables. If we are given values for any three of these variables, we can solve for the fourth.

Example 5   Determining a Simple Interest Rate

You borrow $2500 from a friend and promise to pay back $2655 in six months. What simple interest rate will you pay?

Solution

We use the formula for future value, $A = P(1 + rt)$. You borrow $2500: $P = 2500$. You will pay back $2655, so this is the future value: $A = 2655$. You will do this in six months, which must be expressed in years: $t = \frac{6}{12} = \frac{1}{2} = 0.5$. To determine the simple interest rate you will pay, we solve the future value formula for r.

$$A = P(1 + rt)$$ This is the formula for future value.

$$2655 = 2500[1 + r(0.5)]$$ Substitute the given values.

$$2655 = 2500 + 1250r$$ Use the distributive property.

$$155 = 1250r$$ Subtract 2500 from both sides.

$$\frac{155}{1250} = \frac{1250r}{1250}$$ Divide both sides by 1250.

$$r = 0.124 = 12.4\%$$ Express $\frac{155}{1250}$ as a percent.

You will pay a simple interest rate of 12.4%.

✓ **Checkpoint 5**   You borrow $5000 from a friend and promise to pay back ▶ $6800 in two years. What simple interest rate will you pay?

Example 6   Determining a Present Value

Suppose that you plan to save $2000 for a trip to Europe in two years. You decide to purchase a certificate of deposit (CD) from your bank that pays a simple interest rate of 4%. How much must you put in this CD now in order to have the $2000 in two years?

Solution

We use the formula for future value, $A = P(1 + rt)$. We are interested in finding the principal, P, or the present value.

$$A = P(1 + rt)$$ This is the formula for future value.

$$2000 = P[1 + (0.04)(2)]$$ A(future value) = $2000, r(interest rate) = 0.04, and t = 2 (you want $2000 in two years).

$$2000 = 1.08P$$ Simplify: 1 + (0.04)(2) = 1.08.

$$\frac{2000}{1.08} = \frac{1.08P}{1.08}$$ Divide both sides by 1.08.

$$P \approx 1851.852$$ Simplify.

To make sure you will have enough money for the vacation, let's round this principal *up* to $1851.86. Thus, you should put $1851.86 in the CD now to have $2000 in two years.

✓ **Checkpoint 6**   How much should you put in an investment paying a simple ▶ interest rate of 8% if you need $4000 in six months?

Study Tip

When computing present value, round the principal, or present value, *up*. To round up to the nearest cent, add 1 to the hundredths digit, regardless of the digit to the right. In this way, you'll be sure to have enough money to meet future goals.

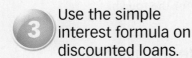

Use the simple interest formula on discounted loans.

Discounted Loans

Some lenders collect the interest from the amount of the loan at the time that the loan is made. This type of loan is called a **discounted loan**. The interest that is deducted from the loan is the **discount**.

Example 7   A Discounted Loan

You borrow $10,000 on a 10% discounted loan for a period of 8 months.

a. What is the loan's discount?

b. Determine the net amount of money you receive.

c. What is the loan's actual interest rate?

Solution

a. Because the loan's discount is the deducted interest, we use the simple interest formula.

$$I = Prt = (10{,}000)(0.10)\left(\tfrac{2}{3}\right) \approx 666.67$$

> Principal is $10,000.   Rate is 10% = 0.10.   Time is 8 months $= \frac{8}{12} = \frac{2}{3}$ year.

The loan's discount is $666.67.

b. The net amount that you receive is the amount of the loan, $10,000, minus the discount, $666.67:

$$10{,}000 - 666.67 = 9333.33.$$

Thus, you receive $9333.33.

c. We can calculate the loan's actual interest rate, rather than the stated 10%, by using the simple interest formula.

$$I = Prt \qquad \text{This is the simple interest formula.}$$

$$666.67 = (9333.33)r\left(\frac{2}{3}\right) \qquad \textit{I}(\text{interest}) = \$666.67,\ P\ (\text{principal, or the net amount received})$$

$$= \$9333.33, \text{ and } t = \frac{2}{3} \text{ year.}$$

$$666.67 = 6222.22r \qquad \text{Simplify: } 9333.33\left(\frac{2}{3}\right) = 6222.22.$$

$$r = \frac{666.67}{6222.22} \approx 0.107 = 10.7\% \qquad \textit{Solve for r and express r to the nearest tenth of a percent.}$$

The actual rate of interest on the 10% discounted loan is approximately 10.7%.

> **Checkpoint 7**   You borrow $5000 on a 12% discounted loan for a period of two years. Determine: **a.** the loan's discount; **b.** the net amount you receive; ▶ **c.** the actual interest rate.

Exercise Set 7.3

Concept and Vocabulary Exercises

In Exercises 1–4, fill in each blank so that the resulting statement is true.

1. The formula for calculating simple interest, I, is _____, where P is the _____, r is the _____, and t is the _____.

2. The future value, A, of P dollars at simple interest rate r for t years is given by the formula _____.

3. A rule that allows financial institutions to calculate simple interest using 360 days in a year is called the _____ rule.

4. Some lenders collect the interest from the amount of a loan at the time the loan is made. This type of loan is called a/an _____. The interest that is deducted from the loan is the _____.

In Exercises 5–7, determine whether each statement is true or false. If the statement is false, make the necessary change(s) to produce a true statement.

5. Interest is the amount of money we get paid for borrowing money or that we pay for investing money.

6. In simple interest, only the original money invested or borrowed generates interest over time

7. If $4000 is borrowed at 7.6% for three months, the loan's future value is $76.

Respond to Exercises 8–10 using verbal or written explanations.

8. Explain how to calculate simple interest.

9. What is the future value of a loan and how is it determined?

10. What is a discounted loan? How is the net amount of money received from such a loan determined?

In Exercises 11–14, determine whether each statement makes sense or does not make sense, and explain your reasoning.

11. After depositing $1500 in an account at a rate of 4%, my balance at the end of the first year was $(1500)(0.04)$.

12. I saved money on my short-term loan for 90 days by finding a financial institution that used the Banker's rule rather than one that calculated interest using 365 days in a year.

13. I planned to save $5000 in four years, computed the present value to be $3846.153, so I rounded the principal to $3846.15.

14. When I borrowed money on a discounted loan, the amount of money I received was less than the amount I wanted to borrow.

Practice Exercises

In Exercises 15–22, the principal P is borrowed at simple interest rate r for a period of time t. Find the simple interest owed for the use of the money. Assume 360 days in a year.

15. $P = \$4000, r = 6\%, t = 1$ year
16. $P = \$7000, r = 5\%, t = 1$ year
17. $P = \$180, r = 3\%, t = 2$ years
18. $P = \$260, r = 4\%, t = 3$ years
19. $P = \$5000, r = 8.5\%, t = 9$ months
20. $P = \$18,000, r = 7.5\%, t = 18$ months
21. $P = \$15,500, r = 11\%, t = 90$ days
22. $P = \$12,600, r = 9\%, t = 60$ days

In Exercises 23–28, the principal P is borrowed at simple interest rate r for a period of time t. Find the loan's future value, A, or the total amount due at time t.

23. $P = \$3000, r = 7\%, t = 2$ years
24. $P = \$2000, r = 6\%, t = 3$ years
25. $P = \$26,000, r = 9.5\%, t = 5$ years
26. $P = \$24,000, r = 8.5\%, t = 6$ years
27. $P = \$9000, r = 6.5\%, t = 8$ months
28. $P = \$6000, r = 4.5\%, t = 9$ months

In Exercises 29–34, the principal P is borrowed and the loan's future value, A, at time t is given. Determine the loan's simple interest rate, r, to the nearest tenth of a percent.

29. $P = \$2000, A = \$2150, t = 1$ year
30. $P = \$3000, A = \$3180, t = 1$ year
31. $P = \$5000, A = \$5900, t = 2$ years
32. $P = \$10,000, A = \$14,060, t = 2$ years
33. $P = \$2300, A = \$2840, t = 9$ months
34. $P = \$1700, A = \$1820, t = 6$ months

In Exercises 35–40, determine the present value, P, you must invest to have the future value, A, at simple interest rate r after time t. Round answers up to the nearest cent.

35. $A = \$6000, r = 8\%, t = 2$ years
36. $A = \$8500, r = 7\%, t = 3$ years
37. $A = \$14,000, r = 9.5\%, t = 6$ years
38. $A = \$16,000, r = 11.5\%, t = 5$ years
39. $A = \$5000, r = 14.5\%, t = 9$ months
40. $A = \$2000, r = 12.6\%, t = 8$ months

Exercises 41–44 involve discounted loans. In each exercise, determine

a. *the loan's discount.*

b. *the net amount of money you receive.*

c. *the loan's actual interest rate, to the nearest tenth of a percent.*

41. You borrow $2000 on a 7% discounted loan for a period of 8 months.

42. You borrow $3000 on an 8% discounted loan for a period of 9 months.

43. You borrow $12,000 on a 6.5% discounted loan for a period of two years.

44. You borrow $20,000 on an 8.5% discounted loan for a period of three years.

Practice Plus

45. Solve for r: $A = P(1 + rt)$.
46. Solve for t: $A = P(1 + rt)$.

47. Solve for P: $A = P(1 + rt)$.

48. Solve for P: $A = P\left(1 + \frac{r}{n}\right)^{nt}$. (We will be using this formula in the next section.)

Application Exercises

49. In order to start a small business, a college student takes out a simple interest loan for $4000 for 9 months at a rate of 8.25%.

   a. How much interest must the student pay?

   b. Find the future value of the loan.

50. In order to pay for baseball uniforms, a college takes out a simple interest loan for $20,000 for 7 months at a rate of 12%.

   a. How much interest must the college pay?

   b. Find the future value of the loan.

51. You borrow $1400 from a friend and promise to pay back $2000 in two years. What simple interest rate, to the nearest tenth of a percent, will you pay?

52. Treasury bills (T-bills) can be purchased from the U.S. Treasury Department. You buy a T-bill for $981.60 that pays $1000 in 13 weeks. What simple interest rate, to the nearest tenth of a percent, does this T-bill earn?

53. A bank offers a CD that pays a simple interest rate of 6.5%. How much must you put in this CD now in order to have $3000 for a class trip to Europe in two years?

54. A bank offers a CD that pays a simple interest rate of 5.5%. How much must you put in this CD now in order to have $8000 for a kitchen remodeling project in two years?

55. In order to pay for dental work, a family borrows $8000 on an 8% discounted loan for a period of three years.

   a. What is the loan's discount?

   b. Determine the net amount of money the family receives.

   c. What is the loan's actual interest rate?

56. In order to pay for a car, your parents borrow $20,000 on a 6% discounted loan for a period of four years.

   a. What is the loan's discount?

   b. Determine the net amount of money your parents receive.

   c. What is the loan's actual interest rate?

Critical Thinking Exercises

57. Use the future value formula to show that the time required for an amount of money P to double in value to $2P$ is given by

$$t = \frac{1}{r}.$$

58. You deposit $5000 in an account that earns 5.5% simple interest.

   a. Express the future value in the account as a linear function of time, t.

   b. Determine the slope of the function in part (a) and describe what this means. Use the phrase "rate of change" in your description.

7.4 Compound Interest

Objectives

1 Use compound interest formulas.

2 Calculate present value.

3 Understand and compute effective annual yield.

So, how did the present value of Manhattan in 1626 — that is, the $24 paid to the native Americans — attain a future value of over $5 billion in 2010, 384 years later, at a mere 5% interest rate? After all, the future value on $24 for 384 years at 5% simple interest is

$$A = P(1 + rt)$$
$$= 24[1 + (0.05)(384)] = 484.8,$$

or a paltry $484.80, compared to over $5 billion. To understand this dramatic difference in future value, we turn to the concept of *compound interest*.

Compound Interest

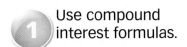

Use compound interest formulas.

Compound interest is interest computed on the original principal as well as on any accumulated interest. Many savings accounts pay compound interest. For example, suppose you deposit $1000 in a savings account at a rate of 5%. **Table 7.3** shows how the investment grows if the interest earned is automatically added on to the principal.

Table 7.3 Calculating the Amount in an Account Subject to Compound Interest

Use $A = P(1 + rt)$ with $r = 0.05$
and $t = 1$, or $A = P(1 + 0.05)$.

Year	Starting Balance	Amount in the Account at Year's End
1	$1000	$A = \$1000(1 + 0.05) = \$1050$
2	$1050 or $1000(1 + 0.05)$	$A = \$1050(1 + 0.05) = \$1102.50$ or $A = \$1000(1 + 0.05)(1 + 0.05) = \$1000(1 + 0.05)^2$
3	$1102.50 or $1000(1 + 0.05)^2$	$A = \$1102.50(1 + 0.05) \approx \$1157.63$ or $A = \$1000(1 + 0.05)^2(1 + 0.05) = \$1000(1 + 0.05)^3$

Using inductive reasoning, the amount, A, in the account after t years is the original principal, $1000, times $(1 + 0.05)^t$: $A = 1000(1 + 0.05)^t$.

If the original principal is P and the interest rate is r, we can use this same approach to determine the amount, A, in an account subject to compound interest.

> ### Calculating the Amount in an Account for Compound Interest Paid Once a Year
>
> If you deposit P dollars at rate r, in decimal form, subject to compound interest, then the amount, A, of money in the account after t years is given by
>
> $$A = P(1 + r)^t.$$
>
> The amount A is called the account's **future value** and the principal P is called its **present value**.

Example 1 Using the Compound Interest Formula

Suppose that you deposit $2000 in a savings account at Hometown Bank, which has a rate of 6%.

a. Find the amount, A, of money in the account after 3 years subject to interest compounded once a year.

b. Find the interest.

Solution

a. The amount deposited, or principal, P, is $2000. The rate, r, is 6%, or 0.06. The time of the deposit, t, is three years. The amount in the account after three years is

$$A = P(1 + r)^t = 2000(1 + 0.06)^3 = 2000(1.06)^3 \approx 2382.03.$$

Rounded to the nearest cent, the amount in the savings account after three years is $2382.03.

b. Because the amount in the account is $2382.03 and the original principal is $2000, the interest is $2382.03 − $2000, or $382.03.

> ✓ **Checkpoint 1** Suppose that you deposit $1000 in a savings account at a bank that has a rate of 4%.
>
> **a.** Find the amount, A, of money in the account after 5 years subject to interest compounded once a year. Round to the nearest cent.
>
> ▶ **b.** Find the interest.

Compound Interest Paid More than Once a Year

The period of time between two interest payments is called the **compounding period**. When compound interest is paid once per year, the compounding period is one year. We say that the interest is **compounded annually**.

Most savings institutions have plans in which interest is paid more than once per year. If compound interest is paid twice per year, the compounding period is six months. We say that the interest is **compounded semiannually**. When compound interest is paid four times per year, the compounding period is three months and the interest is said to be **compounded quarterly**. Some plans allow for monthly compounding or daily compounding.

In general, when compound interest is paid n times per year, we say that there are *n compounding periods per year*. The following formula is used to calculate the amount in an account subject to compound interest with n compounding periods per year:

> **Calculating the Amount in an Account for Compound Interest Paid n Times a Year**
>
> If you deposit P dollars at rate r, in decimal form, subject to compound interest paid n times per year, then the amount, A, of money in the account after t years is given by
>
> $$A = P\left(1 + \frac{r}{n}\right)^{nt}.$$
>
> A is the account's **future value** and the principal P is its **present value**.

Study Tip

If $n = 1$ (interest paid once a year), the formula in the box becomes

$$A = P\left(1 + \frac{r}{1}\right)^{1t}, \text{ or}$$

$$A = P(1 + r)^t.$$

This shows that the amount in an account subject to annual compounding is just one application of the general formula in the box at the right. With this general formula, you no longer need the formula for annual compounding in the box on the previous page.

Example 2   Using the Compound Interest Formula

You deposit $7500 in a savings account that has a rate of 6%. The interest is compounded monthly.

a. How much money will you have after five years?

b. Find the interest after five years.

Solution

a. The amount deposited, or principal, P, is $7500. The rate, r, is 6%, or 0.06. Because interest is compounded monthly, there are 12 compounding periods per year, so $n = 12$. The time of the deposit, t, is five years. The amount in the account after five years is

$$A = P\left(1 + \frac{r}{n}\right)^{nt} = 7500\left(1 + \frac{0.06}{12}\right)^{12 \cdot 5} = 7500(1.005)^{60} \approx 10{,}116.38.$$

Rounded to the nearest cent, you will have $10,116.38 after five years.

b. Because the amount in the account is $10,116.38 and the original principal is $7500, the interest after five years is $10,116.38 − $7500, or $2616.38.

Technology

Here are the calculator keystrokes to compute

$$7500\left(1 + \frac{0.06}{12}\right)^{12 \cdot 5}:$$

Many Scientific Calculators

7500 ⨯ ((1 + .06 ÷ 12)

y^x ((12 ⨯ 5)) =

Many Graphing Calculators

7500 ((1 + .06 ÷ 12)

∧ ((12 ⨯ 5) ENTER

✓ **Checkpoint 2** You deposit $4200 in a savings account that has a rate of 4%. The interest is compounded quarterly.

a. How much money will you have after ten years? Round to the nearest cent.

▶ b. Find the interest after ten years.

Continuous Compounding

Some banks use **continuous compounding**, where the compounding periods increase infinitely (compounding interest every trillionth of a second, every quadrillionth of a second, etc.). As n, the number of compounding periods in a year, increases without bound, the expression $\left(1 + \frac{1}{n}\right)^n$ approaches the irrational number e: $e \approx 2.71828$. This is illustrated in **Table 7.4**. As a result, the formula for the balance in an account with n compounding periods per year, $A = P\left(1 + \frac{r}{n}\right)^{nt}$, becomes $A = Pe^{rt}$ with continuous compounding. Although continuous compounding sounds terrific, it yields only a fraction of a percent more interest over a year than daily compounding.

Table 7.4 As n Takes on Increasingly Large Values, the Expression $\left(1 + \frac{1}{n}\right)^n$ Approaches the Irrational Number e

n	$\left(1 + \frac{1}{n}\right)^n$
1	2
2	2.25
5	2.48832
10	2.59374246
100	2.704813829
1000	2.716923932
10,000	2.718145927
100,000	2.718268237
1,000,000	2.718280469
1,000,000,000	2.718281827

Formulas for Compound Interest

After t years, the balance, A, in an account with principal P and annual interest rate r (in decimal form) is given by the following formulas:

1. For n compounding periods per year: $A = P\left(1 + \frac{r}{n}\right)^{nt}$

2. For continuous compounding: $A = Pe^{rt}$.

Technology

You can compute e to a power using the $\boxed{e^x}$ key on your calculator. Use the key to enter e^1 and verify that e is approximately equal to 2.71828.

Scientific Calculators

1 $\boxed{e^x}$

Graphing Calculators

$\boxed{e^x}$ 1 $\boxed{\text{ENTER}}$

Example 3 Choosing between Investments

You decide to invest $8000 for 6 years and you have a choice between two accounts. The first pays 7% per year, compounded monthly. The second pays 6.85% per year, compounded continuously. Which is the better investment?

Solution

The better investment is the one with the greater balance in the account after 6 years. Let's begin with the account with monthly compounding. We use the compound interest formula with $P = 8000, r = 7\% = 0.07, n = 12$ (monthly compounding means 12 compounding periods per year), and $t = 6$.

$$A = P\left(1 + \frac{r}{n}\right)^{nt} = 8000\left(1 + \frac{0.07}{12}\right)^{12 \cdot 6} \approx 12{,}160.84$$

The balance in this account after 6 years would be $12,160.84.

The second investment option involves $8000 for 6 years paying 6.85% per year, compounded continuously. We use the formula for continuous compounding with $P = 8000$, $r = 6.85\% = 0.0685$, and $t = 6$.

$$A = Pe^{rt} = 8000e^{0.0685(6)} \approx 12{,}066.60$$

The balance in this account after 6 years would be $12,066.60, slightly less than the previous amount, $12,160.84. Thus, the better investment is the 7% monthly compounding option.

✓ **Checkpoint 3**   A sum of $10,000 is invested at an annual rate of 8%. Find the balance in the account after 5 years subject to **a.** quarterly compounding and **b.** continuous compounding.

Planning for the Future with Compound Interest

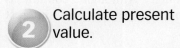

Just as we did in Section 7.3, we can determine P, the principal or present value, that should be deposited now in order to have a certain amount, A, in the future. If an account earns compound interest, the amount of money that should be invested today to obtain a future value of A dollars can be determined by solving the compound interest formula for P:

> **Calculating Present Value**
>
> If A dollars are to be accumulated in t years in an account that pays rate r compounded n times per year, then the present value, P, that needs to be invested now is given by
>
> $$P = \dfrac{A}{\left(1 + \dfrac{r}{n}\right)^{nt}}.$$

Remember to round the principal *up* to the nearest cent when computing present value so there will be enough money to meet future goals.

Example 4   Calculating Present Value

How much money should be deposited today in an account that earns 6% compounded monthly so that it will accumulate to $20,000 in five years?

Solution

The amount we need today, or the present value, is determined by the present value formula. Because the interest is compounded monthly, $n = 12$. Furthermore, A (the future value) = $20,000, r (the rate) = 6% = 0.06, and t (time in years) = 5.

$$P = \dfrac{A}{\left(1 + \dfrac{r}{n}\right)^{nt}} = \dfrac{20{,}000}{\left(1 + \dfrac{0.06}{12}\right)^{12 \cdot 5}} \approx 14{,}827.4439$$

To make sure there will be enough money, we round the principal *up* to $14,827.45. Approximately $14,827.45 should be invested today in order to accumulate to $20,000 in five years.

✓ **Checkpoint 4**   How much money should be deposited today in an account that earns 7% compounded weekly so that it will accumulate to $10,000 in eight years?

Achieving $uccess

The Time Value of Money

When you begin earning your own money, it will be tempting to spend every penny earned. By doing this, you will fail to take advantage of the *time value of money*. The **time value of money** means that a dollar received today is worth more than a dollar received next year or the year after. This is because a sum of money invested today starts earning compound interest sooner than a sum of money invested some time in the future. **A significant way to increase your wealth is to spend less than you earn and invest the difference.** With time on your side, even a small amount of money can be turned into a substantial sum through the power of compounding. Make the time value of money work for you by postponing certain purchases now and investing the savings instead. Do you really need those new shoes? Couldn't you bring lunch from home rather than eating out? Pay close attention to your spending habits as you study the time value of money.

Effective Annual Yield

3 Understand and compute effective annual yield.

As we've seen before, a common problem in personal finance is selecting the best investment from two or more investments. For example, is an investment that pays 8.25% interest compounded quarterly better than one that pays 8.3% interest compounded semiannually? Another way to answer the question is to compare the *effective rates* of the investments, also called their *effective annual yields*.

Blitzer Bonus

Doubling Your Money: The Rule of 72

Here's a shortcut for estimating the number of years it will take for your investment to double: Divide 72 by the effective annual yield without the percent sign. For example, if the effective annual yield is 6%, your money will double in approximately

$$\frac{72}{6}$$

years, or in 12 years.

Effective Annual Yield

The **effective annual yield**, or the **effective rate**, is the simple interest rate that produces the same amount of money in an account at the end of one year as when the account is subjected to compound interest at a stated rate.

Example 5 — Understanding Effective Annual Yield

You deposit $4000 in an account that pays 8% interest compounded monthly.

a. Find the future value after one year.

b. Use the future value formula for simple interest to determine the effective annual yield.

Solution

a. We use the compound interest formula to find the account's future value after one year.

$$A = P\left(1 + \frac{r}{n}\right)^{nt} = 4000\left(1 + \frac{0.08}{12}\right)^{12 \cdot 1} \approx \$4332.00$$

Principal is $4000. Stated rate is 8% = 0.08. Monthly compounding: $n = 12$ Time is one year: $t = 1$.

Rounded to the nearest cent, the future value after one year is $4332.00.

b. **The effective annual yield, or effective rate, is a simple interest rate.** We use the future value formula for simple interest to determine the simple interest rate that produces a future value of $4332 for a $4000 deposit after one year.

$A = P(1 + rt)$ This is the future value formula for simple interest.

$4332 = 4000(1 + r \cdot 1)$ Substitute the given values.

$$4332 = 4000 + 4000r$$ Use the distributive property on the right side of $4332 = 4000(1 + r \cdot 1)$.

$$332 = 4000r$$ Subtract 4000 from both sides.

$$\frac{332}{4000} = \frac{4000r}{4000}$$ Divide both sides by 4000.

$$r = \frac{332}{4000} = 0.083 = 8.3\%$$ Express r as a percent.

The effective annual yield, or effective rate, is 8.3%. This means that money invested at 8.3% simple interest earns the same amount in one year as money invested at 8% interest compounded monthly.

In Example 5, the stated 8% rate is called the **nominal rate**. The 8.3% rate is the effective rate and is a simple interest rate.

✓ **Checkpoint 5** You deposit $6000 in an account that pays 10% interest compounded monthly.

a. Find the future value after one year.

b. Determine the effective annual yield.

Generalizing the procedure of Example 5 and Checkpoint 5 gives a formula for effective annual yield:

Calculating Effective Annual Yield

Suppose that an investment has a nominal interest rate, r, in decimal form, and pays compound interest n times per year. The investment's effective annual yield, Y, in decimal form, is given by

$$Y = \left(1 + \frac{r}{n}\right)^n - 1.$$

The decimal form of Y given by the formula should then be converted to a percent.

Technology

Here are the keystrokes for Example 6:

Many Scientific Calculators

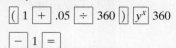

$[(\; 1 \; + \; .05 \; \div \; 360 \;)] \; y^x \; 360$

$- \; 1 \; =$

Many Graphing Calculators

$[(\; 1 \; + \; .05 \; \div \; 360 \;)] \; \wedge \; 360$

$- \; 1 \; \boxed{\text{ENTER}}$.

Given the nominal rate and the number of compounding periods per year, some graphing calculators display the effective annual yield. The screen shows the calculation of the effective rate in Example 6 on the TI-84 Plus.

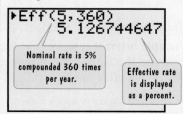

► Eff(5,360)
        5.126744647

Nominal rate is 5% compounded 360 times per year.

Effective rate is displayed as a percent.

Example 6 · Calculating Effective Annual Yield

A savings account has a nominal rate of 5%. The interest is compounded daily. Find the account's effective annual yield. (Assume 360 days in a year.)

Solution

The rate, r, is 5%, or 0.05. Because interest is compounded daily and we assume 360 days in a year, $n = 360$. The account's effective annual yield is

$$Y = \left(1 + \frac{r}{n}\right)^n - 1 = \left(1 + \frac{0.05}{360}\right)^{360} - 1 \approx 0.0513 = 5.13\%.$$

The effective annual yield is 5.13%. Thus, money invested at 5.13% simple interest earns the same amount of interest in one year as money invested at 5% interest, the nominal rate, compounded daily.

✓ **Checkpoint 6** What is the effective annual yield of an account paying 8% compounded quarterly?

The effective annual yield is often included in the information about investments or loans. Because it's the true interest rate you're earning or paying, it's the number you should pay attention to. **If you are selecting the best investment from two or more investments, the best choice is the account with the greatest effective annual yield.** However, there are differences in the types of accounts that you need to take into consideration. Some pay interest from the day of deposit to the day of withdrawal. Other accounts start paying interest the first day of the month that follows the day of deposit. Some savings institutions stop paying interest if the balance in the account falls below a certain amount.

When *borrowing money*, the effective rate or effective annual yield is usually called the **annual percentage rate**. If all other factors are equal and you are borrowing money, select the option with the least annual percentage rate.

Exercise Set 7.4

Concept and Vocabulary Exercises

In Exercises 1–8, fill in each blank so that the resulting statement is true.

1. Compound interest is interest computed on the original _____ as well as on any accumulated _____.

2. The formula $A = P\left(1 + \dfrac{r}{n}\right)^{nt}$ gives the amount of money, A, in an account after _____ years at rate _____ subject to compound interest paid _____ times per year.

3. If interest is compounded once a year, the formula in Exercise 2 becomes _____.

4. If compound interest is paid twice per year, the compounding period is _____ months and the interest is compounded _____.

5. If compound interest is paid four times per year, the compounding period is _____ months and the interest is compounded _____.

6. When the number of compounding periods in a year increases without bound, this is known as _____ compounding.

7. In the formula
$$P = \dfrac{A}{\left(1 + \dfrac{r}{n}\right)^{nt}},$$
the variable _____ represents the amount that needs to be invested now in order to have _____ dollars accumulated in _____ years in an account that pays rate _____ compounded _____ times per year.

8. If you are selecting the best investment from two or more investments, the best choice is the account with the greatest _____, which is the _____ interest rate that produces the same amount of money at the end of one year as when the account is subjected to compound interest at a stated rate.

In Exercises 9–12, determine whether each statement is true or false. If the statement is false, make the necessary change(s) to produce a true statement.

9. Formulas for compound interest show that a dollar invested today is worth more than a dollar invested in the future.

10. Formulas for compound interest show that if you make the decision to postpone certain purchases and save the money instead, small amounts of money can be turned into substantial sums over a period of years.

11. At a given annual interest rate, your money grows faster as the compounding period becomes shorter.

12. According to the Rule of 72, an investment with an effective annual yield of 12% can double in six years.

Respond to Exercises 13–16 using verbal or written explanations.

13. Describe the difference between simple and compound interest.

14. Give two examples that illustrate the difference between a compound interest problem involving future value and a compound interest problem involving present value.

15. What is effective annual yield?

16. Explain how to select the best investment from two or more investments.

In Exercises 17–20, determine whether each statement makes sense or does not make sense, and explain your reasoning.

17. My bank provides simple interest at 3.25% per year, but I can't determine if this is a better deal than a competing bank offering 3.25% compound interest without knowing the compounding period.

18. When choosing between two accounts, the one with the greater annual interest rate is always the better deal.

19. A bank can't increase compounding periods indefinitely without owing its customers an infinite amount of money.

20. My bank advertises a compound interest rate of 2.4%, although, without making deposits or withdrawals, the balance in my account increased by 2.43% in one year.

Practice Exercises

Here is a list of formulas needed to solve the exercises. Be sure you understand what each formula describes and the meaning of the variables in the formulas.

$$A = P\left(1 + \frac{r}{n}\right)^{nt} \qquad P = \frac{A}{\left(1 + \frac{r}{n}\right)^{nt}}$$

$$A = Pe^{rt} \qquad Y = \left(1 + \frac{r}{n}\right)^{n} - 1$$

In Exercises 21–32, the principal represents an amount of money deposited in a savings account subject to compound interest at the given rate.

a. *Find how much money there will be in the account after the given number of years. (Assume 360 days in a year.)*

b. *Find the interest earned.*

Round answers to the nearest cent.

	Principal	Rate	Compounded	Time
21.	$10,000	4%	annually	2 years
22.	$8000	6%	annually	3 years
23.	$3000	5%	semiannually	4 years
24.	$4000	4%	semiannually	5 years
25.	$9500	6%	quarterly	5 years
26.	$2500	8%	quarterly	6 years
27.	$4500	4.5%	monthly	3 years
28.	$2500	6.5%	monthly	4 years
29.	$1500	8.5%	daily	2.5 years
30.	$1200	8.5%	daily	3.5 years
31.	$20,000	4.5%	daily	20 years
32.	$25,000	5.5%	daily	20 years

Solve Exercises 33–36 using appropriate compound interest formulas. Round answers to the nearest cent.

33. Find the accumulated value of an investment of $10,000 for 5 years at an interest rate of 5.5% if the money is **a.** compounded semiannually; **b.** compounded quarterly; **c.** compounded monthly; **d.** compounded continuously.

34. Find the accumulated value of an investment of $5000 for 10 years at an interest rate of 6.5% if the money is **a.** compounded semiannually; **b.** compounded quarterly; **c.** compounded monthly; **d.** compounded continuously.

35. Suppose that you have $12,000 to invest. Which investment yields the greater return over 3 years: 7% compounded monthly or 6.85% compounded continuously?

36. Suppose that you have $6000 to invest. Which investment yields the greater return over 4 years: 8.25% compounded quarterly or 8.3% compounded semiannually?

In Exercises 37–40, round answers up to the nearest cent.

37. How much money should be deposited today in an account that earns 6% compounded semiannually so that it will accumulate to $10,000 in three years?

38. How much money should be deposited today in an account that earns 7% compounded semiannually so that it will accumulate to $12,000 in four years?

39. How much money should be deposited today in an account that earns 9.5% compounded monthly so that it will accumulate to $10,000 in three years?

40. How much money should be deposited today in an account that earns 10.5% compounded monthly so that it will accumulate to $22,000 in four years?

41. Suppose that you deposit $10,000 in an account that pays 4.5% interest compounded quarterly.

a. Find the future value after one year.

b. Use the future value formula for simple interest to determine the effective annual yield.

42. Suppose that you deposit $12,000 in an account that pays 6.5% interest compounded quarterly.

a. Find the future value after one year.

b. Use the future value formula for simple interest to determine the effective annual yield.

In Exercises 43–48, a savings account has a rate of 6%. Find the effective annual yield, rounded to the nearest tenth of a percent, if the interest is compounded

43. semiannually.

44. quarterly.

45. monthly.

46. daily. (Assume 360 days in a year.)

47. 1000 times per year.

48. 100,000 times per year.

In Exercises 49–52, determine the effective annual yield for each investment. Then select the better investment. Assume 360 days in a year. If rounding is required, round to the nearest tenth of a percent.

49. 8% compounded monthly; 8.25% compounded annually

50. 5% compounded monthly; 5.25% compounded quarterly

51. 5.5% compounded semiannually; 5.4% compounded daily

52. 7% compounded annually; 6.85% compounded daily

Practice Plus

In Exercises 53–56, how much more would you earn in the first investment than in the second investment? Round answers to the nearest dollar.

53. • $25,000 invested for 40 years at 12% compounded annually

• $25,000 invested for 40 years at 6% compounded annually

54. • $30,000 invested for 40 years at 10% compounded annually

• $30,000 invested for 40 years at 5% compounded annually

55. • $50,000 invested for 30 years at 10% compounded annually

• $50,000 invested for 30 years at 5% compounded monthly

56. • $20,000 invested for 30 years at 12% compounded annually

• $20,000 invested for 30 years at 6% compounded monthly

Application Exercises

Assume that the accounts described in the exercises have no other deposits or withdrawals except for what is stated. Round all answers to the nearest dollar, rounding up to the nearest dollar in present-value problems. Assume 360 days in a year.

57. At the time of a child's birth, $12,000 was deposited in an account paying 6% interest compounded semiannually. What will be the value of the account at the child's twenty-first birthday?

58. At the time of a child's birth, $10,000 was deposited in an account paying 5% interest compounded semiannually. What will be the value of the account at the child's twenty-first birthday?

59. You deposit $2600 in an account that pays 4% interest compounded once a year. Your friend deposits $2200 in an account that pays 5% interest compounded monthly.

   a. Who will have more money in their account after one year? How much more?

   b. Who will have more money in their account after five years? How much more?

   c. Who will have more money in their account after 20 years? How much more?

60. You deposit $3000 in an account that pays 3.5% interest compounded once a year. Your friend deposits $2500 in an account that pays 4.8% interest compounded monthly.

   a. Who will have more money in their account after one year? How much more?

   b. Who will have more money in their account after five years? How much more?

   c. Who will have more money in their account after 20 years? How much more?

61. You deposit $3000 in an account that pays 7% interest compounded semiannually. After ten years, the interest rate is increased to 7.25% compounded quarterly. What will be the value of the account after 16 years?

62. You deposit $6000 in an account that pays 5.25% interest compounded semiannually. After ten years, the interest rate is increased to 5.4% compounded quarterly. What will be the value of the account after 18 years?

63. In 1626, Peter Minuit convinced the Wappinger Indians to sell him Manhattan Island for $24. If the Native Americans had put the $24 into a bank account paying compound interest at a 5% rate, how much would the investment be worth in the year 2010 ($t = 384$ years) if interest were compounded **a.** monthly? **b.** 360 times per year?

64. In 1777, Jacob DeHaven loaned George Washington's army $450,000 in gold and supplies. Due to a disagreement over the method of repayment (gold versus Continental money), DeHaven was never repaid, dying penniless. In 1989, his descendants sued the U.S. government over the 212-year-old debt. If the DeHavens used an interest rate of 6% and daily compounding (the rate offered by the Continental Congress in 1777), how much money did the DeHaven family demand in their suit? (*Hint:* Use the compound interest formula with $n = 360$ and $t = 212$ years.)

65. Will you earn more interest in one year by depositing $2000 in a simple interest account that pays 6% or in an account that pays 5.9% interest compounded daily? How much more interest will you earn?

66. Will you earn more interest in one year by depositing $1000 in a simple interest account that pays 7% or in an account that pays 6.9% interest compounded daily? How much more interest will you earn?

67. Two accounts each begin with a deposit of $5000. Both accounts have rates of 5.5%, but one account compounds interest once a year while the other account compounds interest continuously. Make a table that shows the amount in each account and the interest earned after one year, five years, ten years, and 20 years.

68. Two accounts each begin with a deposit of $10,000. Both accounts have rates of 6.5%, but one account compounds interest once a year while the other account compounds interest continuously. Make a table that shows the amount in each account and the interest earned after one year, five years, ten years, and 20 years.

69. Parents wish to have $80,000 available for a child's education. If the child is now 5 years old, how much money must be set aside at 6% compounded semiannually to meet their financial goal when the child is 18?

70. A 30-year-old worker plans to retire at age 65. He believes that $500,000 is needed to retire comfortably. How much should be deposited now at 7% compounded monthly to meet the $500,000 retirement goal?

71. You would like to have $75,000 available in 15 years. There are two options. Account A has a rate of 4.5% compounded once a year. Account B has a rate of 4% compounded daily. How much would you have to deposit in each account to reach your goal?

72. You would like to have $150,000 available in 20 years. There are two options. Account A has a rate of 5.5% compounded once a year. Account B has a rate of 5% compounded daily. How much would you have to deposit in each account to reach your goal?

73. You invest $1600 in an account paying 5.4% interest compounded daily. What is the account's effective annual yield? Round to the nearest hundredth of a percent.

74. You invest $3700 in an account paying 3.75% interest compounded daily. What is the account's effective annual yield? Round to the nearest hundredth of a percent.

75. An account has a nominal rate of 4.2%. Find the effective annual yield, rounded to the nearest tenth of a percent, with quarterly compounding, monthly compounding, and daily compounding. How does changing the compounding period affect the effective annual yield?

76. An account has a nominal rate of 4.6%. Find the effective annual yield, rounded to the nearest tenth of a percent, with quarterly compounding, monthly compounding, and daily compounding. How does changing the compounding period affect the effective annual yield?

77. A bank offers a money market account paying 4.5% interest compounded semiannually. A competing bank offers a money market account paying 4.4% interest compounded daily. Which account is the better investment?

78. A bank offers a money market account paying 4.9% interest compounded semiannually. A competing bank offers a money market account paying 4.8% interest compounded daily. Which account is the better investment?

Critical Thinking Exercises

79. A depositor opens a new savings account with $6000 at 5% compounded semiannually. At the beginning of year 3, an additional $4000 is deposited. At the end of six years, what is the balance in the account?

80. A depositor opens a money market account with $5000 at 8% compounded monthly. After two years, $1500 is withdrawn from the account to buy a new computer. A year later, $2000 is put in the account. What will be the ending balance if the money is kept in the account for another three years?

81. Use the future value formulas for simple and compound interest in one year to derive the formula for effective annual yield.

Group Exercise

82. This activity is a group research project intended for four or five people. Present your research in a seminar on the history of interest and banking. The seminar should last about 30 minutes. Address the following questions:

When was interest first charged on loans? How was lending money for a fee opposed historically? What is usury? What connection did banking and interest rates play in the historic European rivalries? When and where were some of the highest interest rates charged? What were the rates? Where does the word *interest* come from? What is the difference between usury and interest in modern times? What is the history of a national bank in the United States?

Chapter 7 Summary

7.1 Percent, Sales Tax, and Discounts

Definitions and Concepts

Percent means per hundred. Thus, $97\% = \frac{97}{100}$.

To express a fraction as a percent, divide the numerator by the denominator, move the decimal point in the quotient two places to the right, and attach a percent sign.

Example

$$0.375 = 0.375\% = 37.5\%$$

$$8\overline{)3.000}$$

Conclusion: $\frac{3}{8} = 37.5\%$

Additional Example to Review

Example 1, page 448

Definitions and Concepts

To express a decimal number as a percent, move the decimal point two places to the right and attach a percent sign.

Examples

• $0.54 = 0.54\% = 54\%$

• $0.012 = 0.012\% = 1.2\%$

Additional Example to Review

Example 2, page 449

Definitions and Concepts

To express a percent as a decimal number, move the decimal point two places to the left and remove the percent sign.

Examples

- $11\% = 11.\% = 0.11\% = 0.11$
- $230\% = 230.\% = 2.30\% = 2.30 \text{ or } 2.3$
- $\frac{3}{4}\% = 0.75\% = .0075\% = 0.0075$

Additional Example to Review

Example 3, page 449

Definitions and Concepts

The percent formula, $A = PB$, means A is P percent of B.

Sales tax amount = tax rate × item's cost

Example

- The local sales tax rate is 6% and you purchase an iPad for $640.

$$\text{Sales tax amount} = 6\% \text{ of } \$640 = 0.06 \times \$640 = \$38.40$$

$$\text{Total cost} = \$640.00 + \$38.40 = \$678.40$$

Additional Example to Review

Example 4, page 450

Definitions and Concepts

Discount amount = discount rate × original price

Example

- A video game with an original price of $24 is on sale at 35% off.

$$\text{Discount amount} = 35\% \text{ of } \$24 = 0.35 \times \$24 = \$8.40$$

$$\text{Sale price} = \$24.00 - \$8.40 = \$15.60$$

Additional Example to Review

Example 5, page 450

Definitions and Concepts

The fraction for percent increase (or decrease) is

$$\frac{\text{amount of increase (or decrease)}}{\text{original amount}}.$$

Find the percent increase (or decrease) by expressing this fraction as a percent.

Example

- Shares of stock were purchased for $5020 and sold for $6777.

$$\text{Percent increase} = \frac{\text{amount of increase}}{\text{original amount}} = \frac{6777 - 5020}{5020} = \frac{1757}{5020} = 0.35 = 35\%$$

Additional Examples to Review
Example 6, page 451; Example 7, page 452; Example 8, page 453; Example 9, page 453

7.2 Income Tax

Definitions and Concepts

Income tax is a percentage of your income collected by the government to fund its services and programs.

Gross income: total income for the year, including interest from investments

Adjusted gross income: Gross income − Adjustments

Adjustments are certain allowable amounts, such as contributions to tax-deferred savings plans.

Taxable income: Adjusted gross income − (Exemptions + Deductions)

An exemption is a fixed amount for each person supported by the taxpayer's income. You are entitled to this fixed amount for yourself. A standard deduction is a lump-sum amount that you can subtract from your adjusted gross income. Itemized deductions are deductions you list separately, including interest on home mortgages, state income taxes, and charitable contributions. Taxpayers should choose the greater of a standard deduction or an itemized deduction.

Example

- Suppose your neighbor earned wages of $34,200, received $470 in interest from a savings account, and contributed $850 to a tax-deferred savings plan. She is entitled to a personal exemption of $3500 and a standard deduction of $5450. The interest on her home mortgage was $6100, she contributed $1100 to charity, and she paid $1265 in state taxes.

Gross income = $34,200 + $470 = $34,670

Wages   Earned interest

Adjusted gross income = Gross income − Adjustments = $34,670 − $850 = $33,820

Contribution to a tax-deferred savings plan

Personal exemption = $3500
Standard deduction = $5450
Itemized deductions = $6100 + $1100 + $1265 = $8465

Interest on home mortgage   Charity   State taxes

Choose the itemized deductions because they are greater than the standard deduction.

Taxable income = Adjusted gross income − (Exemptions + Deductions)
= $33,820 − ($3500 + $8465) = $33,820 − $11,965 = $21,855

Additional Example to Review
Example 1, page 459

Definitions and Concepts

Calculating Income Tax

1. Determine adjusted gross income:

$$\text{Adjusted gross income} = \text{Gross income} - \text{Adjustments.}$$

2. Determine taxable income:

$$\text{Taxable income} = \text{Adjusted gross income} - (\text{Exemptions} + \text{Deductions}).$$

3. Determine the income tax:

$$\text{Income tax} = \text{Tax computation} - \text{Tax credits.}$$

Tax credits are sums of money that reduce the income tax owed by the full dollar-for-dollar amount of the credits.

Example

• Use the 2008 marginal rates in **Table 7.1** on page 462 to calculate the federal income tax owed by the following person:

• Single, no dependents
• Gross income: $34,000
• $2100 paid to a tax-deferred IRA
• $3700 mortgage interest
• $2400 property taxes
• $1000 in tax credits

Tax Rate	Single
10%	up to $8025
15%	$8026 to $32,550
Standard Deduction	$5450
Exemptions (per person)	$3500

A portion of **Table 7.1**

1. Adjusted gross income
 = Gross income − Adjustments
 = $34,000 − $2100 [Tax-deferred IRA]
 = $31,900
2. Personal exemption = $3500
 Standard deduction = $5450
 Itemized deductions
 = $3700 + $2400 = $6100

 [Mortgage interest] [Property taxes]

Choose the itemized deductions.

$$\text{Taxable income} = \text{Adjusted gross income} - (\text{Exemptions} + \text{Deductions})$$
$$= \$31,900 - (\$3500 + \$6100) = \$31,900 - \$9600 = \$22,300$$

3. Tax computation $= 0.10 \times \$8025 + 0.15 \times (\$22,300 - \$8025)$
 $= 0.10 \times \$8025 + 0.15 \times \$14,275 = \$802.50 + \$2141.25 = \$2943.75$

 Income Tax = Tax computation − Tax credits = $2943.75 − $1000 = $1943.75

The federal income tax owed is $1943.75.

Additional Example to Review

Example 2, page 462

Definitions and Concepts

FICA taxes are used for Social Security and Medicare benefits. For people who were not self-employed, the 2008 FICA tax rates were 7.65% on the first $102,000 of income and 1.45% on income in excess of $102,000. People who are self-employed must pay double these rates.

Example

- If you are not self-employed and earn $200,000, what are your FICA taxes?

$$\text{FICA Tax} = 0.0765 \times \$102,000 + 0.0145 \times (\$200,000 - \$102,000)$$
$$= 0.0765 \times \$102,000 + 0.0145 \times \$98,000 = \$7803 + \$1421 = \$9224$$

Additional Examples to Review

Example 3, page 464; Example 4, page 464

Definitions and Concepts

Gross pay is your salary prior to any withheld taxes for a pay period. Net pay is the actual amount of your check after taxes have been withheld.

Example

- You have a part-time job at the local supermarket. The job pays $10 per hour and you work 15 hours per week. Your employer withholds 10% of your gross pay for federal taxes, 7.65% for FICA taxes, and 4% for state taxes.

$$\text{Gross pay} = \cancel{15 \text{ hours}} \times \frac{\$10}{\cancel{\text{hour}}} = \$150$$

Federal taxes withheld = 10% of $150 = $0.10 \times \$150 = \$15$

FICA taxes withheld = 7.65% of $150 = $0.0765 \times \$150 \approx \$11.48$

State taxes withheld = 4% of $150 = $0.04 \times \$150 = \$6$

Total taxes withheld = $15 + $11.48 + $6 = $32.48

Net pay = Gross pay − Taxes withheld = $150.00 − $32.48 = $117.52

$$\text{Percentage of gross pay withheld for taxes} = \frac{\$32.48}{\$150.00} \approx 0.217 = 21.7\%$$

Additional Example to Review

Example 5, page 465

7.3 Simple Interest

Definitions and Concepts

Interest is the amount of money that we get paid for lending or investing money, or that we pay for borrowing money. The amount deposited or borrowed is the principal. The charge for interest, given as a percent, is the rate, assumed to be per year.

Simple interest involves interest calculated only on the principal and is computed using $I = Prt$.

Example

- Simple interest on a loan of $6000 for four months at 20%:

$$I = Prt = (6000)(0.2)\left(\frac{4}{12}\right) = (6000)(0.2)\left(\frac{1}{3}\right) = 400$$

The interest is $400.

Additional Examples to Review

Example 1, page 470; Example 2, page 470

Definitions and Concepts

The future value, A, of P dollars at simple interest rate r for t years is $A = P(1 + rt)$.

Examples

- Future value of a $3000 loan for four years at 14% simple interest:

$$A = P(1 + rt) = 3000[1 + (0.14)(4)] = 3000(1.56) = \$4680$$

- Simple interest rate if you borrow $1200 and pay back $1800 after two years:

$$A = P(1 + rt) \quad \boxed{\text{Solve for } r.}$$
$$1800 = 1200(1 + r \cdot 2)$$
$$1800 = 1200 + 2400r$$
$$600 = 2400r$$
$$r = \frac{600}{2400} = \frac{1}{4} = 0.25 = 25\%$$

- Amount to invest now at 4.5% simple interest to have $3000 in five years:

$$A = P(1 + rt) \quad \boxed{\text{Solve for } P.}$$
$$3000 = P[1 + (0.045)(5)]$$
$$3000 = P(1.225)$$
$$P = \frac{3000}{1.225} \approx \$2448.98$$

Additional Examples to Review

Example 3, page 471; Example 4, page 472; Example 5, page 473; Example 6, page 473

Definitions and Concepts

Discounted loans deduct the interest, called the discount, from the loan amount at the time the loan is made.

Example

- A person borrows $8000 on a 12% discounted loan for a period of 6 months. The loan's discount is the deducted interest.

$$I = Prt = (8000)(0.12)\left(\frac{6}{12}\right) = (8000)(0.12)\left(\frac{1}{2}\right) = \$480$$

The net amount of money that you receive is

$$\$8000 - \$480 = \$7520.$$

The loan's actual interest rate, rather than the stated 12%, can be found using

$$I = Prt.$$
$$480 = (7520)r\left(\frac{1}{2}\right)$$
$$480 = 3760r$$
$$r = \frac{480}{3760} \approx 0.128 = 12.8\%$$

The actual interest rate is approximately 12.8%.

Additional Example to Review

Example 7, page 474

7.4 Compound Interest

Definitions and Concepts

Compound interest involves interest computed on the original principal as well as on any accumulated interest. The amount in an account for one compounding period per year is $A = P(1 + r)^t$. For n compounding periods per year, the amount is $A = P\left(1 + \frac{r}{n}\right)^{nt}$. For continuous compounding, the amount is $A = Pe^{rt}$, where $e \approx 2.72$.

Example

- You receive an inheritance of $50,000 and invest it in an account that pays 12% compounded quarterly. Amount of money in the account after 30 years:

$$A = P\left(1 + \frac{r}{n}\right)^{nt} = 50{,}000\left(1 + \frac{0.12}{4}\right)^{4\cdot30} = 50{,}000(1.03)^{120} \approx \$1{,}735{,}549.36$$

Interest earned after 30 years:

$$\$1{,}735{,}549.36 - \$50{,}000 = \$1{,}685{,}549.36$$

Additional Examples to Review

Example 1, page 477; Example 2, page 478; Example 3, page 479

Definitions and Concepts

Calculating Present Value

If A dollars are to be accumulated in t years in an account that pays rate r compounded n times per year, then the present value, P, that needs to be invested now is given by

$$P = \frac{A}{\left(1 + \frac{r}{n}\right)^{nt}}.$$

Example

- How much should be deposited today in an account that earns 6.5% compounded monthly so that it will accumulate to $50,000 in ten years?

$$P = \frac{A}{\left(1 + \frac{r}{n}\right)^{nt}} = \frac{50{,}000}{\left(1 + \frac{0.065}{12}\right)^{12\cdot10}} \approx \$26{,}148.12$$

Rounding up to the nearest cent, approximately $26,148.12 should be invested today.

Additional Example to Review

Example 4, page 480

Definitions and Concepts

The effective annual yield is the simple interest rate that produces the same amount of money in an account at the end of one year as when the account is subjected to compound interest at a stated rate. The effective annual yield, Y, for an account that pays rate r compounded n times per year is given by

$$Y = \left(1 + \frac{r}{n}\right)^n - 1.$$

Example

Effective annual yield of an account paying 10.5% compounded monthly:

$$Y = \left(1 + \frac{r}{n}\right)^n - 1 = \left(1 + \frac{0.105}{12}\right)^{12} - 1 \approx 0.1102 = 11.02\%.$$

Money invested at 11.02% simple interest earns the same amount of interest in one year as when it is invested at 10.5% interest compounded monthly.

Additional Examples to Review

Example 5, page 481; Example 6, page 482

Summary of Personal Finance Interest Formulas

Simple Interest

$$I = Prt$$
$$A = P(1 + rt)$$

Compound Interest

$$A = P\left(1 + \frac{r}{n}\right)^{nt}$$

$$P = \frac{A}{\left(1 + \frac{r}{n}\right)^{nt}}$$

$$Y = \left(1 + \frac{r}{n}\right)^n - 1$$

Be sure you understand what each formula in the box describes and the meaning of the variables in the formulas. Select the appropriate formula or formulas as you work the exercises in the Review Exercises and the Chapter 7 Tests.

Review Exercises

Section 7.1 Percent, Sales Tax, and Discounts

In Exercises 1–3, express each fraction as a percent.

1. $\frac{4}{5}$ **2.** $\frac{1}{8}$ **3.** $\frac{3}{4}$

In Exercises 4–6, express each decimal as a percent.

4. 0.72   **5.** 0.0035   **6.** 4.756

In Exercises 7–12, express each percent as a decimal.

7. 65%   **8.** 99.7%   **9.** 150%

10. 3%   **11.** 0.65%   **12.** $\frac{1}{4}$%

13. What is 8% of 120?

14. Suppose that the local sales tax rate is 6% and you purchase a backpack for $24.

    a. How much tax is paid?

    b. What is the backpack's total cost?

15. A television with an original price of $850 is on sale at 35% off.

    a. What is the discount amount?

    b. What is the television's sale price?

16. A college that had 40 students for each lecture course increased the number to 45 students. What is the percent increase in the number of students in a lecture course?

17. A dictionary regularly sells for $56.00. The sale price is $36.40. Find the percent decrease of the sale price from the regular price.

18. Consider the following statement:

> My investment portfolio fell 10% last year, but then it rose 10% this year, so at least I recouped my losses.

Is this statement true? In particular, suppose you invested $10,000 in the stock market last year. How much money would be left in your portfolio with a 10% fall and then a 10% rise? If there is a loss, what is the percent decrease, to the nearest tenth of a percent, in your portfolio?

Section 7.2 Income Tax

In Exercises 19–20, find the gross income, the adjusted gross income, and the taxable income. In Exercise 20, base the taxable income on the greater of a standard deduction or an itemized deduction.

19. Your neighbor earned wages of $30,200, received $130 in interest from a savings account, and contributed $1100 to a tax-deferred retirement plan. He was entitled to a personal exemption of $3500 and had deductions totaling $5450.

20. Your neighbor earned wages of $86,400, won $350,000 on a television game show, and contributed $50,000 to a tax-deferred savings plan. She is entitled to a personal exemption of $3500 and a standard deduction of $5450. The interest on her home mortgage was $9200 and she contributed $95,000 to charity.

In Exercises 21–22, use the 2008 marginal tax rates in **Table 7.1** *on page 462 to compute the tax owed by each person or couple.*

21. A single woman with a taxable income of $600,000
22. A married couple filing jointly with a taxable income of $82,000 and a $7500 tax credit
23. Use the 2008 marginal tax rates in **Table 7.1** on page 462 to calculate the income tax owed by the following person:
    - Single, no dependents
    - Gross income: $40,000
    - $2500 paid to a tax-deferred IRA
    - $6500 mortgage interest
    - $1800 property taxes
    - No tax credits

For people who are not self-employed, the 2008 FICA tax rates were as follows:

- *7.65% on the first $102,000 from wages and tips*
- *1.45% on income in excess of $102,000.*

The individual's employer must also pay matching amounts of FICA taxes. People who are self-employed pay double the rates shown above. Use this information to solve Exercises 24–25.

24. If you are not self-employed and earn $86,000, what are your FICA taxes?
25. If you are self-employed and earn $260,000, what are your FICA taxes?
26. You decide to work part-time at a local clothing store. The job pays $8.50 per hour and you work 16 hours per week. Your employer withholds 10% of your gross pay for federal taxes, 7.65% for FICA taxes, and 4% for state taxes.
    a. What is your weekly gross pay?
    b. How much is withheld per week for federal taxes?
    c. How much is withheld per week for FICA taxes?
    d. How much is withheld per week for state taxes?
    e. What is your weekly net pay?
    f. What percentage of your gross pay is withheld for taxes? Round to the nearest tenth of a percent.

Section 7.3 Simple Interest

In Exercises 27–30, find the simple interest. (Assume 360 days in a year.)

	Principal	Rate	Time
27.	$6000	3%	1 year
28.	$8400	5%	6 years
29.	$20,000	8%	9 months
30.	$36,000	15%	60 days

31. In order to pay for tuition and books, a college student borrows $3500 for four months at 10.5% interest.
    a. How much interest must the student pay?
    b. Find the future value of the loan.

In Exercises 32–34, use the formula for future value with simple interest to find the missing quantity. Round dollar amounts to the nearest cent and rates to the nearest tenth of a percent.

32. $A = ?, P = \$12,000, r = 8.2\%, t = 9$ months
33. $A = \$5750, P = \$5000, r = ?, t = 2$ years
34. $A = \$16,000, P = ?, r = 6.5\%, t = 3$ years
35. You plan to buy a $12,000 sailboat in four years. How much should you invest now, at 7.3% simple interest, to have enough for the boat in four years? (Round up to the nearest cent.)
36. You borrow $1500 from a friend and promise to pay back $1800 in six months. What simple interest rate will you pay?
37. You borrow $1800 on a 7% discounted loan for a period of 9 months.
    a. What is the loan's discount?
    b. Determine the net amount of money you will receive.
    c. What is the loan's actual interest rate, to the nearest tenth of a percent?

Section 7.4 Compound Interest

In Exercises 38–40, the principal represents an amount of money deposited in a savings account that provides the lender compound interest at the given rate.

a. *Find how much money, to the nearest cent, there will be in the account after the given number of years.*

b. *Find the interest earned.*

	Principal	Rate	Compounding Periods per Year	Time
38.	$7000	3%	1	5 years
39.	$30,000	2.5%	4	10 years
40.	$2500	4%	12	20 years

41. Suppose that you have $14,000 to invest. Which investment yields the greater return over 10 years: 7% compounded monthly or 6.85% compounded continuously? How much more (to the nearest dollar) is yielded by the better investment?

In Exercises 42–43, round answers up to the nearest cent.

42. How much money should parents deposit today in an account that earns 7% compounded monthly so that it will accumulate to $100,000 in 18 years for their child's college education?
43. How much money should be deposited today in an account that earns 5% compounded quarterly so that it will accumulate to $75,000 in 35 years for retirement?

44. You deposit $2000 in an account that pays 6% interest compounded quarterly.

   a. Find the future value, to the nearest cent, after one year.

   b. Use the future value formula for simple interest to determine the effective annual yield. Round to the nearest tenth of a percent.

45. What is the effective annual yield, to the nearest hundredth of a percent, of an account paying 5.5% compounded quarterly? What does your answer mean?

46. Which investment is the better choice: 6.25% compounded monthly or 6.3% compounded annually?

Chapter 7 Test A

The box on page 493 summarizes the personal finance formulas involving interest. Where applicable, use the appropriate formula to solve an exercise in this test. Unless otherwise stated, round dollar amounts to the nearest cent and rates to the nearest tenth of a percent.

1. A CD player with an original price of $120 is on sale at 15% off.

   a. What is the amount of the discount?

   b. What is the sale price of the CD player?

2. You purchased shares of stock for $2000 and sold them for $3500. Find the percent increase, or your return, on this investment.

3. You earned wages of $46,500, received $790 in interest from a savings account, and contributed $1100 to a tax-deferred savings plan. You are entitled to a personal exemption of $3500 and a standard deduction of $5450. The interest on your home mortgage was $7300, you contributed $350 to charity, and you paid $1395 in state taxes.

   a. Find your gross income.

   b. Find your adjusted gross income.

   c. Find your taxable income. Base your taxable income on the greater of the standard deduction or an itemized deduction.

4. Use the 2008 marginal tax rates in **Table 7.1** on page 462 to calculate the federal income tax owed by the following person:

- Single, no dependents
- Gross income: $36,500
- $2000 paid to a tax-deferred IRA
- $4700 mortgage interest
- $1300 property taxes
- No tax credits

5. Use FICA tax rates for people who are not self-employed, 7.65% on the first $102,000 of income and 1.45% on income in excess of $102,000, to answer this question: If a person is not self-employed and earns $112,000, what are that person's FICA taxes?

6. You decide to work part-time at a local stationery store. The job pays $10 per hour and you work 15 hours per week. Your employer withholds 10% of your gross pay for federal taxes, 7.65% for FICA taxes, and 3% for state taxes.

   a. What is your weekly gross pay?

   b. How much is withheld per week for federal taxes?

   c. How much is withheld per week for FICA taxes?

   d. How much is withheld per week for state taxes?

   e. What is your weekly net pay?

   f. What percentage of your gross pay is withheld for taxes? Round to the nearest tenth of a percent.

7. You borrow $2400 for three months at 12% simple interest. Find the amount of interest paid and the future value of the loan.

8. You borrow $2000 from a friend and promise to pay back $3000 in two years. What simple interest rate will you pay?

9. In six months, your parents want to have $7000 worth of remodeling done to their home. How much should they invest now, at 9% simple interest, to have enough money for the project? (Round up to the nearest cent.)

10. Find the effective annual yield, to the nearest hundredth of a percent, of an account paying 4.5% compounded quarterly. What does your answer mean?

11. You receive an inheritance of $20,000 and invest it in an account that pays 6.5% compounded monthly.

   a. How much, to the nearest dollar, will you have after 40 years?

   b. Find the interest.

12. You would like to have $3000 in four years for a special vacation by making a lump-sum investment in an account that pays 9.5% compounded semiannually. How much should you deposit now? Round up to the nearest dollar.

Chapter 7 Test B

The box on page 493 summarizes the personal finance formulas involving interest. Where applicable, use the appropriate formula to solve an exercise in this test. Unless otherwise stated, round dollar amounts to the nearest cent and rates to the nearest tenth of a percent.

1. A computer with an original price of $1400 is on sale at 45% off.
   a. What is the amount of the discount?
   b. What is the sale price of the computer?

2. You purchased shares of stock for $500 and sold them for $850. Find the percent increase, or your return, on this investment.

3. You earned wages of $61,400, received $6500 in winnings on a television game show, and contributed $4300 to a tax-deferred savings plan. You are entitled to a personal exemption of $3500 and a standard deduction of $5450. The interest on your home mortgage was $8400, you contributed $3750 to charity, and you paid $3280 in state taxes.
   a. Find your gross income.
   b. Find your adjusted gross income.
   c. Find your taxable income. Base your taxable income on the greater of the standard deduction or an itemized deduction.

4. Use the 2008 marginal tax rates in **Table 7.1** on page 462 to calculate the federal income tax owed by the following person:
   • Single, no dependents
   • Gross income: $48,600
   • $3600 paid to a tax-deferred IRA
   • $4300 mortgage interest
   • $1200 property taxes
   • $1000 in tax credits

5. Use FICA tax rates for people who are not self-employed, 7.65% on the first $102,000 of income and 1.45% on income in excess of $102,000, to answer this question: If a person is not self-employed and earns $212,000, what are that person's FICA taxes?

6. You decide to work part-time at a local gift shop. The job pays $9.50 per hour and you work 16 hours per week. Your employer withholds 10% of your gross pay for federal taxes, 7.65% for FICA taxes, and 4% for state taxes.
   a. What is your weekly gross pay?
   b. How much is withheld per week for federal taxes?
   c. How much is withheld per week for FICA taxes?
   d. How much is withheld per week for state taxes?
   e. What is your weekly net pay?
   f. What percentage of your gross pay is withheld for taxes? Round to the nearest tenth of a percent.

7. You borrow $3600 for four months at 15% simple interest. Find the amount of interest paid and the future value of the loan.

8. You borrow $3000 from a friend and promise to pay back $4200 in two years. What simple interest rate will you pay?

9. In six months, you want to save $1200 for a guitar. How much should you invest now, at 12% simple interest, to have enough money for the purchase? (Round up to the nearest cent.)

10. Find the effective annual yield, to the nearest hundredth of a percent, of an account paying 3.5% compounded quarterly. What does your answer mean?

11. You receive an inheritance of $25,000 and invest it in an account that pays 7.5% compounded monthly.
    a. How much, to the nearest dollar, will you have after 40 years?
    b. Find the interest.

12. You would like to have $3500 in five years for a special college graduation vacation by making a lump-sum investment in an account that pays 10.5% compounded semiannually. How much should you deposit now? Round up to the nearest dollar.

Personal Finance: Saving, Investing, and Spending

8

It's a heck of a lot more fun to spend than to save. Of all the lessons taught by America's recent economic crisis, the worst since the Great Depression of the 1930s, the most personal has been that we are not too slick with money. We take out car loans and mortgages we can't afford. We run sky-high credit-card debt. We don't save nearly enough for retirement.

Where do high school students fall on the "money slickness" scale? According to a survey from the Investor Education Foundation, teens display much lower financial literacy than older generations. U.S. Education Secretary Arne Duncan laments, "The reality is that young people don't know the basics of saving and investing. It's a skill they need to be successful in our economy."

The purpose of this chapter is to provide you with an understanding of saving, investing, and the benefits and risks associated with spending. Understanding these topics and the underlying mathematics will enable you to make smarter decisions about money, preparing you for a successful financial future.

8.1 Annuities, Methods of Saving, and Investments

Objectives

1. Determine the value of an annuity.

2. Determine regular annuity payments needed to achieve a financial goal.

3. Understand types of bank accounts.

4. Understand stocks and bonds as investments.

5. Read stock tables.

6. Understand accounts designed for retirement savings.

According to the *Forbes Billionaires List*, in 2009 the two richest Americans were Bill Gates (net worth: $40 billion) and Warren Buffett (net worth: $37 billion). In May 1965, Buffett's new company, Berkshire Hathaway, was selling one share of stock for $18. By the end of 2008, the price of a share had increased to $96,600. If you had purchased one share in May 1965, your **return**, or percent increase, would be

$$\frac{\text{amount of increase}}{\text{original amount}} = \frac{\$96,600 - \$18}{\$18} \approx 5365.67 = 536,567\%.$$

What does a return of nearly 540,000% mean? If you had invested $250 in Warren Buffett's company in May 1965, your shares would have been worth over $1.3 million by December 2008.

Of course, investments that potentially offer outrageous returns come with great risk of losing part or all of the principal. The bottom line: Is there a safe way to save regularly and have an investment worth one million dollars or more? In this section, we consider such savings plans, some of which come with special tax treatment, as well as riskier investments in stocks and bonds.

Annuities

1 Determine the value of an annuity.

The compound interest formula

$$A = P(1 + r)^t$$

gives the future value, A, after t years, when a fixed amount of money, P, the principal, is deposited in an account that pays an annual interest rate r (in decimal form) compounded once a year. However, money is often invested in small amounts at periodic intervals. For example, to save for retirement, you might decide to place $1000 into an Individual Retirement Account (IRA) at the end of each year until you retire. An **annuity** is a sequence of equal payments made at equal time periods. An IRA is an example of an annuity.

The **value of an annuity** is the sum of all deposits plus all interest paid. Our first example illustrates how to find this value.

Example 1   Determining the Value of an Annuity

You deposit $1000 into a savings plan at the end of each year for three years. The interest rate is 8% per year compounded annually.

a. Find the value of the annuity after three years.

b. Find the interest.

Solution

a. The value of the annuity after three years is the sum of all deposits made plus all interest paid over three years.

> This is the $1000 deposit at year's end.

Value at end of year 1   = $1000

> This is the first-year deposit with interest earned for a year.
> This is the $1000 deposit at year's end.

Value at end of year 2   = $1000(1 + 0.08)   + $1000

= $1080   + $1000   = $2080

> Use $A = P(1 + r)^t$ with $r = 0.08$ and $t = 1$, or $A = P(1 + 0.08)$.
> This is the second-year balance, $2080, with interest earned for a year.
> This is the $1000 deposit at year's end.

Value at end of year 3   = $2080(1 + 0.08)   + $1000

= $2246.40 + $1000   = $3246.40

The value of the annuity at the end of three years is $3246.40.

b. You made three payments of $1000 each, depositing a total of 3 × $1000, or $3000. Because the value of the annuity is $3246.40, the interest is $3246.40 − $3000, or $246.40.

Checkpoint 1   You deposit $2000 into a savings plan at the end of each year for three years. The interest rate is 10% per year compounded annually.

a. Find the value of the annuity after three years.

▶ **b.** Find the interest.

Suppose that you deposit P dollars into an account at the end of each year. The account pays an annual interest rate, r, compounded annually. At the end of the first year, the account contains P dollars. At the end of the second year, P dollars is deposited again. At the time of this deposit, the first deposit has received interest earned during the second year. Thus, the value of the annuity after two years is

$$P + P(1 + r).$$

> Deposit of P dollars at end of second year
> First-year deposit of P dollars with interest earned for a year

The value of the annuity after three years is

$$P + P(1 + r) + P(1 + r)^2.$$

> Deposit of P dollars at end of third year
> Second-year deposit of P dollars with interest earned for a year
> First-year deposit of P dollars with interest earned over two years

The value of the annuity after t years is

$$P + P(1 + r) + P(1 + r)^2 + P(1 + r)^3 + \cdots + P(1 + r)^{t-1}.$$

Deposit of P dollars at end of year t

First-year deposit of P dollars with interest earned over $t - 1$ years

Each term in this sum is obtained by multiplying the preceding term by $(1 + r)$. Thus, the terms form a geometric sequence. Using a formula for the sum of the terms of a geometric sequence, we can obtain the following formula that gives the value of this annuity:

Value of an Annuity: Interest Compounded Once a Year

If P is the deposit made at the end of each year for an annuity that pays an annual interest rate r (in decimal form) compounded once a year, the value, A, of the annuity after t years is

$$A = \frac{P[(1 + r)^t - 1]}{r}.$$

> Although you are a long way from retirement, the time to begin retirement savings is when you begin earning a paycheck and can take advantage of the time value of your money.

Example 2 Determining the Value of an Annuity

Suppose that when you are 35, you decide to save for retirement by depositing $1000 into an IRA at the end of each year for 30 years. If you can count on an interest rate of 10% per year compounded annually,

a. How much will you have from the IRA after 30 years?

b. Find the interest.

Round answers to the nearest dollar.

Solution

a. The amount that you will have from the IRA is its value after 30 years.

$$A = \frac{P[(1 + r)^t - 1]}{r}$$

> Use the formula for the value of an annuity.

$$A = \frac{1000[(1 + 0.10)^{30} - 1]}{0.10}$$

> The annuity involves year-end deposits of $1000: $P = 1000$. The interest rate is 10%: $r = 0.10$. The number of years is 30: $t = 30$.
> The Technology box shows how this computation can be done in a single step using parentheses keys.

$$= \frac{1000[(1.10)^{30} - 1]}{0.10}$$

> Add inside parentheses: $1 + 0.10 = 1.10$.

$$\approx \frac{1000(17.4494 - 1)}{0.10}$$

> Use a calculator to find $(1.10)^{30}$: $1.1 \boxed{y^x} \ 30 \ \boxed{=}$.

$$= \frac{1000(16.4494)}{0.10}$$

> Simplify inside parentheses: $17.4494 - 1 = 16.4494$.

$$= 164,494$$

> Use a calculator: $1000 \ \boxed{\times} \ 16.4494 \ \boxed{\div} \ .10 \ \boxed{=}$.

After 30 years, you will have approximately $164,494 from the IRA.

b. You made 30 payments of $1000 each, depositing a total of 30 × $1000, or $30,000. Because the value of the annuity is approximately $164,494, the interest is approximately

$$\$164{,}494 - \$30{,}000, \text{ or } \$134{,}494.$$

The interest is nearly $4\frac{1}{2}$ times the amount of your payments, illustrating the power of compounding.

✔ **Checkpoint 2** Suppose that when you are 25, you deposit $3000 into an IRA at the end of each year for 40 years. If you can count on an interest rate of 8% per year compounded annually,

a. How much will you have from the IRA after 40 years?

b. Find the interest.

▶ Round answers to the nearest dollar.

We can adjust the formula for the value of an annuity if equal payments are made at the end of each of *n* yearly compounding periods.

> **Value of an Annuity: Interest Compounded *n* Times per Year**
>
> If *P* is the deposit made at the end of each compounding period for an annuity that pays an annual interest rate *r* (in decimal form) compounded *n* times per year, the value *A*, of the annuity after *t* years is
>
> $$A = \dfrac{P\left[\left(1 + \dfrac{r}{n}\right)^{nt} - 1\right]}{\left(\dfrac{r}{n}\right)}.$$

Example 3 **Determining the Value of an Annuity**

At age 25, to save for retirement, you decide to deposit $200 at the end of each month into an IRA that pays 7.5% compounded monthly.

a. How much will you have from the IRA when you retire at age 65?

b. Find the interest.

Round answers to the nearest dollar.

Solution

a. Because you are 25, the amount that you will have from the IRA when you retire at 65 is its value after 40 years.

$$A = \dfrac{P\left[\left(1 + \dfrac{r}{n}\right)^{nt} - 1\right]}{\left(\dfrac{r}{n}\right)} \qquad \text{Use the formula for the value of an annuity.}$$

$$A = \frac{P\left[\left(1 + \dfrac{r}{n}\right)^{nt} - 1\right]}{\left(\dfrac{r}{n}\right)}$$

The formula for the value of an annuity (repeated)

$$A = \frac{200\left[\left(1 + \dfrac{0.075}{12}\right)^{12 \cdot 40} - 1\right]}{\left(\dfrac{0.075}{12}\right)}$$

The annuity involves month-end deposits of $200: $P = 200$. The interest rate is 7.5%: $r = 0.075$. The interest is compounded monthly: $n = 12$. The number of years is 40: $t = 40$.

$$= \frac{200\left[(1 + 0.00625)^{480} - 1\right]}{0.00625}$$

Using parentheses keys, these calculations can be performed in a single step on a calculator.

$$= \frac{200\left[(1.00625)^{480} - 1\right]}{0.00625}$$

Add inside parentheses:
$1 + 0.00625 = 1.00625$.

$$\approx \frac{200(19.8989 - 1)}{0.00625}$$

Use a calculator to find $(1.00625)^{480}$:
$1.00625 \boxed{y^x} \, 480 \boxed{=}$.

$$\approx 604{,}765$$

After 40 years, you will have approximately $604,765 when retiring at age 65.

b. Interest = Value of the IRA − Total deposits
$$\approx \$604{,}765 - \$200 \cdot 12 \cdot 40$$

$200 per month × 12 months per year × 40 years

$$= \$604{,}765 - \$96{,}000 = \$508{,}765$$

The interest is approximately $508,765, more than five times the amount of your contributions to the IRA.

Annuities can be categorized by when payments are made. The formula used to solve Example 3 describes **ordinary annuities**, where payments are made at the end of each period. The formula assumes the same number of yearly payments and yearly compounding periods. An annuity plan in which payments are made at the beginning of each period is called an **annuity due**. The formula for the value of this type of annuity is slightly different than the one used in Example 3.

> **Checkpoint 3**   At age 30, to save for retirement, you decide to deposit $100 at the end of each month into an IRA that pays 9.5% compounded monthly.
>
> **a.** How much will you have from the IRA when you retire at age 65?
>
> **b.** Find the interest.
>
> Round answers to the nearest dollar.

Planning for the Future with an Annuity

2 Determine regular annuity payments needed to achieve a financial goal.

By solving the annuity formula for P, we can determine the amount of money that should be deposited at the end of each compounding period so that an annuity has a future value of A dollars. The following formula gives the regular payments, P, needed to reach a financial goal, A:

> **Regular Payments Needed to Achieve a Financial Goal**
>
> The deposit, P, that must be made at the end of each compounding period into an annuity that pays an annual interest rate r (in decimal form) compounded n times per year in order to achieve a value of A dollars after t years is
>
> $$P = \frac{A\left(\dfrac{r}{n}\right)}{\left[\left(1 + \dfrac{r}{n}\right)^{nt} - 1\right]}.$$

When computing regular payments needed to achieve a financial goal, round the deposit made at the end of each compounding period *up*. In this way, you won't fall slightly short of being able to meet future goals. In this section, we will round annuity payments up to the nearest dollar.

Example 4   Using Long-Term Planning to Achieve a Financial Goal

Suppose that once you complete your post–high school education and begin working, you save $20,000 over five years to use as a down payment for a home. You anticipate making regular, end-of-month deposits in an annuity that pays 6% compounded monthly.

  a. How much should you deposit each month? Round up to the nearest dollar.

  b. How much of the $20,000 down payment comes from deposits and how much comes from interest?

Solution

a.
$$P = \dfrac{A\left(\dfrac{r}{n}\right)}{\left[\left(1 + \dfrac{r}{n}\right)^{nt} - 1\right]}$$

Use the formula for regular payments, P, needed to achieve a financial goal, A.

$$P = \dfrac{20{,}000\left(\dfrac{0.06}{12}\right)}{\left[\left(1 + \dfrac{0.06}{12}\right)^{12\cdot5} - 1\right]}$$

Your goal is to accumulate $20,000 (A = 20,000) over five years (t = 5). The interest rate is 6% (r = 0.06) compounded monthly (n = 12).

$$\approx 287$$

Use a calculator and round up to the nearest dollar to be certain you do not fall short of your goal.

You should deposit $287 each month to be certain of having $20,000 for a down payment on a home.

b. Total deposits = $287 · 12 · 5 = $17,220

> $287 per month × 12 months per year × 5 years

      Interest = $20,000 − $17,220 = $2780

We see that $17,220 of the $20,000 comes from your deposits and the remainder, $2780, comes from interest.

Study Tip

In Example 4 of Section 7.4, we saw that a lump-sum deposit of approximately $14,828 at 6% compounded monthly would accumulate to $20,000 in five years. In Example 4 on the right, we see that total deposits of $17,220 are required to reach the same goal. With the same interest rate, compounding period, and time period, a lump-sum deposit will generate more interest than an annuity. If you don't have a large sum of money to open an account, an annuity is a realistic, although more expensive, option to a lump-sum deposit for achieving the same financial goal.

✓ **Checkpoint 4**   Parents of a baby girl are in a financial position to begin saving for her college education. They plan to have $100,000 in a college fund in 18 years by making regular, end-of-month deposits in an annuity that pays 9% compounded monthly.

  a. How much should they deposit each month? Round up to the nearest dollar.

  b. How much of the $100,000 college fund comes from deposits and how much comes from interest?

Methods of Saving and Investing

Financial planning is not something that comes naturally to most people. Although there is no escaping the need to save money, it's always easier to spend than to save. For much of your life, you will face the task of building up your wealth and managing your money. Placing money in various savings options and investments will be critical to success in achieving your financial goals. In the remainder of this section, you will learn about some of the options for saving and investing money, including bank accounts, stocks, bonds, and accounts designed specifically for retirement savings.

Bank Accounts

3 Understand types of bank accounts.

Banks offer a variety of accounts and services. Accounts up to $250,000 are insured by the federal government, so there is no risk of losing the principal you've invested. Because these accounts carry no risk and allow you to take money out on demand, they have the lowest interest rates of all methods of saving. Here are some of the different types of bank accounts and their features:

- **Basic checking accounts** allow immediate access to your money. You can write checks or use an ATM. Basic checking accounts do not pay interest and cannot be used to save money.

- **NOW (Negotiable Order of Withdrawal) accounts** are similar to basic checking accounts with one difference: They pay interest. You are required to maintain a minimum monthly balance, with penalties if your account falls below this amount. Interest rates can be as low as 0.3% compounded annually, so NOW accounts are not a good way to save money.

- **Savings accounts** have low interest rates (slightly higher than NOW accounts) and do not provide check-writing services. They are often used to save money for specific short-term goals, such as purchasing holiday gifts.

- **Money-market accounts** combine the features of checking and savings accounts. They require a minimum balance, pay interest (slightly higher than NOW accounts), and offer limited check-writing privileges. A typical money-market account might require a $2500 minimum balance and allow five checks per month. Fees are imposed if the balance falls below the minimum or if more checks are written than permitted.

- **CDs (Certificates of Deposit)** are accounts that pay fixed interest rates if you keep the principal in the account for a specified amount of time. The most common investment periods for CDs are six months, one year, two years, three years, four years, or five years. If you take the money out of the account before the specified amount of time has expired, you are charged a penalty, which can be quite hefty. You lose not only whatever interest you would have earned, but part of your principal as well. Interest rates vary, but most CDs pay more than savings accounts or money-market accounts. As of April 2010, interest rates were 0.44% on a six-month CD, 0.72% on a one-year CD, and 2.12% on a five-year CD.

Investments: Risk and Return

4 Understand stocks and bonds as investments.

When you deposit money into a bank account, you are making a **cash investment**. Because bank accounts up to $250,000 are insured by the federal government, there is no risk of losing the principal you've invested. The account's interest rate guarantees a certain percent increase in your investment, called its **return**.

All investments involve a tradeoff between risk and return. The different types of bank accounts carry little or no risk, so investors must be willing to accept low returns. There are other kinds of investments that are riskier, meaning that it is possible to lose all or part of your principal. These investments, including *stocks* and *bonds*, give a reasonable expectation of higher returns to attract investors.

Stocks

Investors purchase **stock**, shares of ownership in a company. The shares indicate the percent of ownership. For example, if a company has issued a total of one million shares and an investor owns 20,000 of these shares, that investor owns

$$\frac{20{,}000 \text{ shares}}{1{,}000{,}000 \text{ shares}} = 0.02$$

or 2% of the company. Any investor who owns some percentage of the company is called a **shareholder**.

Buying or selling stock is referred to as **trading**. Shares of stock need both a seller and a buyer to be traded. Stocks are traded on a **stock exchange**. The price of a share of stock is determined by the law of supply and demand. If a company is prospering, investors will be willing to pay a good price for its stock, and so the stock price goes up. If the company does not do well, investors may decide to sell, and the stock price goes down. Stock prices indicate the performance of the companies they represent, as well as the state of the national and global economies.

There are two ways to make money by investing in stock:

- You sell the shares for more money than what you paid for them, in which case you have a **capital gain** on the sale of stock. (There can also be a capital loss by selling for less than what you paid, or if the company goes bankrupt.)

- While you own the stock, the company distributes all or part of its profits to shareholders as **dividends**. Each share is paid the same dividend, so the amount you receive depends on the number of shares owned. (Some companies reinvest all profits and do not distribute dividends.)

When more and more average Americans began investing and making money in stocks in the 1990s, the federal government cut the capital-gains tax rate. Long-term capital gains (profits on items held for more than a year before being sold) and dividends are taxed at lower rates than wages and interest earnings.

Bonds

People who buy stock become part owners in a company. In order to raise money and not dilute the ownership of current stockholders, companies sell **bonds**. People who buy a bond are **lending money** to the company from which they buy the bond. Bonds are a commitment from a company to pay the price an investor pays for the bond at the time it was purchased, called the **face value**, along with interest payments at a given rate.

There are many reasons for issuing bonds. A company might need to raise money for research on a drug that has the potential for curing AIDS, so it issues bonds. The U.S. Treasury Department issues 30-year bonds at a fixed 7% annual rate to borrow money to cover possible federal deficits. Local governments often issue bonds to borrow money to build schools, parks, and libraries.

Bonds are traded like stock, and their price is a function of supply and demand. If a company goes bankrupt, bondholders are the first to claim the company's assets. They make their claims before the stockholders, even though (unlike stockholders) they do not own a share of the company. Generally speaking, investing in bonds is less risky than investing in stocks, although the return is lower.

Mutual Funds

It is not an easy job to determine which stocks and bonds to buy or sell, or when to do so. Even IRAs can be funded by mixing stocks and bonds. Many small investors have decided that they do not have the time to stay informed about the progress of corporations, even with the help of online industry research. Instead, they invest in a **mutual fund**. A mutual fund is a group of stocks and/or bonds managed by a professional investor. When you purchase shares in a mutual fund, you give your money to the **fund manager**. Your money is combined with the money of other investors in the mutual fund. The fund manager invests this pool of money, buying and selling shares of stocks and bonds to obtain the maximum possible returns.

Investors in mutual funds own a small portion of many different companies, which may protect them against the poor performance of a single company. When comparing mutual funds, consider both the fees charged for investing and how well the fund manager is doing with the fund's money. Newspapers publish ratings from 1 (worst) to 5 (best) of mutual fund performance based on whether the manager is doing a good job with the investors' money. Two numbers are given. The first number compares the performance of the mutual fund to a large group of similar funds. The second number compares the performance to funds that are nearly identical. The best rating a fund manager can receive is 5/5; the worst is 1/1.

A listing of all the investments that a person holds is called a **financial portfolio**. Most financial advisors recommend a portfolio with a mixture of low-risk and high-risk investments, called a **diversified portfolio**.

Blitzer Bonus

Risk

The Chinese define *risk* as a combination of danger and opportunity. Investments in which all or part of the principal is in danger offer the opportunity of a greater return. **Table 8.1** includes four types of investments, listed from high risk to low risk, and their average annual returns from 1926 through 2006. Low-risk cash investments barely keep up with inflation, which constantly erodes the buying power of our money. After 20 years, a dollar's value is cut in half even with a seemingly low inflation rate of 3%.

Danger   Opportunity

Once you pay taxes on the interest earned from no-risk cash investments such as CDs and money-market accounts, inflation causes you to inevitably fall behind.

Table 8.1 Investments from High Risk to No Risk and Their Returns: 1926–2006	
Investment Type	**Average Annual Return**
Small-Company Stocks	12.7%
Large-Company Stocks	10.4%
Bonds	5.4%
Cash	3.7%
Average Annual Inflation Rate from 1926–2006: 3%	

Source: Ibbotson Associates

Investing in small-company stocks has the greatest average annual return, but also carries considerable risk. The value of your investment can plunge or even disappear if the company that issues the shares goes bankrupt.

The bottom line: Be weary of low-risk, high-return investment options. If someone promises you an investment that seems too good to be true, it probably is.

Reading Stock Tables

5 Read stock tables.

Daily newspapers and online services give current stock prices and other information about stocks. We will use FedEx (Federal Express) stock to learn how to read these daily stock tables. Look at the following newspaper listing of FedEx stock.

52-Week High	52-Week Low	Stock	SYM	Div	Yld %	PE	Vol 100s	Hi	Lo	Close	Net Chg
99.46	34.02	FedEx	FDX	.44	1.0	19	37701	45	43.47	44.08	−1.60

The headings indicate the meanings of the numbers across the row.

52-Week High
99.46

The heading **52-Week High** refers to the *highest price* at which FedEx stock traded during the past 52 weeks. The highest price was $99.46 per share. This means that during the past 52 weeks at least one investor was willing to pay $99.46 for a share of FedEx stock. Notice that 99.46 represents a quantity in dollars, although the stock table does not show the dollar sign.

52-Week Low
34.02

The heading **52-Week Low** refers to the *lowest price* at which FedEx stock traded during the past 52 weeks. This price was $34.02.

Stock	SYM
FedEx	FDX

The heading **Stock** is the *company name*, FedEx. The heading **SYM** is the *symbol* the company uses for trading. FedEx uses the symbol FDX.

Div
.44

The heading **Div** refers to *dividends* paid per share to stockholders during the past year. FedEx paid a dividend of $0.44 per share. Once again, the dollar symbol does not appear in the table. Thus, if you owned 100 shares, you received a dividend of $0.44 × 100, or $44.00.

Yld %
1.0

The heading **Yld %** stands for *percent yield*. In this case, the percent yield is 1.0%. (The stock table does not show the percent sign.) This means that the dividends alone gave investors an annual return of 1.0%. This is much lower than the average inflation rate. However, this percent does not take into account the fact that FedEx stock prices might rise. If an investor sells shares for more than the purchase price, the gain will probably make FedEx stock a much better investment than a bank account.

In order to understand the meaning of the heading PE, we need to understand some of the other numbers in the table. We will return to this column.

Vol 100s
37701

The heading **Vol 100s** stands for *sales volume in hundreds*. This is the number of shares traded yesterday, in hundreds. The number in the table is 37,701. This means that yesterday, a total of 37,701 × 100, or 3,770,100 shares of FedEx were traded.

Hi
45

The heading **Hi** stands for the *highest price* at which FedEx stock traded *yesterday*. This number is 45. Yesterday, FedEx's highest trading price was $45 a share.

Lo
43.47

The heading **Lo** stands for the *lowest price* at which FedEx stock traded *yesterday*. This number is 43.47. Yesterday, FedEx's lowest trading price was $43.47 a share.

Close
44.08

The heading **Close** stands for the *price* at which shares last traded *when the stock exchange closed yesterday*. This number is 44.08. Thus, the price at which shares of FedEx traded when the stock exchange closed yesterday was $44.08 per share. This is called yesterday's **closing price**.

Net Chg
−1.60

The heading **Net Chg** stands for *net change*. This is the change in price from the market close two days ago to yesterday's market close. This number is −1.60. Thus, the price of a share of FedEx stock went down by $1.60. For some stock listings, the notation ... appears under Net Chg. This means that there was *no change in price* for a share of stock from the market close two days ago to yesterday's market close.

PE
19

Now, we are ready to return to the heading **PE**, standing for the *price-to-earnings ratio*.

$$\text{PE ratio} = \frac{\text{Yesterday's closing price per share}}{\text{Annual earnings per share}}$$

This can also be expressed as

$$\text{Annual earnings per share} = \frac{\text{Yesterday's closing price per share}}{\text{PE ratio}}.$$

The PE ratio for FedEx is given to be 19. Yesterday's closing price per share was 44.08. We can substitute these numbers into the formula to find annual earnings per share:

Close
44.08

PE
19

$$\text{Annual earnings per share} = \frac{44.08}{19} = 2.32.$$

The annual earnings per share for FedEx were $2.32. The PE ratio, 19, tells us that yesterday's closing price per share, $44.08, is 19 times greater than the earnings per share, $2.32.

Example 5 Reading Stock Tables

52-Week High	52-Week Low	Stock	SYM	Div	Yld %	PE	Vol 100s	Hi	Lo	Close	Net Chg
42.38	22.50	Disney	DIS	.21	.6	43	115900	32.50	31.25	32.50	...

Use the stock table for Disney to answer the following questions.

a. What were the high and low prices for the past 52 weeks?

b. If you owned 3000 shares of Disney stock last year, what dividend did you receive?

c. What is the annual return for dividends alone? How does this compare to a bank account offering a 3.5% interest rate?

d. How many shares of Disney were traded yesterday?

e. What were the high and low prices for Disney shares yesterday?

f. What was the price at which Disney shares last traded when the stock exchange closed yesterday?

g. What does the value or symbol in the net change column mean?

h. Compute Disney's annual earnings per share using

$$\text{Annual earnings per share} = \frac{\text{Yesterday's closing price per share}}{\text{PE ratio}}.$$

Solution

a. We find the high price for the past 52 weeks by looking under the heading **High**. The price is listed in dollars, given as 42.38. Thus, the high price for a share of stock for the past 52 weeks was $42.38. We find the low price for the past 52 weeks by looking under the heading **Low**. This price is also listed in dollars, given as 22.50. Thus, the low price for a share of Disney stock for the past 52 weeks was $22.50.

b. We find the dividend paid for a share of Disney stock last year by looking under the heading **Div**. The price is listed in dollars, given as .21. Thus, Disney paid a dividend of $0.21 per share to stockholders last year. If you owned 3000 shares, you received a dividend of $0.21 × 3000, or $630.

c. We find the annual return for dividends alone by looking under the heading **Yld %**, standing for percent yield. The number in the table, .6, is a percent. This means that the dividends alone gave Disney investors an annual return of 0.6%. This is much lower than a bank account paying a 3.5% interest rate. However, if Disney shares increase in value, the gain might make Disney stock a better investment than the bank account.

d. We find the number of shares of Disney traded yesterday by looking under the heading **Vol 100s**, standing for sales volume in hundreds. The number in the table is 115900. This means that yesterday, a total of 115,900 × 100, or 11,590,000 shares, were traded.

e. We find the high and low prices for Disney shares yesterday by looking under the headings **Hi** and **Lo**. Both prices are listed in dollars, given as 32.50 and 31.25. Thus, the high and low prices for Disney shares yesterday were $32.50 and $31.25, respectively.

f. We find the price at which Disney shares last traded when the stock exchange closed yesterday by looking under the heading **Close**. The price is listed in dollars, given as 32.50. Thus, when the stock exchange closed yesterday, the price of a share of Disney stock was $32.50.

g. The ... under **Net Chg** means that there was no change in price in Disney stock from the market close two days ago to yesterday's market close. In part (f), we found that the price of a share of Disney stock at yesterday's close was $32.50, so the price at the market close two days ago was also $32.50.

h. We are now ready to use

$$\text{Annual earnings per share} = \frac{\text{Yesterday's closing price per share}}{\text{PE ratio}}$$

to compute Disney's annual earnings per share. We found that yesterday's closing price per share was $32.50. We find the PE ratio under the heading **PE**. The given number is 43. Thus,

$$\text{Annual earnings per share} = \frac{\$32.50}{43} \approx \$0.76.$$

The annual earnings per share for Disney were $0.76. The PE ratio, 43, tells us that yesterday's closing price per share, $32.50, is 43 times greater than the earnings per share, approximately $0.76.

Checkpoint 5 Use the stock table for Coca Cola to solve parts (a) through (h) in Example 5 for Coca Cola.

52-Week High	52-Week Low	Stock	SYM	Div	Yld %	PE	Vol 100s	Hi	Lo	Close	Net Chg
63.38	42.37	Coca Cola	CocaCl	.72	1.5	37	72032	49.94	48.33	49.50	+0.03

Achieving $uccess

Here are some investment suggestions from financial advisors:

- Do not invest money in the stock market that you will need within ten years. Government bonds, CDs, and money-market accounts are more appropriate options for short-term goals.

- If you are *a* years old, approximately (100 − *a*)% of your investments should be in stocks. For example, at age 25, approximately (100 − 25)%, or 75%, of your portfolio should be invested in stocks.

- Diversify your investments. Invest in a variety of different companies, as well as cushioning stock investments with cash investments and bonds. Diversification enables investors to take advantage of the stock market's superior returns while reducing risk to manageable levels.

Sources: Ralph Frasca, *Personal Finance*, Eighth Edition, Pearson, 2009; Eric Tyson, *Personal Finance for Dummies*, Sixth Edition, Wiley, 2010; Liz Pulliam Weston, *Easy Money*, Pearson, 2008.

Understand accounts designed for retirement savings.

Retirement Savings: Stashing Cash and Making Taxes Less Taxing

As you leave high school and prepare for your future career, retirement probably seems very far away. However, we have seen that you can accumulate wealth much more easily if you have time to make your money work. As soon as you have a job and a paycheck, you should start putting some money away for retirement. Opening a retirement savings account early in your career is a smart way to gain more control over how you will spend a large part of your life.

You can use regular savings and investment accounts to save for retirement. There are also a variety of accounts designed specifically for retirement savings.

- A **traditional individual retirement account (IRA)** is a savings plan that allows you to set aside money for retirement, usually up to $5000 per year. You do not pay taxes on the money you deposit into the IRA. You can start withdrawing from your IRA when you are $59\frac{1}{2}$ years old. The withdrawals are taxed.

- A **Roth IRA** is a type of IRA with slightly different tax benefits. You pay taxes on the money you deposit into the IRA, but then you can withdraw your earnings tax-free when you are $59\frac{1}{2}$ years old. Although your contributions are not tax deductible, your earnings are never taxed, even after withdrawal.

- **Employer-sponsored retirement plans**, including 401(k) and 403(b) plans, are set up by the employer, who often makes some contribution to the plan on your behalf. These plans are not offered by all employers and are used to attract high-quality employees.

All accounts designed specifically for retirement savings have penalties for withdrawals before age $59\frac{1}{2}$.

Example 6   Dollars and Sense of Retirement Plans

a. Suppose that between the ages of 25 and 35, you contribute $4000 per year to a 401(k) and your employer matches this contribution dollar for dollar on your behalf. The interest rate is 8.5% compounded annually. What is the value of the 401(k) at the end of the 10 years?

b. After 10 years of working for this firm, you move on to a new job. However, you keep your accumulated retirement funds in the 401(k). How much money will you have in the plan when you reach age 65?

c. What is the difference between the amount of money you will have accumulated in the 401(k) at age 65 and the amount you contributed to the plan?

Solution

a. We begin by finding the value of your 401(k) after 10 years.

$$A = \frac{P[(1 + r)^t - 1]}{r}$$

Use the formula for the value of an annuity with interest compounded once a year.

$$A = \frac{8000[(1 + 0.085)^{10} - 1]}{0.085}$$

You contribute $4000 and your employer matches this each year: P = 4000 + 4000 = 8000. The interest rate is 8.5%: r = 0.085. The time from age 25 to 35 is 10 years: t = 10.

$$\approx 118{,}681$$

Use a calculator.

The value of the 401(k) at the end of the 10 years is approximately $118,681.

b. Now we find the value of your investment at age 65.

$$A = P(1 + r)^t$$

> Use the formula from Chapter 7 for future value with interest compounded once a year.

$$A = 118{,}681(1 + 0.085)^{30}$$

> The value of the 401(k) is $118,681: P = 118,681. The interest rate is 8.5%: r = 0.085. The time from age 35 to age 65 is 30 years: t = 30.

$$\approx 1{,}371{,}745$$

> Use a calculator.

You will have approximately $1,371,745 in the 401(k) when you reach age 65.

c. You contributed $4000 per year to the 401(k) for 10 years, for a total of $4000 × 10, or $40,000. The difference between the amount you will have accumulated in the plan at age 65, $1,371,745, and the amount you contributed, $40,000, is

$$\$1{,}371{,}745 - \$40{,}000, \text{ or } \$1{,}331{,}745.$$

Even when taxes are taken into consideration, we suspect you'll be quite pleased with your earnings from 10 years of savings.

Checkpoint 6

a. Suppose that between the ages of 25 and 40, you contribute $2000 per year to a 401(k) and your employer contributes $1000 per year on your behalf. The interest rate is 8% compounded annually. What is the value of the 401(k), rounded to the nearest dollar, after 15 years?

b. After 15 years of working for this firm, you move on to a new job. However, you keep your accumulated retirement funds in the 401(k). How much money, to the nearest dollar, will you have in the plan when you reach age 65?

c. What is the difference between the amount of money you will have accumulated in the 401(k) and the amount you contributed to the plan?

Exercise Set 8.1

Concept and Vocabulary Exercises

In Exercises 1–9, fill in each blank so that the resulting statement is true.

1. A sequence of equal payments made at equal time periods is called a/an _____..

2. In the formula

$$A = \frac{P\left[\left(1 + \dfrac{r}{n}\right)^{nt} - 1\right]}{\left(\dfrac{r}{n}\right)},$$

_____ is the deposit made at the end of each compounding period, _____ is the annual interest rate compounded _____ times per year, and A is the _____ after _____ years.

3. A bank account that requires a minimum balance, pays interest, and offers limited check-writing privileges is called a/an _____ account.

4. A bank account that pays a fixed interest rate if the principal is kept in the account for a specified amount of time is called a/an _____, abbreviated _____.

5. Shares of ownership in a company are called _____. If you sell shares for more money than what you paid for them, you have a/an _____ gain on the sale. Some companies distribute all or part of their profits to shareholders as _____.

6. People who buy _____ are lending money to the company from which they buy them.

7. A listing of all the investments that a person holds is called a financial _____. To minimize risk, it should be _____, containing a mixture of low-risk and high-risk investments.

8. A group of investments managed by a professional investor is called a/an _____.

9. With a/an _____ IRA, you pay taxes on the money you deposit, but you can withdraw your earnings tax-free beginning when you are _____ years old.

In Exercises 10–15, determine whether each statement is true or false. If the statement is false, make the necessary change(s) to produce a true statement.

10. With the same interest rate, compounding period, and time period, a lump-sum deposit will generate more interest than an annuity.

11. A CD requires that you keep the principal in the account for a specified amount of time.

12. People who buy bonds are purchasing shares of ownership in a company.

13. Stocks are generally considered higher-risk investments than bonds.

14. The first $250,000 in a bank savings account can plunge in value or even disappear if the bank goes bankrupt.

15. A traditional IRA requires paying taxes when withdrawing money from the account at age $59\frac{1}{2}$ or older.

Respond to Exercises 16–26 using verbal or written explanations.

16. What is an annuity?

17. What is meant by the value of an annuity?

18. What is a money-market account?

19. What is a CD? (Be sure to think finance and not music!)

20. What is stock?

21. Describe how to find the percent ownership that a shareholder has in a company.

22. Describe the two ways that investors make money with stock.

23. What is a bond? Describe the difference between a stock and a bond.

24. If an investor sees that the return from dividends for a stock is lower than the return for a no-risk bank account, should the stock be sold and the money placed in the bank account? Explain your answer.

25. What is a mutual fund?

26. What is the difference between a traditional IRA and a Roth IRA?

In Exercises 27–36, determine whether each statement makes sense or does not make sense, and explain your reasoning.

27. By putting $10 at the end of each month into an annuity that pays 3.5% compounded monthly, I'll be able to retire comfortably in just 30 years.

28. One of the things that I appreciate about my savings account at the bank is that I can write an unlimited number of checks each month.

29. By maintaining a minimum monthly balance, I can earn interest on my checking account.

30. One of the things that I appreciate about my money-market account is that there is no limit to the number of checks I can write each month.

31. When I invest my money, I am making a tradeoff between risk and return.

32. I have little tolerance for risk, so I must be willing to accept lower returns on my investments.

33. Diversification is like saying "don't put all your eggs in one basket."

34. Now that I've purchased bonds, I'm a shareholder in the company.

35. I've been promised a 20% return on an investment without any risk.

36. I appreciate my Roth IRA because I do not pay taxes on my deposits.

Practice Exercises

Here are the formulas needed to solve the exercises. Be sure you understand what each formula describes and the meaning of the variables in the formulas.

$$A = \frac{P[(1+r)^t - 1]}{r} \qquad A = \frac{P\left[\left(1+\frac{r}{n}\right)^{nt} - 1\right]}{\left(\frac{r}{n}\right)}$$

$$P = \frac{A\left(\frac{r}{n}\right)}{\left[\left(1+\frac{r}{n}\right)^{nt} - 1\right]}$$

In Exercises 37–46,

a. *Find the value of each annuity. Round to the nearest dollar.*

b. *Find the interest.*

	Periodic Deposit	Rate	Time
37.	$2000 at the end of each year	5% compounded annually	20 years
38.	$3000 at the end of each year	4% compounded annually	20 years
39.	$4000 at the end of each year	6.5% compounded annually	40 years
40.	$4000 at the end of each year	5.5% compounded annually	40 years
41.	$50 at the end of each month	6% compounded monthly	30 years
42.	$60 at the end of each month	5% compounded monthly	30 years
43.	$100 at the end of every six months	4.5% compounded semiannually	25 years
44.	$150 at the end of every six months	6.5% compounded semiannually	25 years
45.	$1000 at the end of every three months	6.25% compounded quarterly	6 years
46.	$1200 at the end of every three months	3.25% compounded quarterly	6 years

In Exercises 47–54,

a. *Determine the periodic deposit. Round up to the nearest dollar.*

b. *How much of the financial goal comes from deposits and how much comes from interest?*

	Periodic Deposit	Rate	Time	Financial Goal
47.	$? at the end of each year	6% compounded annually	18 years	$140,000
48.	$? at the end of each year	5% compounded annually	18 years	$150,000
49.	$? at the end of each month	4.5% compounded monthly	10 years	$200,000
50.	$? at the end of each month	7.5% compounded monthly	10 years	$250,000
51.	$? at the end of each month	7.25% compounded monthly	40 years	$1,000,000
52.	$? at the end of each month	8.25% compounded monthly	40 years	$1,500,000
53.	$? at the end of every three months	3.5% compounded quarterly	5 years	$20,000
54.	$? at the end of every three months	4.5% compounded quarterly	5 years	$25,000

Exercises 55 and 56 refer to the stock tables for Goodyear (the tire company) and JC Penney (the department store) given below. In each exercise, use the stock table to answer the following questions. Where necessary, round dollar amounts to the nearest cent.

a. *What were the high and low prices for a share for the past 52 weeks?*

b. *If you owned 700 shares of this stock last year, what dividend did you receive?*

c. *What is the annual return for the dividends alone? How does this compare to a bank offering a 3% interest rate?*

d. *How many shares of this company's stock were traded yesterday?*

e. *What were the high and low prices for a share yesterday?*

f. *What was the price at which a share last traded when the stock exchange closed yesterday?*

g. *What was the change in price for a share of stock from the market close two days ago to yesterday's market close?*

h. *Compute the company's annual earnings per share using*

$$\text{Annual earnings per share} = \frac{Yesterday's\ closing\ price\ per\ share}{PE\ ratio}.$$

55.

52-Week High	52-Week Low	Stock	SYM	Div	Yld %	PE	Vol 100s	Hi	Lo	Close	Net Chg
73.25	45.44	Goodyear	GT	1.20	2.2	17	5915	56.38	54.38	55.50	+1.25

56.

52-Week High	52-Week Low	Stock	SYM	Div	Yld %	PE	Vol 100s	Hi	Lo	Close	Net Chg
78.34	35.38	Penney JC	JCP	2.18	4.7	22	7473	48.19	46.63	46.88	−1.31

Practice Plus

Here are additional formulas from Chapter 7 that you will use to solve some of the remaining exercises. Be sure you understand what each formula describes and the meaning of the variables in the formulas.

$$A = P(1 + r)^t \qquad A = P\left(1 + \frac{r}{n}\right)^{nt}$$

In Exercises 57–58, round all answers to the nearest dollar.

57. Here are two ways of investing $30,000 for 20 years.

Lump-Sum Deposit	Rate	Time
$30,000	5% compounded annually	20 years

Periodic Deposit	Rate	Time
$1500 at the end of each year	5% compounded annually	20 years

a. After 20 years, how much more will you have from the lump-sum investment than from the annuity?

b. After 20 years, how much more interest will be earned from the lump-sum investment than from the annuity?

58. Here are two ways of investing $40,000 for 25 years.

Lump-Sum Deposit	Rate	Time
$40,000	6.5% compounded annually	25 years

Periodic Deposit	Rate	Time
$1600 at the end of each year	6.5% compounded annually	25 years

a. After 25 years, how much more will you have from the lump-sum investment than from the annuity?

b. After 25 years, how much more interest will be earned from the lump-sum investment than from the annuity?

59. Solve for P:

$$A = \frac{P[(1 + r)^t - 1]}{r}.$$

What does the resulting formula describe?

60. Solve for P:

$$A = \frac{P\left[\left(1 + \dfrac{r}{n}\right)^{nt} - 1\right]}{\left(\dfrac{r}{n}\right)}.$$

What does the resulting formula describe?

Application Exercises

In Exercises 61–66, round to the nearest dollar.

61. Suppose that you earned a bachelor's degree and now you're teaching high school. The school district offers teachers the opportunity to take a year off to earn a master's degree. To achieve this goal, you deposit $2000 at the end of each year in an annuity that pays 7.5% compounded annually.

  a. How much will you have saved at the end of five years?

  b. Find the interest.

62. Suppose that you earned a bachelor's degree and now you're teaching high school. The school district offers teachers the opportunity to take a year off to earn a master's degree. To achieve this goal, you deposit $2500 at the end of each year in an annuity that pays 6.25% compounded annually.

  a. How much will you have saved at the end of five years?

  b. Find the interest.

63. Suppose that at age 25, you decide to save for retirement by depositing $50 at the end of each month in an IRA that pays 5.5% compounded monthly.

  a. How much will you have from the IRA when you retire at age 65?

  b. Find the interest.

64. Suppose that at age 25, you decide to save for retirement by depositing $75 at the end of each month in an IRA that pays 6.5% compounded monthly.

  a. How much will you have from the IRA when you retire at age 65?

  b. Find the interest.

65. To offer scholarships to children of employees, a company invests $10,000 at the end of every three months in an annuity that pays 10.5% compounded quarterly.

  a. How much will the company have in scholarship funds at the end of ten years?

  b. Find the interest.

66. To offer scholarships to children of employees, a company invests $15,000 at the end of every three months in an annuity that pays 9% compounded quarterly.

  a. How much will the company have in scholarship funds at the end of ten years?

  b. Find the interest.

In Exercises 67–70, round up to the nearest dollar.

67. You would like to have $3500 in four years for a special vacation following college graduation by making deposits at the end of every six months in an annuity that pays 5% compounded semiannually.

  a. How much should you deposit at the end of every six months?

  b. How much of the $3500 comes from deposits and how much comes from interest?

68. You would like to have $4000 in four years for a special vacation following college graduation by making deposits at the end of every six months in an annuity that pays 7% compounded semiannually.

  a. How much should you deposit at the end of every six months?

  b. How much of the $4000 comes from deposits and how much comes from interest?

69. How much should you deposit at the end of each month into an IRA that pays 6.5% compounded monthly to have $2 million when you retire in 45 years? How much of the $2 million comes from interest?

70. How much should you deposit at the end of each month into an IRA that pays 8.5% compounded monthly to have $4 million when you retire in 45 years? How much of the $4 million comes from interest?

71. **a.** Suppose that between the ages of 22 and 40, you contribute $3000 per year to a 401(k) and your employer contributes $1500 per year on your behalf. The interest rate is 8.3% compounded annually. What is the value of the 401(k), rounded to the nearest dollar, after 18 years?

  b. Suppose that after 18 years of working for this firm, you move on to a new job. However, you keep your accumulated retirement funds in the 401(k). How much money, to the nearest dollar, will you have in the plan when you reach age 65?

  c. What is the difference between the amount of money you will have accumulated in the 401(k) and the amount you contributed to the plan?

72. **a.** Suppose that between the ages of 25 and 37, you contribute $3500 per year to a 401(k) and your employer matches this contribution dollar for dollar on your behalf. The interest rate is 8.25% compounded annually. What is the value of the 401(k), rounded to the nearest dollar, after 12 years?

  b. Suppose that after 12 years of working for this firm, you move on to a new job. However, you keep your accumulated retirement funds in the 401(k). How much money, to the nearest dollar, will you have in the plan when you reach age 65?

  c. What is the difference between the amount of money you will have accumulated in the 401(k) and the amount you contributed to the plan?

73. Write a problem involving the formula for regular payments needed to achieve a financial goal. The problem should be similar to Example 4 on page 503. However, the problem should be unique to your situation. Include something for which you would like to save, how much you need to save, and how long you have to achieve your goal. Then solve the problem.

Critical Thinking Exercises

In Exercises 74–75,

  a. *Determine the deposit at the end of each month. Round up to the nearest dollar.*

  b. *Assume that the annuity in part (a) is a tax-deferred IRA belonging to a man whose gross income is $50,000. Use* **Table 7.1** *on page 462 to calculate his taxes first with and then without the IRA. Assume the man is single with no dependents, has no tax credits, and takes the standard deduction.*

c. *What percent of his gross income are the man's federal taxes with and without the IRA? Round to the nearest tenth of a percent.*

	Periodic Deposit	Rate	Time	Financial Goal
74.	$? at the end of each month	8% compounded monthly	40 years	$1,000,000
75.	$? at the end of each month	7% compounded monthly	40 years	$650,000

76. How much should you deposit at the end of each month in an IRA that pays 8% compounded monthly to earn $60,000 per year from interest alone, while leaving the principal untouched, when you retire in 30 years?

Group Exercises

77. Each group should have a newspaper with current stock quotations. Choose nine stocks that group members think would make good investments. Imagine that you invest $10,000 in each of these nine investments. Check the value of your stock each day over the next five weeks and then sell the nine stocks after five weeks. What is the group's profit or loss over the five-week period? Compare this figure with the profit or loss of other groups in your class for this activity.

78. This activity is a group research project intended for four or five people. Use the research to present a seminar on investments. The seminar is intended to last about 30 minutes and should result in an interesting and informative presentation to the entire class. The seminar should include investment considerations, how to read the bond section of the newspaper, how to read the mutual fund section, and higher-risk investments.

79. Group members have inherited $1 million. However, the group cannot spend any of the money for ten years. As a group, determine how to invest this money in order to maximize the money you will make over ten years. The money can be invested in as many ways as the group decides. Explain each investment decision. What are the risks involved in each investment plan?

8.2   Cars

Objectives

1. Compute the monthly payment and interest costs for a car loan.
2. Understand the types of leasing contracts.
3. Understand the pros and cons of leasing versus buying a car.
4. Understand the different kinds of car insurance.
5. Compare monthly payments on new and used cars.
6. Solve problems related to owning and operating a car.

"This car could be automatic, systematic, hydromatic, why it's Greased Lightnin!"

—Kenickie, a senior at Rydell High School, in the musical *Grease!*

To the guys at Rydell High, Kenickie's new car looks like a hunk of junk, but to him it's Greased Lightnin', a hot-rodding work of art on wheels. As with many teens, Kenickie's first car is a rite of passage—a symbol of emerging adulthood.

Our love affair with cars began in the early 1900s when Henry Ford cranked out the first Model T. Since then, we've admired cars to the point of identifying with the vehicles we drive. Cars can serve as status symbols, providing unique insights into a driver's personality.

In this section, we view cars from another vantage point—money. The money pit of owning a car ranges from financing the purchase to escalating costs of everything from fuel to tires to insurance. We open the section with the main reason people spend more money on a car than they can afford: financing.

The Mathematics of Financing a Car

A loan that you pay off with weekly or monthly payments, or payments in some other time period, is called an **installment loan**. The advantage of an installment loan is that the consumer gets to use a product immediately. The disadvantage is that the interest can add a substantial amount to the cost of a purchase.

Let's begin with car loans in which you make regular monthly payments, called **fixed installment loans**. Suppose that you borrow P dollars at interest rate r over t years.

> The lender expects A dollars at the end of t years.

> You save the A dollars in an annuity by paying PMT dollars n times per year.

$$A = P\left(1 + \frac{r}{n}\right)^{nt}$$

$$A = \frac{PMT\left[\left(1 + \frac{r}{n}\right)^{nt} - 1\right]}{\left(\frac{r}{n}\right)}$$

To find your regular payment amount, PMT, we set the amount the lender expects to receive equal to the amount you will save in the annuity:

$$P\left(1 + \frac{r}{n}\right)^{nt} = \frac{PMT\left[\left(1 + \frac{r}{n}\right)^{nt} - 1\right]}{\left(\frac{r}{n}\right)}.$$

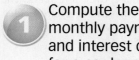

Compute the monthly payment and interest costs for a car loan.

Solving this equation for PMT, we obtain a formula for the loan payment for any installment loan, including payments on car loans.

> **Loan Payment Formula for Fixed Installment Loans**
>
> The regular payment amount, PMT, required to repay a loan of P dollars paid n times per year over t years at an annual rate r is given by
>
> $$PMT = \frac{P\left(\frac{r}{n}\right)}{\left[1 - \left(1 + \frac{r}{n}\right)^{-nt}\right]}.$$

Example 1   Comparing Car Loans

Suppose that you decide to borrow $20,000 for a new car. You can select one of the following loans, each requiring regular monthly payments:

    Installment Loan A: three-year loan at 7%

    Installment Loan B: five-year loan at 9%.

  a. Find the monthly payments and the total interest for Loan A.

  b. Find the monthly payments and the total interest for Loan B.

  c. Compare the monthly payments and total interest for the two loans.

Solution

For each loan, we use the loan payment formula to compute the monthly payments.

  a. We first determine monthly payments and total interest for Loan A.

> P, the loan amount, is $20,000.

> Rate, r, is 7%.

> 12 payments per year

> The loan is for 3 years.

$$PMT = \frac{P\left(\frac{r}{n}\right)}{\left[1 - \left(1 + \frac{r}{n}\right)^{-nt}\right]} = \frac{20{,}000\left(\frac{0.07}{12}\right)}{\left[1 - \left(1 + \frac{0.07}{12}\right)^{-12(3)}\right]} \approx 618$$

The monthly payments are approximately $618.

Now we calculate the interest over 3 years, or 36 months.

Total interest over 3 years		Total of all monthly payments	minus	amount of the loan.
	=	$618 ×	36 −	$20,000
	=	$2248		

The total interest paid over 3 years is approximately $2248.

b. Next, we determine monthly payments and total interest for Loan B.

P, the loan amount, is $20,000. Rate, r, is 9%. 12 payments per year The loan is for 5 years.

$$PMT = \frac{P\left(\frac{r}{n}\right)}{\left[1 - \left(1 + \frac{r}{n}\right)^{-nt}\right]} = \frac{20{,}000\left(\frac{0.09}{12}\right)}{\left[1 - \left(1 + \frac{0.09}{12}\right)^{-12(5)}\right]} \approx 415$$

The monthly payments are approximately $415.
Now we calculate the interest over 5 years, or 60 months.

Total interest over 5 years		Total of all monthly payments	minus	amount of the loan.
	=	$415 ×	60 −	$20,000
	=	$4900		

The total interest paid over 5 years is approximately $4900.

c. Table 8.2 compares the monthly payments and total interest for the two loans.

Table 8.2 Comparing Car Loans

$20,000 loan	Monthly Payment	Total Interest
3-year loan at 7%	$618	$2248
5-year loan at 9%	$415	$4900

Monthly payments are less with the longer-term loan.   Interest is more with the longer-term loan.

Checkpoint 1 Suppose that you decide to borrow $15,000 for a new car. You can select one of the following loans, each requiring regular monthly payments:

Installment Loan A: four-year loan at 8%
Installment Loan B: six-year loan at 10%.

a. Find the monthly payments and the total interest for Loan A.

b. Find the monthly payments and the total interest for Loan B.

▶ **c.** Compare the monthly payments and total interest for the two loans.

Financing Your Car

- Check out financing options. It's a good idea to get preapproved for a car loan through a bank or credit union before going to the dealer. You can then compare the loan offered by the dealer to your preapproved loan. Furthermore, with more money in hand, you'll have more negotiating power.

- Dealer financing often costs 1% or 2% more than a bank or credit union. Shop around for interest rates. Credit unions traditionally offer the best rates on car loans, more than 1.5% less on average than a bank loan.

- Put down as much money as you can. Interest rates generally decrease as the money you put down toward the car increases. Furthermore, you'll be borrowing less money, thereby paying less interest.

- A general rule is that you should spend no more than 20% of your net monthly income on a car payment.

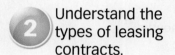

Understand the types of leasing contracts.

The Leasing Alternative

Leasing is the practice of paying a specified amount of money over a specified time for the use of a product. Leasing is essentially a long-term rental agreement.

Leasing a car instead of buying one has become increasingly popular over the past several years. There are two types of leasing contracts:

- **A closed-end lease:** Each month, you make a fixed payment based on estimated usage. When the lease ends, you return the car and pay for mileage in excess of your estimate.

- **An open-end lease:** Each month, you make a fixed payment based on the car's *residual value*. **Residual value** is the estimated resale value of the car at the end of the lease and is determined by the dealer. When the lease ends, you return the car and make a payment based on its appraised value at that time compared to its residual value. If the appraised value is less than the residual value stated in the lease, you pay all or a portion of the difference. If the appraised value is greater than or equal to the residual value, you owe nothing and you may receive a refund.

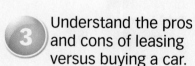

Understand the pros and cons of leasing versus buying a car.

Leasing a car offers both advantages and disadvantages over buying one.

Advantages of Leasing

- Leases require only a small down payment, or no down payment at all.
- Lease payments for a new car are lower than loan payments for the same car. Most people can lease a more expensive car than they would be able to buy.
- When the lease ends, you return the car to the dealer and do not have to be concerned about selling the car.

Disadvantages of Leasing

- When the lease ends, you do not own the car.
- Most lease agreements have mileage limits: 12,000 to 15,000 miles per year is common. If you exceed the number of miles allowed, there can be considerable charges.
- When mileage penalties and other costs at the end of the leasing period are taken into consideration, the total cost of leasing is almost always more expensive than financing a car.

- While leasing the car, you are responsible for keeping it in perfect condition. You are liable for any damage to the car.
- Leasing does not cover maintenance.
- There are penalties for ending the lease early.

Car leases tend to be extremely complicated. It can appear that there are as many lease deals as there are kinds of cars. A helpful pamphlet entitled "Keys to Vehicle Leasing" is published by the Federal Reserve Board. Copies are available on the Internet. Additional information can be found at websites such as autos.msn.com or intellichoice.com.

The Importance of Auto Insurance

4 Understand the different kinds of car insurance.

Who needs auto insurance? The simple answer is that if you own or lease a car, you do.

When you purchase **insurance**, you buy protection against loss associated with unexpected events. Different types of coverage are associated with auto insurance, but the one required by nearly every state is *liability*. There are two components of **liability coverage**:

- **Bodily injury liability** covers the costs of lawsuits if someone is injured or killed in an accident in which you are at fault.
- **Property damage liability** covers damage to other cars and property from negligent operation of your vehicle.

If you have a car loan or lease a car, you will also need *collision* and *comprehensive* coverage:

- **Collision coverage** pays for damage or loss of your car if you're in an accident.
- **Comprehensive coverage** protects your car from perils such as fire, theft, falling objects, acts of nature, and collision with an animal.

There is a big difference in auto insurance rates, so be sure to shop around. Insurance can be very expensive for teens with limited driving experience. A poor driving record dramatically increases your insurance rates. Other factors that impact your insurance premium include where you live, the number of miles you drive each year, whether or not you have taken driver's education, and the value of your car.

New or Used?

5 Compare monthly payments on new and used cars.

Who insists you need a new car? A new car loses an average of 12% of its value the moment it is driven off the dealer's lot. It's already a used car and you haven't even arrived home.

Used cars are a good option for many people. Your best buy is typically a two- to three-year-old car because the annual depreciation in price is greatest over the first few years. Furthermore, many sources of financing for used cars will loan money only on newer models that are less than five years old. Reputable car dealerships offer a good selection of used cars, with extended warranties and other perks.

The two most commonly used sources of pricing information for used cars are the *National Automobile Dealers Association Official Used Car Guide* (www.nada.com) and the *Kelley Blue Book Used Car Guide* (www.kkb.com). They contain the average retail price for many different makes of used cars.

Example 2   Saving Money with a Used Car

Suppose that you are thinking about buying a car and have narrowed down your choices to two options:

- The new-car option: The new car costs $25,000 and can be financed with a four-year loan at 7.9%.
- The used-car option: A three-year-old model of the same car costs $14,000 and can be financed with a four-year loan at 8.45%.

What is the difference in monthly payments between financing the new car and financing the used car?

Solution

We first determine the monthly payments for the new car that costs $25,000, financed with a four-year loan at 7.9%.

P, the loan amount, is $25,000.

Rate, r, is 7.9%.

12 payments per year

The loan is for 4 years.

$$PMT = \frac{P\left(\frac{r}{n}\right)}{\left[1 - \left(1 + \frac{r}{n}\right)^{-nt}\right]} = \frac{25{,}000\left(\frac{0.079}{12}\right)}{\left[1 - \left(1 + \frac{0.079}{12}\right)^{-12(4)}\right]} \approx 609$$

The monthly payments for the new car are approximately $609. Now we determine the monthly payments for the used car that costs $14,000, financed with a four-year loan at 8.45%.

P, the loan amount, is $14,000.

Rate, r, is 8.45%.

12 payments per year

The loan is for 4 years.

$$PMT = \frac{P\left(\frac{r}{n}\right)}{\left[1 - \left(1 + \frac{r}{n}\right)^{-nt}\right]} = \frac{14{,}000\left(\frac{0.0845}{12}\right)}{\left[1 - \left(1 + \frac{0.0845}{12}\right)^{-12(4)}\right]} \approx 345$$

The monthly payments for the used car are approximately $345. The difference in monthly payments between the new-car loan, $609, and the used car loan, $345, is

$$\$609 - \$345, \text{ or } \$264.$$

You save $264 each month over a period of four years with the used-car option.

Checkpoint 2   Suppose that you are thinking about buying a car and have narrowed down your choices to two options:

The new-car option: The new car costs $19,000 and can be financed with a three-year loan at 6.18%.

The used-car option: A two-year-old model of the same car costs $11,500 and can be financed with a three-year loan at 7.5%.

What is the difference in monthly payments between financing the new car and financing the used car?

The Money Pit of Car Ownership

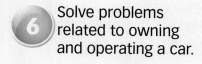

6 Solve problems related to owning and operating a car.

Buying a car is a huge expense. To make matters worse, the car continues costing money after you purchase it. These costs include operating expenses such as fuel, maintenance, tires, tolls, parking, and cleaning. The costs also include ownership expenses such as insurance, license fees, registration fees, taxes, and interest on loans.

The significant expense of owning and operating a car is shown in **Table 8.3**. According to the U.S. Department of Labor Statistics, the average yearly cost of owning and operating a car is just over $8000.

Table 8.3 Average Annual Costs of Owning and Operating a Car, for Selected 2008 Cars

Make and Model	Average Annual Costs		
	Operating	Ownership	Total
Cadillac STS	$5130	$14,407	$19,537
Lexus LS 460	$4750	$14,622	$19,372
Mercury Grand Marquis GS	$4610	$8499	$13,109
Honda Accord LX	$4170	$6820	$10,990
Toyota Camry LE	$3620	$5812	$9432
Ford Focus SE	$3240	$5709	$8949
Chevrolet Cobalt LT	$3450	$5467	$8917
Toyota Corolla CE	$2980	$4953	$7933

Source: Runzheimer International

A large portion of a car's operating expenses involves the cost of gasoline, which averaged $4 per gallon in 2008. As the luster of big gas-guzzlers becomes less appealing, many people are turning to fuel-efficient hybrid cars that use a combination of gasoline and rechargeable batteries as power sources.

Our next example compares fuel expenses for a gas-guzzler and a hybrid. You can estimate the annual fuel expense for a vehicle if you know approximately how many miles the vehicle will be driven each year, how many miles the vehicle can be driven per gallon of gasoline, and how much a gallon of gasoline will cost.

The Cost of Gasoline

$$\text{Annual fuel expense} = \frac{\text{annual miles driven}}{\text{miles per gallon}} \times \text{price per gallon}$$

Example 3 Comparing Fuel Expenses

Suppose that you drive 24,000 miles per year and gas averages $4 per gallon.

a. What will you save in annual fuel expenses by owning a hybrid car averaging 50 miles per gallon rather than an SUV (sport utility vehicle) averaging 12 miles per gallon?

b. If you deposit your monthly fuel savings at the end of each month into an annuity that pays 7.3% compounded monthly, how much will you have saved at the end of six years?

Solution

a. We use the formula for annual fuel expense.

$$\text{Annual fuel expense} = \frac{\text{annual miles driven}}{\text{miles per gallon}} \times \text{price per gallon}$$

$$\text{Annual fuel expense for the hybrid} = \frac{24,000}{50} \times \$4 = 480 \times \$4 = \$1920$$

The hybrid averages 50 miles per gallon.

$$\text{Annual fuel expense} = \frac{\text{annual miles driven}}{\text{miles per gallon}} \times \text{price per gallon}$$

$$\text{Annual fuel expense for the SUV} = \frac{24{,}000}{12} \times \$4 = 2000 \times \$4 = \$8000$$

> The SUV averages 12 miles per gallon.

Your annual fuel expense is $1920 for the hybrid and $8000 for the SUV. By owning the hybrid rather than the SUV, you save

$$\$8000 - \$1920, \text{ or } \$6080$$

in annual fuel expenses.

b. Because you save $6080 per year, you save

$$\frac{\$6080}{12} \approx \$507,$$

or approximately $507 per month. Now you deposit $507 at the end of each month into an annuity that pays 7.3% compounded monthly. We use the formula for the value of an annuity to determine your savings at the end of six years.

$$A = \frac{P\left[\left(1 + \dfrac{r}{n}\right)^{nt} - 1\right]}{\left(\dfrac{r}{n}\right)} \qquad \text{Use the formula for the value of an annuity.}$$

$$A = \frac{507\left[\left(1 + \dfrac{0.073}{12}\right)^{12 \cdot 6} - 1\right]}{\left(\dfrac{0.073}{12}\right)} \qquad \begin{array}{l}\text{The annuity involves month-end deposits of}\\ \$507\text{: } P = 507. \text{ The interest rate is } 7.3\%\text{:}\\ r = 0.073. \text{ The interest is compounded monthly:}\\ n = 12. \text{ The number of years is 6: } t = 6.\end{array}$$

$$\approx 45{,}634 \qquad \text{Use a calculator.}$$

You will have saved approximately $45,634 at the end of six years. This illustrates how driving a car that consumes less gas can yield significant savings for your future.

✓ **Checkpoint 3** Suppose that you drive 36,000 miles per year and gas averages $3.50 per gallon.

a. What will you save in annual fuel expenses by owning a hybrid car averaging 40 miles per gallon rather than an SUV averaging 15 miles per gallon?

b. If you deposit your monthly fuel savings at the end of each month into an annuity that pays 7.25% compounded monthly, how much will you have saved at the end of seven years? Round all computations to the nearest dollar.

Exercise Set 8.2

Concept and Vocabulary Exercises

In Exercises 1–6, fill in each blank so that the resulting statement is true.

1. In the formula

$$PMT = \frac{P\left(\dfrac{r}{n}\right)}{\left[1 - \left(1 + \dfrac{r}{n}\right)^{-nt}\right]},$$

_____ is the regular payment amount required to repay a loan of _____ dollars paid _____ times per year over _____ years at an annual interest rate _____.

2. The two types of contracts involved with leasing a car are called a/an _____ lease and a/an _____ lease.

3. The estimated resale value of a car at the end of its lease is called the car's _____.

4. There are two components of liability insurance. The component that covers costs if someone is injured or killed in an accident in which you are at fault is called _____ liability. The component that covers damage to other cars if you are at fault is called _____ liability.

5. The type of car insurance that pays for damage or loss of your car if you're in an accident is called _____ coverage.

6. The type of insurance that pays for damage to your car due to fire, theft, or falling objects is called _____ coverage.

In Exercises 7–12, determine whether each statement is true or false. If the statement is false, make the necessary change(s) to produce a true statement.

7. The interest on a car loan can be determined by taking the difference between the total of all monthly payments and the amount of the loan.

8. When an open-end lease terminates and the car's appraised value is less than the residual value stated in the lease, you owe nothing.

9. One advantage to leasing a car is that you are not responsible for any damage to the car.

10. One disadvantage to leasing a car is that most lease agreements have mileage limits.

11. Collision coverage pays for damage to another car if you cause an accident.

12. Due to operating and ownership expenses, a car continues costing money after you buy it.

Respond to Exercises 13–20 using verbal or written explanations.

13. If a three-year car loan has the same interest rate as a six-year car loan, how do the monthly payments and the total interest compare for the two loans?

14. What is the difference between a closed-end car lease and an open-end car lease?

15. Describe two advantages of leasing a car over buying one.

16. Describe two disadvantages of leasing a car over buying one.

17. What are the two components of liability coverage and what is covered by each component?

18. What does collision coverage pay for?

19. What does comprehensive coverage pay for?

20. How can you estimate a car's annual fuel expense?

In Exercises 21–26, determine whether each statement makes sense or does not make sense, and explain your reasoning.

21. If I purchase a car using money that I've saved, I can eliminate paying interest on a car loan, but then I have to give up the interest income I could have earned on my savings.

22. The problem with my car lease is that when it ends, I have to be concerned about selling the car.

23. Although lease payments for a new car are lower than loan payments for the same car, once I take mileage penalties and other costs into consideration, the total cost of leasing is more expensive than financing the car.

24. I've paid off my car loan, so I am not required to have liability coverage.

25. Buying a used car or a fuel-efficient car can yield significant savings for my future.

26. Because it is extremely expensive to own and operate a car, I plan to look closely at whether or not a car is essential and consider other modes of transportation.

Practice and Application Exercises

In Exercises 27–36, use

$$PMT = \frac{P\left(\frac{r}{n}\right)}{\left[1 - \left(1 + \frac{r}{n}\right)^{-nt}\right]}.$$

Round answers to the nearest dollar.

27. Suppose that you borrow $10,000 for four years at 8% toward the purchase of a car. Find the monthly payments and the total interest for the loan.

28. Suppose that you borrow $30,000 for four years at 8% for the purchase of a car. Find the monthly payments and the total interest for the loan.

29. Suppose that you decide to borrow $15,000 for a new car. You can select one of the following loans, each requiring regular monthly payments:

    Installment Loan A: three-year loan at 5.1%
    Installment Loan B: five-year loan at 6.4%.

    a. Find the monthly payments and the total interest for Loan A.
    b. Find the monthly payments and the total interest for Loan B.
    c. Compare the monthly payments and the total interest for the two loans.

30. Suppose that you decide to borrow $40,000 for a new car. You can select one of the following loans, each requiring regular monthly payments:

    Installment Loan A: three-year loan at 6.1%
    Installment Loan B: five-year loan at 7.2%.

    a. Find the monthly payments and the total interest for Loan A.
    b. Find the monthly payments and the total interest for Loan B.
    c. Compare the monthly payments and the total interest for the two loans.

31. Suppose that you are thinking about buying a car and have narrowed down your choices to two options:

    The new-car option: The new car costs $28,000 and can be financed with a four-year loan at 6.12%.

    The used-car option: A three-year old model of the same car costs $16,000 and can be financed with a four-year loan at 6.86%.

    What is the difference in monthly payments between financing the new car and financing the used car?

32. Suppose that you are thinking about buying a car and have narrowed down your choices to two options:

> The new-car option: The new car costs $68,000 and can be financed with a four-year loan at 7.14%.
>
> The used-car option: A three-year old model of the same car costs $28,000 and can be financed with a four-year loan at 7.92%.

What is the difference in monthly payments between financing the new car and financing the used car?

33. Suppose that you decide to buy a car for $29,635, including taxes and license fees. You saved $9000 for a down payment and can get a five-year car loan at 6.62%. Find the monthly payment and the total interest for the loan.

34. Suppose that you decide to buy a car for $37,925, including taxes and license fees. You saved $12,000 for a down payment and can get a five-year loan at 6.58%. Find the monthly payment and the total interest for the loan.

35. Suppose that you are buying a car for $60,000, including taxes and license fees. You saved $10,000 for a down payment. The dealer is offering you two incentives:

> Incentive A is $5000 off the price of the car, followed by a five-year loan at 7.34%.
>
> Incentive B does not have a cash rebate, but provides free financing (no interest) over five years.

What is the difference in monthly payments between the two offers? Which incentive is the better deal?

36. Suppose that you are buying a car for $56,000, including taxes and license fees. You saved $8000 for a down payment. The dealer is offering you two incentives:

> Incentive A is $10,000 off the price of the car, followed by a four-year loan at 12.5%.
>
> Incentive B does not have a cash rebate, but provides free financing (no interest) over four years.

What is the difference in monthly payments between the two offers? Which incentive is the better deal?

In Exercises 37–40, use the formula

$$A = \frac{P\left[\left(1 + \frac{r}{n}\right)^{nt} - 1\right]}{\left(\frac{r}{n}\right)}.$$

Round all computations to the nearest dollar.

37. Suppose that you drive 40,000 miles per year and gas averages $4 per gallon.

a. What will you save in annual fuel expenses by owning a hybrid car averaging 40 miles per gallon rather than an SUV averaging 16 miles per gallon?

b. If you deposit your monthly fuel savings at the end of each month into an annuity that pays 5.2% compounded monthly, how much will you have saved at the end of six years?

38. Suppose that you drive 15,000 miles per year and gas averages $3.50 per gallon.

a. What will you save in annual fuel expenses by owning a hybrid car averaging 60 miles per gallon rather than an SUV averaging 15 miles per gallon?

b. If you deposit your monthly fuel savings at the end of each month into an annuity that pays 5.7% compounded monthly, how much will you have saved at the end of six years?

The table shows the expense of operating and owning four selected cars, by average costs per mile. Use the appropriate information in the table to solve Exercises 39–42.

Average Annual Costs of Owning and Operating a Car, for Selected 2008 Cars

Make and Model	Average Costs per Mile		
	Operating	Ownership	Total
Cadillac STS	$0.26	$0.72	$0.98
Mercury Grand Marquis GS	$0.23	$0.42	$0.65
Honda Accord LX	$0.21	$0.34	$0.55
Toyota Corolla CE	$0.15	$0.25	$0.40

Source: Runzheimer International

39. a. If you drive 20,000 miles per year, what is the total annual expense for a Cadillac STS?

b. If the total annual expense for a Cadillac STS is deposited at the end of each year into an IRA paying 8.5% compounded yearly, how much will be saved at the end of six years?

40. a. If you drive 14,000 miles per year, what is the total annual expense for a Toyota Corolla CE?

b. If the total annual expense for a Toyota Corolla CE is deposited at the end of each year into an IRA paying 8.2% compounded yearly, how much will be saved at the end of six years?

41. If you drive 30,000 miles per year, by how much does the total annual expense for a Cadillac STS exceed that of a Toyota Corolla CE over six years?

42. If you drive 25,000 miles per year, by how much does the total annual expense for a Mercury Grand Marquis GS exceed that of a Honda Accord LX over six years?

Critical Thinking Exercises

43. Use the discussion at the top of page 516 to prove the loan payment formula shown in the box in the middle of page 516. Work with the equation in which the amount the lender expects to receive is equal to the amount saved in the annuity. Multiply both sides of this equation by $\frac{r}{n}$ and then solve for *PMT* by dividing both sides by the appropriate expression. Finally, divide the numerator and the denominator of the resulting formula for *PMT* by $\left(1 + \frac{r}{n}\right)^{nt}$ to obtain the form of the loan payment formula shown in the box.

44. The unpaid balance of an installment loan is equal to the present value of the remaining payments. The unpaid balance, P, is given by

$$P = PMT \dfrac{\left[1 - \left(1 + \dfrac{r}{n}\right)^{-nt}\right]}{\left(\dfrac{r}{n}\right)},$$

where PMT is the regular payment amount, r is the annual interest rate, n is the number of payments per year, and t is the number of years remaining in the loan.

a. Use the loan payment formula to derive the unpaid balance formula.

b. The price of a car is $24,000. You have saved 20% of the price as a down payment. After the down payment, the balance is financed with a 5-year loan at 9%. Determine the unpaid balance after three years. Round all calculations to the nearest dollar.

Group Exercises

45. Group members should go to the Internet and select a car that they might like to buy. Price the car and its options. Then find two loans with the best rates, but with different terms. For each loan, calculate the monthly payments and total interest.

46. Student Loans

Group members should present a report on federal loans to finance college costs, including Stafford loans, Perkins loans, and PLUS loans. Also include a discussion of grants that do not have to be repaid, such as Pell Grants and National Merit Scholarships. Refer to *Funding Education Beyond High School*, published by the Department of Education and available at www.studentaid.ed.gov. Use the loan repayment formula that we applied to car loans to determine regular payments and interest on some of the loan options presented in your report.

8.3 The Cost of Home Ownership

Objectives

1. Compute the monthly payment and interest costs for a mortgage.

2. Prepare a partial loan amortization schedule.

3. Solve problems involving what you can afford to spend for a mortgage.

4. Understand the pros and cons of renting versus buying.

The biggest single purchase that most people make in their lives is the purchase of a home. If you choose home ownership at some point in the future, it is likely that you will finance the purchase with an installment loan. Knowing the unique issues surrounding the purchase of a home, and whether or not this aspect of the American dream is right for you, can play a significant role in your financial future.

Mortgages

A **mortgage** is a long-term installment loan (perhaps up to 30, 40, or even 50 years) for the purpose of buying a home, and for which the property is pledged as security for payment. If payments are not made on the loan, the lender may take possession of the property. The **down payment** is the portion of the sale price of the home that the buyer initially pays to the seller. The minimum required down payment is computed as a percentage of the sale price. For example, suppose you decide to buy a $220,000 home. The lender requires you to pay the seller 10% of the sale price. You must pay 10% of $220,000, which is $0.10 \times \$220,000$, or $22,000, to the seller. Thus, $22,000 is the down payment. The **amount of the mortgage** is the difference between the sale price and the down payment. For your $220,000 home, the amount of the mortgage is $220,000 - \$22,000$, or $198,000.

Monthly payments for a mortgage depend on the amount of the mortgage (the principal), the interest rate, and the duration of the mortgage. Mortgages can have a fixed interest rate or a variable interest rate. **Fixed-rate mortgages** have the same monthly principal and interest payment during the entire time of the loan. A loan like this that has a schedule for paying a fixed amount each period is called a **fixed installment loan**. **Variable-rate mortgages**, also known as **adjustable-rate** mortgages (ARMs), have payment amounts that change from time to time depending on changes in the interest rate. ARMS are less predictable than fixed-rate mortgages. They start out at lower rates than fixed-rate mortgages. Caps limit how high rates can go over the term of the loan.

Computations Involved with Buying a Home

1 Compute the monthly payment and interest costs for a mortgage.

Although monthly payments for a mortgage depend on the amount of the mortgage, the duration of the loan, and the interest rate, the interest is not the only cost of a mortgage. Most lending institutions require the buyer to pay one or more **points** at the time of closing—that is, the time at which the mortgage begins. A point is a one-time charge that equals 1% of the loan amount. For example, two points means that the buyer must pay 2% of the loan amount at closing. Often, a buyer can pay fewer points in exchange for a higher interest rate or more points for a lower rate. A document, called the **Truth-in-Lending Disclosure Statement**, shows the buyer the APR, or the annual percentage rate, for the mortgage. The APR takes into account the interest rate and points.

A monthly mortgage payment is used to repay the principal plus interest. In addition, lending institutions can require monthly deposits into an **escrow account**, an account used by the lender to pay real estate taxes and insurance. These deposits increase the amount of the monthly payment.

In the previous section, we used the loan payment formula for fixed installment loans to determine payments on car loans. Because a fixed-rate mortgage is a fixed installment loan, we use the same formula to compute the monthly payment for a mortgage.

Loan Payment Formula for Fixed Installment Loans

The regular payment amount, PMT, required to repay a loan of P dollars paid n times per year over t years at an annual rate r is given by

$$PMT = \frac{P\left(\frac{r}{n}\right)}{\left[1 - \left(1 + \frac{r}{n}\right)^{-nt}\right]}.$$

Example 1   Computing the Monthly Payment and Interest Costs for a Mortgage

The price of a home is $195,000. The bank requires a 10% down payment and two points at the time of closing. The cost of the home is financed with a 30-year fixed-rate mortgage at 7.5%.

a. Find the required down payment.
b. Find the amount of the mortgage.
c. How much must be paid for the two points at closing?
d. Find the monthly payment (excluding escrowed taxes and insurance).
e. Find the total interest paid over 30 years.

Solution

a. The required down payment is 10% of $195,000 or

$$0.10 \times \$195,000 = \$19,500.$$

b. The amount of the mortgage is the difference between the price of the home and the down payment.

Amount of the mortgage = sale price − down payment

= $195,000 − $19,500

= $175,500

c. To find the cost of two points on a mortgage of $175,500, find 2% of $175,500.

0.02 × $175,500 = $3510

The down payment ($19,500) is paid to the seller and the cost of two points ($3510) is paid to the lending institution.

d. We are interested in finding the monthly payment for a $175,500 mortgage at 7.5% for 30 years. We use the loan payment formula for installment loans.

P, the mortgage amount, is $175,500. Fixed rate, r, is 7.5%. 12 payments per year. The mortgage time, t, is 30 years.

$$PMT = \frac{P\left(\frac{r}{n}\right)}{\left[1-\left(1+\frac{r}{n}\right)^{-nt}\right]} = \frac{175,500\left(\frac{0.075}{12}\right)}{\left[1-\left(1+\frac{0.075}{12}\right)^{-12(30)}\right]}$$

$$= \frac{1096.875}{1-(1.00625)^{-360}} \approx 1227$$

The monthly mortgage payment for principal and interest is approximately $1227.00. (Keep in mind that this payment does not include escrowed taxes and insurance.)

e. The total cost of interest over 30 years is equal to the difference between the total of all monthly payments and the amount of the mortgage. The total of all monthly payments is equal to the amount of the monthly payment multiplied by the number of payments. We found the amount of each monthly payment in (d): $1227. The number of payments is equal to the number of months in a year, 12, multiplied by the number of years in the mortgage, 30: 12 × 30 = 360. Thus, the total of all monthly payments = $1227 × 360.
Now we can calculate the interest over 30 years.

Total interest paid = total of all monthly payments minus amount of the mortgage.

= $1227 × 360 − $175,500

= $441,720 − $175,500 = $266,220

The total interest paid over 30 years is approximately $266,220.

Checkpoint 1 In Example 1, the $175,500 mortgage was financed with a 30-year fixed rate at 7.5%. The total interest paid over 30 years was approximately $266,220.

a. Use the loan payment formula for installment loans to find the monthly payment if the time of the mortgage is reduced to 15 years. Round to the nearest dollar.

b. Find the total interest paid over 15 years.

c. How much interest is saved by reducing the mortgage from 30 to 15 years?

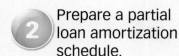

Prepare a partial loan amortization schedule.

Loan Amortization Schedules

When a mortgage loan is paid off through a series of regular payments, it is said to be **amortized**, which literally means "killed off." In working Checkpoint 1(c), were you surprised that nearly $150,000 was saved when the mortgage was amortized over 15 years rather than over 30 years? What adds to the interest cost is the long period over which the loan is financed. **Although each payment is the same, with each successive payment the interest portion decreases and the portion applied toward paying off the principal increases.** The interest is computed using the simple interest formula $I = Prt$. The principal, P, is equal to the balance of the loan, which decreases each month. The rate, r, is the annual interest rate of the mortgage loan. Because a payment is made each month, the time, t, is

$$\frac{1 \text{ month}}{12 \text{ months}} = \frac{1 \cancel{\text{ month}}}{12 \cancel{\text{ months}}}$$

or $\frac{1}{12}$ of a year.

A document showing how the payment each month is split between interest and principal is called a **loan amortization schedule**. Typically, for each payment, this document includes the payment number, the interest for the payment, the amount of the payment applied to the principal, and the balance of the loan after the payment is applied.

Example 2 — Preparing a Loan Amortization Schedule

Prepare a loan amortization schedule for the first two months of the mortgage loan shown in the table below. Round entries to the nearest cent.

Loan Amortization Schedule

Annual % Rate: 9.5%			
Amount of Mortgage: $130,000		Monthly Payment: $1357.50	
Number of Monthly Payments: 180		Term: Years 15, Months 0	
Payment Number	**Interest Payment**	**Principal Payment**	**Balance of Loan**
1			
2			

Solution

We begin with payment number 1.

$$\text{Interest for the month} = Prt = \$130,000 \times 0.095 \times \frac{1}{12} \approx \$1029.17$$

$$\text{Principal payment} = \text{Monthly payment} - \text{Interest payment}$$
$$= \$1357.50 - \$1029.17 = \$328.33$$

$$\text{Balance of loan} = \text{Principal balance} - \text{Principal payment}$$
$$= \$130,000 - \$328.33 = \$129,671.67$$

Now, starting with a loan balance of $129,671.67, we repeat these computations for the second payment.

$$\text{Interest for the month} = Prt = \$129,671.67 \times 0.095 \times \frac{1}{12} \approx \$1026.57$$

$$\text{Principal payment} = \text{Monthly payment} - \text{Interest payment}$$
$$= \$1357.50 - \$1026.57 = \$330.93$$

$$\text{Balance of loan} = \text{Principal balance} - \text{Principal payment}$$
$$= \$129,671.67 - \$330.93 = \$129,340.74$$

The results of these computations are included in **Table 8.4**, a partial loan amortization schedule. By using the simple interest formula month-to-month on the loan's balance, a complete loan amortization schedule for all 180 payments can be calculated.

Table 8.4 Loan Amortization Schedule

Annual % Rate: 9.5%
Amount of Mortgage: $130,000   Monthly Payment: $1357.50
Number of Monthly Payments: 180   Term: Years 15, Months 0

Payment Number	Interest Payment	Principal Payment	Balance of Loan
1	$1029.17	$ 328.33	$129,671.67
2	$1026.57	$ 330.93	$129,340.74
3	$1023.96	$ 333.54	$129,007.22
4	$1021.32	$ 336.18	$128,671.04
30	$ 944.82	$ 412.68	$118,931.35
31	$ 941.55	$ 415.95	$118,515.52
125	$ 484.62	$ 872.88	$ 60,340.84
126	$ 477.71	$ 879.79	$ 59,461.05
179	$ 21.26	$1336.24	$ 1347.74
180	$ 9.76	$1347.74	

Many lenders supply a loan amortization schedule like the one in Example 2 at the time of closing. Such a schedule shows how the buyer pays slightly less in interest and more in principal for each payment over the entire life of the loan.

Checkpoint 2   Prepare a loan amortization schedule for the first two months of the mortgage loan shown in the following table. Round entries to the nearest cent.

Annual % Rate: 7.0%
Amount of Mortgage: $200,000   Monthly Payment: $1550.00
Number of Monthly Payments: 240   Term: Years 20, Months 0

Payment Number	Interest Payment	Principal Payment	Balance of Loan
1			
2			

Blitzer Bonus

Bittersweet Interest

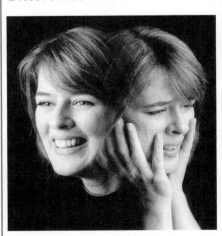

Looking at amortization tables, you could get discouraged by how much of your early mortgage payments goes toward interest and how little goes toward paying off the principal. Although you get socked with tons of interest in the early years of a loan, the one bright side to the staggering cost of a mortgage is the **mortgage interest tax deduction**. To make the cost of owning a home more affordable, the tax code permits deducting all the mortgage interest (but not the principal) that you pay per year on the loan. **Table 8.5** illustrates how this tax loophole reduces the cost of the mortgage.

Table 8.5 Tax Deductions for a $100,000 Mortgage at 7% for a Taxpayer in the 28% Tax Bracket

Year	Interest	Tax Savings	Net Cost of Mortgage
1	$6968	$1951	$5017
2	$6895	$1931	$4964
3	$6816	$1908	$4908
4	$6732	$1885	$4847
5	$6641	$1859	$4782

3 Solve problems involving what you can afford to spend for a mortgage.

Determining What You Can Afford

Here's the bottom line from most financial advisers:

- Spend no more than 28% of your gross monthly income for your mortgage payment.
- Spend no more than 36% of your gross monthly income for your total monthly debt, including mortgage payments, car payments, credit card bills, student loans, and medical debt.

Using these guidelines, **Table 8.6** shows the maximum monthly amount you could afford for mortgage payments and total credit obligations for a variety of income levels.

Table 8.6 Maximum Amount You Can Afford

Gross Annual Income	Monthly Mortgage Payment	Total Monthly Credit Obligations
$20,000	$467	$600
$30,000	$700	$900
$40,000	$933	$1200
$50,000	$1167	$1500
$60,000	$1400	$1800
$70,000	$1633	$2100
$80,000	$1867	$2400
$90,000	$2100	$2700
$100,000	$2333	$3000

Source: Fannie Mae

Example 3 What Can You Afford?

Suppose that your gross annual income is $25,000.

a. What is the maximum amount you should spend each month on a mortgage payment?

b. What is the maximum amount you should spend each month for total credit obligations?

c. If your monthly mortgage payment is 80% of the maximum amount you can afford, what is the maximum amount you should spend each month for all other debt?

Round all computations to the nearest dollar.

Solution

With a gross annual income of $25,000, your gross monthly income is

$$\frac{\$25,000}{12},$$

or approximately $2083.

a. You should spend no more than 28% of your gross monthly income, $2083, on a mortgage payment.

$$28\% \text{ of } \$2083 = 0.28 \times \$2083 \approx \$583.$$

Your monthly mortgage payment should not exceed $583.

b. You should spend no more than 36% of your gross monthly income, $2083, for total monthly debt.

$$36\% \text{ of } \$2083 = 0.36 \times \$2083 \approx \$750.$$

Your total monthly credit obligations should not exceed $750.

c. The problem's conditions state that your monthly mortgage payment is 80% of the maximum you can afford, which is $583. This means that your monthly mortgage payment is 80% of $583.

$$80\% \text{ of } \$583 = 0.8 \times \$583 \approx \$466.$$

In part (b) we saw that your total monthly debt should not exceed $750. Because you are paying $466 for your mortgage payment, this leaves $750 − $466, or $284, for all other debt. Your monthly credit obligations, excluding mortgage payments, should not exceed $284.

> **Checkpoint 3** Suppose that your gross annual income is $240,000. (You've obviously been reading the "Achieving $uccess" essays in this book and following all our suggestions!)
>
> **a.** What is the maximum amount you should spend each month on a mortgage payment?
> **b.** What is the maximum amount you should spend each month for total credit obligations?
> **c.** If your monthly mortgage payment is 90% of the maximum amount you can afford, what is the maximum amount you should spend each month for all other debt?
>
> Round all computations to the nearest dollar.

Renting versus Buying

 4 Understand the pros and cons of renting versus buying.

Nearly everyone is faced at some stage in life with the dilemma "should I rent or should I buy a home?" The rent-or-buy decision can be highly complex and is often based on lifestyle rather than finances. Aside from a changing economic climate, there are many factors to consider. Here are some advantages of both renting and buying to help smooth the way:

Benefits of Renting
- No down payment or points are required. You generally have a security deposit that is returned at the end of your lease.
- Very mobile: You can easily relocate, moving as often as you like and as your lease permits.
- Does not tie up hundreds of thousands of dollars that might be invested more safely and lucratively elsewhere. Most financial advisers agree that you should buy a home because you want to live in it, not because you want to fund your retirement.
- Does not clutter what you can afford for your total monthly debt with mortgage payments.
- May involve lower monthly expenses. You pay rent, whereas a homeowner pays the mortgage, taxes, insurance, and upkeep.
- Can provide amenities like swimming pools, tennis courts, and health clubs.
- Avoids the risk of falling housing prices.
- Does not require home repair, maintenance, and groundskeeping.
- There are no property taxes.
- Generally less costly than buying a home when staying in it for fewer than three years.

Benefits of Home Ownership
- Peace of mind and stability.
- Provides significant tax advantages, including deduction of mortgage interest and property taxes.
- There is no chance of rent increasing over time.

- Allows for freedom to remodel, landscape, and redecorate.
- You can build up **equity**, the difference between the home's value and what you owe on the mortgage, as the mortgage is paid off. The possibility of home appreciation is a potential source of cash in the form of home equity loans.
- When looking at seven-year time frames, the total cost of renting (monthly rent, renter's insurance, loss of potential interest on a security deposit) is more than twice the total cost of buying for home owners who itemize their tax deductions.

Source: Arthur J. Keown, *Personal Finance*, Fourth Edition, Pearson, 2007.

Achieving $uccess

Reducing Rental Costs

Let's assume that one of your long-term financial goals involves home ownership. It's still likely that you'll be renting for a while once you complete your education and begin your first job. Other than living in a tent, here are some realistic suggestions for reducing costs on your first rental:

- Select a lower-cost rental. Who says you should begin your career in a large apartment with fancy amenities, private parking spots, and lakefront views? The less you spend renting, the more you can save for a down payment toward buying your own place. You will ultimately qualify for the most favorable mortgage terms by making a down payment of at least 20% of the purchase price of the property.
- Negotiate rental increases. Landlords do not want to lose good tenants who are respectful of their property and pay rent on time. Filling vacancies can be time consuming and costly.
- Rent a larger place with roommates. By sharing a rental, you will decrease rental costs and get more home for your rental dollars.

Exercise Set 8.3

Concept and Vocabulary Exercises

In Exercises 1–2, fill in each blank so that the resulting statement is true.

1. A long-term installment loan for the purpose of buying a home is called a/an _____. The portion of the sale price of the home that the buyer initially pays to the seller is called the _____.

2. A document showing how each monthly installment payment is split between interest and principal is called a/an _____.

In Exercises 3–6, determine whether each statement is true or false. If the statement is false, make the necessary change(s) to produce a true statement.

3. Over the life of an installment loan, the interest portion increases and the portion applied to paying off the principal decreases with each successive payment.

4. Financial advisors suggest spending no more than 5% of your gross monthly income for your mortgage payment.

5. Renters are not required to pay property taxes.

6. Renters can build up equity as the rent is paid each month.

Respond to Exercises 7–16 using verbal or written explanations.

7. What is a mortgage?

8. What is a down payment?

9. How is the amount of a mortgage determined?

10. Describe why a buyer would select a 30-year fixed-rate mortgage instead of a 15-year fixed-rate mortgage if interest rates are $\frac{1}{4}$% to $\frac{1}{2}$% lower on a 15-year mortgage.

11. Describe one advantage and one disadvantage of an adjustable-rate mortgage over a fixed-rate mortgage.

12. What is a loan amortization schedule?

13. Describe what happens to the portions of payments going to principal and interest over the life of an installment loan.

14. Describe how to determine what you can afford for your monthly mortgage payment.

15. Describe two advantages of renting over home ownership.

16. Describe two advantages of home ownership over renting.

In Exercises 17–20, determine whether each statement makes sense or does not make sense, and explain your reasoning.

17. I use the same formula to determine mortgage payments and payments for car loans.

18. There must be an error in the loan amortization schedule for my mortgage because the annual interest rate is only 3.5%, yet the schedule shows that I'm paying more on interest than on the principal for many of my payments.

19. My landlord required me to pay 2 points when I signed my rental lease.

20. I include rental payments among my itemized tax deductions.

Practice and Application Exercises

In Exercises 21–30, use

$$PMT = \frac{P\left(\dfrac{r}{n}\right)}{\left[1 - \left(1 + \dfrac{r}{n}\right)^{-nt}\right]}$$

to determine the regular payment amount, rounded to the nearest dollar.

21. The price of a home is $220,000. The bank requires a 20% down payment and three points at the time of closing. The cost of the home is financed with a 30-year fixed-rate mortgage at 7%.

   a. Find the required down payment.

   b. Find the amount of the mortgage.

   c. How much must be paid for the three points at closing?

   d. Find the monthly payment (excluding escrowed taxes and insurance).

   e. Find the total cost of interest over 30 years.

22. The price of a condominium is $180,000. The bank requires a 5% down payment and one point at the time of closing. The cost of the condominium is financed with a 30-year fixed-rate mortgage at 8%.

   a. Find the required down payment.

   b. Find the amount of the mortgage.

   c. How much must be paid for the one point at closing?

   d. Find the monthly payment (excluding escrowed taxes and insurance).

   e. Find the total cost of interest over 30 years.

23. The price of a small cabin is $100,000. The bank requires a 5% down payment. The buyer is offered two mortgage options: 20-year fixed at 8% or 30-year fixed at 8%. Calculate the amount of interest paid for each option. How much does the buyer save in interest with the 20-year option?

24. The price of a home is $160,000. The bank requires a 15% down payment. The buyer is offered two mortgage options: 15-year fixed at 8% or 30-year fixed at 8%. Calculate the amount of interest paid for each option. How much does the buyer save in interest with the 15-year option?

25. In terms of paying less in interest, which is more economical for a $150,000 mortgage: a 30-year fixed-rate at 8% or a 20-year fixed-rate at 7.5%? How much is saved in interest?

26. In terms of paying less in interest, which is more economical for a $90,000 mortgage: a 30-year fixed-rate at 8% or a 15-year fixed-rate at 7.5%? How much is saved in interest?

In Exercises 27–28, which mortgage loan has the greater total cost (closing costs + the amount paid for points + total cost of interest)? By how much?

27. A $120,000 mortgage with two loan options:

   Mortgage A: 30-year fixed at 7% with closing costs of $2000 and one point

   Mortgage B: 30-year fixed at 6.5% with closing costs of $1500 and four points

28. A $250,000 mortgage with two loan options:

   Mortgage A: 30-year fixed at 7.25% with closing costs of $2000 and one point

   Mortgage B: 30-year fixed at 6.25% with closing costs of $350 and four points

29. The cost of a home is financed with a $120,000 30-year fixed-rate mortgage at 4.5%.

   a. Find the monthly payments and the total interest for the loan.

   b. Prepare a loan amortization schedule for the first three months of the mortgage. Round entries to the nearest cent.

Payment Number	Interest	Principal	Loan Balance
1			
2			
3			

30. The cost of a home is financed with a $160,000 30-year fixed-rate mortgage at 4.2%.

   a. Find the monthly payments and the total interest for the loan.

   b. Prepare a loan amortization schedule for the first three months of the mortgage. Round entries to the nearest cent.

Payment Number	Interest	Principal	Loan Balance
1			
2			
3			

Use this advice from most financial advisers to solve Exercises 31–34.

- *Spend no more than 28% of your gross monthly income for your mortgage payment.*

- *Spend no more than 36% of your gross monthly income for your total monthly debt.*

Round all computations to the nearest dollar.

31. Suppose that your gross annual income is $36,000.

   a. What is the maximum amount you should spend each month on a mortgage payment?

   b. What is the maximum amount you should spend each month for total credit obligations?

   c. If your monthly mortgage payment is 70% of the maximum you can afford, what is the maximum amount you should spend each month for all other debt?

32. Suppose that your gross annual income is $62,000.

    a. What is the maximum amount you should spend each month on a mortgage payment?

    b. What is the maximum amount you should spend each month for total credit obligations?

    c. If your monthly mortgage payment is 90% of the maximum you can afford, what is the maximum amount you should spend each month for all other debt?

Critical Thinking Exercises

33. If your gross annual income is $75,000, use appropriate computations to determine whether you could afford a $200,000 30-year fixed-rate mortgage at 5.5%.

34. The partial loan amortization schedule shows payments 50–54. Although payment 50 is correct, there are errors in one or more of the payments from 51 through 54. Find the errors and correct them.

Loan Amortization Schedule

Annual % Rate: 6.0% Amount of Mortgage: $120,000 Number of Monthly Payments: 180		Monthly Payment: $1012.63 Term: Years 15, Months 0	
Payment Number	**Interest Payment**	**Principal Payment**	**Balance of Loan**
50	$485.77	$526.86	$96,626.51
51	$483.13	$529.50	$96,097.01
52	$477.82	$534.81	$95,030.06
53	$480.49	$532.14	$95,564.87
54	$495.15	$537.48	$94,492.58

8.4   Credit Cards

Would you like to buy products with a credit card? Although the card will let you use a product while paying for it, the costs associated with such cards, including their high interest rates, fees, and penalties, stack the odds in favor of your getting hurt by them. In 2008, the average credit-card debt per U.S. household was $10,691. One advantage of making a purchase with a credit card is that the consumer gets to use a product immediately. In this section, we will see that a significant disadvantage is that it can add a substantial amount to the cost of a purchase. When it comes to using a credit card, consumer beware!

Open-End Installment Loans

Using a credit card is an example of an open-end installment loan, commonly called **revolving credit**. Open-end loans differ from fixed installment loans such as car loans and mortgages in that there is no schedule for paying a fixed amount each period. Credit card loans require users to make only a minimum monthly payment that depends on the unpaid balance and the interest rate. Credit cards have high interest rates compared to other kinds of loans. The interest on credit cards is computed using the simple interest formula $I = Prt$. However, r represents the *monthly* interest rate and t is time in months rather than in years. A typical interest rate is 1.57% monthly. This is equivalent to a yearly rate of $12 \times 1.57\%$, or 18.84%. With such a high annual percentage rate, credit card balances should be paid off as quickly as possible.

    Most credit card customers are billed every month. A typical billing period is May 1 through May 31, but it can also run from, say, May 5 through June 4. Customers receive a statement, called an **itemized billing**, that includes the unpaid balance on the first day of the billing period, the total balance owed on the last day of the billing period, a list of purchases and cash advances made during the billing period, any finance charges or other fees incurred, the date of the last day of the billing period, the payment due date, and the minimum payment required.

Customers who make a purchase during the billing period and pay the entire amount of the purchase by the payment due date are not charged interest. By contrast, customers who make cash advances using their credit cards must pay interest from the day the money is advanced until the day it is repaid.

Interest on Credit Cards: The Average Daily Balance Method

Methods for calculating interest, or finance charges, on credit cards may vary and the interest can differ on credit cards that show the same annual percentage rate, or APR. The method used for calculating interest on most credit cards is called the *average daily balance method*.

Find the interest, the balance due, and the minimum monthly payment for credit card loans.

> ### The Average Daily Balance Method
>
> Interest is calculated using $I = Prt$, where r is the monthly rate and t is one month. The principal, P, is the average daily balance. The **average daily balance** is the sum of the unpaid balances for each day in the billing period divided by the number of days in the billing period.
>
> Average daily balance
> $$= \frac{\text{Sum of the unpaid balances for each day in the billing period}}{\text{Number of days in the billing period}}$$

In Example 1, we illustrate how to determine the average daily balance. At the conclusion of the example, we summarize the steps used in the computation.

Example 1 Balance Due on a Credit Card

The issuer of a particular VISA card calculates interest using the average daily balance method. The monthly interest rate is 1.3% of the average daily balance. The following transactions occurred during the May 1–May 31 billing period.

Transaction Description	Transaction Amount
Previous balance, $1350.00	
May 1 Billing date	
May 8 Payment	$250.00 credit
May 10 Charge: Airline Tickets	$375.00
May 20 Charge: Books	$ 57.50
May 28 Charge: Restaurant	$ 65.30
May 31 End of billing period	
Payment Due Date: June 9	

a. Find the average daily balance for the billing period. Round to the nearest cent.

b. Find the interest to be paid on June 1, the next billing date. Round to the nearest cent.

c. Find the balance due on June 1.

d. This credit card requires a $10 minimum monthly payment if the balance due at the end of the billing period is less than $360. Otherwise, the minimum monthly payment is $\frac{1}{36}$ of the balance due at the end of the billing period, rounded up to the nearest whole dollar. What is the minimum monthly payment due by June 9?

Solution

a. We begin by finding the average daily balance for the billing period. First make a table that shows the beginning date of the billing period, each transaction date, and the unpaid balance for each date.

Date	Unpaid Balance	
May 1	$1350.00	previous balance
May 8	$1350.00 − $250.00 = $1100.00	$250.00 payment
May 10	$1100.00 + $375.00 = $1475.00	$375.00 charge
May 20	$1475.00 + $57.50 = $1532.50	$57.50 charge
May 28	$1532.50 + $65.30 = $1597.80	$65.30 charge

We now extend our table by adding two columns. One column shows the number of days at each unpaid balance. The final column shows each unpaid balance multiplied by the number of days that the balance is outstanding.

Date	Unpaid Balance	Number of Days at Each Unpaid Balance	$\left(\begin{array}{c}\text{Unpaid}\\\text{Balance}\end{array}\right) \cdot \left(\begin{array}{c}\text{Number}\\\text{of Days}\end{array}\right)$
May 1	$1350.00	7	($1350.00)(7) = $9450.00
May 8	$1100.00	2	($1100.00)(2) = $2200.00
May 10	$1475.00	10	($1475.00)(10) = $14,750.00
May 20	$1532.50	8	($1532.50)(8) = $12,260.00
May 28	$1597.80	4	($1597.80)(4) = $6391.20

Total: 31   Total: $45,051.20

There are 4 days at this unpaid balance, May 28, 29, 30, and 31 before the beginning of the next billing period, June 1.

This is the number of days in the billing period.

This is the sum of the unpaid balances for each day in the billing period.

Notice that we found the sum of the products in the final column of the table. This dollar amount, $45,051.20, gives the sum of the unpaid balances for each day in the billing period.

Now we divide the sum of the unpaid balances for each day in the billing period, $45,051.20, by the number of days in the billing period, 31. This gives the average daily balance.

Average daily balance

$$= \frac{\text{Sum of the unpaid balances for each day in the billing period}}{\text{Number of days in the billing period}}$$

$$= \frac{\$45,051.20}{31} \approx \$1453.26$$

The average daily balance is approximately $1453.26.

b. Now we find the interest to be paid on June 1, the next billing date. The monthly interest rate is 1.3% of the average daily balance. The interest due is computed using $I = Prt$.

$$I = Prt = (\$1453.26)(0.013)(1) \approx \$18.89$$

The average daily balance serves as the principal.

Time, t, is measured in months, and $t = 1$ month.

The interest, or finance charge, for the June 1 billing will be $18.89.

c. The balance due on June 1, the next billing date, is the unpaid balance on May 31 plus the interest.

$$\text{Balance due} = \$1597.80 + \$18.89 = \$1616.69$$

Unpaid balance on May 31, obtained from the second table on the previous page	Interest, or finance charge, obtained from part (b)

The balance due on June 1 is $1616.69.

d. Because the balance due, $1616.69, exceeds $360, the customer must pay a minimum of $\frac{1}{36}$ of the balance due.

$$\text{Minimum monthly payment} = \frac{\text{balance due}}{36} = \frac{\$1616.69}{36} \approx \$45$$

Rounded up to the nearest whole dollar, the minimum monthly payment due by June 9 is $45.

The following box summarizes the steps used in Example 1 to determine the average daily balance. Calculating the average daily balance can be quite tedious when there are numerous transactions during a billing period.

Study Tip

The quotient in part (d) is approximately 44.908, which rounds to 45. Because the minimum monthly payment is rounded up, $45 would still be the payment if the approximate quotient had been 44.098.

Determining the Average Daily Balance

Step 1 Make a table that shows the beginning date of the billing period, each transaction date, and the unpaid balance for each date.

Step 2 Add a column to the table that shows the number of days at each unpaid balance.

Step 3 Add a final column to the table that shows each unpaid balance multiplied by the number of days that the balance is outstanding.

Step 4 Find the sum of the products in the final column of the table. This dollar amount is the sum of the unpaid balances for each day in the billing period.

Step 5 Compute the average daily balance.

Average daily balance

$$= \frac{\text{Sum of the unpaid balances for each day in the billing period}}{\text{Number of days in the billing period}}$$

Checkpoint 1 A credit card company calculates interest using the average daily balance method. The monthly interest rate is 1.6% of the average daily balance. The following transactions occurred during the May 1–May 31 billing period.

Transaction Description	Transaction Amount
Previous balance, $8240.00	
May 1 Billing date	
May 7 Payment	$ 350.00 credit
May 15 Charge: Computer	$1405.00
May 17 Charge: Restaurant	$ 45.20
May 30 Charge: Clothing	$ 180.72
May 31 End of billing period	
Payment Due Date: June 9	

Answer parts (a) through (d) in Example 1 on page 535 using this information.

2 Understand the pros and cons of using credit cards.

Credit Cards: Marvelous Tools or Snakes in Your Wallet?

Credit cards are convenient. Pay the entire balance by the due date for each monthly billing and you avoid interest charges. Carry over a balance and interest charges quickly add up. With this in mind, let's consider the positives and the negatives involved with credit card usage.

Advantages of Using Credit Cards

- Get to use a product before actually paying for it.
- No interest charges by paying the balance due at the end of each billing period.
- Responsible use is an effective way to build a good credit score. (See page 539 for a discussion of credit scores.)
- No need to carry around large amounts of cash.
- More convenient to use than checks.
- Offer consumer protections: If there is a disputed or fraudulent charge on your credit card statement, let the card issuer know and the amount is generally removed.
- Provides a source of temporary emergency funds.
- Extends shopping opportunities to purchases over the phone or the Internet.
- Simple tasks like renting a car or booking a hotel room can be difficult or impossible without a credit card.
- Monthly statements can help keep track of spending. Some card issuers provide an annual statement that aids in tax preparation.
- May provide amenities such as free miles toward air travel.
- Useful as identification when multiple pieces of identification are needed.

Credit Card Woes

- High interest rates on unpaid balances. In 2009, interest rates were as high as 30%.
- No cap on interest rates. In 2009, the U.S. Senate defeated an amendment that would have imposed a 15% cap on credit card interest rates. (The Credit Card Act, passed by Congress in 2009, does restrict when issuers can raise rates on existing unpaid balances.) Your initial credit-card interest rate is unlikely to go down, but it can sure go up.

Average Credit-Card Debt per U.S. Household

A bar graph titled "Average Credit-Card Debt per U.S. Household" with y-axis "Average Credit-Card Balance" ranging from $2000 to $12,000 and x-axis "Year" showing: 1992 = 3803, 1996 = 6912, 2000 = 8308, 2004 = 9577, 2008 = 10,691.

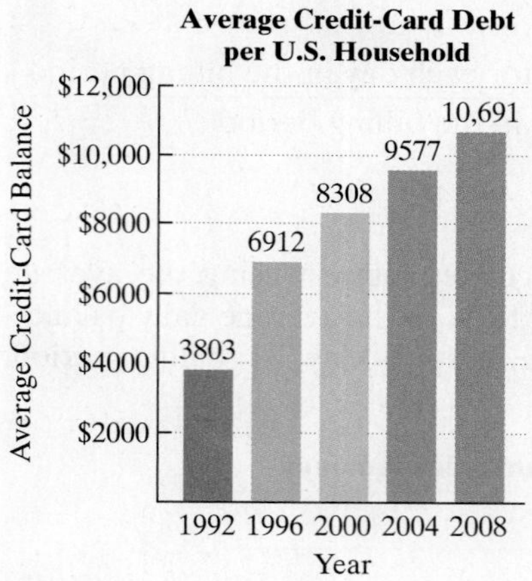

Figure 8.1
Source: CardTrak.com

- No cap on fees. *Consumer Reports* (October 2008) cited a credit card with an enticing 9.9% annual interest rate. But the fine print revealed a $29 account-setup fee, a $95 program fee, a $48 annual fee, and a $7 monthly servicing fee. Nearly 40% of the $40 billion in profits that U.S. card issuers earned in 2008 came from fees. Furthermore, issuers can hike fees at any time, for any reason. Read the fine print of a credit-card agreement before you sign up.
- Easy to overspend. Purchases with credit cards can create the illusion that you are not actually spending money.
- Can serve as a tool for financial trouble. Using a credit card to buy more than you can afford and failing to pay the bill in full each month can result in serious debt. Fees and interest charges are added to the balance, which continues to grow, even if there are no new purchases. **Figure 8.1** indicates that the average credit-card debt per U.S. household nearly tripled from 1992 to 2008.
- The minimum-payment trap: Credit-card debt is made worse by paying only the required minimum, a mistake made by 11% of credit-card debtors. Pay the minimum and most of it goes to interest charges.

3 Understand the difference between credit cards and debit cards.

Debit Cards

Credit cards have been around since the 1950s. Their plastic lookalikes, debit cards, were introduced in the mid-1970s. Although debit cards look like credit cards, the big difference is that debit cards are linked to your bank account. When you use a debit card, the money you spend is deducted electronically from your bank deposit. It's similar to writing an electronic check, but there's no paper involved and the check gets "cashed" instantly.

Debit cards offer the convenience of making purchases with a piece of plastic without the temptation or ability to run up credit-card debt. You can't spend money you don't have because the card won't work if the money isn't in your checking or savings account.

Debit cards have drawbacks. They may not offer the protection a credit card does for disputed purchases. It's easy to rack up overdraft charges if your bank enrolls you in an "overdraft protection" program. That means your card won't be turned down if you do not have sufficient funds in your account to cover your purchase. You are spared the embarrassment of having your card rejected, but it will cost you fees of approximately $27 per overdraft.

Debit card purchases should be treated like those for which you use a check. Record all transactions and their amounts in your checkbook, including any cash received from an ATM. Always keep track of how much money is available in your account. Your balance can be checked at an ATM or online.

Credit Reports and Credit Scores

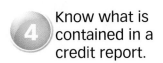

Know what is contained in a credit report.

As a high school junior or senior, it is unlikely that you have a credit history. Once you apply for your first credit card, your personal *credit report* will begin. A **credit report** contains the following information:

- **Identifying Information:** This includes your name, social security number, current address, and previous addresses.
- **Record of Credit Accounts:** This includes details about all open or closed credit accounts, such as when each account was opened, the latest balance, and the payment history.
- **Public Record Information:** Any of your public records, such as bankruptcy information, appears in this section of the credit report.
- **Collection Agency Account Information:** Unpaid accounts are turned over to collection agencies. Information about such actions appear in this section of the credit report.
- **Inquiries:** Companies that have asked for your credit information because you applied for credit are listed here.

Organizations known as **credit bureaus** collect credit information on individual consumers and provide credit reports to potential lenders, employers, and others upon request. The three main credit bureaus are Equifax, Experian, and TransUnion.

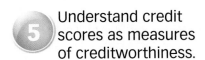

Understand credit scores as measures of creditworthiness.

Credit bureaus use data from your credit report to create a *credit score*, which is used to measure your creditworthiness. **Credit scores**, or **FICO scores**, range from 300 to 850, with a higher score indicating better credit. **Table 8.7** contains ranges of credit scores and their measures of creditworthiness.

Table 8.7 Credit Scores and Their Significance	
Scores	**Creditworthiness**
720–850	Very good to excellent; Best interest rates on loans
650–719	Good; Likely to get credit, but not the best interest rates on loans
630–649	Fair; May get credit, but only at higher rates
580–629	Poor; Likely to be denied credit by all but a high-interest lender
350–579	Bad; Likely to be denied credit

Your credit score will have an enormous effect on your financial life. Individuals with higher credit scores get better interest rates on loans because they are considered to have a lower risk of defaulting. A good credit score can save you thousands of dollars in interest charges over your lifetime.

Blitzer Bonus

Teens, College Students, and Credit Cards

You must be at least 18 to qualify for a credit card on your own. Younger teens can get a card through their parents. Although your name is on a card obtained through your parents, your parents are responsible for paying the bill. The spending limit on most cards issued to teens is $500.

If you have no credit history but are at least 18 years old with a job, you may be able to get a card with a limited amount of credit, usually $500 to $1000. Your chances of getting a credit card increase if you apply through a bank with which you have an account.

Prior to 2010, it was not necessary for college students to have a job to get a credit card. Credit card companies viewed college students as good and responsible customers who would continue to have a lifelong need for credit. The issuers anticipated retaining these students after graduation when their accounts would become more valuable. Many resorted to aggressive marketing tactics, offering everything from T-shirts to iPods to students who signed up.

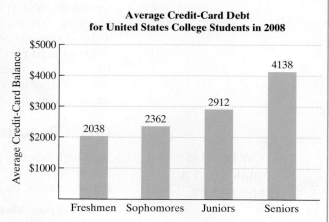

Average Credit-Card Debt for United States College Students in 2008

Source: Nellie Mae

Times have changed. In May 2009, President Obama signed legislation that prohibits issuing credit cards to college students younger than 21 unless they can prove they are able to make payments or get a parent or guardian to co-sign. Because college students do not have much money, most won't be able to get a credit card without permission from their parents. The bill also requires lenders to get permission from the co-signer before increasing the card's credit limit.

Before credit card reform swiped easy plastic from college students, those who fell behind on their credit card bills often left college with blemished credit reports. This made it more difficult for them to rent an apartment, get a car loan, or even find a job. "A lot of kids got themselves in trouble," said Adam Levin, founder of Credit.com, a consumer Web site. "As much as college students are obsessed with GPAs, their credit score is the most important number they're going to have to deal with after graduation."

Achieving $uccess

Responsible Credit Card Use

When you decide to apply for your first credit card, you should look for a card with no annual fee and a low interest rate. You can compare rates and fees at www.creditcards.com, www.bankrate.com, and www.indexcreditcards.com.

A critical component of financial success involves demonstrating that you can handle the responsibility of using a credit card. Responsible credit card use means:

- You pay the entire balance by the due date for each monthly billing.
- You only use the card to make purchases that you can afford.
- You save all receipts for credit-card purchases and check each itemized billing carefully for any errors.
- You use your credit card sometimes, but also use cash, checks, or your debit card.

Some people believe that credit cards are evil by nature. Your author takes issue with this point of view. Whether credit cards become your friends or your foes depends on how you will handle them in the future.

Exercise Set 8.4

Concept and Vocabulary Exercises

In Exercises 1–5, fill in each blank so that the resulting statement is true.

1. Using a credit card is an example of a/an _____ installment loan for which there is no schedule for paying a fixed amount each period.

2. The average daily balance for a credit card's billing period is _____ divided by _____.

3. When you use a/an _____ card, the money you spend is deducted electronically from your bank account.

4. Details about when all your credit accounts were opened, their latest balance, and their payment history are included in a/an _____.

5. Credit scores range from _____ to _____, with a higher score indicating _____.

In Exercises 6–9, determine whether each statement is true or false. If the statement is false, make the necessary change(s) to produce a true statement.

6. Interest on a credit card is calculated using $I = Prt$, where r is the monthly rate, t is one month, and P is the balance due.

7. When using a credit card, the money spent is deducted electronically from the user's bank account.

8. Credit reports contain bankruptcy information.

9. Higher credit scores indicate better credit.

Respond to Exercises 10–17 using verbal or written explanations.

10. Describe the difference between a fixed installment loan and an open-end installment loan.

11. For a credit card billing period, describe how the average daily balance is determined. Why is this computation somewhat tedious when done by hand?

12. Describe two advantages of using credit cards.

13. Describe two disadvantages of using credit cards.

14. What is a debit card?

15. Describe what is contained in a credit report.

16. What are credit scores?

17. Describe two aspects of responsible credit card use.

In Exercises 18–24, determine whether each statement makes sense or does not make sense, and explain your reasoning.

18. I like to keep all my money, so I pay only the minimum required payment on my credit card.

19. One advantage of using credit cards is that there are caps on interest rates and fees.

20. The balance due on my credit card from last month to this month increased even though I made no new purchases.

21. My debit card offers the convenience of making purchases with a piece of plastic without the ability to run up credit-card debt.

22. My credit score is 630, so I anticipate an offer on a car loan at a very low interest rate.

23. In order to achieve financial success, I should consistently pay my entire credit card balance by the due date.

24. Once I begin college, I'll be enticed to sign up for credit cards with offers of free T-shirts or iPods.

Practice and Application Exercises

Exercises 25–26 involve credit cards that calculate interest using the average daily balance method. The monthly interest rate is 1.5% of the average daily balance. Each exercise shows transactions that occurred during the March 1–March 31 billing period. In each exercise,

a. Find the average daily balance for the billing period. Round to the nearest cent.

b. Find the interest to be paid on April 1, the next billing date. Round to the nearest cent.

c. Find the balance due on April 1.

d. This credit card requires a $10 minimum monthly payment if the balance due at the end of the billing period is less than $360. Otherwise, the minimum monthly payment is $\frac{1}{36}$ of the balance due at the end of the billing period, rounded up to the nearest whole dollar. What is the minimum monthly payment due by April 9?

25.

Transaction Description	Transaction Amount
Previous balance, $6240.00	
March 1 Billing date	
March 5 Payment	$300.00 credit
March 7 Charge: Restaurant	$ 40.00
March 12 Charge: Groceries	$ 90.00
March 21 Charge: Car Repairs	$230.00
March 31 End of billing period	
Payment Due Date: April 9	

26.

Transaction Description	Transaction Amount
Previous balance, $7150.00	
March 1 Billing date	
March 4 Payment	$ 400.00 credit
March 6 Charge: Furniture	$1200.00
March 15 Charge: Gas	$ 40.00
March 30 Charge: Groceries	$ 50.00
March 31 End of billing period	
Payment Due Date: April 9	

Exercises 27–28 involve credit cards that calculate interest using the average daily balance method. The monthly interest rate is 1.2% of the average daily balance. Each exercise shows transactions that occurred during the June 1-June 30 billing period. In each exercise,

a. Find the average daily balance for the billing period. Round to the nearest cent.

b. Find the interest to be paid on July 1, the next billing date. Round to the nearest cent.

c. Find the balance due on July 1.

d. This credit card requires a $30 minimum monthly payment if the balance due at the end of the billing period is less than $400. Otherwise, the minimum monthly payment is $\frac{1}{25}$ of the balance due at the end of the billing period, rounded up to the nearest whole dollar. What is the minimum monthly payment due by July 9?

27.

Transaction Description	Transaction Amount
Previous balance, $2653.48	
June 1 Billing date	
June 6 Payment	$1000.00 credit
June 8 Charge: Gas	$ 36.25
June 9 Charge: Groceries	$ 138.43
June 17 Charge: Gas Charge: Groceries	$ 42.36 $ 127.19
June 27 Charge: Clothing	$ 214.83
June 30 End of billing period	
Payment Due Date: July 9	

28.

Transaction Description	Transaction Amount
Previous balance, $4037.93	
June 1 Billing date	
June 5 Payment	$350.00 credit
June 10 Charge: Gas	$ 31.17
June 15 Charge: Prescriptions	$ 42.50
June 22 Charge: Gas Charge: Groceries	$ 43.86 $112.91
June 29 Charge: Clothing	$ 96.73
June 30 End of billing period	
Payment Due Date: July 9	

In Exercises 29–34, use

$$PMT = \frac{P\left(\dfrac{r}{n}\right)}{\left[1 - \left(1 + \dfrac{r}{n}\right)^{-nt}\right]}$$

to determine the regular payment amount, rounded to the nearest dollar.

29. Suppose your credit card has a balance of $4200 and an annual interest rate of 18%. You decide to pay off the balance over two years. If there are no further purchases charged to the card,

a. How much must you pay each month?

b. How much total interest will you pay?

30. Suppose your credit card has a balance of $3600 and an annual interest rate of 16.5%. You decide to pay off the balance over two years. If there are no further purchases charged to the card,

a. How much must you pay each month?

b. How much total interest will you pay?

31. To pay off the $4200 credit-card balance in Exercise 29, suppose that you can get a bank loan at 10.5% with a term of three years.

a. How much will you pay each month? How does this compare with your credit-card payment in Exercise 29?

b. How much total interest will you pay? How does this compare with your total credit-card interest in Exercise 29?

32. To pay off the $3600 credit-card balance in Exercise 30, suppose that you can get a bank loan at 9.5% with a term of three years.

a. How much will you pay each month? How does this compare with your credit-card payment in Exercise 30?

b. How much total interest will you pay? How does this compare with your total credit-card interest in Exercise 30?

33. Rework Exercise 29 assuming you decide to pay off the balance over one year rather than two. How much more must you pay each month and how much less will you pay in total interest?

34. Rework Exercise 30 assuming you decide to pay off the balance over one year rather than two. How much more must you pay each month and how much less will you pay in total interest?

Critical Thinking Exercise

35. A bank bills its credit card holders on the first of each month for each itemized billing. The card provides a 20-day period in which to pay the bill before charging interest. If the card holder wants to buy an expensive gift for a September 30 wedding but can't pay for it until November 5, explain how this can be done without adding an interest charge.

Group Exercises

36. Cellphone Plans

If credit cards can cause financial woes, cellphone plans are not far behind. Group members should present a report on cellphone plans, addressing each of the following questions: What are the monthly fees for these plans and what features are included? What happens if you use the phone more than the plan allows? Are there higher rates for text messaging and Internet access? What additional charges are imposed by the carrier on top of the monthly fee? What are the termination fees if you default on the plan? What can happen to your credit report and your credit score in the event of early termination? Does the carrier use free T-shirts, phones, and other items to entice new subscribers into binding contracts? What suggestions can the group offer to avoid financial difficulties with these plans?

37. Risky Credit Arrangements

Group members should present a report on the characteristics and financial risks associated with payday lending, tax refund loans, and pawn shops.

Chapter 8 Summary

8.1 Annuities, Methods of Saving, and Investments

Definitions and Concepts

An annuity is a sequence of equal payments made at equal time periods. The value of an annuity is the sum of all deposits plus all interest paid.

Value of an Annuity: Interest Compounded Once a Year

If P is the deposit made at the end of each year for an annuity that pays an annual interest rate r (in decimal form) compounded once a year, the value, A, of the annuity after t years is

$$A = \frac{P[(1 + r)^t - 1]}{r}.$$

Value of an Annuity: Interest Compounded n Times per Year

If P is the deposit made at the end of each compounding period for an annuity that pays an annual interest rate r (in decimal form) compounded n times per year, the value, A, of the annuity after t years is

$$A = \frac{P\left[\left(1 + \dfrac{r}{n}\right)^{nt} - 1\right]}{\left(\dfrac{r}{n}\right)}.$$

Example

- You deposit $150 at the end of each month into an account that pays 4.5% compounded monthly. Value after 10 years:

$$A = \frac{P\left[\left(1 + \dfrac{r}{n}\right)^{nt} - 1\right]}{\left(\dfrac{r}{n}\right)} = \frac{150\left[\left(1 + \dfrac{0.045}{12}\right)^{12 \cdot 10} - 1\right]}{\left(\dfrac{0.045}{12}\right)} = \frac{150[(1.00375)^{120} - 1]}{0.00375} \approx \$22{,}680$$

Interest = Value of the annuity − Total deposits ≈ $22,680 − $150 · 12 · 10 = $4680

> $150 per month × 12 months per year × 10 years

Additional Examples to Review

Example 2, page 500; Example 3, page 501

Definitions and Concepts

The formula

$$P = \frac{A\left(\dfrac{r}{n}\right)}{\left[\left(1 + \dfrac{r}{n}\right)^{nt} - 1\right]}$$

gives the deposit, P, that must be made at the end of each compounding period into an annuity that pays an annual interest rate r (in decimal form) compounded n times per year in order to achieve a value of A dollars after t years.

Example

- How much should you deposit at the end of each month in an IRA that pays 6% compounded monthly to have $1,200,000 when you retire in 40 years?

$$P = \frac{A\left(\dfrac{r}{n}\right)}{\left[\left(1 + \dfrac{r}{n}\right)^{nt} - 1\right]} = \frac{1{,}200{,}000\left(\dfrac{0.06}{12}\right)}{\left[\left(1 + \dfrac{0.06}{12}\right)^{12 \cdot 40} - 1\right]} = \frac{1{,}200{,}000(0.005)}{\left[(1.005)^{480} - 1\right]} \approx \$603$$

Interest = Value of the annuity − Total deposits ≈ $1,200,000 − $603 · 12 · 40 = $910,560

> $603 per month × 12 months per year × 40 years

Additional Example to Review
Example 4, page 503

Definitions and Concepts

Bank accounts include basic checking accounts (no interest), NOW accounts (checking with interest; minimum balance required), savings accounts (low interest rates), money-market accounts (combine features of checking and savings; require minimum balance, pay interest, and offer limited check writing), and CDs (certificates of deposit; pay fixed interest if you keep the principal in the account for a specified amount of time; penalties for early withdrawal).

The return on an investment is the percent increase in the investment.

Investors purchase stock, shares of ownership in a company. The shares indicate the percent of ownership. Trading refers to buying and selling stock. Investors make money by selling a stock for more money than they paid for it. They can also make money while they own stock if a company distributes all or part of its profits as dividends. Each share of stock is paid the same dividend.

Investors purchase a bond, lending money to the company from which they purchase the bond. The company commits itself to pay the price an investor pays for the bond at the time it was purchased, called its face value, along with interest payments at a given rate.

A listing of all the investments a person holds is called a financial portfolio. A portfolio with a mixture of low-risk and high-risk investments is called a diversified portfolio.

A mutual fund is a group of stocks and/or bonds managed by a professional investor, called the fund manager.

Example

- Stock Tables

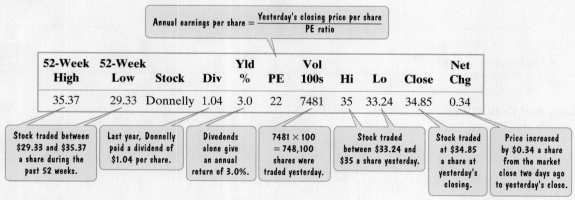

Additional Example to Review
Example 5, page 508

Definitions and Concepts

Accounts designed for retirement savings include traditional IRAs (requires paying taxes when withdrawing money at age $59\frac{1}{2}$ or older), Roth IRAs (requires paying taxes on deposits, but not withdrawals/earnings at age $59\frac{1}{2}$ or older), and employer-sponsored plans, such as 401(k) and 403(b) plans.

Example

You contribute $3500 per year to a 401(k) and your employer matches this contribution. The interest rate is 7.8% compounded annually.

Value after 25 years:

$$A = \frac{P[(1 + r)^t - 1]}{r} = \frac{7000[(1 + 0.078)^{25} - 1]}{0.078} \approx \$497{,}033$$

Additional Example to Review

Example 6, page 510

8.2 Cars

Definitions and Concepts

A fixed installment loan is paid off with a series of equal periodic payments. A car loan is an example of a fixed installment loan.

Loan Payment Formula for Fixed Installment Loans

$$PMT = \frac{P\left(\dfrac{r}{n}\right)}{\left[1 - \left(1 + \dfrac{r}{n}\right)^{-nt}\right]}$$

PMT is the regular payment amount required to repay a loan of P dollars paid n times per year over t years at an annual interest rate r.

Example

- Consider a $15,000 four-year car loan at 6%.
  Monthly payments:

$$PMT = \frac{P\left(\dfrac{r}{n}\right)}{\left[1 - \left(1 + \dfrac{r}{n}\right)^{-nt}\right]} = \frac{15{,}000\left(\dfrac{0.06}{12}\right)}{\left[1 - \left(1 + \dfrac{0.06}{12}\right)^{-12(4)}\right]} = \frac{15{,}000(0.005)}{[1 - (1.005)^{-48}]} \approx \$352$$

Interest = Total payments − Amount of loan = $352 · 12 · 4 − $15,000 = $1896

$352 per month × 12 months per year × 4 years

Additional Examples to Review

Example 1, page 516; Example 2, page 520

Definitions and Concepts

Leasing is a long-term rental agreement. Leasing a car offers some advantages over buying one (small down payment; lower monthly payments; no concerns about selling the car when the lease ends). Leasing also offers some disadvantages over buying (mileage penalties and other costs at the end of the leasing period often make the total cost of leasing more expensive than buying; penalties for ending the lease early; not owning the car when the lease ends; liability for damage to the car).

Auto insurance includes liability coverage: Bodily injury liability covers the costs of lawsuits if someone is injured or killed in an accident in which you are at fault. Property damage liability covers damage to other cars and property from negligent operation of your vehicle. If you have a car loan or lease a car, you also need collision coverage (pays for damage or loss of your car if you're in an accident) and comprehensive coverage (protects your car from fire, theft, and acts of nature).

The Cost of Gasoline

$$\text{Annual fuel expense} = \frac{\text{annual miles driven}}{\text{miles per gallon}} \times \text{price per gallon}$$

Example

- You drive 24,000 miles per year and gas averages $3.50 per gallon.

$$\text{Annual fuel expense for a hybrid averaging 50 miles per gallon} = \frac{24,000}{50} \times \$3.50$$
$$= 480 \times \$3.50 = \$1680$$

$$\text{Annual fuel expense for a truck averaging 16 miles per gallon} = \frac{24,000}{16} \times \$3.50$$
$$= 1500 \times \$3.50 = \$5250$$

$$\text{Annual fuel savings with the hybrid} = \$5250 - \$1680 = \$3570$$

Additional Example to Review

Example 3, page 521

8.3 The Cost of Home Ownership

Definitions and Concepts

A mortgage is a long-term loan for the purpose of buying a home, and for which the property is pledged as security for payment. The term of the mortgage is the number of years until final payoff. The down payment is the portion of the sale price of the home that the buyer initially pays. The amount of the mortgage is the difference between the sale price and the down payment.

Fixed-rate mortgages have the same monthly payment during the entire time of the loan. Variable-rate mortgages, or adjustable-rate mortgages, have payment amounts that change from time to time depending on changes in the interest rate.

A point is a one-time charge that equals 1% of the amount of a mortgage loan.

The loan payment formula for fixed installment loans can be used to determine the monthly payment for a mortgage.

Amortizing a mortgage loan is the process of making regular payments on the principal and interest until the loan is paid off. A loan amortization schedule is a document showing the following information for each mortgage payment: the payment number, the interest paid from the payment, the amount of the payment applied to the principal, and the balance of the loan after the payment. Such a schedule shows how the buyer pays slightly less in interest and more in principal for each payment over the entire life of the loan.

Example

- The price of a home is $450,000. The bank requires a 20% down payment and three points at the time of closing. The cost of the home is financed with a 30-year fixed-rate mortgage at 6.5%.

  Required down payment = 20% of $450,000 = $0.20 \times \$450,000 = \$90,000$

  Mortgage amount = Sale price − Down payment = $\$450,000 - \$90,000 = \$360,000$

  Cost of three points = 3% of $360,000 = $0.03 \times \$360,000 = \$10,800$

  Monthly payment:

  $$PMT = \frac{P\left(\dfrac{r}{n}\right)}{\left[1 - \left(1 + \dfrac{r}{n}\right)^{-nt}\right]} = \frac{360{,}000\left(\dfrac{0.065}{12}\right)}{\left[1 - \left(1 + \dfrac{0.065}{12}\right)^{-12(30)}\right]} \approx \$2275$$

  Interest = Total payments − Mortgage amount = $\$2275 \cdot 12 \cdot 30 - \$360{,}000 = \$459{,}000$

  > $2275 per month × 12 months per year × 30 years

Loan amortization schedule for the first two months:

Payment 1

Interest for one month = $Prt = \$360{,}000 \times 0.065 \times \dfrac{1}{12} = \$1950$

Principal payment = Monthly payment − Interest payment = $\$2275 - \$1950 = \$325$

Loan balance = Principal balance − Principal payment = $\$360{,}000 - \$325 = \$359{,}675$

Payment 2

Interest for one month = $Prt = \$359{,}675 \times 0.065 \times \dfrac{1}{12} \approx \$1948.24$

Principal payment = Monthly payment − Interest payment = $\$2275 - \$1948.24 = \$326.76$

Loan balance = Principal balance − Principal payment = $\$359{,}675 - \$326.76 = \$359{,}348.24$

Additional Examples to Review

Example 1, page 526; Example 2, page 528

Definitions and Concepts

Here are the guidelines for what you can spend for a mortgage:

- Spend no more than 28% of your gross monthly income for your mortgage payment.
- Spend no more than 36% of your gross monthly income for your total monthly debt.

Home ownership provides significant tax advantages, including deduction of mortgage interest and property taxes. Renting does not provide tax benefits, although renters do not pay property taxes.

Example

- Gross annual income is $78,000.

  Gross monthly income = $\dfrac{\$78{,}000}{12} = \$6500$

  Maximum mortgage payment = 28% of $6500 = $0.28 \times \$6500 = \$1820$

  Maximum for total credit obligations = 36% of $6500 = $0.36 \times \$6500 = \$2340$

- Your monthly mortgage payment is 70% of the maximum amount you can afford.

  Mortgage payment = 70% of $1820 = $0.70 \times \$1820 = \$1274$

  Maximum for all other debt = $\$2340 - \$1274 = \$1066$

Additional Example and Concepts to Review

Example 3, page 530; Benefits of Renting, page 531; Benefits of Home Ownership, page 531

8.4 Credit Cards

Definitions and Concepts

An open-end installment loan is paid off with variable monthly payments. Credit card loans are open-end installment loans.

Most credit cards calculate interest using the average daily balance method. Interest is calculated using $I = Prt$, where P is the average daily balance, r is the monthly rate, and t is one month.

$$\text{Average daily balance} = \frac{\text{Sum of the unpaid balances for each day in the billing period}}{\text{Number of days in the billing period}}$$

Determining the Average Daily Balance

Step 1 Make a table that shows the beginning date of the billing period, each transaction date, and the unpaid balance for each date.

Step 2 Add a column to the table that shows the number of days at each unpaid balance.

Step 3 Add a final column to the table that shows each unpaid balance multiplied by the number of days that the balance is outstanding.

Step 4 Find the sum of the products in the final column of the table. This dollar amount is the sum of the unpaid balances for each day in the billing period.

Step 5 Compute the average daily balance.

Example

Transaction Description	Transaction Amount
Previous balance, $4000	
April 1 Billing date	
April 7 Payment	$1000 credit
April 9 Charge: Gas	$ 50
April 18 Charge: Food	$ 100
April 27 Charge: Books	$ 40
April 30 End of billing period	
Payment Due Date: May 9	

- Computing the average daily balance

Date	Unpaid Balance
April 1	$4000
April 7	$4000 − $1000 = $3000
April 9	$3000 + $50 = $3050
April 18	$3050 + $100 = $3150
April 27	$3150 + $40 = $3190

Date	Unpaid Balance	Number of Days at Each Unpaid Balance	$\left(\begin{array}{c}\text{Unpaid}\\\text{Balance}\end{array}\right) \cdot \left(\begin{array}{c}\text{Number}\\\text{of Days}\end{array}\right)$
April 1	$4000	6	($4000)(6) = $24,000
April 7	$3000	2	($3000)(2) = $ 6000
April 9	$3050	9	($3050)(9) = $27,450
April 18	$3150	9	($3150)(9) = $28,350
April 27	$3190	4	($3190)(4) = $12,760
			Total: $98,560

Average daily balance

$$= \frac{\text{Sum of the unpaid balances for each day in the billing period}}{\text{Number of days in the billing period}} = \frac{\$98{,}560}{30} \approx \$3285.33$$

- Monthly interest rate is 1.8% of the average daily balance. Interest to be paid on May 1:

$$I = Prt = (\$3285.33)(0.018)(1) \approx \$59.14$$

Additional Example to Review

Example 1, page 535

Definitions and Concepts

One advantage of using a credit card is that there are no interest charges by paying the balance due at the end of each billing period. A disadvantage is the high interest rate on unpaid balances. Failing to pay the bill in full each month can result in serious debt.

The difference between debit cards and credit cards is that debit cards are linked to your bank account. When you use a debit card, the money you spend is deducted electronically from your account balance.

Credit reports include details about all open or closed credit accounts, such as when each account was opened, the latest balance, and the payment history. They also contain bankruptcy information and information about unpaid accounts that were turned over to collection agencies.

Credit scores range from 300 to 850, with a higher score indicating better credit.

Examples

Determine whether each statement is true or false. If the statement is false, make the necessary change(s) to produce a true statement.

- Money spent using a credit card is deducted from your bank account. False: For a true statement, change *credit card* to *debit card*.
- Credit scores range from 0 to 100. False: For a true statement, change *0 to 100* to *300 to 850*.

Additional Concepts to Review

Advantages of Using Credit Cards, page 538; Credit Card Woes, page 538; Credit Reports and Credit Scores, page 539

Summary of Personal Finance Formulas

Simple Interest

$$I = Prt$$
$$A = P(1 + rt)$$

Compound Interest

$$A = P(1 + r)^t$$
$$A = P\left(1 + \frac{r}{n}\right)^{nt}$$
$$P = \frac{A}{\left(1 + \frac{r}{n}\right)^{nt}}$$
$$Y = \left(1 + \frac{r}{n}\right)^n - 1$$

Annuities

$$A = \frac{P[(1 + r)^t - 1]}{r}$$
$$A = \frac{P\left[\left(1 + \frac{r}{n}\right)^{nt} - 1\right]}{\left(\frac{r}{n}\right)}$$
$$P = \frac{A\left(\frac{r}{n}\right)}{\left[\left(1 + \frac{r}{n}\right)^{nt} - 1\right]}$$

Amortization

$$PMT = \frac{P\left(\frac{r}{n}\right)}{\left[1 - \left(1 + \frac{r}{n}\right)^{-nt}\right]}$$

Be sure you understand what each formula in the box describes and the meaning of the variables in the formulas. Select the appropriate formula or formulas as you work the exercises in the Review Exercises and the Chapter 8 Tests.

Review Exercises

Section 8.1 Annuities, Methods of Saving, and Investments

In Exercises 1–2, round the value of each annuity to the nearest dollar.

1. A person who does not understand probability theory (see Chapter 11) wastes $10 per week on lottery tickets, averaging $520 per year. Instead of buying tickets, if this person deposits the $520 at the end of each year in an annuity paying 6% compounded annually,

   a. How much would he or she have after 20 years?

   b. Find the interest.

2. To save for retirement, you decide to deposit $100 at the end of each month in an IRA that pays 5.5% compounded monthly.

   a. How much will you have from the IRA after 30 years?

   b. Find the interest.

3. Suppose that you would like to have $25,000 to use as a down payment for a home in five years by making regular deposits at the end of every three months in an annuity that pays 7.25% compounded quarterly.

   a. Determine the amount of each deposit. Round up to the nearest dollar.

   b. How much of the $25,000 comes from deposits and how much comes from interest?

4. List three types of bank accounts and describe their features.

For Exercises 5–12, refer to the stock table for Harley Davidson (the motorcycle company). Where necessary, round dollar amounts to the nearest cent.

52-Week High	52-Week Low	Stock	SYM	Div	Yld %	PE
64.06	26.13	Harley Dav	HOG	.16	.3	41

Vol 100s	Hi	Lo	Close	Net Chg
5458	61.25	59.25	61	+1.75

5. What were the high and low prices for a share for the past 52 weeks?

6. If you owned 900 shares of this stock last year, what dividend did you receive?

7. What is the annual return for the dividends alone?

8. How many shares of this company's stock were traded yesterday?

9. What were the high and low prices for a share yesterday?

10. What was the price at which a share last traded when the stock exchange closed yesterday?

11. What was the change in price for a share of stock from the market close two days ago to yesterday's market close?

12. Compute the company's annual earnings per share using

    Annual earnings per share

    $$= \frac{\text{Yesterday's closing price per share}}{\text{PE ratio}}.$$

13. Explain the difference between investing in a stock and investing in a bond.

14. What is the difference between tax benefits for a traditional IRA and a Roth IRA?

Section 8.2 Cars

15. Suppose that you decide to take a $15,000 loan for a new car. You can select one of the following loans, each requiring regular monthly payments:

    Loan A: three-year loan at 7.2%

    Loan B: five-year loan at 8.1%.

    a. Find the monthly payments and the total interest for Loan A.

    b. Find the monthly payments and the total interest for Loan B.

    c. Compare the monthly payments and interest for the longer-term loan to the monthly payments and interest for the shorter-term loan.

16. Describe two advantages of leasing a car.

17. Describe two disadvantages of leasing a car.

18. Two components of auto insurance are property damage liability and collision. What is the difference between these types of coverage?

19. Suppose that you drive 36,000 miles per year and gas averages $3.60 per gallon.

    a. What will you save in annual fuel expenses by owning a hybrid car averaging 40 miles per gallon rather than an SUV averaging 12 miles per gallon?

    b. If you deposit your monthly fuel savings at the end of each month into an annuity that pays 5.2% compounded monthly, how much will you have saved at the end of six years? Round all computations to the nearest dollar.

Section 8.3 The Cost of Home Ownership

In Exercises 20–22, round to the nearest dollar.

20. The price of a home is $240,000. The bank requires a 20% down payment and two points at the time of closing. The cost of the home is financed with a 30-year fixed-rate mortgage at 7%.

    a. Find the required down payment.

    b. Find the amount of the mortgage.

    c. How much must be paid for the two points at closing?

    d. Find the monthly payment (excluding escrowed taxes and insurance).

    e. Find the total cost of interest over 30 years.

21. In terms of paying less in interest, which is more economical for a $70,000 mortgage: a 30-year fixed-rate at 8.5% or a 20-year fixed-rate at 8%? How much is saved in interest? Discuss one advantage and one disadvantage for each mortgage option.

22. Suppose that you need a loan of $100,000 to buy a home. Here are your options:

Option A: 30-year fixed-rate at 8.5% with no closing costs and no points

Option B: 30-year fixed-rate at 7.5% with closing costs of $1300 and three points.

  a. Determine your monthly payments for each option and discuss how you would decide between the two options.

  b. Which mortgage loan has the greater total cost (closing costs + the amount paid for points + total cost of interest)? By how much?

23. The cost of a home is financed with a $300,000 30-year fixed rate mortgage at 6.5%.

  a. Find the monthly payments, rounded to the nearest dollar, for the loan.

  b. Prepare a loan amortization schedule for the first three months of the mortgage. Round entries to the nearest cent.

Payment Number	Interest	Principal	Loan Balance
1			
2			
3			

24. Use these guidelines to solve this exercise: Spend no more than 28% of your gross monthly income for your mortgage payment and no more than 36% for your total monthly debt. Round all computations to the nearest dollar. Suppose that your gross annual income is $54,000.

  a. What is the maximum amount you should spend each month on a mortgage payment?

  b. What is the maximum amount you should spend each month for total credit obligations?

  c. If your monthly mortgage payment is 80% of the maximum amount you can afford, what is the maximum amount you should spend each month for all other debt?

25. Describe three benefits of renting over home ownership.

26. Describe three benefits of home ownership over renting.

Section 8.4 Credit Cards

27. A credit card issuer calculates interest using the average daily balance method. The monthly interest rate is 1.1% of the average daily balance. The following transactions occurred during the November 1–November 30 billing period.

Transaction Description	Transaction Amount
Previous balance, $4620.80	
November 1 Billing date	
November 7 Payment	$650.00 credit
November 11 Charge: Airline Tickets	$350.25
November 25 Charge: Groceries	$125.70
November 28 Charge: Gas	$ 38.25
November 30 End of billing period	
Payment Due Date: December 9	

  a. Find the average daily balance for the billing period. Round to the nearest cent.

  b. Find the interest to be paid on December 1, the next billing date. Round to the nearest cent.

  c. Find the balance due on December 1.

  d. This credit card requires a $10 minimum monthly payment if the balance due at the end of the billing period is less than $360. Otherwise, the minimum monthly payment is $\frac{1}{36}$ of the balance due at the end of the billing period, rounded up to the nearest whole dollar. What is the minimum monthly payment due by December 9?

28. In 2008, the average credit-card debt was $11,211. Suppose your card has this balance and an annual interest rate of 18%. You decide to pay off the balance over two years. If there are no further purchases charged to the card,

  a. How much must you pay each month?

  b. How much total interest will you pay?

  Round answers to the nearest dollar.

29. Describe two advantages of using credit cards.

30. Describe two disadvantages of using credit cards.

31. How does a debit card differ from a credit card?

32. Is a credit report the same thing as a credit score? If not, what is the difference?

33. Describe two ways to demonstrate that you can handle the responsibility of using a credit card.

Chapter 8 Test A

The box on page 549 summarizes the personal finance formulas you have worked with throughout this chapter and the previous chapter. Where applicable, use the appropriate formula to solve an exercise in this test.

1. Suppose that you save money for a down payment to buy a home in five years by depositing $6000 in an account that pays 6.5% compounded monthly.
   a. How much, to the nearest dollar, will you have as a down payment after five years?
   b. Find the interest.

2. Instead of making the lump-sum deposit of $6000 described in Exercise 1, suppose that you decide to deposit $100 at the end of each month in an annuity that pays 6.5% compounded monthly.
   a. How much, to the nearest dollar, will you have as a down payment after five years?
   b. Find the interest.
   c. Why is less interest earned from this annuity than from the lump-sum deposit in Exercise 1? With less interest earned, why would one select the annuity rather than the lump-sum deposit?

3. Suppose that you want to retire in 40 years. How much should you deposit at the end of each month in an IRA that pays 6.25% compounded monthly to have $1,500,000 in 40 years? Round up to the nearest dollar. How much of the $1.5 million comes from interest?

Use the stock table for AT&T to solve Exercises 4–6.

52-Week High	52-Week Low	Stock	SYM	Div	Yld %	PE
26.50	24.25	AT&T	PNS	2.03	7.9	18

Vol 100s	Hi	Lo	Close	Net Chg
961	25.75	25.50	25.75	+0.13

4. What were the high and low prices for a share yesterday?
5. If you owned 1000 shares of this stock last year, what dividend did you receive?
6. Suppose that you bought 600 shares of AT&T, paying the price per share at which a share last traded when the stock exchange closed yesterday. If the broker charges 2.5% of the price paid for all 600 shares, find the broker's commission.
7. Suppose that you drive 30,000 miles per year and gas averages $3.80 per gallon. What will you save in annual fuel expense by owning a hybrid car averaging 50 miles per gallon rather than a pickup truck averaging 15 miles per gallon?

Use this information to solve Exercises 8–13. The price of a home is $120,000. The bank requires a 10% down payment and two points at the time of closing. The cost of the home is financed with a 30-year fixed-rate mortgage at 8.5%.

8. Find the required down payment.
9. Find the amount of the mortgage.

10. How much must be paid for the two points at closing?
11. Find the monthly payment (excluding escrowed taxes and insurance). Round to the nearest dollar.
12. Find the total cost of interest over 30 years.
13. Prepare a loan amortization schedule for the first two months of the mortgage. Round entries to the nearest cent.

Payment Number	Interest	Principal	Loan Balance
1			
2			

14. Use these guidelines to solve this exercise: Spend no more than 28% of your gross monthly income for your mortgage payment and no more than 36% for your total monthly debt. Round all computations to the nearest dollar. Suppose that your gross annual income is $66,000.
   a. What is the maximum amount you should spend each month on a mortgage payment?
   b. What is the maximum amount you should spend each month for total credit obligations?
   c. If your monthly mortgage payment is 90% of the maximum amount you can afford, what is the maximum amount you should spend each month for all other debt?

15. A credit card issuer calculates interest using the average daily balance method. The monthly interest rate is 2% of the average daily balance. The following transactions occurred during the September 1–September 30 billing period.

Transaction Description	Transaction Amount
Previous balance, $3800.00	
September 1 Billing date	
September 5 Payment	$800.00 credit
September 9 Charge: Gas	$ 40.00
September 19 Charge: Clothing	$160.00
September 27 Charge: Airline Ticket	$200.00
September 30 End of billing period	
Payment Due Date: October 9	

   a. Find the average daily balance for the billing period. Round to the nearest cent.
   b. Find the interest to be paid on October 1, the next billing date. Round to the nearest cent.
   c. Find the balance due on October 1.
   d. Terms for the credit card require a $10 minimum monthly payment if the balance due is less than $360. Otherwise, the minimum monthly payment is $\frac{1}{36}$ of the balance due, rounded up to the nearest whole dollar. What is the minimum monthly payment due by October 9?

In Exercises 16–23, determine whether each statement is true or false. If the statement is false, make the necessary change(s) to produce a true statement.

16. If you take money out of a CD before its expiration date, you are charged a penalty.

17. By buying bonds, you purchase shares of ownership in a company.

18. A traditional IRA requires paying taxes when withdrawing money from the account at age $59\frac{1}{2}$ or older.

19. One advantage to leasing a car is that there are no penalties for ending the lease early.

20. If you cause an accident, collision coverage pays for damage to the other car.

21. Home ownership provides significant tax advantages, including deduction of mortgage interest.

22. Money spent using a credit card is deducted electronically from your bank account.

23. Credit scores range from 100 to 1000, with a higher score indicating better credit.

Chapter 8 Test B

The box on page 549 summarizes the personal finance formulas you have worked with throughout this chapter and the previous chapter. Where applicable, use the appropriate formula to solve an exercise in this test.

1. Suppose that you save money for a down payment to buy a home in five years by depositing $9000 in an account that pays 6.2% compounded monthly.

    a. How much, to the nearest dollar, will you have as a down payment after five years?

    b. Find the interest.

2. Instead of making the lump-sum deposit of $9000 described in Exercise 1, you decide to deposit $150 at the end of each month in an annuity that pays 6.2% compounded monthly.

    a. How much, to the nearest dollar, will you have as a down payment after five years?

    b. Find the interest.

    c. Why is less interest earned from this annuity than from the lump-sum deposit in Exercise 1? With less interest earned, why would one select the annuity rather than the lump-sum deposit?

3. Suppose that you want to retire in 35 years. How much should you deposit at the end of each month in an IRA that pays 8.5% compounded monthly to have $1,000,000 in 35 years? Round up to the nearest dollar. How much of the $1 million comes from interest?

Use the stock table for Dow Chemical to solve Exercises 4–6.

52-Week High	52-Week Low	Stock	SYM	Div	Yld %	PE
56.75	37.95	Dow Chemical	DOW	1.34	3.0	12

Vol 100s	Hi	Lo	Close	Net Chg
23997	44.75	44.35	44.69	+0.16

4. What were the high and low prices for a share yesterday?

5. If you owned 1000 shares of this stock last year, what dividend did you receive?

6. Suppose that you bought 600 shares of Dow Chemical, paying the price per share at which a share last traded when the stock exchange closed yesterday. If the broker charges 2.5% of the price paid for all 600 shares, find the broker's commission.

7. Suppose that you drive 18,000 miles per year and gas averages $3.40 per gallon. What will you save in annual fuel expenses by owning a hybrid car averaging 45 miles per gallon rather than a pickup truck averaging 18 miles per gallon?

Use this information to solve Exercises 8–13. The price of a home is $650,000. The bank requires a 20% down payment and three points at the time of closing. The cost of the home is financed with a 30-year fixed-rate mortgage at 6.8%.

8. Find the required down payment.

9. Find the amount of the mortgage.

10. How much must be paid for the three points at closing?

11. Find the monthly payment (excluding escrowed taxes and insurance). Round to the nearest dollar.

12. Find the total cost of interest over 30 years.

13. Prepare a loan amortization schedule for the first two months of the mortgage. Round entries to the nearest cent.

Payment Number	Interest	Principal	Loan Balance
1			
2			

14. Use these guidelines to solve this exercise: Spend no more than 28% of your gross monthly income for your mortgage payment and no more than 36% for your total monthly debt. Round all computations to the nearest dollar. Suppose that your gross annual income is $300,000.

    a. What is the maximum amount you should spend each month on a mortgage payment?

    b. What is the maximum amount you should spend each month for total credit obligations?

    c. If your monthly mortgage payment is 70% of the maximum you can afford, what is the maximum amount you should spend each month for all other debt?

15. A credit card issuer calculates interest using the average daily balance method. The monthly interest rate is 3% of the average daily balance. The following transactions occurred during the September 1–September 30 billing period.

Transaction Description	Transaction Amount
Previous balance, $5200.00	
September 1 Billing date	
September 5 Payment	$700.00 credit
September 10 Charge: Gas	$ 50.00
September 21 Charge: Clothing	$230.00
September 28 Charge: Airline Ticket	$300.00
September 30 End of billing period	
Payment Due Date: October 9	

a. Find the average daily balance for the billing period. Round to the nearest cent.

b. Find the interest to be paid on October 1, the next billing date. Round to the nearest cent.

c. Find the balance due on October 1.

d. Terms for the credit card require a $10 minimum monthly payment if the balance due is less than $360. Otherwise, the minimum monthly payment is $\frac{1}{36}$ of the balance due, rounded up to the nearest whole dollar. What is the minimum monthly payment due by October 9?

In Exercises 16–23, determine whether each statement is true or false. If the statement is false, make the necessary change(s) to produce a true statement.

16. Basic checking accounts do not pay interest.

17. By buying stock, you purchase shares of ownership in a company.

18. A Roth IRA requires paying taxes only when withdrawing money from the account at age $59\frac{1}{2}$ or older.

19. Most car leasing agreements have yearly mileage limits.

20. If you cause an accident, property damage liability pays for damage to the other car.

21. Renting a house or an apartment provides significant tax advantages, including deduction of monthly rent payments.

22. Money spent using a debit card is deducted electronically from your bank account.

23. Credit scores range from 300 to 850, with a higher score indicating better credit.

Measurement

Do you feel energized by crowds or overwhelmed by them? When deciding where to go to college, you might want to consider how crowded the town is. You can look up the population of the town, but that does not take into account the amount of land the population occupies. How is this land measured and how can you use this measure to select a place where wildlife outnumber humans?

In this chapter, we explore ways of measuring things in our English system, as well as in the metric system. We return to finding a place with lots of room to spread out in Example 2 of Section 9.2. Knowing how units of measure are used to describe your world can help you understand issues ranging from population density to the size of viruses (see the Blitzer Bonus on page 561).

9.1 | Measuring Length; The Metric System

Objectives

① Use dimensional analysis to change units of measurement.

② Understand and use metric prefixes.

③ Convert units within the metric system.

④ Use dimensional analysis to change to and from the metric system.

Have you seen either of the *Jurassic Park* films? The popularity of these movies reflects our fascination with dinosaurs and their incredible size. From end to end, the largest dinosaur from the Jurassic period, which lasted from 208 to 146 million years ago, was about 88 feet. To **measure** an object such as a dinosaur is to assign a number to its size. The number representing its measure from end to end is called its **length**. Measurements are used to describe properties of length, area, volume, weight, and temperature. Over the centuries, people have developed systems of measurement that are now accepted in most of the world.

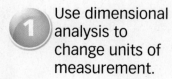

Linear units of measure were originally based on parts of the body. The Egyptians used the palm (equal to four fingers), the span (a handspan), and the cubit (length of forearm).

Length

Every measurement consists of two parts: a number and a unit of measure. For example, if the length of a dinosaur is 88 feet, the number is 88 and the unit of measure is the foot. Many different units are commonly used in measuring length. The foot is from a system of measurement called the **English system**, which is generally used in the United States. In this system of measurement, length is expressed in such units as inches, feet, yards, and miles.

The result obtained from measuring length is called a **linear measurement** and is stated in **linear units**.

> ### Linear Units of Measure: The English System
>
> $$12 \text{ inches (in.)} = 1 \text{ foot (ft)}$$
> $$3 \text{ feet} = 1 \text{ yard (yd)}$$
> $$36 \text{ inches} = 1 \text{ yard}$$
> $$5280 \text{ feet} = 1 \text{ mile (mi)}$$

Because many of us are familiar with the measures in the box, we find it simple to change from one measure to another, say from feet to inches. We know that there are 12 inches in a foot. To convert from 5 feet to a measure in inches, we multiply by 12. Thus, 5 feet = 5 × 12 inches = 60 inches.

Another procedure used to convert from one unit of measurement to another is called **dimensional analysis**. Dimensional analysis uses *unit fractions*. A **unit fraction** has two properties: The numerator and denominator contain different units and the value of the unit fraction is 1. Here are some examples of unit fractions:

$$\frac{12 \text{ in.}}{1 \text{ ft}}, \quad \frac{1 \text{ ft}}{12 \text{ in.}}, \quad \frac{3 \text{ ft}}{1 \text{ yd}}, \quad \frac{1 \text{ yd}}{3 \text{ ft}}, \quad \frac{5280 \text{ ft}}{1 \text{ mi}}, \quad \frac{1 \text{ mi}}{5280 \text{ ft}}.$$

In each unit fraction, the numerator and denominator are equal measures, making the value of the fraction 1.

① Use dimensional analysis to change units of measurement.

Let's see how to convert 5 feet to inches using dimensional analysis.

$$5 \text{ ft} = ? \text{ in.}$$

We need to eliminate feet and introduce inches. The unit we need to introduce, inches, must appear in the numerator of the fraction. The unit we need to eliminate, feet, must appear in the denominator. Therefore, we choose the unit fraction with inches in the numerator and feet in the denominator. The units divide out as follows:

$$5 \text{ ft} = \frac{5 \text{ ft}}{1} \cdot \frac{12 \text{ in.}}{1 \text{ ft}} = 5 \cdot 12 \text{ in.} = 60 \text{ in.}$$

unit fraction

Dimensional Analysis

To convert a measurement to a different unit, multiply by a unit fraction (or by unit fractions). The given unit of measurement should appear in the denominator of the unit fraction so that this unit cancels upon multiplication. The unit of measurement that needs to be introduced should appear in the numerator of the fraction; this unit will be retained upon multiplication.

Example 1 — Using Dimensional Analysis to Change Units of Measurement

Convert:

a. 40 inches to feet   **b.** 13,200 feet to miles   **c.** 9 inches to yards.

Solution

a. Because we want to convert 40 inches to feet, feet should appear in the numerator and inches in the denominator. We use the unit fraction

$$\frac{1 \text{ ft}}{12 \text{ in.}}$$

and proceed as follows:

This period ends the sentence and is not part of the abbreviated unit.

$$40 \text{ in.} = \frac{40 \text{ in.}}{1} \cdot \frac{1 \text{ ft}}{12 \text{ in.}} = \frac{40}{12} \text{ft} = 3\tfrac{1}{3} \text{ ft or } 3.\overline{3} \text{ ft.}$$

b. To convert 13,200 feet to miles, miles should appear in the numerator and feet in the denominator. We use the unit fraction

$$\frac{1 \text{ mi}}{5280 \text{ ft}}$$

and proceed as follows:

$$13{,}200 \text{ ft} = \frac{13{,}200 \text{ ft}}{1} \cdot \frac{1 \text{ mi}}{5280 \text{ ft}} = \frac{13{,}200}{5280} \text{mi} = 2\tfrac{1}{2} \text{ mi or } 2.5 \text{ mi.}$$

c. To convert 9 inches to yards, yards should appear in the numerator and inches in the denominator. We use the unit fraction

$$\frac{1 \text{ yd}}{36 \text{ in.}}$$

and proceed as follows:

$$9 \text{ in.} = \frac{9 \text{ in.}}{1} \cdot \frac{1 \text{ yd}}{36 \text{ in.}} = \frac{9}{36} \text{yd} = \tfrac{1}{4} \text{yd or } 0.25 \text{ yd.}$$

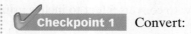

Checkpoint 1   Convert:

a. 78 inches to feet   **b.** 17,160 feet to miles

c. 3 inches to yards.

Length and the Metric System

 Understand and use metric prefixes.

Although the English system of measurement is most commonly used in the United States, most industrialized countries use the metric system of measurement. One of the advantages of the metric system is that units are based on powers of ten, making it much easier than the English system to change from one unit of measure to another.

The basic unit for linear measure in the metric system is the meter (m). A meter is slightly longer than a yard, approximately 39 inches. Prefixes are used to denote a multiple or part of a meter. **Table 9.1** summarizes the more commonly used metric prefixes and their meanings.

kilodollar

hectodollar

dekadollar

dollar

decidollar   centidollar

Table 9.1 Commonly Used Metric Prefixes

Prefix	Symbol	Meaning
kilo	k	$1000 \times$ base unit
hecto	h	$100 \times$ base unit
deka	da	$10 \times$ base unit
deci	d	$\frac{1}{10}$ of base unit
centi	c	$\frac{1}{100}$ of base unit
milli	m	$\frac{1}{1000}$ of base unit

Study Tip

Prefixes ending in "i" (deci, centi, milli) are all fractional parts of one unit.

The prefixes kilo, centi, and milli are used more frequently than hecto, deka, and deci. **Table 9.2** applies all six prefixes to the meter. The first part of the symbol indicates the prefix and the second part (m) indicates meter.

Table 9.2 Commonly Used Units of Linear Measure in the Metric System

Symbol	Unit	Meaning
km	kilometer	1000 meters
hm	hectometer	100 meters
dam	dekameter	10 meters
m	meter	1 meter
dm	decimeter	0.1 meter
cm	centimeter	0.01 meter
mm	millimeter	0.001 meter

Kilometer is pronounced kil'-oh-met-er with the accent on the FIRST syllable. If pronounced correctly, kilometers should sound something like "kill all meters."

In the metric system, the kilometer is used to measure distances comparable to those measured in miles in the English system. One kilometer is approximately 0.6 mile, and one mile is approximately 1.6 kilometers.

Metric units of centimeters and millimeters are used to measure what the English system measures in inches. **Figure 9.1** shows that a centimeter is less than half an inch; there are 2.54 centimeters in an inch. The smaller markings on the bottom scale are millimeters. A millimeter is approximately the thickness of a dime. The length of a bee or a fly may be measured in millimeters.

Figure 9.1

Those of us born in the United States have a good sense of what a length in the English system tells us about an object. An 88-foot dinosaur is huge, about 15 times the height of a 6-foot man. But what sense can we make of knowing that a whale is 25 meters long? The following lengths and the given approximations can help give you a feel for metric units of linear measure.

(1 meter ≈ 39 inches   1 kilometer ≈ 0.6 mile)

Item	Approximate Length
Width of lead in pencil	2 mm or 0.08 in.
Width of an adult's thumb	2 cm or 0.8 in.
Height of adult male	1.8 m or 6 ft
Typical room height	2.5 m or 8.3 ft
Length of medium-size car	5 m or 16.7 ft
Height of Empire State Building	381 m or 1270 ft
Average depth of ocean	4 km or 2.5 mi
Length of Manhattan Island	18 km or 11.25 mi
Distance from New York City to San Francisco	4800 km or 3000 mi
Radius of Earth	6378 km or 3986 mi
Distance from Earth to the moon	384,401 km or 240,251 mi

3 Convert units within the metric system.

Although dimensional analysis can be used to convert from one unit to another within the metric system, there is an easier, faster way to accomplish this conversion. The procedure is based on the observation that successively smaller units involve division by 10 and successively larger units involve multiplication by 10.

Changing Units within the Metric System

Use the following chart to find equivalent measures of length:

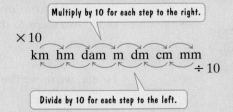

1. To change from a larger unit to a smaller unit (moving to the right in the diagram), multiply by 10 for each step to the right. Thus, move the decimal point in the given quantity one place to the right for each smaller unit until the desired unit is reached.

2. To change from a smaller unit to a larger unit (moving to the left in the diagram), divide by 10 for each step to the left. Thus, move the decimal point in the given quantity one place to the left for each larger unit until the desired unit is reached.

The following sentence provides a way to remember the metric units for length from largest to smallest.

King	Henry	died	Monday	drinking	chocolate	milk.
km	hm	dam	m	dm	cm	mm

Blitzer Bonus

The First Meter

The French first defined the meter in 1791, calculating its length as a romantic one ten-millionth of a line running from the Equator through Paris to the North Pole. Today's meter, officially accepted in 1983, is equal to the length of the path traveled by light in a vacuum during the time interval of 1/299,794,458 of a second.

North Pole

Paris

Equator

1 meter = 1/10,000,000 of the distance from the North Pole to the Equator on the meridian through Paris

Example 2 Changing Units within the Metric System

a. Convert 504.7 meters to kilometers.
b. Convert 27 meters to centimeters.
c. Convert 704 mm to hm.
d. Convert 9.71 dam to dm.

Solution

a. To convert from meters to kilometers, we start at meters and move three steps to the left to obtain kilometers:

km hm dam m dm cm mm.

Hence, we move the decimal point three places to the left:

504.7 m = 0.5047 km.

Thus, 504.7 meters converts to 0.5047 kilometer. Changing from a smaller unit of measurement (meter) to a larger unit of measurement (kilometer) results in an answer with a smaller number of units.

b. To convert from meters to centimeters, we start at meters and move two steps to the right to obtain centimeters:

km hm dam m dm cm mm.

Hence, we move the decimal point two places to the right:

27 m = 2700 cm.

Thus, 27 meters converts to 2700 centimeters. Changing from a larger unit of measurement (meter) to a smaller unit of measurement (centimeter) results in an answer with a larger number of units.

c. To convert from mm (millimeters) to hm (hectometers), we start at mm and move five steps to the left to obtain hm:

km hm dam m dm cm mm.

Hence, we move the decimal point five places to the left:

704 mm = 0.00704 hm.

d. To convert from dam (dekameters) to dm (decimeters), we start at dam and move two places to the right to obtain dm:

km hm dam m dm cm mm.

Hence, we move the decimal point two places to the right:

9.71 dam = 971 dm.

In Example 2(b), we showed that 27 meters converts to 2700 centimeters. This is the average length of the California blue whale, the longest of the great whales. Blue whales can have lengths that exceed 30 meters, making them over 100 feet long.

Blue whale

Checkpoint 2

a. Convert 8000 meters to kilometers.

b. Convert 53 meters to millimeters.

c. Convert 604 cm to hm.

▶ **d.** Convert 6.72 dam to cm.

Blitzer Bonus

Viruses and Metric Prefixes

Viruses are measured in attometers. An attometer is one quintillionth of a meter, or 10^{-18} meter, symbolized am. If a virus measures 1 am, you can place 10^{15} of them across a penciled 1 millimeter dot. If you were to enlarge each of these viruses to the size of the dot, they would stretch far into space, almost reaching Saturn.

Here is a list of all twenty metric prefixes. When applied to the meter, they range from the yottameter (10^{24} meters) to the yoctometer (10^{-24} meters).

Larger Than the Basic Unit

Prefix	Symbol	Power of Ten	English Name
yotta-	Y	+24	septillion
zetta-	Z	+21	sextillion
exa-	E	+18	quintillion
peta-	P	+15	quadrillion
tera-	T	+12	trillion
giga-	G	+9	billion
mega-	M	+6	million
kilo-	k	+3	thousand
hecto-	h	+2	hundred
deca-	da	+1	ten

Smaller Than the Basic Unit

Prefix	Symbol	Power of Ten	English Name
deci-	d	−1	tenth
centi-	c	−2	hundredth
milli-	m	−3	thousandth
micro-	μ	−6	millionth
nano-	n	−9	billionth
pico-	p	−12	trillionth
femto-	f	−15	quadrillionth
atto-	a	−18	quintillionth
zepto-	z	−21	sextillionth
yocto-	y	−24	septillionth

4 Use dimensional analysis to change to and from the metric system.

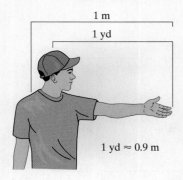

1 m

1 yd

1 yd ≈ 0.9 m

Although dimensional analysis is not necessary when changing units within the metric system, it is a useful tool when converting to and from the metric system. Some conversions are given in **Table 9.3**.

Table 9.3 English and Metric Equivalents	
1 inch (in.) = 2.54 centimeters (cm)	These conversions are exact.
1 foot (ft) = 30.48 centimeters (cm)	
1 yard (yd) ≈ 0.9 meter (m)	These conversions are approximate.
1 mile (mi) ≈ 1.6 kilometers (km)	

Example 3   **Using Dimensional Analysis to Change to and from the Metric System**

a. Convert 8 inches to centimeters.
b. Convert 125 miles to kilometers.
c. Convert 26,800 millimeters to inches.

Solution

a. To convert 8 inches to centimeters, we use a unit fraction with centimeters in the numerator and inches in the denominator:

$$\frac{2.54 \text{ cm}}{1 \text{ in.}}.$$ Table 9.3 shows that 1 in. = 2.54 cm.

We proceed as follows.

$$8 \text{ in.} = \frac{8 \text{ in.}}{1} \cdot \frac{2.54 \text{ cm}}{1 \text{ in.}} = 8(2.54) \text{ cm} = 20.32 \text{ cm}$$

b. To convert 125 miles to kilometers, we use a unit fraction with kilometers in the numerator and miles in the denominator:

$$\frac{1.6 \text{ km}}{1 \text{ mi}}.$$ Table 9.3 shows that 1 mi ≈ 1.6 km.

Thus,

$$125 \text{ mi} \approx \frac{125 \text{ mi}}{1} \cdot \frac{1.6 \text{ km}}{1 \text{ mi}} = 125(1.6) \text{ km} = 200 \text{ km}.$$

c. To convert 26,800 millimeters to inches, we observe that **Table 9.3** has only a conversion factor between inches and centimeters. We begin by changing millimeters to centimeters:

$$26,800 \text{ mm} = 2680.0 \text{ cm}$$

Now we need to convert 2680 centimeters to inches. We use a unit fraction with inches in the numerator and centimeters in the denominator:

$$\frac{1 \text{ in.}}{2.54 \text{ cm}}.$$

Thus,

$$26,800 \text{ mm} = 2680 \text{ cm} = \frac{2680 \text{ cm}}{1} \cdot \frac{1 \text{ in.}}{2.54 \text{ cm}} = \frac{2680}{2.54} \text{ in.} \approx 1055 \text{ in.}$$

Study Tip

Keep in mind that the conversion in part (b) is *approximate*. More accurately,

$$1 \text{ mi} \approx 1.61 \text{ km}.$$

Thus,

$$125 \text{ mi} \approx 125(1.61) \text{ km}$$
$$\approx 201.25 \text{ km}.$$

The measure 1055 inches is equivalent to about 88 feet, the length of the largest dinosaur from the Jurassic period. The diplodocus, a plant eater, was 26.8 meters, approximately 88 feet, long.

Checkpoint 3

a. Convert 8 feet to centimeters.

b. Convert 20 meters to yards.

c. Convert 30 meters to inches.

So far, we have used dimensional analysis to change units of length. Dimensional analysis may also be used to convert other kinds of measures, such as speed.

Example 4 Using Dimensional Analysis

a. The speed limit on many highways in the United States is 55 miles per hour (mi/hr). How many kilometers per hour (km/hr) is this?

b. If a high-speed train in Japan is capable of traveling at 200 kilometers per hour, how many miles per hour is this?

Solution

a. To change miles per hour to kilometers per hour, we need to concentrate on changing miles to kilometers, so we need a unit fraction with kilometers in the numerator and miles in the denominator:

$$\frac{1.6 \text{ km}}{1 \text{ mi}}.\qquad \text{Table 9.3 shows that 1 mi} \approx 1.6 \text{ km.}$$

Thus,

$$\frac{55 \text{ mi}}{\text{hr}} \approx \frac{55 \text{ mi}}{\text{hr}} \cdot \frac{1.6 \text{ km}}{1 \text{ mi}} = 55(1.6)\frac{\text{km}}{\text{hr}} = 88 \text{ km/hr.}$$

This shows that 55 miles per hour is approximately 88 kilometers per hour.

b. To change 200 kilometers per hour to miles per hour, we must convert kilometers to miles. We need a unit fraction with miles in the numerator and kilometers in the denominator:

$$\frac{1 \text{ mi}}{1.6 \text{ km}}.\qquad \text{Table 9.3 shows that 1 mi} \approx 1.6 \text{ km.}$$

Thus,

$$\frac{200 \text{ km}}{\text{hr}} \approx \frac{200 \text{ km}}{\text{hr}} \cdot \frac{1 \text{ mi}}{1.6 \text{ km}} = \frac{200 \text{ mi}}{1.6 \text{ hr}} = 125 \text{ mi/hr.}$$

A train capable of traveling at 200 kilometers per hour can therefore travel at about 125 miles per hour.

Checkpoint 4 A road in Europe has a speed limit of 60 kilometers per hour. Approximately how many miles per hour is this?

Achieving Success

Even if your schedule is jam-packed with everything from homework to sports and friends, getting a sufficient amount of sleep should be one of your priorities. According to the National Sleep Foundation, teens need at least nine hours of sleep per night. Based on this criterion, **Figure 9.2** indicates that approximately 92% of high-school students aren't getting enough sleep. Too little shuteye can affect your concentration at school, your ability to learn new material, and your grades.

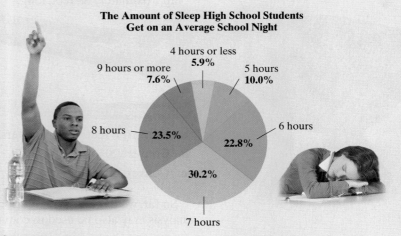

The Amount of Sleep High School Students Get on an Average School Night

4 hours or less **5.9%**
9 hours or more **7.6%**
5 hours **10.0%**
8 hours **23.5%**
6 hours **22.8%**
7 hours **30.2%**

Figure 9.2
Source: Centers for Disease Control and Prevention

Exercise Set 9.1

Concept and Vocabulary Exercises

In Exercises 1–4, fill in each blank so that the resulting statement is true.

1. The result obtained from measuring length is called a/an _____ measurement and is stated in _____ units.

2. In the English system, _____ in. = 1 ft, _____ ft = 1 yd, _____ in. = 1 yd, and _____ ft = 1 mi.

3. Fractions such as $\dfrac{12 \text{ in.}}{1 \text{ ft}}$ and $\dfrac{1 \text{ yd}}{3 \text{ ft}}$ are called _____ fractions. The value of such fractions is _____.

4. In the metric system, 1 km = _____ m, 1 hm = _____ m, 1 dam = _____ m, 1 dm = _____ m, 1 cm = _____ m, and 1 mm = _____ m.

In Exercises 5–8, determine whether each statement is true or false. If the statement is false, make the necessary change(s) to produce a true statement.

5. Dimensional analysis uses powers of ten to convert from one unit of measurement to another.

6. One of the advantages of the English system is that units are based on powers of ten.

7. There are 2.54 inches in a centimeter.

8. The height of an adult male is approximately 6 meters.

Respond to Exercises 9–12 using verbal or written explanations.

9. Describe the two parts of a measurement.

10. Describe how to use dimensional analysis to convert 20 inches to feet.

11. Describe advantages of the metric system over the English system.

12. Explain how to change units within the metric system.

In Exercises 13–16, determine whether each statement makes sense or does not make sense, and explain your reasoning.

13. I can run 4000 meters in approximately one hour.

14. I ran 2000 meters and you ran 2000 yards in the same time, so I ran at a faster rate.

15. The most frequent use of dimensional analysis involves changing units within the metric system.

16. When multiplying by a unit fraction, I put the unit of measure that needs to be introduced in the denominator.

Practice Exercises

In Exercises 17–32, use dimensional analysis to convert the quantity to the indicated unit. If necessary, round the answer to two decimal places.

17. 30 in. to ft
18. 100 in. to ft
19. 30 ft to in.
20. 100 ft to in.
21. 6 in. to yd
22. 21 in. to yd
23. 6 yd to in.
24. 21 yd to in.
25. 6 yd to ft
26. 12 yd to ft
27. 6 ft to yd
28. 12 ft to yd

29. 23,760 ft to mi

30. 19,800 ft to mi

31. 0.75 mi to ft

32. 0.25 mi to ft

In Exercises 33–42, use the diagram in the box on page 559 to convert the given measurement to the unit indicated.

33. 5 m to cm

34. 8 dam to m

35. 16.3 hm to m

36. 0.37 hm to m

37. 317.8 cm to hm

38. 8.64 hm to cm

39. 0.023 mm to m

40. 0.00037 km to cm

41. 2196 mm to dm

42. 71 dm to km

In Exercises 43–60, use the following English and metric equivalents, along with dimensional analysis, to convert the given measurement to the unit indicated.

English and Metric Equivalents

1 in. = 2.54 cm
1 ft = 30.48 cm
1 yd ≈ 0.9 m
1 mi ≈ 1.6 km

43. 14 in. to cm

44. 26 in. to cm

45. 14 cm to in.

46. 26 cm to in.

47. 265 mi to km

48. 776 mi to km

49. 265 km to mi

50. 776 km to mi

51. 12 m to yd

52. 20 m to yd

53. 14 dm to in.

54. 1.2 dam to in.

55. 160 in. to dam

56. 180 in. to hm

57. 5 ft to m

58. 8 ft to m

59. 5 m to ft

60. 8 m to ft

Use 1 mi ≈ 1.6 km to solve Exercises 61–64.

61. Express 96 kilometers per hour in miles per hour.

62. Express 104 kilometers per hour in miles per hour.

63. Express 45 miles per hour in kilometers per hour.

64. Express 50 miles per hour in kilometers per hour.

Practice Plus

In Exercises 65–68, use the unit fractions

$$\frac{36 \text{ in.}}{1 \text{ yd}} \quad \text{and} \quad \frac{2.54 \text{ cm}}{1 \text{ in.}}.$$

65. Convert 5 yd to cm.

66. Convert 8 yd to cm.

67. Convert 762 cm to yd.

68. Convert 1016 cm to yd.

In Exercises 69–72, use the unit fractions

$$\frac{5280 \text{ ft}}{1 \text{ mi}}, \quad \frac{12 \text{ in.}}{1 \text{ ft}}, \quad \text{and} \quad \frac{2.54 \text{ cm}}{1 \text{ in.}}.$$

69. Convert 30 mi to km.

70. Convert 50 mi to km.

71. Use unit fractions to express 120 miles per hour in feet per second.

72. Use unit fractions to express 100 miles per hour in feet per second.

Application Exercises

In Exercises 73–82, selecting from millimeter, meter, and kilometer, determine the best unit of measure to express the given length.

73. A person's height

74. The length of a football field

75. The length of a bee

76. The distance from New York City to Washington, D.C.

77. The distance around a one-acre lot

78. The length of a car

79. The width of a book

80. The altitude of an airplane

81. The diameter of a screw

82. The width of a human foot

In Exercises 83–90, select the best estimate for the measure of the given item.

83. The length of a pen

   a. 30 cm   **b.** 19 cm   **c.** 19 mm

84. The length of this page

   a. 2.5 mm   **b.** 25 mm   **c.** 250 mm

85. The height of a skyscraper

   a. 325 m   **b.** 32.5 km   **c.** 325 km   **d.** 3250 km

86. The length of a pair of pants

   a. 700 cm   **b.** 70 cm   **c.** 7 cm

87. The height of a room

   a. 4 mm   **b.** 4 cm   **c.** 4 m   **d.** 4 dm

88. The length of a rowboat

   a. 4 cm   **b.** 4 dm   **c.** 4 m   **d.** 4 dam

89. The width of an electric cord

   a. 4 mm   **b.** 4 cm   **c.** 4 dm   **d.** 4 m

90. The dimensions of a piece of typing paper

   a. 22 mm by 28 mm   **b.** 22 cm by 28 cm

   c. 22 dm by 28 dm   **d.** 22 m by 28 m

91. A baseball diamond measures 27 meters along each side. If a batter scored two home runs in a game, how many kilometers did the batter run?

92. If you jog six times around a track that is 700 meters long, how many kilometers have you covered?

93. The distance from the Earth to the sun is about 93 million miles. What is this distance in kilometers?

94. The distance from New York City to Los Angeles is 4690 kilometers. What is the distance in miles?

Exercises 95–96 give the approximate length of some of the world's longest rivers. In each exercise, determine which is the longer river and by how many kilometers.

95. Nile: 4130 miles; Amazon: 6400 kilometers

96. Yangtze: 3940 miles; Mississippi: 6275 kilometers

Exercises 97–98 give the approximate height of some of the world's tallest mountains. In each exercise, determine which is the taller mountain and by how many meters. Round to the nearest meter.

97. K2: 8611 meters; Everest: 29,035 feet

98. Lhotse: 8516 meters; Kangchenjunga: 28,170 feet

Exercises 99–100 give the average rainfall of some of the world's wettest places. In each exercise, determine which location has the greater average rainfall and by how many inches. Round to the nearest inch.

99. Debundscha (Cameroon): 10,280 millimeters; Waialeale (Hawaii): 451 inches

100. Mawsynram (India): 11,870 millimeters; Cherrapunji (India): 498 inches

(Source for Exercises 95–100: Russell Ash, *The Top 10 of Everything 2009*)

Critical Thinking Exercises

101. You jog 500 meters in a given period of time. The next day, you jog 500 yards over the same time period. On which day was your speed faster? Explain your answer.

102. What kind of difficulties might arise if the United States immediately eliminated all units of measure in the English system and replaced the system by the metric system?

103. The United States is the only Westernized country that does not use the metric system as its primary system of measurement. What reasons might be given for continuing to use the English system?

In Exercises 104–108, convert to an appropriate metric unit so that the numerical expression in the given measure does not contain any zeros.

104. 6000 cm

105. 900 m

106. 7000 dm

107. 11,000 mm

108. 0.0002 km

9.2 : Measuring Area and Volume

Objectives

1 Use square units to measure area.

2 Use dimensional analysis to change units for area.

3 Use cubic units to measure volume.

4 Use English and metric units to measure capacity.

Are you feeling a bit crowded in? Although there are more people on the East Coast of the United States than there are bears, there are places in the Northwest where bears outnumber humans. The most densely populated state is New Jersey, averaging 1171.1 people per square mile. The least densely populated state is Alaska, averaging 1.2 persons per square mile.

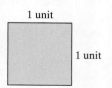

A square mile is one way of measuring the **area** of a state. A state's area is the region within its boundaries. Its **population density** is its population divided by its area. In this section, we discuss methods for measuring both area and volume.

Measuring Area

1 Use square units to measure area.

In order to measure a region that is enclosed by boundaries, we begin by selecting a *square unit*. A **square unit** is a square, each of whose sides is one unit in length, illustrated in **Figure 9.3**. The region in **Figure 9.3** is said to have an area of **one square unit**. The side of the square can be 1 inch, 1 centimeter, 1 meter, 1 foot, or one of any linear unit of measure. The corresponding

Figure 9.3
One square unit

units of area are the square inch (in.²), the square centimeter (cm²), the square meter (m²), the square foot (ft²), and so on. **Figure 9.4** illustrates 1 square inch and 1 square centimeter, drawn to actual size.

1 in.

1 in.

1 cm

1 cm

1 square inch, symbolized 1 in.²

1 square centimeter, symbolized 1 cm²

Figure 9.4 Common units of measurement for area, drawn to actual size

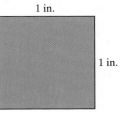

Square Unit

Figure 9.5

Example 1   Measuring Area

What is the area of the region shown in **Figure 9.5**?

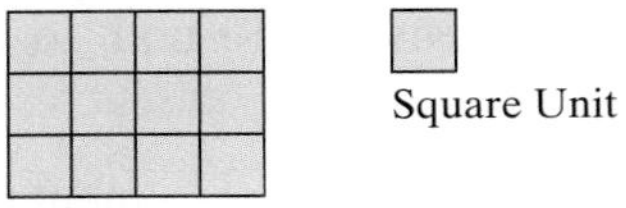

Solution

We can determine the area of the region by counting the number of square units contained within the region. There are 12 such units. Therefore, the area of the region is 12 square units.

 Checkpoint 1   What is the area of the region represented by the first two rows in **Figure 9.5**?

Although there are 12 inches in one foot and 3 feet in one yard, these numerical relationships are not the same for square units.

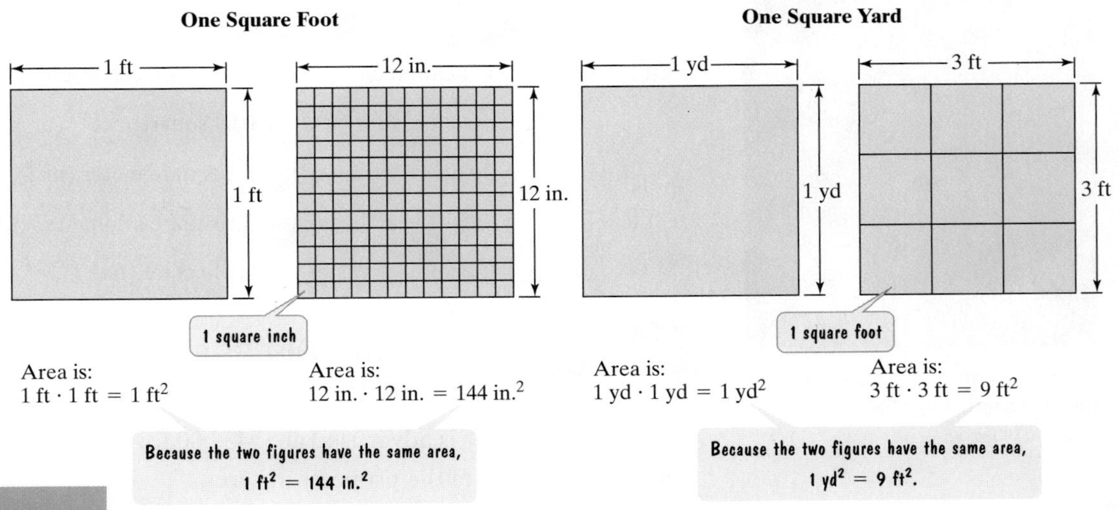

One Square Foot

1 ft

1 ft

12 in.

12 in.

1 square inch

Area is:
1 ft · 1 ft = 1 ft²

Area is:
12 in. · 12 in. = 144 in.²

Because the two figures have the same area,
1 ft² = 144 in.²

One Square Yard

1 yd

1 yd

3 ft

3 ft

1 square foot

Area is:
1 yd · 1 yd = 1 yd²

Area is:
3 ft · 3 ft = 9 ft²

Because the two figures have the same area,
1 yd² = 9 ft².

Study Tip

A small plot of land is usually measured in square feet, rather than a small fraction of an acre. Curiously, square yards are rarely used in measuring land. However, square yards are commonly used for carpet and flooring measures.

Square Units of Measure: The English System

1 square foot (ft²) = 144 square inches (in.²)
1 square yard (yd²) = 9 square feet (ft²)
1 acre (a)   = 43,560 ft² or 4840 yd²
1 square mile (mi²) = 640 acres

Using Square Units to Compute Population Density

After Alaska, Wyoming is the least densely populated state. The population of Wyoming is 515,004 and its area is 97,814 square miles. What is Wyoming's population density?

Solution

We compute the population density by dividing Wyoming's population by its area:

$$\text{population density} = \frac{\text{population}}{\text{area}} = \frac{515{,}004 \text{ people}}{97{,}814 \text{ square miles}}$$

Using a calculator and rounding to the nearest tenth, we obtain a population density of 5.3 people per square mile. This means that there is an average of only 5.3 people for each square mile of area.

Checkpoint 2 The population of California is 36,457,549 and its area is 158,633 square miles. What is California's population density? Round to the nearest tenth.

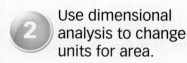

2 **Use dimensional analysis to change units for area.**

One theory is that using square units to measure area has its origin in the weaving of fabric.

In Section 9.1, we saw that in most other countries, the system of measurement that is used is the metric system. In the metric system, the square centimeter is used instead of the square inch. The square meter replaces the square foot and the square yard.

The English system uses the acre and the square mile to measure large land areas. The metric system uses the hectare (symbolized ha and pronounced "hectair"). A hectare is about the area of two football fields placed side by side, approximately 2.5 acres. One square mile of land consists of approximately 260 hectares. Just as the hectare replaces the acre, the square kilometer is used instead of the square mile. One square kilometer is approximately 0.38 square mile.

Some basic approximate conversions for units of area are given in **Table 9.4**.

Table 9.4 English and Metric Equivalents for Area		
1 square inch (in.^{2})	$\approx$ 6.5 square centimeters (cm^2)	
1 square foot (ft^2)	$\approx$ 0.09 square meter (m^2)	
1 square yard (yd^2)	$\approx$ 0.8 square meter (m^2)	
1 square mile (mi^2)	$\approx$ 2.6 square kilometers (km^2)	
1 acre	$\approx$ 0.4 hectare (ha)	

Using Dimensional Analysis on Units of Area

A property in Italy is advertised at $545,000 for 6.8 hectares.
a. Find the area of the property in acres.
b. What is the price per acre?

Solution

a. Using **Table 9.4**, we see that 1 acre $\approx$ 0.4 hectare. To convert 6.8 hectares to acres, we use a unit fraction with acres in the numerator and hectares in the denominator.

$$6.8 \text{ ha} \approx \frac{6.8 \text{ ha}}{1} \cdot \frac{1 \text{ acre}}{0.4 \text{ ha}} = \frac{6.8}{0.4} \text{ acres} = 17 \text{ acres}$$

The area of the property is approximately 17 acres.

b. The price per acre is the total price, $545,000, divided by the number of acres, 17.

$$\text{price per acre} = \frac{\$545,000}{17 \text{ acres}} \approx \$32,059/\text{acre}$$

The price is approximately $32,059 per acre.

✓ **Checkpoint 3**   A property in Northern California is on the market at $415,000 for 1.8 acres.

a. Find the area of the property in hectares.

▶ **b.** What is price per hectare?

Measuring Volume

③ Use cubic units to measure volume.

A shoe box and a basketball are examples of three-dimensional figures. **Volume** refers to the amount of space occupied by such figures. In order to measure this space, we begin by selecting a *cubic unit*. Two such cubic units are shown in **Figure 9.6**.

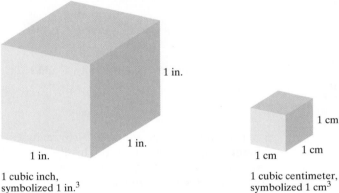

Figure 9.6 Common units of measurement for volume

1 cubic inch, symbolized 1 in.³

1 cubic centimeter, symbolized 1 cm³

The edges of a cube all have the same length. Other cubic units used to measure volume include 1 cubic foot (1 ft^3) and 1 cubic meter (1 m^3). One way to measure the volume of a solid is to calculate the number of cubic units contained in its interior.

Example 4   Measuring Volume

What is the volume of the solid shown in **Figure 9.7**?

Solution

We can determine the volume of the solid by counting the number of cubic units contained within the region. Because we have drawn a solid three-dimensional figure on a flat two-dimensional page, some of the small cubic units in the back, right are hidden. The figures below show how the cubic units are used to fill the inside of the solid.

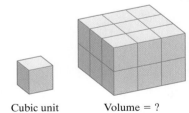

Cubic unit   Volume = ?

Figure 9.7

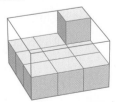 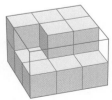

Do these figures help you to see that there are 18 cubic units inside the solid? The volume of the solid is 18 cubic units.

✓ **Checkpoint 4**   What is the volume of the region represented by the bottom row of blocks in **Figure 9.7**?

We have seen that there are 3 feet in a yard, but 9 square feet in a square yard. Neither of these relationships holds for cubic units. **Figure 9.8** illustrates that there are 27 cubic feet in a cubic yard. Furthermore, there are 1728 cubic inches in a cubic foot.

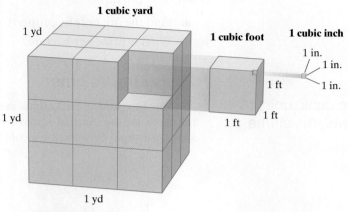

Figure 9.8 $27 \text{ ft}^3 = 1 \text{ yd}^3$ and $1728 \text{ in.}^3 = 1 \text{ ft}^3$

The measure of volume also includes the amount of fluid that a three-dimensional object can hold. This is often called the object's **capacity**. For example, we often refer to the capacity, in gallons, of a gas tank. A cubic yard has a capacity of about 200 gallons and a cubic foot has a capacity of about 7.48 gallons.

Table 9.5 contains information about standard units of capacity in the English system.

④ Use English and metric units to measure capacity.

Table 9.5 English Units for Capacity	
2 pints (pt)	= 1 quart (qt)
4 quarts	= 1 gallon (gal)
1 gallon	= 128 ounces (oz)
1 cup (c)	= 8 ounces

Volume in Cubic Units	Capacity
1 cubic yard	about 200 gallons
1 cubic foot	about 7.48 gallons
231 cubic inches	about 1 gallon

Example 5 **Volume and Capacity in the English System**

A swimming pool has a volume of 22,500 cubic feet. How many gallons of water does the pool hold?

Solution

We use the fact that 1 cubic foot has a capacity of about 7.48 gallons to set up our unit fraction:

$$\frac{7.48 \text{ gal}}{1 \text{ ft}^3}.$$

We use this unit fraction to find the capacity of the 22,500 cubic feet.

$$22{,}500 \text{ ft}^3 \approx \frac{22{,}500 \text{ ft}^3}{1} \cdot \frac{7.48 \text{ gal}}{1 \text{ ft}^3} = 22{,}500(7.48) \text{ gal} = 168{,}300 \text{ gal}$$

The pool holds approximately 168,300 gallons of water.

> ✓ **Checkpoint 5** A pool has a volume of 10,000 cubic feet. How many gallons of water does the pool hold?

As we have come to expect, things are simpler when the metric system is used to measure capacity. The basic unit is the **liter**, symbolized by L. A liter is slightly larger than a quart.

$$1 \text{ liter} \approx 1.0567 \text{ quarts}$$

The standard metric prefixes are used to denote a multiple or part of a liter. **Table 9.6** applies these prefixes to the liter.

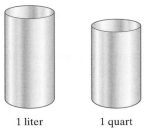

1 liter   1 quart

Table 9.6 Units of Capacity in the Metric System		
Symbol	**Unit**	**Meaning**
kL	kiloliter	1000 liters
hL	hectoliter	100 liters
daL	dekaliter	10 liters
L	liter	1 liter ≈ 1.06 quarts
dL	deciliter	0.1 liter
cL	centiliter	0.01 liter
mL	milliliter	0.001 liter

The following list should help give you a feel for capacity in the metric system.

Item	**Capacity**
Average cup of coffee	250 mL
12-ounce can of soda	355 mL
One quart of fruit juice	0.95 L
One gallon of milk	3.78 L
Average gas tank capacity of a car (about 18.5 gallons)	70 L

Figure 9.9 shows a 1-liter container filled with water. The water in the liter container will fit exactly into the box shown to its right. The volume of this box is 1000 cubic centimeters, or equivalently, 1 cubic decimeter. Thus,

$$1000 \text{ cm}^3 = 1 \text{ dm}^3 = 1 \text{ L}.$$

1 L = 1000 mL   1000 cm³ = 1 dm³

10 cm = 1 dm
10 cm = 1 dm
10 cm = 1 dm

Figure 9.9

Table 9.7 expands on the relationship between volume and capacity in the metric system.

Table 9.7 Volume and Capacity in the Metric System		
Volume in Cubic Units		**Capacity**
1 cm^3	=	1 mL
$1 \text{ dm}^3 = 1000 \text{ cm}^3$	=	1 L
1 m^3	=	1 kL

Blitzer Bonus

Measuring Dosages of Medicine

Table 9.7 indicates that

$$1 \text{ cm}^3 = 1 \text{ mL}.$$

Dosages of medicine are measured using cubic centimeters, or milliliters as they are also called. (In the United States, cc, rather than cm^3, denotes cubic centimeters.) When taking medication by mouth, such as cough syrup, there is an easy conversion between the English and metric systems:

$$1 \text{ teaspoon} = 5 \text{ cc}.$$

Example 6   Volume and Capacity in the Metric System

An aquarium has a volume of 36,000 cubic centimeters. How many liters of water does the aquarium hold?

Solution

We use the fact that 1000 cubic centimeters corresponds to a capacity of 1 liter to set up our unit fraction:

$$\frac{1 \text{ L}}{1000 \text{ cm}^3}.$$

We use this unit fraction to find the capacity of the 36,000 cubic centimeters.

$$36,000 \text{ cm}^3 = \frac{36,000 \text{ cm}^3}{1} \cdot \frac{1 \text{ L}}{1000 \text{ cm}^3} = \frac{36,000}{1000} \text{L} = 36 \text{ L}$$

The aquarium holds 36 liters of water.

Checkpoint 6   A fish pond has a volume of 220,000 cubic centimeters. How many liters of water does the pond hold?

Achieving Success

Make it your own.

Think back on the courses you've taken in high school.

- Give examples of things you learned in courses that still have importance in your thinking or in your life.
- Give examples of things you learned in courses that you believe will have no long-term influence on your life.

How can you make the subject matter of this math course more like the first examples? How can you use this course to bring value to your life?

Exercise Set 9.2

Concept and Vocabulary Exercises

In Exercises 1–5, fill in each blank so that the resulting statement is true.

1. Area is measured in _____ units and volume is measured in _____ units.

2. In the English system, $1 \text{ ft}^2 = $ _____ in.^2 and $1 \text{ yd}^2 = $ _____ ft^2.

3. In the English system, _____ pt = 1 qt and _____ qt = 1 gal.

4. The amount of fluid that a three-dimensional object can hold is called the object's _____, measured in the metric system using a basic unit called the _____.

5. A state's population density is its population divided by its _____.

In Exercises 6–8, determine whether each statement is true or false. If the statement is false, make the necessary change(s) to produce a true statement.

6. Because there are 3 feet in one yard, there are also 3 square feet in one square yard.

7. The English system uses in.² to measure large land areas.

8. One quart is approximately 1.06 liters.

Respond to Exercises 9–14 using verbal or written explanations.

9. Describe how area is measured. Explain why linear units cannot be used.

10. New Mexico has a population density of 16 people per square mile. Describe what this means. Use an almanac or the Internet to compare this population density to that of the state in which you are now living.

11. Describe the difference between the following problems: How much fencing is needed to enclose a garden? How much fertilizer is needed for the garden?

12. Describe how volume is measured. Explain why linear or square units cannot be used.

13. For a swimming pool, what is the difference between the following units of measure: cubic feet and gallons? For each unit, write a sentence about the pool that makes the use of the unit appropriate.

14. If there are 10 decimeters in a meter, explain why there are not 10 cubic decimeters in a cubic meter.

In Exercises 15–18, determine whether each statement makes sense or does not make sense, and explain your reasoning.

15. The gas capacity of my car is approximately 40 meters.

16. The population density of Montana is approximately 6.5 people per mile.

17. Yesterday I drank 2000 cm³ of water.

18. I read that one quart is approximately 0.94635295 liter.

Practice Exercises

In Exercises 19–22, use the given figure to find its area in square units.

19.

20.

21.

22.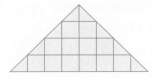

*In Exercises 23–30, use **Table 9.4** on page 568, along with dimensional analysis, to convert the given area measurement to the square unit indicated. Where necessary, round answers to two decimal places.*

23. 14 cm² to in.²   24. 20 m² to ft²

25. 30 m² to yd²   26. 14 mi² to km²

27. 10.2 ha to acres   28. 20.6 ha to acres

29. 14 in.² to cm²   30. 20 in.² to cm²

In Exercises 31–32, use the given figure to find its volume in cubic units.

31.    32.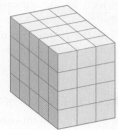

*In Exercises 33–40, use **Table 9.5** on page 570, along with dimensional analysis, to convert the given measurement of volume or capacity to the unit indicated. Where necessary, round answers to two decimal places.*

33. 10,000 ft³ to gal

34. 25,000 ft³ to gal

35. 8 yd³ to gal   36. 35 yd³ to gal

37. 2079 in.³ to gal   38. 6237 in.³ to gal

39. 2700 gal to yd³   40. 1496 gal to ft³

*In Exercises 41–50, use **Table 9.7** on page 572, along with dimensional analysis, to convert the given measurement of volume or capacity to the unit indicated.*

41. 45,000 cm³ to L   42. 75,000 cm³ to L

43. 17 cm³ to mL   44. 19 cm³ to mL

45. 1.5 L to cm³   46. 4.5 L to cm³

47. 150 mL to cm³   48. 250 mL to cm³

49. 12 kL to dm³   50. 16 kL to dm³

Practice Plus

The bar graph shows the resident population and the land area of the United States for selected years from 1800 through 2000. Use the information shown by the graph to solve Exercises 51–54.

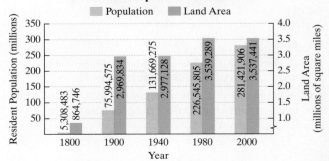

U.S. Resident Population and Land Area

Source: U.S. Bureau of the Census

(In Exercises 51–54, refer to the bar graph on the previous page.)

51. a. Find the population density of the United States, to the nearest tenth, in 1900 and in 2000.

   b. Find the percent increase in population density, to the nearest tenth of a percent, from 1900 to 2000.

52. a. Find the population density of the United States, to the nearest tenth, in 1800 and in 2000.

   b. Find the percent increase in population density, to the nearest tenth of a percent, from 1800 to 2000.

53. Find the population density of the United States, to the nearest tenth, expressed in people per square kilometer, in 1940.

54. Find the population density of the United States, to the nearest tenth, expressed in people per square kilometer, in 1980.

Application Exercises

55. A property that measures 8 hectares is for sale.

   a. How large is the property in acres?

   b. If the property is selling for $250,000, what is the price per acre?

56. A property that measures 100 hectares is for sale.

   a. How large is the property in acres?

   b. If the property is selling for $350,000, what is the price per acre?

In Exercises 57–60, selecting from square centimeters, square meters, or square kilometers, determine the best unit of measure to express the area of the object described.

57. The top of a desk

58. A dollar bill

59. A national park

60. The wall of a room

In Exercises 61–64, select the best estimate for the measure of the area of the object described.

61. The area of the floor of a room

   a. 25 cm^2 **b.** 25 m^2 **c.** 25 km^2

62. The area of a television screen

   a. 2050 mm^2 **b.** 2050 cm^2 **c.** 2050 dm^2

63. The area of the face of a small coin

   a. 6 mm^2 **b.** 6 cm^2 **c.** 6 dm^2

64. The area of a parcel of land in a large metropolitan area on which a house can be built

   a. 900 cm^2 **b.** 900 m^2 **c.** 900 ha

65. A swimming pool has a volume of 45,000 cubic feet. How many gallons of water does the pool hold?

66. A swimming pool has a volume of 66,000 cubic feet. How many gallons of water does the pool hold?

67. A container of grapefruit juice has a volume of 4000 cubic centimeters. How many liters of juice does the container hold?

68. An aquarium has a volume of 17,500 cubic centimeters. How many liters of water does the aquarium hold?

Exercises 69–70 give the approximate area of some of the world's largest island countries. In each exercise, determine which country has the greater area and by how many square kilometers.

69. Philippines: $300,000 \text{ km}^2$; Japan: $145,900 \text{ mi}^2$

70. Iceland: $103,000 \text{ km}^2$; Cuba: $42,800 \text{ mi}^2$

Exercises 71–72 give the approximate area of some of the world's largest islands. In each exercise, determine which island has the greater area and by how many square miles. Round to the nearest square mile.

71. Baffin Island, Canada: $194,574 \text{ mi}^2$; Sumatra, Indonesia: $443,070 \text{ km}^2$

72. Honshu, Japan: $87,805 \text{ mi}^2$; Victoria Island, Canada: $217,300 \text{ km}^2$

(Source for Exercises 69–72: Russell Ash, *The Top 10 of Everything 2009*)

Critical Thinking Exercises

73. Singapore has the highest population density of any country: 46,690 people per 1000 hectares. How many people are there per square mile?

74. Nebraska has a population density of 22.9 people per square mile and a population of 1,768,331. What is the area of Nebraska? Round to the nearest square mile.

75. A high population density is a condition common to extremely poor and extremely rich locales. Explain why this is so.

76. Although Alaska is the least densely populated state, over 90% of its land is protected federal property that is off-limits to settlement. A resident of Anchorage, Alaska, might feel hemmed in. In terms of "elbow room," what other important factor must be considered when calculating a state's population density?

77. Does an adult's body contain approximately 6.5 liters, kiloliters, or milliliters of blood? Explain your answer.

78. Is the volume of a coin approximately 1 cubic centimeter, 1 cubic millimeter, or 1 cubic decimeter? Explain your answer.

Group Exercise

79. If you could select any place in the world, where would you like to live? Look up the population and land area of your ideal place, and compute its population density. Group members should share places to live and population densities. What trend, if any, does the group observe?

9.3 Measuring Weight and Temperature

Objectives

1. Apply metric prefixes to units of weight.

2. Convert units of weight within the metric system.

3. Use relationships between volume and weight within the metric system.

4. Use dimensional analysis to change units of weight to and from the metric system.

5. Understand temperature scales.

You are watching CNN International on cable television. The temperature in Honolulu, Hawaii, is reported as 30°C. Are Honolulu's tourists running around in winter jackets? In this section, we will make sense of Celsius temperature readings, as we discuss methods for measuring temperature and weight.

Measuring Weight

You step on the scale at the doctor's office to check your weight, discovering that you are 150 pounds. Compare this to your weight on the moon: 25 pounds. Why the difference? **Weight** is the measure of the gravitational pull on an object. The gravitational pull on the moon is only about one-sixth the gravitational pull on Earth. Although your weight varies depending on the force of gravity, your mass is exactly the same in all locations. **Mass** is a measure of the quantity of matter in an object, determined by its molecular structure. On Earth, as your weight increases, so does your mass. In this section, measurements are assumed to involve everyday situations on the surface of Earth. Thus, we will treat weight and mass as equivalent, and refer strictly to weight.

> **Units of Weight: The English System**
>
> 16 ounces (oz) = 1 pound (lb)
>
> 2000 pounds (lb) = 1 ton (T)

1 Apply metric prefixes to units of weight.

The basic metric unit of weight is the **gram** (g), used for very small objects such as a coin, a candy bar, or a teaspoon of salt. A nickel has a weight of about 5 grams.

As with meters, prefixes are used to denote a multiple or part of a gram. **Table 9.8** applies the common metric prefixes to the gram. The first part of the symbol indicates the prefix and the second part (g) indicates gram.

Table 9.8 Commonly Used Units of Weight in the Metric System		
Symbol	**Unit**	**Meaning**
kg	kilogram	1000 grams
hg	hectogram	100 grams
dag	dekagram	10 grams
g	gram	1 gram
dg	decigram	0.1 gram
cg	centigram	0.01 gram
mg	milligram	0.001 gram

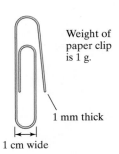

Weight of paper clip is 1 g.

1 mm thick

1 cm wide

In the metric system, the kilogram is the comparable unit to the pound in the English system. **A weight of 1 kilogram is approximately 2.2 pounds.** Thus, an average man has a weight of about 75 kilograms. Objects that we measure in pounds are measured in kilograms in most countries.

A milligram, equivalent to 0.001 gram, is an extremely small unit of weight and is used extensively in the pharmaceutical industry. If you look at the label on a bottle of tablets, you will see that the amounts of different substances in each tablet are expressed in milligrams.

The weight of a very heavy object is expressed in terms of the metric tonne (t), which is equivalent to 1000 kilograms, or about 2200 pounds. This is 10 percent more than the English ton (T) of 2000 pounds.

We change units of weight within the metric system exactly the same way that we changed units of length.

② Convert units of weight within the metric system.

Changing Units of Weight within the Metric System
Use the following diagram to find equivalent measures of weight:

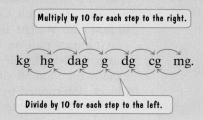

Multiply by 10 for each step to the right.

kg hg dag g dg cg mg.

Divide by 10 for each step to the left.

Example 1 **Changing Units within the Metric System**

a. Convert 8.7 dg to mg.
b. Convert 950 mg to g.

Solution

a. To convert from dg (decigrams) to mg (milligrams), we start at dg and move two steps to the right:

kg hg dag g dg cg mg.

Hence, we move the decimal point two places to the right:

$$8.7 \text{ dg} = 870 \text{ mg.}$$

b. To convert from mg (milligrams) to g (grams), we start at mg and move three steps to the left:

kg hg dag g dg cg mg.

Hence, we move the decimal point three places to the left:

$$950 \text{ mg} = 0.950 \text{ g.}$$

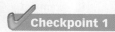 **Checkpoint 1**

a. Convert 4.2 dg to mg.
b. Convert 620 cg to g.

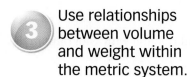 Use relationships between volume and weight within the metric system.

We have seen a convenient relationship in the metric system between volume and capacity:

$$1000 \text{ cm}^3 = 1 \text{ dm}^3 = 1 \text{ L}.$$

This relationship can be extended to include weight based on the following:
One kilogram of water has a volume of 1 liter.
Thus,

$$1000 \text{ cm}^3 = 1 \text{ dm}^3 = 1 \text{ L} = 1 \text{ kg}.$$

Table 9.9 shows the relationships between volume and weight of water in the metric system.

Table 9.9 Volume and Weight of Water in the Metric System				
Volume		**Capacity**		**Weight**
1 cm^3	=	1 mL	=	1 g
$1 \text{ dm}^3 = 1000 \text{ cm}^3$	=	1 L	=	1 kg
1 m^3	=	1 kL	=	$1000 \text{ kg} = 1 \text{ t}$

Example 2   Volume and Weight in the Metric System

An aquarium holds 0.25 m^3 of water. How much does the water weigh?

Solution

We use the fact that 1 m^3 of water $= 1000$ kg of water to set up our unit fraction:

$$\frac{1000 \text{ kg}}{1 \text{ m}^3}.$$

Thus,

$$0.25 \text{ m}^3 = \frac{0.25 \text{ m}^3}{1} \cdot \frac{1000 \text{ kg}}{1 \text{ m}^3} = 250 \text{ kg}.$$

The water weighs 250 kilograms.

✓ **Checkpoint 2**   An aquarium holds 0.145 m^3 of water. How much does the water weigh?

 Use dimensional analysis to change units of weight to and from the metric system.

A problem like Example 2 involves more awkward computation in the English system. For example, if you know the aquarium's volume in cubic feet, you must also know that 1 cubic foot of water weighs about 62.5 pounds to determine the water's weight.

Dimensional analysis is a useful tool when converting units of weight between the English and metric systems. Some basic approximate conversions are given in **Table 9.10**.

Table 9.10 Weight: English and Metric Equivalents
1 ounce (oz) $\approx$ 28 grams (g)
1 pound (lb) $\approx$ 0.45 kilogram (kg)
1 ton (T) $\approx$ 0.9 tonne (t)

Example 3   Using Dimensional Analysis

a. Convert 160 pounds to kilograms.
b. Convert 300 grams to ounces.

Solution

a. To convert 160 pounds to kilograms, we use a unit fraction with kilograms in the numerator and pounds in the denominator:

$$\frac{0.45 \text{ kg}}{1 \text{ lb}} \cdot$$ Table 9.10 on the previous page shows that 1 lb ≈ 0.45 kg.

Thus,

$$160 \text{ lb} \approx \frac{160 \text{ lb}}{1} \cdot \frac{0.45 \text{ kg}}{1 \text{ lb}} = 160(0.45) \text{ kg} = 72 \text{ kg}.$$

b. To convert 300 grams to ounces, we use a unit fraction with ounces in the numerator and grams in the denominator:

$$\frac{1 \text{ oz}}{28 \text{ g}} \cdot$$ Table 9.10 on the previous page shows that 1 oz ≈ 28 g.

Thus,

$$300 \text{ g} \approx \frac{300 \text{ g}}{1} \cdot \frac{1 \text{ oz}}{28 \text{ g}} = \frac{300}{28} \text{ oz} \approx 10.7 \text{ oz}.$$

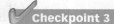 **Checkpoint 3**

a. A man weighs 186 pounds. Convert his weight to kilograms.

b. For each kilogram of weight, 1.2 milligrams of a drug is to be given. What dosage should a 186-pound man be given?

Blitzer Bonus

Using the Metric System to Measure Blood Alcohol Concentration

Although it is not legal for people under age 21 to consume alcoholic beverages in the United States, it is never too early to learn about the dangers of alcohol. Mathematics allows us to predict and quantify the concentration of alcohol in the blood for a person of given weight who consumes a certain amount of alcohol. Knowing this information helps people make responsible choices.

Blood alcohol concentration (BAC) is measured in grams of alcohol per 100 milliliters of blood. To put this measurement into perspective, a 175-pound man has approximately 5 liters (5000 milliliters) of blood and a 12-ounce can of 6%-alcohol-by-volume beer contains about 15 grams of alcohol. Based on the time it takes for alcohol to be absorbed into the bloodstream, as well as its elimination at a rate of 10–15 grams per hour, **Figure 9.10** shows the effect on BAC of a number of drinks on individuals within weight classes. It is illegal to drive with a BAC at 0.08 g/100 mL, or 0.8 gram of alcohol per 100 milliliters of blood, or greater.

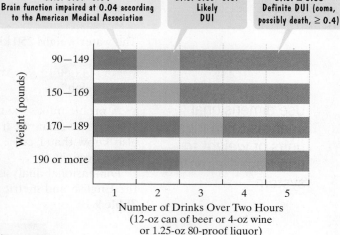

Blood Alcohol Concentration by Number of Drinks and Weight

| BAC: 0.01–0.04 Brain function impaired at 0.04 according to the American Medical Association | BAC: 0.05–0.07 Likely DUI | BAC: ≥ 0.08 Definite DUI (coma, possibly death, ≥ 0.4) |

Figure 9.10
Source: Patrick McSharry, *Everyday Numbers*, Random House, 2002.

Measuring Temperature

5 Understand temperature scales.

You'll be leaving the cold of winter for a vacation to Hawaii. CNN International reports a temperature in Hawaii of 30°C. Should you pack a winter coat?

The idea of changing from Celsius readings—or is it Centigrade?—to familiar Fahrenheit readings can be disorienting. Reporting a temperature of 30°C doesn't have the same impact as the Fahrenheit equivalent of 86 degrees (don't pack the coat). Why these annoying temperature scales?

The **Fahrenheit temperature scale**, the one we are accustomed to, was established in 1714 by the German physicist Gabriel Daniel Fahrenheit. He took a mixture of salt and ice, then thought to be the coldest possible temperature, and called it 0 degrees. He called the temperature of the human body 96 degrees, dividing the space between 0 and 96 into 96 parts. Fahrenheit was wrong about body temperature. It was later found to be 98.6 degrees. On his scale, water froze (without salt) at 32 degrees and boiled at 212 degrees. The symbol ° was used to replace the word *degree*.

Twenty years later, the Swedish scientist Anders Celsius introduced another temperature scale. He set the freezing point of water at 0° and its boiling point at 100°, dividing the space into 100 parts. Degrees were called centigrade until 1948, when the name was officially changed to honor its inventor. However, *centigrade* is still commonly used in the United States.

Figure 9.11 shows a thermometer that measures temperatures in both degrees Celsius (°C is the scale on the left) and degrees Fahrenheit (°F is the scale on the right). The thermometer should help orient you if you need to know what a temperature in °C means. For example, if it is 40°C, find the horizontal line representing this temperature on the left. Now read across to the °F scale on the right. The reading is above 100°, indicating heat wave conditions.

The following formulas can be used to convert from one temperature scale to the other:

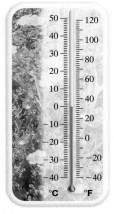

Figure 9.11 The Celsius scale is on the left and the Fahrenheit scale is on the right.

From Celsius to Fahrenheit

$$F = \frac{9}{5}C + 32$$

From Fahrenheit to Celsius

$$C = \frac{5}{9}(F - 32)$$

Study Tip

The formula used to convert from Celsius to Fahrenheit can be expressed without the use of fractions:

$$F = 1.8C + 32.$$

Some students find this form of the formula easier to memorize.

Example 4 **Converting from Celsius to Fahrenheit**

The bills from your European vacation have you feeling a bit feverish, so you decide to take your temperature. The thermometer reads 37°C. Should you panic?

Solution

Use the formula

$$F = \frac{9}{5}C + 32$$

to convert 37°C from °C to °F. Substitute 37 for C in the formula and find the value of F.

$$F = \frac{9}{5}(37) + 32 = 66.6 + 32 = 98.6$$

No need to panic! Your temperature is 98.6° F, which is perfectly normal.

 **Checkpoint 4** Convert 50°C from °C to °F.

Example 5 | Converting from Fahrenheit to Celsius

The temperature on a warm spring day is 77°F. Find the equivalent temperature on the Celsius scale.

Solution

Use the formula

$$C = \frac{5}{9}(F - 32)$$

to convert 77°F from °F to °C. Substitute 77 for F in the formula and find the value of C.

$$C = \frac{5}{9}(77 - 32) = \frac{5}{9}(45) = \frac{5}{\overset{1}{9}}\left(\frac{\overset{5}{45}}{1}\right) = 25$$

Thus, 77°F is equivalent to 25°C.

> ✓ **Checkpoint 5** Convert 59°F from °F to °C.

Because temperature is a measure of heat, scientists do not find negative temperatures meaningful in their work. In 1948, the British physicist Lord Kelvin introduced a third temperature scale. He put 0 degrees at absolute zero, the coldest possible temperature, at which there is no heat and molecules stop moving. **Figure 9.12** illustrates the three temperature scales.

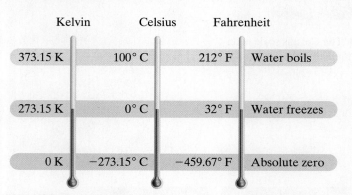

Figure 9.12 The three temperature scales

Lake Baikal, Siberia, is one of the coldest places on Earth, reaching −76°F (−60°C) in winter. The lowest temperature possible is absolute zero. Scientists have cooled atoms to a few millionths of a degree above absolute zero.

Figure 9.12 shows that water freezes at 273.15 K (read "K" or "Kelvins," not "degrees Kelvin") and boils at 373.15 K. The Kelvin scale is the same as the Celsius scale, except in its starting (zero) point. This makes it easy to go back and forth from Celsius to Kelvin.

From Celsius to Kelvin

$$K = C + 273.15$$

From Kelvin to Celsius

$$C = K - 273.15$$

Kelvin's scale was embraced by the scientific community. Today, it is the final authority, as scientists define Celsius and Fahrenheit in terms of Kelvins.

Achieving Success

Research postsecondary opportunities online, in your school guidance center, and at the library. Most careers require education you can get only after you graduate from high school. There are more than 3500 postsecondary schools in the United States, ranging from technical colleges that award certification for a specific job or career, to universities that grant bachelors' and graduate degrees. As you research various programs, here are some helpful questions to consider:

- Does the school offer a degree in your field of interest?
- Where is the school located? Will you live at home or on campus? Do you prefer living in a city or a small town?
- How many students are there? Are you more comfortable in a small school or a large school?
- What is the annual tuition? How much is room and board? Are there additional fees?
- Are you qualified for admission? Does your grade point average meet the school's standards? Are your ACT or SAT scores high enough? Do you have the right extracurricular activities?

Source for helpful questions: Suzanne Weixel and Faithe Wempen, *Life Skills for the 21st Century*, Pearson, 2010.

Exercise Set 9.3

Concept and Vocabulary Exercises

In Exercises 1–6, fill in each blank so that the resulting statement is true.

1. In the English system, _____ oz = 1 lb and _____ lb = 1 T.

2. The basic metric unit of weight is the _____, used for very small objects such as a coin, a candy bar, or a teaspoon of salt.

3. A weight of 1 kilogram is approximately _____ pounds.

4. One kilogram of water has a volume of _____ liter(s).

5. On the Fahrenheit temperature scale, water freezes at _____ degrees and boils at _____ degrees.

6. On the Celsius temperature scale, water freezes at _____ degrees and boils at _____ degrees.

In Exercises 7–10, determine whether each statement is true or false. If the statement is false, make the necessary change(s) to produce a true statement.

7. 1 gram = 1000 kilograms

8. 1 gram ≈ 28 ounces

9. The formula $F = \frac{9}{5}C + 32$ is used to convert from Fahrenheit to Celsius.

10. The formula $C = \frac{1}{9}(F - 32)$ is used to convert from Fahrenheit to Celsius.

Respond to Exercises 11–16 using verbal or written explanations.

11. Describe the difference between weight and mass.

12. Explain how to use dimensional analysis to convert 200 pounds to kilograms.

13. Why do you think that countries using the metric system prefer the Celsius scale over the Fahrenheit scale?

14. Describe in words how to convert from Celsius to Fahrenheit.

15. Describe in words how to convert from Fahrenheit to Celsius.

16. If you decide to travel outside the United States, which one of the two temperature conversion formulas should you take? Explain your answer.

In Exercises 17–20, determine whether each statement makes sense or does not make sense, and explain your reasoning.

17. When I played high school basketball, I weighed 250 kilograms.

18. It's not realistic to convert 500 mg to g because measurements should not involve fractional units.

19. Both the English system and the metric system enable me to easily see relationships among volume, capacity, and weight.

20. A comfortable classroom temperature is 20°C.

Practice Exercises

In Exercises 21–30, convert the given weight to the unit indicated.

21. 7.4 dg to mg
22. 6.9 dg to mg
23. 870 mg to g
24. 640 mg to g
25. 8 g to cg
26. 7 g to cg
27. 18.6 kg to g
28. 0.37 kg to g
29. 0.018 mg to g
30. 0.029 mg to g

*In Exercises 31–38, use **Table 9.9** on page 577 to convert the given measurement of water to the unit indicated.*

31. 0.05 m³ to kg
32. 0.02 m³ to kg

33. 4.2 kg to cm³   **34.** 5.8 kg to cm³

35. 1100 m³ to t   **36.** 1500 t to m³

37. 0.04 kL to g   **38.** 0.03 kL to g

In Exercises 39–50, use the following equivalents, along with dimensional analysis, to convert the given weight to the unit indicated. When necessary, round answers to two decimal places.

$$16 \text{ oz} = 1 \text{ lb}$$
$$2000 \text{ lb} = 1 \text{ T}$$
$$1 \text{ oz} \approx 28 \text{ g}$$
$$1 \text{ lb} \approx 0.45 \text{ kg}$$
$$1 \text{ T} \approx 0.9 \text{ t}$$

39. 36 oz to lb   **40.** 26 oz to lb

41. 36 oz to g   **42.** 26 oz to g

43. 540 lb to kg   **44.** 220 lb to kg

45. 80 lb to g   **46.** 150 lb to g

47. 540 kg to lb   **48.** 220 kg to lb

49. 200 t to T   **50.** 100 t to T

In Exercises 51–58, convert the given Celsius temperature to its equivalent temperature on the Fahrenheit scale. Where appropriate, round to the nearest tenth of a degree.

51. 10°C   **52.** 20°C

53. 35°C   **54.** 45°C

55. 57°C   **56.** 98°C

57. −5°C   **58.** −10°C

In Exercises 59–70, convert the given Fahrenheit temperature to its equivalent temperature on the Celsius scale. Where appropriate, round to the nearest tenth of a degree.

59. 68°F   **60.** 86°F

61. 41°F   **62.** 50°F

63. 72°F   **64.** 90°F

65. 23°F   **66.** 14°F

67. 350°F   **68.** 475°F

69. −22°F   **70.** −31°F

Practice Plus

71. The nine points shown below represent Celsius temperatures and their equivalent Fahrenheit temperatures. Also shown is a line that passes through the points.

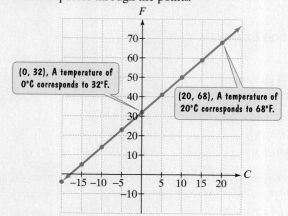

(0, 32), A temperature of 0°C corresponds to 32°F.

(20, 68), A temperature of 20°C corresponds to 68°F.

a. Use the coordinates of the two points identified by the voice balloons to compute the line's slope. Express the answer as a fraction reduced to lowest terms. What does this mean about the change in Fahrenheit temperature for each degree change in Celsius temperature?

b. Use the slope-intercept form of the equation of a line, $y = mx + b$, the slope from part (a), and the y-intercept shown by the graph to derive the formula used to convert from Celsius to Fahrenheit.

72. Solve the formula used to convert from Celsius to Fahrenheit for C and derive the other temperature conversion formula.

Application Exercises

In Exercises 73–79, selecting from milligram, gram, kilogram, and tonne, determine the best unit of measure to express the given item's weight.

73. A bee   **74.** This book

75. A tablespoon of salt   **76.** A Boeing 747

77. A stacked washer-dryer   **78.** A pen

79. An adult male

In Exercises 80–86, select the best estimate for the weight of the given item.

80. A newborn infant's weight
   a. 3000 kg   **b.** 300 kg
   c. 30 kg   **d.** 3 kg

81. The weight of a nickel
   a. 5 kg   **b.** 5 g   **c.** 5 mg

82. A person's weight
   a. 60 kg   **b.** 60 g   **c.** 60 dag

83. The weight of a box of cereal
   a. 0.5 kg   **b.** 0.5 g   **c.** 0.5 t

84. The weight of a glass of water
   a. 400 dg   **b.** 400 g   **c.** 400 dag   **d.** 400 hg

85. The weight of a regular-size car
   a. 1500 dag   **b.** 1500 hg   **c.** 1500 kg   **d.** 15,000 kg

86. The weight of a bicycle
   a. 140 kg   **b.** 140 hg   **c.** 140 dag   **d.** 140 g

87. Six items purchased at a grocery store weigh 14 kilograms. One of the items is detergent weighing 720 grams. What is the total weight, in kilograms, of the other five items?

88. If a nickel weighs 5 grams, how many nickels are there in 4 kilograms of nickels?

89. If the cost to mail a letter is 44 cents for mail weighing up to one ounce and 24 cents for each additional ounce or fraction of an ounce, find the cost of mailing a letter that weighs 85 grams.

90. Using the information given below the pictured finback whale, estimate the weight, in tons and kilograms, of the killer whale.

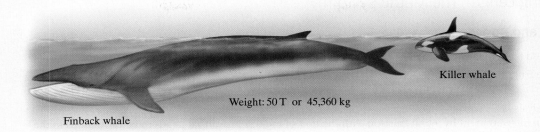

Killer whale

Weight: 50 T or 45,360 kg

Finback whale

91. Which is more economical: purchasing the economy size of a detergent at 3 kilograms for \$3.15 or purchasing the regular size at 720 grams for 60¢?

Exercises 92–93 ask you to determine drug dosage by a patient's weight. Use the fact that 1 lb ≈ 0.45 kg.

92. For each kilogram of a person's weight, 20 milligrams of a drug is to be given. What dosage should be given to a person who weighs 200 pounds?

93. For each kilogram of a person's weight, 2.5 milligrams of a drug is to be given. What dosage should be given to a child who weighs 80 pounds?

The label on a bottle of Emetrol ("for food or drink indiscretions") reads

> *Each 5 mL teaspoonful contains glucose, 1.87 g; levulose, 1.87 g; and phosphoric acid, 21.5 mg.*

Use this information to solve Exercises 94–95.

94. a. Find the amount of glucose in the recommended dosage of two teaspoons.

b. If the bottle contains 4 ounces, find the quantity of glucose in the bottle. (1 oz ≈ 30 mL)

95. a. Find the amount of phosphoric acid in the recommended dosage of two teaspoons.

b. If the bottle contains 4 ounces, find the quantity of phosphoric acid in the bottle. (1 oz ≈ 30 mL)

In Exercises 96–99, select the best estimate of the Celsius temperature of

96. A very hot day.
  a. 85°C   **b.** 65°C   **c.** 35°C   **d.** 20°C

97. A warm winter day in Washington, D.C.
  a. 10°C   **b.** 30°C   **c.** 50°C   **d.** 70°C

98. A setting for a home thermostat.
  a. 20°C   **b.** 40°C   **c.** 60°C   **d.** 80°C

99. The oven temperature for cooking a roast.
  a. 80°C   **b.** 100°C   **c.** 175°C   **d.** 350°C

Exercises 100–101 give the average temperature of some of the world's hottest places. In each exercise, determine which location has the hotter average temperature and by how many degrees Fahrenheit. Round to the nearest tenth of a degree.

100. Assab, Eritrea: 86.8°F; Dalol, Ethiopia: 34.6°C

101. Berbera, Somalia: 86.2°F; Néma, Mauritania: 30.3°C

Exercises 102–103 give the average temperature of some of the world's coldest places. In each exercise, determine which location has the colder average temperature and by how many degrees Celsius. Round to the nearest tenth of a degree.

102. Plateau, Antarctica: −56.7°C; Amundsen-Scott, Antarctica: −56.2°F

103. Eismitte, Greenland: −29.2°C; Resolute, Canada: −11.6°F

(Source for Exercises 100–103: Russell Ash, *The Top 10 of Everything* 2009)

Critical Thinking Exercises

In Exercises 104–111, determine whether each statement is true or false. If the statement is false, make the necessary change(s) to produce a true statement.

104. A 4-pound object weighs more than a 2000-gram object.

105. A 100-milligram object weighs more than a 2-ounce object.

106. A 50-gram object weighs more than a 2-ounce object.

107. A 10-pound object weighs more than a 4-kilogram object.

108. Flour selling at 3¢ per gram is a better buy than flour selling at 55¢ per pound.

109. The measures

$$32,600 \text{ g}, \quad 32.1 \text{ kg}, \quad 4 \text{ lb}, \quad 36 \text{ oz}$$

are arranged in order, from greatest to least weight.

110. If you are taking aspirin to relieve cold symptoms, a reasonable dose is 2 kilograms four times a day.

111. A large dog weighs about 350 kilograms.

Group Exercise

112. Present a group report on the current status of the metric system in the United States. At present, does it appear that the United States will convert to the metric system? Who supports the conversion and who opposes it? Summarize each side's position. Give examples of how our current system of weights and measures is an economic liability. What are the current obstacles to metric conversion?

Chapter 9 Summary

9.1 Measuring Length; The Metric System

Definitions and Concepts

The result obtained from measuring length is called a linear measurement, stated in linear units.

Linear Units: The English System

$$12 \text{ in.} = 1 \text{ ft}, 3 \text{ ft} = 1 \text{ yd}, 36 \text{ in.} = 1 \text{ yd}, 5280 \text{ ft} = 1 \text{ mi}$$

Dimensional Analysis
Multiply the given measurement by a unit fraction with the unit of measurement that needs to be introduced in the numerator and the unit of measurement that needs to be eliminated in the denominator.

Examples

- Convert 50 inches to feet.

$$50 \text{ in.} = \frac{50 \text{ in.}}{1} \cdot \frac{1 \text{ ft}}{12 \text{ in.}} = \frac{50}{12} \text{ ft} = 4\frac{1}{6} \text{ ft or } 4.1\overline{6} \text{ ft}$$

- Convert 4 inches to yards.

$$4 \text{ in.} = \frac{4 \text{ in.}}{1} \cdot \frac{1 \text{ yd}}{36 \text{ in.}} = \frac{4}{36} \text{ yd} = \frac{1}{9} \text{ yd or } 0.\overline{1} \text{ yd}$$

Additional Example to Review
Example 1, page 557

Definitions and Concepts

Linear Units: The Metric System
The basic unit is the meter (m), approximately 39 inches.

$$1 \text{ km} = 1000 \text{ m}, 1 \text{ hm} = 100 \text{ m}, 1 \text{ dam} = 10 \text{ m},$$
$$1 \text{ dm} = 0.1 \text{ m}, 1 \text{ cm} = 0.01 \text{ m}, 1 \text{ mm} = 0.001 \text{ m}$$

Changing Linear Units within the Metric System

$$\times 10$$

km hm dam m dm cm mm

$$\div 10$$

Examples

- 905 mm = 0.00905 hm

- 86 m = 8600 cm

- 57 m = 0.057 km

Additional Concept and Example to Review
Approximate Metric Lengths, table on page 559; Example 2, page 560

Definitions and Concepts

English and Metric Equivalents for Length

1 inch (in.) = 2.54 centimeters (cm)
1 foot (ft) = 30.48 centimeters (cm)
1 yard (yd) ≈ 0.9 meter (m)
1 mile (mi) ≈ 1.6 kilometers (km)

Examples

- Convert 381 centimeters to inches.

$$381 \text{ cm} = \frac{381 \text{ cm}}{1} \cdot \frac{1 \text{ in.}}{2.54 \text{ cm}} = \frac{381}{2.54} \text{ in.} = 150 \text{ in.}$$

- Convert 50 miles per hour to kilometers per hour.

$$\frac{50 \text{ mi}}{\text{hr}} \approx \frac{50 \text{ mi}}{\text{hr}} \cdot \frac{1.6 \text{ km}}{1 \text{ mi}} = 50(1.6)\frac{\text{km}}{\text{hr}} = 80 \text{ km/hr}$$

Additional Examples to Review

Example 3, page 562; Example 4, page 563

9.2 Measuring Area and Volume

Definitions and Concepts

The area measure of a plane region is the number of square units contained in the given region.

Square Units of Measure: The English System

$$1 \text{ square foot (ft}^2) = 144 \text{ square inches (in.}^2)$$
$$1 \text{ square yard (yd}^2) = 9 \text{ square feet (ft}^2)$$
$$1 \text{ acre (a)} = 43{,}560 \text{ ft}^2 \text{ or } 4840 \text{ yd}^2$$
$$1 \text{ square mile (mi}^2) = 640 \text{ acres}$$

$$\text{Population density} = \frac{\text{population}}{\text{area}}$$

Example

- Jamaica's population: 2,825,928   Jamaica's area: 4244 mi^2

$$\text{Jamaica's population density} = \frac{\text{population}}{\text{area}} = \frac{2{,}825{,}928}{4244} \approx 665.9$$

In Jamaica, there is an average of approximately 665.9 people for each square mile of land.

Additional Examples to Review

Example 1, page 567; Example 2, page 568

Definitions and Concepts

English and Metric Equivalents for Area

1 square inch (in.^{2}) $\approx$ 6.5 square centimeters (cm^2)
1 square foot (ft^2) $\approx$ 0.09 square meter (m^2)
1 square yard (yd^2) $\approx$ 0.8 square meter (m^2)
1 square mile (mi^2) $\approx$ 2.6 square kilometers (km^2)
1 acre $\approx$ 0.4 hectare (ha)

Example

- Convert 26 hectares to acres.

$$26 \text{ ha} \approx \frac{26 \text{ ha}}{1} \cdot \frac{1 \text{ acre}}{0.4 \text{ ha}} = \frac{26}{0.4} \text{ acres} = 65 \text{ acres}$$

Additional Example to Review

Example 3, page 568

Definitions and Concepts

The volume measure of a three-dimensional figure is the number of cubic units contained in its interior.

Capacity refers to the amount of fluid that a three-dimensional object can hold. English units for capacity include pints, quarts, and gallons: 2 pt = 1 qt; 4 qt = 1 gal; one cubic yard has a capacity of about 200 gallons.

English Units for Capacity
2 pints (pt) = 1 quart (qt)
4 quarts = 1 gallon (gal)
1 gallon = 128 ounces (oz)
1 cup (c) = 8 ounces

Volume in Cubic Units	Capacity
1 cubic yard	about 200 gallons
1 cubic foot	about 7.48 gallons
231 cubic inches	about 1 gallon

Example

- A pool has a volume of 20,000 cubic feet. How many gallons of water does the pool hold?

$$20,000 \text{ ft}^3 \approx \frac{20,000 \text{ ft}^3}{1} \cdot \frac{7.48 \text{ gal}}{1 \text{ ft}^3} = 20,000(7.48) \text{ gal} = 149,600 \text{ gal}$$

Additional Examples to Review
Example 4, page 569; Example 5, page 570

Definitions and Concepts

The basic unit for capacity in the metric system is the liter (L). One liter is about 1.06 quarts. Prefixes for the liter are the same as throughout the metric system.

Units of Capacity in the Metric System

Symbol	Unit	Meaning
kL	kiloliter	1000 liters
hL	hectoliter	100 liters
daL	dekaliter	10 liters
L	liter	1 liter ≈ 1.06 quarts
dL	deciliter	0.1 liter
cL	centiliter	0.01 liter
mL	milliliter	0.001 liter

Item	Capacity
Average cup of coffee	250 mL
12-ounce can of soda	355 mL
One quart of fruit juice	0.95 L
One gallon of milk	3.78 L
Average gas tank capacity of a car (about 18.5 gallons)	70 L

Volume and Capacity in the Metric System

Volume in Cubic Units		Capacity
1 cm³	=	1 mL
1 dm³ = 1000 cm³	=	1 L
1 m³	=	1 kL

Example

- A pond has a volume of 300,000 cubic centimeters. How many liters of water does the pond hold?

$$300,000 \text{ cm}^3 = \frac{300,000 \text{ cm}^3}{1} \cdot \frac{1 \text{ L}}{1000 \text{ cm}^3} = \frac{300,000}{1000} \text{ L} = 300 \text{ L}$$

Additional Example to Review
Example 6, page 572

9.3 Measuring Weight and Temperature

Definitions and Concepts

Weight: The English System

$$16 \text{ oz} = 1 \text{ lb}, \quad 2000 \text{ lb} = 1 \text{ T}$$

Units of Weight: The Metric System
The basic unit is the gram (g).

$$1000 \text{ grams } (1 \text{ kg}) \approx 2.2 \text{ lb},$$

$$1 \text{ kg} = 1000 \text{ g}, 1 \text{ hg} = 100 \text{ g}, 1 \text{ dag} = 10 \text{ g},$$

$$1 \text{ dg} = 0.1 \text{ g}, 1 \text{ cg} = 0.01 \text{ g}, 1 \text{ mg} = 0.001 \text{ g}$$

Changing Units of Weight within the Metric System

$$\times 10$$

kg hg dag g dg cg mg

$$\div 10$$

Examples

- $1285 \text{ g} = 1.285 \text{ kg}$
- $4.3 \text{ dg} = 430 \text{ mg}$

Additional Example to Review
Example 1, page 576

Definitions and Concepts

One kilogram of water has a volume of 1 liter.

Volume and Weight of Water in the Metric System

Volume		Capacity		Weight
1 cm^3	=	1 mL	=	1 g
$1 \text{ dm}^3 = 1000 \text{ cm}^3$	=	1 L	=	1 kg
1 m^3	=	1 kL	=	1000 kg = 1 t

Example

- An aquarium holds 0.35 m^3 of water. How much does the water weigh?

$$0.35 \text{ m}^3 = \frac{0.35 \text{ m}^3}{1} \cdot \frac{1000 \text{ kg}}{1 \text{ m}^3} = 0.35 (1000) \text{ kg} = 350 \text{ kg}$$

Additional Example to Review
Example 2, page 577

Definitions and Concepts

English and Metric Equivalents for Weight

1 ounce (oz)	$\approx$ 28 grams (g)
1 pound (lb)	$\approx$ 0.45 kilogram (kg)
1 ton (T)	$\approx$ 0.9 tonne (t)

Examples

• Convert 300 pounds to kilograms.

$$300 \text{ lb} \approx \frac{300 \text{ lb}}{1} \cdot \frac{0.45 \text{ kg}}{1 \text{ lb}} = 300(0.45) \text{ kg} = 135 \text{ kg}$$

• Convert 196 grams to ounces.

$$196 \text{ g} \approx \frac{196 \text{ g}}{1} \cdot \frac{1 \text{ oz}}{28 \text{ g}} = \frac{196}{28} \text{ oz} = 7 \text{ oz}$$

Additional Example to Review

Example 3, page 577

Definitions and Concepts

Converting between Temperature Scales

Celsius to Fahrenheit: $F = \frac{9}{5}C + 32$ Fahrenheit to Celsius: $C = \frac{5}{9}(F - 32)$

Examples

• Convert 20°C from °C to °F.

$$F = \frac{9}{5}C + 32 = \frac{9}{5}(20) + 32 = 36 + 32 = 68: \quad 20°C = 68°F$$

• Convert −49°F from °F to °C.

$$C = \frac{5}{9}(F - 32) = \frac{5}{9}(-49 - 32) = \frac{5}{9}(-81) = \frac{5}{\cancel{9}_{1}}\left(\frac{\overset{9}{\cancel{-81}}}{1}\right) = -45: \quad -49°F = -45°C$$

Additional Examples to Review

Example 4, page 579; Example 5, page 580

Review Exercises

Section 9.1 Measuring Length; The Metric System

In Exercises 1–4, use dimensional analysis to convert the quantity to the indicated unit.

1. 69 in. to ft
2. 9 in. to yd
3. 21 ft to yd
4. 13,200 ft to mi

In Exercises 5–10, convert the given linear measurement to the metric unit indicated.

5. 22.8 m to cm
6. 7 dam to m
7. 19.2 hm to m
8. 144 cm to hm
9. 0.5 mm to m
10. 18 cm to mm

In Exercises 11–16, use the given English and metric equivalents, along with dimensional analysis, to convert the given measurement to the unit indicated. Where necessary, round answers to two decimal places.

11. 23 in. to cm
12. 19 cm to in.
13. 330 mi to km
14. 600 km to mi
15. 14 m to yd
16. 12 m to ft

1 in.	= 2.54 cm
1 ft	= 30.48 cm
1 yd	$\approx$ 0.9 m
1 mi	$\approx$ 1.6 km

17. Express 45 kilometers per hour in miles per hour.

18. Express 60 miles per hour in kilometers per hour.

19. Arrange from smallest to largest: 0.024 km, 2400 m, 24,000 cm.

20. If you jog six times around a track that is 800 meters long, how many kilometers have you covered?

Section 9.2 Measuring Area and Volume

21. Use the given figure to find its area in square units.

22. In April 2006, the United States, with an area of 3,537,441 square miles, had a population of 298,923,319. Find the population density, to the nearest tenth, at that time. Describe what this means.

23. Given 1 acre ≈ 0.4 hectare, use dimensional analysis to find the size of a property in acres measured at 7.2 hectares.

24. Using $1 \text{ ft}^2 \approx 0.09 \text{ m}^2$, convert 30 m^2 to ft^2.

25. Using $1 \text{ mi}^2 \approx 2.6 \text{ km}^2$, convert 12 mi^2 to km^2.

26. Which one of the following is a reasonable measure for the area of a flower garden in a person's yard?

    a. 100 m^2 **b.** 0.4 ha   **c.** 0.01 km^2

27. Use the given figure to find its volume in cubic units.

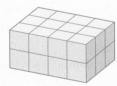

28. A swimming pool has a volume of 33,600 cubic feet. Given that 1 cubic foot has a capacity of about 7.48 gallons, how many gallons of water does the pool hold?

29. An aquarium has a volume of 76,000 cubic centimeters. How many liters of water does the aquarium hold?

30. The capacity of a one-quart container of juice is approximately

    a. 0.1 kL   **b.** 0.5 L   **c.** 1 L   **d.** 1 mL

31. There are 3 feet in a yard. Explain why there are not 3 square feet in a square yard. If helpful, illustrate your explanation with a diagram.

32. Explain why the area of Texas could not be measured in cubic miles.

Section 9.3 Measuring Weight and Temperature

In Exercises 33–36, convert the given weight to the unit indicated.

33. 12.4 dg to mg   **34.** 12 g to cg

35. 0.012 mg to g   **36.** 450 mg to kg

*In Exercises 37–38, use **Table 9.9** on page 577 to convert the given measurement of water to the unit or units indicated.*

37. 50 kg to cm^3

38. $4 \text{ kL to dm}^3 \text{ to g}$

39. Using 1 lb ≈ 0.45 kg, convert 210 pounds to kilograms.

40. Using 1 oz ≈ 28 g, convert 392 grams to ounces.

41. If you are interested in your weight in the metric system, would it be best to report it in milligrams, grams, or kilograms? Explain why the unit you selected would be most appropriate. Explain why each of the other two units is not the best choice for reporting your weight.

42. Given 16 oz = 1 lb, use dimensional analysis to convert 36 ounces to pounds.

In Exercises 43–44, select the best estimate for the weight of the given item.

43. A dollar bill:

    a. 1 g   **b.** 10 g   **c.** 1 kg   **d.** 4 kg

44. A hamburger:

    a. 3 kg   **b.** 1 kg   **c.** 200 g   **d.** 5 g

In Exercises 45–49, convert the given Celsius temperature to its equivalent temperature on the Fahrenheit scale.

45. 15°C   **46.** 100°C

47. 5°C   **48.** 0°C

49. −25°C

In Exercises 50–55, convert the given Fahrenheit temperature to its equivalent temperature on the Celsius scale.

50. 59°F   **51.** 41°F

52. 212°F   **53.** 98.6°F

54. 0°F   **55.** 14°F

56. Is a decrease of 15° Celsius more or less than a decrease of 15° Fahrenheit? Explain your answer.

Chapter 9 Test A

1. Change 807 mm to hm.

2. Given 1 inch = 2.54 centimeters, use dimensional analysis to change 635 centimeters to inches.

3. If you jog eight times around a track that is 600 meters long, how many kilometers have you covered?

In Exercises 4–6, write the most reasonable metric unit for length in each blank. Select from mm, cm, m, and km.

4. A human thumb is 20 _____ wide.

5. The height of the table is 45 _____.

6. The towns are 60 _____ apart.

7. If 1 mile ≈ 1.6 kilometers, express 80 miles per hour in kilometers per hour.

8. How many times greater is a square yard than a square foot?

9. Spain has a population of 40,491,051 and an area of 194,896 square miles. Find Spain's population density to the nearest tenth. Describe what this means.

10. Given 1 acre ≈ 0.4 hectare, use dimensional analysis to find the area of a property measured at 18 hectares.

11. The area of a dollar bill is approximately

    a. 10 cm² **b.** 100 cm² **c.** 1000 cm² **d.** 1 m²

12. There are 10 decimeters in a meter. Explain why there are not 10 cubic decimeters in a cubic meter. How many times greater is a cubic meter than a cubic decimeter?

13. A swimming pool has a volume of 10,000 cubic feet. Given that 1 cubic foot has a capacity of about 7.48 gallons, how many gallons of water does the pool hold?

14. The capacity of a pail used to wash floors is approximately

    a. 3 L **b.** 12 L **c.** 80 L **d.** 2 kL

15. Change 137 g to kg.

16. Using 1 lb ≈ 0.45 kg, convert 90 pounds to kilograms.

In Exercises 17–18, write the most reasonable metric unit for weight in each blank. Select from mg, g, and kg.

17. My suitcase weighs 20 _____.

18. I took a 350 _____ aspirin.

19. Convert 30°C to Fahrenheit.

20. Convert 176°F to Celsius.

21. Comfortable room temperature is approximately

    a. 70°C **b.** 50°C **c.** 30°C **d.** 20°C

 ## Chapter 9 Test B

1. Change 403 cm to hm.

2. Given 1 inch = 2.54 centimeters, use dimensional analysis to change 889 centimeters to inches.

3. If you jog nine times around a track that is 700 meters long, how many kilometers have you covered?

In Exercises 4–6, write the most reasonable metric unit for length in each blank. Select from mm, cm, m, and km.

4. A human thumb is 2 _____ wide.

5. The height of the room is 2.5 _____.

6. The average depth of the ocean is 4_____.

7. If 1 mile ≈ 1.6 kilometers, express 50 kilometers per hour in miles per hour.

8. How many times greater is a cubic yard than a cubic foot?

9. Singapore has a population of 4,657,542 and an area of 269 square miles. Find Singapore's population density to the nearest tenth. Describe what this means.

10. Given 1 acre ≈ 0.4 hectare, use dimensional analysis to find the area of a property measured at 16 hectares.

11. The area of a sheet of paper is approximately

    a. 60 cm² **b.** 588 cm² **c.** 6000 cm³ **d.** 588 m²

12. There are 10 decimeters in a meter. Explain why there are not 10 square decimeters in a square meter. How many times greater is a square meter than a square decimeter?

13. A swimming pool has a volume of 8000 cubic feet. Given that 1 cubic foot has a capacity of about 7.48 gallons, how many gallons of water does the pool hold?

14. The capacity of a soda can is approximately

    a. 35 mL **b.** 35 L **c.** 350 mL **d.** 350 L

15. Change 2600 g to kg.

16. Using 1 lb ≈ 0.45 kg, convert 200 pounds to kilograms.

In Exercises 17–18, write the most reasonable metric unit for weight in each blank. Select from mg, g, and kg.

17. A pair of running shoes weighs approximately 1.1_____.

18. Each day I take an 81_____ aspirin.

19. Convert 120°C to Fahrenheit.

20. Convert −40°F to Celsius.

21. It's 35°C in the room. Are you comfortable, hot, or cold?

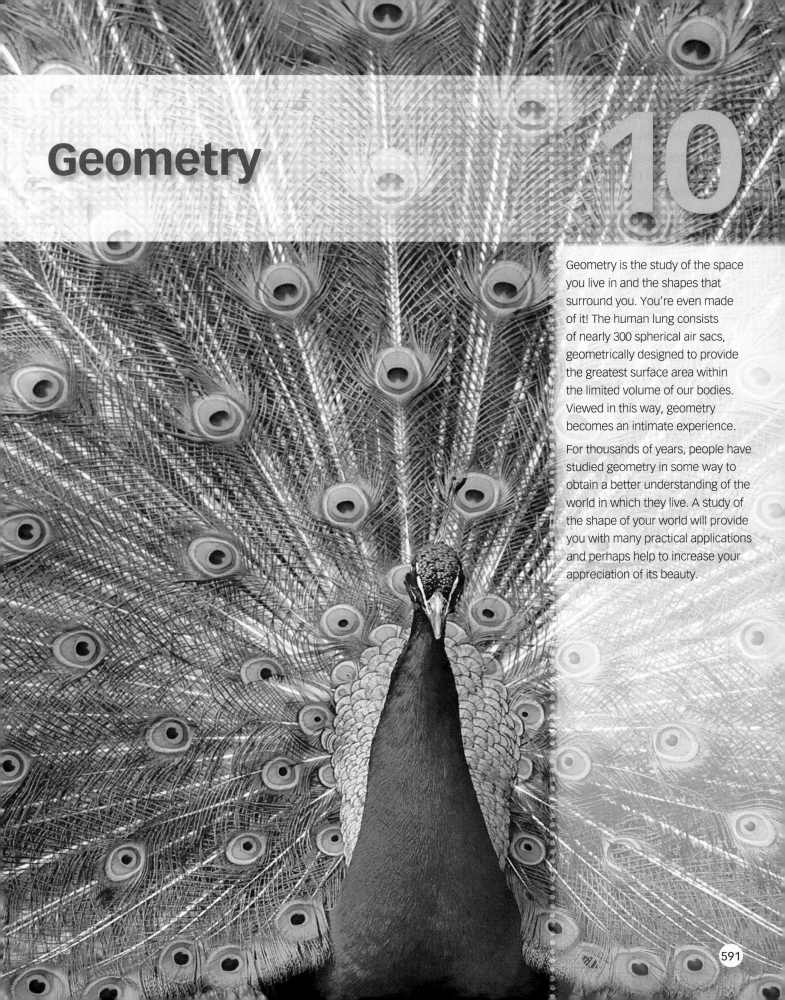

Geometry

10

Geometry is the study of the space you live in and the shapes that surround you. You're even made of it! The human lung consists of nearly 300 spherical air sacs, geometrically designed to provide the greatest surface area within the limited volume of our bodies. Viewed in this way, geometry becomes an intimate experience.

For thousands of years, people have studied geometry in some way to obtain a better understanding of the world in which they live. A study of the shape of your world will provide you with many practical applications and perhaps help to increase your appreciation of its beauty.

10.1 Points, Lines, Planes, and Angles

Objectives

1. Understand points, lines, and planes as the basis of geometry.

2. Solve problems involving angle measures.

3. Solve problems involving angles formed by parallel lines and transversals.

The San Francisco Museum of Modern Art was constructed in 1995 to illustrate how art and architecture can enrich one another. The exterior involves geometric shapes, symmetry, and unusual facades. Although there are no windows, natural light streams in through a truncated cylindrical skylight that crowns the building. The architect worked with a scale model of the museum at the site and observed how light hit it during different times of the day. These observations were used to cut the cylindrical skylight at an angle that maximizes sunlight entering the interior.

Angles play a critical role in creating modern architecture. They are also fundamental in the study of geometry. The word "geometry" means "earth measure." Because it involves the mathematics of shapes, geometry connects mathematics to art and architecture. It also has many practical applications. You can use geometry at home when you buy carpet, build a fence, tile a floor, or determine whether a piece of furniture will fit through your doorway. In this chapter, we look at the shapes that surround us and their applications.

Points, Lines, and Planes

1. **Understand points, lines, and planes as the basis of geometry.**

Points, lines, and planes make up the basis of all geometry. Stars in the night sky look like points of light. Long stretches of overhead power lines that appear to extend endlessly look like lines. The top of a flat table resembles part of a plane. However, stars, power lines, and tabletops only approximate points, lines, and planes. Points, lines, and planes do not exist in the physical world. Representations of these forms are shown in **Figure 10.1**. A **point**, represented as a small dot, has no length, width, or thickness. No object in the real world has zero size. A **line**, connecting two points along the shortest possible path, has no thickness and extends infinitely in both directions. However, no familiar everyday object is infinite in length. A **plane** is a flat surface with no thickness and no boundaries. This page resembles a plane, although it does not extend indefinitely and it does have thickness.

A line may be named using any two of its points. In **Figure 10.2(a)**, line AB can be symbolized $\overleftrightarrow{AB}$ or $\overleftrightarrow{BA}$. Any point on the line divides the line into three parts— the point and two **half-lines**. **Figure 10.2(b)** illustrates half-line AB, symbolized $\overset{\circ}{AB}$. The open circle above the A in the symbol and in the diagram indicates that point A is not included in the half-line. A **ray** is a half-line with its endpoint included. **Figure 10.2(c)** illustrates ray AB, symbolized $\overrightarrow{AB}$. The closed dot above the A in the diagram shows that point A is included in the ray. A portion of a line joining two points and including the endpoints is called a **line segment**. **Figure 10.2(d)** illustrates line segment AB, symbolized $\overline{AB}$ or $\overline{BA}$.

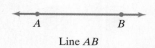

Point A

Line AB

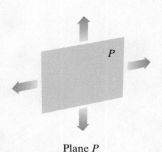

Plane P

Figure 10.1 Representing a point, a line, and a plane

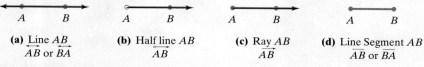

| (a) Line AB $\overleftrightarrow{AB}$ or $\overleftrightarrow{BA}$ | (b) Half line AB $\overset{\circ}{AB}$ | (c) Ray AB $\overrightarrow{AB}$ | (d) Line Segment AB $\overline{AB}$ or $\overline{BA}$ |

Figure 10.2 Lines, half-lines, rays, and line segments

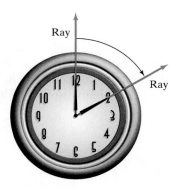

Figure 10.3 Clock with hands forming an angle

Angles

An **angle**, symbolized ∡, is formed by the union of two rays that have a common endpoint. One ray is called the **initial side** and the other the **terminal side**.

A rotating ray is often a useful way to think about angles. The ray in **Figure 10.3** rotates from 12 to 2. The ray pointing to 12 is the initial side and the ray pointing to 2 is the terminal side. The common endpoint of an angle's initial side and terminal side is the **vertex** of the angle.

Figure 10.4 shows an angle. The common endpoint of the two rays, B, is the vertex. The two rays that form the angle, $\overrightarrow{BA}$ and $\overrightarrow{BC}$, are the sides. Four ways of naming the angle are shown to the right of **Figure 10.4**.

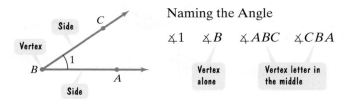

Figure 10.4 An angle: two rays with a common endpoint

Measuring Angles Using Degrees

Figure 10.5 A complete 360° rotation

Solve problems involving angle measures.

Angles are measured by determining the amount of rotation from the initial side to the terminal side. One way to measure angles is in **degrees**, symbolized by a small, raised circle °. Think of the hour hand of a clock. From 12 noon to 12 midnight, the hour hand moves around in a complete circle. By definition, the ray has rotated through 360 degrees, or 360°, shown in **Figure 10.5**. Using 360° as the amount of rotation of a ray back onto itself, **a degree, 1°, is $\frac{1}{360}$ of a complete rotation**.

Example 1 **Using Degree Measure**

The hand of a clock moves from 12 to 2 o'clock, shown in **Figure 10.3**. Through how many degrees does it move?

Solution

We know that one complete rotation is 360°. Moving from 12 to 2 o'clock is $\frac{2}{12}$, or $\frac{1}{6}$, of a complete revolution. Thus, the hour hand moves

$$\frac{1}{6} \times 360° = \frac{360°}{6} = 60°$$

in going from 12 to 2 o'clock.

Checkpoint 1 The hand of a clock moves from 12 to 1 o'clock. Through how many degrees does it move?

Figure 10.6 shows angles classified by their degree measurement. An **acute angle** measures less than 90° [see **Figure 10.6(a)**]. A **right angle**, one-quarter of a complete rotation, measures 90° [**Figure 10.6(b)**]. Examine the right angle—do you see a small square at the vertex? This symbol is used to indicate a right angle. An **obtuse angle** measures more than 90°, but less than 180° [**Figure 10.6(c)**]. Finally, a **straight angle**, one-half a complete rotation, measures 180° [**Figure 10.6(d)**]. The two rays in a straight angle form a straight line.

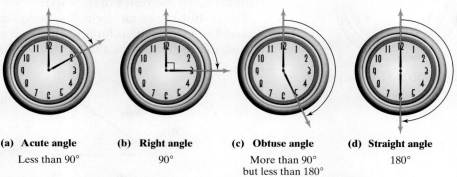

(a) **Acute angle**	(b) **Right angle**	(c) **Obtuse angle**	(d) **Straight angle**
Less than 90°	90°	More than 90° but less than 180°	180°

Figure 10.6 Classifying angles by their degree measurements

Figure 10.7 illustrates a **protractor**, used for finding the degree measure of an angle. As shown in the figure, we measure an angle by placing the center point of the protractor on the vertex of the angle and the straight side of the protractor along one side of the angle. The measure of ∡ABC is then read as 50°. Observe that the measure is not 130° because the angle is obviously less than 90°. We indicate the angle's measure by writing m∡ABC = 50°, read "the measure of angle ABC is 50°."

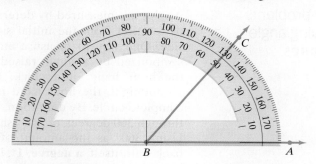

Figure 10.7 Using a protractor to measure an angle: m∡ABC = 50°

Two angles whose measures have a sum of 90° are called **complementary angles**. For example, angles measuring 70° and 20° are complementary angles because 70° + 20° = 90°. For angles such as those measuring 70° and 20°, each angle is the **complement** of the other: The 70° angle is the complement of the 20° angle and the 20° angle is the complement of the 70° angle. **The measure of the complement can be found by subtracting the angle's measure from 90°.** For example, we can find the complement of a 25° angle by subtracting 25° from 90°: 90° − 25° = 65°. Thus, an angle measuring 65° is the complement of one measuring 25°.

Example 2   Angle Measures and Complements

Use **Figure 10.8(a)** to find m∡DBC.

Solution

The measure of ∡DBC is not yet known. It is shown as ?° in **Figure 10.8(a)**. The acute angles ∡ABD, which measures 62°, and ∡DBC form a right angle, indicated by the square at the vertex. This means that the measures of the acute angles add up

Figure 10.8(a)

to 90°. Thus, ∡*DBC* is the complement of the angle measuring 62°. The measure of ∡*DBC* is found by subtracting 62° from 90°:

$$m\angle DBC = 90° - 62° = 28°.$$

The measure of ∡*DBC* can also be found using an algebraic approach.

$m\angle ABD + m\angle DBC = 90°$	The sum of the measures of complementary angles is 90°.
$62° + m\angle DBC = 90°$	We are given $m\angle ABD = 62°$.
$m\angle DBC = 90° - 62° = 28°$	Subtract 62° from both sides of the equation.

 **Checkpoint 2** In **Figure 10.8(b)**, let $m\angle DBC = 19°$. Find $m\angle DBA$.

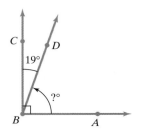

Figure 10.8(b)

Two angles whose measures have a sum of 180° are called **supplementary angles**. For example, angles measuring 110° and 70° are supplementary angles because 110° + 70° = 180°. For angles such as those measuring 110° and 70°, each angle is the **supplement** of the other: The 110° angle is the supplement of the 70° angle, and the 70° angle is the supplement of the 110° angle. **The measure of the supplement can be found by subtracting the angle's measure from 180°.** For example, we can find the supplement of a 25° angle by subtracting 25° from 180°: 180° − 25° = 155°. Thus, an angle measuring 155° is the supplement of one measuring 25°.

Example 3 Angle Measures and Supplements

Figure 10.9 shows that ∡*ABD* and ∡*DBC* are supplementary angles. If $m\angle ABD$ is 66° greater than $m\angle DBC$, find the measure of each angle.

Solution

Let $m\angle DBC = x$. Because $m\angle ABD$ is 66° greater than $m\angle DBC$, then $m\angle ABD = x + 66°$. We are given that these angles are supplementary.

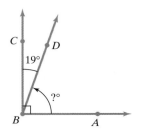

Figure 10.9

$m\angle DBC + m\angle ABD = 180°$	The sum of the measures of supplementary angles is 180°.
$x + (x + 66°) = 180°$	Substitute the variable expressions for the measures.
$2x + 66° = 180°$	Combine like terms: $x + x = 2x$.
$2x = 114°$	Subtract 66° from both sides.
$x = 57°$	Divide both sides by 2.

Thus, $m\angle DBC = 57°$ and $m\angle ABD = 57° + 66° = 123°$.

 **Checkpoint 3** In **Figure 10.9**, if $m\angle ABD$ is 88° greater than $m\angle DBC$, find the measure of each angle.

Figure 10.10 illustrates a highway sign that warns of a railroad crossing. When two lines intersect, the opposite angles formed are called **vertical angles**.

In **Figure 10.11**, there are two pairs of vertical angles. Angles 1 and 3 are vertical angles. Angles 2 and 4 are also vertical angles.

Figure 10.10

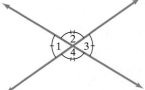

Figure 10.11

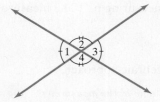

Figure 10.11 (repeated)

We can use **Figure 10.11** to show that vertical angles have the same measure. Let's concentrate on angles 1 and 3, each denoted by one tick mark. Can you see that each of these angles is supplementary to angle 2?

$$m\angle 1 + m\angle 2 = 180°$$ The sum of the measures of

$$m\angle 2 + m\angle 3 = 180°$$ supplementary angles is 180°.

$$m\angle 1 + m\angle 2 = m\angle 2 + m\angle 3$$ Substitute $m\angle 2 + m\angle 3$ for 180° in the first equation.

$$m\angle 1 = m\angle 3$$ Subtract $m\angle 2$ from both sides.

Using a similar approach, we can show that $m\angle 2 = m\angle 4$, each denoted by two tick marks in **Figure 10.11**.

> Vertical angles have the same measure.

Example 4   Using Vertical Angles

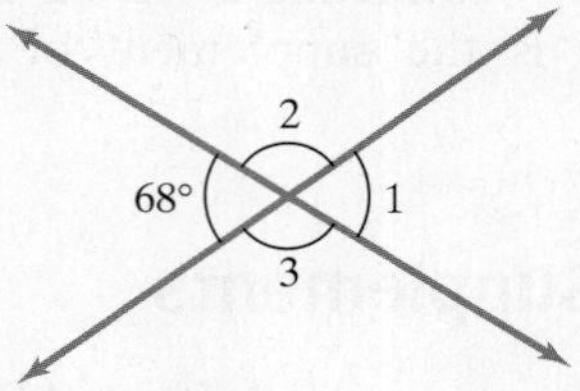

Figure 10.12 shows that the angle on the left measures 68°. Find the measures of the other three angles.

Solution

Angle 1 and the angle measuring 68° are vertical angles. Because vertical angles have the same measure,

$$m\angle 1 = 68°.$$

Angle 2 and the angle measuring 68° form a straight angle and are supplementary. Because their measures add up to 180°,

$$m\angle 2 = 180° - 68° = 112°.$$

Angle 2 and angle 3 are also vertical angles, so they have the same measure. Because the measure of angle 2 is 112°,

$$m\angle 3 = 112°.$$

✓ **Checkpoint 4**   In **Figure 10.12**, assume that the angle on the left measures 57°. ► Find the measures of the other three angles.

Figure 10.12

Parallel Lines

Parallel lines are lines that lie in the same plane and have no points in common. If two different lines in the same plane are not parallel, they have a single point in common and are called **intersecting lines**. If the lines intersect at an angle of 90°, they are called **perpendicular lines**.

If we intersect a pair of parallel lines with a third line, called a **transversal**, eight angles are formed, as shown in **Figure 10.13**. Certain pairs of these angles have special names, as well as special properties. These names and properties are summarized in **Table 10.1**.

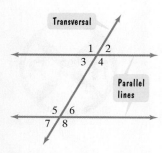

Figure 10.13

Table 10.1 Names of Angle Pairs Formed by a Transversal Intersecting Parallel Lines

Name	Description	Sketch	Angle Pairs Described	Property
Alternate interior angles	Interior angles that do not have a common vertex on alternate sides of the transversal		$\angle 3$ and $\angle 6$ $\angle 4$ and $\angle 5$	Alternate interior angles have the same measure. $m\angle 3 = m\angle 6$ $m\angle 4 = m\angle 5$
Alternate exterior angles	Exterior angles that do not have a common vertex on alternate sides of the transversal		$\angle 1$ and $\angle 8$ $\angle 2$ and $\angle 7$	Alternate exterior angles have the same measure. $m\angle 1 = m\angle 8$ $m\angle 2 = m\angle 7$
Corresponding angles	One interior and one exterior angle on the same side of the transversal		$\angle 1$ and $\angle 5$ $\angle 2$ and $\angle 6$ $\angle 3$ and $\angle 7$ $\angle 4$ and $\angle 8$	Corresponding angles have the same measure. $m\angle 1 = m\angle 5$ $m\angle 2 = m\angle 6$ $m\angle 3 = m\angle 7$ $m\angle 4 = m\angle 8$

3 Solve problems involving angles formed by parallel lines and transversals.

When two parallel lines are intersected by a transversal, the following relationships are true:

> **Parallel Lines and Angle Pairs**
>
> If parallel lines are intersected by a transversal,
>
> - alternate interior angles have the same measure,
> - alternate exterior angles have the same measure, and
> - corresponding angles have the same measure.
>
> Conversely, if two lines are intersected by a third line and a pair of alternate interior angles or a pair of alternate exterior angles or a pair of corresponding angles have the same measure, then the two lines are parallel.

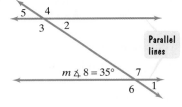

Figure 10.14

Example 5   **Finding Angle Measures when Parallel Lines Are Intersected by a Transversal**

In **Figure 10.14**, two parallel lines are intersected by a transversal. One of the angles ($\angle 8$) has a measure of 35°. Find the measure of each of the other seven angles.

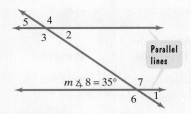

Figure 10.14 (repeated)

Study Tip

There is more than one way to solve Example 5. For example, once you know that $m\angle 2 = 35°$, $m\angle 5$ is also 35° because $\angle 2$ and $\angle 5$ are vertical angles.

Solution

Look carefully at **Figure 10.14** and fill in the angle measures as you read each line in this solution.

$m\angle 1 = 35°$ — $\angle 8$ and $\angle 1$ are vertical angles and vertical angles have the same measure.

$m\angle 6 = 180° - 35° = 145°$ — $\angle 8$ and $\angle 6$ are supplementary.

$m\angle 7 = 145°$ — $\angle 6$ and $\angle 7$ are vertical angles, so they have the same measure.

$m\angle 2 = 35°$ — $\angle 8$ and $\angle 2$ are alternate interior angles, so they have the same measure.

$m\angle 3 = 145°$ — $\angle 7$ and $\angle 3$ are alternate interior angles. Thus, they have the same measure.

$m\angle 5 = 35°$ — $\angle 8$ and $\angle 5$ are corresponding angles. Thus, they have the same measure.

$m\angle 4 = 180° - 35° = 145°$ — $\angle 4$ and $\angle 5$ are supplementary.

Checkpoint 5   In **Figure 10.14**, assume that $m\angle 8 = 29°$. Find the measure of each of the other seven angles.

Achieving Success

If you plan to attend college, selecting your classes will be important in determining what type of education you receive. Our tips in this chapter include advice for students by students on choosing college classes.

Sample a variety of subject areas.

"There are many college courses that you will have never even heard of while in high school. Sociology, psychology, political science, communication, art history . . . the list is endless. You will be cheating yourself if you don't try out some of these new areas of study."

—Katie MacDonald, University of Washington

Exercise Set 10.1

Concept and Vocabulary Exercises

In Exercises 1–6, fill in each blank so that the resulting statement is true.

1. $\overleftrightarrow{AB}$ symbolizes _____ AB, $\overrightarrow{AB}$ symbolizes _____ AB, $\vec{AB}$ symbolizes _____ AB, and $\overline{AB}$ symbolizes _____ AB.

2. A/an _____ angle measures less than 90°, a/an _____ angle measures 90°, a/an _____ angle measures more than 90° and less than 180°, and a/an _____ angle measures 180°.

3. Two angles whose measures have a sum of 90° are called _____ angles. Two angles whose measures have a sum of 180° are called _____ angles.

4. When two lines intersect, the opposite angles are called _____ angles.

5. Lines that lie in the same plane and have no points in common are called _____ lines. If these lines are intersected by a third line, called a _____ , eight angles are formed.

6. Lines that intersect at an angle of 90° are called _____ lines.

In Exercises 7–12, determine whether each statement is true or false. If the statement is false, make the necessary change(s) to produce a true statement.

7. A ray extends infinitely in both directions.

8. An angle is formed by the union of two lines that have a common endpoint.

9. A degree, 1°, is $\frac{1}{90}$ of a complete rotation.

10. A ruler is used for finding the degree measure of an angle.

11. The measure of an angle's complement is found by subtracting the angle's measure from 90°.

12. Vertical angles have the same measure.

Respond to Exercises 13–20 using verbal or written explanations.

13. Describe the differences among lines, half-lines, rays, and line segments.

14. What is an angle and what determines its size?

15. Describe each type of angle: acute, right, obtuse, and straight.

16. What are complementary angles? Describe how to find the measure of an angle's complement.

17. What are supplementary angles? Describe how to find the measure of an angle's supplement.

18. Describe the difference between perpendicular and parallel lines.

19. If two parallel lines are intersected by a transversal, describe the location of the alternate interior angles, the alternate exterior angles, and the corresponding angles.

20. If a transversal is perpendicular to one of two parallel lines, must it be perpendicular to the other parallel line as well? Explain your answer.

In Exercises 21–24, determine whether each statement makes sense or does not make sense, and explain your reasoning.

21. I drew two lines that are not parallel and that intersect twice.

22. I used the length of an angle's sides to determine that it was obtuse.

23. I'm working with two angles that are not complementary, so I can conclude that they are supplementary.

24. The rungs of a ladder are perpendicular to each side, so the rungs are parallel to each other.

Practice Exercises

25. The hour hand of a clock moves from 12 to 5 o'clock. Through how many degrees does it move?

26. The hour hand of a clock moves from 12 to 4 o'clock. Through how many degrees does it move?

27. The hour hand of a clock moves from 1 to 4 o'clock. Through how many degrees does it move?

28. The hour hand of a clock moves from 1 to 7 o'clock. Through how many degrees does it move?

In Exercises 29–34, use the protractor to find the measure of each angle. Then classify the angle as acute, right, straight, or obtuse.

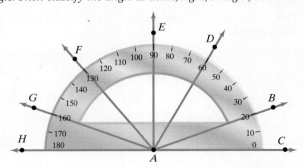

29. $\angle CAB$ **30.** $\angle CAF$

31. $\angle HAB$ **32.** $\angle HAF$

33. $\angle CAH$ **34.** $\angle HAE$

In Exercises 35–38, find the measure of the angle in which a question mark with a degree symbol appears.

35.   **36.**

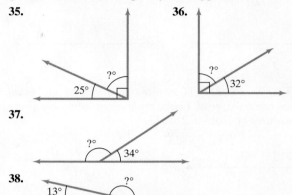

37.

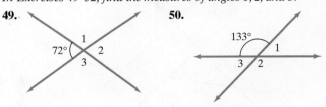

38.

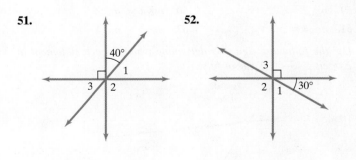

In Exercises 39–44, find the measure of the complement and the supplement of each angle.

39. 48°   **40.** 52°   **41.** 89°

42. 1°   **43.** 37.4°   **44.** $15\frac{1}{3}°$

In Exercises 45–48, use an algebraic equation to find the measures of the two angles described. Begin by letting x represent the degree measure of the angle's complement or its supplement.

45. The measure of the angle is 12° greater than its complement.

46. The measure of the angle is 56° greater than its complement.

47. The measure of the angle is three times greater than its supplement.

48. The measure of the angle is 81° more than twice that of its supplement.

In Exercises 49–52, find the measures of angles 1, 2, and 3.

49.   **50.**

51.   **52.**

The figures for Exercises 53–54 show two parallel lines intersected by a transversal. One of the angle measures is given. Find the measure of each of the other seven angles.

53.
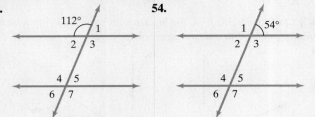

54.

The figures for Exercises 55–58 show two parallel lines intersected by more than one transversal. Two of the angle measures are given. Find the measures of angles 1, 2, and 3.

55.
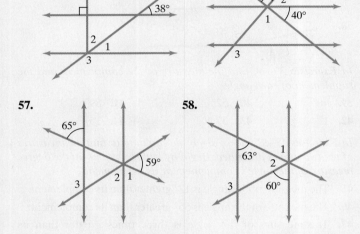

56.

57.

58.

Use the following figure to determine whether each statement in Exercises 59–62 is true or false.

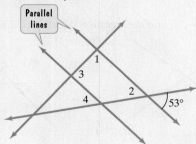

59. $m\angle 2 = 37°$

60. $m\angle 1 = m\angle 2$

61. $m\angle 4 = 53°$

62. $m\angle 3 = m\angle 4$

Use the following figure to determine whether each statement in Exercises 63–66 is true or false.

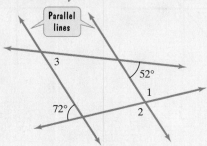

63. $m\angle 1 = 38°$

64. $m\angle 1 = 108°$

65. $m\angle 2 = 52°$

66. $m\angle 3 = 72°$

Practice Plus

In Exercises 67–70, use an algebraic equation to find the measure of each angle that is represented in terms of x.

67.
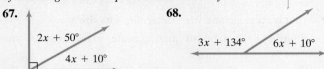

68.
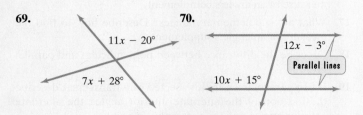

69.
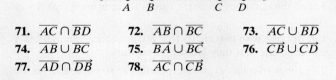

70.

Because geometric figures consist of sets of points, we can apply set operations to obtain the union, ∪, or the intersection, ∩, of such figures. The union of two geometric figures is the set of points that belongs to either of the figures or to both figures. The intersection of two geometric figures is the set of points common to both figures. In Exercises 71–78, use the line shown to find each set of points.

71. $\overline{AC} \cap \overline{BD}$

72. $\overline{AB} \cap \overline{BC}$

73. $\overline{AC} \cup \overline{BD}$

74. $\overline{AB} \cup \overline{BC}$

75. $\overrightarrow{BA} \cup \overrightarrow{BC}$

76. $\overrightarrow{CB} \cup \overrightarrow{CD}$

77. $\overrightarrow{AD} \cap \overrightarrow{DB}$

78. $\overrightarrow{AC} \cap \overrightarrow{CB}$

Application Exercises

79. The picture shows the top of an umbrella in which all the angles formed by the spokes have the same measure. Find the measure of each angle.

80. Use the definition of parallel lines to explain the joke in this B.C. cartoon.

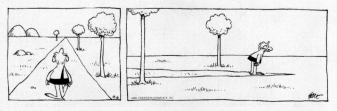

Reproduced by permission of John L. Hart FLP and Creators Syndicate, Inc.

81. The picture shows a window with parallel framing in which snow has collected in the corners. What property of parallel lines is illustrated by where the snow has collected?

In Exercises 82–83, consider the following uppercase letters from the English alphabet:

A E F H N T X Z.

82. Which letters contain parallel line segments?

83. Which letters contain perpendicular line segments?

Angles play an important role in custom bikes that are properly fitted to the biking needs of cyclists. One of the angles to help find the perfect fit is called the **hip angle**. *The figure indicates that the hip angle is created when you're sitting on the bike, gripping the handlebars, and your leg is fully extended. Your hip is the vertex, with one ray extending to your shoulder and the other ray extending to the front-bottom of your foot.*

The table indicates hip angles for various biking needs. Use this information to pedal through Exercises 84–87.

Hip Angle	Used For
85° ≤ hip angle ≤ 89°	short-distance aggressive racing
91° ≤ hip angle ≤ 115°	long-distance riding
116° ≤ hip angle ≤ 130°	mountain biking

84. Which type or types of biking require an acute hip angle?

85. Which type or types of biking require an obtuse hip angle?

86. A racer who had an 89° hip angle decides to switch to long-distance riding. What is the maximum difference in hip angle for the two types of biking?

87. A racer who had an 89° hip angle decides to switch to mountain biking. What is the minimum difference in hip angle for the two types of biking?

(Source for Exercises 84–87: *Scholastic Math*, January 11, 2010)

Critical Thinking Exercises

88. Use the figure to select a pair of complementary angles.

  a. ∡1 and ∡4   **b.** ∡3 and ∡6

  c. ∡2 and ∡5   **d.** ∡1 and ∡5

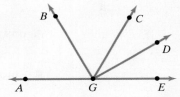

89. If $m\angle AGB = m\angle BGC$, and $m\angle CGD = m\angle DGE$, find $m\angle BGD$.

10.2 | Triangles

Objectives

1 Solve problems involving angle relationships in triangles.

2 Solve problems involving similar triangles.

3 Solve problems using the Pythagorean Theorem.

In Chapter 1, we defined *deductive reasoning* as the process of proving a specific conclusion from one or more general statements. A conclusion that is proved to be true through deductive reasoning is called a **theorem**. The Greek mathematician Euclid, who lived more than 2000 years ago, used deductive reasoning. In his 13-volume book, the *Elements*, Euclid proved over 465 theorems about geometric figures. Euclid's work established deductive reasoning as a fundamental tool of mathematics. Here's looking at Euclid!

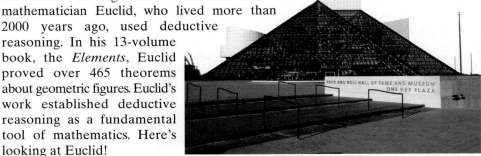

A **triangle** is a geometric figure that has three sides, all of which lie on a flat surface or plane. If you start at any point along the triangle and trace along the entire figure exactly once, you will end at the same point at which you started. Because the beginning point and ending point are the same, the triangle is called a **closed** geometric figure. Euclid used parallel lines to prove one of the most important properties of triangles: The sum of the measures of the three angles of any triangle is 180°. Here is how he did it. He began with the following general statement:

Euclid's Assumption about Parallel Lines

Given a line and a point not on the line, one and only one line can be drawn through the given point parallel to the given line.

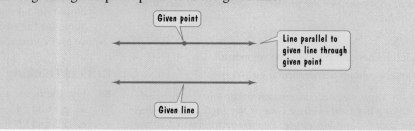

In **Figure 10.15**, triangle *ABC* represents any triangle. Using the general assumption given above, we draw a line through point *B* parallel to line *AC*.

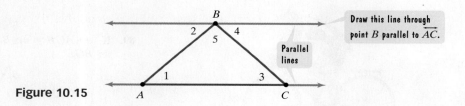

Figure 10.15

Because the lines are parallel, alternate interior angles have the same measure.

$$m\angle 1 = m\angle 2 \quad \text{and} \quad m\angle 3 = m\angle 4$$

Also observe that angles 2, 5, and 4 form a straight angle.

$$m\angle 2 + m\angle 5 + m\angle 4 = 180°$$

Because $m\angle 1 = m\angle 2$, replace $m\angle 2$ with $m\angle 1$.   Because $m\angle 3 = m\angle 4$, replace $m\angle 4$ with $m\angle 3$.

$$m\angle 1 + m\angle 5 + m\angle 3 = 180°$$

Because $\angle 1, \angle 5$, and $\angle 3$ are the three angles of the triangle, this last equation shows that the measures of the triangle's three angles have a sum of 180°.

The Angles of a Triangle

The sum of the measures of the three angles of any triangle is 180°.

Solve problems involving angle relationships in triangles.

Example 1   Using Angle Relationships in Triangles

Find the measure of angle A for triangle ABC in **Figure 10.16**.

Solution

Because $m \angle A + m \angle B + m \angle C = 180°$, we obtain

$$m \angle A + 120° + 17° = 180°.$$ The sum of the measures of a triangle's three angles is 180°.

$$m \angle A + 137° = 180°$$ Simplify: $120° + 17° = 137°$.

$$m \angle A = 180° - 137°$$ Find the measure of A by subtracting 137° from both sides of the equation.

$$m \angle A = 43°$$ Simplify.

Checkpoint 1   In **Figure 10.16**, suppose that $m \angle B = 116°$ and $m \angle C = 15°$. Find $m \angle A$.

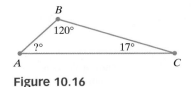

Figure 10.16

Example 2   Using Angle Relationships in Triangles

Find the measures of angles 1 through 5 in **Figure 10.17**.

Solution

Because $\angle 1$ is supplementary to the right angle, $m \angle 1 = 90°$.

$m \angle 2$ can be found using the fact that the sum of the measures of the angles of a triangle is 180°.

$$m \angle 1 + m \angle 2 + 43° = 180°$$ The sum of the measures of a triangle's three angles is 180°.

$$90° + m \angle 2 + 43° = 180°$$ We previously found that $m \angle 1 = 90°$.

$$m \angle 2 + 133° = 180°$$ Simplify: $90° + 43° = 133°$.

$$m \angle 2 = 180° - 133°$$ Subtract 133° from both sides.

$$m \angle 2 = 47°$$ Simplify.

$m \angle 3$ can be found using the fact that vertical angles have equal measures: $m \angle 3 = m \angle 2$. Thus, $m \angle 3 = 47°$.

$m \angle 4$ can be found using the fact that the sum of the measures of the angles of a triangle is 180°. Refer to the triangle at the top of **Figure 10.17**.

$$m \angle 3 + m \angle 4 + 60° = 180°$$ The sum of the measures of a triangle's three angles is 180°.

$$47° + m \angle 4 + 60° = 180°$$ We previously found that $m \angle 3 = 47°$.

$$m \angle 4 + 107° = 180°$$ Simplify: $47° + 60° = 107°$.

$$m \angle 4 = 180° - 107°$$ Subtract 107° from both sides.

$$m \angle 4 = 73°$$ Simplify.

Finally, we can find $m \angle 5$ by observing that angles 4 and 5 form a straight angle.

$$m \angle 4 + m \angle 5 = 180°$$ A straight angle measures 180°.

$$73° + m \angle 5 = 180°$$ We previously found that $m \angle 4 = 73°$.

$$m \angle 5 = 180° - 73°$$ Subtract 73° from both sides.

$$m \angle 5 = 107°$$ Simplify.

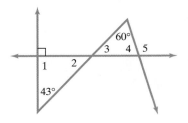

Figure 10.17

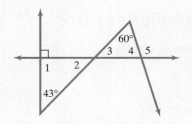

Figure 10.17 (repeated)

Study Tip

We checked a variety of math textbooks and found two slightly different definitions for an isosceles triangle. One definition states that an isosceles triangle has *exactly* two sides of equal length. A second definition asserts that an isosceles triangle has *at least* two sides of equal length. (This makes an equilateral triangle, with three sides of the same length, an isosceles triangle.) In this book, we assume that an isosceles triangle has exactly two sides, but not three sides, of the same length.

✓ **Checkpoint 2**   In **Figure 10.17**, suppose that the angle shown to measure 43° measures, instead, 36°. Further suppose that the angle shown to measure 60° measures, instead, 58°. Under these new conditions, find the measures of angles 1 through 5 in the figure.

Triangles can be described using characteristics of their angles or their sides.

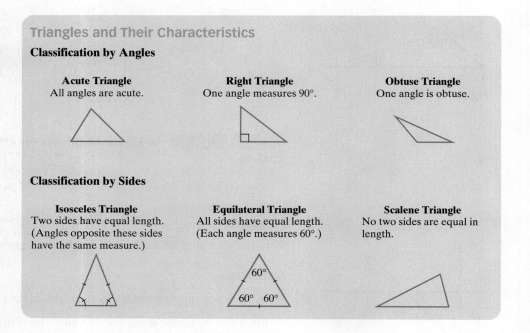

Triangles and Their Characteristics

Classification by Angles

Acute Triangle
All angles are acute.

Right Triangle
One angle measures 90°.

Obtuse Triangle
One angle is obtuse.

Classification by Sides

Isosceles Triangle
Two sides have equal length. (Angles opposite these sides have the same measure.)

Equilateral Triangle
All sides have equal length. (Each angle measures 60°.)

Scalene Triangle
No two sides are equal in length.

② Solve problems involving similar triangles.

Similar Triangles

Shown in the margin is an international road sign. This sign is shaped just like the actual sign, although its size is smaller. Figures that have the same shape, but not the same size, are used in **scale drawings**. A scale drawing always pictures the exact shape of the object that the drawing represents. Architects, engineers, landscape gardeners, and interior decorators use scale drawings in planning their work.

Figures that have the same shape, but not necessarily the same size, are called **similar figures**. In **Figure 10.18**, triangles ABC and DEF are similar. Angles A and D measure the same number of degrees and are called **corresponding angles**. Angles C and F are corresponding angles, as are angles B and E. Angles with the same number of tick marks in **Figure 10.18** are the corresponding angles.

Pedestrian crossing

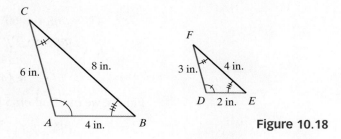

Figure 10.18

The sides opposite the corresponding angles are called **corresponding sides**. Thus, $\overline{CB}$ and $\overline{FE}$ are corresponding sides. $\overline{AB}$ and $\overline{DE}$ are also corresponding sides, as are $\overline{AC}$ and $\overline{DF}$. Corresponding angles measure the same number of degrees, but corresponding sides may or may not be the same length. For the triangles in **Figure 10.18**, each side in the smaller triangle is half the length of the corresponding side in the larger triangle.

The triangles in **Figure 10.18** illustrate what it means to be **similar triangles. Corresponding angles have the same measure and the ratios of the lengths of the corresponding sides are equal.**

$$\frac{\text{length of}}{\text{length of}} = \frac{6 \text{ in.}}{3 \text{ in.}} = \frac{2}{1}; \quad \frac{\text{length of } \overline{CB}}{\text{length of } \overline{FE}} = \frac{8 \text{ in.}}{4 \text{ in.}} = \frac{2}{1}; \quad \frac{\text{length of } \overline{AB}}{\text{length of } \overline{DE}} = \frac{4 \text{ in.}}{2 \text{ in.}} = \frac{2}{1}$$

In similar triangles, the lengths of the corresponding sides are proportional. Thus, in **Figure 10.18**,

AC represents the length of $\overline{AC}$, DF the length of $\overline{DF}$, and so on. $\quad \frac{AC}{DF} = \frac{CB}{FE} = \frac{AB}{DE}.$

How can we quickly determine if two triangles are similar? **If the measures of two angles of one triangle are equal to those of two angles of a second triangle, then the two triangles are similar.** If the triangles are similar, then their corresponding sides are proportional.

Example 3 Using Similar Triangles

Explain why the triangles in **Figure 10.19** are similar. Then find the missing length, x.

Solution

Figure 10.20 shows that two angles of the small triangle are equal in measure to two angles of the large triangle. One angle pair is given to have the same measure. Another angle pair consists of vertical angles with the same measure. Thus, the triangles are similar and their corresponding sides are proportional.

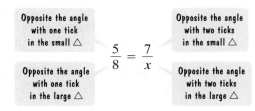

$$\frac{5}{8} = \frac{7}{x}$$

We solve $\frac{5}{8} = \frac{7}{x}$ for x by applying the cross-products principle for proportions that we discussed in Section 4.4: If $\frac{a}{b} = \frac{c}{d}$, then $ad = bc$.

$5x = 8 \cdot 7$ Apply the cross-products principle.

$5x = 56$ Multiply: $8 \cdot 7 = 56$.

$\dfrac{5x}{5} = \dfrac{56}{5}$ Divide both sides by 5.

$x = 11.2$ Simplify.

The missing length, x, is 11.2 inches.

Checkpoint 3 Explain why the triangles in **Figure 10.21** are similar. Then find the missing length, x.

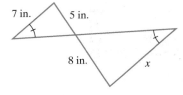

Figure 10.19

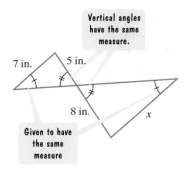

Vertical angles have the same measure.

Given to have the same measure

Figure 10.20

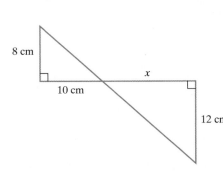

Figure 10.21

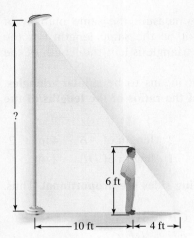

Figure 10.22

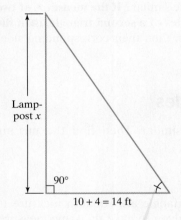

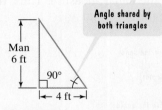

Figure 10.23

Example 4 Problem Solving Using Similar Triangles

A man who is 6 feet tall is standing 10 feet from the base of a lamppost (see **Figure 10.22**). The man's shadow has a length of 4 feet. How tall is the lamppost?

Solution

The drawing in **Figure 10.23** makes the similarity of the triangles easier to see. The large triangle with the lamppost on the left and the small triangle with the man on the left both contain 90° angles. They also share an angle. Thus, two angles of the large triangle are equal in measure to two angles of the small triangle. This means that the triangles are similar and their corresponding sides are proportional. We begin by letting x represent the height of the lamppost, in feet. Because corresponding sides of the two similar triangles are proportional,

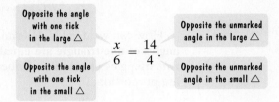

We solve for x by applying the cross-products principle.

$$4x = 6 \cdot 14 \quad \text{Apply the cross-products principle.}$$
$$4x = 84 \quad \text{Multiply: } 6 \cdot 14 = 84.$$
$$\frac{4x}{4} = \frac{84}{4} \quad \text{Divide both sides by 4.}$$
$$x = 21 \quad \text{Simplify.}$$

The lamppost is 21 feet tall.

Checkpoint 4 Find the height of the lookout tower shown in **Figure 10.24** using the figure that lines up the top of the tower with the top of a stick that is 2 yards long.

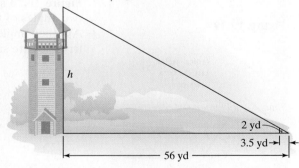

Figure 10.24

The Pythagorean Theorem

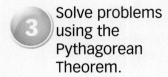

3 Solve problems using the Pythagorean Theorem.

The ancient Greek philosopher and mathematician Pythagoras (approximately 582–500 B.C.) founded a school whose motto was "All is number." Pythagoras is best remembered for his work with the **right triangle**, a triangle with one angle measuring 90°. The side opposite the 90° angle is called the **hypotenuse**. The other sides are called **legs**. Pythagoras found that if he constructed squares on each of the legs, as well as a larger square

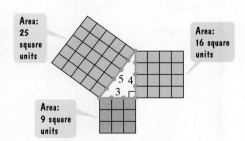

Figure 10.25 The area of the large square equals the sum of the areas of the smaller squares.

on the hypotenuse, the sum of the areas of the smaller squares is equal to the area of the larger square. This is illustrated in **Figure 10.25**.

This relationship is usually stated in terms of the lengths of the three sides of a right triangle and is called the **Pythagorean Theorem**.

> **The Pythagorean Theorem**
>
> The sum of the squares of the lengths of the legs of a right triangle equals the square of the length of the hypotenuse.
>
> If the legs have lengths a and b and the hypotenuse has length c, then
>
> $$a^2 + b^2 = c^2.$$

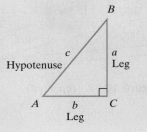

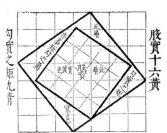
Example 5   Using the Pythagorean Theorem

Find the length of the hypotenuse c in the right triangle shown in **Figure 10.26**.

Figure 10.26

Solution

Let $a = 9$ and $b = 12$. Substituting these values into $c^2 = a^2 + b^2$ enables us to solve for c.

$$c^2 = a^2 + b^2 \quad \text{Use the symbolic statement of the Pythagorean Theorem.}$$
$$c^2 = 9^2 + 12^2 \quad \text{Let } a = 9 \text{ and } b = 12.$$
$$c^2 = 81 + 144 \quad 9^2 = 9 \cdot 9 = 81 \text{ and } 12^2 = 12 \cdot 12 = 144.$$
$$c^2 = 225 \quad \text{Add.}$$
$$c = \sqrt{225} = 15 \quad \text{Solve for } c \text{ by taking the positive square root of 225.}$$

The length of the hypotenuse is 15 feet.

Checkpoint 5   Find the length of the hypotenuse in a right triangle whose legs have lengths 7 feet and 24 feet.

Example 6   Using the Pythagorean Theorem

a. A wheelchair ramp with a length of 122 inches has a horizontal distance of 120 inches. How high is the ramp at the top?

b. Construction laws are very specific when it comes to access ramps for the disabled. Every vertical rise of 1 inch requires a horizontal run of 12 inches. Does this ramp satisfy the requirement?

Solution

a. The problem's conditions state that the wheelchair ramp has a length of 122 inches and a horizontal distance of 120 inches. **Figure 10.27** shows the right triangle that is formed by the ramp, the wall, and the ground. We can find x, the ramp's height, using the Pythagorean Theorem.

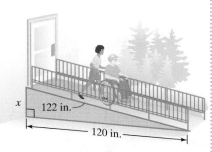

Figure 10.27

(leg)²	plus	(leg)²	equals	(hypotenuse)²
x^2	$+$	120^2	$=$	122^2

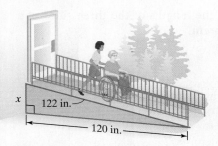

x 122 in.

120 in.

Figure 10.27 (repeated)

$$x^2 + 120^2 = 122^2$$ This is the equation resulting from the Pythagorean Theorem.

$$x^2 + 14,400 = 14,884$$ Square 120 and 122.

$$x^2 = 484$$ Isolate x^2 by subtracting 14,400 from both sides.

$$x = \sqrt{484} = 22$$ Solve for x by taking the positive square root of 484.

The ramp is 22 inches high at the top.

b. Every vertical rise of 1 inch requires a horizontal run of 12 inches. Because the ramp has a height of 22 inches, it requires a horizontal distance of 22(12) inches, or 264 inches. The horizontal distance is only 120 inches, so this ramp does not satisfy construction laws for access ramps for the disabled.

Checkpoint 6 A radio tower is supported by two wires that are each 130 yards long and attached to the ground 50 yards from the base of the tower. How far from the ground are the wires attached to the tower?

Exercise Set 10.2

Concept and Vocabulary Exercises

In Exercises 1–8, fill in each blank so that the resulting statement is true.

1. The sum of the measures of the three angles of any triangle is _____.

2. A triangle in which each angle measures less than 90° is called a/an _____ triangle.

3. A triangle in which one angle measures more than 90° is called a/an _____ triangle.

4. A triangle with exactly two sides of the same length is called a/an _____ triangle.

5. A triangle whose sides are all the same length is called a/an _____ triangle.

6. A triangle that has no sides of the same length is called a/an _____ triangle.

7. Triangles that have the same shape, but not necessarily the same size, are called _____ triangles. For such triangles, corresponding angles have _____ and the lengths of the corresponding sides are _____.

8. The Pythagorean Theorem states that in any _____ triangle, the sum of the squares of the lengths of the _____ equals _____.

In Exercises 9–13, determine whether each statement is true or false. If the statement is false, make the necessary change(s) to produce a true statement.

9. Euclid's assumption about parallel lines states that given a line and a point not on the line, one and only one line can be drawn through the given point parallel to the given line.

10. A triangle cannot have a right angle and an obtuse angle.

11. Each angle of an isosceles triangle measures 60°.

12. If the measures of two angles of one triangle are equal to those of two angles of a second triangle, then the two triangles have the same size and shape.

13. In any triangle, the sum of the squares of the two shorter sides equals the square of the length of the longest side.

Respond to Exercises 14–22 using verbal or written explanations.

14. If the measures of two angles of a triangle are known, explain how to find the measure of the third angle.

15. Can a triangle contain two right angles? Explain your answer.

16. What general assumption did Euclid make about a point and a line in order to prove that the sum of the measures of the angles of a triangle is 180°?

17. What are similar triangles?

18. If the ratio of the corresponding sides of two similar triangles is 1 to $1\left(\frac{1}{1}\right)$, what must be true about the triangles?

19. What are corresponding angles in similar triangles?

20. Describe how to identify the corresponding sides in similar triangles.

21. In your own words, state the Pythagorean Theorem.

22. In the 1939 movie *The Wizard of Oz*, upon being presented with a Th.D. (Doctor of Thinkology), the Scarecrow proudly exclaims, "The sum of the square roots of any two sides of an isosceles triangle is equal to the square root of the remaining side." Did the Scarecrow get the Pythagorean Theorem right? In particular, describe four errors in the Scarecrow's statement.

In Exercises 23–26, determine whether each statement makes sense or does not make sense, and explain your reasoning.

23. I'm fencing off a triangular plot of land that has two right angles.

24. Triangle I is equilateral, as is triangle II, so the triangles are similar.

25. Triangle I is a right triangle, as is triangle II, so the triangles are similar.

26. If I am given the lengths of two sides of a triangle, I can use the Pythagorean Theorem to find the length of the third side.

Practice Exercises

In Exercises 27–30, find the measure of angle A for the triangle shown.

27.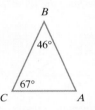

28.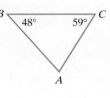

29.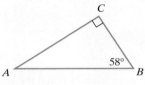

30.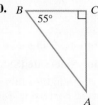

In Exercises 31–32, find the measures of angles 1 through 5 in the figure shown.

31.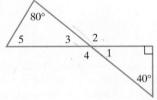

32.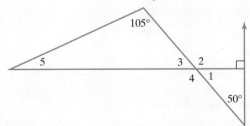

We have seen that isosceles triangles have two sides of equal length. The angles opposite these sides have the same measure. In Exercises 33–34, use this information to help find the measure of each numbered angle.

33.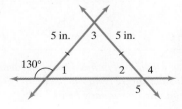

34.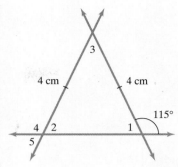

In Exercises 35–36, lines l and m are parallel. Find the measure of each numbered angle.

35.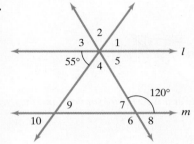

36.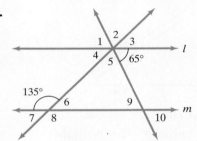

In Exercises 37–42, explain why the triangles are similar. Then find the missing length, x.

37.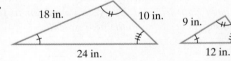

38.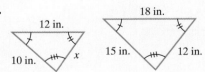

39.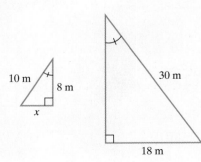

40.

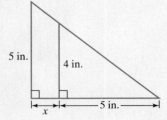

41.

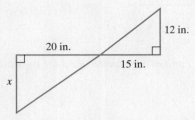

42.

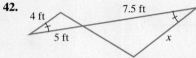

In Exercises 43–45, △ABC and △ADE are similar. Find the length of the indicated side.

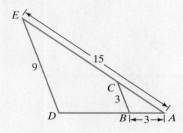

43. $\overline{CA}$ **44.** $\overline{DB}$ **45.** $\overline{DA}$

46. In the diagram for Exercises 43–45, suppose that you are not told that △ABC and △ADE are similar. Instead, you are given that $\overleftrightarrow{ED}$ and $\overleftrightarrow{CB}$ are parallel. Under these conditions, explain why the triangles must be similar.

In Exercises 47–52, use the Pythagorean Theorem to find the missing length in each right triangle. Use your calculator to find square roots, rounding, if necessary, to the nearest tenth.

47.

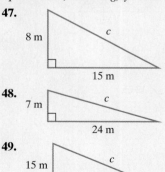

48.

49.

50.

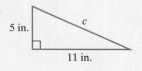

51.

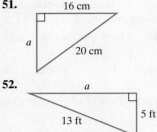

52.

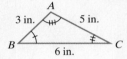 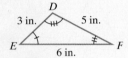

Practice Plus

Two triangles are **congruent** if they have the same shape and the same size. In congruent triangles, the measures of corresponding angles are equal and the corresponding sides have the same length. The following triangles are congruent:

Any one of the following may be used to determine if two triangles are congruent.

Determining Congruent Triangles

1. Side-Side-Side (SSS)

If the lengths of three sides of one triangle equal the lengths of the corresponding sides of a second triangle, then the two triangles are congruent.

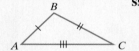 **SSS**

2. Side-Angle-Side (SAS)

If the lengths of two sides of one triangle equal the lengths of the corresponding sides of a second triangle, and the measures of the angles between each pair of sides are equal, then the two triangles are congruent.

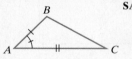 **SAS**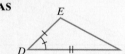

3. Angle-Side-Angle (ASA)

If the measures of two angles of one triangle equal the measures of two angles of a second triangle, and the lengths of the sides between each pair of angles are equal, then the two triangles are congruent.

ASA

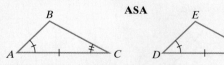

In Exercises 53–62, determine whether △I and △II are congruent. If the triangles are congruent, state the reason why, selecting from SSS, SAS, or ASA. (More than one reason may be possible.)

53.

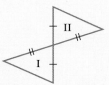

54.

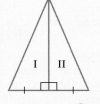

55.

56.

57.

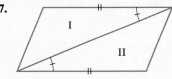

58.

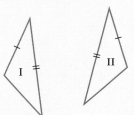

59.

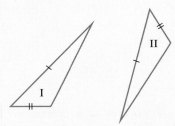

60.

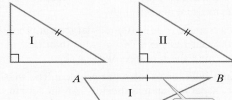

61.

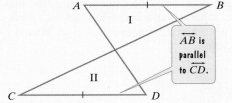

62.

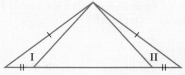

Application Exercises

Use similar triangles to solve Exercises 63–64.

63. A person who is 5 feet tall is standing 80 feet from the base of a tree and the tree casts an 86-foot shadow. The person's shadow is 6 feet in length. What is the tree's height? Round to the nearest tenth of a foot.

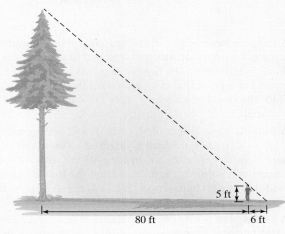

64. A tree casts a shadow 12 feet long. At the same time, a vertical rod 8 feet high casts a shadow that is 6 feet long. How tall is the tree?

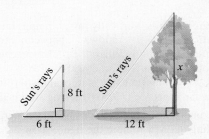

Use the Pythagorean Theorem to solve Exercises 65–72. Use your calculator to find square roots, rounding, if necessary, to the nearest tenth.

65. A baseball diamond is actually a square with 90-foot sides. What is the distance from home plate to second base?

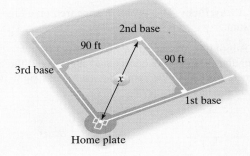

66. The base of a 20-foot ladder is 15 feet from the house. How far up the house does the ladder reach?

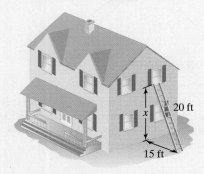

67. A flagpole has a height of 16 yards. It will be supported by three cables, each of which is attached to the flagpole at a point 4 yards below the top of the pole and attached to the ground at a point that is 9 yards from the base of the pole. Find the total number of feet of cable that will be required.

68. A flagpole has a height of 10 yards. It will be supported by three cables, each of which is attached to the flagpole at a point 4 yards below the top of the pole and attached to the ground at a point that is 8 yards from the base of the pole. Find the total number of feet of cable that will be required.

69. A rectangular garden bed measures 5 feet by 12 feet. A water faucet is located at one corner of the garden bed. A hose will be connected to the water faucet. The hose must be long enough to reach the opposite corner of the garden bed when stretched straight. Find the required length of hose.

70. A rocket ascends vertically after being launched from a location that is midway between two ground-based tracking stations. When the rocket reaches an altitude of 4 kilometers, it is 5 kilometers from each of the tracking stations. Assuming that this is a locale where the terrain is flat, how far apart are the two tracking stations?

71. If construction costs are $150,000 per *kilometer*, find the cost of building the new road in the figure shown.

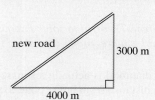

72. Picky, Picky, Picky This problem appeared on a high school exit exam:

Alex is building a ramp for a bike competition. He has two rectangular boards. One board is six meters and the other is five meters long. If the ramp has to form a right triangle, what should its height be?

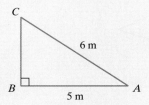

Students were asked to select the correct answer from the following options:

3 meters; 4 meters; 3.3 meters; 7.8 meters.

a. Among the available choices, which option best expresses the ramp's height? How many feet, to the nearest tenth of a foot, is this? Does a bike competition that requires riders to jump off these heights seem realistic? (ouch!)

b. Express the ramp's height to the nearest hundredth of a meter. By how many centimeters does this differ from the "correct" answer on the test? How many inches, to the nearest half inch, is this? Is it likely that a carpenter with a tape measure would make this error?

c. According to the problem, Alex has boards that measure 5 meters and 6 meters. A 6-meter board? How many feet, to the nearest tenth of a foot, is this? When was the last time you found a board of this length at Home Depot? (*Source: The New York Times,* April 24, 2005)

Critical Thinking Exercises

73. Find the measure of angle *R*.

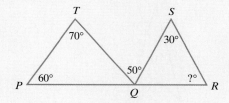

74. What is the length of $\overline{AB}$ in the accompanying figure?

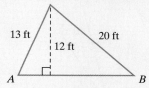

75. One angle of a triangle is twice as large as another. The measure of the third angle is 20° more than that of the smallest angle. Find the measure of each angle.

76. One angle of a triangle is three times as large as another. The measure of the third angle is 30° greater than that of the smallest angle. Find the measure of each angle.

77. Use a quadratic equation to solve this problem. The width of a rectangular carpet is 7 meters shorter than the length, and the diagonal is 1 meter longer than the length. What are the carpet's dimensions?

10.3 : Polygons, Perimeter, and Tessellations

You have just purchased a beautiful plot of land in the country, shown in **Figure 10.28**. In order to have more privacy, you decide to put fencing along each of its four sides. The cost of this project depends on the distance around the four outside edges of the plot, called its **perimeter**, as well as the cost for each foot of fencing.

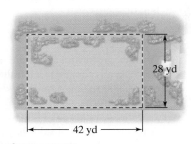

Figure 10.28

Your plot of land is a geometric figure: It has four straight sides that are line segments. The plot is on level ground, so that the four sides all lie on a flat surface, or plane. The plot is an example of a *polygon*. Any closed shape in the plane formed by three or more line segments that intersect only at their endpoints is a **polygon**.

A polygon is named according to the number of sides it has. We know that a three-sided polygon is called a **triangle**. A four-sided polygon is called a **quadrilateral**.

A polygon whose sides are all the same length and whose angles all have the same measure is called a **regular polygon**. **Table 10.2** provides the names of six polygons. Also shown are illustrations of regular polygons.

① **Name certain polygons according to the number of sides.**

Table 10.2 Illustrations of Regular Polygons			
Name	**Picture**	**Name**	**Picture**
Triangle 3 sides	△	Hexagon 6 sides	⬡
Quadrilateral 4 sides	□	Heptagon 7 sides	⬡
Pentagon 5 sides	⬠	Octagon 8 sides	⯃

Quadrilaterals

② **Recognize the characteristics of certain quadrilaterals.**

The plot of land in **Figure 10.28** is a four-sided polygon, or a quadrilateral. However, when you first looked at the figure, perhaps you thought of the plot as a rectangular field. A **rectangle** is a special kind of quadrilateral in which both pairs of opposite sides are parallel, have the same measure, and whose angles are right angles. **Table 10.3** on the next page presents some special quadrilaterals and their characteristics.

Table 10.3 Types of Quadrilaterals

Name	Characteristics	Representation
Parallelogram	Quadrilateral in which both pairs of opposite sides are parallel and have the same measure. Opposite angles have the same measure.	
Rhombus	Parallelogram with all sides having equal length.	
Rectangle	Parallelogram with four right angles. Because a rectangle is a parallelogram, opposite sides are parallel and have the same measure.	
Square	A rectangle with all sides having equal length. Each angle measures 90°, and the square is a regular quadrilateral.	
Trapezoid	A quadrilateral with exactly one pair of parallel sides.	

Perimeter

The **perimeter**, P, of a polygon is the sum of the lengths of its sides. Perimeter is measured in linear units, such as inches, feet, yards, meters, or kilometers.

Example 1 involves the perimeter of a rectangle. Because perimeter is the sum of the lengths of the sides, the perimeter of the rectangle shown in **Figure 10.29** is $l + w + l + w$. This can be expressed as

$$P = 2l + 2w.$$

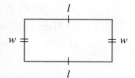

Figure 10.29
A rectangle with length l and width w

3 | Solve problems involving a polygon's perimeter.

Example 1 An Application of Perimeter

The rectangular field we discussed at the beginning of this section (see **Figure 10.28** on page 613) has a length of 42 yards and a width of 28 yards. If fencing costs $5.25 per foot, find the cost to enclose the field with fencing.

Solution

We begin by finding the perimeter of the rectangle in yards. Using 3 ft = 1 yd and dimensional analysis, we express the perimeter in feet. Finally, we multiply the perimeter, in feet, by $5.25 because the fencing costs $5.25 per foot.

The length, l, is 42 yards and the width, w, is 28 yards. The perimeter of the rectangle is determined using the formula $P = 2l + 2w$.

$$P = 2l + 2w = 2 \cdot 42 \text{ yd} + 2 \cdot 28 \text{ yd} = 84 \text{ yd} + 56 \text{ yd} = 140 \text{ yd}$$

Because 3 ft = 1 yd, we use the unit fraction $\frac{3 \text{ ft}}{1 \text{ yd}}$ to convert from yards to feet.

$$140 \text{ yd} = \frac{140 \text{ yd}}{1} \cdot \frac{3 \text{ ft}}{1 \text{ yd}} = 140 \cdot 3 \text{ ft} = 420 \text{ ft}$$

The perimeter of the rectangle is 420 feet. Now we are ready to find the cost of the fencing. We multiply 420 feet by $5.25, the cost per foot.

$$\text{Cost} = \frac{420 \text{ feet}}{1} \cdot \frac{\$5.25}{\text{foot}} = 420(\$5.25) = \$2205$$

The cost to enclose the field with fencing is $2205.

Checkpoint 1   A rectangular field has a length of 50 yards and a width of 30 yards. If fencing costs $6.50 per foot, find the cost to enclose the field with fencing.

The Sum of the Measures of a Polygon's Angles

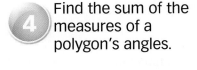
Find the sum of the measures of a polygon's angles.

We know that the sum of the measures of the three angles of any triangle is 180°. We can use inductive reasoning to find the sum of the measures of the angles of any polygon. Start by drawing line segments from a single point where two sides meet so that nonoverlapping triangles are formed. This is done in **Figure 10.30**.

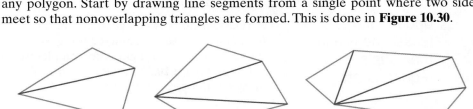

4 sides	5 sides	6 sides
2 triangles	3 triangles	4 triangles
Angle sum: 2(180°) = 360°	Angle sum: 3(180°) = 540°	Angle sum: 4(180°) = 720°

Figure 10.30

In each case, the number of triangles is two less than the number of sides of the polygon. Thus, for an n-sided polygon, there are $n - 2$ triangles. Because each triangle has an angle-measure sum of 180°, the sum of the measures for the angles in the $n - 2$ triangles is $(n - 2)180°$. Thus, the sum of the measures of the angles of an n-sided polygon is $(n - 2)180°$.

> **The Angles of a Polygon**
>
> The sum of the measures of the angles of a polygon with n sides is
>
> $$(n - 2)180°.$$

Example 2   Using the Formula for the Angles of a Polygon

a. Find the sum of the measures of the angles of an octagon.

b. **Figure 10.31** shows a regular octagon. Find the measure of angle A.

c. Find the measure of exterior angle B.

Solution

a. An octagon has eight sides. Using the formula $(n - 2)180°$ with $n = 8$, we can find the sum of the measures of its eight angles.
The sum of the measures of an octagon's angles is

$$(n - 2)180° = (8 - 2)180°$$
$$= 6 \cdot 180°$$
$$= 1080°.$$

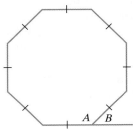

Figure 10.31 A regular octagon

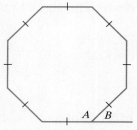

Figure 10.31 (repeated)
A regular octagon

b. Examine the regular octagon in **Figure 10.31**. Note that all eight sides have the same length. Likewise, all eight angles have the same degree measure. Angle A is one of these eight angles. We find its measure by taking the sum of the measures of all eight angles, 1080°, and dividing by 8.

$$m \angle A = \frac{1080°}{8} = 135°$$

c. Because $\angle B$ is the supplement of $\angle A$,

$$m \angle B = 180° - 135° = 45°.$$

a. Find the sum of the measures of the angles of a 12-sided polygon.
b. Find the measure of an angle of a regular 12-sided polygon.

Tessellations

⑤ Understand tessellations and their angle requirements.

A relationship between geometry and the visual arts is found in an art form called *tessellations*. A **tessellation**, or **tiling**, is a pattern consisting of the repeated use of the same geometric figures to completely cover a plane, leaving no gaps and no overlaps. **Figure 10.32** shows eight tessellations, each consisting of the repeated use of two or more regular polygons.

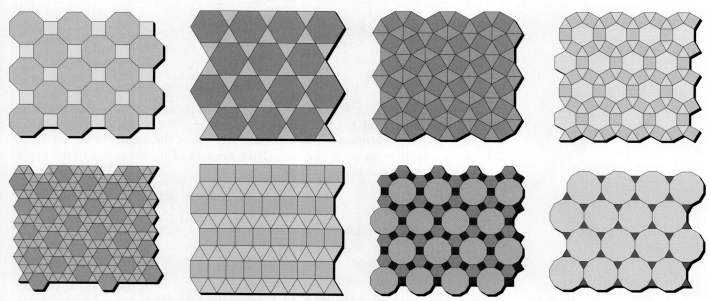

Figure 10.32 Eight tessellations formed by two or more regular polygons

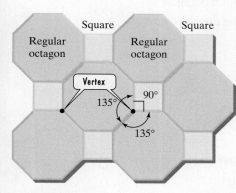

Figure 10.33

In each tessellation in **Figure 10.32**, the same types of regular polygons surround every vertex (intersection point). Furthermore, **the sum of the measures of the angles that come together at each vertex is 360°, a requirement for the formation of a tessellation.** This is illustrated in the enlarged version of the tessellation in **Figure 10.33**. If you select any vertex and count the sides of the polygons that touch it, you'll see that vertices are surrounded by two regular octagons and a square. Can you see why the sum of the angle measures at every vertex is 360°?

$$135° + 135° + 90° = 360°$$

In Example 2, we found that each angle of a regular octagon measures 135°.

Each angle of a square measures 90°.

The most restrictive condition in creating tessellations is that just *one type of regular polygon* may be used. With this restriction, there are only three possible tessellations, made from equilateral triangles or squares or regular hexagons, as shown in **Figure 10.34**.

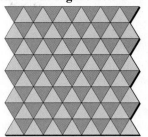 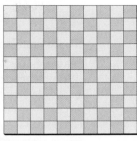 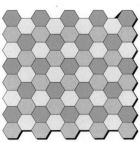

Figure 10.34 The three tessellations formed using one regular polygon

Each tessellation is possible because the angle sum at every vertex is 360°:

> **Six equilateral triangles at each vertex** $60° + 60° + 60° + 60° + 60° + 60° = 360°$

> **Four squares at each vertex** $90° + 90° + 90° + 90° = 360°$

> **Three regular hexagons at each vertex** $120° + 120° + 120° = 360°$.

> Each angle measures $\dfrac{(n-2)180°}{n} = \dfrac{(6-2)180°}{6} = 120°$.

Example 3   Angle Requirements of Tessellations

Explain why a tessellation cannot be created using only regular pentagons.

Solution

Let's begin by applying $(n - 2)180°$ to find the measure of each angle of a regular pentagon. Each angle measures

$$\frac{(5-2)180°}{5} = \frac{3(180°)}{5} = 108°.$$

A requirement for the formation of a tessellation is that the sum of the measures of the angles that come together at each vertex is 360°. With each angle of a regular pentagon measuring 108°, **Figure 10.35** shows that three regular pentagons fill in $3 \cdot 108° = 324°$ and leave a $360° - 324°$, or a 36°, gap. Because the 360° angle requirement cannot be met, no tessellation by regular pentagons is possible.

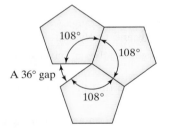

Figure 10.35

> **Checkpoint 3**   Explain why a tessellation cannot be created using only regular octagons.

"Drawings showing how a surface can be regularly filled with similar-shaped figures which are contiguous to one another, without leaving any open spaces, are my richest source of inspiration."

—M. C. Escher

Tessellations that are not restricted to the repeated use of regular polygons are endless in number. They are prominent in Islamic art, Italian mosaics, quilts, and ceramics. The Dutch artist M. C. Escher (1898–1972) created a dazzling array of prints, drawings, and paintings using tessellations composed of stylized interlocking animals. Escher's art reflects the mathematics that underlies all things, while creating surreal manipulations of space and perspective that make gentle fun of consensus reality.

M. C. Escher, *Symmetry Drawing, E85 (Lizard, Fish, Bat)*, drawn at Baarn, April, 1952, ink, pencil, watercolor. © 2006 The M. C. Escher Company, Holland. All rights reserved.

Exercise Set 10.3

Concept and Vocabulary Exercises

In Exercises 1–11, fill in each blank so that the resulting statement is true.

1. The distance around the sides of a polygon is called its _____.

2. A four-sided polygon is a/an _____, a five-sided polygon is a/an _____, a six-sided polygon is a/an _____, a seven-sided polygon is a/an _____, and an eight-sided polygon is a/an _____.

3. A polygon whose sides are all the same length and whose angles all have the same measure is called a/an _____ polygon.

4. Opposite sides of a parallelogram are _____ and _____.

5. A parallelogram with all sides of equal length without any right angles is called a/an _____.

6. A parallelogram with four right angles without all sides of equal length is called a/an _____.

7. A parallelogram with four right angles and all sides of equal length is called a/an _____.

8. A four-sided figure with exactly one pair of parallel sides is called a/an _____.

9. The perimeter, P, of a rectangle with length l and width w is given by the formula _____.

10. The sum of the measures of the angles of a polygon with n sides is _____.

11. A pattern consisting of the repeated use of the same geometric figures to completely cover a plane, leaving no gaps and no overlaps, is called a/an _____.

In Exercises 12–18, determine whether each statement is true or false. If the statement is false, make the necessary change(s) to produce a true statement.

12. Every parallelogram is a rhombus.

13. Every rhombus is a parallelogram.

14. All squares are rectangles.

15. Some rectangles are not squares.

16. No triangles are polygons.

17. Every rhombus is a regular polygon.

18. A requirement for the formation of a tessellation is that the sum of the measures of the angles that come together at each vertex is 360°.

Respond to Exercises 19–24 using verbal or written explanations.

19. What is a polygon?

20. Explain why rectangles and rhombuses are also parallelograms.

21. Explain why every square is a rectangle, a rhombus, a parallelogram, a quadrilateral, and a polygon.

22. Explain why a square is a regular polygon, but a rhombus is not.

23. Using words only, describe how to find the perimeter of a rectangle.

24. Describe how to find the measure of an angle of a regular pentagon.

In Exercises 25–28, determine whether each statement makes sense or does not make sense, and explain your reasoning.

25. I drew a polygon having two sides that intersect at their midpoints.

26. I find it helpful to think of a polygon's perimeter as the length of its boundary.

27. If a polygon is not regular, I can determine the sum of the measures of its angles, but not the measure of any one of its angles.

28. I used floor tiles in the shape of regular pentagons to completely cover my kitchen floor.

Practice Exercises

In Exercises 29–32, use the number of sides to name the polygon.

29.

30.

31. 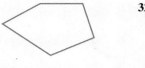 **32.**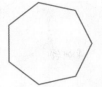

Use these quadrilaterals to solve Exercises 33–38.

a.  **b.**

c. 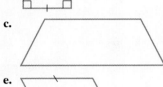 **d.**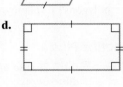

e.

33. Which of these quadrilaterals have opposite sides that are parallel? Name these quadrilaterals.

34. Which of these quadrilaterals have sides of equal length that meet at a vertex? Name these quadrilaterals.

35. Which of these quadrilaterals have right angles? Name these quadrilaterals.

36. Which of these quadrilaterals do not necessarily have four sides of equal length? Name these quadrilaterals.

37. Which of these quadrilaterals is not a parallelogram? Name this quadrilateral.

38. Which of these quadrilaterals is/are a regular polygon? Name this/these quadrilateral(s).

In Exercises 39–48, find the perimeter of the figure named and shown. Express the perimeter using the same unit of measure that appears on the given side or sides.

39. Rectangle

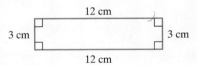

40. Parallelogram

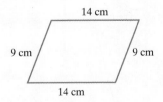

41. Rectangle

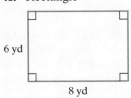

42. Rectangle

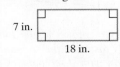

43. Square

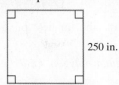

44. Square

45. Triangle

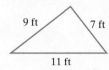

46. Triangle

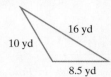

47. Equilateral triangle

48. Regular hexagon

In Exercises 49–52, find the perimeter of the figure shown. Express the perimeter using the same unit of measure that appears on the given side or sides.

49.

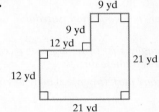

50.

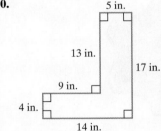

51.

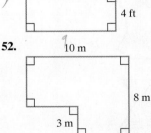

52.

53. Find the sum of the measures of the angles of a five-sided polygon.

54. Find the sum of the measures of the angles of a six-sided polygon.

55. Find the sum of the measures of the angles of a quadrilateral.

56. Find the sum of the measures of the angles of a heptagon.

In Exercises 57–58, each figure shows a regular polygon. Find the measures of angle A and angle B.

57.   **58.**

In Exercises 59–60, **a.** *Find the sum of the measures of the angles for the figure given;* **b.** *Find the measures of angle A and angle B.*

59. 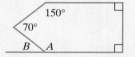   **60.**

In Exercises 61–64, tessellations formed by two or more regular polygons are shown.

a. *Name the types of regular polygons that surround each vertex.*

b. *Determine the number of angles that come together at each vertex, as well as the measures of these angles.*

c. *Use the angle measures from part (b) to explain why the tessellation is possible.*

61.

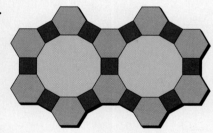

62.

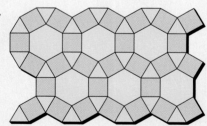

63.

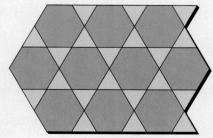

64.

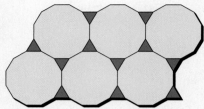

65. Can a tessellation be created using only regular nine-sided polygons? Explain your answer.

66. Can a tessellation be created using only regular ten-sided polygons? Explain your answer.

Practice Plus

In Exercises 67–70, use an algebraic equation to determine each rectangle's dimensions.

67. A rectangular field is four times as long as it is wide. If the perimeter of the field is 500 yards, what are the field's dimensions?

68. A rectangular field is five times as long as it is wide. If the perimeter of the field is 288 yards, what are the field's dimensions?

69. An American football field is a rectangle with a perimeter of 1040 feet. The length is 200 feet more than the width. Find the width and length of the rectangular field.

70. A basketball court is a rectangle with a perimeter of 86 meters. The length is 13 meters more than the width. Find the width and length of the basketball court.

In Exercises 71–72, use algebraic equations to find the measure of each angle that is represented in terms of x.

71.

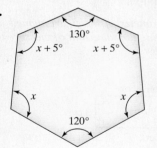

72.

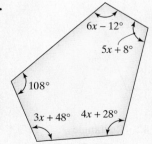

In the figure shown at the top of the next page, the artist has cunningly distorted the "regular" polygons to create a fraudulent tessellation with discrepancies that are too subtle for the eye to notice. In Exercises 73–74, you will use mathematics, not your eyes, to observe the irregularities.

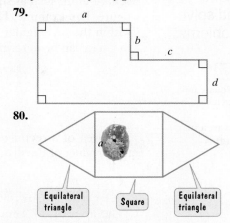

73. Find the sum of the angle measures at vertex *A*. Then explain why the tessellation is a fake.

74. Find the sum of the angle measures at vertex *B*. Then explain why the tessellation is a fake.

Application Exercises

75. A school playground is in the shape of a rectangle 400 feet long and 200 feet wide. If fencing costs $14 per yard, what will it cost to place fencing around the playground?

76. A rectangular field is 70 feet long and 30 feet wide. If fencing costs $8 per yard, how much will it cost to enclose the field?

77. One side of a square flower bed is 8 feet long. How many plants are needed if they are to be spaced 8 inches apart around the outside of the bed?

78. What will it cost to place baseboard around the region shown if the baseboard costs $0.25 per foot? No baseboard is needed for the 2-foot doorway.

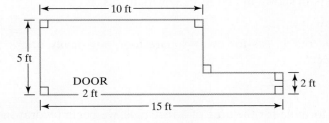

Critical Thinking Exercises

In Exercises 79–80, write an algebraic expression that represents the perimeter of the figure shown.

79.

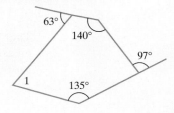

80.

81. Find *m*∡1 in the figure shown.

Group Exercise

82. Group members should consult sites on the Internet devoted to tessellations, or tilings, and present a report that expands upon the information in this section. Include a discussion of cultures that have used tessellations on fabrics, wall coverings, baskets, rugs, and pottery, with examples. Include the Alhambra, a fourteenth-century palace in Granada, Spain, in the presentation, as well as works by the artist M. C. Escher. Discuss the various symmetries (translations, rotations, reflections) associated with tessellations. Demonstrate how to create unique tessellations, including Escher-type patterns. Other than creating beautiful works of art, are there any practical applications of tessellations?

10.4 Area and Circumference

Objectives

1 Use area formulas to compute the areas of plane regions and solve applied problems.

2 Use formulas for a circle's circumference and area.

The size of a house is described in square feet. But how do you know from the real estate ad whether a 1200-square-foot home with the backyard pool is large enough to warrant a visit? Faced with hundreds of ads, you need some way to sort out the best bets. What does 1200 square feet mean and how is this area determined? In this section, we discuss how to compute the areas of plane regions.

Use area formulas to compute the areas of plane regions and solve applied problems.

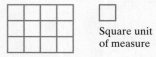

Square unit of measure

Figure 10.36 The area of the region on the left is 12 square units.

Formulas for Area

In Section 9.2, we saw that the area of a two-dimensional figure is the number of square units, such as square inches or square miles, it takes to fill the interior of the figure. For example, **Figure 10.36** shows that there are 12 square units contained within the rectangular region. The area of the region is 12 square units. Notice that the area can be determined in the following manner:

Distance across   Distance down

$$4 \text{ units} \times 3 \text{ units} = 4 \times 3 \times \text{units} \times \text{units}$$
$$= 12 \text{ square units.}$$

The area of a rectangular region, usually referred to as the area of a rectangle, is the product of the distance across (length) and the distance down (width).

Area of a Rectangle and a Square

The area, A, of a rectangle with length l and width w is given by the formula

$$A = lw.$$

The area, A, of a square with one side measuring s linear units is given by the formula

$$A = s^2.$$

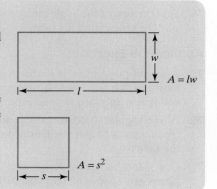

Example 1 Solving an Area Problem

You decide to cover the path shown in **Figure 10.37** with bricks.

a. Find the area of the path.

b. If the path requires four bricks for every square foot, how many bricks are needed for the project?

Solution

a. Because we have a formula for the area of a rectangle, we begin by drawing a dashed line that divides the path into two rectangles. One way of doing this is shown below. We then use the length and width of each rectangle to find its area. The computations for area are shown in the green and blue voice balloons.

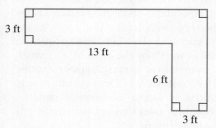

Figure 10.37

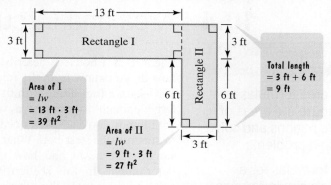

The area of the path is found by adding the areas of the two rectangles.

$$\text{Area of path} = 39 \text{ ft}^2 + 27 \text{ ft}^2 = 66 \text{ ft}^2$$

b. The path requires 4 bricks per square foot. The number of bricks needed for the project is the number of square feet in the path, its area, times 4.

$$\text{Number of bricks needed} = 66 \text{ ft}^2 \cdot \frac{4 \text{ bricks}}{\text{ft}^2} = 66 \cdot 4 \text{ bricks} = 264 \text{ bricks}$$

Thus, 264 bricks are needed for the project.

> **Checkpoint 1** Find the area of the path described in Example 1, rendered below as a green region, by first measuring off a large rectangle as shown. The area of the path is the area of the large rectangle (the blue and green regions combined) minus the area of the blue rectangle. Do you get the same answer as we did in Example 1(a)?

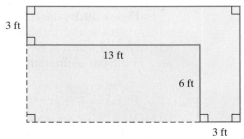

In Section 9.2, we saw that although there are 3 linear feet in a linear yard, there are 9 square feet in a square yard. If a problem requires measurement of area in square yards and the linear measures are given in feet, to avoid errors, first convert feet to yards. Then apply the area formula. This idea is illustrated in Example 2.

Example 2   Solving an Area Problem

What will it cost to carpet a rectangular floor measuring 12 feet by 15 feet if the carpet costs $18.50 per square yard?

Solution

We begin by converting the linear measures from feet to yards.

$$12 \text{ ft} = \frac{12 \text{ ft}}{1} \cdot \frac{1 \text{ yd}}{3 \text{ ft}} = \frac{12}{3} \text{ yd} = 4 \text{ yd}$$

$$15 \text{ ft} = \frac{15 \text{ ft}}{1} \cdot \frac{1 \text{ yd}}{3 \text{ ft}} = \frac{15}{3} \text{ yd} = 5 \text{ yd}$$

Next, we find the area of the rectangular floor in square yards.

$$A = lw = 5 \text{ yd} \cdot 4 \text{ yd} = 20 \text{ yd}^2$$

Finally, we find the cost of the carpet by multiplying the cost per square yard, $18.50, by the number of square yards in the floor, 20.

$$\text{Cost of carpet} = \frac{\$18.50}{\text{yd}^2} \cdot \frac{20 \text{ yd}^2}{1} = \$18.50(20) = \$370$$

It will cost $370 to carpet the floor.

> **Checkpoint 2** What will it cost to carpet a rectangular floor measuring 18 feet by 21 feet if the carpet costs $16 per square yard?

We can use the formula for the area of a rectangle to develop formulas for areas of other polygons. We begin with a parallelogram, a quadrilateral with opposite sides equal and parallel. The **height** of a parallelogram is the perpendicular distance between two of the parallel sides. Height is denoted by h in **Figure 10.38**. The **base**, denoted by b, is the length of either of these parallel sides.

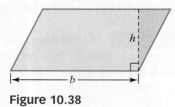

Figure 10.38

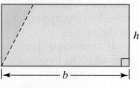

Figure 10.39

In **Figure 10.39**, the red triangular region has been cut off from the right of the parallelogram and attached to the left. The resulting figure is a rectangle with length b and width h. Because bh is the area of the rectangle, it also represents the area of the parallelogram.

Area of a Parallelogram

The area, A, of a parallelogram with height h and base b is given by the formula

$$A = bh.$$

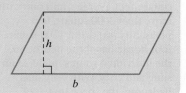

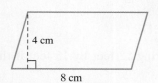

Figure 10.40

Example 3   Using the Formula for a Parallelogram's Area

Find the area of the parallelogram in **Figure 10.40**.

Solution

As shown in the figure, the base is 8 centimeters and the height is 4 centimeters. Thus, $b = 8$ and $h = 4$.

$$A = bh$$
$$A = 8 \text{ cm} \cdot 4 \text{ cm} = 32 \text{ cm}^2$$

The area is 32 cm².

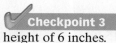 **Checkpoint 3**   Find the area of a parallelogram with a base of 10 inches and a height of 6 inches.

Figure 10.41 demonstrates how we can use the formula for the area of a parallelogram to obtain a formula for the area of a triangle. The area of the parallelogram in **Figure 10.41(a)** is given by $A = bh$. The diagonal shown in the parallelogram divides it into two triangles with the same size and shape. This means that the area of each triangle is one-half that of the parallelogram. Thus, the area of the triangle in **Figure 10.41(b)** is given by $A = \frac{1}{2}bh$.

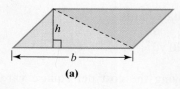

(a)

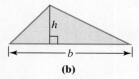

(b)

Figure 10.41

Area of a Triangle

The area, A, of a triangle with height h and base b is given by the formula

$$A = \frac{1}{2}bh.$$

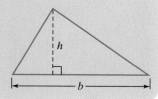

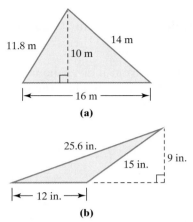

(a)

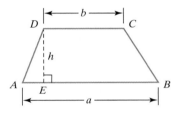

(b)

Figure 10.42

Example 4 **Using the Formula for a Triangle's Area**

Find the area of each triangle in **Figure 10.42**.

Solution

a. In **Figure 10.42(a)**, the base is 16 meters and the height is 10 meters, so $b = 16$ and $h = 10$. We do not need the 11.8 meters or the 14 meters to find the area. The area of the triangle is

$$A = \frac{1}{2}bh = \frac{1}{2} \cdot 16 \text{ m} \cdot 10 \text{ m} = 80 \text{ m}^2.$$

The area is 80 square meters.

b. In **Figure 10.42(b)**, the base is 12 inches. The base line needs to be extended to draw the height. However, we still use 12 inches for b in the area formula. The height, h, is given to be 9 inches. The area of the triangle is

$$A = \frac{1}{2}bh = \frac{1}{2} \cdot 12 \text{ in.} \cdot 9 \text{ in.} = 54 \text{ in.}^2.$$

The area of the triangle is 54 square inches.

Checkpoint 4 A sailboat has a triangular sail with a base of 12 feet and a height of 5 feet. Find the area of the sail.

The formula for the area of a triangle can be used to obtain a formula for the area of a trapezoid. Consider the trapezoid shown in **Figure 10.43**. The lengths of the two parallel sides, called the **bases**, are represented by a (the lower base) and b (the upper base). The trapezoid's height, denoted by h, is the perpendicular distance between the two parallel sides.

In **Figure 10.44**, we have drawn line segment BD, dividing the trapezoid into two triangles, shown in yellow and red. The area of the trapezoid is the sum of the areas of these triangles.

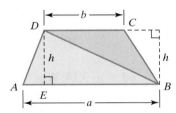

Figure 10.43

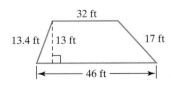

Figure 10.44

Area of trapezoid	=	Area of yellow △	plus	Area of red △
A	=	$\frac{1}{2}ah$	$+$	$\frac{1}{2}bh$
	=	$\frac{1}{2}h(a + b)$		

Area of a Trapezoid

The area, A, of a trapezoid with parallel bases a and b and height h is given by the formula

$$A = \frac{1}{2}h(a + b).$$

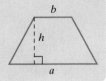

Example 5 **Finding the Area of a Trapezoid**

Find the area of the trapezoid in **Figure 10.45**.

Solution

The height, h, is 13 feet. The lower base, a, is 46 feet, and the upper base, b, is 32 feet. We do not use the 17-foot and 13.4-foot sides in finding the trapezoid's area.

$$A = \frac{1}{2}h(a + b) = \frac{1}{2} \cdot 13 \text{ ft} \cdot (46 \text{ ft} + 32 \text{ ft})$$

$$= \frac{1}{2} \cdot 13 \text{ ft} \cdot 78 \text{ ft} = 507 \text{ ft}^2$$

The area of the trapezoid is 507 square feet.

Figure 10.45

2 Use formulas for a circle's circumference and area.

The point at which a pebble hits a flat surface of water becomes the center of a number of circular ripples.

✔ **Checkpoint 5** Find the area of a trapezoid with bases of length 20 feet and 10 feet and height 7 feet.

A **circle** is a set of points in the plane equally distant from a given point, its **center**. **Figure 10.46** shows two circles. The **radius** (plural: radii), r, is a line segment from the center to any point on the circle. For a given circle, all radii have the same length. The **diameter**, d, is a line segment through the center whose endpoints both lie on the circle. For a given circle, all diameters have the same length. In any circle, the **length of the diameter is twice the length of the radius**.

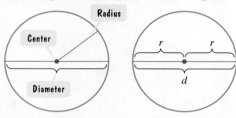

Figure 10.46

The words *radius* and *diameter* refer to both the line segments in **Figure 10.46**, as well as to their linear measures. The distance around a circle (its perimeter) is called its **circumference**, C. For all circles, if you divide the circumference by the diameter, or by twice the radius, you will get the same number. This ratio is the irrational number π and is approximately equal to 3.14:

$$\frac{C}{d} = \pi \quad \text{or} \quad \frac{C}{2r} = \pi.$$

Thus,

$$C = \pi d \quad \text{or} \quad C = 2\pi r.$$

Finding the Distance Around a Circle

The circumference, C, of a circle with diameter d and radius r is

$$C = \pi d \quad \text{or} \quad C = 2\pi r.$$

When computing a circle's circumference by hand, round π to 3.14. When using a calculator, use the $\boxed{\pi}$ key, which gives the value of π rounded to approximately 11 decimal places. In either case, calculations involving π give approximate answers. These answers can vary slightly depending on how π is rounded. The symbol $\approx$ (is approximately equal to) will be written in these calculations.

Example 6 **Finding a Circle's Circumference**

Find the circumference of the circle in **Figure 10.47**.

Solution

The diameter is 40 yards, so we use the formula for circumference with d in it.

$$C = \pi d = \pi(40 \text{ yd}) = 40\pi \text{ yd} \approx 125.7 \text{ yd}$$

The distance around the circle is approximately 125.7 yards.

40 yd

Figure 10.47

✔ **Checkpoint 6** Find the circumference of a circle whose diameter measures 10 inches. Express the answer in terms of π and then round to the nearest tenth of an inch.

Figure 10.48

Example 7 Using the Circumference Formula

How much trim, to the nearest tenth of a foot, is needed to go around the window shown in **Figure 10.48**?

Solution

The trim covers the 6-foot bottom of the window, the two 8-foot sides, and the half-circle (called a semicircle) on top. The length needed is

6 ft + 8 ft + 8 ft + circumference of the semicircle.

The circumference of the semicircle is half the circumference of a circle whose diameter is 6 feet.

Circumference of semicircle

$d = 6$ ft

$$\text{Circumference of semicircle} = \frac{1}{2}\pi d$$

$$= \frac{1}{2}\pi(6 \text{ ft}) = 3\pi \text{ ft} \approx 9.4 \text{ ft}$$

Rounding the circumference to the nearest tenth (9.4 feet), the length of trim that is needed is approximately

6 ft + 8 ft + 8 ft + 9.4 ft,

or 31.4 feet.

✓ **Checkpoint 7** In **Figure 10.48**, suppose that the dimensions are 10 feet and 12 feet for the window's bottom and side, respectively. How much trim, to the nearest tenth of a foot, is needed to go around the window?

We also use π to find the area of a circle in square units.

> **Finding the Area of a Circle**
> The area, A, of a circle with radius r is
> $$A = \pi r^2.$$

Example 8 Problem Solving Using the Formula for a Circle's Area

Which one of the following is the better buy: a large pizza with a 16-inch diameter for $15.00 or a medium pizza with an 8-inch diameter for $7.50?

Solution

The better buy is the pizza with the lower price per square inch. The radius of the large pizza is $\frac{1}{2} \cdot 16$ inches, or 8 inches, and the radius of the medium pizza is $\frac{1}{2} \cdot 8$ inches, or 4 inches. The area of the surface of each circular pizza is determined using the formula for the area of a circle.

Large pizza: $A = \pi r^2 = \pi(8 \text{ in.})^2 = 64\pi \text{ in.}^2 \approx 201 \text{ in.}^2$

Medium pizza: $A = \pi r^2 = \pi(4 \text{ in.})^2 = 16\pi \text{ in.}^2 \approx 50 \text{ in.}^2$

The area of the large pizza is 64π square inches, or approximately 201 square inches. The area of the small pizza is 16π square inches, or approximately 50 square inches. For each pizza, the price per square inch is found by dividing the price by the area:

$$\text{Price per square inch for large pizza} = \frac{\$15.00}{64\pi \text{ in.}^2} \approx \frac{\$15.00}{201 \text{ in.}^2} \approx \frac{\$0.07}{\text{in.}^2}$$

$$\text{Price per square inch for medium pizza} = \frac{\$7.50}{16\pi \text{ in.}^2} \approx \frac{\$7.50}{50 \text{ in.}^2} = \frac{\$0.15}{\text{in.}^2}$$

The large pizza costs approximately \$0.07 per square inch and the medium pizza costs approximately \$0.15 per square inch. Thus, the large pizza is the better buy.

In Example 8, did you at first think that the price per square inch would be the same for the large and the medium pizzas? After all, the radius of the large pizza is twice that of the medium pizza, and the cost of the large is twice that of the medium. However, the large pizza's area, 64π square inches, is *four times the area* of the medium pizza's, 16π square inches. Doubling the radius of a circle increases its area by a factor of 2^2, or 4. In general, if the radius of a circle is increased by k times its original linear measure, the area is multiplied by k^2. The same principle is true for any two-dimensional figure: If the shape of the figure is kept the same while linear dimensions are increased k times, the area of the larger figure is k^2 times greater than the area of the original figure.

> **Checkpoint 8** Which one of the following is the better buy: a large pizza with an 18-inch diameter for \$20.00 or a medium pizza with a 14-inch diameter for \$14.00?

Exercise Set 10.4

Concept and Vocabulary Exercises

In Exercises 1–8, fill in each blank so that the resulting statement is true.

1. The area, A, of a rectangle with length l and width w is given by the formula _____.

2. The area, A, of a square with one side measuring s linear units is given by the formula _____.

3. The area, A, of a parallelogram with height h and base b is given by the formula _____.

4. The area, A, of a triangle with height h and base b is given by the formula _____.

5. The area, A, of a trapezoid with parallel bases a and b and height h is given by the formula _____ .

6. The circumference, C, of a circle with diameter d is given by the formula _____.

7. The circumference, C, of a circle with radius r is given by the formula _____.

8. The area, A, of a circle with radius r is given by the formula _____.

In Exercises 9–13, determine whether each statement is true or false. If the statement is false, make the necessary change(s) to produce a true statement.

9. The area, A, of a rectangle with length l and width w is given by the formula $A = 2l + 2w$.

10. The height of a parallelogram is the perpendicular distance between two of the parallel sides.

11. The area of either triangle formed by drawing a diagonal in a parallelogram is one-half that of the parallelogram.

12. In any circle, the length of the radius is twice the length of the diameter.

13. The ratio of a circle's circumference to its diameter is the irrational number π.

Respond to Exercises 14–18 using verbal or written explanations.

14. Using the formula for the area of a rectangle, explain how the formula for the area of a parallelogram ($A = bh$) is obtained.

15. Using the formula for the area of a parallelogram ($A = bh$), explain how the formula for the area of a triangle ($A = \frac{1}{2}bh$) is obtained.

16. Using the formula for the area of a triangle, explain how the formula for the area of a trapezoid is obtained.

17. Explain why a circle is not a polygon.

18. Describe the difference between the formulas needed to solve the following problems: How much fencing is needed to enclose a circular garden? How much fertilizer is needed for a circular garden?

In Exercises 19–22, determine whether each statement makes sense or does not make sense, and explain your reasoning.

19. The house is a 1500-square-foot mansion with six bedrooms.

20. Because a parallelogram can be divided into two triangles with the same size and shape, the area of a triangle is one-half that of a parallelogram.

21. I used $A = \pi r^2$ to determine the amount of fencing needed to enclose my circular garden.

22. I paid \$10 for a pizza, so I would expect to pay approximately \$20 for the same kind of pizza with twice the radius.

Practice Exercises

In Exercises 23–36, use the formulas developed in this section to find the area of each figure.

23.
3 m
6 m

24.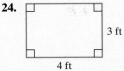
3 ft
4 ft

25.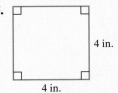
4 in.
4 in.

26.
3 cm
3 cm

27.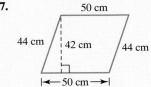
50 cm
44 cm   42 cm   44 cm
50 cm

28.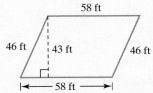
58 ft
46 ft   43 ft   46 ft
58 ft

29.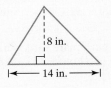
8 in.
14 in.

30.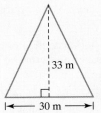
33 m
30 m

31.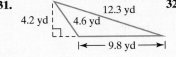
12.3 yd
4.2 yd   4.6 yd
9.8 yd

32.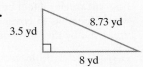
8.73 yd
3.5 yd
8 yd

33.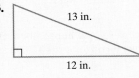
13 in.
12 in.

34.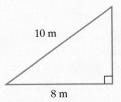
10 m
8 m

35.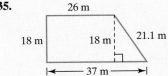
26 m
18 m   18 m   21.1 m
37 m

36.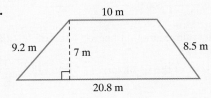
10 m
9.2 m   7 m   8.5 m
20.8 m

In Exercises 37–40, find the circumference and area of each circle. Express answers in terms of π and then round to the nearest tenth.

37.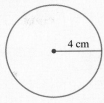
4 cm

38.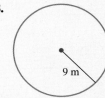
9 m

39.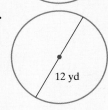
12 yd

40.
40 ft

Find the area of each figure in Exercises 41–46. Where necessary, express answers in terms of π and then round to the nearest tenth.

41.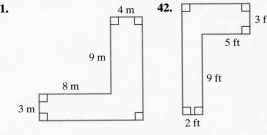
4 m
9 m
8 m
3 m

42.
3 ft
5 ft
9 ft
2 ft

43.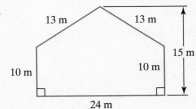
13 m   13 m
10 m   15 m   10 m
24 m

44.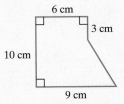
6 cm
3 cm
10 cm
9 cm

45.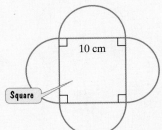
10 cm
Square

46.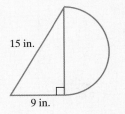
15 in.
9 in.

In Exercises 47–50, find a formula for the total area, A, of each figure in terms of the variable(s) shown. Where necessary, use π in the formula.

47.

48.

49.

50.

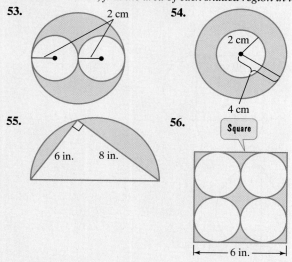

Practice Plus

In Exercises 51–52, find the area of each shaded region.

51.

52.

In Exercises 53–56, find the area of each shaded region in terms of π.

53.   2 cm

54.   2 cm   4 cm

55.   6 in.   8 in.

56.   Square   6 in.

In Exercises 57–58, find the perimeter and the area of each figure. Where necessary, express answers in terms of π and round to the nearest tenth.

57.

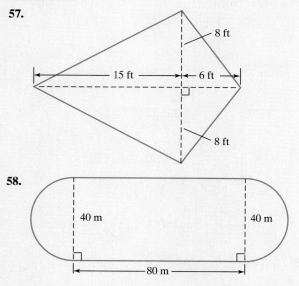

8 ft   15 ft   6 ft   8 ft

58.

40 m   40 m   80 m

Application Exercises

59. What will it cost to carpet a rectangular floor measuring 9 feet by 21 feet if the carpet costs $26.50 per square yard?

60. A plastering contractor charges $18 per square yard. What is the cost of plastering 60 feet of wall in a house with a 9-foot ceiling?

61. A rectangular kitchen floor measures 12 feet by 15 feet. A stove on the floor has a rectangular base measuring 3 feet by 4 feet, and a refrigerator covers a rectangular area of the floor measuring 4 feet by 5 feet. How many square feet of tile will be needed to cover the kitchen floor not counting the area used by the stove and the refrigerator?

62. A rectangular room measures 12 feet by 15 feet. The entire room is to be covered with rectangular tiles that measure 3 inches by 2 inches. If the tiles are sold at ten for 30¢, what will it cost to tile the room?

63. The lot in the figure shown, except for the house, shed, and driveway, is lawn. One bag of lawn fertilizer costs $25.00 and covers 4000 square feet.

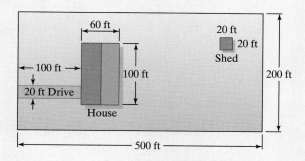

60 ft   20 ft   20 ft   Shed   100 ft   100 ft   200 ft   20 ft Drive   House   500 ft

a. Determine the minimum number of bags of fertilizer needed for the lawn.

b. Find the total cost of the fertilizer.

64. Taxpayers with an office in their home may deduct a percentage of their home-related expenses. This percentage is based on the ratio of the office's area to the area of the home. A taxpayer with an office in a 2200-square-foot home maintains a 20 foot by 16 foot office. If the yearly electric bills for the home come to $4800, how much of this is deductible?

65. You are planning to paint the house whose dimensions are shown in the figure.

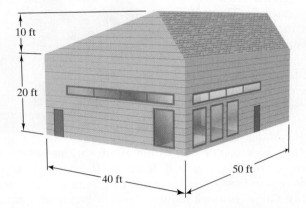

    a. How many square feet will you need to paint? (There are four windows, each 8 ft by 5 ft; two windows, each 30 ft by 2 ft; and two doors, each 80 in. by 36 in., that do not require paint.)

    b. The paint that you have chosen is available in gallon cans only. Each can covers 500 square feet. If you want to use two coats of paint, how many cans will you need for the project?

    c. If the paint you have chosen sells for $26.95 per gallon, what will it cost to paint the house?

The diagram shows the floor plan for a one-story home. Use the given measurements to solve Exercises 66–68. (A calculator will be helpful in performing the necessary computations.)

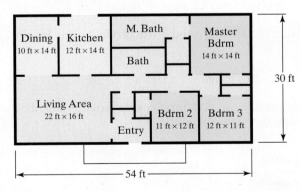

66. If construction costs $95 per square foot, find the cost of building the home.

67. If carpet costs $17.95 per square yard and is available in whole square yards only, find the cost of carpeting the three bedroom floors.

68. If ceramic tile costs $26.95 per square yard and is available in whole square yards only, find the cost of installing ceramic tile on the kitchen and dining room floors.

In Exercises 69–70, express the required calculation in terms of π and then round to the nearest tenth.

69. How much fencing is required to enclose a circular garden whose radius is 20 meters?

70. A circular rug is 6 feet in diameter. How many feet of fringe is required to edge this rug?

71. How many plants spaced every 6 inches are needed to surround a circular garden with a 30-foot radius?

72. A stained glass window is to be placed in a house. The window consists of a rectangle, 6 feet high by 3 feet wide, with a semicircle at the top. Approximately how many feet of stripping, to the nearest tenth of a foot, will be needed to frame the window?

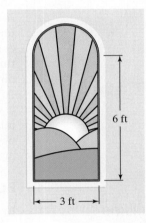

73. Which one of the following is a better buy: a large pizza with a 14-inch diameter for $12.00 or a medium pizza with a 7-inch diameter for $5.00?

74. Which one of the following is a better buy: a large pizza with a 16-inch diameter for $12.00 or two small pizzas, each with a 10-inch diameter, for $12.00?

Critical Thinking Exercises

75. You need to enclose a rectangular region with 200 feet of fencing. Experiment with different lengths and widths to determine the maximum area you can enclose. Which quadrilateral encloses the most area?

76. Suppose you know the cost for building a rectangular deck measuring 8 feet by 10 feet. If you decide to increase the dimensions to 12 feet by 15 feet, by how much will the cost increase?

77. A rectangular swimming pool measures 14 feet by 30 feet. The pool is surrounded on all four sides by a path that is 3 feet wide. If the cost to resurface the path is $2 per square foot, what is the total cost of resurfacing the path?

78. A proposed oil pipeline will cross 16.8 miles of national forest. The width of the land needed for the pipeline is 200 feet. If the U.S. Forest Service charges the oil company $32 per acre, calculate the total cost. (1 mile = 5280 feet and 1 acre = 43,560 square feet.)

10.5 : Volume

Objectives

1. Use volume formulas to compute the volumes of three-dimensional figures and solve applied problems.

2. Compute the surface area of a three-dimensional figure.

You are considering going to Judge Judy's Web site and filling out a case submission form to appear on her TV show. The case involves your contractor, who promised to install a water tank that holds 500 gallons of water. Upon delivery, you noticed that the capacity was not printed anywhere, so you decided to do some measuring. The tank is shaped like a giant tuna can, with a circular top and bottom. You measured the radius of each circle to be 3 feet and you measured the tank's height to be 2 feet 4 inches. You know that 500 gallons is the capacity of a solid figure with a volume of about 67 cubic feet. Now you need some sort of method to compute the volume of the water tank. In this section, we discuss how to compute the volumes of various solid, three-dimensional figures. Using a formula you will learn in the section, you can determine whether the evidence indicates you can win a case against the contractor if you appear on *Judge Judy*. Or do you risk joining a cast of bozos who entertain television viewers by being loudly castigated by the judge? (Before a possible ear-piercing "Baloney, sir, you're a geometric idiot!", we suggest working Exercise 65 in Exercise Set 10.5.)

Formulas for Volume

In Section 9.2, we saw that **volume** refers to the amount of space occupied by a solid object, determined by the number of cubic units it takes to fill the interior of that object. For example, **Figure 10.49** shows that there are 18 cubic units contained within the box. The volume of the box, called a **rectangular solid**, is 18 cubic units. The box has a length of 3 units, a width of 3 units, and a height of 2 units. The volume, 18 cubic units, may be determined by finding the product of the length, the width, and the height:

$$\text{Volume} = 3 \text{ units} \cdot 3 \text{ units} \cdot 2 \text{ units} = 18 \text{ units}^3.$$

In general, the volume, V, of a rectangular solid is the product of its length, l, its width, w, and its height, h:

$$V = lwh.$$

If the length, width, and height are the same, the rectangular solid is called a **cube**. Formulas for these boxlike shapes are given below.

① Use volume formulas to compute the volumes of three-dimensional figures and solve applied problems.

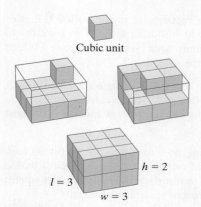

Cubic unit

Figure 10.49
Volume = 18 cubic units

$l = 3$
$w = 3$
$h = 2$

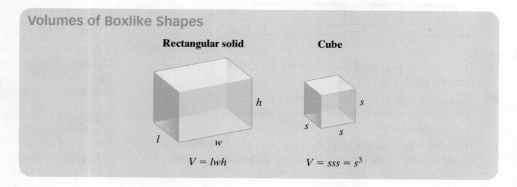

Volumes of Boxlike Shapes

Rectangular solid   **Cube**

h s

l w s s

$V = lwh$ $V = sss = s^3$

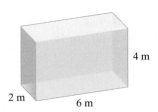

Figure 10.50

Example 1   Finding the Volume of a Rectangular Solid

Find the volume of the rectangular solid in **Figure 10.50**.

Solution

As shown in the figure, the length is 6 meters, the width is 2 meters, and the height is 4 meters. Thus, $l = 6$, $w = 2$, and $h = 4$.

$$V = lwh = 6\text{ m} \cdot 2\text{ m} \cdot 4\text{ m} = 48\text{ m}^3$$

The volume of the rectangular solid is 48 cubic meters.

Checkpoint 1   Find the volume of a rectangular solid with length 5 feet, width 3 feet, and height 7 feet.

In Section 9.2, we saw that although there are 3 feet in a yard, there are 27 cubic feet in a cubic yard. If a problem requires measurement of volume in cubic yards and the linear measures are given in feet, to avoid errors, first convert feet to yards. Then apply the volume formula. This idea is illustrated in Example 2.

Example 2   Solving a Volume Problem

You are about to begin work on a swimming pool in your yard. The first step is to dig a hole that is 90 feet long, 60 feet wide, and 6 feet deep. You will use a truck that can carry 10 cubic yards of dirt and charges $12 per load. How much will it cost you to have all the dirt hauled away?

Solution

We begin by converting feet to yards:

$$90\text{ ft} = \frac{90\text{ ft}}{1} \cdot \frac{1\text{ yd}}{3\text{ ft}} = \frac{90}{3}\text{ yd} = 30\text{ yd}.$$

Similarly, 60 ft = 20 yd and 6 ft = 2 yd. Next, we find the volume of dirt that needs to be dug out and hauled off.

$$V = lwh = 30\text{ yd} \cdot 20\text{ yd} \cdot 2\text{ yd} = 1200\text{ yd}^3$$

Now, we find the number of loads necessary for the truck to haul off all the dirt. Because the truck carries 10 cubic yards, divide the number of cubic yards of dirt by 10.

$$\text{Number of truckloads} = \frac{1200\text{ yd}^3}{\dfrac{10\text{ yd}^3}{\text{load}}} = \frac{1200\text{ yd}^3}{1} \cdot \frac{\text{load}}{10\text{ yd}^3} = \frac{1200}{10}\text{ loads} = 120\text{ loads}$$

Because the truck charges $12 per load, the cost to have all the dirt hauled away is the number of loads, 120, times the cost per load, $12.

$$\text{Cost to haul all dirt away} = \frac{120\text{ loads}}{1} \cdot \frac{\$12}{\text{load}} = 120(\$12) = \$1440$$

The dirt-hauling phase of the pool project will cost you $1440.

Checkpoint 2   Find the volume, in cubic yards, of a cube whose edges each measure 6 feet.

Figure 10.51 The volume of a pyramid is $\frac{1}{3}$ the volume of a rectangular solid having the same base and the same height.

A rectangular solid is an example of a **polyhedron**, a solid figure bounded by polygons. A rectangular solid is bounded by six rectangles, called faces. By contrast, a **pyramid** is a polyhedron whose base is a polygon and whose sides are triangles. **Figure 10.51** shows a pyramid with a rectangular base drawn inside a rectangular solid. The contents of three pyramids with rectangular bases exactly fill a rectangular solid of the same base and height. Thus, the formula for the volume of the pyramid is $\frac{1}{3}$ that of the rectangular solid.

Volume of a Pyramid

Pyramid

The volume, V, of a pyramid is given by the formula

$$V = \frac{1}{3}Bh,$$

where B is the area of the base and h is the height (the perpendicular distance from the top to the base).

The Transamerica Tower's 3678 windows take cleaners one month to wash. Its foundation is sunk 15.5 m (52 ft) into the ground and is designed to move with earth tremors.

Example 3 Using the Formula for a Pyramid's Volume

Capped with a pointed spire on top of its 48 stories, the Transamerica Tower in San Francisco is a pyramid with a square base. The pyramid is 256 meters (853 feet) tall. Each side of the square base has a length of 52 meters. Although San Franciscans disliked it when it opened in 1972, they have since accepted it as part of the skyline. Find the volume of the building.

Solution

First find the area of the base, represented as B in the volume formula. Because each side of the square base is 52 meters, the area of the base is

$$B = 52 \text{ m} \cdot 52 \text{ m} = 2704 \text{ m}^2.$$

The area of the base is 2704 square meters. Because the pyramid is 256 meters tall, its height, h, is 256 meters. Now we apply the formula for the volume of a pyramid:

$$V = \frac{1}{3}Bh = \frac{1}{3} \cdot \frac{2704 \text{ m}^2}{1} \cdot \frac{256 \text{ m}}{1} = \frac{2704 \cdot 256}{3} \text{ m}^3 \approx 230{,}741 \text{ m}^3.$$

The volume of the building is approximately 230,741 cubic meters.

The San Francisco pyramid is relatively small compared to the Great Pyramid outside Cairo, Egypt. Built about 2550 B.C. by a labor force of 100,000, the Great Pyramid is approximately 11 times the volume of San Francisco's pyramid.

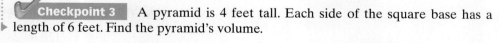

Checkpoint 3 A pyramid is 4 feet tall. Each side of the square base has a length of 6 feet. Find the pyramid's volume.

Not every three-dimensional figure is a polyhedron. Take, for example, the right circular cylinder shown in **Figure 10.52**. Its shape should remind you of a soup can or a stack of coins. The right circular cylinder is so named because the top and bottom are circles, and the side forms a right angle with the top and bottom. The formula for the volume of a right circular cylinder is given at the top of the next page.

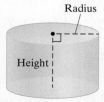

Radius

Height

Figure 10.52

Volume of a Right Circular Cylinder

The volume, V, of a right circular cylinder is given by the formula

$$V = \pi r^2 h,$$

where r is the radius of the circle at either end and h is the height.

Right circular cylinder

Example 4   Finding the Volume of a Cylinder

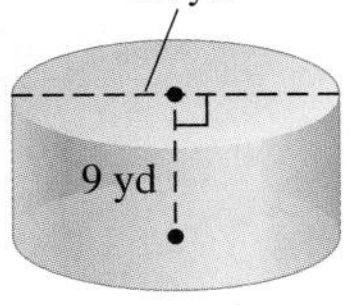

Find the volume of the cylinder in **Figure 10.53**.

Solution

In order to find the cylinder's volume, we need both its radius and its height. Because the diameter is 20 yards, the radius is half this length, or 10 yards. The height of the cylinder is given to be 9 yards. Thus, $r = 10$ and $h = 9$. Now we apply the formula for the volume of a cylinder.

$$V = \pi r^2 h = \pi(10 \text{ yd})^2 \cdot 9 \text{ yd} = 900\pi \text{ yd}^3 \approx 2827 \text{ yd}^3$$

The volume of the cylinder is approximately 2827 cubic yards.

20 yd

9 yd

Figure 10.53

✓ **Checkpoint 4**   Find the volume, to the nearest cubic inch, of a cylinder with a diameter of 8 inches and a height of 6 inches.

Figure 10.54 shows a **right circular cone** inside a cylinder, sharing the same circular base as the cylinder. The height of the cone, the perpendicular distance from the top to the circular base, is the same as that of the cylinder. Three such cones can occupy the same amount of space as the cylinder. Therefore, the formula for the volume of the cone is $\frac{1}{3}$ the volume of the cylinder.

Figure 10.54

Volume of a Cone

The volume, V, of a right circular cone that has height h and radius r is given by the formula

$$V = \frac{1}{3}\pi r^2 h.$$

Cone

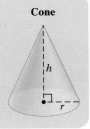

Example 5   Finding the Volume of a Cone

Find the volume of the cone in **Figure 10.55**.

Solution

The radius of the cone is 7 meters and the height is 10 meters. Thus, $r = 7$ and $h = 10$. Now we apply the formula for the volume of a cone.

$$V = \frac{1}{3}\pi r^2 h = \frac{1}{3}\pi(7 \text{ m})^2 \cdot 10 \text{ m} = \frac{490\pi}{3} \text{ m}^3 \approx 513 \text{ m}^3$$

The volume of the cone is approximately 513 cubic meters.

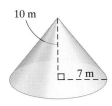

10 m

7 m

Figure 10.55

✓ **Checkpoint 5**   Find the volume, to the nearest cubic inch, of a cone with a radius of 4 inches and a height of 6 inches.

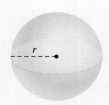

Figure 10.56

Figure 10.56 shows a *sphere*. Its shape may remind you of a basketball. The Earth is not a perfect sphere, but it's close. A **sphere** is the set of points in space equally distant from a given point, its **center**. Any line segment from the center to a point on the sphere is a **radius** of the sphere. The word *radius* is also used to refer to the length of this line segment. A sphere's volume can be found by using π and its radius.

Volume of a Sphere

The volume, V, of a sphere of radius r is given by the formula

$$V = \frac{4}{3}\pi r^3.$$

Sphere

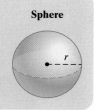

Example 6 Applying Volume Formulas

An ice cream cone is 5 inches deep and has a radius of 1 inch. A spherical scoop of ice cream also has a radius of 1 inch. (See **Figure 10.57**.) If the ice cream melts into the cone, will it overflow?

Solution

The ice cream will overflow if the volume of the ice cream, a sphere, is greater than the volume of the cone. Find the volume of each.

$$V_{\text{cone}} = \frac{1}{3}\pi r^2 h = \frac{1}{3}\pi(1 \text{ in.})^2 \cdot 5 \text{ in.} = \frac{5\pi}{3} \text{ in.}^3 \approx 5 \text{ in.}^3$$

$$V_{\text{sphere}} = \frac{4}{3}\pi r^3 = \frac{4}{3}\pi(1 \text{ in.})^3 = \frac{4\pi}{3} \text{ in.}^3 \approx 4 \text{ in.}^3$$

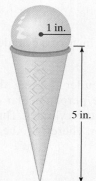

Figure 10.57

The volume of the spherical scoop of ice cream is less than the volume of the cone, so there will be no overflow.

✓ **Checkpoint 6** A basketball has a radius of 4.5 inches. If the ball is filled with 350 cubic inches of air, is this enough air to fill it completely?

Surface Area

② Compute the surface area of a three-dimensional figure.

In addition to volume, we can also measure the area of the outer surface of a three-dimensional object, called its **surface area**. Like area, surface area is measured in square units. For example, the surface area of the rectangular solid in **Figure 10.58** is the sum of the areas of the six outside rectangles of the solid.

Figure 10.58

$$\text{Surface Area} = lw + lw + lh + lh + wh + wh$$

Areas of top and bottom rectangles	Areas of front and back rectangles	Areas of rectangles on left and right sides

$$= \quad 2lw \quad + \quad 2lh \quad + \quad 2wh$$

Formulas for the surface area, abbreviated *SA*, of three-dimensional figures are given in **Table 10.4**.

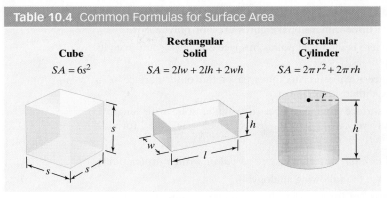

Table 10.4 Common Formulas for Surface Area

Cube	Rectangular Solid	Circular Cylinder
$SA = 6s^2$	$SA = 2lw + 2lh + 2wh$	$SA = 2\pi r^2 + 2\pi rh$

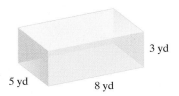

Figure 10.59

5 yd   8 yd   3 yd

Example 7   Finding the Surface Area of a Solid

Find the surface area of the rectangular solid in **Figure 10.59**.

Solution

As shown in the figure, the length is 8 yards, the width is 5 yards, and the height is 3 yards. Thus, $l = 8$, $w = 5$, and $h = 3$.

$$SA = 2lw + 2lh + 2wh$$
$$= 2 \cdot 8 \text{ yd} \cdot 5 \text{ yd} + 2 \cdot 8 \text{ yd} \cdot 3 \text{ yd} + 2 \cdot 5 \text{ yd} \cdot 3 \text{ yd}$$
$$= 80 \text{ yd}^2 + 48 \text{ yd}^2 + 30 \text{ yd}^2 = 158 \text{ yd}^2$$

The surface area is 158 square yards.

Checkpoint 7   If the length, width, and height shown in **Figure 10.59** are each doubled, find the surface area of the resulting rectangular solid.

Achieving Success

College courses require much more reading than courses in high school. It is not unusual for a professor to assign 500 pages or more of reading per week. Balance your class load between courses that require extensive reading (usually humanities courses) and courses that primarily have problem sets for homework (usually math and sciences).

Exercise Set 10.5

Concept and Vocabulary Exercises

In Exercises 1–7, fill in each blank so that the resulting statement is true.

1. The volume, *V*, of a rectangular solid with length *l*, width *w*, and height *h* is given by the formula _____.

2. The volume, *V*, of a cube with an edge that measures *s* linear units is given by the formula _____.

3. A solid figure bounded by polygons is called a _____.

4. The volume, *V*, of a pyramid with base area *B* and height *h* is given by the formula _____.

5. The volume, *V*, of a right circular cylinder with height *h* and radius *r* is given by the formula _____.

6. The volume, *V*, of a right circular cone with height *h* and radius *r* is given by the formula _____.

7. The volume, *V*, of a sphere of radius *r* is given by the formula _____.

In Exercises 8–14, determine whether each statement is true or false. If the statement is false, make the necessary change(s) to produce a true statement.

8. A cube is a rectangular solid with the same length, width, and height.

9. A cube is an example of a polyhedron.

10. The volume of a pyramid is $\frac{1}{2}$ the volume of a rectangular solid having the same base and the same height.

11. Some three-dimensional figures are not polyhedrons.

12. A sphere is the set of points in space equally distant from its center.

13. Surface area refers to the area of the outer surface of a three-dimensional object.

14. The surface area, SA, of a rectangular solid with length l, width w, and height h is given by the formula $SA = lw + lh + wh$.

Respond to Exercises 15–16 using verbal or written explanations.

15. Explain the following analogy:

    In terms of formulas used to compute volume, a pyramid is to a rectangular solid just as a cone is to a cylinder.

16. Explain why a cylinder is not a polyhedron.

In Exercises 17–20, determine whether each statement makes sense or does not make sense, and explain your reasoning.

17. The physical education department ordered new basketballs in the shape of right circular cylinders.

18. When completely full, a cylindrical soup can with a diameter of 3 inches and a height of 4 inches holds more soup than a cylindrical can with a diameter of 4 inches and a height of 3 inches.

19. I found the volume of a rectangular solid in cubic inches and then divided by 12 to convert the volume to cubic feet.

20. Because a cylinder is a solid figure, I use cubic units to express its surface area.

Practice Exercises

In Exercises 21–40, find the volume of each figure. If necessary, express answers in terms of π and then round to the nearest whole number.

21.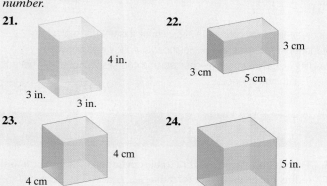

4 in.

3 in.

3 in.

22.

3 cm

3 cm

5 cm

23.

4 cm

4 cm

4 cm

24.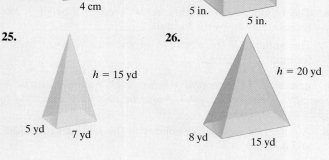

5 in.

5 in.

5 in.

25.

$h = 15$ yd

5 yd   7 yd

26.

$h = 20$ yd

8 yd   15 yd

27.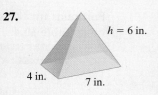

$h = 6$ in.

4 in.   7 in.

28.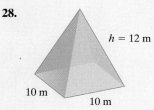

$h = 12$ m

10 m   10 m

29.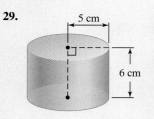

5 cm

6 cm

30.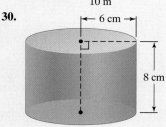

6 cm

8 cm

31.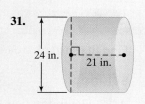

24 in.   21 in.

32.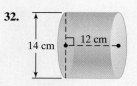

14 cm   12 cm

33.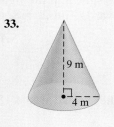

9 m

4 m

34.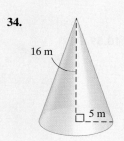

16 m

5 m

35.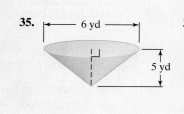

6 yd

5 yd

36.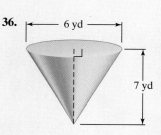

6 yd

7 yd

37.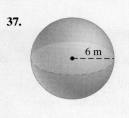

6 m

38.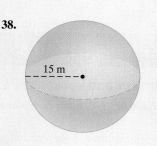

15 m

39.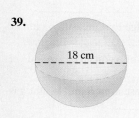

18 cm

40.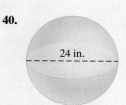

24 in.

In Exercises 41–44, find the surface area of each figure.

41. **42.**

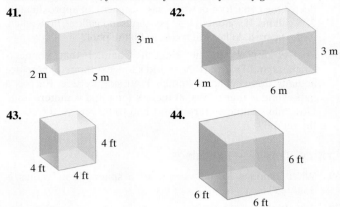

43. **44.**

In Exercises 45–50, use two formulas for volume to find the volume of each figure. Express answers in terms of π and then round to the nearest whole number.

45. **46.**

47. **48.**

49. **50.**

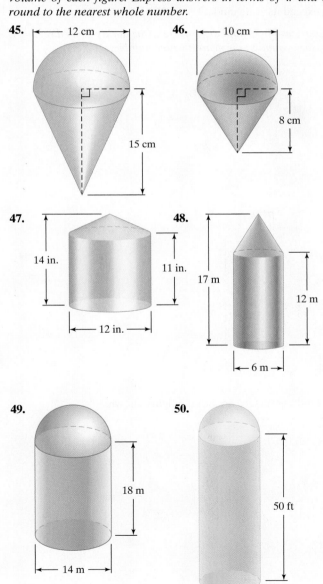

Practice Plus

51. Find the surface area and the volume of the figure shown.

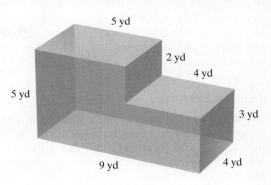

52. Find the surface area and the volume of the cement block in the figure shown.

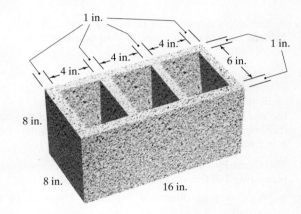

53. Find the surface area of the figure shown.

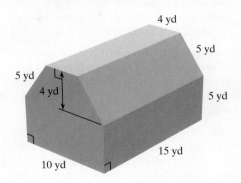

54. A machine produces open boxes using square sheets of metal measuring 12 inches on each side. The machine cuts equal-sized squares whose sides measure 2 inches from each corner. Then it shapes the metal into an open box by turning up the sides. Find the volume of the box.

55. Find the ratio, reduced to lowest terms, of the volume of a sphere with a radius of 3 inches to the volume of a sphere with a radius of 6 inches.

56. Find the ratio, reduced to lowest terms, of the volume of a sphere with a radius of 3 inches to the volume of a sphere with a radius of 9 inches.

57. A cylinder with radius 3 inches and height 4 inches has its radius tripled. How many times greater is the volume of the larger cylinder than the smaller cylinder?

58. A cylinder with radius 2 inches and height 3 inches has its radius quadrupled. How many times greater is the volume of the larger cylinder than the smaller cylinder?

Application Exercises

59. A building contractor is to dig a foundation 12 feet long, 9 feet wide, and 6 feet deep for a toll booth. The contractor pays $10 per load for trucks to remove the dirt. Each truck holds 6 cubic yards. What is the cost to the contractor to have all the dirt hauled away?

60. What is the cost of concrete for a walkway that is 15 feet long, 8 feet wide, and 9 inches deep if the concrete costs $30 per cubic yard?

61. A furnace is designed to heat 10,000 cubic feet. Will this furnace be adequate for a 1400-square-foot house with a 9-foot ceiling?

62. A water reservoir is shaped like a rectangular solid with a base that is 50 yards by 30 yards, and a vertical height of 20 yards. At the start of a three-month period of no rain, the reservoir was completely full. At the end of this period, the height of the water was down to 6 yards. How much water was used in the three-month period?

63. The Great Pyramid outside Cairo, Egypt, has a square base measuring 756 feet on a side and a height of 480 feet.
   a. What is the volume of the Great Pyramid, in cubic yards?
   b. The stones used to build the Great Pyramid were limestone blocks with an average volume of 1.5 cubic yards. How many of these blocks were needed to construct the Great Pyramid?

64. Although the Eiffel Tower in Paris is not a solid pyramid, its shape approximates that of a pyramid with a square base measuring 120 feet on a side and a height of 980 feet. If it were a solid pyramid, what would be the Eiffel Tower's volume, in cubic yards?

65. You are about to sue your contractor who promised to install a water tank that holds 500 gallons of water. You know that 500 gallons is the capacity of a tank that holds 67 cubic feet. The cylindrical tank has a radius of 3 feet and a height of 2 feet 4 inches. Does the evidence indicate you can win the case against the contractor if it goes to court?

66. Two cylindrical cans of soup sell for the same price. One can has a diameter of 6 inches and a height of 5 inches. The other has a diameter of 5 inches and a height of 6 inches. Which can contains more soup and, therefore, is the better buy?

67. A circular backyard pool has a diameter of 24 feet and is 4 feet deep. One cubic foot of water has a capacity of approximately 7.48 gallons. If water costs $2 per thousand gallons, how much, to the nearest dollar, will it cost to fill the pool?

68. The tunnel under the English Channel that connects England and France is the world's longest tunnel. There are actually three separate tunnels built side by side. Each is a half-cylinder that is 50,000 meters long and 4 meters high. How many cubic meters of dirt had to be removed to build the tunnel?

Critical Thinking Exercises

69. What happens to the volume of a sphere if its radius is doubled?

70. A scale model of a car is constructed so that its length, width, and height are each $\frac{1}{10}$ the length, width, and height of the actual car. By how many times does the volume of the car exceed its scale model?

In Exercises 71–72, find the volume of the darkly shaded region. If necessary, round to the nearest whole number.

71.

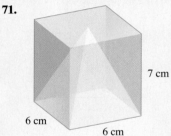

7 cm
6 cm
6 cm

72.

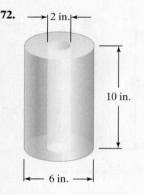

2 in.
10 in.
6 in.

73. Find the surface area of the figure shown.

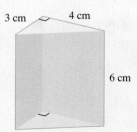

3 cm   4 cm
6 cm

10.6 : Right Triangle Trigonometry

Objectives

1. Use the lengths of the sides of a right triangle to find trigonometric ratios.

2. Use trigonometric ratios to find missing parts of right triangles.

3. Use trigonometric ratios to solve applied problems.

Mountain climbers have forever been fascinated by reaching the top of Mount Everest, sometimes with tragic results. The mountain, on Asia's Tibet-Nepal border, is Earth's highest, peaking at an incredible 29,035 feet. The heights of mountains can be found using *trigonometry*. The word **trigonometry** means *measurement of triangles*. Trigonometry is used in navigation, building, and engineering. Historically, people have used trigonometry and the stars to navigate deserts and oceans. The ancient Greeks used trigonometry to record the locations of thousands of stars and worked out the motion of the Moon relative to Earth. Today, trigonometry is used to study the structure of DNA, the master molecule that determines how we grow from a single cell to a complex, fully developed adult.

In the last century, Ang Rita Sherpa climbed Mount Everest ten times, all without the use of bottled oxygen.

Ratios in Right Triangles

The right triangle forms the basis of trigonometry. If either acute angle of a right triangle stays the same size, the shape of the triangle does not change even if it is made larger or smaller. Because of properties of similar triangles, this means that the ratios of certain lengths stay the same regardless of the right triangle's size. These ratios have special names and are defined in terms of the **side opposite** an acute angle, the **side adjacent** to the acute angle, and the **hypotenuse**. In **Figure 10.60**, the length of the hypotenuse, the side opposite the 90° angle, is represented by c. The length of the side opposite angle A is represented by a. The length of the side adjacent to angle A is represented by b.

The three fundamental trigonometric ratios, **sine** (abbreviated sin), **cosine** (abbreviated cos), and **tangent** (abbreviated tan), are defined as ratios of the lengths of the sides of a right triangle. In the box that follows, when a side of a triangle is mentioned, we are referring to the *length* of that side.

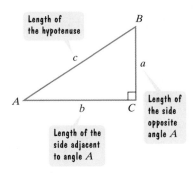

Figure 10.60 Naming a right triangle's sides from the point of view of an acute angle

1. **Use the lengths of the sides of a right triangle to find trigonometric ratios**

Trigonometric Ratios

Let A represent an acute angle of a right triangle, with right angle C, shown in **Figure 10.60**. For angle A, the trigonometric ratios are defined as follows:

sine of A

$$\sin A = \frac{\text{side opposite angle } A}{\text{hypotenuse}} = \frac{a}{c}$$

cosine of A

$$\cos A = \frac{\text{side adjacent to angle } A}{\text{hypotenuse}} = \frac{b}{c}$$

tangent of A

$$\tan A = \frac{\text{side opposite angle } A}{\text{side adjacent to angle } A} = \frac{a}{b}.$$

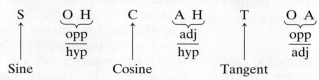
Example 1 — Becoming Familiar with the Trigonometric Ratios

Find the sine, cosine, and tangent of A in **Figure 10.61**.

Solution

We begin by finding the measure of the hypotenuse c using the Pythagorean Theorem.

$$c^2 = a^2 + b^2 = 5^2 + 12^2 = 25 + 144 = 169$$
$$c = \sqrt{169} = 13$$

Now, we apply the definitions of the trigonometric ratios.

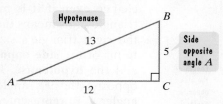

$$\sin A = \frac{\text{side opposite angle } A}{\text{hypotenuse}} = \frac{5}{13}$$

$$\cos A = \frac{\text{side adjacent to angle } A}{\text{hypotenuse}} = \frac{12}{13}$$

$$\tan A = \frac{\text{side opposite angle } A}{\text{side adjacent to angle } A} = \frac{5}{12}$$

Figure 10.61

(Figure: right triangle with vertices A, B, C; side $BC = 5$, side $AC = 12$, right angle at C.)

Study Tip

Having difficulties identifying opposite and adjacent sides? We know that the longest side of the right triangle is the hypotenuse. The two legs of the triangle are described by their relationship to each acute angle. The adjacent leg "touches" the acute angle by forming one of the angle's sides. The opposite leg is not a side of the acute angle because it lies directly opposite the angle.

Checkpoint 1 Find the sine, cosine, and tangent of A in the figure shown.

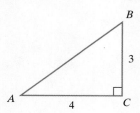

A Brief Review • Equations with Fractions

Finding missing parts of right triangles involves solving equations with fractions.

- The variable can appear in a fraction's numerator.

  Example

  Solve for a: $0.3 = \dfrac{a}{150}$

  $$150(0.3) = 150\left(\dfrac{a}{150}\right) \quad \text{Clear fractions by multiplying both sides by 150.}$$

  $$45 = a \qquad \text{Simplify: } 150(0.3) = 45 \text{ and } 150\left(\dfrac{a}{150}\right) = \dfrac{\overset{1}{\cancel{150}}}{1} \cdot \dfrac{a}{\underset{1}{\cancel{150}}} = a.$$

- The variable can appear in a fraction's denominator.

  Example

  Solve for c: $0.8 = \dfrac{150}{c}$

  $$0.8c = \left(\dfrac{150}{c}\right)c \qquad \text{Clear fractions by multiplying both sides by } c.$$

  $$0.8c = 150 \qquad \text{Simplify: } \left(\dfrac{150}{c}\right)c = \dfrac{150}{\cancel{c}} \cdot \dfrac{\overset{1}{\cancel{c}}}{1} = 150.$$

  > We're still not done. We have not isolated c.

  $$\dfrac{0.8c}{0.8} = \dfrac{150}{0.8} \qquad \text{Divide both sides by 0.8.}$$

  $$c = 187.5 \qquad \text{Simplify.}$$

---

② Use trigonometric ratios to find missing parts of right triangles.

A scientific or graphing calculator in the degree mode will give you decimal approximations for the trigonometric ratios of any angle. For example, to find an approximation for tan 37°, the tangent of an angle measuring 37°, a keystroke sequence similar to one of the following can be used:

Many Scientific Calculators: 37 |TAN|

Many Graphing Calculators: |TAN| 37 |ENTER| .

The tangent of 37°, rounded to four decimal places, is 0.7536.

If we are given the length of one side and the measure of an acute angle of a right triangle, we can use trigonometry to solve for the length of either of the other two sides. Example 2 illustrates how this is done.

Example 2 ⋮ Finding a Missing Leg of a Right Triangle

Find a in the right triangle in **Figure 10.62**.

Solution

We need to identify a trigonometric ratio that will make it possible to find a. Because we have a known angle, 40°, an unknown opposite side, a, and a known adjacent side, 150 cm, we use the tangent ratio.

$$\tan 40° = \dfrac{a}{150}$$

Side opposite the 40° angle

Side adjacent to the 40° angle

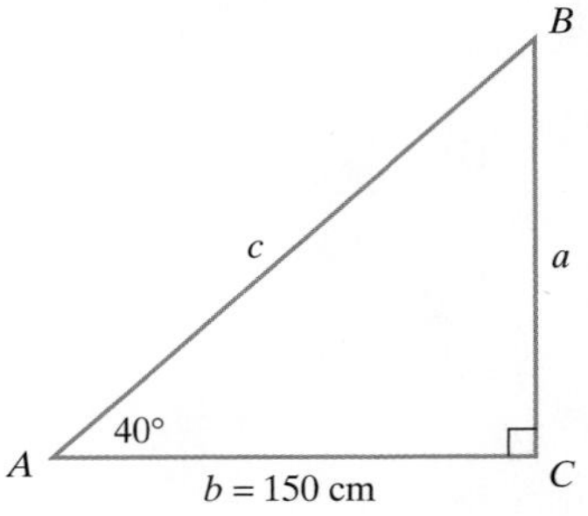

Figure 10.62

Now we solve for a by multiplying both sides by 150.

$$a = 150 \tan 40° \approx 126$$

The tangent ratio reveals that a is approximately 126 centimeters.

Technology

Here is the keystroke sequence for 150 tan 40°:

Many Scientific Calculators

150 |×| 40 |TAN| |=|

Many Graphing Calculators

150 |TAN| 40 |ENTER|

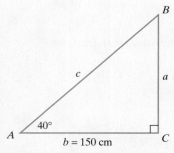

Figure 10.62 (repeated)

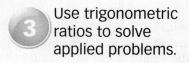

3 Use trigonometric ratios to solve applied problems.

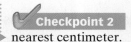

Checkpoint 2 In **Figure 10.62**, let $m\angle A = 62°$ and $b = 140$ cm. Find a to the nearest centimeter.

Example 3 ### Finding a Missing Hypotenuse of a Right Triangle

Find c in the right triangle in **Figure 10.62**.

Solution

In Example 2, we found a: $a \approx 126$. Because we are given that $b = 150$, it is possible to find c using the Pythagorean Theorem: $c^2 = a^2 + b^2$. However, if we made an error in computing a, we will perpetuate our mistake using this approach.

Instead, we will use the quantities given and identify a trigonometric ratio that will make it possible to find c without using a. Refer to **Figure 10.62**. Because we have a known angle, 40°, a known adjacent side, 150 cm, and an unknown hypotenuse, c, we use the cosine ratio.

$$\cos 40° = \frac{150}{c}$$

Side adjacent to the 40° angle

Hypotenuse

$$c \cos 40° = 150 \qquad \text{Multiply both sides by } c.$$

$$c = \frac{150}{\cos 40°} \qquad \text{Divide both sides by } \cos 40°.$$

$$c \approx 196 \qquad \text{Use a calculator.}$$

The cosine ratio reveals that the hypotenuse is approximately 196 centimeters.

Checkpoint 3 In **Figure 10.62**, let $m\angle A = 62°$ and $b = 140$ cm. Find c to the nearest centimeter.

Applications of the Trigonometric Ratios

Trigonometry was first developed to determine heights and distances that are inconvenient or impossible to measure. These applications often involve the angle made with an imaginary horizontal line. As shown in **Figure 10.63**, an angle formed by a horizontal line and the line of sight to an object that is above the horizontal line is called the **angle of elevation**. The angle formed by a horizontal line and the line of sight to an object that is below the horizontal line is called the **angle of depression**. Transits and sextants are instruments used to measure such angles.

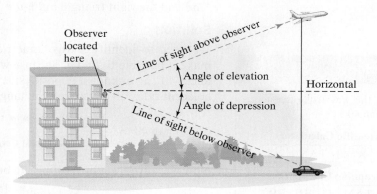

Observer located here

Line of sight above observer

Angle of elevation

Horizontal

Angle of depression

Line of sight below observer

Figure 10.63

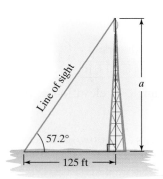

Figure 10.64 Determining height without using direct measurement

Example 4   Problem Solving Using an Angle of Elevation

From a point on level ground 125 feet from the base of a tower, the angle of elevation to the top of the tower is 57.2°. Approximate the height of the tower to the nearest foot.

Solution

A sketch is shown in **Figure 10.64**, where a represents the height of the tower. In the right triangle, we have a known angle, an unknown opposite side, and a known adjacent side. Therefore, we use the tangent ratio.

$$\tan 57.2° = \frac{a}{125}$$

Side opposite the **57.2°** angle

Side adjacent to the **57.2°** angle

We solve for a by multiplying both sides of this equation by 125:

$$a = 125 \tan 57.2° \approx 194.$$

The tower is approximately 194 feet high.

✓ **Checkpoint 4**   From a point on level ground 80 feet from the base of the Eiffel Tower, the angle of elevation to the top of the tower is 85.4°. Approximate the height of the Eiffel Tower to the nearest foot.

If the measures of two sides of a right triangle are known, the measures of the two acute angles can be found using the **inverse trigonometric keys** on a calculator. For example, suppose that $\sin A = 0.866$. We can find the measure of angle A by using the *inverse sine* key, usually labeled $\boxed{\text{SIN}^{-1}}$. The key $\boxed{\text{SIN}^{-1}}$ is not a button you will actually press; it is the secondary function for the button labeled $\boxed{\text{SIN}}$.

Many Scientific Calculators:

.866 $\boxed{\text{2nd}}$ $\boxed{\text{SIN}}$

Pressing $\boxed{\text{2nd}}$ $\boxed{\text{SIN}}$ accesses the inverse sine key, $\boxed{\text{SIN}^{-1}}$.

Many Graphing Calculators:

$\boxed{\text{2nd}}$ $\boxed{\text{SIN}}$ .866 $\boxed{\text{ENTER}}$

The display should show approximately 59.99°, which can be rounded to 60°. Thus, if $\sin A = 0.866$, then $m\angle A \approx 60°$.

Example 5   Determining the Angle of Elevation

A building that is 21 meters tall casts a shadow 25 meters long. Find the angle of elevation of the sun.

Solution

The situation is illustrated in **Figure 10.65**. We are asked to find $m\angle A$. We begin with the tangent ratio.

$$\tan A = \frac{\text{side opposite } A}{\text{side adjacent to } A} = \frac{21}{25}$$

We use the **inverse tangent** key, $\boxed{\text{TAN}^{-1}}$, to find A.

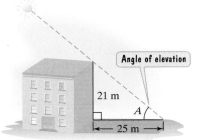

Figure 10.65

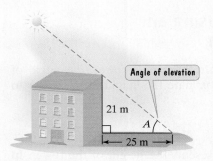

Figure 10.65 (repeated)

We use tan $A = \dfrac{21}{25}$ and $\boxed{\text{TAN}^{-1}}$ to find A.

Many Scientific Calculators:

$\boxed{(}\,\boxed{21}\,\boxed{\div}\,\boxed{25}\,\boxed{)}\,\boxed{\text{2nd}}\,\boxed{\text{TAN}}$

Pressing $\boxed{\text{2nd}}$ $\boxed{\text{TAN}}$ accesses the inverse tangent key, $\boxed{\text{TAN}^{-1}}$.

Many Graphing Calculators:

$\boxed{\text{2nd}}\,\boxed{\text{TAN}}\,\boxed{(}\,\boxed{21}\,\boxed{\div}\,\boxed{25}\,\boxed{)}\,\boxed{\text{ENTER}}$

The display should show approximately 40. Thus, the angle of elevation of the sun is approximately 40°.

✓ **Checkpoint 5** A flagpole that is 14 meters tall casts a shadow 10 meters long. Find the angle of elevation of the sun to the nearest degree.

Blitzer Bonus

The Mountain Man

In the 1930s, a *National Geographic* team headed by Brad Washburn used trigonometry to create a map of the 5000-square-mile region of the Yukon, near the Canadian border. The team started with aerial photography. By drawing a network of angles on the photographs, the approximate locations of the major mountains and their rough heights were determined. The expedition then spent three months on foot to find the exact heights. Team members established two base points a known distance apart, one directly under the mountain's peak. By measuring the angle of elevation from one of the base points to the peak, the tangent ratio was used to determine the peak's height. The Yukon expedition was a major advance in the way maps are made.

Exercise Set 10.6

Concept and Vocabulary Exercises

In Exercises 1–5, fill in each blank so that the resulting statement is true. Exercises 1–3 are based on the following right triangle:

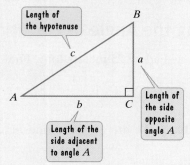

1. sin A, which represents the _____ of angle A, is defined by the length of the side _____ angle A divided by the length of the _____. This is represented by _____ in the figure shown.

2. cos A, which represents the _____ of angle A, is defined by the length of the side _____ angle A divided by the length of the _____. This is represented by _____ in the figure shown.

3. tan A, which represents the _____ of angle A, is defined by the length of the side _____ angle A divided by the length of the side _____ angle A. This is represented by _____ in the figure shown.

4. An angle formed by a horizontal line and the line of sight to an object that is above the horizontal line is called the angle of _____.

5. An angle formed by a horizontal line and the line of sight to an object that is below the horizontal line is called the angle of _____.

In Exercises 6–9, determine whether each statement is true or false. If the statement is false, make the necessary change(s) to produce a true statement.

6. sin 30° increases as the size of a right triangle with an acute angle of 30° grows larger.

7. The side of a right triangle opposite an acute angle forms one of the acute angle's sides.

8. On a scientific calculator, the keystroke 47 $\boxed{\text{TAN}}$ gives a decimal approximation for tan 47°.

9. The equation $\cos 40° = \frac{150}{c}$ is solved for c by multiplying both sides by 150.

Respond to Exercises 10–16 using verbal or written explanations.

10. If you are given the lengths of the sides of a right triangle, describe how to find the sine of either acute angle.

11. Describe one similarity and one difference between the sine ratio and the cosine ratio in terms of the sides of a right triangle.

12. If one of the acute angles of a right triangle is 37°, explain why the sine ratio does not increase as the size of the triangle increases.

13. If the measure of one of the acute angles and the hypotenuse of a right triangle are known, describe how to find the measures of the remaining parts of the triangle.

14. Describe what is meant by an angle of elevation and an angle of depression.

15. Give an example of an applied problem that can be solved using one or more trigonometric ratios. Be as specific as possible.

16. Stonehenge, the famous "stone circle" in England, was built between 2750 B.C. and 1300 B.C. using solid stone blocks weighing over 99,000 pounds each. It required 550 people to pull a single stone up a ramp inclined at a 9° angle. Describe how right triangle trigonometry can be used to determine the distance the 550 workers had to drag a stone in order to raise it to a height of 30 feet.

In Exercises 17–20, determine whether each statement makes sense or does not make sense, and explain your reasoning.

17. For a given angle A, I found a slight increase in sin A as the size of the triangle increased.

18. I'm working with a right triangle in which the hypotenuse is the side adjacent to the acute angle.

19. Standing under this arch, I can determine its height by measuring the angle of elevation to the top of the arch and my distance to a point directly under the arch.

Delicate Arch in Arches National Park, Utah

20. A wheelchair ramp must be constructed so that the slope is not more than 1 inch of rise for every 1 foot of run, so I used the tangent ratio to determine the maximum angle that the ramp can make with the ground.

Practice Exercises

In Exercises 21–28, use the given right triangles to find ratios, in reduced form, for sin A, cos A, and tan A.

21.
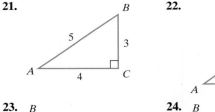

22.

23.
35

24.

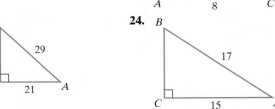

25.
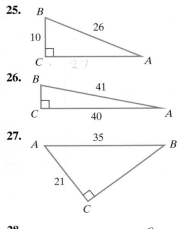

26.

27.

28.
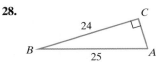

In Exercises 29–38, find the measure of the side of the right triangle whose length is designated by a lowercase letter. Round answers to the nearest whole number.

29.

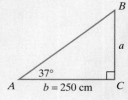

30.

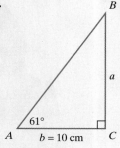

31.

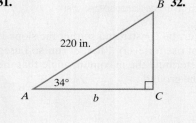

32.

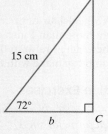

33.

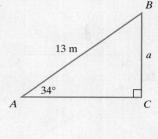

34.

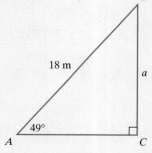

35.

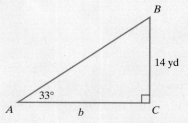

36.

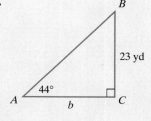

37.

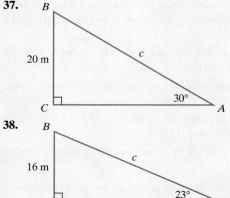

38.

In Exercises 39–42, find the measures of the parts of the right triangle that are not given. Round all answers to the nearest whole number.

39.

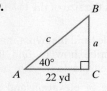

40.

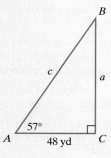

41.

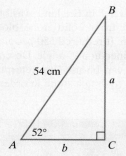

42.

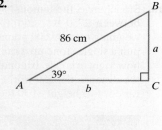

In Exercises 43–46, use the inverse trigonometric keys on a calculator to find the measure of angle A, rounded to the nearest whole degree.

43.

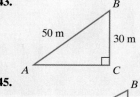

44.

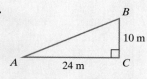

45.

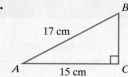

46.

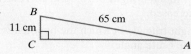

Practice Plus

In Exercises 47–54, find the length x to the nearest whole number.

47.

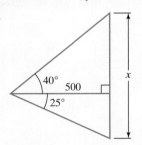

48.

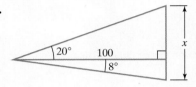

49.

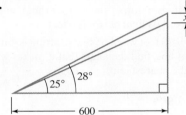

50.

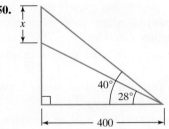

51.

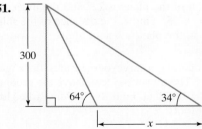

52.

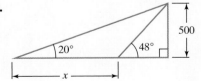

53.

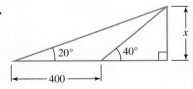

54.

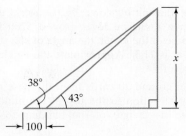

Application Exercises

55. To find the distance across a lake, a surveyor took the measurements shown in the figure. Use these measurements to determine how far it is across the lake. Round to the nearest yard.

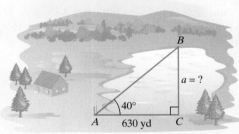

56. A tree casts a shadow 35 feet long when the angle of elevation of the sun is 40°. To the nearest foot, find the height of the tree.

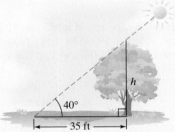

57. A plane rises from take-off and flies at an angle of 10° with the horizontal runway. When it has gained 500 feet in altitude, find the distance, to the nearest foot, the plane has flown.

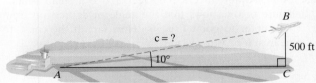

58. A road is inclined at an angle of 5°. After driving 5000 feet along this road, find the driver's increase in altitude. Round to the nearest foot.

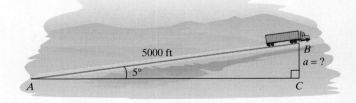

59. The tallest television transmitting tower in the world is in North Dakota. From a point on level ground 5280 feet (one mile) from the base of the tower, the angle of elevation to the top of the tower is 21.3°. Approximate the height of the tower to the nearest foot.

60. From a point on level ground 30 yards from the base of a building, the angle of elevation to the top of the building is 38.7°. Approximate the height of the building to the nearest foot.

61. The Statue of Liberty is approximately 305 feet tall. If the angle of elevation from a ship to the top of the statue is 23.7°, how far, to the nearest foot, is the ship from the statue's base?

62. A 200-foot cliff drops vertically into the ocean. If the angle of elevation from a ship to the top of the cliff is 22.3°, how far off shore, to the nearest foot, is the ship?

63. A tower that is 125 feet tall casts a shadow 172 feet long. Find the angle of elevation of the sun to the nearest degree.

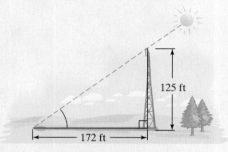

64. The Washington Monument is 555 feet high. If you stand one quarter of a mile, or 1320 feet, from the base of the monument and look to the top, find the angle of elevation to the nearest degree.

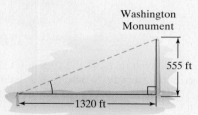

65. A helicopter hovers 1000 feet above a small island. The figure below shows that the angle of depression from the helicopter to point *P* is 36°. How far off the coast, to the nearest foot, is the island?

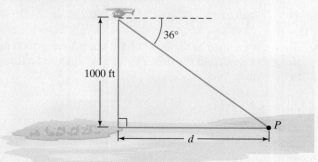

66. A police helicopter is flying at 800 feet. A stolen car is sighted at an angle of depression of 72°. Find the distance of the stolen car, to the nearest foot, from a point directly below the helicopter.

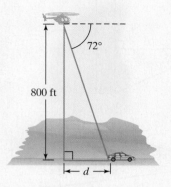

67. A wheelchair ramp is to be built beside the steps to the campus library. Find the angle of elevation of the 23-foot ramp, to the nearest tenth of a degree, if its final height is 6 feet.

68. A kite flies at a height of 30 feet when 65 feet of string is out. If the string is in a straight line, find the angle that it makes with the ground. Round to the nearest tenth of a degree.

Critical Thinking Exercises

69. Use a calculator to find each of the following: sin 32° and cos 58°; sin 17° and cos 73°; sin 50° and cos 40°; sin 88° and cos 2°. Describe what you observe. Based on your observations, what do you think the *co* in *cosine* stands for?

70. Explain why the sine or cosine of an acute angle cannot be greater than or equal to 1.

71. Describe what happens to the tangent of an angle as the measure of the angle gets close to 90°.

72. From the top of a 250-foot lighthouse, a plane is sighted overhead and a ship is observed directly below the plane. The angle of elevation of the plane is 22° and the angle of depression of the ship is 35°. Find **a.** the distance of the ship from the lighthouse; **b.** the plane's height above the water. Round to the nearest foot.

73. Sighting the top of a building, a surveyor measured the angle of elevation to be 22°. The transit is 5 feet above the ground and 300 feet from the building. Find the building's height. Round to the nearest foot.

10.7 : Distance and Midpoint Formulas; Circles

1. Find the distance between two points.

2. Find the midpoint of a line segment.

3. Write the standard form of a circle's equation.

4. Give the center and radius of a circle whose equation is in standard form.

It's a good idea to know your way around a circle. Clocks, angles, maps, and compasses are based on circles. Circles occur everywhere in nature: in ripples on water, patterns on a moth's wings, and cross sections of trees. Some consider the circle to be the most pleasing of all shapes.

The rectangular coordinate system gives us a unique way of knowing a circle. It enables us to translate a circle's geometric definition into an algebraic equation. To do this, we must first develop a formula for the distance between any two points in rectangular coordinates.

① **Find the distance between two points.**

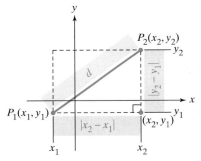

Figure 10.66

The Distance Formula

Using the Pythagorean Theorem, we can find the distance between the two points $P_1(x_1, y_1)$ and $P_2(x_2, y_2)$ in the rectangular coordinate system. The two points are illustrated in **Figure 10.66**.

The distance that we need to find is represented by d and shown in blue. Notice that the distance between the two points on the dashed horizontal line is the absolute value of the difference between the x-coordinates of the points. This distance, $|x_2 - x_1|$, is shown in pink. Similarly, the distance between the two points on the dashed vertical line is the absolute value of the difference between the y-coordinates of the points. This distance, $|y_2 - y_1|$, is also shown in pink.

Because the dashed lines are horizontal and vertical, a right triangle is formed. Thus, we can use the Pythagorean Theorem to find the distance d. Squaring the lengths of the triangle's sides results in positive numbers, so absolute value notation is not necessary.

$$d^2 = (x_2 - x_1)^2 + (y_2 - y_1)^2$$

Apply the Pythagorean Theorem to the right triangle in Figure 10.66.

$$d = \sqrt{(x_2 - x_1)^2 + (y_2 - y_1)^2}$$

Take the square root of both sides of the equation: $\sqrt{d^2} = d$. Because distance is nonnegative, write only the principal square root.

This result is called the **distance formula**.

The Distance Formula

The distance, d, between the points (x_1, y_1) and (x_2, y_2) in the rectangular coordinate system is

$$d = \sqrt{(x_2 - x_1)^2 + (y_2 - y_1)^2}.$$

To compute the distance between two points, find the square of the difference between the x-coordinates plus the square of the difference between the y-coordinates. The principal square root of this sum is the distance.

When using the distance formula to find the distance between two given points, it does not matter which point you call (x_1, y_1) and which you call (x_2, y_2).

Example 1 · Using the Distance Formula

Find the distance between $(-1, 4)$ and $(3, -2)$.

Solution

We will let $(x_1, y_1) = (-1, 4)$ and $(x_2, y_2) = (3, -2)$.

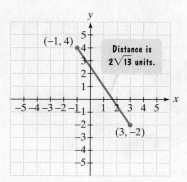

Figure 10.67 Finding the distance between two points

$$d = \sqrt{(x_2 - x_1)^2 + (y_2 - y_1)^2}$$ Use the distance formula.

$$= \sqrt{[3 - (-1)]^2 + (-2 - 4)^2}$$ Substitute the given values.

$$= \sqrt{4^2 + (-6)^2}$$ Perform operations inside grouping symbols: $3 - (-1) = 3 + 1 = 4$ and $-2 - 4 = -6$.

$$= \sqrt{16 + 36}$$ **Caution:** This does not equal $\sqrt{16} + \sqrt{36}$.   Square 4 and -6.

$$= \sqrt{52}$$ Add.

$$= \sqrt{4 \cdot 13} = 2\sqrt{13} \approx 7.21$$ $\sqrt{52} = \sqrt{4 \cdot 13} = \sqrt{4}\sqrt{13} = 2\sqrt{13}$

The distance between the given points is $2\sqrt{13}$ units, or approximately 7.21 units. The situation is illustrated in **Figure 10.67**.

> ✓ **Checkpoint 1** Find the distance between $(-4, 9)$ and $(1, -3)$.

The Midpoint Formula

② Find the midpoint of a line segment.

The distance formula can be used to derive a formula for finding the midpoint of a line segment between two given points. The formula is given as follows:

The Midpoint Formula

Consider a line segment whose endpoints are (x_1, y_1) and (x_2, y_2). The coordinates of the segment's midpoint are

$$\left(\frac{x_1 + x_2}{2}, \frac{y_1 + y_2}{2} \right).$$

To find the midpoint, take the average of the two x-coordinates and the average of the two y-coordinates.

Study Tip

The midpoint formula requires finding the *sum* of coordinates. By contrast, the distance formula requires finding the *difference* of coordinates:

Midpoint: Sum of coordinates
$$\left(\frac{x_1 + x_2}{2}, \frac{y_1 + y_2}{2} \right)$$

Distance: Difference of coordinates
$$\sqrt{(x_2 - x_1)^2 + (y_2 - y_1)^2}$$

It's easy to confuse the two formulas. Be sure to use addition, not subtraction, when applying the midpoint formula.

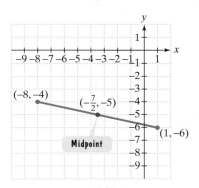

Figure 10.68 Finding a line segment's midpoint

Example 2   Using the Midpoint Formula

Find the midpoint of the line segment with endpoints $(1, -6)$ and $(-8, -4)$.

Solution

To find the coordinates of the midpoint, we average the coordinates of the endpoints.

$$\text{Midpoint} = \left(\frac{1 + (-8)}{2}, \frac{-6 + (-4)}{2} \right) = \left(\frac{-7}{2}, \frac{-10}{2} \right) = \left(-\frac{7}{2}, -5 \right)$$

Average the
x-coordinates.

Average the
y-coordinates.

Figure 10.68 illustrates that the point $\left(-\frac{7}{2}, -5 \right)$ is midway between the points $(1, -6)$ and $(-8, -4)$.

✓ **Checkpoint 2**   Find the midpoint of the line segment with endpoints $(1, 2)$ and $(7, -3)$.

Circles

Our goal is to translate a circle's geometric definition into an equation. We begin by restating this geometric definition.

> ### Definition of a Circle
> A **circle** is the set of all points in a plane that are equidistant from a fixed point, called the **center**. The fixed distance from the circle's center to any point on the circle is called the **radius**.

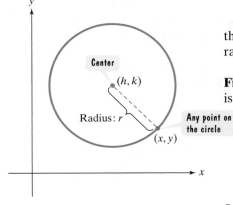

Figure 10.69 A circle centered at (h, k) with radius r

Figure 10.69 is our starting point for obtaining a circle's equation. We've placed the circle into a rectangular coordinate system. The circle's center is (h, k) and its radius is r. We let (x, y) represent the coordinates of any point on the circle.

What does the geometric definition of a circle tell us about the point (x, y) in **Figure 10.69**? The point is on the circle if and only if its distance from the center is r. We can use the distance formula to express this idea algebraically:

The distance between
(x, y) and (h, k) is always r.

$$\sqrt{(x - h)^2 + (y - k)^2} = r.$$

Squaring both sides of $\sqrt{(x - h)^2 + (y - k)^2} = r$ yields the *standard form of the equation of a circle*.

3 Write the standard form of a circle's equation.

> ### The Standard Form of the Equation of a Circle
> The **standard form of the equation of a circle** with center (h, k) and radius r is
> $$(x - h)^2 + (y - k)^2 = r^2.$$

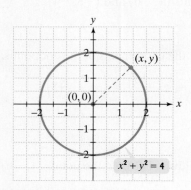

Figure 10.70 The graph of $x^2 + y^2 = 4$

Example 3   Finding the Standard Form of a Circle's Equation

Write the standard form of the equation of the circle with center $(0, 0)$ and radius 2. Graph the circle.

Solution

The center is $(0, 0)$. Because the center is represented as (h, k) in the standard form of the equation, $h = 0$ and $k = 0$. The radius is 2, so we will let $r = 2$ in the equation.

$(x - h)^2 + (y - k)^2 = r^2$ This is the standard form of a circle's equation.
$(x - 0)^2 + (y - 0)^2 = 2^2$ Substitute 0 for h, 0 for k, and 2 for r.
$x^2 + y^2 = 4$ Simplify.

The standard form of the equation of the circle is $x^2 + y^2 = 4$. **Figure 10.70** shows the graph.

✓ **Checkpoint 3**   Write the standard form of the equation of the circle with center $(0, 0)$ and radius 4.

Example 3 and Checkpoint 3 involved circles centered at the origin. The standard form of the equation of all such circles is $x^2 + y^2 = r^2$, where r is the circle's radius. Now, let's consider a circle whose center is not at the origin.

Example 4   Finding the Standard Form of a Circle's Equation

Write the standard form of the equation of the circle with center $(-2, 3)$ and radius 4.

Solution

The center is $(-2, 3)$. Because the center is represented as (h, k) in the standard form of the equation, $h = -2$ and $k = 3$. The radius is 4, so we will let $r = 4$ in the equation.

$(x - h)^2 + (y - k)^2 = r^2$ This is the standard form of a circle's equation.
$[x - (-2)]^2 + (y - 3)^2 = 4^2$ Substitute −2 for h, 3 for k, and 4 for r.
$(x + 2)^2 + (y - 3)^2 = 16$ Simplify.

The standard form of the equation of the circle is $(x + 2)^2 + (y - 3)^2 = 16$.

✓ **Checkpoint 4**   Write the standard form of the equation of the circle with center $(0, -6)$ and radius 10.

 Give the center and radius of a circle whose equation is in standard form.

Example 5   Using the Standard Form of a Circle's Equation to Graph the Circle

a. Find the center and radius of the circle whose equation is
$$(x - 2)^2 + (y + 4)^2 = 9.$$

b. Graph the equation.

Solution

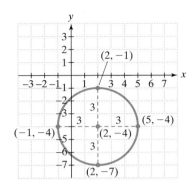

Figure 10.71 The graph of $(x - 2)^2 + (y + 4)^2 = 9$

a. We begin by finding the circle's center, (h, k), and its radius, r. We can find the values for h, k, and r by comparing the given equation to the standard form of the equation of a circle, $(x - h)^2 + (y - k)^2 = r^2$.

$$(x - 2)^2 + (y + 4)^2 = 9$$
$$(x - 2)^2 + (y - (-4))^2 = 3^2$$

| This is $(x - h)^2$, with $h = 2$. | This is $(y - k)^2$, with $k = -4$. | This is r^2, with $r = 3$. |

We see that $h = 2$, $k = -4$, and $r = 3$. Thus, the circle has center $(h, k) = (2, -4)$ and a radius of 3 units.

b. To graph this circle, first locate the center $(2, -4)$. Because the radius is 3, you can locate at least four points on the circle by going out three units to the right, to the left, up, and down from the center.

The points three units to the right and to the left of $(2, -4)$ are $(5, -4)$ and $(-1, -4)$, respectively. The points three units up and down from $(2, -4)$ are $(2, -1)$ and $(2, -7)$, respectively.

Using these points, we obtain the graph in **Figure 10.71**.

Study Tip

It's easy to make sign errors when finding h and k, the coordinates of a circle's center, (h, k). Keep in mind that h and k are the numbers that *follow the subtraction signs* in a circle's equation:

$$(x - 2)^2 + (y + 4)^2 = 9$$
$$(x - 2)^2 + (y - (-4))^2 = 9.$$

| The number *after* the subtraction is 2: $h = 2$. | The number *after* the subtraction is −4: $k = -4$. |

✓ **Checkpoint 5**

a. Find the center and radius of the circle whose equation is
$$(x + 3)^2 + (y - 1)^2 = 4.$$

b. Graph the equation.

Achieving Success

Most colleges have requirements that make it necessary for students to take courses in a variety of areas. Take these required courses early in your college education, primarily in your freshman year. These requirements can turn out to be interesting courses, which may lead your education in unexpected directions.

Exercise Set 10.7

Concept and Vocabulary Exercises

In Exercises 1–4, fill in each blank so that the resulting statement is true.

1. The distance, d, between the points (x_1, y_1) and (x_2, y_2) in the rectangular coordinate system is given by the formula _____.

2. The midpoint of a line segment whose endpoints are (x_1, y_1) and (x_2, y_2) is _____.

3. The standard form of the equation of a circle with center (h, k) and radius r is _____.

4. The standard form of the equation of a circle centered at the origin with radius r is _____.

In Exercises 5–8, determine whether each statement is true or false. If the statement is false, make the necessary change(s) to produce a true statement.

5. The midpoint formula requires finding the sum of coordinates, whereas the distance formula requires finding the difference of coordinates.

6. The equation of the circle whose center is at the origin with radius 16 is $x^2 + y^2 = 16$.

7. The graph of $(x - 3)^2 + (y + 5)^2 = 36$ is a circle with radius 6 centered at $(-3, 5)$.

8. The graph of $(x - 4) + (y + 6) = 25$ is a circle with radius 5 centered at $(4, -6)$.

Respond to Exercises 9–12 using verbal or written explanations.

9. In your own words, describe how to find the distance between two points in the rectangular coordinate system.

10. In your own words, describe how to find the midpoint of a line segment if its endpoints are known.

11. What is a circle? Without using variables, describe how the definition of a circle can be used to obtain the standard form of its equation.

12. Give an example of a circle's equation in standard form. Describe how to find the center and radius for this circle.

In Exercises 13–16, determine whether each statement makes sense or does not make sense, and explain your reasoning.

13. I've noticed that in mathematics, one topic often leads logically to a new topic:

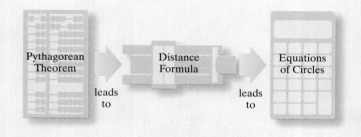

Pythagorean Theorem → leads to → Distance Formula → leads to → Equations of Circles

14. To avoid sign errors when finding h and k, I place parentheses around the numbers that follow the subtraction signs in a circle's equation.

15. I found the distance between $(-1, 4)$ and $(9, 6)$ by simplifying $\sqrt{8^2 + 2^2}$.

16. My graph of $(x - 2)^2 + (y + 1)^2 = 16$ is my graph of $x^2 + y^2 = 16$ translated two units right and one unit down.

Practice Exercises

In Exercises 17–28, find the distance between each pair of points. If necessary, round answers to two decimals places.

17. $(2, 3)$ and $(14, 8)$
18. $(5, 1)$ and $(8, 5)$
19. $(4, -1)$ and $(-6, 3)$
20. $(2, -3)$ and $(-1, 5)$
21. $(0, 0)$ and $(-3, 4)$
22. $(0, 0)$ and $(3, -4)$
23. $(-2, -6)$ and $(3, -4)$
24. $(-4, -1)$ and $(2, -3)$
25. $(0, -3)$ and $(4, 1)$
26. $(0, -2)$ and $(4, 3)$
27. $(3.5, 8.2)$ and $(-0.5, 6.2)$
28. $(2.6, 1.3)$ and $(1.6, -5.7)$

In Exercises 29–36, find the midpoint of each line segment with the given endpoints.

29. $(6, 8)$ and $(2, 4)$
30. $(10, 4)$ and $(2, 6)$
31. $(-2, -8)$ and $(-6, -2)$
32. $(-4, -7)$ and $(-1, -3)$
33. $(-3, -4)$ and $(6, -8)$
34. $(-2, -1)$ and $(-8, 6)$
35. $\left(-\frac{7}{2}, \frac{3}{2}\right)$ and $\left(-\frac{5}{2}, -\frac{11}{2}\right)$
36. $\left(-\frac{2}{5}, \frac{7}{15}\right)$ and $\left(-\frac{2}{5}, -\frac{4}{15}\right)$

In Exercises 37–46, write the standard form of the equation of the circle with the given center and radius.

37. Center $(0, 0)$, $r = 7$
38. Center $(0, 0)$, $r = 8$
39. Center $(3, 2)$, $r = 5$
40. Center $(2, -1)$, $r = 4$
41. Center $(-1, 4)$, $r = 2$
42. Center $(-3, 5)$, $r = 3$
43. Center $(-3, -1)$, $r = \sqrt{3}$
44. Center $(-5, -3)$, $r = \sqrt{5}$
45. Center $(-4, 0)$, $r = 10$
46. Center $(-2, 0)$, $r = 6$

In Exercises 47–58, give the center and radius of the circle described by the equation and graph each equation.

47. $x^2 + y^2 = 16$
48. $x^2 + y^2 = 49$
49. $(x - 3)^2 + (y - 1)^2 = 36$
50. $(x - 2)^2 + (y - 3)^2 = 16$
51. $(x + 3)^2 + (y - 2)^2 = 4$
52. $(x + 1)^2 + (y - 4)^2 = 25$
53. $(x + 2)^2 + (y + 2)^2 = 4$
54. $(x + 4)^2 + (y + 5)^2 = 36$
55. $x^2 + (y - 1)^2 = 1$
56. $x^2 + (y - 2)^2 = 4$
57. $(x + 1)^2 + y^2 = 25$
58. $(x + 2)^2 + y^2 = 16$

Practice Plus

In Exercises 59–62, the standard form of a circle's equation is given in a rectangular coordinate system in which each unit represents 1 inch. Find the circumference and the area of each circle in terms of π.

59. $x^2 + y^2 = 25$
60. $x^2 + y^2 = 49$
61. $(x - 2)^2 + y^2 = 100$
62. $x^2 + (y - 4)^2 = 81$

In Exercises 63–66, graph both equations in the same rectangular coordinate system and find all points of intersection. Then show that these ordered pairs satisfy the equations.

63. $x^2 + y^2 = 16$
$x - y = 4$

64. $x^2 + y^2 = 9$
$x - y = 3$

65. $(x - 2)^2 + (y + 3)^2 = 4$
$$y = x - 3$$

66. $(x - 3)^2 + (y + 1)^2 = 9$
$$y = x - 1$$

Application Exercises

The cellphone screen shows coordinates of six cities from a rectangular coordinate system placed on North America by long-distance telephone companies. Each unit in this system represents $\sqrt{0.1}$ mile.

Source: Peter H. Dana

In Exercises 67–68, use this information to find the distance, to the nearest mile, between each pair of cities.

67. Boston and San Francisco

68. New Orleans and Houston

69. A rectangular coordinate system with coordinates in miles is placed with the origin at the center of Los Angeles. The figure indicates that the University of Southern California is located 2.4 miles west and 2.7 miles south of central Los Angeles. A seismograph on the campus shows that a small earthquake occurred. The quake's epicenter is estimated to be approximately 30 miles from the university. Write the standard form of the equation for the set of points that could be the epicenter of the quake.

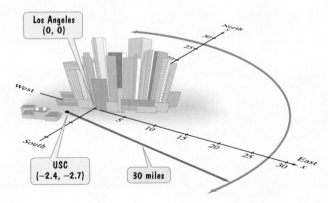

70. The Ferris wheel in the figure at the top of the next column has a radius of 68 feet. The clearance between the wheel and the ground is 14 feet. The rectangular coordinate system shown has its origin on the ground directly below the center of the wheel. Use the coordinate system to write the equation of the circular wheel.

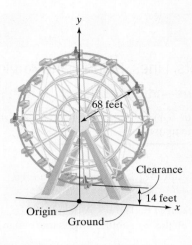

Critical Thinking Exercises

71. Find the area of the donut-shaped region bounded by the graphs of $(x - 2)^2 + (y + 3)^2 = 25$ and $(x - 2)^2 + (y + 3)^2 = 36$. Assume that all units are expressed in centimeters. Write the area in terms of π.

72. Write and solve a problem about the flying time between a pair of cities shown on the cellphone screen for Exercises 67–68. Do not use the pairs in Exercise 67 or Exercise 68. Begin by determining a reasonable average speed, in miles per hour, for a jet flying between the cities.

In Exercises 73–74, a line segment through the center of each circle intersects the circle at the points shown.

  a. *Find the coordinates of the circle's center.*

  b. *Find the radius of the circle.*

  c. *Use your answers from parts (a) and (b) to write the standard form of the circle's equation.*

73.

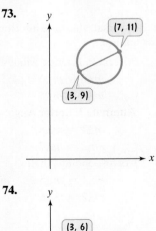

74.

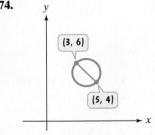

Chapter 10 Summary

10.1 Points, Lines, Planes, and Angles

Definitions and Concepts

Line AB ($\overleftrightarrow{AB}$ or $\overleftrightarrow{BA}$), half-line AB ($\overset{\circ}{\rightarrow}{AB}$), ray AB ($\overrightarrow{AB}$), and line segment AB ($\overline{AB}$ or $\overline{BA}$) are represented in the following figure:

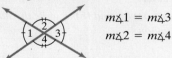

Line AB	Half line AB	Ray AB	Line Segment AB
$\overleftrightarrow{AB}$ or $\overleftrightarrow{BA}$	$\overset{\circ}{\rightarrow}{AB}$	$\overrightarrow{AB}$	$\overline{AB}$ or $\overline{BA}$

Angles are measured in degrees. A degree, 1°, is $\frac{1}{360}$ of a complete rotation. Acute angles measure less than 90°, right angles 90°, obtuse angles more than 90° but less than 180°, and straight angles 180°.

Complementary angles are two angles whose measures have a sum of 90°. The measure of an angle's complement is found by subtracting the angle's measure from 90°.

Supplementary angles are two angles whose measures have a sum of 180°. The measure of an angle's supplement is found by subtracting the angle's measure from 180°.

Example

- An angle measures 4°.

  measure of its complement $= 90° - 4° = 86°$

  measure of its supplement $= 180° - 4° = 176°$

Additional Examples to Review

Example 1, page 593; Example 2, page 594; Example 3, page 595

Definitions and Concepts

Vertical angles have the same measure.

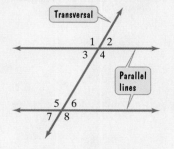

$$m\angle 1 = m\angle 3$$
$$m\angle 2 = m\angle 4$$

If parallel lines are intersected by a transversal, alternate interior angles, alternate exterior angles, and corresponding angles have the same measure.

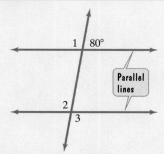

Alternate Interior Angles
$$m\angle 3 = m\angle 6$$
$$m\angle 4 = m\angle 5$$

Alternate Exterior Angles
$$m\angle 1 = m\angle 8$$
$$m\angle 2 = m\angle 7$$

Corresponding Angles
$$m\angle 1 = m\angle 5 \qquad m\angle 3 = m\angle 7$$
$$m\angle 2 = m\angle 6 \qquad m\angle 4 = m\angle 8$$

Example

- $m\angle 1 = 180° - 80° = 100°$ ∡1 and the 80° angle are supplementary.
- $m\angle 2 = 100°$ ∡1 and ∡2 are corresponding angles with the same measure.
- $m\angle 3 = 100°$ ∡2 and ∡3 are vertical angles with the same measure.

Additional Examples to Review

Example 4, page 596; Example 5, page 597

10.2 Triangles

Definitions and Concepts

The sum of the measures of the three angles of any triangle is 180°.

Example

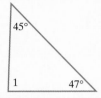

- $m\angle 1 + 45° + 47° = 180°$

$$m\angle 1 + 92° = 180°$$

$$m\angle 1 = 180° - 92° = 88°$$

Additional Examples to Review

Example 1, page 603; Example 2, page 603

Definitions and Concepts

Triangles can be classified by angles (acute, right, obtuse) or by sides (isosceles, equilateral, scalene).

Classification by Angles

Acute Triangle
All angles are acute.

Right Triangle
One angle measures 90°.

Obtuse Triangle
One angle is obtuse.

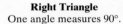

Classification by Sides

Isosceles Triangle
Two sides have equal length.
(Angles opposite these sides
have the same measure.)

Equilateral Triangle
All sides have equal length.
(Each angle measures 60°.)

Scalene Triangle
No two sides are equal in
length.

Similar triangles have the same shape, but not necessarily the same size. Corresponding angles have the same measure and corresponding sides are proportional. If the measures of two angles of one triangle are equal to those of two angles of a second triangle, then the two triangles are similar.

Example

- The triangles are similar.

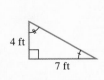
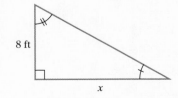

$$\frac{4}{8} = \frac{7}{x}$$

$$4x = 8 \cdot 7$$

$$4x = 56$$

$$\frac{4x}{4} = \frac{56}{4}$$

$$x = 14$$

The missing length, x, is 14 feet.

Additional Examples to Review

Example 3, page 605; Example 4, page 606

Definitions and Concepts

The Pythagorean Theorem

The sum of the squares of the lengths of the legs
of a right triangle equals the square of the length
of the hypotenuse.

If the legs have lengths a and b and the hypotenuse
has length c, then

$$a^2 + b^2 = c^2.$$

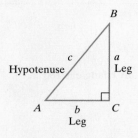

Example

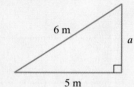

- $a^2 + 5^2 = 6^2$
  $a^2 + 25 = 36$
  $a^2 = 11$
  $a = \sqrt{11} \approx 3.3\text{ m}$

Additional Examples to Review

Example 5, page 607; Example 6, page 607

10.3 Polygons, Perimeter, and Tessellations

Definitions and Concepts

A polygon is a closed geometric figure in a plane formed by three or more line segments. Names of
some polygons include triangles (three sides), quadrilaterals (four sides), pentagons (five sides),
hexagons (six sides), heptagons (seven sides), and octagons (eight sides). A regular polygon is one
whose sides are all the same length and whose angles all have the same measure. The perimeter of a
polygon is the sum of the lengths of its sides.

Types of Quadrilaterals		
Name	**Characteristics**	**Representation**
Parallelogram	Quadrilateral in which both pairs of opposite sides are parallel and have the same measure. Opposite angles have the same measure.	
Rhombus	Parallelogram with all sides having equal length.	
Rectangle	Parallelogram with four right angles. Because a rectangle is a parallelogram, opposite sides are parallel and have the same measure.	
Square	A rectangle with all sides having equal length. Each angle measures 90°, and the square is a regular quadrilateral.	
Trapezoid	A quadrilateral with exactly one pair of parallel sides.	

Example

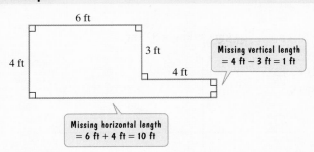

- Perimeter

$$P = 6 + 3 + 4 + 1 + 10 + 4$$
$$= 28 \text{ ft}$$

Additional Example to Review

Example 1, page 614

Definitions and Concepts

The sum of the measures of the angles of an n-sided polygon is $(n - 2)180°$. If the n-sided polygon is a regular polygon, then each angle measures $\dfrac{(n - 2)180°}{n}$.

Examples

- The sum of the measures of the angles of a hexagon is
$$(n - 2)180° = (6 - 2)180° = 4 \cdot 180° = 720°.$$
- Each angle of a regular hexagon measures
$$\frac{720°}{6} = 120°.$$

Additional Example to Review

Example 2, page 615

Definitions and Concepts

A tessellation is a pattern consisting of the repeated use of the same geometric figures to completely cover a plane, leaving no gaps and having no overlaps. The angle requirement for the formation of a tessellation is that the sum of the measures of the angles at each vertex must be 360°.

Example

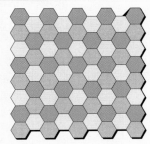

- Regular hexagons surround each vertex.
- Three angles come together at each vertex, measuring $120° + 120° + 120° = 360°$.
- The tessellation is possible because the sum of the measures of the angles at each vertex is 360°.

Additional Example to Review

Example 3, page 617

10.4 Area and Circumference

Definitions and Concepts

Formulas for Area

Rectangle: $A = lw$;   Square: $A = s^2$;   Parallelogram: $A = bh$;   Triangle: $A = \frac{1}{2}bh$;
Trapezoid: $A = \frac{1}{2}h(a + b)$

Examples

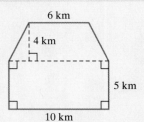

- Area of Trapezoid
  $A = \frac{1}{2}h(a+b) = \frac{1}{2} \cdot 4\,\text{km} \cdot (6\,\text{km} + 10\,\text{km}) = \frac{1}{2} \cdot 4\,\text{km} \cdot 16\,\text{km} = 32\,\text{km}^2$
- Area of Rectangle
  $A = lw = 10\,\text{km} \cdot 5\,\text{km} = 50\,\text{km}^2$
- Total Area $= 32\,\text{km}^2 + 50\,\text{km}^2 = 82\,\text{km}^2$

Additional Examples to Review

Example 1, page 622; Example 2, page 623; Example 3, page 624; Example 4, page 625; Example 5, page 625

Definitions and Concepts

A circle is a set of points in the plane equally distant from a given point, its center. The radius (plural: radii), r, is a line segment from the center to any point on the circle. The diameter, d, is a line segment through the center whose endpoints both lie on the circle. The length of the diameter is twice the length of the radius.

$$\text{Circumference: } C = 2\pi r \quad \text{or} \quad C = \pi d$$
$$\text{Area: } A = \pi r^2$$

Examples

A circle has a diameter of 20 meters.

- Circumference: $C = \pi d = \pi(20\,\text{m}) = 20\pi\,\text{m} \approx 62.8\,\text{m}$
- Area: $A = \pi r^2 = \pi(10\,\text{m})^2 = 100\pi\,\text{m}^2 \approx 314.2\,\text{m}^2$

> The diameter measures 20 m, so the radius measures $\frac{1}{2} \cdot 20\,\text{m} = 10\,\text{m}$.

Additional Examples to Review

Example 6, page 626; Example 7, page 627; Example 8, page 627

10.5 Volume

Definitions and Concepts

Formulas for Volume

Rectangular Solid: $V = lwh$;   Cube: $V = s^3$;   Pyramid: $V = \frac{1}{3}Bh$;   Cylinder: $V = \pi r^2 h$;
Cone: $V = \frac{1}{3}\pi r^2 h$ Sphere: $V = \frac{4}{3}\pi r^3$

Examples

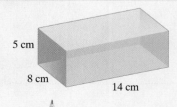

- Volume of the Rectangular Solid
  $V = lwh = 14\,\text{cm} \cdot 8\,\text{cm} \cdot 5\,\text{cm} = 560\,\text{cm}^3$

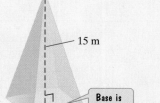

- Area of the Square Base
  $B = 7\,\text{m} \cdot 7\,\text{m} = 49\,\text{m}^2$
- Volume of the Pyramid
  $V = \frac{1}{3}Bh = \frac{1}{3} \cdot 49\,\text{m}^2 \cdot 15\,\text{m} = \frac{1}{3} \cdot 49\,\text{m}^2 \cdot \overset{5}{\underset{1}{15}}\,\text{m} = 245\,\text{m}^3$

> Base is a square.

- Volume of the Cylinder

$$V = \pi r^2 h = \pi(3\text{ in.})^2 \cdot 5\text{ in.} = 45\pi \text{ in.}^3 \approx 141 \text{ in.}^3$$

Additional Examples to Review

Example 1, page 633; Example 2, page 633; Example 3, page 634; Example 4, page 635; Example 5, page 635; Example 6, page 636

Definitions and Concepts

The measure of the area of the outer surface of a three-dimensional object is called its surface area and is measured in square units.

Common Formulas for Surface Area

Cube
$SA = 6s^2$

Rectangular Solid
$SA = 2lw + 2lh + 2wh$

Circular Cylinder
$SA = 2\pi r^2 + 2\pi rh$

Example

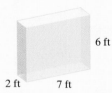

Surface Area of the Rectangular Solid

$$SA = 2lw + 2lh + 2wh$$
$$= 2 \cdot 7\text{ ft} \cdot 2\text{ ft} + 2 \cdot 7\text{ ft} \cdot 6\text{ ft} + 2 \cdot 2\text{ ft} \cdot 6\text{ ft}$$
$$= 28\text{ ft}^2 + 84\text{ ft}^2 + 24\text{ ft}^2 = 136\text{ ft}^2$$

6 ft, 2 ft, 7 ft

Additional Example to Review

Example 7, page 637

10.6 Right Triangle Trigonometry

Definitions and Concepts

Definitions of the Trigonometric Ratios

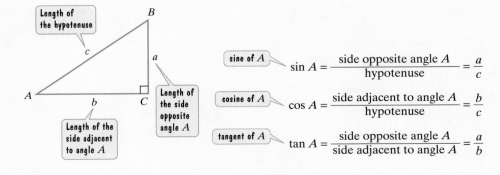

$$\sin A = \frac{\text{side opposite angle } A}{\text{hypotenuse}} = \frac{a}{c}$$

$$\cos A = \frac{\text{side adjacent to angle } A}{\text{hypotenuse}} = \frac{b}{c}$$

$$\tan A = \frac{\text{side opposite angle } A}{\text{side adjacent to angle } A} = \frac{a}{b}$$

Examples

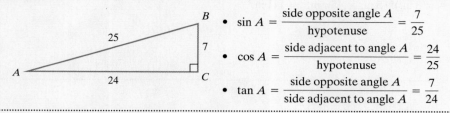

- $\sin A = \dfrac{\text{side opposite angle } A}{\text{hypotenuse}} = \dfrac{7}{25}$
- $\cos A = \dfrac{\text{side adjacent to angle } A}{\text{hypotenuse}} = \dfrac{24}{25}$
- $\tan A = \dfrac{\text{side opposite angle } A}{\text{side adjacent to angle } A} = \dfrac{7}{24}$

Additional Example to Review

Example 1, page 642

Definitions and Concepts

Given one side and the measure of an acute angle of a right triangle, the trigonometric ratios can be used to find the measures of the other parts of the right triangle.

An angle formed by a horizontal line and the line of sight to an object that is above the horizontal line is called the angle of elevation. The angle formed by a horizontal line and the line of sight to an object that is below the horizontal line is called the angle of depression.

Examples

- Find c.

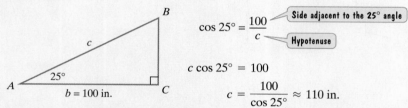

$\cos 25° = \dfrac{100}{c}$ Side adjacent to the 25° angle   Hypotenuse

$c \cos 25° = 100$

$c = \dfrac{100}{\cos 25°} \approx 110 \text{ in.}$

- When the sun's angle of elevation is 38°, a building casts a shadow of 92 feet. Find the building's height.

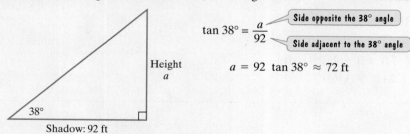

$\tan 38° = \dfrac{a}{92}$ Side opposite the 38° angle   Side adjacent to the 38° angle

$a = 92 \ \tan 38° \approx 72 \text{ ft}$

Additional Examples to Review

Example 2, page 643; Example 3, page 644; Example 4, page 645

Definitions and Concepts

Given the measures of two sides of a right triangle, the measures of the acute angles can be found using the inverse trigonometric ratio keys $\boxed{\text{SIN}^{-1}}$, $\boxed{\text{COS}^{-1}}$, and $\boxed{\text{TAN}^{-1}}$, on a calculator.

Example

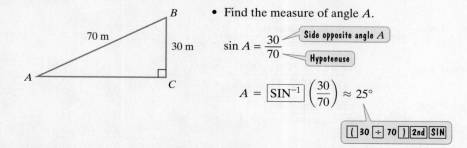

- Find the measure of angle A.

$\sin A = \dfrac{30}{70}$ Side opposite angle A Hypotenuse

$A = \boxed{\text{SIN}^{-1}}\left(\dfrac{30}{70}\right) \approx 25°$

$\boxed{(}\ \boxed{30}\ \boxed{\div}\ \boxed{70}\ \boxed{)}\ \boxed{\text{2nd}}\ \boxed{\text{SIN}}$

Additional Example to Review

Example 5, page 645

10.7 Distance and Midpoint Formulas; Circles

Definitions and Concepts

The distance, d, between the points (x_1, y_1) and (x_2, y_2) is given by $d = \sqrt{(x_2 - x_1)^2 + (y_2 - y_1)^2}$.

Example

The distance between $(x_1, y_1) = (-1, -4)$ and $(x_2, y_2) = (5, -2)$ is

$$d = \sqrt{(x_2 - x_1)^2 + (y_2 - y_1)^2} = \sqrt{[5 - (-1)]^2 + [-2 - (-4)]^2} = \sqrt{(5 + 1)^2 + (-2 + 4)^2} = \sqrt{6^2 + 2^2}$$
$$= \sqrt{36 + 4} = \sqrt{40} = \sqrt{4 \cdot 10} = 2\sqrt{10} \approx 6.32$$

Additional Example to Review

Example 1, page 652

Definitions and Concepts

The midpoint of the line segment whose endpoints are (x_1, y_1) and (x_2, y_2) is the point with

coordinates $\left(\dfrac{x_1 + x_2}{2}, \dfrac{y_1 + y_2}{2} \right)$.

Example

The midpoint of the line segment with endpoints $(-8, 4)$ and $(2, 11)$ is

$$\left(\frac{-8 + 2}{2}, \frac{4 + 11}{2} \right) = \left(\frac{-6}{2}, \frac{15}{2} \right) = \left(-3, \frac{15}{2} \right).$$

Additional Example to Review

Example 2, page 653

Definitions and Concepts

The standard form of the equation of a circle with center (h, k) and radius r is
$(x - h)^2 + (y - k)^2 = r^2$.

Examples

- The standard form of the equation of the circle with center $(0, -6)$ and radius 8 is

$$(x - 0)^2 + [y - (-6)]^2 = 8^2, \text{ or}$$
$$x^2 + (y + 6)^2 = 64.$$

- Give the center and radius of the circle whose standard equation is
$$(x + 4)^2 + (y - 5)^2 = 100.$$

$$(x - (-4))^2 \;+\; (y - 5)^2 \;=\; 10^2$$

$\boxed{h = -4}$ $\boxed{k = 5}$ $\boxed{r = 10}$

center: $(h, k) = (-4, 5)$

radius: $r = 10$

Additional Examples to Review

Example 3, page 654; Example 4, page 654; Example 5, page 654

Review Exercises

Section 10.1 Points, Lines, Planes, and Angles

In the figure shown, lines l and m are parallel. In Exercises 1–7, match each term with the numbered angle or angles in the figure.

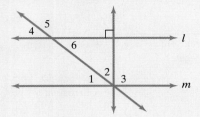

1. right angle

2. obtuse angle

3. vertical angles

4. alternate interior angles

5. corresponding angles

6. the complement of ∡1

7. the supplement of ∡6

In Exercises 8–9, find the measure of the angle in which a question mark with a degree symbol appears.

8.

9.

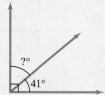

10. If an angle measures 73°, find the measure of its complement.

11. If an angle measures 46°, find the measure of its supplement.

12. In the figure shown, find the measures of angles 1, 2, and 3.

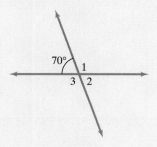

13. In the figure shown, two parallel lines are intersected by a transversal. One of the angle measures is given. Find the measure of each of the other seven angles.

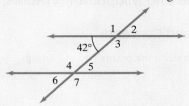

Section 10.2 Triangles

In Exercises 14–15, find the measure of angle A for the triangle shown.

14.   15.

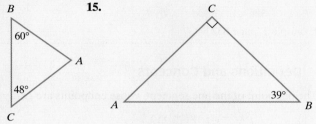

16. Find the measures of angles 1 through 5 in the figure shown.

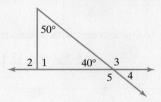

17. In the figure shown, lines *l* and *m* are parallel. Find the measure of each numbered angle.

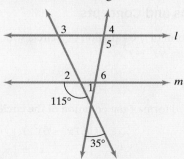

In Exercises 18–19, use similar triangles and the fact that corresponding sides are proportional to find the length of each side marked with an x.

18.

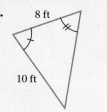

19.

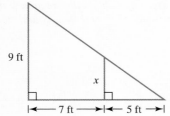

9 ft

x

|← 7 ft →|← 5 ft →|

In Exercises 20–22, use the Pythagorean Theorem to find the missing length in each right triangle. Round, if necessary, to the nearest tenth.

20.

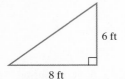

6 ft

8 ft

21.

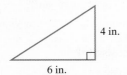

4 in.

6 in.

22.

15 cm

11 cm

23. Find the height of the lamppost in the figure.

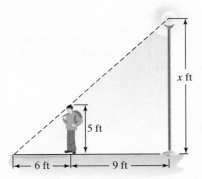

x ft

5 ft

|← 6 ft →|← 9 ft →|

24. How far away from the building in the figure shown is the bottom of the ladder?

20 ft   25 ft

|← ? ft →|

25. A vertical pole is to be supported by three wires. Each wire is 13 yards long and is anchored 5 yards from the base of the pole. How far up the pole will the wires be attached?

Section 10.3 Polygons, Perimeter, and Tessellations

26. Write the names of all quadrilaterals that always have four right angles.

27. Write the names of all quadrilaterals with four sides always having the same measure.

28. Write the names of all quadrilaterals that do not always have four angles with the same measure.

In Exercises 29–31, find the perimeter of the figure shown. Express the perimeter using the same unit of measure that appears in the figure.

29.

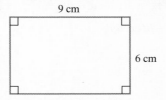

9 cm

6 cm

30.

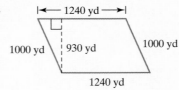

|← 1240 yd →|

1000 yd   930 yd   1000 yd

1240 yd

31.

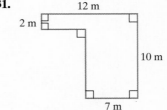

12 m

2 m

10 m

7 m

32. Find the sum of the measures of the angles of a 12-sided polygon.

33. Find the sum of the measures of the angles of an octagon.

34. The figure shown is a regular polygon. Find the measures of angle 1 and angle 2.

2 1

35. A carpenter is installing a baseboard around a room that has a length of 35 feet and a width of 15 feet. The room has four doorways and each doorway is 3 feet wide. If no baseboard is to be put across the doorways and the cost of the baseboard is $1.50 per foot, what is the cost of installing the baseboard around the room?

36. Use the following tessellation to solve this exercise.

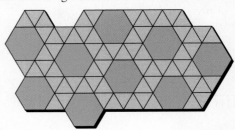

a. Name the types of regular polygons that surround each vertex.

b. Determine the number of angles that come together at each vertex, as well as the measures of these angles.

c. Use the angle measures from part (b) to explain why the tessellation is possible.

37. Can a tessellation be created using only regular hexagons? Explain your answer.

Section 10.4 Area and Circumference

In Exercises 38–41, find the area of each figure.

38.

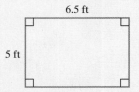

6.5 ft

5 ft

39.

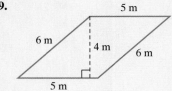

5 m
6 m   4 m   6 m
5 m

40.

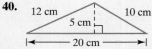

12 cm   10 cm
5 cm
← 20 cm →

41.

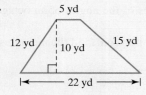

5 yd
12 yd   10 yd   15 yd
← 22 yd →

42. Find the circumference and the area of a circle with a diameter of 20 meters. Express answers in terms of π and then round to the nearest tenth.

In Exercises 43–44, find the area of each figure.

43.

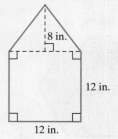

8 in.
12 in.
12 in.

44.

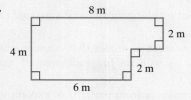

8 m
2 m
4 m
2 m
6 m

In Exercises 45–46, find the area of each shaded region. Where necessary, express answers in terms of π and then round to the nearest tenth.

45.

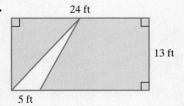

24 ft
13 ft
5 ft

46.

4 in.

47. What will it cost to carpet a rectangular floor measuring 15 feet by 21 feet if the carpet costs $22.50 per square yard?

48. What will it cost to cover a rectangular floor measuring 40 feet by 50 feet with square tiles that measure 2 feet on each side if a package of 10 tiles costs $13?

49. How much fencing, to the nearest whole yard, is needed to enclose a circular garden that measures 10 yards across?

Section 10.5 Volume

In Exercises 50–54, find the volume of each figure. Where necessary, express answers in terms of π and then round to the nearest whole number.

50.

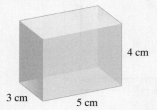

4 cm
3 cm   5 cm

51.

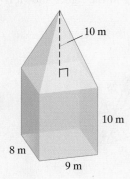

10 m
10 m
8 m
9 m

52.

8 yd
4 yd

53.

28 in.   40 in.

54.

6 m

55. Find the surface area of the figure shown.

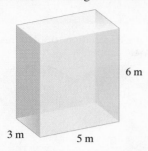

6 m

3 m   5 m

56. A train is being loaded with shipping boxes. Each box is 8 meters long, 4 meters wide, and 3 meters high. If there are 50 shipping boxes, how much space is needed?

57. An Egyptian pyramid has a square base measuring 145 meters on each side. If the height of the pyramid is 93 meters, find its volume.

58. What is the cost of concrete for a walkway that is 27 feet long, 4 feet wide, and 6 inches deep if the concrete is $40 per cubic yard?

Section 10.6 Right Triangle Trigonometry

59. Use the right triangle shown to find ratios, in reduced form, for sin A, cos A, and tan A.

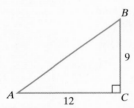

B

9

A   12   *C*

In Exercises 60–62, find the measure of the side of the right triangle whose length is designated by a lowercase letter. Round answers to the nearest whole number.

60.

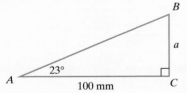

B

a

23°

A   100 mm   *C*

61.

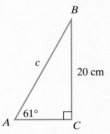

B

c   20 cm

61°

A   *C*

62.

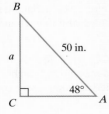

B

50 in.

a

48°

C   *A*

63. Find the measure of angle A in the right triangle shown. Round to the nearest whole degree.

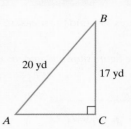

B

20 yd   17 yd

A   *C*

64. A hiker climbs for a half mile (2640 feet) up a slope whose inclination is 17°. How many feet of altitude, to the nearest foot, does the hiker gain?

65. To find the distance across a lake, a surveyor took the measurements in the figure shown. What is the distance across the lake? Round to the nearest meter.

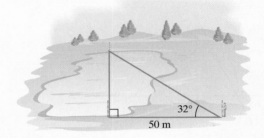

32°

50 m

66. When a six-foot pole casts a four-foot shadow, what is the angle of elevation of the sun? Round to the nearest whole degree.

Section 10.7 Distance and Midpoint Formulas; Circles

In Exercises 67–68, find the distance between each pair of points. If necessary, round answers to two decimal places.

67. $(-2, 3)$ and $(3, -9)$

68. $(-4, 3)$ and $(-2, 5)$

In Exercises 69–70, find the midpoint of each line segment with the given endpoints.

69. $(2, 6)$ and $(-12, 4)$

70. $(4, -6)$ and $(-15, 2)$

In Exercises 71–72, write the standard form of the equation of the circle with the given center and radius.

71. Center $(0, 0)$, $r = 3$

72. Center $(-2, 4)$, $r = 6$

In Exercises 73–74, give the center and radius of each circle and graph its equation.

73. $x^2 + y^2 = 1$

74. $(x + 2)^2 + (y - 3)^2 = 9$

Chapter 10 Test A

1. If an angle measures 54°, find the measure of its complement and supplement.

In Exercises 2–4, use the figure shown to find the measure of angle 1.

2.

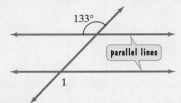

3.

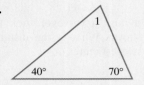

4.

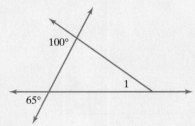

5. The triangles in the figure are similar. Find the length of the side marked with an *x*.

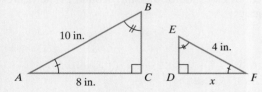

6. A vertical pole is to be supported by three wires. Each wire is 26 feet long and is anchored 24 feet from the base of the pole. How far up the pole should the wires be attached?

7. Find the sum of the measures of the angles of a ten-sided polygon.

8. Find the perimeter of the figure shown.

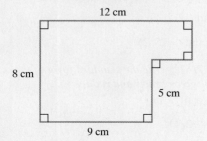

9. Which one of the following names a quadrilateral in which the sides that meet at each vertex have the same measure?

   a. rectangle
   b. parallelogram
   c. trapezoid
   d. rhombus

10. Use the following tessellation to solve this exercise.

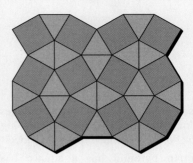

   a. Name the types of regular polygons that surround each vertex.

   b. Determine the number of angles that come together at each vertex, as well as the measures of these angles.

   c. Use the angle measures from part (b) to explain why the tessellation is possible.

In Exercises 11–12, find the area of each figure.

11.

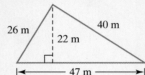

12.

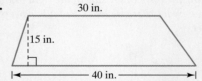

13. The right triangle shown has one leg of length 5 centimeters and a hypotenuse of length 13 centimeters.

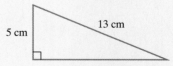

   a. Find the length of the other leg.

   b. What is the perimeter of the triangle?

   c. What is the area of the triangle?

14. Find the circumference and area of a circle with a diameter of 40 meters. Express answers in terms of π and then round to the nearest tenth.

15. A rectangular floor measuring 8 feet by 6 feet is to be completely covered with square tiles measuring 8 inches on each side. How many tiles are needed to completely cover the floor?

In Exercises 16–18, find the volume of each figure. If necessary, express the answer in terms of π and then round to the nearest whole number.

16.

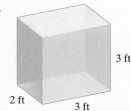

3 ft

2 ft   3 ft

17.

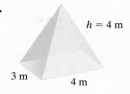

h = 4 m

3 m   4 m

18.

5 cm

7 cm

19. Find the measure, to the nearest whole number, of the side of the right triangle whose length is designated by *c*.

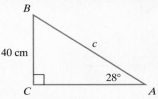

B

40 cm   c

C   28°   A

20. If a building casts a shadow measuring 104 feet when the angle of elevation of the sun is 34°, find the height of the building to the nearest foot.

21. Find the length and the midpoint of the line segment whose endpoints are $(2, -2)$ and $(5, 2)$

22. Write the standard form of the equation of the circle with center $(0, -2)$ and radius 5.

23. Give the center and radius of the circle whose standard equation is $(x - 2)^2 + (y + 1)^2 = 4$. Then graph the equation.

Chapter 10 Test B

1. If an angle measures 88°, find the measure of its complement and supplement.   ;

In Exercises 2–4, use the figure shown to find the measure of angle 1.

2.

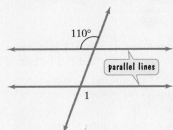

110°

parallel lines

1

3.

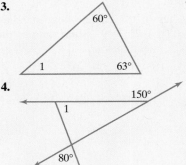

60°

1   63°

4.

150°

1

80°

5. The triangles in the figure are similar. Find the length of the side marked with an *x*.

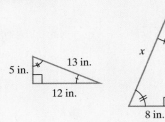

13 in.

5 in.

12 in.

x

8 in.

6. A vertical pole is to be supported by three wires. Each wire is 39 feet long and is anchored 36 feet from the base of the pole. How far up the pole should the wires be attached?

7. Find the sum of the measures of the angles of a 12-sided polygon.

8. Find the perimeter of the figure shown.

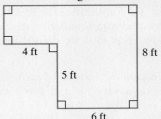

4 ft   8 ft

5 ft

6 ft

9. Which one of the following names a quadrilateral in which the angles have the same measure?
  a. rectangle   **b.** parallelogram
  c. trapezoid   **d.** rhombus

10. Use the following tessellation to solve this exercise.

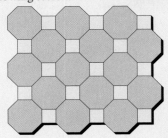

  a. Name the types of regular polygons that surround each vertex.
  b. Determine the number of angles that come together at each vertex, as well as the measures of these angles.
  c. Use the angle measures from part (b) to explain why the tessellation is possible.

In Exercises 11–12, find the area of each figure.

11.

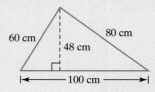

12.

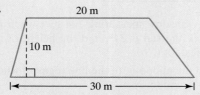

13. The right triangle shown has one leg of length 6 yards and a hypotenuse of length 10 yards.

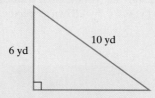

   a. Find the length of the other leg.
   b. What is the perimeter of the triangle?
   c. What is the area of the triangle?

14. Find the circumference and area of a circle with a diameter of 30 meters. Express answers in terms of π and then round to the nearest tenth.

15. A rectangular floor measuring 10 feet by 6 feet is to be completely covered with square tiles measuring 4 inches on each side. How many tiles are needed to completely cover the floor?

In Exercises 16–18, find the volume of each figure. If necessary, express the answer in terms of π and then round to the nearest whole number.

16.

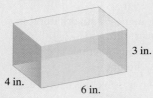

17.

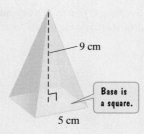

18.

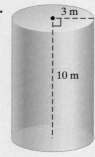

19. Find the measure, to the nearest whole number, of the side of the right triangle whose length is designated by c.

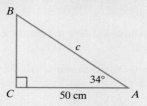

20. If a building casts a shadow measuring 110 feet when the angle of elevation of the sun is 26°, find the height of the building to the nearest foot.

21. Find the length and the midpoint of the line segment whose endpoints are $(4, -3)$ and $(7, 1)$.

22. Write the standard form of the equation of the circle with center $(-4, 0)$ and radius 6.

23. Give the center and radius of the circle whose standard equation is $(x - 3)^2 + (y + 2)^2 = 9$. Then graph the equation.

Counting Methods and Probability Theory

Two of America's best-loved presidents, Abraham Lincoln and John F. Kennedy, are linked by a bizarre series of coincidences:

- Lincoln was elected president in 1860. Kennedy was elected president in 1960.

- Lincoln's assassin, John Wilkes Booth, was born in 1839. Kennedy's assassin, Lee Harvey Oswald, was born in 1939.

- Lincoln's secretary, named Kennedy, warned him not to go to the theater on the night he was shot. Kennedy's secretary, named Lincoln, warned him not to go to Dallas on the day he was shot.

- Booth shot Lincoln in a theater and ran into a warehouse. Oswald shot Kennedy from a warehouse and ran into a theater.

- Both Lincoln and Kennedy were shot from behind, with their wives present.

- Andrew Johnson, who succeeded Lincoln, was born in 1808. Lyndon Johnson, who succeeded Kennedy, was born in 1908.

Source: Edward Burger and Michael Starbird, *Coincidences, Chaos, and All That Math Jazz*, W. W. Norton and Company, 2005.

Amazing coincidences? A cosmic conspiracy? Not really. In this chapter, you will see how the mathematics of uncertainty and risk, called probability theory, numerically describes situations in which to expect the unexpected. By assigning numbers to things that are extraordinarily unlikely, we can logically analyze coincidences without erroneous beliefs that strange and mystical events are occurring. We'll even see how wildly inaccurate our intuition can be about the likelihood of an event by examining an "amazing" coincidence that is nearly certain.

Coincidences are discussed in the Blitzer Bonus on page 728. Coincidences that are nearly certain are developed in Exercise 89 of Exercise Set 11.7.

11.1 : The Fundamental Counting Principle

Objective

1 Use the Fundamental Counting Principle to determine the number of possible outcomes in a given situation.

1 Use the Fundamental Counting Principle to determine the number of possible outcomes in a given situation.

Have you ever imagined what your life would be like if you won the lottery? What changes would you make? Before you fantasize about becoming a person of leisure with a staff of obedient elves, think about this: The probability of winning top prize in the lottery is about the same as the probability of being struck by lightning. There are millions of possible number combinations in lottery games and only one way of winning the grand prize. Determining the probability of winning involves calculating the chance of getting the winning combination from all possible outcomes. In this section, we begin preparing for the surprising world of probability by looking at methods for counting possible outcomes.

The Fundamental Counting Principle with Two Groups of Items

It's early morning, you're groggy, and you have to select something to wear for your 8 A.M. math class. Fortunately, your "school wardrobe" is rather limited—just two pairs of jeans to choose from (one blue, one black) and three T-shirts to choose from (one beige, one yellow, and one blue). Your early-morning dilemma is illustrated in **Figure 11.1**.

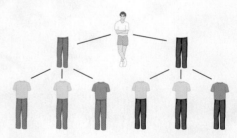

Figure 11.1 Selecting a wardrobe

The **tree diagram**, so named because of its branches, shows that you can form six different outfits from your two pairs of jeans and three T-shirts. Each pair of jeans can be combined with one of three T-shirts. Notice that the total number of outfits can be obtained by multiplying the number of choices for the jeans, 2, by the number of choices for the T-shirts, 3:

$$2 \cdot 3 = 6.$$

We can generalize this idea to any two groups of items—not just jeans and T-shirts—with the **Fundamental Counting Principle**.

The Fundamental Counting Principle

If you can choose one item from a group of M items and a second item from a group of N items, then the total number of two-item choices is $M \cdot N$.

Example 1 Applying the Fundamental Counting Principle

The Greasy Spoon Restaurant offers 6 appetizers and 14 main courses. In how many ways can a person order a two-course meal?

Solution

Choosing from one of 6 appetizers and one of 14 main courses, the total number of two-course meals is

$$6 \cdot 14 = 84.$$

A person can order a two-course meal in 84 different ways.

Checkpoint 1 A restaurant offers 10 appetizers and 15 main courses. In how many ways can you order a two-course meal?

Example 2 Applying the Fundamental Counting Principle

Suppose that you are taking two classes, one in psychology and one in social science, at your local community college. There are 15 sections of psychology from which you can choose. Furthermore, there are 9 sections of social science that are available at times that do not conflict with those for psychology. In how many ways can you create two-course schedules?

Solution

The number of ways that you can satisfy the requirement is found by multiplying the number of choices for each course. You can choose your psychology course from 15 sections and your social science course from 9 sections. For both courses you have

$$15 \cdot 9, \text{ or } 135$$

choices. Thus, you can satisfy the psychology–social science requirement in 135 ways.

Checkpoint 2 Rework Example 2 given that the number of sections of psychology and nonconflicting sections of social science each decrease by 5.

The Fundamental Counting Principle with More Than Two Groups of Items

Whoops! You forgot something in choosing your lecture wardrobe—shoes! You have two pairs of sneakers to choose from—one black and one red, for that extra fashion flair! Your possible outfits including sneakers are shown in **Figure 11.2**.

The number of possible ways of playing the first four moves on each side in a game of chess is 318,979,564,000.

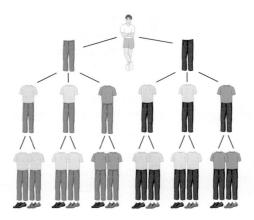

Figure 11.2
Increasing wardrobe selections

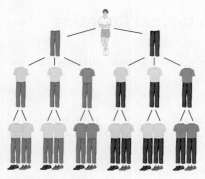

Figure 11.2 (repeated)

The tree diagram shows that you can form 12 outfits from your two pairs of jeans, three T-shirts, and two pairs of sneakers. Notice that the number of outfits can be obtained by multiplying the number of choices for jeans, 2, the number of choices for T-shirts, 3, and the number of choices for sneakers, 2:

$$2 \cdot 3 \cdot 2 = 12.$$

Unlike your earlier dilemma, you are now dealing with *three* groups of items. The Fundamental Counting Principle can be extended to determine the number of possible outcomes in situations in which there are three or more groups of items.

> **The Fundamental Counting Principle**
>
> The number of ways in which a series of successive things can occur is found by multiplying the number of ways in which each thing can occur.

For example, if you own 30 pairs of jeans, 20 T-shirts, and 12 pairs of sneakers, you have

$$30 \cdot 20 \cdot 12 = 7200$$

choices for your wardrobe.

Example 3   Options in Planning a College Course Schedule

Suppose that you are planning to take three college courses—math, English, and humanities. Based on time blocks and highly recommended professors, there are 8 sections of math, 5 of English, and 4 of humanities that you find suitable. Assuming no scheduling conflicts, how many different three-course schedules are possible?

Solution

This situation involves making choices with three groups of items.

Math	English	Humanities
8 choices	5 choices	4 choices

We use the Fundamental Counting Principle to find the number of three-course schedules. Multiply the number of choices for each of the three groups.

$$8 \cdot 5 \cdot 4 = 160$$

Thus, there are 160 different three-course schedules.

Checkpoint 3   A pizza can be ordered with two choices of size (medium or large), three choices of crust (thin, thick, or regular), and five choices of toppings (ground beef, sausage, pepperoni, bacon, or mushrooms). How many different one-topping pizzas can be ordered?

Example 4   Car of the Future

Car manufacturers are now experimenting with lightweight three-wheel cars, designed for one person, and considered ideal for city driving. Intrigued? Suppose you could order such a car with a choice of 9 possible colors, with or without air conditioning, electric or gas powered, and with or without an onboard computer. In how many ways can this car be ordered with regard to these options?

Solution

This situation involves making choices with four groups of items.

Color	Air conditioning	Power	Computer
9 choices	2 choices: with or without	2 choices: electric or gas	2 choices: with or without

We use the Fundamental Counting Principle to find the number of ordering options. Multiply the number of choices for each of the four groups.

$$9 \cdot 2 \cdot 2 \cdot 2 = 72$$

Thus, the car can be ordered in 72 different ways.

✓ **Checkpoint 4** The car in Example 4 is now available in 10 possible colors. The options involving air conditioning, power, and an onboard computer still apply. Furthermore, the car is available with or without a global positioning system (for pinpointing your location at every moment). In how many ways can this car be ordered in terms of these options?

Example 5 A Multiple-Choice Test

You are taking a multiple-choice test that has ten questions. Each of the questions has four answer choices, with one correct answer per question. If you select one of these four choices for each question and leave nothing blank, in how many ways can you answer the questions?

Solution

This situation involves making choices with ten questions.

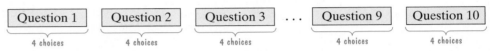

| Question 1 | Question 2 | Question 3 | ... | Question 9 | Question 10 |
| 4 choices | 4 choices | 4 choices | | 4 choices | 4 choices |

We use the Fundamental Counting Principle to determine the number of ways that you can answer the questions on the test. Multiply the number of choices, 4, for each of the ten questions.

$$4 \cdot 4 \cdot 4 \cdot 4 \cdot 4 \cdot 4 \cdot 4 \cdot 4 \cdot 4 \cdot 4 = 4^{10} = 1,048,576 \quad \text{Use a calculator: 4 } \boxed{y^x} \text{ 10 } \boxed{=}.$$

Thus, you can answer the questions in 1,048,576 different ways.

Are you surprised that there are over one million ways of answering a ten-question multiple-choice test? Of course, there is only one way to answer the test and receive a perfect score. The probability of guessing your way into a perfect score involves calculating the chance of getting a perfect score, just one way, from all 1,048,576 possible outcomes. In short, prepare for the test and do not rely on guessing!

✓ **Checkpoint 5** You are taking a multiple-choice test that has six questions. Each of the questions has three answer choices, with one correct answer per question. If you select one of these three choices for each question and leave nothing blank, in how many ways can you answer the questions?

Example 6 Telephone Numbers in the United States

Telephone numbers in the United States begin with three-digit area codes followed by seven-digit local telephone numbers. Area codes and local telephone numbers cannot begin with 0 or 1. How many different telephone numbers are possible?

Solution

This situation involves making choices with ten groups of items.

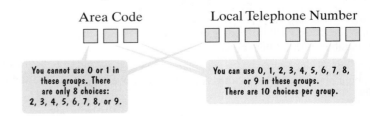

Area Code   Local Telephone Number

You cannot use 0 or 1 in these groups. There are only 8 choices: 2, 3, 4, 5, 6, 7, 8, or 9.

You can use 0, 1, 2, 3, 4, 5, 6, 7, 8, or 9 in these groups. There are 10 choices per group.

Blitzer Bonus

Running Out of Telephone Numbers

By the year 2020, portable telephones used for business and pleasure will all be videophones. At that time, U.S. population is expected to be 323 million. Faxes, beepers, cellphones, computer phone lines, and business lines may result in certain areas running out of phone numbers. Solution: Add more digits!

With or without extra digits, we expect that the 2020 videophone greeting will still be "hello," a word created by Thomas Edison in 1877. Phone inventor Alexander Graham Bell preferred "ahoy," but "hello" won out, appearing in the *Oxford English Dictionary* in 1883.

(*Source: New York Times*)

Here are the numbers of choices for each of the ten groups of items:

Area Code | Local Telephone Number
8 10 10 | 8 10 10   10 10 10 10

We use the Fundamental Counting Principle to determine the number of different telephone numbers that are possible. The total number of telephone numbers possible is

$$8 \cdot 10 \cdot 10 \cdot 8 \cdot 10 \cdot 10 \cdot 10 \cdot 10 \cdot 10 \cdot 10 = 6,400,000,000.$$

There are six billion four hundred million different telephone numbers that are possible.

✔ **Checkpoint 6**   An electronic gate can be opened by entering five digits on a keypad containing the digits $0, 1, 2, 3, \ldots, 8, 9$. How many different keypad sequences are possible if the digit 0 cannot be used as the first digit?

Achieving Success

Don't overburden yourself in your freshman year in college.

Because you will be bombarded with new acquaintances, new activities, and new responsibilities, it's not a good idea to overburden yourself with a gigantic course load.

Example 3 on page 676 involved taking three college courses—math, English, and humanities. If you plan to be a full-time student, forget about such a light class load. Perhaps the best idea in each semester of your freshman year is to take an average course load, which consists of 15 units or credits, at most colleges.

You will have less class time in college than you do in high school. The fact that you are taking 25 hours of classes per week in high school does not mean that you will be slacking off by taking 15 hours per week in your freshman year.

Exercise Set 11.1

Concept and Vocabulary Exercises

In Exercises 1–2, fill in each blank so that the resulting statement is true.

1. If you can choose one item from a group of M items and a second item from a group of N items, then the total number of two-item choices is _____.

2. The number of ways in which a series of successive things can occur is found by _____ the number of ways in which each thing can occur. This is called the _____ Principle.

In Exercises 3–4, determine whether each statement is true or false. If the statement is false, make the necessary change(s) to produce a true statement.

3. If one item is chosen from M items, a second item is chosen from N items, and a third item is chosen from P items, the total number of three-item choices is $M + N + P$.

4. Regardless of the United States population, we will not run out of telephone numbers as long as we continue to add new digits.

Respond to Exercises 5–6 using verbal or written explanations.

5. Explain the Fundamental Counting Principle.

6. **Figure 11.2** on page 675 shows that a tree diagram can be used to find the total number of outfits. Describe one advantage of using the Fundamental Counting Principle rather than a tree diagram.

In Exercises 7–10, determine whether each statement makes sense or does not make sense, and explain your reasoning.

7. I used the Fundamental Counting Principle to determine the number of five-digit ZIP codes that are available to the U.S. Postal Service.

8. The Fundamental Counting Principle can be used to determine the number of ways of arranging the numbers $1, 2, 3, 4, 5, \ldots, 98, 99, 100$.

9. I estimate there are approximately 10,000 ways to arrange the letters $A, B, C, D, E, \ldots, X, Y, Z$.

10. The statement in the last frame of this cartoon is highly unlikely because there are 2048 possible sets of answers on an eleven-item true/false test.

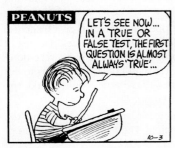

Peanuts: © United Feature Syndicate, Inc.

Practice and Application Exercises

11. A restaurant offers 8 appetizers and 10 main courses. In how many ways can a person order a two-course meal?

12. The model of the car you are thinking of buying is available in nine different colors and three different styles (hatchback, sedan, or station wagon). In how many ways can you order the car?

13. A popular brand of pen is available in three colors (red, green, or blue) and four writing tips (bold, medium, fine, or micro). How many different choices of pens do you have with this brand?

14. In how many ways can a casting director choose a female lead and a male lead from five female actors and six male actors?

15. A family is planning a two-part trip. The first leg of the trip is from San Francisco to New York, and the second leg is from New York to Paris. From San Francisco to New York, travel options include airplane, train, or bus. From New York to Paris, the options are limited to airplane or ship. In how many ways can the two-part trip be made?

16. For a summer job, you are painting the parking spaces for a new shopping mall with a letter of the alphabet and a single digit from 1 to 9. The first parking space is A1 and the last parking space is Z9. How many parking spaces can you paint with distinct labels?

17. An ice cream store sells two drinks (sodas or milk shakes), in four sizes (small, medium, large, or jumbo), and five flavors (vanilla, strawberry, chocolate, coffee, or pistachio). In how many ways can a customer order a drink?

18. A pizza can be ordered with three choices of size (small, medium, or large), four choices of crust (thin, thick, crispy, or regular), and six choices of toppings (ground beef, sausage, pepperoni, bacon, mushrooms, or onions). How many one-topping pizzas can be ordered?

19. A restaurant offers the following limited lunch menu.

Main Course	Vegetables	Beverages	Desserts
Ham	Potatoes	Coffee	Cake
Chicken	Peas	Tea	Pie
Fish	Green beans	Milk	Ice cream
Beef		Soda	

If one item is selected from each of the four groups, in how many ways can a meal be ordered? Describe two such orders.

20. An apartment complex offers apartments with four different options, designated by A through D.

A	B	C	D
one bedroom	one bathroom	first floor	lake view
two bedrooms	two bathrooms	second floor	golf course view
three bedrooms			no special view

How many apartment options are available? Describe two such options.

21. Shoppers in a large shopping mall are categorized as male or female, over 30 or 30 and under, and cash or credit card shoppers. In how many ways can the shoppers be categorized?

22. There are three highways from city A to city B, two highways from city B to city C, and four highways from city C to city D. How many different highway routes are there from city A to city D?

23. A person can order a new car with a choice of six possible colors, with or without air conditioning, with or without automatic transmission, with or without power windows, and with or without a CD player. In how many different ways can a new car be ordered with regard to these options?

24. A car model comes in nine colors, with or without air conditioning, with or without a sun roof, with or without automatic transmission, and with or without antilock brakes. In how many ways can the car be ordered with regard to these options?

25. You are taking a multiple-choice test that has five questions. Each of the questions has three answer choices, with one correct answer per question. If you select one of these three choices for each question and leave nothing blank, in how many ways can you answer the questions?

26. You are taking a multiple-choice test that has eight questions. Each of the questions has three answer choices, with one correct answer per question. If you select one of these three choices for each question and leave nothing blank, in how many ways can you answer the questions?

27. In the original plan for area codes in 1945, the first digit could be any number from 2 through 9, the second digit was either 0 or 1, and the third digit could be any number except 0. With this plan, how many different area codes are possible?

28. The local seven-digit telephone numbers in Inverness, California, have 669 as the first three digits. How many different telephone numbers are possible in Inverness?

29. License plates in a particular state display two letters followed by three numbers, such as AT-887 or BB-013. How many different license plates can be manufactured for this state?

30. How many different four-letter radio station call letters can be formed if the first letter must be W or K?

31. A stock can go up, go down, or stay unchanged. How many possibilities are there if you own seven stocks?

32. A social security number contains nine digits, such as 074-66-7795. How many different social security numbers can be formed?

Critical Thinking Exercises

33. How many four-digit odd numbers are there? Assume that the digit on the left cannot be 0.

34. In order to develop a more appealing hamburger, a franchise uses taste tests with 12 different buns, 30 sauces, 4 types of lettuce, and 3 types of tomatoes. If the taste tests were done at one restaurant by one tester who takes 10 minutes to eat each hamburger, approximately how long would it take the tester to eat all possible hamburgers?

35. Write an original problem that can be solved using the Fundamental Counting Principle. Then solve the problem.

11.2 Permutations

Objectives

1. Use the Fundamental Counting Principle to count permutations.

2. Evaluate factorial expressions.

3. Use the permutations formula.

4. Find the number of permutations of duplicate items.

We open this section with six jokes about books. (Stay with us on this one.)

- *"Outside of a dog, a book is man's best friend. Inside of a dog, it's too dark to read."—Groucho Marx*

- *"I recently bought a book of free verse. For $12."—George Carlin*

- *"If a word in the dictionary was misspelled, how would we know?"—Steven Wright*

- *"I wrote a book under a pen name: Bic."'—Henny Youngman*

- *"A bit of advice: Never read a pop-up book about giraffes."—Jerry Seinfeld*

- *"Fang just got out of the hospital. He was in a speed-reading contest. He hit a bookmark."—Phyllis Diller*

"Read the book!"   *"See the movie!"*

© The New Yorker Collection 1992 Arnie Levin from cartoonbank.com. All Rights Reserved.

1 Use the Fundamental Counting Principle to count permutations.

We can use the Fundamental Counting Principle to determine the number of ways these jokes can be delivered. You can choose any one of the six jokes as the first one told. Once this joke is delivered, you'll have five jokes to choose from for the second joke told. You'll then have four jokes left to choose from for the third delivery. Continuing in this manner, the situation can be shown as follows:

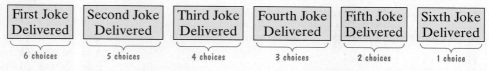

First Joke Delivered	Second Joke Delivered	Third Joke Delivered	Fourth Joke Delivered	Fifth Joke Delivered	Sixth Joke Delivered
6 choices	5 choices	4 choices	3 choices	2 choices	1 choice

Using the Fundamental Counting Principle, we multiply the choices:

$$6 \cdot 5 \cdot 4 \cdot 3 \cdot 2 \cdot 1 = 720.$$

Thus, there are 720 different ways to deliver the six jokes about books. One of the 720 possible arrangements is

Seinfeld joke—Youngman joke—Carlin joke—Marx joke—Wright joke—Diller joke.

Such an ordered arrangement is called a *permutation* of the six jokes.

A **permutation** is an ordered arrangement of items that occurs when

- No item is used more than once. (Each joke is told exactly once.)
- The order of arrangement makes a difference. (The order in which these jokes are told makes a difference in terms of how they are received.)

Example 1 · Counting Permutations

How many ways can the six jokes about books be delivered if George Carlin's joke is delivered first and Jerry Seinfeld's joke is told last?

Solution

The conditions of Carlin's joke first and Seinfeld's joke last can be shown as follows:

First Joke Delivered	Second Joke Delivered	Third Joke Delivered	Fourth Joke Delivered	Fifth Joke Delivered	Sixth Joke Delivered
1 choice: Carlin					1 choice: Seinfeld

Now let's fill in the number of choices for positions two through five. You can choose any one of the four remaining jokes for the second delivery. Once you've chosen this joke, you'll have three jokes to choose from for the third delivery. This leaves only two jokes to choose from for the fourth delivery. Once this choice is made, there is just one joke left to choose for the fifth delivery.

First Joke Delivered	Second Joke Delivered	Third Joke Delivered	Fourth Joke Delivered	Fifth Joke Delivered	Sixth Joke Delivered
1 choice: Carlin	4 choices	3 choices	2 choices	1 choice	1 choice: Seinfeld

We use the Fundamental Counting Principle to find the number of ways the six jokes can be delivered. Multiply the choices:

$$1 \cdot 4 \cdot 3 \cdot 2 \cdot 1 \cdot 1 = 24.$$

Thus, there are 24 different ways the jokes can be delivered if Carlin's joke is told first and Seinfeld's joke is told last.

✓ **Checkpoint 1**   How many ways can the six jokes about books be delivered if a man's joke is told first?

Example 2 · Counting Permutations

You need to arrange seven of your favorite books along a small shelf. How many different ways can you arrange the books, assuming that the order of the books makes a difference to you?

Solution

You may choose any of the seven books for the first position on the shelf. This leaves six choices for second position. After the first two positions are filled, there are five books to choose from for third position, four choices left for the fourth position, three choices left for the fifth position, then two choices for the sixth position, and only one choice for the last position. This situation can be shown as follows:

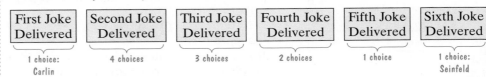

First Shelf Position	Second Shelf Position	Third Shelf Position	Fourth Shelf Position	Fifth Shelf Position	Sixth Shelf Position	Seventh Shelf Position
7 choices	6 choices	5 choices	4 choices	3 choices	2 choices	1 choice

We use the Fundamental Counting Principle to find the number of ways you can arrange the seven books along the shelf. Multiply the choices:

$$7 \cdot 6 \cdot 5 \cdot 4 \cdot 3 \cdot 2 \cdot 1 = 5040.$$

Thus, you can arrange the books in 5040 ways. There are 5040 different possible permutations.

> ✔ **Checkpoint 2**   In how many ways can you arrange five books along a shelf, assuming that the order of the books makes a difference?

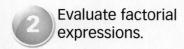

Evaluate factorial expressions.

Factorial Notation

The product in Example 2,

$$7 \cdot 6 \cdot 5 \cdot 4 \cdot 3 \cdot 2 \cdot 1$$

is given a special name and symbol. It is called 7 **factorial**, and written 7!. Thus,

$$7! = 7 \cdot 6 \cdot 5 \cdot 4 \cdot 3 \cdot 2 \cdot 1.$$

In general, if n is a positive integer, then $n!$ (n *factorial*) is the product of all positive integers from n down through 1. For example,

$$1! = 1$$
$$2! = 2 \cdot 1 = 2$$
$$3! = 3 \cdot 2 \cdot 1 = 6$$
$$4! = 4 \cdot 3 \cdot 2 \cdot 1 = 24$$
$$5! = 5 \cdot 4 \cdot 3 \cdot 2 \cdot 1 = 120$$
$$6! = 6 \cdot 5 \cdot 4 \cdot 3 \cdot 2 \cdot 1 = 720.$$

Factorials: 0! Through 20!

0!	1
1!	1
2!	2
3!	6
4!	24
5!	120
6!	720
7!	5040
8!	40,320
9!	362,880
10!	3,628,800
11!	39,916,800
12!	479,001,600
13!	6,227,020,800
14!	87,178,291,200
15!	1,307,674,368,000
16!	20,922,789,888,000
17!	355,687,428,096,000
18!	6,402,373,705,728,000
19!	121,645,100,408,832,000
20!	2,432,902,008,176,640,000

> **Factorial Notation**
>
> If n is a positive integer, the notation $n!$ (read "n factorial") is the product of all positive integers from n down through 1.
> $$n! = n(n-1)(n-2)\cdots(3)(2)(1)$$
> 0! (zero factorial), by definition, is 1.
> $$0! = 1$$

Example 3   Using Factorial Notation

Evaluate the following factorial expressions without using the factorial key on your calculator:

a. $\dfrac{8!}{5!}$ **b.** $\dfrac{26!}{21!}$ **c.** $\dfrac{500!}{499!}.$

Solution

a. We can evaluate the numerator and the denominator of $\frac{8!}{5!}$. However, it is easier to use the following simplification:

$$\frac{8!}{5!} = \frac{8 \cdot 7 \cdot 6 \cdot \boxed{5 \cdot 4 \cdot 3 \cdot 2 \cdot 1}}{\boxed{5 \cdot 4 \cdot 3 \cdot 2 \cdot 1}} = \frac{8 \cdot 7 \cdot 6 \cdot \boxed{5!}}{\boxed{5!}} = \frac{8 \cdot 7 \cdot 6 \cdot \cancel{5!}}{\cancel{5!}} = 8 \cdot 7 \cdot 6 = 336.$$

b. Rather than write out 26!, the numerator of $\frac{26!}{21!}$, as the product of all integers from 26 down to 1, we can express 26! as

$$26! = 26 \cdot 25 \cdot 24 \cdot 23 \cdot 22 \cdot 21!.$$

In this way, we can cancel 21! in the numerator and the denominator of the given expression.

$$\frac{26!}{21!} = \frac{26 \cdot 25 \cdot 24 \cdot 23 \cdot 22 \cdot 21!}{21!} = \frac{26 \cdot 25 \cdot 24 \cdot 23 \cdot 22 \cdot \cancel{21!}}{\cancel{21!}}$$
$$= 26 \cdot 25 \cdot 24 \cdot 23 \cdot 22 = 7{,}893{,}600$$

c. In order to cancel identical factorials in the numerator and the denominator of $\frac{500!}{499!}$, we can express 500! as $500 \cdot 499!$.

$$\frac{500!}{499!} = \frac{500 \cdot 499!}{499!} = \frac{500 \cdot \cancel{499!}}{\cancel{499!}} = 500$$

> ✓ **Checkpoint 3** Evaluate without using a calculator's factorial key:
>
> **a.** $\dfrac{9!}{6!}$ **b.** $\dfrac{16!}{11!}$ **c.** $\dfrac{100!}{99!}$.

A Formula for Permutations

Suppose that you are the coach of a little league baseball team. There are 13 players on the team (and lots of parents hovering in the background, dreaming of stardom for their little "Derek Jeter"). You need to choose a batting order having 9 players. The order makes a difference, because, for instance, if bases are loaded and "Little Derek" is fourth or fifth at bat, his possible home run will drive in three additional runs. How many batting orders can you form?

You can choose any of 13 players for the first person at bat. Then you will have 12 players from which to choose the second batter, then 11 from which to choose the third batter, and so on. The situation can be shown as follows:

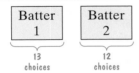
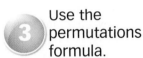
3 Use the permutations formula.

The total number of batting orders is

$$13 \cdot 12 \cdot 11 \cdot 10 \cdot 9 \cdot 8 \cdot 7 \cdot 6 \cdot 5 = 259{,}459{,}200.$$

Nearly 260 million batting orders are possible for your 13-player little league team. Each batting order is a permutation because the order of the batters makes a difference. The number of permutations of 13 players taken 9 at a time is 259,459,200.

We can obtain a formula for finding the number of permutations by rewriting our computation:

$13 \cdot 12 \cdot 11 \cdot 10 \cdot 9 \cdot 8 \cdot 7 \cdot 6 \cdot 5$

$$= \frac{13 \cdot 12 \cdot 11 \cdot 10 \cdot 9 \cdot 8 \cdot 7 \cdot 6 \cdot 5 \cdot \boxed{4 \cdot 3 \cdot 2 \cdot 1}}{\boxed{4 \cdot 3 \cdot 2 \cdot 1}} = \frac{13!}{4!} = \frac{13!}{(13-9)!}.$$

Thus, the number of permutations of 13 things taken 9 at a time is $\frac{13!}{(13-9)!}$. The special notation $_{13}P_9$ is used to replace the phrase "the number of permutations of 13 things taken 9 at a time." Using this new notation, we can write

$$_{13}P_9 = \frac{13!}{(13-9)!}.$$

The numerator of this expression is the factorial of the number of items, 13 team members: 13!. The denominator is also a factorial. It is the factorial of the difference between the number of items, 13, and the number of items in each permutation, 9 batters: $(13-9)!$.

The notation $_nP_r$ means the **number of permutations of n things taken r at a time**. We can generalize from the situation in which 9 batters were taken from 13 players. By generalizing, we obtain the following formula for the number of permutations if r items are taken from n items:

> **Permutations of n Things Taken r at a Time**
>
> The number of possible permutations if r items are taken from n items is
>
> $$_nP_r = \frac{n!}{(n-r)!}.$$

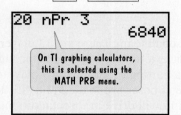
Example 4 Using the Formula for Permutations

Suppose that you and 19 of your friends have decided to form an Internet marketing consulting firm. The group needs to choose three officers—a CEO, an operating manager, and a treasurer. In how many ways can those offices be filled?

Solution

Your group is choosing $r = 3$ officers from a group of $n = 20$ people (you and 19 friends). The order in which the officers are chosen matters because the CEO, the operating manager, and the treasurer each have different responsibilities. Thus, we are looking for the number of permutations of 20 things taken 3 at a time. We use the formula

$$ {}_nP_r = \frac{n!}{(n-r)!} $$

with $n = 20$ and $r = 3$.

$$ {}_{20}P_3 = \frac{20!}{(20-3)!} = \frac{20!}{17!} = \frac{20 \cdot 19 \cdot 18 \cdot 17!}{17!} = \frac{20 \cdot 19 \cdot 18 \cdot \cancel{17!}}{\cancel{17!}} = 20 \cdot 19 \cdot 18 = 6840 $$

Thus, there are 6840 different ways of filling the three offices.

✓ **Checkpoint 4** A corporation has seven members on its board of directors. In how many different ways can it elect a president, vice-president, secretary, and treasurer?

Example 5 Using the Formula for Permutations

Suppose that you are working for The Sitcom Television Network. Your assignment is to help set up the television schedule for Monday evenings between 7 and 10 P.M. You need to schedule a show in each of six 30-minute time blocks, beginning with 7 to 7:30 and ending with 9:30 to 10:00. You can select from among the following situation comedies: *Home Improvement, Seinfeld, Mad About You, Cheers, Friends, Frasier, All in the Family, I Love Lucy, M*A*S*H, The Larry Sanders Show, The Jeffersons, Married with Children,* and *Happy Days.* How many different programming schedules can be arranged?

Solution

You are choosing $r = 6$ situation comedies from a collection of $n = 13$ classic sitcoms. The order in which the programs are aired matters. Family-oriented comedies have higher ratings when aired in earlier time blocks, such as 7 to 7:30. By contrast, comedies with adult themes do better in later time blocks. In short, we are looking for the number of permutations of 13 things taken 6 at a time. We use the formula

$$ {}_nP_r = \frac{n!}{(n-r)!} $$

with $n = 13$ and $r = 6$.

$$ {}_{13}P_6 = \frac{13!}{(13-6)!} = \frac{13!}{7!} = \frac{13 \cdot 12 \cdot 11 \cdot 10 \cdot 9 \cdot 8 \cdot \cancel{7!}}{\cancel{7!}} = 13 \cdot 12 \cdot 11 \cdot 10 \cdot 9 \cdot 8 = 1{,}235{,}520 $$

There are 1,235,520 different programming schedules that can be arranged.

✓ **Checkpoint 5** How many different programming schedules can be arranged by choosing 5 situation comedies from a collection of 9 classic sitcoms?

Permutations of Duplicate Items

4 Find the number of permutations of duplicate items.

The number of permutations of the letters in the word SET is 3!, or 6. The six permutations are

SET, STE, EST, ETS, TES, TSE.

Are there also six permutations of the letters in the name ANA? The answer is no. Unlike SET, with three distinct letters, ANA contains three letters, of which the two As are duplicates. If we rearrange the letters just as we did with SET, we obtain

ANA, AAN, NAA, NAA, ANA, AAN.

Without the use of color to distinguish between the two As, there are only three distinct permutations: ANA, AAN, NAA.

There is a formula for finding the number of distinct permutations when duplicate items exist:

Permutations of Duplicate Items

The number of permutations of n items, where p items are identical, q items are identical, r items are identical, and so on, is given by

$$\frac{n!}{p!\, q!\, r! \dots}.$$

For example, ANA contains three letters ($n = 3$), where two of the letters are identical ($p = 2$). The number of distinct permutations is

$$\frac{n!}{p!} = \frac{3!}{2!} = \frac{3 \cdot 2!}{2!} = 3.$$

We saw that the three distinct permutations are ANA, AAN, and NAA.

Technology

Parentheses are necessary to enclose the factorials in the denominator when using a calculator to find

$$\frac{11!}{4!\, 4!\, 2!}.$$

```
11!/(4!4!2!)
        34650
```

Example 6 — Using the Formula for Permutations of Duplicate Items

In how many distinct ways can the letters of the word MISSISSIPPI be arranged?

Solution

The word contains 11 letters ($n = 11$), where four Is are identical ($p = 4$), four Ss are identical ($q = 4$), and 2 Ps are identical ($r = 2$). The number of distinct permutations is

$$\frac{n!}{p!\, q!\, r!} = \frac{11!}{4!\, 4!\, 2!} = \frac{11 \cdot 10 \cdot 9 \cdot 8 \cdot 7 \cdot 6 \cdot 5 \cdot 4!}{4! \cdot 4 \cdot 3 \cdot 2 \cdot 1 \cdot 2 \cdot 1} = 34{,}650$$

There are 34,650 distinct ways the letters in the word MISSISSIPPI can be arranged.

✓ **Checkpoint 6** In how many ways can the letters of the word OSMOSIS be arranged?

Exercise Set 11.2

Concept and Vocabulary Exercises

In Exercises 1–3, fill in each blank so that the resulting statement is true.

1. 5!, called 5 _____, is the product of all positive integers from _____ down through _____. By definition, 0! = _____.

2. The number of possible permutations if r objects are taken from n items is $_nP_r$ = _____.

3. The number of permutations of n items, where p items are identical and q items are identical, is given by _____.

In Exercises 4–7, determine whether each statement is true or false. If the statement is false, make the necessary change(s) to produce a true statement.

4. A permutation occurs when the order of arrangement does not matter.

5. $8! = 8 + 7 + 6 + 5 + 4 + 3 + 2 + 1$

6. Because all permutation problems are also Fundamental Counting problems, they can be solved using the formula for $_nP_r$ or using the Fundamental Counting Principle.

7. Because the words BET and BEE both contain three letters, the number of permutations of the letters in each word is 3!, or 6.

Respond to Exercises 8–12 using verbal or written explanations.

8. What is a permutation?

9. Explain how to find $n!$, where n is a positive integer.

10. Explain the best way to evaluate $\frac{900!}{899!}$ without a calculator.

11. Describe what $_nP_r$ represents.

12. If 24 permutations can be formed using the letters in the word BAKE, why can't 24 permutations also be formed using the letters in the word BABE? How is the number of permutations in BABE determined?

In Exercises 13–16, determine whether each statement makes sense or does not make sense, and explain your reasoning.

13. I used the formula for $_nP_r$ to determine the number of ways the manager of a baseball team can form a 9-player batting order from a team of 25 players.

14. I used the Fundamental Counting Principle to determine the number of permutations of the letters of the word ENGLISH.

15. I used the formula for $_nP_r$ to determine the number of ways people can select their 9 favorite baseball players from a team of 25 players.

16. I used the Fundamental Counting Principle to determine the number of permutations of the letters of the word SUCCESS.

Practice and Application Exercises

Use the Fundamental Counting Principle to solve Exercises 17–28.

17. Six performers are to present their comedy acts on a weekend evening at a comedy club. How many different ways are there to schedule their appearances?

18. Five singers are to perform on a weekend evening at a night club. How many different ways are there to schedule their appearances?

19. In the *Cambridge Encyclopedia of Language* (Cambridge University Press, 1987), author David Crystal presents five sentences that make a reasonable paragraph regardless of their order. The sentences are as follows:

   • Mark had told him about the foxes.
   • John looked out of the window.
   • Could it be a fox?
   • However, nobody had seen one for months.
   • He thought he saw a shape in the bushes.

   In how many different orders can the five sentences be arranged?

20. In how many different ways can a police department arrange eight suspects in a police lineup if each lineup contains all eight people?

21. As in Exercise 17, six performers are to present their comedy acts on a weekend evening at a comedy club. One of the performers insists on being the last stand-up comic of the evening. If this performer's request is granted, how many different ways are there to schedule the appearances?

22. As in Exercise 18, five singers are to perform at a night club. One of the singers insists on being the last performer of the evening. If this singer's request is granted, how many different ways are there to schedule the appearances?

23. You need to arrange nine of your favorite books along a small shelf. How many different ways can you arrange the books, assuming that the order of the books makes a difference to you?

24. You need to arrange ten of your favorite photographs on the mantle above a fireplace. How many ways can you arrange the photographs, assuming that the order of the pictures makes a difference to you?

In Exercises 25–26, use the five sentences that are given in Exercise 19.

25. How many different five-sentence paragraphs can be formed if the paragraph begins with "He thought he saw a shape in the bushes" and ends with "John looked out of the window"?

26. How many different five-sentence paragraphs can be formed if the paragraph begins with "He thought he saw a shape in the bushes" followed by "Mark had told him about the foxes"?

27. A television programmer is arranging the order that five movies will be seen between the hours of 6 P.M. and 4 A.M. Two of the movies have a G rating, and they are to be shown in the first two time blocks. One of the movies is rated NC-17, and it is to be shown in the last of the time blocks, from 2 A.M. until 4 A.M. Given these restrictions, in how many ways can the five movies be arranged during the indicated time blocks?

28. A camp counselor and six campers are to be seated along a picnic bench. In how many ways can this be done if the counselor must be seated in the middle and a camper who has a tendency to engage in food fights must sit to the counselor's immediate left?

In Exercises 29–48, evaluate each factorial expression.

29. $\frac{9!}{6!}$

30. $\frac{12!}{10!}$

31. $\frac{29!}{25!}$

32. $\frac{31!}{28!}$

33. $\frac{19!}{11!}$

34. $\frac{17!}{9!}$

35. $\frac{600!}{599!}$

36. $\frac{700!}{699!}$

37. $\frac{104!}{102!}$

38. $\frac{106!}{104!}$

39. $7! - 3!$

40. $6! - 3!$

41. $(7 - 3)!$

42. $(6 - 3)!$

43. $\left(\frac{12}{4}\right)!$

44. $\left(\frac{45}{9}\right)!$

45. $\frac{7!}{(7 - 2)!}$

46. $\frac{8!}{(8 - 5)!}$

47. $\frac{13!}{(13 - 3)!}$

48. $\frac{17!}{(17 - 3)!}$

In Exercises 49–56, use the formula for $_nP_r$ to evaluate each expression.

49. $_9P_4$

50. $_7P_3$

51. $_8P_5$

52. $_{10}P_4$

53. $_6P_6$

54. $_9P_9$

55. $_8P_0$

56. $_6P_0$

Use an appropriate permutations formula to solve Exercises 57–72.

57. A club with ten members is to choose three officers—president, vice-president, and secretary-treasurer. If each office is to be held by one person and no person can hold more than one office, in how many ways can those offices be filled?

58. A corporation has six members on its board of directors. In how many different ways can it elect a president, vice-president, secretary, and treasurer?

59. For a segment of a radio show, a disc jockey can play 7 records. If there are 13 records to select from, in how many ways can the program for this segment be arranged?

60. Suppose you are asked to list, in order of preference, the three best movies you have seen this year. If you saw 20 movies during the year, in how many ways can the three best be chosen and ranked?

61. In a race in which six automobiles are entered and there are no ties, in how many ways can the first three finishers come in?

62. In a production of *West Side Story*, eight actors are considered for the male roles of Tony, Riff, and Bernardo. In how many ways can the director cast the male roles?

63. Nine bands have volunteered to perform at a benefit concert, but there is only enough time for five of the bands to play. How many lineups are possible?

64. How many arrangements can be made using four of the letters of the word COMBINE if no letter is to be used more than once?

65. In how many distinct ways can the letters of the word DALLAS be arranged?

66. In how many distinct ways can the letters of the word SCIENCE be arranged?

67. How many distinct permutations can be formed using the letters of the word TALLAHASSEE?

68. How many distinct permutations can be formed using the letters of the word TENNESSEE?

69. In how many ways can the digits in the number 5,446,666 be arranged?

70. In how many ways can the digits in the number 5,432,435 be arranged?

In Exercises 71–72, a signal can be formed by running different colored flags up a pole, one above the other.

71. Find the number of different signals consisting of eight flags that can be made using three white flags, four red flags, and one blue flag.

72. Find the number of different signals consisting of nine flags that can be made using three white flags, five red flags, and one blue flag.

Critical Thinking Exercises

73. Write a word problem that can be solved by evaluating 5!.

74. Write a word problem that can be solved by evaluating $_7P_3$.

75. Ten people board an airplane that has 12 aisle seats. In how many ways can they be seated if they all select aisle seats?

76. Six horses are entered in a race. If two horses are tied for first place and there are no ties among the other four horses, in how many ways can the six horses cross the finish line?

77. Five men and five women line up at a checkout counter in a store. In how many ways can they line up if the first person in line is a woman, and the people in line alternate woman, man, woman, man, and so on?

78. How many four-digit odd numbers less than 6000 can be formed using the digits 2, 4, 6, 7, 8, and 9?

79. Express $_nP_{n-2}$ without using factorials.

11.3 Combinations

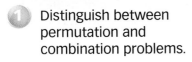 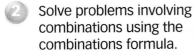

Objectives

① Distinguish between permutation and combination problems.

② Solve problems involving combinations using the combinations formula.

Throughout the history of entertainment, performers have featured choreography in their acts. Singers who are known for their serious moves include Beyoncé, Lady Gaga, Shakira, Justin Timberlake, and Usher.

Imagine that you ask your friends the following question: "Of these five entertainers, which three would you select to be included in a documentary on singers and choreography?" You are not asking your friends to rank their three favorite artists in any kind of order—they should merely select the three to be included in the documentary.

One friend answers, "Beyoncé, Lady Gaga, and Usher." Another responds, "Usher, Lady Gaga, and Beyoncé." These two people have the same artists in their group of selections, even if they are named in a different order. We are interested *in which artists are named, not the order in which they are named,* for the documentary. Because the items are taken without regard to order, this is not a permutation problem. No ranking of any sort is involved.

Later on, you ask your sister which three artists she would select for the documentary. She names Justin Timberlake, Beyoncé, and Usher. Her selection is different from those of your two other friends because different entertainers are cited.

Mathematicians describe the group of artists given by your sister as a *combination*. A **combination** of items occurs when

- The items are selected from the same group (the five entertainers who are known for their choreography).

- No item is used more than once. (You may view Beyoncé as a phenomenal performer, but your three selections cannot be Beyoncé, Beyoncé, and Beyoncé.)

- The order of the items makes no difference. (Beyoncé, Lady Gaga, Usher is the same group in the documentary as Usher, Lady Gaga, Beyoncé.)

① Distinguish between permutation and combination problems.

Do you see the difference between a permutation and a combination? A permutation is an ordered arrangement of a given group of items. A combination is a group of items taken without regard to their order. **Permutation** problems involve situations in which **order matters**. **Combination** problems involve situations in which the **order** of the items **makes no difference**.

Example 1   **Distinguishing between Permutations and Combinations**

For each of the following problems, determine whether the problem is one involving permutations or combinations. (It is not necessary to solve the problem.)

a. Six students are running for student government president, vice-president, and treasurer. The student with the greatest number of votes becomes the president, the second highest vote-getter becomes vice-president, and the student who gets the third largest number of votes will be treasurer. How many different outcomes are possible for these three positions?

b. Six people are on the board of supervisors for your neighborhood park. A three-person committee is needed to study the possibility of expanding the park. How many different committees could be formed from the six people?

c. Baskin-Robbins offers 31 different flavors of ice cream. One of its items is a bowl consisting of three scoops of ice cream, each a different flavor. How many such bowls are possible?

Solution

a. Students are choosing three student government officers from six candidates. The order in which the officers are chosen makes a difference because each of the offices (president, vice-president, treasurer) is different. Order matters. This is a problem involving permutations.

b. A three-person committee is to be formed from the six-person board of supervisors. The order in which the three people are selected does not matter because they are not filling different roles on the committee. Because order makes no difference, this is a problem involving combinations.

c. A three-scoop bowl of three different flavors is to be formed from Baskin-Robbins's 31 flavors. The order in which the three scoops of ice cream are put into the bowl is irrelevant. A bowl with chocolate, vanilla, and strawberry is exactly the same as a bowl with vanilla, strawberry, and chocolate. Different orderings do not change things, and so this is a problem involving combinations.

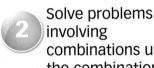 **Checkpoint 1** For each of the following problems, determine whether the problem is one involving permutations or combinations. (It is not necessary to solve the problem.)

a. How many ways can you select 6 free DVDs from a list of 200 DVDs?

b. In a race in which there are 50 runners and no ties, in how many ways can the first three finishers come in?

A Formula for Combinations

Solve problems involving combinations using the combinations formula.

We have seen that the notation $_nP_r$ means the number of permutations of n things taken r at a time. Similarly, the notation $_nC_r$ **means the number of combinations of n things taken r at a time**.

We can develop a formula for $_nC_r$ by comparing permutations and combinations. Consider the letters A, B, C, and D. The number of permutations of these four letters taken three at a time is

$$_4P_3 = \frac{4!}{(4-3)!} = \frac{4!}{1!} = \frac{4 \cdot 3 \cdot 2 \cdot 1}{1} = 24.$$

Here are the 24 permutations:

ABC,	ABD,	ACD,	BCD,
ACB,	ADB,	ADC,	BDC,
BAC,	BAD,	CAD,	CBD,
BCA,	BDA,	CDA,	CDB,
CAB,	DAB,	DAC,	DBC,
CBA,	DBA,	DCA,	DCB.

This column contains only one combination, ABC.	This column contains only one combination, ABD.	This column contains only one combination, ACD.	This column contains only one combination, BCD.

Because the order of items makes no difference in determining combinations, each column of six permutations represents one combination. There is a total of four combinations:

ABC, ABD, ACD, BCD.

Thus, $_4C_3 = 4$: The number of combinations of 4 things taken 3 at a time is 4. With 24 permutations and only four combinations, there are 6, or 3!, times as many permutations as there are combinations.

In general, there are $r!$ times as many permutations of n things taken r at a time as there are combinations of n things taken r at a time. Thus, we find the number of combinations of n things taken r at a time by dividing the number of permutations of n things taken r at a time by $r!$.

$$_nC_r = \frac{_nP_r}{r!} = \frac{\frac{n!}{(n-r)!}}{r!} = \frac{n!}{(n-r)!\, r!}$$

Study Tip

The number of combinations if r items are taken from n items cannot be found using the Fundamental Counting Principle and requires the use of the formula shown in the orange box.

Combinations of n Things Taken r at a Time

The number of possible combinations if r items are taken from n items is

$$_nC_r = \frac{n!}{(n-r)!\, r!}.$$

Technology

Graphing calculators have a menu item for calculating combinations, usually labeled $_nC_r$ (On TI graphing calculators, $_nC_r$ is selected using the MATH PRB menu.) For example, to find $_8C_3$, the keystrokes on most graphing calculators are

8 $\boxed{_nC_r}$ 3 $\boxed{\text{ENTER}}$.

If you are using a scientific calculator, check your manual to see whether there is a menu item for calculating combinations.

If you use your calculator's factorial key to find $\frac{8!}{5!3!}$, be sure to enclose the factorials in the denominator with parentheses

8 $\boxed{!}$ $\boxed{\div}$ $\boxed{(}$ 5 $\boxed{!}$ $\boxed{\times}$ 3 $\boxed{!}$ $\boxed{)}$

pressing $\boxed{=}$ or $\boxed{\text{ENTER}}$ to obtain the answer.

Example 2 **Using the Formula for Combinations**

A three-person committee is needed to study ways of improving public transportation. How many committees could be formed from the eight people on the board of supervisors?

Solution

The order in which the three people are selected does not matter. This is a problem of selecting $r = 3$ people from a group of $n = 8$ people. We are looking for the number of combinations of eight things taken three at a time. We use the formula

$$_nC_r = \frac{n!}{(n-r)!\, r!}$$

with $n = 8$ and $r = 3$.

$$_8C_3 = \frac{8!}{(8-3)!\,3!} = \frac{8!}{5!\,3!} = \frac{8\cdot7\cdot6\cdot5!}{5!\cdot3\cdot2\cdot1} = \frac{8\cdot7\cdot6\cdot5!}{5!\cdot3\cdot2\cdot1} = 56$$

Thus, 56 committees of three people each can be formed from the eight people on the board of supervisors.

Checkpoint 2 You volunteer to pet-sit for your friend who has seven different animals. How many different pet combinations are possible if you take three of the seven pets?

Example 3 **Using the Formula for Combinations**

In poker, a person is dealt 5 cards from a standard 52-card deck. The order in which the 5 cards are received does not matter. How many different 5-card poker hands are possible?

Solution

Because the order in which the 5 cards are received does not matter, this is a problem involving combinations. We are looking for the number of combinations of $n = 52$ cards drawn $r = 5$ at a time. We use the formula

$$_nC_r = \frac{n!}{(n-r)!\, r!}$$

with $n = 52$ and $r = 5$.

$$_{52}C_5 = \frac{52!}{(52 - 5)!\,5!} = \frac{52!}{47!\,5!} = \frac{52 \cdot 51 \cdot 50 \cdot 49 \cdot 48 \cdot \cancel{47!}}{\cancel{47!} \cdot 5 \cdot 4 \cdot 3 \cdot 2 \cdot 1} = 2{,}598{,}960$$

Thus, there are 2,598,960 different 5-card poker hands possible. It surprises many people that more than 2.5 million 5-card hands can be dealt from a mere 52 cards.

Figure 11.3 A royal flush

If you are a card player, it does not get any better than to be dealt the 5-card poker hand shown in **Figure 11.3**. This hand is called a *royal flush*. It consists of an ace, king, queen, jack, and 10, all of the same suit: all hearts, all diamonds, all clubs, or all spades. The probability of being dealt a royal flush involves calculating the number of ways of being dealt such a hand: just 4 of all 2,598,960 possible hands. In the next section, we move from counting possibilities to computing probabilities.

> **Checkpoint 3** How many different 4-card hands can be dealt from a deck that has 16 different cards?

Example 4 Using the Formula for Combinations and the Fundamental Counting Principle

In December 2009, the U.S. Senate consisted of 60 Democrats and 40 Republicans. How many distinct five-person committees can be formed if each committee must have 3 Democrats and 2 Republicans?

Solution

The order in which the members are selected does not matter. Thus, this is a problem involving combinations.

We begin with the number of ways of selecting 3 Democrats out of 60 Democrats without regard to order. We are looking for the number of combinations of $n = 60$ people taken $r = 3$ people at a time. We use the formula

$$_nC_r = \frac{n!}{(n - r)!\,r!}$$

with $n = 60$ and $r = 3$.

> We are picking 3 Democrats out of 60 Democrats.

$$_{60}C_3 = \frac{60!}{(60 - 3)!\,3!} = \frac{60!}{57!\,3!} = \frac{60 \cdot 59 \cdot 58 \cdot \cancel{57!}}{\cancel{57!}\,3 \cdot 2 \cdot 1} = \frac{60 \cdot 59 \cdot 58}{3 \cdot 2 \cdot 1} = 34{,}220$$

There are 34,220 ways to choose three Democrats for a committee.

Next, we find the number of ways of selecting 2 Republicans out of 40 Republicans without regard to order. We are looking for the number of combinations of $n = 40$ people taken $r = 2$ people at a time. Once again, we use the formula

$$_nC_r = \frac{n!}{(n - r)!\,r!}.$$

This time, $n = 40$ and $r = 2$.

> We are picking 2 Republicans out of 40 Republicans.

$$_{40}C_2 = \frac{40!}{(40 - 2)!\,2!} = \frac{40!}{38!\,2!} = \frac{40 \cdot 39 \cdot \cancel{38!}}{\cancel{38!}\,2 \cdot 1} = \frac{40 \cdot 39}{2 \cdot 1} = 780$$

There are 780 ways to choose two Republicans for a committee.

There are 34,220 ways to choose three Democrats for the committee and 780 ways to choose two Republicans for the committee. We use the Fundamental Counting Principle to find the number of committees that can be formed:

$$_{60}C_3 \cdot {}_{40}C_2 = 34,220 \cdot 780 = 26,691,600.$$

▶ Thus, 26,691,600 distinct committees can be formed.

✓ **Checkpoint 4** A zoo has six male bears and seven female bears. Two male bears and three female bears will be selected for an animal exchange program with another zoo. How many five-bear collections are possible?

Achieving Success

Many colleges put a student on academic probation if he or she takes nine units/credits or less. For this reason, it is a good idea to take a class load of at least 15 units. In this way, you can drop a course (perhaps one that you are failing) and not go on academic probation.

Exercise Set 11.3

Concept and Vocabulary Exercises

In Exercises 1–2, fill in each blank so that the resulting statement is true.

1. The number of possible combinations if r objects are taken from n items is $_nC_r =$ _____.

2. The formula for $_nC_r$ has the same numerator as the formula for $_nP_r$, but contains an extra factor of _____ in the denominator.

In Exercises 3–4, determine whether each statement is true or false. If the statement is false, make the necessary change(s) to produce a true statement.

3. Combination problems involve situations in which the order of the items makes a difference.

4. Permutation problems involve situations in which the order of the items does not matter.

Respond to Exercises 5–6 using verbal or written explanations.

5. What is a combination?

6. Explain how to distinguish between permutation and combination problems.

In Exercises 7–10, determine whether each statement makes sense or does not make sense, and explain your reasoning.

7. I used the formula for $_nC_r$ to determine the number of possible outcomes when ranking five politicians from most admired to least admired.

8. I used the formula for $_nC_r$ to determine how many four-letter passwords with no repeated letters can be formed using a, d, h, n, p, and w.

9. I solved a problem involving the number of possible outcomes when selecting from two groups, which required me to use both the formula for $_nC_r$ and the Fundamental Counting Principle.

10. I used the formula for $_nC_r$ to determine how many outcomes are possible when choosing four letters from a, d, h, n, p, and w.

Practice Exercises

In Exercises 11–14, does the problem involve permutations or combinations? Explain your answer. (It is not necessary to solve the problem.)

11. A medical researcher needs 6 people to test the effectiveness of an experimental drug. If 13 people have volunteered for the test, in how many ways can 6 people be selected?

12. Fifty people purchase raffle tickets. Three winning tickets are selected at random. If first prize is $1000, second prize is $500, and third prize is $100, in how many different ways can the prizes be awarded?

13. How many different four-letter passwords can be formed from the letters A, B, C, D, E, F, and G if no repetition of letters is allowed?

14. Fifty people purchase raffle tickets. Three winning tickets are selected at random. If each prize is $500, in how many different ways can the prizes be awarded?

In Exercises 15–30, use the formula for $_nC_r$ to evaluate each expression.

15. $_6C_5$

16. $_8C_7$

17. $_9C_5$

18. $_{10}C_6$

19. $_{11}C_4$

20. $_{12}C_5$

21. $_8C_1$

22. $_7C_1$

23. $_7C_7$

24. $_4C_4$

25. $_{30}C_3$

26. $_{25}C_4$

27. $_5C_0$

28. $_6C_0$

29. $\dfrac{_7C_3}{_5C_4}$

30. $\dfrac{_{10}C_3}{_6C_4}$

Practice Plus

In Exercises 31–38, evaluate each expression.

31. $\dfrac{_7P_3}{3!} - {_7C_3}$ **32.** $\dfrac{_{20}P_2}{2!} - {_{20}C_2}$ **33.** $1 - \dfrac{_3P_2}{_4P_3}$

34. $1 - \dfrac{_5P_3}{_{10}P_4}$ **35.** $\dfrac{_7C_3}{_5C_4} - \dfrac{98!}{96!}$ **36.** $\dfrac{_{10}C_3}{_6C_4} - \dfrac{46!}{44!}$

37. $\dfrac{_4C_2 \cdot {_6C_1}}{_{18}C_3}$ **38.** $\dfrac{_5C_1 \cdot {_7C_2}}{_{12}C_3}$

Application Exercises

Use the formula for $_nC_r$ to solve Exercises 39–50.

39. An election ballot asks voters to select three city commissioners from a group of six candidates. In how many ways can this be done?

40. A four-person committee is to be elected from an organization's membership of 11 people. How many different committees are possible?

41. Of 12 possible books, you plan to take 4 with you on vacation. How many different collections of 4 books can you take?

42. There are 14 standbys who hope to get seats on a flight, but only 6 seats are available on the plane. How many different ways can the 6 people be selected?

43. You volunteer to help drive children at a charity event to the zoo, but you can fit only 8 of the 17 children present in your van. How many different groups of 8 children can you drive?

44. Of the 100 people in the U.S. Senate, 18 serve on the Foreign Relations Committee. How many ways are there to select Senate members for this committee (assuming party affiliation is not a factor in the selection)?

45. To win at LOTTO in the state of Florida, one must correctly select 6 numbers from a collection of 53 numbers (1 through 53). The order in which the selection is made does not matter. How many different selections are possible?

46. To win in the New York State lottery, one must correctly select 6 numbers from 59 numbers. The order in which the selection is made does not matter. How many different selections are possible?

47. In how many ways can a committee of four men and five women be formed from a group of seven men and seven women?

48. How many different committees can be formed from 5 professors and 15 students if each committee is made up of 2 professors and 10 students?

49. The U.S. Senate of the 109th Congress consisted of 55 Republicans, 44 Democrats, and 1 Independent. How many committees can be formed if each committee must have 4 Republicans and 3 Democrats?

50. A mathematics exam consists of 10 multiple-choice questions and 5 open-ended problems in which all work must be shown. If an examinee must answer 8 of the multiple-choice questions and 3 of the open-ended problems, in how many ways can the questions and problems be chosen?

In Exercises 51–70, solve by the method of your choice.

51. In a race in which six automobiles are entered and there are no ties, in how many ways can the first four finishers come in?

52. A book club offers a choice of 8 books from a list of 40. In how many ways can a member make a selection?

53. A medical researcher needs 6 people to test the effectiveness of an experimental drug. If 13 people have volunteered for the test, in how many ways can 6 people be selected?

54. Fifty people purchase raffle tickets. Three winning tickets are selected at random. If first prize is $1000, second prize is $500, and third prize is $100, in how many different ways can the prizes be awarded?

55. From a club of 20 people, in how many ways can a group of three members be selected to attend a conference?

56. Fifty people purchase raffle tickets. Three winning tickets are selected at random. If each prize is $500, in how many different ways can the prizes be awarded?

57. How many different four-letter passwords can be formed from the letters A, B, C, D, E, F, and G if no repetition of letters is allowed?

58. Nine comedy acts will perform over two evenings. Five of the acts will perform on the first evening. How many ways can the schedule for the first evening be made?

59. Using 15 flavors of ice cream, how many cones with three different flavors can you create if it is important to you which flavor goes on the top, middle, and bottom?

60. Baskin-Robbins offers 31 different flavors of ice cream. One of its items is a bowl consisting of three scoops of ice cream, each a different flavor. How many such bowls are possible?

61. A restaurant lunch special allows the customer to choose two vegetables from this list: okra, corn, peas, carrots, and squash. How many outcomes are possible if the customer chooses two different vegetables?

62. There are six employees in the stock room at an appliance retail store. The manager will choose three of them to deliver a refrigerator. How many three-person groups are possible?

63. Suppose that you have three dress shirts, two ties, and two jackets. You need to select a dress shirt, a tie, and a jacket for work today. How many outcomes are possible?

64. Suppose that you have four flannel shirts. You are going to choose two of them to take on a camping trip. How many outcomes are possible?

65. A chef has five brands of hot sauce. Three of the brands will be chosen to mix into gumbo. How many outcomes are possible?

66. In the Mathematics Department, there are four female professors and six male professors. Three female professors will be chosen to serve as mentors for a special program designed to encourage female students to pursue careers in mathematics. In how many ways can the professors be chosen?

67. There are four Democrats and five Republicans on the county commission. From among their group they will choose a committee of two Democrats and two Republicans to examine a proposal to purchase land for a new county park. How many four-person groups are possible?

68. An office employs six customer service representatives. Each day, two of them are randomly selected and their customer interactions are monitored for the purposes of improving customer relations. In how many ways can the representatives be chosen?

69. A group of campers is going to occupy five campsites at a campground. There are 12 campsites from which to choose. In how many ways can the campsites be chosen?

70. Two teachers have driven to school in separate cars. When they arrive, there are seven empty spaces in the parking lot. They each choose a parking space. How many outcomes are possible?

(*Source* for Exercises 61–70: James Wooland, *CLAST Manual for Thinking Mathematically*)

Thousands of jokes have been told about marriage and divorce. Exercises 71–78 are based on the following observations:

- *"By all means, marry; if you get a good wife, you'll be happy. If you get a bad one, you'll become a philosopher."—Socrates*

- *"My wife and I were happy for 20 years. Then we met."—Rodney Dangerfield*

- *"Whatever you may look like, marry a man your own age. As your beauty fades, so will his eyesight."—Phyllis Diller*

- *"Why do Jewish divorces cost so much? Because they're worth it."—Henny Youngman*

- *"I think men who have a pierced ear are better prepared for marriage. They've experienced pain and bought jewelry."—Rita Rudner*

- *"For a while we pondered whether to take a vacation or get a divorce. We decided that a trip to Bermuda is over in two weeks, but a divorce is something you always have."—Woody Allen*

71. In how many ways can these six jokes be ranked from best to worst?

72. If Socrates's thoughts about marriage are excluded, in how many ways can the remaining five jokes be ranked from best to worst?

73. In how many ways can people select their three favorite jokes from these thoughts about marriage and divorce?

74. In how many ways can people select their two favorite jokes from these thoughts about marriage and divorce?

75. If the order in which these jokes are told makes a difference in terms of how they are received, how many ways can they be delivered if Socrates's comments are scheduled first and Dangerfield's joke is told last?

76. If the order in which these jokes are told makes a difference in terms of how they are received, how many ways can they be delivered if a joke by a woman (Rudner or Diller) is told first?

77. In how many ways can people select their favorite joke told by a woman (Rudner or Diller) and their two favorite jokes told by a man?

78. In how many ways can people select their favorite joke told by a woman (Rudner or Diller) and their three favorite jokes told by a man?

Critical Thinking Exercises

79. Write a word problem that can be solved by evaluating $_7C_3$.

80. Write a word problem that can be solved by evaluating $_{10}C_3 \cdot {}_7C_2$.

81. A 6/53 lottery involves choosing 6 of the numbers from 1 through 53 and a 5/36 lottery involves choosing 5 of the numbers from 1 through 36. The order in which the numbers are chosen does not matter. Which lottery is easier to win? Explain your answer.

82. If the number of permutations of n objects taken r at a time is six times the number of combinations of n objects taken r at a time, determine the value of r. Is there enough information to determine the value of n? Why or why not?

83. In a group of 20 people, how long will it take each person to shake hands with each of the other persons in the group, assuming that it takes three seconds for each shake and only 2 people can shake hands at a time? What if the group is increased to 40 people?

84. A sample of 4 telephones is selected from a shipment of 20 phones. There are 5 defective telephones in the shipment. How many of the possible samples of 4 phones do not include any of the defective ones?

11.4 : Fundamentals of Probability

① Compute theoretical probability.

② Compute empirical probability.

Table 11.1 The Hours of Sleep Americans Get on a Typical Night	
Hours of Sleep	**Number of Americans, in millions**
4 or less	12
5	27
6	75
7	90
8	81
9	9
10 or more	6
	Total: 300

Source: Discovery Health Media

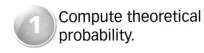

① **Compute theoretical probability.**

How many hours of sleep do you typically get each night? **Table 11.1** indicates that 75 million out of 300 million Americans are getting six hours of sleep on a typical night. The *probability* of an American getting six hours of sleep on a typical night is $\frac{75}{300}$. This fraction can be reduced to $\frac{1}{4}$, or expressed as 0.25, or 25%. Thus, 25% of Americans get six hours of sleep each night.

We find a probability by dividing one number by another. Probabilities are assigned to an *event*, such as getting six hours of sleep on a typical night. Events that are certain to occur are assigned probabilities of 1, or 100%. For example, the probability that a given individual will eventually die is 1. Although Woody Allen whined, "I don't want to achieve immortality through my work. I want to achieve it through not dying," death (and taxes) are always certain. By contrast, if an event cannot occur, its probability is 0. Regrettably, the probability that Elvis will return and serenade us with one final reprise of "Don't Be Cruel" (and we hope we're not) is 0.

Probabilities of events are expressed as numbers ranging from 0 to 1, or 0% to 100%. The closer the probability of a given event is to 1, the more likely it is that the event will occur. The closer the probability of a given event is to 0, the less likely it is that the event will occur.

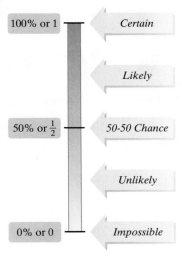

Possible Values for Probabilities

Theoretical Probability

You toss a coin. Although it is equally likely to land either heads up, denoted by H, or tails up, denoted by T, the actual outcome is uncertain. Any occurrence for which the outcome is uncertain is called an **experiment**. Thus, tossing a coin is an example of an experiment. The set of all possible outcomes of an experiment is the **sample space** of the experiment, denoted by S. The sample space for the coin-tossing experiment is

$$S = \{H, T\}.$$

Lands heads up   Lands tails up

An **event**, denoted by E, is any subset of a sample space. For example, the subset $E = \{T\}$ is the event of landing tails up when a coin is tossed.

Theoretical probability applies to situations like this, in which the sample space only contains equally likely outcomes, all of which are known. To calculate the theoretical probability of an event, we divide the number of outcomes resulting in the event by the total number of outcomes in the sample space.

Computing Theoretical Probability

If an event E has $n(E)$ equally likely outcomes and its sample space S has $n(S)$ equally likely outcomes, the **theoretical probability** of event E, denoted by $P(E)$, is

$$P(E) = \frac{\text{number of outcomes in event } E}{\text{total number of possible outcomes}} = \frac{n(E)}{n(S)}.$$

How can we use the formula $P(E) = \dfrac{n(E)}{n(S)}$ to compute the probability of a coin landing tails up? We use the following sets:

$$E = \{T\} \qquad S = \{H, T\}.$$

This is the event of landing tails up.

This is the sample space with all equally likely outcomes.

The probability of a coin landing tails up is

$$P(E) = \frac{\text{number of outcomes that result in tails up}}{\text{total number of possible outcomes}} = \frac{n(E)}{n(S)} = \frac{1}{2}.$$

Theoretical probability applies to many games of chance, including dice rolling, lotteries, and card games. We begin with rolling a die. **Figure 11.4** illustrates that when a die is rolled, there are six equally likely possible outcomes. The sample space can be shown as

$$S = \{1, 2, 3, 4, 5, 6\}.$$

Figure 11.4 Outcomes when a die is rolled

Example 1   Computing Theoretical Probability

A die is rolled once. Find the probability of rolling

a. a 3.   **b.** an even number.   **c.** a number less than 5.

d. a number less than 10.   **e.** a number greater than 6.

Solution

The sample space is $S = \{1, 2, 3, 4, 5, 6\}$ with $n(S) = 6$. We will use 6, the total number of possible outcomes, in the denominator of each probability fraction.

a. The phrase "rolling a 3" describes the event $E = \{3\}$. This event can occur in one way: $n(E) = 1$.

$$P(3) = \frac{\text{number of outcomes that result in 3}}{\text{total number of possible outcomes}} = \frac{n(E)}{n(S)} = \frac{1}{6}$$

The probability of rolling a 3 is $\frac{1}{6}$.

b. The phrase "rolling an even number" describes the event $E = \{2, 4, 6\}$. This event can occur in three ways: $n(E) = 3$.

$$P(\text{even number}) = \frac{\text{number of outcomes that result in an even number}}{\text{total number of possible outcomes}} = \frac{n(E)}{n(S)} = \frac{3}{6} = \frac{1}{2}$$

The probability of rolling an even number is $\frac{1}{2}$.

c. The phrase "rolling a number less than 5" describes the event $E = \{1, 2, 3, 4\}$. This event can occur in four ways: $n(E) = 4$.

$$P(\text{less than 5}) = \frac{\text{number of outcomes that are less than 5}}{\text{total number of possible outcomes}} = \frac{n(E)}{n(S)} = \frac{4}{6} = \frac{2}{3}$$

The probability of rolling a number less than 5 is $\frac{2}{3}$.

d. The phrase "rolling a number less than 10" describes the event $E = \{1, 2, 3, 4, 5, 6\}$. This event can occur in six ways: $n(E) = 6$. Can you see that all of the possible outcomes are less than 10? This event is certain to occur.

$$P(\text{less than }10) = \frac{\text{number of outcomes that are less than }10}{\text{total number of possible outcomes}} = \frac{n(E)}{n(S)} = \frac{6}{6} = 1$$

The probability of any certain event is 1.

e. The phrase "rolling a number greater than 6" describes an event that cannot occur, or the empty set. Thus, $E = \varnothing$ and $n(E) = 0$.

$$P(\text{greater than }6) = \frac{\text{number of outcomes that are greater than }6}{\text{total number of possible outcomes}} = \frac{n(E)}{n(S)} = \frac{0}{6} = 0$$

The probability of an event that cannot occur is 0.

In Example 1, there are six possible outcomes, each with a probability of $\frac{1}{6}$:

$$P(1) = \frac{1}{6} \quad P(2) = \frac{1}{6} \quad P(3) = \frac{1}{6} \quad P(4) = \frac{1}{6} \quad P(5) = \frac{1}{6} \quad P(6) = \frac{1}{6}.$$

The sum of these probabilities is 1: $\frac{1}{6} + \frac{1}{6} + \frac{1}{6} + \frac{1}{6} + \frac{1}{6} + \frac{1}{6} = 1$. In general, **the sum of the theoretical probabilities of all possible outcomes in the sample space is 1**.

Checkpoint 1   A die is rolled once. Find the probability of rolling

a. a 2.

b. a number less than 4.

c. a number greater than 7.

d. a number less than 7.

Our next example involves a standard 52-card bridge deck, illustrated in **Figure 11.5**. The deck has four suits: Hearts and diamonds are red, and clubs and spades are black. Each suit has 13 different face values—A(ace), 2, 3, 4, 5, 6, 7, 8, 9, 10, J(jack), Q(queen), and K(king). Jacks, queens, and kings are called **picture cards** or **face cards**.

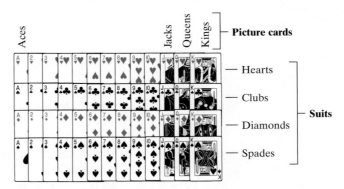

Figure 11.5 A standard 52-card bridge deck

Example 2   Probability and a Deck of 52 Cards

You are dealt one card from a standard 52-card deck. Find the probability of being dealt

a. a king.

b. a heart.

c. the king of hearts.

Solution

Because there are 52 cards in the deck, the total number of possible ways of being dealt a single card is 52. The number of outcomes in the sample space is 52: $n(S) = 52$. We use 52 as the denominator of each probability fraction.

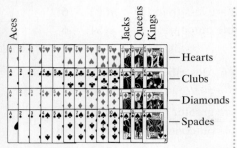

Figure 11.5 (repeated)
A standard 52-card bridge deck

—Hearts
—Clubs
—Diamonds
—Spades

Aces · Jacks · Queens · Kings

a. Let E be the event of being dealt a king. Because there are four kings in the deck, this event can occur in four ways: $n(E) = 4$.

$$P(\text{king}) = \frac{\text{number of outcomes that result in a king}}{\text{total number of possible outcomes}} = \frac{n(E)}{n(S)} = \frac{4}{52} = \frac{1}{13}$$

The probability of being dealt a king is $\frac{1}{13}$.

b. Let E be the event of being dealt a heart. Because there are 13 hearts in the deck, this event can occur in 13 ways: $n(E) = 13$.

$$P(\text{heart}) = \frac{\text{number of outcomes that result in a heart}}{\text{total number of possible outcomes}} = \frac{n(E)}{n(S)} = \frac{13}{52} = \frac{1}{4}$$

The probability of being dealt a heart is $\frac{1}{4}$.

c. Let E be the event of being dealt the king of hearts. Because there is only one card in the deck that is the king of hearts, this event can occur in just one way: $n(E) = 1$.

$$P(\text{king of hearts}) = \frac{\text{number of outcomes that result in the king of hearts}}{\text{total number of possible outcomes}} = \frac{n(E)}{n(S)} = \frac{1}{52}$$

The probability of being dealt the king of hearts is $\frac{1}{52}$.

✓ **Checkpoint 2**   You are dealt one card from a standard 52-card deck. Find the probability of being dealt

a. an ace.   **b.** a red card.   **c.** a red king.

Probabilities play a valuable role in the science of genetics. Example 3 deals with cystic fibrosis, an inherited lung disease occurring in about 1 out of every 2000 births among Caucasians and in about 1 out of every 250,000 births among non-Caucasians.

Blitzer Bonus

Try to Bench Press this at the Gym

You have a deck of cards for every permutation of the 52 cards. If each deck weighed only as much as a single hydrogen atom, all the decks together would weigh a billion times as much as the sun.

Source: Isaac Asimov's Book of Facts

Example 3   Probabilities in Genetics

Each person carries two genes that are related to the absence or presence of the disease cystic fibrosis. Most Americans have two normal genes for this trait and are unaffected by cystic fibrosis. However, 1 in 25 Americans carries one normal gene and one defective gene. If we use c to represent a defective gene and C a normal gene, such a carrier can be designated as Cc. Thus, CC is a person who neither carries nor has cystic fibrosis, Cc is a carrier who is not actually sick, and cc is a person sick with the disease. **Table 11.2** shows the four equally likely outcomes for a child's genetic inheritance from two parents who are both carrying one cystic fibrosis gene. One copy of each gene is passed on to the child from the parents.

Table 11.2 Cystic Fibrosis and Genetic Inheritance

		Second Parent	
		C	c
First	C	CC	Cc
Parent	c	cC	cc

Shown in the table are the four possibilities for a child whose parents each carry one cystic fibrosis gene.

If each parent carries one cystic fibrosis gene, what is the probability that their child will have cystic fibrosis?

Solution

Table 11.2 shows that there are four equally likely outcomes. The sample space is $S = \{CC, Cc, cC, cc\}$ and $n(S) = 4$. The phrase "will have cystic fibrosis" describes only the cc child. Thus, $E = \{cc\}$ and $n(E) = 1$.

$$P(\text{cystic fibrosis}) = \frac{\text{number of outcomes that result in cystic fibrosis}}{\text{total number of possible outcomes}} = \frac{n(E)}{n(S)} = \frac{1}{4}$$

If each parent carries one cystic fibrosis gene, the probability that their child will have cystic fibrosis is $\frac{1}{4}$.

✓ **Checkpoint 3**   Use **Table 11.2** in Example 3 to solve this exercise. If each parent carries one cystic fibrosis gene, find the probability that their child will be a carrier of the disease who is not actually sick.

Empirical Probability

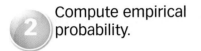

2 Compute empirical probability.

Theoretical probability is based on a set of equally likely outcomes and the number of elements in the set. By contrast, *empirical probability* applies to situations in which we observe how frequently an event occurs. We use the following formula to compute the empirical probability of an event:

> ### Computing Empirical Probability
>
> The empirical probability of event E is
>
> $$P(E) = \frac{\text{observed number of times } E \text{ occurs}}{\text{total number of observed occurrences}}.$$

Example 4   ## Computing Empirical Probability

In 2007, there were approximately 235.8 million Americans ages 18 or older. **Table 11.3** shows the distribution, by marital status and gender, of this population.

Table 11.3 Marital Status of the U.S. Population, Ages 18 or Older, in Millions

	Never Married	Married	Widowed	Divorced	Total
Male	37.5	64.7	2.7	9.6	114.5
Female	31.7	65.2	11.2	13.2	121.3
Total	69.2	129.9	13.9	22.8	235.8

Total male: 37.5 + 64.7 + 2.7 + 9.6 = 114.5
Total female: 31.7 + 65.2 + 11.2 + 13.2 = 121.3
Total never married: 37.5 + 31.7 = 69.2
Total widowed: 2.7 + 11.2 = 13.9
Total adult population: 114.5 + 121.3 = 235.8
Total married: 64.7 + 65.2 = 129.9
Total divorced: 9.6 + 13.2 = 22.8

Source: U.S. Census Bureau

If one person is randomly selected from the population described in **Table 11.3**, find the probability, to the nearest hundreth, that the person

a. is divorced.   **b.** is female.

Solution

As you read this solution, turn back a page to verify the numbers in the numerator and the denominator of each empirical probability fraction.

a. The probability of selecting a divorced person is the observed number of divorced people, 22.8 (million), divided by the total number of U.S. adults, 235.8 (million).

P(selecting a divorced person from the U.S. adult population)

$$= \frac{\text{number of divorced people}}{\text{total number of U.S. adults}} = \frac{22.8}{235.8} \approx 0.10$$

The empirical probability of selecting a divorced person from the U.S. adult population is approximately 0.10.

b. The probability of selecting a female is the observed number of females, 121.3 (million), divided by the total number of U.S. adults, 235.8 (million).

P(selecting a female from the U.S. adult population)

$$= \frac{\text{number of females}}{\text{total number of U.S. adults}} = \frac{121.3}{235.8} \approx 0.51$$

The empirical probability of selecting a female from the U.S. adult population is approximately 0.51.

Checkpoint 4 If one person is randomly selected from the population described in **Table 11.3** on the previous page, find the probability, expressed as a decimal rounded to the nearest hundredth, that the person

a. has never been married.

b. is male.

In certain situations, we can establish a relationship between the two kinds of probability. Consider, for example, a coin that is equally likely to land heads or tails. Such a coin is called a **fair coin**. Empirical probability can be used to determine whether a coin is fair. Suppose we toss a coin 10, 50, 100, 1000, 10,000, and 100,000 times. We record the number of heads observed, shown in **Table 11.4**. For each of the six cases in the table, the empirical probability of heads is determined by dividing the number of heads observed by the number of tosses.

Table 11.4 Empirical Probabilities of Heads as the Number of Tosses Increases		
Number of Tosses	**Number of Heads Observed**	**Empirical Probability of Heads, or *P(H)***
10	4	$P(H) = \frac{4}{10} = 0.4$
50	27	$P(H) = \frac{27}{50} = 0.54$
100	44	$P(H) = \frac{44}{100} = 0.44$
1000	530	$P(H) = \frac{530}{1000} = 0.53$
10,000	4851	$P(H) = \frac{4851}{10,000} = 0.4851$
100,000	49,880	$P(H) = \frac{49,880}{100,000} = 0.4988$

A pattern is exhibited by the empirical probabilities in the right-hand column of **Table 11.4**. As the number of tosses increases, the empirical probabilities tend to get closer to 0.5, the theoretical probability. These results give us no reason to suspect that the coin is not fair.

Table 11.4 illustrates an important principle when observing uncertain outcomes such as the event of a coin landing on heads. As an experiment is repeated more and more times, the empirical probablity of an event tends to get closer to the theoretical probability of that event. This principle is known as the **law of large numbers**.

Exercise Set 11.4

Concept and Vocabulary Exercises

In Exercises 1–4, fill in each blank so that the resulting statement is true.

1. The set of all possible outcomes of an experiment is called the _____ of the experiment.

2. The theoretical probability of event *E*, denoted by _____, is the _____ divided by the _____.

3. A standard bridge deck has _____ cards with four suits: _____ and _____ are red, and _____ and _____ are black.

4. Probability that is based on situations in which we observe how frequently an event occurs is called _____ probability.

In Exercises 5–8, determine whether each statement is true or false. If the statement is false, make the necessary change(s) to produce a true statement.

5. If an event is certain to occur, its probability is 1.

6. If an event cannot occur, its probability is −1.

7. The sum of the probabilities of all possible outcomes in an experiment is 1.

8. If an experiment is repeated more and more times, the theoretical probability of an event tends to get closer to the empirical probability of that event.

Respond to Exercises 9–14 using verbal or written explanations.

9. What is the sample space of an experiment? What is an event?

10. How is the theoretical probability of an event computed?

11. Describe the difference between theoretical probability and empirical probability.

12. Give an example of an event whose probability must be determined empirically rather than theoretically.

13. Use the definition of theoretical probability to explain why the probability of an event that cannot occur is 0.

14. Use the definition of theoretical probability to explain why the probability of an event that is certain to occur is 1.

In Exercises 15–18, determine whether each statement makes sense or does not make sense, and explain your reasoning.

15. Assuming the next U.S. president will be a Democrat or a Republican, the probability of a Republican president is 0.5.

16. The probability that I will go to college is 1.5.

17. When I toss a coin, the probability of getting heads *or* tails is 1, but the probability of getting heads *and* tails is 0.

18. When I am dealt one card from a standard 52-card deck, the probability of getting a red card *or* a black card is 1, but the probability of getting a red card *and* a black card is 0.

Practice and Application Exercises

In Exercises 19–72, express each probability as a fraction reduced to lowest terms.

In Exercises 19–28, a die is rolled. The set of equally likely outcomes is {1, 2, 3, 4, 5, 6}. Find the probability of rolling

19. a 4.

20. a 5.

21. an odd number.

22. a number greater than 3.

23. a number less than 3.

24. a number greater than 4.

25. a number less than 20.

26. a number less than 8.

27. a number greater than 20.

28. a number greater than 8.

In Exercises 29–38, you are dealt one card from a standard 52-card deck. Find the probability of being dealt

29. a queen.

30. a jack.

31. a club.

32. a diamond.

33. a picture card.

34. a card greater than 3 and less than 7.

35. the queen of spades.

36. the ace of clubs.

37. a diamond and a spade.

38. a card with a green heart.

In Exercises 39–44, a fair coin is tossed two times in succession. The set of equally likely outcomes is {HH, HT, TH, TT}. Find the probability of getting

39. two heads.

40. two tails.

41. the same outcome on each toss.

42. different outcomes on each toss.

43. a head on the second toss.

44. at least one head.

In Exercises 45–52, you select a family with three children. If M represents a male child and F a female child, the set of equally likely outcomes for the children's genders is {MMM, MMF, MFM, MFF, FMM, FMF, FFM, FFF}. Find the probability of selecting a family with

45. exactly one female child.

46. exactly one male child.

47. exactly two male children.

48. exactly two female children.

49. at least one male child.

50. at least two female children.

51. four male children.

52. fewer than four female children.

In Exercises 53–58, a single die is rolled twice. The 36 equally likely outcomes are shown as follows:

		Second Roll					
		⚀	⚁	⚂	⚃	⚄	⚅
First Roll	⚀	(1, 1)	(1, 2)	(1, 3)	(1, 4)	(1, 5)	(1, 6)
	⚁	(2, 1)	(2, 2)	(2, 3)	(2, 4)	(2, 5)	(2, 6)
	⚂	(3, 1)	(3, 2)	(3, 3)	(3, 4)	(3, 5)	(3, 6)
	⚃	(4, 1)	(4, 2)	(4, 3)	(4, 4)	(4, 5)	(4, 6)
	⚄	(5, 1)	(5, 2)	(5, 3)	(5, 4)	(5, 5)	(5, 6)
	⚅	(6, 1)	(6, 2)	(6, 3)	(6, 4)	(6, 5)	(6, 6)

Find the probability of getting

53. two even numbers.

54. two odd numbers.

55. two numbers whose sum is 5.

56. two numbers whose sum is 6.

57. two numbers whose sum exceeds 12.

58. two numbers whose sum is less than 13.

Use the spinner shown to answer Exercises 59–66. Assume that it is equally probable that the pointer will land on any one of the ten colored regions. If the pointer lands on a borderline, spin again.

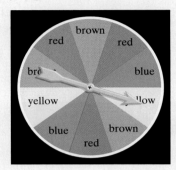

Find the probability that the spinner lands in

59. a red region.

60. a yellow region.

61. a blue region.

62. a brown region.

63. a region that is red or blue.

64. a region that is yellow or brown.

65. a region that is red and blue.

66. a region that is yellow and brown.

Exercises 67–72 deal with sickle cell anemia, an inherited disease in which red blood cells become distorted and deprived of oxygen. Approximately 1 in every 500 African-American infants is born with the disease; only 1 in 160,000 white infants has the disease. A person with two sickle cell genes will have the disease, but a person with only one sickle cell gene will have a mild, nonfatal anemia called sickle cell trait. (Approximately 8%–10% of the African-American population has this trait.)

		Second Parent	
		S	s
First	S	SS	Ss
Parent	s	sS	ss

If we use s to represent a sickle cell gene and S a healthy gene, the table above shows the four possibilities for the children of two Ss parents. Each parent has only one sickle cell gene, so each has the relatively mild sickle cell trait. Find the probability that these parents give birth to a child who

67. has sickle cell anemia.

68. has sickle cell trait.   69. is healthy.

In Exercises 70–72, use the following table that shows the four possibilities for the children of one healthy, SS, parent, and one parent with sickle cell trait, Ss.

		Second Parent (with Sickle Cell Trait)	
		S	s
Healthy	S	SS	Ss
First Parent	S	SS	Ss

Find the probability that these parents give birth to a child who

70. has sickle cell anemia.

71. has sickle cell trait.

72. is healthy.

The table shows the distribution, by age and gender, of the 29.3 million Americans who live alone. Use the data in the table to solve Exercises 73–78.

Number of People in the United States Living Alone, in Millions

	Ages 15–24	Ages 25–34	Ages 35–44	Ages 45–64	Ages 65–74	Ages ≥75	Total
Male	0.7	2.2	2.6	4.3	1.3	1.4	12.5
Female	0.8	1.6	1.6	5.0	2.9	4.9	16.8
Total	1.5	3.8	4.2	9.3	4.2	6.3	29.3

Source: U.S. Census Bureau

Find the probability, expressed as a decimal rounded to the nearest hundredth, that a randomly selected American living alone is

73. male.

74. female.

75. in the 25–34 age range.

76. in the 35–44 age range.

77. a woman in the 15–24 age range.

78. a man in the 45–64 age range.

The table shows the number of Americans who moved in a recent year, categorized by where they moved and whether they were an owner or a renter. Use the data in the table, expressed in millions, to solve Exercises 79–84.

Number of People in the United States Who Moved, in Millions

	Moved to Same State	Moved to Different State	Moved to Different Country
Owner	11.7	2.8	0.3
Renter	18.7	4.5	1.0

Source: U.S. Census Bureau

Use the table to find the probability, expressed as a decimal rounded to the nearest hundredth, that a randomly selected American who moved was

79. an owner.

80. a renter.

81. a person who moved within the same state.

82. a person who moved to a different country.

83. a renter who moved to a different state.

84. an owner who moved to a different state.

The table shows the educational attainment of the U.S. population, ages 65 and over, in 2007. Use the data in the table, expressed in millions, to solve Exercises 85–88.

Educational Attainment of the U.S. Population, Ages 65 and Over, in Millions

	Less Than 4 Years High School	4 Years High School Only	Some College (Less Than 4 years)	4 Years College (or More)
Male	3.6	5.0	2.6	3.9
Female	5.2	8.0	4.1	3.0

Source: U.S. Census Bureau

Find the probability, expressed as a decimal rounded to the nearest hundredth, that a randomly selected American in 2007, aged 65 or over,

85. had less than 4 years of high school.

86. had 4 years of high school only.

87. was a woman with 4 years of college or more.

88. was a man with 4 years of college or more.

Critical Thinking Exercises

89. Write a probability word problem whose answer is one of the following fractions: $\frac{1}{6}$ or $\frac{1}{4}$ or $\frac{1}{3}$.

90. The target in the figure shown contains four squares. If a dart thrown at random hits the target, find the probability that it will land in an orange region.

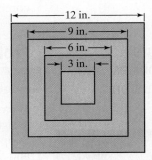

91. Some three-digit numbers, such as 101 and 313, read the same forward and backward. If you select a number from all three-digit numbers, find the probability that it will read the same forward and backward.

Probability with the Fundamental Counting Principle, Permutations, and Combinations

11.5

1 Compute probabilities with permutations.

2 Compute probabilities with combinations.

Probability of Dying at Any Given Age

Age	Probability of Male Death	Probability of Female Death
10	0.00013	0.00010
20	0.00140	0.00050
30	0.00153	0.00050
40	0.00193	0.00095
50	0.00567	0.00305
60	0.01299	0.00792
70	0.03473	0.01764
80	0.07644	0.03966
90	0.15787	0.11250
100	0.26876	0.23969
110	0.39770	0.39043

Source: George Shaffner, *The Arithmetic of Life and Death*

George Tooker (b. 1920) *"Mirror II"* 1963, egg tempera on gesso panel, 20 × 20 in., 1968. Gift of R. H. Donnelley Erdman (PA 1956). Addison Gallery of American Art, Phillips Academy, Andover, Massachusetts. All Rights Reserved © George Tooker.

According to actuarial tables, there is no year in which death is as likely as continued life, at least until the age of 115. Until that age, the probability of dying in any one year ranges from a low of 0.00009 for a girl at age 11 to a high of 0.465 for either gender at age 114. For a healthy 30-year-old, how does the probability of dying this year compare to the probability of winning the top prize in a state lottery game? In this section, we provide the surprising answer to this question, as we study probability with the Fundamental Counting Principle, permutations, and combinations.

 Compute probabilities with permutations.

Probability with Permutations

Groucho Marx

Example 1   Probability and Permutations

We return to the six jokes about books by Groucho Marx, George Carlin, Steven Wright, Henny Youngman, Jerry Seinfeld, and Phyllis Diller that opened Section 11.2. Suppose that each joke is written on one of six cards. The cards are placed in a hat and then six cards are drawn, one at a time. The order in which the cards are drawn determines the order in which the jokes are delivered. What is the probability that a man's joke will be delivered first and a man's joke will be delivered last?

Solution

We begin by applying the definition of probability to this situation.

P(man's joke first, man's joke last)

$= \dfrac{\text{number of permutations with man's joke first, man's joke last}}{\text{total number of possible permutations}}$

George Carlin

Steven Wright

Henny Youngman

Jerry Seinfeld

Phyllis Diller

We can use the Fundamental Counting Principle to find the total number of possible permutations. This represents the number of ways the six jokes can be delivered.

There are $6 \cdot 5 \cdot 4 \cdot 3 \cdot 2 \cdot 1$, or 720 possible permutations. Equivalently, there are 720 different ways to deliver the six jokes about books.

We can also use the Fundamental Counting Principle to find the number of permutations with a man's joke delivered first and a man's joke delivered last. These conditions can be shown as follows:

Now let's fill in the number of choices for positions two through five.

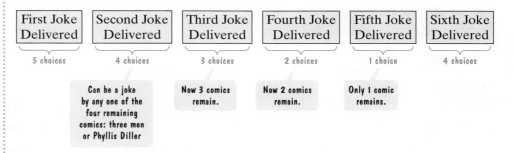

Thus, there are $5 \cdot 4 \cdot 3 \cdot 2 \cdot 1 \cdot 4$, or 480 possible permutations. Equivalently, there are 480 ways to deliver the jokes with a man's joke told first and a man's joke delivered last.

Now we can return to our probability fraction.

P (man's joke first, man's joke last)

$$= \frac{\text{number of permutations with man's joke first, man's joke last}}{\text{total number of possible permutations}}$$

$$= \frac{480}{720} = \frac{2}{3}$$

The probability of a man's joke delivered first and a man's joke told last is $\frac{2}{3}$.

✔ **Checkpoint 1**   Consider the six jokes about books by Groucho Marx, George Carlin, Steven Wright, Henny Youngman, Jerry Seinfeld, and Phyllis Diller. As in Example 1, each joke is written on one of six cards which are randomly drawn one card at a time. The order in which the cards are drawn determines the order in which the jokes are delivered. What is the probability that a joke by a comic whose first name begins with G is told first and a man's joke is delivered last?

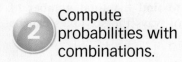

Compute probabilities with combinations.

Probability with Combinations

In 2008, Americans spent approximately $60 billion on lotteries set up by 42 revenue-hungry states and Washington, D.C. If your state has a lottery drawing each week, the probability that someone will win the top prize is relatively high. If there is no winner this week, it is virtually certain that eventually someone will be graced with millions of dollars. So, why are you so unlucky compared to this undisclosed someone? In Example 2, we provide an answer to this question.

Example 2   Probability and Combinations: Winning the Lottery

Florida's lottery game, LOTTO, is set up so that each player chooses six different numbers from 1 to 53. If the six numbers chosen match the six numbers drawn randomly, the player wins (or shares) the top cash prize. (As of this writing, the top cash prize has ranged from $7 million to $106.5 million.) With one LOTTO ticket, what is the probability of winning this prize?

Solution

Because the order of the six numbers does not matter, this is a situation involving combinations. We begin with the formula for probability.

$$P(\text{winning}) = \frac{\text{number of ways of winning}}{\text{total number of possible combinations}}$$

We can use the combinations formula

$$_nC_r = \frac{n!}{(n-r)!\,r!}$$

to find the total number of possible combinations. We are selecting $r = 6$ numbers from a collection of $n = 53$ numbers.

$$_{53}C_6 = \frac{53!}{(53-6)!\,6!} = \frac{53!}{47!\,6!} = \frac{53 \cdot 52 \cdot 51 \cdot 50 \cdot 49 \cdot 48 \cdot \cancel{47!}}{\cancel{47!} \cdot 6 \cdot 5 \cdot 4 \cdot 3 \cdot 2 \cdot 1 \cdot} = 22{,}957{,}480$$

There are nearly 23 million number combinations possible in LOTTO. If a person buys one LOTTO ticket, that person has selected only one combination of the six numbers. With one LOTTO ticket, there is only one way of winning.

Now we can return to our probability fraction.

$$P(\text{winning}) = \frac{\text{number of ways of winning}}{\text{total number of possible combinations}} = \frac{1}{22{,}957{,}480} \approx 0.0000000436$$

The probability of winning the top prize with one LOTTO ticket is $\frac{1}{22{,}957{,}480}$, or about 1 in 23 million.

Suppose that a person buys 5000 different tickets in Florida's LOTTO. Because that person has selected 5000 different combinations of the six numbers, the probability of winning is

$$\frac{5000}{22{,}957{,}480} \approx 0.000218.$$

The chances of winning the top prize are about 218 in a million. At $1 per LOTTO ticket, it is highly probable that our LOTTO player will be $5000 poorer. Knowing a little probability helps a lotto.

Checkpoint 2   People lose interest when they do not win at games of chance, including Florida's LOTTO. With drawings twice weekly instead of once, the game

---

Blitzer Bonus

Comparing the Probability of Dying to the Probability of Winning Florida's Lotto

As a healthy nonsmoking 30-year-old, your probability of dying this year is approximately 0.001. Divide this probability by the probability of winning LOTTO with one ticket:

$$\frac{0.001}{0.0000000436} \approx 22{,}936.$$

A healthy 30-year-old is nearly 23,000 times more likely to die this year than to win Florida's lottery.

described in Example 2 was brought in to bring back lost players and increase ticket sales. The original LOTTO was set up so that each player chose six different numbers from 1 to 49, rather than from 1 to 53, with a lottery drawing only once a week. With one LOTTO ticket, what was the probability of winning the top cash prize in Florida's original LOTTO? Express the answer as a fraction and as a decimal correct to ten places.

Example 3   Probability and Combinations

A club consists of five men and seven women. Three members are selected at random to attend a conference. Find the probability that the selected group consists of

a. three men.

b. one man and two women.

Solution

The order in which the three people are selected does not matter, so this is a problem involving combinations.

a. We begin with the probability of selecting three men.

$$P(3 \text{ men}) = \frac{\text{number of ways of selecting 3 men}}{\text{total number of possible combinations}}$$

First, we consider the denominator of the probability fraction. We are selecting $r = 3$ people from a total group of $n = 12$ people (five men and seven women). The total number of possible combinations is

$$_{12}C_3 = \frac{12!}{(12-3)!\,3!} = \frac{12!}{9!\,3!} = \frac{12 \cdot 11 \cdot 10 \cdot \cancel{9!}}{\cancel{9!} \cdot 3 \cdot 2 \cdot 1} = 220.$$

Thus, there are 220 possible three-person selections.

   Next, we consider the numerator of the probability fraction. We are interested in the number of ways of selecting three men from five men. We are selecting $r = 3$ men from a total group of $n = 5$ men. The number of possible combinations of three men is

$$_5C_3 = \frac{5!}{(5-3)!\,3!} = \frac{5!}{2!\,3!} = \frac{5 \cdot 4 \cdot \cancel{3!}}{2 \cdot 1 \cdot \cancel{3!}} = 10.$$

Thus, there are 10 ways of selecting three men from five men. Now we can fill in the numbers in the numerator and the denominator of our probability fraction.

$$P(3 \text{ men}) = \frac{\text{number of ways of selecting 3 men}}{\text{total number of possible combinations}} = \frac{10}{220} = \frac{1}{22}$$

The probability that the group selected to attend the conference consists of three men is $\frac{1}{22}$.

b. We set up the fraction for the probability that the selected group consists of one man and two women.

$$P(1 \text{ man, 2 women}) = \frac{\text{number of ways of selecting 1 man and 2 women}}{\text{total number of possible combinations}}$$

The denominator of this fraction is the same as the denominator in part (a). The total number of possible combinations is found by selecting $r = 3$ people from $n = 12$ people: $_{12}C_3 = 220$.

   Next, we move to the numerator of the probability fraction. The number of ways of selecting $r = 1$ man from $n = 5$ men is

$$_5C_1 = \frac{5!}{(5-1)!\,1!} = \frac{5!}{4!\,1!} = \frac{5 \cdot \cancel{4!}}{\cancel{4!} \cdot 1} = 5.$$

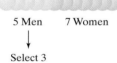

12 Club Members

5 Men   7 Women

Select 3

12 Club Members

5 Men   7 Women

Select 1   Select 2

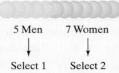

12 Club Members

5 Men   7 Women

↓   ↓

Select 1   Select 2

The number of ways of selecting $r = 2$ women from $n = 7$ women is

$$_7C_2 = \frac{7!}{(7-2)!\,2!} = \frac{7!}{5!\,2!} = \frac{7 \cdot 6 \cdot \cancel{5!}}{\cancel{5!} \cdot 2 \cdot 1} = 21.$$

By the Fundamental Counting Principle, the number of ways of selecting 1 man and 2 women is

$$_5C_1 \cdot {}_7C_2 = 5 \cdot 21 = 105.$$

Now we can fill in the numbers in the numerator and the denominator of our probability fraction.

$$P(1 \text{ man, 2 women}) = \frac{\text{number of ways of selecting 1 man and 2 women}}{\text{total number of possible combinations}}$$

$$= \frac{_5C_1 \cdot {}_7C_2}{_{12}C_3} = \frac{105}{220} = \frac{21}{44}$$

The probability that the group selected to attend the conference consists of one man and two women is $\frac{21}{44}$.

✓ **Checkpoint 3**   A club consists of six men and four women. Three members are selected at random to attend a conference. Find the probability that the selected group consists of

a. three men.

▶ **b.** two men and one woman.

Achieving Success

Some of the difficulties facing students who take a college math class for the first time are a result of the differences between a high school math class and a college math course.
It is helpful to understand these differences before entering a college level math class.

High School Math Class	College Math Course
Attendance is required.	Attendance may be optional.
Teachers monitor progress and performance closely.	Students receive grades, but may not be informed by the professor if they are in trouble.
There are frequent tests, as well as make-up tests if grades are poor.	There are usually no more than three or four tests per semester. Make-up tests are rarely allowed.
Grades are often based on participation and effort.	Grades are usually based exclusively on test grades.
Students have contact with their instructor every day.	Students usually meet with their instructor two or three times per week.
Teachers cover all material for tests in class through lectures and activities.	Students are responsible for information whether or not it is covered in class.
A course is covered over the school year, usually ten months.	A course is covered in a semester, usually four months.
Extra credit is often available for struggling students.	Extra credit is almost never offered.

Source: Alan Bass, *Math Study Skills*, Pearson, 2008.

Exercise Set 11.5

Concept and Vocabulary Exercises

In Exercises 1–2, fill in each blank so that the resulting statement is true.

1. Six stand-up comics, A, B, C, D, E, and F, are to perform on a single evening at a comedy club. The order of performance is determined by random selection. The probability that comic E will perform first is the number of ———————— with comic E performing first divided by ———————— ————————.

2. The probability of winning a lottery with one lottery ticket is the number of ways of winning, which is precisely ————————, divided by the total number of possible ————————.

In Exercises 3–4, determine whether each statement is true or false. If the statement is false, make the necessary change(s) to produce a true statement.

3. When working problems involving probability with permutations, the denominators of the probability fractions consist of the total number of possible permutations.

4. When working problems involving probability with combinations, the numerators of the probability fractions consist of the total number of possible combinations.

Respond to Exercises 5–6 using verbal or written explanations.

5. If you purchase ten lottery tickets, explain how you determine the probability of winning the lottery. How do you find the number in the numerator of the probability fraction and the number in the denominator of the probability fraction?

6. If people understood the mathematics involving probabilities and lotteries, as you now do, do you think they would continue to spend hundreds of dollars per year on lottery tickets? Explain your answer.

In Exercises 7–10, determine whether each statement makes sense or does not make sense, and explain your reasoning.

7. When solving probability problems using the Fundamental Counting Principle, I find it easier to reduce the probability fraction if I leave the numerator and denominator as products of numbers.

8. I would never choose the lottery numbers 1, 2, 3, 4, 5, 6 because the probability of winning with six numbers in a row is less than winning with six random numbers.

9. From an investment point of view, a state lottery is a very poor place to put my money.

10. When finding the probability of randomly selecting two men and one woman from a group of ten men and ten women, I used the formula for $_nC_r$ three times.

Practice and Application Exercises

11. Martha, Lee, Nancy, Paul, and Armando have all been invited to a dinner party. They arrive randomly and each person arrives at a different time.
    a. In how many ways can they arrive?
    b. In how many ways can Martha arrive first and Armando last?
    c. Find the probability that Martha will arrive first and Armando last.

12. Three men and three women line up at a checkout counter in a store.
    a. In how many ways can they line up?
    b. In how many ways can they line up if the first person in line is a woman, and then the line alternates by gender—that is a woman, a man, a woman, a man, and so on?
    c. Find the probability that the first person in line is a woman and the line alternates by gender.

13. Six stand-up comics, A, B, C, D, E, and F, are to perform on a single evening at a comedy club. The order of performance is determined by random selection. Find the probability that
    a. Comic E will perform first.
    b. Comic C will perform fifth and comic B will perform last.
    c. The comedians will perform in the following order: D, E, C, A, B, F.
    d. Comic A or comic B will perform first.

14. Seven performers, A, B, C, D, E, F, and G, are to appear in a fund raiser. The order of performance is determined by random selection. Find the probability that
    a. D will perform first.
    b. E will perform sixth and B will perform last.
    c. They will perform in the following order: C, D, B, A, G, F, E.
    d. F or G will perform first.

15. A group consists of four men and five women. Three people are selected to attend a conference.
    a. In how many ways can three people be selected from this group of nine?
    b. In how many ways can three women be selected from the five women?
    c. Find the probability that the selected group will consist of all women.

16. A political discussion group consists of five Democrats and six Republicans. Four people are selected to attend a conference.
    a. In how many ways can four people be selected from this group of eleven?
    b. In how many ways can four Republicans be selected from the six Republicans?
    c. Find the probability that the selected group will consist of all Republicans.

17. To play the California lottery, a person has to correctly select 6 out of 51 numbers, paying $1 for each six-number selection. If the six numbers picked are the same as the ones drawn by the lottery, mountains of money are bestowed. What is the probability that a person with one combination of six numbers will win? What is the probability of winning if 100 different lottery tickets are purchased?

18. A state lottery is designed so that a player chooses five numbers from 1 to 30 on one lottery ticket. What is the probability that a player with one lottery ticket will win? What is the probability of winning if 100 different lottery tickets are purchased?

19. A box contains 25 transistors, 6 of which are defective. If 6 are selected at random, find the probability that

    a. all are defective.   b. none are defective.

20. A committee of five people is to be formed from six lawyers and seven teachers. Find the probability that

    a. all are lawyers.   b. none are lawyers.

21. A city council consists of six Democrats and four Republicans. If a committee of three people is selected, find the probability of selecting one Democrat and two Republicans.

22. A parent-teacher committee consisting of four people is to be selected from fifteen parents and five teachers. Find the probability of selecting two parents and two teachers.

*Exercises 23–28 involve a deck of 52 cards. If necessary, refer to the picture of a deck of cards, **Figure 11.5** on page 697.*

23. A poker hand consists of five cards.

    a. Find the total number of possible five-card poker hands.

    b. A diamond flush is a five-card hand consisting of all diamonds. Find the number of possible diamond flushes.

    c. Find the probability of being dealt a diamond flush.

24. A poker hand consists of five cards.

    a. Find the total number of possible five-card poker hands.

    b. Find the number of ways in which four aces can be selected.

    c. Find the number of ways in which one king can be selected.

    d. Use the Fundamental Counting Principle and your answers from parts (b) and (c) to find the number of ways of getting four aces and one king.

    e. Find the probability of getting a poker hand consisting of four aces and one king.

25. If you are dealt 3 cards from a shuffled deck of 52 cards, find the probability that all 3 cards are picture cards.

26. If you are dealt 4 cards from a shuffled deck of 52 cards, find the probability that all 4 are hearts.

27. If you are dealt 4 cards from a shuffled deck of 52 cards, find the probability of getting two queens and two kings.

28. If you are dealt 4 cards from a shuffled deck of 52 cards, find the probability of getting three jacks and one queen.

Critical Thinking Exercises

29. Write and solve an original problem involving probability and permutations.

30. Write and solve an original problem involving probability and combinations whose solution requires $\dfrac{_{14}C_{10}}{_{20}C_{10}}$.

31. An apartment complex offers apartments with four different options, designated by A through D. There are an equal number of apartments with each combination of options.

A	B	C	D
one bedroom	one bathroom	first floor	lake view
two bedrooms	two bathrooms	second floor	golf course view
three bedrooms			no special view

If there is only one apartment left, what is the probability that it is precisely what a person is looking for, namely two bedrooms, two bathrooms, first floor, and a lake or golf course view?

32. Reread Exercise 17. How much must a person spend so that the probability of winning the California lottery is $\frac{1}{2}$?

33. Suppose that it is a week in which the cash prize in Florida's LOTTO is promised to exceed $50 million. If a person purchases 22,957,480 tickets in LOTTO at $1 per ticket (all possible combinations), isn't this a guarantee of winning the lottery? Because the probability in this situation is 1, what's wrong with doing this?

34. The digits 1, 2, 3, 4, and 5 are randomly arranged to form a three-digit number. (Digits are not repeated.) Find the probability that the number is even and greater than 500.

35. In a five-card poker hand, what is the probability of being dealt exactly one ace and no picture cards?

Group Exercise

36. Research and present a group report on state lotteries. Include answers to some or all of the following questions. As always, make the report interesting and informative. Which states do not have lotteries? Why not? How much is spent per capita on lotteries? What are some of the lottery games? What is the probability of winning top prize in these games? What income groups spend the greatest amount of money on lotteries? If your state has a lottery, what does it do with the money it makes? Is the way the money is spent what was promised when the lottery first began?

11.6 : Events Involving *Not* and *Or*; Odds

Objectives

1 Find the probability that an event will not occur.

2 Find the probability of one event or a second event occurring.

3 Understand and use odds.

What are you most afraid of? A shark attack? An airplane crash? The Harvard Center for Risk Analysis helps to put these fears in perspective. According to the Harvard Center, the odds in favor of a deadly shark attack are 1 to 280 million and the odds against a deadly airplane accident are 3 million to 1.

There are several ways to express the likelihood of an event. For example, we can determine the probability of a deadly shark attack or a deadly airplane accident. We can also determine the *odds in favor* and the *odds against* these events. In this section, we expand our knowledge of probability and explain the meaning of odds.

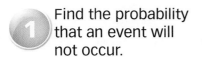

Probability of an Event Not Occurring

1 Find the probability that an event will not occur.

If we know $P(E)$, the probability of an event E, we can determine the probability that the event will not occur, denoted by $P(\text{not } E)$. The event *not E* is the **complement** of E because it is the set of all outcomes in the sample space S that are not outcomes in the event E. In any experiment, an event must occur or its complement must occur. Thus, the sum of the probability that an event will occur and the probability that it will not occur is 1:

$$P(E) + P(\text{not } E) = 1.$$

Solving for $P(E)$ or for $P(\text{not } E)$, we obtain the following formulas:

Complement Rules of Probability

• The probability that an event E will not occur is equal to 1 minus the probability that it will occur.

$$P(\text{not } E) = 1 - P(E)$$

• The probability that an event E will occur is equal to 1 minus the probability that it will not occur.

$$P(E) = 1 - P(\text{not } E)$$

Using set notation, if E' is the complement of E, then $P(E') = 1 - P(E)$ and $P(E) = 1 - P(E')$.

Example 1 : The Probability of an Event Not Occurring

If you are dealt one card from a standard 52-card deck, find the probability that you are not dealt a queen.

Solution

Because

$$P(\text{not } E) = 1 - P(E)$$

then

$$P(\text{not a queen}) = 1 - P(\text{queen}).$$

There are four queens in a deck of 52 cards. The probability of being dealt a queen is $\frac{4}{52} = \frac{1}{13}$. Thus,

$$P(\text{not a queen}) = 1 - P(\text{queen}) = 1 - \frac{1}{13} = \frac{13}{13} - \frac{1}{13} = \frac{12}{13}.$$

The probability that you are not dealt a queen is $\frac{12}{13}$.

Checkpoint 1 If you are dealt one card from a standard 52-card deck, find the probability that you are not dealt a diamond.

Example 2 Using the Complement Rules

The circle graph in **Figure 11.6** shows the distribution, by age group, of the 191 million car drivers in the United States, with all numbers rounded to the nearest million. If one driver is randomly selected from this population, find the probability that the person

a. is not in the 20–29 age group.

b. is less than 80 years old.

Express probabilities as simplified fractions.

Number of U.S. Car Drivers, by Age Group

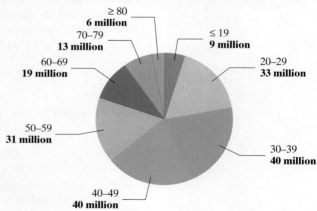

Figure 11.6
Source: U.S. Census Bureau

Solution

a. We begin with the probability that a randomly selected driver is not in the 20–29 age group.

$$P(\text{not in 20–29 age group}) = 1 - P(\text{in 20–29 age group})$$

$$= 1 - \frac{33}{191}$$

The graph shows 33 million drivers in the 20–29 age group.

This number, 191 million drivers, was given, but can be obtained by adding the numbers in the eight sectors.

$$= \frac{191}{191} - \frac{33}{191} = \frac{158}{191}$$

The probability that a randomly selected driver is not in the 20–29 age group is $\frac{158}{191}$.

b. We could compute the probability that a randomly selected driver is less than 80 years old by adding the numbers in each of the seven sectors representing drivers less than 80 and dividing this sum by 191 (million). However, it is easier to use complements. The complement of selecting a driver less than 80 years old is selecting a driver 80 or older.

$$P(\text{less than 80 years old}) = 1 - P(\text{80 or older})$$

$$= 1 - \frac{6}{191} \quad \boxed{\begin{array}{l}\text{The graph shows 6 million}\\\text{drivers 80 or older.}\end{array}}$$

$$= \frac{191}{191} - \frac{6}{191} = \frac{185}{191}$$

The probability that a randomly selected driver is less than 80 years old is $\frac{185}{191}$.

✓ **Checkpoint 2**   If one driver is randomly selected from the population represented in **Figure 11.6**, find the probability, expressed as a simplified fraction, that the person

a. is not in the 50–59 age group.

▶ **b.** is at least 20 years old.

Or Probabilities with Mutually Exclusive Events

 Find the probability of one event or a second event occurring.

Suppose that you randomly select one card from a deck of 52 cards. Let A be the event of selecting a king and B be the event of selecting a queen. Only one card is selected, so it is impossible to get both a king and a queen. The events of selecting a king and a queen cannot occur simultaneously. They are called *mutually exclusive events*.

> **Mutually Exclusive Events**
>
> If it is impossible for events A and B to occur simultaneously, the events are said to be **mutually exclusive**.

In general, if A and B are mutually exclusive events, the probability that either A or B will occur is determined by adding their individual probabilities.

> **Or Probabilities with Mutually Exclusive Events**
>
> If A and B are mutually exclusive events, then
> $$P(A \text{ or } B) = P(A) + P(B).$$
> Using set notation, $P(A \cup B) = P(A) + P(B)$.

Example 3   The Probability of Either of Two Mutually Exclusive Events Occurring

If one card is randomly selected from a deck of cards, what is the probability of selecting a king or a queen?

Solution

We find the probability that either of these mutually exclusive events will occur by adding their individual probabilities.

$$P(\text{king or queen}) = P(\text{king}) + P(\text{queen}) = \frac{4}{52} + \frac{4}{52} = \frac{8}{52} = \frac{2}{13}$$

The probability of selecting a king or a queen is $\frac{2}{13}$.

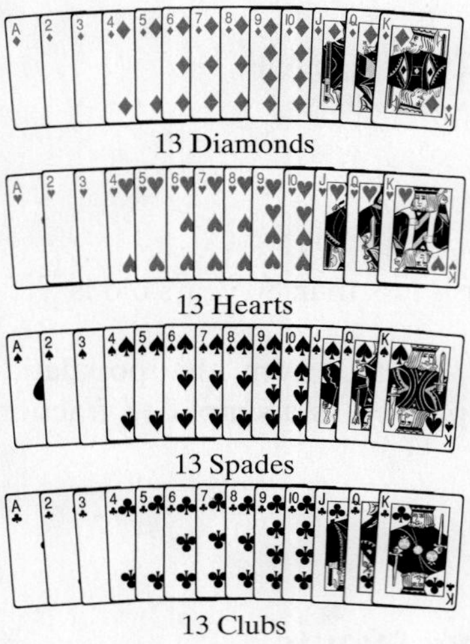

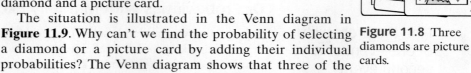

Figure 11.7 A deck of 52 cards

13 Diamonds

13 Hearts

13 Spades

13 Clubs

Figure 11.9

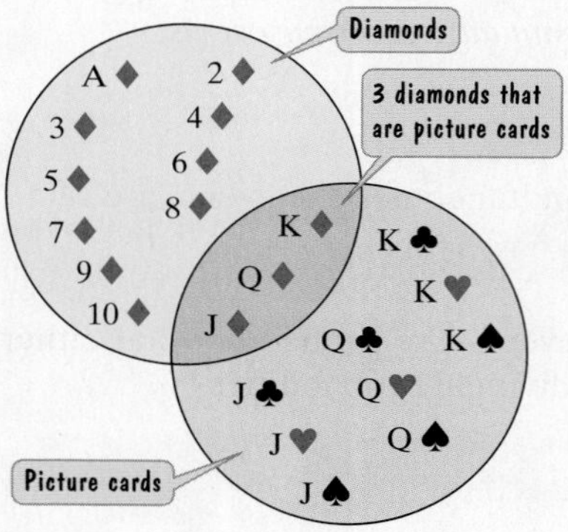

Diamonds

3 diamonds that
are picture cards

Picture cards

> **✓ Checkpoint 3** If you roll a single, six-sided die, what is the probability of getting either a 4 or a 5?

Or Probabilities with Events that are not Mutually Exclusive

Consider the deck of 52 cards shown in **Figure 11.7**. Suppose that these cards are shuffled and you randomly select one card from the deck. What is the probability of selecting a diamond or a picture card (jack, queen, king)? Begin by adding their individual probabilities.

$$P(\text{diamond}) + P(\text{picture card}) = \frac{13}{52} + \frac{12}{52}$$

There are 13 diamonds in the deck of 52 cards.

There are 12 picture cards in the deck of 52 cards.

However, this sum is not the probability of selecting a diamond or a picture card. The problem is that there are three cards that are *simultaneously* diamonds and picture cards, shown in **Figure 11.8**. The events of selecting a diamond and selecting a picture card are not mutually exclusive. It is possible to select a card that is both a diamond and a picture card.

The situation is illustrated in the Venn diagram in **Figure 11.9**. Why can't we find the probability of selecting a diamond or a picture card by adding their individual probabilities? The Venn diagram shows that three of the cards, the three diamonds that are picture cards, get counted twice when we add the individual probabilities. First the three cards get counted as diamonds and then they get counted as picture cards. In order to avoid the error of counting the three cards twice, we need to subtract the probability of getting a diamond and a picture card, $\frac{3}{52}$, as follows:

$$P(\text{diamond or picture card})$$
$$= P(\text{diamond}) + P(\text{picture card}) - P(\text{diamond and picture card})$$
$$= \frac{13}{52} + \frac{12}{52} - \frac{3}{52} = \frac{13 + 12 - 3}{52} = \frac{22}{52} = \frac{11}{26}.$$

Thus, the probability of selecting a diamond or a picture card is $\frac{11}{26}$.

In general, if A and B are events that are not mutually exclusive, the probability that A or B will occur is determined by adding their individual probabilities and then subtracting the probability that A and B occur simultaneously.

Figure 11.8 Three diamonds are picture cards.

> #### *Or* Probabilities with Events that are not Mutually Exclusive
>
> If A and B are not mutually exclusive events, then
> $$P(A \text{ or } B) = P(A) + P(B) - P(A \text{ and } B).$$
> Using set notation,
> $$P(A \cup B) = P(A) + P(B) - P(A \cap B).$$

Example 4 An *Or* Probability with Events That Are Not Mutually Exclusive

In a group of 25 baboons, 18 enjoy grooming their neighbors, 16 enjoy screeching wildly, while 10 enjoy grooming their neighbors and screeching wildly. If one baboon is selected at random from the group, find the probability that it enjoys grooming its neighbors or screeching wildly.

Solution

It is possible for a baboon in the group to enjoy both grooming its neighbors and screeching wildly. Ten of the brutes are given to engage in both activities. These events are not mutually exclusive.

$$P\left(\begin{array}{c}\text{grooming}\\\text{or screeching}\end{array}\right) = P(\text{grooming}) + P(\text{screeching}) - P\left(\begin{array}{c}\text{grooming}\\\text{and screeching}\end{array}\right)$$

$$= \frac{18}{25} + \frac{16}{25} - \frac{10}{25}$$

| 18 of the 25 baboons enjoy grooming. | 16 of the 25 baboons enjoy screeching. | 10 of the 25 baboons enjoy both. |

$$= \frac{18 + 16 - 10}{25} = \frac{24}{25}$$

The probability that a baboon in the group enjoys grooming its neighbors or screeching wildly is $\frac{24}{25}$.

> **Checkpoint 4** In a group of 50 students, 23 take math, 11 take psychology, and 7 take both math and psychology. If one student is selected at random, find the probability that the student takes math or psychology.

Example 5   An *Or* Probability with Events That Are Not Mutually Exclusive

Figure 11.10 illustrates a spinner. It is equally probable that the pointer will land on any one of the eight regions, numbered 1 through 8. If the pointer lands on a borderline, spin again. Find the probability that the pointer will stop on an even number or on a number greater than 5.

Solution

It is possible for the pointer to land on a number that is both even and greater than 5. Two of the numbers, 6 and 8, are even and greater than 5. These events are not mutually exclusive. The probability of landing on a number that is even or greater than 5 is calculated as follows:

$$P\left(\begin{array}{c}\text{even or}\\\text{greater than 5}\end{array}\right) = P(\text{even}) + P(\text{greater than 5}) - P\left(\begin{array}{c}\text{even and}\\\text{greater than 5}\end{array}\right)$$

$$= \frac{4}{8} + \frac{3}{8} - \frac{2}{8}$$

| Four of the eight numbers, 2, 4, 6, and 8, are even. | Three of the eight numbers, 6, 7, and 8, are greater than 5. | Two of the eight numbers, 6 and 8, are even and greater than 5. |

$$= \frac{4 + 3 - 2}{8} = \frac{5}{8}.$$

The probability that the pointer will stop on an even number or a number greater than 5 is $\frac{5}{8}$.

> **Checkpoint 5** Use **Figure 11.10** to find the probability that the pointer will stop on an odd number or a number less than 5.

Figure 11.10 It is equally probable that the pointer will land on any one of the eight regions.

Example 6   *Or* Probabilities with Real-World Data

Each year the Internal Revenue Service audits a sample of tax forms to verify their accuracy. **Table 11.5** shows the number of tax returns filed and audited in 2006, by taxable income.

Table 11.5 Tax Returns Filed and Audited, by Taxable Income, 2006

	< $25,000	$25,000–$49,999	$50,000–$99,999	≥ $100,000	Total
Audit	461,729	191,150	163,711	166,839	983,429
No audit	51,509,900	30,637,782	26,300,262	12,726,963	121,174,907
Total	51,971,629	30,828,932	26,463,973	12,893,802	122,158,336

Source: Internal Revenue Service

If one person is randomly selected from the population represented in **Table 11.5**, find the probability that

a. the taxpayer had a taxable income less than $25,000 or was audited.
b. the taxpayer had a taxable income less than $25,000 or at least $100,000.

Express probabilities as decimals rounded to the nearest hundredth.

Solution

a. It is possible to select a taxpayer who both earned less than $25,000 and was audited. Thus, these events are not mutually exclusive.

P(less than $25,000 or audited)

$= P$(less than $25,000) $+ P$(audited) $- P$(less than $25,000 and audited)

$$= \frac{51,971,629}{122,158,336} + \frac{983,429}{122,158,336} - \frac{461,729}{122,158,336}$$

Of the 122,158,336 taxpayers, 51,971,629 had taxable incomes less than $25,000.	Of the 122,158,336 taxpayers, 983,429 were audited.	Of the 122,158,336 taxpayers, 461,729 earned less than $25,000 and were audited.

$$= \frac{52,493,329}{122,158,336} \approx 0.43$$

The probability that a taxpayer had a taxable income less than $25,000 or was audited is approximately 0.43.

b. A taxable income of *at least* $100,000 means $100,000 or more. Thus, it is not possible to select a taxpayer with both a taxable income of less than $25,000 and at least $100,000. These events are mutually exclusive.

P(less than $25,000 or at least $100,000)

$$= P(\text{less than } \$25,000) + P(\text{at least } \$100,000)$$

$$= \frac{51,971,629}{122,158,336} + \frac{12,893,802}{122,158,336}$$

Of the 122,158,336 taxpayers, 51,971,629 had taxable incomes less than $25,000.	Of the 122,158,336 taxpayers, 12,893,802 had taxable incomes of $100,000 or more.

$$= \frac{64,865,431}{122,158,336} \approx 0.53$$

The probability that a taxpayer had a taxable income less than $25,000 or at least $100,000 is approximately 0.53.

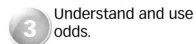 **Checkpoint 6**   If one person is randomly selected from the population represented in **Table 11.5**, find the probability, expressed as a decimal rounded to the nearest hundredth, that

a. the taxpayer had a taxable income of at least $100,000 or was not audited.

b. the taxpayer had a taxable income less than $25,000 or between $50,000 and $99,999, inclusive.

Odds

3 Understand and use odds.

If we know the probability of an event E, we can also speak of the *odds in favor*, or the *odds against*, the event. The following definitions link together the concepts of odds and probabilities:

Probability to Odds

If $P(E)$ is the probability of an event E occurring, then

1. The **odds in favor of E** are found by taking the probability that E will occur and dividing by the probability that E will not occur.

$$\text{Odds in favor of } E = \frac{P(E)}{P(\text{not } E)}$$

2. The **odds against E** are found by taking the probability that E will not occur and dividing by the probability that E will occur.

$$\text{Odds against } E = \frac{P(\text{not } E)}{P(E)}$$

The odds against E can also be found by reversing the ratio representing the odds in favor of E.

Example 7   From Probability to Odds

You roll a single, six-sided die.

a. Find the odds in favor of rolling a 2.

b. Find the odds against rolling a 2.

Solution

Let E represent the event of rolling a 2. In order to determine odds, we must first find the probability of E occurring and the probability of E not occurring. With $S = \{1, 2, 3, 4, 5, 6\}$ and $E = \{2\}$, we see that

$$P(E) = \frac{1}{6}$$

$$\text{and } P(\text{not } E) = 1 - \frac{1}{6} = \frac{6}{6} - \frac{1}{6} = \frac{5}{6}.$$

Now we are ready to construct the ratios for the odds in favor of E and the odds against E.

a. Odds in favor of E (rolling a 2) $= \dfrac{P(E)}{P(\text{not } E)} = \dfrac{\frac{1}{6}}{\frac{5}{6}} = \dfrac{1}{6} \cdot \dfrac{6}{5} = \dfrac{1}{5}$

The odds in favor of rolling a 2 are $\frac{1}{5}$. The ratio $\frac{1}{5}$ is usually written 1:5 and is read "1 to 5." Thus, the odds in favor of rolling a 2 are 1 to 5.

b. Now that we have the odds in favor of rolling a 2, namely $\frac{1}{5}$ or 1:5, we can find the odds against rolling a 2 by reversing this ratio. Thus,

$$\text{Odds against } E \text{ (rolling a 2)} = \frac{5}{1} \text{ or } 5:1.$$

The odds against rolling a 2 are 5 to 1.

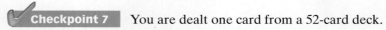 **Checkpoint 7** You are dealt one card from a 52-card deck.

a. Find the odds in favor of getting a red queen.
b. Find the odds against getting a red queen.

Example 8 From Probability to Odds

The winner of a raffle will receive a new sports utility vehicle. If 500 raffle tickets were sold and you purchased ten tickets, what are the odds against your winning the car?

Solution

Let E represent the event of winning the SUV. Because you purchased ten tickets and 500 tickets were sold,

$$P(E) = \frac{10}{500} = \frac{1}{50} \quad \text{and} \quad P(\text{not } E) = 1 - \frac{1}{50} = \frac{50}{50} - \frac{1}{50} = \frac{49}{50}.$$

Now we are ready to construct the ratio for the odds against E (winning the SUV).

$$\text{Odds against } E = \frac{P(\text{not } E)}{P(E)} = \frac{\frac{49}{50}}{\frac{1}{50}} = \frac{49}{50} \cdot \frac{50}{1} = \frac{49}{1}$$

The odds against winning the SUV are 49 to 1, written 49:1.

Checkpoint 8 The winner of a raffle will receive a two-year scholarship to the college of his or her choice. If 1000 raffle tickets were sold and you purchased five tickets, what are the odds against your winning the scholarship?

Now that we know how to convert from probability to odds, let's see how to convert from odds to probability. Suppose that the odds in favor of event E occurring are a to b. This means that

$$\frac{P(E)}{P(\text{not } E)} = \frac{a}{b} \quad \text{or} \quad \frac{P(E)}{1 - P(E)} = \frac{a}{b}.$$

By solving the equation on the right for $P(E)$, we obtain a formula for converting from odds to probability.

> **Odds to Probability**
>
> If the odds in favor of event E are a to b, then the probability of the event is given by
>
> $$P(E) = \frac{a}{a + b}.$$

Example 9   From Odds to Probability

The odds in favor of a particular candidate winning an election are 2 to 5. What is the probability that this candidate will win the election?

Solution

Because odds in favor, a to b, means a probability of $\frac{a}{a + b}$, then odds in favor, 2 to 5, means a probability of

$$\frac{2}{2 + 5} = \frac{2}{7}.$$

The probability that this candidate will win the election is $\frac{2}{7}$.

✓ **Checkpoint 9**   The odds against a particular candidate winning an election are 15 to 1. Find the odds in favor of the candidate winning the election and the probability of the candidate winning the election.

Exercise Set 11.6

Concept and Vocabulary Exercises

In Exercises 1–5, fill in each blank so that the resulting statement is true.

1. Because $P(E) + P(\text{not } E) = 1$, then $P(\text{not } E) = $ _____ and $P(E) = $ _____.

2. If it is impossible for events A and B to occur simultaneously, the events are said to be _____. For such events, $P(A \text{ or } B) = $ _____.

3. If it is possible for events A and B to occur simultaneously, then $P(A \text{ or } B) = $ _____.

4. The odds in favor of E can be found by taking the probability that _____ and dividing by the probability that _____.

5. If the odds in favor of event E are a to b, then the probability of the event, $P(E)$ is given by the formula _____.

In Exercises 6–9, determine whether each statement is true or false. If the statement is false, make the necessary change(s) to produce a true statement.

6. The probability that an event will not occur is equal to the probability that it will occur minus 1.

7. The probability of A or B can always be found by adding the probability of A and the probability of B.

8. The odds against E can always be found by reversing the ratio representing the odds in favor of E.

9. According to the National Center for Health Statistics, the lifetime odds in favor of dying from heart disease are 1 to 5, so the probability of dying from heart disease is $\frac{1}{5}$.

Respond to Exercises 10–16 using verbal or written explanations.

10. Explain how to find the probability of an event not occurring. Give an example.

11. What are mutually exclusive events? Give an example of two events that are mutually exclusive.

12. Explain how to find *or* probabilities with mutually exclusive events. Give an example.

13. Give an example of two events that are not mutually exclusive.

14. Explain how to find *or* probabilities with events that are not mutually exclusive. Give an example.

15. Explain how to find the odds in favor of an event if you know the probability that the event will occur.

16. Explain how to find the probability of an event if you know the odds in favor of that event.

In Exercises 17–20, determine whether each statement makes sense or does not make sense, and explain your reasoning.

17. The probability that Jill will win the election is 0.7 and the probability that she will not win is 0.4.

18. The probability of selecting a king from a deck of 52 cards is $\frac{4}{52}$ and the probability of selecting a heart is $\frac{13}{52}$, so the probability of selecting a king or a heart is $\frac{4}{52} + \frac{13}{52}$.

19. The probability of selecting a king or a heart from a deck of 52 cards is the same as the probability of selecting the king of hearts.

20. I estimate that the odds in favor of most college students getting married before receiving a bachelor's degree are 9:1.

Practice and Application Exercises

In Exercises 21–26, you are dealt one card from a 52-card deck. Find the probability that you are not dealt

21. an ace.   **22.** a 3.   **23.** a heart.

24. a club.   **25.** a picture card.

26. a red picture card.

The graph shows the probability of cardiovascular disease, by age and gender. Use the information in the graph to solve Exercises 27–28. Express all probabilities as decimals, estimated to two decimal places.

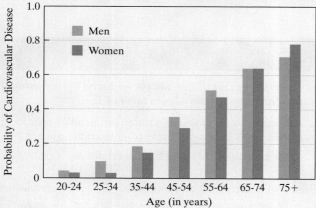

Probability of Cardiovascular Disease, by Age and Gender

Source: American Heart Association

27. a. What is the probability that a randomly selected man between the ages of 25 and 34 has cardiovascular disease?

b. What is the probability that a randomly selected man between the ages of 25 and 34 does not have cardiovascular disease?

28. a. What is the probability that a randomly selected woman, 75 or older, has cardiovascular disease?

b. What is the probability that a randomly selected woman, 75 or older, does not have cardiovascular disease?

The table shows the distribution, by annual income, of the 118 million households in the United States in 2007, with all numbers rounded to the nearest million. Use this distribution to solve Exercises 29–32.

Income Distribution of U.S. Households, in Millions

Annual Income	Number
Less than $10,000	8
$10,000–$14,999	7
$15,000–$24,999	14
$25,000–$34,999	13
$35,000–$49,999	17
$50,000–$74,999	21
$75,000–$99,999	14
$100,000 or more	24

Source: U.S. Census Bureau

If one household is randomly selected from this population, find the probability, expressed as a simplified fraction, that

29. the household income is not in the $50,000–$74,999 range.

30. the household income is not in the $15,000–$24,999 range.

31. the household income is less than $100,000.

32. the household income is at least $10,000.

In Exercises 33–38, you randomly select one card from a 52-card deck. Find the probability of selecting

33. a 2 or a 3.

34. a 7 or an 8.

35. a red 2 or a black 3.

36. a red 7 or a black 8.

37. the 2 of hearts or the 3 of spades.

38. the 7 of hearts or the 8 of spades.

39. The mathematics faculty at a college consists of 8 professors, 12 associate professors, 14 assistant professors, and 10 instructors. If one faculty member is randomly selected, find the probability of choosing a professor or an instructor.

40. A political discussion group consists of 30 Republicans, 25 Democrats, 8 Independents, and 4 members of the Green party. If one person is randomly selected from the group, find the probability of choosing an Independent or a member of the Green party.

In Exercises 41–42, a single die is rolled. Find the probability of rolling

41. an even number or a number less than 5.

42. an odd number or a number less than 4.

In Exercises 43–46, you are dealt one card from a 52-card deck. Find the probability that you are dealt

43. a 7 or a red card.

44. a 5 or a black card.

45. a heart or a picture card.

46. a card greater than 2 and less than 7, or a diamond.

In Exercises 47–50, it is equally probable that the pointer on the spinner shown will land on any one of the eight regions, numbered 1 through 8. If the pointer lands on a borderline, spin again.

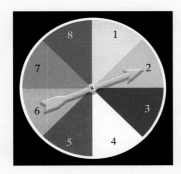

Find the probability that the pointer will stop on

47. an odd number or a number less than 6.

48. an odd number or a number greater than 3.

49. an even number or a number greater than 5.

50. an even number or a number less than 4.

Use this information to solve Exercises 51–54. The mathematics department of a college has 8 male professors, 11 female professors, 14 male teaching assistants, and 7 female teaching assistants. If a person is selected at random from the group, find the probability that the selected person is

51. a professor or a male.

52. a professor or a female.

53. a teaching assistant or a female.

54. a teaching assistant or a male.

55. In a class of 50 students, 29 are Democrats, 11 are business majors, and 5 of the business majors are Democrats. If one student is randomly selected from the class, find the probability of choosing a Democrat or a business major.

56. A student is selected at random from a group of 200 students in which 135 take math, 85 take English, and 65 take both math and English. Find the probability that the selected student takes math or English.

The table shows the educational attainment of the U.S. population, ages 25 and over, in 2007. Use the data in the table, expressed in millions, to solve Exercises 57–64.

Educational Attainment of The U.S. Population, Ages 25 and Over, in Millions

	Less Than 4 Years High School	4 Years High School Only	Some College (Less than 4 years)	4 Years College (or More)	Total
Male	14	25	20	23	82
Female	15	31	24	22	92
Total	29	56	44	45	174

Source: U.S. Census Bureau

Find the probability, expressed as a simplified fraction, that a randomly selected American, aged 25 or over,

57. has not completed four years (or more) of college.

58. has not completed four years of high school.

59. has completed four years of high school only or less than four years of college.

60. has completed less than four years of high school or four years of high school only.

61. has completed four years of high school only or is a man.

62. has completed four years of high school only or is a woman.

Find the odds in favor and the odds against a randomly selected American, aged 25 and over, with

63. four years (or more) of college.

64. less than four years of high school.

The graph shows the distribution, by branch and gender, of the 1.38 million, or 1380 thousand, active-duty personnel in the U.S. military in 2007. Numbers are given in thousands and rounded to the nearest ten thousand. Use the data to solve Exercises 65–76.

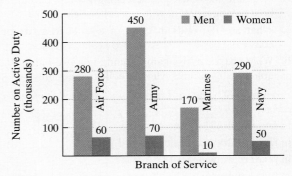

Active Duty U.S. Military Personnel

Source: U.S. Defense Department

If one person is randomly selected from the population represented in the bar graph at the bottom of the previous page, find the probability, expressed as a simplified fraction, that the person

65. is not in the Army.

66. is not in the Marines.

67. is in the Navy or is a man.

68. is in the Army or is a woman.

69. is in the Air Force or the Marines.

70. is in the Army or the Navy.

Find the odds in favor and the odds against a randomly selected person from the population represented in the bar graph on the previous page being

71. in the Navy.

72. in the Army.

73. a woman in the Marines.

74. a woman in the Air Force.

75. a man.

76. a woman.

In Exercises 77–80, a single die is rolled. Find the odds

77. in favor of rolling a number greater than 2.

78. in favor of rolling a number less than 5.

79. against rolling a number greater than 2.

80. against rolling a number less than 5.

The circle graphs show the percentage of children in the United States whose parents are college graduates in one-parent households and two-parent households. Use the information shown to solve Exercises 81–82.

**Percentage of U.S. Children
Whose Parents Are College Graduates**

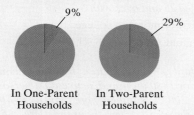

9%   29%

In One-Parent   In Two-Parent
Households   Households

Source: U.S. Census Bureau

81. a. What are the odds in favor of a child in a one-parent household having a parent who is a college graduate?

   b. What are the odds against a child in a one-parent household having a parent who is a college graduate?

82. a. What are the odds in favor of a child in a two-parent household having parents who are college graduates?

   b. What are the odds against a child in a two-parent household having parents who are college graduates?

In Exercises 83–92, one card is randomly selected from a deck of cards. Find the odds

83. in favor of drawing a heart.

84. in favor of drawing a picture card.

85. in favor of drawing a red card.

86. in favor of drawing a black card.

87. against drawing a 9.

88. against drawing a 5.

89. against drawing a black king.

90. against drawing a red jack.

91. against drawing a spade greater than 3 and less than 9.

92. against drawing a club greater than 4 and less than 10.

93. The winner of a raffle will receive a 21-foot outboard boat. If 1000 raffle tickets were sold and you purchased 20 tickets, what are the odds against your winning the boat?

94. The winner of a raffle will receive a 30-day all-expense-paid trip throughout Europe. If 5000 raffle tickets were sold and you purchased 30 tickets, what are the odds against your winning the trip?

Of the 38 plays attributed to Shakespeare, 18 are comedies, 10 are tragedies, and 10 are histories. In Exercises 95–102, one play is randomly selected from Shakespeare's 38 plays. Find the odds

95. in favor of selecting a comedy.

96. in favor of selecting a tragedy.

97. against selecting a history.

98. against selecting a comedy.

99. in favor of selecting a comedy or a tragedy.

100. in favor of selecting a tragedy or a history.

101. against selecting a tragedy or a history.

102. against selecting a comedy or a history.

103. If you are given odds of 3 to 4 in favor of winning a bet, what is the probability of winning the bet?

104. If you are given odds of 3 to 7 in favor of winning a bet, what is the probability of winning the bet?

105. Based on his skills in basketball, it was computed that when Michael Jordan shot a free throw, the odds in favor of his making it were 21 to 4. Find the probability that when Michael Jordan shot a free throw, he missed it. Out of every 100 free throws he attempted, on the average how many did he make?

106. The odds in favor of a person who is alive at age 20 still being alive at age 70 are 193 to 270. Find the probability that a person who is alive at age 20 will still be alive at age 70.

Exercises 107–108 give the odds against various flight risks. Use these odds to determine the probability of the underlined event for those in flight. (Source: Men's Health)

107. odds against <u>contracting an airborne disease</u>: 999 to 1

108. odds against <u>deep-vein thrombosis</u> (blood clot in the leg): 28 to 1

Critical Thinking Exercises

109. In Exercise 55, find the probability of choosing

   a. a Democrat who is not a business major;

   b. a student who is neither a Democrat nor a business major.

110. On New Year's Eve, the probability of a person driving while intoxicated or having a driving accident is 0.35. If the probability of driving while intoxicated is 0.32 and the probability of having a driving accident is 0.09, find the probability of a person having a driving accident while intoxicated.

111. The formula for converting from odds to probability is given in the box on page 719. Read the paragraph on the bottom of page 718 that precedes this box and derive the formula.

Events Involving *And*; Conditional Probability

1. Find the probability of one event and a second event occurring.

2. Compute conditional probabilities.

You are considering a job offer in South Florida. You were thrilled by images of Miami on MTV's *The Real World*. The job offer is just what you wanted and you are excited about living in the midst of Miami's tropical diversity. However, there is just one thing: the risk of hurricanes. You expect to stay in Miami ten years and buy a home. What is the probability that South Florida will be hit by a hurricane at least once in the next ten years?

In this section, we look at the probability that an event occurs at least once by expanding our discussion of probability to events involving *and*.

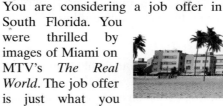

1. Find the probability of one event and a second event occurring.

And Probabilities with Independent Events

Consider tossing a fair coin two times in succession. The outcome of the first toss, heads or tails, does not affect what happens when you toss the coin a second time. For example, the occurrence of tails on the first toss does not make tails more likely or less likely to occur on the second toss. The repeated toss of a coin produces *independent events* because the outcome of one toss does not affect the outcome of others.

Study Tip

Do not confuse *independent events* and *mutually exclusive events*. Mutually exclusive events cannot occur at the same time. Independent events occur at different times, although they have no effect on each other.

Independent Events

Two events are **independent events** if the occurrence of either of them has no effect on the probability of the other.

When a fair coin is tossed two times in succession, the set of equally likely outcomes is

{heads heads, heads tails, tails heads, tails tails}.

We can use this set to find the probability of getting heads on the first toss and heads on the second toss:

$$P(\text{heads and heads}) = \frac{\text{number of ways two heads can occur}}{\text{total number of possible outcomes}} = \frac{1}{4}.$$

We can also determine the probability of two heads, $\frac{1}{4}$, without having to list all the equally likely outcomes. The probability of heads on the first toss is $\frac{1}{2}$. The probability of heads on the second toss is also $\frac{1}{2}$. The product of these probabilities, $\frac{1}{2} \cdot \frac{1}{2}$, results in the probability of two heads, namely $\frac{1}{4}$. Thus,

$$P(\text{heads and heads}) = P(\text{heads}) \cdot P(\text{heads}).$$

In general, if two events are independent, we can calculate the probability of the first occurring and the second occurring by multiplying their probabilities.

And Probabilities with Independent Events

If A and B are independent events, then

$$P(A \text{ and } B) = P(A) \cdot P(B).$$

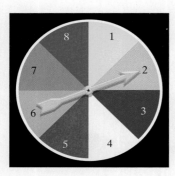

Figure 11.11 It is equally probable that the pointer will land on any one of the eight regions.

Example 1 — Independent Events on a Spinner

Figure 11.11 illustrates a spinner. It is equally probable that the pointer will land on any one of the eight regions, numbered 1 through 8. If the pointer lands on a borderline, spin again. Find the probability that the pointer will stop on a number less than 3 on two consecutive spins.

Solution

The spinner has eight equally likely outcomes and two of the outcomes, 1 and 2, are less than 3. Thus, the probability of landing on a number less than 3 on a spin is $\frac{2}{8}$, or $\frac{1}{4}$. The result that occurs on each spin is independent of all previous results. Thus,

$$P(\text{less than 3 and less than 3}) = P(\text{less than 3}) \cdot P(\text{less than 3}) = \frac{1}{4} \cdot \frac{1}{4} = \frac{1}{16}.$$

The probability that the pointer will stop on a number less than 3 on two consecutive spins is $\frac{1}{16}$.

Some people incorrectly believe that if a number less than 3 occurs on two consecutive spins, then a number greater than or equal to 3 is "due." Because the events are independent, the outcomes of previous spins have no effect on any other spins.

Checkpoint 1 Use **Figure 11.11** to find the probability that the pointer will stop on a number greater than 5 on two consecutive spins.

The *and* rule for independent events can be extended to cover three or more independent events. Thus, if A, B, and C are independent events, then

$$P(A \text{ and } B \text{ and } C) = P(A) \cdot P(B) \cdot P(C).$$

Example 2 — Independent Events in a Family

The picture in the margin shows a family that had nine girls in a row. Find the probability of this occurrence.

Solution

If two or more events are independent, we can find the probability of them all occurring by multiplying their probabilities. The probability of a baby girl is $\frac{1}{2}$, so the probability of nine girls in a row is $\frac{1}{2}$ used as a factor nine times.

$$P(\text{nine girls in a row}) = \frac{1}{2} \cdot \frac{1}{2} \cdot \frac{1}{2} \cdot \frac{1}{2} \cdot \frac{1}{2} \cdot \frac{1}{2} \cdot \frac{1}{2} \cdot \frac{1}{2} \cdot \frac{1}{2}$$

$$= \left(\frac{1}{2}\right)^9 = \frac{1}{512}$$

The probability of a run of nine girls in a row is $\frac{1}{512}$. (If another child is born into the family, this event is independent of the other nine and the probability of a girl is still $\frac{1}{2}$.)

Checkpoint 2 Find the probability of a family having four boys in a row.

Now let us return to the hurricane problem that opened this section. The Saffir/Simpson scale assigns numbers 1 through 5 to measure the disaster potential of a hurricane's winds. **Table 11.6** describes the scale. According to the National Hurricane Center, the probability that South Florida will be hit by a category 1 hurricane or higher in any single year is $\frac{5}{19}$, or approximately 0.26. In Example 3, we explore the risks of living in "Hurricane Alley."

Table 11.6 The Saffir/Simpson Hurricane Scale

Category	Winds (Miles per Hour)
1	74–95
2	96–110
3	111–130
4	131–155
5	> 155

| Example 3 | **Hurricanes and Probabilities** |

If the probability that South Florida will be hit by a hurricane in any single year is $\frac{5}{19}$,

a. What is the probability that South Florida will be hit by a hurricane in three consecutive years?

b. What is the probability that South Florida will not be hit by a hurricane in the next ten years?

Solution

a. The probability that South Florida will be hit by a hurricane in three consecutive years is

$$P(\text{hurricane and hurricane and hurricane})$$

$$= P(\text{hurricane}) \cdot P(\text{hurricane}) \cdot P(\text{hurricane}) = \frac{5}{19} \cdot \frac{5}{19} \cdot \frac{5}{19} = \frac{125}{6859} \approx 0.018.$$

b. We will first find the probability that South Florida will not be hit by a hurricane in any single year.

$$P(\text{no hurricane}) = 1 - P(\text{hurricane}) = 1 - \frac{5}{19} = \frac{19}{19} - \frac{5}{19} = \frac{14}{19} \approx 0.737$$

The probability of not being hit by a hurricane in a single year is $\frac{14}{19}$. Therefore, the probability of not being hit by a hurricane ten years in a row is $\frac{14}{19}$ used as a factor ten times.

$$P(\text{no hurricanes for ten years})$$

$$= P\left(\begin{array}{c}\text{no hurricane} \\ \text{for year 1}\end{array}\right) \cdot P\left(\begin{array}{c}\text{no hurricane} \\ \text{for year 2}\end{array}\right) \cdot P\left(\begin{array}{c}\text{no hurricane} \\ \text{for year 3}\end{array}\right) \cdot \cdots \cdot P\left(\begin{array}{c}\text{no hurricane} \\ \text{for year 10}\end{array}\right)$$

$$= \frac{14}{19} \quad \cdot \quad \frac{14}{19} \quad \cdot \quad \frac{14}{19} \quad \cdot \cdots \cdot \quad \frac{14}{19}$$

$$= \left(\frac{14}{19}\right)^{10} \approx (0.737)^{10} \approx 0.047$$

The probability that South Florida will not be hit by a hurricane in the next ten years is approximately 0.047.

Now we are ready to answer your question from the section opener:

What is the probability that South Florida will be hit by a hurricane at least once in the next ten years?

Because $P(\text{not } E) = 1 - P(E)$,

$$P(\text{no hurricane for ten years}) = 1 - P(\text{at least one hurricane in ten years}).$$

Equivalently,

$$P(\text{at least one hurricane in ten years}) = 1 - P(\text{no hurricane for ten years})$$
$$= 1 - 0.047 = 0.953.$$

With a probability of 0.953, it is nearly certain that South Florida will be hit by a hurricane at least once in the next ten years.

The Probability of an Event Happening at Least Once

$$P(\text{event happening at least once}) = 1 - P(\text{event does not happen})$$

✓ **Checkpoint 3** If the probability that South Florida will be hit by a hurricane in any single year is $\frac{5}{19}$,

a. What is the probability that South Florida will be hit by a hurricane in four consecutive years?

b. What is the probability that South Florida will not be hit by a hurricane in the next four years?

c. What is the probability that South Florida will be hit by a hurricane at least once in the next four years?

▶ Express all probabilities as fractions and as decimals rounded to three places.

And Probabilities with Dependent Events

5 chocolate-covered cherries lie within the 20 pieces.

Once a chocolate-covered cherry is selected, only 4 chocolate-covered cherries lie within the remaining 19 pieces.

Chocolate lovers, please help yourselves! There are 20 mouth-watering tidbits to select from. What's that? You want 2? And you prefer chocolate-covered cherries? The problem is that there are only 5 chocolate-covered cherries and it's impossible to tell what is inside each piece. They're all shaped exactly alike. At any rate, reach in, select a piece, enjoy, choose another piece, eat, and be well. There is nothing like savoring a good piece of chocolate in the midst of all this chit-chat about probability and hurricanes.

Another question? You want to know what your chances are of selecting 2 chocolate-covered cherries? Well, let's see. Five of the 20 pieces are chocolate-covered cherries, so the probability of getting one of them on your first selection is $\frac{5}{20}$, or $\frac{1}{4}$. Now, suppose that you did choose a chocolate-covered cherry on your first pick. Eat it slowly; there's no guarantee that you'll select your favorite on the second selection. There are now only 19 pieces of chocolate left. Only 4 are chocolate-covered cherries. The probability of getting a chocolate-covered cherry on your second try is 4 out of 19, or $\frac{4}{19}$. This is a different probability than the $\frac{1}{4}$ probability on your first selection. Selecting a chocolate-covered cherry the first time changes what is in the candy box. The probability of what you select the second time *is* affected by the outcome of the first event. For this reason, we say that these are *dependent events*.

> ### Dependent Events
>
> Two events are **dependent events** if the occurrence of one of them has an effect on the probability of the other.

The probability of selecting two chocolate-covered cherries in a row can be found by multiplying the $\frac{1}{4}$ probability on the first selection by the $\frac{4}{19}$ probability on the second selection:

P(chocolate-covered cherry and chocolate-covered cherry)

$$= P(\text{chocolate-covered cherry}) \cdot P\left(\begin{array}{c}\text{chocolate-covered cherry}\\\text{given that one was selected}\end{array}\right)$$

$$= \frac{1}{4} \cdot \frac{4}{19} = \frac{1}{19}.$$

The probability of selecting two chocolate-covered cherries in a row is $\frac{1}{19}$. This is a special case of finding the probability that each of two dependent events occurs.

> ### *And* Probabilities with Dependent Events
>
> If A and B are dependent events, then
>
> $$P(A \text{ and } B) = P(A) \cdot P(B \text{ given that } A \text{ has occurred}).$$

Example 4 An *And* Probability with Dependent Events

Good news: You won a free trip to Madrid and can take two people with you, all expenses paid. Bad news: Ten of your cousins have appeared out of nowhere and are begging you to take them. You write each cousin's name on a card, place the cards in a hat, and select one name. Then you select a second name without replacing the first card. If three of your ten cousins speak Spanish, find the probability of selecting two Spanish-speaking cousins.

Solution

Because $P(A \text{ and } B) = P(A) \cdot P(B \text{ given that } A \text{ has occurred})$, then

$P(\text{two Spanish-speaking cousins})$

$= P(\text{speaks Spanish and speaks Spanish})$

$= P(\text{speaks Spanish}) \cdot P\left(\begin{array}{c}\text{speaks Spanish given that a Spanish-speaking}\\ \text{cousin was selected first}\end{array}\right)$

$= \dfrac{3}{10} \quad \cdot \quad \dfrac{2}{9}$

> There are ten cousins, three of whom speak Spanish.

> After picking a Spanish-speaking cousin, there are nine cousins left, two of whom speak Spanish.

$= \dfrac{6}{90} = \dfrac{1}{15} \approx 0.067.$

The probability of selecting two Spanish-speaking cousins is $\frac{1}{15}$.

> **Checkpoint 4** You are dealt two cards from a 52-card deck. Find the probability of getting two kings.

The multiplication rule for dependent events can be extended to cover three or more dependent events. For example, in the case of three such events,

$P(A \text{ and } B \text{ and } C)$

$= P(A) \cdot P(B \text{ given that } A \text{ occurred}) \cdot P(C \text{ given that } A \text{ and } B \text{ occurred}).$

Study Tip

Example 4 can also be solved using the combinations formula.

$P(\text{two Spanish speakers})$

$= \dfrac{\begin{array}{c}\text{number of ways of}\\ \text{selecting 2 Spanish-}\\ \text{speaking cousins}\end{array}}{\begin{array}{c}\text{number of ways of}\\ \text{selecting 2 cousins}\end{array}}$

$= \dfrac{{}_3C_2}{{}_{10}C_2}$

> 2 Spanish speakers selected from 3 Spanish-speaking cousins

> 2 cousins selected from 10 cousins

$= \dfrac{3}{45} = \dfrac{1}{15}$

Example 5 An *And* Probability with Three Dependent Events

Three people are randomly selected, one person at a time, from five freshmen, two sophomores, and four juniors. Find the probability that the first two people selected are freshmen and the third is a junior.

Solution

$P(\text{first two are freshmen and the third is a junior})$

$= P(\text{freshman}) \cdot P\left(\begin{array}{c}\text{freshman given that a}\\ \text{freshman was selected first}\end{array}\right) \cdot P\left(\begin{array}{c}\text{junior given that a freshman was}\\ \text{selected first and a freshman was}\\ \text{selected second}\end{array}\right)$

$= \dfrac{5}{11} \quad \cdot \quad \dfrac{4}{10} \quad \cdot \quad \dfrac{4}{9}$

> There are 11 people, five of whom are freshmen.

> After picking a freshman, there are 10 people left, four of whom are freshmen.

> After the first two selections, 9 people are left, four of whom are juniors.

$= \dfrac{8}{99}$

The probability that the first two people selected are freshmen and the third is a junior is $\frac{8}{99}$.

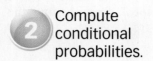

Compute conditional probabilities.

Checkpoint 5 You are dealt three cards from a 52-card deck. Find the probability of getting three hearts.

Conditional Probability

We have seen that for any two dependent events A and B,

$$P(A \text{ and } B) = P(A) \cdot P(B \text{ given that } A \text{ occurs}).$$

The probability of B given that A occurs is called *conditional probability*, denoted by $P(B \mid A)$.

Conditional Probability

The probability of event B, assuming that event A has already occurred, is called the **conditional probability** of B, given A. This probability is denoted by $P(B \mid A)$.

It is helpful to think of the conditional probability $P(B \mid A)$ as the **probability that event B occurs if the sample space is restricted to the outcomes associated with event A.**

Example 6 **Finding Conditional Probability**

A letter is randomly selected from the letters of the English alphabet. Find the probability of selecting a vowel, given that the outcome is a letter that precedes h.

Solution

We are looking for

$$P(\text{vowel}|\text{letter precedes h}).$$

This is the probability of a vowel if the sample space is restricted to the set of letters that precede h. Thus, the sample space is given by

$$S = \{a, b, c, d, e, f, g\}.$$

There are seven possible outcomes in the sample space. We can select a vowel from this set in one of two ways: a or e. Therefore, the probability of selecting a vowel, given that the outcome is a letter that precedes h, is $\frac{2}{7}$.

$$P(\text{vowel}|\text{letter precedes h}) = \tfrac{2}{7}$$

Checkpoint 6 A letter is randomly selected from the letters of the English alphabet. Find the probability of selecting a letter that precedes h, given that the outcome is a vowel. (Do not include the letter y among the vowels.)

Example 7 **Finding Conditional Probability**

You are dealt one card from a 52-card deck.

a. Find the probability of getting a heart, given that the card you were dealt is a red card.

b. Find the probability of getting a red card, given that the card you were dealt is a heart.

Solution

a. We begin with

$$P(\text{heart}\,|\,\text{red card}).$$

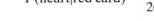

Probability of getting a heart if the sample space is restricted to the set of red cards

The sample space is shown in **Figure 11.12**. There are 26 outcomes in the sample space. We can get a heart from this set in 13 ways. Thus,

$$P(\text{heart}|\text{red card}) = \frac{13}{26} = \frac{1}{2}.$$

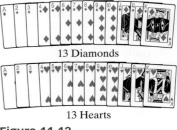

13 Diamonds

13 Hearts

Figure 11.12

b. We now find

$$P(\text{red card}\,|\,\text{heart}).$$

Probability of getting a red card if the sample space is restricted to the set of hearts

The sample space is shown in **Figure 11.13**. There are 13 outcomes in the sample space. All of the outcomes are red. We can get a red card from this set in 13 ways. Thus,

$$P(\text{red card}|\text{heart}) = \frac{13}{13} = 1.$$

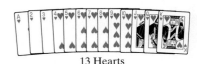

13 Hearts

Figure 11.13

Example 7 illustrates that $P(\text{heart}|\text{red card})$ is not equal to $P(\text{red card}|\text{heart})$. In general, $P(B|A) \neq P(A|B)$.

 Checkpoint 7 You are dealt one card from a 52-card deck.

a. Find the probability of getting a black card, given the card you were dealt is a spade.

b. Find the probability of getting a spade, given the card you were dealt is a black card.

Example 8   Conditional Probabilities with Real-World Data

When women turn 40, their gynecologists typically remind them that it is time to undergo mammography screening for breast cancer. The data in **Table 11.7** are based on 100,000 U.S. women, ages 40 to 50, who participated in mammography screening.

Table 11.7 Mammography Screening on 100,000 U.S. Women, Ages 40 to 50			
	Breast Cancer	**No Breast Cancer**	**Total**
Positive Mammogram	720	6944	7664
Negative Mammogram	80	92,256	92,336
Total	800	99,200	100,000

Source: Gerd Gigerenzer, *Calculated Risks*. Simon and Schuster, 2002.

	Breast Cancer	No Breast Cancer
Positive Mammogram	720	6944
Negative Mammogram	80	92,256

Table 11.7 (partly repeated)

Assuming that these numbers are representative of all U.S. women ages 40 to 50, find the probability that a woman in this age range

a. has a positive mammogram, given that she does not have breast cancer.

b. does not have breast cancer, given that she has a positive mammogram.

Solution

a. We begin with the probability that a U.S. woman aged 40 to 50 has a positive mammogram, given that she does not have breast cancer:

$$P(\text{positive mammogram}|\text{no breast cancer}).$$

This is the probability of a positive mammogram if the data are restricted to women without breast cancer:

	No Breast Cancer
Positive Mammogram	6944
Negative Mammogram	92,256
Total	99,200

Within the restricted data, there are 6944 women with positive mammograms and $6944 + 92{,}256$, or 99,200 women without breast cancer. Thus,

$$P(\text{positive mammogram}|\text{no breast cancer}) = \frac{6944}{99{,}200} = 0.07.$$

Among women without breast cancer, the probability of a positive mammogram is 0.07.

b. Now, we find the probability that a U.S. woman aged 40 to 50 does not have breast cancer, given that she has a positive mammogram:

$$P(\text{no breast cancer}|\text{positive mammogram}).$$

This is the probability of not having breast cancer if the data are restricted to women with positive mammograms:

	Breast Cancer	No Breast Cancer	Total
Positive Mammogram	720	6944	7664

Within the restricted data, there are 6944 women without breast cancer and $720 + 6944$, or 7664 women with positive mammograms. Thus,

$$P(\text{no breast cancer} \mid \text{positive mammogram}) = \frac{6944}{7664} \approx 0.906.$$

Among women with positive mammograms, the probability of not having breast cancer is $\frac{6944}{7664}$, or approximately 0.906.

Study Tip

Example 8 shows that the probability of a positive mammogram among women without breast cancer, 0.07, is not the same as the probability of not having breast cancer among women with positive mammograms, approximately 0.9. We have seen that the conditional probability of B, given A, is not the same as the conditional probability of A, given B:

$$P(B|A) \neq P(A|B).$$

Checkpoint 8 Use the data in **Table 11.7** at the bottom of the previous page to find the probability that a U.S. woman aged 40 to 50

a. has a positive mammogram, given that she has breast cancer.

b. has breast cancer, given that she has a positive mammogram.

▶ Express probabilities as decimals and, if necessary, round to three decimal places.

We have seen that $P(B|A)$ is the probability that event B occurs if the sample space is restricted to event A. Thus,

$$P(B|A) = \frac{\text{number of outcomes of } B \text{ that are in the restricted sample space } A}{\text{number of outcomes in the restricted sample space } A}.$$

This can be stated in terms of the following formula:

A Formula for Conditional Probability

$$P(B|A) = \frac{n(B \cap A)}{n(A)} = \frac{\text{number of outcomes common to } B \text{ and } A}{\text{number of outcomes in } A}$$

Achieving Success

FOXTROT © 2008 Bill Amend

A test-taking tip: Live for partial credit.

Always show your work. If worse comes to worst, write *something* down, anything, even if it's a formula that you think might solve a problem or a possible idea or procedure for solving the problem. As Bill Amend's *FoxTrot* cartoon indicates, partial credit has salvaged more than a few test scores.

Exercise Set 11.7

Concept and Vocabulary Exercises

In Exercises 1–4, fill in each blank so that the resulting statement is true.

1. If the occurrence of one event has no effect on the probability of another event, the events are said to be _____. For such events, $P(A \text{ and } B) = $ _____.

2. The probability of an event occurring at least once is equal to 1 minus the probability that _____.

3. If the occurrence of one event has an effect on the probability of another event, the events are said to be _____. For such events, $P(A \text{ and } B) = $ _____.

4. The probability of event B, assuming that event A has already occurred, is called the _____ probability of B, given A. This probability is denoted by _____.

In Exercises 5–8, determine whether each statement is true or false. If the statement is false, make the necessary change(s) to produce a true statement.

5. *And* probabilities can always be determined using the formula $P(A \text{ and } B) = P(A) \cdot P(B)$.

6. Probability problems with the word *and* involve more than one selection.

7. The probability that an event happens at least once can be found by subtracting the probability that the event does not happen from 1.

8. $P(B|A)$ is the probability that event B occurs if the sample space is restricted to the outcomes associated with event A.

Respond to Exercises 9–10 using verbal or written explanations.

9. Explain how to find *and* probabilities with independent events. Give an example.

10. Explain how to find *and* probabilities with dependent events. Give an example.

In Exercises 11–14, determine whether each statement makes sense or does not make sense, and explain your reasoning.

11. If a fourth child is born into a family with three boys, the odds in favor of a girl are better than 1:1.

12. In a group of five men and five women, the probability of randomly selecting a man is $\frac{1}{2}$, so if I select two people from the group, the probability that both are men is $\frac{1}{2} \cdot \frac{1}{2}$.

13. I found the probability of getting rain at least once in ten days by calculating the probability that none of the days have rain and subtracting this probability from 1.

14. I must have made an error calculating probabilities because $P(A|B)$ is not the same as $P(B|A)$.

Practice and Application Exercises

Exercises 15–40 involve probabilities with independent events.

Use the spinner shown to solve Exercises 15–24. It is equally probable that the pointer will land on any one of the six regions. If the pointer lands on a borderline, spin again. If the pointer is spun twice, find the probability it will land on

15. green and then red.

16. yellow and then green.

17. yellow and then yellow.

18. red and then red.

19. a color other than red each time.

20. a color other than green each time.

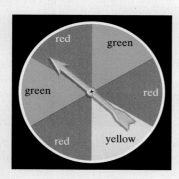

If the pointer is spun three times, find the probability it will land on

21. green and then red and then yellow.

22. red and then red and then green.

23. red every time.   24. green every time.

In Exercises 25–28, a single die is rolled twice. Find the probability of rolling

25. a 2 the first time and a 3 the second time.

26. a 5 the first time and a 1 the second time.

27. an even number the first time and a number greater than 2 the second time.

28. an odd number the first time and a number less than 3 the second time.

In Exercises 29–34, you draw one card from a 52-card deck. Then the card is replaced in the deck, the deck is shuffled, and you draw again. Find the probability of drawing

29. a picture card the first time and a heart the second time.

30. a jack the first time and a club the second time.

31. a king each time.   32. a 3 each time.

33. a red card each time.   34. a black card each time.

35. If you toss a fair coin six times, what is the probability of getting all heads?

36. If you toss a fair coin seven times, what is the probability of getting all tails?

In Exercises 37–38, a coin is tossed and a die is rolled. Find the probability of getting

37. a head and a number greater than 4.

38. a tail and a number less than 5.

39. The probability that South Florida will be hit by a major hurricane (category 4 or 5) in any single year is $\frac{1}{16}$. (*Source:* National Hurricane Center)
    a. What is the probability that South Florida will be hit by a major hurricane two years in a row?
    b. What is the probability that South Florida will be hit by a major hurricane in three consecutive years?
    c. What is the probability that South Florida will not be hit by a major hurricane in the next ten years?
    d. What is the probability that South Florida will be hit by a major hurricane at least once in the next ten years?

40. The probability that a region prone to flooding will flood in any single year is $\frac{1}{10}$.
    a. What is the probability of a flood two years in a row?
    b. What is the probability of flooding in three consecutive years?
    c. What is the probability of no flooding for ten consecutive years?
    d. What is the probability of flooding at least once in the next ten years?

The graph shows that U.S. adults dependent on tobacco have a greater probability of suffering from some ailments than the general adult population. When making two or more selections from populations with large numbers, such as the U.S. adult population or the population dependent on tobacco, we assume that each selection is independent of every other selection. In Exercises 41–46, assume that the selections are independent events.

Probability That U.S. Adults Suffer from Various Ailments

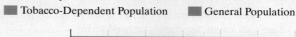

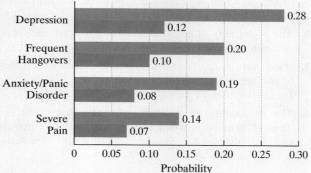

Source: MARS 2005 OTC/DTC

41. If two adults are randomly selected from the general population, what is the probability that they both suffer from depression?

42. It two adults are randomly selected from the population of cigarette smokers, what is the probability that they both suffer from depression?

43. If three adults are randomly selected from the population of cigarette smokers, what is the probability that they all suffer from frequent hangovers?

44. If three adults are randomly selected from the general population, what is the probability that they all suffer from frequent hangovers?

45. If three adults are randomly selected from the population of cigarette smokers, what is the probability, expressed as a decimal correct to four places, that at least one person suffers from anxiety/panic disorder?

46. If three adults are randomly selected from the population of cigarette smokers, what is the probability, expressed as a decimal correct to four places, that at least one person suffers from severe pain?

Exercises 47–62 involve probabilities with dependent events.

In Exercises 47–50, we return to our box of chocolates. There are 30 chocolates in the box, all identically shaped. Five are filled with coconut, 10 with caramel, and 15 are solid chocolate. You randomly select one piece, eat it, and then select a second piece. Find the probability of selecting

47. two solid chocolates in a row.

48. two caramel-filled chocolates in a row.

49. a coconut-filled chocolate followed by a caramel-filled chocolate.

50. a coconut-filled chocolate followed by a solid chocolate.

In Exercises 51–56, consider a political discussion group consisting of 5 Democrats, 6 Republicans, and 4 Independents. Suppose that two group members are randomly selected, in succession, to attend a political convention. Find the probability of selecting

51. two Democrats.

52. two Republicans.

53. an Independent and then a Republican.

54. an Independent and then a Democrat.

55. no Independents.

56. no Democrats.

In Exercises 57–62, an ice chest contains six cans of apple juice, eight cans of grape juice, four cans of orange juice, and two cans of mango juice. Suppose that you reach into the container and randomly select three cans in succession. Find the probability of selecting

57. three cans of apple juice.

58. three cans of grape juice.

59. a can of grape juice, then a can of orange juice, then a can of mango juice.

60. a can of apple juice, then a can of grape juice, then a can of orange juice.

61. no grape juice.

62. no apple juice.

In Exercises 63–70, the numbered disks shown are placed in a box and one disk is selected at random.

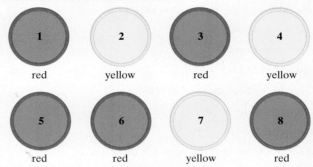

Find the probability of selecting

63. a 3, given that a red disk is selected.

64. a 7, given that a yellow disk is selected.

65. an even number, given that a yellow disk is selected.

66. an odd number, given that a red disk is selected.

67. a red disk, given that an odd number is selected.

68. a yellow disk, given that an odd number is selected.

69. a red disk, given that the number selected is at least 5.

70. a yellow disk, given that the number selected is at most 3.

The table shows the outcome of car accidents in Florida for a recent year by whether or not the driver wore a seat belt. Use the data to solve Exercises 71–74. Express probabilities as fractions and as decimals rounded to three places.

Car Accidents in Florida

	Wore Seat Belt	**No Seat Belt**	**Total**
Driver Survived	412,368	162,527	574,895
Driver Died	510	1601	2111
Total	412,878	164,128	577,006

Source: Alan Agresti and Christine Franklin, *Statistics,* Prentice Hall, 2007.

71. Find the probability of surviving a car accident, given that the driver wore a seat belt.

72. Find the probability of not surviving a car accident, given that the driver did not wear a seat belt.

73. Find the probability of wearing a seat belt, given that a driver survived a car accident.

74. Find the probability of not wearing a seat belt, given that a driver did not survive a car accident.

In Exercises 75–86, we return to the table showing the distribution, by marital status and gender, of the 235.8 million Americans ages 18 or older.

Marital Status of The U.S. Population, Ages 18 or Older, in Millions

	Never Married	**Married**	**Widowed**	**Divorced**	**Total**
Male	37.5	64.7	2.7	9.6	114.5
Female	31.7	65.2	11.2	13.2	121.3
Total	69.2	129.9	13.9	22.8	235.8

Source: U.S. Census Bureau

If one person is selected from the population described in the table at the bottom of the previous page, find the probability, expressed as a decimal rounded to three places, that the person

75. is not divorced.

76. is not widowed.

77. is widowed or divorced.

78. has never been married or is divorced.

79. is male or divorced.

80. is female or divorced.

81. is male, given that this person is divorced.

82. is female, given that this person is divorced.

83. is widowed, given that this person is a woman.

84. is divorced, given that this person is a man.

85. has never been married or is married, given that this person is a man.

86. has never been married or is married, given that this person is a woman.

Critical Thinking Exercises

87. If the probability of being hospitalized during a year is 0.1, find the probability that no one in a family of five will be hospitalized in a year.

88. If a single die is rolled five times, what is the probability it lands on 2 on the first, third, and fourth rolls, but not on either of the other rolls?

89. **Probabilities and Coincidence of Shared Birthdays**

    a. If two people are selected at random, the probability that they do not have the same birthday (day and month) is $\frac{365}{365} \cdot \frac{364}{365}$. Explain why this is so. (Ignore leap years and assume 365 days in a year.)

    b. If three people are selected at random, find the probability that they all have different birthdays.

    c. If three people are selected at random, find the probability that at least two of them have the same birthday.

    d. If 20 people are selected at random, find the probability that at least 2 of them have the same birthday.

    e. How large a group is needed to give a 0.5 chance of at least two people having the same birthday?

90. Nine cards numbered from 1 through 9 are placed into a box and two cards are selected without replacement. Find the probability that both numbers selected are odd, given that their sum is even.

91. If a single die is rolled twice, find the probability of rolling an odd number and a number greater than 4 in either order.

In Exercises 92–96, write a probability problem involving the word "and" whose solution results in the probability fractions shown.

92. $\frac{1}{2} \cdot \frac{1}{2}$

93. $\frac{1}{6} \cdot \frac{1}{6} \cdot \frac{1}{6}$

94. $\frac{1}{2} \cdot \frac{1}{6}$

95. $\frac{13}{52} \cdot \frac{12}{51}$

96. $\frac{1}{4} \cdot \frac{3}{5}$

Group Exercises

97. Do you live in an area prone to catastrophes, such as earthquakes, fires, tornados, hurricanes, or floods? If so, research the probability of this catastrophe occurring in a single year. Group members should then use this probability to write and solve a problem similar to Exercise 39 in this exercise set.

98. Group members should use the table for Exercises 75–86 to write and solve four probability problems different than those in the exercises. Two should involve *or* (one with events that are mutually exclusive and one with events that are not), one should involve *and*—that is, events in succession—and one should involve conditional probability.

Chapter 11 Summary

11.1 The Fundamental Counting Principle

Definitions and Concepts

The Fundamental Counting Principle

The number of ways in which a series of successive things can occur is found by multiplying the number of ways in which each thing can occur.

Example

- A car is available in 6 possible colors, with or without air conditioning, with or without a global positioning system, and electric, gas powered, or hybrid.

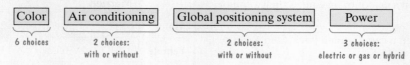

Color	Air conditioning	Global positioning system	Power
6 choices	2 choices: with or without	2 choices: with or without	3 choices: electric or gas or hybrid

The car can be ordered in $6 \cdot 2 \cdot 2 \cdot 3 = 72$ different ways.

Additional Examples to Review

Example 1, page 675; Example 2, page 675; Example 3, page 676; Example 4, page 676;
Example 5, page 677; Example 6, page 677

11.2 Permutations

Definitions and Concepts

A permutation from a group of items occurs when no item is used more than once and the order of arrangement makes a difference. The Fundamental Counting Principle can be used to determine the number of permutations possible.

Example

- Seven acts (A, B, C, D, E, F, and G) perform in a variety show. B performs first, F performs next-to-last, and A performs last.

Act 1	Act 2	Act 3	Act 4	Act 5	Act 6	Act 7
1 choice: B	4 choices: C, D, E or G	3 choices	2 choices	1 choice	1 choice: F	1 choice: A

There are $1 \cdot 4 \cdot 3 \cdot 2 \cdot 1 \cdot 1 \cdot 1 = 24$ different ways to schedule their appearances.

Additional Examples to Review

Example 1, page 681; Example 2, page 681

Definitions and Concepts

Factorial Notation

$$n! = n(n-1)(n-2)\cdots(3)(2)(1) \text{ and } 0! = 1$$

Example

$$\frac{13!}{7!} = \frac{13 \cdot 12 \cdot 11 \cdot 10 \cdot 9 \cdot 8 \cdot \cancel{7!}}{\cancel{7!}} = 13 \cdot 12 \cdot 11 \cdot 10 \cdot 9 \cdot 8 = 1{,}235{,}520$$

Additional Example to Review

Example 3, page 682

Definitions and Concepts

Permutations Formula

The number of permutations possible if r items are taken from n items is $_nP_r = \dfrac{n!}{(n-r)!}$.

Example

If 16 equally qualified applicants fill five different positions, this can be done in

$$_{16}P_5 = \frac{16!}{(16-5)!} = \frac{16!}{11!} = \frac{16 \cdot 15 \cdot 14 \cdot 13 \cdot 12 \cdot \cancel{11!}}{\cancel{11!}} = 524{,}160 \text{ ways}$$

Additional Examples to Review

Example 4, page 684; Example 5, page 684

Definitions and Concepts

Permutations of Duplicate Items

The number of permutations of n items, where p items are identical, q items are identical, r items are identical, and so on, is

$$\frac{n!}{p!\, q!\, r! \ldots}.$$

Example

• The number of distinct ways the letters of the word ERRATA can be arranged is

6 letters:
$n = 6$

2 identical Rs
and 2 identical As

$$\frac{6!}{2!\, 2!} = \frac{6 \cdot 5 \cdot 4 \cdot 3 \cdot 2 \cdot 1}{2 \cdot 1 \cdot 2 \cdot 1} = 180.$$

Additional Example to Review

Example 6, page 685

11.3 Combinations

Definitions and Concepts

A combination from a group of items occurs when no item is used more than once and the order of items makes no difference.

Combinations Formula

The number of combinations possible if r items are taken from n items is $_nC_r = \dfrac{n!}{(n-r)!\, r!}$.

Example

• The number of five-person committees that can be formed from nine people is

$$_9C_5 = \frac{9!}{(9-5)!5!} = \frac{9!}{4!5!} = \frac{9 \cdot 8 \cdot 7 \cdot 6 \cdot 5!}{4 \cdot 3 \cdot 2 \cdot 1 \cdot 5!} = 126.$$

Additional Examples to Review

Example 1, page 688; Example 2, page 690; Example 3, page 690; Example 4, page 691

11.4 Fundamentals of Probability

Definitions and Concepts

Theoretical probability applies to experiments in which the set of all equally likely outcomes, called the sample space, is known. An event is any subset of the sample space.

The theoretical probability of event E with sample space S is

$$P(E) = \frac{\text{number of outcomes in } E}{\text{total number of possible outcomes}} = \frac{n(E)}{n(S)}.$$

Example

• 3 freshmen, 4 sophomores, 2 juniors, 1 senior

Probability of selecting a junior:

$$P(\text{junior}) = \frac{\text{number of outcomes that result in a junior}}{\text{total number of possible outcomes}} = \frac{2}{3+4+2+1} = \frac{2}{10} = \frac{1}{5}$$

Additional Examples to Review

Example 1, page 696; Example 2, page 697; Example 3, page 698

Definitions and Concepts

Empirical probability applies to situations in which we observe the frequency of the occurrence of an event.

The empirical probability of event E is

$$P(E) = \frac{\text{observed number of times } E \text{ occurs}}{\text{total number of observed occurrences}}.$$

Example

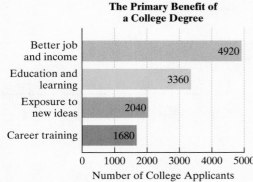

The Primary Benefit of a College Degree

Better job and income — 4920
Education and learning — 3360
Exposure to new ideas — 2040
Career training — 1680

Number of College Applicants

Source: Princeton Review

- The bar graph is based on a survey of 12,000 college applicants and shows how the applicants responded to the question

    What is the primary benefit of a college degree?

P(selecting an applicant who believes the primary benefit is career training)

$$= \frac{\text{number who stated the primary benefit is career training}}{\text{total number of applicants}}$$

$$= \frac{1680}{12,000} = \frac{7}{50} = 0.14$$

Additional Example to Review
Example 4, page 699

11.5 Probability with the Fundamental Counting Principle, Permutations, and Combinations

Definitions and Concepts

Probability of a permutation

$$= \frac{\text{the number of ways the permutation can occur}}{\text{total number of possible permutations}}$$

Example

- Seven acts (A, B, C, D, E, F, and G) perform in a variety show. The order of performance is determined by random selection. What is the probability that C performs first and G performs last?

P(C performs first and G performs last)

$$= \frac{\text{number of permutations with C first and G last}}{\text{total number of possible permutations}}$$

C first   G last

$$= \frac{1\ 5\ 4\ 3\ 2\ 1\ 1}{7\ 6\ 5\ 4\ 3\ 2\ 1} = \frac{1 \cdot 5 \cdot 4 \cdot 3 \cdot 2 \cdot 1 \cdot 1}{7 \cdot 6 \cdot 5 \cdot 4 \cdot 3 \cdot 2 \cdot 1} = \frac{1}{7 \cdot 6} = \frac{1}{42}$$

Additional Example to Review
Example 1, page 704

Definitions and Concepts

Probability of a combination

$$= \frac{\text{the number of ways the combination can occur}}{\text{total number of possible combinations}}$$

Example

A lottery game is set up so that each player chooses five different numbers from 1 to 20. If the five numbers match the five numbers drawn in the lottery, the player wins (or shares) top cash prize. The probability of winning with 100 lottery tickets is

$$\frac{\text{number of ways of winning}}{\text{total number of possible combinations}} = \frac{100}{_{20}C_5} = \frac{100}{15{,}504} \approx 0.0064$$

$$_{20}C_5 = \frac{20!}{(20-5)!\ 5!} = \frac{20!}{15!\ 5!} = \frac{20 \cdot 19 \cdot 18 \cdot 17 \cdot 16}{5 \cdot 4 \cdot 3 \cdot 2 \cdot 1} = 15{,}504$$

Additional Examples to Review
Example 2, page 706; Example 3, page 707

11.6 Events Involving *Not* and *Or*; Odds

Definitions and Concepts

Complement Rules of Probability

$$P(\text{not } E) = 1 - P(E) \quad \text{and} \quad P(E) = 1 - P(\text{not } E)$$

Example

- 3 freshmen, 4 sophomores, 2 juniors, 1 senior
  Probability of not selecting a freshman:

$$P(\text{not freshman}) = 1 - P(\text{freshman}) = 1 - \frac{3}{10} = \frac{10}{10} - \frac{3}{10} = \frac{7}{10}$$

Additional Examples to Review
Example 1, page 711; Example 2, page 712

Definitions and Concepts

If it is impossible for events A and B to occur simultaneously, the events are mutually exclusive.

If A and B are mutually exclusive events, then $P(A \text{ or } B) = P(A) + P(B)$.

Example

- 3 freshmen, 4 sophomores, 2 juniors, 1 senior
  Probability of selecting a sophomore or a senior:

$$P(\text{sophomore or senior}) = P(\text{sophomore}) + P(\text{senior}) = \frac{4}{10} + \frac{1}{10} = \frac{5}{10} = \frac{1}{2}$$

Additional Examples to Review
Example 3, page 713; Example 6(b), page 716

Definitions and Concepts

If A and B are not mutually exclusive events, then

$$P(A \text{ or } B) = P(A) + P(B) - P(A \text{ and } B).$$

Example

- 8 male juniors, 4 female juniors, 3 male seniors, 5 female seniors

  Probability of selecting a junior or a male:

  $$P(\text{junior or male}) = P(\text{junior}) + P(\text{male}) - P(\text{junior and male})$$

$8 + 4 = 12$ of the 20 students are juniors.	$8 + 3 = 11$ of the 20 students are male.	8 of the 20 students are male juniors.

  $$= \frac{12}{20} + \frac{11}{20} - \frac{8}{20} = \frac{15}{20} = \frac{3}{4}$$

Additional Examples to Review

Example 4, page 714; Example 5, page 715; Example 6(a), page 716

Definitions and Concepts

Probability to Odds

1. Odds in favor of $E = \dfrac{P(E)}{P(\text{not } E)}$ **2.** Odds against $E = \dfrac{P(\text{not } E)}{P(E)}$

Example

You roll a single, six-sided die. $S = \{1, 2, 3, 4, 5, 6\}$

- Probability of an outcome greater than 4:   Two outcomes, 5 and 6, are greater than 4.

  $$P(E) = \frac{2}{6} = \frac{1}{3}$$

- Odds in favor of an outcome greater than 4:

  $$\frac{P(E)}{P(\text{not } E)} = \frac{\dfrac{1}{3}}{1 - \dfrac{1}{3}} = \frac{\dfrac{1}{3}}{\dfrac{2}{3}} = \frac{1}{3} \cdot \frac{3}{2} = \frac{1}{2} = 1 \text{ to } 2$$

- Odds against an outcome greater than 4:

  $$\frac{P(\text{not } E)}{P(E)} = \frac{1 - \dfrac{1}{3}}{\dfrac{1}{3}} = \frac{\dfrac{2}{3}}{\dfrac{1}{3}} = \frac{2}{3} \cdot \frac{3}{1} = \frac{2}{1} = 2 \text{ to } 1$$

  You can quickly obtain these odds by reversing the ratio for the odds in favor of E.

Additional Examples to Review

Example 7, page 717; Example 8, page 718

Definitions and Concepts

Odds to Probability

If odds in favor of E are a to b, then $P(E) = \dfrac{a}{a + b}$.

Example

Odds against E are 7 to 10.

- Odds in favor of E are 10 to 7.

- $P(E) = \dfrac{10}{10 + 7} = \dfrac{10}{17}$

Additional Example to Review
Example 9, page 719

11.7 Events Involving *And*; Conditional Probability

Definitions and Concepts

Two events are independent if the occurrence of either of them has no effect on the probability of the other.

If A and B are independent events,

$$P(A \text{ and } B) = P(A) \cdot P(B).$$

The probability of a succession of independent events is the product of each of their probabilities.

Examples

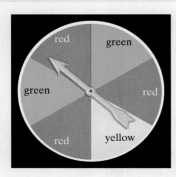

- The spinner is spun twice.

  $$P(\text{red and then green}) = P(\text{red}) \cdot P(\text{green})$$
  $$= \frac{3}{6} \cdot \frac{2}{6} = \frac{1}{2} \cdot \frac{1}{3} = \frac{1}{6}$$

- The spinner is spun three times.

  $$P(\text{yellow on three consecutive spins})$$
  $$= P(\text{yellow}) \cdot P(\text{yellow}) \cdot P(\text{yellow})$$
  $$= \frac{1}{6} \cdot \frac{1}{6} \cdot \frac{1}{6} = \frac{1}{216}$$

Additional Examples to Review
Example 1, page 724; Example 2, page 724; Example 3, page 725

Definitions and Concepts

Two events are dependent if the occurrence of one of them has an effect on the probability of the other.

If A and B are dependent events,

$$P(A \text{ and } B) = P(A) \cdot P(B \text{ given that } A \text{ has occurred}).$$

The multiplication rule for dependent events can be extended to cover three or more dependent events. In the case of three such events,

$$P(A \text{ and } B \text{ and } C)$$
$$= P(A) \cdot P(B \text{ given } A \text{ occurred}) \cdot P(C \text{ given } A \text{ and } B \text{ occurred}).$$

Example

- Two balls are selected without replacement. Probability that both are red:

  $$P(\text{red and red})$$
  $$= P(\text{red}) \cdot P(\text{red given that a red was selected first})$$
  $$= \frac{4}{9} \cdot \frac{3}{8} = \frac{12}{72} = \frac{1}{6}$$

  > There are 9 balls, 4 of which are red.

  > After picking a red ball, there are 8 balls left, 3 of which are red.

Additional Examples to Review
Example 4, page 727; Example 5, page 727

Definitions and Concepts

The conditional probability of B, given A, written $P(B|A)$, is the probability of B if the sample space is restricted to A.

Examples

Numbers in the table represent the number of juniors and seniors ages 16, 17, and 18, by gender, in a group of 100 juniors and seniors.

	Age 16	Age 17	Age 18
Male	18	17	10
Female	20	23	12

- Probability that a randomly selected student is male, given that the student is 16 years old:

	Age 16
Male	18
Female	20

$P(\text{male}|\text{age is 16})$ — Sample space is restricted to students who are 16.

$$= \frac{\text{number of males}}{\text{number of 16-year-olds}} = \frac{18}{18 + 20} = \frac{18}{38} = \frac{9}{19}$$

- Probability that a randomly selected student is 18, given that the student is female:

	Age 16	Age 17	Age 18
Female	20	23	12

$P(\text{age 18}|\text{female})$ — Sample space is restricted to females.

$$= \frac{\text{number of 18-year-olds}}{\text{number of females}} = \frac{12}{20 + 23 + 12} = \frac{12}{55}$$

Additional Examples to Review

Example 6, page 728; Example 7, page 728; Example 8, page 729

Review Exercises

Section 11.1 The Fundamental Counting Principle

1. A restaurant offers 20 appetizers and 40 main courses. In how many ways can a person order a two-course meal?

2. A popular brand of pen comes in red, green, blue, or black ink. The writing tip can be chosen from extra bold, bold, regular, fine, or micro. How many different choices of pens do you have with this brand?

3. In how many ways can first and second prize be awarded in a contest with 100 people, assuming that each prize is awarded to a different person?

4. You are answering three multiple-choice questions. Each question has five answer choices, with one correct answer per question. If you select one of these five choices for each question and leave nothing blank, in how many ways can you answer the questions?

5. A stock can go up, go down, or stay unchanged. How many possibilities are there if you own five stocks?

6. A person can purchase a condominium with a choice of five kinds of carpeting, with or without a pool, with or without a porch, and with one, two, or three bedrooms. How many different options are there for the condominium?

Section 11.2 Permutations and Section 11.3 Combinations

In Exercises 7–10, evaluate each factorial expression.

7. $\dfrac{16!}{14!}$

8. $\dfrac{800!}{799!}$

9. $5! - 3!$

10. $\dfrac{11!}{(11 - 3)!}$

In Exercises 11–12, use the formula for $_nP_r$ to evaluate each expression.

11. $_{10}P_6$

12. $_{100}P_2$

In Exercises 13–14, use the formula for $_nC_r$ to evaluate each expression.

13. $_{11}C_7$

14. $_{14}C_5$

In Exercises 15–17, does the problem involve permutations or combinations? Explain your answer. (It is not necessary to solve the problem.)

15. How many different 4-card hands can be dealt from a 52-card deck?

16. How many different ways can a director select from 20 male actors to cast the roles of Mark, Roger, Angel, and Collins in the musical *Rent*?

17. How many different ways can a director select 4 actors from a group of 20 actors to attend a workshop on performing in rock musicals?

In Exercises 18–28, solve each problem using an appropriate method.

18. Six acts are scheduled to perform in a variety show. How many different ways are there to schedule their appearances?

19. A club with 15 members is to choose four officers—president, vice-president, secretary, and treasurer. In how many ways can these offices be filled?

20. An election ballot asks voters to select four city commissioners from a group of ten candidates. In how many ways can this be done?

21. In how many distinct ways can the letters of the word TORONTO be arranged?

22. From the 20 CDs that you've bought during the past year, you plan to take 3 with you on vacation. How many different sets of three CDs can you take?

23. You need to arrange seven of your favorite books along a small shelf. Although you are not arranging the books by height, the tallest of the books is to be placed at the left end and the shortest of the books at the right end. How many different ways can you arrange the books?

24. Suppose you are asked to list, in order of preference, the five favorite CDs you purchased in the past 12 months. If you bought 20 CDs over this time period, in how many ways can the five favorite be ranked?

25. In how many ways can five airplanes line up for departure on a runway?

26. How many different 5-card hands can be dealt from a deck that has only hearts (13 different cards)?

27. A political discussion group consists of 12 Republicans and 8 Democrats. In how many ways can 5 Republicans and 4 Democrats be selected to attend a conference on politics and social issues?

28. In how many ways can the digits in the number 335,557 be arranged?

Section 11.4 Fundamentals of Probability

In Exercises 29–32, a die is rolled. Find the probability of rolling

29. a 6.   **30.** a number less than 5.

31. a number less than 7.   **32.** a number greater than 6.

In Exercises 33–37, you are dealt one card from a 52-card deck. Find the probability of being dealt

33. a 5.   **34.** a picture card.

35. a card greater than 4 and less than 8.

36. a 4 of diamonds.   **37.** a red ace.

In Exercises 38–40, suppose that you reach into a bag and randomly select one piece of candy from 15 chocolates, 10 caramels, and 5 peppermints. Find the probability of selecting

38. a chocolate.   **39.** a caramel.

40. a peppermint.

41. Tay-Sachs disease occurs in 1 of every 3600 births among Jewish people from central and eastern Europe, and in 1 in 600,000 births in other populations. The disease causes abnormal accumulation of certain fat compounds in the spinal cord and brain, resulting in paralysis, blindness, and mental impairment. Death generally occurs before the age of five. If we use t to represent a Tay-Sachs gene and T a healthy gene, the table at the top of the next column shows the four possibilities for the children of one healthy, TT, parent, and one parent who carries the disease, Tt, but is not sick.

| | | Second Parent | |
		T	t
First	T	TT	Tt
Parent	T	TT	Tt

a. Find the probability that a child of these parents will be a carrier without the disease.

b. Find the probability that a child of these parents will have the disease.

The table shows the employment status of the U.S. civilian labor force, by gender, for a recent year. Use the data in the table, expressed in millions, to solve Exercises 42–44.

Employment Status of the U.S. Labor Force, in Millions

	Employed	**Unemployed**	**Total**
Male	74.5	33.2	107.7
Female	64.7	51.0	115.7
Total	139.2	84.2	223.4

Source: U.S. Bureau of Labor Statistics

Find the probability, expressed as a decimal rounded to three places, that a randomly selected person from the civilian labor force represented in the table

42. is employed.   **43.** is female.

44. is an unemployed male.

Section 11.5 Probability with the Fundamental Counting Principle, Permutations, and Combinations

45. If cities A, B, C, and D are visited in random order, each city visited once, find the probability that city D will be visited first, city B second, city A third, and city C last.

In Exercises 46–49, suppose that six singers are being lined up to perform at a charity fundraiser. Call the singers A, B, C, D, E, and F. The order of performance is determined by writing each singer's name on one of six cards, placing the cards in a hat, and then drawing one card at a time. The order in which the cards are drawn determines the order in which the singers perform. Find the probability that

46. singer C will perform last.

47. singer B will perform first and singer A will perform last.

48. the singers will perform in the following order: F, E, A, D, C, B.

49. the performance will begin with singer A or C.

50. A lottery game is set up so that each player chooses five different numbers from 1 to 20. If the five numbers match the five numbers drawn in the lottery, the player wins (or shares) the top cash prize. What is the probability of winning the prize

a. with one lottery ticket?

b. with 100 different lottery tickets?

51. A committee of four people is to be selected from six Democrats and four Republicans. Find the probability that

a. all are Democrats.

b. two are Democrats and two are Republicans.

52. If you are dealt 3 cards from a shuffled deck of red cards (26 different cards), find the probability of getting exactly 2 picture cards.

11.6 Events Involving *Not* and *Or*; Odds

In Exercises 53–57, a die is rolled. Find the probability of

53. not rolling a 5.

54. not rolling a number less than 4.

55. rolling a 3 or a 5.

56. rolling a number less than 3 or greater than 4.

57. rolling a number less than 5 or greater than 2.

In Exercises 58–63, you draw one card from a 52-card deck. Find the probability of

58. not drawing a picture card.

59. not drawing a diamond.

60. drawing an ace or a king.

61. drawing a black 6 or a red 7.

62. drawing a queen or a red card.

63. drawing a club or a picture card.

In Exercises 64–69, it is equally probable that the pointer on the spinner shown will land on any one of the six regions, numbered 1 through 6, and colored as shown. If the pointer lands on a borderline, spin again. Find the probability of

64. not stopping on 4.

65. not stopping on yellow.

66. not stopping on red.

67. stopping on red or yellow.

68. stopping on red or an even number.

69. stopping on red or a number greater than 3.

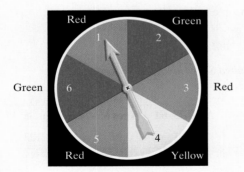

Use this information to solve Exercises 70–71. At a workshop on police work and the African-American community, there are 50 African-American male police officers, 20 African-American female police officers, 90 white male police officers, and 40 white female police officers. If one police officer is selected at random from the people at the workshop, find the probability that the selected person is

70. African American or male.

71. female or white.

Suppose that a survey of 350 college students is taken. Each student is asked the type of college attended (public or private) and the family's income level (low, middle, high). Use the data in the table to solve Exercises 72–75. Express probabilities as simplified fractions.

	Public	Private	Total
Low	120	20	140
Middle	110	50	160
High	22	28	50
Total	252	98	350

Find the probability that a randomly selected student in the survey

72. attends a public college.

73. is not from a high-income family.

74. is from a middle-income or a high-income family.

75. attends a private college or is from a high-income family.

76. One card is randomly selected from a deck of 52 cards. Find the odds in favor and the odds against getting a queen.

77. The winner of a raffle will receive a two-year scholarship to any college of the winner's choice. If 2000 raffle tickets were sold and you purchased 20 tickets, what are the odds against your winning the scholarship?

78. The odds in favor of a candidate winning an election are given at 3 to 1. What is the probability that this candidate will win the election?

Section 11.7 Events Involving *And*; Conditional Probability

Use the spinner shown to solve Exercises 79–83. It is equally likely that the pointer will land on any one of the six regions, numbered 1 through 6, and colored as shown. If the pointer lands on a borderline, spin again. If the pointer is spun twice, find the probability it will land on

79. yellow and then red.

80. 1 and then 3.

81. yellow both times.

If the pointer is spun three times, find the probability it will land on

82. yellow and then 4 and then an odd number.

83. red every time.

84. What is the probability of a family having five boys born in a row?

85. The probability of a flood in any given year in a region prone to flooding is 0.2.

   a. What is the probability of a flood two years in a row?

   b. What is the probability of a flood for three consecutive years?

   c. What is the probability of no flooding for four consecutive years?

   d. What is the probability of a flood at least once in the next four years?

In Exercises 86–87, two students are selected from a group of four psychology majors, three business majors, and two music majors. The two students are to meet with the campus cafeteria manager to voice the group's concerns about food prices and quality. One student is randomly selected and leaves for the cafeteria manager's office. Then, a second student is selected. Find the probability of selecting

86. a music major and then a psychology major.

87. two business majors.

88. A final visit to the box of chocolates: It's now grown to a box of 50, of which 30 are solid chocolate, 15 are filled with jelly, and 5 are filled with cherries. The story is still the same: They all look alike. You select a piece, eat it, select a second piece, eat it, and help yourself to a final sugar rush. Find the probability of selecting a solid chocolate followed by two cherry-filled chocolates.

89. A single die is tossed. Find the probability that the tossed die shows 5, given that the outcome is an odd number.

90. A letter is randomly selected from the letters of the English alphabet. Find the probability of selecting a vowel, given that the outcome is a letter that precedes k.

91. The numbers shown below are each written on a colored chip. The chips are placed into a bag and one chip is selected at random. Find the probability of selecting

   a. an odd number, given that a red chip is selected.

   b. a yellow chip, given that the number selected is at least 3.

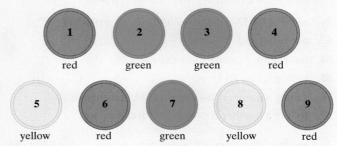

The data in the table are based on 145 Americans tested for tuberculosis. Use the data to solve Exercises 92–99. Express probabilities as simplified fractions.

	TB	**No TB**
Positive Screening Test	9	11
Negative Screening Test	1	124

Source: Deborah J. Bennett, *Randomness,* Harvard University Press, 1998.

Find the probability that a randomly selected person from this group

92. does not have TB.

93. tests positive.

94. does not have TB or tests positive.

95. does not have TB, given a positive test.

96. has a positive test, given no TB.

97. has TB, given a negative test.

Suppose that two people are randomly selected, in succession, from this group. Find the probability of selecting

98. two people with TB.

99. two people with positive screening tests.

Exercises 100–106 involve a combination of topics from Section 11.4, Section 11.6, and Section 11.7.

The table shows the distribution, by age and gender, of the 29,625 deaths in the United States involving firearms in 2004.

Deaths in the United States Involving Firearms, 2004

	Under Age 5	**Ages 5–14**	**Ages 15–19**	**Ages 20–24**	**Ages 25–44**	**Ages 45–64**	**Ages 65–74**	**Age ≥ 75**	**Total**
Male	31	231	2206	3700	9463	6127	1665	2123	25,546
Female	27	69	293	353	1581	1323	222	211	4079
Total	58	300	2499	4053	11,044	7450	1887	2334	29,625

Source: National Safety Council

In Exercises 100–106, use the data in the table to find the probability, expressed as a fraction and as a decimal rounded to three places, that a firearm death in the United States

100. involved a male.

101. involved a person in the 25–44 age range.

102. involved a person less than 75 years old.

103. involved a person in the 20–24 age range or in the 25–44 age range.

104. involved a female or a person younger than 5.

105. involved a person in the 20–24 age range, given that this person was a male.

106. involved a male, given that this person was at least 75.

Chapter 11 Test A

1. A person can purchase a particular model of a new car with a choice of ten colors, with or without automatic transmission, with or without four-wheel drive, with or without air conditioning, and with two, three, or four radio-CD speakers. How many different options are there for this model of the car?

2. Four acts are scheduled to perform in a variety show. How many different ways are there to schedule their appearances?

3. In how many ways can seven airplanes line up for a departure on a runway if the plane with the greatest number of passengers must depart first?

4. A human resource manager has 11 applicants to fill three different positions. Assuming that all applicants are equally qualified for any of the three positions, in how many ways can this be done?

5. From the ten books that you've recently bought but not read, you plan to take four with you on vacation. How many different sets of four books can you take?

6. In how many distinct ways can the letters of the word ATLANTA be arranged?

In Exercises 7–9, one student is selected at random from a group of 12 freshmen, 16 sophomores, 20 juniors, and 2 seniors. Find the probability that the person selected is

7. a freshman.

8. not a sophomore.

9. a junior or a senior.

10. If you are dealt one card from a 52-card deck, find the probability of being dealt a card greater than 4 and less than 10.

11. Seven movies (A, B, C, D, E, F, and G) are being scheduled for showing. The order of showing is determined by random selection. Find the probability that film C will be shown first, film A next-to-last, and film E last.

12. A lottery game is set up so that each player chooses six different numbers from 1 to 15. If the six numbers match the six numbers drawn in the lottery, the player wins (or shares) the top cash prize. What is the probability of winning the prize with 50 different lottery tickets?

In Exercises 13–14, it is equally probable that the pointer on the spinner shown will land on any one of the eight colored regions. If the pointer lands on a borderline, spin again.

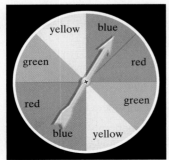

13. If the spinner is spun once, find the probability that the pointer will land on red or blue.

14. If the spinner is spun twice, find the probability that the pointer lands on red on the first spin and blue on the second spin.

15. A region is prone to flooding once every 20 years. The probability of flooding in any one year is $\frac{1}{20}$. What is the probability of flooding for three consecutive years?

16. One card is randomly selected from a deck of 52 cards. Find the probability of selecting a black card or a picture card.

17. A group of students consists of 10 male freshmen, 15 female freshmen, 20 male sophomores, and 5 female sophomores. If one person is randomly selected from the group, find the probability of selecting a freshman or a female.

18. A box contains five red balls, six green balls, and nine yellow balls. Suppose you select one ball at random from the box and do not replace it. Then you randomly select a second ball. Find the probability that both balls selected are red.

19. A quiz consisting of four multiple-choice questions has four available options (a, b, c, or d) for each question. If a person guesses at every question, what is the probability of answering *all* questions correctly?

20. A group is comprised of 20 men and 15 women. If one person is randomly selected from the group, find the odds against the person being a man.

21. The odds against a candidate winning an election are given at 1 to 4.
    a. What are the odds in favor of the candidate winning?
    b. What is the probability that the candidate will win the election?

A class is collecting data on eye color and gender. They organize the data they collected into the table shown. Numbers in the table represent the number of students from the class that belong to each of the categories. Use the data to solve Exercises 22–26. Express probabilities as simplified fractions.

	Brown	Blue	Green
Male	22	18	10
Female	18	20	12

Find the probability that a randomly selected student from this class

22. does not have brown eyes.

23. has brown eyes or blue eyes.

24. is female or has green eyes.

25. is male, given the student has blue eyes.

26. If two people are randomly selected, in succession, from the students in this class, find the probability that they both have green eyes.

Chapter 11 Test B

1. A person can purchase a particular model of a new car with a choice of nine colors, with or without automatic transmission, with or without four-wheel drive, with or without air conditioning, and with two, three, four, or five radio-CD speakers. How many different options are there for this model of the car?

2. Six acts are scheduled to perform in a variety show. How many different ways are there to schedule their appearances?

3. In how many ways can five airplanes line up for a departure on a runway if the plane with the greatest number of passengers must depart first?

4. A human resource manager has ten applicants to fill three different positions. Assuming that all applicants are equally qualified for any of the three positions, in how many ways can this be done?

5. From the nine books that you've recently bought but not read, you plan to take four with you on vacation. How many different sets of four books can you take?

6. In how many distinct ways can the letters of the word NEWSWEEK be arranged?

In Exercises 7–9, one student is selected at random from a group of 10 freshmen, 18 sophomores, 16 juniors, and 6 seniors. Find the probability that the person selected is

7. a senior.

8. not a junior.

9. a freshman or a sophomore.

10. If you are dealt one card from a 52-card deck, find the probability of being dealt a card greater than 5 and less than 9.

11. Six movies (A, B, C, D, E, and F) are being scheduled for showing. The order of showing is determined by random selection. Find the probability that film B will be shown first, film C next-to-last, and film E last.

12. A lottery game is set up so that each player chooses six different numbers from 1 to 18. If the six numbers match the six numbers drawn in the lottery, the player wins (or shares) the top cash prize. What is the probability of winning the prize with 30 different lottery tickets?

In Exercises 13–14, it is equally probable that the pointer on the spinner shown will land on any one of the eight colored regions. If the pointer lands on a borderline, spin again.

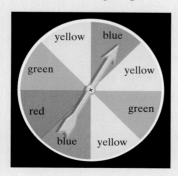

13. If the spinner is spun once, find the probability that the pointer will land on yellow or blue.

14. If the spinner is spun twice, find the probability that the pointer lands on yellow on the first spin and blue on the second spin.

15. A region is prone to flooding once every 10 years. The probability of flooding in any one year is $\frac{1}{10}$. What is the probability of flooding for four consecutive years?

16. One card is randomly selected from a deck of 52 cards. Find the probability of selecting a red card or a picture card.

17. A group of students consists of 10 male freshmen, 20 female freshmen, 25 male sophomores, and 5 female sophomores. If one person is randomly selected from the group, find the probability of selecting a freshman or a female.

18. A box contains four red balls, seven green balls, and eight yellow balls. Suppose you select one ball at random from the box and do not replace it. Then you randomly select a second ball. Find the probability that both balls selected are red.

19. A quiz consisting of four multiple-choice questions has five available options (a, b, c, d, or e) for each question. If a person guesses at every question, what is the probability of answering *all* questions correctly?

20. A group is comprised of 30 men and 10 women. If one person is randomly selected from the group, find the odds against the person being a man.

21. The odds against a candidate winning an election are given at 1 to 6.

    a. What are the odds in favor of the candidate winning?

    b. What is the probability that the candidate will win the election?

A class is collecting data on eye color and gender. They organize the data they collected into the table shown. Numbers in the table represent the number of students from the class that belong to each of the categories. Use the data to solve Exercises 22–26. Express probabilities as simplified fractions.

	Brown	**Blue**	**Green**
Male	24	16	10
Female	16	20	14

Find the probability that a randomly selected student from this class

22. does not have blue eyes.

23. has blue eyes or green eyes.

24. is male or has brown eyes.

25. is female, given the student has blue eyes.

26. If two people are randomly selected, in succession, from the students in this class, find the probability that they both have brown eyes.

Statistics

12

Some random statistical factoids:

- 70% of high school teens who use the Internet visit social networking sites, 30% have blogs, and 20% create their own art with online content. (Pew)

- 16% of American teens report they have been a victim of cyber-bullying. (Horatio Alger Assoc.)

- 49% of Americans cite a "lot" of stress at age 22, compared with 45% at 42, 35% at 58, 29% at 62, and 20% by 70. (*Proceedings of the National Academy of Sciences*)

- 97% of American teens play video games. (*Scholastic Scope*)

- 34% of American adults believe in ghosts. (AP/Ipsos)

- 48% of the world's population lives on less than $2 a day. (Population Reference Bureau)

Statisticians collect numerical data from subgroups of populations to find out everything imaginable about the population as a whole, including whom they support in an election, what they watch on TV, how much money they make, or what worries them. Comedians and statisticians joke that 62.38% of all statistics are made up on the spot. Because statisticians both record and influence our behavior, it is important to distinguish between good and bad methods for collecting, presenting, and interpreting data. In this chapter, you will gain an understanding of where data come from and how these numbers are used to make decisions.

12.1 Sampling, Frequency Distributions, and Graphs

Objectives

1 Describe the population whose properties are to be analyzed.

2 Select an appropriate sampling technique.

3 Organize and present data.

4 Identify deceptions in visual displays of data.

At the end of the twentieth century, there were 94 million households in the United States with television sets. The television program viewed by the greatest percentage of such households in that century was the final episode of $M*A*S*H$. Over 50 million American households watched this program.

Numerical information, such as the information about the top three TV shows of the twentieth century, shown in **Table 12.1**, is called **data**. The word **statistics** is often used when referring to data. However, statistics has a second meaning: Statistics is also a method for collecting, organizing, analyzing, and interpreting data, as well as drawing conclusions based on the data. This methodology divides statistics into two main areas. **Descriptive statistics** is concerned with collecting, organizing, summarizing, and presenting data. **Inferential statistics** has to do with making generalizations about and drawing conclusions from the data collected.

*M*A*S*H* took place in the early 1950s, during the Korean War. By the final episode, the show had lasted four times as long as the Korean War.

Table 12.1 TV Programs with the Greatest U.S. Audience Viewing Percentage of the Twentieth Century		
Program	**Total Households**	**Viewing Percentage**
1. $M*A*S*H$ Feb. 28, 1983	50,150,000	60.2%
2. *Dallas* Nov. 21, 1980	41,470,000	53.3%
3. *Roots Part 8* Jan. 30, 1977	36,380,000	51.1%

Source: Nielsen Media Research

Populations and Samples

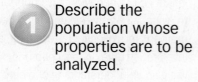

Describe the population whose properties are to be analyzed.

Consider the set of all American TV households. Such a set is called the *population*. In general, a **population** is the set containing all the people or objects whose properties are to be described and analyzed by the data collector.

The population of American TV households is huge. At the time of the $M*A*S*H$ conclusion, there were nearly 84 million such households. Did over 50 million American TV households really watch the final episode of $M*A*S*H$? A friendly phone call to each household ("So, how are you? What's new? Watch any good television last night? If so, what?") is, of course, absurd. A **sample**, which is a subset or subgroup of the population, is needed. In this case, it would be appropriate to have a sample of a few thousand TV households to draw conclusions about the population of all TV households.

Cover of the *Literary Digest,* October 31, 1936. General Research Division, The New York Public Library, Astor, Lenox and Tilden Foundations. The New York Public Library/Art Resource, NY.

In 1936, the *Literary Digest* mailed out over ten million ballots to voters throughout the country. The results poured in, and the magazine predicted a landslide victory for Republican Alf Landon over Democrat Franklin Roosevelt. However, the prediction of the *Literary Digest* was wrong. Why? The mailing lists the editors used included people from their own subscriber list, directories of automobile owners, and telephone books. As a result, its sample was anything but random. It excluded most of the poor, who were unlikely to subscribe to the *Literary Digest*, or to own a car or telephone in the heart of the Depression. Prosperous people in 1936 were more likely to be Republican than the poor. Thus, although the sample was massive, it included a higher percentage of affluent individuals than the population as a whole did. A victim of both the Depression and the 1936 sampling fiasco, the *Literary Digest* folded in 1937.

Example 1   Populations and Samples

The board of supervisors in a large city decides to conduct a survey among citizens of the city to determine their opinions about restricting land use in order to protect the environment.

a. Describe the population.

b. A supervisor suggests obtaining a sample by surveying people who hike along one of the city's nature trails. Each person will be asked to express his or her opinion on land use restriction to protect the environment. Does this seem like a good idea?

Solution

a. The population is the set containing all the citizens of the city.

b. Questioning people hiking along one of the city's nature trails is a terrible idea. The subset of hikers is likely to have a more positive attitude about environmental issues, thereby favoring land use restriction, than the population of all the city's citizens.

Checkpoint 1   A city government wants to conduct a survey among the city's homeless to discover their opinions about required residence in city shelters from midnight until 6 A.M.

a. Describe the population.

b. A city commissioner suggests obtaining a sample by surveying all the homeless people at the city's largest shelter on a Sunday night. Does this seem like a good idea? Explain your answer.

Random Sampling

There is a way to use a small sample to make generalizations about a large population: Guarantee that every member of the population has an equal chance to be selected for the sample. Surveying hikers along one of the city's nature trails does not provide this guarantee. Unless it can be established that all citizens of the city hike these trails, which seems unlikely, this sampling scheme does not permit each citizen an equal chance of selection.

> ### Random Samples
>
> A **random sample** is a sample obtained in such a way that every element in the population has an equal chance of being selected for the sample.

Suppose that in your first year of college, you are elated with the quality of one of your courses. Although it's an auditorium section with 120 students, you feel that the professor is lecturing right to you. During a wonderful lecture, you look around the auditorium to see if any of the other students are sharing your enthusiasm. Based on body language, it's hard to tell. You really want to know the opinion of the population of 120 students taking this course. You think about asking students to grade the course on an A to F scale, anticipating a unanimous A. You cannot survey everyone. Eureka! Suddenly you have an idea on how to take a sample. Place cards numbered from 1 through 120, one number per card, in a box. Because the course has assigned seating by number, each numbered card corresponds to a student in the class. Reach in and randomly select six cards. Each card, and therefore each student, has an equal chance of being selected. Then use the opinions about the course from the six randomly selected students to generalize about the course opinion for the entire 120-student population.

Your idea is precisely how random samples are obtained. In random sampling, each element in the population must be identified and assigned a number. The numbers are generally assigned in order. The way to sample from the larger numbered population is to generate random numbers using a computer or calculator. Each numbered element from the population that corresponds to one of the generated random numbers is selected for the sample.

Call-in polls on radio and television are not reliable because those polled do not represent the larger population. A person who calls in is likely to have feelings about an issue that are consistent with the politics of the show's host. For a poll to be accurate, the sample must be chosen randomly from the larger population. The A. C. Nielsen Company uses a random sample of approximately 5000 TV households to measure the percentage of households tuned in to a television program.

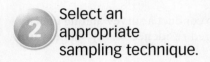

Select an appropriate sampling technique.

Example 2   Selecting an Appropriate Sampling Technique

We return to the board of supervisors in the large city and their interest in how the city's citizens feel about restricting land use in order to protect the environment. Which of the following would be the most appropriate way to select a random sample?

a. Randomly survey construction workers who live in the city.

b. Survey the first 200 people whose names appear in the city's telephone directory.

c. Randomly select neighborhoods of the city and then randomly survey people within the selected neighborhoods.

Solution

Keep in mind that the population is the set containing all the city's citizens. A random sample must give each citizen an equal chance of being selected.

a. Randomly selecting construction workers who live in the city is not a good idea. The salaries of many construction workers depend on new construction, which would diminish with land use restriction to protect the environment. Furthermore, this sample does not give each citizen of the city an equal chance of being selected.

b. If the supervisors survey the first 200 people whose names appear in the city's telephone directory, all citizens do not have an equal chance of selection. For example, an individual whose last name begins with a letter toward the end of the alphabet has no chance of being selected.

c. Randomly selecting neighborhoods of the city and then randomly surveying people within the selected neighborhoods is an appropriate technique. Using this method, each citizen has an equal chance of being selected.

In summary, given the three options, the sampling technique in part (c) is the most appropriate.

Surveys and polls involve data from a sample of some population. Regardless of the sampling technique used, the sample should exhibit characteristics typical of those possessed by the target population. This type of sample is called a **representative sample**.

✓ **Checkpoint 2** Explain why the sampling technique described in Checkpoint 1(b) on page 749 is not a random sample. Then describe an appropriate way to select a random sample of the city's homeless.

Blitzer Bonus
••••••••••••••••••
The United States Census

A census is a survey that attempts to include the entire population. The U.S. Constitution requires a census of the American population every ten years. When the Founding Fathers invented American democracy, they realized that if you are going to have government by the people, you need to know who and where they are. Nowadays about $400 billion per year in federal aid is distributed based on the Census numbers, for everything from jobs to bridges to schools. For every 100 people not counted, states and communities could lose as much as $130,000 annually, or $1300 per person each year, so this really matters.

Although the Census generates volumes of statistics, its main purpose is to give the government block-by-block population figures. The U.S. Census is not foolproof. The 1990 Census missed 1.6% of the American population, including an estimated 4.4% of the African-American population, largely in inner cities. Only 67% of households responded to the 2000 Census, even after door-to-door canvassing. About 6.4 million people were missed and 3.1 million were counted twice. Although the 2010 Census was one of the shortest forms in history, counting each person was not an easy task, particularly with concerns about immigration status and privacy of data.

Of course, there would be more than $400 billion to spread around if it didn't cost so much to count us in the first place: about $15 billion for the 2010 Census. That included $338 million for ads in 28 languages, a Census-sponsored NASCAR entry, and $2.5 million for a Super Bowl ad. The ads were meant to boost the response rate, since any household that did not mail back its form got visited by a Census worker, another pricey item. In all, the cost of the 2010 Census worked out to approximately $49 per person.

Frequency Distributions

3 Organize and present data.

After data have been collected from a sample of the population, the next task facing the statistician is to present the data in a condensed and manageable form. In this way, the data can be more easily interpreted.

Suppose, for example, that researchers are interested in determining the age at which adolescent males show the greatest rate of physical growth. A random sample of 35 ten-year-old boys is measured for height and then remeasured each year until they reach 18. The age of maximum yearly growth for each subject is as follows:

12, 14, 13, 14, 16, 14, 14, 17, 13, 10, 13, 18, 12, 15, 14, 15, 15, 14, 14, 13, 15, 16, 15, 12, 13, 16, 11, 15, 12, 13, 12, 11, 13, 14, 14.

A piece of data is called a **data item**. This list of data has 35 data items. Some of the data items are identical. Two of the data items are 11 and 11. Thus, we can say that the **data value** 11 occurs twice. Similarly, because five of the data items are 12, 12, 12, 12, and 12, the data value 12 occurs five times.

Collected data can be presented using a **frequency distribution**. Such a distribution consists of two columns. The data values are listed in one column. Numerical data are generally listed from smallest to largest. The adjacent column is labeled **frequency** and indicates the number of times each value occurs.

Table 12.2 A Frequency Distribution for a Boy's Age of Maximum Yearly Growth

Age of Maximum Growth	Number of Boys (Frequency)
10	1
11	2
12	5
13	7
14	9
15	6
16	3
17	1
18	1
Total:	$n = 35$

35 is the sum of the frequencies.

Example 3　Constructing a Frequency Distribution

Construct a frequency distribution for the data of the age of maximum yearly growth for 35 boys:

12, 14, 13, 14, 16, 14, 14, 17, 13, 10, 13, 18, 12, 15, 14, 15, 15, 14, 14, 13, 15, 16, 15, 12, 13, 16, 11, 15, 12, 13, 12, 11, 13, 14, 14.

Solution

It is difficult to determine trends in the data above in their current format. Perhaps we can make sense of the data by organizing them into a frequency distribution. Let us create two columns. One lists all possible data values, from smallest (10) to largest (18). The other column indicates the number of times the value occurs in the sample. The frequency distribution is shown in **Table 12.2**.

The frequency distribution indicates that one subject had maximum growth at age 10, two at age 11, five at age 12, seven at age 13, and so on. The maximum growth for most of the subjects occurred between the ages of 12 and 15. Nine boys experienced maximum growth at age 14, more than at any other age within the sample. The sum of the frequencies, 35, is equal to the original number of data items.

The trend shown by the frequency distribution in **Table 12.2** indicates that the number of boys who attain their maximum yearly growth at a given age increases until age 14 and decreases after that. This trend is not evident in the data in their original format.

Checkpoint 3　Construct a frequency distribution for the data showing final course grades for students in a college math course, listed alphabetically by student name in a grade book:

F, A, B, B, C, C, B, C, A, A, C, C, D, C, B, D, C, C, B, C.

A frequency distribution that lists all possible data items can be quite cumbersome when there are many such items. For example, consider the following data items. These are statistics test scores for a class of 40 college students.

82	47	75	64	57	82	63	93
76	68	84	54	88	77	79	80
94	92	94	80	94	66	81	67
75	73	66	87	76	45	43	56
57	74	50	78	71	84	59	76

It's difficult to determine how well the group did when the grades are displayed like this. Because there are so many data items, one way to organize these data so that the results are more meaningful is to arrange the grades into groups, or **classes**, based on something that interests us. Many grading systems assign an A to grades in the 90–100 class, B to grades in the 80–89 class, C to grades in the 70–79 class, and so on. These classes provide one way to organize the data.

Looking at the 40 statistics test scores, we see that they range from a low of 43 to a high of 94. We can use classes that run from 40 through 49, 50 through 59, 60 through 69, and so on up to 90 through 99, to organize the scores. In Example 4, we go through the data and tally each item into the appropriate class. This method for organizing data is called a **grouped frequency distribution**.

Example 4   **Constructing a Grouped Frequency Distribution**

Use the classes 40–49, 50–59, 60–69, 70–79, 80–89, and 90–99 to construct a grouped frequency distribution for the 40 test scores on the previous page.

Solution

We use the 40 given scores and tally the number of scores in each class.

Tallying Statistics Test Scores

Test Scores (Class)	Tally	Number of Students (Frequency)
40–49	|||	3
50–59	⊤⊤⊤⊤		6
60–69	⊤⊤⊤⊤		6
70–79	⊤⊤⊤⊤ ⊤⊤⊤⊤		11
80–89	⊤⊤⊤⊤ ||||	9
90–99	⊤⊤⊤⊤	5

The second score in the list, 47, is shown as the first tally in this row.

The first score in the list, 82, is shown as the first tally in this row.

Table 12.3 A Grouped Frequency Distribution for Statistics Test Scores

Class	Frequency
40–49	3
50–59	6
60–69	6
70–79	11
80–89	9
90–99	5
Total:	$n = 40$

40, the sum of the frequencies, is the number of data items.

Omitting the tally column results in the grouped frequency distribution in **Table 12.3**. The distribution shows that the greatest frequency of students scored in the 70–79 class. The number of students decreases in classes that contain successively lower and higher scores. The sum of the frequencies, 40, is equal to the original number of data items.

The leftmost number in each class of a grouped frequency distribution is called the **lower class limit**. For example, in **Table 12.3**, the lower limit of the first class is 40 and the lower limit of the third class is 60. The rightmost number in each class is called the **upper class limit**. In **Table 12.3**, 49 and 69 are the upper limits for the first and third classes, respectively. Notice that if we take the difference between any two consecutive lower class limits, we get the same number:

$$50 - 40 = 10, \ 60 - 50 = 10, \ 70 - 60 = 10, \ 80 - 70 = 10, \ 90 - 80 = 10.$$

The number 10 is called the **class width**.

When setting up class limits, each class, with the possible exception of the first or last, should have the same width. Because each data item must fall into exactly one class, it is sometimes helpful to vary the width of the first or last class to allow for items that fall far above or below most of the data.

Checkpoint 4   Use the classes in **Table 12.3** to construct a grouped frequency distribution for the following 37 exam scores:

73	58	68	75	94	79	96	79
87	83	89	52	99	97	89	58
95	77	75	81	75	73	73	62
69	76	77	71	50	57	41	98
77	71	69	90	75.			

Table 12.2 (repeated) A Frequency Distribution for a Boy's Age of Maximum Yearly Growth	
Age of Maximum Growth	**Number of Boys (Frequency)**
10	1
11	2
12	5
13	7
14	9
15	6
16	3
17	1
18	1
Total:	$n = 35$

Histograms and Frequency Polygons

Take a second look at the frequency distribution for the age of a boy's maximum yearly growth, repeated in **Table 12.2**. A bar graph with bars that touch can be used to visually display the data. Such a graph is called a **histogram**. **Figure 12.1** illustrates a histogram that was constructed using the frequency distribution in **Table 12.2**. A series of rectangles whose heights represent the frequencies are placed next to each other. For example, the height of the bar for the data value 10, shown in **Figure 12.1**, is 1. This corresponds to the frequency for 10 given in **Table 12.2**. The higher the bar, the more frequent the age of maximum yearly growth. The break along the horizontal axis, symbolized by ∿, eliminates listing the ages 1 through 9.

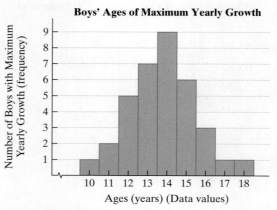

Figure 12.1 A histogram for a boy's age of maximum yearly growth

A line graph called a **frequency polygon** can also be used to visually convey the information shown in **Figure 12.1**. The axes are labeled just like those in a histogram. Thus, the horizontal axis shows data values and the vertical axis shows frequencies. Once a histogram has been constructed, it's fairly easy to draw a frequency polygon. **Figure 12.2** shows a histogram with a dot at the top of each rectangle at its midpoint. Connect each of these midpoints with a straight line. To complete the frequency polygon at both ends, the lines should be drawn down to touch the horizontal axis. The completed frequency polygon is shown in **Figure 12.3**.

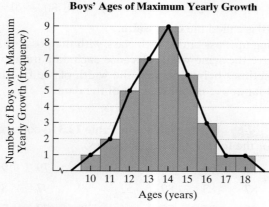

Figure 12.2 A histogram with a superimposed frequency polygon

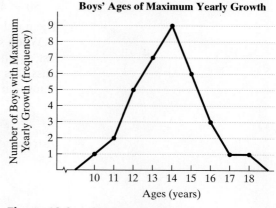

Figure 12.3 A frequency polygon

Stem-and-Leaf Plots

A unique way of displaying data uses a tool called a **stem-and-leaf plot**. Example 5 illustrates how we sort the data, revealing the same visual impression created by a histogram.

Example 5   Constructing a Stem-and-Leaf Plot

Use the data showing statistics test scores for 40 college students to construct a stem-and-leaf plot:

82	47	75	64	57	82	63	93
76	68	84	54	88	77	79	80
94	92	94	80	94	66	81	67
75	73	66	87	76	45	43	56
57	74	50	78	71	84	59	76.

Solution

The plot is constructed by separating each data item into two parts. The first part is the *stem*. The **stem** consists of the tens digit. For example, the stem for the score of 82 is 8. The second part is the *leaf*. The **leaf** consists of the units digit for a given value. For the score of 82, the leaf is 2. The possible stems for the 40 scores are 4, 5, 6, 7, 8, and 9, entered in the left column of the plot.

Begin by entering each data item in the first row:

82   47   75   64   57   82   63   93.

Entering 8 2:

Stems	Leaves
4	
5	
6	
7	
8	2
9	

Adding 4 7:

Stems	Leaves
4	7
5	
6	
7	
8	2
9	

Adding 7 5:

Stems	Leaves
4	7
5	
6	
7	5
8	2
9	

Adding 6 4:

Stems	Leaves
4	7
5	
6	4
7	5
8	2
9	

Adding 5 7:

Stems	Leaves
4	7
5	7
6	4
7	5
8	2
9	

Adding 8 2:

Stems	Leaves
4	7
5	7
6	4
7	5
8	2 2
9	

Adding 6 3:

Stems	Leaves
4	7
5	7
6	4 3
7	5
8	2 2
9	

Adding 9 3:

Stems	Leaves
4	7
5	7
6	4 3
7	5
8	2 2
9	3

We continue in this manner and enter all the data items. **Figure 12.4** at the top of the next page shows the completed stem-and-leaf plot.

If you turn the page so that the left margin is on the bottom and facing you, the visual impression created by the enclosed leaves is the same as that created by a histogram. An advantage over the histogram is that the stem-and-leaf plot preserves exact data items. The enclosed leaves extend farthest to the right when the stem is 7. This shows that the greatest frequency of students scored in the 70s.

A Stem-and-Leaf Plot for 40 Test Scores

Tens digit	Units digit
Stems	**Leaves**
4	7 5 3
5	7 4 6 7 0 9
6	4 3 8 6 7 6
7	5 6 7 9 5 3 6 4 8 1 6
8	2 2 4 8 0 0 1 7 4
9	3 4 2 4 4

Figure 12.4 A stem-and-leaf plot displaying 40 test scores

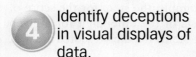 **Checkpoint 5** Construct a stem-and-leaf plot for the data in Checkpoint 4 on page 753.

Deceptions in Visual Displays of Data

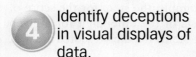 **4** Identify deceptions in visual displays of data.

Benjamin Disraeli, Queen Victoria's prime minister, stated that there are "lies, damned lies, and statistics." The problem is not that statistics lie, but rather that liars use statistics. Graphs can be used to distort the underlying data, making it difficult for the viewer to learn the truth. One potential source of misunderstanding is the scale on the vertical axis used to draw the graph. This scale is important because it lets a researcher "inflate" or "deflate" a trend. For example, both graphs in **Figure 12.5** present identical data for the percentage of people in the United States living below the poverty level from 2001 through 2005. The graph on the left stretches the scale on the vertical axis to create an overall impression of a poverty rate increasing rapidly over time. The graph on the right compresses the scale on the vertical axis to create an impression of a poverty rate that is slowly increasing, and beginning to level off, over time.

Percentage of People in the United States Living below the Poverty Level, 2001–2005

Table 12.4 U.S. Poverty Rate from 2001 to 2008	
Year	**Poverty Rate**
2001	11.3%
2002	11.7%
2003	12.1%
2004	12.5%
2005	12.7%
2006	12.3%
2007	12.5%
2008	12.4%

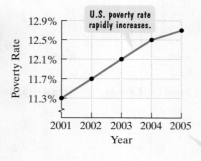

Year	Poverty Rate
2001	11.3%
2002	11.7%
2003	12.1%
2004	12.5%
2005	12.7%

The graph in both figures present this data.

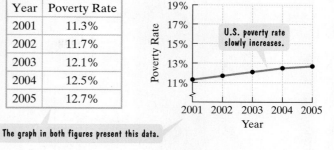

Figure 12.5
Source: U.S. Census Bureau

There is another problem with the data in **Figure 12.5**. Look at **Table 12.4**, which shows the poverty rate from 2001 through 2008. Depending on the time frame chosen, the data can be interpreted in various ways. Carefully choosing a time frame can help represent data trends in the most positive or negative light.

Things to Watch for in Visual Displays of Data

1. Is there a title that explains what is being displayed?

2. Are numbers lined up with tick marks on the vertical axis that clearly indicate the scale? Has the scale been varied to create a more or less dramatic impression than shown by the actual data?

3. Do too many design and cosmetic effects draw attention from or distort the data?

4. Has the wrong impression been created about how the data are changing because equally spaced time intervals are not used on the horizontal axis? Furthermore, has a time interval been chosen that allows the data to be interpreted in various ways?

5. Are bar sizes scaled proportionately in terms of the data they represent?

6. Is there a source that indicates where the data in the display came from? Do the data come from an entire population or a sample? Was a random sample used and, if so, are there possible differences between what is displayed in the graph and what is occurring in the entire population? (We'll discuss these *margins of error* in Section 12.4.) Who is presenting the visual display, and does that person have a special case to make for or against the trend shown by the graph?

Table 12.5 contains two examples of misleading visual displays.

Table 12.5 Examples of Misleading Visual Displays

Graphic Display	Presentation Problems
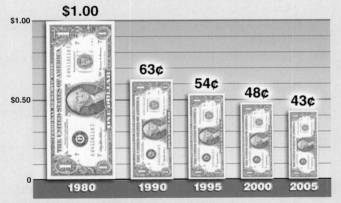 Purchasing Power of the Diminishing Dollar *Source:* Bureau of Labor Statistics	Although the length of each dollar bill is proportional to its spending power, the visual display varies both the length *and width* of the bills to show the diminishing power of the dollar over time. Because our eyes focus on the *areas* of the dollar-shaped bars, this creates the impression that the purchasing power of the dollar diminished even more than it really did. If the area of the dollar were drawn to reflect its purchasing power, the 2005 dollar would be approximately twice as large as the one shown in the graphic display.
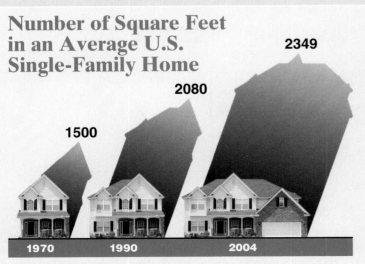 Number of Square Feet in an Average U.S. Single-Family Home *Source:* National Association of Home Builders	Cosmetic effects of homes with equal heights, but different frontal additions and shadow lengths, make it impossible to tell if they proportionately depict the given areas. Time intervals on the horizontal axis are not uniform in size, making it appear that dwelling swelling has been linear from 1970 through 2004. The data indicate that this is not the case. There was a greater increase in area from 1970 through 1990, averaging 29 square feet per year, than from 1990 through 2004, averaging approximately 19.2 square feet per year.

Exercise Set 12.1

Concept and Vocabulary Exercises

In Exercises 1–6, fill in each blank so that the resulting statement is true.

1. A sample obtained in such a way that every element in the population has an equal chance of being selected is called a/an _____ sample.

2. If data values are listed in one column and the adjacent column indicates the number of times each value occurs, the data presentation is called a/an _____.

3. If the data presentation in Exercise 2 is varied by organizing the data into classes, the data presentation is called a/an _____. If one class in such a distribution is 80–89, the lower class limit is _____ and the upper class limit is _____.

4. Data can be displayed using a bar graph with bars that touch each other. This visual presentation of the data is called a/an _____. The heights of the bars represent the _____ of the data values.

5. If the midpoints of the tops of the bars for the data presentation in Exercise 4 are connected with straight lines, the resulting line graph is a data presentation called a/an _____. To complete such a graph at both ends, the lines are drawn down to touch the _____.

6. A data presentation that separates each data item into two parts is called a/an _____.

In Exercises 7–10, determine whether each statement is true or false. If the statement is false, make the necessary change(s) to produce a true statement.

7. A sample is the set of all the people or objects whose properties are to be described and analyzed by the data collector.

8. A call-in poll on radio or television is not reliable because the sample is not chosen randomly from a larger population.

9. One disadvantage of a stem-and-leaf plot is that it does not display the data items.

10. A deception in the visual display of data can result by stretching or compressing the scale on a graph's vertical axis.

Respond to Exercises 11–20 using verbal or written explanations.

11. What is a population? What is a sample?

12. Describe what is meant by a random sample.

13. Suppose you are interested in whether or not the students at your high school would favor a grading system in which students may receive final grades of A+, A, A−, B+, B, B−, C+, C, C−, and so on. Describe how you might obtain a random sample of 100 students from the entire student population.

14. For Exercise 13, would questioning every fifth student as he or she is leaving the school library until 100 students are interviewed be a good way to obtain a random sample? Explain your answer.

15. What is a frequency distribution?

16. What is a histogram?

17. What is a frequency polygon?

18. Describe how to construct a frequency polygon from a histogram.

19. Describe how to construct a stem-and-leaf plot from a set of data.

20. Describe two ways that graphs can be misleading.

In Exercises 21–24, determine whether each statement makes sense or does not make sense, and explain your reasoning.

21. The death rate from this new strain of flu is catastrophic because 25% of the people hospitalized with the disease have died.

22. The graph indicates that for the period from 2000 through 2007, the percentage of English majors among incoming college freshmen has rapidly increased.

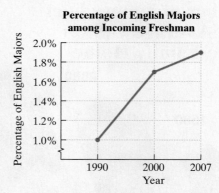

Percentage of English Majors among Incoming Freshman

Source: Higher Education Research Institute

23. A public radio station needs to survey its contributors to determine their programming interests, so they should select a random sample of 100 of their largest contributors.

24. The joke in this cartoon is based on the observation that improperly worded questions can steer respondents toward answers that are not their own.

THE WIZARD OF ID   parker and hart

Creators Syndicate/John Hart Studios/Patti Hart-Pomeroy

Practice and Application Exercises

25. The government of a large city needs to determine whether the city's residents will support the construction of a new jail. The government decides to conduct a survey of a sample of the city's residents. Which one of the following procedures would be most appropriate for obtaining a sample of the city's residents?

   a. Survey a random sample of the employees and inmates at the old jail.

   b. Survey every fifth person who walks into City Hall on a given day.

   c. Survey a random sample of persons within each geographic region of the city.

   d. Survey the first 200 people listed in the city's telephone directory.

26. The city council of a large city needs to know whether its residents will support the building of three new schools. The council decides to conduct a survey of a sample of the city's residents. Which procedure would be most appropriate for obtaining a sample of the city's residents?

   a. Survey a random sample of teachers who live in the city.

   b. Survey 100 individuals who are randomly selected from a list of all people living in the state in which the city in question is located.

   c. Survey a random sample of persons within each neighborhood of the city.

   d. Survey every tenth person who enters City Hall on a randomly selected day.

A questionnaire was given to college students in an introductory statistics class during the first week of the course. One question asked, "How stressed have you been in the last $2\frac{1}{2}$ weeks, on a scale of 0 to 10, with 0 being not at all stressed and 10 being as stressed as possible?" The students' responses are shown in the frequency distribution in the next column. Use this frequency distribution to solve Exercises 27–30.

Stress Rating	Frequency
0	2
1	1
2	3
3	12
4	16
5	18
6	13
7	31
8	26
9	15
10	14

Source: Journal of Personality and Social Psychology, 69, 1102–1112

27. Which stress rating describes the greatest number of students? How many students responded with this rating?

28. Which stress rating describes the least number of students? How many responded with this rating?

29. How many students were involved in this study?

30. How many students had a stress rating of 8 or more?

31. A random sample of 30 high school students is selected. Each student is asked how much time he or she spent on homework during the previous week. The following times (in hours) are obtained:

16, 24, 18, 21, 18, 16, 18, 17, 15, 21, 19, 17, 17, 16, 19, 18, 15, 15, 20, 17, 15, 17, 24, 19, 16, 20, 16, 19, 18, 17.

Construct a frequency distribution for the data.

32. A random sample of 30 male high school seniors is selected. Each student is asked his height (to the nearest inch). The heights are as follows:

72, 70, 68, 72, 71, 71, 71, 69, 73, 71, 73, 75, 66, 67, 75, 74, 73, 71, 72, 67, 72, 68, 67, 71, 73, 71, 72, 70, 73, 70.

Construct a frequency distribution for the data.

A college professor had students keep a diary of their social interactions for a week. Excluding family and work situations, the number of social interactions of ten minutes or longer over the week is shown in the following grouped frequency distribution. Use this information to solve Exercises 33–40.

Number of Social Interactions	Frequency
0–4	12
5–9	16
10–14	16
15–19	16
20–24	10
25–29	11
30–34	4
35–39	3
40–44	3
45–49	3

Source: Society for Personality and Social Psychology

33. Identify the lower class limit for each class.

34. Identify the upper class limit for each class.

35. What is the class width?

36. How many students were involved in this study?

37. How many students had at least 30 social interactions for the week?

38. How many students had at most 14 social interactions for the week?

39. Among the classes with the greatest frequency, which class has the least number of social interactions?

40. Among the classes with the smallest frequency, which class has the least number of social interactions?

41. As of 2011, the following are the ages, in chronological order, at which U.S. presidents were inaugurated:

    57, 61, 57, 57, 58, 57, 61, 54, 68, 51, 49, 64, 50, 48, 65, 52, 56, 46, 54, 49, 50, 47, 55, 55, 54, 42, 51, 56, 55, 51, 54, 51, 60, 62, 43, 55, 56, 61, 52, 69, 64, 46, 54, 47.

    Source: Time Almanac

    Construct a grouped frequency distribution for the data. Use 41–45 for the first class and use the same width for each subsequent class.

42. The IQ scores of 70 students enrolled in a liberal arts course at a college are as follows:

    102, 100, 103, 86, 120, 117, 111, 101, 93, 97, 99, 95, 95, 104, 104, 105, 106, 109, 109, 89, 94, 95, 99, 99, 103, 104, 105, 109, 110, 114, 124, 123, 118, 117, 116, 110, 114, 114, 96, 99, 103, 103, 104, 107, 107, 110, 111, 112, 113, 117, 115, 116, 100, 104, 102, 94, 93, 93, 96, 96, 111, 116, 107, 109, 105, 106, 97, 106, 107, 108.

    Construct a grouped frequency distribution for the data. Use 85–89 for the first class and use the same width for each subsequent class.

43. Construct a histogram and a frequency polygon for the data involving stress ratings in Exercises 27–30.

44. Construct a histogram and a frequency polygon for the data in Exercise 31.

45. Construct a histogram and a frequency polygon for the data in Exercise 32.

The histogram shows the distribution of starting salaries (rounded to the nearest thousand dollars) for college graduates based on a random sample of recent graduates.

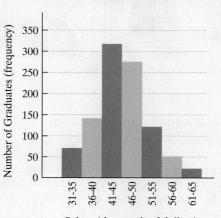

Starting Salaries of Recent College Graduates

In Exercises 46–49, determine whether each statement is true or false according to the graph.

46. The graph is based on a sample of approximately 500 recent college graduates.

47. More college graduates had starting salaries in the $51,000–$55,000 range than in the $36,000–$40,000 range.

48. If the sample is truly representative, then for a group of 400 college graduates, we can expect about 28 of them to have starting salaries in the $31,000–$35,000 range.

49. The percentage of starting salaries falling above those shown by any rectangular bar is equal to the percentage of starting salaries falling below that bar.

The frequency polygon shows a distribution of IQ scores.

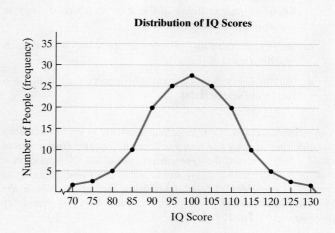

Distribution of IQ Scores

In Exercises 50–53, determine whether each statement is true or false according to the graph at the bottom of the previous page.

50. The graph is based on a sample of approximately 50 people.

51. More people had an IQ score of 100 than any other IQ score, and as the deviation from 100 increases or decreases, the scores fall off in a symmetrical manner.

52. More people had an IQ score of 110 than a score of 90.

53. The percentage of scores above any IQ score is equal to the percentage of scores below that score.

54. Construct a stem-and-leaf plot for the data in Exercise 41 showing the ages at which U.S. presidents were inaugurated.

55. A random sample of 40 high school teachers is selected from all high school teachers in a school district. The following list gives their ages:

63, 48, 42, 42, 38, 59, 41, 44, 45, 28, 54, 62, 51, 44, 63, 66, 59, 46, 51, 28, 37, 66, 42, 40, 30, 31, 48, 32, 29, 42, 63, 37, 36, 47, 25, 34, 49, 30, 35, 50.

Construct a stem-and-leaf plot for the data. What does the shape of the display reveal about the ages of the teachers?

56. In "Ages of Oscar-Winning Best Actors and Actresses" (*Mathematics Teacher* magazine) by Richard Brown and Gretchen Davis, the stem-and-leaf plots shown compare the ages of 30 actors and 30 actresses at the time they won the award.

Actors	Stems	Actresses
	2	146667
98753221	3	00113344455778
88776543322100	4	11129
6651	5	
210	6	011
6	7	4
	8	0

a. What is the age of the youngest actor to win an Oscar?

b. What is the age difference between the oldest and the youngest actress to win an Oscar?

c. What is the oldest age shared by two actors to win an Oscar?

d. What differences do you observe between the two stem-and-leaf plots? What explanations can you offer for these differences?

In Exercises 57–61, describe what is misleading in each visual display of data.

57.

World Population, in Billions

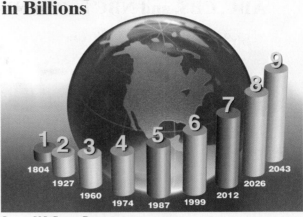

Source: U.S. Census Bureau

58.

Book Title Output in the United States

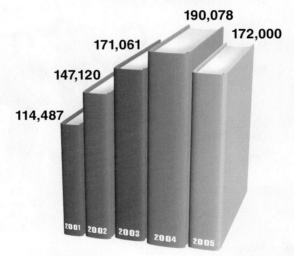

Source: R. R. Bowker

59.

Percentage of the World's Computers in Use, by Country

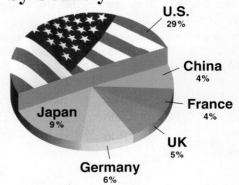

Source: Computer Industry Almanac

60.

Percentage of U.S. Households Watching ABC, CBS, and NBC in Prime Time

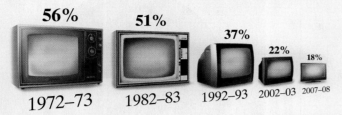

56% 1972–73
51% 1982–83
37% 1992–93
22% 2002–03
18% 2007–08

Source: Nielsen Media Research

61.

Domestic Box-Office Receipts for Musical Films

Box-Office Receipts (millions of dollars)

$180
$160
$140
$120
$100
$80
$60
$40
$20

| Chicago (2002) $170.7 | The Phantom of the Opera (2004) $51.3 | Rent (2005) $29.1 | The Producers (2005) $19.4 | Dreamgirls (2006) $103.4 | Hairspray (2007) $118.9 | Sweeney Todd (2007) $52.9 |

Source: Entertainment Weekly

Critical Thinking Exercises

62. Create a graph that shows a more accurate way of presenting the data from one of the visual displays in Exercises 57–61.

63. Construct a grouped frequency distribution for the following data, showing the length, in miles, of the 25 longest rivers in the United States. Use five classes that have the same width.

2540	2340	1980	1900	1900
1460	1450	1420	1310	1290
1280	1240	1040	990	926
906	886	862	800	774
743	724	692	659	649

Source: U.S. Department of the Interior

Group Exercises

64. The classic book on distortion using statistics is *How to Lie with Statistics* by Darrell Huff. This activity is designed for five people. Each person should select two chapters from Huff's book and then present to the class the common methods of statistical manipulation and distortion that Huff discusses.

65. Each group member should find one example of a graph that presents data with integrity and one example of a graph that is misleading. Use newspapers, magazines, the Internet, books, and so forth. Once graphs have been collected, each member should share his or her graphs with the entire group. Be sure to explain why one graph depicts data in a forthright manner and how the other graph misleads the viewer.

12.2 Measures of Central Tendency

1 Determine the mean for a data set.

2 Determine the median for a data set.

3 Determine the mode for a data set.

4 Determine the midrange for a data set.

CROCK   by Bill Rechin & Brant Parker

© NAS. Reprinted with permission of North America Syndicate.

According to researchers, "Robert," the average American guy, is 31 years old, 5 feet 10 inches, 172 pounds, works 6.1 hours daily, and sleeps 7.7 hours. These numbers represent what is "average" or "typical" of American men. In statistics, such values are known as **measures of central tendency** because they are generally located toward the center of a distribution. Four such measures are discussed in this section: the mean, the median, the mode, and the midrange. Each measure of central tendency is calculated in a different way. Thus, it is better to use a specific term (mean, median, mode, or midrange) than to use the generic descriptive term "average."

The Mean

1 Determine the mean for a data set.

By far the most commonly used measure of central tendency is the *mean*. The **mean** is obtained by adding all the data items and then dividing the sum by the number of items. The Greek letter sigma, Σ, called a **symbol of summation**, is used to indicate the sum of data items. The notation Σx, read "the sum of x," means to add all the data items in a given data set. We can use this symbol to give a formula for calculating the mean.

> **The Mean**
>
> The **mean** is the sum of the data items divided by the number of items.
>
> $$\text{Mean} = \frac{\Sigma x}{n},$$
>
> where Σx represents the sum of all the data items and n represents the number of items.

The mean of a sample is symbolized by $\bar{x}$ (read "x bar"), while the mean of an entire population is symbolized by μ (the lowercase Greek letter *mu*). Unless otherwise indicated, the data sets throughout this chapter represent samples, so we will use $\bar{x}$ for the mean: $\bar{x} = \frac{\Sigma x}{n}$.

Example 1   Calculating the Mean

Table 12.6 shows the ten youngest male singers in the United States to have a number 1 single. Find the mean age of these male singers at the time of their number 1 single.

Table 12.6 Youngest U.S. Male Singers to Have a Number 1 Single		
Artist/Year	**Title**	**Age**
Stevie Wonder, 1963	"Fingertips"	13
Donny Osmond, 1971	"Go Away Little Girl"	13
Michael Jackson, 1972	"Ben"	14
Laurie London, 1958	"He's Got the Whole World in His Hands"	14
Chris Brown, 2005	"Run It!"	15
Paul Anka, 1957	"Diana"	16
Brian Hyland, 1960	"Itsy Bitsy Teenie Weenie Yellow Polkadot Bikini"	16
Shaun Cassidy, 1977	"Da Doo Ron Ron"	17
Soulja Boy, 2007	"Crank That Soulja Boy"	17
Sean Kingston, 2007	"Beautiful Girls"	17

Source: Russell Ash, *The Top 10 of Everything*

Solution

We find the mean, $\overline{x}$, by adding the ages and dividing this sum by 10, the number of data items.

$$\overline{x} = \frac{\Sigma x}{n} = \frac{13 + 13 + 14 + 14 + 15 + 16 + 16 + 17 + 17 + 17}{10} = \frac{152}{10} = 15.2$$

The mean age of the ten youngest male singers to have a number 1 single is 15.2.

One and only one mean can be calculated for any group of numerical data. The mean may or may not be one of the actual data items. In Example 1, the mean was 15.2, although no data item is 15.2.

✓ **Checkpoint 1**   Find the mean, $\overline{x}$, for each group of data items:

a. 10, 20, 30, 40, 50   **b.** 3, 10, 10, 10, 117.

In Example 1, some of the data items were identical. We can use multiplication when computing the mean for these identical items:

$$\overline{x} = \frac{13 + 13 + 14 + 14 + 15 + 16 + 16 + 17 + 17 + 17}{10}$$

$$= \frac{13 \cdot 2 + 14 \cdot 2 + 15 \cdot 1 + 16 \cdot 2 + 17 \cdot 3}{10}$$

The data values 13, 14, and 16 each have a frequency of 2.
The data value 17 has a frequency of 3.

When many data values occur more than once and a frequency distribution is used to organize the data, we can use the following formula to calculate the mean:

> **Calculating the Mean for a Frequency Distribution**
>
> $$\text{Mean} = \bar{x} = \frac{\Sigma x f}{n},$$
>
> where
> x represents a data value.
> f represents the frequency of that data value.
> $\Sigma x f$ represents the sum of all the products obtained by multiplying each data value by its frequency.
> n represents the *total frequency* of the distribution.

Table 12.7 Students' Stress-Level Ratings

Stress Rating x	Frequency f
0	2
1	1
2	3
3	12
4	16
5	18
6	13
7	31
8	26
9	15
10	14

Source: Journal of Personality and Social Psychology, 69, 1102–1112

Example 2   Calculating the Mean for a Frequency Distribution

In the previous exercise set, we mentioned a questionnaire given to college students in an introductory statistics class during the first week of the course. One question asked, "How stressed have you been in the last $2\frac{1}{2}$ weeks, on a scale of 0 to 10, with 0 being not at all stressed and 10 being as stressed as possible?" **Table 12.7** shows the students' responses. Use this frequency distribution to find the mean of the stress-level ratings.

Solution

We use the formula for the mean, $\bar{x}$:

$$\bar{x} = \frac{\Sigma x f}{n}.$$

First, we must find xf, obtained by multiplying each data value, x, by its frequency, f. Then, we need to find the sum of these products, $\Sigma x f$. We can use the frequency distribution to organize these computations. Add a third column in which each data value is multiplied by its frequency. This column, shown on the right, is headed xf. Then, find the sum of the values, $\Sigma x f$, in this column.

x	f	xf
0	2	$0 \cdot 2 = 0$
1	1	$1 \cdot 1 = 1$
2	3	$2 \cdot 3 = 6$
3	12	$3 \cdot 12 = 36$
4	16	$4 \cdot 16 = 64$
5	18	$5 \cdot 18 = 90$
6	13	$6 \cdot 13 = 78$
7	31	$7 \cdot 31 = 217$
8	26	$8 \cdot 26 = 208$
9	15	$9 \cdot 15 = 135$
10	14	$10 \cdot 14 = 140$

Totals: $n = 151$ $\Sigma x f = 975$

$\Sigma x f$ is the sum of the numbers in the third column.

This value, the sum of the numbers in the second column, is the total frequency of the distribution.

Now, substitute these values into the formula for the mean, $\bar{x}$. Remember that n is the *total frequency* of the distribution, or 151.

$$\bar{x} = \frac{\Sigma x f}{n} = \frac{975}{151} \approx 6.46$$

The mean of the 0 to 10 stress-level ratings is approximately 6.46. Notice that the mean is greater than 5, the middle of the 0 to 10 scale.

Checkpoint 2 Find the mean, $\bar{x}$, for the data items in the frequency distribution. (In order to save space, we've written the frequency distribution horizontally.)

Score, x	30	33	40	50
Frequency, f	3	4	4	1

2 Determine the median for a data set.

The Median

The *median* age in the United States is 35.3. The oldest state by median age is Florida (38.7) and the youngest state is Utah (27.1). To find these values, researchers begin with appropriate random samples. The data items—that is, the ages—are arranged from youngest to oldest. The median age is the data item in the middle of each set of ranked, or ordered, data.

> ### The Median
>
> To find the **median** of a group of data items,
>
> 1. Arrange the data items in order, from smallest to largest.
> 2. If the number of data items is odd, the median is the data item in the middle of the list.
> 3. If the number of data items is even, the median is the mean of the two middle data items.

Example 3 Finding the Median

Find the median for each of the following groups of data:

a. 84, 90, 98, 95, 88   **b.** 68, 74, 7, 13, 15, 25, 28, 59, 34, 47.

Solution

a. Arrange the data items in order, from smallest to largest. The number of data items in the list, five, is odd. Thus, the median is the middle number.

$$84, 88, 90, 95, 98$$

Middle data item

The median is 90. Notice that two data items lie above 90 and two data items lie below 90.

b. Arrange the data items in order, from smallest to largest. The number of data items in the list, ten, is even. Thus, the median is the mean of the two middle data items.

$$7, 13, 15, 25, 28, 34, 47, 59, 68, 74$$

Middle data items are 28 and 34.

$$\text{Median} = \frac{28 + 34}{2} = \frac{62}{2} = 31$$

The median is 31. Five data items lie above 31 and five data items lie below 31.

7 13 15 25 28 | 34 47 59 68 74

Five data items lie below 31.   Five data items lie above 31.

Median is 31.

Study Tip

The median splits the data items down the middle, like the median strip in a road.

✓ **Checkpoint 3** Find the median for each of the following groups of data:

a. 28, 42, 40, 25, 35 **b.** 72, 61, 85, 93, 79, 87.

If a relatively long list of data items is arranged in order, it may be difficult to identify the item or items in the middle. In cases like this, the median can be found by determining its position in the list of items.

Position of the Median

If n data items are arranged in order, from smallest to largest, the median is the value in the

$$\frac{n + 1}{2}$$

position.

Example 4 Finding the Median Using the Position Formula

Listed below are the points scored per season by the 13 top point scorers in the National Football League. Find the median points scored per season for the top 13 scorers.

144, 144, 145, 145, 145, 146, 147, 149, 150, 155, 161, 164, 176

Solution

The data items are arranged from smallest to largest. There are 13 data items, so $n = 13$. The median is the value in the

$$\frac{n + 1}{2} \text{ position} = \frac{13 + 1}{2} \text{ position} = \frac{14}{2} \text{ position} = \text{seventh position.}$$

We find the median by selecting the data item in the seventh position.

| Position 3 | Position 4 | | Position 7 |

144, 144, 145, 145, 145, 146, 147, 149, 150, 155, 161, 164, 176

| Position 1 | Position 2 | | Position 5 | Position 6 |

The median is 147. Notice that six data items lie above 147 and six data items lie below it. The median points scored per season for the top 13 scorers in the National Football League is 147.

✓ **Checkpoint 4** Find the median for the following group of data items:

1, 2, 2, 2, 3, 3, 3, 3, 3, 5, 6, 7, 7, 10, 11, 13, 19, 24, 26.

Table 12.8 Time per Day, in Hours and Minutes, Spent Sleeping and Eating in Selected Countries

Country	Sleeping	Eating
France	8:50	2:15
U.S.	8:38	1:14
Spain	8:34	1:46
New Zealand	8:33	2:10
Australia	8:32	1:29
Turkey	8:32	1:29
Canada	8:29	1:09
Poland	8:28	1:34
Finland	8:27	1:21
Belgium	8:25	1:49
United Kingdom	8:23	1:25
Mexico	8:21	1:06
Italy	8:18	1:54
Germany	8:12	1:45
Sweden	8:06	1:34
Norway	8:03	1:22
Japan	7:50	1:57
S. Korea	7:49	1:36

Source: Organization for Economic Cooperation and Development

Example 5 **Finding the Median Using the Position Formula**

Table 12.8 gives the mean amount time per day, in hours and minutes, spent sleeping and eating in 18 selected countries. Find the median amount of time per day, in hours and minutes, spent sleeping for these countries.

Solution

Reading from the bottom to the top of **Table 12.8**, the data items for sleeping appear from smallest to largest. There are 18 data items, so $n = 18$. The median is the value in the

$$\frac{n+1}{2} \text{ position} = \frac{18+1}{2} \text{ position} = \frac{19}{2} \text{ position} = 9.5 \text{ position}.$$

This means that the median is the mean of the data items in positions 9 and 10.

Position 3 Position 4   Position 7 Position 8

7:49, 7:50, 8:03, 8:06, 8:12, 8:18, 8:21, 8:23, 8:25, 8:27, 8:28, 8:29, 8:32, 8:32, 8:33, 8:34, 8:38, 8:50

Position 1 Position 2   Position 5 Position 6   Position 9 Position 10

$$\text{Median} = \frac{8:25 + 8:27}{2} = \frac{16:52}{2} = 8:26$$

The median amount of time per day spent sleeping for the 18 countries is 8 hours, 26 minutes.

✓ **Checkpoint 5** Arrange the data items for eating in **Table 12.8** from smallest to largest. Then find the median amount of time per day, in hours and minutes, spent eating for the 18 countries.

When individual data items are listed from smallest to largest, you can find the median by identifying the item or items in the middle or by using the $\frac{n+1}{2}$ formula for its position. However, the formula for the position of the median is more useful when data items are organized in a frequency distribution.

Example 6 **Finding the Median for a Frequency Distribution**

The frequency distribution for the stress-level ratings of 151 students is repeated below using a horizontal format. Find the median stress-level rating.

Stress rating →

x	0	1	2	3	4	5	6	7	8	9	10
f	2	1	3	12	16	18	13	31	26	15	14

← Number of college students   Total: $n = 151$

Solution

There are 151 data items, so $n = 151$. The median is the value in the

$$\frac{n+1}{2} \text{ position} = \frac{151+1}{2} \text{ position} = \frac{152}{2} \text{ position} = 76\text{th position}.$$

We find the median by selecting the data item in the 76th position. The frequency distribution indicates that the data items begin with

$$0, 0, 1, 2, 2, 2, \ldots .$$

We can write the data items all out and then select the median, the 76th data item. A more efficient way to proceed is to count down the frequency column in the distribution until we identify the 76th data item:

x	f
0	2
1	1
2	3
3	12
4	16
5	18
6	13
7	31
8	26
9	15
10	14

> We count down the frequency column.
> 1, 2
> 3
> 4, 5, 6
> 7, 8, 9, 10, 11, 12, 13, 14, 15, 16, 17, 18
> 19, 20, 21, 22, 23, 24, 25, 26, 27, 28, 29, 30, 31, 32, 33, 34
> 35, 36, 37, 38, 39, 40, 41, 42, 43, 44, 45, 46, 47, 48, 49, 50, 51, 52
> 53, 54, 55, 56, 57, 58, 59, 60, 61, 62, 63, 64, 65
> 66, 67, 68, 69, 70, 71, 72, 73, 74, 75, 76

Stop counting. We've reached the 76th data item.

The 76th data item is 7. The median stress-level rating is 7.

Checkpoint 6   Find the median for the following frequency distribution.

Age at presidential inauguration

Number of U.S. presidents assuming office in the 20th century with the given age

x	42	43	46	51	52	54	55	56	60	61	64	69
f	1	1	1	3	1	2	2	2	1	2	1	1

Statisticians generally use the median, rather than the mean, when reporting income. Why? Our next example will help to answer this question.

Example 7   Comparing the Median and the Mean

Five employees in the assembly section of a television manufacturing company earn salaries of $19,700, $20,400, $21,500, $22,600, and $23,000 annually. The section manager has an annual salary of $95,000.

a. Find the median annual salary for the six people.

b. Find the mean annual salary for the six people.

Solution

a. To compute the median, first arrange the salaries in order:

$$\$19,700, \quad \$20,400, \quad \$21,500, \quad \$22,600, \quad \$23,000, \quad \$95,000.$$

Because the list contains an even number of data items, six, the median is the mean of the two middle items.

The median is the mean of the two middle items of the salaries in order:

$$\$19{,}700, \quad \$20{,}400, \quad \$21{,}500, \quad \$22{,}600, \quad \$23{,}000, \quad \$95{,}000.$$

$$\text{Median} = \frac{\$21{,}500 + \$22{,}600}{2} = \frac{\$44{,}100}{2} = \$22{,}050$$

The median annual salary is $22,050.

b. We find the mean annual salary by adding the six annual salaries and dividing by 6.

$$\text{Mean} = \frac{\$19{,}700 + \$20{,}400 + \$21{,}500 + \$22{,}600 + \$23{,}000 + \$95{,}000}{6}$$

$$= \frac{\$202{,}200}{6} = \$33{,}700$$

The mean annual salary is $33,700.

In Example 7, the median annual salary is $22,050 and the mean annual salary is $33,700. Why such a big difference between these two measures of central tendency? The relatively high annual salary of the section manager, $95,000, pulls the mean salary to a value considerably higher than the median salary. When one or more data items are much greater than the other items, these extreme values can greatly influence the mean. In cases like this, the median is often more representative of the data.

This is why the median, rather than the mean, is used to summarize the incomes, by gender and race, shown in **Figure 12.6**. Because no one can earn less than $0, the distribution of income must come to an end at $0 for each of these eight groups. By contrast, there is no upper limit on income on the high side. In the United States, the wealthiest 5% of the population earn about 21% of the total income. The relatively few people with very high annual incomes tend to pull the mean income to a value considerably greater than the median income. Reporting mean incomes in **Figure 12.6** would inflate the numbers shown, making them nonrepresentative of the millions of workers in each of the eight groups.

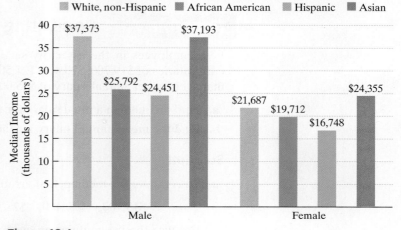

Figure 12.6
Source: U.S. Census Bureau

Checkpoint 7 **Table 12.9** shows 2008 compensation (sum of salary, bonus, perks, stock and option awards) for CEOs of six U.S. companies.

Table 12.9 Executive Compensation for Six U.S. Companies

Company	Executive	Total Compensation (millions of dollars)
Apple	Steve Jobs	$0
Coca-Cola	Muhtar Kent	$19.6
IBM	Samuel Palmisano	$21.0
Exxon Mobil	Rex Tillerson	$23.9
Comcast	Brian Roberts	$24.7
Abbott Laboratories	Miles White	$25.1

Source: USA TODAY

a. Find the mean compensation, in millions of dollars, for the six CEOs.
b. Find the median compensation, in millions of dollars, for the six CEOs.
c. Describe why one of the measures of central tendency is greater than the other.

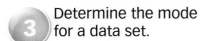
Determine the mode for a data set.

The Mode

Let's take one final look at the frequency distribution for the stress-level ratings of 151 college students.

Stress rating

Number of college students

x	0	1	2	3	4	5	6	7	8	9	10
f	2	1	3	12	16	18	13	31	26	15	14

7 is the stress rating with the greatest frequency.

The data value that occurs most often in this distribution is 7, the stress rating for 31 of the 151 students. We call 7 the *mode* of this distribution.

> **The Mode**
>
> The **mode** is the data value that occurs most often in a data set. If more than one data value has the highest frequency, then each of these data values is a mode. If there is no data value that occurs most often, then the data set has no mode.

Example 8 **Finding the Mode**

Find the mode for the following group of data:

$$7, 2, 4, 7, 8, 10.$$

Solution

The number 7 occurs more often than any other. Therefore, 7 is the mode.

Checkpoint 8 Find the mode for the following group of data:

$$8, 6, 2, 4, 6, 8, 10, 8.$$

Be aware that a data set might not have a mode. For example, no data item in 2, 1, 4, 5, 3 occurs most often, so this data group has no mode. By contrast, 3, 3, 4, 5, 6, 6 has two data values with the highest frequency, namely 3 and 6. Each of these data values is a mode and the data set is said to be **bimodal**.

<table>
<tr><td>④</td><td>Determine the midrange for a data set.</td></tr>
</table>

Table 12.10 Ten Hottest U.S. Cities

City	Mean Temperature
Key West, FL	77.8°
Miami, FL	75.9°
West Palm Beach, FL	74.7°
Fort Myers, FL	74.4°
Yuma, AZ	74.2°
Brownsville, TX	73.8°
Phoenix, AZ	72.6°
Vero Beach, FL	72.4°
Orlando, FL	72.3°
Tampa, FL	72.3°

Source: National Oceanic and Atmospheric Administration

The Midrange

Table 12.10 shows the ten hottest cities in the United States. Because temperature is constantly changing, you might wonder how the mean temperatures shown in the table are obtained.

First, we need to find a representative daily temperature. This is obtained by adding the lowest and highest temperatures for the day and then dividing this sum by 2. Next, we take the representative daily temperatures for all 365 days, add them, and divide the sum by 365. These are the mean temperatures that appear in **Table 12.10**.

Representative daily temperature,

$$\frac{\text{lowest daily temperature} + \text{highest daily temperature}}{2},$$

is an example of a measure of central tendency called the *midrange*.

The Midrange

The **midrange** is found by adding the lowest and highest data values and dividing the sum by 2.

$$\text{Midrange} = \frac{\text{lowest data value} + \text{highest data value}}{2}$$

Example 9 Finding the Midrange

One criticism of Major League baseball is that the discrepancy between team payrolls hampers fair competition. In 2009, the New York Yankees had the greatest payroll, a record $201,449,181 (median salary: $5,200,000). The Florida Marlins were the worst paid team, with a payroll of $36,834,000 (median salary: $470,000). Find the midrange for the annual payroll of Major League baseball teams in 2009. (*Source:* usatoday.com)

Solution

$$\text{Midrange} = \frac{\text{lowest annual payroll} + \text{highest annual payroll}}{2}$$

$$= \frac{\$36,834,000 + \$201,449,181}{2} = \frac{\$238,283,181}{2} = \$119,141,590.50$$

The midrange for the annual payroll of Major League baseball teams in 2009 was ▶ $119,141,590.50.

We could find the mean annual payroll of the 30 professional baseball teams in 2009 by adding up the payrolls of all 30 teams and then dividing the sum by 30. It is much faster to calculate the midrange, which is often used as an estimate for the mean.

✓ **Checkpoint 9** In 2008, the Oakland Raiders had the greatest payroll of any team in the National Football League, a record $152,389,371 (median salary: $967,700). The Kansas City Chiefs were the worst paid team, with a payroll of $83,623,776 (median salary: $695,000). Find the midrange for the annual payroll of ▶ National Football League teams in 2008. (*Source*: usatoday.com)

Example 10 ### Finding the Four Measures of Central Tendency

Suppose your six exam grades in a class are

$$52, 69, 75, 86, 86, \text{ and } 92.$$

Compute your final grade $(90-100 = A, 80-89 = B, 70-79 = C, 60-69 = D,$ below $60 = F)$ using the

a. mean.   **b.** median.   **c.** mode.   **d.** midrange.

Solution

a. The mean is the sum of the data items divided by the number of items, 6.

$$\text{Mean} = \frac{52 + 69 + 75 + 86 + 86 + 92}{6} = \frac{460}{6} \approx 76.67$$

Using the mean, your final course grade is C.

b. The six data items, 52, 69, 75, 86, 86, and 92, are arranged in order. Because the number of data items is even, the median is the mean of the two middle items.

$$\text{Median} = \frac{75 + 86}{2} = \frac{161}{2} = 80.5$$

Using the median, your final course grade is B.

c. The mode is the data value that occurs most frequently. Because 86 occurs most often, the mode is 86. Using the mode, your final course grade is B.

d. The midrange is the mean of the lowest and highest data values.

$$\text{Midrange} = \frac{52 + 92}{2} = \frac{144}{2} = 72$$

Using the midrange, your final course grade is C.

✓ **Checkpoint 10** *Consumer Reports* magazine gave the following data for the number of calories in a meat hot dog for each of 17 brands:

173, 191, 182, 190, 172, 147, 146, 138, 175, 136, 179, 153, 107, 195, 135, 140, 138.

Find the mean, median, mode, and midrange for the number of calories in a meat hot dog for the 17 brands. If necessary, round answers to the nearest tenth of a ▶ calorie.

Exercise Set 12.2

Concept and Vocabulary Exercises

In Exercises 1–5, fill in each blank so that the resulting statement is true.

1. $\frac{\Sigma x}{n}$, the sum of all the data items divided by the number of data items, is the measure of central tendency called the _____.

2. The measure of central tendency that is the data item in the middle of ranked, or ordered, data is called the _____.

3. If n data items are arranged in order, from smallest to largest, the data item in the middle is the value in _____ position.

4. A data value that occurs most often in a data set is the measure of central tendency called the _____.

5. The measure of central tendency that is found by adding the lowest and highest data values and dividing the sum by 2 is called the _____.

In Exercises 6–9, determine whether each statement is true or false. If the statement is false, make the necessary change(s) to produce a true statement.

6. Numbers representing what is average or typical about a data set are called measures of central tendency.

7. When finding the mean, it is necessary to arrange the data items in order.

8. If one or more data items are much greater than the other items, the mean, rather than the median, is more representative of the data.

9. A data set can contain more than one median, or no median at all.

Respond to Exercises 10–18 using verbal or written explanations.

10. What is the mean and how is it obtained?

11. What is the median and how is it obtained?

12. What is the mode and how is it obtained?

13. What is the midrange and how is it obtained?

14. The "average" income in the United States can be given by the mean or the median.
   a. Which measure would be used in anti-U.S. propaganda? Explain your answer.
   b. Which measure would be used in pro-U.S. propaganda? Explain your answer.

15. In a class of 40 students, 21 have examination scores of 77%. Which measure or measures of central tendency can you immediately determine? Explain your answer.

16. You read an article that states. "Of the 411 players in the National Basketball Association, only 138 make more than the average salary of $3.12 million." Is $3.12 million the mean or the median salary? Explain your answer.

17. A college student's parents promise to pay for next semester's tuition if an A average is earned in chemistry. With examination grades of 97%, 97%, 75%, 70%, and 55%, the student reports that an A average has been earned. Which measure of central tendency is the student reporting as the average? How is this student misrepresenting the course performance with statistics?

18. According to the National Oceanic and Atmospheric Administration, the coldest city in the United States is International Falls, Minnesota, with a mean Fahrenheit temperature of 36.8°. Explain how this mean is obtained.

In Exercises 19–22, determine whether each statement makes sense or does not make sense, and explain your reasoning.

19. I'm working with a data set for which neither the mean nor the median is one of the data items.

20. I made a distribution of the heights of the 12 players on our basketball team. Because one player is much taller than the others, the team's median height is greater than its mean height.

21. Although the data set 1, 1, 2, 3, 3, 3, 4, 4 has a number of repeated items, there is only one mode.

22. If teachers use the same test scores for a particular student and calculate measures of central tendency correctly, they will always agree on the student's final course grade.

Practice Exercises

In Exercises 23–30, find the mean for each group of data items.

23. 7, 4, 3, 2, 8, 5, 1, 3

24. 11, 6, 4, 0, 2, 1, 12, 0, 0

25. 91, 95, 99, 97, 93, 95

26. 100, 100, 90, 30, 70, 100

27. 100, 40, 70, 40, 60

28. 1, 3, 5, 10, 8, 5, 6, 8

29. 1.6, 3.8, 5.0, 2.7, 4.2, 4.2, 3.2, 4.7, 3.6, 2.5, 2.5

30. 1.4, 2.1, 1.6, 3.0, 1.4, 2.2, 1.4, 9.0, 9.0, 1.8

In Exercises 31–34, find the mean for the data items in the given frequency distribution.

31.

Score x	Frequency f
1	1
2	3
3	4
4	4
5	6
6	5
7	3
8	2

32.

Score x	Frequency f
1	2
2	4
3	5
4	7
5	6
6	4
7	3

33.

Score x	Frequency f
1	1
2	1
3	2
4	5
5	7
6	9
7	8
8	6
9	4
10	3

34.

Score x	Frequency f
1	3
2	4
3	6
4	8
5	9
6	7
7	5
8	2
9	1
10	1

In Exercises 35–42, find the median for each group of data items.

35. 7, 4, 3, 2, 8, 5, 1, 3

36. 11, 6, 4, 0, 2, 1, 12, 0, 0

37. 91, 95, 99, 97, 93, 95

38. 100, 100, 90, 30, 70, 100

39. 100, 40, 70, 40, 60

40. 1, 3, 5, 10, 8, 5, 6, 8

41. 1.6, 3.8, 5.0, 2.7, 4.2, 4.2, 3.2, 4.7, 3.6, 2.5, 2.5

42. 1.4, 2.1, 1.6, 3.0, 1.4, 2.2, 1.4, 9.0, 9.0, 1.8

Find the median for the data items in the frequency distribution in

43. Exercise 31.
44. Exercise 32.
45. Exercise 33.
46. Exercise 34.

In Exercises 47–54, find the mode for each group of data items. If there is no mode, so state.

47. 7, 4, 3, 2, 8, 5, 1, 3
48. 11, 6, 4, 0, 2, 1, 12, 0, 0
49. 91, 95, 99, 97, 93, 95
50. 100, 100, 90, 30, 70, 100
51. 100, 40, 70, 40, 60
52. 1, 3, 5, 10, 8, 5, 6, 8
53. 1.6, 3.8, 5.0, 2.7, 4.2, 4.2, 3.2, 4.7, 3.6, 2.5, 2.5
54. 1.4, 2.1, 1.6, 3.0, 1.4, 2.2, 1.4, 9.0, 9.0, 1.8

Find the mode for the data items in the frequency distribution in

55. Exercise 31.
56. Exercise 32.
57. Exercise 33.
58. Exercise 34.

In Exercises 59–66, find the midrange for each group of data items.

59. 7, 4, 3, 2, 8, 5, 1, 3
60. 11, 6, 4, 0, 2, 1, 12, 0, 0
61. 91, 95, 99, 97, 93, 95
62. 100, 100, 90, 30, 70, 100
63. 100, 40, 70, 40, 60
64. 1, 3, 5, 10, 8, 5, 6, 8
65. 1.6, 3.8, 5.0, 2.7, 4.2, 4.2, 3.2, 4.7, 3.6, 2.5, 2.5
66. 1.4, 2.1, 1.6, 3.0, 1.4, 2.2, 1.4, 9.0, 9.0, 1.8

Find the midrange for the data items in the frequency distribution in

67. Exercise 31.
68. Exercise 32.
69. Exercise 33.
70. Exercise 34.

Practice Plus

In Exercises 71–76, use each display of data items to find the mean, median, mode, and midrange.

71.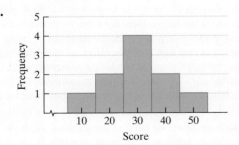

72.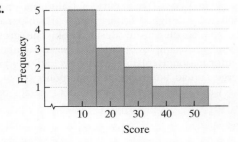

73.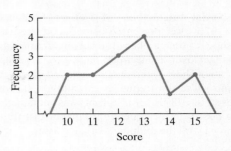

74.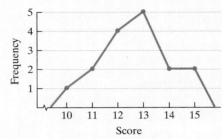

75.

Stems	Leaves			
2	1	4	5	
3	0	1	1	3
4	2	5		

76.

Stems	Leaves			
2	8			
3	2	4	4	9
4	0	1	5	7

Application Exercises

Exercises 77–79 present data on a variety of topics. For each data set described in boldface, find the

a. *mean.*

b. *median.*

c. *mode (or state that there is no mode).*

d. *midrange.*

77. Values (in millions of dollars) of the 30 Major League Baseball Teams

1500, 912, 833, 722, 700, 509, 496, 486, 471, 450, 446, 445, 426, 406, 406, 401, 400, 399, 390, 373, 371, 356, 353, 347, 342, 320, 319, 314, 288, 277

Source: Forbes (May 11, 2009)

78. Revenues (in millions of dollars) of the 30 Major League Baseball Teams

375, 269, 261, 241, 239, 216, 212, 196, 196, 195, 194, 189, 186, 186, 184, 181, 178, 177, 176, 174, 174, 173, 172, 171, 160, 160, 158, 144, 143, 139

Source: Forbes (May 11, 2009)

79. **Number of Social Interactions of College Students** In Exercise Set 12.1, we presented a grouped frequency distribution showing the number of social interactions of ten minutes or longer over a one-week period for a group of college students. (These interactions excluded family and work situations.) Use the frequency distribution shown to solve this exercise. (This distribution was obtained by replacing the classes in the grouped frequency distribution previously shown with the midpoints of the classes.)

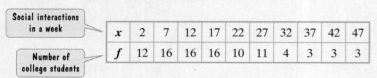

Social interactions in a week	x	2	7	12	17	22	27	32	37	42	47
Number of college students	f	12	16	16	16	10	11	4	3	3	3

The weights (to the nearest five pounds) of 40 randomly selected male high school seniors are organized in a histogram with a superimposed frequency polygon. Use the graph to answer Exercises 80–83.

Weights of 40 Male High School Seniors

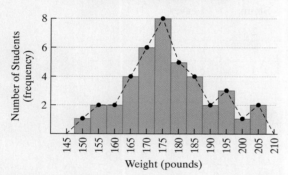

80. Find the mean weight.
81. Find the median weight.
82. Find the modal weight.
83. Find the midrange weight.
84. An advertisement for a speed-reading course claimed that the "average" reading speed for people completing the course was 1000 words per minute. Shown below are the actual data for the reading speeds per minute for a sample of 24 people who completed the course.

1000	900	800	1000	900	850
650	1000	1050	800	1000	850
700	750	800	850	900	950
600	1100	950	700	750	650

a. Find the mean, median, mode, and midrange. (If you prefer, first organize the data in a frequency distribution.)

b. Which measure of central tendency was given in the advertisement?

c. Which measure of central tendency is the best indicator of the "average" reading speed in this situation? Explain your answer.

85. In one common system for finding a grade-point average, or GPA,

$$A = 4, B = 3, C = 2, D = 1, F = 0.$$

The GPA is calculated by multiplying the number of credit hours for a course and the number assigned to each grade, and then adding these products. Then divide this sum by the total number of credit hours. Because each course grade is weighted according to the number of credits of the course, GPA is called a *weighted mean*. Calculate the GPA for this transcript:

Sociology: 3 cr. A; Biology: 3.5 cr. C; Music: 1 cr. B; Math: 4 cr. B; English: 3 cr. C.

Critical Thinking Exercises

86. Give an example of a set of six examination grades (from 0 to 100) with each of the following characteristics:

a. The mean and the median have the same value, but the mode has a different value.

b. The mean and the mode have the same value, but the median has a different value.

c. The mean is greater than the median.

d. The mode is greater than the mean.

e. The mean, median, and mode have the same value.

f. The mean and mode have values of 72.

87. On an examination given to 30 students, no student scored below the mean. Describe how this occurred.

Group Exercises

88. Select a characteristic, such as shoe size or height, for which each member of the group can provide a number. Choose a characteristic of genuine interest to the group. For this characteristic, organize the data collected into a frequency distribution and a graph. Compute the mean, median, mode, and midrange. Discuss any differences among these values. What happens if the group is divided (men and women, or people under a certain age and people over a certain age) and these measures of central tendency are computed for each of the subgroups? Attempt to use measures of central tendency to discover something interesting about the entire group or the subgroups.

89. A recent book on spotting bad statistics and learning to think critically about these influential numbers is *Damn Lies and Statistics* by Joel Best (University of California Press, 2001). This activity is designed for six people. Each person should select one chapter from Best's book. The group report should include examples of the use, misuse, and abuse of statistical information. Explain exactly how and why bad statistics emerge, spread, and come to shape policy debates. What specific ways does Best recommend to detect bad statistics?

12.3   Measures of Dispersion

① Determine the range for a data set.

② Determine the standard deviation for a data set.

When you think of Houston, Texas and Honolulu, Hawaii, do balmy temperatures come to mind? Both cities have a mean temperature of 75°. However, the mean temperature does not tell the whole story. The temperature in Houston differs seasonally from a low of about 40° in January to a high of close to 100° in July and August. By contrast, Honolulu's temperature varies less throughout the year, usually ranging between 60° and 90°.

Measures of dispersion are used to describe the spread of data items in a data set. Two of the most common measures of dispersion, the *range* and the *standard deviation*, are discussed in this section.

① Determine the range for a data set.

The Range

A quick but rough measure of dispersion is the **range**, the difference between the highest and lowest data values in a data set. For example, if Houston's hottest annual temperature is 103° and its coldest annual temperature is 33°, the range in temperature is

$$103° - 33°, \quad \text{or} \quad 70°.$$

If Honolulu's hottest day is 89° and its coldest day 61°, the range in temperature is

$$89° - 61°, \quad \text{or} \quad 28°.$$

> **The Range**
>
> The **range**, the difference between the highest and lowest data values in a data set, indicates the total spread of the data.
>
> Range = highest data value − lowest data value

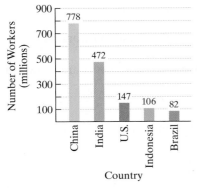

Countries with the Most Workers

Number of Workers (millions): 900, 700, 500, 300, 100

China 778, India 472, U.S. 147, Indonesia 106, Brazil 82

Country

Figure 12.7
Source: Central Intelligence Agency

Example 1   Computing the Range

Figure 12.7 shows the number of workers, in millions, for the five countries with the largest labor forces. Find the range of workers, in millions, for these five countries.

Solution

$$\text{Range} = \text{highest data value} - \text{lowest data value}$$
$$= 778 - 82 = 696$$

The range is 696 million workers.

> ✓ **Checkpoint 1** Find the range for the following group of data items:
>
> $$4, 2, 11, 7.$$

The Standard Deviation

A second measure of dispersion, and one that is dependent on *all* of the data items, is called the **standard deviation**. The standard deviation is found by determining how much each data item differs from the mean.

In order to compute the standard deviation, it is necessary to find by how much each data item deviates from the mean. First compute the mean, $\bar{x}$. Then subtract the mean from each data item, $x - \bar{x}$. Example 2 shows how this is done. In Example 3, we will use this skill to actually find the standard deviation.

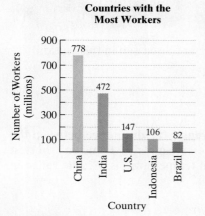

Countries with the Most Workers

Number of Workers (millions)

China 778
India 472
U.S. 147
Indonesia 106
Brazil 82

Country

Figure 12.7 (repeated)

Example 2 **Preparing to Find the Standard Deviation; Finding Deviations from the Mean**

Find the deviations from the mean for the five data items 778, 472, 147, 106, and 82, shown in **Figure 12.7**.

Solution

First, calculate the mean, $\bar{x}$.

$$\bar{x} = \frac{\Sigma x}{n} = \frac{778 + 472 + 147 + 106 + 82}{5} = \frac{1585}{5} = 317$$

The mean for the five countries with the largest labor forces is 317 million workers. Now, let's find by how much each of the five data items in **Figure 12.7** differs from 317, the mean. For China, with 778 million workers, the computation is shown as follows:

$$\begin{aligned} \text{Deviation from mean} &= \text{data item} - \text{mean} \\ &= x - \bar{x} \\ &= 778 - 317 = 461. \end{aligned}$$

This indicates that the labor force in China exceeds the mean by 461 million workers.

The computation for the United States, with 147 million workers, is given by

$$\begin{aligned} \text{Deviation from mean} &= \text{data item} - \text{mean} \\ &= x - \bar{x} \\ &= 147 - 317 = -170. \end{aligned}$$

This indicates that the labor force in the United States is 170 million workers below the mean.

The deviations from the mean for each of the five given data items are shown in **Table 12.11**.

Table 12.11 Deviations from the Mean

Data Item x	Deviation: data item − mean $x - \bar{x}$
778	$778 - 317 = 461$
472	$472 - 317 = 155$
147	$147 - 317 = -170$
106	$106 - 317 = -211$
82	$82 - 317 = -235$

> ✓ **Checkpoint 2** Compute the mean for the following group of data items:
>
> $$2, 4, 7, 11.$$

Then find the deviations from the mean for the four data items. Organize your work in table form just like **Table 12.11**. Keep track of these computations. You will be using them in Checkpoint 3.

The sum of the deviations from the mean for a set of data is always zero: $\Sigma(x - \bar{x}) = 0$. For the deviations from the mean shown in **Table 12.11**,

$$461 + 155 + (-170) + (-211) + (-235) = 616 + (-616) = 0.$$

This shows that we cannot find a measure of dispersion by finding the mean of the deviations, because this value is always zero. However, a kind of average of the deviations from the mean, called the **standard deviation**, can be computed. We do so by squaring each deviation and later introducing a square root in the computation. Here are the details on how to find the standard deviation for a set of data:

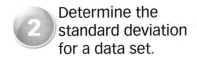

Determine the standard deviation for a data set.

Computing the Standard Deviation for a Data Set

1. Find the mean of the data items.
2. Find the deviation of each data item from the mean:

    data item − mean.

3. Square each deviation:

    $$(\text{data item} - \text{mean})^2.$$

4. Sum the squared deviations:

    $$\Sigma(\text{data item} - \text{mean})^2.$$

5. Divide the sum in step 4 by $n - 1$, where n represents the number of data items:

    $$\frac{\Sigma(\text{data item} - \text{mean})^2}{n - 1}.$$

6. Take the square root of the quotient in step 5. This value is the standard deviation for the data set.

    $$\text{Standard deviation} = \sqrt{\frac{\Sigma(\text{data item} - \text{mean})^2}{n - 1}}$$

The standard deviation of a sample is symbolized by s, while the standard deviation of an entire population is symbolized by σ (the lowercase Greek letter *sigma*). Unless otherwise indicated, data sets represent samples, so we will use s for the standard deviation:

$$s = \sqrt{\frac{\Sigma(x - \bar{x})^2}{n - 1}}.$$

The computation of the standard deviation can be organized using a table with three columns:

Data item x	Deviation: $x - \bar{x}$ Data item − mean	(Deviation)^2: $(x - \bar{x})^2$ (Data item − mean)^2

In Example 2, we worked out the first two columns of such a table. Let's continue working with the data for the countries with the most workers and compute the standard deviation.

Example 3 Computing the Standard Deviation

Figure 12.7, showing the number of workers, in millions, for the five countries with the largest labor forces, is repeated in the margin on the facing page. Find the standard deviation, in millions, for these five countries.

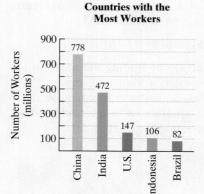

Countries with the Most Workers

Number of Workers (millions)

China 778
India 472
U.S. 147
Indonesia 106
Brazil 82

Country

Figure 12.7 (repeated)

Almost all scientific and graphing calculators compute the standard deviation of a set of data. Using the data items in Example 3,

778, 472, 147, 106, 82,

the keystrokes for obtaining the standard deviation on many scientific calculators are as follows:

778 $\boxed{\Sigma+}$ 472 $\boxed{\Sigma+}$ 147 $\boxed{\Sigma+}$

106 $\boxed{\Sigma+}$ 82 $\boxed{\Sigma+}$ $\boxed{\text{2nd}}$ $\boxed{\sigma n-1}$.

Graphing calculators require that you specify if data items are from an entire population or a sample of the population.

Solution

Step 1 Find the mean. From our work in Example 2, the mean is 317: $\bar{x} = 317$.

Step 2 Find the deviation of each data item from the mean: data item − mean or $x - \bar{x}$. This, too, was done in Example 2 for each of the five data items.

Step 3 Square each deviation: (data item − mean)^2 or $(x - \bar{x})^2$. We square each of the numbers in the (data item − mean) column, shown in **Table 12.12**. Notice that squaring the difference always results in a nonnegative number.

Table 12.12 Computing the Standard Deviation

Data item x	Deviation: data item − mean $x - \bar{x}$	(Deviation)^2: (data item − mean)^2 $(x - \bar{x})^2$
778	$778 - 317 = 461$	$461^2 = 461 \cdot 461 = 212{,}521$
472	$472 - 317 = 155$	$155^2 = 155 \cdot 155 = 24{,}025$
147	$147 - 317 = -170$	$(-170)^2 = (-170) \cdot (-170) = 28{,}900$
106	$106 - 317 = -211$	$(-211)^2 = (-211) \cdot (-211) = 44{,}521$
82	$82 - 317 = -235$	$(-235)^2 = (-235) \cdot (-235) = 55{,}225$
Totals:	$\Sigma(x - \bar{x}) = 0$	$\Sigma(x - \bar{x})^2 = 365{,}192$

The sum of the deviations for a set of data is always zero.

Adding the five numbers in the third column gives the sum of the squared deviations: $\Sigma(\text{data item} - \text{mean})^2$.

Step 4 Sum the squared deviations: $\Sigma(\text{data item} - \text{mean})^2$. This step is shown in **Table 12.12**. The squares in the third column were added, resulting in a sum of 365,192: $\Sigma(x - \bar{x})^2 = 365{,}192$.

Step 5 Divide the sum in step 4 by $n - 1$, where n represents the number of data items. The number of data items is 5 so we divide by 4.

$$\frac{\Sigma(x - \bar{x})^2}{n - 1} = \frac{\Sigma(\text{data item} - \text{mean})^2}{n - 1} = \frac{365{,}192}{5 - 1} = \frac{365{,}192}{4} = 91{,}298$$

Step 6 The standard deviation, s, is the square root of the quotient in step 5.

$$s = \sqrt{\frac{\Sigma(x - \bar{x})^2}{n - 1}} = \sqrt{\frac{\Sigma(\text{data item} - \text{mean})^2}{n - 1}} = \sqrt{91{,}298} \approx 302.16$$

The standard deviation for the five countries with the largest labor forces is approximately 302.16 million workers.

✓ **Checkpoint 3** Find the standard deviation for the group of data items in Checkpoint 2 on page 778. Round to two decimal places.

Example 4 illustrates that as the spread of data items increases, the standard deviation gets larger.

Example 4 Computing the Standard Deviation

Find the standard deviation of the data items in each of the samples shown below.

Sample A

17, 18, 19, 20, 21, 22, 23

Sample B

5, 10, 15, 20, 25, 30, 35

Solution

Begin by finding the mean for each sample.

Sample A:

$$\text{Mean} = \frac{17 + 18 + 19 + 20 + 21 + 22 + 23}{7} = \frac{140}{7} = 20$$

Sample B:

$$\text{Mean} = \frac{5 + 10 + 15 + 20 + 25 + 30 + 35}{7} = \frac{140}{7} = 20$$

Although both samples have the same mean, the data items in sample B are more spread out. Thus, we would expect sample B to have the greater standard deviation. The computation of the standard deviation requires that we find $\Sigma(\text{data item} - \text{mean})^2$, shown in **Table 12.13**.

Table 12.13 Computing Standard Deviations for Two Samples

	Sample A			Sample B	
Data Item x	Deviation: data item − mean $x - \bar{x}$	(Deviation)^2: (data item − mean)^2 $(x - \bar{x})^2$	Data Item x	Deviation: data item − mean $x - \bar{x}$	(Deviation)^2: (data item − mean)^2 $(x - \bar{x})^2$
17	$17 - 20 = -3$	$(-3)^2 = 9$	5	$5 - 20 = -15$	$(-15)^2 = 225$
18	$18 - 20 = -2$	$(-2)^2 = 4$	10	$10 - 20 = -10$	$(-10)^2 = 100$
19	$19 - 20 = -1$	$(-1)^2 = 1$	15	$15 - 20 = -5$	$(-5)^2 = 25$
20	$20 - 20 = 0$	$0^2 = 0$	20	$20 - 20 = 0$	$0^2 = 0$
21	$21 - 20 = 1$	$1^2 = 1$	25	$25 - 20 = 5$	$5^2 = 25$
22	$22 - 20 = 2$	$2^2 = 4$	30	$30 - 20 = 10$	$10^2 = 100$
23	$23 - 20 = 3$	$3^2 = 9$	35	$35 - 20 = 15$	$15^2 = 225$
Totals:		$\Sigma(x - \bar{x})^2 = 28$			$\Sigma(x - \bar{x})^2 = 700$

Each sample contains seven data items, so we compute the standard deviation by dividing the sums in **Table 12.13**, 28 and 700, by $7 - 1$, or 6. Then we take the square root of each quotient.

$$\text{Standard deviation} = \sqrt{\frac{\Sigma(x - \bar{x})^2}{n - 1}} = \sqrt{\frac{\Sigma(\text{data item} - \text{mean})^2}{n - 1}}$$

Sample A: Sample B:

$$\text{Standard deviation} = \sqrt{\frac{28}{6}} \approx 2.16 \qquad \text{Standard deviation} = \sqrt{\frac{700}{6}} \approx 10.80$$

Sample A has a standard deviation of approximately 2.16 and sample B has a standard deviation of approximately 10.80. The data in sample B are more spread out than those in sample A.

Checkpoint 4 Find the standard deviation of the data items in each of the samples shown below. Round to two decimal places.

Sample A: 73, 75, 77, 79, 81, 83

Sample B: 40, 44, 92, 94, 98, 100

Figure 12.8 illustrates four sets of data items organized in histograms. From left to right, the data items are

Figure 12.8(a): 4, 4, 4, 4, 4, 4, 4
Figure 12.8(b): 3, 3, 4, 4, 4, 5, 5
Figure 12.8(c): 3, 3, 3, 4, 5, 5, 5
Figure 12.8(d): 1, 1, 1, 4, 7, 7, 7.

Each data set has a mean of 4. However, as the spread of the data items increases, the standard deviation gets larger. Observe that when all the data items are the same, the standard deviation is 0.

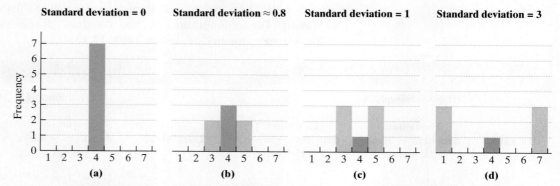

Figure 12.8 The standard deviation gets larger with increased dispersion among data items. In each case, the mean is 4.

Example 5   Interpreting Standard Deviation

Two fifth-grade classes have nearly identical mean scores on an aptitude test, but one class has a standard deviation three times that of the other. All other factors being equal, which class is easier to teach, and why?

Solution

The class with the smaller standard deviation is easier to teach because there is less variation among student aptitudes. Course work can be aimed at the average student without too much concern that the work will be too easy for some or too difficult for others. By contrast, the class with greater dispersion poses a greater challenge. By teaching to the average student, the students whose scores are significantly above the mean will be bored; students whose scores are significantly below the mean will be confused.

Checkpoint 5   Shown below are the means and standard deviations of the yearly returns on two investments from 1926 through 2004.

Investment	Mean Yearly Return	Standard Deviation
Small-Company Stocks	17.5%	33.3%
Large-Company Stocks	12.4%	20.4%

Source: Summary Statistics of Annual Total Returns 1926 to 2004 Yearbook, Ibbotson Associates, Chicago

a. Use the means to determine which investment provided the greater yearly return.

b. Use the standard deviations to determine which investment had the greater risk. Explain your answer.

Exercise Set 12.3

Concept and Vocabulary Exercises

In Exercises 1–2, fill in each blank so that the resulting statement is true.

1. The difference between the highest and lowest data values in a data set is called the _____.

2. The formula

$$\sqrt{\frac{\Sigma(\text{data item } - \text{ mean})^2}{n - 1}}$$

gives the value of the _____ for a data set.

In Exercises 3–5, determine whether each statement is true or false. If the statement is false, make the necessary change(s) to produce a true statement.

3. Measures of dispersion are used to describe the spread of data items in a data set.

4. The sum of the deviations from the mean for a data set is always zero.

5. Measures of dispersion get smaller as the spread of data items increases.

Respond to Exercises 6–12 using verbal or written explanations.

6. Describe how to find the range of a data set.

7. Describe why the range might not be the best measure of dispersion.

8. Describe how the standard deviation is computed.

9. Describe what the standard deviation reveals about a data set.

10. If a set of test scores has a standard deviation of zero, what does this mean about the scores?

11. Two classes took a statistics test. Both classes had a mean score of 73. The scores of class A had a standard deviation of 5 and those of class B had a standard deviation of 10. Discuss the difference between the two classes' performance on the test.

12. A sample of cereals indicates a mean potassium content per serving of 93 milligrams and a standard deviation of 2 milligrams. Write a description of what this means for a person who knows nothing about statistics.

In Exercises 13–16, determine whether each statement makes sense or does not make sense, and explain your reasoning.

13. The joke in this cartoon is based on the observation that the mean can be misleading if you don't know the spread of data items.

© 1990 Creators Syndicate Inc. By permission of Mell Lazarus and Creators Syndicate, Inc.

14. The standard deviation for the weights of high school students is greater than the standard deviation for the weights of 3-year-old children.

15. I'm working with data sets with different means and the same standard deviation.

16. I'm working with data sets with the same mean and different standard deviations.

Practice Exercises

In Exercises 17–22, find the range for each group of data items.

17. 1, 2, 3, 4, 5

18. 16, 17, 18, 19, 20

19. 7, 9, 9, 15

20. 11, 13, 14, 15, 17

21. 3, 3, 4, 4, 5, 5

22. 3, 3, 3, 4, 5, 5, 5

In Exercises 23–26, a group of data items and their mean are given.

a. *Find the deviation from the mean for each of the data items.*

b. *Find the sum of the deviations in part (a).*

23. 3, 5, 7, 12, 18, 27; Mean = 12

24. 84, 88, 90, 95, 98; Mean = 91

25. 29, 38, 48, 49, 53, 77; Mean = 49

26. 60, 60, 62, 65, 65, 65, 66, 67, 70, 70; Mean = 65

In Exercises 27–32, find **a.** *the mean;* **b.** *the deviation from the mean for each data item; and* **c.** *the sum of the deviations in part (b).*

27. 85, 95, 90, 85, 100

28. 94, 62, 88, 85, 91

29. 146, 153, 155, 160, 161

30. 150, 132, 144, 122

31. 2.25, 3.50, 2.75, 3.10, 1.90

32. 0.35, 0.37, 0.41, 0.39, 0.43

In Exercises 33–42, find the standard deviation for each group of data items. Round answers to two decimal places.

33. 1, 2, 3, 4, 5

34. 16, 17, 18, 19, 20

35. 7, 9, 9, 15

36. 11, 13, 14, 15, 17

37. 3, 3, 4, 4, 5, 5

38. 3, 3, 3, 4, 5, 5, 5

39. 1, 1, 1, 4, 7, 7, 7

40. 6, 6, 6, 6, 7, 7, 7, 4, 8, 3

41. 9, 5, 9, 5, 9, 5, 9, 5

42. 6, 10, 6, 10, 6, 10, 6, 10

In Exercises 43–44, compute the mean, range, and standard deviation for the data items in each of the three samples. Then describe one way in which the samples are alike and one way in which they are different.

43. Sample A: 6, 8, 10, 12, 14, 16, 18

Sample B: 6, 7, 8, 12, 16, 17, 18

Sample C: 6, 6, 6, 12, 18, 18, 18

44. Sample A: 8, 10, 12, 14, 16, 18, 20

Sample B: 8, 9, 10, 14, 18, 19, 20

Sample C: 8, 8, 8, 14, 20, 20, 20

Practice Plus

In Exercises 45–52, use each display of data items to find the standard deviation. Where necessary, round answers to two decimal places.

45.

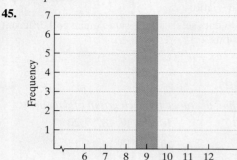

46.

47.

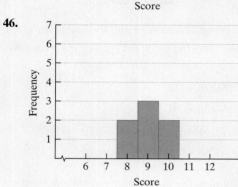

48.

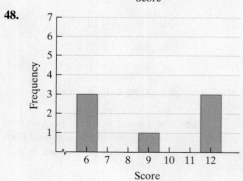

49.

Stems	Leaves
0	5
1	0 5
2	0 5

50.

Stems	Leaves
0	4 8
1	2 6
2	0

51.

Stems	Leaves
1	8 9 9 8 7 8
2	0 1 0 2

52.

Stems	Leaves
1	3 5 3 8 3 4
2	3 0 0 4

Application Exercises

53. The data sets give the number of platinum albums for the five male artists and the five female artists in the United States with the most platinum albums. (Platinum albums sell one million units or more.)

Male Artists with the Most Platinum Albums

Artist	Platinum Albums
Garth Brooks	106
Elvis Presley	89
Billy Joel	68
Elton John	63
Michael Jackson	59

Female Artists with the Most Platinum Albums

Artist	Platinum Albums
Madonna	62
Barbra Streisand	61
Mariah Carey	60
Whitney Houston	54
Celine Dion	48

Source: RIAA

a. Without calculating, which data set has the greater mean number of platinum albums? Explain your answer.

b. Verify your conjecture from part (a) by calculating the mean number of platinum albums for each data set.

c. Without calculating, which data set has the greater standard deviation? Explain your answer.

d. Verify your conjecture from part (c) by calculating the standard deviation for each data set. Round answers to two decimal places.

54. The data sets give the ages of the first six U.S. presidents and the six most recent U.S. presidents (through Barack Obama).

Age of First Six U.S. Presidents at Inauguration

President	Age
Washington	57
J. Adams	61
Jefferson	57
Madison	57
Monroe	58
J. Q. Adams	57

Age of Six Most Recent U.S. Presidents at Inauguration

President	Age
Carter	52
Reagan	69
G. H. W. Bush	64
Clinton	46
G. W. Bush	54
Obama	47

Source: Time Almanac

a. Without calculating, which set has the greater standard deviation? Explain your answer.

b. Verify your conjecture from part (b) by calculating the standard deviation for each data set. Round answers to two decimal places.

Critical Thinking Exercises

55. Describe a situation in which a relatively large standard deviation is desirable.

56. If a set of test scores has a large range but a small standard deviation, describe what this means about students' performance on the test.

57. Use the data 1, 2, 3, 5, 6, 7. Without actually computing the standard deviation, which of the following best approximates the standard deviation?

    a. 2   **b.** 6   **c.** 10   **d.** 20

58. Use the data 0, 1, 3, 4, 4, 6. Add 2 to each of the numbers. How does this affect the mean? How does this affect the standard deviation?

Group Exercises

59. As a follow-up to Group Exercise 88 on page 776, the group should reassemble and compute the standard deviation for each data set whose mean you previously determined. Does the standard deviation tell you anything new or interesting about the entire group or subgroups that you did not discover during the previous group activity?

60. Group members should consult a current almanac or the Internet and select intriguing data. The group's function is to use statistics to tell a story. Once "intriguing" data are identified, as a group

    a. Summarize the data. Use words, frequency distributions, and graphic displays.

    b. Compute measures of central tendency and dispersion, using these statistics to discuss the data.

12.4 : The Normal Distribution

Objectives

1. Recognize characteristics of normal distributions.

2. Understand the 68–95–99.7 Rule.

3. Find scores at a specified standard deviation from the mean.

4. Use the 68–95–99.7 Rule.

5. Convert a data item to a z-score.

6. Understand percentiles and quartiles.

7. Use and interpret margins of error.

8. Recognize distributions that are not normal.

Our heights are on the rise! In one million B.C., the mean height for men was 4 feet 6 inches. The mean height for women was 4 feet 2 inches. Because of improved diets and medical care, the mean height for men is now 5 feet 10 inches and for women it is 5 feet 5 inches. Mean adult heights are expected to plateau by 2050.

Mean Adult Heights

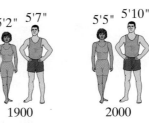

5'2" 5'7"
1900

5'5" 5'10"
2000

5'7" 6'0"
2050

Source: National Center for Health Statistics

1 Recognize characteristics of normal distributions.

Suppose that a researcher selects a random sample of 100 adult men, measures their heights, and constructs a histogram for the data. The graph is shown in **Figure 12.9(a)** below. **Figure 12.9(a)** and **(b)** illustrate what happens as the sample size increases. In **Figure 12.9(c)**, if you were to fold the graph down the middle, the left side would fit the right side. As we move out from the middle, the heights of the bars are the same to the left and right. Such a histogram is called **symmetric**. As the sample size increases, so does the graph's symmetry. If it were possible to measure the heights of all adult males, the entire population, the histogram would approach what is called the **normal distribution**, shown in **Figure 12.9(d)**. This distribution is also called the **bell curve** or the **Gaussian distribution**, named for the German mathematician Carl Friedrich Gauss (1777–1855).

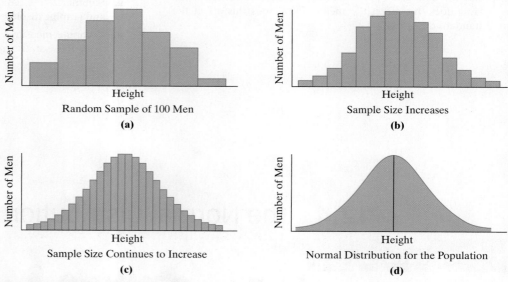

Figure 12.9 Heights of adult males

Figure 12.9(d) illustrates that the normal distribution is bell shaped and symmetric about a vertical line through its center. Furthermore, **the mean, median, and mode** of a normal distribution **are all equal** and located at the center of the distribution.

The shape of the normal distribution depends on the mean and the standard deviation. **Figure 12.10** illustrates three normal distributions with the same mean, but different standard deviations. As the standard deviation increases, the distribution becomes more dispersed, or spread out, but retains its symmetric bell shape.

The normal distribution provides a wonderful model for all kinds of phenomena because many sets of data items closely resemble this population distribution. Examples include heights and weights of adult males, intelligence quotients, SAT scores, prices paid for a new car model, and life spans of light bulbs. In these distributions, the data items tend to cluster around the mean. The more an item differs from the mean, the less likely it is to occur.

The normal distribution is used to make predictions about an entire population using data from a sample. In this section, we focus on the characteristics and applications of the normal distribution.

The Standard Deviation and z-Scores in Normal Distributions

2 Understand the 68–95–99.7 Rule.

The standard deviation plays a crucial role in the normal distribution, summarized by the **68–95–99.7 Rule**. This rule is illustrated in **Figure 12.11**.

Well-Worn Steps and the Normal Distribution

These ancient steps each take on the shape of a normal distribution when the picture is viewed upside down. The center of each step is more worn than the outer edges. The greatest number of people have walked in the center, making this the mean, median, and mode for where people have walked.

The 68–95–99.7 Rule for the Normal Distribution

1. Approximately 68% of the data items fall within 1 standard deviation of the mean (in both directions).

2. Approximately 95% of the data items fall within 2 standard deviations of the mean.

3. Approximately 99.7% of the data items fall within 3 standard deviations of the mean.

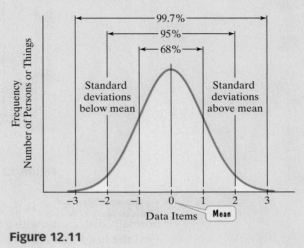

Figure 12.11

Figure 12.11 illustrates that a very small percentage of the data in a normal distribution lies more than 3 standard deviations above or below the mean. As we move from the mean, the curve falls rapidly, and then more and more gradually, toward the horizontal axis. The tails of the curve approach, but never touch, the horizontal axis, although they are quite close to the axis at 3 standard deviations from the mean. The range of the normal distribution is infinite. No matter how far out from the mean we move, there is always the probability (although very small) of a data item occurring even farther out.

  Find scores at a specified standard deviation from the mean.

Example 1   **Finding Scores at a Specified Standard Deviation from the Mean**

Male adult heights in North America are approximately normally distributed with a mean of 70 inches and a standard deviation of 4 inches. Find the height that is

a. 2 standard deviations above the mean.

b. 3 standard deviations below the mean.

Solution

a. First, let us find the height that is 2 standard deviations above the mean.

$$\text{Height} = \text{mean} + 2 \cdot \text{standard deviation}$$
$$= 70 + 2 \cdot 4 = 70 + 8 = 78$$

A height of 78 inches is 2 standard deviations above the mean.

b. Next, let us find the height that is 3 standard deviations below the mean.

$$\text{Height} = \text{mean} - 3 \cdot \text{standard deviation}$$
$$= 70 - 3 \cdot 4 = 70 - 12 = 58$$

A height of 58 inches is 3 standard deviations below the mean.

The normal distribution of male adult heights in North America, with a mean of 70 inches and a standard deviation of 4 inches, is illustrated in **Figure 12.12**.

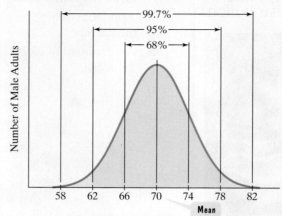

Normal Distribution of Male Adult Heights

Male Adult Heights in North America

Figure 12.12

> ✓ **Checkpoint 1** Female adult heights in North America are approximately normally distributed with a mean of 65 inches and a standard deviation of 3.5 inches. Find the height that is
>
> **a.** 3 standard deviations above the mean.
> **b.** 2 standard deviations below the mean.

4 Use the 68–95–99.7 Rule.

Example 2   **Using the 68–95–99.7 Rule**

Use the distribution of male adult heights in **Figure 12.12** to find the percentage of men in North America with heights

a. between 66 inches and 74 inches.   **b.** between 70 inches and 74 inches.
c. above 78 inches.

Solution

a. The 68–95–99.7 Rule states that approximately 68% of the data items fall within 1 standard deviation, 4, of the mean, 70.

$$\text{mean} - 1 \cdot \text{standard deviation} = 70 - 1 \cdot 4 = 70 - 4 = 66$$
$$\text{mean} + 1 \cdot \text{standard deviation} = 70 + 1 \cdot 4 = 70 + 4 = 74$$

Figure 12.12 shows that 68% of male adults have heights between 66 inches and 74 inches.

b. The percentage of men with heights between 70 inches and 74 inches is not given directly in **Figure 12.12**. Because of the distribution's symmetry, the percentage with heights between 66 inches and 70 inches is the same as the percentage with heights between 70 and 74 inches. **Figure 12.13** indicates that 68% have heights between 66 inches and 74 inches. Thus, half of 68%, or 34%, of men have heights between 70 inches and 74 inches.

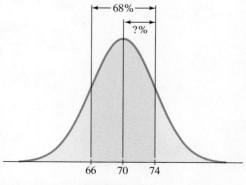

Figure 12.13 What percentage have heights between 70 inches and 74 inches?

c. The percentage of men with heights above 78 inches is not given directly in **Figure 12.12**. A height of 78 inches is 2 standard deviations, $2 \cdot 4$, or 8 inches, above the mean, 70 inches. The 68–95–99.7 Rule states that approximately 95% of the data items fall within 2 standard deviations of the mean. Thus, approximately 100% − 95%, or 5%, of the data items are farther than 2 standard deviations from the mean. The 5% of the data items are represented by the two shaded green regions in **Figure 12.14**. Because of the distribution's symmetry, half of 5%, or 2.5%, of the data items are more than 2 standard deviations above the mean. This means that 2.5% of men have heights above 78 inches.

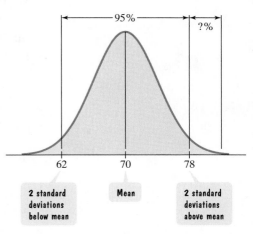

Figure 12.14 What percentage have heights above 78 inches?

✓ **Checkpoint 2**   Use the distribution of male adult heights in North America in **Figure 12.12** to find the percentage of men with heights

a. between 62 inches and 78 inches.

b. between 70 inches and 78 inches.

▶ c. above 74 inches.

Because the normal distribution of male adult heights in North America has a mean of 70 inches and a standard deviation of 4 inches, a height of 78 inches lies 2 standard deviations above the mean. In a normal distribution, a **z-score** describes how many standard deviations a particular data item lies above or below the mean. Thus, the z-score for the data item 78 is 2.

The following formula can be used to express a data item in a normal distribution as a z-score:

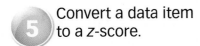

Convert a data item to a z-score.

Computing z-Scores

A z-score describes how many standard deviations a data item in a normal distribution lies above or below the mean. The z-score can be obtained using

$$z\text{-score} = \frac{\text{data item} - \text{mean}}{\text{standard deviation}}.$$

Data items above the mean have positive z-scores. Data items below the mean have negative z-scores. The z-score for the mean is 0.

Example 3 ⋮ **Computing z-Scores**

The mean weight of newborn infants is 7 pounds and the standard deviation is 0.8 pound. The weights of newborn infants are normally distributed. Find the z-score for a weight of

a. 9 pounds.   b. 7 pounds.   c. 6 pounds.

Solution

We compute the z-score for each weight by using the z-score formula. The mean is 7 and the standard deviation is 0.8.

a. The z-score for a weight of 9 pounds, written z_9, is

$$z_9 = \frac{\text{data item} - \text{mean}}{\text{standard deviation}} = \frac{9 - 7}{0.8} = \frac{2}{0.8} = 2.5.$$

The z-score of a data item greater than the mean is always positive. A 9-pound infant is a chubby little tyke, with a weight that is 2.5 standard deviations above the mean.

b. The z-score for a weight of 7 pounds is

$$z_7 = \frac{\text{data item} - \text{mean}}{\text{standard deviation}} = \frac{7 - 7}{0.8} = \frac{0}{0.8} = 0.$$

The z-score for the mean is always 0. A 7-pound infant is right at the mean, deviating 0 pounds above or below it.

c. The z-score for a weight of 6 pounds is

$$z_6 = \frac{\text{data item} - \text{mean}}{\text{standard deviation}} = \frac{6 - 7}{0.8} = \frac{-1}{0.8} = -1.25.$$

The z-score of a data item less than the mean is always negative. A 6-pound infant's weight is 1.25 standard deviations below the mean.

Figure 12.15 shows the normal distribution of weights of newborn infants. The horizontal axis is labeled in terms of weights and z-scores.

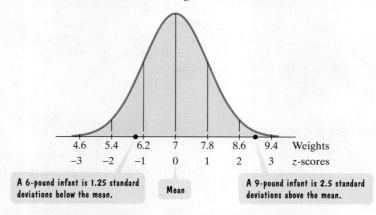

Figure 12.15 Infants' weights are normally distributed.

Checkpoint 3   The length of horse pregnancies from conception to birth is normally distributed with a mean of 336 days and a standard deviation of 3 days. Find the z-score for a horse pregnancy of

a. 342 days.   **b.** 336 days.   **c.** 333 days.

In Example 4, we consider two normally distributed sets of test scores, in which a higher score generally indicates a better result. To compare scores on two different tests in relation to the mean on each test, we can use z-scores. The better score is the item with the greater z-score.

Example 4   Using and Interpreting *z*-Scores

A student scores 70 on an arithmetic test and 66 on a vocabulary test. The scores for both tests are normally distributed. The arithmetic test has a mean of 60 and a standard deviation of 20. The vocabulary test has a mean of 60 and a standard deviation of 2. On which test did the student have the better score?

Solution

To answer the question, we need to find the student's *z*-score on each test, using

$$z = \frac{\text{data item} - \text{mean}}{\text{standard deviation}}.$$

The arithmetic test has a mean of 60 and a standard deviation of 20.

$$z\text{-score for } 70 = z_{70} = \frac{70 - 60}{20} = \frac{10}{20} = 0.5$$

The vocabulary test has a mean of 60 and a standard deviation of 2.

$$z\text{-score for } 66 = z_{66} = \frac{66 - 60}{2} = \frac{6}{2} = 3$$

The arithmetic score, 70, is half a standard deviation above the mean, whereas the vocabulary score, 66, is 3 standard deviations above the mean. The student did much better than the mean on the vocabulary test.

✓ **Checkpoint 4**   The SAT (Scholastic Aptitude Test) has a mean of 500 and a standard deviation of 100. The ACT (American College Test) has a mean of 18 and a standard deviation of 6. Both tests measure the same kind of ability, with scores that are normally distributed. Suppose that you score 550 on the SAT and 24 on the ACT. On which test did you have the better score?

Example 5   Understanding *z*-Scores

Intelligence quotients (IQs) on the Stanford-Binet intelligence test are normally distributed with a mean of 100 and a standard deviation of 16.

  a. What is the IQ corresponding to a *z*-score of -1.5?

  b. Mensa is a group of people with high IQs whose members have *z*-scores of 2.05 or greater on the Stanford-Binet intelligence test. What is the IQ corresponding to a *z*-score of 2.05?

Solution

  a. We begin with the IQ corresponding to a *z*-score of -1.5. The negative sign in -1.5 tells us that the IQ is $1\frac{1}{2}$ standard deviations below the mean.

$$\text{IQ} = \text{mean} - 1.5 \cdot \text{standard deviation}$$
$$= 100 - 1.5(16) = 100 - 24 = 76$$

The IQ corresponding to a *z*-score of -1.5 is 76.

  b. Next, we find the IQ corresponding to a *z*-score of 2.05. The positive sign implied in 2.05 tells us that the IQ is 2.05 standard deviations above the mean.

$$\text{IQ} = \text{mean} + 2.05 \cdot \text{standard deviation}$$
$$= 100 + 2.05(16) = 100 + 32.8 = 132.8$$

The IQ corresponding to a *z*-score of 2.05 is 132.8. (An IQ score of at least 133 is required to join Mensa.)

✓ **Checkpoint 5**   Use the information in Example 5 to find the IQ corresponding to a *z*-score of

  a. -2.25.   **b.** 1.75.

Blitzer Bonus
••••••••••••••
The IQ Controversy

Is intelligence something we are born with or is it a quality that can be manipulated through education? Can it be measured accurately and is IQ the way to measure it? There are no clear answers to these questions.

In a study by Carolyn Bird (*Pygmalion in the Classroom*), a group of third-grade teachers was told that they had classes of students with IQs well above the mean. These classes made incredible progress throughout the year. In reality, these were not gifted kids, but, rather, a random sample of all third-graders. It was the teachers' expectations, and not the IQs of the students, that resulted in increased performance.

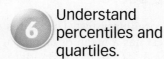

6 Understand percentiles and quartiles.

Percentiles and Quartiles

A *z*-score measures a data item's position in a normal distribution. Another measure of a data item's position is its **percentile**. Percentiles are often associated with scores on standardized tests. If a score is in the 45th percentile, this means that 45% of the scores are less than this score. If a score is in the 95th percentile, this indicates that 95% of the scores are less than this score.

> **Percentiles**
>
> If *n*% of the items in a distribution are less than a particular data item, we say that the data item is in the ***nth* percentile** of the distribution.

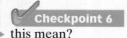 **Example 6   Interpreting Percentile**

The cutoff IQ score for Mensa membership, 132.8, is in the 98th percentile. What does this mean?

Solution

Because 132.8 is in the 98th percentile, this means that 98% of IQ scores fall below 132.8.

✔ **Checkpoint 6**   A student scored in the 75th percentile on the SAT. What does this mean?

Three commonly encountered percentiles are the *quartiles*. **Quartiles** divide data sets into four equal parts. The 25th percentile is the **first quartile**: 25% of the data fall below the first quartile. The 50th percentile is the **second quartile**: 50% of the data fall below the second quartile, so the second quartile is equivalent to the median. The 75th percentile is the **third quartile**: 75% of the data fall below the third quartile. **Figure 12.16** illustrates the concept of quartiles for the normal distribution.

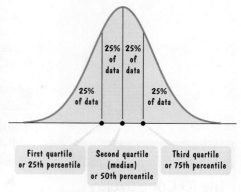

Figure 12.16 Quartiles

Polls and Margins of Error

7 Use and interpret margins of error.

When you were between the ages of 6 and 14, how would you have responded to this question:

> What is bad about being a kid?

In a random sample of 1172 children ages 6 through 14, 17% of the children responded, "Getting bossed around." The problem is that this is a single random sample. Do 17% of kids in the entire population of children ages 6 through 14 think that getting bossed around is a bad thing?

Statisticians use properties of the normal distribution to estimate the probability that a result obtained from a single sample reflects what is truly happening in the population. If you look at the results of a poll like the one shown in the margin, you will observe that a *margin of error* is reported. Surveys and opinion polls often give a margin of error. Let's use our understanding of the normal distribution to see how to calculate and interpret margins of error.

Suppose that *p*% of the population of children ages 6 through 14 hold the opinion that getting bossed around is a bad thing about being a kid. Instead of taking only one random sample of 1172 children, we repeat the process of selecting a random sample of 1172 children hundreds of times. Then, we calculate the percentage of

What Is Bad about Being a Kid?

Kids Say	
Getting bossed around	17%
School, homework	15%
Can't do everything I want	11%
Chores	9%
Being grounded	9%

Source: Penn, Schoen, and Berland using 1172 interviews with children ages 6 to 14 from May 14 to June 1, 1999, Margin of error: ±2.9%.

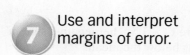

Note the margin of error.

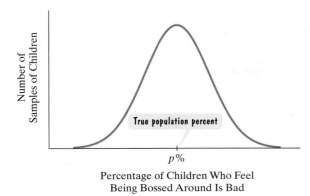

Number of Samples of Children

True population percent

$p\%$

Percentage of Children Who Feel Being Bossed Around Is Bad

Figure 12.17

children for each sample who think being bossed around is bad. With random sampling, we expect to find the percentage in many of the samples close to $p\%$, with relatively few samples having percentages far from $p\%$. **Figure 12.17** shows that the percentages of children from the hundreds of samples can be modeled by a normal distribution. The mean of this distribution is the actual population percent, $p\%$, and is the most frequent result from the samples.

Mathematicians have shown that the standard deviation of a normal distribution of samples like the one in **Figure 12.17** is approximately $\frac{1}{2\sqrt{n}} \times 100\%$, where n is the sample size. Using the 68–95–99.7 Rule, approximately 95% of the samples have a percentage within 2 standard deviations of the true population percentage, $p\%$:

$$2 \text{ standard deviations } = 2 \cdot \frac{1}{2\sqrt{n}} \times 100\% = \frac{1}{\sqrt{n}} \times 100\%.$$

If we use a single random sample of size n, there is a 95% probability that the percent obtained will lie within two standard deviations, or $\frac{1}{\sqrt{n}} \times 100\%$, of the true population percent. We can be 95% confident that the true population percent lies between

$$\text{the sample percent } - \frac{1}{\sqrt{n}} \times 100\%$$

and

$$\text{the sample percent } + \frac{1}{\sqrt{n}} \times 100\%.$$

We call $\pm \frac{1}{\sqrt{n}} \times 100\%$ the **margin of error**.

Margin of Error in a Survey

If a statistic is obtained from a random sample of size n, there is a 95% probability that it lies within $\frac{1}{\sqrt{n}} \times 100\%$ of the true population percent, where $\pm \frac{1}{\sqrt{n}} \times 100\%$ is called the **margin of error**.

Example 7 Using and Interpreting Margin of Error

In a random sample of 1172 children ages 6 through 14, 17% of the children said getting bossed around is a bad thing about being a kid.

a. Verify the margin of error that was given for this survey.

b. Write a statement about the percentage of children in the population who feel that getting bossed around is a bad thing about being a kid.

Solution

a. The sample size is $n = 1172$. The margin of error is

$$\pm \frac{1}{\sqrt{n}} \times 100\% = \pm \frac{1}{\sqrt{1172}} \times 100\% \approx \pm 0.029 \times 100\% = \pm 2.9\%.$$

b. There is a 95% probability that the true population percentage lies between

$$\text{the sample percent } - \frac{1}{\sqrt{n}} \times 100\% = 17\% - 2.9\% = 14.1\%$$

and

$$\text{the sample percent } + \frac{1}{\sqrt{n}} \times 100\% = 17\% + 2.9\% = 19.9\%.$$

We can be 95% confident that between 14.1% and 19.9% of all children feel that getting bossed around is a bad thing about being a kid.

What Is Bad about Being a Kid?

Kids Say	
Getting bossed around	17%
School, homework	15%
Can't do everything I want	11%
Chores	9%
Being grounded	9%

Source: Penn, Schoen, and Berland using 1172 interviews with children ages 6 to 14 from May 14 to June 1, 1999, Margin of error: ±2.9%.

✓ **Checkpoint 7** A Harris Poll of 2513 U.S. adults ages 18 and older asked the question

> How many books do you typically read in a year?

The results of the poll are shown in **Figure 12.18**.

a. Find the margin of error for this survey. Round to the nearest tenth of a percent.

b. Write a statement about the percentage of U.S. adults who read more than ten books per year.

c. Why might some people not respond honestly and accurately to the question in this poll?

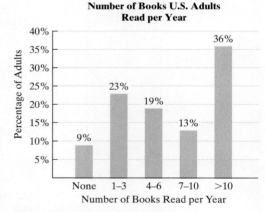

Figure 12.18
Source: Harris Poll of 2513 U.S. adults ages 18 and older conducted March 11 and 18, 2008

8 Recognize distributions that are not normal.

Other Kinds of Distributions

Although the normal distribution is the most important of all distributions in terms of analyzing data, not all data can be approximated by this symmetric distribution with its mean, median, and mode all having the same value.

In our discussion of measures of central tendency, we mentioned that the median, rather than the mean, is used to summarize income. **Figure 12.19** illustrates the population distribution of weekly earnings in the United States. There is no upper limit on weekly earnings. The relatively few people with very high weekly incomes tend to pull the mean income to a value greater than the median. The most frequent income, the mode, occurs toward the low end of the data items. The mean, median, and mode do not have the same value, and a normal distribution is not an appropriate model for describing weekly earnings in the United States.

The distribution in **Figure 12.19** is called a *skewed distribution*. A distribution of data is **skewed** if a large number of data items are piled up at one end or the other, with a "tail" at the opposite end. In the distribution of weekly earnings in **Figure 12.19**, the tail is to the right. Such a distribution is said to be **skewed to the right**.

In contrast to the distribution of weekly earnings, the distribution in **Figure 12.20** has more data items at the high end of the scale than at the low end. The tail of this distribution is to the left. The distribution is said to be **skewed to the left**. An example of a distribution skewed to the left is

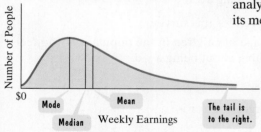

Figure 12.19 Skewed to the right

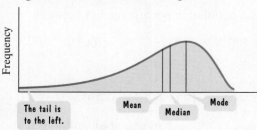

Figure 12.20 Skewed to the left

based on the student ratings of faculty teaching performance in many colleges. Most professors are given rather high ratings, while only a few are rated as terrible. These low ratings pull the value of the mean lower than the median.

Achieving Success

"If you want to learn something, read about it.
If you want to understand something, write about it.
If you want to master something, teach it."

—Yogi Bhajan

This sentiment has appeared in almost every culture of the world. Yogi Bhajan was a Sikh spiritual leader who moved from Pakistan to the United States in the 1960s. In an earlier version of the same idea, the nineteenth-century American theologian Tryon Edwards wrote, "If you would thoroughly know anything, teach it to others."

In the near future, you may be interested in specific jobs available to college students. Consider tutoring positions sponsored by the college. Attractive benefits of this college-sponsored job include reasonable pay, flexible hours, and the unique opportunity to master something.

Exercise Set 12.4

Concept and Vocabulary Exercises

In Exercises 1–4, fill in each blank so that the resulting statement is true.

1. In a normal distribution, approximately _____% of the data items fall within 1 standard deviation of the mean, approximately _____% of the data items fall within 2 standard deviations of the mean, and approximately _____% of the data items fall within 3 standard deviations of the mean.

2. A z-score describes how many standard deviations a data item in a normal distribution lies above or below the _____.

3. If n% of the items in a distribution are less than a particular data item, we say that the data item is in the nth _____ of the distribution.

4. If a statistic is obtained from a random sample of size n, there is a 95% probability that it lies within $\frac{1}{\sqrt{n}} \times 100\%$ of the true population percent, where $\pm\frac{1}{\sqrt{n}} \times 100\%$ is called the _____.

In Exercises 5–8, determine whether each statement is true or false. If the statement is false, make the necessary change(s) to produce a true statement.

5. The mean, median, and mode of a normal distribution are all equal.

6. In a normal distribution, the z-score for the mean is 0.

7. The z-score for a data item in a normal distribution is obtained using

$$z\text{-score} = \frac{\text{data item} - \text{standard deviation}}{\text{mean}}.$$

8. A score in the 50th percentile on a standardized test is the median.

Respond to Exercises 9–18 using verbal or written explanations.

9. What is a symmetric histogram?

10. Describe the normal distribution and discuss some of its properties.

11. Describe the 68–95–99.7 Rule.

12. Describe how to determine the z-score for a data item in a normal distribution.

13. What does a z-score measure?

14. Give an example of both a commonly occurring and an infrequently occurring z-score. Explain how you arrived at these examples.

15. Describe when a z-score is negative.

16. If you score in the 83rd percentile, what does this mean?

17. If your weight is in the third quartile, what does this mean?

18. Two students have scores with the same percentile, but for different administrations of the SAT. Does this mean that the students have the same score on the SAT? Explain your answer.

In Exercises 19–22, determine whether each statement makes sense or does not make sense, and explain your reasoning.

19. The heights of the men on our high school basketball team are normally distributed with a mean of 6 feet 3 inches and a standard deviation of 1 foot 2 inches.

20. A poll administered to a random sample of 1150 voters shows 51% in favor of candidate A, so I'm 95% confident that candidate A will win the election.

21. My math teacher gave a very difficult exam for which the distribution of scores was skewed to the right.

22. Although the grade distributions in this *FoxTrot* cartoon are not normal, the first two graphs correctly model Jason's descriptions in the voice balloons.

FOXTROT © 2010 Bill Amend

Practice and Application Exercises

The scores on a test are normally distributed with a mean of 100 and a standard deviation of 20. In Exercises 23–32, find the score that is

23. 1 standard deviation above the mean.

24. 2 standard deviations above the mean.

25. 3 standard deviations above the mean.

26. $1\frac{1}{2}$ standard deviations above the mean.

27. $2\frac{1}{2}$ standard deviations above the mean.

28. 1 standard deviation below the mean.

29. 2 standard deviations below the mean.

30. 3 standard deviations below the mean.

31. one-half a standard deviation below the mean.

32. $2\frac{1}{2}$ standard deviations below the mean.

Not everyone pays the same price for the same model of a car. The figure illustrates a normal distribution for the prices paid for a particular model of a new car. The mean is $17,000 and the standard deviation is $500.

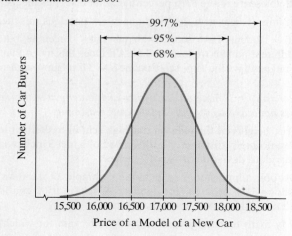

In Exercises 33–44, use the 68–95–99.7 Rule, illustrated in the figure at the bottom of the previous column, to find the percentage of buyers who paid

33. between $16,500 and $17,500.

34. between $16,000 and $18,000.

35. between $17,000 and $17,500.

36. between $17,000 and $18,000.

37. between $16,000 and $17,000.

38. between $16,500 and $17,000.

39. between $15,500 and $17,000.

40. between $17,000 and $18,500.

41. more than $17,500.   **42.** more than $18,000.

43. less than $16,000.   **44.** less than $16,500.

Intelligence quotients (IQs) on the Stanford-Binet intelligence test are normally distributed with a mean of 100 and a standard deviation of 16. In Exercises 45–54, use the 68–95–99.7 Rule to find the percentage of people with IQs

45. between 68 and 132.   **46.** between 84 and 116.

47. between 68 and 100.   **48.** between 84 and 100.

49. above 116.   **50.** above 132.

51. below 68.   **52.** below 84.

53. above 148.   **54.** below 52.

A set of data items is normally distributed with a mean of 60 and a standard deviation of 8. In Exercises 55–70, convert each data item to a z-score.

55. 68   **56.** 76   **57.** 84

58. 92   **59.** 64   **60.** 72

61. 74   **62.** 78   **63.** 60

64. 100   **65.** 52   **66.** 44

67. 48   **68.** 40   **69.** 34

70. 30

Scores on a dental anxiety scale range from 0 (no anxiety) to 20 (extreme anxiety). The scores are normally distributed with a mean of 11 and a standard deviation of 4. In Exercises 71–78, find the z-score for the given score on this dental anxiety scale.

71. 17   **72.** 18   **73.** 20

74. 12   **75.** 6   **76.** 8

77. 5   **78.** 1

Intelligence quotients on the Stanford-Binet intelligence test are normally distributed with a mean of 100 and a standard deviation of 16. Intelligence quotients on the Wechsler intelligence test are normally distributed with a mean of 100 and a standard deviation of 15. Use this information to solve Exercises 79–80.

79. Use z-scores to determine which person has the higher IQ: an individual who scores 128 on the Stanford-Binet or an individual who scores 127 on the Wechsler.

80. Use z-scores to determine which person has the higher IQ: an individual who scores 150 on the Stanford-Binet or an individual who scores 148 on the Wechsler.

A set of data items is normally distributed with a mean of 400 and a standard deviation of 50. In Exercises 81–88, find the data item in this distribution that corresponds to the given z-score.

81. $z = 2$ **82.** $z = 3$

83. $z = 1.5$ **84.** $z = 2.5$

85. $z = -3$ **86.** $z = -2$

87. $z = -2.5$ **88.** $z = -1.5$

89. Using a random sample of 1023 high school students 14 to 18 years old, a Ridgid survey asked respondents to name their career choice. The top five career choices and the percentage of students who named each of these careers are shown in the bar graph.

Career Choices for U.S. High School Students

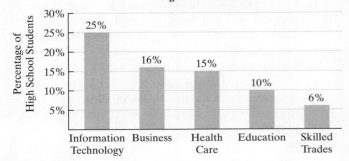

Source: Ridgid Survey

a. Find the margin of error, to the nearest tenth of a percent, for this survey.

b. Write a statement about the percentage of high school students in the population whose career choice is information technology.

90. Using a random sample of 2774 college students, an Experience.com survey asked respondents the following question:

> If you had two job offers and one company was "green," would that have an impact on your decision?

The responses are shown in the circle graph.

Would a "Green" Company Impact Your Decision?

Yes 79%

No 21%

Source: Experience.com Survey

a. Find the margin of error, to the nearest tenth of a percent, for this survey.

b. Write a statement about the percentage of college students in the population for whom a "green" company would have an impact on their decision.

91. Using a random sample of 4000 TV households, Nielsen Media Research found that 60.2% watched the final episode of $M*A*S*H$.

a. Find the margin of error in this percent.

b. Write a statement about the percentage of TV households in the population who tuned into the final episode of $M*A*S*H$.

92. Using a random sample of 4000 TV households, Nielsen Media Research found that 51.1% watched *Roots, Part 8.*

a. Find the margin of error in this percent.

b. Write a statement about the percentage of TV households in the population who tuned into *Roots, Part 8.*

93. In 1997, Nielsen Media Research increased its random sample to 5000 TV households. By how much, to the nearest tenth of a percent, did this improve the margin of error over that in Exercises 91 and 92?

94. If Nielsen Media Research were to increase its random sample from 5000 to 10,000 TV households, by how much, to the nearest tenth of a percent, would this improve the margin of error?

The histogram shows murder rates per 100,000 residents and the number of U.S. states that had these rates in 2006. Use this histogram to solve Exercises 95–96.

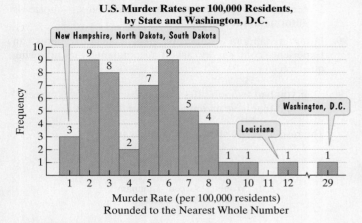

U.S. Murder Rates per 100,000 Residents, by State and Washington, D.C.

Murder Rate (per 100,000 residents)
Rounded to the Nearest Whole Number

Source: FBI, *Crime in the United States*

95. a. Is the shape of this distribution best classified as normal, skewed to the right, or skewed to the left?

b. Calculate the mean murder rate per 100,000 residents for the 50 states and Washington, D.C.

c. Find the median murder rate per 100,000 residents for the 50 states and Washington, D.C.

d. Are the mean and median murder rates consistent with the shape of the distribution that you described in part (a)? Explain your answer.

e. The standard deviation for the data is approximately 4.2. If the distribution were roughly normal, what would be the z-score, rounded to one decimal place, for Washington, D.C.? Does this seem unusually high? Explain your answer.

96. a. Find the median murder rate per 100,000 residents for the 50 states and Washington, D.C.

b. Find the first quartile by determining the median of the lower half of the data. (This is the median of the items that lie below the median that you found in part (a).)

c. Find the third quartile by determining the median of the upper half of the data. (This is the median of the items that lie above the median that you found in part (a).)

d. Use the following numerical scale:

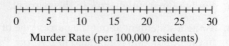

Murder Rate (per 100,000 residents)

Above this scale, show five points, each at the same height. (The height is arbitrary.) Each point should represent one of the following numbers:

   lowest data value, first quartile, median, third quartile, highest data value.

e. A **box-and-whisker plot** consists of a rectangular box extending from the first quartile to the third quartile, with a dashed line representing the median, and line segments (or whiskers) extending outward from the box to the lowest and highest data values:

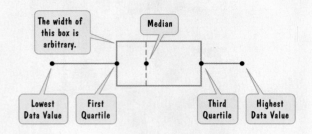

Use your graph from part (d) to create a box-and-whisker plot for U.S. murder rates per 100,000 residents.

f. If one of the whiskers in a box-and-whisker plot is clearly longer, the distribution is usually skewed in the direction of the longer whisker. Based on this observation, does your box-and-whisker plot in part (e) indicate that the distribution is skewed to the right or skewed to the left?

g. Is the shape of the distribution of scores shown by the given histogram consistent with your observation in part (f)?

Critical Thinking Exercises

97. Give an example of a phenomenon that is normally distributed. Explain why. (Try to be creative and not use one of the distributions discussed in this section.) Estimate what the mean and the standard deviation might be and describe how you determined these estimates.

98. Give an example of a phenomenon that is not normally distributed and explain why.

Group Exercise

99. For this activity, group members will conduct interviews with a random sample of students in your high school. Each student is to be asked. "What is the worst thing about being a high school student?" One response should be recorded for each student.

a. Each member should interview enough students so that there are at least 50 randomly selected students in the sample.

b. After all responses have been recorded, the group should organize the four most common answers. For each answer, compute the percentage of students in the sample who felt that this is the worst thing about being a student.

c. Find the margin of error for your survey.

d. For each of the four most common answers, write a statement about the percentage of all students in your school who feel that this is the worst thing about being a student.

Chapter 12 Summary

12.1 Sampling, Frequency Distributions, and Graphs

Definitions and Concepts

A population is the set containing all objects whose properties are to be described and analyzed. A sample is a subset of the population.

Random samples are obtained in such a way that each member of the population has an equal chance of being selected.

Example

- A newspaper wants to find out which of its features are popular with its readers and decides to conduct a survey.
  Appropriate for obtaining a sample: Survey a random sample of readers from a list of all subscribers.
  Not appropriate for obtaining a sample: Survey a random sample of people from the telephone directory (This does not address the target population.); Survey the first 100 subscribers from an alphabetical list of all subscribers (This is not sufficiently random. A subscriber whose last name begins with a letter toward the end of the alphabet has no chance of being selected.)

Additional Examples to Review
Example 1, page 749; Example 2, page 750

Definitions and Concepts

Data can be organized and presented in frequency distributions (data values in one column with frequencies in the adjacent column), grouped frequency distributions (data values arranged into classes), histograms (bar graphs with data values on a horizontal axis and frequencies on a vertical axis), frequency polygons (line graphs with data values on a horizontal axis and frequencies on a vertical axis), and stem-and-leaf plots (data items separated into two parts).

Stretching or compressing the scale on a graph's vertical axis can create a more or less dramatic impression than shown by the actual data.

Examples

- 4, 5, 2, 2, 1, 3, 2, 3, 5, 5, 2, 1, 3, 3, 3

Frequency Distribution

Score	Frequency
1	2
2	4
3	5
4	1
5	3

Histogram

Frequency Polygon

- 62, 63, 71, 74, 78, 79, 81, 81, 82, 83, 90, 92

Grouped Frequency Distribution

Class	Frequency
60–69	2
70–79	4
80–89	4
90–99	2

Stem-and-Leaf Plot

Stems	Leaves
6	2 3
7	1 4 8 9
8	1 1 2 3
9	0 2

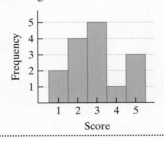
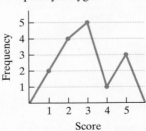

Additional Examples and Graphs to Review
Example 3, page 752; Example 4, page 753; Figure 12.2, page 754; Figure 12.3, page 754;
Example 5, page 755; Figure 12.5, page 756; Table 12.5, page 757

12.2 Measures of Central Tendency

Definitions and Concepts

The mean, $\bar{x}$, is the sum of the data items divided by the number of items: $\bar{x} = \dfrac{\Sigma x}{n}$.

Example

• 2, 7, 8, 8, 5

$$\bar{x} = \frac{\Sigma x}{n} = \frac{2 + 7 + 8 + 8 + 5}{5} = \frac{30}{5} = 6$$

Additional Example to Review
Example 1, page 764

Definitions and Concepts

The mean, $\bar{x}$, of a frequency distribution is computed using

$$\bar{x} = \frac{\Sigma xf}{n},$$

where x is a data value, f is its frequency, and n is the total frequency of the distribution.

Example

•

Score, x	Frequency, f	xf
5	10	$5 \cdot 10 = 50$
6	15	$6 \cdot 15 = 90$
7	30	$7 \cdot 30 = 210$
8	35	$8 \cdot 35 = 280$
9	5	$9 \cdot 5 = 45$
10	5	$10 \cdot 5 = 50$
Totals:	$n = 100$	$\Sigma xf = 725$

$$\bar{x} = \frac{\Sigma xf}{n} = \frac{725}{100} = 7.25$$

Additional Example to Review
Example 2, page 765

Definitions and Concepts

The median of ranked data is the item in the middle or the mean of the two middlemost items.
The median is the value in the $\dfrac{n + 1}{2}$ position in the list of ranked data.

Examples

• 5, 8, 1, 8, 4, 3, 6, 7

Numbers in order: 1, 3, 4, 5, 6, 7, 8, 8

Middle items are 5 and 6.

$$\text{Median} = \frac{5 + 6}{2} = \frac{11}{2} = 5.5$$

•

Score, x	Frequency, f
5	10
6	15
7	30
8	35
9	5
10	5
	$n = 100$

Median is in $\dfrac{n+1}{2}$ position $= \dfrac{100+1}{2} = \dfrac{101}{2} = 50.5$

The median is the mean of data items in position 50 (which is 7) and position 51 (which is also 7).

Median $= \dfrac{7+7}{2} = \dfrac{14}{2} = 7$

Additional Examples to Review

Example 3, page 766; Example 4, page 767; Example 5, page 768; Example 6, page 768

Definitions and Concepts

When one or more data items are much greater than or much less than the other items, these extreme values greatly influence the mean, often making the median more representative of the data.

Example

• $1, 2, 2, 5, 7, 8, 96$

Mean $= \bar{x} = \dfrac{\Sigma x}{n} = \dfrac{1+2+2+5+7+8+96}{7} = \dfrac{121}{7} \approx 17.3$

Median (middlemost item) $= 5$

The median is more representative of the data than the mean.

Additional Example to Review

Example 7, page 769

Definitions and Concepts

The mode of a data set is the value that occurs most often. If there is no such value, there is no mode. If more than one data value has the highest frequency, then each of these data values is a mode.

Examples

• $4, 3, 7, 9, 3$
Mode is 3.

• $5, 8, 4, 3, 4, 8$
Modes are 4 and 8.

• $2, 5, 4, 4, 2, 5$
No mode

•

Score, x	Frequency, f
5	10
6	15
7	30
8	35
9	5
10	5

The mode is 8 because it occurs most often (35 times).

Additional Example to Review

Example 8, page 771

Definitions and Concepts

The midrange is computed using

$$\frac{\text{lowest data value } + \text{ highest data value}}{2}.$$

Examples

- $4, 3, 7, 9, 3$

  Midrange $= \dfrac{3 + 9}{2} = \dfrac{12}{2} = 6$

- $2, 5, 4, 4, 2, 5$

  Midrange $= \dfrac{2 + 5}{2} = \dfrac{7}{2} = 3.5$

Additional Examples to Review

Example 9, page 772; Example 10, page 773

12.3 Measures of Dispersion

Definitions and Concepts

Range $=$ highest data value $-$ lowest data value

Examples

- $4, 3, 7, 9, 3$

  Range $= 9 - 3 = 6$

- $2, 5, 4, 4, 2, 5$

  Range $= 5 - 2 = 3$

Additional Example to Review

Example 1, page 777

Definitions and Concepts

Standard deviation $= \sqrt{\dfrac{\Sigma(\text{data item } - \text{ mean})^2}{n - 1}}$

This is symbolized by $s = \sqrt{\dfrac{\Sigma(x - \bar{x})^2}{n - 1}}$.

As the spread of data items increases, the standard deviation gets larger.

Example

- $5, 7, 8, 12$ Find the standard deviation.

  First find the mean. $\bar{x} = \dfrac{\Sigma x}{n} = \dfrac{5 + 7 + 8 + 12}{4} = \dfrac{32}{4} = 8$

x	$x - \bar{x}$	$(x - \bar{x})^2$
5	$5 - 8 = -3$	$(-3)^2 = 9$
7	$7 - 8 = -1$	$(-1)^2 = 1$
8	$8 - 8 = 0$	$0^2 = 0$
12	$12 - 8 = 4$	$4^2 = 16$
		$\Sigma(x - \bar{x})^2 = 26$

Standard deviation:

$$s = \sqrt{\frac{\Sigma(x - \bar{x})^2}{n - 1}} = \sqrt{\frac{26}{4 - 1}} = \sqrt{\frac{26}{3}} \approx 2.94$$

Additional Examples to Review

Example 2, page 778; Example 3, page 779; Example 4, page 780; Example 5, page 782

12.4 The Normal Distribution

Definitions and Concepts

The normal distribution is a theoretical distribution for the entire population. The distribution is bell shaped and symmetric about a vertical line through its center, where the mean, median, and mode are located.

The 68–95–99.7 Rule

Approximately 68% of the data items fall within 1 standard deviation of the mean.
Approximately 95% of the data items fall within 2 standard deviations of the mean.
Approximately 99.7% of the data items fall within 3 standard deviations of the mean.

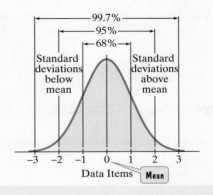

Examples

Cholesterol levels for U.S. men are normally distributed with a mean of 200 and a standard deviation of 15.

- mean ± 1 st. dev. = $200 \pm 1 \cdot 15 = 200 \pm 15 = 185$ or 215
- mean ± 2 st. dev. = $200 \pm 2 \cdot 15 = 200 \pm 30 = 170$ or 230
- mean ± 3 st. dev. = $200 \pm 3 \cdot 15 = 200 \pm 45 = 155$ or 245
- percentage of men with levels between 200 and 215: $\frac{1}{2} \cdot 68\% = 34\%$
- percentage of men with levels below 170: $\frac{1}{2}(100\% - 95\%) = \frac{1}{2}(5\%) = 2.5\%$

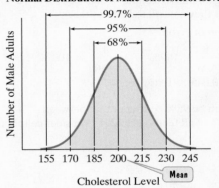

Normal Distribution of Male Cholesterol Levels

Additional Examples to Review

Example 1, page 787; Example 2, page 788

Definitions and Concepts

A *z*-score describes how many standard deviations a data item in a normal distribution lies above or below the mean.

$$z\text{-score} = \frac{\text{data item} - \text{mean}}{\text{standard deviation}}$$

Examples

Scores on both tests are normally distributed.

- Test 1: Mean is 100 and standard deviation is 20. Your score is 140.

$$z\text{-score for } 140 = \frac{140 - 100}{20} = \frac{40}{20} = 2$$

- Test 2: Mean is 95 and standard deviation is 16. Your score is 135.

$$z\text{-score for } 135 = \frac{135 - 95}{16} = \frac{40}{16} = 2.5$$

- Your score on test 2 is farther above the mean (2.5 standard deviations) than your score on test 1 (2 standard deviations), so it is the better score.

Additional Examples to Review

Example 3, page 789; Example 4, page 791; Example 5, page 791

Definitions and Concepts

If $n\%$ of the items in a distribution are less than a particular data item, that data item is in the nth percentile of the distribution. The 25th percentile is the first quartile, the 50th percentile, or the median, is the second quartile, and the 75th percentile is the third quartile.

Example

- A student scored in the 80th percentile on the SAT. This means that 80% of the scores on the SAT are less than this student's score.

Additional Example to Review

Example 6, page 792

Definitions and Concepts

If a statistic is obtained from a random sample of size n, there is a 95% probability that it lies within $\frac{1}{\sqrt{n}} \times 100\%$ of the true population statistic. $\pm\frac{1}{\sqrt{n}} \times 100\%$ is called the margin of error.

Examples

- Using a random sample of 500 teens ages 12–17, a survey asked respondents to identify things they considered necessities.

  Margin of error

  $$= \pm\frac{1}{\sqrt{n}} \times 100\% = \pm\frac{1}{\sqrt{500}} \times 100\%$$
  $$\approx \pm0.045 \times 100\% = \pm4.5\%$$

  We can be 95% confident that between 47% − 4.5%, or 42.5%, and 47% + 4.5%, or 51.5%, of all teens ages 12–17 consider cellphones a necessity.

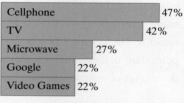

Teens Ages 12–17 Who Consider These Necessities

Cellphone	47%
TV	42%
Microwave	27%
Google	22%
Video Games	22%

Source: Lemelson-MIT Invention Index Survey of 500 teens

Additional Example to Review

Example 7, page 793

Definitions and Concepts

A distribution of data is skewed if a large number of data items are piled up at one end or the other, with a "tail" at the opposite end.

Example

- A very easy exam has many high scores. The distribution of scores has a tail to the left and is said to be skewed to the left.

Additional Graphs to Review

Figure 12.19, page 794; Figure 12.20, page 794

Review Exercises

Section 12.1 Sampling, Frequency Distributions, and Graphs

1. The government of a large city wants to know if its citizens will support a three-year tax increase to provide additional support to the city's community college system. The government decides to conduct a survey of the city's residents before placing a tax increase initiative on the ballot. Which one of the following is most appropriate for obtaining a sample of the city's residents?

   a. Survey a random sample of persons within each geographic region of the city.

   b. Survey a random sample of community college professors living in the city.

   c. Survey every tenth person who walks into the city's government center on two randomly selected days of the week.

   d. Survey a random sample of persons within each geographic region of the state in which the city is located.

A random sample of ten high school students is selected and each student is asked how much time he or she spent on homework during the previous weekend. The following times, in hours, are obtained:

$$8, 10, 9, 7, 9, 8, 7, 6, 8, 7.$$

Use these data items to solve Exercises 2–4.

2. Construct a frequency distribution for the data.

3. Construct a histogram for the data.

4. Construct a frequency polygon for the data.

The 50 grades on a chemistry test are shown. Use the data to solve Exercises 5–6.

44	24	54	81	18
34	39	63	67	60
72	36	91	47	75
57	74	87	49	86
59	14	26	41	90
13	29	13	31	68
63	35	29	70	22
95	17	50	42	27
73	11	42	31	69
56	40	31	45	51

5. Construct a grouped frequency distribution for the data. Use 0–39 for the first class, 40–49 for the second class, and make each subsequent class width the same as the second class.

6. Construct a stem-and-leaf plot for the data.

7. Describe what is misleading about the size of the barrels in the following visual display.

Average Daily Price per Barrel of Oil

Source: U.S. Department of Energy

Section 12.2 Measures of Central Tendency

In Exercises 8–9, find the mean for each group of data items.

8. $84, 90, 95, 89, 98$

9. $33, 27, 9, 10, 6, 7, 11, 23, 27$

10. Find the mean for the data items in the given frequency distribution.

Score x	Frequency f
1	2
2	4
3	3
4	1

In Exercises 11–12, find the median for each group of data items.

11. $33, 27, 9, 10, 6, 7, 11, 23, 27$

12. $28, 16, 22, 28, 34$

13. Find the median for the data items in the frequency distribution in Exercise 10.

In Exercises 14–15, find the mode for each group of data items. If there is no mode, so state.

14. $33, 27, 9, 10, 6, 7, 11, 23, 27$

15. $582, 585, 583, 585, 587, 587, 589$

16. Find the mode for the data items in the frequency distribution in Exercise 10.

In Exercises 17–18, find the midrange for each group of data items.

17. $84, 90, 95, 88, 98$

18. $33, 27, 9, 10, 6, 7, 11, 23, 27$

19. Find the midrange for the data items in the frequency distribution in Exercise 10.

20. A student took seven tests in a course, scoring between 90% and 95% on three of the tests, between 80% and 89% on three of the tests, and below 40% on one of the tests. In this distribution, is the mean or the median more representative of the student's overall performance in the course? Explain your answer.

21. The data items below are the ages of U.S. presidents at the time of their first inauguration.

   57 61 57 57 58 57 61 54 68 51 49 64 50 48

   65 52 56 46 54 49 51 47 55 55 54 42 51 56

   55 51 54 51 60 62 43 55 56 61 52 69 64 46 54 47

   a. Organize the data in a frequency distribution.

   b. Use the frequency distribution to find the mean age, median age, modal age, and midrange age of the presidents when they were inaugurated.

Section 12.3 Measures of Dispersion

In Exercises 22–23, find the range for each group of data items.

22. $28, 34, 16, 22, 28$

23. $312, 783, 219, 312, 426, 219$

24. The mean for the data items $29, 9, 8, 22, 46, 51, 48, 42, 53, 42$ is 35. Find **a.** the deviation from the mean for each data item and **b.** the sum of the deviations in part (a).

25. Use the data items $36, 26, 24, 90,$ and 74 to find **a.** the mean, **b.** the deviation from the mean for each data item, and **c.** the sum of the deviations in part (b).

In Exercises 26–27, find the standard deviation for each group of data items.

26. $3, 3, 5, 8, 10, 13$

27. $20, 27, 23, 26, 28, 32, 33, 35$

28. A test measuring anxiety levels is administered to a sample of ten high school students with the following results. (High scores indicate high anxiety.)

   $$10, 30, 37, 40, 43, 44, 45, 69, 86, 86$$

   Find the mean, range, and standard deviation for the data.

29. Compute the mean and the standard deviation for each of the following data sets. Then, write a brief description of similarities and differences between the two sets based on each of your computations.

Set A: 80, 80, 80, 80   Set B: 70, 70, 90, 90

30. Describe how you would determine

    a. which of the two groups, boys or girls, at your high school has a higher mean grade point average.

    b. which of the groups is more consistently close to its mean grade point average.

Section 12.4 The Normal Distribution

The scores on a test are normally distributed with a mean of 70 and a standard deviation of 8. In Exercises 31–33, find the score that is

31. 2 standard deviations above the mean.

32. $3\frac{1}{2}$ standard deviations above the mean.

33. $1\frac{1}{4}$ standard deviations below the mean.

The ages of people living in a retirement community are normally distributed with a mean age of 68 years and a standard deviation of 4 years. In Exercises 34–40, use the 68–95–99.7 Rule to find the percentage of people in the community whose ages

34. are between 64 and 72.

35. are between 60 and 76.

36. are between 68 and 72.

37. are between 56 and 80.

38. exceed 72.   **39.** are less than 72.

40. exceed 76.

A set of data items is normally distributed with a mean of 50 and a standard deviation of 5. In Exercises 41–45, convert each data item to a z-score.

41. 50   **42.** 60

43. 58   **44.** 35

45. 44

46. A student scores 60 on a vocabulary test and 80 on a grammar test. The data items for both tests are normally distributed. The vocabulary test has a mean of 50 and a standard deviation of 5. The grammar test has a mean of 72 and a standard deviation of 6. On which test did the student have the better score? Explain why this is so.

The number of miles that a particular brand of car tires lasts is normally distributed with a mean of 32,000 miles and a standard deviation of 4000 miles. In Exercises 47–49, find the data item in this distribution that corresponds to the given z-score.

47. $z = 1.5$

48. $z = 2.25$

49. $z = -2.5$

50. Using a random sample of 2281 American adults ages 18 and older, an Adecco survey asked respondents if they would be willing to sacrifice a percentage of their salary in order to work for an environmentally friendly company. The poll indicated that 31% of the respondents said "yes," 39% said "no," and 30% declined to answer.

    a. Find the margin of error, to the nearest tenth of a percent, for this survey.

    b. Write a statement about the percentage of American adults who would be willing to sacrifice a percentage of their salary in order to work for an environmentally friendly company.

51. The histogram indicates the frequencies of the number of syllables per word for 100 randomly selected words in Japanese.

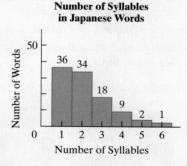

Number of Syllables in Japanese Words

    a. Is the shape of this distribution best classified as normal, skewed to the right, or skewed to the left?

    b. Find the mean, median, and mode for the number of syllables in the sample of Japanese words.

    c. Are the measures of central tendency from part (b) consistent with the shape of the distribution that you described in part (a)? Explain your answer.

Chapter 12 Test A

1. Politicians in the Florida Keys need to know if the residents of Key Largo think the amount of money charged for water is reasonable. The politicians decide to conduct a survey of a sample of Key Largo's residents. Which procedure would be most appropriate for a sample of Key Largo's residents?

    a. Survey all water customers who pay their water bills at Key Largo City Hall on the third day of the month.

    b. Survey a random sample of executives who work for the water company in Key Largo.

    c. Survey 5000 individuals who are randomly selected from a list of all people living in Georgia and Florida.

    d. Survey a random sample of persons within each neighborhood of Key Largo.

Use these scores on a ten-point quiz to solve Exercises 2–4.

8, 5, 3, 6, 5, 10, 6, 9, 4, 5, 7, 9, 7, 4, 8, 8

2. Construct a frequency distribution for the data.

3. Construct a histogram for the data.

4. Construct a frequency polygon for the data.

Use the 30 test scores listed below to solve Exercises 5–6.

79	51	67	50	78
62	89	83	73	80
88	48	60	71	79
89	63	55	93	71
41	81	46	50	61
59	50	90	75	61

5. Construct a grouped frequency distribution for the data. Use 40–49 for the first class and use the same width for each subsequent class.

6. Construct a stem-and-leaf display for the data.

7. The graph shows the percentage of students in the United States through grade 12 who were home-schooled in 1999 and 2007. What impression does the roofline in the visual display imply about what occurred in 2000 through 2006? How might this be misleading?

Source: National Center for Education Statistics

Use the data items listed below to solve Exercises 8–11.

9, 4, 6, 4, 2

8. Find the mean.

9. Find the median.

10. Find the midrange,

11. Find the standard deviation. Round to two decimal places.

Use the frequency distribution shown to solve Exercises 12–14.

Score x	Frequency f
1	3
2	5
3	2
4	2

12. Find the mean.

13. Find the median.

14. Find the mode.

15. The annual salaries of four salespeople and the owner of a bookstore are

$17,500, $19,000, $22,000, $27,500, $98,500.

Is the mean or the median more representative of the five annual salaries? Briefly explain your answer.

Monthly charges for cellphone plans in the United States are normally distributed with a mean of $62 and a standard deviation of $18. In Exercises 16–17, use the 68–95–99.7 Rule to find the percentage of plans that have monthly charges

16. between $62 and $80.

17. that exceed $98.

18. IQ scores are normally distributed in the population. Who has a higher IQ: a student with a 120 IQ on a scale where 100 is the mean and 10 is the standard deviation, or a teacher with a 128 IQ on a scale where 100 is the mean and 15 is the standard deviation? Briefly explain your answer.

19. Using a random sample of 1000 American adults ages 18 and older, a Tiller social action survey asked respondents if they would be willing to pay 20% more in taxes annually if that eliminated poverty and hunger in the United States. The poll indicated that 25% of the respondents said "yes," 40% said "no," and 35% were not sure.

   a. Find the margin of error, to the nearest tenth of a percent, for this survey.

   b. Write a statement about the percentage of American adults who would be willing to pay 20% more in taxes if that eliminated poverty and hunger.

Chapter 12 Test B

1. A jewelry maker wants to survey a sample of Davis County high school seniors in order to determine preferences for class rings. Which procedure would be most appropriate for obtaining a sample of Davis County high school seniors?

   a. Mail surveys to a random sample of Davis County high school students.

   b. Randomly select an assortment of classes for seniors at each Davis County high school and distribute surveys to the students in those classes.

   c. Randomly select an assortment of Davis County high school teachers and distribute surveys to them.

   d. Distribute surveys to all Davis County high school seniors on the honor roll.

Use these scores on a ten-point quiz to solve Exercises 2–4.

7, 4, 2, 5, 4, 10, 5, 8, 3, 4, 6, 8, 6, 3, 7, 7

2. Construct a frequency distribution for the data.

3. Construct a histogram for the data.

4. Construct a frequency polygon for the data.

Use the 30 test scores listed below to solve Exercises 5–6.

78	50	66	51	78
61	88	84	72	80
87	47	60	70	78
88	62	54	92	70
40	80	47	50	60
58	50	90	74	60

5. Construct a grouped frequency distribution for the data. Use 40–49 for the first class and use the same width for each subsequent class.

6. Construct a stem-and-leaf display for the data.

7. The bar graph shows where teen women meet the guys they date. Describe what is misleading in this visual display of data.

Where Teen Women Meet the Guys They Date

1% After-School Jobs
2% After-School Activities
6% Online
9% Shopping Mall
12% Through Friends
70% School

Source: J–14

Use the data items listed below to solve Exercises 8–11.

23, 25, 18, 15, 24

8. Find the mean.

9. Find the median.

10. Find the midrange.

11. Find the standard deviation. Round to two decimal places.

Use the frequency distribution shown to solve Exercises 12–14.

Score, x	Frequency, f
1	10
2	10
3	5
4	35
5	40

12. Find the mean.   13. Find the median.

14. Find the mode.

15. In a family of seven people, six family members are under the age of 40 and one is 96 years old. Is the mean or the median more representative of the family's average age? Briefly explain your answer.

Lengths of pregnancies of women are normally distributed with a mean of 266 days and a standard deviation of 16 days. In Exercises 16–17, use the 68–95–99.7 Rule to find the percentage of pregnancies that last

16. between 266 days and 282 days.

17. more than 298 days.

18. IQ scores are normally distributed in the population. Who has a higher IQ: a teacher with a 125 IQ on a scale where 100 is the mean and 10 is the standard deviation, or a student with a 130 IQ on a scale where 100 is the mean and 15 is the standard deviation? Briefly explain your answer.

19. Using a random sample of 938 Americans ages 16 and older, a survey asked respondents to identify the word, phrase, or sentence they found most annoying in conversation. The bar graph summarizes the survey's results.

Which Is Most Annoying In Conversation?

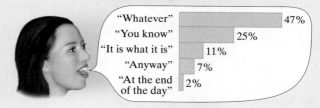

"Whatever" 47%
"You know" 25%
"It is what it is" 11%
"Anyway" 7%
"At the end of the day" 2%

Source: Marist Poll Survey of 938 Americans ages 16 and older, August 3–6, 2010

a. Find the margin of error, to the nearest tenth of a percent, for this survey.

b. Write a statement about the percentage of Americans ages 16 and older who find "whatever" most annoying in conversation?

Logic

We are inundated with arguments that attempt to convince us of a variety of claims. P. T. Barnum (1810–1891), cofounder of the circus called "the Greatest Show on Earth," shamelessly engaged in the art of ballyhoo and humbug, feeding the public "bonafide baloney, with no truth in it." His philosophy: There is a sucker born every minute.

Logic is a kind of self-defense to avoid being suckered in by the Barnums of the world. It will enable you to apply deductive reasoning to arrive at valid conclusions in complicated situations and to avoid being fooled into believing things for which insufficient reasons are available. The rules of logic will help you to evaluate the vast array of claims facing you as a consumer, a citizen, a student, and a human being. Understanding logic will also allow you to construct better and more convincing arguments of your own, thereby becoming a more effective advocate for your beliefs.

13.1 Statements, Negations, and Quantified Statements

History is filled with bad predictions. Here are examples of statements that turned out to be notoriously false:

"The actual building of roads devoted to motor cars will not occur in the future."

Harper's Weekly, August 2, 1902

"Everything that can be invented has been invented."

CHARLES H. DUELL, Commissioner, U.S. Office of Patents, 1899

"The abdomen, the chest, and the brain will forever be shut from the intrusion of the wise and humane surgeon."

JOHN ERICKSEN, Queen Victoria's surgeon, 1873

"Television won't be able to hold onto any market. People will soon get tired of staring at a plywood box every night."

DARRYL F. ZANUCK, 1949

"Whatever happens in Vietnam, I can conceive of nothing except military victory."

LYNDON B. JOHNSON, in a speech at West Point, 1967

"When the President does it, that means that it is not illegal."

RICHARD M. NIXON, TV interview with David Frost, May 20, 1977

Understanding that these statements are false enables us to negate each statement mentally and, with the assistance of historical perspective, obtain a true statement. We begin our study of logic by looking at statements and their negations.

Statements and Using Symbols to Represent Statements

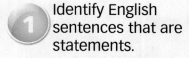

Identify English sentences that are statements.

In everyday English, we use many different kinds of sentences. Some of these sentences are clearly true or false. Others are opinions, questions, and exclamations such as *Help!* or *Fire!* However, in logic we are concerned solely with statements, and not all English sentences are statements.

> **Definition of a Statement**
> A **statement** is a sentence that is either true or false, but not both simultaneously.

Here are two examples of statements:

1. London is the capital of England.

2. William Shakespeare wrote the television series *Glee*.

Statement 1 is true and statement 2 is false. Shakespeare had nothing to do with *Glee* (perhaps writer's block after *Macbeth*).

As long as a sentence is either true or false, *even if we do not know which it is*, then that sentence is a statement. For example, the sentence

The United States has the world's highest divorce rate

is a statement. It's clearly either true or false, and it's not necessary to know which it is.

Some sentences, such as commands, questions, and opinions, are not statements because they are not either true or false. The following sentences are not statements:

1. Tweet me.   (This is an order or command.)

2. Do you have friends who seldom come through?   (This is a question.)

3. *Titanic* is the greatest movie of all time.   (This is an opinion.)

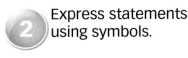

Express statements using symbols.

In symbolic logic, we use lowercase letters such as p, q, r, and s to represent statements. Here are two examples:

p: London is the capital of England.

q: William Shakespeare wrote the television series *Glee*.

The letter p represents the first statement.
The letter q represents the second statement.

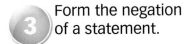

Form the negation of a statement.

Negating Statements

The sentence "London is the capital of England" is a true statement. The *negation* of this statement, "London is not the capital of England," is a false statement. The **negation** of a true statement is a false statement and the negation of a false statement is a true statement.

Example 1   Forming Negations

Form the negation of each statement:

a. Shakespeare wrote the television series *Glee*.

b. Today is not Monday.

Solution

a. The most common way to negate this statement is to introduce *not* into the sentence. The negation is

Shakespeare did not write the television series *Glee*.

The English language provides many ways of expressing a statement's meaning. Here is another way to express the negation:

It is not true that Shakespeare wrote the television series *Glee*.

b. The negation of "Today is not Monday" is

It is not true that today is not Monday.

The negation is more naturally expressed in English as

Today is Monday.

 Checkpoint 1 Form the negation of each statement:

a. Paris is the capital of Spain.

▶ **b.** July is not a month.

4 **Express negations using symbols.**

The negation of statement p is expressed by writing $\sim p$. We read this as "not p" or "It is not true that p."

 **Example 2** **Expressing Negations Symbolically**

Let p and q represent the following statements:

p: Shakespeare wrote the television series *Glee*.

q: Today is not Monday.

Express each of the following statements symbolically:

a. Shakespeare did not write the television series *Glee*.

b. Today is Monday.

Solution

a. Shakespeare did not write the television series *Glee* is the negation of statement p. Therefore, it is expressed symbolically as $\sim p$.

b. Today is Monday is the negation of statement q. Therefore, it is expressed symbolically as $\sim q$.

 Checkpoint 2 Let p and q represent the following statements:

p: Paris is the capital of Spain.

q: July is not a month.

Express each of the following statements symbolically:

a. Paris is not the capital of Spain.

▶ **b.** July is a month.

5 **Translate a negation represented by symbols into English.**

In Example 2, we translated English statements into symbolic statements. In Example 3, we reverse the direction of our translation.

Example 3 **Translating a Symbolic Statement into Words**

Let p represent the following statement:

p: The United States has the world's highest divorce rate.

Express the symbolic statement $\sim p$ in words.

Solution

The symbol $\sim$ is translated as "not." Therefore, $\sim p$ represents

The United States does not have the world's highest divorce rate.

This can also be expressed as

It is not true that the United States has the world's highest divorce rate.

✓ **Checkpoint 3**   Let q represent the following statement:

    q: Chicago O'Hare is the world's busiest airport.

▶ Express the symbolic statement $\sim q$ in words.

Quantified Statements

6 **Express quantified statements in two ways.**

In English, we frequently encounter statements containing the words **all, some**, and **no** (or **none**). These words are called **quantifiers**. A statement that contains one of these words is a **quantified statement**. Here are some examples:

    All poets are writers.

    Some people are bigots.

    No common colds are fatal.

    Some students do not work hard.

Using our knowledge of the English language, we can express each of these quantified statements in two equivalent ways, that is, in two ways that have exactly the same meaning. These equivalent statements are shown in **Table 13.1**.

Table 13.1 Equivalent Ways of Expressing Quantified Statements

Statement	An Equivalent Way to Express the Statement	Example (Two Equivalent Quantified Statements)
All A are B.	There are no A that are not B.	All poets are writers. There are no poets that are not writers.
Some A are B.	There exists at least one A that is a B.	Some people are bigots. At least one person is a bigot.
No A are B.	All A are not B.	No common colds are fatal. All common colds are not fatal.
Some A are not B.	Not all A are B.	Some students do not work hard. Not all students work hard.

7 **Write negations of quantified statements.**

Forming the negation of a quantified statement can be a bit tricky. Suppose we want to negate the statement "All writers are poets." Because this statement is false, its negation must be true. The negation is "Not all writers are poets." This means the same thing as "Some writers are not poets." Notice that the negation is a true statement.

Study Tip

The negation of "All writers are poets" cannot be "No writers are poets" because both statements are false. The negation of a false statement must be a true statement. In general, the negation of "All A are B" is *not* "No A are B."

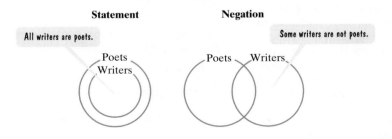

In general, the negation of "All A are B" is "Some A are not B." Likewise, the negation of "Some A are not B" is "All A are B."

Now let's investigate how to negate a statement with the word *some*. Consider the statement "Some canaries weigh 50 pounds." Because *some* means "there exists at least one," the negation is "It is not true that there is at least one canary that weighs 50 pounds." Because it is not true that there is even one such critter, we can express the negation as "No canary weighs 50 pounds."

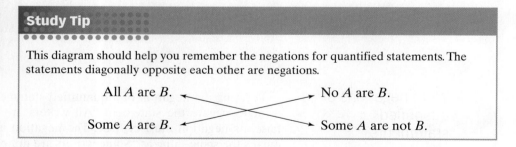

Statement — Some canaries weigh 50 pounds. — Canaries / 50-pound objects

Negation — No canary weighs 50 pounds. — Canaries / 50-pound objects

In general, the negation of "Some *A* are *B*" is "No *A* are *B*." Likewise, the negation of "No *A* are *B*" is "Some *A* are *B*."

Negations of quantified statements are summarized in **Table 13.2**.

Table 13.2 Negations of Quantified Statements

Statement	Negation	Example (A Quantified Statement and Its Negation)
All *A* are *B*.	Some *A* are not *B*.	All people take exams honestly. Negation: Some people do not take exams honestly.
Some *A* are *B*.	No *A* are *B*.	Some roads are open. Negation: No roads are open.

(The negations of the statements in the second column are the statements in the first column.)

Study Tip

This diagram should help you remember the negations for quantified statements. The statements diagonally opposite each other are negations.

All *A* are *B*. ⟷ No *A* are *B*.

Some *A* are *B*. ⟷ Some *A* are not *B*.

Table 13.3 contains examples of negations for each of the four kinds of quantified statements.

Table 13.3 Examples of Negations of Quantified Statements

Statement	Negation
All humans are mortal.	Some humans are not mortal.
Some students do not come to class prepared.	All students come to class prepared.
Some psychotherapists are in therapy.	No psychotherapists are in therapy.
No well-behaved dogs shred couches.	Some well-behaved dogs shred couches.

Example 4   **Negating a Quantified Statement**

The mechanic told me, "All piston rings were replaced." I later learned that the mechanic never tells the truth. What can I conclude?

Solution

Let's begin with the mechanic's statement:

All piston rings were replaced.

Because the mechanic never tells the truth, I can conclude that the truth is the negation of what I was told. The negation of "All A are B" is "Some A are not B." Thus, I can conclude that

Some piston rings were not replaced.

Because *some* means *at least one*, I can also correctly conclude that

At least one piston ring was not replaced.

Checkpoint 4   The board of supervisors told us, "All new tax dollars will be used to improve education." I later learned that the board of supervisors never tells the truth. What can I conclude? Express the conclusion in two equivalent ways.

Exercise Set 13.1

Concept and Vocabulary Exercises

In Exercises 1–10, fill in each blank so that the resulting statement is true.

1. A statement is a sentence that is either _____ or _____, but not both simultaneously.

2. The negation of a true statement is a _____ statement and the negation of a false statement is a _____ statement.

3. The negation of statement p is expressed by writing _____. We read this as _____.

4. Statements that contain the words *all, some,* and *no* are called _____ statements.

5. The statement "All A are B" can be expressed equivalently as _____.

6. The statement "Some A are B" can be expressed equivalently as _____.

7. The statement "No A are B" can be expressed equivalently as _____.

8. The statement "Some A are not B" can be expressed equivalently as _____.

9. The negation of "All A are B" is _____.

10. The negation of "Some A are B" is _____.

In Exercises 11–14, determine whether each statement is true or false. If the statement is false, make the necessary change(s) to produce a true statement.

11. A false sentence is a statement.

12. The negation of a false statement must be a true statement.

13. The negation of "All A are B" is "No A are B."

14. The negation of "Some A are B" is "Some A are not B."

Respond to Exercises 15–20 using verbal or written explanations.

15. What is a statement? Explain why commands, questions, and opinions are not statements.

16. Explain how to form the negation of a given English statement. Give an example.

17. Describe how the negation of statement p is expressed using symbols.

18. List the words identified as quantifiers. Give an example of a statement that uses each of these quantifiers.

19. Explain how to write the negation of a quantified statement in the form "All A are B." Give an example.

20. Explain how to write the negation of a quantified statement in the form "Some A are B." Give an example.

In Exercises 21–24, determine whether each statement makes sense or does not make sense, and explain your reasoning.

21. I have no idea if a particular sentence is true or false, so I cannot determine whether or not that sentence is a statement.

22. "All beagles are dogs" is true and "no beagles are dogs" is false, so the second statement must be the negation of the first statement.

23. Little Richard's "A-wop-bop-a-lula-a-wop-bam-boom!" is an exclamation and not a statement.

24. Researchers at Cambridge University made the following comments on how we read:

   > It deson't mtater waht oerdr the ltteres in a wrod are, so lnog as the frist and lsat ltteer are in the crorcet pclae. Tihs is bcuseae we dno't raed ervey lteter but the wrod as a wlohe.

   Because of the incorrect spellings in these sentences, neither sentence is a statement.

Practice Exercises

In Exercises 25–38, determine whether or not each sentence is a statement.

25. René Descartes came up with the theory of analytic geometry by watching a fly walk across a ceiling.

26. As a young and struggling artist, Pablo Picasso kept warm by burning his own paintings.

27. On January 20, 2009, John McCain became America's 44th president.

28. On January 20, 2009, Barack Obama became America's first Harvard-educated president.

29. Take the most interesting classes you can find.

30. Don't text me.

31. The average human brain contains 100 billion neurons.

32. There are 2,500,000 rivets in the Eiffel Tower.

33. Is the unexamined life worth living?

34. Is this the best of all possible worlds?

35. All U.S. presidents with beards have been Republicans.

36. No U.S. president was an only child.

37. The shortest sentence in the English language is "Go!"

38. Go!

In Exercises 39–44, form the negation of each statement.

39. It is raining.

40. It is snowing.

41. The Dallas Cowboys are not the team with the most Super Bowl wins.

42. The New York Yankees are not the team with the most World Series wins.

43. It is not true that chocolate in moderation is good for the heart.

44. It is not true that Albert Einstein was offered the presidency of Israel.

In Exercises 45–48, let p, q, r, and s represent the following statements:

   p: A student works hard.
   q: A student succeeds.
   r: The temperature outside is not freezing.
   s: It is not true that the heater is working.

Express each of the following statements symbolically.

45. A student does not work hard.

46. A student does not succeed.

47. The temperature outside is freezing.

48. The heater is working.

According to Condensed Knowledge: A Deliciously Irreverent Guide to Feeling Smart Again *(Harper Collins, 2004), each statement listed below is false (who knew!).*

   p: Listening to classical music makes infants smarter.
   q: Eating chocolate causes acne breakouts.
   r: Sigmund Freud's father was not 20 years older than his mother.
   s: Humans and bananas do not share approximately 60% of the same DNA structure.

In Exercises 49–52, use the representations shown above to express each symbolic statement in words. What can you conclude about the resulting verbal statement?

49. $\sim p$ 50. $\sim q$ 51. $\sim r$ 52. $\sim s$

In Exercises 53–66,

   a. *Express the quantified statement in an equivalent way, that is, in a way that has exactly the same meaning.*

   b. *Write the negation of the quantified statement. (The negation should begin with "all," "some," or "no.")*

53. All whales are mammals.

54. All journalists are writers.

55. Some students are business majors.

56. Some movies are comedies.

57. Some thieves are not criminals.

58. Some pianists are not keyboard players.

59. No Democratic presidents have been impeached.

60. No women have served as Supreme Court justices.

61. There are no seniors who did not graduate.

62. There are no applicants who were not hired.

63. Not all parrots are pets.

64. Not all dogs are playful.

65. All atheists are not churchgoers.

66. All burnt muffins are not edible.

Here's another list of false statements from Condensed Knowledge.

   p: No Africans have Jewish ancestry.
   q: Some sleepwalkers have died as a result of being woken up.
   r: All rap is hip-hop.
   s: Some hip-hop is not rap.

In Exercises 67–70, use the representations shown to express each symbolic statement in words. Verbal statements should begin with "all," "some," or "no." What can you conclude about each resulting verbal statement?

67. $\sim p$ 68. $\sim q$ 69. $\sim r$ 70. $\sim s$

Practice Plus

Exercises 71–74 contain diagrams that show relationships between two sets. (These diagrams are just like the Venn diagrams studied in Chapter 2. However, the circles are not enclosed in a rectangle representing a universal set.)

   a. *Use each diagram to write a statement beginning with the word "all," "some," or "no" that illustrates the relationship between the sets.*

   b. *Determine if the statement in part (a) is true or false. If it is false, write its negation.*

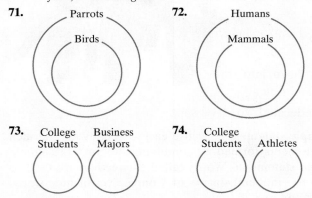

71. Parrots / Birds   **72.** Humans / Mammals

73. College Students / Business Majors   **74.** College Students / Athletes

In Exercises 75–80,

   a. *Express each statement in an equivalent way that begins with "all," "some," or "no."*

   b. *Write the negation of the statement in part (a).*

75. Nobody doesn't like Sara Lee.

76. A problem well stated is a problem half solved.

77. Nothing is both safe and exciting.

78. Many a person has lived to regret a misspent youth.

79. Not every great actor is a Tom Hanks.

80. Not every generous philanthropist is a Bill Gates.

Application Exercises

In Exercises 81 and 82, choose the correct statement.

81. The City Council of a large northern metropolis promised its citizens that in the event of snow, all major roads connecting the city to its airport would remain open. The City Council did not keep its promise during the first blizzard of the season. Therefore, during the first blizzard:

   a. No major roads connecting the city to the airport were open.

   b. At least one major road connecting the city to the airport was not open.

   c. At least one major road connecting the city to the airport was open.

   d. The airport was forced to close.

82. During the Watergate scandal in 1974, President Richard Nixon assured the American people that "In all my years of public service, I have never obstructed justice." Later, events indicated that the president was not telling the truth. Therefore, in his years of public service:

   a. Nixon always obstructed justice.

   b. Nixon sometimes did not obstruct justice.

   c. Nixon sometimes obstructed justice.

   d. Nixon never obstructed justice.

Our culture celebrates romantic love—affection and passion for another person—as the basis for marriage. However, the bar graph illustrates that in some countries, romantic love plays a less important role in marriage.

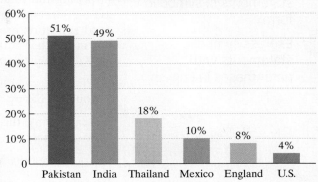

Percentage of College Students Willing to Marry without Romantic Love

Source: Robert Levine, "Is Love a Luxury?" *American Demographics*

In Exercises 83–90, use the graph shown above to determine whether each statement is true or false. If the statement is false, write its negation.

83. A majority of college students in Pakistan are willing to marry without romantic love.

84. Nearly half of India's college students are willing to marry without romantic love.

85. No college students in the United States are willing to marry without romantic love.

86. All college students in Pakistan are willing to marry without romantic love.

87. Not all college students in Mexico are willing to marry without romantic love.

88. Not all college students in England are willing to marry without romantic love.

89. The sentence "5% of college students in Australia are willing to marry without romantic love" is not a statement.

90. The sentence "12% of college students in the Philippines are willing to marry without romantic love" is not a statement.

Critical Thinking Exercises

91. Give an example of a sentence that is not a statement because it is true and false simultaneously.

92. Give an example in which the statement "Some A are not B" is true, but the statement "Some B are not A" is false.

93. The statement

   She isn't dating him because he is muscular

is confusing because it can mean two different things. Describe the two different meanings that make this statement ambiguous.

94. If the ancient Greek god Zeus could do anything, could he create a rock so huge that he could not move it? Explain your answer.

13.2 Compound Statements and Connectives

What conditions enable us to flourish and our hearts to sing? Researchers in the new science of happiness have learned some surprising things about what it doesn't take to ring our inner chimes. Neither wealth nor a good education are sufficient for happiness. Put in another way, we should not rely on the following statement:

> If you're wealthy or well educated, then you'll be happy.

We can break this statement down into three basic sentences:

> You're wealthy. You're well educated. You'll be happy.

These sentences are called **simple statements** because each one conveys one idea with no connecting words. Statements formed by combining two or more simple statements are called **compound statements**. Words called **connectives** are used to join simple statements to form a compound statement. Connectives include words such as **and, or, if . . . then**, and **if and only if**.

Compound statements appear throughout written and spoken language. We need to be able to understand the logic of such statements to analyze information objectively. In this section, we analyze four kinds of compound statements.

And Statements

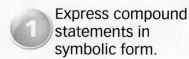
Express compound statements in symbolic form.

If p and q represent two simple statements, then **the compound statement "p and q" is symbolized by $p \wedge q$.** The compound statement formed by connecting statements with the word **and** is called a **conjunction**. The symbol for *and* is $\wedge$.

Example 1 Translating from English to Symbolic Form

Let p and q represent the following simple statements:

> p: It is after 5 P.M.
>
> q: They are working.

Write each compound statement below in symbolic form:

a. It is after 5 P.M. and they are working.

b. It is after 5 P.M. and they are not working.

Solution

a.

It is after 5 P.M.	and	they are working.
p	$\wedge$	q

The symbolic form is $p \wedge q$.

b.

It is after 5 P.M.	and	they are not working.
p	$\wedge$	$\sim q$

The symbolic form is $p \wedge \sim q$.

✓ **Checkpoint 1** Use the representations in Example 1 on the previous page to write each compound statement below in symbolic form:

a. They are working and it is after 5 P.M.

▶ **b.** It is not after 5 P.M. and they are working.

The English language has a variety of ways to express the connectives that appear in compound statements. **Table 13.4** shows a number of ways to translate $p \wedge q$ into English.

Table 13.4 Common English Expressions for $p \wedge q$

Symbolic Statement	English Statement	Example p: It is after 5 P.M. q: They are working.
$p \wedge q$	p and q.	It is after 5 P.M. and they are working.
$p \wedge q$	p but q.	It is after 5 P.M., but they are working.
$p \wedge q$	p yet q.	It is after 5 P.M., yet they are working.
$p \wedge q$	p nevertheless q.	It is after 5 P.M.; nevertheless, they are working.

Study Tip

Not every English statement with the word *and* is a conjunction.

- Not a conjunction:

  "Nonviolence and truth are inseparable."
  —GANDHI

  This statement cannot be broken down into two simple statements. It is itself a simple statement.

- Conjunction:

  Pizza and soda are not recommended for people with ulcers.

  Can be broken down as follows: Pizza is not recommended for people with ulcers and soda is not recommended for people with ulcers.

Or Statements

The connective *or* can mean two different things. For example, consider this statement:

I visited London or Paris.

The statement can mean

I visited London or Paris, but not both.

This is an example of the **exclusive or**, which means "one or the other, but not both." By contrast, the statement can mean

I visited London or Paris or both.

This is an example of the **inclusive or**, which means "either or both."

In this chapter and in mathematics in general, when the connective *or* appears, it means the *inclusive or*. If p and q represent two simple statements, then the compound statement "p or q" means p or q or both. The compound statement formed by connecting statements with the word *or* is called a **disjunction**. **The symbol for *or* is $\vee$. Thus, we can symbolize the compound statement "p or q or both" by $p \vee q$.**

| Example 2 | Translating from English to Symbolic Form |

Let p and q represent the following simple statements:

  $p:$ The bill receives majority approval.

  $q:$ The bill becomes a law.

Write each compound statement below in symbolic form:

a. The bill receives majority approval or the bill becomes a law.

b. The bill receives majority approval or the bill does not become a law.

Solution

a.

The symbolic form is $p \lor q$.

b.

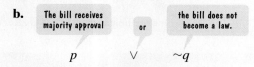

The symbolic form is $p \lor \sim q$.

| ✓ Checkpoint 2 | Let p and q represent the following simple statements:

  $p:$ You graduate.

  $q:$ You satisfy the math requirement.

Write each compound statement below in symbolic form:

a. You graduate or you satisfy the math requirement.

▶ **b.** You satisfy the math requirement or you do not graduate.

If–Then Statements

The diagram in **Figure 13.1** shows that

  All poets are writers.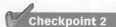

In Section 13.1, we saw that this can be expressed as

  There are no poets that are not writers.

Another way of expressing this statement is

  If a person is a poet, then that person is a writer.

The form of this statement is "If p, then q." **The compound statement "If p, then q" is symbolized by $p \rightarrow q$.** The compound statement formed by connecting statements with "if–then" is called a **conditional statement**. The symbol for "if–then" is $\rightarrow$.

Figure 13.1

In a conditional statement, the statement before the → connective is called the **antecedent**. The statement after the → connective is called the **consequent**:

$$antecedent \rightarrow consequent.$$

Example 3 Translating from English to Symbolic Form

Let *p* and *q* represent the following simple statements:

p: A person is a father.

q: A person is a male.

Write each compound statement below in symbolic form:

a. If a person is a father, then that person is a male.

b. If a person is a male, then that person is a father.

c. If a person is not a male, then that person is not a father.

Solution

We use *p*: A person is a father and *q*: A person is a male.

a.

The symbolic form is $p \rightarrow q$.

b.

The symbolic form is $q \rightarrow p$.

c.

The symbolic form is $\sim q \rightarrow \sim p$.

✓ **Checkpoint 3** Use the representations in Example 3 to write each compound statement below in symbolic form:

a. If a person is not a father, then that person is not a male.

▶ **b.** If a person is a male, then that person is not a father.

Conditional statements in English often omit the word *then* and simply use a comma. When *then* is included, the comma can be included or omitted. Here are some examples:

If a person is a father, then that person is a male.

If a person is a father then that person is a male.

If a person is a father, that person is a male.

Table 13.5 shows some of the common ways to translate $p \rightarrow q$ into English.

Table 13.5 Common English Expressions for $p \rightarrow q$		
Symbolic Statement	**English Statement**	**Example** p: A person is a father. q: A person is a male.
$p \rightarrow q$	If p then q.	If a person is a father, then that person is a male.
$p \rightarrow q$	q if p.	A person is a male if that person is a father.
$p \rightarrow q$	p is sufficient for q.	Being a father is sufficient for being a male.
$p \rightarrow q$	q is necessary for p.	Being a male is necessary for being a father.
$p \rightarrow q$	p only if q.	A person is a father only if that person is a male.
$p \rightarrow q$	Only if q, p.	Only if a person is a male is that person a father.

Study Tip

The sufficient condition of a conditional statement is the part that precedes $\rightarrow$, or the antecedent. The necessary condition of a conditional statement is the part that follows $\rightarrow$, or the consequent.

$$\square \rightarrow \triangle$$

The *if* part or the *sufficient* part   The *only if* part or the *necessary* part

Example 4   Translating from English to Symbolic Form

Let p and q represent the following simple statements:

   p: We suffer huge budget deficits.

   q: We control military spending.

Write the following compound statement in symbolic form:

   Controlling military spending is necessary for not suffering huge budget deficits.

Solution

The necessary part of a conditional statement follows the *if–then* connective. Because "controlling military spending" is the necessary part, we can rewrite the compound statement as follows:

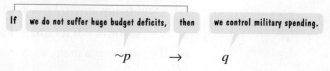

If   we do not suffer huge budget deficits,   then   we control military spending.

$$\sim p \qquad \rightarrow \qquad q$$

The symbolic form is $\sim p \rightarrow q$.

✓ **Checkpoint 4**   Use the representations in Example 4 to write the following compound statement in symbolic form.

   Suffering huge budget deficits is necessary for not controlling military spending.

If and Only If Statements

If a conditional statement is true, reversing the antecedent and consequent may result in a statement that is not necessarily true:

- If a person is a father, then that person is a male.

  true

- If a person is a male, then that person is a father.

  not necessarily true

However, some true conditional statements are still true when the antecedent and consequent are reversed:

- If a person is an unmarried male, then that person is a bachelor.

  true

- If a person is a bachelor, then that person is an unmarried male.

  also true

Rather than deal with two separate conditionals, we can combine them into one *biconditional statement:*

> A person is an unmarried male if and only if that person is a bachelor.

If p and q represent two simple statements, then **the compound statement "p if and only if q" is symbolized by $p \leftrightarrow q$.** The compound statement formed by connecting statements with *if and only if* is called a **biconditional**. The symbol for *if and only if* is $\leftrightarrow$. The phrase *if and only if* can be abbreviated as *iff*.

Table 13.6 Words with the Most Meanings in the *Oxford English Dictionary*

Word	Meanings
Set	464
Run	396
Go	368
Take	343
Stand	334

Example 5 Translating from English to Symbolic Form

Table 13.6 shows that the word *set* has 464 meanings, making it the word with the most meanings in the English language. Let p and q represent the following simple statements:

p: The word is *set*.

q: The word has 464 meanings.

Write each of the compound statements below in its symbolic form:

a. The word is *set* if and only if the word has 464 meanings.

b. The word does not have 464 meanings if and only if the word is not *set*.

Solution

a. The word is *set* | if and only if | the word has 464 meanings.

$$p \qquad \leftrightarrow \qquad q$$

The symbolic form is $p \leftrightarrow q$. Observe that each of the following statements is true:

> If the word is *set*, then it has 464 meanings.
> If the word has 464 meanings, then it is *set*.

b. The word does not have 464 meanings | if and only if | the word is not *set*.

$$\sim q \qquad \leftrightarrow \qquad \sim p$$

The symbolic form is $\sim q \leftrightarrow \sim p$.

> ✔ **Checkpoint 5**   Let p and q represent the following simple statements:
>
> p: The word is *run*.
>
> q: The word has 396 meanings.
>
> Write each of the compound statements below in its symbolic form:
>
> **a.** The word has 396 meanings if and only if the word is *run*.
>
> ▶ **b.** The word is not *run* if and only if the word does not have 396 meanings.

Table 13.7 shows some of the common ways to translate $p \leftrightarrow q$ into English.

Table 13.7 Common English Expressions for $p \leftrightarrow q$

Symbolic Statement	English Statement	Example p: A person is an unmarried male. q: A person is a bachelor.
$p \leftrightarrow q$	p if and only if q.	A person is an unmarried male if and only if that person is a bachelor.
$p \leftrightarrow q$	q if and only if p.	A person is a bachelor if and only if that person is an unmarried male.
$p \leftrightarrow q$	If p then q, and if q then p.	If a person is an unmarried male then that person is a bachelor, and if a person is a bachelor then that person is an unmarried male.
$p \leftrightarrow q$	p is necessary and sufficient for q.	Being an unmarried male is necessary and sufficient for being a bachelor.
$p \leftrightarrow q$	q is necessary and sufficient for p.	Being a bachelor is necessary and sufficient for being an unmarried male.

Table 13.8 summarizes the statements discussed in the first two sections of this chapter.

Table 13.8 Statements of Symbolic Logic

Name	Symbolic Form	Common English Translations
Negation	$\sim p$	Not p. It is not true that p.
Conjunction	$p \wedge q$	p and q. p but q.
Disjunction	$p \vee q$	p or q.
Conditional	$p \rightarrow q$	If p, then q. p is sufficient for q. q is necessary for p.
Biconditional	$p \leftrightarrow q$	p if and only if q. p is necessary and sufficient for q.

Symbolic Statements with Parentheses

 2 Express symbolic statements with parentheses in English.

Parentheses in symbolic statements indicate which statements are to be grouped together. For example, $\sim(p \wedge q)$ means the negation of the entire statement $p \wedge q$. By contrast, $\sim p \wedge q$ means that only statement p is negated. We read $\sim(p \wedge q)$ as "it is not true that p and q." We read $\sim p \wedge q$ as "not p and q." Unless parentheses appear in a symbolic statement, the symbol $\sim$ negates only the statement that immediately follows it.

Example 6 Expressing Symbolic Statements with and without Parentheses in English

Let p and q represent the following simple statements:

p: She is wealthy.

q: She is happy.

Write each of the following symbolic statements in words:

a. $\sim(p \wedge q)$ **b.** $\sim p \wedge q$ **c.** $\sim(p \vee q)$.

Solution

The voice balloons illustrate the differences among the three statements.

- $\sim(p \wedge q)$

  It is not true that | she is wealthy and she is happy.

- $\sim p \wedge q$

  She is not wealthy | and | she is happy.

- $\sim(p \vee q)$

  It is not true that | she is wealthy or she is happy.

a. The symbolic statement $\sim(p \wedge q)$ means the negation of the entire statement $p \wedge q$. A translation of $\sim(p \wedge q)$ is

  It is not true that she is wealthy and happy.

  We can also express this statement as

  It is not true that she is both wealthy and happy.

b. A translation of $\sim p \wedge q$ is

  She is not wealthy and she is happy.

c. The symbolic statement $\sim(p \vee q)$ means the negation of the entire statement $p \vee q$. A translation of $\sim(p \vee q)$ is

  It is not true that she is wealthy or happy.

  We can express this statement as

  She is neither wealthy nor happy.

 Checkpoint 6 Let p and q represent the following simple statements:

p: He earns $105,000 yearly.

q: He is often happy.

Write each of the following symbolic statements in words:

a. $\sim(p \wedge q)$ **b.** $\sim q \wedge p$ **c.** $\sim(q \rightarrow p)$.

Many compound statements contain more than one connective. When expressed symbolically, parentheses are used to indicate which simple statements are grouped together. When expressed in English, commas are used to indicate the groupings. Here is a table that illustrates groupings using parentheses in symbolic statements and commas in English statements:

Symbolic Statement	Statements to Group Together	English Translation
$(q \wedge \sim p) \rightarrow \sim r$	$q \wedge \sim p$	If q and not p, then not r.
$q \wedge (\sim p \rightarrow \sim r)$	$\sim p \rightarrow \sim r$	q, and if not p then not r.

The statement in the first row is an *if–then* conditional statement. Notice that the symbol $\rightarrow$ is outside the parentheses. By contrast, the statement in the second row is an *and* conjunction. In this case, the symbol $\wedge$ is outside the parentheses. Notice that when we translate the symbolic statement into English, **the simple statements in parentheses appear on the same side of the comma.**

Example 7 Expressing Symbolic Statements with Parentheses in English

Let p, q, and r represent the following simple statements:

p: A student misses class.

q: A student studies.

r: A student fails.

Write each of the symbolic statements below in words:

a. $(q \land \sim p) \to \sim r$ **b.** $q \land (\sim p \to \sim r)$.

Solution

a.

$$(q \quad \land \quad \sim p) \quad \to \quad \sim r$$

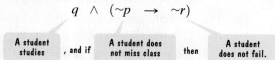

If [A student studies] and [A student does not miss class], then [A student does not fail.]

One possible English translation for the symbolic statement is

If a student studies and does not miss class, then the student does not fail.

Observe how the symbolic statements in parentheses appear on the same side of the comma in the English translation.

b.

$$q \quad \land \quad (\sim p \quad \to \quad \sim r)$$

[A student studies], and if [A student does not miss class] then [A student does not fail.]

One possible English translation for the symbolic statement is

A student studies, and if a student does not miss class then the student does not fail.

Once again, the symbolic statements in parentheses appear on the same side of the comma in the English statement.

✓ **Checkpoint 7** Let p, q, and r represent the following simple statements:

p: The plant is fertilized.

q: The plant is not watered.

r: The plant wilts.

Write each of the symbolic statements in words:

a. $(p \land \sim q) \to \sim r$

▶ **b.** $p \land (\sim q \to \sim r)$.

Exercise Set 13.2

Concept and Vocabulary Exercises

In Exercises 1–4, fill in each blank so that the resulting statement is true.

1. The compound statement "*p* and *q*" is symbolized by _____ and is called a/an _____.

2. The compound statement "*p* or *q*" is symbolized by _____ and is called a/an _____.

3. The compound statement "If *p*, then *q*" is symbolized by _____ and is called a/an _____ statement.

4. The compound statement "*p* if and only if *q*" is symbolized by _____ and is called a/an _____.

In Exercises 5–12, determine whether each statement is true or false. If the statement is false, make the necessary change(s) to produce a true statement.

5. $p \wedge q$ can be translated as "*p* but *q*."

6. $p \vee q$ means *p* or *q*, but not both.

7. $p \rightarrow q$ can be translated as "*p* is sufficient for *q*."

8. $p \rightarrow q$ can be translated as "*p* is necessary for *q*."

9. The consequent is the necessary condition in a conditional statement.

10. $p \leftrightarrow q$ can be translated as "If *p* then *q*, and if *q* then *p*."

11. $p \leftrightarrow q$ can be translated as "*p* is necessary and sufficient for *q*."

12. When symbolic statements are translated into English, the simple statements in parentheses appear on the same side of the comma.

Respond to Exercises 13–18 using verbal or written explanations.

13. Describe what is meant by a compound statement.

14. What is a conjunction? Describe the symbol that forms a conjunction.

15. What is a disjunction? Describe the symbol that forms a disjunction.

16. What is a conditional statement? Describe the symbol that forms a conditional statement.

17. What is a biconditional statement? Describe the symbol that forms a biconditional statement.

18. Discuss the difference between the symbolic statements $\sim(p \wedge q)$ and $\sim p \wedge q$.

In Exercises 19–22, determine whether each statement makes sense or does not make sense, and explain your reasoning.

19. When the waiter asked if I would like soup or salad, he used the exclusive *or*. However when he asked if I would like coffee or dessert, he used the inclusive *or*.

20. When you wrote me that you planned to enroll in math or chemistry, I knew that if you enrolled in math, you would not be taking chemistry.

21. In China, the bride wears red, so wearing red is sufficient for being a Chinese bride.

22. Earth is the only planet not named after a god, so not being named after a god is both necessary and sufficient for a planet being Earth.

Practice Exercises

In Exercises 23–28, let p and q represent the following simple statements:

   p: I'm leaving
   q: You're staying.

Write each compound statement in symbolic form.

23. I'm leaving and you're staying.

24. You're staying and I'm leaving.

25. You're staying and I'm not leaving.

26. I'm leaving and you're not staying.

27. You're not staying, but I'm leaving.

28. I'm not leaving, but you're staying.

In Exercises 29–32, let p and q represent the following simple statements:

   p: I study.
   q: I pass the course.

Write each compound statement in symbolic form.

29. I study or I pass the course.

30. I pass the course or I study.

31. I study or I do not pass the course.

32. I do not study or I do not pass the course.

In Exercises 33–40, let p and q represent the following simple statements:

   p: This is an alligator.
   q: This is a reptile.

Write each compound statement in symbolic form.

33. If this is an alligator, then this is a reptile.

34. If this is a reptile, then this is an alligator.

35. If this is not an alligator, then this is not a reptile.

36. If this is not a reptile, then this is not an alligator.

37. This is not an alligator if it's not a reptile.

38. This is a reptile if it's an alligator.

39. Being a reptile is necessary for being an alligator.

40. Being an alligator is sufficient for being a reptile.

In Exercises 41–48, let p and q represent the following simple statements:

   p: You are human.
   q: You have feathers.

Write each compound statement in symbolic form.

41. You do not have feathers if you are human.

42. You are not human if you have feathers.

43. Not being human is necessary for having feathers.

44. Not having feathers is necessary for being human.

45. Being human is sufficient for not having feathers.

46. Having feathers is sufficient for not being human.

47. You have feathers only if you're not human.

48. You're human only if you do not have feathers.

In Exercises 49–54, let p and q represent the following simple statements:

   p: The school is closed.

   q: It is Sunday.

Write each compound statement in symbolic form.

49. The school is closed if and only if it is Sunday.

50. It is Sunday if and only if the school is closed.

51. It is not Sunday if and only if the school is not closed.

52. The school is not closed if and only if it is not Sunday.

53. Being Sunday is necessary and sufficient for the school being closed.

54. The school being closed is necessary and sufficient for being Sunday.

In Exercises 55–62, let p and q represent the following simple statements:

   p: The heater is working.

   q: The house is cold.

Write each symbolic statement in words.

55. $\sim p \wedge q$ 56. $p \wedge \sim q$ 57. $p \vee \sim q$

58. $\sim p \vee q$ 59. $p \rightarrow \sim q$ 60. $q \rightarrow \sim p$

61. $p \leftrightarrow \sim q$ 62. $\sim p \leftrightarrow q$

In Exercises 63–70, let q and r represent the following simple statements:

   q: It is July 4th.

   r: We are having a barbecue.

Write each symbolic statement in words.

63. $q \wedge \sim r$ 64. $\sim q \wedge r$ 65. $\sim q \vee r$

66. $q \vee \sim r$ 67. $r \rightarrow \sim q$ 68. $q \rightarrow \sim r$

69. $\sim q \leftrightarrow r$ 70. $q \leftrightarrow \sim r$

In Exercises 71–80, let p and q represent the following simple statements:

   p: Romeo loves Juliet.

   q: Juliet loves Romeo.

Write each symbolic statement in words.

71. $\sim(p \wedge q)$ 72. $\sim(q \wedge p)$ 73. $\sim p \wedge q$

74. $\sim q \wedge p$ 75. $\sim(q \vee p)$ 76. $\sim(p \vee q)$

77. $\sim q \vee p$ 78. $\sim p \vee q$ 79. $\sim p \wedge \sim q$

80. $\sim q \wedge \sim p$

In Exercises 81–88, let p, q, and r represent the following simple statements:

   p: The temperature outside is freezing.

   q: The heater is working.

   r: The house is cold.

Write each compound statement in symbolic form.

81. The temperature outside is freezing and the heater is working, or the house is cold.

82. If the temperature outside is freezing, then the heater is working or the house is not cold.

83. If the temperature outside is freezing or the heater is not working, then the house is cold.

84. It is not the case that if the house is cold then the heater is not working.

85. The house is cold, if and only if the temperature outside is freezing and the heater isn't working.

86. If the heater is working, then the temperature outside is freezing if and only if the house is cold.

87. Sufficient conditions for the house being cold are freezing outside temperatures and a heater not working.

88. A freezing outside temperature is both necessary and sufficient for a cold house if the heater is not working.

In Exercises 89–102, let p, q, and r represent the following simple statements:

   p: The temperature is above 85°.

   q: We finished studying.

   r: We go to the beach.

Write each symbolic statement in words.

89. $(p \wedge q) \rightarrow r$ 90. $(q \wedge r) \rightarrow p$

91. $p \wedge (q \rightarrow r)$ 92. $p \wedge (r \rightarrow q)$

93. $\sim r \rightarrow (\sim p \vee \sim q)$ 94. $\sim p \rightarrow (q \vee r)$

95. $(\sim r \rightarrow \sim q) \vee p$ 96. $(\sim p \rightarrow \sim r) \vee q$

97. $r \leftrightarrow (p \wedge q)$ 98. $r \leftrightarrow (q \wedge p)$

99. $(p \leftrightarrow q) \wedge r$ 100. $q \rightarrow (r \leftrightarrow p)$

101. $\sim r \rightarrow \sim(p \wedge q)$ 102. $\sim(p \wedge q) \rightarrow \sim r$

In Exercises 103–112, write each compound statement in symbolic form. Let letters assigned to the simple statements represent English sentences that are not negated.

103. If I like the teacher or the course is interesting then I do not miss class.

104. If the lines go down or the transformer blows then we do not have power.

105. I like the teacher, or if the course is interesting then I do not miss class.

106. The lines go down, or if the transformer blows then we do not have power.

107. I miss class if and only if it's not true that both I like the teacher and the course is interesting.

108. We have power if and only if it's not true that both the lines go down and the transformer blows.

109. If I like the teacher I do not miss class if and only if the course is interesting.

110. If the lines go down we do not have power if and only if the transformer blows.

111. If I do not like the teacher and I miss class then the course is not interesting or I spend extra time reading the textbook.

112. If the lines do not go down and we have power then the transformer does not blow or there is an increase in the cost of electricity.

Practice Plus

In Exercises 113–117, write each compound statement in symbolic form. Assign letters to simple statements that are not negated and show grouping symbols in symbolic statements.

113. If it's not true that being French is necessary for being a Parisian then it's not true that being German is necessary for being a Berliner.

114. If it's not true that being English is necessary for being a Londoner then it's not true that being American is necessary for being a New Yorker.

115. Filing an income tax report and a complete statement of earnings is necessary for each taxpayer or an authorized tax preparer.

116. It is not true that being wealthy is a sufficient condition for being happy and living contentedly.

117. It is not true that being happy and living contentedly are necessary conditions for being wealthy.

Application Exercises

Exercises 118–123 contain statements made by well-known people. Use letters to represent each non-negated simple statement and rewrite the given compound statement in symbolic form.

118. "If you cannot get rid of the family skeleton, you may as well make it dance." (George Bernard Shaw)

119. "I wouldn't turn out the way I did if I didn't have all the old-fashioned values to rebel against." (Madonna)

120. "If you know what you believe then it makes it a lot easier to answer questions, and I can't answer your question." (George W. Bush)

121. "If you don't like what you're doing, you can always pick up your needle and move to another groove." (Timothy Leary)

122. "If I were an intellectual, I would be pessimistic about America, but since I'm not an intellectual, I am optimistic about America." (General Lewis B. Hershey, Director of the Selective Service during the Vietnam war) (For simplicity, regard "optimistic" as "not pessimistic.")

123. "You cannot be both a good socializer and a good writer." (Erskine Caldwell)

Critical Thinking Exercises

124. Use letters to represent each simple statement in the compound statement that follows. Then express the compound statement in symbolic form.

     Shooting unarmed civilians is morally justifiable if and only if bombing them is morally justifiable, and as the former is not morally justifiable, neither is the latter.

125. Using a topic on which you have strong opinions, write a compound statement that contains at least two different connectives. Then express the statement in symbolic form.

Group Exercise

126. Each group member should find a legal document that contains at least six connectives in one paragraph. The connectives should include at least three different kinds of connectives, such as *and, or, if–then*, and *if and only if*. Share your example with other members of the group and see if the group can explain what some of the more complicated statements actually mean.

Truth Tables for Negation, Conjunction, and Disjunction

13.3

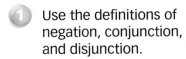

Objectives

1. Use the definitions of negation, conjunction, and disjunction.

2. Construct truth tables.

3. Determine the truth value of a compound statement for a specific case.

In 2006, *USA Today* analyzed patterns in the deaths of four-year college students since January 2000. Their most dominant finding was that freshmen emerged as the class most likely to make a fatal mistake. Freshmen accounted for more than one-third of all undergraduate deaths, even though they made up only 24% of the population enrolled in four-year institutions.

In this section, you will work with two circle graphs based on data from the *USA Today* study. By determining when statements involving negation, ~ (not), conjunction, ∧ (and), and disjunction, ∨ (or), are true and when they are false, you will be able to draw conclusions from the data. Classifying a statement as true or false is called **assigning a truth value to the statement**.

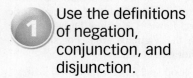

Use the definitions of negation, conjunction, and disjunction.

Negation, ~

The negation of a true statement is a false statement. We can express this in a table in which T represents true and F represents false.

p	~*p*
T	F

The negation of a false statement is a true statement. This, too, can be shown in table form.

p	~*p*
F	T

Combining the two tables results in **Table 13.9**, called the **truth table for negation**. This truth table expresses the idea that ~*p* has the opposite truth value from *p*.

Table 13.9 Negation	
p	~*p*
T	F
F	T

~*p* has the opposite truth value from *p*.

Conjunction, ∧

A friend tells you, "I visited London and I visited Paris." In order to understand the truth values for this statement, let's break it down into its two simple statements:

p: I visited London.

q: I visited Paris.

There are four possible cases to consider.

Case 1 Your friend actually visited both cities, so *p* is true and *q* is true. The conjunction "I visited London and I visited Paris" is true because your friend did both things. If both *p* and *q* are true, the conjunction *p* ∧ *q* is true. We can show this in truth table form:

p	*q*	*p* ∧ *q*
T	T	T

Case 2   Your friend actually visited London, but did not tell the truth about visiting Paris. In this case, p is true and q is false. Your friend didn't do what was stated, namely visit both cities, so $p \land q$ is false. If p is true and q is false, the conjunction $p \land q$ is false.

p	q	$p \land q$
T	F	F

Case 3   This time, London was not visited, but Paris was. This makes p false and q true. As in case 2, your friend didn't do what was stated, namely visit both cities, so $p \land q$ is false. If p is false and q is true, the conjunction $p \land q$ is false.

p	q	$p \land q$
F	T	F

Case 4   This time your friend visited neither city, so p is false and q is false. The statement that both were visited, $p \land q$, is false.

p	q	$p \land q$
F	F	F

Let's use a truth table to summarize all four cases. Only in the case that your friend visited London and visited Paris is the conjunction true. Each of the four cases appears in **Table 13.10**, the truth table for conjunction, $\land$. The definition of conjunction is given in words to the left of the table.

I visited London and I visited Paris.

The Definition of Conjunction

A conjunction is true only when both simple statements are true.

Table 13.10 Conjunction

p	q	$p \land q$
T	T	T
T	F	F
F	T	F
F	F	F

Table 13.11 contains an example of each of the four cases in the conjunction truth table.

Table 13.11 Statements of Conjunction and Their Truth Values

Statement	Truth Value	Reason
3 + 2 = 5 and London is in England.	T	Both simple statements are true.
3 + 2 = 5 and London is in France.	F	The second simple statement is false.
3 + 2 = 6 and London is in England.	F	The first simple statement is false.
3 + 2 = 6 and London is in France.	F	Both simple statements are false.

The statements that come before and after the main connective in a compound statement do not have to be simple statements. Consider, for example, the compound statement

$$(\sim p \lor q) \land \sim q.$$

The statements that make up this conjunction are $\sim p \lor q$ and $\sim q$. The conjunction is true only when both $\sim p \lor q$ and $\sim q$ are true. Notice that $\sim p \lor q$ is not a simple statement. We call $\sim p \lor q$ and $\sim q$ the *component statements* of the conjunction. The statements making up a compound statement are called **component statements**.

Disjunction, $\lor$

Now your friend states, "I will visit London or I will visit Paris." Because we assume that this is the inclusive "or," if your friend visits either or both of these cities, the truth has been told. The disjunction is false only in the event that neither city is visited. An *or* statement is true in every case, except when both component statements are false.

The truth table for disjunction, $\lor$, is shown in **Table 13.12**. The definition of disjunction is given in the words above the table.

The Definition of Disjunction

A disjunction is false only when both component statements are false.

Table 13.12 Disjunction

p	q	$p \lor q$
T	T	T
T	F	T
F	T	T
F	F	F

Table 13.13 contains an example of each of the four cases in the disjunction truth table.

Table 13.13 Statements of Disjunction and Their Truth Values

Statement	Truth Value	Reason
3 + 2 = 5 or London is in England.	T	Both component statements are true.
3 + 2 = 5 or London is in France.	T	The first component statement is true.
3 + 2 = 6 or London is in England.	T	The second component statement is true.
3 + 2 = 6 or London is in France.	F	Both component statements are false.

Example 1	Using the Definitions of Negation, Conjunction, and Disjunction

Let p and q represent the following statements:

p: $10 > 4$

q: $3 < 5$.

Determine the truth value for each statement:

a. $p \wedge q$ **b.** $\sim p \wedge q$ **c.** $p \vee \sim q$ **d.** $\sim p \vee \sim q$.

Solution

a. $p \wedge q$ translates as

$$10 > 4 \qquad \text{and} \qquad 3 < 5.$$

> 10 is greater than 4 is true.   3 is less than 5 is true.

By definition, a conjunction, $\wedge$, is true only when both component statements are true. Thus, $p \wedge q$ is a true statement.

b. $\sim p \wedge q$ translates as

$$10 \not> 4 \qquad \text{and} \qquad 3 < 5.$$

> 10 is not greater than 4 is false.   3 is less than 5 is true.

By definition, a conjunction, $\wedge$, is true only when both component statements are true. In this conjunction, only one of the two component statements is true. Thus, $\sim p \wedge q$ is a false statement.

c. $p \vee \sim q$ translates as

$$10 > 4 \qquad \text{or} \qquad 3 \not< 5.$$

> 10 is greater than 4 is true.   3 is not less than 5 is false.

By definition, a disjunction, $\vee$, is false only when both component statements are false. In this disjunction, only one of the two component statements is false. Thus, $p \vee \sim q$ is a true statement.

d. $\sim p \vee \sim q$ translates as

$$10 \not> 4 \qquad \text{or} \qquad 3 \not< 5.$$

> 10 is not greater than 4 is false.   3 is not less than 5 is false.

By definition, a disjunction, $\vee$, is false only when both component statements are false. Thus, $\sim p \vee \sim q$ is a false statement.

✓ **Checkpoint 1**   Let p and q represent the following statements:

p: $3 + 5 = 8$

q: $2 \times 7 = 20$.

Determine the truth value for each statement:

a. $p \wedge q$ **b.** $p \wedge \sim q$

c. $\sim p \vee q$ **d.** $\sim p \vee \sim q$.

Constructing Truth Tables

② Construct truth tables.

Truth tables can be used to gain a better understanding of English statements. The truth tables in this section are based on the definitions of negation, $\sim$, conjunction, $\wedge$, and disjunction, $\vee$. It is helpful to remember these definitions in words.

Definitions of Negation, Conjunction, and Disjunction

1. **Negation ~: not**
   The negation of a statement has the opposite truth value from the statement.
2. **Conjunction ∧: and**
   The only case in which a conjunction is true is when both component statements are true.
3. **Disjunction ∨: or**
   The only case in which a disjunction is false is when both component statements are false.

Breaking compound statements into component statements and applying these definitions will enable you to construct truth tables.

Constructing Truth Tables for Compound Statements Containing Only the Simple Statements *p* and *q*

• List the four possible combinations of truth values for *p* and *q*.

p	*q*	
T	T	
T	F	
F	T	
F	F	

> We will always list the combinations in this order. Although any order can be used, this standard order makes for a consistent presentation.

• Determine each column heading by reconstructing the given compound statement one component statement at a time. The final column heading should be the given compound statement.

• Use each column heading to fill in the four truth values.
   → If a column heading involves negation, ~ (not), fill in the column by looking back at the column that contains the statement that must be negated. Take the opposite of the truth values in this column.
   → If a column heading involves the symbol for conjunction, ∧ (and), fill in the truth values in the column by looking back at two columns—the column for the statement before the ∧ connective and the column for the statement after the ∧ connective. Fill in the column by applying the definition of conjunction, writing T only when both component statements are true.
   → If a column heading involves the symbol for disjunction, ∨ (or), fill in the truth values in the column by looking back at two columns—the column for the statement before the ∨ connective and the column for the statement after the ∨ connective. Fill in the column by applying the definition of disjunction, writing F only when both component statements are false.

Example 2 Constructing a Truth Table

Construct a truth table for

$$\sim(p \wedge q)$$

to determine when the statement is true and when the statement is false.

Solution

The parentheses in ~(p ∧ q) indicate that we must first determine the truth values for the conjunction p ∧ q. After this, we determine the truth values for the negation ~(p ∧ q) by taking the opposite of the truth values for p ∧ q.

Step 1 As with all truth tables, first list the simple statements on top. Then show all the possible truth values for these statements. In this case there are two simple statements and four possible combinations, or cases.

p	q	
T	T	
T	F	
F	T	
F	F	

Step 2 Make a column for p ∧ q, the statement within the parentheses in ~(p ∧ q). Use p ∧ q as the heading for the column, and then fill in the truth values for the conjunction by looking back at the p and q columns. A conjunction is true only when both component statements are true.

p	q	p ∧ q
T	T	T
T	F	F
F	T	F
F	F	F

p and q are true, so p ∧ q is true.

Step 3 Construct one more column for ~(p ∧ q). Fill in this column by negating the values in the p ∧ q column. Using the negation definition, take the opposite of the truth values in the third column.

p	q	p ∧ q	~(p ∧ q)
T	T	T	F
T	F	F	T
F	T	F	T
F	F	F	T

Opposite truth values because we are negating column 3

This completes the truth table for ~(p ∧ q).

The final column in the truth table for ~(p ∧ q) tells us that the statement is false only when both p and q are true. For example, using

p: Harvard is a college (true)
q: Yale is a college (true),

the statement ~(p ∧ q) translates as

It is not true that Harvard and Yale are colleges.

This compound statement is false. It *is* true that Harvard and Yale are colleges.

Checkpoint 2 Construct a truth table for ~(p ∨ q) to determine when the statement is true and when the statement is false.

Example 3　Constructing a Truth Table

Construct a truth table for

$$\sim p \vee \sim q$$

to determine when the statement is true and when the statement is false.

Solution

Without parentheses, the negation symbol, $\sim$, negates only the statement that immediately follows it. Therefore, we first determine the truth values for $\sim p$ and for $\sim q$. Then we determine the truth values for the *or* disjunction, $\sim p \vee \sim q$.

Step 1　List the simple statements on top and show the four possible cases for the truth values.

p	q	
T	T	
T	F	
F	T	
F	F	

Step 2　Make columns for $\sim p$ and for $\sim q$. Fill in the $\sim p$ column by looking back at the p column, the first column, and taking the opposite of the truth values in that column. Fill in the $\sim q$ column by taking the opposite of the truth values in the second column, the q column.

Opposite truth values

p	q	$\sim p$	$\sim q$
T	T	F	F
T	F	F	T
F	T	T	F
F	F	T	T

Opposite truth values

Step 3　Construct one more column for $\sim p \vee \sim q$. To determine the truth values of $\sim p \vee \sim q$, look back at the $\sim p$ column, column 3, and the $\sim q$ column, column 4. Now use the disjunction definition on the entries in columns 3 and 4. Disjunction definition: An *or* statement is false only when both component statements are false. This occurs only in the first row.

p	q	$\sim p$	$\sim q$	$\sim p \vee \sim q$
T	T	F	F	F
T	F	F	T	T
F	T	T	F	T
F	F	T	T	T

$\sim p$ is false and $\sim q$ is false, so $\sim p \vee \sim q$ is false.

column 3 $\vee$ column 4

Checkpoint 3　Construct a truth table for $\sim p \wedge \sim q$ to determine when the statement is true and when the statement is false.

| Example 4 | Constructing a Truth Table |

Construct a truth table for

$$(\sim p \vee q) \wedge \sim q$$

to determine when the statement is true and when the statement is false.

Solution

The statement $(\sim p \vee q) \wedge \sim q$ is an *and* conjunction because the conjunction symbol, $\wedge$, is outside the parentheses. We cannot determine the truth values for the statement until we first determine the truth values for $\sim p \vee q$ and for $\sim q$, the component statements before and after the $\wedge$ connective:

$$\boxed{(\sim p \ \vee \ q)} \ \wedge \ \boxed{\sim q}.$$

We'll need a column with truth values for this component statement.

We'll need a column with truth values for this component statement.

Step 1 The compound statement involves two simple statements and four possible cases.

p	*q*	
T	T	
T	F	
F	T	
F	F	

$\boxed{(\sim p \ \vee \ q)} \ \wedge \ \boxed{\sim q}$

Column needed   Column needed

Step 2 Because we need a column with truth values for $\sim p \vee q$, begin with $\sim p$. Use $\sim p$ as the heading. Fill in the column by looking back at the *p* column, column 1, and take the opposite of the truth values in that column.

p	*q*	$\sim p$
T	T	F
T	F	F
F	T	T
F	F	T

Opposite truth values

Step 3 Now add a $\sim p \vee q$ column. To determine the truth values of $\sim p \vee q$, look back at the $\sim p$ column, column 3, and the *q* column, column 2. Now use the disjunction definition on the entries in columns 3 and 2. Disjunction definition: An *or* statement is false only when both component statements are false. This occurs only in the second row.

p	*q*	$\sim p$	$\sim p \vee q$
T	T	F	T
T	F	F	F
F	T	T	T
F	F	T	T

$\sim p$ is false and *q* is false, so $\sim p \vee q$ is false.

column 3 $\vee$ column 2

Step 4   The statement following the ∧ connective in $(\sim p \lor q) \land \sim q$ is $\sim q$, so add a $\sim q$ column. Fill in the column by looking back at the q column, column 2, and take the opposite of the truth values in that column.

p	q	$\sim p$	$\sim p \lor q$	$\sim q$
T	T	F	T	F
T	F	F	F	T
F	T	T	T	F
F	F	T	T	T

Opposite truth values

Step 5   The final column heading is

$$(\sim p \lor q) \land \sim q,$$

which is our given statement. To determine its truth values, look back at the $\sim p \lor q$ column, column 4, and the $\sim q$ column, column 5. Now use the conjunction definition on the entries in columns 4 and 5. Conjunction definition: An *and* statement is true only when both component statements are true. This occurs only in the last row.

p	q	$\sim p$	$\sim p \lor q$	$\sim q$	$(\sim p \lor q) \land \sim q$
T	T	F	T	F	F
T	F	F	F	T	F
F	T	T	T	F	F
F	F	T	T	T	T

$\sim p \lor q$ is true and $\sim q$ is true, so $(\sim p \lor q) \land \sim q$ is true.

column 4 ∧ column 5

The truth table is now complete. By looking at the truth values in the last column, we can see that the compound statement

$$(\sim p \lor q) \land \sim q$$

is true only in the fourth row, that is, when p is false and q is false.

✓ **Checkpoint 4**   Construct a truth table for $(p \land \sim q) \lor \sim p$ to determine when the statement is true and when the statement is false.

Some compound statements, such as $p \lor \sim p$, consist of only one simple statement. In cases like this, there are only two true–false possibilities: p can be true or p can be false.

Example 5   Constructing a Truth Table

Construct a truth table for

$$p \lor \sim p$$

to determine when the statement is true and when the statement is false.

Solution

We first determine the truth values for $\sim p$. Then we determine the truth values for the *or* disjunction, $p \lor \sim p$.

Step 1   The compound statement involves one simple statement and two possible cases.

p	
T	
F	

Step 2   Add a column for $\sim p$.

p	$\sim p$
T	F
F	T

Take the opposite of the truth values in the first column.

Step 3   Construct one more column for $p \vee \sim p$.

p	$\sim p$	$p \vee \sim p$
T	F	T
F	T	T

Look back at columns 1 and 2 and apply the disjunction definition: An *or* statement is false only when both component statements are false. This does not occur in either row.

The truth table is now complete. By looking at the truth values in the last column, we can see that the compound statement $p \vee \sim p$ is true in all cases.

A compound statement that is always true is called a **tautology**. Example 5 proves that $p \vee \sim p$ is a tautology.

 **Checkpoint 5**   Construct a truth table for

$$p \wedge \sim p$$

to determine when the statement is true and when the statement is false.

Some compound statements involve three simple statements, usually represented by p, q, and r. In this situation, there are eight different true–false combinations, shown in **Table 13.14**. The first column has four Ts followed by four Fs. The second column has two Ts, two Fs, two Ts, and two Fs. Under the third statement, r, T alternates with F. It is not necessary to list the eight cases in this order, but this systematic method ensures that no case is repeated and that all cases are included.

Blitzer Bonus

Tautologies and Commercials

Think about the logic in the following statement from a commercial for a high-fiber cereal:

> Some studies suggest a high-fiber, low-fat diet may reduce the risk of some kinds of cancer.

From a strictly logical point of view, "may reduce" means "It may reduce the risk of cancer or it may not." The symbolic translation of this sentence, $p \vee \sim p$, is a tautology. The claim made in the cereal commercial can be made about any substance that is not known to actually cause cancer and, like $p \vee \sim p$, will always be true.

Table 13.14

	p	q	r
Case 1	T	T	T
Case 2	T	T	F
Case 3	T	F	T
Case 4	T	F	F
Case 5	F	T	T
Case 6	F	T	F
Case 7	F	F	T
Case 8	F	F	F

There are eight different true–false combinations for compound statements consisting of three simple statements.

Example 6   Constructing a Truth Table with Eight Cases

a. Construct a truth table for the following statement:

I study hard and ace the final, or I fail the course.

b. Suppose that you study hard, you do not ace the final, and you fail the course. Under these conditions, is the compound statement in part (a) true or false?

Solution

a. We begin by assigning letters to the simple statements. Use the following representations:

p: I study hard.

q: I ace the final.

r: I fail the course.

Now we can write the given statement in symbolic form.

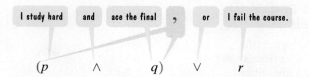

The statement $(p \wedge q) \vee r$ is a disjunction because the *or* symbol, $\vee$, is outside the parentheses. We cannot determine the truth values for this disjunction until we have determined the truth values for $p \wedge q$ and for r, the statements before and after the $\vee$ connective. The completed truth table appears as follows:

> The conjunction is true only when p, the first column, is true and q, the second column, is true.

> The disjunction is false only when $p \wedge q$, column 4, is false and r, column 3, is false.

> Show eight possible cases.

> These are the conditions in part (b).

p	q	r	$p \wedge q$	$(p \wedge q) \vee r$
T	T	T	T	T
T	T	F	T	T
T	F	T	F	T
T	F	F	F	F
F	T	T	F	T
F	T	F	F	F
F	F	T	F	T
F	F	F	F	F

b. We are given the following:

p: I study hard. — This is true. We are told you study hard.

q: I ace the final. — This is false. We are told you do not ace the final.

r: I fail the course. — This is true. We are told you fail the course.

The given conditions, T F T, correspond to case 3 of the truth table, indicated by the voice balloon on the far left. Under these conditions, the original compound statement is true, shown by the red T in the truth table.

 **Checkpoint 6**

a. Construct a truth table for the following statement:

I study hard, and I ace the final or fail the course.

b. Suppose that you do not study hard, you ace the final, and you fail the course. Under these conditions, is the compound statement in part (a) true or false?

Determining Truth Values for Specific Cases

A truth table shows the truth values of a compound statement for every possible case. In our next example, we will determine the truth value of a compound statement for a specific case in which the truth values of the simple statements are known. This does not require constructing an entire truth table. By substituting the truth values of the simple statements into the symbolic form of the compound statement and using the appropriate definitions, we can determine the truth value of the compound statement.

③ Determine the truth value of a compound statement for a specific case.

| **Example 7** | **Determining the Truth Value of a Compound Statement** |

College administrators, public health officials, and parents increasingly have become concerned about the safety of college students after highly publicized deaths on campus from alcohol abuse and other causes. The data in **Figure 13.2** indicate that freshmen are particularly vulnerable.

Undergraduate Enrollment and Deaths in the United States

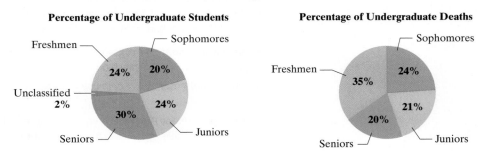

Figure 13.2
Source: USA Today

Use the information in the circle graphs to determine the truth value of the following statement:

It is not true that freshmen make up 24% of the undergraduate college population and account for more than one-third of the undergraduate deaths, or seniors do not account for 30% of the undergraduate deaths.

Solution

We begin by assigning letters to the simple statements, using the graphs to determine whether each simple statement is true or false. As always, we let these letters represent statements that are not negated.

p: Freshmen make up 24% of the undergraduate college population.   *This statement is true.*

q: Freshmen account for more than one-third of the undergraduate deaths.   *This statement is true. They account for 35% of the deaths, which is greater than $\frac{1}{3}$: $\frac{1}{3} = 33\frac{1}{3}\%$.*

r: Seniors account for 30% of the undergraduate deaths.   *This statement is false. They account for 20% of the deaths.*

Using these representations, the given compound statement can be expressed in symbolic form as

$$\sim(p \wedge q) \vee \sim r.$$

Now we substitute the truth values for p, q, and r that we obtained from the circle graphs to determine the truth value for the given compound statement.

$\sim(p \wedge q) \vee \sim r$ This is the given compound statement in symbolic form.

$\sim(\text{T} \wedge \text{T}) \vee \sim\text{F}$ Substitute the truth values obtained from the graph.

$\sim\text{T} \vee \sim\text{F}$ Replace T $\wedge$ T with T. Conjunction is true when both parts are true.

$\text{F} \vee \text{T}$ Replace $\sim$T with F and $\sim$F with T. Negation gives the opposite truth value.

T Replace F $\vee$ T with T. Disjunction is true when at least one part is true.

We conclude that the given statement is true.

Checkpoint 7 Use the information in the circle graphs in **Figure 13.2** on the previous page to determine the truth value for the following statement:

Freshmen and juniors make up the same percentage of the undergraduate college population or seniors make up 35% of the population, and freshmen and juniors do not account for the same percentage of the undergraduate deaths.

Achieving Success

After leaving high school and continuing your higher education, **consider the community college option**. Half of undergraduate college students in the United States are in community colleges. Most community colleges have at least three kinds of programs that you can select from:

- **College Preparatory Programs.** Community colleges test incoming students in writing, reading, and math. If you need a brush-up or review, you can get it through a variety of college-prep courses.

- **Vocational Programs.** Community colleges have two-year vocational programs that lead directly into the job market.

- **College Transfer Programs.** Many students enroll in community colleges to get the first two years of their college career in a less stressful and more economical environment.

Exercise Set 13.3

Concept and Vocabulary Exercises

In Exercises 1–3, fill in each blank so that the resulting statement is true.

1. $\sim p$ has the _____ truth value from p.
2. A conjunction, $p \wedge q$, is true only when _____.
3. A disjunction, $p \vee q$, is false only when _____.

In Exercises 4–7, determine whether each statement is true or false. If the statement is false, make the necessary change(s) to produce a true statement.

4. If one component statement in a conjunction is false, the conjunction is false.
5. If one component statement in a disjunction is true, the disjunction is true.
6. A truth table for $p \vee \sim q$ requires four possible combinations of truth values.
7. A truth table for $p \vee \sim p$ requires four possible combinations of truth values.

Respond to Exercises 8–14 using verbal or written explanations.

8. Under which conditions is a conjunction true?
9. Under which conditions is a conjunction false?
10. Under which conditions is a disjunction true?
11. Under which conditions is a disjunction false?

12. Describe how to construct a truth table for a compound statement.
13. Describe the information given by the truth values in the final column of a truth table.
14. Describe how to set up the eight different true–false combinations for a compound statement consisting of three simple statements.

In Exercises 15–18, determine whether each statement makes sense or does not make sense, and explain your reasoning.

15. I'm filling in the truth values for a column in a truth table that requires me to look back at three columns.
16. If I know that p is true, q is false, and r is false, the most efficient way to determine the truth value of $(p \wedge \sim q) \vee r$ is to construct a truth table.
17. My truth table for $\sim(\sim p)$ has four possible combinations of truth values.
18. Using inductive reasoning, I conjecture that a truth table for a compound statement consisting of n simple statements has 2^n true–false combinations.

Practice Exercises

In Exercises 19–34, let p and q represent the following statements:

p: $4 + 6 = 10$
q: $5 \times 8 = 80$

Determine the truth value for each statement.

19. $\sim q$ **20.** $\sim p$

21. $p \wedge q$ **22.** $q \wedge p$

23. $\sim p \wedge q$ **24.** $p \wedge \sim q$

25. $\sim p \wedge \sim q$ **26.** $q \wedge \sim q$

27. $q \vee p$ **28.** $p \vee q$

29. $p \vee \sim q$ **30.** $\sim p \vee q$

31. $p \vee \sim p$ **32.** $q \vee \sim q$

33. $\sim p \vee \sim q$ **34.** $\sim q \vee \sim p$

In Exercises 35–42, complete the truth table for the given statement by filling in the required columns.

35. $\sim p \wedge p$

p	$\sim p$	$\sim p \wedge p$
T		
F		

36. $\sim(\sim p)$

p	$\sim p$	$\sim(\sim p)$
T		
F		

37. $\sim p \wedge q$

p	q	$\sim p$	$\sim p \wedge q$
T	T		
T	F		
F	T		
F	F		

38. $\sim p \vee q$

p	q	$\sim p$	$\sim p \vee q$
T	T		
T	F		
F	T		
F	F		

39. $\sim(p \vee q)$

p	q	$p \vee q$	$\sim(p \vee q)$
T	T		
T	F		
F	T		
F	F		

40. $\sim(p \vee \sim q)$

p	q	$\sim q$	$p \vee \sim q$	$\sim(p \vee \sim q)$
T	T			
T	F			
F	T			
F	F			

41. $\sim p \wedge \sim q$

p	q	$\sim p$	$\sim q$	$\sim p \wedge \sim q$
T	T			
T	F			
F	T			
F	F			

42. $p \wedge \sim q$

p	q	$\sim q$	$p \wedge \sim q$
T	T		
T	F		
F	T		
F	F		

In Exercises 43–60, construct a truth table for the given statement.

43. $p \vee \sim q$ **44.** $\sim q \wedge p$

45. $\sim(\sim p \vee q)$ **46.** $\sim(p \wedge \sim q)$

47. $(p \vee q) \wedge \sim p$ **48.** $(p \wedge q) \vee \sim p$

49. $\sim p \vee (p \wedge \sim q)$ **50.** $\sim p \wedge (p \vee \sim q)$

51. $(p \vee q) \wedge (\sim p \vee \sim q)$ **52.** $(p \wedge \sim q) \vee (\sim p \wedge q)$

53. $(p \wedge \sim q) \vee (p \wedge q)$ **54.** $(p \vee \sim q) \wedge (p \vee q)$

55. $p \wedge (\sim q \vee r)$ **56.** $p \vee (\sim q \wedge r)$

57. $(r \wedge \sim p) \vee \sim q$ **58.** $(r \vee \sim p) \wedge \sim q$

59. $\sim(p \vee q) \wedge \sim r$ **60.** $\sim(p \wedge q) \vee \sim r$

In Exercises 61–70,

a. *Write each statement in symbolic form. Assign letters to simple statements that are not negated.*

b. *Construct a truth table for the symbolic statement in part (a).*

c. *Use the truth table to indicate one set of conditions that makes the compound statement true, or state that no such conditions exist.*

61. You did not do the dishes and you left the room a mess.

62. You did not do the dishes, but you did not leave the room a mess.

63. It is not true that I bought a ticket to the concert and did not use it.

64. It is not true that I snacked while doing homework and did not gain weight.

65. The student is intelligent or an overachiever, and not an overachiever.

66. You're blushing or sunburned, and you're not sunburned.

67. Married people are healthier than single people and more economically stable than single people, and children of married people do better on a variety of indicators.

68. You walk or jog, or engage in something physical.

69. I go to office hours and ask questions, or my teacher does not know that I am trying to improve my grade.

70. You marry the person you love, but you do not always love that person or do not always have a successful marriage.

In Exercises 71–80, determine the truth value for each statement when p is false, q is true, and r is false.

71. $p \wedge (q \vee r)$

72. $p \vee (q \wedge r)$

73. $\sim p \vee (q \wedge \sim r)$

74. $\sim p \wedge (\sim q \wedge r)$

75. $\sim (p \wedge q) \vee r$

76. $\sim (p \vee q) \wedge r$

77. $\sim (p \vee q) \wedge \sim (p \wedge r)$

78. $\sim (p \wedge q) \vee \sim (p \vee r)$

79. $(\sim p \wedge q) \vee (\sim r \wedge p)$

80. $(\sim p \vee q) \wedge (\sim r \vee p)$

Practice Plus

In Exercises 81–84, construct a truth table for each statement.

81. $\sim [\sim (p \wedge \sim q) \vee \sim (\sim p \vee q)]$

82. $\sim [\sim (p \vee \sim q) \wedge \sim (\sim p \wedge q)]$

83. $[(p \wedge \sim r) \vee (q \wedge \sim r)] \wedge \sim (\sim p \vee r)$

84. $[(p \vee \sim r) \wedge (q \vee \sim r)] \vee \sim (\sim p \vee r)$

In Exercises 85–88, write each statement in symbolic form and construct a truth table. Then indicate under what conditions, if any, the compound statement is true.

85. You notice this notice or you do not, and you notice this notice is not worth noticing.

86. You notice this notice and you notice this notice is not worth noticing, or you do not notice this notice.

87. It is not true that $x \leq 3$ or $x \geq 7$, but $x > 3$ and $x < 7$.

88. It is not true that $x < 5$ or $x > 8$, but $x \geq 5$ and $x \leq 8$.

Application Exercises

With aging, body fat increases and muscle mass declines. The line graphs show the percent body fat in adult women and men as they age from 25 to 75 years.

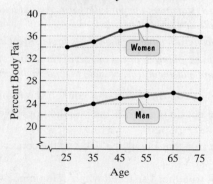

Percent Body Fat in Adults

Source: Thompson et al., *The Science of Nutrition*, Benjamin Cummings, 2008.

In Exercises 89–98, let p, q, and r represent the following simple statements:

    p: The percent body fat in women peaks at age 55.

    q: The percent body fat in men peaks at age 65.

    r: Men have more than 24% body fat at age 25.

Write each symbolic statement in words. Then use the information given by the graph to determine the truth value of the statement.

89. $p \wedge \sim q$

90. $p \wedge \sim r$

91. $\sim p \wedge r$

92. $q \wedge \sim p$

93. $p \vee \sim q$

94. $p \vee \sim r$

95. $\sim p \vee r$

96. $q \vee \sim p$

97. $(p \wedge q) \vee r$

98. $p \wedge (q \vee r)$

Completing the transition to adulthood is measured by one or more of the following: leaving home, finishing school, getting married, having a child, or being financially independent. The bar graph shows the percentage of Americans, ages 20 and 30, who had completed the transition to adulthood in 1960 and in 2000.

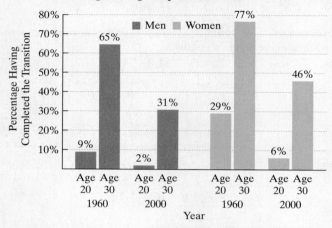

Percentage Having Completed the Transition to Adulthood

Source: James M. Henslin, *Sociology*, Eighth Edition, Allyn and Bacon, 2007.

In Exercises 99–102, write each statement in symbolic form. Then use the information in the graph above to determine the truth value of the compound statement.

99. It is not true that in 2000, 2% of 20-year-old men and 46% of 30-year-old men had completed the transition to adulthood.

100. It is not true that in 2000, 2% of 20-year-old women and 46% of 30-year-old women had completed the transition to adulthood.

101. Between 1960 and 2000, the percentage of 20-year-old women making the transition to adulthood decreased or the percentage of 30-year-old women making the transition increased, and the percentage of 30-year-old men making the transition did not decrease.

102. Between 1960 and 2000, the percentage of 20-year-old men making the transition to adulthood decreased, but the percentage of 30-year-old men making the transition did not decrease or the percentage of 20-year-old women making the transition increased.

103. To qualify for the position of music professor, an applicant must have a master's degree in music, and be able to play at least three musical instruments or have at least five years' experience playing with a symphony orchestra.

There are three applicants for the position:

- *Bolero Mozart* has a master's degree in journalism and plays the piano, tuba, and violin.

- *Cha-Cha Bach* has a master's degree in music, plays the piano and the harp, and has two years' experience with a symphony orchestra.

- *Hora Gershwin* has a master's degree in music, plays 14 instruments, and has two years' experience in a symphony orchestra.

a.   Which of the applicants qualifies for the position?

b.   Explain why each of the other applicants does not qualify for the position.

104. To qualify for the position of art professor, an applicant must have a master's degree in art, and a body of work judged as excellent by at least two working artists or at least two works on public display in the United States.

There are three applicants for the position:

- *Adagio Picasso* needs two more courses to complete a master's degree in art, has a body of work judged as excellent by ten working artists, and has over 50 works on public display in a number of different American cities.

- *Rondo Seurat* has a master's degree in art, a body of work judged as excellent by a well-known working artist, and two works on public display in New York City.

- *Yodel Van Gogh* has a master's degree in art, is about to complete a doctorate in art history, has 20 works on public display in Paris, France, and has a body of work judged as excellent by a working artist.

a.   Which of the applicants qualifies for the position?

b.   Explain why each of the other applicants does not qualify for the position.

Critical Thinking Exercises

105. The circle graphs in **Figure 13.2** on page 841 indicate that college freshmen account for more than one-third of undergraduate deaths, although they comprise only 24% of the undergraduate population. What explanations can you offer for these grim statistics?

106. Use the bar graph for Exercises 99–102 to write a true compound statement with each of the following characteristics. Do not use any of the simple statements that appear in Exercises 99–102.

a.   The statement contains three different simple statements.

b.   The statement contains two different connectives.

c.   The statement contains one simple statement with the word *not*.

107. If $\sim(p \vee q)$ is true, determine the truth values for p and q.

108. The truth table that defines $\vee$, the *inclusive or*, indicates that the compound statement is true if one or both of its component statements are true. The symbol for the *exclusive or* is $\underline{\vee}$. The *exclusive or* means *either p or q, but not both*. Use this meaning to construct the truth table that defines $p \underline{\vee} q$.

Group Exercise

109. Each member of the group should find a graph that is of particular interest to that person. Share the graphs. The group should select the three graphs that it finds most intriguing. For the graphs selected, group members should write four compound statements. Two of the statements should be true and two should be false. One of the statements should contain three different simple statements and two different connectives.

13.4   Truth Tables for the Conditional and the Biconditional

Objectives

1   Understand the logic behind the definition of the conditional.

2   Construct truth tables for conditional statements.

3   Understand the definition of the biconditional.

4   Construct truth tables for biconditional statements.

5   Determine the truth value of a compound statement for a specific case.

Your author received junk mail with this claim:

> If your Super Million Dollar Prize Entry Number matches the winning preselected number and you return the number before the deadline stated below, you will win $1,000,000.00.

Should he obediently return the number before the deadline or trash the whole thing?

In this section, we will use logic to analyze the claim in the junk mail. By understanding when statements involving the conditional, $\rightarrow$ (if–then), and the biconditional, $\leftrightarrow$ (if and only if), are true and when they are false, you will be able to determine the truth value of the claim.

Understand the logic behind the definition of the conditional.

Conditional Statements, →

We begin by looking at the truth table for conditional statements. Suppose that your teacher promises you the following:

If you pass the final, then you pass the course.

Break the statement down into its two component statements:

p: You pass the final.

q: You pass the course.

Translated into symbolic form, your teacher's statement is $p \to q$. We now look at the four cases shown in **Table 13.15**, the truth table for the conditional.

Case 1 (T, T) You do pass the final and you do pass the course. Your teacher did what was promised, so the conditional statement is true.

Case 2 (T, F) You pass the final, but you do not pass the course. Your teacher did not do what was promised, so the conditional statement is false.

Case 3 (F, T) You do not pass the final, but you do pass the course. Your teacher's original statement talks about only what would happen if you passed the final. It says nothing about what would happen if you did not pass the final. Your teacher did not break the promise of the original statement, so the conditional statement is true.

Case 4 (F, F) You do not pass the final and you do not pass the course. As with case 3, your teacher's original statement talks about only what would happen if you passed the final. The promise of the original statement has not been broken. Therefore, the conditional statement is true.

Table 13.15 illustrates that a conditional statement is false only when the antecedent, the statement before the → connective, is true and the consequent, the statement after the → connective, is false. A conditional statement is true in all other cases.

Table 13.15 Conditional			
	p	*q*	$p \to q$
Case 1	T	T	T
Case 2	T	F	F
Case 3	F	T	T
Case 4	F	F	T

The Definition of the Conditional

A conditional is false only when the antecedent is true and the consequent is false.

p	*q*	$p \to q$
T	T	T
T	F	F
F	T	T
F	F	T

Construct truth tables for conditional statements.

Constructing Truth Tables

Our first example shows how truth tables can be used to gain a better understanding of conditional statements.

Example 1 **Constructing a Truth Table**

Construct a truth table for

$$\sim q \to \sim p$$

to determine when the statement is true and when the statement is false.

Solution

Remember that without parentheses, the symbol ~ negates only the statement that immediately follows it. Therefore, we cannot determine the truth values for this conditional statement until we first determine the truth values for $\sim q$ and for $\sim p$, the statements before and after the $\rightarrow$ connective.

Step 1   List the simple statements on top and show the four possible cases for the truth values.

p	q	
T	T	
T	F	
F	T	
F	F	

Step 2   Make columns for $\sim q$ and for $\sim p$. Fill in the $\sim q$ column by looking back at the q column, the second column, and taking the opposite of the truth values in this column. Fill in the $\sim p$ column by taking the opposite of the truth values in the first column, the p column.

Opposite truth values

p	q	$\sim q$	$\sim p$
T	T	F	F
T	F	T	F
F	T	F	T
F	F	T	T

Opposite truth values

Step 3   Construct one more column for $\sim q \rightarrow \sim p$. Look back at the $\sim q$ column, column 3, and the $\sim p$ column, column 4. Now use the conditional definition to determine the truth values for $\sim q \rightarrow \sim p$ based on columns 3 and 4. Conditional definition: An *if–then* statement is false only when the antecedent is true and the consequent is false. This occurs only in the second row.

p	q	$\sim q$	$\sim p$	$\sim q \rightarrow \sim p$
T	T	F	F	T
T	F	T	F	F
F	T	F	T	T
F	F	T	T	T

$\sim q$ is true and $\sim p$ is false, so $\sim q \rightarrow \sim p$ is false.

column 3 → column 4

Table 13.16

p	q	$p \rightarrow q$	$\sim q \rightarrow \sim p$
T	T	T	T
T	F	F	F
F	T	T	T
F	F	T	T

$p \rightarrow q$ and $\sim q \rightarrow \sim p$ have the same truth values.

Checkpoint 1   Construct a truth table for $\sim p \rightarrow \sim q$ to determine when the statement is true and when the statement is false.

The truth values for $p \rightarrow q$, as well as those for $\sim q \rightarrow \sim p$ from Example 1, are shown in **Table 13.16**. Notice that $p \rightarrow q$ and $\sim q \rightarrow \sim p$ have the same truth value in each of the four cases. What does this mean? **Every time you hear or utter a conditional statement, you can reverse and negate the antecedent and consequent,**

and the statement's truth value will not change. Here's an example from a student providing advice on high school fashion:

- If you're cool, you won't wear clothing with a brand name on it.
- If you wear clothing with a brand name on it, you're not cool.

> If the fashion tip above is true then so is this, and if it's false then this is false as well.

We'll have lots more to say about this (that is, variations of conditional statements, not tips on dressing up and down around school) in the next section.

Example 2   Constructing a Truth Table

Construct a truth table for

$$[(p \lor q) \land \sim p] \to q$$

to determine when the statement is true and when the statement is false.

Solution

The statement is a conditional statement because the *if–then* symbol, $\to$, is outside the grouping symbols. We cannot determine the truth values for this conditional until we first determine the truth values for the statements before and after the $\to$ connective.

$$\boxed{[(p \lor q) \land \sim p]} \quad \to \quad \boxed{q}$$

> We'll need a column with truth values for this statement. Prior to this column, we'll need columns for $p \lor q$ and for $\sim p$.

> We'll need a column with truth values for this statement. This will be the second column of the truth table.

> The disjunction, $\lor$, is false only when both component statements are false.

> The truth value of $\sim p$ is opposite that of p.

> The conjunction, $\land$, is true only when both $p \lor q$ and $\sim p$ are true.

> The conditional, $\to$, is false only when $(p \lor q) \land \sim p$ is true and q is false.

Show four possible cases.

p	q	$p \lor q$	$\sim p$	$(p \lor q) \land \sim p$	$[(p \lor q) \land \sim p] \to q$
T	T	T	F	F	T
T	F	T	F	F	T
F	T	T	T	T	T
F	F	F	T	F	T

The completed truth table shows that the conditional statement in the last column, $[(p \lor q) \land \sim p] \to q$, is true in all cases.

In Section 13.3, we defined a **tautology** as a compound statement that is always true. Example 2 proves that the conditional statement $[(p \lor q) \land \sim p] \to q$ is a tautology.

Conditional statements that are tautologies are called **implications**. For the conditional statement

$$[(p \lor q) \land \sim p] \to q$$

we can say that

$$(p \lor q) \land \sim p \text{ implies } q.$$

Using *p:* I am visiting London and *q:* I am visiting Paris, we can say that

I am visiting London or Paris, and I am not visiting London, implies that I am visiting Paris.

✓ **Checkpoint 2**   Construct a truth table for $[(p \rightarrow q) \wedge \sim q] \rightarrow \sim p$ and show
▶ that the compound statement is a tautology.

Some compound statements are false in all possible cases. Such statements are called **self-contradictions**. An example of a self-contradiction is the statement $p \wedge \sim p$:

p	~*p*	*p* ∧ ~*p*
T	F	F
F	T	F

$p \wedge \sim p$ is always false.

If *p* represents "I am going," then $p \wedge \sim p$ translates as "I am going and I am not going." Such a translation sounds like a contradiction.

Example 3   Constructing a Truth Table with Eight Cases

The following is from an editorial that appeared in *The New York Times*:

Our entire tax system depends upon the vast majority of taxpayers who attempt to pay the taxes they owe having confidence that they're being treated fairly and that their competitors and neighbors are also paying what is due. <u>If the public concludes that the IRS cannot meet these basic expectations, the risk to the tax system will become very high and the effects very difficult to reverse.</u>

The New York Times, February 13, 2000

a. Construct a truth table for the underlined statement.

b. Suppose that the public concludes that the IRS cannot meet basic expectations, the risk to the tax system becomes very high, but the effects are not very difficult to reverse. Under these conditions, is the underlined statement true or false?

Solution

a. We begin by assigning letters to the simple statements. Use the following representations:

> *p:* The public concludes that the IRS *can* meet basic expectations (of fair treatment and others paying what is due).
>
> *q:* The risk to the tax system will become very high.
>
> *r:* The effects will be very difficult to reverse.

The underlined statement in symbolic form is

$$\sim p \quad \rightarrow \quad (q \quad \wedge \quad r).$$

... cannot meet basic expectations   ... high risk   ... difficult to reverse effects

The statement $\sim p \rightarrow (q \wedge r)$ is a conditional statement because the *if–then* symbol, $\rightarrow$, is outside the parentheses. We cannot determine the truth values for this conditional until we have determined the truth values for $\sim p$ and for $q \wedge r$, the statements before and after the $\rightarrow$ connective. Because the compound statement $\sim p \to (q \wedge r)$ consists of three simple statements, represented by p, q, and r, the truth table must contain eight cases. The completed truth table appears as follows:

The conjunction is true only when q, column 2, is true and r, column 3, is true.

The conditional is false only when the $\sim p$ column is true and the $q \wedge r$ column is false.

Take the opposite of the truth values in column 1.

Show eight possible cases.

p	q	r	$\sim p$	$q \wedge r$	$\sim p \to (q \wedge r)$
T	T	T	F	T	T
T	T	F	F	F	T
T	F	T	F	F	T
T	F	F	F	F	T
F	T	T	T	T	T
F	T	F	T	F	F
F	F	T	T	F	F
F	F	F	T	F	F

These are the conditions in part (b).

b. We are given that p (. . . can meet basic expectations) is false, q (. . . high risk) is true, and r (. . . difficult to reverse effects) is false. The given conditions, F T F, correspond to case 6 of the truth table, indicated by the voice balloon on the far left. Under these conditions, the original compound statement is false, shown by the red F in the truth table.

Checkpoint 3   An advertisement makes the following claim:

If you use Hair Grow and apply it daily, then you will not go bald.

a. Construct a truth table for the claim.

b. Suppose you use Hair Grow, forget to apply it every day, and you go bald. Under these conditions, is the claim in the advertisement false?

Biconditional Statements

3 Understand the definition of the biconditional.

In Section 13.2, we introduced the biconditional connective, $\leftrightarrow$, translated as "if and only if." The biconditional statement $p \leftrightarrow q$ means that $p \to q$ and $q \to p$. We write this symbolically as

$$(p \to q) \wedge (q \to p).$$

To create the truth table for $p \leftrightarrow q$, we will first make a truth table for the conjunction of the two conditionals $p \to q$ and $q \to p$. The truth table for $(p \to q) \wedge (q \to p)$ is shown as follows:

The conditional is false only when p is true and q is false.

The conditional is false only when q is true and p is false.

The conjunction is true only when both $p \to q$ and $q \to p$ are true.

Show four possible cases.

p	q	$p \to q$	$q \to p$	$(p \to q) \wedge (q \to p)$
T	T	T	T	T
T	F	F	T	F
F	T	T	F	F
F	F	T	T	T

col. 1 $\to$ col. 2   col. 2 $\to$ col. 1   col. 3 $\wedge$ col. 4

The truth values in the column for $(p \rightarrow q) \wedge (q \rightarrow p)$ show the truth values for the biconditional statement $p \leftrightarrow q$.

The Definition of the Biconditional			
A biconditional is true only when the component statements have the same truth value.	**p**	**q**	**p ↔ q**
	T	T	T
	T	F	F
	F	T	F
	F	F	T

Before we continue our work with truth tables, let's take a moment to summarize the basic definitions of symbolic logic.

The Definitions of Symbolic Logic

1. Negation ~: not
   The negation of a statement has the opposite meaning, as well as the opposite truth value, from the statement.

2. Conjunction ∧: and
   The only case in which a conjunction is true is when both component statements are true.

3. Disjunction ∨: or
   The only case in which a disjunction is false is when both component statements are false.

4. Conditional →: if–then
   The only case in which a conditional is false is when the first component statement, the antecedent, is true and the second component statement, the consequent, is false.

5. Biconditional ↔: if and only if
   A biconditional is true only when the component statements have the same truth value.

4 Construct truth tables for biconditional statements.

Example 4   Constructing a Truth Table

Construct a truth table for

$$(p \vee q) \leftrightarrow (\sim q \rightarrow p)$$

to determine whether the statement is a tautology.

Solution

The statement is a biconditional because the biconditional symbol, ↔, is outside the parentheses. We cannot determine the truth values for this biconditional until we determine the truth values for the statements in parentheses.

$$\boxed{(p \vee q)} \quad \leftrightarrow \quad \boxed{(\sim q \rightarrow p)}$$

We need a column with truth values for this statement.   We need a column with truth values for this statement.

The completed truth table for $(p \vee q) \leftrightarrow (\sim q \rightarrow p)$ appears as follows:

p	q	$p \vee q$	$\sim q$	$\sim q \rightarrow p$	$(p \vee q) \leftrightarrow (\sim q \rightarrow p)$
T	T	T	F	T	T
T	F	T	T	T	T
F	T	T	F	T	T
F	F	F	T	F	T

col. 1 $\vee$ col. 2 $\sim$ col. 2   col. 4 $\rightarrow$ col. 1   col. 3 $\leftrightarrow$ col. 5

We applied the definition of the biconditional to fill in the last column. In each case, the truth values of $p \vee q$ and $\sim q \rightarrow p$ are the same. Therefore, the biconditional $(p \vee q) \leftrightarrow (\sim q \rightarrow p)$ is true in each case. Because all cases are true, the biconditional is a tautology.

✓ **Checkpoint 4**   Construct a truth table for $(p \vee q) \leftrightarrow (\sim p \rightarrow q)$ to determine whether the statement is a tautology.

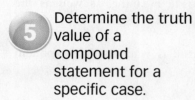
Determine the truth value of a compound statement for a specific case.

Example 5   Determining the Truth Value of a Compound Statement

Your author recently received a letter from his credit card company that began as follows:

Dear Mr. Bob Blitzer,

I am pleased to inform you that a personal Super Million Dollar Prize Entry Number—665567010—has been assigned in your name as indicated above. <u>If your Super Million Dollar Prize Entry Number matches the winning preselected number and you return the number before the deadline stated below, you will win $1,000,000.00</u>. It's as simple as that.

Consider the claim in the underlined conditional statement. Suppose that your Super Million Dollar Prize Entry Number does not match the winning preselected number (those dirty rotten scoundrels!), you obediently return the number before the deadline, and you win only a free issue of a magazine (with the remaining 11 issues billed to your credit card). Under these conditions, can you sue the credit card company for making a false claim?

Solution

Let's begin by assigning letters to the simple statements in the claim. We'll also indicate the truth value of each simple statement.

p: Your Super Million Dollar Prize Entry Number   *false*
matches the winning preselected number.

q: You return the number before the stated deadline.   *true*

r: You win $1,000,000.00.   *false; To make matters worse, you were duped into buying a magazine subscription.*

Now we can write the underlined claim in the letter to Bob in symbolic form.

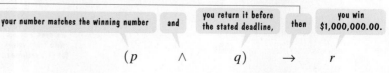

We substitute the truth values for p, q, and r to determine the truth value for the credit card company's claim.

$(p \wedge q) \rightarrow r$ This is the claim in symbolic form.

$(F \wedge T) \rightarrow F$ Substitute the truth values for the simple statements.

$F \rightarrow F$ Replace F $\wedge$ T with F. Conjunction is false when one part is false.

T   Replace F $\rightarrow$ F with T. The conditional is true with a false antecedent and a false consequent.

Our truth-value analysis indicates that you cannot sue the credit card company for making a false claim. Call it conditional trickery, but the company's claim is true.

✓ **Checkpoint 5**   Consider the underlined claim in the letter in Example 5. Suppose that your number actually matches the winning preselected number, you do not return the number, and you win nothing. Under these conditions, determine ▶ the claim's truth value.

Blitzer Bonus
Conditional Wishful Thinking

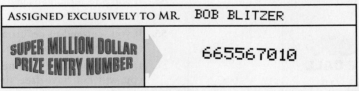

Bob's credit card company is too kind. It even offers options as to how he wants to receive his million-dollar winnings. With this lure, people who do not think carefully might interpret the conditional claim in the letter to read as follows:

If you return your winning number and do so before the stated deadline, you win $1,000,000.00.

This misreading is wishful thinking. There is no winning number to return. What there is, of course, is a deceptive attempt to sell magazines.

Exercise Set 13.4

Concept and Vocabulary Exercises

In Exercises 1–3, fill in each blank so that the resulting statement is true.

1. A conditional statement, $p \rightarrow q$, is false only when _____.

2. A compound statement that is always true is called a/an _____. Conditional statements that are always true are called _____. A compound statement that is always false is called a/an _____.

3. A biconditional statement, $p \leftrightarrow q$, is true only when _____.

In Exercises 4–7, determine whether each statement is true or false. If the statement is false, make the necessary change(s) to produce a true statement.

4. A conditional statement is false only when the consequent is true and the antecedent is false.

5. Some implications are not tautologies.

6. An equivalent form for a conditional statement is obtained by reversing and negating the antecedent and consequent.

7. A compound statement consisting of two simple statements that are both false can be true.

Respond to Exercises 8–10 using verbal or written explanations.

8. Explain when conditional statements are true and when they are false.

9. Explain when biconditional statements are true and when they are false.

10. What is the difference between a tautology and a self-contradiction?

In Exercises 11–14, determine whether each statement makes sense or does not make sense, and explain your reasoning.

11. The statement "If $2 + 2 = 5$, then the moon is made of green cheese" is true in logic, but does not make much sense in everyday speech.

12. I'm working with a true conditional statement, but when I reverse the antecedent and the consequent, my new conditional statement is no longer true.

13. I'm working with a true conditional statement, but when I reverse and negate the antecedent and the consequent, my new conditional statement is no longer true.

14. This ad was submitted to the Editorial Department of *Consumer Reports.* Using

p: There is 10% off on service calls,

the ad represents the self-contradiction $p \wedge \sim p$.

Practice Exercises

In Exercises 15–38, construct a truth table for the given statement.

15. $p \rightarrow \sim q$
16. $\sim p \rightarrow q$
17. $\sim(q \rightarrow p)$
18. $\sim(p \rightarrow q)$
19. $(p \wedge q) \rightarrow (p \vee q)$
20. $(p \vee q) \rightarrow (p \wedge q)$
21. $(p \rightarrow q) \wedge \sim q$
22. $(p \rightarrow q) \wedge \sim p$
23. $(p \vee q) \rightarrow r$
24. $p \rightarrow (q \vee r)$
25. $\sim r \wedge (\sim q \rightarrow p)$
26. $\sim r \wedge (q \rightarrow \sim p)$
27. $p \leftrightarrow \sim q$
28. $\sim p \leftrightarrow q$
29. $\sim(p \leftrightarrow q)$
30. $\sim(q \leftrightarrow p)$
31. $(p \leftrightarrow q) \rightarrow p$
32. $(p \leftrightarrow q) \rightarrow q$
33. $(\sim p \leftrightarrow q) \rightarrow (\sim p \rightarrow q)$
34. $(p \leftrightarrow \sim q) \rightarrow (q \rightarrow \sim p)$
35. $(p \leftrightarrow q) \rightarrow \sim r$
36. $(p \rightarrow q) \leftrightarrow \sim r$
37. $(p \wedge r) \leftrightarrow \sim(q \vee r)$
38. $(p \vee r) \leftrightarrow \sim(q \wedge r)$

In Exercises 39–52, use a truth table to determine whether each statement is a tautology, a self-contradiction, or neither.

39. $[(p \rightarrow q) \wedge q] \rightarrow p$
40. $[(p \rightarrow q) \wedge p] \rightarrow q$
41. $[(p \rightarrow q) \wedge \sim q] \rightarrow \sim p$
42. $[(p \rightarrow q) \wedge \sim p] \rightarrow \sim q$
43. $[(p \vee q) \wedge p] \rightarrow \sim q$
44. $[(p \vee q) \wedge \sim q] \rightarrow p$
45. $(p \rightarrow q) \leftrightarrow (q \rightarrow p)$
46. $(p \rightarrow q) \leftrightarrow (\sim p \rightarrow \sim q)$
47. $(p \rightarrow q) \leftrightarrow (\sim p \vee q)$
48. $(p \rightarrow q) \leftrightarrow (p \vee \sim q)$
49. $(p \leftrightarrow q) \leftrightarrow [(q \rightarrow p) \wedge (p \rightarrow q)]$
50. $(q \leftrightarrow p) \leftrightarrow [(p \rightarrow q) \wedge (q \rightarrow p)]$
51. $[(p \rightarrow q) \wedge (q \rightarrow r)] \rightarrow (p \rightarrow r)$
52. $[(p \rightarrow q) \wedge (q \rightarrow r)] \rightarrow (\sim r \rightarrow \sim p)$

Exercises 53–58 involve statements related to college. In each exercise,

a. *Write each statement in symbolic form. Assign letters to simple statements that are not negated.*

b. *Construct a truth table for the symbolic statement in part (a).*

c. *Use the truth table to indicate one set of conditions that makes the compound statement false, or state that no such conditions exist.*

53. If you do homework right after class then you will not fall behind, and if you do not do homework right after class then you will.

54. If you do a little bit each day then you'll get by, and if you do not do a little bit each day then you won't.

55. If you "cut-and-paste" from the Internet and do not cite the source, then you will be charged with plagiarism.

56. If you take more than one class with a lot of reading, then you will not have free time and you'll be in the library until 1 A.M.

57. You'll be comfortable in your room if and only if you're honest with your roommate, or you won't enjoy the college experience.

58. I enjoy the course if and only if I choose the class based on the teacher and not the course description.

In Exercises 59–66, determine the truth value for each statement when p is false, q is true, and r is false.

59. $\sim(p \rightarrow q)$
60. $\sim(p \leftrightarrow q)$
61. $\sim p \leftrightarrow q$
62. $\sim p \rightarrow q$
63. $q \rightarrow (p \wedge r)$
64. $(p \wedge r) \rightarrow q$
65. $(\sim p \wedge q) \leftrightarrow \sim r$
66. $\sim p \leftrightarrow (\sim q \wedge r)$

Practice Plus

In Exercises 67–70, construct a truth table for each statement. Then use the table to indicate one set of conditions that make the compound statement false, or state that no such conditions exist.

67. Loving a person is necessary for marrying that person, but not loving someone is sufficient for not marrying that person.

68. Studying hard is necessary for getting an A, but not studying hard is sufficient for not getting an A.

69. It is not true that being happy and living contentedly are necessary conditions for being wealthy.

70. It is not true that being wealthy is a sufficient condition for being happy and living contentedly.

Application Exercises

The bar graph shows the percentage of American adults who believed in God, Heaven, the devil, and Hell in 2008 compared with 2003.

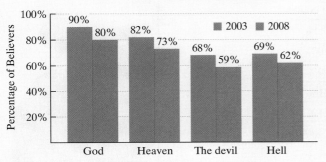

Percentage of American Adults Believing in God, Heaven, the Devil, and Hell

Source: Harris Interactive poll of 2201 adults Jan. 21–27, 2003, and 2126 adults Nov. 10–17, 2008

In Exercises 71–74, write each statement in symbolic form. (Increases or decreases in each simple statement refer to 2008 compared with 2003.) Then use the information in the graph to determine the truth value of each compound statement.

71. If there was an increase in the percentage who believed in God and a decrease in the percentage who believed in Heaven, then there was an increase in the percentage who believed in the devil.

72. If there was a decrease in the percentage who believed in God, then it is not the case that there was an increase in the percentage who believed in the devil or in Hell.

73. There was a decrease in the percentage who believed in God if and only if there was an increase in the percentage who believed in Heaven, or the percentage believing in the devil decreased.

74. There was an increase in the percentage who believed in God if and only if there was an increase in the percentage who believed in Heaven, and the percentage believing in Hell decreased.

Sociologists Joseph Kahl and Dennis Gilbert developed a six-tier model to portray the class structure of the United States. The bar graph gives the percentage of Americans who are members of each of the six social classes.

The United States Social Class Ladder

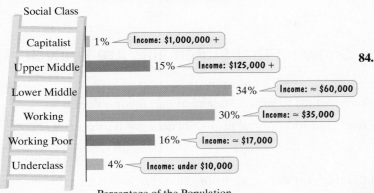

Source: James Henslin, *Sociology*, Eighth Edition, Allyn and Bacon, 2007.

In Exercises 75–78, write each statement in symbolic form. Then use the information given by the graph at the bottom of the previous column to determine the truth value of each compound statement.

75. Fifteen percent are capitalists or 34% are not members of the upper-middle class, if and only if the number of working poor exceeds the number belonging to the working class.

76. Fifteen percent are capitalists and 34% are not members of the upper-middle class, if and only if the number of working poor exceeds the number belonging to the working class.

77. If there are more people in the lower-middle class than in the capitalist and upper-middle classes combined, then 1% are capitalists and 34% belong to the upper-middle class.

78. If there are more people in the lower-middle class than in the capitalist and upper-middle classes combined, then 1% are capitalists or 34% belong to the upper-middle class.

79. When asked the question "What is time?", the fourth-century Christian philosopher St. Augustine replied,

"If you don't ask me, I know, but if you ask me, I don't know."

Construct a truth table for St. Augustine's statement and determine whether or not it is a tautology.

80. In "Computing Machines and Intelligence," the English mathematician Alan Turing (1912–1954) wrote,

"If each man had a definite set of rules of conduct by which he regulated his life, he would be a machine, but there are no such rules, so men cannot be machines."

Construct a truth table for Turing's statement and determine whether or not it is a tautology.

Critical Thinking Exercises

81. Based on the meaning of the inclusive *or*, explain why it is reasonable that if $p \lor q$ is true, then $\sim p \to q$ must also be true.

82. Based on the meaning of the inclusive *or*, explain why if $p \lor q$ is true, then $p \to \sim q$ is not necessarily true.

In Exercises 83–84, the headings for the columns in the truth tables are missing. Fill in the statements to replace the missing headings. (More than one correct statement may be possible.)

83.

> Do not repeat the statement from the third column.

p	q				
T	T	T	F	T	T
T	F	F	F	F	T
F	T	T	T	T	T
F	F	T	T	T	T

84.

> Do not repeat the previous statement.

p	q				
T	T	T	T	F	F
T	F	T	F	T	T
F	T	T	F	T	T
F	F	F	F	T	T

13.5 Equivalent Statements and Variations of Conditional Statements

Objectives

1. Use a truth table to show that statements are equivalent.

2. Write the contrapositive for a conditional statement.

3. Write the converse and inverse of a conditional statement.

Top U.S. Last Names

Name	% of All Names
Smith	1.006%
Johnson	0.810%
Williams	0.699%
Brown	0.621%
Jones	0.621%

Source: Russell Ash, *The Top 10 of Everything*

The top U.S. last names shown above make up more than $3\frac{3}{4}\%$ of the entire population. One American in every 26 bears one of the names in the table. The list indicates that the following statement is true:

> If your last name is Brown, your name is shared by 0.621% of the population.

Venus and Serena Williams

Does this mean that if your name is shared by 0.621% of the population, then it's Brown? If your last name isn't Brown, can we conclude that it's not shared by 0.621% of the population? Furthermore, if your name is not shared by 0.621% of the population, can we conclude that it's not Brown? In this section, we will use truth tables and logic to unravel this verbal morass of conditional statements.

Equivalent Statements

Equivalent compound statements are made up of the same simple statements and have the same corresponding truth values for all true–false combinations of these simple statements. If a compound statement is true, then its equivalent statement must also be true. Likewise, if a compound statement is false, its equivalent statement must also be false.

Truth tables are used to show that two statements are equivalent. When translated into English, equivalencies can be used to gain a better understanding of English statements.

1. Use a truth table to show that statements are equivalent.

Example 1   Showing That Statements Are Equivalent

a. Show that $p \lor \sim q$ and $\sim p \to \sim q$ are equivalent.

b. Use the result from part (a) to write a statement that is equivalent to

> The bill receives majority approval or the bill does not become law.

Solution

a. Construct a truth table that shows the truth values for $p \lor \sim q$ and $\sim p \to \sim q$. The truth values for each statement are shown at the top of the next page.

p	q	$\sim q$	$p \vee \sim q$	$\sim p$	$\sim p \to \sim q$
T	T	F	T	F	T
T	F	T	T	F	T
F	T	F	F	T	F
F	F	T	T	T	T

Corresponding truth
values are the same.

The table shows that the truth values for $p \vee \sim q$ and $\sim p \to \sim q$ are the same. Therefore, the statements are equivalent.

b. The statement

The bill receives majority approval or the bill does not become law

can be expressed in symbolic form using the following representations:

p: The bill receives majority approval.

q: The bill becomes law.

In symbolic form, the statement is $p \vee \sim q$. Based on the truth table in part (a), we know that an equivalent statement is $\sim p \to \sim q$. The equivalent statement can be expressed in words as

If the bill does not receive majority approval, then the bill does not become law.

Notice that the given statement and its equivalent are both true.

Checkpoint 1

a. Show that $p \vee q$ and $\sim q \to p$ are equivalent.

b. Use the result from part (a) to write a statement that is equivalent to

I attend classes or I fail the course.

A special symbol, $\equiv$, is used to show that two statements are equivalent. Because $p \vee \sim q$ and $\sim p \to \sim q$ are equivalent, we can write

$$p \vee \sim q \equiv \sim p \to \sim q \quad \text{or} \quad \sim p \to \sim q \equiv p \vee \sim q.$$

Example 2 Showing That Statements Are Equivalent

Show that $\sim(\sim p) \equiv p$.

Solution

Determine the truth values for $\sim(\sim p)$ and p. These are shown in the truth table at the left.

The truth values for $\sim(\sim p)$ were obtained by taking the opposite of each truth value for $\sim p$. The table shows that the truth values for $\sim(\sim p)$ and p are the same. Therefore, the statements are equivalent:

$$\sim(\sim p) \equiv p.$$

The equivalence in Example 2 illustrates that **the double negation of a statement is equivalent to the statement.** For example, the statement "It is not true that Ernest Hemingway was not a writer" means the same thing as "Ernest Hemingway was a writer."

p	$\sim p$	$\sim(\sim p)$
T	F	T
F	T	F

Corresponding truth
values are the same.

Study Tip

A familiar algebraic statement should help you remember the logical equivalency

$$\sim(\sim p) \equiv p.$$

If a represents a number, then

$$-(-a) = a.$$

For example, $-(-4) = 4.$

Checkpoint 2 Show that $\sim[\sim(\sim p)] \equiv \sim p$.

Example 3   Equivalencies and Truth Tables

Select the statement that is not equivalent to

Miguel is blushing or sunburned.

a. If Miguel is blushing, then he is not sunburned.
b. Miguel is sunburned or blushing.
c. If Miguel is not blushing, then he is sunburned.
d. If Miguel is not sunburned, then he is blushing.

Solution

To determine which of the choices is not equivalent to the given statement, begin by writing the given statement and the choices in symbolic form. Then construct a truth table and compare each statement's truth values to those of the given statement. The nonequivalent statement is the one that does not have exactly the same truth values as the given statement.

The simple statements that make up "Miguel is blushing or sunburned" can be represented as follows:

p: Miguel is blushing.

q: Miguel is sunburned.

Here are the symbolic representations for the given statement and the four choices:

Miguel is blushing or sunburned: $p \vee q$.

a. If Miguel is blushing, then he is not sunburned: $p \rightarrow \sim q$.
b. Miguel is sunburned or blushing: $q \vee p$.
c. If Miguel is not blushing, then he is sunburned: $\sim p \rightarrow q$.
d. If Miguel is not sunburned, then he is blushing: $\sim q \rightarrow p$.

Next, construct a truth table that contains the truth values for the given statement, $p \vee q$, as well as those for the four options. The truth table is shown as follows:

Equivalent (same corresponding truth values)

		Given		(a)	(b)		(c)	(d)
p	q	$p \vee q$	$\sim q$	$p \rightarrow \sim q$	$q \vee p$	$\sim p$	$\sim p \rightarrow q$	$\sim q \rightarrow p$
T	T	T	F	F	T	F	T	T
T	F	T	T	T	T	F	T	T
F	T	T	F	T	T	T	T	T
F	F	F	T	T	F	T	F	F

Not equivalent

The statement in option (a) does not have the same corresponding truth values as those for $p \vee q$. Therefore, this statement is not equivalent to the given statement.

In Example 3, we used a truth table to show that $p \vee q$ and $p \rightarrow \sim q$ are not equivalent. We can use our understanding of the inclusive *or* to see why the following English translations for these symbolic statements are not equivalent:

Miguel is blushing or sunburned.

If Miguel is blushing, then he is not sunburned.

Let us assume that the first statement is true. The inclusive *or* tells us that Miguel might be both blushing and sunburned. This means that the second statement might not be true. The fact that Miguel is blushing does not indicate that he is not sunburned; he might be both.

> ✓ **Checkpoint 3** Select the statement that is not equivalent to
>
> If it's raining, then I need a jacket.
>
> **a.** It's not raining or I need a jacket.
> **b.** I need a jacket or it's not raining.
> **c.** If I need a jacket, then it's raining.
> **d.** If I do not need a jacket, then it's not raining.

 2 Write the contrapositive for a conditional statement.

Variations of the Conditional Statement $p \to q$

In Section 13.4, we learned that $p \to q$ is equivalent to $\sim q \to \sim p$. The truth value of a conditional statement does not change if the antecedent and the consequent are reversed and then both of them are negated. The **contrapositive** of a conditional statement is a statement obtained by reversing and negating the antecedent and the consequent.

p	q	$p \to q$	$\sim q \to \sim p$
T	T	T	T
T	F	F	F
F	T	T	T
F	F	T	T

$p \to q$ and $\sim q \to \sim p$ are equivalent.

> **A Conditional Statement and Its Equivalent Contrapositive**
>
> $$p \to q \equiv \sim q \to \sim p$$
>
> The truth value of a conditional statement does not change if the antecedent and consequent are reversed and both are negated. The statement $\sim q \to \sim p$ is called the **contrapositive** of the conditional $p \to q$.

Example 4   Writing Equivalent Contrapositives

Write the contrapositive for each of the following statements:

a. If you live in Los Angeles, then you live in California.
b. If the patient is not breathing, then the patient is dead.
c. If all people obey the law, then prisons are not needed.
d. $\sim(p \wedge q) \to r$

Solution

In parts (a)–(c), we write each statement in symbolic form. Then we form the contrapositive by reversing and negating the antecedent and the consequent. Finally, we translate the symbolic form of the contrapositive back into English.

a. Use the following representations:

p: You live in Los Angeles.

q: You live in California.

If you live in Los Angeles, then you live in California.
$\quad\quad\quad\quad p \quad\quad\quad\quad\quad\quad \to \quad\quad\quad q$

$p \to q$ This is the statement's symbolic form.

$\sim q \to \sim p$ Form the contrapositive: Reverse and negate the components.

Translating $\sim q \to \sim p$ into English, the contrapositive is

If you do not live in California, then you do not live in Los Angeles.

Notice that the given conditional statement and its contrapositive are both true.

b. Use the following representations for "If the patient is not breathing, then the patient is dead."

p: The patient is breathing.

q: The patient is dead.

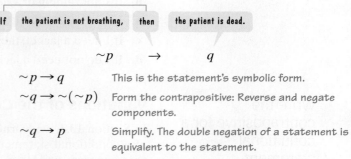

$\sim p \to q$	This is the statement's symbolic form.
$\sim q \to \sim(\sim p)$	Form the contrapositive: Reverse and negate components.
$\sim q \to p$	Simplify. The double negation of a statement is equivalent to the statement.

Translating $\sim q \to p$ into English, the contrapositive is

If the patient is not dead, then the patient is breathing.

c. Use the following representations for "If all people obey the law, then prisons are not needed."

p: All people obey the law.

q: Prisons are needed.

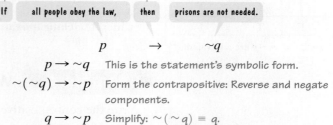

$p \to \sim q$	This is the statement's symbolic form.
$\sim(\sim q) \to \sim p$	Form the contrapositive: Reverse and negate components.
$q \to \sim p$	Simplify: $\sim(\sim q) \equiv q$.

Recall, as shown in the margin, that the negation of *all* is *some . . . not*. Using this negation and translating $q \to \sim p$ into English, the contrapositive is

If prisons are needed, then some people do not obey the law.

d.

$\sim(p \wedge q) \to r$	This is the given symbolic statement.
$\sim r \to \sim[\sim(p \wedge q)]$	Form the contrapositive: Reverse and negate the components.
$\sim r \to (p \wedge q)$	Simplify: $\sim[\sim(p \wedge q)] \equiv p \wedge q$.

The contrapositive of $\sim(p \wedge q) \to r$ is $\sim r \to (p \wedge q)$.

> ✓ **Checkpoint 4**   Write the contrapositive for each of the following statements:
>
> **a.** If you can read this, then you're driving too closely.
> **b.** If you do not have clean underwear, it's time to do the laundry.
> **c.** If all students are honest, then supervision during exams is not required.
> **d.** $\sim(p \vee r) \to \sim q$

Negations of Quantified Statements

All   No

Some   Some ... not

③ Write the converse and inverse of a conditional statement.

The truth value of a conditional statement does not change if the antecedent and the consequent are reversed and then both of them are negated. But what happens to the conditional's truth value if just one, but not both, of these changes is made? If the antecedent and the consequent are reversed but not negated, the resulting statement is called the **converse** of the conditional statement. By negating both the antecedent and the consequent but not reversing them, we obtain the **inverse** of the conditional statement.

Variations of the Conditional Statement		
Name	**Symbolic Form**	**English Translation**
Conditional	$p \rightarrow q$	If p, then q.
Converse	$q \rightarrow p$	If q, then p.
Inverse	$\sim p \rightarrow \sim q$	If not p, then not q.
Contrapositive	$\sim q \rightarrow \sim p$	If not q, then not p.

Let's see what happens to the truth value of a true conditional statement when we form its converse and its inverse.

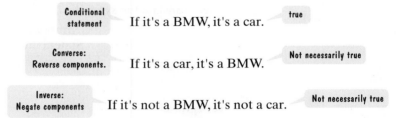

Conditional statement — If it's a BMW, it's a car. — true

Converse: Reverse components. — If it's a car, it's a BMW. — Not necessarily true

Inverse: Negate components — If it's not a BMW, it's not a car. — Not necessarily true

These statements illustrate that if a conditional statement is true, its converse and inverse are not necessarily true. Because the equivalent of a true statement must be true, we see that a conditional statement is not equivalent to its converse or its inverse.

The relationships among the truth values for a conditional statement, its converse, its inverse, and its contrapositive are shown in the truth table that follows:

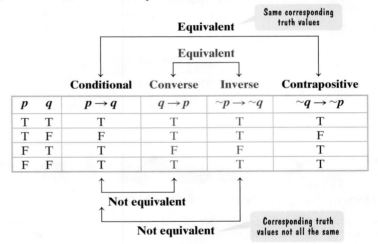

Equivalent — Same corresponding truth values

Equivalent

p	q	Conditional $p \rightarrow q$	Converse $q \rightarrow p$	Inverse $\sim p \rightarrow \sim q$	Contrapositive $\sim q \rightarrow \sim p$
T	T	T	T	T	T
T	F	F	T	T	F
F	T	T	F	F	T
F	F	T	T	T	T

Not equivalent

Not equivalent — Corresponding truth values not all the same

The truth table confirms that a conditional statement is equivalent to its contrapositive. The table also shows that a conditional statement is not equivalent to its converse; in some cases they have the same truth value, but in other cases they have opposite truth values. Also, a conditional statement is not equivalent to its inverse. By contrast, the converse and the inverse are equivalent to each other.

Example 5　Writing Variations of a Conditional Statement

The following conditional statement regarding state and national elections in the United States is true:

If you are 17, then you are not eligible to vote.

(In 1971, the voting age was lowered from 21 to 18 with the ratification of the 26th Amendment.) Write the statement's converse, inverse, and contrapositive.

Solution

Use the following representations for "If you are 17, then you are not eligible to vote."

p: You are 17.

q: You are eligible to vote.

$$p \quad \rightarrow \quad \sim q$$

We now work with $p \rightarrow \sim q$ to form the converse, inverse, and contrapositive. We then translate the symbolic form of each statement back into English.

	Symbolic Statement	English Translation
Given Conditional Statement	$p \rightarrow \sim q$	If you are 17, then you are not eligible to vote. *(true)*
Converse: Reverse the components of $p \rightarrow \sim q$.	$\sim q \rightarrow p$	If you are not eligible to vote, then you are 17. *(not necessarily true)*
Inverse: Negate the components of $p \rightarrow \sim q$.	$\sim p \rightarrow \sim(\sim q)$ simplifies to $\sim p \rightarrow q$	If you are not 17, then you are eligible to vote. *(not necessarily true)*
Contrapositive: Reverse and negate the components of $p \rightarrow \sim q$.	$\sim(\sim q) \rightarrow \sim p$ simplifies to $q \rightarrow \sim p$	If you are eligible to vote, then you are not 17. *(true)*

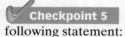 **Checkpoint 5**   Write the converse, inverse, and contrapositive of the following statement:

▶   If you are in New Orleans, then you are not in New England.

Exercise Set 13.5

Concept and Vocabulary Exercises

In Exercises 1–5, fill in each blank so that the resulting statement is true.

1. Compound statements that are made up of the same simple statements and have the same corresponding truth values for all true–false combinations of these simple statements are said to be _____, connected by the symbol _____.

2. The contrapositive of $p \rightarrow q$ is _____.

3. The converse of $p \rightarrow q$ is _____.

4. The inverse of $p \rightarrow q$ is _____.

5. A conditional statement is _____ to its contrapositive, but not to its _____ or its _____.

In Exercises 6–8, determine whether each statement is true or false. If the statement is false, make the necessary change(s) to produce a true statement.

6. Truth tables are used to show that two statements are equivalent.

7. The double negation of a statement is equivalent to the statement's negation.

8. If a conditional statement is true, its inverse must be false.

Respond to Exercises 9–12 using verbal or written explanations.

9. What are equivalent statements?

10. Describe how to determine if two statements are equivalent.

11. Describe how to obtain the contrapositive of a conditional statement.

12. Describe how to obtain the converse and the inverse of a conditional statement.

In Exercises 13–16, determine whether each statement makes sense or does not make sense, and explain your reasoning.

13. A conditional statement can sometimes be true if its contrapositive is false.

14. A conditional statement can never be false if its converse is true.

15. The inverse of a statement's converse is the statement's contrapositive.

16. Groucho Marx stated, "I cannot say that I do not disagree with you," which is equivalent to asserting that I disagree with you.

Practice Exercises

17. **a.** Use a truth table to show that $\sim p \rightarrow q$ and $p \vee q$ are equivalent.

    b. Use the result from part (a) to write a statement that is equivalent to

    If the United States does not energetically support the development of solar-powered cars, then it will suffer increasing atmospheric pollution.

18. **a.** Use a truth table to show that $p \rightarrow q$ and $\sim p \vee q$ are equivalent.

    b. Use the result from part (a) to write a statement that is equivalent to

    If a number is even, then it is divisible by 2.

In Exercises 19–26, use a truth table to determine whether the two statements are equivalent.

19. $\sim p \rightarrow q, q \rightarrow \sim p$ 20. $\sim p \rightarrow q, p \rightarrow \sim q$

21. $(p \rightarrow \sim q) \wedge (\sim q \rightarrow p), p \leftrightarrow \sim q$

22. $(\sim p \rightarrow q) \wedge (q \rightarrow \sim p), \sim p \leftrightarrow q$

23. $(p \vee r) \rightarrow \sim q, (\sim p \wedge \sim r) \rightarrow q$

24. $(p \wedge \sim r) \rightarrow q, (\sim p \vee r) \rightarrow \sim q$

25. $\sim p \rightarrow (q \vee \sim r), (r \wedge \sim q) \rightarrow p$

26. $\sim p \rightarrow (\sim q \wedge r), (\sim r \vee q) \rightarrow p$

27. Select the statement that is equivalent to

    I saw the original *King Kong* or the 2005 version.

    a. If I did not see the original *King Kong*, I saw the 2005 version.

    b. I saw both the original *King Kong* and the 2005 version.

    c. If I saw the original *King Kong*, I did not see the 2005 version.

    d. If I saw the 2005 version, I did not see the original *King Kong*.

28. Select the statement that is equivalent to

    Citizen Kane or *Howard the Duck* appears in a list of greatest U.S. movies.

    a. If *Citizen Kane* appears in the list of greatest U.S. movies, *Howard the Duck* does not.

    b. If *Howard the Duck* does not appear in the list of greatest U.S. movies, then *Citizen Kane* does.

    c. Both *Citizen Kane* and *Howard the Duck* appear in a list of greatest U.S. movies.

    d. If *Howard the Duck* appears in the list of greatest U.S. movies, *Citizen Kane* does not.

29. Select the statement that is *not* equivalent to

    It is not true that Taylor Lautner and Justin Bieber are both actors.

    a. Taylor Lautner is not an actor or Justin Bieber is not an actor.

    b. If Taylor Lautner is an actor, then Justin Bieber is not an actor.

    c. Taylor Lautner is not an actor and Justin Bieber is not an actor.

    d. If Justin Bieber is an actor, then Taylor Lautner is not an actor.

30. Select the statement that is *not* equivalent to

    It is not true that England and Africa are both countries.

    a. If England is a country, then Africa is not a country.

    b. England is not a country and Africa is not a country.

    c. England is not a country or Africa is not a country.

    d. If Africa is a country, then England is not a country.

In Exercises 31–42, write the converse, inverse, and contrapositive of each statement.

31. If I am in Chicago, then I am in Illinois.

32. If I am in Birmingham, then I am in the South.

33. If the stereo is playing, then I cannot hear you.

34. If it is blue, then it is not an apple.

35. "If we don't change, we don't grow." (Gail Sheehy)

36. "If the phone doesn't ring, it's me." (Jimmy Buffett)

37. "If slavery is not wrong, then no things are wrong." (Abraham Lincoln)

38. "If you think your way is best, all new ideas will pass you by." (Akio Morita)

39. "If all people think alike, then some people are not thinking." (George S. Patton)

40. "If all your thoughts interest you, at least one person is pleased." (Katharine Hepburn)

41. $\sim q \rightarrow \sim r$ 42. $\sim p \rightarrow r$

Practice Plus

In Exercises 43–50, express each statement in "if . . . then" form. (More than one correct wording in "if . . . then" form may be possible.) Then write the statement's converse, inverse, and contrapositive.

43. All people who diet lose weight.

44. All senators are politicians.

45. No vehicle that has no flashing light on top is an ambulance.

46. All people who are not fearful are crazy.

47. Passing the bar exam is a necessary condition for being an attorney.

48. Being a citizen is a necessary condition for voting.

49. Being a pacifist is sufficient for not being a warmonger.

50. Being a writer is sufficient for not being illiterate.

Application Exercises

The Corruption Perceptions Index uses perceptions of the general public, business people, and risk analysts to rate countries by how likely they are to accept bribes. The ratings are on a scale from 0 to 10, where higher scores represent less corruption. The graph shows the corruption ratings for the world's least corrupt and most corrupt countries. (The rating for the United States is 7.6.) Use the graph to solve Exercises 51–52.

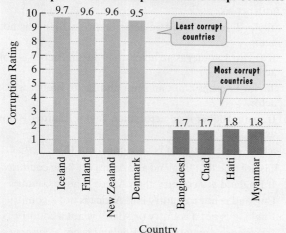

Top Four Least Corrupt and Most Corrupt Countries

Source: Transparency International, *Corruption Perceptions Index*

51. **a.** Consider the statement

    If the country is Finland, then the corruption rating is 9.6.

    Use the information given by the graph at the bottom of the previous column to determine the truth value of this conditional statement.

    b. Write the converse, inverse, and contrapositive of the statement in part (a). Then use the information given by the graph to determine whether each statement is true or not necessarily true.

52. **a.** Consider the statement

    If the country is Haiti, then the corruption rating is 1.8.

    Use the information given by the graph at the bottom of the previous column to determine the truth value of this conditional statement.

    b. Write the converse, inverse, and contrapositive of the statement in part (a). Then use the information given by the graph to determine whether each statement is true or not necessarily true.

Critical Thinking Exercises

53. Give an example of a conditional statement that is true, but whose converse and inverse are not necessarily true. Try to make the statement somewhat different from the conditional statements that you have encountered throughout this section. Explain why the converse and the inverse that you wrote are not necessarily true.

54. Read the Blitzer Bonus on page 862. The Dormouse's last statement is the setup for a joke. The punchline, delivered by the Hatter to the Dormouse, is, "For you, it's the same thing." Explain the joke. What does this punchline have to do with the difference between a conditional and a biconditional statement?

Group Exercise

55. Can you think of an advertisement in which the person using a product is extremely attractive or famous? It is true that if you are this attractive or famous person, then you use the product. (Or at least pretend, for monetary gain, that you use the product!) In order to get you to buy the product, here is what the advertisers would *like* you to believe: If I use this product, then I will be just like this attractive or famous person. This, the converse, is not necessarily true and, for most of us, is unfortunately false. Each group member should find an example of this kind of deceptive advertising to share with the other group members.

13.6 Arguments and Truth Tables

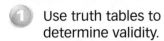

1. Use truth tables to determine validity.

2. Recognize and use forms of valid and invalid arguments.

You're having a bit of a hissy fit. Just because you turned in a report one day late, your teacher lowered the grade from B to C. She reminded you that the course policy states that work turned in late will be marked down a grade. Is your annoyance justified?

An **argument** consists of two parts: the given statements, called the **premises**, and a **conclusion**. Here's your teacher's argument for changing your grade from B to C:

Premise 1: If you turn in your work late, it will be marked down a grade.

Premise 2: You turned in your work late.

Conclusion: Therefore, your work was marked down a grade.

It appears that if the premises are true, then your teacher must change the grade on your report from B to C. The true premises force the conclusion to be true, making this an example of a ***valid argument***.

Definition of a Valid Argument

An argument is **valid** if the conclusion is true whenever the premises are assumed to be true. An argument that is not valid is said to be an **invalid argument**, also called a **fallacy**.

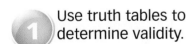

1 Use truth tables to determine validity.

Truth tables can be used to test validity. We begin by writing the argument in symbolic form. Let's do this for the scenario involving your lowered grade. Represent each simple statement with a letter:

p: You turn in your work late.

q: Your work will be marked down a grade.

Now we write the two premises and the conclusion in symbolic form:

Premise 1: $p \rightarrow q$ If you turn in your work late, it will be marked down a grade.

Premise 2: p____   You turned in your work late.

Conclusion: $\therefore q$ Therefore, your work was marked down a grade.

(The three-dot triangle, $\therefore$, is read "therefore.")

To decide whether this argument is valid, we rewrite it as a conditional statement that has the following form:

$$[(p \rightarrow q) \wedge p] \rightarrow q.$$

If premise 1 and premise 2,   then   conclusion.

At this point, we can determine whether the conjunction of the premises, $p \rightarrow q$ and p, implies that the conclusion, q, is true for all possible truth values for p and q. We construct a truth table for the statement

$$[(p \rightarrow q) \wedge p] \rightarrow q.$$

If the final column in the truth table for $[(p \rightarrow q) \wedge p] \rightarrow q$ is true in every case, then the statement is a tautology and the argument is valid. If the conditional statement in the last column is false in at least one case, then the statement is not a tautology, and the argument is invalid. The truth table is shown below.

p	q	$p \rightarrow q$	$(p \rightarrow q) \wedge p$	$[(p \rightarrow q) \wedge p] \rightarrow q$
T	T	T	T	T
T	F	F	F	T
F	T	T	F	T
F	F	T	F	T

The final column in the table is true in every case. The conditional statement is a tautology. This means that the premises imply the conclusion. The conclusion necessarily follows from the premises. Therefore, the argument is valid.

The form of your teacher's argument for changing your grade from B to C

$$p \rightarrow q$$
$$\underline{p \qquad\quad}$$
$$\therefore q$$

is called **direct reasoning**. All arguments that have the direct reasoning form are valid regardless of the English statements that p and q represent.

Here's a step-by-step procedure to test the validity of an argument using truth tables:

Testing the Validity of an Argument with a Truth Table

1. Use a letter to represent each simple statement in the argument.
2. Express the premises and the conclusion symbolically.
3. Write a symbolic conditional statement of the form

   $$[(\text{premise 1}) \wedge (\text{premise 2}) \wedge \cdots \wedge (\text{premise } n)] \rightarrow \text{conclusion},$$

   where n is the number of premises.
4. Construct a truth table for the conditional statement in step 3.
5. If the final column of the truth table has all trues, the conditional statement is a tautology and the argument is valid. If the final column does not have all trues, the conditional statement is not a tautology and the argument is invalid.

 **Example 1**   **Did the Pickiest Logician in the Galaxy Foul Up?**

In an episode of the television series *Star Trek*, the starship *Enterprise* is hit by an ion storm, causing the power to go out. Captain Kirk wonders if Mr. Scott, the engineer, is aware of the problem. Mr. Spock, the paragon of extraterrestrial intelligence, replies, "If Mr. Scott is still with us, the power should be on momentarily." Moments later, the ship's power comes on and Spock arches his Vulcan brow: "Ah, Mr. Scott is still with us."

Spock's logic can be expressed in the form of an argument:

If Mr. Scott is still with us, then the power will come on.

The power comes on.

Therefore, Mr. Scott is still with us.

Determine whether this argument is valid or invalid.

Solution

Step 1 Use a letter to represent each simple statement in the argument. We introduce the following representations:

p: Mr. Scott is still with us.

q: The power will come on.

Step 2 Express the premises and the conclusion symbolically.

$p \rightarrow q$ If Mr. Scott is still with us, then the power will come on.

q The power comes on.

$\therefore p$ Mr. Scott is still with us.

Step 3 Write a symbolic statement of the form

$$[(\text{premise 1}) \wedge (\text{premise 2})] \rightarrow \text{conclusion}.$$

The symbolic statement is

$$[(p \rightarrow q) \wedge q] \rightarrow p.$$

Step 4 Construct a truth table for the conditional statement in step 3.

p	q	$p \rightarrow q$	$(p \rightarrow q) \wedge q$	$[(p \rightarrow q) \wedge q] \rightarrow p$
T	T	T	T	T
T	F	F	F	T
F	T	T	T	F
F	F	T	F	T

Step 5 Use the truth values in the final column to determine if the argument is valid or invalid. The entries in the final column of the truth table are not all true, so the conditional statement is not a tautology. Spock's argument is invalid, or a fallacy.

The form of the argument in Spock's logical foul-up

$$p \rightarrow q$$
$$q$$
$$\therefore p$$

is called the **fallacy of the converse**. It should remind you that a conditional statement is not equivalent to its converse. All arguments that have this form are invalid regardless of the English statements that p and q represent.

You may also recall that a conditional statement is not equivalent to its inverse. Another common invalid form of argument is called the **fallacy of the inverse**:

$$p \rightarrow q$$
$$\sim p$$
$$\therefore \sim q.$$

An example of the fallacy of the inverse is "If I study, I pass. I do not study. Therefore, I do not pass." For most students, the conclusion is true, but it does not have to be. If an argument is invalid, then the conclusion is not necessarily true. This, however, does not mean that the conclusion must be false.

Study Tip

If Spock's fallacious reasoning represents television writers' best efforts at being logical, do not expect much from their commercials.

> ✓ **Checkpoint 1** Use a truth table to determine whether the following argument is valid or invalid:
>
> The United States must energetically support the development of solar-powered cars or suffer increasing atmospheric pollution.
>
> <u>The United States must not suffer increasing atmospheric pollution.</u>
>
> ▶ Therefore, the United States must energetically support the development of solar-powered cars.

Example 2   Determining Validity with a Truth Table

Determine whether the following argument is valid or invalid:

"I can't have anything more to do with the operation. If I did, I'd have to lie to the Ambassador. And I can't do that."

—Henry Bromell, "I Know Your Heart, Marco Polo," *The New Yorker*

Solution

We can express the argument as follows:

If I had anything more to do with the operation, I'd have to lie to the Ambassador.

<u>I can't lie to the Ambassador.</u>

Therefore, I can't have anything more to do with the operation.

Step 1   Use a letter to represent each statement in the argument. We introduce the following representations:

p: I have more to do with the operation.

q: I have to lie to the Ambassador.

Step 2   Express the premises and the conclusion symbolically.

$p \rightarrow q$ If I had anything more to do with the operation, I'd have to lie to the Ambassador.

<u>$\sim q$ I can't lie to the Ambassador.</u>

$\therefore \sim p$ Therefore, I can't have anything more to do with the operation.

Step 3   Write a symbolic statement of the form

$$[(\text{premise 1}) \wedge (\text{premise 2})] \rightarrow \text{conclusion}.$$

The symbolic statement is

$$[(p \rightarrow q) \wedge \sim q] \rightarrow \sim p.$$

Step 4   Construct a truth table for the conditional statement in step 3.

p	q	$p \rightarrow q$	$\sim q$	$(p \rightarrow q) \wedge \sim q$	$\sim p$	$[(p \rightarrow q) \wedge \sim q] \rightarrow \sim p$
T	T	T	F	F	F	T
T	F	F	T	F	F	T
F	T	T	F	F	T	T
F	F	T	T	T	T	T

Step 5 Use the truth values in the final column to determine if the argument is valid or invalid. The entries in the final column of the truth table at the bottom of the previous page are all true, so the conditional statement is a tautology. The given argument is valid.

The form of the argument in Example 2

$$p \to q$$
$$\sim q$$
$$\therefore \sim p$$

should remind you that a conditional statement is equivalent to its contrapositive:

$$p \to q \equiv \sim q \to \sim p.$$

The form of this argument is called **contrapositive reasoning**.

> ✓ **Checkpoint 2**   Use a truth table to determine whether the following argument is valid or invalid:
>
> I study for 5 hours or I fail.
> I did not study for 5 hours.
> Therefore, I failed.

Example 3   An Argument for Not Lowering Your Grade

We opened this section with a course policy that stated work turned in late is marked down a grade. However, you believe it was unfair for your teacher to lower the grade on your report from B to C. Although you turned in the report a day late, it was because you were absent on the due date because of illness. You explain to your teacher that you never miss a deadline unless you're sick. Here's your argument:

If I turn in my work late, I am sick.
If I am sick, my work should not be marked down a grade.
Therefore, if I turn in my work late, it should not be marked down a grade.

Determine whether this argument is valid or invalid.

Solution

Step 1 Use a letter to represent each statement in the argument. We introduce the following representations:

p: I turn in my work late.

q: I am sick.

r: My work should not be marked down a grade.

Step 2 Express the premises and the conclusion symbolically.

$p \to q$	If I turn in my work late, I am sick.
$q \to r$	If I am sick, my work should not be marked down a grade.
$\therefore p \to r$	Therefore, if I turn in my work late, it should not be marked down a grade.

Step 3 Write a symbolic statement of the form

[(premise 1) ∧ (premise 2)] → conclusion.

The symbolic statement is

$$[(p \to q) \land (q \to r)] \to (p \to r).$$

Step 4 Construct a truth table for the conditional statement in step 3, $[(p \to q) \land (q \to r)] \to (p \to r)$.

p	q	r	$p \to q$	$q \to r$	$p \to r$	$(p \to q) \land (q \to r)$	$[(p \to q) \land (q \to r)] \to (p \to r)$
T	T	T	T	T	T	T	T
T	T	F	T	F	F	F	T
T	F	T	F	T	T	F	T
T	F	F	F	T	F	F	T
F	T	T	T	T	T	T	T
F	T	F	T	F	T	F	T
F	F	T	T	T	T	T	T
F	F	F	T	T	T	T	T

Step 5 Use the truth values in the final column to determine if the argument is valid or invalid. The entry in each of the eight rows in the final column of the truth table is true, so the conditional statement is a tautology. The argument for not lowering your grade is valid.

The form of the argument in Example 3

$$p \to q$$
$$q \to r$$
$$\therefore \; p \to r$$

is called **transitive reasoning**. If p implies q and q implies r, then p must imply r. Because $p \to r$ is a valid conclusion, the contrapositive, $\sim r \to \sim p$, is also a valid conclusion. Not necessarily true are the converse, $r \to p$, and the inverse, $\sim p \to \sim r$.

> ✓ **Checkpoint 3** Use a truth table to determine whether the following argument is valid or invalid:

If you lower the fat in your diet, you lower your cholesterol.

If you lower your cholesterol, you reduce the risk of heart disease.

Therefore, if you do not lower the fat in your diet, you do not reduce the risk of heart disease.

We have seen two valid arguments that resulted in very different conclusions. The argument that opened this section concluded that your work needed to be marked down a grade. The argument in Example 3 concluded that your work should not be marked down a grade. This illustrates that the conclusion of a valid argument is true *relative to the premises*. The conclusion may follow from the premises, although one or more of the premises may not be true.

A valid argument with true premises is called a **sound argument**. The conclusion of a sound argument is true relative to the premises, but it is also true as a separate statement removed from the premises. When an argument is sound, its conclusion represents perfect certainty. Knowing how to assess the validity and soundness of arguments is a very important skill that will enable you to avoid being fooled into thinking that something is proven with certainty when it is not.

Table 13.17 at the top of the next page contains the standard forms of commonly used valid and invalid arguments. If an English argument translates into one of these forms, you can immediately determine whether or not it is valid without using a truth table.

2 Recognize and use forms of valid and invalid arguments.

Table 13.17 Standard Forms of Arguments

Valid Arguments

Direct Reasoning	Contrapositive Reasoning	Disjunctive Reasoning		Transitive Reasoning
$p \rightarrow q$	$p \rightarrow q$	$p \vee q$ $p \vee q$		$p \rightarrow q$
p	$\sim q$	$\sim p$ $\sim q$		$q \rightarrow r$
$\therefore q$	$\therefore \sim p$	$\therefore q$ $\therefore p$		$\therefore p \rightarrow r$
				$\therefore \sim r \rightarrow \sim p$

Invalid Arguments

Fallacy of the Converse	Fallacy of the Inverse	Misuse of Disjunctive Reasoning		Misuse of Transitive Reasoning
$p \rightarrow q$	$p \rightarrow q$	$p \vee q$ $p \vee q$		$p \rightarrow q$
q	$\sim p$	p q		$q \rightarrow r$
$\therefore p$	$\therefore \sim q$	$\therefore \sim q$ $\therefore \sim p$		$\therefore r \rightarrow p$
				$\therefore \sim p \rightarrow \sim r$

Example 4   Determining Validity without Truth Tables

Determine whether each argument is valid or invalid. Identify any sound arguments.

a. There is no need for surgery. I know this because if there is a tumor then there is need for surgery, but there is no tumor.

b. The emergence of democracy is a cause for hope or environmental problems will overshadow any promise of a bright future. Because environmental problems will overshadow any promise of a bright future, it follows that the emergence of democracy is not a cause for hope.

c. If evidence of the defendant's DNA is found at the crime scene, we can connect him with the crime. If we can connect him with the crime, we can have him stand trial. Therefore, if the defendant's DNA is found at the crime scene, we can have him stand trial.

Solution

a. We introduce the following representations:

p: There is a tumor.

q: There is need for surgery.

We express the premises and conclusion symbolically.

If there is a tumor then there is need for surgery.	$p \rightarrow q$
There is no tumor.	$\sim p$
Therefore, there is no need for surgery.	$\therefore \sim q$

The argument is in the form of the fallacy of the inverse. Therefore, the argument is invalid.

b. We introduce the following representations:

p: The emergence of democracy is a cause for hope.

q: Environmental problems will overshadow any promise of a bright future.

We express the premises and conclusion symbolically.

The emergence of democracy is a cause for hope or environmental problems will overshadow any promise of a bright future.	$p \lor q$
Environmental problems will overshadow any promise of a bright future.	q
Therefore, the emergence of democracy is not a cause for hope.	$\therefore \sim p$

The argument is in a form that represents a misuse of disjunctive reasoning. Therefore, the argument is invalid.

c. We introduce the following representations:

p: Evidence of the defendant's DNA is found at the crime scene.

q: We can connect him with the crime.

r: We can have him stand trial.

The argument can now be expressed symbolically.

If evidence of the defendant's DNA is found at the crime scene, we can connect him with the crime.	$p \rightarrow q$
If we can connect him with the crime, we can have him stand trial.	$q \rightarrow r$
Therefore, if the defendant's DNA is found at the crime scene, we can have him stand trial.	$\therefore \ p \rightarrow r$

The argument is in the form of transitive reasoning. Therefore, the argument is valid. Furthermore, the premises appear to be true statements, so this is a sound argument.

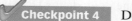 **Checkpoint 4** Determine whether each argument is valid or invalid.

a. The emergence of democracy is a cause for hope or environmental problems will overshadow any promise of a bright future. Environmental problems will not overshadow any promise of a bright future. Therefore, the emergence of democracy is a cause for hope.

b. If the defendant's DNA is found at the crime scene, then we can have him stand trial. He is standing trial. Consequently, we found evidence of his DNA at the crime scene.

c. If you mess up, your self-esteem goes down. If your self-esteem goes down, everything else falls apart. So, if you mess up, everything else falls apart.

Example 5 **Nixon's Resignation**

"The decision of the Supreme Court in U.S. v. Nixon *(1974), handed down the first day of the Judiciary Committee's final debate, was critical. If the President defied the order, he would be impeached. If he obeyed the order, it was increasingly apparent he would be impeached on the evidence."*

—Victoria Schuck, "Watergate," *The Key Reporter*

Richard Nixon's resignation on August 8, 1974, was the sixth anniversary of the day he had triumphantly accepted his party's nomination for his first term as president.

Based on the paragraph at the bottom of the previous page, we can formulate the following argument:

If Nixon did not obey the Supreme Court order, he would be impeached.

If Nixon obeyed the Supreme Court order, he would be impeached.

Therefore, Nixon's impeachment was certain.

Determine whether this argument is valid or invalid.

Solution

Step 1   Use a letter to represent each simple statement in the argument. We introduce the following representations:

p: Nixon obeys the Supreme Court order.

q: Nixon is impeached.

Step 2   Express the premises and the conclusion symbolically.

$\sim p \rightarrow q$ If Nixon did not obey the Supreme Court order, he would be impeached.

$p \rightarrow q$ If Nixon obeyed the Supreme Court order, he would be impeached.

$\therefore q$ Therefore, Nixon's impeachment was certain.

Because this argument is not in the form of a recognizable valid or invalid argument, we will use a truth table to determine validity.

Step 3   Write a symbolic statement of the form

$$[(\textbf{premise 1}) \wedge (\textbf{premise 2})] \rightarrow \textbf{conclusion}.$$

The symbolic statement is

$$[(\sim p \rightarrow q) \wedge (p \rightarrow q)] \rightarrow q.$$

Step 4   Construct a truth table for the conditional statement in step 3.

p	q	$\sim p$	$\sim p \rightarrow q$	$p \rightarrow q$	$(\sim p \rightarrow q) \wedge (p \rightarrow q)$	$[(\sim p \rightarrow q) \wedge (p \rightarrow q)] \rightarrow q$
T	T	F	T	T	T	T
T	F	F	T	F	F	T
F	T	T	T	T	T	T
F	F	T	F	T	F	T

Step 5   Use the truth values in the final column to determine if the argument is valid or invalid. The entries in the final column of the truth table are all true, so the conditional statement is a tautology. Thus, the given argument is valid. Because the premises are true statements, this is a sound argument, with impeachment a certainty. In a 16-minute broadcast on August 8, 1974, Richard Nixon yielded to the inevitability of the argument's conclusion and, staring sadly into the cameras, announced his resignation.

Checkpoint 5   Determine whether the following argument is valid or invalid:

If people are good, laws are not needed to prevent wrongdoing.

If people are not good, laws will not succeed in preventing wrongdoing.

Therefore, laws are not needed to prevent wrongdoing or laws will not succeed in preventing wrongdoing.

A **logical** or **valid conclusion** is one that forms a valid argument when it follows a given set of premises. Suppose that the premises of an English argument translate into any one of the symbolic forms of premises for the valid arguments in **Table 13.17** on page 871. The symbolic conclusion can be used to find a valid English conclusion. Example 6 shows how this is done.

Example 6 **Drawing a Logical Conclusion**

Draw a valid conclusion from the following premises:

If all college students get requirements out of the way early, then no college students take required courses in their last semester. Some college students take required courses in their last semester.

Solution

Let p be: All college students get requirements out of the way early.

Let q be: No college students take required courses in their last semester.

The form of the premises is

$p \rightarrow q$ If all college students get requirements out of the way early, then no college students take required courses in their last semester.

$\underline{\sim q}$ Some college students take required courses in their last semester. (Recall that the negation of *no* is *some*.)

$\therefore$?

The conclusion $\sim p$ is valid because it forms the contrapositive reasoning of a valid argument when it follows the given premises. The conclusion $\sim p$ translates as

Not all college students get requirements out of the way early.

Because the negation of *all* is *some . . . not*, we can equivalently conclude that

Some college students do not get requirements out of the way early.

Checkpoint 6   Draw a valid conclusion from the following premises:

If all people lead, then no people follow. Some people follow.

Achieving Success

Example 6 is based on the fact that most colleges have requirements that make it mandatory for students to take courses in a variety of areas. **Get your college requirements out of the way early.** Sample various disciplines during your freshman year by taking classes that fulfill college requirements.

Exercise Set 13.6

Concept and Vocabulary Exercises

In Exercises 1–7, fill in each blank so that the resulting statement is true.

1. An argument is _____ if the conclusion is true whenever the premises are assumed to be true.

2. The argument

$$p \rightarrow q$$
$$\underline{p}$$
$$\therefore \underline{}$$

is called direct reasoning and is _____ because _____ is a tautology.

3. The argument

$$p \rightarrow q$$
$$\underline{\sim q}$$
$$\therefore \underline{}$$

is called contrapositive reasoning and is _____ because _____ is a tautology.

4. The argument

$$p \rightarrow q$$
$$\underline{q \rightarrow r}$$
$$\therefore \underline{}$$

is called transitive reasoning and is _____ because _____ is a tautology.

5. The argument

$$p \lor q$$
$$\underline{\sim p}$$
$$\therefore \underline{}$$

is called disjunctive reasoning and is _____ because _____ is a tautology.

6. The fallacy of the converse has the form

$$p \to q$$
$$\underline{q}$$
$$\therefore \underline{}.$$

7. The fallacy of the inverse has the form

$$p \to q$$
$$\underline{\sim p}$$
$$\therefore \underline{}.$$

In Exercises 8–10, determine whether each statement is true or false. If the statement is false, make the necessary change(s) to produce a true statement.

8. Any argument with true premises is valid.

9. The conclusion of a sound argument is true relative to the premises, but it is also true as a separate statement removed from the premises.

10. Any argument whose premises are $p \to q$ and $q \to r$ is valid regardless of the conclusion.

Respond to Exercises 11–16 using verbal or written explanations.

11. Describe what is meant by a valid argument.

12. If you are given an argument in words that contains two premises and a conclusion, describe how to determine if the argument is valid or invalid.

13. Write an original argument in words for the direct reasoning form.

14. Write an original argument in words for the contrapositive reasoning form.

15. Write an original argument in words for the transitive reasoning form.

16. What is a valid conclusion?

Exercises 17–20 are based on the following argument by conservative radio talk show host Rush Limbaugh and directed at former vice president Al Gore.

You would think that if Al Gore and company believe so passionately in their environmental crusading that they would first put these ideas to work in their own lives, right? . . . Al Gore thinks the automobile is one of the greatest threats to the planet, but he sure as heck still travels in one of them —a gas guzzler too.

(*See, I Told You So*, p. 168)

Limbaugh's passage can be expressed in the form of an argument:

If Gore really believed that the automobile were a threat to the planet, he would not travel in a gas guzzler.

Gore does travel in a gas guzzler.
_____

Therefore, Gore does not really believe that the automobile is a threat to the planet.

In Exercises 17–20, use Limbaugh's argument to determine whether each statement makes sense or does not make sense, and explain your reasoning.

17. I know for a fact that Al Gore does not travel in a gas guzzler, so Limbaugh's argument is invalid.

18. I think Limbaugh is a fanatic and all his arguments are invalid.

19. In order to avoid a long truth table and instead use a standard form of an argument, I tested the validity of Limbaugh's argument using the following representations:

p: Gore really believes that the automobile is a threat to the planet.

q: He does not travel in a gas guzzler.

20. Using my representations in Exercise 19, I determined that Limbaugh's argument is invalid.

Practice Exercises

In Exercises 21–32, use a truth table to determine whether the symbolic form of the argument is valid or invalid.

21.
$$p \to q$$
$$\underline{\sim p}$$
$$\therefore \sim q$$

22.
$$p \to q$$
$$\underline{\sim p}$$
$$\therefore q$$

23.
$$p \to \sim q$$
$$\underline{q}$$
$$\therefore \sim p$$

24.
$$\sim p \to q$$
$$\underline{\sim q}$$
$$\therefore p$$

25.
$$p \land \sim q$$
$$\underline{p}$$
$$\therefore \sim q$$

26.
$$\sim p \lor q$$
$$\underline{p}$$
$$\therefore q$$

27.
$$p \to q$$
$$\underline{q \to p}$$
$$\therefore p \land q$$

28.
$$(p \to q) \land (q \to p)$$
$$\underline{p}$$
$$\therefore p \lor q$$

29.
$$p \to q$$
$$\underline{q \to r}$$
$$\therefore r \to p$$

30.
$$p \to q$$
$$\underline{q \to r}$$
$$\therefore \sim p \to \sim r$$

31.
$$p \to q$$
$$\underline{q \land r}$$
$$\therefore p \lor r$$

32.
$$\sim p \land q$$
$$\underline{p \leftrightarrow r}$$
$$\therefore p \land r$$

In Exercises 33–58, translate each argument into symbolic form. Then determine whether the argument is valid or invalid. You may use a truth table or, if applicable, compare the argument's symbolic form to a standard valid or invalid form. (You can ignore differences in past, present, and future tense.)

33. If it is cold, my motorcycle will not start.

My motorcycle started.
_____
$\therefore$ It is not cold.

34. If a metrorail system is not in operation, there are traffic delays.

Over the past year there have been no traffic delays.
_____
$\therefore$ Over the past year a metrorail system has been in operation.

35. There must be a dam or there is flooding.

This year there is flooding.
_____
$\therefore$ This year there is no dam.

36. You must eat well or you will not be healthy.
I eat well.

∴ I am healthy.

37. If we close the door, then there is less noise.
There is less noise.

∴ We closed the door.

38. If an argument is in the form of the fallacy of the inverse, then it is invalid.
This argument is invalid.

∴ This argument is in the form of the fallacy of the inverse.

39. If he was disloyal, his dismissal was justified.
If he was loyal, his dismissal was justified.

∴ His dismissal was justified.

40. If I tell you I cheated, I'm miserable.
If I don't tell you I cheated, I'm miserable.

∴ I'm miserable.

41. We criminalize drugs or we damage the future of young people.
We will not damage the future of young people.

∴ We criminalize drugs.

42. He is intelligent or an overachiever.
He is not intelligent.

∴ He is an overachiever.

43. If all people obey the law, then no jails are needed.
Some people do not obey the law.

∴ Some jails are needed.

44. If all people obey the law, then no jails are needed.
Some jails are needed.

∴ Some people do not obey the law.

45. If I'm tired, I'm edgy.
If I'm edgy, I'm moody.

∴ If I'm tired, I'm moody.

46. If I am at the beach, then I swim in the ocean.
If I swim in the ocean, then I feel refreshed.

∴ If I am at the beach, then I feel refreshed.

47. If I'm tired, I'm edgy.
If I'm edgy, I'm moody.

∴ If I'm moody, I'm tired.

48. If I'm at the beach, then I swim in the ocean.
If I swim in the ocean, then I feel refreshed.

∴ If I'm not at the beach, then I don't feel refreshed.

49. If Tim and Janet play, then the team wins.
Tim played and the team did not win.

∴ Janet did not play.

50. If you have a garden and a library, you have everything you need. (Cicero, 1st century B.C.)
You have a library and you do not have everything you need.

∴ You do not have a garden.

51. If it rains or snows, then I read.
I am not reading.

∴ It is neither raining nor snowing.

52. If I am tired or hungry, I cannot concentrate.
I can concentrate.

∴ I am neither tired nor hungry.

53. If it rains or snows, then I read.
I am reading.

∴ It is raining or snowing.

54. If I am tired or hungry, I cannot concentrate.
I cannot concentrate.

∴ I am tired or hungry.

55. If it is hot and humid, I complain.
It is not hot or it is not humid.

∴ I am not complaining.

56. If the young knew and the old could, everything can be done. (Italian proverb)
The young do not know and the old cannot.

∴ Some things cannot be done.

57. If some journalists learn about the invasion, the newspapers will print the news.

If the newspapers print the news, the invasion will not be a secret.
The invasion was a secret.

∴ No journalists learned about the invasion.

58. If some journalists learn about the invasion, the newspapers will print the news.

If the newspapers print the news, the invasion will not be a secret.
No journalists learned about the invasion.

∴ The invasion was a secret.

In Exercises 59–66, use the standard forms of valid arguments to draw a valid conclusion from the given premises.

59. If a person is a chemist, then that person has a college degree.
My best friend does not have a college degree.
Therefore, . . .

60. If the Westway Expressway is not in operation, automobile traffic makes the East Side Highway look like a parking lot.
On June 2, the Westway Expressway was completely shut down because of an overturned truck.
Therefore, . . .

61. The writers of *My Mother the Car* were told by the network to improve their scripts or be dropped from prime time.
The writers of *My Mother the Car* did not improve their scripts.
Therefore, . . .

62. You exercise or you do not feel energized.
I do not exercise.
Therefore, . . .

63. If all electricity is off, then no lights work.
Some lights work.
Therefore, . . .

64. If all houses meet the hurricane code, then none of them are destroyed by a category 4 hurricane.
Some houses were destroyed by Andrew, a category 4 hurricane.
Therefore, . . .

65. If I vacation in Paris, I eat French pastries.
If I eat French pastries, I gain weight.
Therefore, . . .

66. If we don't change, we don't grow.
If we don't grow, we are not really living.
Therefore, . . .

Practice Plus

In Exercises 67–74, translate each argument into symbolic form. Then determine whether the argument is valid or invalid.

67. If it was any of your business, I would have invited you. It is not, and so I did not.

68. If it was any of your business, I would have invited you. I did, and so it is.

69. It is the case that $x < 5$ or $x > 8$, but $x \geq 5$, so $x > 8$.

70. It is the case that $x < 3$ or $x > 10$, but $x \leq 10$, so $x < 3$.

71. Having a college degree is necessary for obtaining a teaching position. You have a college degree, so you have a teaching position.

72. Having a college degree is necessary for obtaining a teaching position. You do not obtain a teaching position, so you do not have a college degree.

73. "I do know that this pencil exists; but I could not know this if Hume's principles were true. Therefore, Hume's principles, one or both of them, are false."
— G. E. Moore, *Some Main Problems of Philosophy*

74. (In this exercise, determine if the argument is sound, valid but not sound, or invalid.)

If an argument is invalid, it does not produce truth, whereas a valid unsound argument also does not produce truth. Arguments are invalid or they are valid but unsound. Therefore, no arguments produce truth.

Application Exercises

Exercises 75–76 illustrate arguments that have appeared in cartoons. Each argument is restated following the cartoon. Translate the argument into symbolic form and then determine whether it is valid or invalid.

75.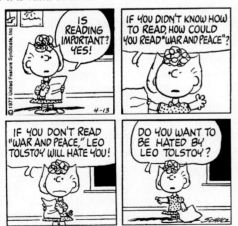

Peanuts reprinted by permission of United Feature Syndicate, Inc.

If you do not know how to read, you cannot read *War and Peace*. If you cannot read *War and Peace*, then Leo Tolstoy will hate you. Therefore, if you do not know how to read, Leo Tolstoy will hate you.

76.

Betty by Gary Delainey and Gery Rasmussen. Reprinted by permission of Newspaper Enterprise Association, Inc.

If I say I'm in denial, then I'm not in denial. I am saying that I'm in denial, so I'm not in denial at all.

77. Conservative commentator Rush Limbaugh directed this passage at liberals and the way they think about crime.

Of course, liberals will argue that these actions [contemporary youth crime] can be laid at the foot of socioeconomic inequities, or poverty. However, the Great Depression caused a level of poverty unknown to exist in America today, and yet I have been unable to find any accounts of crime waves sweeping our large cities. Let the liberals chew on that.

(*See, I Told You So*, p. 83)

Limbaugh's passage can be expressed in the form of an argument:

If poverty causes crime, then crime waves would have swept American cities during the Great Depression.

Crime waves did not sweep American cities during the Great Depression.

∴ Poverty does not cause crime. (Liberals are wrong.)

Translate this argument into symbolic form and determine whether it is valid or invalid.

78. In the following passage, Martin Luther King, Jr. presents an argument with the conclusion "Segregation statutes are unjust." Use two premises, one a conditional statement and the other a simple statement, to rewrite King's argument in the format used throughout this section. Then determine if the argument is sound, valid but not sound, or invalid.

"Any law that uplifts human personality is just. Any law that degrades human personality is unjust. All segregation statutes are unjust because segregation distorts the soul and damages the personality. It gives the segregator a false sense of superiority, and the segregated a false sense of inferiority."
— Martin Luther King, Jr., "Letter from a Birmingham Jail"

In addition to the forms of invalid arguments, fallacious reasoning occurs in everyday logic. Some people use the fallacies described below to intentionally deceive. Others use fallacies innocently; they are not even aware they are using them. Match each description below with the example from Exercises 79–90 that illustrates the fallacy. The matching is one-to-one.

Common Fallacies in Everyday Reasoning

a. *The* **fallacy of emotion** *consists of appealing to emotion (pity, force, etc.) in an argument.*

b. *The* **fallacy of inappropriate authority** *consists of claiming that a statement is true because a person cited as an authority says it's true or because most people believe it's true.*

c. *The* **fallacy of hasty generalization** *occurs when an inductive generalization is made on the basis of a few observations or an unrepresentative sample.*

d. *The* **fallacy of questionable cause** *consists of claiming that A caused B when it is known that A occurred before B.*

e. *The* **fallacy of ambiguity** *occurs when the conclusion of an argument is based on a word or phrase that is used in two different ways in the premises.*

f. *The* **fallacy of ignorance** *consists of claiming that a statement is true simply because it has not been proven false, or vice versa.*

g. *The* **mischaracterization fallacy** *consists of misrepresenting an opponent's position or attacking the opponent rather than that person's ideas in order to refute his or her argument.*

h. *The* **slippery slope fallacy** *occurs when an argument reasons without justification that an event will set off a series of events leading to an undesirable consequence.*

i. *The* **either/or fallacy** *mistakenly presents only two solutions to a problem, negates one of these either/or alternatives, and concludes that the other must be true.*

j. *The* **fallacy of begging the question** *assumes that the conclusion is true within the premises.*

k. *The* **fallacy of composition** *occurs when an argument moves from premises about the parts of a group to a conclusion about the whole group. It also occurs when characteristics of an entire group are mistakenly applied to parts of the group.*

l. *The* **fallacy of the complex question** *consists of drawing a conclusion from a self-incriminating question.*

79. If we allow physician-assisted suicide for those who are terminally ill and request it, it won't be long before society begins pressuring the old and infirm to get out of the way and make room for the young. Before long the government will be deciding who should live and who should die.

80. Of course there are extraterrestrials. Haven't you read that article in the *National Enquirer* about those UFOs spotted in Texas last month?

81. Either you go to college and make something of yourself, or you'll end up as an unhappy street person. You cannot be an unhappy street person, so you should go to college.

82. Scientists have not proved that AIDS cannot be transmitted through casual contact. Therefore, we should avoid casual contact with suspected AIDS carriers.

83. Each of my three uncles smoked two packs of cigarettes every day and they all lived into their 90s. Smoking can't be that bad for your health.

84. You once cheated on tests. I know this because when I asked you if you had stopped cheating on tests, you said yes.

85. My paper is late, but I know you'll accept it because I've been sick and my parents will kill me if I flunk this course.

86. We've all heard Professor Jones tell us about how economic systems should place human need above profit. But I'm not surprised that he neglected to tell you that he's a communist who has visited Cuba twice. How can he possibly speak the truth about economic systems?

87. It's easy to see that suicide is wrong. After all, no one is ever justified in taking his or her own life.

88. The reason I hurt your arm is because you hurt me just as much by telling Dad.

89. During the last full moon, YouTube featured hundreds of videos showing people behaving weirdly. It's reasonable to conclude that a full moon is the cause of a lot of peculiar behavior.

90. I know, without even looking, that question #17 on this test is difficult. This is the case because the test was made up by Professor Flunkem and Flunkem's exams are always difficult.

Critical Thinking Exercises

91. Write a valid argument on one of the following questions. If you can, write valid arguments on both sides.

    a. Should the death penalty be abolished?

    b. Should women be allowed combat roles in the military?

    c. Should grades be abolished?

    d. Should there be restrictions on who can speak or what can be said at your high school graduation ceremony?

92. Write an original argument in words that has a true conclusion, yet is invalid.

93. Draw a valid conclusion from the given premises. Then use a truth table to verify your answer.

    If you only spoke when spoken to and I only spoke when spoken to, then nobody would ever say anything. Some people do say things. Therefore, . . .

94. Translate the argument below into symbolic form. Then use a truth table to determine if the argument is valid or invalid.

    It's wrong to smoke in public if secondary cigarette smoke is a health threat. If secondary cigarette smoke were not a health threat, the American Lung Association would not say that it is. The American Lung Association says that secondary cigarette smoke is a health threat. Therefore, it's wrong to smoke in public.

95. Draw what you believe is a valid conclusion in the form of a disjunction for the following argument. Then verify that the argument is valid for your conclusion.

> "Inevitably, the use of the placebo involved built-in contradictions. A good patient–doctor relationship is essential to the process, but what happens to that relationship when one of the partners conceals important information from the other? If the doctor tells the truth, he destroys the base on which the placebo rests. If he doesn't tell the truth, he jeopardizes a relationship built on trust."
>
> —Norman Cousins, *Anatomy of an Illness*

Group Exercise

96. In this section, we used a variety of examples, including arguments about marking down your work a grade, the inevitability of Nixon's impeachment, Spock's (fallacious) logic on *Star Trek*, and even two cartoons, to illustrate symbolic arguments.

a. From any source that is of particular interest to you (these can be the words of someone you truly admire or a person who really gets under your skin), select a paragraph or two in which the writer argues a particular point. (An intriguing source is *What Is Your Dangerous Idea?*, edited by John Brockman, published by Harper Perennial, 2007.) Rewrite the reasoning in the form of an argument using words. Then translate the argument into symbolic form and use a truth table to determine if it is valid or invalid.

b. Each group member should share the selected passage with other people in the group. Explain how it was expressed in argument form. Then tell why the argument is valid or invalid.

13.7 Arguments and Euler Diagrams

Objective

1 Use Euler diagrams to determine validity.

William Shakespeare

Leonhard Euler

He is the Shakespeare of mathematics, yet he is unknown by the general public. Most people cannot even correctly pronounce his name. The Swiss mathematician Leonhard Euler (1707–1783), whose last name rhymes with *boiler*, not *ruler*, is the most prolific mathematician in history. His collected books and papers fill some 80 volumes; Euler published an average of 800 pages of new mathematics per year over a career that spanned six decades. Euler was also an astronomer, botanist, chemist, physicist, and linguist. His productivity was not at all slowed down by the total blindness he experienced the last 17 years of his life. An equation discovered by Euler, $e^{\pi i} + 1 = 0$, connected five of the most important numbers in mathematics in a totally unexpected way.

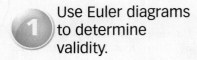

Use Euler diagrams to determine validity.

Euler invented an elegant way to determine the validity of arguments whose premises contain the words *all, some,* and *no.* The technique for doing this uses geometric ideas and involves four basic diagrams, known as **Euler diagrams**. **Figure 13.3** illustrates how Euler diagrams represent four quantified statements.

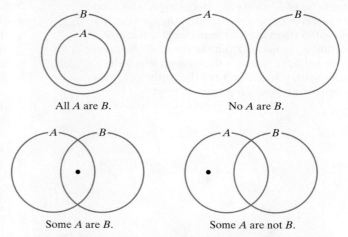

Figure 13.3   Euler diagrams for quantified statements

The Euler diagrams in **Figure 13.3** are just like the Venn diagrams that we used in studying sets. However, there is no need to enclose the circles inside a rectangle representing a universal set. In these diagrams, circles are used to indicate relationships of premises to conclusions.

Here's a step-by-step procedure for using Euler diagrams to determine whether or not an argument is valid:

> **Euler Diagrams and Arguments**
>
> 1. Make an Euler diagram for the first premise.
> 2. Make an Euler diagram for the second premise on top of the one for the first premise.
> 3. The argument is valid if and only if every possible diagram illustrates the conclusion of the argument. If there is even *one* possible diagram that contradicts the conclusion, this indicates that the conclusion is not true in every case, so the argument is invalid.

The goal of this procedure is to produce, if possible, *a diagram that does* **not** *illustrate the argument's conclusion.* The method of Euler diagrams boils down to determining whether such a diagram is possible. If it is, this serves as a counterexample to the argument's conclusion, and the argument is immediately declared invalid. By contrast, if no such counterexample can be drawn, the argument is valid.

The technique of using Euler diagrams is illustrated in Examples 1–6.

Example 1   Arguments and Euler Diagrams

Use Euler diagrams to determine whether the following argument is valid or invalid:

All people who arrive late cannot perform.

All people who cannot perform are ineligible for scholarships.

Therefore, all people who arrive late are ineligible for scholarships.

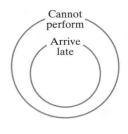

Solution

Step 1   Make an Euler diagram for the first premise. We begin by diagramming the premise

All people who arrive late cannot perform.

The region inside the smaller circle represents people who arrive late. The region inside the larger circle represents people who cannot perform.

Step 2   Make an Euler diagram for the second premise on top of the one for the first premise. We add to our previous figure the diagram for the second premise:

All people who cannot perform are ineligible for scholarships.

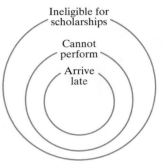

A third, larger, circle representing people who are ineligible for scholarships is drawn surrounding the circle representing people who cannot perform.

Step 3   The argument is valid if and only if every possible diagram illustrates the argument's conclusion. There is only one possible diagram. Let's see if this diagram illustrates the argument's conclusion, namely

All people who arrive late are ineligible for scholarships.

This is indeed the case because the Euler diagram shows the circle representing the people who arrive late contained within the circle of people who are ineligible for scholarships. The Euler diagram supports the conclusion and the given argument is valid.

✓ **Checkpoint 1**   Use Euler diagrams to determine whether the following argument is valid or invalid:

> All U.S. voters must register.
> All people who register must be U.S. citizens.
> Therefore, all U.S. voters are U.S. citizens.

Example 2   Arguments and Euler Diagrams

Use Euler diagrams to determine whether the following argument is valid or invalid:

> All poets appreciate language.
> All writers appreciate language.
> Therefore, all poets are writers.

Solution

Step 1   Make an Euler diagram for the first premise. We begin by diagramming the premise

All poets appreciate language.

Up to this point, our work is similar to what we did in Example 1.

Step 2 Make an Euler diagram for the second premise on top of the one for the first premise. We add to our previous figure the diagram for the second premise:

All writers appreciate language.

A third circle representing writers must be drawn inside the circle representing people who appreciate language. There are four ways to do this.

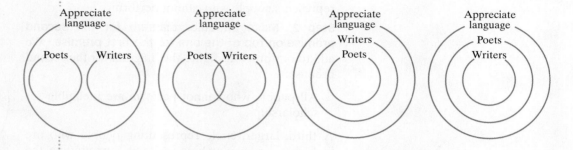

Step 3 The argument is valid if and only if every possible diagram illustrates the argument's conclusion. The argument's conclusion is

All poets are writers.

This conclusion is not illustrated by every possible diagram shown above. One of these diagrams is repeated in the margin. This diagram shows "no poets are writers." There is no need to examine the other three diagrams.

The diagram on the left serves as a counterexample to the argument's conclusion. This means that the given argument is invalid. It would have sufficed to draw only the counterexample on the left to determine that the argument is invalid.

Checkpoint 2 Use Euler diagrams to determine whether the following argument is valid or invalid:

> All baseball players are athletes.
> All ballet dancers are athletes.
> Therefore, no baseball players are ballet dancers.

Example 3   Arguments and Euler Diagrams

Use Euler diagrams to determine whether the following argument is valid or invalid:

> All college freshmen live on campus.
> No people who live on campus can own cars.
> Therefore, no college freshmen can own cars.

Solution

Step 1 Make an Euler diagram for the first premise. The diagram for

All college freshmen live on campus

is shown on the right. The region inside the smaller circle represents college freshmen. The region inside the larger circle represents people who live on campus.

Step 2 Make an Euler diagram for the second premise on top of the one for the first premise. We add to our previous figure the diagram for the second premise:

No people who live on campus can own cars.

A third circle representing people who own cars is drawn outside the circle representing people who live on campus.

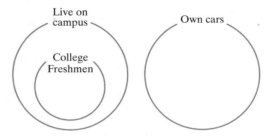

Step 3 The argument is valid if and only if every possible diagram illustrates the argument's conclusion. There is only one possible diagram. The argument's conclusion is

No college freshmen can own cars.

This is supported by the diagram shown above because it shows the circle representing college freshmen drawn outside the circle representing people who own cars. The Euler diagram supports the conclusion, and it is impossible to find a counterexample that does not. The given argument is valid.

Checkpoint 3 Use Euler diagrams to determine whether the following argument is valid or invalid:

> All mathematicians are logical.
> No poets are logical.
> Therefore, no poets are mathematicians.

Let's see what happens to the validity if we reverse the second premise and the conclusion of the argument in Example 3.

Example 4 Euler Diagrams and Validity

Use Euler diagrams to determine whether the following argument is valid or invalid:

> All college freshmen live on campus.
> No college freshmen can own cars.
> Therefore, no people who live on campus can own cars.

Solution

Step 1 Make an Euler diagram for the first premise. We once again begin with the diagram for

All college freshmen live on campus.

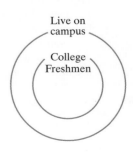

So far, our work is exactly the same as in the previous example.

Step 2   Make an Euler diagram for the second premise on top of the one for the first premise. We add to our previous figure the diagram for the second premise:

No college freshmen can own cars.

The circle representing people who own cars is drawn outside the college freshmen circle. At least two Euler diagrams are possible.

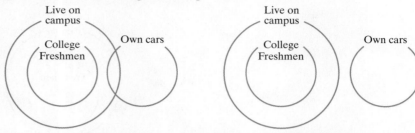

Step 3   The argument is valid if and only if every possible diagram illustrates the argument's conclusion. The argument's conclusion is

No people who live on campus can own cars.

This conclusion is not supported by both diagrams shown above. The diagram that does not support the conclusion is repeated in the margin. Notice that the "live on campus" circle and the "own cars" circle intersect. This diagram serves as a counterexample to the argument's conclusion. This means that the argument is invalid. Once again, only the counterexample on the left is needed to conclude that the argument is invalid.

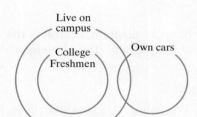

✓ **Checkpoint 4**   Use Euler diagrams to determine whether the following argument is valid or invalid:

> All mathematicians are logical.
> No poets are mathematicians.
> Therefore, no poets are logical.

So far, the arguments that we have looked at have contained "all" or "no" in the premises and conclusions. The quantifier "some" is a bit trickier to work with. Because the statement "Some *A* are *B*" means there exists at least one *A* that is a *B*, we diagram this existence by showing a dot in the region where *A* and *B* intersect, illustrated in **Figure 13.4**.

Suppose that it is true that "Some *A* are not *B*," illustrated by the dot in **Figure 13.5**. This Euler diagram does not let us conclude that "Some *B* are not *A*" because there is not a dot in the part of the *B* circle that is not in the *A* circle. Conclusions with the word "some" must be shown by existence of at least one element represented by a dot in an Euler diagram.

Here is an example that shows the premise "Some *A* are not *B*" does not enable us to logically conclude that "Some *B* are not *A*."

> Some U.S. citizens are not U.S. senators. (true)
> ∴ Some U.S. senators are not U.S. citizens. (false)

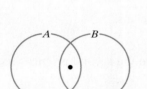

Figure 13.4   Some *A* are *B*.

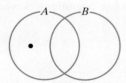

Figure 13.5
Illustrated by the dot is some *A* are not *B*. We cannot validly conclude that some *B* are not *A*.

Example 5   Euler Diagrams and the Quantifier "Some"

Use Euler diagrams to determine whether the following argument is valid or invalid:

> All people are mortal.
> Some mortals are students.
> Therefore, some people are students.

Solution

Step 1 Make an Euler diagram for the first premise. Begin with the premise

All people are mortal.

The Euler diagram is shown on the right.

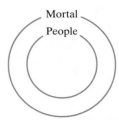

Step 2 Make an Euler diagram for the second premise on top of the one for the first premise. We add to our previous figure the diagram for the second premise:

Some mortals are students.

The circle representing students intersects the circle representing mortals. The dot in the region of intersection shows that at least one mortal is a student. Another diagram is possible, but if the diagram on the right serves as a counterexample then it is all we need. Let's check if it is a counterexample.

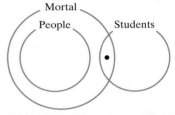

Step 3 The argument is valid if and only if every possible diagram illustrates the conclusion of the argument. The argument's conclusion is

Some people are students.

This conclusion is not supported by the Euler diagram. The diagram does not show the "people" circle and the "students" circle intersecting with a dot in the region of intersection. Although this conclusion is true in the real world, the Euler diagram serves as a counterexample that shows it does not follow from the premises. Therefore, the argument is invalid.

✓ **Checkpoint 5** Use Euler diagrams to determine whether the following argument is valid or invalid:

> All mathematicians are logical.
> Some poets are logical.
> Therefore, some poets are mathematicians.

Some arguments show existence without using the word "some." Instead, a particular person or thing is mentioned in one of the premises. This particular person or thing can be represented by a dot. Here is an example:

> All men are mortal.
> Aristotle is a man.
> Therefore, Aristotle is mortal.

The two premises can be represented by the following Euler diagrams:

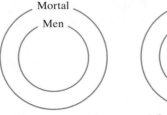

All men are mortal.

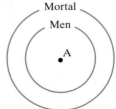

Aristotle (•) is a man.

The Euler diagram on the right uses a dot labeled A (for Aristotle). The diagram shows Aristotle (•) winding up in the "mortal" circle. The diagram supports the conclusion that Aristotle is mortal. This argument is valid.

Example 6 **An Argument Mentioning One Person**

Use Euler diagrams to determine whether the following argument is valid or invalid:

> All children love to swim.
> Michael Phelps loves to swim.
> Therefore, Michael Phelps is a child.

Solution

Step 1   Make an Euler diagram for the first premise. Begin with the premise

   All children love to swim.

The Euler diagram is shown on the right.

Step 2   Make an Euler diagram for the second premise on top of the one for the first premise. We add to our previous figure the diagram for the second premise:

   Michael Phelps loves to swim.

Michael Phelps is represented by a dot labeled M. The dot must be placed in the "love to swim" circle. At least two Euler diagrams are possible.

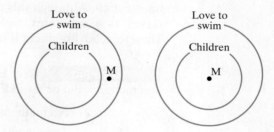

Step 3   The argument is valid if and only if every possible diagram illustrates the conclusion of the argument. The argument's conclusion is

   Michael Phelps is a child.

This conclusion is not supported by the Euler diagram shown above on the left. The dot representing Michael Phelps is outside the "children" circle. Michael Phelps might not be a child. This diagram serves as a counterexample to the argument's conclusion. The argument is invalid.

✓ **Checkpoint 6**   Use Euler diagrams to determine whether the following argument is valid or invalid:

> All mathematicians are logical.
> Euclid was logical.
> Therefore, Euclid was a mathematician.

Blitzer Bonus
●●●●●●●●●●●●●●●●
Aristotle 384–322 B.C.

The first systematic attempt to describe the logical rules that may be used to arrive at a valid conclusion was made by the ancient Greeks, in particular Aristotle. Aristotelian forms of valid arguments are built into the ways that Westerners think and view the world. In this detail of Raphael's painting *The School of Athens*, Aristotle (on the left) is debating with his teacher and mentor, Plato.

Raphael (1483–1520). *School of Athens*, Stanza della Segnatura, Vatican Palace, Vatican State. © Scala/Art Resource.

Achieving Success

A recent government study cited in *Math: A Rich Heritage* (Globe Fearon Educational Publisher) found this simple fact: **The more college mathematics courses you take, the greater your earning potential will be.** Even jobs that do not require a college degree require mathematical thinking that involves attending to precision, making sense of complex problems, and persevering in solving them. No other discipline comes close to math in offering a more extensive set of tools for application and intellectual development. Take as much math as possible as you pursue your journey into higher education.

Exercise Set 13.7

Concept and Vocabulary Exercises

In Exercises 1–4, fill in each blank so that the resulting statement is true. Refer to parts (a) through (d) in the following figure.

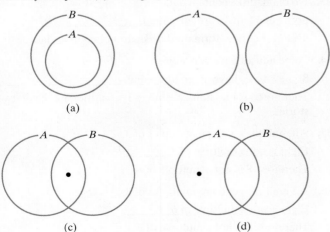

(a)

(b)

(c)

(d)

1. The figure in part (a) illustrates the quantified statement _____.

2. The figure in part (b) illustrates the quantified statement _____.

3. The figure in part (c) illustrates the quantified statement _____.

4. The figure in part (d) illustrates the quantified statement _____.

In Exercises 5–6, determine whether each statement is true or false. If the statement is false, make the necessary change(s) to produce a true statement.

5. Truth tables are used to represent quantified statements.

6. The most important part in a quantified statement's representation is the size of each circle.

Respond to Exercises 7–8 using verbal or written explanations.

7. Explain how to use Euler diagrams to determine whether or not an argument is valid.

8. Under what circumstances should Euler diagrams rather than truth tables be used to determine whether or not an argument is valid?

In Exercises 9–12, determine whether each statement makes sense or does not make sense, and explain your reasoning.

9. I made Euler diagrams for the premises of an argument and one of my possible diagrams illustrated the conclusion, so the argument is valid.

10. I made Euler diagrams for the premises of an argument and one of my possible diagrams did not illustrate the conclusion, so the argument is invalid.

11. I used Euler diagrams to determine that an argument is valid, but when I reverse one of the premises and the conclusion, this new argument is invalid.

12. I can't use Euler diagrams to determine the validity of an argument if one of the premises is false.

Practice Exercises

In Exercises 13–36, use Euler diagrams to determine whether each argument is valid or invalid.

13. All writers appreciate language.
    All poets are writers.
    Therefore, all poets appreciate language.

14. All physicists are scientists.
    All scientists attended college.
    Therefore, all physicists attended college.

15. All clocks keep time accurately.
    All time-measuring devices keep time accurately.
    Therefore, all clocks are time-measuring devices.

16. All cowboys live on ranches.
    All cowherders live on ranches.
    Therefore, all cowboys are cowherders.

17. All insects have six legs.
    No spiders have six legs.
    Therefore, no spiders are insects.

18. All humans are warm-blooded.
    No reptiles are warm-blooded.
    Therefore, no reptiles are human.

19. All insects have six legs.
    No spiders are insects.
    Therefore, no spiders have six legs.

20. All humans are warm-blooded.
    No reptiles are human.
    Therefore, no reptiles are warm-blooded.

21. All teachers are wise people.
    Some wise people are actors.
    Therefore, some teachers are actors.

22. All comedians are funny people.
    Some funny people are teachers.
    Therefore, some comedians are teachers.

23. All teachers are wise people.
    Some teachers are actors.
    Therefore, some wise people are actors.

24. All comedians are funny people.
    Some comedians are teachers.
    Therefore, some funny people are teachers.

25. All dancers are athletes.
    Savion Glover is a dancer.
    Therefore, Savion Glover is an athlete.

26. All actors are artists.
    Tom Hanks is an actor.
    Therefore, Tom Hanks is an artist.

27. All dancers are athletes.
    Savion Glover is an athlete.
    Therefore, Savion Glover is a dancer.

28. All actors are artists.
    Tom Hanks is an artist.
    Therefore, Tom Hanks is an actor.

29. Some people enjoy reading.
    Some people enjoy TV.
    Therefore, some people who enjoy reading enjoy TV.

30. All thefts are immoral acts.
    Some thefts are justifiable.
    Therefore, some immoral acts are justifiable.

31. All dogs have fleas.
    Some dogs have rabies.
    Therefore, all dogs with rabies have fleas.

32. All logic problems make sense.
    Some jokes make sense.
    Therefore, some logic problems are jokes.

33. No blank disks contain data.
    Some blank disks are formatted.
    Therefore, some formatted disks do not contain data.

34. Some houses have two stories.
    Some houses have air conditioning.
    Therefore, some houses with air conditioning have two stories.

35. All multiples of 6 are multiples of 3.
    Eight is not a multiple of 3.
    Therefore, 8 is not a multiple of 6.

36. All multiples of 6 are multiples of 3.
    Eight is not a multiple of 6.
    Therefore, 8 is not a multiple of 3.

Practice Plus

In Exercises 37–48, determine whether each argument is valid or invalid.

37. All natural numbers are whole numbers, all whole numbers are integers, and −4006 is not a whole number. Thus, −4006 is not an integer.

38. Some natural numbers are even, all natural numbers are whole numbers, and all whole numbers are integers. Thus, some integers are even.

39. All natural numbers are real numbers, all real numbers are complex numbers, but some complex numbers are not real numbers. The number $19 + 0i$ is a complex number, so it is not a natural number.

40. All rational numbers are real numbers, all real numbers are complex numbers, but some complex numbers are not real numbers. The number $\frac{1}{2} + 0i$ is a complex number, so it is not a rational number.

41. All A are B, all B are C, and all C are D. Thus, all A are D.

42. All A are B, no C are B, and all D are C. Thus, no A are D.

43. No A are B, some A are C, and all C are D. Thus, some D are B.

44. No A are B, some A are C, and all C are D. Thus, some D are C.

45. No A are B, no B are C, and no C are D. Thus, no A are D.

46. Some A are B, some B are C, and some C are D. Thus, some A are D.

47. All A are B, all A are C, and some B are D. Thus, some A are D.

48. Some A are B, all B are C, and some C are D. Thus, some A are D.

Application Exercises

49. This is an excerpt from a 1967 speech in the U.S. House of Representatives by Representative Adam Clayton Powell:

He who is without sin should cast the first stone. There is no one here who does not have a skeleton in his closet. I know, and I know them by name.

Powell's argument can be expressed as follows:

No sinner is one who should cast the first stone.

All people here are sinners.

Therefore, no person here is one who should cast the first stone.

Use an Euler diagram to determine whether the argument is valid or invalid.

50. In the *Sixth Meditation*, Descartes writes

I first take notice here that there is a great difference between the mind and the body, in that the body, from its nature, is always divisible and the mind is completely indivisible.

Descartes's argument can be expressed as follows:

All bodies are divisible.

No minds are divisible.

Therefore, no minds are bodies.

Use an Euler diagram to determine whether the argument is valid or invalid.

51. In *Symbolic Logic*, Lewis Carroll presents the following argument:

Babies are illogical. (All babies are illogical persons.)

Illogical persons are despised. (All illogical persons are despised persons.)

Nobody is despised who can manage a crocodile. (No persons who can manage crocodiles are despised persons.)

Therefore, babies cannot manage crocodiles.

Use an Euler diagram to determine whether the argument is valid or invalid.

Critical Thinking Exercises

52. Write an example of an argument with two quantified premises that is invalid but that has a true conclusion.

53. No animals that eat meat are vegetarians.

No cat is a vegetarian.

Felix is a cat.

Therefore, ...

   a. Felix is a vegetarian.

   b. Felix is not a vegetarian.

   c. Felix eats meat.

   d. All animals that do not eat meat are vegetarians.

54. Supply the missing first premise that will make this argument valid.

Some opera singers are terrible actors.

Therefore, some people who take voice lessons are terrible actors.

55. Supply the missing first premise that will make this argument valid.

All amusing people are entertaining.

Therefore, some teachers are entertaining.

Chapter 13 Summary

13.1 Statements, Negations, and Quantified Statements

13.2 Compound Statements and Connectives

Definitions and Concepts

A statement is a sentence that is either true or false, but not both simultaneously.

Negations and equivalences of quantified statements are given in the following diagram.
Each quantified statement's equivalent is written in parentheses below the statement. The statements
diagonally opposite each other are negations.

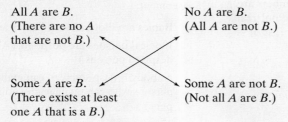

All A are B.
(There are no A
that are not B.)

No A are B.
(All A are not B.)

Some A are B.
(There exists at least
one A that is a B.)

Some A are not B.
(Not all A are B.)

Examples

• All students are hard workers.

  Equivalent: There are no students who are not hard workers.

  Negation: Some students are not hard workers.

• Some writers are poets.

  Equivalent: There exists at least one writer who is a poet.

  Negation: No writers are poets.

Additional Examples to Review

Table 13.1, page 813; Table 13.2, page 814; Example 4, page 815

Definitions and Concepts

The statements of symbolic logic and their translations are given as follows:

• Negation
  $\sim p$: Not p. It is not true that p.
• Conjunction
  $p \wedge q$: p and q. p but q. p yet q. p nevertheless q.
• Disjunction
  $p \vee q$: p or q.
• Conditional
  $p \rightarrow q$: If p, then q. q if p. p is sufficient for q. q is necessary for p. p only if q. Only if q, p.
• Biconditional
  $p \leftrightarrow q$: p if and only if q. q if and only if p. If p then q, and if q then p. p is necessary and sufficient for q.
  q is necessary and sufficient for p.

Groupings in symbolic statements are determined as follows:

• Unless parentheses follow the negation symbol, $\sim$, only the statement that immediately follows it is
  negated.
• When translating symbolic statements into English, the simple statements in parentheses appear on
  the same side of the comma.

Examples

Expressing Symbolic Statements in English

- $(q \vee r) \leftrightarrow \sim p$
  p: There are clouds. q: I see Venus. r: You see Saturn.
  I see Venus or you see Saturn, if and only if there are no clouds.

- $\sim(p \wedge q)$
  p: *American Idol* is a reality show. q: *Glee* is a reality show.
  It is not true that *American Idol* and *Glee* are both reality shows.

Expressing English Statements in Symbolic Form

- If I am your sister's son then I am your nephew, and I am not your sister's son.
  p: I am your sister's son. q: I am your nephew.

  $$(p \rightarrow q) \wedge \sim p$$

- Not being in ill health is necessary for being a successful athlete.

  > This means: If you are a successful athlete, then you are not in ill health.

  p: You are a successful athlete. q: You are in ill health.

  $$p \rightarrow \sim q$$

Additional Examples to Review

Example 1, page 818; Example 2, page 820; Example 3, page 821; Example 4, page 822;
Example 5, page 823; Example 6, page 825; Example 7, page 826

13.3 Truth Tables for Negation, Conjunction, and Disjunction

13.4 Truth Tables for the Conditional and the Biconditional

Definitions and Concepts

The definitions of symbolic logic are given by the truth values in the following table:

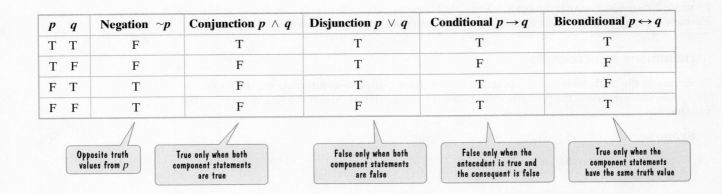

p q	Negation $\sim p$	Conjunction $p \wedge q$	Disjunction $p \vee q$	Conditional $p \rightarrow q$	Biconditional $p \leftrightarrow q$
T T	F	T	T	T	T
T F	F	F	T	F	F
F T	T	F	T	T	F
F F	T	F	F	T	T

Opposite truth values from p | True only when both component statements are true | False only when both component statements are false | False only when the antecedent is true and the consequent is false | True only when the component statements have the same truth value

A truth table for a compound statement shows when the statement is true and when it is false. The first few columns show the simple statements that comprise the compound statement and their possible truth values. The final column heading is the given compound statement. The truth values in each column are determined by looking back at appropriate columns and using one of the five definitions of symbolic logic. If a compound statement is always true, it is called a tautology.

Examples

* $\sim p \wedge (p \vee \sim q)$

p q	$\sim p$	$\sim q$	$p \vee \sim q$	$\sim p \wedge (p \vee \sim q)$
T T	F	F	T	F
T F	F	T	T	F
F T	T	F	F	F
F F	T	T	T	T

Opposite truth values

column 1 ∨ column 4
False when both components are false.

column 3 ∧ column 5
True when both components are true.

$\sim p \wedge (p \vee \sim q)$ is not a tautology. The statement is true only when p is false and q is false.

* $p \rightarrow (q \wedge r)$

p q r	$q \wedge r$	$p \rightarrow (q \wedge r)$
T T T	T	T
T T F	F	F
T F T	F	F
T F F	F	F
F T T	T	T
F T F	F	T
F F T	F	T
F F F	F	T

column 2 ∧ column 3
True when both components are true.

column 1 → column 4
False when column 1 is true and column 4 is false.

Additional Examples to Review

Example 2, page 834; Example 3, page 836; Example 4, page 837; Example 5, page 838; Example 6, page 839; Example 1, page 846; Example 2, page 848; Example 3, page 849; Example 4, page 851

Definitions and Concepts

To determine the truth value of a compound statement for a specific case, substitute the truth values of the simple statements into the symbolic form of the compound statement and then use the appropriate definitions.

Example

* Determine the truth value of
$$(p \wedge \sim r) \rightarrow (r \leftrightarrow \sim q)$$
when p is true, q is false, and r is false.

$(p \wedge \sim r) \rightarrow (r \leftrightarrow \sim q)$	This is the given statement.
$(T \wedge \sim F) \rightarrow (F \leftrightarrow \sim F)$	Substitute the given truth values.
$(T \wedge T) \rightarrow (F \leftrightarrow T)$	Replace $\sim F$ with T.
$T \rightarrow F$	Replace $T \wedge T$ with T: Conjunction is true when both parts are true. Replace $F \leftrightarrow T$ with F: A biconditional is false when both parts have different truth values.
F	Replace $T \rightarrow F$ with F: A conditional is false with a true antecedent and a false consequent.

Additional Examples to Review

Example 7, page 841; Example 5, page 852

13.5 Equivalent Statements and Variations of Conditional Statements

Definitions and Concepts

Two statements are equivalent, symbolized by ≡, if they have the same truth value in every possible case.

Examples

- Select the statement that is equivalent to

$$\underbrace{\text{It's hot}}_{p} \quad \text{or} \quad \underbrace{\text{it's humid.}}_{q}$$

  a. If it's hot, then it's not humid. $p \rightarrow \sim q$
  b. It's humid and it's hot. $q \wedge p$
  c. It is false that it is not hot or not humid. $\sim(\sim p \vee \sim q)$
  d. If it's not humid, then it's hot. $\sim q \rightarrow p$

		Given		(a)	(b)		(c)	(d)
p	q	$p \vee q$	$\sim q$	$p \rightarrow \sim q$	$q \wedge p$	$\sim p$	$\sim p \vee \sim q$	$\sim(\sim p \vee \sim q)$	$\sim q \rightarrow p$
T	T	T	F	F	T	F	F	T	T
T	F	T	T	T	F	F	T	F	T
F	T	T	F	T	F	T	T	F	T
F	F	F	T	T	F	T	T	F	F

equivalent

  Statement (d), If it's not humid, then it's hot, is equivalent to It's hot or it's humid:
  $$(\sim q \rightarrow p) \equiv (p \vee q).$$

- Determine which, if any, of the three statements are equivalent.
  a. If you love what you're doing, you're successful. $p \rightarrow q$

$$\underbrace{\text{If you love what you're doing,}}_{p} \underbrace{\text{you're successful.}}_{q}$$

  b. If you're successful, you love what you're doing. $q \rightarrow p$

  c. You're successful or you don't love what you're doing. $q \vee \sim p$

		(a)	(b)		(c)
p	q	$p \rightarrow q$	$q \rightarrow p$	$\sim p$	$q \vee \sim p$
T	T	T	T	F	T
T	F	F	T	F	F
F	T	T	F	T	T
F	F	T	T	T	T

Statements (a) and (c) are equivalent.

Additional Examples to Review

Example 1, page 856; Example 2, page 857; Example 3, page 858

Definitions and Concepts

Variations of the Conditional Statement $p \rightarrow q$

- $p \rightarrow q$ is equivalent to $\sim q \rightarrow \sim p$, the contrapositive: $p \rightarrow q \equiv \sim q \rightarrow \sim p$.
- $p \rightarrow q$ is not equivalent to $q \rightarrow p$, the converse.
- $p \rightarrow q$ is not equivalent to $\sim p \rightarrow \sim q$, the inverse.

Example

- If we use the pool, it's not snowing. $p \rightarrow \sim q$

  (p: We use the pool. q: It's snowing.)

  Converse: $\sim q \rightarrow p$

  If it's not snowing, we use the pool.

  Inverse: $\sim p \rightarrow \sim (\sim q)$

  $\sim p \rightarrow q$

  If we do not use the pool, it's snowing.

  Contrapositive: $\sim (\sim q) \rightarrow \sim p$

  $q \rightarrow \sim p$

  It it's snowing, we do not use the pool.

Additional Examples to Review

Example 4, page 859; Example 5, page 861

13.6 Arguments and Truth Tables

Definitions and Concepts

An argument consists of two parts: the given statements, called the premises, and a conclusion. An argument is valid if the conclusion is true whenever the premises are assumed to be true. An argument that is not valid is called an invalid argument or a fallacy. A valid argument with true premises is called a sound argument.

Testing the Validity of an Argument with a Truth Table

If the argument contains n premises, write a conditional statement of the form

$$[(\text{premise 1}) \wedge (\text{premise 2}) \wedge \cdots \wedge (\text{premise } n)] \rightarrow \text{conclusion}$$

and construct a truth table. If the conditional statement is a tautology, the argument is valid; if not, the argument is invalid.

Example

- You don't own a car or you need insurance.

  You need insurance.

  $\therefore$ You own a car.

  Is this argument valid or invalid?

  p: You own a car. q: You need insurance.

  $\sim p \vee q$ Construct a truth table for

  q $[(\sim p \vee q) \wedge q] \rightarrow p$.

  $\therefore p$

p q	$\sim p$	$\sim p \vee q$	$(\sim p \vee q) \wedge q$	$[(\sim p \vee q) \wedge q] \rightarrow p$
T T	F	T	T	T
T F	F	F	F	T
F T	T	T	T	F
F F	T	T	F	T

Not a tautology

The argument is invalid.

Additional Examples to Review

Example 1, page 866; Example 2, page 868; Example 3, page 869; Example 5, page 872

Definitions and Concepts

Standard Forms of Arguments

Valid Arguments			
Direct Reasoning	**Contrapositive Reasoning**	**Disjunctive Reasoning**	**Transitive Reasoning**
$p \rightarrow q$ p $\therefore q$	$p \rightarrow q$ $\sim q$ $\therefore \sim p$	$p \vee q \qquad p \vee q$ $\sim p \qquad\quad \sim q$ $\therefore q \qquad\quad \therefore p$	$p \rightarrow q$ $q \rightarrow r$ $\therefore p \rightarrow r$ $\therefore \sim r \rightarrow \sim p$

Invalid Arguments			
Fallacy of the Converse	**Fallacy of the Inverse**	**Misuse of Disjunctive Reasoning**	**Misuse of Transitive Reasoning**
$p \rightarrow q$ q $\therefore p$	$p \rightarrow q$ $\sim p$ $\therefore \sim q$	$p \vee q \qquad p \vee q$ $p \qquad\quad q$ $\therefore \sim q \qquad \therefore \sim p$	$p \rightarrow q$ $q \rightarrow r$ $\therefore r \rightarrow p$ $\therefore \sim p \rightarrow \sim r$

Examples

Determine whether each argument is valid or invalid.

- Class is cancelled or some people are at their desks.
  No people are at their desks.
  $\therefore$ Class is cancelled.
  p: Class is cancelled.
  q: Some people are at their desks.
  Symbolic Form of the Argument

  $p \vee q$
  $\sim q$ The negation of *some* is *no*.
  $\therefore p$ The argument is in the form of disjunctive reasoning and is valid.

- If stock prices rise, then bond values fall.
  Bond values are falling.
  $\therefore$ Stock prices are rising.
  p: Stock prices rise.
  q: Bond values fall.
  Symbolic Form of the Argument

  $p \rightarrow q$
  q
  $\therefore p$ The argument is in the form of the fallacy of the converse and is invalid

Additional Examples to Review
Example 4, page 871; Example 6, page 874

13.7 Arguments and Euler Diagrams

Definitions and Concepts

Euler diagrams for quantified statements are given as follows:

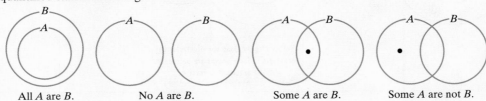

All A are B.   No A are B.   Some A are B.   Some A are not B.

To test the validity of an argument with an Euler diagram,

1. Make an Euler diagram for the first premise.
2. Make an Euler diagram for the second premise on top of the one for the first premise.
3. The argument is valid if and only if every possible diagram illustrates the conclusion of the argument.

Examples

Determine whether each argument is valid or invalid.

- All mice live in holes.
  <u>All rodents live in holes.</u>
  ∴ All mice are rodents.

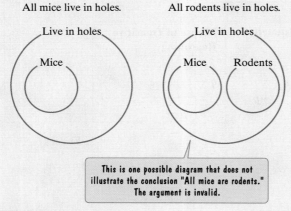

This is one possible diagram that does not illustrate the conclusion "All mice are rodents." The argument is invalid.

- All bears are dangerous animals
  <u>No poodles are dangerous animals.</u>
  ∴ No poodles are bears.

All bears are dangerous animals.

Dangerous animals

Bears

No poodles are dangerous animals.

Dangerous animals

Bears   Poodles

This one possible diagram illustrates the conclusion "No poodles are bears." The argument is valid.

- All snakes are vipers.
  <u>Some vipers are poisonous.</u>
  ∴ Some snakes are poisonous.

All snakes are vipers.

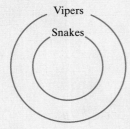

Vipers

Snakes

Some vipers are poisonous.

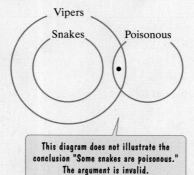

Vipers

Snakes   Poisonous

This diagram does not illustrate the conclusion "Some snakes are poisonous." The argument is invalid.

Additional Examples to Review

Example 1, page 880; Example 2, page 881; Example 3, page 882; Example 4, page 883;
Example 5, page 884; Example 6, page 886

Review Exercises

Section 13.1 Statements, Negations, and Quantified Statements and Section 13.2 Compound Statements and Connectives

In Exercises 1–6, let p, q, and r represent the following simple statements:

 p: The temperature is below 32°.
 q: We finished studying.
 r: We go to the movies.

Express each symbolic compound statement in English.

1. $(p \wedge q) \rightarrow r$

2. $\sim r \rightarrow (\sim p \vee \sim q)$

3. $p \wedge (q \rightarrow r)$

4. $r \leftrightarrow (p \wedge q)$

5. $\sim(p \wedge q)$

6. $\sim r \leftrightarrow (\sim p \vee \sim q)$

In Exercises 7–12, let p, q, and r represent the following simple statements:

 p: The outside temperature is at least 80°.
 q: The air conditioner is working.
 r: The house is hot.

Express each English statement in symbolic form.

7. The outside temperature is at least 80° and the air conditioner is working, or the house is hot.

8. If the outside temperature is at least 80° or the air conditioner is not working, then the house is hot.

9. If the air conditioner is working, then the outside temperature is at least 80° if and only if the house is hot.

10. The house is hot, if and only if the outside temperature is at least 80° and the air conditioner is not working.

11. Having an outside temperature of at least 80° is sufficient for having a hot house.

12. Not having a hot house is necessary for the air conditioner to be working.

In Exercises 13–16, write the negation of each statement.

13. All houses are made with wood.

14. No students major in business.

15. Some crimes are motivated by passion.

16. Some Democrats are not registered voters.

17. The speaker stated that, "All new taxes are for education." We later learned that the speaker was not telling the truth. What can we conclude about new taxes and education?

Section 13.3 Truth Tables for Negation, Conjunction, and Disjunction and Section 13.4 Truth Tables for the Conditional and the Biconditional

In Exercises 18–24, construct a truth table for each statement. Then indicate whether the statement is a tautology, a self-contradiction, or neither.

18. $p \vee (\sim p \wedge q)$

19. $\sim p \vee \sim q$

20. $p \rightarrow (\sim p \vee q)$

21. $p \leftrightarrow \sim q$

22. $\sim(p \vee q) \rightarrow (\sim p \wedge \sim q)$

23. $(p \vee q) \rightarrow \sim r$

24. $(p \wedge q) \leftrightarrow (p \wedge r)$

In Exercises 25–26,

 a. Write each statement in symbolic form. Assign letters to simple statements that are not negated.

 b. Construct a truth table for the symbolic statement in part (a).

 c. Use the truth table to indicate one set of conditions that make the compound statement true, or state that no such conditions exist.

25. I'm in class or I'm studying, and I'm not in class.

26. If you spit from a truck then it's legal, but if you spit from a car then it's not. (This law is still on the books in Georgia!)

In Exercises 27–29, determine the truth value for each statement when p is true, q is false, and r is false.

27. $\sim(q \leftrightarrow r)$

28. $(p \wedge q) \rightarrow (p \vee r)$

29. $(\sim q \rightarrow p) \vee (r \wedge \sim p)$

The circle graph shows the percentage of American adults who consider various subjects the most taboo to discuss at work.

What Is the Most Taboo Topic to Discuss at Work?

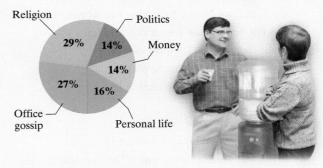

Source: Adecco survey of 1807 workers

In Exercises 30–32, write each statement in symbolic form. Then use the information displayed by the graph above to determine the truth value of the compound statement.

30. Twenty-nine percent consider religion the most taboo topic and it is not true that 14% consider politics the most taboo topic.

31. If a greater percentage of people consider money a more taboo topic than personal life, then 16% consider money the most taboo topic and 14% consider personal life the most taboo topic.

32. Fourteen percent consider money the most taboo topic if and only if 14% consider politics the most taboo topic, or it is not true that the greatest percentage consider religion the most taboo topic.

Section 13.5 Equivalent Statements and Variations of Equivalent Statements

33. a. Use a truth table to show that $\sim p \vee q$ and $p \rightarrow q$ are equivalent.

   b. Use the result from part (a) to write a statement that is equivalent to

   The triangle is not isosceles or it has two equal sides.

34. Select the statement that is equivalent to

   Joe grows mangos or oranges.

   a. If Joe grows mangos, he does not grow oranges.

   b. If Joe grows oranges, he does not grow mangos.

   c. If Joe does not grow mangos, he grows oranges.

   d. Joe grows both mangos and oranges.

In Exercises 35–36, use a truth table to determine whether the two statements are equivalent.

35. $\sim(p \leftrightarrow q), \sim p \vee \sim q$

36. $\sim p \wedge (q \vee r), (\sim p \wedge q) \vee (\sim p \wedge r)$

In Exercises 37–40, write the converse, inverse, and contrapositive of each statement.

37. If I am in Atlanta, then I am in the South.

38. If I am in class, then today is not a holiday.

39. If I work hard, then I pass all courses.

40. $\sim p \rightarrow \sim q$

In Exercises 41–44, determine which, if any, of the three given statements are equivalent.

41. a. If it is hot, then I use the air conditioner.

   b. If it is not hot, then I do not use the air conditioner.

   c. It is not hot or I use the air conditioner.

42. a. If she did not play, then we lost.

   b. If we did not lose, then she played.

   c. She did not play and we did not lose.

43. a. He is here or I'm not.

   b. If I'm not here, he is.

   c. It is not true that he isn't here and I am.

44. a. If the class interests me and I like the teacher, then I enjoy studying.

   b. If the class interests me, then I like the teacher and I enjoy studying.

   c. The class interests me, or I like the teacher and I enjoy studying.

Section 13.6 Arguments and Truth Tables

In Exercises 45–46, use a truth table to determine whether the symbolic form of the argument is valid or invalid.

45. $p \rightarrow q$
   $\underline{\sim q}$
   $\therefore p$

46. $p \wedge q$
   $\underline{q \rightarrow r}$
   $\therefore p \rightarrow r$

In Exercises 47–52, translate each argument into symbolic form. Then determine whether the argument is valid or invalid. You may use a truth table or, if applicable, compare the argument's symbolic form to a standard valid or invalid form.

47. If Tony plays, the team wins.
   The team won.
   $\therefore$ Tony played.

48. My plant is fertilized or it turns yellow.
   My plant is turning yellow.
   $\therefore$ My plant is not fertilized.

49. A majority of legislators vote for a bill or that bill does not become law.
   A majority of legislators did not vote for bill x.
   $\therefore$ Bill x did not become law.

50. Having good eye–hand coordination is necessary for being a good baseball player.
   Todd does not have good eye–hand coordination.
   $\therefore$ Todd is not a good baseball player.

51. If you love the person you marry, you can fall out of love with that person.
   If you do not love the person you marry, you can fall in love with that person.
   $\therefore$ You love the person you marry if and only if you can fall out of love with that person.

52. If I purchase season tickets to the football games, then I do not attend all classes.
   If I do well in school, then I attend all classes.
   $\therefore$ If I do not do well in school, then I purchased season tickets to the football games.

Section 13.7 Arguments and Euler Diagrams

In Exercises 53–58, use Euler diagrams to determine whether each argument is valid or invalid.

53. All birds have feathers.
   All parrots have feathers.
   $\therefore$ All parrots are birds.

54. All botanists are scientists.
   All scientists have college degrees.
   $\therefore$ All botanists have college degrees.

55. All native desert plants can withstand severe drought.
   No tree ferns can withstand severe drought.
   $\therefore$ No tree ferns are native desert plants.

56. All native desert plants can withstand severe drought.
   No tree ferns are native desert plants.
   $\therefore$ No tree ferns can withstand severe drought.

57. All poets are writers.
   Some writers are wealthy.
   $\therefore$ Some poets are wealthy.

58. Some people enjoy reading.
   All people who enjoy reading appreciate language.
   $\therefore$ Some people appreciate language.

Chapter 13 Test A

Use the following representations in Exercises 1–6:

   p: I'm registered.

   q: I'm a citizen.

   r: I vote.

Express each compound statement in English.

1. $(p \wedge q) \to r$ **2.** $\sim r \leftrightarrow (\sim p \vee \sim q)$

3. $\sim (p \vee q)$

Express each English statement in symbolic form.

4. I am registered and a citizen, or I do not vote.

5. If I am not registered or not a citizen, then I do not vote.

6. Being a citizen is necessary for voting.

In Exercises 7–8, write the negation of the statement.

7. All numbers are divisible by 5.

8. Some people wear glasses.

In Exercises 9–11, construct a truth table for the statement.

9. $p \wedge (\sim p \vee q)$ **10.** $\sim (p \wedge q) \leftrightarrow (\sim p \vee \sim q)$

11. $p \leftrightarrow (q \vee r)$

12. Write the following statement in symbolic form and construct a truth table. Then indicate one set of conditions that makes the compound statement false.

   If you break the law and change the law, then you have not broken the law.

In Exercises 13–14, determine the truth value for each statement when p is false, q is true, and r is false.

13. $\sim (q \to r)$ **14.** $(p \vee r) \leftrightarrow (\sim r \wedge p)$

15. The following statement pertains to support for the death penalty in the United States in 2008 compared to 1997:

   There was no increase in the percentage of Americans who supported the death penalty, or there was an increase in the percentage who opposed the death penalty and those who were not sure about the death penalty.

   a. Write the statement in symbolic form.

   b. Use the information in the following graph to determine the truth value of the statement in part (a).

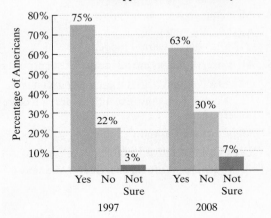

**Americans and Capital Punishment:
Do You Support the Death Penalty?**

Source: Harris Interactive

16. Select the statement below that is equivalent to

   Gene is an actor or a musician.

   a. If Gene is an actor, then he is not a musician.

   b. If Gene is not an actor, then he is a musician.

   c. It is false that Gene is not an actor or not a musician.

   d. If Gene is an actor, then he is a musician.

17. Write the contrapositive of

   If it is August, it does not snow.

18. Write the converse and the inverse of the following statement:

   If the radio is playing, then I cannot concentrate.

In Exercises 19–20, determine which, if any, of the three given statements are equivalent.

19.   **a.** If I'm not feeling well, I'm grouchy.

   b. I'm feeling well or I'm grouchy.

   c. If I'm feeling well, I'm not grouchy.

20.   **a.** It is not true that today is a holiday or tomorrow is a holiday.

   b. If today is not a holiday, then tomorrow is not a holiday.

   c. Today is not a holiday and tomorrow is not a holiday.

Determine whether each argument in Exercises 21–26 is valid or invalid.

21. If a parrot talks, it is intelligent.

   This parrot is intelligent.

   ∴ This parrot talks.

22. I am sick or I am tired.

   I am not tired.

   ∴ I am sick.

23. I am going if and only if you are not.

   You are going.

   ∴ I'm going.

24. All mammals are warm-blooded.

   All dogs are warm-blooded.

   ∴ All dogs are mammals.

25. All conservationists are advocates of solar-powered cars.

   No oil company executives are advocates of solar-powered cars.

   ∴ No conservationists are oil company executives.

26. All rabbis are Jewish.

   Some Jewish people observe kosher dietary traditions.

   ∴ Some rabbis observe kosher dietary traditions.

Chapter 13 Test B

Use the following representations in Exercises 1–6:

    p: I live at home.

    q: I eat at home.

    r: I pay for my meals.

Express each compound statement in English.

1. $(p \wedge q) \rightarrow {\sim}r$ **2.** $r \leftrightarrow ({\sim}p \vee {\sim}q)$

3. ${\sim}(p \wedge q)$

Express each English statement in symbolic form.

4. I live at home and eat at home, or I pay for my meals.

5. If I do not live at home or do not eat at home, then I pay for my meals.

6. Eating at home is necessary for not paying for my meals.

In Exercises 7–8, write the negation of the statement.

7. All natural numbers are even.

8. No students are hard workers.

In Exercises 9–11, construct a truth table for the statement.

9. $p \vee ({\sim}p \wedge q)$ **10.** ${\sim}(p \vee q) \leftrightarrow ({\sim}p \wedge {\sim}q)$

11. $p \leftrightarrow (q \wedge r)$

12. Write the following statement in symbolic form and construct a truth table. Then indicate one set of conditions that makes the compound statement false.

    "If you have a job and do not have aggravations, then you do not have a job."

                              (Malcolm Forbes)

In Exercises 13–14, determine the truth value for each statement when p is true, q is false, and r is false.

13. ${\sim}(q \leftrightarrow r)$

14. $(p \wedge r) \leftrightarrow ({\sim}r \vee p)$

15. The following statement pertains to book buyers in the United States:

    The average age of book buyers is 20 and a higher percentage of women buy books than men, or the average price paid for a book exceeds $20.

    a. Write the statement in symbolic form.

    b. Use the following information to determine the truth value of the statement in part (a).

Book Buyers in the United States

Average Age: 42

Gender: Male 42%   Female 58%

Average Price Paid: $12.19

Source: Bowker's 2009 U.S. Book Consumer Annual Review

16. Select the statement below that is equivalent to

    Hugh appears on television or in film.

    a. If Hugh appears on television, then he does not appear in film.

    b. If Hugh appears in film, then he appears on television.

    c. If Hugh does not appear in film, then he appears on television.

    d. It is false that Hugh does not appear on television or does not appear in film.

17. Write the contrapositive of

    If it is a holiday, I am not in class.

18. Write the converse and the inverse of the following statement:

    If I know the answer, then I don't guess.

In Exercises 19–20, determine which, if any, of the three given statements are equivalent.

19. **a.** If it tastes good, it's not good for you.

    b. If it does not taste good, it's good for you.

    c. It does not taste good or it's good for you.

20. **a.** It is not true that I got a better job or a better apartment.

    b. I did not get a better job and I did not get a better apartment.

    c. If I did not get a better job, then I did not get a better apartment.

Determine whether each argument in Exercises 21–26 is valid or invalid.

21. If a parrot talks, it is intelligent.

    <u>This parrot is not intelligent.</u>

    ∴ This parrot does not talk.

22. Joe is a plumber or a carpenter.

    <u>Joe is a carpenter.</u>

    ∴ Joe is not a plumber.

23. You graduate if and only if you do not fail this course.

    <u>You failed this course.</u>

    ∴ You do not graduate.

24. All mammals are warm-blooded.

    <u>All dogs are mammals.</u>

    ∴ All dogs are warm-blooded.

25. All conservationists are advocates of solar-powered cars.

    <u>No conservationists are oil company executives.</u>

    ∴ No oil company executives are advocates of solar-powered cars.

26. All priests are Catholic.

    <u>Some Catholics attend church every Sunday.</u>

    ∴ Some priests attend church every Sunday.

Answers to Selected Exercises

CHAPTER 1

Section 1.1

Checkpoint Exercises

1. Answers will vary; an example is $40 \times 40 = 1600$. **2. a.** Each number in the list is obtained by adding 6 to the previous number.; 33 **b.** Each number in the list is obtained by multiplying the previous number by 5.; 1250 **c.** To get the second number, multiply the previous number by 2. Then multiply by 3 and then by 4. Then multiply by 2, then by 3, and then by 4, repeatedly.; 3456 **d.** To get the second number, add 8 to the previous number. Then add 8 and then subtract 14. Then add 8, then add 8, and then subtract 14, repeatedly.; 7 **3. a.** Starting with the third number, each number is the sum of the previous two numbers.; 76 **b.** Starting with the second number, each number is one less than twice the previous number.; 257

4. The figures alternate between rectangles and triangles, and the number of appendages follows the pattern: one, two, three, one, two, three, etc.;   **5. a.** The result of the process is two times the original number selected. **b.** Using n to represent the original number, we have

Select a number:	n
Multiply the number by 4:	$4n$
Add 6 to the product:	$4n + 6$
Divide this sum by 2:	$\dfrac{4n + 6}{2} = 2n + 3$
Subtract 3 from the quotient:	$2n + 3 - 3 = 2n.$

Exercise Set 1.1

1. counterexample **2.** deductive **3.** inductive **4.** false **5.** true **6.** true **9.** does not make sense **11.** does not make sense **13.** Answers will vary; an example is: Barack Obama was younger than 65 at the time of his inauguration. **15.** Answers will vary; an example is: 3 multiplied by itself is 9, which is not even. **17.** Answers will vary; an example is: Adding 1 to the numerator and denominator of $\frac{1}{2}$ results in $\frac{2}{3}$, which is not equal to $\frac{1}{2}$. **19.** Answers will vary; an example is: When -1 is added to itself, the result is -2, which is less than -1. **21.** Each number in the list is obtained by adding 4 to the previous number.; 28 **23.** Each number in the list is obtained by subtracting 5 from the previous number.; 12 **25.** Each number in the list is obtained by multiplying the previous number by 3.; 729 **27.** Each number in the list is obtained by multiplying the previous number by 2.; 32 **29.** The numbers in the list alternate between 1 and numbers obtained by multiplying the number prior to the previous number by 2.; 32 **31.** Each number in the list is obtained by subtracting 2 from the previous number.; -6 **33.** Each number in the list is obtained by adding 4 to the denominator of the previous fraction.; $\frac{1}{22}$ **35.** Each number in the list is obtained by multiplying the previous number by $\frac{1}{3}$.; $\frac{1}{81}$ **37.** The second number is obtained by adding 4 to the first number. The third number is obtained by adding 5 to the second number. The number being added to the previous number increases by 1 each time.; 42 **39.** The second number is obtained by adding 3 to the first number. The third number is obtained by adding 5 to the second number. The number being added to the previous number increases by 2 each time.; 51 **41.** Starting with the third number, each number is the sum of the previous two numbers.; 71 **43.** To get the second number, add 5 to the previous number. Then add 5 and then subtract 7. Then add 5, then add 5, and then subtract 7, repeatedly.; 18 **45.** The second number is obtained by multiplying the first number by 2. The third number is obtained by subtracting 1 from the second number. Then multiply by 2 and then subtract 1, repeatedly.; 33 **47.** Each number in the list is obtained by multiplying the previous number by $-\frac{1}{4}$.; $\frac{1}{4}$ **49.** For each pair in the list, the second number is obtained by subtracting 4 from the first number.; -1

51. The pattern is: square, triangle, circle, square, triangle, circle, etc.;

53. Each figure contains the letter of the alphabet following the letter in the previous figure with one more occurrence than in the previous figure.;

d	d	d
d	d	

55. a. The result of the process is two times the original number selected. **b.** Using n to represent the original number, we have

Select a number:	n
Multiply the number by 4:	$4n$
Add 8 to the product:	$4n + 8$
Divide this sum by 2:	$\dfrac{4n + 8}{2} = 2n + 4$
Subtract 4 from the quotient:	$2n + 4 - 4 = 2n.$

57. a. The result of the process is 3. **b.** Using n to represent the original number, we have

Select a number:	n
Add 5:	$n + 5$
Double the result:	$2(n + 5) = 2n + 10$
Subtract 4:	$2n + 10 - 4 = 2n + 6$
Divide by 2:	$\dfrac{2n + 6}{2} = n + 3$
Subtract n:	$n + 3 - n = 3.$

59. $1 + 2 + 3 + 4 + 5 + 6 = \dfrac{6 \times 7}{2}$; $21 = 21$ **61.** $1 + 3 + 5 + 7 + 9 + 11 = 6 \times 6$; $36 = 36$ **63.** $98,765 \times 9 + 3 = 888,888$; correct **65.** $165 \times 3367 = 555,555$; correct **67.** b **69.** deductive reasoning; Answers will vary. **71.** inductive reasoning; Answers will vary. **73. a.** $28, 36, 45, 55, 66$ **b.** $36, 49, 64, 81, 100$ **c.** $35, 51, 70, 92, 117$ **d.** square **75.** $(11 - 7)^2 = 121 - 154 + 49$

77. Answers will vary; an example is: 1, 5, 1, 10, 1, …. **79. a.** 36; 4356; 443,556; 44,435,556 **b.** In each multiplication, a number is multiplied by itself; this number has one more digit than the number in the previous multiplication; all such digits are 6.; Each product has one more digit of 4 than the previous product, followed by a digit of 3, followed by one more digit of 5 than the previous product, ending in a digit of 6; the first product has zero digits of 4 and zero digits of 5. **c.** $66,666 \times 66,666 = 4,444,355,556; 666,666 \times 666,666 = 444,443,555,556$ **d.** inductive reasoning; Answers will vary.

Section 1.2

Checkpoint Exercises

1. a. 7,000,000,000 **b.** 6,800,000,000 **c.** 6,751,590,000 **2. a.** 3.1 **b.** 3.1416 **3. a.** $2, $1, $5, $4, $1, $2, and $3; ≈$18 **b.** no **4. a.** $≈40 \times 50 = 2000 per wk **b.** $≈50 \times 2000 = $100,000$ per yr **5. a.** 0.48×2148.72 **b.** $0.5 \times 2100 = 1050$; Your family spent approximately $1050 on heating and cooling last year. **6. a.** ≈0.16 year for each subsequent birth year **b.** ≈87.1 yr **7. a.** ≈66% **b.** 1990 through 1995 **c.** 1990 **8. a.** $1069 **b.** $T = 15,518 + 1069x$ **c.** $28,346

Exercise Set 1.2

1. estimation **2.** circle graph **3.** mathematical model **4.** true **5.** true **6.** false **19.** makes sense **21.** makes sense **23. a.** 19,490,300 **b.** 19,490,000 **c.** 19,490,000 **d.** 19,500,000 **e.** 19,000,000 **f.** 20,000,000 **25.** 2.718 **27.** 2.71828 **29.** 2.718281828 **31.** $350 + 600 = 950$; 955; reasonably well **33.** $9 + 1 + 19 = 29$; 29.23; quite well **35.** $32 - 11 = 21$; 20.911; quite well **37.** $40 \times 6 = 240$; 218.185; not so well **39.** $0.8 \times 400 = 320$; 327.06; reasonably well **41.** $48 \div 3 = 16$; ≈16.49; quite well **43.** 30% of 200,000 is 60,000.; 59,920.96; quite well **45.** ≈3 + 6 + 20 + 2 + 12 + 0 = $43 **47.** $≈50 \times 40 \times 20 = $40,000$ per yr **49.** 3 yr ≈40 months; total cost: $≈40 \times 600 = $24,000 **51.** $≈300,000 \div 300 = $1000 **53.** $≈40 \times 50 = 2000$ hr per yr; $≈60,000 \div 2000 = 30 per hr

55. $≈80 \times 365 \times 24 = 700,800$ hr **57.** $\frac{0.2 \times 100}{0.5} = 40$; ≈42.03; quite reasonable **59.** b **61.** c **63.** ≈3 hr **65.** ≈100,000,000 American adults **67. a.** ≈85 people per 100 **b.** ≈5400 more people **69. a.** ≈0.9% per yr **b.** ≈60.1% **71. a.** 1986; 3.3 million **b.** from 1991 through 1996 **c.** 1981 **d.** 1976 **73. a.** ≈1.4 million per yr **b.** $F = 122 + 1.4x$ **c.** 171 million **75.** a **77.** b **79.** ≈667 days; ≈1.8 yr **81.** Since there are infinitely many digits, the digits of these numbers cannot be reversed.

Section 1.3

Checkpoint Exercises

1. the amount of money given to the cashier **2.** The 128-ounce bottle at approximately 4¢ an ounce is the better value. **3.** 14 months **4.** 5 combinations **5.** 6 outfits **6.** A route that will cost less than $1460 is A, D, E, C, B, A. **7.** Answers will vary.

Trick Questions

1. 12 **2.** 12 **3.** sister and brother **4.** match

Exercise Set 1.3

1. understand **2.** devise a plan **3.** false **4.** false **9.** does not make sense **11.** makes sense **13.** the price of the computer **15.** the number of words on the page **17.** unnecessary information: weekly salary of $350; extra pay: $180 **19.** unnecessary information: $20 given to the parking attendant; charge: $4.50 **21. a.** 24-ounce box for $4.59 **b.** 22¢ per ounce for the 15.3-ounce box and $3.06 per pound for the 24-ounce box **c.** no; Answers will vary. **23.** $3000 **25.** $50 **27.** $90 **29.** $4525 **31.** $565 **33.** 4 mi **35.** $104 **37.** $14,300 **39.** 5 ways **41.** 6 ways **43.** 9 ways **45.** 10 different total scores **47.** B owes $18 and C owes $2.; A is owed $14, D is owed $4, and E is owed $2.; B should give A $14 and D $4, while C should give E $2. **49.** 4 ways **51.** Andy, Darnell, Caleb, Beth, Ella **53.** Home, Bank, Post Office, Dry Cleaners, Home

55. Sample answer: CO, WY, UT, AZ, NM, CO, UT **57.** Bob's major is psychology.

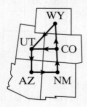

59. a.

5	22	18
28	15	2
12	8	25

b.

4	9	8
11	7	3
6	5	10

61.

9	6	7
8	1	4
2	2	5

63.
```
        156
   28)4368
        28
       156
       140
       168
       168
         0
```

65. the dentist with poor dental work **67.** Friday **69.** Sample answer:

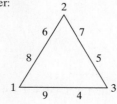

71. There is no missing dollar; in the end, the customers paid a total of $27 of which $25 went to the restaurant and $2 was stolen by the waiter.

73. Answers will vary; an example is

State	A	B	C	D
Congressional Seats	4	6	8	12

Chapter 1 Review Exercises

1. deductive reasoning: Answers will vary. **2.** inductive reasoning; Answers will vary. **3.** Each number in the list is obtained by adding 5 to the previous number.; 24 **4.** Each number in the list is obtained by multiplying the previous number by 2.; 112 **5.** The successive differences are consecutive counting numbers beginning with 2.; 21   **6.** Each number in the list is obtained by writing a fraction with a denominator that is one more than the denominator of the previous fraction before it is reduced to lowest terms.; $\frac{3}{8}$ **7.** Each number in the list is obtained by multiplying the previous number by $-\frac{1}{2}$; $\frac{5}{2}$ **8.** Each number in the list is obtained by subtracting 60 from the previous number; -200.

9. Each number beginning with the third number is the sum of the two previous numbers.; 42   **10.** To get the second number, multiply the first number by 3. Then multiply the second number by 2 to get the third number. Then multiply by 3 and then by 2, repeatedly.; 432   **11.** The figures alternate between squares and circles, and in each figure the tick mark has been rotated 90° clockwise from its position in the previous figure.;
12. $2 + 4 + 8 + 16 + 32 = 64 - 2$; correct   **13.** $444 \div 12 = 37$; correct
14. a. The result is the original number.
   b. Using n to represent the original number, we have
   Select a number: n
   Double the number: $2n$
   Add 4 to the product: $2n + 4$
   Divide the sum by 2: $\frac{2n + 4}{2} = n + 2$
   Subtract 2 from the quotient: $n + 2 - 2 = n$.

15. a. 4,257,500   **b.** 4,257,000   **c.** 4,300,000   **d.** 4,000,000   **16. a.** 1.5   **b.** 1.51   **c.** 1.507   **d.** 1.5065917   **17.** $\approx 2 + 4 + 10 = 16$; 15.71; quite well   **18.** $\approx 9 \times 50 = 450$; 432.67; somewhat reasonable   **19.** $\approx 20 \div 4 = 5$; ≈ 4.76; quite reasonable   **20.** $\approx 0.6 \times 4000 = 2400$; 2397.0548; quite reasonable   **21.** $\approx 8 + 1 + 3 + 0 + 1 + 5 = \$18$ **22.** $\approx 80 \times 8 = \$640$ **23.** books: $\approx 21 \times 1 = \$21$; chairs: $2 \times 12 = \$24$; plate: \$15; total: $\approx 21 + 24 + 15 = \$60$ **24.** $\approx 0.3 \times 1{,}600{,}000 = 480{,}000$ students   **25.** b   **26.** c   **27. a.** Asian; ≈ 122
b. $\approx 33 \times 30 = 990$ million people   **28. a.** $\approx 0.4\%$ per yr   **b.** 34%   **29. a.** 115 beats per min; 10 min   **b.** 64 beats per min; 8 min
c. between 9 and 10 minutes   **d.** 9 min   **30. a.** 0.5% per yr   **b.** $C = 42.4 - 0.5x$ **c.** 17.4%   **31.** the weight of the child
32. unnecessary information: \$20 given to driver; cost of trip: \$8.00   **33.** 8 lb   **34.** \$885   **35.** Plan A; \$90   **36.** 5.75 hr or 5 hr 45 min
37. \$15,500   **38.** 6 combinations

Chapter 1 Test A

1. deductive   **2.** inductive   **3.** 20   **4.** $\frac{1}{96}$ **5.** $3367 \times 15 = 50{,}505$ **6.**

7. a. The original number is doubled.
   b. Using n to represent the original number, we have
   Select a number: n
   Multiply the number by 4: $4n$
   Add 8 to the product: $4n + 8$
   Divide the sum by 2: $\frac{4n + 8}{2} = 2n + 4$
   Subtract 4 from the quotient: $2n + 4 - 4 = 2n$.
8. 3,300,000   **9.** 706.38   **10.** total expenses: $\approx 50 + 310 + 410 = \$770$; amount needed: $770 - 680 = \$90$ **11.** $\approx 500{,}000 \div 20 = \$25{,}000$ per person   **12.** $\approx 0.5 \times 120 = 60$ **13.** $\approx 0.1 \times 500 = 50$ billion pounds   **14.** a   **15. a.** 36 yr old; 31%   **b.** 16 yr old; 22%
c. between 56 yr old and 66 yr old   **d.** 46 yr old   **16. a.** 1.26 ppm per yr   **b.** $C = 310 + 1.26x$ **c.** 436 ppm   **17.** Estes Rental; \$12
18. \$14,080   **19.** 26 weeks   **20.** Belgium; 160,000 people

Chapter 1 Test B

1. inductive   **2.** deductive   **3.** 19   **4.** $\frac{1}{162}$ **5.** $1 + 3 + 5 + 7 + 9 + 11 = 6^2$ **6.**

7. a. The result is 5 times the original number.
   b. Using n to represent the original number, we have
   Select a number: n
   Multiply the number by 20: $20n$
   Add 12 to the product: $20n + 12$
   Divide this sum by 4: $\frac{20n + 12}{4} = 5n + 3$
   Subtract 3 from the quotient: $5n + 3 - 3 = 5n$.
8. 3,000,000   **9.** 706.385   **10.** total purchase: $\approx 5 + 2 + 1 + 7 = \$15$; change: $\approx 20 - 15 = \$5$ **11.** $\approx 40 \times 20 \times 50 = \$40{,}000$
12. $\approx \frac{500}{20} = 25$ **13.** $\approx 0.1 \times 17{,}000{,}000 = 1{,}7000{,}000$ students   **14.** $\approx 100{,}000 \div 25 = 4000$ gal   **15. a.** 8 yr old; 1 awakening per night
b. 65 yr old; 8 awakenings per night   **c.** 2.9 awakenings per night   **16. a.** $\approx 0.01°$F per yr   **b.** $T = 56.58 + 0.01x$ **c.** 58.08°F
17. \$246.40   **18.** allowing partial packs: \$13; considering only whole packs: \$15   **19.** 18 months or $1\frac{1}{2}$ yr   **20.** Company A; \$600

CHAPTER 2

Section 2.1

Checkpoint Exercises

1. L is the set of the first six lowercase letters of the alphabet.   **2.** $M = \{\text{April, August}\}$ **3.** $O = \{1, 3, 5, 7, 9\}$ **4.** b   **5. a.** true
b. true   **c.** false   **6. a.** $A = \{1, 2, 3\}$ **b.** $B = \{15, 16, 17, 18, \ldots\}$ **c.** $O = \{1, 3, 5, 7, \ldots\}$ **7. a.** $\{1, 2, 3, 4, \ldots, 199\}$

b. $\{51, 52, 53, 54, \ldots, 200\}$ **8. a.** $n(A) = 5$ **b.** $n(B) = 1$ **c.** $n(C) = 8$ **d.** $n(D) = 0$ **9.** No; the sets do not contain the same number of distinct elements. **10. a.** true **b.** false

Exercise Set 2.1

1. empty/null; $\{\ \}$ or $\varnothing$. **2.** an object is an element of a set **3.** natural numbers **4.** cardinality **5.** equivalent **6.** equal **7.** false
8. false **9.** false **10.** true **11.** false **12.** false **19.** makes sense **21.** does not make sense **23.** well defined; set
25. not well defined; not a set **27.** well defined; set **29.** the set of planets in our solar system **31.** the set of months that begin with J
33. the set of natural numbers greater than 5 **35.** the set of natural numbers between 6 and 20, inclusive **37.** $\{$winter, spring, summer, fall$\}$
39. $\{$September, October, November, December$\}$ **41.** $\{1, 2, 3\}$ **43.** $\{1, 3, 5, 7, 9, 11\}$ **45.** $\{1, 2, 3, 4, 5\}$ **47.** $\{6, 7, 8, 9, \ldots\}$
49. $\{7, 8, 9, 10\}$ **51.** $\{10, 11, 12, 13, \ldots, 79\}$ **53.** $\{2\}$ **55.** not the empty set **57.** empty set **59.** not the empty set **61.** empty set
63. empty set **65.** not the empty set **67.** not the empty set **69.** true **71.** true **73.** false **75.** true **77.** false **79.** false
81. true **83.** false **85.** false **87.** true **89.** 5 **91.** 15 **93.** 0 **95.** 1 **97.** 4 **99.** 5 **101.** 0 **103. a.** not equivalent;
Answers will vary. **b.** not equal; Answers will vary. **105. a.** equivalent; Answers will vary. **b.** not equal; Answers will vary.
107. a. equivalent; Answers will vary **b.** equal; Answers will vary. **109. a.** equivalent; Answers will vary. **b.** not equal; Answers will vary.
111. a. equivalent; Answers will vary. **b.** equal; Answers will vary. **113.** infinite **115.** finite **117.** finite **119.** $\{x \mid x \in \mathbf{N} \text{ and } x \geq 61\}$
121. $\{x \mid x \in \mathbf{N} \text{ and } 61 \leq x \leq 89\}$ **123.** $\{$Caracas, Mexico City, Nairobi$\}$ **125.** $\{$Beijing, Mumbai, Caracas, Mexico City$\}$
127. $\{$Tokyo, New York, Paris$\}$ **129.** $\{\ \}$ or $\varnothing$ **131.** $\{12, 19\}$ **133.** $\{20, 21\}$ **135.** no one-to-one correspondence; not equivalent
137. Answers will vary; an example is: $\{x \mid x \in \mathbf{N} \text{ and } x < 5\}$ and $\{1, 2, 3, 4\}$ **139.** Answers will vary; an example is: $\{1\}$ and $\{1, 2\}$

Section 2.2

Checkpoint Exercises

1. a. $\nsubseteq$ **b.** $\subseteq$ **c.** $\subseteq$ **2. a.** $\subseteq$, $\subset$ **b.** $\subseteq$, $\subset$ **3.** yes **4. a.** 16; 15 **b.** 64; 63 **5. a.** 8; There are eight different ways to choose some, all, or none of the movies to watch. **b.** $\varnothing$, $\{Citizen\ Kane\}$, $\{The\ Godfather\}$, $\{Casablanca\}$, $\{Citizen\ Kane, The\ Godfather\}$, $\{Citizen\ Kane, Casablanca\}$, $\{The\ Godfather, Casablanca\}$, $\{Citizen\ Kane, The\ Godfather, Casablanca\}$; These sets represent the eight different viewing options. **c.** 7

Exercise Set 2.2

1. $A \subseteq B$; every element in set A is also an element in set B **2.** $A \subset B$; sets A and B are not equal **3.** 2^n **4.** $2^n - 1$ **5.** false **6.** true
7. false **8.** false **15.** does not make sense **17.** does not make sense **19.** $\subseteq$ **21.** $\nsubseteq$ **23.** $\nsubseteq$ **25.** $\nsubseteq$ **27.** $\subseteq$ **29.** $\nsubseteq$
31. $\subseteq$ **33.** $\nsubseteq$ **35.** $\subseteq$ **37.** both **39.** $\subseteq$ **41.** neither **43.** both **45.** $\subseteq$ **47.** $\subseteq$ **49.** both **51.** both **53.** both
55. neither **57.** $\subseteq$ **59.** true **61.** false; Answers will vary. **63.** true **65.** false; Answers will vary. **67.** true **69.** false; Answers
will vary. **71.** true **73.** $\varnothing$, $\{$border collie$\}$, $\{$poodle$\}$, $\{$border collie, poodle$\}$ **75.** $\varnothing$, $\{t\}$, $\{a\}$, $\{b\}$, $\{t, a\}$, $\{t, b\}$, $\{a, b\}$, $\{t, a, b\}$
77. $\varnothing$ and $\{0\}$ **79.** 16; 15 **81.** 64; 63 **83.** 128; 127 **85.** 8; 7 **87.** false; The set $\{1, 2, 3, \ldots, 1000\}$ has $2^{1000} - 1$ proper subsets.
89. true **91.** false; $\{\varnothing\} \subseteq \{\varnothing, \{\varnothing\}\}$ **93.** true **95.** true **97.** 32 **99.** 64 **101.** 256 **103.** \$0.00, \$0.05, \$0.10, \$0.15, \$0.25, \$0.30,
\$0.35, \$0.40

Section 2.3

Checkpoint Exercises

1. a. $\{1, 5, 6, 7, 9\}$ **b.** $\{1, 5, 6\}$ **c.** $\{7, 9\}$ **2. a.** $\{a, b, c, d\}$ **b.** $\{e\}$ **c.** $\{e, f, g\}$ **d.** $\{f, g\}$ **3.** $\{b, c, e\}$ **4. a.** $\{7, 10\}$
b. $\varnothing$ **c.** $\varnothing$ **5. a.** $\{1, 3, 5, 6, 7, 10, 11\}$ **b.** $\{1, 2, 3, 4, 5, 6, 7\}$ **c.** $\{1, 2, 3\}$ **6. a.** $\{a, d\}$ **b.** $\{a, d\}$ **7. a.** $\{5\}$
b. $\{2, 3, 7, 11, 13, 17, 19\}$ **c.** $\{2, 3, 5, 7, 11, 13\}$ **d.** $\{17, 19\}$ **e.** $\{5, 7, 11, 13, 17, 19\}$ **f.** $\{2, 3\}$

Exercise Set 2.3

1. Venn diagrams **2.** complement; A' **3.** intersection; $A \cap B$ **4.** union; $A \cup B$ **5.** true **6.** false
7. true **8.** false **19.** does not make sense **21.** does not make sense **23.** the set of all composers **25.** the set of all brands of soft
drinks **27.** $\{c, d, e\}$ **29.** $\{b, c, d, e, f\}$ **31.** $\{6, 7, 8, 9, \ldots, 20\}$ **33.** $\{2, 4, 6, 8, \ldots, 20\}$ **35.** $\{21, 22, 23, 24, \ldots\}$ **37.** $\{1, 3, 5, 7, \ldots\}$
39. $\{1, 3\}$ **41.** $\{1, 2, 3, 5, 7\}$ **43.** $\{2, 4, 6\}$ **45.** $\{4, 6\}$ **47.** $\{1, 3, 5, 7\}$ or A **49.** $\{1, 2, 4, 6, 7\}$ **51.** $\{1, 2, 4, 6, 7\}$ **53.** $\{4, 6\}$
55. $\{1, 3, 5, 7\}$ or A **57.** $\varnothing$ **59.** $\{1, 2, 3, 4, 5, 6, 7\}$ or U **61.** $\{1, 3, 5, 7\}$ or A **63.** $\{g, h\}$ **65.** $\{a, b, g, h\}$ **67.** $\{b, c, d, e, f\}$ or C
69. $\{c, d, e, f\}$ **71.** $\{a, g, h\}$ or A **73.** $\{a, b, c, d, e, f, g, h\}$ or U **75.** $\{a, b, c, d, e, f, g, h\}$ or U **77.** $\{c, d, e, f\}$ **79.** $\{a, g, h\}$ or A
81. $\varnothing$ **83.** $\{a, b, c, d, e, f, g, h\}$ or U **85.** $\{a, g, h\}$ or A **87.** $\{a, c, d, e, f, g, h\}$ **89.** $\{1, 3, 4, 7\}$ **91.** $\{1, 2, 3, 4, 5, 6, 7, 8, 9\}$ **93.** $\{3, 7\}$
95. $\{1, 4, 8, 9\}$ **97.** $\{8, 9\}$ **99.** $\{1, 4\}$ **101.** $\{\Delta, \text{two, four, six}\}$ **103.** $\{\Delta, \#, \$, \text{two, four, six}\}$ **105.** 6 **107.** 5
109. $\{\#, \$, \text{two, four, six, 10, 01}\}$ **111.** $\{\text{two, four, six}\}$ **113.** 4 **115.** $\{1, 2, 3, 4, 5, 6, 7, 8\}$ or U **117.** $\{1, 3, 5, 7\}$ or A **119.** $\{1, 7\}$
121. $\{1, 3, 5, 6, 7, 8\}$ **123.** $\{$Ashley, Mike, Josh$\}$ **125.** $\{$Ashley, Mike, Josh, Emily, Hanna, Ethan$\}$ **127.** $\{$Ashley$\}$ **129.** $\{$Jacob$\}$
131. III **133.** I **135.** II **137.** I **139.** IV **141.** II **143.** III **145.** I **147.** true **149.** false **151.** false **153.** true

155. 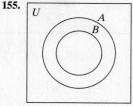 **157.**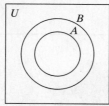

Section 2.4

Checkpoint Exercises

1. a. {a, b, c, d, f}   **b.** {a, b, c, d, f}   **c.** {a, b, d}   **2. a.** {5, 6, 7, 8, 9}   **b.** {1, 2, 5, 6, 7, 8, 9, 10, 12}   **c.** {5, 6, 7}   **d.** {3, 4, 6, 8, 11}
e. {1, 2, 3, 5, 6, 7, 8, 9, 10, 11, 12}   **3. a.** IV   **b.** IV   **c.** $(A \cup B)' = A' \cap B'$ **4. a.** II, IV, and V   **b.** II, IV, and V
c. $A \cap (B \cup C) = (A \cap B) \cup (A \cap C)$

Exercise Set 2.4

1. inside parentheses   **2.** eight   **3.** false   **4.** true   **7.** makes sense   **9.** does not make sense   **11.** {1, 2, 3, 5, 7}   **13.** {1, 2, 3, 5, 7}
15. {2}   **17.** {2}   **19.** $\varnothing$ **21.** {4, 6}   **23.** {a, b, g, h}   **25.** {a, b, g, h}   **27.** {b}   **29.** {b}   **31.** $\varnothing$ **33.** {c, d, e, f}
35. II, III, V, and VI   **37.** I, II, IV, V, VI, and VII   **39.** II and V   **41.** I, IV, VII, and VIII   **43.** {1, 2, 3, 4, 5, 6, 7, 8}
45. {1, 2, 3, 4, 5, 6, 7, 8, 9, 10, 11}   **47.** {12, 13}   **49.** {4, 5, 6}   **51.** {6}   **53.** {1, 2, 3, 4, 5, 7, 8, 9, 10, 11, 12, 13}   **55. a.** II   **b.** II
c. $A \cap B = B \cap A$ **57. a.** I, III, IV   **b.** IV   **c.** no; Answers will vary.   **59.** not equal   **61.** not equal   **63.** equal
65. a. II, IV, V, VI, and VII   **b.** II, IV, V, VI, and VII   **c.** $(A \cap B) \cup C = (A \cup C) \cap (B \cup C)$ **67. a.** II, IV, and V   **b.** I, II, IV, V, and VI
c. no; Answers will vary.   **69.** not true   **71.** true; theorem   **73.** true; theorem   **75. a.** {c, e, f}; {c, e, f}   **b.** {1, 3, 5, 7, 8}; {1, 3, 5, 7, 8}
c. $A \cup (B' \cap C') = (A \cup B') \cap (A \cup C')$ **d.** theorem   **77.** {Ann, Jose, Al, Gavin, Amy, Ron, Grace}   **79.** {Jose}   **81.** {Lily, Emma}
83. {Lily, Emma, Ann, Jose, Lee, Maria, Fred, Ben, Sheila, Ellen, Gary}   **85.** {Lily, Emma, Al, Gavin, Amy, Lee, Maria}
87. {Al, Gavin, Amy}   **89.** The set of students who scored 90% or above on exam 1 and exam 3 but not on exam 2   **91.** VI   **93.** V   **95.** I
97. III   **99.** V   **101.** VII   **103.** AB$^+$ **105.** no   **107.** $(A \cap B)'$ **109.** $A' \cup B$ **111.** $(A \cap B) \cup C$ **113.** $A' \cap (B \cup C)$

Section 2.5

Checkpoint Exercises

1. a. 75   **b.** 90   **c.** 20   **d.** 145   **e.** 55   **f.** 70   **g.** 30   **h.** 175   **2. a.**   **b.** 7   **c.** 3
3.   **4. a.** 63   **b.** 3   **c.** 136   **d.** 30
e. 228   **f.** 22

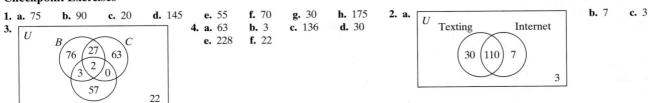

Exercise Set 2.5

1. and/but   **2.** or   **3.** not   **4.** false   **5.** true   **7.** makes sense   **9.** 26   **11.** 17   **13.** 37   **15.** 7   **17.** I: 14; III: 22; IV: 5
19. 17   **21.** 6   **23.** 28   **25.** 9   **27.** 3   **29.** 19   **31.** 21   **33.** 34   **35.** I: 5; II: 1; III: 4; IV: 3; VI: 1; VII: 8; VIII: 6   **37.** I: 5; II: 10;
III: 3; IV: 4; V: 7; VI: 1; VII: 6; VIII: 2   **39.** impossible; There are only 10 elements in set A but there are 13 elements in set A that are also in sets B or
C. A similar problem exists for set C.   **33.** 18   **35.** 9   **37.** 9   **41.**   **a.** 22   **b.** 36   **c.** 65   **d.** 10

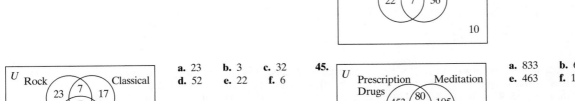

43.   **a.** 23   **b.** 3   **c.** 32   **45.**   **a.** 833   **b.** 675   **c.** 8   **d.** 80
d. 52   **e.** 22   **f.** 6   **e.** 463   **f.** 102   **g.** 683

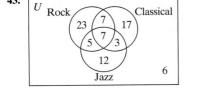

47. a. 0   **b.** 30   **c.** 60

Chapter 2 Review Exercises

1. the set of the days of the week beginning with the letter T   **2.** the set of natural numbers between 1 and 10, inclusive   **3.** {m, i, s}
4. {8, 9, 10, 11, 12}   **5.** {1, 2, 3, ..., 30}   **6.** not empty   **7.** empty set   **8.** $\in$ **9.** $\notin$ **10.** 12   **11.** 15   **12.** $\neq$ **13.** $\neq$
14. equivalent   **15.** both   **16.** finite   **17.** infinite   **18.** $\subseteq$ **19.** $\not\subseteq$ **20.** $\subseteq$ **21.** $\subseteq$ **22.** both   **23.** false; Answers will vary.
24. false; Answers will vary.   **25.** true   **26.** false; Answers will vary.   **27.** true   **28.** false; It has $2^1 = 2$ subsets.   **29.** true
30. $\varnothing$, {1}, {5}, {1, 5}; {1, 5}   **31.** 32; 31   **32.** 8; 7   **33.** {1, 2, 4}   **34.** {1, 2, 3, 4, 6, 7, 8}   **35.** {5}   **36.** {6, 7, 8}   **37.** {6, 7, 8}
38. {4, 5, 6}   **39.** {2, 3, 6, 7}   **40.** {1, 4, 5, 6, 8, 9}   **41.** {4, 5}   **42.** {1, 2, 3, 6, 7, 8, 9}   **43.** {2, 3, 7}   **44.** {6}
45. {1, 2, 3, 4, 5, 6, 7, 8, 9}   **46.** {1, 2, 3, 4, 5}   **47.** {1, 2, 3, 4, 5, 6, 7, 8} or U **48.** {c, d, e, f, k, p, r}   **49.** {f, p}   **50.** {c, d, f, k, p, r}
51. {c, d, e}   **52.** {a, b, c, d, e, g, h, p, r}   **53.** {f}

54. Use Figure 2.21 on page 89.

Set	Regions in the Venn Diagram
A	I, II
B	II, III
$A \cup B$	I, II, III
$(A \cup B)'$	IV

Set	Regions in the Venn Diagram
A'	III, IV
B'	I, IV
$A' \cap B'$	IV

Since $(A \cup B)'$ and $A' \cap B'$ are represented by the same region, $(A \cup B)' = A' \cap B'$.

55. Use Figure 2.22 on page 90.

Set	Regions in the Venn Diagram
A	I, II, IV, V
B	II, III, V, VI
C	IV, V, VI, VII
$B \cup C$	II, III, IV, V, VI, VII
$A \cap (B \cup C)$	II, IV, V

Set	Regions in the Venn Diagram
A	I, II, IV, V
B	II, III, V, VI
C	IV, V, VI, VII
$B \cap C$	V, VI
$A \cup (B \cap C)$	I, II, IV, V, VI

Since $A \cap (B \cup C)$ and $A \cup (B \cap C)$ are not represented by the same regions, $A \cap (B \cup C) = A \cup (B \cap C)$ is false.

56. United States: V; Italy: IV; Turkey: VIII; Norway: V; Pakistan: VIII; Iceland: V; Mexico: I

57.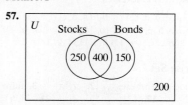

a. 250   **b.** 800   **c.** 200

58.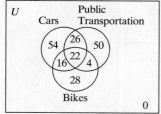

a. 50   **b.** 26   **c.** 130   **d.** 46   **e.** 0

Chapter 2 Test A

1. {18, 19, 20, 21, 22, 23, 24}   **2.** false; Answers will vary.   **3.** true   **4.** true   **5.** false; Answers will vary.   **6.** true   **7.** false; Answers will vary.   **8.** false; Answers will vary.   **9.** false; Answers will vary.   **10.** ∅, {6}, {9}, {6, 9}; {6, 9}   **11.** {a, b, c, d, e, f}   **12.** {a, b, c, d, f, g}   **13.** {b, c, d}   **14.** {a, e}   **15.** 5   **16.** {b, c, d, i, j, k}   **17.** {a}   **18.** {a, f, h}   **19.** theorem   **20. a.** V   **b.** VII   **c.** IV   **d.** I   **e.** VI   **21. a.**   **b.** 263   **c.** 25   **d.** 62   **e.** 0   **f.** 147   **g.** 116

Chapter 2 Test B

1. {8, 9, 10, 11, 12, 13}   **2.** false; Answers will vary.   **3.** true   **4.** true   **5.** false; Answers will vary.   **6.** true   **7.** false; Answers will vary.   **8.** true   **9.** true   **10.** ∅, {6}, {8}, {10}, {6, 8}, {6, 10}, {8, 10}, {6, 8, 10}; {6, 8, 10}   **11.** {1, 2, 3, 4, 5, 7}   **12.** {1, 2, 5, 6, 7}   **13.** {3, 4, 6}   **14.** {1, 3, 4}   **15.** 4   **16.** {2, 3, 5, 6, 7, 9}   **17.** {2, 3, 4, 5, 6, 7, 8, 9}   **18.** {1, 5, 6, 7, 8, 10, 11}   **19.** not a theorem   **20. a.** II   **b.** VI   **c.** VIII   **d.** VII   **e.** V   **21. a.**   **b.** 42   **c.** 17   **d.** 61   **e.** 40   **f.** 72   **g.** 14

CHAPTER 3

Section 3.1

Checkpoint Exercises

1. b **2.** $2^3 \cdot 3 \cdot 5$ **3.** 75 **4.** 96 **5.** 90 **6.** 5:00 P. M.

Exercise Set 3.1

1. prime **2.** composite **3.** greatest common divisor **4.** least common multiple **5.** false **6.** true **7.** false **8.** false **19.** does not make sense **21.** does not make sense **23. a.** yes **b.** no **c.** yes **d.** no **e.** no **f.** yes **g.** no **h.** no **i.** no **25. a.** yes **b.** yes **c.** yes **d.** no **e.** yes **f.** yes **g.** no **h.** no **i.** yes **27. a.** yes **b.** no **c.** yes **d.** no **e.** no **f.** no **g.** no **h.** no **i.** no **29. a.** yes **b.** yes **c.** yes **d.** no **e.** yes **f.** yes **g.** yes **h.** no **i.** yes **31. a.** yes **b.** yes **c.** yes **d.** yes **e.** yes **f.** yes **g.** yes **h.** yes **i.** yes **33.** true; $5 + 9 + 5 + 8 = 27$ which is divisible by 3. **35.** true; the last two digits of 10,612 form the number 12 and 12 is divisible by 4. **37.** false **39.** true; 104,538 is an even number so it is divisible by 2. $1 + 0 + 4 + 5 + 3 + 8 = 21$ which is divisible by 3 so it is divisible by 3. Any number divisible by 2 and 3 is divisible by 6. **41.** true; the last three digits of 20,104 form the number 104 and 104 is divisible by 8. **43.** false **45.** true; $5 + 1 + 7 + 8 + 7 + 2 = 30$ which is divisible by 3 so it is divisible by 3. The last two digits of 517,872 form the number 72 and 72 is divisible by 4 so it is divisible by 4. Any number divisible by 3 and 4 is divisible by 12. **47.** $3 \cdot 5^2$ **49.** $2^3 \cdot 7$ **51.** $3 \cdot 5 \cdot 7$ **53.** $2^2 \cdot 5^3$ **55.** $3 \cdot 13 \cdot 17$ **57.** $3 \cdot 5 \cdot 59$ **59.** $2^5 \cdot 3^2 \cdot 5$ **61.** $2^2 \cdot 499$ **63.** $3 \cdot 5^2 \cdot 7^2$ **65.** $2^3 \cdot 3 \cdot 5^2 \cdot 11 \cdot 13$ **67.** 14 **69.** 2 **71.** 12 **73.** 24 **75.** 38 **77.** 15 **79.** 168 **81.** 336 **83.** 540 **85.** 360 **87.** 3420 **89.** 4560 **91.** 8 **93.** 6 **95.** 2, 6 **97.** perfect **99.** not perfect **101.** not an emirp **103.** emirp **105.** not a Germain prime **107.** not a Germain prime **109.** 648; 648; Answers will vary. An example is: The product of the greatest common divisor and least common multiple of two numbers equals the product of the two numbers. **111.** The numbers are the prime numbers between 2 and 97. **113.** 12 **115.** 10 **117.** July 1 **119.** 90 min or $1\frac{1}{2}$ hr **121.** Answers will vary; an example is 1020. **123.** 53 **125.** 2 and 3 **127.** no

Section 3.2

Checkpoint Exercises

1. (number line with points at $-4, -1, 3$; labeled (a), (b), (c); range -5 to 5) **2. a.** $>$ **b.** $<$ **c.** $<$ **d.** $<$ **3. a.** 8 **b.** 6 **c.** -8 **4. a.** 37 **b.** -4 **c.** -24 **5.** \$297 billion **6. a.** 25 **b.** -25 **c.** -64 **d.** 81 **7.** 36 **8.** 82

Exercise Set 3.2

1. $\{\ldots, -3, -2, -1, 0, 1, 2, 3, \ldots\}$ **2.** left **3.** the distance from 0 to a **4.** additive inverses **5.** false **6.** true **7.** true **8.** false **17.** makes sense **19.** does not make sense **21.** (number line, point at 3, range -5 to 5) **23.** (number line, point at -4, range -5 to 5) **25.** $<$ **27.** $<$ **29.** $>$ **31.** $<$ **33.** 14 **35.** 14 **37.** 300,000 **39.** -12 **41.** 4 **43.** -3 **45.** -5 **47.** -18 **49.** 0 **51.** 5 **53.** -7 **55.** 14 **57.** 11 **59.** -9 **61.** -28 **63.** -54 **65.** 21 **67.** -12 **69.** 13 **71.** 0 **73.** 25 **75.** 25 **77.** 64 **79.** -125 **81.** 625 **83.** -81 **85.** 81 **87.** -3 **89.** -7 **91.** 30 **93.** 0 **95.** undefined **97.** -20 **99.** -31 **101.** 25 **103.** 13 **105.** 13 **107.** 33 **109.** 32 **111.** -24 **113.** 0 **115.** -32 **117.** 45 **119.** 14 **121.** 14 **123.** -36 **125.** 88 **127.** -29 **129.** -10 **131.** $-10 - (-2)^3; -2$ **133.** $[2(7 - 10)]^2; 36$ **135.** 20,602 ft **137.** shrink by 5 years **139.** shrink by 11 years **141.** 10 years **143.** no change **145.** 9 years **147.** $-\$375$ billion; deficit **149.** \$587 billion **151.** $7\,°F$ **153.** $-2\,°F$ **155.** $8 - 2 \cdot (3 - 4) = 10$ **157.** -7 **159.** 150

Section 3.3

Checkpoint Exercises

1. $\frac{4}{5}$ **2.** $\frac{21}{8}$ **3.** $1\frac{2}{3}$ **4. a.** 0.375 **b.** $0.\overline{45}$ **5. a.** $\frac{9}{10}$ **b.** $\frac{43}{50}$ **c.** $\frac{53}{1000}$ **6.** $\frac{2}{9}$ **7.** $\frac{79}{99}$ **8. a.** $\frac{8}{33}$ **b.** $\frac{3}{2}$ or $1\frac{1}{2}$ **c.** $\frac{51}{10}$ or $5\frac{1}{10}$ **9. a.** $\frac{36}{55}$ **b.** $-\frac{4}{3}$ or $-1\frac{1}{3}$ **c.** $\frac{3}{2}$ or $1\frac{1}{2}$ **10. a.** $\frac{2}{3}$ **b.** $\frac{3}{2}$ or $1\frac{1}{2}$ **c.** $-\frac{9}{4}$ or $-2\frac{1}{4}$ **11.** $\frac{19}{20}$ **12.** $-\frac{17}{60}$ **13.** $\frac{3}{4}$ **14.** $\frac{5}{12}$ **15.** $2\frac{4}{5}$ eggs; 3 eggs

Exercise Set 3.3

1. rational numbers; integers; zero **2.** an improper fraction; the numerator is greater than the denominator **3.** terminate/stop; have repeating digits **4.** reciprocal/multiplicative inverse **5.** false **6.** false **7.** true **8.** false **19.** does not make sense **21.** does not make sense **23.** $\frac{2}{3}$ **25.** $\frac{5}{6}$ **27.** $\frac{4}{7}$ **29.** $\frac{5}{9}$ **31.** $\frac{9}{10}$ **33.** $\frac{14}{19}$ **35.** $\frac{19}{8}$ **37.** $-\frac{38}{5}$ **39.** $\frac{199}{16}$ **41.** $4\frac{3}{5}$ **43.** $-8\frac{4}{9}$ **45.** $35\frac{11}{20}$ **47.** 0.75 **49.** 0.35 **51.** 0.875 **53.** $0.\overline{81}$ **55.** $3.\overline{142857}$ **57.** $0.\overline{285714}$ **59.** $\frac{3}{10}$ **61.** $\frac{2}{5}$ **63.** $\frac{39}{100}$ **65.** $\frac{41}{50}$ **67.** $\frac{29}{40}$ **69.** $\frac{5399}{10,000}$ **71.** $\frac{7}{9}$ **73.** 1 **75.** $\frac{4}{11}$ **77.** $\frac{257}{999}$ **79.** $\frac{21}{88}$ **81.** $-\frac{7}{120}$ **83.** $\frac{3}{2}$ **85.** 6 **87.** $\frac{10}{3}$ **89.** $-\frac{14}{15}$ **91.** 6 **93.** $\frac{5}{11}$ **95.** $\frac{2}{3}$ **97.** $\frac{2}{3}$ **99.** $\frac{7}{10}$ **101.** $\frac{9}{10}$ **103.** $\frac{53}{120}$ **105.** $\frac{1}{2}$ **107.** $\frac{7}{12}$ **109.** $-\frac{71}{150}$ **111.** $\frac{53}{12}$ or $4\frac{5}{12}$ **113.** $\frac{7}{6}$ or $1\frac{1}{6}$ **115.** $-\frac{5}{2}$ or $-2\frac{1}{2}$

117. $\dfrac{11}{14}$ **119.** $\dfrac{4}{15}$ **121.** $-\dfrac{9}{40}$ **123.** $-1\dfrac{1}{36}$ **125.** $-19\dfrac{3}{4}$ **127.** $\dfrac{7}{24}$ **129.** $\dfrac{7}{12}$ **131.** $-\dfrac{3}{4}$ **133.** Both are equal to $\dfrac{169}{36}$.

135. $\dfrac{1}{2^2 \cdot 3}$ **137.** $-\dfrac{289}{2^4 \cdot 5^4 \cdot 7}$ **139.** $0.\overline{54}; 0.58\overline{3}; <$ **141.** $-0.8\overline{3}; -0.\overline{8}; >$ **143. a.** $\dfrac{560}{1600} = \dfrac{7}{20}$ **b.** $0.35 = 35\%$ **c.** $\dfrac{720}{1600} = \dfrac{9}{20}$

d. $0.45 = 45\%$ **e.** 10% **145.** $\dfrac{1}{3}$ cup butter, $\dfrac{5}{2} = 2.5$ ounces unsweetened chocolate, $\dfrac{3}{4}$ cup sugar, 1 teaspoon vanilla, 1 egg, $\dfrac{1}{2}$ cup flour

147. $\dfrac{5}{6}$ cup butter, $\dfrac{25}{4} = 6.25$ ounces unsweetened chocolate, $\dfrac{15}{8} = 1\dfrac{7}{8}$ cups sugar, $\dfrac{5}{2} = 2\dfrac{1}{2}$ teaspoons vanilla, $\dfrac{5}{2} = 2\dfrac{1}{2}$ eggs, $\dfrac{5}{4} = 1\dfrac{1}{4}$ cups flour

149. 24 brownies **151.** $3\dfrac{2}{3}$ c **153. a.** D, E, G, A, B **b.** There are black keys to the left of the keys for the notes D, E, G, A, and B.

155. $16\dfrac{7}{16}$ in. **157.** $\dfrac{1}{3}$ of the business **159.** $1\dfrac{3}{20}$ mi; $\dfrac{7}{20}$ mi **161.** $\dfrac{1}{10}$

163.

say does that   Star-span-gled Ban-ner   yet   wave O'er the

165. a. 0.24625 **b.** 1.45248 **c.** 0.00112

Section 3.4

Checkpoint Exercises

1. a. $2\sqrt{3}$ **b.** $2\sqrt{15}$ **c.** cannot be simplified **2. a.** $\sqrt{30}$ **b.** 10 **c.** $2\sqrt{3}$ **3. a.** 4 **b.** $2\sqrt{2}$ **4. a.** $18\sqrt{3}$ **b.** $-5\sqrt{13}$

c. $8\sqrt{10}$ **5. a.** $3\sqrt{3}$ **b.** $-13\sqrt{2}$ **6. a.** $\dfrac{5\sqrt{10}}{2}$ **b.** $\dfrac{\sqrt{14}}{7}$ **c.** $\dfrac{5\sqrt{2}}{6}$

Exercise Set 3.4

1. terminating; repeating **2.** π **3.** $\sqrt{n}; n$ **4.** coefficient **5.** false **6.** false **7.** true **8.** false **15.** makes sense **17.** does not make sense **19.** 3 **21.** 5 **23.** 8 **25.** 11 **27.** 13 **29. a.** 13.2 **b.** 13.15 **c.** 13.153 **31. a.** 133.3 **b.** 133.27 **c.** 133.270 **33. a.** 1.8 **b.** 1.77 **c.** 1.772 **35.** $2\sqrt{5}$ **37.** $4\sqrt{5}$ **39.** $5\sqrt{10}$ **41.** $14\sqrt{7}$ **43.** $\sqrt{42}$ **45.** 6 **47.** $3\sqrt{2}$ **49.** $2\sqrt{13}$ **51.** 3 **53.** $3\sqrt{5}$ **55.** $-4\sqrt{3}$ **57.** $13\sqrt{3}$ **59.** $-2\sqrt{13}$ **61.** $2\sqrt{5}$ **63.** $7\sqrt{2}$ **65.** $3\sqrt{5}$ **67.** $2\sqrt{2}$ **69.** $34\sqrt{2}$

71. $-\dfrac{3}{2}\sqrt{3}$ **73.** $11\sqrt{3}$ **75.** $\dfrac{5\sqrt{3}}{3}$ **77.** $3\sqrt{7}$ **79.** $\dfrac{2\sqrt{30}}{5}$ **81.** $\dfrac{5\sqrt{3}}{2}$ **83.** $\dfrac{\sqrt{10}}{5}$ **85.** $20\sqrt{2} - 5\sqrt{3}$ **87.** $-7\sqrt{7}$ **89.** $\dfrac{43\sqrt{2}}{35}$

91. $\dfrac{5\sqrt{6}}{6}$ **93.** $6\sqrt{3}$ miles; 10.4 miles **95.** 70 mph; He was speeding. **97. a.** 41 in. **b.** 40.6 in.; quite well **99. a.** 36 cm **b.** 44.7 cm

c. 46.9 cm **d.** The model describes healthy children. **101.** $0.6R_f$; 60 weeks **103.** A = 11; B = 8; C = 1; D = 17; E = 10; F = 2; G = 5; H = 19; I = 13; J = 7; K = 26; L = 6; M = 22; N = 20; O = 15; P = 16; Q = 24; R = 9; S = 23; T = 12; U = 3; V = 25; W = 4; X = 18; Y = 21; Z = 14; Paige Fox is bad at math. **105.** $<$ **107.** $>$ **109.** -7 and -6 **111.** Answers will vary.; An example is $(3 + \sqrt{2}) - (1 + \sqrt{2}) = 2$.

Section 3.5

Checkpoint Exercises

1. a. $\sqrt{9}$ **b.** $0, \sqrt{9}$ **c.** $-9, 0, \sqrt{9}$ **d.** $-9, -1.3, 0, 0.\overline{3}, \sqrt{9}$ **e.** $\dfrac{\pi}{2}, \sqrt{10}$ **f.** $-9, -1.3, 0, 0.\overline{3}, \dfrac{\pi}{2}, \sqrt{9}, \sqrt{10}$ **2. a.** associative property of multiplication **b.** commutative property of addition **c.** distributive property of multiplication over addition **d.** commutative property of multiplication **e.** identity property of addition **f.** inverse property of multiplication **3. a.** yes **b.** no

Exercise Set 3.5

1. rational; irrational **2.** closure **3.** $ab = ba$ **4.** $(a + b) + c = a + (b + c)$ **5.** $a(b + c) = ab + ac$ **6.** identity **7.** identity; 1 **8.** multiplicative inverse; multiplicative identity **9.** false **10.** false **11.** true **12.** false **13.** false **14.** false **15.** true **16.** false **25.** makes sense **27.** does not make sense **29. a.** $\sqrt{100}$ **b.** $0, \sqrt{100}$ **c.** $-9, 0, \sqrt{100}$ **d.** $-9, -\dfrac{4}{5}, 0, 0.25, 9.2, \sqrt{100}$

e. $\sqrt{3}$ **f.** $-9, -\dfrac{4}{5}, 0, 0.25, \sqrt{3}, 9.2, \sqrt{100}$ **31. a.** $\sqrt{64}$ **b.** $0, \sqrt{64}$ **c.** $-11, 0, \sqrt{64}$ **d.** $-11, -\dfrac{5}{6}, 0, 0.75, \sqrt{64}$ **e.** $\sqrt{5}, \pi$

f. $-11, -\dfrac{5}{6}, 0, 0.75, \sqrt{5}, \pi, \sqrt{64}$ **33.** 0 **35.** Answers will vary; an example is: $\dfrac{1}{2}$. **37.** Answers will vary; an example is: 1. **39.** Answers will vary; an example is: $\sqrt{2}$. **41.** 4 **43.** 6 **45.** 4; 3 **47.** 7 **49.** $30 + 5\sqrt{2}$ **51.** $3\sqrt{7} + \sqrt{14}$ **53.** $5\sqrt{3} + 3$ **55.** $2\sqrt{3} + 6$ **57.** commutative property of addition **59.** associative property of addition **61.** commutative property of addition **63.** distributive property of multiplication over addition **65.** associative property of multiplication **67.** identity property of multiplication **69.** inverse property of addition **71.** inverse property of multiplication **73.** Answers will vary; an example is: $1 - 2 = -1$. **75.** Answers will vary; an example is: $4 \div 8 = \dfrac{1}{2}$. **77.** Answers will vary; an example is: $\sqrt{2} \cdot \sqrt{2} = 2$. **79.** true **81.** false **83.** vampire **85.** vampire **87.** narcissistic **89.** narcissistic **91.** distributive property; commutative property of addition; associative property of addition; commutative property of addition **93.** Answers will vary. An example is: In a group of at least 367 people, at least two people will have the same birthday, but a subset of the group containing 50 people would not necessarily have two people with the same birthday.

Section 3.6

Checkpoint Exercises

1. a. 1 **b.** 1 **c.** 1 **d.** -1 **2. a.** $\dfrac{1}{81}$ **b.** $\dfrac{1}{216}$ **c.** $\dfrac{1}{12}$ **3. a.** 7,400,000,000 **b.** 0.000003017 **4. a.** 7.41×10^9

b. 9.2×10^{-8} **5. a.** 3.07×10^8 **b.** billion **6.** 520,000 **7.** 0.0000023 **8. a.** 1.872×10^4 **b.** 18,720 **9.** $2560

Exercise Set 3.6

1. add **2.** multiply **3.** subtract **4.** one **5.** a number greater than or equal to 1 and less than 10; 10 to an integer power **6.** false
7. false **8.** false **9.** true **10.** false **21.** makes sense **23.** does not make sense **25.** $2^5 = 32$ **27.** $4^3 = 64$ **29.** $2^6 = 64$
31. $1^{20} = 1$ **33.** $4^2 = 16$ **35.** $2^4 = 16$ **37.** 1 **39.** 1 **41.** -1 **43.** $\frac{1}{4}$ **45.** $\frac{1}{64}$ **47.** $\frac{1}{32}$ **49.** $3^2 = 9$ **51.** $3^{-2} = \frac{1}{9}$
53. $2^{-4} = \frac{1}{16}$ **55.** 270 **57.** 912,000 **59.** 80,000,000 **61.** 100,000 **63.** 0.79 **65.** 0.0215 **67.** 0.000786 **69.** 0.00000318
71. 3.7×10^2 **73.** 3.6×10^3 **75.** 3.2×10^4 **77.** 2.2×10^8 **79.** 2.7×10^{-2} **81.** 3.7×10^{-3} **83.** 2.93×10^{-6} **85.** 8.2×10^7
87. 4.1×10^5 **89.** 2.1×10^{-6} **91.** 600,000 **93.** 60,000 **95.** 0.123 **97.** 30,000 **99.** 3,000,000 **101.** 0.03 **103.** 0.0021
105. $(8.2 \times 10^7)(3.0 \times 10^9) = 2.46 \times 10^{17}$ **107.** $(5.0 \times 10^{-4})(6.0 \times 10^6) = 3 \times 10^3$ **109.** $\frac{9.5 \times 10^6}{5 \times 10^2} = 1.9 \times 10^4$
111. $\frac{8 \times 10^{-5}}{2 \times 10^2} = 4 \times 10^{-7}$ **113.** $\frac{4.8 \times 10^{11}}{1.2 \times 10^{-4}} = 4 \times 10^{15}$ **115.** $\frac{11}{18}$ **117.** $\frac{99}{25} = 3\frac{24}{25}$ **119.** 2.5×10^{-3} **121.** 8×10^{-5}
123. a. 1.35×10^{12} **b.** 3.07×10^8 **c.** $\$4.40 \times 10^3$; $\$4400$ **125.** 42,000 years **127. a.** 5.32×10^7 **b.** 5.554×10^{11}
c. $\$1.04 \times 10^4$; $\$10,400$ **129.** $\$9.792 \times 10^9$ **131.** 1.06×10^{-18} g **133.** false **135.** false **137.** false **139.** true **141.** $\frac{1}{4}$

Section 3.7

Checkpoint Exercises

1. 100, 120, 140, 160, 180, and 200 **2.** 8, 5, 2, -1, -4, and -7 **3.** -34 **4. a.** $a_n = -0.08n + 50.28$ **b.** 49.48%
5. 12, -6, 3, $-\frac{3}{2}$, $\frac{3}{4}$, and $-\frac{3}{8}$ **6.** 3645 **7.** $a_n = 3(2)^{n-1}$; 384

Exercise Set 3.7

1. arithmetic; common difference **2.** $a_n = a_1 + (n-1)d$; the first term; the common difference **3.** geometric; common ratio
4. $a_n = a_1 r^{n-1}$; the first term; the common ratio **5.** false **6.** false **7.** false **8.** false **9.** false **10.** false **11.** true
19. does not make sense **21.** makes sense **23.** 8, 10, 12, 14, 16, and 18 **25.** 200, 220, 240, 260, 280, and 300 **27.** $-7, -3, 1, 5,$
9, and 13 **29.** $-400, -100, 200, 500, 800,$ and 1100 **31.** 7, 4, 1, -2, -5, and -8 **33.** 200, 140, 80, 20, -40, and -100 **35.** $\frac{5}{2}, 3, \frac{7}{2}, 4, \frac{9}{2},$ and 5
37. $\frac{3}{2}, \frac{7}{4}, 2, \frac{9}{4}, \frac{5}{2},$ and $\frac{11}{4}$ **39.** 4.25, 4.55, 4.85, 5.15, 5.45, and 5.75 **41.** 4.5, 3.75, 3, 2.25, 1.5, and 0.75 **43.** 33 **45.** 252 **47.** 67 **49.** 955
51. -82 **53.** -142 **55.** -43 **57.** -248 **59.** $\frac{23}{2}$ **61.** 1.75 **63.** $a_n = 1 + (n-1)4$; 77 **65.** $a_n = 7 + (n-1)(-4)$; -69
67. $a_n = 9 + (n-1)2$; 47 **69.** $a_n = -20 + (n-1)(-4)$; -96 **71.** 4, 8, 16, 32, 64, and 128 **73.** 1000, 1000, 1000, 1000, 1000, and 1000
75. 3, -6, 12, -24, 48, and -96 **77.** 10, -40, 160, -640, 2560, and $-10,240$ **79.** 2000, -2000, 2000, -2000, 2000, and -2000
81. -2, 6, -18, 54, -162, and 486 **83.** -6, 30, -150, 750, -3750, and 18,750 **85.** $\frac{1}{4}, \frac{1}{2}, 1, 2, 4,$ and 8 **87.** $\frac{1}{4}, \frac{1}{8}, \frac{1}{16}, \frac{1}{32}, \frac{1}{64},$ and $\frac{1}{128}$
89. $-\frac{1}{16}, \frac{1}{4}, -1, 4, -16,$ and 64 **91.** 2, 0.2, 0.02, 0.002, 0.0002, and 0.00002 **93.** 256 **95.** 2,324,522,934 $\approx 2.32 \times 10^9$ **97.** 50 **99.** 320
101. -2 **103.** 486 **105.** $\frac{3}{64}$ **107.** $-\frac{2}{27}$ **109.** $\approx -1.82 \times 10^{-9}$ **111.** 0.1 **113.** $a_n = 3(4)^{n-1}$; 12,288 **115.** $a_n = 18\left(\frac{1}{3}\right)^{n-1}$; $\frac{2}{81}$
117. $a_n = 1.5(-2)^{n-1}$; 96 **119.** $a_n = 0.0004(-10)^{n-1}$; 400 **121.** arithmetic; 18 and 22 **123.** geometric; 405 and 1215 **125.** arithmetic;
13 and 18 **127.** geometric; $\frac{3}{16}$ and $\frac{3}{32}$ **129.** arithmetic; $\frac{5}{2}$ and 3 **131.** geometric; 7 and -7 **133.** arithmetic; -49 and -63
135. geometric; $25\sqrt{5}$ and 125 **137.** arithmetic; 310 **139.** geometric; 59,048 **141.** geometric; -1023 **143.** arithmetic; 80 **145.** 5050
147. a. $a_n = 0.5n + 10.5$ **b.** 35.5% **149.** Company A; $\$1400$ **151.** $\$16,384$ **153.** $\$3,795,957$ **155. a.** approximately 1.01 for each
division **b.** $a_n = 35.48(1.01)^{n-1}$ **c.** 38.04 million **157.** Company B

Chapter 3 Review Exercises

1. 2: yes; 3: yes; 4: yes; 5: no; 6: yes; 8: yes; 9: no; 10: no; 12: yes **2.** 2: yes; 3: yes; 4: no; 5: yes; 6: yes; 8: no; 9: yes; 10: yes; 12: no **3.** $3 \cdot 5 \cdot 47$
4. $2^6 \cdot 3 \cdot 5$ **5.** $3 \cdot 5^2 \cdot 7 \cdot 13$ **6.** GCD: 6; LCM: 240 **7.** GCD: 6; LCM: 900 **8.** GCD: 2; LCM: 27,432 **9.** 12 **10.** 11:48 A.M.
11. $<$ **12.** $>$ **13.** 860 **14.** 53 **15.** 0 **16.** -3 **17.** -11 **18.** -15 **19.** 1 **20.** 99 **21.** -15 **22.** 9 **23.** -4
24. -16 **25.** -16 **26.** 10 **27.** -2 **28.** 17 **29. a.** $-\$162$ billion **b.** $\$735$ billion **30.** $\frac{8}{15}$ **31.** $\frac{6}{25}$ **32.** $\frac{11}{12}$ **33.** $\frac{64}{11}$
34. $-\frac{23}{7}$ **35.** $5\frac{2}{5}$ **36.** $-1\frac{8}{9}$ **37.** 0.8 **38.** $0.\overline{428571}$ **39.** 0.625 **40.** 0.5625 **41.** $\frac{3}{5}$ **42.** $\frac{17}{25}$ **43.** $\frac{147}{250}$ **44.** $\frac{21}{2500}$ **45.** $\frac{5}{9}$
46. $\frac{34}{99}$ **47.** $\frac{113}{999}$ **48.** $\frac{21}{50}$ **49.** $\frac{35}{6}$ **50.** $\frac{8}{3}$ **51.** $-\frac{1}{4}$ **52.** $\frac{2}{3}$ **53.** $\frac{43}{36}$ **54.** $\frac{37}{60}$ **55.** $\frac{11}{15}$ **56.** $\frac{5}{16}$ **57.** $-\frac{2}{5}$ **58.** $-\frac{20}{3}$ or
$-6\frac{2}{3}$ **59.** $\frac{15}{112}$ **60.** $\frac{27}{40}$ **61.** $11\frac{1}{4}$ or about 11 pounds **62.** $\frac{5}{12}$ of the tank **63.** $2\sqrt{7}$ **64.** $6\sqrt{2}$ **65.** $5\sqrt{6}$ **66.** $10\sqrt{3}$
67. $4\sqrt{3}$ **68.** $5\sqrt{2}$ **69.** $2\sqrt{3}$ **70.** 3 **71.** $5\sqrt{5}$ **72.** $-6\sqrt{11}$ **73.** $7\sqrt{2}$ **74.** $-17\sqrt{3}$ **75.** $12\sqrt{2}$ **76.** $6\sqrt{5}$ **77.** $\frac{\sqrt{6}}{3}$
78. $8\sqrt{3} \approx 13.9$ feet per second **79. a.** $\sqrt{81}$ **b.** 0, $\sqrt{81}$ **c.** $-17, 0, \sqrt{81}$ **d.** $-17, -\frac{9}{13}, 0, 0.75, \sqrt{81}$ **e.** $\sqrt{2}, \pi$
f. $-17, -\frac{9}{13}, 0, 0.75, \sqrt{2}, \pi, \sqrt{81}$ **80.** Answers will vary; an example is: 0. **81.** Answers will vary; an example is: $\frac{1}{2}$.

82. Answers will vary; an example is: $\sqrt{2}$.   **83.** commutative property of addition   **84.** associative property of multiplication   **85.** distributive property of multiplication over addition   **86.** commutative property of multiplication   **87.** commutative property of multiplication   **88.** commutative property of addition   **89.** inverse property of multiplication   **90.** identity property of multiplication   **91.** Answers will vary; an example is: $2 \div 6 = \frac{1}{3}$.   **92.** Answers will vary; an example is: $0 - 2 = -2$.   **93.** 216   **94.** 64   **95.** 16   **96.** 729   **97.** 25   **98.** 1   **99.** 1

100. $\frac{1}{216}$ **101.** $\frac{1}{16}$ **102.** $\frac{1}{49}$ **103.** 27   **104.** 460   **105.** 37,400   **106.** 0.00255   **107.** 0.0000745   **108.** 7.52×10^3 **109.** 3.59×10^6 **110.** 7.25×10^{-3} **111.** 4.09×10^{-7} **112.** 4.2×10^{13} **113.** 9.7×10^{-5} **114.** 390   **115.** 1,150,000   **116.** 0.023

117. 40   **118.** $(6.0 \times 10^4)(5.4 \times 10^5) = 3.24 \times 10^{10}$ **119.** $(9.1 \times 10^4)(4 \times 10^{-4}) = 3.64 \times 10^1$ **120.** $\frac{8.4 \times 10^6}{4 \times 10^3} = 2.1 \times 10^3$

121. $\frac{3 \times 10^{-6}}{6 \times 10^{-8}} = 5 \times 10^1$ **122.** $\$5.36 \times 10^{10}$ **123.** 3.07×10^8 **124.** approximately $\$175$ **125.** 2.88×10^{13} **126.** 7, 11, 15, 19, 23, and 27

127. $-4, -9, -14, -19, -24,$ and -29 **128.** $\frac{3}{2}, 1, \frac{1}{2}, 0, -\frac{1}{2},$ and -1 **129.** 20   **130.** -30 **131.** -38 **132.** $a_n = -7 + (n-1)4; 69$

133. $a_n = 200 + (n-1)(-20); -180$ **134.** 3, 6, 12, 24, 48, and 96   **135.** $\frac{1}{2}, \frac{1}{4}, \frac{1}{8}, \frac{1}{16}, \frac{1}{32},$ and $\frac{1}{64}$ **136.** $16, -8, 4, -2, 1,$ and $-\frac{1}{2}$ **137.** 54

138. $\frac{1}{2}$ **139.** -48 **140.** $a_n = 1(2)^{n-1}; 128$ **141.** $a_n = 100\left(\frac{1}{10}\right)^{n-1}; \frac{1}{100,000}$ **142.** arithmetic; 24 and 29   **143.** geometric; 162 and 486   **144.** geometric; $\frac{1}{256}$ and $\frac{1}{1024}$ **145.** arithmetic; -28 and -35 **146. a.** $a_n = 4.75n + 34.25$ **b.** 96%

147. a. $\frac{5.9}{4.2} \approx 1.4; \frac{8.3}{5.9} \approx 1.4; \frac{11.6}{8.3} \approx 1.4; \frac{16.2}{11.6} \approx 1.4; \frac{22.7}{16.2} \approx 1.4$ **b.** $4.2(1.4)^{n-1}$ **c.** 62.0 million

Chapter 3 Test A

1. 2, 3, 4, 6, 8, 9, and 12   **2.** $2^2 \cdot 3^2 \cdot 7$ **3.** GCD: 24; LCM: 144   **4.** 1   **5.** -4 **6.** -32 **7.** $0.58\overline{3}$ **8.** $\frac{64}{99}$ **9.** $\frac{1}{5}$ **10.** $\frac{37}{60}$

11. $-\frac{19}{2}$ **12.** $\frac{7}{12}$ **13.** $5\sqrt{2}$ **14.** $9\sqrt{2}$ **15.** $3\sqrt{2}$ **16.** $-7, -\frac{4}{5}, 0, 0.25, \sqrt{4},$ and $\frac{22}{7}$ **17.** commutative property of addition

18. distributive property of multiplication over addition   **19.** 243   **20.** 64   **21.** $\frac{1}{64}$ **22.** 7500   **23.** $\frac{4.9 \times 10^4}{7 \times 10^{-3}} = 7 \times 10^6$ **24. a.** $\$1.44 \times 10^{13}$ **b.** approximately $\$46,900$ **25.** 1.86×10^{12} ounces   **26.** $1, -4, -9, -14, -19,$ and -24 **27.** 22   **28.** $16, 8, 4, 2, 1,$ and $\frac{1}{2}$ **29.** 320

Chapter 3 Test B

1. 2, 3, 4, 6, 8, 9, and 12   **2.** $2^3 \cdot 3^2 \cdot 5$ **3.** GCD: 36; LCM: 216   **4.** -2 **5.** -6 **6.** -9000 **7.** $0.91\overline{6}$ **8.** $\frac{51}{99}$ **9.** $\frac{9}{44}$ **10.** $\frac{41}{72}$

11. $-\frac{43}{6}$ or $-7\frac{1}{6}$ **12.** $\frac{19}{40}$ **13.** $3\sqrt{10}$ **14.** $7\sqrt{3}$ **15.** $3\sqrt{5}$ **16.** $-100, -\frac{2}{3}, 0, 1.57283, \sqrt{100}, 12.\overline{7}$ **17.** commutative property of multiplication   **18.** associative property of multiplication   **19.** 64   **20.** 243   **21.** $\frac{1}{125}$ **22.** 2,800,000,000   **23.** $\frac{3.6 \times 10^7}{6 \times 10^{-4}} = 6 \times 10^{10}$ **24. a.** $\$7.8 \times 10^{12}$ **b.** $\$6000$ **25.** 4.34×10^{12} e-mails   **26.** $7, 4, 1, -2, -5,$ and -8 **27.** 36   **28.** $25, -5, 1, -\frac{1}{5}, \frac{1}{25},$ and $-\frac{1}{125}$ **29.** 1280

CHAPTER 4

Section 4.1

Checkpoint Exercises

1. 608   **2.** -2 **3.** -94 **4.** 2662 calories; underestimates by 38 calories   **5.** $3x - 21$ **6.** $38x^2 + 23x$ **7.** $2x + 36$

Exercise Set 4.1

1. evaluating   **2.** equation   **3.** terms   **4.** coefficient   **5.** factors   **6.** like terms   **7.** false   **8.** false   **9.** false   **10.** true   **11.** false   **12.** false   **13.** true   **14.** false   **21.** makes sense   **23.** makes sense   **25.** 27   **27.** 23   **29.** 29   **31.** -2 **33.** -21 **35.** -10 **37.** 140   **39.** 217   **41.** 176   **43.** 30   **45.** 44   **47.** 22   **49.** 27   **51.** -12 **53.** 69   **55.** -33 **57.** -8 **59.** $10°C$ **61.** 60 ft   **63.** $17x$ **65.** $-3x^2$ **67.** $3x + 15$ **69.** $8x - 12$ **71.** $15x + 16$ **73.** $27x - 10$ **75.** $29y - 29$ **77.** $8y - 12$ **79.** $-8x^2 + 10x$ **81.** $16y - 25$ **83.** $-10x + 6$ **85.** $12x^2 + 11$ **87.** $-2x^2 - 9$ **89. a.** 140 beats per minute   **b.** 160 beats per minute   **91.** 9.5%; underestimates by 0.5   **93. a.** $\$52,000$ **b.** $\$52,723$; reasonably well   **c.** $\$52,628$; reasonably well   **95.** Model 1; overestimates by $\$63$ **97. a.** $\$50.50, \$5.50,$ and $\$1.00$ per clock, respectively   **b.** no; Answer will very.

Section 4.2

Checkpoint Exercises

1. $\{6\}$ **2.** $\{-2\}$ **3.** $\{2\}$ **4.** $\{5\}$ **5.** $\{6\}$ **6.** $\frac{37}{10}$ or 3.7; If a horizontal line is drawn from 10 on the scale for level of depression until it touches the blue line graph for the low-humor group and then a vertical line is drawn from that point on the blue line graph to the scale for the intensity of the negative life event, the vertical line will touch the scale at 3.7   **7.** $\varnothing$ **8.** $\{x | x \text{ is a real number}\}$

Exercise Set 4.2

1. real numbers; linear equation **2.** $b + c$; addition **3.** ac; multiplication **4.** simplified; solved **5.** false **6.** false **7.** false

8. false **15.** makes sense **17.** does not make sense **19.** $\{10\}$ **21.** $\{-17\}$ **23.** $\{12\}$ **25.** $\{9\}$ **27.** $\{-3\}$ **29.** $\left\{-\dfrac{1}{4}\right\}$

31. $\{3\}$ **33.** $\{11\}$ **35.** $\{-17\}$ **37.** $\{11\}$ **39.** $\left\{\dfrac{7}{5}\right\}$ **41.** $\left\{\dfrac{25}{3}\right\}$ **43.** $\{8\}$ **45.** $\{-3\}$ **47.** $\{2\}$ **49.** $\{-4\}$ **51.** $\left\{-\dfrac{1}{5}\right\}$

53. $\{-4\}$ **55.** $\{5\}$ **57.** $\{6\}$ **59.** $\{1\}$ **61.** $\{-57\}$ **63.** $\{18\}$ **65.** $\{1\}$ **67.** $\{24\}$ **69.** $\{-6\}$ **71.** $\{20\}$ **73.** $\{5\}$

75. $\{-7\}$ **77.** $\varnothing$ **79.** $\{x | x \text{ is a real number}\}$ **81.** $\left\{\dfrac{2}{3}\right\}$ **83.** $\{x | x \text{ is a real number}\}$ **85.** $\varnothing$ **87.** $\{0\}$ **89.** $\varnothing$ **91.** $\{0\}$

93. $\left\{\dfrac{7}{2}\right\}$ **95.** $\{0\}$ **97.** 2 **99.** 161 **101.** $\{-2\}$ **103.** $\varnothing$ **105.** $\{10\}$ **107.** $\{-2\}$ **109.** 142 lb; 13 lb **111.** 409.2 ft

113. The student made two mistakes, not changing the sign of the second term when distributing the -3 and subtracting 18 on the right when adding 18 on the left. Answers will vary. **115.** Answers will vary; examples are $x = 5, x + 3 = 8,$ and $2x = 10$. **117.** Yes, according to the formula, the femur belongs to a woman who was approximately 61.2 inches tall.

Section 4.3

Checkpoint Exercises

1. some college: \$37,000; associate degree: \$41,000; bachelor's degree: \$58,000 **2.** by 67 years after 1969, or in 2036 **3.** 720 miles **4.** \$1200

5. $w = \dfrac{P - 2l}{2}$ **6.** $m = \dfrac{T - D}{p}$

Exercise Set 4.3

1. $x + 120$ **2.** $55 + 6x$ **3.** $125 + 0.15x$ **4.** $x - 0.35x$ or $0.65x$ **5.** incorrect; $x + (x + 2.4) = 19.2$ **6.** correct
7. incorrect; $125 + 0.20x = 335$ **8.** incorrect; $x - 0.4x = 12$ **11.** does not make sense **13.** does not make sense **15.** 6 **17.** 25
19. 120 **21.** 320 **23.** 19 and 45 **25.** $x - (x + 4); -4$ **27.** $6(-5x); -30x$ **29.** $5x - 2x; 3x$ **31.** $8x - (3x + 6); 5x - 6$
33. computer science; \$56,000; psychology: \$36,000 **35.** computer science: \$96,000; political science: \$78,000; economics; \$99,000
37. less than 5 years; \$43,000; 10 to 20 years; \$72,000 **39. a.** births; 367,000; deaths: 153,000 **b.** 78 million **c.** approximately 4 years
41. by 2015 **43.** after 5 years **45.** after 5 months; \$165 **47.** 30 times **49.** \$600 of merchandise; \$580 **51.** 2017; 22,300 students

53. \$420 **55.** \$150 **57.** \$467.20 **59.** $L = \dfrac{A}{W}$ **61.** $b = \dfrac{2A}{h}$ **63.** $P = \dfrac{1}{rt}$ **65.** $m = \dfrac{E}{c^2}$ **67.** $m = \dfrac{y - b}{x}$ **69.** $a = \dfrac{2A}{h} - b$

71. $r = \dfrac{S - P}{Pt}$ **73.** $x = \dfrac{C - By}{A}$ **75.** $n = \dfrac{a_n - a_1}{d} + 1$ or $n = \dfrac{a_n - a_1 + d}{d}$ **77.** \$200 **79.** 10 problems **81.** 36 plants

Section 4.4

Checkpoint Exercises

1. a. $\{15\}$ **b.** $\{5\}$ **2.** \$5880 **3.** 720 deer

Exercise Set 4.4

1. proportion **2.** $ad = bc$ **3.** false **4.** true **9.** makes sense **11.** does not make sense **13.** $\{14\}$ **15.** $\{27\}$ **17.** $\{-15\}$

19. $\{-9\}$ **21.** $\{34\}$ **23.** $\{-49\}$ **25.** $\{10\}$ **27.** $x = \dfrac{ab}{c}$ **29.** $x = \dfrac{ad + bd}{c}$ **31.** $x = \dfrac{ab}{c}$ **33.** \$9100 **35.** 5.4 feet

37. 20,489 seal pups **39.** Either change the rear sprocket to 12 teeth or change the front sprocket to 100 teeth. **41.** $\{-10.77\}$

Section 4.5

Checkpoint Exercises

1. 66 gal **2.** 9375 lb **3.** 512 cycles per second **4.** 24 min

Exercise Set 4.5

1. $y = kx$ **2.** $y = kx^n$ **3.** $y = \dfrac{k}{x}$ **4.** $y = \dfrac{kx}{z}$ **5.** false **6.** false **7.** false **8.** true **13.** does not make sense

15. does not make sense **17.** 156 **19.** 150 **21.** 30 **23.** 2 **25.** 10 **27.** $\dfrac{5}{6}$ **29.** $x = ky; y = \dfrac{x}{k}$ **31.** $x = \dfrac{k}{y}; y = \dfrac{k}{x}$

33. $x = \dfrac{ky}{z}; y = \dfrac{xz}{k}$ **35.** 80 in. **37.** about 607 lb **39.** 32° **41.** 90 milliroentgens per hour

43. This person has a BMI of 24.4 and is not overweight. **45. a.** $L = \dfrac{1890}{R}$ **b.** an approximate model **c.** 70 years

47. $\dfrac{1}{3}$ of what it was originally **49.** The wind pressure is 4 times more destructive. **51.** Distance is increased by $\sqrt{50}$, or about 7.07, for the space telescope.

Section 4.6

Checkpoint Exercises

1. a. (number line: $-2\ -1\ 0\ 1\ 2\ 3\ 4\ 5\ 6\ 7$) **b.** (number line: $-4\ -3\ -2\ -1\ 0\ 1\ 2\ 3\ 4\ 5$) **c.** (number line: $-5\ -4\ -3\ -2\ -1\ 0\ 1\ 2\ 3\ 4$)

2. $\{x \mid x \le 4\}$; (number line: $-2\ -1\ 0\ 1\ 2\ 3\ 4\ 5\ 6\ 7$) **3. a.** $\{x \mid x < 8\}$; (number line: $0\ 1\ 2\ 3\ 4\ 5\ 6\ 7\ 8\ 9$)

b. $\{x \mid x > -3\}$; (number line: $-7\ -6\ -5\ -4\ -3\ -2\ -1\ 0\ 1\ 2$) **4.** $\{x \mid x < -6\}$; (number line: $-10\ -9\ -8\ -7\ -6\ -5\ -4\ -3\ -2\ -1$)

5. $\{x \mid x \ge 1\}$; (number line: $-4\ -3\ -2\ -1\ 0\ 1\ 2\ 3\ 4\ 5$) **6.** $\{x \mid -1 \le x < 4\}$; (number line: $-4\ -3\ -2\ -1\ 0\ 1\ 2\ 3\ 4\ 5$) **7.** $\varnothing$

8. $\{x \mid x \text{ is a real number}\}$ **9.** at least 83%

Exercise Set 4.6

1. is not included; is included **2.** $\{x \mid x < a\}$ **3.** $\{x \mid a < x \le b\}$ **4.** reverse the direction or change the sense **5.** $\varnothing$ **6.** $\{x \mid x \text{ is a real number}\}$
7. false **8.** false **9.** false **10.** true **15.** makes sense **17.** makes sense **19.** (number line: $-1\ 0\ 1\ 2\ 3\ 4\ 5\ 6\ 7\ 8$)

21. (number line: $-6\ -5\ -4\ -3\ -2\ -1\ 0\ 1\ 2\ 3$) **23.** (number line: $-10\ -9\ -8\ -7\ -6\ -5\ -4\ -3\ -2\ -1$) **25.** (number line: $2\ 3\ 4\ 5\ 6\ 7\ 8\ 9\ 10\ 11$)

27. (number line: $-3\ -2\ -1\ 0\ 1\ 2\ 3\ 4\ 5\ 6$) **29.** (number line: $-3\ -2\ -1\ 0\ 1\ 2\ 3\ 4\ 5\ 6$) **31.** $\{x \mid x > 5\}$; (number line: $-2\ -1\ 0\ 1\ 2\ 3\ 4\ 5\ 6\ 7$)

33. $\{x \mid x \le 5\}$; (number line: $3\ 4\ 5\ 6\ 7\ 8\ 9\ 10\ 11\ 12$) **35.** $\{x \mid x < 3\}$; (number line: $1\ 2\ 3\ 4\ 5\ 6\ 7\ 8\ 9\ 10$)

37. $\{x \mid x < 5\}$; (number line: $3\ 4\ 5\ 6\ 7\ 8\ 9\ 10\ 11\ 12$) **39.** $\{x \mid x \ge -5\}$; (number line: $-11\ -10\ -9\ -8\ -7\ -6\ -5\ -4\ -3\ -2$)

41. $\{x \mid x > 5\}$; (number line: $-2\ -1\ 0\ 1\ 2\ 3\ 4\ 5\ 6\ 7$) **43.** $\{x \mid x < 5\}$; (number line: $3\ 4\ 5\ 6\ 7\ 8\ 9\ 10\ 11\ 12$)

45. $\{x \mid x < 8\}$; (number line: $6\ 7\ 8\ 9\ 10\ 11\ 12\ 13\ 14\ 15$) **47.** $\{x \mid x > -6\}$; (number line: $-13\ -12\ -11\ -10\ -9\ -8\ -7\ -6\ -5\ -4$)

49. $\{x \mid x > -5\}$; (number line: $-13\ -12\ -11\ -10\ -9\ -8\ -7\ -6\ -5\ -4$) **51.** $\{x \mid x \le 5\}$; (number line: $3\ 4\ 5\ 6\ 7\ 8\ 9\ 10\ 11\ 12$)

53. $\{x \mid x \le 3\}$; (number line: $1\ 2\ 3\ 4\ 5\ 6\ 7\ 8\ 9\ 10$) **55.** $\{x \mid x < 16\}$; (number line: $14\ 15\ 16\ 17\ 18\ 19\ 20\ 21\ 22\ 23$)

57. $\{x \mid x > -3\}$; (number line: $-11\ -10\ -9\ -8\ -7\ -6\ -5\ -4\ -3\ -2$) **59.** $\{x \mid x \ge -2\}$; (number line: $-9\ -8\ -7\ -6\ -5\ -4\ -3\ -2\ -1\ 0$)

61. $\{x \mid x > -4\}$; (number line: $-11\ -10\ -9\ -8\ -7\ -6\ -5\ -4\ -3\ -2$) **63.** $\{x \mid x \ge 4\}$; (number line: $-2\ -1\ 0\ 1\ 2\ 3\ 4\ 5\ 6\ 7$)

65. $\left\{ x \mid x > \dfrac{11}{3} \right\}$; (number line with $\frac{11}{3}$ marked: $-2\ -1\ 0\ 1\ 2\ 3\ 4\ 5\ 6\ 7$) **67.** $\{x \mid x > 2\}$ (number line: $-5\ -4\ -3\ -2\ -1\ 0\ 1\ 2\ 3\ 4$)

69. $\{x \mid x < 3\}$; (number line: $1\ 2\ 3\ 4\ 5\ 6\ 7\ 8\ 9\ 10$) **71.** $\left\{ x \mid x > \dfrac{5}{3} \right\}$; (number line with $\frac{5}{3}$ marked: $-5\ -4\ -3\ -2\ -1\ 0\ 1\ 2\ 3\ 4$)

73. $\{x \mid x \ge -10\}$; (number line: $-17\ -16\ -15\ -14\ -13\ -12\ -11\ -10\ -9\ -8$) **75.** $\{x \mid x < -6\}$; (number line: $-8\ -7\ -6\ -5\ -4\ -3\ -2\ -1\ 0\ 2$)

77. $\{x \mid 3 < x < 5\}$; (number line: $-1\ 0\ 1\ 2\ 3\ 4\ 5\ 6\ 7\ 8$) **79.** $\{x \mid -1 \le x < 3\}$; (number line: $-4\ -3\ -2\ -1\ 0\ 1\ 2\ 3\ 4\ 5$)

81. $\{x \mid -5 < x \le -2\}$; (number line: $-8\ -7\ -6\ -5\ -4\ -3\ -2\ -1\ 0\ 1$) **83.** $\{x \mid 3 \le x < 6\}$; (number line: $0\ 1\ 2\ 3\ 4\ 5\ 6\ 7\ 8\ 9$)

85. $\varnothing$ **87.** $\{x \mid x \text{ is real number}\}$ **89.** $\varnothing$ **91.** $\{x \mid x \text{ is real number}\}$ **93.** $\{x \mid x \le 0\}$ **95.** $x > \dfrac{C - By}{A}$ **97.** $x < \dfrac{C - By}{A}$

99. $\{x \mid x + 5 \ge 2x\}$ **101.** $\{x \mid 2(4 + x) \le 36\}$ **103.** $\left\{ x \mid \dfrac{3x}{5} + 4 \le 34 \right\}$ **105.** $\{x \mid 0 < x < 4\}$

107. intimacy $\ge$ passion or passion $\le$ intimacy **109.** commitment $>$ passion or passion $<$ commitment **111.** 9; after 3 years
113. a. overestimates by $334 **b.** more than 22 years after 1990; years after 2012 **115. a.** at least 96 **b.** less than 66
117. at most 1280 mi **119.** at most 29 bags **121.** between 80 and 110 minutes, inclusive **123.** more than 720 miles
125. more then 1200 packages

Chapter 4 Review Exercises

1. 33 **2.** 15 **3.** -48 **4.** 38.55% overestimates by 3.55 **5.** $17x - 15$ **6.** $5y - 13$ **7.** $14x^2 + x$ **8.** $\{6\}$ **9.** $\{2\}$
10. $\{12\}$ **11.** $\left\{ -\dfrac{13}{3} \right\}$ **12.** $\{2\}$ **13.** $\varnothing$ **14.** $\{x \mid x \text{ is a real number}\}$ **15. a.** Model 2 **b.** underestimates by 3 channels
c. 15 years after 2000, or in 2015 **16.** watching TV: 270 minutes; listening to music: 151 minutes; text messaging: 90 minutes
17. by 12 years after 2007, or in 2019 **18.** 500 min **19.** $60 **20.** $10,000 in sales **21.** $x = \dfrac{By + C}{A}$ **22.** $h = \dfrac{2A}{b}$ **23.** $B = 2A - C$

24. $g = \dfrac{s - vt}{t^2}$ **25.** $\{5\}$ **26.** $\{-65\}$ **27.** $\left\{\dfrac{2}{5}\right\}$ **28.** $\{3\}$ **29.** 324 teachers   **30.** 287 trout   **31.** 134.4 cubic centimeters

32. 1600 feet   **33.** 440 vibrations per second   **34.** 112 decibels   **35.** 16 hours   **36.** $\{x \mid x < 4\}$;

37. $\{x \mid x > -8\}$; ![number line]

38. $\{x \mid x \geq -3\}$; ![number line]

39. $\{x \mid x > 6\}$; ![number line]

40. $\{x \mid x \geq 4\}$; ![number line]

41. $\{x \mid x \leq 2\}$; ![number line]

42. $\left\{x \mid -\dfrac{3}{4} < x \leq 1\right\}$; ![number line]

43. $\{x \mid x \text{ is a real number}\}$ **44.** $\varnothing$ **45.** at least 64

Chapter 4 Test A

1. -44 **2.** $14x - 4$ **3.** \$25,075   **4.** $\{-5\}$ **5.** $\{2\}$ **6.** $\varnothing$ **7.** $\{-15\}$ **8.** $y = \dfrac{Ax + A}{B}$ **9.** 14 years after 2000, or 2014

10. $\left\{\dfrac{15}{2}\right\}$ **11.** $\{3\}$ **12.** watching TV: 9 years; sleeping: 28 years; working: 25 years   **13.** after 7 years   **14.** 20 prints; \$3.80   **15.** \$50

16. 6000 tule elk   **17.** 2442 miles per hour   **18.** 45 foot-candles   **19.** $\{x \mid x \leq -3\}$; ![number line]

20. $\{x \mid x > 7\}$; ![number line]   **21.** $\left\{x \mid -2 \leq x < \dfrac{5}{2}\right\}$; ![number line]   **22.** at least 92

Chapter 4 Test B

1. -244 **2.** $13x - 11$ **3.** overestimates by 0.4   **4.** $\{7\}$ **5.** $\{0\}$ **6.** $\{x \mid x \text{ is a real number}\}$ **7.** $\{12\}$ **8.** $x = \dfrac{B - C}{A}$

9. 20 years after 2000, or 2020   **10.** $\left\{\dfrac{15}{2}\right\}$ **11.** $\{-19\}$ **12.** Bachelor's: \$55 thousand; Master's: \$61 thousand; Doctorate: \$71 thousand

13. after 6 years   **14.** 18 times   **15.** \$860   **16.** 954 deer   **17.** 1296 feet   **18.** 52.5 amperes
19. $\{x \mid x \geq -5\}$; ![number line]   **20.** $\{x \mid x < -1\}$; ![number line]

21. $\{x \mid -1 < x \leq 5\}$; ![number line]   **22.** at least 98

CHAPTER 5

Section 5.1

Checkpoint Exercises

1. 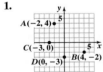 $A(-2, 4)$ $C(-3, 0)$ $B(4, -2)$ $D(0, -3)$

2.

3. a.

$y = 2x$		$y = 10 + x$	
x	(x, y)	x	(x, y)
0	$(0, 0)$	0	$(0, 10)$
2	$(2, 4)$	2	$(2, 12)$
4	$(4, 8)$	4	$(4, 14)$
6	$(6, 12)$	6	$(6, 16)$
8	$(8, 16)$	8	$(8, 18)$
10	$(10, 20)$	10	$(10, 20)$
12	$(12, 24)$	12	$(12, 22)$

b.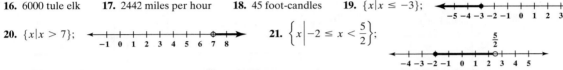

Number of Times the Bridge Is Used Each Month

c. $(10, 20)$; When the bridge is used 10 times during a month, the cost is \$20 with or without the discount pass.

4. a. 29   **b.** 65   **c.** 46
5. a. 190 ft   **b.** 191 ft
6. ![graph] The graph of g is the graph of f shifted down 3 units.

7. a. function   **b.** function   **c.** not a function   **8. a.** 0 to 3 hours   **b.** 3 to 13 hours   **c.** 0.05 mg per 100 ml; after 3 hours
d. None of the drug is left in the body.   **e.** No vertical line intersects the graph in more than one point.
9. a. 20; With 40 calling minutes, the monthly cost is \$20.; $(40, 20)$.   **b.** 28; With 80 calling minutes, the monthly cost is \$28.; $(80, 28)$.

Exercise Set 5.1

1. x-coordinate; y-coordinate   **2.** solution; satisfies   **3.** y; x; function   **4.** more than once; function   **5.** false   **6.** false   **7.** true
8. false   **9.** false   **10.** true   **17.** makes sense   **19.** makes sense

21.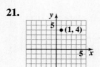

23.

25.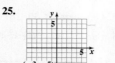

27.

29.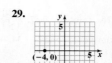

31.

33.

35.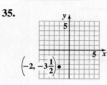

37.

39.

41.

43.

45.

47.

49.

51.

53. a. 4   **b.** -3 **55. a.** 19   **b.** -2
57. a. 5   **b.** 5   **59. a.** -14 **b.** -7
61. a. 53   **b.** 8   **63. a.** 26   **b.** 19
65. a. 1   **b.** -1

67.

x	$f(x) = x^2 - 1$
-2	3
-1	0
0	-1
1	0
2	3

69.

x	$f(x) = x - 1$
-2	-3
-1	-2
0	-1
1	0
2	1

71.

x	$f(x) = (x - 2)^2$
0	4
1	1
2	0
3	1
4	4

73.

x	$f(x) = x^3 + 1$
-3	-26
-2	-7
-1	0
0	1
1	2

75. function   **77.** function   **79.** not a function   **81.** function   **83.** $-2; 10$ **85.** -38

87. $y = 2x + 4$ **89.** $y = 3 - x^2$

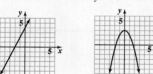

91. $(2, 7)$; The football is 7 feet above the ground when it is 2 yards from the quarterback.
93. $(6, 9.25)$ **95.** 12 ft; 15 yd   **97. a.** 707; Approximately 707,000 bachelor's degrees were awarded to women in 2000.; $(20, 707)$ **b.** overestimates by 2 thousand
99. a. 68; Approximately 68,000 more bachelor's degrees were awarded to women than to men in 1990.; The points on the graphs with first coordinate 10 are 68 units apart.   **b.** overestimates by 13 thousand
101. 440; For 20-year-old drivers, there are 440 accidents per 50 million miles driven.; $(20, 440)$
103. $x = 45; y = 190$; The minimum number of accidents is 190 per 50 million miles driven and is attributed to 45-year-old drivers.

105. $0.30t - 6$

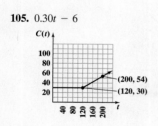

107. $C(x) = \begin{cases} 30 & \text{if } 0 \le t \le 500 \\ 30 + 0.25(t - 500) & \text{if } t > 500 \end{cases}$;

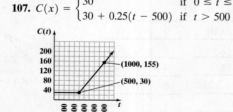

109. b   **111.** c   **113.** Answers will vary; an example is: At the beginning, there is a positive balance in the account. Then a check is written, lowering the balance. After that, a deposit is made, raising the balance above the beginning balance. Finally, a check is written for an amount greater than the new balance, resulting in a negative balance.   **115.** -2 **117.** 1

Section 5.2

Checkpoint Exercises

1.

2. a. $m = 6$ **b.** $m = -\dfrac{7}{5}$

3.

4.

5.

6.

7. 51.4; The slope indicates that for the period from 2007 through 2016, the cost of Medicare is projected to increase by approximately \$51.4 billion per year. The rate of change is approximately \$51.4 billion per year.   **8. a.** $F(x) = 765x + 5791$ **b.** \$14,971

Exercise Set 5.2

1. x-intercept   **2.** y-intercept   **3.** $\dfrac{y_2 - y_1}{x_2 - x_1}$ **4.** $y = mx + b$; slope; the y-intercept   **5.** horizontal   **6.** vertical   **7.** false   **8.** false

9. true   **10.** false   **19.** does not make sense   **21.** does not make sense

23. 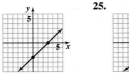   **25.**   **27.**   **29.**   **31.** -1; falls   **33.** $\dfrac{1}{4}$; rises   **35.** -5; falls

37. undefined; vertical   **39.** -4; falls
41. 0; horizontal

43.   **45.**   **47.**   **49.**   **51.**   **53.**

55. a. $y = -3x$ or $y = -3x + 0$ **57. a.** $y = \dfrac{4}{3}x$ or $y = \dfrac{4}{3}x + 0$ **59. a.** $y = -2x + 3$ **61. a.** $y = -\dfrac{7}{2}x + 7$

b. slope $= -3$; y-intercept $= 0$ **b.** slope $= \dfrac{4}{3}$; y-intercept $= 0$ **b.** slope $= -2$; y-intercept $= 3$ **b.** slope $= -\dfrac{7}{2}$; y-intercept $= 7$

c. 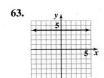   **c.**   **c.**   **c.**

63.   **65.**   **67.**   **69.**   **71.** $m = -\dfrac{a}{b}$; falls

73. undefined slope; vertical

75. $m = -\dfrac{A}{B}$; $b = \dfrac{C}{B}$ **77.** -2

79. m_1, m_3, m_2, m_4

81. -0.275; The slope indicates that for the period from 2000 through 2008, the percentage of total sales of rap-hip hop decreased by approximately 0.275 each year. The rate of change is -0.275 per year.   **83.** $P(x) = -1.2x + 47$ **85. a.** $p(x) = -0.52x + 68$ **b.** 16%   **87.** $-6; 3$

89. Find the slope: $m = \dfrac{b - 0}{0 - a} = -\dfrac{b}{a}$. The y-intercept is b.

Write the slope-intercept equation: $y = -\dfrac{b}{a}x + b$.

Add $\dfrac{b}{a}x$ to each side: $\dfrac{b}{a}x + y = b$.

Divide each side by b: $\dfrac{x}{a} + \dfrac{y}{b} = 1$.

This is called the *intercept form* because the number dividing x is the x-intercept and the number dividing y is the y-intercept.

Section 5.3

Checkpoint Exercises

1. $y + 5 = 6(x - 2)$; $y = 6x - 17$ **2. a.** $y + 1 = -5(x + 2)$ or $y + 6 = -5(x + 1)$ **b.** $y = -5x - 11$ **3.** $y = 0.28x + 27.2$; 41.2
4. There is a moderately strong negative relationship between annual income and hours per week watching TV.

Exercise Set 5.3

1. $y - y_1 = m(x - x_1)$; a point on the line; slope   **2.** scatter plot   **3.** regression line   **4.** correlation coefficient; -1; 1   **5.** false
6. false   **7.** true   **8.** false   **15.** makes sense   **17.** does not make sense   **19.** $y - 5 = 3(x - 2)$; $y = 3x - 1$
21. $y - 6 = 5(x + 2)$; $y = 5x + 16$ **23.** $y + 2 = -8(x + 3)$; $y = -8x - 26$ **25.** $y - 0 = -12(x + 8)$; $y = -12x - 96$
27. $y + 2 = -1\left(x + \dfrac{1}{2}\right)$; $y = -x - \dfrac{5}{2}$ **29.** $y - 0 = \dfrac{1}{2}(x - 0)$; $y = \dfrac{1}{2}x$ **31.** $y + 2 = -\dfrac{2}{3}(x - 6)$; $y = -\dfrac{2}{3}x + 2$
33. $y - 2 = 2(x - 1)$ or $y - 10 = 2(x - 5)$; $y = 2x$ **35.** $y - 0 = 1(x + 3)$ or $y - 3 = 1(x - 0)$; $y = x + 3$
37. $y + 1 = 1(x + 3)$ or $y - 4 = 1(x - 2)$; $y = x + 2$ **39.** $y + 1 = \dfrac{5}{7}(x + 4)$ or $y - 4 = \dfrac{5}{7}(x - 3)$; $y = \dfrac{5}{7}x + \dfrac{13}{7}$
41. $y + 1 = 0(x + 3)$ or $y + 1 = 0(x - 4)$; $y = -1$ **43.** $y - 4 = 1(x - 2)$ or $y - 0 = 1(x + 2)$; $y = x + 2$
45. $y - 0 = 8\left(x + \dfrac{1}{2}\right)$ or $y - 4 = 8(x - 0)$; $y = 8x + 4$

47.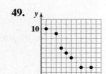

There appears to be a positive correlation.

49.

There appears to be a negative correlation.

51. a   **53.** d   **55.** $y - 2 = 4(x + 3); y = 4x + 14$
57. $y + 5 = -3(x + 1); y = -3x - 8$
59. $y - 0 = -5(x + 4); y = -5x - 20$
61. a. $y - 31.1 = 0.78(x - 10)$ or $y - 38.9 = 0.78(x - 20)$
b. $f(x) = 0.78x + 23.3$ **c.** 54.5%   **63.** false
65. true   **67.** false   **69.** false
71. positive; Answers will vary.
73. no correlation; Answers will vary.

Section 5.4

Checkpoint Exercises

1. not a solution   **2.** 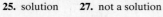   ; $\{(-3, 4)\}$ **3.** $\{(6, 11)\}$ **4.** $\{(1, -2)\}$ **5.** $\{(2, -1)\}$ **6.** $\{(2, -1)\}$ **7.** $\varnothing$
8. $\{(x,y) \mid y = 4x - 4\}$ or $\{(x,y) \mid 8x - 2y = 8\}$ **9. a.** $C(x) = 300,000 + 30x$ **b.** $R(x) = 80x$
c. $(6000, 480,000)$; The company will break even if it produces and sells 6000 pairs of shoes. At this level, the money coming in is equal to the money going out: $480,000.

Exercise Set 5.4

1. satisfies both equations in the system   **2.** the intersection point   **3.** substitution; addition   **4.** no solution; infinitely many solutions
5. revenue; profit   **6.** break-even point   **7.** false   **8.** false   **9.** false   **10.** false   **21.** makes sense   **23.** does not make sense
25. solution   **27.** not a solution

29.   **31.**   **33.** 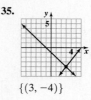   **35.**

$\{(4, 2)\}$ $\{(3, 0)\}$ $\{(-1, 4)\}$ $\{(3, -4)\}$

37. $\{(1, 3)\}$ **39.** $\{(5, 1)\}$ **41.** $\{(2, 1)\}$ **43.** $\{(-1, 3)\}$ **45.** $\{(4, 5)\}$ **47.** $\left\{\left(-4, \dfrac{5}{4}\right)\right\}$ **49.** $\{(2, -1)\}$ **51.** $\{(3, 0)\}$
53. $\{(-4, 3)\}$ **55.** $\{(3, 1)\}$ **57.** $\{(1, -2)\}$ **59.** $\{(-5, -2)\}$ **61.** $\varnothing$ **63.** $\{(x,y) \mid y = 3x - 5\}$ or $\{(x,y) \mid 21x - 35 = 7y\}$
65. $\{(1, 4)\}$ **67.** $\{(x,y) \mid x + 3y = 2\}$ or $\{(x,y) \mid 3x + 9y = 6\}$ **69.** $\begin{cases} y = x - 4 \\ y = -\dfrac{1}{3}x + 4 \end{cases}$ **71.** $\left\{\left(\dfrac{1}{a}, 3\right)\right\}$ **73.** $m = -4, b = 3$

75. 500 radios   **77.** -6000; When the company produces and sells 200 radios, the loss is $6000.   **79. a.** $P(x) = 20x - 10,000$ **b.** $190,000
81. a. $C(x) = 18,000 + 20x$ **b.** $R(x) = 80x$ **c.** $(300, 24,000)$; When 300 canoes are produced and sold, both revenue and cost are $24,000.
83. a. $C(x) = 30,000 + 2500x$ **b.** $R(x) = 3125x$ **c.** $(48, 150,000)$; For 48 sold-out performances, both cost and revenue are $150,000.
85. a. $y = 0.45x + 0.8$ **b.** $y = 0.15x + 2.6$ **c.** week 6; 3.5 symptoms; by the intersection point $(6, 3.5)$
87. Mr. Goodbar: 264 cal; Mounds bar: 258 cal   **89.** 3 Mr. Goodbars and 2 Mounds bars   **91.** 50 rooms with kitchen facilities and 150 rooms without kitchen facilities   **95.** the twin who always lies

Section 5.5

Checkpoint Exercises

1.   **2.**   **3. a.**   **b.**

4. Point $B = (66, 130)$; $4.9(66) - 130 \geq 165$, or $193.4 \geq 165$, is true; $3.7(66) - 130 \leq 125$, or $114.2 \leq 125$, is true.

5.   **6.**

Exercise Set 5.5

1. solution; x; y; $9 > 2$ **2.** graph   **3.** half-plane   **4.** system   **5.** false   **6.** false   **7.** false   **8.** true   **15.** does not make sense
17. makes sense

19.   **21.**   **23.**   **25.**   **27.**   **29.**

31.   **33.**   **35.**   **37.**   **39.**   **41.**

43.   **45.**   **47.**   **49.**   **51.**   **53.**

55.   **57.** $y \geq -2x + 4$ **59.** $\begin{cases} x + y \leq 4 \\ 3x + y \leq 6 \end{cases}$ **61.**

63. Point $A = (66, 160)$; $5.3(66) - 160 \geq 180$, or $189.8 \geq 180$, is true; $4.1(66) - 160 \leq 140$, or $110.6 \leq 140$, is true.   **65.** no
67. a. $50x + 150y > 2000$

b.   **69. a.** 27.1   **b.** overweight   **71.** $\begin{cases} x \geq -2 \\ y > -1 \end{cases}$ **73.** no solution   **75.** infinitely many solutions

Chapter 5 Review Exercises

1.   **2.**   **3.**   **4.**   **5.**   **6.**

7.   **8.** 3   **9.** 26   **10.** 30   **11.** -64 **12.**

x	$f(x) = \frac{1}{2}|x|$
-6	3
-4	2
-2	1
0	0
2	1
4	2
6	3

13.

x	$f(x) = x^2 - 2$
-2	2
-1	-1
0	-2
1	-1
2	2

14. y is a function of x.   **15.** y is not a function of x.   **16. a.** $(1985, 50\%)$; In 1985, the top marginal tax rate in the U.S. was 50%.   **b.** 35%   **c.** 1945; 94%   **d.** 1990; 28%   **e.** 1950–1960; 91%   **f.** 1930–1935; 38%

17.   **18.**   **19.**   **20.** $-\dfrac{1}{2}$; falls **21.** 3; rises **22.** 0; horizontal **23.** undefined; vertical

24.   **25.**   **26.**   **27.**

28. a. $y = -2x$ **29. a.** $y = \dfrac{5}{3}x$ **30. a.** $y = -\dfrac{3}{2}x + 2$

b. slope: -2; y-intercept: 0 **b.** slope: $\dfrac{5}{3}$; y-intercept: 0 **b.** slope: $-\dfrac{3}{2}$; y-intercept: 2

c.   **c.** 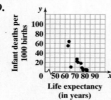 **c.**

31.   **32.**   **33.**

34. a. 254; If no women in a country are literate, the mortality rate of children under 5 is 254 per thousand. **b.** -2.4; For each 1% of adult females who are literate, the mortality rate of children under 5 decreases by 2.4 per thousand. **c.** $f(x) = -2.4x + 254$ **d.** 134 per thousand
35. $y - 2 = -6(x + 3)$; $y = -6x - 16$ **36.** $y - 6 = 2(x - 1)$ or $y - 2 = 2(x + 1)$; $y = 2x + 4$
37. a. $y - 2.3 = 0.116(x - 15)$ or $y - 11 = 0.116(x - 90)$ **b.** $f(x) = 0.116x + 0.56$ **c.** approximately 5 deaths per 100,000 persons
d. 4.3 deaths per 100,000 persons; underestimates by 0.7 death per 100,000 persons; The line passes below the point for France.

38.

There appears to be a positive correlation.

39.

There appears to be a negative correlation

40. false **41.** true **42.** false **43.** false
44. true **45.** false **46.** true **47.** c
48. $\{(2, 3)\}$ **49.** $\{(-2, -3)\}$ **50.** $\{(3, 2)\}$
51. $\{(4, -2)\}$ **52.** $\{(1, 5)\}$ **53.** $\{(2, 3)\}$
54. $\{(-9, 3)\}$ **55.** $\{(3, 4)\}$ **56.** $\{(2, -3)\}$
57. $\varnothing$ **58.** $\{(3, -1)\}$

59. $\{(x,y) \mid 3x - 2y = 6\}$ or $\{(x,y) \mid 6x - 4y = 12\}$ **60. a.** $C(x) = 60{,}000 + 200x$ **b.** $R(x) = 450x$ **c.** $(240, 108{,}000)$; This means the company will break even if it produces and sells 240 desks. At this level, both revenue and cost will be $108,000. **61. a.** Answer will vary.; approximately $(2004, 180)$; 2004; 180 million **b.** $f(x) = 19.8x + 98$ **c.** 2004; 180 million **d.** Answers will vary from "quite well" to "extremely well."

62.   **63.** **64.** **65.** **66.** **67.**

68. **69.** **70.** **71.** **72.** **73.**

74.

Chapter 5 Test A

1.   **2.** 21   **3.** y is not a function of x.   **4.** y is a function of x.

5. a. Yes; no vertical line intersects the graph in more than one point.   **b.** 0; The eagle was on the ground after 15 seconds.
c. 45 meters   **d.** between 3 and 12 seconds

6.   **7.** $m = 3$ **8.**   **9.**

10. a. 2.2; In 2003, the cost per 30 seconds was $2.2 million.
b. 0.1; The cost of a 30-second commercial increased approximately $0.1 million each year.; The rate of change is $0.1 million per year.   **c.** $C(x) = 0.1x + 2.2$
d. $3.2 million
11. $y - 1 = 3(x - 2)$ or $y + 8 = 3(x + 1)$; $y = 3x - 5$

12. a. $y = 0.08x + 2.1$ **b.** 8.1 billion   **13.**

There appears to be a strong negative correlation.
14. false   **15.** false   **16.** true   **17.** $\{(2, 4)\}$ **18.** $\{(1, -3)\}$
19. $\{(6, -5)\}$ **20. a.** $C(x) = 360{,}000 + 850x$ **b.** $R(x) = 1150x$
c. $(1200, 1{,}380{,}000)$; The company will break even if it produces and sells 1200 computers. At this level, both revenue and cost are $1,380,000.

21.   **22.**   **23.**   **24.**

Chapter 5 Test B

1.   **2.** 127   **3.** y is not a function of x.   **4.** y is a function of x.

5. a. Yes; no vertical line intersects the graph in more than one point.   **b.** between 25 and 55   **c.** 65; 26%
d. 24; At age 35, the average body fat in men is 24%.

6.   **7.** $m = 4$ **8.**   **9.**

10. a. 30; In 1989, there were 30 million viewers.   **b.** -1.5; The number of viewers is decreasing by approximately 1.5 million each year.; The rate of change is -1.5 million viewers per year.
c. $V(x) = -1.5x + 30$ **d.** 0; underestimates by approximately 4 million viewers
11. $y - 4 = -3(x - 2)$ or $y + 2 = -3(x - 4)$; $y = -3x + 10$

12. a. $y = 433x + 4138$ **b.** 8468 deaths
13.

There appears to be a strong negative correlation.   **14.** true   **15.** false   **16.** false   **17.** $\{(3, 2)\}$ **18.** $\{(-1, 3)\}$ **19.** $\{(-1, 2)\}$
20. a. $C(x) = 400{,}000 + 20x$ **b.** $R(x) = 100x$ **c.** $(5000, 500{,}000)$; The company will break even when it produces and sells 5000 PDAs. At this level, both revenue and cost are $500,000.

21.   **22.**   **23.**   **24.**

CHAPTER 6

Section 6.1

Checkpoint Exercises

1. $-3x^3 + 10x^2 - 10$ **2.** $10x^3 - 2x^2 + 8x - 10$ **3.** $12x^9 - 18x^6 + 24x^4$ **4.** $15x^3 - 31x^2 + 30x - 8$ **5. a.** $28x^2 - x - 15$
b. $4x^6 - 12x^4 - 5x^3 + 15x$ **6. a.** $49x^2 - 64$ **b.** $4y^6 - 25$ **7. a.** $x^2 + 20x + 100$ **b.** $25x^2 + 40x + 16$ **8. a.** $x^2 - 18x + 81$
b. $49x^2 - 42x + 9$ **9. a.** $T(x) = 3.9x^2 + 5x + 6451$ **b.** 6573.5 thousand   **c.** underestimates by 1.5 thousand

Exercise Set 6.1

1. whole **2.** monomial; binomial; trinomial **3.** n **4.** the greatest degree of all its terms **5.** false **6.** false **7.** false **8.** true
21. makes sense **23.** does not make sense **25.** yes; $3x^2 + 2x - 5$ **27.** yes; $2x^2 + \frac{1}{3}x - 5$ **29.** 2 **31.** 4 **33.** $11x^3 + 7x^2 - 12x - 4; 3$
35. $12x^3 + 4x^2 + 12x - 14; 3$ **37.** $6x^2 - 6x + 2; 2$ **39.** $12x^3 + 8x^2$ **41.** $2y^3 - 10y^2$ **43.** $10x^8 - 20x^5 + 45x^3$ **45.** $x^3 + 1$
47. $2x^3 - 9x^2 + 19x - 15$ **49.** $x^2 + 10x + 21$ **51.** $x^2 - 2x - 15$ **53.** $6x^2 + 13x + 5$ **55.** $10x^2 - 9x - 9$ **57.** $15x^4 - 47x^2 + 28$
59. $8x^5 - 40x^3 + 3x^2 - 15$ **61.** $x^2 - 9$ **63.** $9x^2 - 4$ **65.** $25 - 49x^2$ **67.** $16x^4 - 25x^2$ **69.** $1 - y^{10}$ **71.** $x^2 + 4x + 4$
73. $4x^2 + 12x + 9$ **75.** $x^2 - 6x + 9$ **77.** $16x^4 - 8x^2 + 1$ **79.** $4x^2 - 28x + 49$ **81.** $48x$ **83.** $-9x^2 + 3x + 9$ **85.** $16x^4 - 625$
87. $4x^2 - 28x + 49$ **89. a.** $f(x) = -0.3x^2 + 3.4x + 35.4$ **b.** 44.5% **c.** overestimates by 0.6 **91. a.** $F(x) = -0.12x^2 + 2.2x + 11.7$
b. 21.6% **c.** overestimates by 0.5 **93.** $x^2 + 2x + 1$ **95.** $4x^2 - 9$ **97.** $6x + 22$ **99. a.** $V(x) = -2x^3 + 10x^2 + 300x$
b. $V(10) = 2000$; Carry-on luggage with a depth of 10 inches has a volume of 2000 cubic inches. **c.** $(10, 2000)$
d. $\{x|0 < x < 15\}$, although answers may vary. **101.** $49x^4 + 54x^2 + 25$ **103.** $6x^n - 13$

Section 6.2

Checkpoint Exercises

1. $10x(2x + 3)$ **2. a.** $2x^2(5x - 2)$ **b.** $(x - 7)(2x + 3)$ **3.** $(x + 5)(x^2 + 2)$ **4.** $(x + 2)(x + 3)$ **5.** $(x + 5)(x - 2)$
6. $(5x - 4)(x - 2)$ **7.** $(3y - 1)(2y + 7)$ **8. a.** $(x + 9)(x - 9)$ **b.** $(6x + 5)(6x - 5)$ **9.** $(9x^2 + 4)(3x + 2)(3x - 2)$
10. a. $(x + 7)^2$ **b.** $(4x - 7)^2$ **11.** $4(x - 6)(x + 2)$ **12.** $4x(x^2 + 4)(x + 2)(x - 2)$ **13.** $(x - 4)(x + 3)(x - 3)$ **14.** $3x(x - 5)^2$

Exercise Set 6.2

1. $10x$; greatest common factor **2.** factoring by grouping **3.** $7x^2; 6$ **4.** sum; difference **5.** perfect square; x; 3; $3x$
6. $(A + B)^2; (A - B)^2$ **7.** false **8.** false **9.** true **10.** false **11.** false **12.** true **19.** makes sense **21.** makes sense
23. $9(2x + 3)$ **25.** $3x(x + 2)$ **27.** $9x^2(x^2 - 2x + 3)$ **29.** $(x + 5)(x + 3)$ **31.** $(x - 3)(x^2 + 12)$ **33.** $(x - 2)(x^2 + 5)$
35. $(x - 1)(x^2 + 2)$ **37.** $(3x - 2)(x^2 - 2)$ **39.** $(x + 2)(x + 3)$ **41.** $(x - 5)(x + 3)$ **43.** $(x - 5)(x - 3)$ **45.** $(3x + 2)(x - 1)$
47. $(3x - 28)(x + 1)$ **49.** $(2x - 1)(3x - 4)$ **51.** $(2x + 3)(2x + 5)$ **53.** $(3x - 2)(3x - 1)$ **55.** $(5x + 8)(4x - 1)$
57. $(x + 10)(x - 10)$ **59.** $(6x + 7)(x - 7)$ **61.** $(3 + 5y)(3 - 5y)$ **63.** $(x^2 + 4)(x + 2)(x - 2)$ **65.** $(4x^2 + 9)(2x + 3)(2x - 3)$
67. $(x + 1)^2$ **69.** $(x - 7)^2$ **71.** $(2x + 1)^2$ **73.** $(3x - 1)^2$ **75.** $5x(x + 2)(x - 2)$ **77.** $7x(x^2 + 1)$ **79.** $5(x - 3)(x + 2)$
81. $2(x^2 + 9)(x + 3)(x - 3)$ **83.** $(x + 2)(x + 3)(x - 3)$ **85.** $3x(x - 4)^2$ **87.** $2x(3x + 4)$ **89.** $2(y - 8)(y + 7)$ **91.** $7y^2(y + 1)^2$
93. prime **95.** $2(4y + 1)(2y - 1)$ **97.** $r(r - 25)$ **99.** $(2w + 5)(2w - 1)$ **101.** $x(x + 2)(x - 2)$ **103.** prime
105. $(9y + 4)(y + 1)$ **107.** $(y + 2)(y + 2)(y - 2)$ **109.** $(4y + 3)^2$ **111.** $4y(y - 5)(y - 2)$ **113.** $y(y^2 + 9)(y + 3)(y - 3)$
115. $5a^2(2a + 3)(2a - 3)$ **117.** $3x^2(3x^2 + 6x + 2)$ **119.** $(4y - 1)(3y - 2)$ **121.** $(3y + 8)(3y - 8)$ **123.** prime
125. $(2y + 3)(y + 5)(y - 5)$ **127.** $2r(r + 17)(r - 2)$ **129.** $2x^3(2x + 1)(2x - 1)$ **131.** $3(x^2 + 81)$ **133.** $(x + 1)(5x - 6)(2x + 1)$
135. $(x^2 + 6)(6x^2 - 1)$ **137.** $(x - 7 + 2a)(x - 7 - 2a)$ **139.** $(x + 4 + 5a)(x + 4 - 5a)$
141. a. $(x - 0.4x)(1 - 0.4) = (0.6x)(0.6) = 0.36x$ **b.** no; 36% **143. a.** $x^2 - 25$ **b.** $(x + 5)(x - 5)$ **145. a.** $9x^2 - 16$
b. $(3x + 4)(3x - 4)$ **147.** $a - b = 0$ and division by 0 is not permitted. **149.** $(x^n + 5y^n)(x^n - 5y^n)$ **151.** $[(x + 3) - 1]^2$ or $(x + 2)^2$
153. 3 and 4

Section 6.3

Checkpoint Exercises

1. $\left\{-\frac{1}{2}, 5\right\}$ **2. a.** $\left\{0, \frac{2}{3}\right\}$ **b.** $\{5\}$ **c.** $\{-4, 3\}$ **3.** $\left\{-2, -\frac{3}{2}, 2\right\}$ **4.** after 3 seconds; $(3, 336)$ **5.** 2 ft

Exercise Set 6.3

1. $ax^2 + bx + c = 0$ **2.** $A = 0$ or $B = 0$ **3.** x-intercepts **4.** descending **5.** false **6.** false **7.** false **8.** true
17. does not make sense **19.** makes sense **21.** $\{-4, 3\}$ **23.** $\{-7, 1\}$ **25.** $\left\{-4, \frac{2}{3}\right\}$ **27.** $\left\{\frac{3}{5}, 1\right\}$ **29.** $\left\{-2, \frac{1}{3}\right\}$ **31.** $\{0, 8\}$
33. $\left\{0, \frac{5}{3}\right\}$ **35.** $\{-2\}$ **37.** $\{7\}$ **39.** $\left\{\frac{5}{3}\right\}$ **41.** $\{-5, 5\}$ **43.** $\left\{-\frac{10}{3}, \frac{10}{3}\right\}$ **45.** $\{-3, 6\}$ **47.** $\{-3, -2\}$ **49.** $\{4\}$
51. $\{-2, 8\}$ **53.** $\left\{-1, \frac{2}{5}\right\}$ **55.** $\{-6, -3\}$ **57.** $\{-5, -4, 5\}$ **59.** $\{-5, 1, 5\}$ **61.** $\{-4, 0, 4\}$ **63.** $\{0, 2\}$ **65.** $\{-5, -3, 0\}$
67. 2 and 4; d **69.** -4 and -2; c **71.** $\{-7, -1, 6\}$ **73.** $\left\{-\frac{4}{3}, 0, \frac{4}{5}, 2\right\}$ **75.** -4 and 8 **77.** $-2, -\frac{1}{2},$ and 2 **79.** -7 and 4
81. 2 and 3 **83.** 1 second; 0.25, 0.5, 0.75, 1, 1.25 **85.** 7 **87.** $(7, 21)$ **89.** length: 9 ft; width: 6ft **91.** 5 in. **93.** 5 m **95. a.** $4x^2 + 44x$
b. 3 ft **97.** length: 10 in.; width: 10 in. **99.** $\{4, 5\}$ **101.** $x^2 - 2x - 15 = 0$ **103.** $\{-3, 2\}$ **105.** $\{1\}$

Section 6.4

Checkpoint Exercises

1. $\{\pm \sqrt{7}\}$ **2.** $\left\{\pm\frac{\sqrt{33}}{3}\right\}$ **3.** $\{3 \pm \sqrt{10}\}$ **4.** $\left\{-\frac{1}{2}, \frac{1}{4}\right\}$ **5.** $\left\{\frac{3 + \sqrt{7}}{2}, \frac{3 - \sqrt{7}}{2}\right\}$
6. approximately 26 years old; The point $(26, 115)$ lies approximately on the blue graph.

Exercise Set 6.4

1. $u = \pm\sqrt{d}$ **2.** $\dfrac{-b \pm \sqrt{b^2 - 4ac}}{2a}$; quadratic formula   **3.** the square root property   **4.** the quadratic formula   **5.** false   **6.** false

7. false   **8.** false   **13.** does not make sense   **15.** does not make sense   **17.** $\{\pm 4\}$ **19.** $\{\pm 9\}$ **21.** $\{\pm\sqrt{7}\}$ **23.** $\{\pm 5\sqrt{2}\}$

25. $\{\pm 2\}$ **27.** $\left\{\pm\dfrac{7}{2}\right\}$ **29.** $\{\pm 5\}$ **31.** $\left\{\pm\dfrac{\sqrt{6}}{3}\right\}$ **33.** $\left\{\pm\dfrac{\sqrt{35}}{5}\right\}$ **35.** $\{-1, 7\}$ **37.** $\{-16, 6\}$ **39.** $\left\{-\dfrac{5}{3}, \dfrac{1}{3}\right\}$ **41.** $\{5 \pm \sqrt{3}\}$

43. $\{-8 \pm \sqrt{11}\}$ **45.** $\{4 \pm 3\sqrt{2}\}$ **47.** $\{-5, -3\}$ **49.** $\left\{\dfrac{-5 + \sqrt{13}}{2}, \dfrac{-5 - \sqrt{13}}{2}\right\}$ **51.** $\{-2 + \sqrt{10}, -2 - \sqrt{10}\}$

53. $\{-2 + \sqrt{11}, -2 - \sqrt{11}\}$ **55.** $\{-3, 6\}$ **57.** $\left\{-\dfrac{2}{3}, \dfrac{3}{2}\right\}$ **59.** $\{1 + \sqrt{11}, 1 - \sqrt{11}\}$ **61.** $\left\{\dfrac{1 + \sqrt{57}}{2}, \dfrac{1 - \sqrt{57}}{2}\right\}$

63. $\left\{\dfrac{-3 + \sqrt{3}}{6}, \dfrac{-3 - \sqrt{3}}{6}\right\}$ **65.** $\left\{\dfrac{3}{2}\right\}$ **67.** $\left\{-\dfrac{1}{2}, 1\right\}$ **69.** $\left\{\dfrac{1}{5}, 2\right\}$ **71.** $\{\pm 2\sqrt{5}\}$ **73.** $\{1 \pm \sqrt{2}\}$ **75.** $\left\{\dfrac{-11 \pm \sqrt{33}}{4}\right\}$

77. $\left\{0, \dfrac{8}{3}\right\}$ **79.** $\{2\}$ **81.** $\{-2, 2\}$ **83.** $\{-9, 0\}$ **85.** $\left\{\dfrac{2 \pm \sqrt{10}}{3}\right\}$ **87.** $\left\{-\dfrac{2}{3}, 4\right\}$ **89.** $\{-3, 1\}$ **91.** $\left\{1, \dfrac{5}{2}\right\}$ **93.** $\{\pm\sqrt{6}\}$

95. $1 + \sqrt{7}$ **97.** $10\sqrt{3}$ seconds ≈ 17.3 seconds   **99. a.** 10.1%; underestimates by 0.3   **b.** 2036   **101.** width: 4.7 meters; length: 7.7 meters

103. a. $\dfrac{1}{\Phi - 1}$ **b.** $\Phi = \dfrac{1 + \sqrt{5}}{2}$ **c.** $\dfrac{1 + \sqrt{5}}{2}$ **105.** $\{-3\sqrt{3}, \sqrt{3}\}$

Section 6.5

Checkpoint Exercises

1. a. upward   **b.** $f(x) = x^2 - 6x + 8$ **2.** 2 and 4   **3.** 8   **4.** $(3, -1)$ **5.** $f(x) = x^2 + 6x + 5$

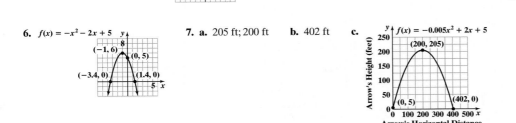

6. $f(x) = -x^2 - 2x + 5$ **7. a.** 205 ft; 200 ft   **b.** 402 ft   **c.**

Exercise Set 6.5

1. $a > 0; a < 0$ **2.** $ax^2 + bx + c = 0$ **3.** $0; f(0)$ **4.** $-\dfrac{b}{2a}; -\dfrac{b}{2a}$ or the x-coordinate   **5.** false   **6.** true   **7.** false   **8.** false

15. does not make sense   **17.** makes sense   **19.** upward   **21.** downward   **23.** 1 and 3   **25.** 2 and 6   **27.** 1.2 and -3.2 **29.** 3

31. -12 **33.** -4 **35.** 0   **37.** $(2, -1)$ **39.** $(-1, -8)$ **41.** $(-3, -9)$

43. $f(x) = x^2 + 8x + 7$ **45.** $f(x) = x^2 - 2x - 8$ **47.** $f(x) = -x^2 + 4x - 3$

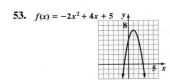

49. $f(x) = x^2 - 1$ **51.** $f(x) = x^2 + 2x + 1$ **53.** $f(x) = -2x^2 + 4x + 5$

55. $(3, 2)$ **57.** $(-5, -4)$ **59.** $(1, -3)$ **61.** $(-2, 5)$ **63. a.** 18.35 ft; 35 ft   **b.** 77.8 ft   **c.** 6.1 ft   **65. a.** 2 sec; 224 ft   **b.** 5.7 sec

c. 160; 160 feet is the height of the building.   **d.**

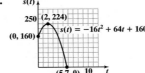

67. length: 60 ft; width: 30 ft; 1800 sq ft   **69.** 8 and 8; 64

71.

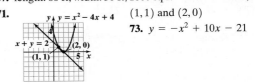

$(1, 1)$ and $(2, 0)$ **73.** $y = -x^2 + 10x - 21$

Chapter 6 Review Exercises

1. $8x^3 + 10x^2 - 20x - 4$; degree 3 **2.** $8x^4 - 5x^3 + 6$; degree 4 **3.** $40x^5 - 16x^3 - 8x^2$ **4.** $12x^3 + x^2 - 21x + 10$ **5.** $6x^2 - 7x - 5$
6. $16x^2 - 25$ **7.** $4x^2 + 20x + 25$ **8.** $9x^2 - 24x + 16$ **9. a.** $c(x) = 0.005x^2 + 0.5x + 29.9$ **b.** 35.4% **c.** underestimates by 1.3
d. $d(x) = -0.013x^2 + 1.7x - 7.5$ **e.** -7.5% **f.** underestimates by 0.3 **10.** $4x(3x^2 + 4x - 100)$ **11.** $(x - 5)(x^2 + 13)$
12. $(x^2 + 2)(x + 3)$ **13.** $(x^2 + 5)(x + 1)$ **14.** $(x - 2)(x - 1)$ **15.** $(x - 5)(x + 4)$ **16.** $(x + 3)(x + 16)$ **17.** $(x - 4)(x - 2)$
18. prime **19.** $(x + 5)(3x + 2)$ **20.** $(y - 3)(5y - 2)$ **21.** $(2x + 5)(2x - 3)$ **22.** prime **23.** $2(2x + 3)(2x - 1)$

24. $x(2x - 9)(x + 8)$ **25.** $4y(3y + 1)(y + 2)$ **26.** $(2x + 1)(2x - 1)$ **27.** $(9 + 10y)(9 - 10y)$ **28.** $(z^2 + 4)(z + 2)(z - 2)$
29. $(x + 11)^2$ **30.** $(x - 8)^2$ **31.** $(3y + 8)^2$ **32.** $(4x - 5)^2$ **33.** prime **34.** $x(x - 7)(x - 1)$ **35.** $(5y + 2)(2y + 1)$
36. $2(8 + y)(8 - y)$ **37.** $(3x + 1)^2$ **38.** $4x^3(5x^4 - 9)$ **39.** $(x - 3)^2(x + 3)$ **40.** prime **41.** $x(2x + 5)(x + 7)$ **42.** $3x(x - 5)^2$
43. $4y^2(y + 3)(y - 3)$ **44.** $5(x + 7)(x - 3)$ **45.** prime **46.** $2x^3(5x - 2)(x - 4)$ **47.** $(10y + 7)(10y - 7)$ **48.** $9x^4(x - 2)$
49. $(x^2 + 1)(x + 1)(x - 1)$ **50.** $(3x - 2)(2x + 5)$ **51.** $3x^2(x + 2)(x - 2)$ **52.** $(x - 10)(x + 9)$ **53.** $2y(4y + 3)(4y + 1)$

54. $2(y - 4)^2$ **55.** $\{-5, -1\}$ **56.** $\left\{\dfrac{1}{3}, 7\right\}$ **57.** $\{-8, 7\}$ **58.** $\{0, 4\}$ **59.** $\{-5, -3, 3\}$ **60.** length: 9 ft; width: 6 ft **61.** 2 in.

62. $\{-8, 8\}$ **63.** $\{\pm\sqrt{17}\}$ **64.** $\{\pm 5\sqrt{3}\}$ **65.** $\{0, 6\}$ **66.** $\{-4 \pm \sqrt{5}\}$ **67.** $\left\{\pm\dfrac{\sqrt{15}}{3}\right\}$ **68.** $\{1, 6\}$ **69.** $\{-5 \pm 2\sqrt{3}\}$

70. $\left\{-3, \dfrac{1}{2}\right\}$ **71.** $\{1 \pm \sqrt{5}\}$ **72.** $\left\{\dfrac{9 \pm \sqrt{21}}{6}\right\}$ **73.** $\left\{\dfrac{1}{2}, 5\right\}$ **74.** $\left\{-2, \dfrac{10}{3}\right\}$ **75.** $\left\{\dfrac{7 \pm \sqrt{37}}{6}\right\}$ **76.** $\{-3, 3\}$ **77.** $\left\{\dfrac{3 \pm \sqrt{5}}{2}\right\}$
78. a. 261 ft; overestimates by 1 ft **b.** 40 miles per hour **79. a.** by the point $(35, 261)$ **b.** by the point $(40, 267)$ **80.** about 8.8 seconds

81. 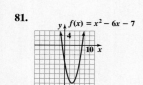 **82.**   **83.** 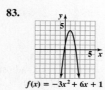 **84.**

85. 6.25 sec; 629 ft **86. a.** 20 yd; 16 ft **b.** 7.9 ft **c.** 45.3 yd **d.**

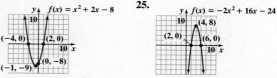

Chapter 6 Test A

1. $13x^3 + x^2 - x - 24$ **2.** $5x^3 + 2x^2 + 2x - 9$ **3.** $48x^5 - 30x^3 - 12x^2$ **4.** $3x^3 - 10x^2 - 17x - 6$ **5.** $6y^2 - 13y - 63$ **6.** $49x^2 - 25$
7. $x^4 + 6x^2 + 9$ **8.** $25x^2 - 30x + 9$ **9.** $(x - 3)(x - 6)$ **10.** $(x^2 + 3)(x + 2)$ **11.** $5y^2(3y - 1)(y - 2)$ **12.** $(5x + 3)(5x - 3)$
13. prime **14.** $(6x - 7)^2$ **15.** $(7x - 1)(x - 7)$ **16.** $4(y + 3)(y - 3)$ **17.** $(2x + 3)^2$ **18.** $\left\{\dfrac{-5 \pm \sqrt{13}}{6}\right\}$ **19.** $\left\{-\dfrac{4}{3}, 1\right\}$
20. $\left\{-\dfrac{7}{2}, \dfrac{5}{2}\right\}$ **21.** $\left\{\dfrac{3 \pm \sqrt{7}}{2}\right\}$ **22.** $\left\{-5, \dfrac{1}{2}\right\}$ **23.** $\{-1, 1, 4\}$ **24.**   **25.**
26. 2 sec; 69 ft **27.** 4.1 sec **28.** length: 8 ft; width: 5 ft

Chapter 6 Test B

1. $18x^3 + x^2 - x - 20$ **2.** $-2x^3 - 5x^2 + x - 24$ **3.** $40x^5 - 48x^3 - 24x^2$ **4.** $2x^3 - 7x^2 - 23x - 12$ **5.** $12y^2 - y - 35$ **6.** $25x^2 - 49$
7. $x^4 + 8x^2 + 16$ **8.** $49x^2 - 28x + 4$ **9.** $(x - 4)(x - 5)$ **10.** $(x^2 + 5)(x + 4)$ **11.** $3y^2(4y - 1)(y - 2)$ **12.** $(3x + 5)(3x - 5)$
13. prime **14.** $(5x - 9)^2$ **15.** $(5x - 9)(x - 8)$ **16.** $5(y + 4)(y - 4)$ **17.** $(3x + 2)^2$ **18.** $\left\{\dfrac{7 \pm \sqrt{17}}{4}\right\}$ **19.** $\left\{-1, \dfrac{6}{5}\right\}$
20. $\{-2, 3\}$ **21.** $\left\{\dfrac{3 \pm \sqrt{3}}{3}\right\}$ **22.** $\left\{-\dfrac{1}{3}, 2\right\}$ **23.** $\{-3, -1, 1\}$ **24.**   **25.**
26. after 12.5 seconds; 2540 feet **27.** 25.1 seconds
28. length: 5 yd; width: 3 yd

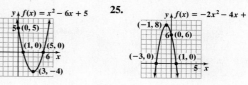

CHAPTER 7

Section 7.1

Checkpoint Exercises

1. 12.5%   **2.** 2.3%   **3. a.** 0.67   **b.** 2.5 ·   **4. a.** $75.60   **b.** $1335.60   **5. a.** $133   **b.** $247   **6. a.** $66\frac{2}{3}$%   **b.** 40%   **7.** 35%

8. 20%   **9. a.** $1152   **b.** 4% decrease

Exercise Set 7.1

1. 100   **2.** 7; 8; 100; a percent sign   **3.** two; right; a percent sign   **4.** two; left; the percent sign   **5.** tax rate; item's cost   **6.** discount rate; original price   **7.** the amount of increase; the original amount   **8.** the amount of decrease; the original amount   **9.** false   **10.** false
17. does not make sense   **19.** does not make sense   **21.** 40%   **23.** 25%   **25.** 37.5%   **27.** 2.5%   **29.** 11.25%   **31.** 59%
33. 38.44%   **35.** 287%   **37.** 1487%   **39.** 10,000%   **41.** 0.72   **43.** 0.436   **45.** 1.3   **47.** 0.02   **49.** 0.005   **51.** 0.00625
53. 0.625   **55.** 6   **57.** 7.2   **59.** 5   **61.** 170   **63.** 20%   **65.** 12%   **67. a.** $1968   **b.** $34,768   **69. a.** $103.20   **b.** $756.80
71. 19.2%   **73.** 100%   **75.** 15%   **77.** no; 2% loss   **79.** $648.72   **81.** $2400 increase

Section 7.2

Checkpoint Exercises

1. a. $89,880   **b.** $86,680   **c.** $46,960   **2.** $4106.25   **3.** $9224   **4.** $16,273   **5. a.** $180   **b.** $18   **c.** $13.77   **d.** $7.20
e. $141.03   **f.** 21.7%

Exercise Set 7.2

1. gross   **2.** adjusted gross; adjustments   **3.** exemption   **4.** adjusted gross; exemptions; deductions   **5.** credits   **6.** FICA
7. gross   **8.** net   **9.** false   **10.** true   **11.** true   **12.** true   **25.** does not make sense   **27.** makes sense   **29.** gross income: $53,320; adjusted gross income: $50,120; taxable income: $39,370   **31.** gross income: $87,490; adjusted gross income: $85,290; taxable income: $68,765
33. $6343.75   **35.** $28,964.50   **37.** −$651.25   **39.** $53,729   **41.** $3062.50   **43.** $4443.75   **45.** $2152.50   **47.** $8064   **49.** $16,998
51. a. $1530   **b.** $1256.25   **c.** 13.9%   **53. a.** $170   **b.** $17   **c.** $13.01   **d.** $5.10   **e.** $134.89   **f.** 20.7%   **55.** $2500 credit; $2100
57. yes; Answers will vary.

Section 7.3

Checkpoint Exercises

1. $150   **2.** $336   **3.** $2091   **4. a.** $1820   **b.** $1892.80   **5.** 18%   **6.** $3846.16   **7. a.** $1200   **b.** $3800   **c.** 15.8%

Exercise Set 7.3

1. $I = Prt$; principal; rate; time   **2.** $A = P(1 + rt)$ **3.** Banker's   **4.** discounted loan; discount   **5.** false   **6.** true   **7.** false
11. does not make sense   **13.** does not make sense   **15.** $240   **17.** $10.80   **19.** $318.75   **21.** $426.25   **23.** $3420   **25.** $38,350
27. $9390   **29.** 7.5%   **31.** 9%   **33.** 31.3%   **35.** $5172.42   **37.** $8917.20   **39.** $4509.59   **41. a.** $93.33   **b.** $1906.67   **c.** 7.3%

43. a. $1560   **b.** $10,440   **c.** 7.5%   **45.** $r = \dfrac{A - P}{Pt}$ **47.** $P = \dfrac{A}{1 + rt}$ **49. a.** $247.50   **b.** $4247.50   **51.** 21.4%

53. $2654.87   **55. a.** $1920   **b.** $6080   **c.** 10.5%
57. $2P = P(1 + rt)$

$$2P = P + Prt \qquad \text{Distribute } P.$$
$$P = Prt \qquad \text{Subtract } P \text{ from each side.}$$
$$1 = rt \qquad \text{Divide each side by } P.$$
$$\frac{1}{r} = t \qquad \text{Divide each side by } r.$$
$$t = \frac{1}{r} \qquad \text{Interchange the sides.}$$

Section 7.4

Checkpoint Exercises

1. a. $1216.65   **b.** $216.65   **2. a.** $6253.23   **b.** $2053.23   **3. a.** $14,859.47   **b.** $14,918.25   **4.** $5714.25   **5. a.** $6628.28
b. 10.5%   **6.** ≈ 8.24%

Exercise Set 7.4

1. principal; interest   **2.** $t; r; n$ **3.** $A = P(1 + r)^t$ **4.** six; semiannually   **5.** three; quarterly   **6.** continuous   **7.** $P; A; t; r; n$
8. effective annual yield; simple   **9.** true   **10.** true   **11.** true   **12.** true   **17.** does not make sense   **19.** does not make sense
21. a. $10,816   **b.** $816   **23. a.** $3655.21   **b.** $655.21   **25. a.** $12,795.12   **b.** $3295.12   **27. a.** $5149.12   **b.** $649.12
29. a. $1855.10   **b.** $355.10   **31. a.** $49,189.30   **b.** $29,189.30   **33. a.** $13,116.51   **b.** $13,140.67   **c.** $13,157.04   **d.** $13,165.31
35. 7% compounded monthly   **37.** $8374.85   **39.** $7528.59   **41. a.** $10,457.65   **b.** 4.6%   **43.** 6.1%   **45.** 6.2%   **47.** 6.2%

49. 8.3%; 8.25%; 8% compounded monthly   **51.** 5.6%; 5.5%; 5.5% compounded semiannually   **53.** $2,069,131   **55.** $649,083   **57.** $41,528
59. a. you; $391   **b.** you; $340   **c.** your friend; $271   **61.** $9187   **63. a.** $\approx$ $5,027,400,000   **b.** $\approx$ $5,225,000,000
65. 5.9% compounded daily; $2
67.

Years	Once a Year		Continuous	
	Amount	Interest	Amount	Interest
1	$5275	$275	$5283	$283
5	$6535	$1535	$6583	$1583
10	$8541	$3541	$8666	$3666
20	$14,589	$9589	$15,021	$10,021

69. $37,096   **71.** A: $38,755; B: $41,163   **73.** 5.55%   **75.** 4.3%; 4.3%; 4.3%; As the number of compounding periods increases, the effective
annual yield increases slightly. However, with the rates rounded to the nearest tenth of a percent, this increase is not evident.
77. 4.5% compounded semiannually   **79.** $12,942.94

81. $A = P\left(1 + \dfrac{r}{n}\right)^{nt}$

Substitute this value of A into the future value formula for simple interest. Since the interest rates are two different values, use Y, the effective yield, for the interest rate on the right side of the equation.

$P\left(1 + \dfrac{r}{n}\right)^{nt} = P(1 + Yt)$

$\left(1 + \dfrac{r}{n}\right)^{nt} = (1 + Yt)$ Divide each side by P.

$\left(1 + \dfrac{r}{n}\right)^{n} = 1 + Y$ Let $t = 1$.

$Y = \left(1 + \dfrac{r}{n}\right)^{n} - 1$ Subtract 1 from each side and interchange the sides.

Chapter 7 Review Exercises

1. 80%   **2.** 12.5%   **3.** 75%   **4.** 72%   **5.** 0.35%   **6.** 475.6%   **7.** 0.65   **8.** 0.997   **9.** 1.50   **10.** 0.03   **11.** 0.0065
12. 0.0025   **13.** 9.6   **14. a.** $1.44   **b.** $25.44   **15. a.** $297.50   **b.** $552.50   **16.** 12.5% increase   **17.** 35% decrease
18. no; $9900; 1% decrease   **19.** gross income: $30,330; adjusted gross income: $29,230; taxable income: $20,280   **20.** gross income: $436,400;
adjusted gross income: $386,400; taxable income: $278,700   **21.** $188,596.75   **22.** $5687.50   **23.** $3453.75   **24.** $6579   **25.** $20,188
26. a. $136   **b.** $13.60   **c.** $10.40   **d.** $5.44   **e.** $106.56   **f.** 21.6%   **27.** $180   **28.** $2520   **29.** $1200   **30.** $900
31. a. $122.50   **b.** $3622.50   **32.** $12,738   **33.** 7.5%   **34.** $13,389.12   **35.** $9287.93   **36.** 40%   **37. a.** $94.50   **b.** $1705.50
c. 7.4%   **38. a.** $8114.92   **b.** $1114.92   **39. a.** $38,490.80   **b.** $8490.80   **40. a.** $5556.46   **b.** $3056.46   **41.** 7% compounded
monthly; $362   **42.** $28,469.44   **43.** $13,175.19   **44. a.** $2122.73   **b.** 6.1%   **45.** 5.61%; Answers will vary; an example is: The same
amount of interest would be earned at 5.61% in a simple interest account for a year.   **46.** 6.25% compounded monthly

Chapter 7 Test A

1. a. $18   **b.** $102   **2.** 75%   **3. a.** $47,290   **b.** $46,190   **c.** $33,645   **4.** $3348.75   **5.** $7948   **6. a.** $150   **b.** $15
c. $11.48   **d.** $4.50   **e.** $119.02   **f.** 20.7%   **7.** $72; $2472   **8.** 25%   **9.** $6698.57   **10.** 4.58%; Answers will vary; an example is: The
same amount of interest would be earned at 4.58% in a simple interest account for a year.   **11. a.** $267,392   **b.** $247,392   **12.** $2070

Chapter 7 Test B

1. a. $630   **b.** $770   **2.** 70%   **3. a.** $67,900   **b.** $63,600   **c.** $44,670   **4.** $4343.75   **5.** $9398   **6. a.** $152   **b.** $15.20
c. $11.63   **d.** $6.08   **e.** $119.09   **f.** 21.7%   **7.** $180; $3780   **8.** 20%   **9.** $1132.08   **10.** 3.55%; Answers will vary; an example is:
The same amount of interest would be earned at 3.55% in a simple interest account for a year.   **11. a.** $497,472   **b.** $472,472   **c.** $2099

CHAPTER 8

Section 8.1

Checkpoint Exercises

1. a. $6620   **b.** $620   **2. a.** $777,170   **b.** $657,170   **3. a.** $333,946   **b.** $291,946   **4. a.** $187   **b.** deposits: $40,392; interest: $59,608
5. a. high: $63.38; low: $42.37   **b.** $2160   **c.** 1.5%; This is much lower than a bank account paying 3.5%.   **d.** 7,203,200 shares
e. high: $49.94; low: $48.33   **f.** $49.50   **g.** The closing price is up $0.03 from the previous day's closing price.   **h.** $\approx$ $1.34   **6. a.** $81,456
b. $557,849   **c.** $527,849

Exercise Set 8.1

1. annuity   **2.** P; r; n; value of the annuity; t **3.** money-market   **4.** certificate of deposit; CD   **5.** stock; capital; dividends
6. bonds   **7.** portfolio; diversified   **8.** mutual fund   **9.** Roth; $59\frac{1}{2}$ **10.** true   **11.** true   **12.** false   **13.** true   **14.** false
15. true   **27.** does not make sense   **29.** makes sense   **31.** makes sense   **33.** makes sense   **35.** does not make sense   **37. a.** $66,132
b. $26,132   **39. a.** $702,528   **b.** $542,528   **41. a.** $50,226   **b.** $32,226   **43. a.** $9076   **b.** $4076   **45. a.** $28,850   **b.** $4850
47. a. $4530   **b.** deposits: $81,540; interest: $58,460   **49. a.** $1323   **b.** deposits: $158,760; interest: $41,240   **51. a.** $356
b. deposits: $170,880; interest: $829,120   **53. a.** $920   **b.** deposits: $18,400; interest: $1600   **55. a.** high: $73.25; low: $45.44   **b.** $840
c. 2.2%; Answers will vary.   **d.** 591,500 shares   **e.** high: $56.38; low: $54.38   **f.** $55.50   **g.** $1.25 increase   **h.** $\approx$ $3.26   **57. a.** $30,000

b. $30,000   **59.** $P = \dfrac{Ar}{[(1 + r)^t - 1]}$; the deposit at the end of each year that yields A dollars after t years with interest rate r compounded annually

61. a. $11,617   **b.** $1617   **63. a.** $87,052   **b.** $63,052   **65. a.** $693,031   **b.** $293,031   **67. a.** $401   **b.** deposits: $3208; interest: $292
69. $620; $1,665,200   **71. a.** $173,527   **b.** $1,273,733   **c.** $1,219,733   **75. a.** $248   **b.** with: $5862.25; without: $6606.25

Section 8.2

Checkpoint Exercises

1. a. payment: $366; interest: $2568   **b.** payment: $278; interest: $5016   **c.** Monthly payments are less with the longer-term loan, but there is
more interest with this loan.   **2.** $222   **3. a.** $5250   **b.** $47,746

Exercise Set 8.2

1. PMT; P; n; t; r **2.** closed-end; open-end   **3.** residual value   **4.** bodily injury; property damage   **5.** collision   **6.** comprehensive
7. true   **8.** false   **9.** false   **10.** true   **11.** false   **12.** true   **21.** makes sense   **23.** makes sense   **25.** makes sense
27. $244; $1712   **29. a.** $450; $1200   **b.** $293; $2580   **c.** The monthly payment is lower with the longer-term loan, but the total interest is less
with the shorter-term loan.   **31.** $277   **33.** $405; $3665   **35.** $65; Incentive B   **37. a.** $6000   **b.** $42,142   **39. a.** $19,600   **b.** $145,609
41. $104,400

43.
$$P\left(1 + \frac{r}{n}\right)^{nt} = \frac{PMT\left[\left(1 + \frac{r}{n}\right)^{nt} - 1\right]}{\left(\frac{r}{n}\right)}$$

$$P\left(\frac{r}{n}\right)\left(1 + \frac{r}{n}\right)^{nt} = PMT\left[\left(1 + \frac{r}{n}\right)^{nt} - 1\right]$$

$$\frac{P\left(\frac{r}{n}\right)\left(1 + \frac{r}{n}\right)^{nt}}{\left[\left(1 + \frac{r}{n}\right)^{nt} - 1\right]} = PMT$$

$$\frac{P\left(\frac{r}{n}\right)}{\left[1 - \left(1 + \frac{r}{n}\right)^{-nt}\right]} = PMT$$

Section 8.3

Checkpoint Exercises

1. a. $1627   **b.** $117,360   **c.** $148,860
2.

Payment Number	Interest Payment	Principal Payment	Balance of Loan
1	$1166.67	$383.33	$199,616.67
2	$1164.43	$385.57	$199,231.10

3. a. $5600   **b.** $7200   **c.** $2160

Exercise Set 8.3

1. mortgage; down payment   **2.** loan amortization schedule   **3.** false   **4.** false   **5.** true   **6.** false   **17.** makes sense
19. does not make sense   **21. a.** $44,000   **b.** $176,000   **c.** $5280   **d.** $1171   **e.** $245,560   **23.** $60,120
25. 20-year at 7.5%; $106,400   **27.** Mortgage A; $11,300
29 a. $608; $98,880   **b.**

Payment Number	Interest	Principal	Loan Balance
1	$450.00	$158.00	$119,842.00
2	$449.41	$158.59	$119,683.41
3	$448.81	$159.19	$119,524.22

31. a. $840   **b.** $1080   **c.** $492   **33.** yes

Section 8.4

Checkpoint Exercise

1. a. $8761.76   **b.** $140.19   **c.** $9661.11   **d.** $269

Exercise Set 8.4

1. open-end   **2.** the sum of the unpaid balances for each day in the billing period; the number of days in the billing period   **3.** debit
4. credit report   **5.** 300; 850; better credit (worthiness)   **6.** false   **7.** false   **8.** true   **9.** true   **19.** does not make sense
21. makes sense   **23.** makes sense   **25. a.** $6150.65   **b.** $92.26   **c.** $6392.26   **d.** $178   **27. a.** $2057.22   **b.** $24.69
c. $2237.23   **d.** $90   **29. a.** $210   **b.** $840   **31. a.** $137; lower monthly payment   **b.** $732; less interest   **33.** monthly payment: $385;
total interest: $420; $175 more each month; $420 less interest

Chapter 8 Review Exercises

1. a. $19,129   **b.** $8729   **2. a.** $91,361   **b.** $55,361   **3. a.** $1049   **b.** $20,980; $4020   **5.** high: $64.06; low: $26.13   **6.** $144
7. 0.3%   **8.** 545,800 shares   **9.** high: $61.25; low: $59.25   **10.** $61.00   **11.** $1.75 increase   **12.** $1.49   **15. a.** monthly payment: $465;
total interest: $1740   **b.** monthly payment: $305; total interest: $3300   **c.** Longer term has lower monthly payment but greater total interest.
19. a. $7560   **b.** $53,099   **20. a.** $48,000   **b.** $192,000   **c.** $3840   **d.** $1277   **e.** $267,720   **21.** 20-year at 8%; $53,040; Answers
will vary.   **22. a.** option A: $769; option B: $699; Answers will vary.   **b.** Mortgage A; $20,900
23. a. $1896   **b.**

Payment Number	Interest	Principal	Loan Balance
1	$1625.00	$271.00	$299,729.00
2	$1623.53	$272.47	$299,456.53
3	$1622.06	$273.94	$299,182.59

24. a. $1260   **b.** $1620   **c.** $612   **27. a.** $4363.27   **b.** $48.00   **c.** $4533.00   **d.** $126   **28. a.** $560   **b.** $2229

Chapter 8 Test A

1. a. $8297   **b.** $2297   **2. a.** $7067   **b.** $1067   **c.** Only part of the $6000 is invested for the entire five years.; Answers will vary.
3. $704 per month; $1,162,080 interest   **4.** high: $25.75; low: $25.50   **5.** $2030   **6.** $386.25   **7.** $5320   **8.** $12,000   **9.** $108,000
10. $2160   **11.** $830   **12.** $190,800   **13.**

Payment Number	Interest	Principal	Loan Balance
1	$765.00	$65.00	$107,935.00
2	$764.54	$65.46	$107,869.54

14. a. $1540   **b.** $1980   **c.** $594   **15. a.** $3226.67   **b.** $64.53   **c.** $3464.53   **d.** $97   **16.** true   **17.** false   **18.** true   **19.** false
20. false   **21.** true   **22.** false   **23.** false

Chapter 8 Test B

1. a. $12,261   **b.** $3261   **2. a.** $10,519   **b.** $1519   **c.** Only part of the $9000 is invested for the entire five years.; Answers will vary.
3. $386 per month; $837,880 interest   **4.** high: $44.75; low: $44.35   **5.** $1340   **6.** $670.35   **7.** $2040   **8.** $130,000   **9.** $520,000
10. $15,600   **11.** $3390   **12.** $700,400
13.

Payment Number	Interest	Principal	Loan Balance
1	$2946.67	$443.33	$519,556.67
2	$2944.15	$445.85	$519,110.82

14. a. $7000   **b.** $9000   **c.** $4100   **15. a.** $4375   **b.** $142.05   **c.** $5222.05   **d.** $146   **16.** true   **17.** true   **18.** false
19. true   **20.** true   **21.** false   **22.** true   **23.** true

CHAPTER 9

Section 9.1

Checkpoint Exercises

1. a. 6.5 ft   **b.** 3.25 mi   **c.** $\frac{1}{12}$ yd   **2. a.** 8 km   **b.** 53,000 mm   **c.** 0.0604 hm   **d.** 6720 cm   **3. a.** 243.84 cm   **b.** $\approx$22.22 yd
c. $\approx$1181.1 in.   **4.** $\approx$37.5 mi/hr

Exercise Set 9.1

1. linear; linear   **2.** 12; 3; 36; 5280   **3.** unit; 1   **4.** 1000; 100; 10; 0.1; 0.01; 0.001   **5.** false   **6.** false   **7.** false   **8.** false
13. makes sense   **15.** does not makes sense   **17.** 2.5 ft   **19.** 360 in.   **21.** $\frac{1}{6}$ yd $\approx$ 0.17 yd   **23.** 216 in.   **25.** 18 ft   **27.** 2 yd

29. 4.5 mi   **31.** 3960 ft   **33.** 500 cm   **35.** 1630 m   **37.** 0.03178 hm   **39.** 0.000023 m   **41.** 21.96 dm   **43.** 35.56 cm   **45.** ≈5.51 in.
47. ≈424 km   **49.** ≈165.625 mi   **51.** ≈13.33 yd   **53.** ≈55.12 in.   **55.** 0.4064 dam   **57.** 1.524 m   **59.** ≈16.40 ft   **61.** ≈60 mi/hr
63. ≈72 km/hr   **65.** 457.2 cm   **67.** $8\frac{1}{3}$ yd   **69.** 48.28032 km   **71.** 176 ft/sec   **73.** meter   **75.** millimeter   **77.** meter   **79.** millimeter
81. millimeter   **83.** b   **85.** a   **87.** c   **89.** a   **91.** 0.216 km   **93.** ≈148.8 million km   **95.** Nile; ≈208 km   **97.** Everest; ≈239 m
99. Waialeale; ≈46 in.   **101.** the first day; Answers will vary.   **105.** 9 hm   **107.** 11 m or 1.1 dam

Section 9.2

Checkpoint Exercises

1. 8 square units   **2.** 229.8 people per square mile   **3. a.** ≈0.72 ha   **b.** ≈$576,389 per hectare   **4.** 9 cubic units   **5.** ≈74,800 gal   **6.** 220 L

Exercise Set 9.2

1. square; cubic   **2.** 144; 9   **3.** 2; 4   **4.** capacity; liter   **5.** area   **6.** false   **7.** false   **8.** false   **15.** does not makes sense
17. makes sense   **19.** 16 square units   **21.** 8 square units   **23.** ≈2.15 in.^{2} **25.** ≈37.5 yd^2 **27.** ≈25.5 acres   **29.** ≈91 cm^2
31. 24 cubic units   **33.** ≈74,800 gal   **35.** ≈1600 gal   **37.** ≈9 gal   **39.** ≈13.5 yd^3 **41.** 45 L   **43.** 17 mL   **45.** 1500 cm^3 **47.** 150 cm^3
49. 12,000 dm^3 **51. a.** 1900: 25.6 people per square mile; 2000: 79.6 people per square mile   **b.** 210.9%   **53.** 17.0 people per square kilometer
55. a. ≈20 acres   **b.** ≈$12,500 per acre   **57.** square centimeters   **59.** square kilometers   **61.** b   **63.** b   **65.** ≈336,600 gal   **67.** 4 L
69. Japan; ≈79,000 km^2 **71.** Baffin Island; ≈24,162 mi^2 **73.** ≈11,952.64 people per square mile   **75.** Answers will vary.   **77.** 6.5 liters;
Answers will vary.

Section 9.3

Checkpoint Exercises

1. a. 420 mg   **b.** 6.2 g   **2.** 145 kg   **3. a.** ≈83.7 kg   **b.** ≈100.44 mg   **4.** 122°F   **5.** 15°C

Exercise Set 9.3

1. 16; 2000   **2.** gram   **3.** 2.2   **4.** 1   **5.** 32; 212   **6.** 0; 100   **7.** false   **8.** false   **9.** false   **10.** false   **17.** does not makes sense
19. does not makes sense   **21.** 740 mg   **23.** 0.87 g   **25.** 800 cg   **27.** 18,600 g   **29.** 0.000018 g   **31.** 50 kg   **33.** 4200 cm^3
35. 1100 t   **37.** 40,000 g   **39.** 2.25 lb   **41.** ≈1008 g   **43.** ≈243 kg   **45.** ≈36,000 g   **47.** ≈1200 lb   **49.** ≈222.22 T   **51.** 50° F
53. 95°F   **55.** 134.6°F   **57.** 23°F   **59.** 20°C   **61.** 5°C   **63.** 22.2°C   **65.** −5°C   **67.** 176.7°C   **69.** −30°C   **71. a.** $\frac{9}{5}$;
Fahrenheit temperature increases by $\frac{9}{5}$° for each 1° change in Celsius temperature.   **b.** $F = \frac{9}{5}C + 32$ **73.** milligram   **75.** gram
77. kilogram   **79.** kilogram   **81.** b   **83.** a   **85.** c   **87.** 13.28 kg   **89.** $1.16   **91.** Purchasing the regular size is more economical.
93. ≈90 mg   **95. a.** 43 mg   **b.** ≈516 mg   **97.** a   **99.** c   **101.** Néma; ≈0.3°F   **103.** Eismitte; ≈5°C   **105.** false   **107.** true
109. true   **111.** false

Chapter 9 Review Exercises

1. 5.75 ft   **2.** 0.25 yd   **3.** 7 yd   **4.** 2.5 mi   **5.** 2280 cm   **6.** 70 m   **7.** 1920 m   **8.** 0.0144 hm   **9.** 0.0005 m   **10.** 180 mm
11. 58.42 cm   **12.** ≈7.48 in.   **13.** ≈528 km   **14.** ≈375 mi   **15.** ≈15.56 yd   **16.** ≈39.37 ft   **17.** ≈28.13 mi/hr   **18.** ≈96 km/hr
19. 0.024 km, 24,000 cm, 2400 m   **20.** 4.8 km   **21.** 24 square units   **22.** 84.5 people per square mile; In the United States, there was an average
of 84.5 people for each square mile in 2006.   **23.** ≈18 acres   **24.** ≈333.33 ft^2 **25.** ≈31.2 km^2 **26.** a   **27.** 24 cubic units
28. ≈251,328 gal   **29.** 76 L   **30.** c   **31–32.** Answers will vary.   **33.** 1240 mg   **34.** 1200 cg   **35.** 0.000012 g   **36.** 0.00045 kg
37. 50,000 cm^3 **38.** 4000 dm^3, 4,000,000 g   **39.** 94.5 kg   **40.** 14 oz   **41.** kilograms; Answers will vary.   **42.** 2.25 lb   **43.** a   **44.** c
45. 59°F   **46.** 212°F   **47.** 41°F   **48.** 32°F   **49.** −13°F   **50.** 15°C   **51.** 5°C   **52.** 100° C   **53.** 37°C   **54.** ≈−17.8°C
55. −10°C   **56.** more; Answers will vary.

Chapter 9 Test A

1. 0.00807 hm   **2.** 250 in.   **3.** 4.8 km   **4.** mm   **5.** cm   **6.** km   **7.** ≈128 km/hr   **8.** 9 times   **9.** 207.8 people per square mile; In
Spain, there is an average of 207.8 people for each square mile.   **10.** ≈45 acres   **11.** b   **12.** Answers will vary.; 1000 times   **13.** ≈74,800 gal
14. b   **15.** 0.137 kg   **16.** ≈40.5 kg   **17.** kg   **18.** mg   **19.** 86° F   **20.** 80° C   **21.** d

Chapter 9 Test B

1. 0.0403 hm   **2.** 350 in.   **3.** 6.3 km   **4.** cm   **5.** m   **6.** km   **7.** ≈31.25 mi/hr   **8.** 27 times   **9.** 17,314.3 people per square mile; In
Singapore, there is an average of 17,314.3 people for each square mile.   **10.** ≈40 acres   **11.** b   **12.** Answers will vary.; 100 times   **13.** ≈59,840 gal
14. c   **15.** 2.6 kg   **16.** ≈90 kg   **17.** kg   **18.** mg   **19.** 248°F   **20.** −40°C   **21.** hot

CHAPTER 10

Section 10.1

Checkpoint Exercises

1. 30°   **2.** 71°   **3.** 46°, 134°   **4.** $m\angle 1 = 57°, m\angle 2 = 123°,$ and $m\angle 3 = 123°$ **5.** $m\angle 1 = 29°, m\angle 2 = 29°, m\angle 3 = 151°, m\angle 4 = 151°,$ $m\angle 5 = 29°, m\angle 6 = 151°,$ and $m\angle 7 = 151°$

Exercise Set 10.1

1. line; half-line; ray; line segment   **2.** acute; right; obtuse; straight   **3.** complementary; supplementary   **4.** vertical   **5.** parallel; transversal   **6.** perpendicular   **7.** false   **8.** false   **9.** false   **10.** false   **11.** true   **12.** true   **21.** does not make sense   **23.** does not make sense   **25.** 150°   **27.** 90°   **29.** 20°; acute   **31.** 160°; obtuse   **33.** 180°; straight   **35.** 65°   **37.** 146°   **39.** 42°; 132°   **41.** 1°; 91°   **43.** 52.6°; 142.6°   **45.** $x + x + 12° = 90°; 39°, 51°$ **47.** $x + 3x = 180°; 45°, 135°$ **49.** $m\angle 1 = 108°; m\angle 2 = 72°; m\angle 3 = 108°$ **51.** $m\angle 1 = 50°$; $m\angle 2 = 90°; m\angle 3 = 50°$ **53.** $m\angle 1 = 68°; m\angle 2 = 68°; m\angle 3 = 112°; m\angle 4 = 112°; m\angle 5 = 68°; m\angle 6 = 68°; m\angle 7 = 112°$ **55.** $m\angle 1 = 38°$; $m\angle 2 = 52°; m\angle 3 = 142°$ **57.** $m\angle 1 = 65°; m\angle 2 = 56°; m\angle 3 = 124°$ **59.** false   **61.** true   **63.** false   **65.** false   **67.** 60°, 30°   **69.** 112°, 112°   **71.** $\overline{BC}$ **73.** $\overline{AD}$ **75.** $\overrightarrow{AD}$ **77.** $\overleftrightarrow{AD}$ **79.** 45°   **81.** When two parallel lines are intersected by a transversal, corresponding angles have the same measure.   **83.** E, F, H, and T   **85.** long-distance riding and mountain biking   **87.** 27°   **89.** 90°

Section 10.2

Checkpoint Exercises

1. 49°   **2.** $m\angle 1 = 90°; m\angle 2 = 54°; m\angle 3 = 54°; m\angle 4 = 68°; m\angle 5 = 112°$ **3.** Two angles of the small triangle are equal to two angles of the large triangle. One angle pair is given to have the same measure (right angles). Another angle pair consists of vertical angles with the same measure.; 15 cm   **4.** 32 yd   **5.** 25 ft   **6.** 120 yd

Exercise Set 10.2

1. 180°   **2.** acute   **3.** obtuse   **4.** isosceles   **5.** equilateral   **6.** scalene   **7.** similar; the same measure; proportional   **8.** right; legs; the square of the length of the hypotenuse   **9.** true   **10.** true   **11.** false   **12.** false   **13.** false   **23.** does not make sense   **25.** does not make sense   **27.** 67°   **29.** 32°   **31.** $m\angle 1 = 50°; m\angle 2 = 130°; m\angle 3 = 50°; m\angle 4 = 130°; m\angle 5 = 50°$ **33.** $m\angle 1 = 50°; m\angle 2 = 50°; m\angle 3 = 80°; m\angle 4 = 130°; m\angle 5 = 130°$ **35.** $m\angle 1 = 55°; m\angle 2 = 65°; m\angle 3 = 60°; m\angle 4 = 65°; m\angle 5 = 60°; m\angle 6 = 120°; m\angle 7 = 60°; m\angle 8 = 60°; m\angle 9 = 55°; m\angle 10 = 55°$ **37.** The three angles of the large triangle are given to have the same measures as the three angles of the small triangle.; 5 in.   **39.** Two angles of the large triangle are given to have the same measures as two angles of the small triangle.; 6 m   **41.** One angle pair is given to have the same measure (right angles). Another angle pair consists of vertical angles with the same measure.; 16 in.   **43.** 5   **45.** 9   **47.** 17 m   **49.** 39 m   **51.** 12 cm   **53.** congruent; SAS   **55.** congruent; SSS   **57.** congruent; SAS   **59.** not necessarily congruent   **61.** congruent; ASA   **63.** 71.7 ft   **65.** $90\sqrt{2}$ ft ≈ 127.3 ft   **67.** 45 yd   **69.** 13 ft   **71.** $750,000   **73.** 70°   **75.** 40°, 80°, 60°   **77.** 5 ft by 12 ft

Section 10.3

Checkpoint Exercises

1. $3120   **2. a.** 1800°   **b.** 150°   **3.** Each angle of a regular octagon measures 135°. 360° is not a multiple of 135°.

Exercise Set 10.3

1. perimeter   **2.** quadrilateral; pentagon; hexagon; heptagon; octagon   **3.** regular   **4.** equal in measure; parallel   **5.** rhombus   **6.** rectangle   **7.** square   **8.** trapezoid   **9.** $P = 2l + 2w$ **10.** $(n - 2)180°$ **11.** tessellation   **12.** false   **13.** true   **14.** true   **15.** true   **16.** false   **17.** false   **18.** true   **25.** does not make sense   **27.** makes sense   **29.** quadrilateral   **31.** pentagon   **33.** a: square; b: rhombus; d: rectangle; e: parallelogram   **35.** a: square; d: rectangle   **37.** c: trapezoid   **39.** 30 cm   **41.** 28 yd   **43.** 1000 in.   **45.** 27 ft   **47.** 18 yd   **49.** 84 yd   **51.** 32 ft   **53.** 540°   **55.** 360°   **57.** 108°; 72°   **59. a.** 540°   **b.** $m\angle A = 140°; m\angle B = 40°$ **61. a.** squares, hexagons, dodecagons (12-sided polygons)   **b.** 3 angles; 90°, 120°, 150°   **c.** The sum of the measures is 360°.   **63. a.** triangles, hexagons   **b.** 4 angles; 60°, 60°, 120°, 120°   **c.** The sum of the measures is 360°.   **65.** no; Each angle of the polygon measures 140°. 360° is not a multiple of 140°.   **67.** 50 yd by 200 yd   **69.** 160 ft by 360 ft   **71.** 115°, 115°, 120°, 120°   **73.** If the polygons were all regular polygons, the sum would be 363°. The sum is not 360°.   **75.** $5600   **77.** 48   **79.** $2(a + b + c + d)$ **81.** 65°

Section 10.4

Checkpoint Exercises

1. 66 ft²; yes   **2.** $672   **3.** 60 in.²   **4.** 30 ft²   **5.** 105 ft²   **6.** 10π in.; 31.4 in.   **7.** 49.7 ft   **8.** large pizza

Exercise Set 10.4

1. $A = lw$ **2.** $A = s^2$ **3.** $A = bh$ **4.** $A = \frac{1}{2}bh$ **5.** $A = \frac{1}{2}h(a + b)$ **6.** $C = \pi d$ **7.** $C = 2\pi r$ **8.** $A = \pi r^2$ **9.** false   **10.** true   **11.** true   **12.** false   **13.** true   **19.** does not make sense   **21.** does not make sense   **23.** 18 m²   **25.** 16 in.²   **27.** 2100 cm²   **29.** 56 in.²   **31.** 20.58 yd²   **33.** 30 in.²   **35.** 567 m²   **37.** C: 8π cm, ≈ 25.1 cm; A: 16π cm², ≈ 50.3 cm²   **39.** C: 12π yd, ≈ 37.7 yd; A: 36π yd², ≈ 113.1 yd²   **41.** 72 m²   **43.** 300 m²   **45.** $100 + 50\pi$; 257.1 cm²   **47.** $A = ab + \frac{1}{2}(c - a)b$ or $A = \frac{1}{2}(a + c)b$ **49.** $A = 2a^2 + ab$ **51.** 192 cm²   **53.** 8π cm²   **55.** $(12.5\pi - 24)$ in.²

57. perimeter: 54 ft; area: 168 ft^2 **59.** $556.50   **61.** 148 ft^2 **63. a.** 23 bags   **b.** $575   **65. a.** 3680 ft^2 **b.** 15 cans of paint
c. $404.25   **67.** $933.40   **69.** 40π m; 125.7 m   **71.** 377 plants   **73.** large pizza   **75.** the square with sides of 50 ft; 2500 ft^2 **77.** $600

Section 10.5

Checkpoint Exercises

1. 105 ft^3 **2.** 8 yd^3 **3.** 48 ft^3 **4.** 302 in.^{3} **5.** 101 in.^{3} **6.** no   **7.** 632 yd^2

Exercise Set 10.5

1. $V = lwh$ **2.** $V = s^3$ **3.** polyhedron   **4.** $V = \frac{1}{3}Bh$ **5.** $V = \pi r^2 h$ **6.** $V = \frac{1}{3}\pi r^2 h$ **7.** $V = \frac{4}{3}\pi r^3$ **8.** true   **9.** true
10. false   **11.** true   **12.** true   **13.** true   **14.** false   **17.** does not make sense   **19.** does not make sense   **21.** 36 in.^{3} **23.** 64 cm^3
25. 175 yd^3 **27.** 56 in.^{3} **29.** 150π cm^3, 471 cm^3 **31.** 3024π in.^{3}, 9500 in.^{3} **33.** 48π m^3, 151 m^3 **35.** 15π yd^3, 47 yd^3
37. 288π m^3, 905 m^3 **39.** 972π cm^3, 3054 cm^3 **41.** 62 m^2 **43.** 96 ft^2 **45.** 324π cm^3, 1018 cm^3 **47.** 432π in.^{3}, 1357 in.^{3}
49. $\frac{3332}{3}\pi$ m^3, 3489 m^3 **51.** surface area: 186 yd^2; volume: 148 yd^3 **53.** 666 yd^2 **55.** $\frac{1}{8}$ **57.** 9 times   **59.** $40   **61.** no
63. a. 3,386,880 yd^3 **b.** 2,257,920 blocks   **65.** yes   **67.** $27   **69.** The volume is multiplied by 8.   **71.** 168 cm^3 **73.** 84 cm^2

Section 10.6

Checkpoint Exercises

1. $\sin A = \frac{3}{5}$; $\cos A = \frac{4}{5}$; $\tan A = \frac{3}{4}$ **2.** 263 cm   **3.** 298 cm   **4.** 994 ft   **5.** 54°

Exercise Set 10.6

1. sine; opposite; hypotenuse; $\frac{a}{c}$ **2.** cosine; adjacent to; hypotenuse; $\frac{b}{c}$ **3.** tangent; opposite; adjacent to; $\frac{a}{b}$ **4.** elevation   **5.** depression
6. false   **7.** false   **8.** true   **9.** false   **17.** does not make sense   **19.** makes sense   **21.** $\sin A = \frac{3}{5}$; $\cos A = \frac{4}{5}$; $\tan A = \frac{3}{4}$
23. $\sin A = \frac{20}{29}$; $\cos A = \frac{21}{29}$; $\tan A = \frac{20}{21}$ **25.** $\sin A = \frac{5}{13}$; $\cos A = \frac{12}{13}$; $\tan A = \frac{5}{12}$ **27.** $\sin A = \frac{4}{5}$; $\cos A = \frac{3}{5}$; $\tan A = \frac{4}{3}$ **29.** 188 cm
31. 182 in.   **33.** 7 m   **35.** 22 yd   **37.** 40 m   **39.** $m\measuredangle B = 50°$, $a = 18$ yd, $c = 29$ yd   **41.** $m\measuredangle B = 38°$, $a = 43$ cm, $b = 33$ cm   **43.** 37°
45. 28°   **47.** 653 units   **49.** 39 units   **51.** 298 units   **53.** 257 units   **55.** 529 yd   **57.** 2879 ft   **59.** 2059 ft   **61.** 695 ft   **63.** 36°
65. 1376 ft   **67.** 15.1°   **69.** 0.5299 and 0.5299; 0.2924 and 0.2924; 0.7660 and 0.7660; 0.9994 and 0.9994; $\sin A = \cos(90° - A)$; Answers will vary.
71. As the measure of an angle gets closer to 90°, the tangent of the angle grows without bound (gets very large).   **73.** 126 ft

Section 10.7

Checkpoint Exercises

1. 13   **2.** $\left(4, -\frac{1}{2}\right)$ **3.** $x^2 + y^2 = 16$ **4.** $x^2 + (y + 6)^2 = 100$
5. a. center: $(-3, 1)$; radius: 2   **b.**

$(x + 3)^2 + (y - 1)^2 = 4$

Exercise Set 10.7

1. $\sqrt{(x_2 - x_1)^2 + (y_2 - y_1)^2}$ **2.** $\left(\frac{x_1 + x_2}{2}, \frac{y_1 + y_2}{2}\right)$ **3.** $(x - h)^2 + (y - k)^2 = r^2$ **4.** $x^2 + y^2 = r^2$ **5.** true   **6.** false   **7.** false
8. false   **13.** makes sense   **15.** does not make sense   **17.** 13   **19.** $2\sqrt{29} \approx 10.77$ **21.** 5   **23.** $\sqrt{29} \approx 5.39$ **25.** $4\sqrt{2} \approx 5.66$
27. $2\sqrt{5} \approx 4.47$ **29.** $(4, 6)$ **31.** $(-4, -5)$ **33.** $\left(\frac{3}{2}, -6\right)$ **35.** $(-3, -2)$ **37.** $x^2 + y^2 = 49$ **39.** $(x - 3)^2 + (y - 2)^2 = 25$
41. $(x + 1)^2 + (y - 4)^2 = 4$ **43.** $(x + 3)^2 + (y + 1)^2 = 3$ **45.** $(x + 4)^2 + y^2 = 100$
47. center: $(0, 0)$ **49.** center: $(3, 1)$ **51.** center: $(-3, 2)$ **53.** center: $(-2, -2)$ **55.** center: $(0, 1)$
      radius: 4   radius: 6   radius: 2   radius: 2   radius: 1

$x^2 + y^2 = 16$

$(x - 3)^2 + (y - 1)^2 = 36$

$(x + 3)^2 + (y - 2)^2 = 4$

$(x + 2)^2 + (y + 2)^2 = 4$

$x^2 + (y - 1)^2 = 1$

57. center: $(-1, 0)$ radius: 5

$(x + 1)^2 + y^2 = 25$

59. circumference: 10π in.; area: 25π in.^{2}
61. circumference: 20π in.; area: 100π in.^{2}

63. $(0, -4), (4, 0)$

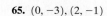
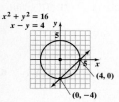

$x^2 + y^2 = 16$
$x - y = 4$

$(4, 0)$
$(0, -4)$

65. $(0, -3), (2, -1)$

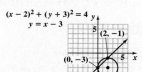

$(x - 2)^2 + (y + 3)^2 = 4$
$y = x - 3$
$(2, -1)$
$(0, -3)$

67. 2693 mi   **69.** $(x + 2.4)^2 + (y + 2.7)^2 = 900$ **71.** 11π cm^2 **73. a.** $(5, 10)$ **b.** $\sqrt{5}$ **c.** $(x - 5)^2 + (y - 10)^2 = 5$

Chapter 10 Review Exercises

1. $\angle 3$ **2.** $\angle 5$ **3.** $\angle 4$ and $\angle 6$ **4.** $\angle 1$ and $\angle 6$ **5.** $\angle 4$ and $\angle 1$ **6.** $\angle 2$ **7.** $\angle 5$ **8.** $65°$ **9.** $49°$ **10.** $17°$ **11.** $134°$
12. $m\angle 1 = 110°$; $m\angle 2 = 70°$; $m\angle 3 = 110°$ **13.** $m\angle 1 = 138°$; $m\angle 2 = 42°$; $m\angle 3 = 138°$; $m\angle 4 = 138°$; $m\angle 5 = 42°$; $m\angle 6 = 42°$; $m\angle 7 = 138°$
14. $72°$ **15.** $51°$ **16.** $m\angle 1 = 90°$; $m\angle 2 = 90°$; $m\angle 3 = 140°$; $m\angle 4 = 40°$; $m\angle 5 = 140°$ **17.** $m\angle 1 = 80°$; $m\angle 2 = 65°$; $m\angle 3 = 115°$;
$m\angle 4 = 80°$; $m\angle 5 = 100°$; $m\angle 6 = 80°$ **18.** 5 ft   **19.** 3.75 ft   **20.** 10 ft   **21.** 7.2 in.   **22.** 10.2 cm   **23.** 12.5 ft   **24.** 15 ft   **25.** 12 yd
26. rectangle, square   **27.** rhombus, square   **28.** parallelogram, rhombus, trapezoid   **29.** 30 cm   **30.** 4480 yd   **31.** 44 m   **32.** $1800°$
33. $1080°$ **34.** $m\angle 1 = 135°$; $m\angle 2 = 45°$ **35.** \$132   **36. a.** triangles, hexagons   **b.** 5 angles; $60°, 60°, 60°, 60°, 120°$ **c.** The sum of the
measures is $360°$.   **37.** yes; Each angle of a regular hexagon measures $120°$, and $360°$ is a multiple of $120°$.   **38.** 32.5 ft^2 **39.** 20 m^2 **40.** 50 cm^2
41. 135 yd^2 **42.** $C = 20\pi$ m ≈ 62.8 m; $A = 100\pi$ m$^2 \approx 314.2$ m^2 **43.** 192 in.^{2} **44.** 28 m^2 **45.** 279.5 ft^2 **46.** $(128 - 32\pi)$ in.^{2}, 27.5 in.^{2}
47. \$787.50   **48.** \$650   **49.** 31 yd   **50.** 60 cm^3 **51.** 960 m^3 **52.** 128π yd^3, 402 yd^3 **53.** $\dfrac{44{,}800\pi}{3}$ in.^{3}, 46,914 in.^{3} **54.** 288π m^3; 905 m^3
55. 126 m^2 **56.** 4800 m^3 **57.** 651,775 m^3 **58.** \$80   **59.** $\sin A = \dfrac{3}{5}$; $\cos A = \dfrac{4}{5}$; $\tan A = \dfrac{3}{4}$ **60.** 42 mm   **61.** 23 cm   **62.** 37 in.
63. $58°$ **64.** 772 ft   **65.** 31 m   **66.** $56°$ **67.** 13   **68.** $2\sqrt{2} \approx 2.83$ **69.** $(-5, 5)$ **70.** $\left(-\dfrac{11}{2}, -2\right)$ **71.** $x^2 + y^2 = 9$
72. $(x + 2)^2 + (y - 4)^2 = 36$
73. center: $(0, 0)$; radius: 1   **74.** center: $(-2, 3)$; radius: 3

$x^2 + y^2 = 1$

$(x + 2)^2 + (y - 3)^2 = 9$

Chapter 10 Test A

1. $36°$; $126°$ **2.** $133°$ **3.** $70°$ **4.** $35°$ **5.** 3.2 in.   **6.** 10 ft   **7.** $1440°$ **8.** 40 cm   **9.** d   **10. a.** triangles, squares
b. 5 angles; $60°, 60°, 60°, 90°, 90°$ **c.** The sum of the measures is $360°$.   **11.** 517 m^2 **12.** 525 in.^{2} **13. a.** 12 cm   **b.** 30 cm   **c.** 30 cm^2
14. $C = 40\pi$ m ≈ 125.7 m; $A = 400\pi$ m$^2 \approx 1256.6$ m^2 **15.** 108 tiles   **16.** 18 ft^3 **17.** 16 m^3 **18.** 175π cm$^3 \approx 550$ cm^3 **19.** 85 cm
20. 70 ft   **21.** 5; $(3.5, 0)$ **22.** $x^2 + (y + 2)^2 = 25$
23. center: $(2, -1)$; radius: 2;

$(2, -1)$

Chapter 10 Test B

1. $2°$; $92°$ **2.** $110°$ **3.** $57°$ **4.** $70°$ **5.** 20.8   **6.** 15 ft   **7.** $1800°$ **8.** 36 ft   **9.** a   **10. a.** squares, octagons
b. 3 angles; $90°, 135°, 135°$ **c.** The sum of the measures is $360°$.   **11.** 2400 cm^2 **12.** 250 m^2 **13. a.** 8 yd   **b.** 24 yd   **c.** 24 yd^2
14. $C = 30\pi \approx 94.2$ m; $A = 225\pi \approx 706.9$ m^2 **15.** 540 tiles   **16.** 72 in.^{3} **17.** 75 cm^3 **18.** 90π m$^3 \approx 283$ m^3 **19.** 60 cm   **20.** 54 ft
21. 5; $(5.5, -1)$ **22.** $(x + 4)^2 + y^2 = 36$ **23.** center: $(3, -2)$; radius: 3;

$(3, -2)$

CHAPTER 11

Section 11.1

Checkpoint Exercises

1. 150   **2.** 40   **3.** 30   **4.** 160   **5.** 729   **6.** 90,000

Exercise Set 11.1

1. $M \cdot N$ **2.** multiplying; Fundamental Counting   **3.** false   **4.** true   **7.** makes sense   **9.** does not make sense   **11.** 80   **13.** 12
15. 6   **17.** 40   **19.** 144; Answers will vary.   **21.** 8   **23.** 96   **25.** 243   **27.** 144   **29.** 676,000   **31.** 2187   **33.** 4500

Section 11.2

Checkpoint Exercises

1. 600   **2.** 120   **3. a.** 504   **b.** 524,160   **c.** 100   **4.** 840   **5.** 15,120   **6.** 420

Exercise Set 11.2

1. factorial; 5; 1; 1   **2.** $\dfrac{n!}{(n-r)!}$ **3.** $\dfrac{n!}{p!q!}$ **4.** false   **5.** false   **6.** true   **7.** false   **13.** makes sense   **15.** does not make sense
17. 720   **19.** 120   **21.** 120   **23.** 362,880   **25.** 6   **27.** 4   **29.** 504   **31.** 570,024   **33.** 3,047,466,240   **35.** 600   **37.** 10,712
39. 5034   **41.** 24   **43.** 6   **45.** 42   **47.** 1716   **49.** 3024   **51.** 6720   **53.** 720   **55.** 1   **57.** 720   **59.** 8,648,640   **61.** 120
63. 15,120   **65.** 180   **67.** 831,600   **69.** 105   **71.** 280   **75.** 239,500,800   **77.** 14,400   **79.** $\dfrac{n(n-1)\cdots 3 \cdot 2 \cdot 1}{2} = n(n-1)\cdots 3$

Section 11.3

Checkpoint Exercises

1. a. combinations   **b.** permutations   **2.** 35   **3.** 1820   **4.** 525

Exercise Set 11.3

1. $\dfrac{n!}{(n-r)!\,r!}$ **2.** $r!$ **3.** false   **4.** false   **7.** does not make sense   **9.** makes sense   **11.** combinations   **13.** permutations
15. 6   **17.** 126   **19.** 330   **21.** 8   **23.** 1   **25.** 4060   **27.** 1   **29.** 7   **31.** 0   **33.** $\dfrac{3}{4}$ **35.** -9499 **37.** $\dfrac{3}{68}$ **39.** 20
41. 495   **43.** 24,310   **45.** 22,957,480   **47.** 735   **49.** 4,516,932,420   **51.** 360   **53.** 1716   **55.** 1140   **57.** 840   **59.** 2730   **61.** 10
63. 12   **65.** 10   **67.** 60   **69.** 792   **71.** 720   **73.** 20   **75.** 24   **77.** 12   **81.** The 5/36 lottery is easier to win.; Answers will vary.
83. 570 sec or 9.5 min; 2340 sec or 39 min

Section 11.4

Checkpoint Exercises

1. a. $\dfrac{1}{6}$ **b.** $\dfrac{1}{2}$ **c.** 0   **d.** 1   **2. a.** $\dfrac{1}{13}$ **b.** $\dfrac{1}{2}$ **c.** $\dfrac{1}{26}$ **3.** $\dfrac{1}{2}$ **4. a.** 0.29   **b.** 0.49

Exercise Set 11.4

1. sample space   **2.** $P(E)$; number of outcomes in E; total number of possible outcomes   **3.** 52; hearts; diamonds; clubs; spades
4. empirical   **5.** true   **6.** false   **7.** true   **8.** false   **15.** does not make sense   **17.** makes sense   **19.** $\dfrac{1}{6}$ **21.** $\dfrac{1}{2}$ **23.** $\dfrac{1}{3}$ **25.** 1
27. 0   **29.** $\dfrac{1}{13}$ **31.** $\dfrac{1}{4}$ **33.** $\dfrac{3}{13}$ **35.** $\dfrac{1}{52}$ **37.** 0   **39.** $\dfrac{1}{4}$ **41.** $\dfrac{1}{2}$ **43.** $\dfrac{1}{2}$ **45.** $\dfrac{3}{8}$ **47.** $\dfrac{3}{8}$ **49.** $\dfrac{7}{8}$ **51.** 0   **53.** $\dfrac{1}{4}$ **55.** $\dfrac{1}{9}$
57. 0   **59.** $\dfrac{3}{10}$ **61.** $\dfrac{1}{5}$ **63.** $\dfrac{1}{2}$ **65.** 0   **67.** $\dfrac{1}{4}$ **69.** $\dfrac{1}{4}$ **71.** $\dfrac{1}{2}$ **73.** 0.43   **75.** 0.13   **77.** 0.03   **79.** 0.38   **81.** 0.78
83. 0.12   **85.** 0.25   **87.** 0.08   **91.** $\dfrac{1}{10}$

Section 11.5

Checkpoint Exercises

1. $\dfrac{4}{15}$ **1.** $\dfrac{1}{13,983,816} \approx 0.0000000715$ **3. a.** $\dfrac{1}{6}$ **b.** $\dfrac{1}{2}$

Exercise Set 11.5

1. permutations; the total number of possible permutations **2.** 1; combinations **3.** true **4.** false **7.** makes sense **9.** makes sense
11. a. 120 **b.** 6 **c.** $\frac{1}{20}$ **13. a.** $\frac{1}{6}$ **b.** $\frac{1}{30}$ **c.** $\frac{1}{720}$ **d.** $\frac{1}{3}$ **15. a.** 84 **b.** 10 **c.** $\frac{5}{42}$ **17.** $\frac{1}{18,009,460} \approx 0.0000000555$;
$\frac{100}{18,009,460} \approx 0.00000555$ **19. a.** $\frac{1}{177,100} \approx 0.00000565$ **b.** $\frac{27,132}{177,100} \approx 0.153$ **21.** $\frac{3}{10} = 0.3$ **23. a.** 2,598,960 **b.** 1287
c. $\frac{1287}{2,598,960} \approx 0.000495$ **25.** $\frac{11}{1105} \approx 0.00995$ **27.** $\frac{36}{270,725} \approx 0.000133$ **31.** $\frac{1}{18}$ **33.** The prize is shared among all winners. You are
guaranteed to win but not to win \$50 million. **35.** $\frac{235,620}{2,598,960} \approx 0.0907$

Section 11.6

Checkpoint Exercises

1. $\frac{3}{4}$ **2. a.** $\frac{160}{191}$ **b.** $\frac{182}{191}$ **3.** $\frac{1}{3}$ **4.** $\frac{27}{50}$ **5.** $\frac{3}{4}$ **6. a.** 0.99 **b.** 0.64 **7. a.** 2:50 or 1:25 **b.** 50:2 or 25:1 **8.** 199:1
9. 1:15; $\frac{1}{16}$

Exercise Set 11.6

1. $1 - P(E)$; $1 - P(\text{not } E)$ **2.** mutually exclusive; $P(A) + P(B)$ **3.** $P(A) + P(B) - P(A \text{ and } B)$ **4.** E will occur; E will not occur
5. $P(E) = \frac{a}{a+b}$ **6.** false **7.** false **8.** true **9.** false **17.** does not make sense **19.** does not make sense **21.** $\frac{12}{13}$ **23.** $\frac{3}{4}$
25. $\frac{10}{13}$ **27. a.** 0.10 **b.** 0.90 **29.** $\frac{97}{118}$ **31.** $\frac{47}{59}$ **33.** $\frac{2}{13}$ **35.** $\frac{1}{13}$ **37.** $\frac{1}{26}$ **39.** $\frac{9}{22}$ **41.** $\frac{5}{6}$ **43.** $\frac{7}{13}$ **45.** $\frac{11}{26}$ **47.** $\frac{3}{4}$
49. $\frac{5}{8}$ **51.** $\frac{33}{40}$ **53.** $\frac{4}{5}$ **55.** $\frac{7}{10}$ **57.** $\frac{43}{58}$ **59.** $\frac{50}{87}$ **61.** $\frac{113}{174}$ **63.** 15:43; 43:15 **65.** $\frac{43}{69}$ **67.** $\frac{62}{69}$ **69.** $\frac{26}{69}$ **71.** 17:52; 52:17
73. 1:137; 137:1 **75.** 119:19; 19:119 **77.** 2:1 **79.** 1:2 **81. a.** 9:91 **b.** 91:9 **83.** 1:3 **85.** 1:1 **87.** 12:1 **89.** 25:1 **91.** 47:5
93. 49:1 **95.** 9:10 **97.** 14:5 **99.** 14:5 **101.** 9:10 **103.** $\frac{3}{7}$ **105.** $\frac{4}{25}$; 84 **107.** $\frac{1}{1000}$ **109. a.** $\frac{12}{25}$ **b.** $\frac{3}{10}$
111.
$$\frac{P(E)}{1 - P(E)} = \frac{a}{b}$$
$$bP(E) = a(1 - P(E))$$
$$bP(E) = a - aP(E)$$
$$aP(E) + bP(E) = a$$
$$P(E)(a + b) = a$$
$$P(E) = \frac{a}{a + b}$$

Section 11.7

Checkpoint Exercises

1. $\frac{9}{64}$ **2.** $\frac{1}{16}$ **3. a.** $\frac{625}{130,321} \approx 0.005$ **b.** $\frac{38,416}{130,321} \approx 0.295$ **c.** $\frac{91,905}{130,321} \approx 0.705$ **4.** $\frac{1}{221} \approx 0.00452$ **5.** $\frac{11}{850} \approx 0.0129$
6. $\frac{2}{5}$ **7. a.** 1 **b.** $\frac{1}{2}$ **8. a.** $\frac{9}{10} = 0.9$ **b.** $\frac{45}{479} \approx 0.094$

Exercise Set 11.7

1. independent; $P(A) \cdot P(B)$ **2.** the event does not occur **3.** dependent; $P(A) \cdot P(B \text{ given that } A \text{ occurred})$ **4.** conditional; $P(B|A)$ **5.** false
6. true **7.** true **8.** true **11.** does not make sense **13.** makes sense **15.** $\frac{1}{6}$ **17.** $\frac{1}{36}$ **19.** $\frac{1}{4}$ **21.** $\frac{1}{36}$ **23.** $\frac{1}{8}$ **25.** $\frac{1}{36}$
27. $\frac{1}{3}$ **29.** $\frac{3}{52}$ **31.** $\frac{1}{169}$ **33.** $\frac{1}{4}$ **35.** $\frac{1}{64}$ **37.** $\frac{1}{6}$ **39. a.** $\frac{1}{256} \approx 0.00391$ **b.** $\frac{1}{4096} \approx 0.000244$ **c.** ≈ 0.524 **d.** ≈ 0.476
41. 0.0144 **43.** 0.008 **45.** 0.4686 **47.** $\frac{7}{29}$ **49.** $\frac{5}{87}$ **51.** $\frac{2}{21}$ **53.** $\frac{4}{35}$ **55.** $\frac{11}{21}$ **57.** $\frac{1}{57}$ **59.** $\frac{8}{855}$ **61.** $\frac{11}{57}$ **63.** $\frac{1}{5}$

65. $\frac{2}{3}$ **67.** $\frac{3}{4}$ **69.** $\frac{3}{4}$ **71.** $\frac{412,368}{412,878} \approx 0.999$ **73.** $\frac{412,368}{574,895} \approx 0.717$ **75.** 0.903 **77.** 0.156 **79.** 0.542 **81.** 0.421 **83.** 0.092

85. 0.893 **87.** 0.59049 **89. a.** Answers will vary. **b.** $\frac{365}{365} \cdot \frac{364}{365} \cdot \frac{363}{365} \approx 0.992$ **c.** ≈ 0.008 **d.** 0.411 **e.** 23 people **91.** $\frac{11}{36}$

Chapter 11 Review Exercises

1. 800 **2.** 20 **3.** 9900 **4.** 125 **5.** 243 **6.** 60 **7.** 240 **8.** 800 **9.** 114 **10.** 990 **11.** 151,200 **12.** 9900 **13.** 330
14. 2002 **15.** combinations **16.** permutations **17.** combinations **18.** 720 **19.** 32,760 **20.** 210 **21.** 420 **22.** 1140
23. 120 **24.** 1,860,480 **25.** 120 **26.** 1287 **27.** 55,440 **28.** 60 **29.** $\frac{1}{6}$ **30.** $\frac{2}{3}$ **31.** 1 **32.** 0 **33.** $\frac{1}{13}$ **34.** $\frac{3}{13}$ **35.** $\frac{3}{13}$
36. $\frac{1}{52}$ **37.** $\frac{1}{26}$ **38.** $\frac{1}{2}$ **39.** $\frac{1}{3}$ **40.** $\frac{1}{6}$ **41. a.** $\frac{1}{2}$ **b.** 0 **42.** 0.623 **43.** 0.518 **44.** 0.149 **45.** $\frac{1}{24}$ **46.** $\frac{1}{6}$ **47.** $\frac{1}{30}$
48. $\frac{1}{720}$ **49.** $\frac{1}{3}$ **50. a.** $\frac{1}{15,504} \approx 0.0000645$ **b.** $\frac{100}{15,504} \approx 0.00645$ **51. a.** $\frac{1}{14}$ **b.** $\frac{3}{7}$ **52.** $\frac{3}{26}$ **53.** $\frac{5}{6}$ **54.** $\frac{1}{2}$ **55.** $\frac{1}{3}$ **56.** $\frac{2}{3}$
57. 1 **58.** $\frac{10}{13}$ **59.** $\frac{3}{4}$ **60.** $\frac{2}{13}$ **61.** $\frac{1}{13}$ **62.** $\frac{7}{13}$ **63.** $\frac{11}{26}$ **64.** $\frac{5}{6}$ **65.** $\frac{5}{6}$ **66.** $\frac{1}{2}$ **67.** $\frac{2}{3}$ **68.** 1 **69.** $\frac{5}{6}$ **70.** $\frac{4}{5}$
71. $\frac{3}{4}$ **72.** $\frac{18}{25}$ **73.** $\frac{6}{7}$ **74.** $\frac{3}{5}$ **75.** $\frac{12}{35}$ **76.** in favor: 1:12; against: 12:1 **77.** 99:1 **78.** $\frac{3}{4}$ **79.** $\frac{2}{9}$ **80.** $\frac{1}{36}$ **81.** $\frac{1}{9}$ **82.** $\frac{1}{36}$
83. $\frac{8}{27}$ **84.** $\frac{1}{32}$ **85. a.** 0.04 **b.** 0.008 **c.** 0.4096 **d.** 0.5904 **86.** $\frac{1}{9}$ **87.** $\frac{1}{12}$ **88.** $\frac{1}{196}$ **89.** $\frac{1}{3}$ **90.** $\frac{3}{10}$ **91. a.** $\frac{1}{2}$
b. $\frac{2}{7}$ **92.** $\frac{27}{29}$ **93.** $\frac{4}{29}$ **94.** $\frac{144}{145}$ **95.** $\frac{11}{20}$ **96.** $\frac{11}{135}$ **97.** $\frac{1}{125}$ **98.** $\frac{1}{232}$ **99.** $\frac{19}{1044}$ **100.** $\frac{25,546}{29,625} \approx 0.862$
101. $\frac{11,044}{29,625} \approx 0.373$ **102.** $\frac{27,291}{29,625} \approx 0.921$ **103.** $\frac{15,097}{29,625} \approx 0.510$ **104.** $\frac{4110}{29,625} \approx 0.139$ **105.** $\frac{3700}{25,546} \approx 0.145$ **106.** $\frac{2123}{2334} \approx 0.910$

Chapter 11 Test A

1. 240 **2.** 24 **3.** 720 **4.** 990 **5.** 210 **6.** 420 **7.** $\frac{6}{25}$ **8.** $\frac{17}{25}$ **9.** $\frac{11}{25}$ **10.** $\frac{5}{13}$ **11.** $\frac{1}{210}$ **12.** $\frac{10}{1001} \approx 0.00999$ **13.** $\frac{1}{2}$
14. $\frac{1}{16}$ **15.** $\frac{1}{8000} = 0.000125$ **16.** $\frac{8}{13}$ **17.** $\frac{3}{5}$ **18.** $\frac{1}{19}$ **19.** $\frac{1}{256}$ **20.** 3:4 **21. a.** 4:1 **b.** $\frac{4}{5}$ **22.** $\frac{3}{5}$ **23.** $\frac{39}{50}$ **24.** $\frac{3}{5}$
25. $\frac{9}{19}$ **26.** $\frac{7}{150}$

Chapter 11 Test B

1. 288 **2.** 720 **3.** 24 **4.** 720 **5.** 126 **6.** 3360 **7.** $\frac{3}{25}$ **8.** $\frac{17}{25}$ **9.** $\frac{14}{25}$ **10.** $\frac{3}{13}$ **11.** $\frac{1}{120}$ **12.** $\frac{5}{3094} \approx 0.00162$ **13.** $\frac{5}{8}$
14. $\frac{3}{32}$ **15.** $\frac{1}{10,000} = 0.0001$ **16.** $\frac{8}{13}$ **17.** $\frac{7}{12}$ **18.** $\frac{2}{57}$ **19.** $\frac{1}{625}$ **20.** 1:3 **21. a.** 6:1 **b.** $\frac{6}{7}$ **22.** $\frac{16}{25}$ **23.** $\frac{3}{5}$ **24.** $\frac{33}{50}$
25. $\frac{5}{9}$ **26.** $\frac{26}{165}$

CHAPTER 12

Section 12.1

Checkpoint Exercises

1. a. the set containing all the city's homeless **b.** no: People already in the shelters are probably less likely to be against mandatory residence in the shelters. **2.** By selecting people from a shelter, homeless people who do not go to the shelters have no chance of being selected. An appropriate method would be to randomly select neighborhoods of the city and then randomly survey homeless people within the selected neighborhoods.

3.

Grade	Frequency
A	3
B	5
C	9
D	2
F	1
	20

4.

Class	Frequency
40–49	1
50–59	5
60–69	4
70–79	15
80–89	5
90–99	7
	37

5.

Stem	Leaves
4	1
5	8 2 8 0 7
6	8 2 9 9
7	3 5 9 9 7 5 5 3 3 6 7 1 7 1 5
8	7 3 9 9 1
9	4 6 9 7 5 8 0

Exercise Set 12.1

1. random **2.** frequency distribution **3.** grouped frequency distribution; 80; 89 **4.** histogram; frequencies **5.** frequency polygon; horizontal axis **6.** stem-and-leaf plot **7.** false **8.** true **9.** false **10.** true **21.** does not make sense **23.** does not make sense **25.** c **27.** 7; 31 **29.** 151

31.

Time Spent on Homework (in hours)	Number of Students
15	4
16	5
17	6
18	5
19	4
20	2
21	2
22	0
23	0
24	2
	30

33. $0, 5, 10, \ldots, 40, 45$ **35.** 5 **37.** 13 **39.** the 5–9 class **41.**

Age at Inauguration	Number of Presidents
41–45	2
46–50	9
51–55	15
56–60	9
61–65	7
66–70	2
	44

43. a.  **b.** 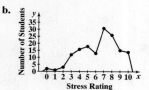 **45.** 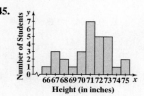 **b.**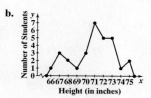

47. false **49.** false **51.** true **53.** false **55.**

Stem	Leaves
2	8 8 9 5
3	8 7 0 1 2 7 6 4 0 5
4	8 2 2 1 4 5 4 6 2 0 8 2 7 9
5	9 4 1 9 1 0
6	3 2 3 6 6 3

The greatest number of college professors are in their 40s.

57. Time intervals on the horizontal axis do not represent equal amounts of time. **59.** Percentages do not add up to 100%. **61.** It is not clear whether the bars or the actors represent the box-office receipts. **63.** Sample answer:

Length (miles)	Number of Rivers
501–1000	12
1001–1500	8
1501–2000	3
2001–2500	1
2501–3000	1
	25

Section 12.2

Checkpoint Exercises

1. a. 30 **b.** 30 **2.** 36 **3. a.** 35 **b.** 82 **4.** 5 **5.** 1:06, 1:09, 1:14, 1:21, 1:22, 1:25, 1:29, 1:29, 1:34, 1:34, 1:36, 1:45, 1:46, 1:49, 1:54, 1:57, 2:10, 2:15; median: 1 hour, 34 minutes **6.** 54.5 **7. a.** $19.05 million **b.** $22.45 million **c.** One salary, $0 million, is much smaller than the other salaries. **8.** 8 **9.** $118,006,573.50 **10.** mean: 158.6 cal; median: 153 cal; mode: 138 cal; midrange: 151 cal

Exercise Set 12.2

1. mean **2.** median **3.** $\frac{n+1}{2}$ **4.** mode **5.** midrange **6.** true **7.** false **8.** false **9.** false **19.** makes sense **21.** makes sense **23.** 4.125 **25.** 95 **27.** 62 **29.** ≈ 3.45 **31.** ≈ 4.71 **33.** ≈ 6.26 **35.** 3.5 **37.** 95 **39.** 60 **41.** 3.6 **43.** 5 **45.** 6 **47.** 3 **49.** 95 **51.** 40 **53.** 2.5, 4.2 (bimodal) **55.** 5 **57.** 6 **59.** 4.5 **61.** 95 **63.** 70 **65.** 3.3 **67.** 4.5 **69.** 5.5 **71.** mean: 30; median: 30; mode: 30; midrange: 30 **73.** mean: approximately 12.4; median: 12.5; mode: 13; midrange: 12.5 **75.** mean: approximately 31.3; median: 31; mode: 31; midrange: 33 **77. a.** approximately $481.9 million **b.** $403.5 million **c.** $406 million **d.** $888.5 million **79. a.** ≈ 17.27 **b.** 17 **c.** 7, 12, 17 **d.** 24.5 **81.** 175 lb **83.** 177.5 lb **85.** ≈ 2.76 **87.** All 30 students had the same grade.

Section 12.3

Checkpoint Exercises

1. 9 **2.** mean: 6;

Data item	Deviation
2	−4
4	−2
7	1
11	5

3. ≈ 3.92 **4.** sample A: 3.74; sample B: 28.06 **5. a.** small-company stocks **b.** small-company stocks; Answers will vary.

Exercise Set 12.3

1. range   **2.** standard deviation   **3.** true   **4.** true   **5.** false   **13.** makes sense   **15.** makes sense   **17.** 4   **19.** 8   **21.** 2

23. a.

Data item	Deviation
3	−9
5	−7
7	−5
12	0
18	6
27	15

b. 0

25. a.

Data item	Deviation
29	−20
38	−11
48	−1
49	0
53	4
77	28

b. 0

27. a. 91
b.

Data item	Deviation
85	−6
95	4
90	−1
85	−6
100	9

c. 0

29. a. 155
b.

Data item	
146	−9
153	−2
155	0
160	5
161	6

c. 0

31. a. 2.70
b.

Data item	Deviation
2.25	−0.45
3.50	0.80
2.75	0.05
3.10	0.40
1.90	−0.80

c. 0

33. ≈1.58   **35.** ≈3.46   **37.** ≈0.89   **39.** 3   **41.** ≈2.14
43. *Sample A*: mean: 12; range: 12; standard deviation: ≈4.32
Sample B: mean: 12; range: 12; standard deviation: ≈5.07
Sample C: mean: 12; range: 12; standard deviation: 6
The samples have the same mean and range, but different standard deviations.
45. 0   **47.** 1   **49.** 7.91   **51.** 1.55   **53. a.** male artists; All but one of the data items for the men are greater than the greatest data item for the women.   **b.** male artists: 77; female artists: 57   **c.** male artists; There is greater spread in the data for the men.   **d.** male artists: 19.91; female artists: 5.92   **57.** a

Section 12.4

Checkpoint Exercises

1. a. 75.5 in.   **b.** 58 in.   **2. a.** 95%   **b.** 47.5%   **c.** 16%   **3. a.** 2   **b.** 0   **c.** −1   **4.** ACT   **5. a.** 64   **b.** 128
6. 75% of the scores on the SAT are less than this student's score.   **7. a.** ± 2.0%   **b.** We can be 95% confident that between 34% and 38% of Americans read more than ten books per year.   **c.** Sample answer: Some people may be embarrassed to admit that they read few or no books per year.

Exercise Set 12.4

1. 68; 95; 99.7   **2.** mean   **3.** percentile   **4.** margin of error   **5.** true   **6.** true   **7.** false   **8.** true   **19.** does not make sense
21. makes sense   **23.** 120   **25.** 160   **27.** 150   **29.** 60   **31.** 90   **33.** 68%   **35.** 34%   **37.** 47.5%   **39.** 49.85%   **41.** 16%
43. 2.5%   **45.** 95%   **47.** 47.5%   **49.** 16%   **51.** 2.5%   **53.** 0.15%   **55.** 1   **57.** 3   **59.** 0.5   **61.** 1.75   **63.** 0   **65.** −1
67. −1.5   **69.** −3.25   **71.** 1.5   **73.** 2.25   **75.** −1.25   **77.** −1.5   **79.** The person who scores 127 on the Wechsler has the higher IQ.
81. 500   **83.** 475   **85.** 250   **87.** 275   **89. a.** ± 3.1%   **b.** We can be 95% confident that between 21.9% and 28.1% of high school students have information technology as their career choice.   **91. a.** ± 1.6%   **b.** We can be 95% confident that between 58.6% and 61.8% of all TV households watched the final episode of *M*A*S*H*.   **93.** 0.2%   **95. a.** skewed to the right   **b.** 5.3 murders per 100,000 residents
c. 5 murders per 100,000 residents   **d.** yes; The mean is greater than the median, which is consistent with a distribution skewed to the right.
e. 5.6; yes; For a normal distribution, almost 100% of the *z*-scores are between −3 and 3.

Chapter 12 Review Exercises

1. a

2.

Time Spent on Homework (in hours)	Number of Students
6	1
7	3
8	3
9	2
10	1
	10

3.

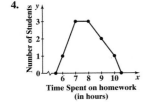

4.

5.

Grades	Number of Students
0–39	19
40–49	8
50–59	6
60–69	6
70–79	5
80–89	3
90–99	3
	50

6.

Stem	Leaves
1	8 4 3 3 7 1
2	4 6 9 9 2 7
3	4 9 6 1 5 1 1
4	4 7 9 1 2 2 0 5
5	4 7 9 0 6 1
6	3 7 0 8 3 9
7	2 5 4 0 3
8	1 7 6
9	1 0 5

7. Sizes of barrels are not scaled proportionally in terms of the data they represent.   **8.** 91.2   **9.** 17   **10.** 2.3   **11.** 11   **12.** 28   **13.** 2
14. 27   **15.** 585, 587 (bimodal)   **16.** 2   **17.** 91   **18.** 19.5   **19.** 2.5
21. a.

Age at First Inauguration	Number of Presidents
42	1
43	1
44	0
45	0
46	2
47	2
48	1
49	2
50	1
51	5
52	2
53	0
54	5
55	4
56	3
57	4
58	1
59	0
60	1
61	3
62	1
63	0
64	2
65	1
66	0
67	0
68	1
69	1
	44

b. mean: ≈54.66 yr; median: 54.5 yr; mode: 51 yr, 54 yr (bimodal); midrange: 55.5 yr   **22.** 18   **23.** 564

24. a.

Data item	Deviation
29	−6
9	−26
8	−27
22	−13
46	11
51	16
48	13
42	7
53	18
42	7

b. 0

25. a. 50
b.

Data item	Deviation
36	−14
26	−24
24	−26
90	40
74	24

c. 0

26. ≈4.05   **27.** ≈5.13   **28.** mean: 49; range: 76; standard deviation: ≈24.32   **29.** Set A: mean: 80; standard deviation: 0; Set B: mean: 80;
standard deviation: ≈ 11.55; Answers will vary.   **30.** Answers will vary.   **31.** 86   **32.** 98   **33.** 60   **34.** 68%   **35.** 95%   **36.** 34%
37. 99.7%   **38.** 16%   **39.** 84%   **40.** 2.5%   **41.** 0   **42.** 2   **43.** 1.6   **44.** −3   **45.** −1.2   **46.** vocabulary test
47. 38,000 miles   **48.** 41,000 miles   **49.** 22,000 miles   **50. a.** ±2.1%   **b.** We can be 95% confident that between 28.9% and 33.1% of
American adults would be willing to sacrifice a percentage of their salary to work for an environmentally friendly company.   **51. a.** skewed to the
right   **b.** mean: 2.1 syllables; median: 2 syllables; mode: 1 syllable   **c.** yes; The mean is greater than the median, which is consistent with a
distribution skewed to the right.

Chapter 12 Test A

1. d   **2.**

Score	Frequency
3	1
4	2
5	3
6	2
7	2
8	3
9	2
10	1
	16

3.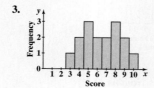

4.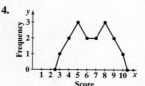

5.

Class	Frequency
40–49	3
50–59	6
60–69	6
70–79	7
80–89	6
90–99	2
	30

6.

Stem	Leaves
4	8 1 6
5	1 0 5 0 9 0
6	7 2 0 3 1 1
7	9 8 3 1 9 1 5
8	9 3 0 8 9 1
9	3 0

7. The roofline gives the impression that the percentage of home-schooled students grew at the same rate each year between the years shown which might not have happened.   **8.** 5   **9.** 4   **10.** 5.5   **11.** 2.65   **12.** 2.25   **13.** 2   **14.** 2   **15.** Answers will vary.   **16.** 34%
17. 2.5%   **18.** student   **19. a.** ± 3.2%   **b.** We can be 95% confident that between 21.8% and 28.2% of American adults would be willing to pay 20% more in taxes if that eliminated poverty and hunger.

Chapter 12 Test B
1. b   **2.**

Score	Frequency
2	1
3	2
4	3
5	2
6	2
7	3
8	2
9	0
10	1
	16

3.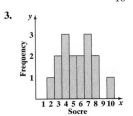

4.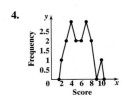

5.

Class	Frequency
40–49	3
50–59	6
60–69	6
70–79	7
80–89	6
90–99	2
	30

6.

Stem	Leaves
4	7 0 7
5	0 1 4 0 8 0
6	6 1 0 2 0 0
7	8 8 2 0 8 0 4
8	8 4 0 7 8 0
9	2 0

7. The heights of the bars are not proportional to the percents they represent.   **8.** 21   **9.** 23   **10.** 20   **11.** 4.30   **12.** 3.85
13. 4   **14.** 5   **15.** Answers will vary.   **16.** 34%   **17.** 2.5%
18. teacher   **19. a.** ±3.3%   **b.** We can be 95% confident that between 43.7% and 50.3% of Americans ages 16 and older find "whatever" most annoying in conversation.

CHAPTER 13

Section 13.1

Checkpoint Exercises

1. a. Paris is not the capital of Spain.   **b.** July is a month.   **2. a.** $\sim p$ **b.** $\sim q$ **3.** Chicago O'Hare is not the world's busiest airport.
4. Some new tax dollars will not be used to improve education.; At least one new tax dollar will not be used to improve education.

Exercise Set 13.1

1. true; false   **2.** false; true   **3.** $\sim p$; not p **4.** quantified   **5.** There are no A that are not B.   **6.** There exists at least one A that is a B.
7. All A are not B.   **8.** Not all A are B.   **9.** Some A are not B.   **10.** No A are B.   **11.** true   **12.** true   **13.** false   **14.** false
21. does not make sense   **23.** makes sense   **25.** statement   **27.** statement   **29.** not a statement   **31.** statement   **33.** not a statement
35. statement   **37.** statement   **39.** It is not raining.   **41.** The Dallas Cowboys are the team with the most Super Bowl wins.   **43.** Chocolate in moderation is good for the heart.   **45.** $\sim p$ **47.** $\sim r$ **49.** Listening to classical music does not make infants smarter.   **51.** Sigmund Freud's father was 20 years older than his mother.   **53. a.** There are no whales that are not mammals.   **b.** Some whales are not mammals.
55. a. At least one student is a business major.   **b.** No students are business majors.   **57. a.** At least one thief is not a criminal.   **b.** All thieves are criminals.   **59. a.** All Democratic presidents have not been impeached.   **b.** Some Democratic presidents have been impeached.
61. a. All seniors graduated.   **b.** Some seniors did not graduate.   **63. a.** Some parrots are not pets.   **b.** All parrots are pets.

65. a. No atheist is a churchgoer.   **b.** Some atheists are churchgoers.   **67.** Some Africans have Jewish ancestry.   **69.** Some rap is not hip-hop.
71. a. All birds are parrots.   **b.** false; Some birds are not parrots.   **73. a.** No college students are business majors.   **b.** false; Some college students are business majors.   **75. a.** All people like Sara Lee.   **b.** Some people don't like Sara Lee.   **77. a.** No safe thing is exciting.
b. Some safe things are exciting.   **79. a.** Some great actors are not a Tom Hanks.   **b.** All great actors are a Tom Hanks.   **81.** b   **83.** true
85. false; Some college students in the United States are willing to marry without romantic love.   **87.** true   **89.** false; The sentence "5% of college students in Australia are willing to marry without romantic love" is a statement.   **93.** One interpretation is the reason she is not dating him is because he is muscular. Another interpretation is that she is dating him but it is not because he is muscular.

Section 13.2

Checkpoint Exercises

1. a. $q \wedge p$ **b.** $\sim p \wedge q$ **2. a.** $p \vee q$ **b.** $q \vee \sim p$ **3. a.** $\sim p \to \sim q$ **b.** $q \to \sim p$ **4.** $\sim q \to p$ **5. a.** $q \leftrightarrow p$ **b.** $\sim p \leftrightarrow \sim q$
6. a. It is not true that he earns \$105,000 yearly and that he is often happy.   **b.** He is not often happy and he earns \$105,000 yearly.
c. It is not true that if he is often happy then he earns \$105,000 yearly.   **7. a.** If the plant is fertilized and watered, then it will not wilt.
b. The plant is fertilized, and if it is watered then it will not wilt.

Exercise Set 13.2

1. $p \wedge q$; conjunction   **2.** $p \vee q$; disjunction   **3.** $p \to q$; conditional   **4.** $p \leftrightarrow q$; biconditional   **5.** true   **6.** false   **7.** true   **8.** false
9. true   **10.** true   **11.** true   **12.** true   **19.** makes sense   **21.** does not make sense   **23.** $p \wedge q$ **25.** $q \wedge \sim p$ **27.** $\sim q \wedge p$
29. $p \vee q$ **31.** $p \vee \sim q$ **33.** $p \to q$ **35.** $\sim p \to \sim q$ **37.** $\sim q \to \sim p$ **39.** $p \to q$ **41.** $p \to \sim q$ **43.** $q \to \sim p$ **45.** $p \to \sim q$
47. $q \to \sim p$ **49.** $p \leftrightarrow q$ **51.** $\sim q \leftrightarrow \sim p$ **53.** $q \leftrightarrow p$ **55.** The heater is not working and the house is cold.   **57.** The heater is working or the house is not cold.   **59.** If the heater is working, then the house is not cold.   **61.** The heater is working if and only if the house is not cold.
63. It is July 4th and we are not having a barbecue.   **65.** It is not July 4th or we are having a barbecue.   **67.** If we are having a barbecue, then it is not July 4th.   **69.** It is not July 4th if and only if we are having a barbecue.   **71.** It is not true that Romeo loves Juliet and Juliet loves Romeo.
73. Romeo does not love Juliet and Juliet loves Romeo.   **75.** It is not true that Juliet loves Romeo or Romeo loves Juliet; Neither Juliet loves Romeo nor Romeo loves Juliet.   **77.** Juliet does not love Romeo or Romeo loves Juliet.   **79.** Romeo does not love Juliet and Juliet does not love Romeo.   **81.** $(p \wedge q) \vee r$ **83.** $(p \vee \sim q) \to r$ **85.** $r \leftrightarrow (p \wedge \sim q)$ **87.** $(p \wedge \sim q) \to r$ **89.** If the temperature is above 85° and we have finished studying, then we will go to the beach.   **91.** The temperature is above 85°, and if we have finished studying then we will go to the beach.   **93.** If we do not go to the beach, then the temperature is not above 85° or we have not finished studying.   **95.** If we do not go to the beach then we have not finished studying, or the temperature is above 85°.   **97.** We will go to the beach if and only if the temperature is above 85° and we have finished studying.   **99.** The temperature is above 85° if and only if we have finished studying, and we go to the beach.   **101.** If we do not go to the beach, then it is not true that both the temperature is above 85° and we have finished studying.   **103.** p: I like the teacher.; q: The course is interesting.; r: I miss class.; $(p \vee q) \to \sim r$ **105.** p: I like the teacher.; q: The course is interesting.; r: I miss class.; $p \vee (q \to \sim r)$
107. p: I like the teacher.; q: The course is interesting.; r: I miss class.; $r \leftrightarrow \sim (p \wedge q)$ **109.** p: I like the teacher.; q: The course is interesting.; r: I miss class.; $(p \to \sim r) \leftrightarrow q$ **111.** p: I like the teacher.; q: The course is interesting.; r: I miss class.; s: I spend extra time reading the book.; $(\sim p \wedge r) \to (\sim q \vee s)$ **113.** p: Being French is necessary for being a Parisian.; q: Being German is necessary for being a Berliner.; $\sim p \to \sim q$
115. p: You file an income tax report.; q: You file a complete statement of earnings.; r: You are a taxpayer.; s: You are an authorized tax preparer.; $(r \vee s) \to (p \wedge q)$ **117.** p: You are wealthy.; q: You are happy.; r: You live contentedly.; $\sim (p \to (q \wedge r))$ **119.** $\sim p \to \sim q$
121. $\sim p \to (q \wedge r)$ **123.** $\sim (p \wedge q)$

Section 13.3

Checkpoint Exercises

1. a. false   **b.** true   **c.** false   **d.** true

2.

p	q	$p \vee q$	$\sim(p \vee q)$
T	T	T	F
T	F	T	F
F	T	T	F
F	F	F	T

$\sim(p \vee q)$ is true when both p and q are false; otherwise, $\sim(p \vee q)$ is false.

3.

p	q	$\sim p$	$\sim q$	$\sim p \wedge \sim q$
T	T	F	F	F
T	F	F	T	F
F	T	T	F	F
F	F	T	T	T

$\sim p \wedge \sim q$ is true when both p and q are false; otherwise, $\sim p \wedge \sim q$ is false.

4.

p	q	$\sim q$	$p \wedge \sim q$	$\sim p$	$(p \wedge \sim q) \vee \sim p$
T	T	F	F	F	F
T	F	T	T	F	T
F	T	F	F	T	T
F	F	T	F	T	T

$(p \wedge \sim q) \vee \sim p$ is false when both p and q are true; otherwise, $(p \wedge \sim q) \vee \sim p$ is true.

5.

p	$\sim p$	$p \wedge \sim p$
T	F	F
F	T	F

$p \wedge \sim p$ is false in all cases.

6. *p*: I study hard.; *q*: I ace the final.; *r*: I fail the course.

a.

p	*q*	*r*	*q* ∨ *r*	*p* ∧ (*q* ∨ *r*)
T	T	T	T	T
T	T	F	T	T
T	F	T	T	T
T	F	F	F	F
F	T	T	T	F
F	T	F	T	F
F	F	T	T	F
F	F	F	F	F

b. false   **7.** true

Exercise Set 13.3

1. opposite   **2.** both *p* and *q* are true   **3.** both *p* and *q* are false   **4.** true   **5.** true   **6.** true   **7.** false   **15.** does not make sense
17. does not make sense   **19.** true   **21.** false   **23.** false   **25.** false   **27.** true   **29.** true   **31.** true   **33.** true

35.

p	~*p*	~*p* ∧ *p*
T	F	F
F	T	F

37.

p	*q*	~*p*	~*p* ∧ *q*
T	T	F	F
T	F	F	F
F	T	T	T
F	F	T	F

39.

p	*q*	*p* ∨ *q*	~(*p* ∨ *q*)
T	T	T	F
T	F	T	F
F	T	T	F
F	F	F	T

41.

p	*q*	~*p*	~*q*	~*p* ∧ ~*q*
T	T	F	F	F
T	F	F	T	F
F	T	T	F	F
F	F	T	T	T

43.

p	*q*	~*q*	*p* ∨ ~*q*
T	T	F	T
T	F	T	T
F	T	F	F
F	F	T	T

45.

p	*q*	~*p*	~*p* ∨ *q*	~(~*p* ∨ *q*)
T	T	F	T	F
T	F	F	F	T
F	T	T	T	F
F	F	T	T	F

47.

p	*q*	~*p*	*p* ∨ *q*	(*p* ∨ *q*) ∧ ~*p*
T	T	F	T	F
T	F	F	T	F
F	T	T	T	T
F	F	T	F	F

49.

p	*q*	~*p*	~*q*	*p* ∧ ~*q*	~*p* ∨ (*p* ∧ ~*q*)
T	T	F	F	F	F
T	F	F	T	T	T
F	T	T	F	F	T
F	F	T	T	F	T

51.

p	*q*	~*p*	~*q*	*p* ∨ *q*	~*p* ∨ ~*q*	(*p* ∨ *q*) ∧ (~*p* ∨ ~*q*)
T	T	F	F	T	F	F
T	F	F	T	T	T	T
F	T	T	F	T	T	T
F	F	T	T	F	T	F

53.

p	*q*	~*q*	*p* ∧ ~*q*	*p* ∧ *q*	(*p* ∧ ~*q*) ∨ (*p* ∧ *q*)
T	T	F	F	T	T
T	F	T	T	F	T
F	T	F	F	F	F
F	F	T	F	F	F

55.

p	*q*	*r*	~*q*	~*q* ∨ *r*	*p* ∧ (~*q* ∨ *r*)
T	T	T	F	T	T
T	T	F	F	F	F
T	F	T	T	T	T
T	F	F	T	T	T
F	T	T	F	T	F
F	T	F	F	F	F
F	F	T	T	T	F
F	F	F	T	T	F

57.

p	*q*	*r*	~*p*	~*q*	*r* ∧ ~*p*	(*r* ∧ ~*p*) ∨ ~*q*
T	T	T	F	F	F	F
T	T	F	F	F	F	F
T	F	T	F	T	F	T
T	F	F	F	T	F	T
F	T	T	T	F	T	T
F	T	F	T	F	F	F
F	F	T	T	T	T	T
F	F	F	T	T	F	T

59.

p	*q*	*r*	*p* ∨ *q*	~(*p* ∨ *q*)	~*r*	~(*p* ∨ *q*) ∧ ~*r*
T	T	T	T	F	F	F
T	T	F	T	F	T	F
T	F	T	T	F	F	F
T	F	F	T	F	T	F
F	T	T	T	F	F	F
F	T	F	T	F	T	F
F	F	T	F	T	F	F
F	F	F	F	T	T	T

61. a. *p*: You did the dishes.; *q*: You left the room a mess.; ~*p* ∧ *q*
b. See truth table for Exercise 37.
c. The statement is true when *p* is false and *q* is true.

63. a. p: I bought a ticket to the concert.; q: I used it.; $\sim(p \wedge \sim q)$

b.

p	q	$\sim q$	$p \wedge \sim q$	$\sim(p \wedge \sim q)$
T	T	F	F	T
T	F	T	T	F
F	T	F	F	T
F	F	T	F	T

c. Answers will vary; an example is: The statement is true when p and q are true.

65. a. p: The student is intelligent.; q: The student is an overachiever.; $(p \vee q) \wedge \sim q$

b.

p	q	$\sim q$	$p \vee q$	$(p \vee q) \wedge \sim q$
T	T	F	T	F
T	F	T	T	T
F	T	F	T	F
F	F	T	F	F

c. The statement is true when p is true and q is false.

67. a. p: Married people are healthier than single people.; q: Married people are more economically stable than single people.; r: Children of married people do better on a variety of indicators.; $(p \wedge q) \wedge r$

b.

p	q	r	$p \wedge q$	$(p \wedge q) \wedge r$
T	T	T	T	T
T	T	F	T	F
T	F	T	F	F
T	F	F	F	F
F	T	T	F	F
F	T	F	F	F
F	F	T	F	F
F	F	F	F	F

c. The statement is true when p, q, and r are all true.

69. a. p: I go to office hours.; q: I ask questions.; r: My teacher does not know that I am trying to improve my grade.; $(p \wedge q) \vee \sim r$

b.

p	q	r	$\sim r$	$p \wedge q$	$(p \wedge q) \vee \sim r$
T	T	T	F	T	T
T	T	F	T	T	T
T	F	T	F	F	F
T	F	F	T	F	T
F	T	T	F	F	F
F	T	F	T	F	T
F	F	T	F	F	F
F	F	F	T	F	T

c. Answers will vary; an example is: The statement is true when p, q, and r are all true.   **71.** false   **73.** true   **75.** true   **77.** false

79. true   **81.**

p	q	$\sim[\sim(p \wedge \sim q) \vee \sim(\sim p \vee q)]$
T	T	F
T	F	F
F	T	F
F	F	F

83.

p	q	r	$[(p \wedge \sim r) \vee (q \wedge \sim r)] \wedge \sim(\sim p \vee r)$
T	T	T	F
T	T	F	T
T	F	T	F
T	F	F	T
F	T	T	F
F	T	F	F
F	F	T	F
F	F	F	F

85. p: You notice this notice.; q: You notice this notice is not worth noticing.; $(p \vee \sim p) \wedge q$

p	q	$\sim p$	$p \vee \sim p$	$(p \vee \sim p) \wedge q$
T	T	F	T	T
T	F	F	T	F
F	T	T	T	T
F	F	T	T	F

The statement is true when q is true.

87. p: $x \leq 3$; q: $x \geq 7$; $\sim(p \vee q) \wedge (\sim p \wedge \sim q)$

p	q	$\sim p$	$\sim q$	$p \vee q$	$\sim(p \vee q)$	$\sim p \wedge \sim q$	$\sim(p \vee q) \wedge (\sim p \wedge \sim q)$
T	T	F	F	T	F	F	F
T	F	F	T	T	F	F	F
F	T	T	F	T	F	F	F
F	F	T	T	F	T	T	T

The statement is true when both p and q are false.

89. The percent body fat in women peaks at age 55 and the percent body fat in men does not peak at age 65.; false   **91.** The percent body fat in women does not peak at age 55 and men have more than 24% body fat at age 25.; false   **93.** The percent body fat in women peaks at age 55 or the percent body fat in men does not peak at age 65.; true   **95.** The percent body fat in women does not peak at age 55 or men have more than 24% body fat at age 25.; false   **97.** The percent body fat in women peaks at age 55 and the percent body fat in men peaks at age 65, or men have more than 24% body fat at age 25.; true   **99.** p: In 2000, 2% of 20-year-old men had completed the transition to adulthood.; q: In 2000, 46% of 30-year-old men had completed the transition to adulthood.; $\sim(p \wedge q)$; true   **101.** p: From 1960 through 2000, the percentage of 20-year-old women making the transition to adulthood decreased.; q: From 1960 through 2000, the percentage of 30-year-old women making the transition increased.; r: From 1960 through 2000, the percentage of 30-year-old men making the transition decreased.; $(p \vee q) \wedge \sim r$; false   **103. a.** Hora Gershwin   **b.** Bolera Mozart does not have a master's degree in music. Cha-Cha Bach does not either play three instruments or have five years experience playing with a symphony orchestra.   **107.** p is false and q is false.

Section 13.4

Checkpoint Exercises

1.

p	q	$\sim p$	$\sim q$	$\sim p \rightarrow \sim q$
T	T	F	F	T
T	F	F	T	T
F	T	T	F	F
F	F	T	T	T

2. The table shows that $[(p \rightarrow q) \wedge \sim q] \rightarrow \sim p$ is always true; therefore, it is a tautology.

p	q	$\sim p$	$\sim q$	$p \rightarrow q$	$(p \rightarrow q) \wedge \sim q$	$[(p \rightarrow q) \wedge \sim q] \rightarrow \sim p$
T	T	F	F	T	F	T
T	F	F	T	F	F	T
F	T	T	F	T	F	T
F	F	T	T	T	T	T

3. a. p: You use Hair Grow.; q: You apply it daily.; r: You go bald.

p	q	r	$\sim r$	$p \wedge q$	$(p \wedge q) \rightarrow \sim r$
T	T	T	F	T	F
T	T	F	T	T	T
T	F	T	F	F	T
T	F	F	T	F	T
F	T	T	F	F	T
F	T	F	T	F	T
F	F	T	F	F	T
F	F	F	T	F	T

4.

p	q	$p \vee q$	$\sim p$	$\sim p \rightarrow q$	$(p \vee q) \leftrightarrow (\sim p \rightarrow q)$
T	T	T	F	T	T
T	F	T	F	T	T
F	T	T	T	T	T
F	F	F	T	F	T

Because all cases are true, the statement is a tautology.

b. no

5. true

Exercise Set 13.4

1. p is true and q is false   **2.** tautology; implications; self-contradiction   **3.** p and q have the same truth value   **4.** false   **5.** false   **6.** true   **7.** true   **11.** makes sense   **13.** does not make sense

15.

p	q	$\sim q$	$p \rightarrow \sim q$
T	T	F	F
T	F	T	T
F	T	F	T
F	F	T	T

17.

p	q	$q \rightarrow p$	$\sim(q \rightarrow p)$
T	T	T	F
T	F	T	F
F	T	F	T
F	F	T	F

19.

p	q	$p \wedge q$	$p \vee q$	$(p \wedge q) \rightarrow (p \vee q)$
T	T	T	T	T
T	F	F	T	T
F	T	F	T	T
F	F	F	F	T

21.

p	q	$p \rightarrow q$	$\sim q$	$(p \rightarrow q) \wedge \sim q$
T	T	T	F	F
T	F	F	T	F
F	T	T	F	F
F	F	T	T	T

23.

p	q	r	$p \vee q$	$(p \vee q) \rightarrow r$
T	T	T	T	T
T	T	F	T	F
T	F	T	T	T
T	F	F	T	F
F	T	T	T	T
F	T	F	T	F
F	F	T	F	T
F	F	F	F	T

25.

p	q	r	$\sim q$	$\sim r$	$\sim q \rightarrow p$	$\sim r \wedge (\sim q \rightarrow p)$
T	T	T	F	F	T	F
T	T	F	F	T	T	T
T	F	T	T	F	T	F
T	F	F	T	T	T	T
F	T	T	F	F	T	F
F	T	F	F	T	T	T
F	F	T	T	F	F	F
F	F	F	T	T	F	F

27.

p	q	$\sim q$	$p \leftrightarrow \sim q$
T	T	F	F
T	F	T	T
F	T	F	T
F	F	T	F

29.

p	q	$p \leftrightarrow q$	$\sim(p \leftrightarrow q)$
T	T	T	F
T	F	F	T
F	T	F	T
F	F	T	F

31.

p	q	$p \leftrightarrow q$	$(p \leftrightarrow q) \to p$
T	T	T	T
T	F	F	T
F	T	F	T
F	F	T	F

33.

p	q	$\sim p$	$\sim p \leftrightarrow q$	$\sim p \to q$	$(\sim p \leftrightarrow q) \to (\sim p \to q)$
T	T	F	F	T	T
T	F	F	T	T	T
F	T	T	T	T	T
F	F	T	F	F	T

35.

p	q	r	$\sim r$	$p \leftrightarrow q$	$(p \leftrightarrow q) \to \sim r$
T	T	T	F	T	F
T	T	F	T	T	T
T	F	T	F	F	T
T	F	F	T	F	T
F	T	T	F	F	T
F	T	F	T	F	T
F	F	T	F	T	F
F	F	F	T	T	T

37.

p	q	r	$p \wedge r$	$q \vee r$	$\sim(q \vee r)$	$(p \wedge r) \leftrightarrow \sim(q \vee r)$
T	T	T	T	T	F	F
T	T	F	F	T	F	T
T	F	T	T	T	F	F
T	F	F	F	F	T	F
F	T	T	F	T	F	T
F	T	F	F	T	F	T
F	F	T	F	T	F	T
F	F	F	F	F	T	F

39. neither

p	q	$p \to q$	$(p \to q) \wedge q$	$[(p \to q) \wedge q] \to p$
T	T	T	T	T
T	F	F	F	T
F	T	T	T	F
F	F	T	F	T

41. tautology

p	q	$\sim p$	$\sim q$	$p \to q$	$(p \to q) \wedge \sim q$	$[(p \to q) \wedge \sim q] \to \sim p$
T	T	F	F	T	F	T
T	F	F	T	F	F	T
F	T	T	F	T	F	T
F	F	T	T	T	T	T

43. neither

p	q	$\sim q$	$p \vee q$	$(p \vee q) \wedge p$	$[(p \vee q) \wedge p] \to \sim q$
T	T	F	T	T	F
T	F	T	T	T	T
F	T	F	T	F	T
F	F	T	F	F	T

45. neither

p	q	$p \to q$	$q \to p$	$(p \to q) \leftrightarrow (q \to p)$
T	T	T	T	T
T	F	F	T	F
F	T	T	F	F
F	F	T	T	T

47. tautology

p	q	$\sim p$	$p \to q$	$\sim p \vee q$	$(p \to q) \leftrightarrow (\sim p \vee q)$
T	T	F	T	T	T
T	F	F	F	F	T
F	T	T	T	T	T
F	F	T	T	T	T

49. tautology

p	q	$p \leftrightarrow q$	$p \to q$	$q \to p$	$(q \to p) \wedge (p \to q)$	$(p \leftrightarrow q) \leftrightarrow [(q \to p) \wedge (p \to q)]$
T	T	T	T	T	T	T
T	F	F	F	T	F	T
F	T	F	T	F	F	T
F	F	T	T	T	T	T

51. tautology

p	q	r	$p \to q$	$q \to r$	$(p \to q) \wedge (q \to r)$	$p \to r$	$[(p \to q) \wedge (q \to r)] \to (p \to r)$
T	T	T	T	T	T	T	T
T	T	F	T	F	F	F	T
T	F	T	F	T	F	T	T
T	F	F	F	T	F	F	T
F	T	T	T	T	T	T	T
F	T	F	T	F	F	T	T
F	F	T	T	T	T	T	T
F	F	F	T	T	T	T	T

53. a. p: You do homework right after class.; q: You fall behind.; $(p \rightarrow \sim q) \wedge (\sim p \rightarrow q)$

b.

p	q	$\sim p$	$\sim q$	$p \rightarrow \sim q$	$\sim p \rightarrow q$	$(p \rightarrow \sim q) \wedge (\sim p \rightarrow q)$
T	T	F	F	F	T	F
T	F	F	T	T	T	T
F	T	T	F	T	T	T
F	F	T	T	T	F	F

c. Answers will vary; an example is: The statement is true when p and q have opposite truth values.

55. a. p: You cut and paste from the Internet.; q: You cite the source.; r: You are charged with plagiarism.; $(p \wedge \sim q) \rightarrow r$

b.

p	q	r	$\sim q$	$p \wedge \sim q$	$(p \wedge \sim q) \rightarrow r$
T	T	T	F	F	T
T	T	F	F	F	T
T	F	T	T	T	T
T	F	F	T	T	F
F	T	T	F	F	T
F	T	F	F	F	T
F	F	T	T	F	T
F	F	F	T	F	T

c. Answers will vary; an example is: The statement is true when p, q, and r are all true.

57. a. p: You are comfortable in your room.; q: You are honest with your roommate.; r: You enjoy the college experience.; $(p \leftrightarrow q) \vee \sim r$

b.

p	q	r	$\sim r$	$p \leftrightarrow q$	$(p \leftrightarrow q) \vee \sim r$
T	T	T	F	T	T
T	T	F	T	T	T
T	F	T	F	F	F
T	F	F	T	F	T
F	T	T	F	F	F
F	T	F	T	F	T
F	F	T	F	T	T
F	F	F	T	T	T

c. Answers will vary; an example is: The statement is true when p, q, and r are all true.

59. false   **61.** true   **63.** false   **65.** true

67. p: You love a person.; q: You marry that person.; $(q \rightarrow p) \wedge (\sim p \rightarrow \sim q)$

p	q	$\sim p$	$\sim q$	$q \rightarrow p$	$\sim p \rightarrow \sim q$	$(q \rightarrow p) \wedge (\sim p \rightarrow \sim q)$
T	T	F	F	T	T	T
T	F	F	T	T	T	T
F	T	T	F	F	F	F
F	F	T	T	T	T	T

Answers will vary; an example is: The statement is false when p is false and q is true.

69. p: You are happy.; q: You live contentedly.; r: You are wealthy.; $\sim[r \rightarrow (p \wedge q)]$

p	q	r	$p \wedge q$	$r \rightarrow (p \wedge q)$	$\sim[r \rightarrow (p \wedge q)]$
T	T	T	T	T	F
T	T	F	T	T	F
T	F	T	F	F	T
T	F	F	F	T	F
F	T	T	F	F	T
F	T	F	F	T	F
F	F	T	F	F	T
F	F	F	F	T	F

Answers will vary; an example is: The statement is false when p, q, and r are all true.

71. p: There was an increase in the percentage who believed in God.; q: There was a decrease in the percentage who believed in Heaven.; r: There was an increase in the percentage who believed in the devil.; $(p \wedge q) \rightarrow r$; true   **73.** p: There was a decrease in the percentage who believed in God.; q: There was an increase in the percentage who believed in Heaven.; r: The percentage believing in the devil decreased.; $(p \leftrightarrow q) \vee r$; true

75. p: Fifteen percent are capitalists.; q: Thirty-four percent are members of the upper-middle class.; r: The number of working poor exceeds the number belonging to the working class.; $(p \vee \sim q) \leftrightarrow r$; false   **77.** p: There are more people in the lower-middle class than in the capitalist and upper-middle classes combined.; q: One percent are capitalists.; r: Thirty-four percent belong to the upper-middle class.; $p \rightarrow (q \wedge r)$; false

79. not a tautlogy; p: You ask me.; q: I know.

p	q	$\sim p$	$\sim q$	$\sim p \rightarrow q$	$p \rightarrow \sim q$	$(\sim p \rightarrow q) \wedge (p \rightarrow \sim q)$
T	T	F	F	T	F	F
T	F	F	T	T	T	T
F	T	T	F	T	T	T
F	F	T	T	F	T	F

81. Sample answer: Assuming $p \vee q$ is true, at least one of p and q is true, so if p is false it follows that q must be true, $\sim p \rightarrow q$.   **83.** Answers will vary; examples are $p \rightarrow q$, $\sim p$, $(p \rightarrow q) \vee \sim p$, and $\sim p \rightarrow [(p \rightarrow q) \vee \sim p]$.

Section 13.5

Checkpoint Exercises

1. a.

p	q	$\sim q$	$p \vee q$	$\sim q \to p$
T	T	F	T	T
T	F	T	T	T
F	T	F	T	T
F	F	T	F	F

The statements are equivalent since their truth values are the same.

2.

p	$\sim p$	$\sim(\sim p)$	$\sim[\sim(\sim p)]$
T	F	T	F
F	T	F	T

Since the truth values are the same, $\sim[\sim(\sim p)] \equiv \sim p$.

b. If I don't fail the course, then I attend classes.

3. c   **4. a.** If you're not driving too closely, then you can't read this.   **b.** If it's not time to do the laundry, then you have clean underwear.
c. If supervision during exams is required, then some students are not honest.   **d.** $q \to (p \vee r)$ **5.** Converse: If you are not in New England, then you are in New Orleans.; Inverse: If you are not in New Orleans, then you are in New England.; Contrapositive: If you are in New England, then you are not in New Orleans.

Exercise Set 13.5

1. equivalent; $\equiv$ **2.** $\sim q \to \sim p$ **3.** $q \to p$ **4.** $\sim p \to \sim q$ **5.** equivalent; converse; inverse   **6.** true   **7.** false   **8.** false
13. does not make sense   **15.** makes sense
17. a.

p	q	$\sim p$	$p \vee q$	$\sim p \to q$
T	T	F	T	T
T	F	F	T	T
F	T	T	T	T
F	F	T	F	F

b. The United States supports the development of solar-powered cars or it will suffer increasing atmospheric pollution.   **19.** not equivalent
21. equivalent   **23.** not equivalent   **25.** equivalent   **27.** a   **29.** c   **31.** Converse: If I am in Illinois, then I am in Chicago.; Inverse: If I am not in Chicago, then I am not in Illinois.; Contrapositive: If I am not in Illinois, then I am not in Chicago.   **33.** Converse: If I cannot hear you, then the stereo is playing.; Inverse: If the stereo is not playing, then I can hear you.; Contrapositive: If I can hear you, then the stereo is not playing.
35. Converse: If we don't grow, we don't change.; Inverse: If we change, we grow.; Contrapositive: If we grow, we change.   **37.** Converse: If no things are wrong, then slavery is not wrong.; Inverse: If slavery is wrong, then some things are wrong.; Contrapositive: If some things are not wrong, then slavery is wrong.   **39.** Converse: If some people are not thinking, then all people think alike.; Inverse: If some people do not think alike, then all people are thinking.; Contrapositive: If all people are thinking, then some people do not think alike.   **41.** Converse: $\sim r \to \sim q$; Inverse: $q \to r$;
Contrapositive: $r \to q$ **43.** If a person diets, then he or she loses weight.; Converse: If a person loses weight, then he or she is dieting.; Inverse: If a person is not dieting, then he or she is not losing weight.; Contrapositive: If a person is not losing weight, then he or she is not dieting.   **45.** If a vehicle has no flashing light on top, then it is not an ambulance.; Converse: If a vehicle is not an ambulance, then it has no flashing light on top.; Inverse: If a vehicle has a flashing light on top, then it is an ambulance.; Contrapositive: If a vehicle is an ambulance, then it has a flashing light on top.
47. If a person is an attorney, then he or she has passed the bar exam.; Converse: If a person has passed the bar exam, then he or she is an attorney.; Inverse: If a person is not an attorney, then he or she has not passed the bar exam.; Contrapositive: If a person has not passed the bar exam, then he or she is not an attorney.   **49.** If a person is a pacifist, then he or she is not a warmonger.; Converse: If a person is not a warmonger, then he or she is a pacifist.; Inverse: If a person is not a pacifist, then he or she is a warmonger.; Contrapositive: If a person is a warmonger, then he or she is not a pacifist.
51. a. true   **b.** Converse: If the corruption rating is 9.6, then the country is Finland.; Inverse: If the country is not Finland, then the corruption rating is not 9.6.; Contrapositive: If the corruption rating is not 9.6, then the country is not Finland.; The contrapositive is true. The converse and inverse are not necessarily true.

Section 13.6

Checkpoint Exercises

1. valid   **2.** valid   **3.** invalid   **4. a.** valid   **b.** invalid   **c.** valid   **5.** valid   **6.** Some people do not lead.

Exercise Set 13.6

1. valid   **2.** q; valid; $[(p \to q) \wedge p] \to q$ **3.** $\sim p$; valid; $[(p \to q) \wedge \sim q] \to \sim p$ **4.** $p \to r$; valid; $[(p \to q) \wedge (q \to r)] \to (p \to r)$
5. q; valid; $[(p \vee q) \wedge \sim p] \to q$ **6.** p **7.** $\sim q$ **8.** false   **9.** true   **10.** false   **17.** does not make sense   **19.** makes sense
21. invalid   **23.** valid   **25.** valid   **27.** invalid   **29.** invalid   **31.** valid

33. $\dfrac{\begin{array}{c} p \to \sim q \\ q \end{array}}{\therefore \sim p}$ valid   **35.** $\dfrac{\begin{array}{c} p \vee q \\ q \end{array}}{\therefore \sim p}$ invalid   **37.** $\dfrac{\begin{array}{c} p \to q \\ q \end{array}}{\therefore p}$ invalid   **39.** $\dfrac{\begin{array}{c} p \to q \\ \sim p \to q \end{array}}{\therefore q}$ valid   **41.** $\dfrac{\begin{array}{c} p \vee q \\ \sim q \end{array}}{\therefore p}$ valid   **43.** $\dfrac{\begin{array}{c} p \to q \\ \sim p \end{array}}{\therefore \sim q}$ invalid   **45.** $\dfrac{\begin{array}{c} p \to q \\ q \to r \end{array}}{\therefore p \to r}$ valid   **47.** $\dfrac{\begin{array}{c} p \to q \\ q \to r \end{array}}{\therefore r \to p}$ invalid

49. $\dfrac{\begin{array}{c} (p \wedge q) \to r \\ p \wedge \sim r \end{array}}{\therefore \sim q}$ valid   **51.** $\dfrac{\begin{array}{c} (p \vee q) \to r \\ \sim r \end{array}}{\therefore \sim p \wedge \sim q}$ valid   **53.** $\dfrac{\begin{array}{c} (p \vee q) \to r \\ r \end{array}}{\therefore p \vee q}$ invalid   **55.** $\dfrac{\begin{array}{c} (p \wedge q) \to r \\ \sim p \vee \sim q \end{array}}{\therefore \sim r}$ invalid   **57.** $\dfrac{\begin{array}{c} p \to q \\ q \to \sim r \\ r \end{array}}{\therefore \sim p}$ valid

59. My best friend is not a chemist.   **61.** They were dropped from prime time.   **63.** Some electricity is not off.   **65.** If I vacation in Paris, I gain weight.

67.
$$\frac{p \to q}{\sim p}$$
$$\overline{\therefore \sim q}$$
invalid
69.
$$\frac{p \vee q}{\sim p}$$
$$\overline{\therefore q}$$
valid
71.
$$\frac{p \to q}{q}$$
$$\overline{\therefore p}$$
invalid
73.
$$\frac{p \to q}{\sim q}$$
$$\overline{\therefore \sim p}$$
valid
75.
$$\frac{p \to q}{q \to r}$$
$$\overline{\therefore p \to r}$$
valid
77.
$$\frac{p \to q}{\sim q}$$
$$\overline{\therefore \sim p}$$
valid
79. h   **81.** i   **83.** c   **85.** a   **87.** j

89. d   **93.** People sometimes speak without being spoken to.   **95.** The doctor either destroys the base on which the placebo rests or jeopardizes a relationship built on trust.

Section 13.7

Checkpoint Exercises

1. valid   **2.** invalid   **3.** valid   **4.** invalid   **5.** invalid   **6.** invalid

Exercise Set 3.7

1. All A are B.   **2.** No A are B.   **3.** Some A are B.   **4.** Some A are not B.   **5.** false   **6.** false   **9.** does not make sense   **11.** makes sense   **13.** valid   **15.** invalid   **17.** valid   **19.** invalid   **21.** invalid   **23.** valid   **25.** valid   **27.** invalid   **29.** invalid   **31.** valid   **33.** valid   **35.** valid   **37.** invalid   **39.** invalid   **41.** valid   **43.** invalid   **45.** invalid   **47.** invalid   **49.** valid   **51.** valid   **53.** b   **55.** Some teachers are amusing people.

Chapter 13 Review Exercises

1. If the temperature is below 32° and we have finished studying, we will go to the movies.   **2.** If we do not go to the movies, then the temperature is not below 32° or we have not finished studying.   **3.** The temperature is below 32°, and if we finish studying, we will go to the movies.   **4.** We will go to the movies if and only if the temperature is below 32° and we have finished studying.   **5.** It is not true that both the temperature is below 32° and we have finished studying.   **6.** We will not go to the movies if and only if the temperature is not below 32° or we have not finished studying.   **7.** $(p \wedge q) \vee r$ **8.** $(p \vee \sim q) \to r$ **9.** $q \to (p \leftrightarrow r)$ **10.** $r \leftrightarrow (p \wedge \sim q)$ **11.** $p \to r$ **12.** $q \to \sim r$ **13.** Some houses are not made with wood.   **14.** Some students major in business.   **15.** No crimes are motivated by passion.   **16.** All Democrats are registered voters.   **17.** Some new taxes will not be used for education.

18. neither

p	q	$\sim p$	$\sim p \wedge q$	$p \wedge (\sim p \wedge q)$
T	T	F	F	T
T	F	F	F	T
F	T	T	T	T
F	F	T	F	F

19. neither

p	q	$\sim p$	$\sim q$	$\sim p \vee \sim q$
T	T	F	F	F
T	F	F	T	T
F	T	T	F	T
F	F	T	T	T

20. neither

p	q	$\sim p$	$\sim p \vee q$	$p \to (\sim p \vee q)$
T	T	F	T	T
T	F	F	F	F
F	T	T	T	T
F	F	T	T	T

21. neither

p	q	$\sim q$	$p \leftrightarrow \sim q$
T	T	F	F
T	F	T	T
F	T	F	T
F	F	T	F

22. tautology

p	q	$\sim p$	$\sim q$	$p \vee q$	$\sim(p \vee q)$	$\sim p \wedge \sim q$	$\sim(p \vee q) \to (\sim p \wedge \sim q)$
T	T	F	F	T	F	F	T
T	F	F	T	T	F	F	T
F	T	T	F	T	F	F	T
F	F	T	T	F	T	T	T

23. neither

p	q	r	$\sim r$	$p \vee q$	$(p \vee q) \to \sim r$
T	T	T	F	T	F
T	T	F	T	T	T
T	F	T	F	T	F
T	F	F	T	T	T
F	T	T	F	T	F
F	T	F	T	T	T
F	F	T	F	F	T
F	F	F	T	F	T

24. neither

p	q	r	$p \wedge q$	$p \wedge r$	$(p \wedge q) \leftrightarrow (p \wedge r)$
T	T	T	T	T	T
T	T	F	T	F	F
T	F	T	F	T	F
T	F	F	F	F	T
F	T	T	F	F	T
F	T	F	F	F	T
F	F	T	F	F	T
F	F	F	F	F	T

25. a. p: I'm in class.; q: I'm studying.; $(p \vee q) \wedge \sim p$

b.

p	q	$\sim p$	$p \vee q$	$(p \vee q) \wedge \sim p$
T	T	F	T	F
T	F	F	T	F
F	T	T	T	T
F	F	T	F	F

c. The statement is true when p is false and q is true.

26. a. *p*: You spit from a truck.; *q*: It's legal.; *r*: You spit from a car.; $(p \rightarrow q) \wedge (r \rightarrow \sim q)$

b.

p	*q*	*r*	~*q*	*p* → *q*	*r* → ~*q*	(p → q) ∧ (r → ~q)
T	T	T	F	T	F	F
T	T	F	F	T	T	T
T	F	T	T	F	T	F
T	F	F	T	F	T	F
F	T	T	F	T	F	F
F	T	F	F	T	T	T
F	F	T	T	T	T	T
F	F	F	T	T	T	T

c. The statement is true when *p* and *q* are both false.

27. false   **28.** true   **29.** true   **30.** *p*: Twenty-nine percent consider religion the most taboo topic.; *q*: Fourteen percent consider politics the most taboo topic.; $p \wedge \sim q$; false   **31.** *p*: A greater percentage of people consider money a more taboo topic than personal life.; *q*: Sixteen percent consider money the most taboo topic.; *r*: Fourteen percent consider personal life the most taboo topic.; $p \rightarrow (q \wedge r)$; true

32. *p*: Fourteen percent consider money the most taboo topic.; *q*: Fourteen percent consider politics the most taboo topic.; *r*: The greatest percentage consider religion the most taboo topic.; $(p \leftrightarrow q) \vee \sim r$; true

33. a.

p	*q*	~*p*	~*p* ∨ *q*	*p* → *q*
T	T	F	T	T
T	F	F	F	F
F	T	T	T	T
F	F	T	T	T

b. If the triangle is isosceles, then it has two equal sides.

34. c   **35.** not equivalent   **36.** equivalent   **37.** Converse: If I am in the South, then I am in Atlanta.; Inverse: If I am not in Atlanta, then I am not in the South.; Contrapositive: If I am not in the South, then I am not in Atlanta.   **38.** Converse: If today is not a holiday, then I am in class.; Inverse: If I am not in class, then today is a holiday.; Contrapositive: If today is a holiday, then I am not in class.   **39.** Converse: If I pass all courses, then I worked hard.; Inverse: If I don't work hard, then I don't pass some courses.; Contrapositive: If I do not pass some course, then I did not work hard.   **40.** Converse: $\sim q \rightarrow \sim p$; Inverse: $p \rightarrow q$.; Contrapositive: $q \rightarrow p$ **41.** a and c   **42.** a and b   **43.** a and c   **44.** none   **45.** invalid

46. valid   **47.** $\dfrac{p \rightarrow q}{\therefore p}$ invalid   **48.** $\dfrac{p \vee q}{\therefore \sim p}$ invalid   **49.** $\dfrac{p \vee q}{\sim p}{\therefore q}$ valid   **50.** $\dfrac{p \rightarrow q}{\sim q}{\therefore \sim p}$ valid   **51.** $\dfrac{p \rightarrow \sim q}{\sim p \rightarrow q}{\therefore p \leftrightarrow \sim q}$ valid   **52.** $\dfrac{p \rightarrow q}{r \rightarrow q}{\therefore \sim r \rightarrow p}$ invalid   **53.** invalid   **54.** valid

55. valid   **56.** invalid   **57.** invalid   **58.** valid

Chapter 13 Test A

1. If I'm registered and I'm a citizen, then I vote.   **2.** I don't vote if and only if I'm not registered or I'm not a citizen.   **3.** I'm neither registered nor a citizen.   **4.** $(p \wedge q) \vee \sim r$ **5.** $(\sim p \vee \sim q) \rightarrow \sim r$ **6.** $r \rightarrow q$ **7.** Some numbers are not divisible by 5.   **8.** No people wear glasses.

9.

p	*q*	~*p*	~*p* ∨ *q*	*p* ∧ (~*p* ∨ *q*)
T	T	F	T	T
T	F	F	F	F
F	T	T	T	F
F	F	T	T	F

10.

p	*q*	~*p*	~*q*	*p* ∧ *q*	~(*p* ∧ *q*)	(~*p* ∨ ~*q*)	~(*p* ∧ *q*) ↔ (~*p* ∨ ~*q*)
T	T	F	F	T	F	F	T
T	F	F	T	F	T	T	T
F	T	T	F	F	T	T	T
F	F	T	T	F	T	T	T

11.

p	*q*	*r*	*q* ∨ *r*	*p* ↔ (*q* ∨ *r*)
T	T	T	T	T
T	T	F	T	T
T	F	T	T	T
T	F	F	F	F
F	T	T	T	F
F	T	F	T	F
F	F	T	T	F
F	F	F	F	T

12. *p*: You break the law.; *q*: You change the law.; $(p \wedge q) \rightarrow \sim p$

p	*q*	~*p*	*p* ∧ *q*	(*p* ∧ *q*) → ~*p*
T	T	F	T	F
T	F	F	F	T
F	T	T	F	T
F	F	T	F	T

The statement is false when *p* and *q* are both true.

13. true   **14.** true   **15. a.** *p*: There was an increase in the percentage of Americans who supported the death penalty.; *q*: There was an increase in the percentage who opposed the death penalty.; *r*: There was an increase in the percentage who were not sure about the death penalty.; ~*p* ∨ (*q* ∧ *r*)   **b.** true   **16.** b   **17.** If it snows, then it is not August.   **18.** Converse: If I cannot concentrate, then the radio is playing.; Inverse: If the radio is not playing, then I can concentrate.   **19.** a and b   **20.** a and c   **21.** invalid   **22.** valid   **23.** invalid   **24.** invalid **25.** valid   **26.** invalid

Chapter 13 Test B

1. If I live at home and eat at home, then I do not pay for meals.   **2.** I pay for my meals, if and only if I do not live at home or do not eat at home. **3.** It is not the case that I live at home and eat at home.   **4.** $(p \wedge q) \vee r$ **5.** $(\sim p \vee \sim q) \rightarrow r$ **6.** $\sim r \rightarrow q$ **7.** Some natural numbers are not even.   **8.** Some students are hard workers.

9.

p	*q*	~*p*	~*p* ∧ *q*	*p* ∨ (~*p* ∧ *q*)
T	T	F	F	T
T	F	F	F	T
F	T	T	T	T
F	F	T	F	F

10.

p	*q*	~*p*	~*q*	*p* ∨ *q*	~(*p* ∨ *q*)	~*p* ∧ ~*q*	~(*p* ∨ *q*) ↔ (~*p* ∧ ~*q*)
T	T	F	F	T	F	F	T
T	F	F	T	T	F	F	T
F	T	T	F	T	F	F	T
F	F	T	T	F	T	T	T

11.

p	*q*	*r*	*q* ∧ *r*	*p* ↔ (*q* ∧ *r*)
T	T	T	T	T
T	T	F	F	F
T	F	T	F	F
T	F	F	F	F
F	T	T	T	F
F	T	F	F	T
F	F	T	F	T
F	F	F	F	T

12. *p*: You have a job.; *q*: You have aggravations.; $(p \vee \sim q) \rightarrow \sim p$

p	*q*	~*p*	~*q*	*p* ∧ ~*q*	(*p* ∧ ~*q*) → ~*p*
T	T	F	F	F	T
T	F	F	T	T	F
F	T	T	F	F	T
F	F	T	T	F	T

The statement is false when *p* is true and *q* is false.

13. false   **14.** false   **15. a.** *p*: The average age of book buyers is 20.; *q*: A higher percentage of women buy books than men.; *r*: The average price paid for a book exceeds $20.; $(p \wedge q) \vee r$ **b.** false   **16.** c   **17.** If I am in class, then it is not a holiday.   **18.** Converse: If I don't guess, then I know the answer.; Inverse: If I don't know the answer, then I guess.   **19.** none   **20.** a and b   **21.** valid   **22.** invalid   **23.** valid **24.** valid   **25.** invalid   **26.** invalid

Subject Index

Application Index

Photo Credits

Chapter 1

1 Ron Buskirk/Alamy; **2** R.F. Voss/IBM Research; **7** Stone Allstock/Getty Images; **13** Joe Belanger/iStockphoto; **16** Stockbyte/Thinkstock; **23** Iofoto/Shutterstock; **23** Stephen Coburn/Shutterstock; **29 (top)** Nikada/iStockphoto; **29 (bottom)** Ohiophoto/iStockphoto; **34** Ragne Kabanova/Shutterstock; **40** Clifford A. Pickover, A Passion for Mathematics, John Wiley & Sons, Inc., 2005. Reproduced with permission of John Wiley & Sons, Inc.

Chapter 2

51 Michael Svoboda/iStockphoto; **52** Moodboard/SuperStock; **53** SHNS Photo/MTV Games/Newscom; **54** John Tenniel/National Archives; **58** Crysrob/Dreamstime; **65** WilleeCole/Shutterstock; **70 (top)** Bob London/Getty Images; **70 (bottom)** Jerry Young/Dorling Kindersley; **71 (top)** Labat J. M. Jacana/Scientific Control/Photo Researchers; **71 (bottom)** Marina Jay/Shutterstock; **77** Mero/Labat J.M. Jacana/Scientific Control/PhotoResearchers; **74** Courtesy of Si TV; **86** Tatniz/Shutterstock; **96 (left)** RubberBall/Alamy; **96 (center)** Karimala/Dreamstime; **96 (right)** Thomas Northcut/Photodisc/Thinkstock

Chapter 3

117 Domen Colja/Alamy; **118** SuperStock RF/SuperStock; **129** Tom Mareschal/The Image Bank/Getty Images; **132** American Images Inc./Taxi/Getty Images; **143** John Burwell/iStockphoto; **155** Robert Voets/CBS Television/Newscom; **159** *School of Athens* (detail of left half showing Pythagoras, seated) (1509), Raphael. Stanza della Segnatura, Vatican Palace, Vatican State/Scala/Art Resource, New York; **160** Alan Schein/Alamy; **165** John G. Ross/Photo Researchers, Inc.; **166** *The Persistence of Memory* (1931) Salvador Dali. Oil on canvas, 9 1/2 × 13 in. (24.1 × 33 cm). (162.1934). Given anonymously to the Museum of Modern Art/Digital image copyright © the Museum of Modern Art/Licensed by SCALA/Art Resource, NY/Artists Rights Society, NY; **167 (top)** *Foxtrot* copyright © 2003, 2009 by Bill Amend/Distributed by Universal Uclick; **167 (center)** *Hagar the Horrible* by Chris Browne. Copyright © 1995 by King Features Syndicate; **167 (bottom left)** *Peanuts* copyright © 2010 by Peanuts Worldwide LLC./Distributed by United Feature Syndicate; **167 (bottom right)** *Foxtrot* copyright © 2003, 2009 by Bill Amend/Distributed by Universal Uclick; **170** *Foxtrot* copyright © 2003, 2009 by Bill Amend/Distributed by Universal Uclick;

171 Unimedia Europe/SND Films/Newscom; **172** Raisa Kanareva/Shutterstock; **175** Copyright © 2010 Roz Chast/The New Yorker Collection/CartoonBank; **177** Gordon Caulkins; **181 (top)** *Invoking the Googulplex* (1981) John Scott. Watercolor on paper, 24 × 18 in. A Space Gallery, Toronto; **181 (bottom)** David Weintraub/Photo Researchers, Inc.; **186** Charles Rex Arbogast/AP Images; **190** iStockphoto/Thinkstock; **196** Jokerpro/Shutterstock; **197** Harry Blair

Chapter 4

219 Melissa Moseley/Columbia/Sony Pictures/The Kobal Collection/The Picture Desk, Inc.; **220** Sergiy Bykhunenko/iStockphoto; **222 (left)** Kris Ubach and Quim Roser/Cultura/Alamy; **222 (right)** Erik Isakson/Tetra Images/Alamy; **229** SHNS Photo/Comedy Central/Newscom; **240** Tammy616/iStockphotos; **246 (left)** Pictorial Press Ltd./Alamy; **246 (right)** Moviestore Collection Lt./Alamy; **241 (top left)** Stuart Ramson/AP Images; **241 (top center left)** Frederick M. Brown/Getty Images; **241 (top center right)** Piyal Hosain/Fotos International/Getty Images; **241 (top right)** Lucy Nicholson/Reuters/Corbis; **241 (bottom left)** PictureIndia/Getty Images; **241 (bottom center left)** Red Chopstocks/Getty Images; **241 (bottom center right)** Moodboard/Corbis; **241 (bottom right)** Creatas Images/Jupiter Images; **255** John Woodcock/Getty Images; **259** *Real Life Adventures* copyright © 2010 by Gary Wise and Lance Aldrich/Distributed by Universal Uclick; **258** Chris Burt/Shutterstock; **260 (left)** Alamy; **260 (right)** Alamy Royalty Free; **262** Stan Fellerman/Corbis; **263** Superstock Royalty Free; **267** Bettmann/Corbis; **269** Greg Nicholas/iStockphoto

Chapter 5

293 Big Cheese Photo LLC/Alamy; **294** Iofoto/Dreamstime; **299** Copyright © 2010 Warren Miller/The New Yorker Collection/CartoonBank; **309** Toynutz/Dreamstime.com; **322** Paul Prescott/Shutterstock; **330** *Foxtrot* copyright ©2003, 2009 by Bill Amend/Distributed by Universal Uclick; **334** RubberBall/Alamy; **343** Roman Milert/Alamy; **348** Terry J. Alcorn/iStockphoto

Chapter 6

373 Greg Baker/AP Images; **374** Mutlu/iStockphotos; **384** Robert F. Blitzer; **387** Steve Shott/Dorling Kindersley; **399** Christophe Testi/iStockphotos; **402** Lise Gagne/iStockphotos; **413** David Young-Wolff/PhotoEdit, Inc.;

424 (top) Benjamin K. Frederick; **424 (bottom)** Simon Bruty; **421** *Bathers at Asnières* (1884), Georges Seurat. Oil on canvas, 201 × 300 cm. National Gallery, London/Art Resource, New York

Chapter 7

447 Oksana Perkins/Shutterstock; **448** Florea Marius Catalin/iStockphotos; **457 (left)** Matt Groening/20th Century Fox/The Kobal Collection/The Picture Desk, Inc.; **457 (right)** Alex Slobodkin/iStockphotos; **458** David Breslauer/AP Images; **460 (right)** TinCup/Alamy; **460 (left)** Thomas J. Peterson/Alamy; **460 (bottom)** Morgan Lane Photography/Shutterstock; **465** James Steidl/iStockphotos; **469** North Wind Picture Archives/Alamy; **476** Donald Swartz/iStockphoto

Chapter 8

497 Heng Kong Chen/iStockphoto; **498** Jeff Christensen/WireImage/Getty Images; **515** Moviestore Collection Ltd./Alamy; **525** Grafissimo/iStockphotos; **529** Mel Yates/Getty Images; **534** Eyebyte/Alamy

Chapter 9

555 Nico Smit/iStockphotos; **556** Dorling Kindersley; **558 (top-bottom)** U.S. Bureau of Engraving and Printing; **558 (left-right)** Frank LaBua/Pearson; **566** Oksana Perkins/iStockphotos; **568** Peter Langoné; **575** Ben Goode/Dreamstime; **575 (center)** Vicki Reid/iStockphotos; **580** DAJ/Getty

Chapter 10

591 Alla Shcherbak/iStockphoto; **592** Hemis/Alamy; **600** *B.C.* copyright © 2010 by John L. Hart FLP/Distributed by Creators Syndicate; **601** Visions of America, LLC/Alamy; **608** Lowe's Incorporated/The Kobal Collection/The Picture Desk, Inc.; **613** Omni-Photo Communications, Inc.; **614** Florin Tirlea/iStockphoto; **617** *Symmetry Drawing (Lizard/Fish/Bat)* (1952), M.C. Escher. Ink, pencil, watercolor. Copyright © 2010 The M.C. Escher Company, Holland. All rights reserved.; **621** Red Cover/Alamy; **626** Benjamin Shearn/Taxi/Getty; **632** KRT Photos/Newscom; **634** Uyen Le/Getty; **641** STR/AFP/Getty Images; **646** Janet Foster/Masterfile; **647 (bottom)** Hugh Sitton/Getty Images; **647 (top)** Wesley Hitt/Photographer's Choice/Getty Images; **651** Scott Berner/Index Stock/Jupiter